ISBN 978-1-5279-2030-9
PIBN 10903608

1 MONTH OF
FREE
READING

at
www.ForgottenBooks.com

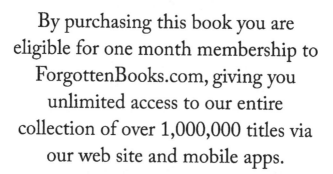

By purchasing this book you are eligible for one month membership to ForgottenBooks.com, giving you unlimited access to our entire collection of over 1,000,000 titles via our web site and mobile apps.

To claim your free month visit:
www.forgottenbooks.com/free903608

English
Français
Deutsche
Italiano
Español
Português

www.forgottenbooks.com

Mythology Photography **Fiction**
Fishing Christianity **Art** Cooking
Essays Buddhism Freemasonry
Medicine **Biology** Music **Ancient
Egypt** Evolution Carpentry Physics
Dance Geology **Mathematics** Fitness
Shakespeare **Folklore** Yoga Marketing
Confidence Immortality Biographies
Poetry **Psychology** Witchcraft
Electronics Chemistry History **Law**
Accounting **Philosophy** Anthropology
Alchemy Drama Quantum Mechanics
Atheism Sexual Health **Ancient History**
Entrepreneurship Languages Sport
Paleontology Needlework Islam
Metaphysics Investment Archaeology
Parenting Statistics Criminology
Motivational

A GENERAL

SYSTEM OF BOTANY

DESCRIPTIVE AND ANALYTICAL.

IN TWO PARTS.

PART I.—OUTLINES OF ORGANOGRAPHY, ANATOMY, AND PHYSIOLOGY.

PART II.—DESCRIPTIONS AND ILLUSTRATIONS OF THE ORDERS.

BY

Emm. LE MAOUT,

DOCTOR OF MEDICINE;
MEMBER OF THE SOCIÉTÉ PHILOMATHIQUE OF PARIS.

Jos. DECAISNE,

MEMBER OF THE INSTITUTE OF FRANCE;
PROFESSOR OF CULTIVATION, JARDIN DES PLANTES, PARIS.

WITH 5500 FIGURES BY L. STEINHEIL AND A. RIOCREUX.

TRANSLATED FROM THE ORIGINAL BY

MRS. HOOKER.

THE ORDERS ARRANGED AFTER THE METHOD FOLLOWED IN THE UNIVERSITIES AND SCHOOLS OF
GREAT BRITAIN, ITS COLONIES, AMERICA, AND INDIA; WITH ADDITIONS, AN APPENDIX
ON THE NATURAL METHOD, AND A SYNOPSIS OF THE ORDERS, BY

J. D. HOOKER, C.B.,

F.R.S. L.S. & G.S., M.D., D.C.L. Oxon., LL.D. Cantab.

DIRECTOR OF THE ROYAL BOTANICAL GARDENS, KEW;
CORRESPONDENT OF THE INSTITUTE OF FRANCE.

SECOND THOUSAND.

LONDON:

LONGMANS, GREEN, AND CO.

1876.

EDITOR'S PREFACE.

THIS English reproduction of LE MAOUT and DECAISNE's work differs from the original published in Paris in 1868, first and mainly in the Natural Orders of Flowering Plants being arranged more nearly in the sequence followed in England and its dependencies, in the United States, and over the greater part of the Continent :—a course necessary to adapt it to the use of schools, universities, and the keepers of herbariums, botanical museums and gardens, in all English-speaking countries. This sequence, which is that originally proposed by De Candolle, and adopted with modifications by himself and by most classifiers, is further, in the opinion of the Editor, on the whole, the best linear arrangement hitherto devised.

The sequence of the Orders followed in the original is that of the late accomplished Professor Adrien de Jussieu, son of Antoine-Laurent de Jussieu, the establisher of the Natural Orders of Plants upon the principles his uncle Bernard had devised. This sequence has been but partially adopted, even in Paris, where, although the lectures on the Natural Orders given at the Jardin des Plantes are conducted in accordance with it, the plants in the garden itself are arranged according to that of Professor Adolphe Brongniart (see p. 165).

To render this part of the work complete, and to facilitate its use, I have added in an Appendix—what is a great desideratum in the original —a Conspectus of the Orders arranged under groups (cohorts), according to their affinities, in so far as this is practicable in a linear series. These groups are analogous to the 'alliances' devised by Lindley for his

' Vegetable Kingdom,' though widely differing from them; they more nearly approach the 'groups' of Asa Gray's 'Introduction to Botany,' and are identical with the 'cohorts' of Mr. Bentham's and my 'Genera Plantarum' in so far as these have been published, namely, to the end of *Polypetalæ*. The remaining Dicotyledonous Orders are grouped approximately by Mr. Bentham and myself, and are subject to rectification as we advance with our analyses of the genera for that work; for it must be borne in mind that no Natural Order or higher group can be accurately limited till all the genera belonging to itself and its allied groups have been thoroughly investigated, compared, and contrasted. For the grouping of the Monocotyledons I am alone responsible.

The next considerable deviation from the original consists in the intro-duction of various omitted Orders, and of much additional matter under the others, especially the tribes, sub-tribes, etc., of the large Orders, and in the increased numbers of genera (the selection of which is necessarily to some extent arbitrary) which have been cited. This will render the English edition more useful to voyagers and travellers, and to dwellers in America, India and the Colonies, whose requirements in this respect have been especially regarded.

The twenty-four Orders omitted in the original, and supplied here, are chiefly small ones; but some, as XX., XXXIV., LVII., LXXIV., XCII., CLXVII., and CCXXI., are either of considerable extent, or of importance under other points of view. They are as follow :—

XX. CANELLACEÆ.	CLXXXIX. HERNANDIEÆ.
XXIV. VOCHYSIACEÆ.	CCIV. PENÆACEÆ.
XXXIV. DIPTEROCARPEÆ.	CCV. GEISSOLOMEÆ.
XXXV. CHLÆNACEÆ.	CCVI. LACISTEMACEÆ.
LVII. BURSERACEÆ.	CCXV. GRUBBIACEÆ.
LIX. CHAILLETIACEÆ.	CCXXI. PODOSTEMACEÆ.
LXV. STACKHOUSIEÆ.	
LXXII. SABIACEÆ.	*MONOCOTYLEDONS.*
LXXIV. CONNARACEÆ.	VII. APOSTASIACEÆ.
XCII. RHIZOPHOREÆ.	XVI. TRIURIDEÆ.
CII. SAMYDACEÆ.	XLII. ROXBURGHIACEÆ.
CX. FICOIDEÆ.	XLV. RAPATEEÆ.
CLXVII. CRESCENTIEÆ.	XLIX. MAYACEÆ.

The number of Orders adopted in this work greatly exceeds that which will be adopted by Mr. Bentham and myself in the 'Genera Plantarum,' or than is accepted by Professor Asa Gray, and most modern systematists: many of them are not in our opinion entitled to that rank, being rather to be regarded as tribes or aberrant genera of larger Orders. A multiplication of these is, however, in a work of this description far from a great evil: it enables the student to form a clearer idea of the essential characters of the more important Orders, from which the lesser are departures; and it affords the opportunity of illustrating more copiously many structural and physiological matters of high importance. It will be observed that the authors have been scrupulously careful in indicating the very slender pretensions that many of these lesser groups have to ordinal rank, and in pointing out their affinities.

In dealing with the Introduction to Botany, p. 1, the Translator has had much difficulty. In point of style, a literal translation of the original was inadmissible; its copiousness of expression and repetitions of adjective terms, however suited to French, are obstacles to English students, who associate clearness with a concise, rather than with a more diffuse method of exposition. The Translator has therefore condensed the matter of this part of the work—it is to be hoped, without loss of sense or substance; and the space gained has been devoted to those additions to the Systematic portion which are enumerated above.

It will be obvious to the English reader, that it has not been the aim of the Authors to give an exhaustive history of the Natural Orders: what they have given is a clear and precise structural and morphological account of each, with a sketch of its affinities, geographical distribution, and principal uses in medicine and the arts; and in this, I think, they have succeeded to a degree not attained in any previous work of the kind. On the extent and utility of the Illustrations there is no need to dwell; but it is only my duty to one of the Authors to state (which does not appear in the original) that their great value is due to the use made of my friend M. Decaisne's unique collection of analytical drawings, the fruits of his life-long botanical labours, and which for scientific accuracy and artistic excellence have never been surpassed.

Nearly thirty years have elapsed since I first had the privilege of inspecting those portfoliòs, the contents of which have, with a rare liberality, been ever since placed at my disposal when desired.

Few or no allusions are made under the Natural Orders to histological characters ; to the differentiation and development of the organs ; to the phenomena of fertilization; to the functions of nutrition, circulation, and respiration ; nor to the structural characters of Fossil Types, which, in the case of a few Orders (chiefly *Cycadeæ* and *Lycopodiaceæ*), are of great significance in reference to these and their allies. To have introduced all these subjects to any useful purpose was beyond the scope of this work. A companion volume devoted to them—that is, one completing the Life-history of the Natural Orders—is the great desideratum of Botanical Science.

It remains for the Editor and Translator to thank the Authors for their confidence, both in entrusting them with the task, and in liberally permitting the re-arrangement of the Orders according to the Editor's judgment of the requirements of those for whose use the English version is made.

Jos. D. Hooker.

Royal Gardens, Kew:
December 1872.

PREFACE.

———∘◦❁◦∘———

THE FIRST PART of this work, together with the woodcuts illustrating the Natural Orders, is a reprint of the 'Atlas Élémentaire de la Botanique,' edited by one of the Authors some years ago, and which has been favourably received by the scientific public. This, however, being devoted to European Orders, and confined to brief systematic descriptions only of these, could not illustrate the affinities of all the known types of the Vegetable Kingdom. To supply this deficiency, we have here added nearly all the exotic Orders, with detailed descriptions of their affinities and uses ; so as to give such a general view of the Vegetable Kingdom as may be advantageously consulted by students and professed botanists.

For the sequence of the Orders we have followed the classification of A. de Jussieu[1] in the valuable article on Taxonomy in the 'Dictionnaire Universel,' simply inverting the series, so as to commence with the most highly organized, and end with the Families of lowest organization, whose history is still obscure.

The reader will observe that we have treated the Monocotyledons and Cryptogams with greater fulness than the Dicotyledons : this is because the two first, and especially the Cryptogams, having hitherto been much less fully studied than the Dicotyledons, required much more careful illustration.

We have also thought it best to detach from the larger groups many monotypic Orders, so as to give them greater prominence ; following in this

---

[1] In this English edition the Editor has, with the approval of the authors, adopted that modification of the elder Jussieu's system known as De Candolle's, in order to suit the convenience of the Universities, Medical Schools, and other educational bodies of Great Britain, as well as of working botanists, herbarium keepers, &c. A sketch of A. de Jussieu's sequence of the Families will be found in the chapter devoted to Taxonomy (p. 167 ; see also p. 988).

descend, and they together constitute the *vegetable axis*. In its early stage this axis is simple, but by successive growths it usually gives off branches, which form *secondary axes*; each branch may thus be regarded as an independent axis.

The point of junction of the stem and root is the *neck* (*collum*, c). It is from this point, which may be thickened, shortened, or obscure, that the ascending fibres of the stem and the descending fibres of the root diverge.

The stem, which alone possesses the power of emitting lateral expansions, develops from its sides more or less flattened bodies, the *leaves* (F). The point at which the leaves issue from the stem is generally thickened, and is termed a *node* (*nodus*); the intervals between the nodes are termed *internodes* (*internodium*, *merithallus*). When the nodes develop leaves only, the stem remains perfectly *simple* or unbranched; but at each node a *bud* (*gemma*, B, B) may spring from the axil of the leaf; and this bud, which appears at first as a small protuberance, afterwards becomes a *branch* (*ramus*), which lengthens, develops leaves, and ramifies in its turn. The buds springing from the axils of the leaves on the primitive axis thus give origin to as many fresh axes, whence it results that the mother-plant is repeated by every bud which it produces. Hence it is more logical to say that a plant *multiplies*, than that it *divides* by branching; and a vegetable may thus be looked upon, not as an individual, but as a collective being, or an aggregation of individuals nourished in common, like the zoophytes of a coral.

The node does not always produce a leaf and bud; the bud may be absent or scarcely visible, or the leaf may be imperfectly developed: but the latter is rarely entirely suppressed; and when the bud is undeveloped, it is owing to the rigour of the climate or the short duration of the plant.

Leaves are not developed promiscuously on the stem; they may be given off singly, when they are *alternate* (*alterna*, fig. 2); or two may be placed *opposite* to each other (*opposita*, fig. 3); or they may be *whorled* around the stem (*verticillata*,

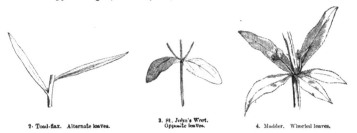

2. Toad-flax. Alternate leaves.   3. St. John's Wort.   4. Madder. Whorled leaves.
                                        Opposite leaves.

fig. 4). Stem-leaves are rarely whorled, but floral leaves are arranged in several superimposed *whorls* (*verticilli*).

Alternate leaves, though apparently scattered without order on the axis, are really arranged in a spiral (fig. 5); so that, in starting from any one leaf (1), we arrive, after one or more turns of the spiral, at another leaf (6), placed directly above the first; whence it results that, if the leaves completing the spiral (1, 2, 3, 4, 5) were all placed on a level with the first, they would form a whorl around the stem. This arrangement is more easily traced on young branches of trees than on herbaceous stems.

The fibro-vascular bundle connecting the green expansion of the leaf with the stem is the *petiole* (*petiolus*, fig. 6). It extends from the axis to the *blade* or *limb*

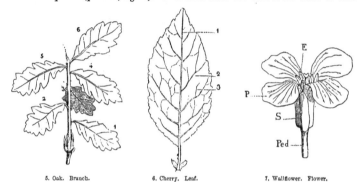

5. Oak. Branch. 6. Cherry. Leaf. 7. Wallflower. Flower.

(*limbus, lamina*), which is composed of parenchyma and fibro-vascular bundles, which latter form the *nerves* (*nervi*, 1, 2, 3). The middle nerve of the limb, which is continuous with the petiole, is the *median nerve* or *midrib* (*n. medius, costa media*). The bundles which rise from each side of the midrib are the *lateral nerves* (*n. laterales*); and these again give rise to *secondary* (2), *tertiary* (3), &c. nerves, according to their subdivision.

A leaf springing directly from the stem without a petiole is *sessile* (*f. sessile*, figs. 2–4), and that with a petiole is *petiolate* (*f. petiolatum*, figs. 5, 6). The leaf-blade is protected on both surfaces by a thin, colourless, and transparent skin (*epidermis*), which covers almost the entire plant, and will be described later.

The coloured leaves, arranged in whorls at the extremities of the ultimate branches of the axis, together form the *flower* (*flos*, fig. 7). The branch which immediately bears a flower, and forms the axis of its component whorls, is its *peduncle* or *pedicel* (*pedunculus, pedicellus*, fig. 7, Ped). Its more or less swollen extremity, upon which the whorls of the flower are grouped, is the *receptacle* (*receptaculum*, fig. 10, R).

In the most fully developed plants the flower is usually composed of four successive whorls (fig. 7), of which the internodes are suppressed. The outer or lower whorl is the *calyx* (*calyx*, figs. 7, 8, and 8), the leaves of which are *sepals* (*sepala*, fig. 8). The whorl within or above the calyx is the *corolla* (*corolla*, fig. 7, P), and its leaves are *petals* (*petala*, fig. 9). When a petal is not sessile, but has its blade (L) borne on a petiole (O), this petiole is called the *claw* (*unguis*).

The whorl within or above the corolla is the *andrœcium* (*andrœcium*, figs. 7, E, and 10), and its leaves are *stamens* (*stamina*, figs. 10, E, and 11). The petiole of the stamen is the *filament*

8. Wallflower. 9. Wallflower.
Calyx (mag.). Petal.

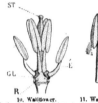

10. Wallflower. 11. Wallflower.
Andrœcium and pistil (mag.). Stamen (mag.).

(*filamentum*), and its blade is the *anther* (*anthera*). The dusty parenchyma contained in the anther is called *pollen* (*pollen*, P). This pollen leaves the anther at a certain period, and, falling on the central organ of the flower, assists in the formation of the seed.

There frequently occur on the receptacle (fig. 10, R) small bodies (GL) which secrete a sweet juice, named *nectariferous glands* or *nectaries* (*glandulæ nectariferæ, nectaria*).

The whorl within or above the andrœcium is the *pistil* (*pistillum*, fig. 12). This,

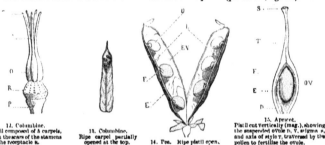

12. Columbine.
Pistil composed of 5 carpels, with the scars of the stamens on the receptacle R.

13. Columbine.
Ripe carpel partially opened at the top.

14. Pea. Ripe pistil open.

15. Apricot.
Pistil cut vertically (mag.), showing the suspended ovule D, V. stigma S, and axis of style T, traversed by the pollen to fertilise the ovule.

the central or last of the floral whorls, is composed of one or more leaves called *carpels* (*carpidia, carpella*, fig. 13), bearing on their edges small bodies called *ovules* (*ovula*), destined to reproduce the plant when fertilized by the pollen. The blade of the carpel, which encloses and protects the ovules, is the *ovary* (*ovarium*, fig. 12, O); its prolongation upwards into a longer or shorter neck is the *style* (*stylus*, T); and the *stigma* (*stigma*, S) is an organ of variable form, spongy and viscous when young, usually placed on the top of the style, and destined to receive the pollen, which adheres to its surface.

The substance of an ordinary leaf, however thin, consists of three parts: (1) an upper and (2) an under surface, enclosing (3) a network of fibres and parenchyma; and a slight inspection will show that a carpellary leaf is constructed on the same plan. Thus in the Pea, the pistil of which is composed of a single carpel, which splits into halves when ripe (fig. 14), the outer portion of the leaf (E) is a thin skin, easily torn away, named *epicarp* (*epicarpium*). The inner portion (EN) consists of a thicker and paler membrane than the first, named *endocarp* (*endocarpium*). The intermediate portion consists of a more or less succulent tissue (according to the proportions of fibre and parenchyma), named *mesocarp* (*mesocarpium*). In the solitary carpel which forms the pistil of a Cherry (fig. 16), Peach, or Apricot (fig. 15), the epicarp (F) is a thin skin, the mesocarp (figs. 16, ME, and 15, E) is very thick and succulent when ripe, and the very hard endocarp (figs. 16, N, and 15, D) forms the stone.

The fibro-vascular bundles (fig. 14, L) which are found on the edges of the blade of the carpellary leaf, and which both bear the ovules (O) and transmit nourishment to them, are called the *placentæ* (*placentæ, trophospermia*). Each placenta produces lateral branches or cords, called *funicles* (*funiculi*, F), which are sometimes very short, and through which the nourishing juices are conveyed to the seeds. When the funicles are absent (fig. 13), nourishment is transmitted directly to the seed from the placenta.

The *seed* or *plant-egg* (*semen*, fig. 17) is the ovule fertilized by the pollen. It is composed, (1) of a very small body, destined to reproduce the plant, the *embryo* (or *plantula*); (2) of an envelope or integument surrounding and protecting the embryo. This integument (in the ovule) either originates from the top of the funicle (fig. 17, F), or directly from the placenta (fig. 13). It usually consists of two layers or coats, an external *testa* (fig. 17, I), and an internal *endopleura* (E). The point of union of the seed and funicle, and at which its nourishment enters, is called the *hilum* or *umbilicus* (fig. 18, H), and is a part of the testa. The *chalaza* (fig. 17, H) marks the spot where the juices penetrate the internal coat and reach the

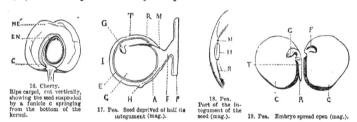

16. Cherry. Ripe carpel, cut vertically, showing the seed suspended by a funicle c springing from the bottom of the kernel.

17. Pea. Seed deprived of half its integument (mag.).

18. Pea. Part of the integument of the seed (mag.).

19. Pea. Embryo spread open (mag.).

embryo, and is usually indicated on the outside by a projection or thickening or discoloration. When the hilum and chalaza are superimposed, the juices reach the embryo directly; when they are at opposite ends, they are connected by a small cord, *raphe* (figs. 17, A, and 18, R), which runs between the two coats. The small opening through which the ovule is acted on by the pollen is the *micropyle* (figs. 17, and 18, M).

' The embryo (figs. 17, 19) is a complete plant in miniature, composed of a stem, *caulicle* (*tigellus, cauliculus*, T), a root, *radicle* (*radicula*, R), one or two leaves, *cotyledons* (*cotyledones*, C), and a bud, *plumule* (*gemmula, plumula*, G), usually occupying a small pit (F) sunk in the thickness of the cotyledons. The young plant, after having been nourished by the juices transmitted through the funicle, detaches itself from the latter with its integuments; and when placed under favourable circumstances, it sheds or leaves its coats, and becomes developed into a plant similar to its parent.

The *caulicle* (T) is a small cylindric or conical body, bearing the first leaves of the plant (fig. 19, c), which ascends to form the stem. The radicle (R), or organ destined to develop the roots, is at first merely a transparent point terminating the free end of the caulicle, and tending downwards; it usually corresponds in the seed to the position of the micropyle (figs. 17, 18). The cotyledons (figs. 17, and 19, c), which are the first leaves of the young plant, spring laterally from the caulicle, and protect the plumule, or first shoot of the future plant; they are usually thick and succulent, and nourish the young plant until it is able to support itself. Within the integuments of the seed there always exists, at an early period, a peculiar form of cellular tissue, the study of which is important, and to which we shall recur; it is sometimes rapidly absorbed by the embryo, but at others it is retained in the seed until germination, in which case it is called albumen, and supplies the young plant with its first food.

Considering the embryo as the plant in its simplest form, let us follow the growth and lateral development of its primitive axis. The two first leaves (*cotyledons*) are attached to the small stem (*caulicle*), as may be seen in the Pea (fig. 19), or, better, in a germinating Bean (fig. 20, c, c). The *radicle*, which terminates the free end of the caulicle (fig. 20, T), sends out many descending branches, and forms the root (R). Sometimes the cotyledon is solitary, as in the Maize (fig. 21, c), when the rootlets usually spring from various points of the caulicle (t), and branch very little. At the point of union of the cotyledons or cotyledon with the caulicle is the plumule (fig. 20, G, G, and fig. 21, g). Each cotyledon and each leaf of the plumule is produced from a node, but the internodes are scarcely visible.

20. Germination of Kidney-bean.

21. Germination of Maize.

Soon after germination, as the plant grows and the axis lengthens, the nodes, and consequently the leaves, become separated. Near the flower the internodes of the axis shorten, the leaves usually become smaller and changed in form and colour; finally, at the termination of the axis, the leaves (*flower*), instead of forming a spiral or being placed in pairs, are arranged in superimposed whorls of different structure, the leaves in each whorl usually alternating with those of the next within or above it; which results in the blades of the different leaves composing the flower being separated as far as is compatible with being crowded in a very small space.

The leaves of the three first floral whorls (*sepals, petals, stamens*) have no buds in their axils or on their edges; those of the pistil alone (*carpels*) produce and protect buds; each edge of the carpel (*placenta*) giving origin to cords, which convey nourishment to one or more seed-buds (*ovules*), which eventually become seeds, consisting of an embryo with its integuments, which is destined to produce a plant similar to its parent. Though so dissimilar, the ovule or seed-bud presents a remarkable analogy to an ordinary bud : both spring from a node, and are protected by a leaf; both are destined to reproduce the plant. They only differ in the conditions of their existence : the *seed-bud* needing for its development the fertilizing action of the pollen ; the *branch-bud* needing only the nourishment contributed by the node. To this must be added, that the *branch-bud* multiplies the plant without separating from it; whilst the *seed-bud* is destined to leave its parent, and reproduce at a distance the plant which gave it birth.

In some cases the *branch-bud* may be separated from its parent, and made to germinate, which is due to the power which the stem possesses of emitting from its surface supplementary or *adventitious* roots (*r. adventitiæ*). Sometimes a young branch, with its buds, may be detached from the stem, and planted, when the buried portion speedily sends forth roots, and the new individual becomes an independent organism : this is called propagating by *slips* or *cuttings* (*talea*). Or the branch, still attached to the trunk, may be surrounded with damp mould, into which it emits roots, which soon become sufficiently strong to nourish the branch, and to permit of its removal from its parent stem : this is termed propagation by *layers* (*malleoli*). Or again, the branch, with its buds, may be separated from its parent, and so attached to another plant, whose sap resembles its own, as to bring into contact the parts in which the sap circulates ; the branch then grows as if on its parent : this is called propagation by grafts, and the plant on which the branch is grafted is called the *stock*. Lastly, the branch-bud may separate spontaneously from the parent like a seed-bud, and falling to the ground, may strike root and become a separate individual, as in the Tiger-lily (fig. 22, B) : such branch-buds are called *bulbils* (*bulbilli*).

The power of producing (naturally or artificially) buds and adventitious roots is not confined to the stem, the branches of many plants having also this power. The physiologist Duhamel, having planted a tree with its branches in the ground, saw the roots become covered with buds, while the buried branches produced roots. In some cases the divided root will reproduce the plant, as in

22. Bulbiferous Lily. Portion of stem.

23. *Bryophyllum.* Leaf giving off embryos at each crenature.

the Japan *Quince*, the *Osage Orange* (*Maclura*), and especially the *Paulownia*, the roots of which may be cut into small sections, each of which, if planted, will produce a perfect tree.

In some plants the leaf itself possesses this reproductive power, as in the *Watercress*, *Cardamine pratensis*, and *Malaxis*, &c., amongst native plants; and amongst exotics, *Bryophyllum calycinum* (fig. 23), a succulent tropical plant, whose leaf produces buds furnished with root, stem, and leaves, at the extremities of its lateral nerves; these buds, which spontaneously fall off, and root in the earth, may be likened to embryos that do not need to be fertilized before developing; and the leaf of *Bryophyllum* may be regarded as an open carpel, on which the seeds have been developed by nutritive action alone. This fecundity of *Bryophyllum* completes the analogy between the true bud and the fertilized embryo.

Amongst the examples of reproduction by leaves, the *Begonias* hold the first rank; for if a Begonia leaf be placed on damp soil, and incisions made across its nerves, roots and buds will spring from every incision, and as many fresh plants will be obtained as the leaf has received wounds. The same vitality is observable in some woody plants: thus, if a fresh-cut Orange leaf be placed under suitable conditions of heat and moisture, a small swelling will be formed on the broken petiole, from which will shortly spring roots and shoots, that will eventually form a tree, capable of growing, flowering, and fruiting, like an Orange-tree raised from seed.

In this brief summary we have only spoken of the structure of the higher plants, whose organs of fructification being obvious, and their seeds provided with mono- or di-cotyledonous embryos, are called *cotyledonous or phænogamous* (*p. cotyledoneæ* v. *phænerogamæ*). Other plants, which have no obvious stamens or pistils, and seeds without embryos, are called *cryptogamous* or *Acotyledonous* (*p. cryptogameæ* v. *acotyledoneæ*), and are of much simpler organization.

ORGANOGRAPHY AND GLOSSOLOGY.

THE ROOT.

The *root* (*radix*) is that part of the plant which tends towards the centre of the earth; it is not coloured green, even when exposed to light, and rarely produces leaves or shoots. It serves to fix the plant in the earth, and to draw thence the nourishment necessary to its growth.

The root is absent in certain plants, which, from growing upon and drawing their nourishment from others, are called *parasites* (*p. parasiticæ*). Such is the Mistleto, which fixes itself beneath the bark of certain trees by the dilated base of its stem.

The root may be *simple*, or irregularly branched. Its axis or branches terminate in delicate fibrils, which together are termed the *root-fibres* (*fibrillæ*); the

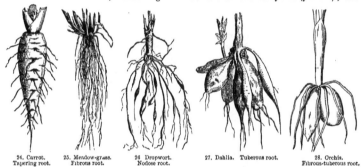

24. Carrot.   25. Meadow-grass.   26 Dropwort.   27. Dahlia. Tuberous root.   28. Orchis.
Tapering root.   Fibrous root.   Nodose root.   Fibrous-tuberous root.

tips of these fibres, being soft, loose, and cellular, are named *spongioles* (*spongiolæ*). The individual fibrils die annually, like leaves, and fresh ones spring from the youngest parts of the root.

Roots with a single, descending, vertical stock, are called *tap-roots* (*r. perpendicularis*); their main trunk or tap may branch (*Stock*, fig. 1), or remain nearly simple (*Carrot*, fig. 24). Sometimes the original, usually simple, tap-root perishes soon after germination, and is replaced by a bundle of fibrils, which spring from the neck or crown of the root. A root is *fibrous* (*r. fibrosa*) when its fibrils form a bundle of fine, long, scarcely-branched threads (*Meadow-grass*, fig. 25); *nodose* (*r. nodosa*), when the fibres are swollen at intervals (*Dropwort*, fig. 26)*; *tuberous* (*r. tuberosa*), when the fibres are much swollen in the middle, thus becoming stores of nourishment destined to sustain the plant (*Dahlia*, fig. 27). The Orchis root (fig. 28) is both fibrous and tuberous, the ovoid or palmate tubers being reservoirs of nutritious matters, and the cylindric fibres being organs of absorption. The fibres of young Crocus roots are similarly swollen.

We have said that the stem has the power of emitting adventitious roots; these are sometimes artificially induced (as on *slips* or *layers*), sometimes spontaneously developed on the nodes of the stem: when these emerge at a considerable height, and descend to enter the earth, they are termed *aërial roots* (many tropical climbers and epiphytal orchids); when they spring from the lower branches of creeping plants, they are called *accessory roots* (*Strawberry*, *Ground-ivy*).

THE STEM.

The *stem* (*caulis*) is that portion of the vegetable axis which grows in an opposite direction to the root. It branches by means of shoots, which originate in the axils of the leaves. The stem exists in all *phænogams*, but is sometimes scarcely developed, when the leaves and flowering branches appear to spring from the root, and the plant is termed *stemless* (*p. acaulis*), and its leaves *radical* (*f. radicalia*, Hyacinth, Dandelion, fig. 29). The stem is *perennial* (*c. perennis*) when it lives many years (*Strawberry*); *annual* (*c. annuus*), when it only lives one (*Wheat*); *biennial* (*c. biennis*), when it lives two years (*Carrot*); a biennial stem usually produces leaves only the first year, and in the

29. Dandelion, with root.

second it flowers, fruits, and dies. The stem is *herbaceous* (*c. herbaceus*) when soft and easily broken; such are annual, biennial, and many perennial stems; it is *woody* (*c. lignosus*, *fruticosus*) when it forms a solid, more or less durable wood (*Oak*); it is *suffruticose* (*c. suffruticosus*) when the lower part is hard, and remains above ground for many years, while the branches and twigs die, and are annually renewed (*Rue*, *Thyme*, *Sage*, *Bitter-sweet*). The woody stem of trees is called a *trunk*.

The stem is *indefinite* (*c. indeterminatus*) when the flowers are borne only on the secondary axes (those springing from the axils of the leaves), thus appearing to elongate indefinitely (*Periwinkle, Pimpernel,* fig. 30). .

30. Pimpernel. Indefinite stem. 31. Columbine. Definite stem.

The stem is *definite* (*c. determinatus*) when each axis terminates in a flower, and cannot therefore be indefinitely prolonged (*Campanula,* fig. 159 ; *Columbine,* fig. 31) ; it is aërial when it grows entirely above ground (*Stock,* fig. 1).

. . The *rhizome* or *rootstock* (*rhizoma*) is a stem which extends obliquely or horizontally below or on the surface of the ground, the advancing portion emitting fibrous roots, leaves, and shoots, the posterior gradually dying. The rootstock is *indefinite* (*rh. indeterminatum*) when it grows by means of a terminal shoot, which lengthens indefinitely, and never itself flowers, but gives off lateral flowering shoots. Thus in the *Primrose* (fig. 32) the extremity of the rootstock bears a bundle of leaves, in the centre of which is the shoot by which it is indefinitely prolonged, whilst the flowering shoot is developed in the axil of one of the leaves (B). After flowering, the aërial portion of the leaves dies, but the subterranean portion survives, and from its axil spring accessory roots.

The rootstock is *definite* (*rh. determinatum*) when, after producing laterally one or more creeping branches, it rises above the earth, and terminates its existence by a flowering branch. In the Iris (fig. 33) and *Arum* (figs. 34, 35) the leaf-bases

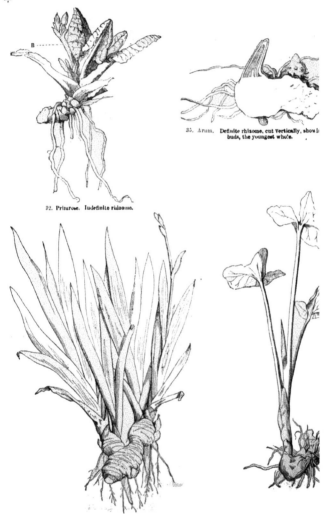

35. Arum. Definite rhizome, cut vertically, showin buds, the youngest whole.

32. Primrose. Indefinite rhizome.

persist as dry scales on the fleshy mass of the rootstock, after the decay of the aërial portions.

In *Carex* (fig. 36) each shoot remains under ground during the first year of its existence; it rises in the spring of the second year, makes a tuft of leaves, and emits from the axil of the lowest of these a shoot, which lengthens during its first year, as its predecessor did. In the autumn the two-year-old shoot loses its leaves, but the axis, sheltered by their persistent bases, lengthens, and sends up flowers and leaves in the spring of the third year, when it dies. During the following autumn the

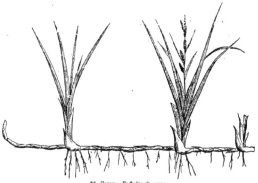

36. Carex. Definite rhizome.

flowering stem fruits and dies, together with the old shoot that produced it, but the second year's shoot, which has now produced a tuft of leaves, will in its turn flower in the following year. A shoot of Carex thus requires three years for its full development.

The stem is *stoloniferous* (*c. stolonifer*) when creeping shoots (*flagellum*) spring from the axils of its lower leaves, develop terminal tufts of leaves, then rise, and produce root-fibres below the tufts (*Creeping Buttercup; Strawberry,* fig. 37). The *rosette* (*propagulum*) is the tuft of

37. Strawberry. Creeping stem.

leaves produced on the lateral shoots of succulent plants (*Houseleek*).

The stem may present both *stolons* and *rootstock* when some of the lower branches are underground, and others aërial and creeping (*Clubmoss*).

The *bulb* (*bulbus*, Lily, fig. 38) is a subterranean swollen stock, consisting, firstly, of a more or less convex fleshy disk (*lecus*, L), which below gives rise to the roots; secondly, of fleshy, closely-appressed coats or scales (E) borne on the disk; thirdly of a more or less central shoot (T), equally borne by the disk, protected by the coats, and formed of rudimentary leaves and flowers; fourthly, of one or more lateral shoots, called *cloves* (*bulbuli*), destined to reproduce the plant.

A bulb is *coated* (*b. tunicatus*) when the outer leaves overlap each other so as completely to sheathe the base of the stem (*Narcissus*, fig. 39 ; *Onion*, fig. 40) it is *scaly* (*b. squamosus*) when the leaves are narrow, almost flat, and imbricated in many

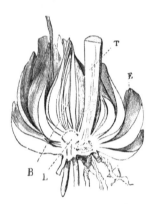

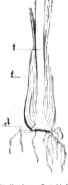

38. Lily. Scaly bulb, cut vertically.

39. Narcissus. Coated bulb.
l, disk ; *t*, stem ; *f*, leaves.

41. Colchicum. Solid bulb.

rows (*Lily*, fig. 38) ; solid (*b. solidus*) when the leaf-bases are very close and confluent with the disk, so that the latter appears to form the entire stock (*Colchicum*, fig. 41).

In the *Crocus* (fig. 42), the underground stock is formed of two or three solid bulbs, superimposed like the beads of a chaplet. The primitive bulb (1), which terminates in a flower, pushes out a lateral shoot, which perpetuates the plant. After flowering, it swells considerably, to nourish the shoot which is to succeed it ; this latter flowers in its turn the following year, and emits a shoot like its predecessor ; to nourish this it

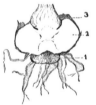

40. Onion. Coated bulb.

42. Crocus. Superimposed bulbs.

swells and forms a bulb (2) above the original one, which then gradually decays. At the flowering of the third shoot (3) adventitious roots grow from the base of the second bulb, which soon withers and dries like the first. At the side of the middle bulb a lateral bulbil often springs, which separates from the parent, and becomes a fresh plant.

In comparing rootstocks with bulbs, it is easy to perceive that they differ only by the greater or less length of the disk, and the more or less fleshy texture of their underground leaves. The rootstock may thus be regarded as a bulb with a horizontally lengthened disk, and the bulb as a short rootstock with fleshy leaves.

The superimposed rootstock of the Crocus presents a transition from the bulb to the rootstock proper, for it may equally be regarded as a vertical rootstock or as a series of superimposed bulbs.

The roots of *Orchis*, which are both fibrous and tuberous, are classed with true bulbs, differing from ordinary bulbs only in the swelling of some of the root-fibres. The two tubers are ovoid (fig. 43) or palmate (fig. 44), and are unequal; one (T 1) is dark-coloured, wrinkled, flabby, and empty, and gives off the flowering stem; the

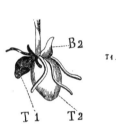

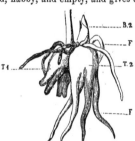

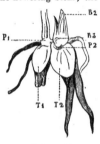

43. Orchis. Tuberous root.   44. Orchis maculata. Palmate tuberous root.   45. Orchis maculata. Palma.e tuberous root, cut vertically.

other (T 2) is larger, whiter, and more succulent, often ending in well-developed fibres (F), and bears a shoot (B 2), from the base of which grow fibrous roots. The two tubers (fig. 45) are united above by a very short neck (P 1). This neck connects the old tuber (T 1) with the new one (T 2), and from it the latter descends, and a leafy shoot (B 2) ascends, which in the following year will bear a flowering stem; between this large shoot and the old stem a vertical cut reveals a third tiny shoot (B 3), developed from the young tuber, and destined to succeed it in the third year. There are thus three generations in the rootstock of an Orchis, each of which requires two years for its perfect development, and dies at the end of the third, after having flowered; and the same may be seen in *Carex, Crocus*, and ordinary bulbs.

The term *tubers (tubera, tubercula)* has been given to the dilated extremities of underground roots,

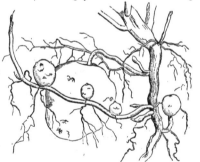

46. Potato. Subterranean branches bearing tubercles.

usually containing starch. These swellings bear rudimentary leaves, in the axils of which are *eyes* or buds, which develop into stems (*Jerusalem Artichoke; Potato,* fig. 46). The conversion of shoots into tubers can be encouraged by earthing up the lower portion of the stem; if the covering is slight, the tuber swells but little; if the light can penetrate to the stem, the tuber becomes green, and produces rosettes of leaves.

Props (fulcra) are a kind of aërial roots which spring from the axils of leaves, or from various points of the stem in certain climbing plants (*Ivy*, fig. 47), which are attached by them to walls or trees; these organs are non-absorbent, but under suitable conditions they behave like ordinary roots, as is seen with ivy cultivated for edgings.

Suckers (haustoria) are small warts upon certain parasitic stems (*Cuscuta*, fig. 48), whence issue true supplementary roots, which attach themselves to the neighbouring plants, and draw nourishment from their juices.

The stem is *cylindric* or *terete* (*c. cylindricus, teres*), when a transverse cut presents a circular outline (*Cabbage*);—*compressed* (*c. compressus*), when an elliptic

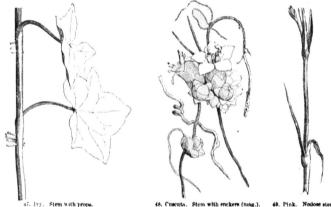

47. Ivy. Stem with props.   48. Cuscuta. Stem with suckers (nag.).   49. Pink. Nodose stem.

one, as if squeezed from opposite sides (*St. John's Wort, Tutsan*);—*triangular* or *trigonous* (*c. triangularis, trigonus*), when a cut shows three sides (*Carex*);—*square* (*c. quadrangularis, tetragonus*), when it shows four right angles (*Lamium*);—*pentagonal* (*c. quinquangularis, pentagonus*), when it shows five faces and five angles (*Bramble*).

The stem is *glabrous* (*c. glaber*), when there are no hairs on it (*Horse-tail*);—*smooth* (*lævis*), when, being glabrous, it presents no roughness, and its surface is quite even (*Tulip*);—*scabrous* (*c. scaber, asper*), when its surface presents little inequalities (*Carrot*);—*striate* (*c. striatus*), when it is marked with small raised longitudinal lines or *striæ* (*Sorrel*);—*winged* (*c. alatus*), when furnished with foliaceous expansions (*Comfrey*, fig. 66);—*nodose* (*c. nodosus*), when its nodes are tumid (*Pink*, fig. 49);—*pilose* (*c. pilosus*), when it is furnished with long scattered hairs (*Herb-Robert*);—

pubescent (*c. pubescens*), when it is covered with more or less appressed short hairs (*Henbane*) ;—*woolly* (*c. lanatus*), when the hairs are long, close, appressed, and curly (*Thistle*) ;—*tomentose* (*c. tomentosus*), when the hairs are short, soft, and matted (*Mullein*) ;—*villous* (*c. villosus*), when the hairs are long, soft, and close-set (*Forget-me-not*) ;—*hirsute* (*c. hirsutus*), when it bears straight, stiff hairs (*Borage*) ;—*hispid* (*c. hispidus*), when the hairs are straight, stiff and very long (*Poppy*). The ana-tomical structure of hairs will be described hereafter.

The stem is *prickly* (*c. aculeatus*), when the hairs which clothe it thicken, harden, and end in a sharp point; the *prickles* (*aculei*) always belong to the epidermis, and come away with it (*Rose*, fig. 50) ;—it is *spinous* (*c. spinosus*), when the woody tissue of the stem is elongated into a hard point. *Spines* (*spinæ*) are usually partially developed or arrested branches (*Blackthorn*, fig. 51), which, under favourable circumstances, produce leaves and shoots.

The stem is *erect* (*c. erectus*), when vertical (*Stock*, fig. 1) ;—*procumbent* or *prostrate* (*c. procumbens, prostratus*), when, too weak to support itself, it trails

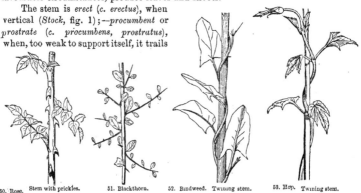

50. Rose. Stem with prickles.   51. Blackthorn.   52. Bindweed. Twining stem.   53. Hop. Twining stem.

along the ground (*Knot-grass*) ;—*spreading* (*c. patulus*), when many branches start from the neck, and spread on all sides horizontally (*Pimpernel*) ; —*ascending* (*c. ascendens*), when, after being horizontal or oblique at its commencement, its tip becomes upright (*Speedwell*) ;—*creeping* (*c. repens*), when a prostrate stem gives off adventitious roots from the nodes (*Strawberry*, fig. 37) ;—*scandent* (*c. scandens*), when it raises itself by aid of neighbouring bodies, and attaches itself to them, either by *props* (*Ivy*, fig. 47), *suckers* (*Cuscuta*, fig. 48), or *tendrils* (*Vine*, fig. 130 ; *Melon*, fig. 61) ;—the climbing stem is termed *twining* (*c. volubilis*), when it coils spirally round other bodies, rising either from left to right (*c. dextrorsum volubilis*, *Bindweed*, fig. 52), or from right to left (*c. sinistrorsum volubilis*, *Hop*, fig. 53) of the spectator placed opposite its convexity.

The direction of the branches depends on that of the leaves from the axils of which they spring; and they are *alternate* (*r. alterni*, *Rose*), *opposite* (*r. oppositi*, *Valerian*), or *whorled* (*r. verticillati*, *Pine*).

c

The stem bearing opposite branches is *dichotomous* (*c. dichotomus*, Lamb's Lettuce) ; and *trichotomous* (*c. trichotomus*, *Oleander*), when it continually forks or **trifurcates to** the extremities of its branches.

THE LEAVES.

Leaves (*folia*) are usually flat, green, horizontal expansions, arising from the nodes, and are the result of the spreading out of a bundle of fibres, the interstices between which are filled with parenchyma. The point of the stem constituting the base of the leaf, and of which the latter is a continuation, forms a small swelling (*pulvinus*, fig. 54, c), which, when the leaf has fallen, is clearly indicated by a scar (F).

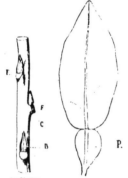

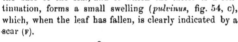

54. Glycine
Branch showing the buds
after the leaves have fallen.

55. Orange.
Leaf with a winged petiole.

57. Ranunculus.
Leaf with amplexicaul petiole.

56. Acacia heterophylla.
Phyllode.

The leaves and roots are the principal organs of nutrition, absorbing from the atmosphere gases and liquids suited for the nutrition of the vegetable : they also act as respirators, and as exhalers of useless matters ; and it is in their tissues that the sap, absorbed by the root, and conducted upwards by the stem, parts with its surplus fluids, and acquires all its nutritious properties.

Of all plant-organs, the leaves are those which present the greatest variety, and which supply most specific characters.

When the vascular bundle which enters the leaf is prolonged for a certain length before branching to form the skeleton of the *blade* (*limbus*), it takes the name of *petiole* (*petiolus*), and the leaf is called *petiolate* (*f. petiolatum*, Cherry, **fig. 6**) ;— when it expands immediately after leaving the node, the leaf is reduced **to its blade**, and is called *sessile* (*f. sessile*, St. John's Wort). When the blade merely **narrows so** as to form an obscure petiole, it is called *sub-petiolate* (*f. sub-petiolatum*).

The petiole may be *cylindric* (*p. cylindricus*) ; longitudinally grooved or *channelled* (*p. canaliculatus*); flattened horizontally, or *depressed* (*p. depressus*);—flattened laterally or *compressed* (*p. compressus*); in this case it is usually flexible, and the pendulous blade trembles with every breath of wind (*Aspen*).

The petiole is usually of tolerably uniform diameter throughout its length (*p. continuus, Ivy*, fig. 47); but it may be much dilated in the middle, and thus resemble a blade separated from the true blade by a constriction, when it is *winged* (*p. alatus, Orange*, fig. 55, P; *Acacia heterophylla*, fig. 56). Lastly, a dilated petiole

58. Wheat. Sheathing leaf. 59. Clematis. Twining petiole.

may replace the true blade, when it is called a *phyllode* (*phyllodium*), as in most Australian *Acacias*. When the enlarged base of the petiole, and the node from which it issues, occupy a large portion of the circumference of the stem, the petiole is called *amplexicaul* (*p. amplexicaulis, Ranunculus*, fig. 57); if the entire petiole is enlarged, and sheathes the stem, the leaf is called *sheathing* (*vaginans, Carex, Wheat*, fig. 58).

The direction of the petiole is usually straight, but in some plants it twines round neighbouring objects (*Clematis*, fig. 59).

Stipules.—A leaf is *stipulate* (*f. stipulatum*), when provided at its base with appendages more or less analogous to leaves, named *stipules* (*stipulæ, Heartsease*, fig. 60). These may be *persistent* (*s. persistentes*), when they persist as long as the leaf which they accompany (*Heartsease*, fig. 60); or *caducous* (*s. caducæ*), when they fall before the leaf, or as soon as the shoot lengthens (*Willow, Oak*).

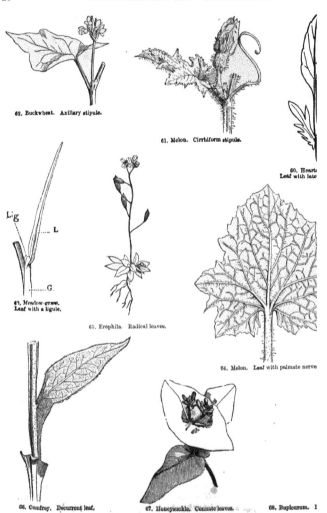

62. Buckwheat. Axillary stipule.

61. Melon. Cirrhiform stipule.

60. Heart.
Leaf with late

63. Meadow-grass.
Leaf with a ligule.

65. Erophila. Radical leaves.

64. Melon. Leaf with palmate nerve

66. Comfrey. Decurrent leaf.   67. Honeysuckle. Connate leaves.   68. Buplourum. I

Stipules are *foliaceous* (*s. foliaceæ*), when of the colour and texture of leaves (*Heartsease*, fig. 60) ;—*scale-like* (*s. squamiformes*), when thin like scales ;—*membranous* (*s. membranaceæ*), when thin, flexible, and almost transparent ;—*scarious* (*s. scariosæ*), when dry and coriaceous (*Beech, Willow, Hornbeam*) ;—*spinous* (*s. spinosæ*), when contracted and hardened into spines (*Robinia*, fig. 114) ; *cirrhose* (*s. cirrhiformes*), when they lengthen into twining tendrils (*Melon*, fig. 61). (We retain the name of stipules for the tendrils of the Melon and other *Cucurbitaceæ*, in deference to the glossology adopted by botanists; but we shall return to this subject when discussing tendrils.

Stipules are *lateral* (*s. laterales*), when inserted left and right of the leaf (*Heartsease*, fig. 60 ; *Robinia*, fig. 114) ;—*axillary* (*s. axillares*), when in the axil of the leaf; they are then usually consolidated into one. Such axillary stipules may cover only a part of the circumference of the stem (*Drosera*), or may completely surround it (*Buckwheat*, fig. 62), in which latter case it bears the name of *ochrea*.

The *ligule* of grasses (*ligula*, *Meadow-grass*, fig. 63) is simply an axillary stipule (Lig.), situated at the separation of the blade (L) from the sheathing petiole (G) ; it may be *entire, emarginate, laciniate, pilose*, &c.

Of the whorled leaves of *Madder* (fig. 4) and other *Rubiaceæ*, the two opposite ones are alone considered as true leaves, and bear each a bud in its axil ; the others are regarded as stipules, sometimes multiplied, when there are more than four, or confluent, when fewer than four.

The **Nerves** of the leaf are said to be *parallel* (*n. paralleli*), when they run free and parallel to the edge of the leaf and to each other (*Iris*, figs. 33, 79) ;—*branching* or *anastomosing* (*n. ramosi, anastomosantes*), when they subdivide and join each other (*Cherry*, fig. 6).

Branching nerves are *pinnate* (*n. pinnati*), and the leaves *penni-nerved* (*f. penni-nervia*), when lateral nerves, like the plumes of a feather, spring from the midrib (*Cherry*, fig. 6) ;—*palmate* (*n. palmati, palmatinervia*), when several primary nerves diverge from the base of the blade like the fingers of a hand (*Melon*, fig. 64). The primary nerves only are palmate ; the secondary, tertiary, &c., are always pinnately arranged.

As regards *position*, leaves are *radical* (*f. radicalia*), when they spring from near the neck, and hence appear to rise from the root (*Dandelion*, fig. 29 ; *Plantain, Erophila*, fig. 65) ;—*cauline* (*f. caulina*), when they spring from the stem and branches (*Rose*, fig. 50). Leaves are *clasping* or *amplexicaul* (*f. amplexicaulia*), when the base of their petiole or blade surrounds the stem (*Buttercup, Henbane*) ;—*decurrent* (*f. decurrentia*), when their blade is continued down the stem, forming a sort of foliaceous *wing* ;

69. Yew. Distichous leaves.

70. Fascicled leaves, Weymouth Pine.

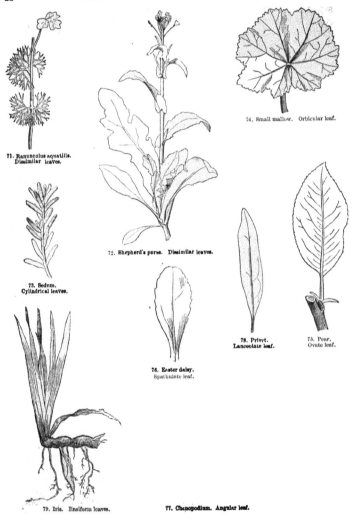

71. Ranunculus aquatilis.
Dissimilar leaves.

74. Small mallow. Orbicular leaf.

72. Shepherd's purse. Dissimilar leaves.

73. Sedum.
Cylindrical leaves.

78. Privet.
Lanceolate leaf.

75. Pear.
Ovate leaf.

76. Easter daisy.
Spathulate leaf.

79. Iris. Ensiform leaves.

77. Chenopodium. Angular leaf.

the stem is then *winged* (*caulis alatus*, Comfrey, fig. 66) ;—*confluent* or *connate* (*f. connata*), when the bases of two opposite leaves join around the stem (*Honeysuckle*, fig. 67, *Chlora*) ;—when the base of a single leaf spreads completely round the stem, the stem and leaves are *perfoliate* (c. *f. perfoliatus*, *Bupleurum*, fig. 68). Leaves are *alternate* (*f. alterna*; *Stock*, fig. 1; *Toad-flax*, fig. 2; *Oak*, fig. 5) ;—*opposite* (*f. opposita*, *St. John's Wort*, fig. 3) ;—*whorled* (*f. verticillata*, *Oleander*, fig. 82; *Madder*, fig. 4) ;—*distichous* (*f. disticha*), when they spring from alternate nodes placed on two lines to right and left (*Yew*, fig. 69) ;—*fascicled* (*f. fasciculata*), when crowded into a bundle on very short branches (*Weymouth Pine*, fig. 70). In true Pines this bundle is persistent; in larches the leaves become solitary and scattered, in consequence of the elongation of the axis. *Imbricated* leaves (*f. imbricata*) overlap like roof-tiles (*Houseleek*, *Cypress*, *Thuja*).

Colour of Leaves.—Leaves are *green* when of the usual colour;—*glaucous* (*f. glauca*) when of a whitish dusty green or blue (*Poppy*, *Cabbage*) ;—*spotted* (*f. maculata*) when they have spots of a different colour from the ground (*Arum*) ;—*variegated* (*f. variegata*) when they are of many colours arranged without order (variegated *Holly*, tricoloured *Amaranth*) ;—*hoary* (*f. incana*), when they owe their colour to short and close hairs (*Ten-week-stock*).

Forms of Leaves.—Without being precisely alike, yet the leaves of any one plant are usually very similar; but in some species they are obviously dissimilar (*Paper Mulberry*, *Calthrop*, *Water Crowfoot*, fig. 71 ; *Shepherd's purse*, fig. 72) ; the plant is then said to be *heterophyllous* (*pl. heterophylla*).

Leaves are *plane* (*f. plana*), when their blade is much flattened, as is usually the case (*Lime*, fig. 86) ;—*cylindric* or *terete* (*f. teretia*), when the blade is rounded throughout its length (*Sedum*, fig. 73) ;—*orbicular* (*f. orbiculata*), when the circumference of the blade is more or less circular (small *Mallow*, fig. 74) ;—*ovate* (*f. ovata*), when the blade resembles the longitudinal section of an egg, with the larger end at the base (*Pear*, fig. 75) ;—*obovate* (*f. obovata*), when ovate, with the smaller end at the base (*Meadow-sweet*, *St. John's Wort*) ;—*oblong* (*f. oblonga*), when the width is about a third of the length (small *Centaury*) ;—*elliptic* (*f. elliptica*), when the two ends of the blade are rounded and equal, like an ellipse (*St. John's Wort*, fig. 3) ;—*spathulate* (*f. spathulata*), when the blade is narrow at the base, and large and rounded at the end, like a spatula (*Easter Daisy*, fig. 76) ;—*angular* (*f. angulata*), when the circumference of the leaf presents three or more angles;—*deltoid* (*f. deltoidea*), if it presents three nearly equal angles, like a delta, Δ (*Chenopodium*, fig. 77).

Leaves are *lanceolate* (*f. lanceolata*), when the blade is largest in the middle, and diminishes insensibly towards the extremities (*Privet*, fig. 78) ;—*linear* (*f. linearia*), when the sides of the blade are nearly parallel, and the space between them narrow (*Toad-flax*, fig. 2);—*ensiform* (*f. ensiformia*), when of the shape of a sword; in this case, the two surfaces are in apposition and consolidated in the upper part (*Iris*, fig. 79);—*subulate* (*f. subulata*), when the cylindrical blade terminates somewhat like an awl (*Ledum reflexum*) ;—*needle-shaped* (*f. acerosa*), when the blade is hard, narrow, and pointed like a needle (*Pine*, fig. 70, *Juniper*, fig. 80) ;—

apillary (*f. capillacea*), when slender and flexible like hairs (*Water Crowfoot*, fig. 71);—*filiform* (*f. filiformis*), when thin and slender like threads (*Asparagus*, fig. 81). The false leaves of Asparagus here alluded to, and which have been described as leaves by most botanists, ought to be considered as branches springing from the axils of small scarious scales, which are the true leaves.

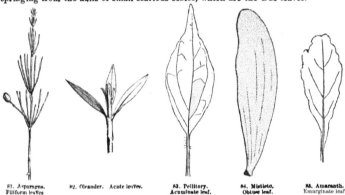

81. Asparagus.   82. Oleander. Acute leaves.   83. Pellitory.   84. Mistleto.   85. Amaranth.
Filiform leaves   Acuminate leaf.   Obtuse leaf.   Emarginate leaf.

Leaves are *acute* (*f. acuta*) when they terminate in a sharp angle (*Oleander*, fig. 82);—*acuminate* (*f. acuminata*), when the tip narrows rapidly and lengthens into a point (*Pellitory*, fig. 83);—*obtuse* (*f. obtusa*), when the tip is rounded (*Mistleto*,

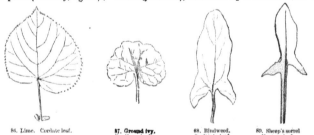

86. Lime. Cordate leaf.   87. Ground ivy.   88. Bindweed.   89. Sheep's sorrel
                  Reniform leaf.   Sagittate leaf.   Hastate leaf.

fig. 84);—*emarginate* (*f. emarginata*), when it terminates in a shallow sinus (*Amaranth*, fig. 85).

Leaves are *cordate* (*f. cordata*), when the base forms two rounded lobes and the tip is pointed, somewhat like an ace of hearts (*Lime*, fig. 86);—*reniform* (*f. reniformia*), when the base is cordate but the tip rounded, like a kidney (*Ground ivy*, fig. 87);—*sagittate* (*f. sagittata*), when the base is lengthened into two sharp lobes, which are oblique or parallel to the petiole, like an arrow (*Bindweed*, fig. 88);—

hastate (*f. hastata*), when 'the two lobes are nearly perpendicular to the petiole, like a halbert (*Sheep's sorrel*, fig. 89) ;—*peltate* (*f. peltatum*), when the petiole is joined to the centre of the under surface of the blade (*Nasturtium*, fig. 90), in which case the primary nerves diverge symmetrically from the petiole, like the spokes of a wheel. A peltate leaf may be compared with the orbicular palmately-nerved leaves of *Mallows*, for if the two

90. Nasturtium, Peltate leaves,   91. Curled Mallow.

edges nearest the petiole of the leaf of the small Mallow (fig. 74) were joined, a peltate leaf would be the result.

Surface of Leaves.—Leaves are *smooth* (*f. lævia*), when their surface presents neither hairs nor inequalities (*Orange*);—*scabrid* (*f. scabra*), when rough or harsh to the touch (*Carex*);—*glabrous* (*f. glabra*), when, whether smooth or not, they have no hairs (*Tulip*) ;—*silky* (*f. sericea*), when clothed with long, even, shining hairs (*Silverweed*) ;—*pubescent* or *downy* (*f. pubescentia*), when they are clothed with soft short hairs (*Strawberry*) ;—*pilose* (*f. pilosa*), when the hairs are long and scattered (*Herb-Robert*) ;—*villous* (*f. villosa*), when the hairs are rather long, soft, white, and close (*Forget-me-not*) ;—*hirsute* (*f. hirsuta*), when the hairs are long and numerous (*Rose-campion*);—*hispid* (*f. hispida*), when they are erect and stiff (*Borage*) ;—*setose* (*f. setosa*), when they are long, spreading, and bristly (*Poppy*) ;—*tomentose* (*f. tomentosa*), when they are rather short, soft, and matted (*Quince*) ;—*woolly* (*f. lanata*), when long, appressed, curly, but not matted (*Corn-centaury*) ;—*velvety* (*f. velutina, holosericea*), when the pubescence is short and soft to the touch (*Foxglove*) ;—*cobwebby* (*f. arachnoidea*), when the hairs are long, very fine, and interlaced like a cobweb (*Thistle, Cobwebby Houseleek*).

Leaves are *wrinkled* or *rugose* (*f. rugosa*), when their surface presents inequalities, due to there being more parenchyma than is enough to fill the spaces between the nerves (*Sage*) ;—*bullate* (*f. bullata*), when this excess of parenchyma renders the inequalities more visible, and, the whole blade is swollen between the nerves (*Cabbage*) ;—*crisped* (*f. crispa*), when the extra parenchyma only appears at the edge of the blade, which appears crimped (*Curled Mallow*, fig. 91) ;—*waved* (*f. undulata*), when for the same reason the edges are in rounded folds (*Tulip*).

Hairs and Spines on the Margins of Leaves.—The leaf is *ciliate* (*f. ciliatum*), when its margins bear long hairs like eyelashes (*Sundew*, fig. 92) ;—*spinous* (*f. spinosum*), when the nerves lengthen and harden into thorns (*Holly*, fig. 93 ; **Barberry**, fig. 94) ; in *Berberis*, the leaves which first appear after germination are provided with parenchyma like ordinary leaves, and the base of their petiole is furnished with two little stipules ; but on the subsequent branches the stipules harden, lengthen into spines, and the leaf itself is reduced to one

91. Barberry. Spiny leaves.   92. Sundew. Ciliate leaf.   93. Holly. Spiny leaf.   95. Gooseberry. Spines.

or three thickened and spinous nerves, from the axils of which short branches, bearing ordinary leaves, are developed. In the Gooseberry (fig. 95) the three or five spines (c) which spring below the leaves (f) may be considered as a development of the pulvinus of the leaf.

Divisions of Leaves.—The leaf is *entire* (*f. integrum*) when its blade is quite undivided (*Oleander*, fig. 82) ;—*cut* when its edge, instead of being a continuous line, presents a series of broken

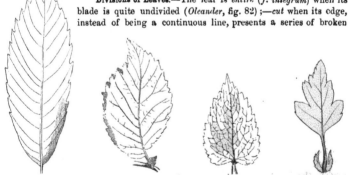

96. Chestnut. Dentate leaf.   97. Archangel. Serrated leaf.   98. Elm. Bidentate leaf.   99. Hawthorn. Incised leaf.

lines, owing to the parenchyma not accompanying the nerves to their extremities (*Chestnut, Oak, Hawthorn*) ;—*dentate* or *toothed* (*f. dentatum*) when it has sharp teeth with a rounded sinus; it is the slightest way in which a leaf can be cut (*Chestnut*, fig. 96);—*crenate* (*f. crenatum*) when it has rounded teeth and a sharp sinus (*Ground ivy*, fig. 87);—*serrate* (*f. serratum*) when the sinus and teeth are sharp and turned towards the tip of the leaf, like the teeth of a saw (*White Archangel*, fig. 97);—*doubly-dentate* or -*crenate* or -*serrate* (*f. duplicato-dentatum*, &c.) when the teeth or crenatures are themselves toothed or crenate (*Elm*, fig. 98);—*incised* (*f. incisum*) when the teeth are very unequal, and the sinus sharp and deep (*Hawthorn*, fig. 99);—*sinuate* (*f. sinuatum*), when the divisions (deeper than teeth) and the sinus are large and obtuse (*Oak*, fig. 100). The divisions of the leaf are called *laciniæ*[1] (*laciniæ*), when acute, and separated by an acute sinus, which reaches half-way to the middle of the blade. If the nerves are pinnate, the laciniæ are so also, and

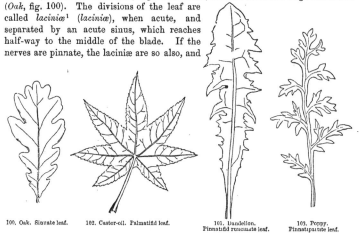

100. Oak. Sinuate leaf. 102. Castor-oil. Palmatifid leaf. 101. Dandelion. Pinnatifid runcinate leaf. 103. Poppy. Pinnatipartite leaf.

the leaf is pinnatifid (*f. pinnatifidum, Artichoke*);—if palmate, so also are the laciniæ, and the leaf is *palmate* (*f. palmatifidum, Castor-oil*, fig. 102). A pinnatifid leaf of which the laciniæ point downwards, is called *runcinate* (*f. runcinatum, Dandelion*, fig. 101).

The divisions of the leaf are called *partitions* (*partitiones*), when the sinuses extend beyond the middle, and nearly reach the midrib or the base of the blade; according to the nervation the leaf is then *pinnatipartite* (*f. pinnatipartitum, Poppy*, fig. 103), or *palmatipartite* (*f. palmatipartitum, Aconite*, fig. 104). The divisions of the leaf are called *segments* (*segmenta*), when the sinuses extend to the midrib or to the base of the blade; then, according to the nervation, the leaf is *pinnatisect* (*f. pinnatisectum, Watercress*, fig. 105) or *palmatisect* (*f. palmatisectum, Cinq-foil*, fig. 106;[2] *Strawberry*, fig. 107). The divisions are termed *lobes* (*lobi*) when the sinuses

---

[1] There are no current exact equivalents for the substantive terms *laciniæ, partitions, segments,* and *lobes* of this work; though when rendered into adjectives we usually apply lobes to divisions which descend to or about the middle of the leaf, and segments to divisions to or near the base.—ED.

[2] The Strawberry and Cinq-foil have undoubted *compound* leaves.—ED.

are as long as those of the laciniæ or partitions or segments, and when the divisions, of indefinite depth, are rounded; according to the arrangement of its nerves the leaf is then said to be *pinnately lobed* (*f. pinnatilobatum, Coronopus*, fig. 108), or *palmately lobed* (*f. palmatilobatum, Maple*, fig. 109).

104. Aconite. Palmipartite leaf.   105. Watercress. Pinnatisect leaf.   106, Cinq-foil. Palmatisect leaf.

The leaf is *lyrate* (*f. lyratum*), when, being pinnati -fid, -partite, -sect, or pinnately lobed, it terminates in a rounded division, much larger than the others

(*Turnip*, fig. 110);—*pedate* (*f. pedatum*), when its lobes, segments, partitions, or laciniæ diverge from the base; this occurs when three palmate divisions spring from the petiole, their midrib remaining undivided, whilst the two lateral produce on each side one or two parallel divisions, which are perpendicular to that from which they spring (*Hellebore*, fig. 111).

The same leaf is often variously divided; thus the segments of the lower pinnatisect leaves of *Chelidonium* (fig. 112) are lobed, sinuate, crenulate, and dentate; the lower leaves of *Aconite* (fig. 104) are palmi-partite, with bifid or trifid partitions, and incised and toothed laciniæ; the lower leaves of *Herb-Robert* (fig. 113) are palmatisect, with trifid

107. Strawberry. Palmatisect leaf.

108. Coronopus. Pennilobed leaf.

segments and incised and toothed laciniæ; the laciniæ being rounded and abruptly terminated by a small point, and said to be *apiculate* (*l. apiculatæ*). The *Castor-oil* (fig. 102), *Poppy* (fig. 103), *Cinq-foil* (fig. 106), and *Maple* (fig. 109), have toothed divisions.

Compound Leaves.—A leaf is *simple* (*f. simplex*), however deeply cut its divisions may be, when these cannot be separated from each other without tearing, as in most of the leaves mentioned above. It is *compound* (*f. compositum*), when its component divisions can be separated without tearing; and its divisions are named *leaflets* (*foliola*). The petiole of a compound leaf is the *common petiole* (*p. communis*), and that of each leaflet is a *petiolule* (*petiolulus*).

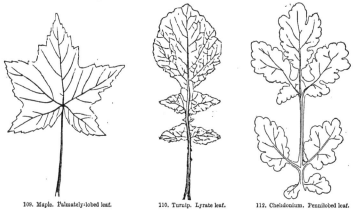

109. Maple. Palmately-lobed leaf.   110. Turnip. Lyrate leaf.   112. Chelidonium. Pennilobed leaf.

A leaf is *simply compound*, when the leaflets, whether petiolulate or not, spring directly from the common petiole; and, according to its nervation, the leaf is

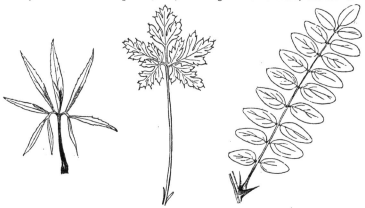

111. Hellebore Pedate leaf.   113. Herb-Robert. Palmatisect leaf.   114. Robinia. Pinnate leaf.

pinnate (*f. pinnatum*, *Robinia*, fig. 114) or *digitate* (*f. digitatum*, *Horse-chestnut*, fig. 115; *Lupin*, fig. 116). When there are but few leaflets, their insertion must

115. Horse chestnut. Digitate leaf.

116. Lupin. Digitate leaf.

117. Melilotus. Pinnate leaf.   118. Trefoil. Digitate leaf.

119. Gleditschia triacanthos. Bipinnate leaf.

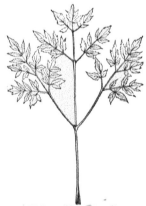

120. Actæa spicata. Tripinnate leaf.

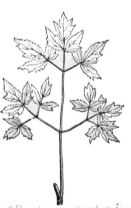

121. Actæa racemosa. Triternate leaf.

be carefully observed: thus the *Melilot* (fig. 117) has a pinnately tri-foliolate leaf, but the *Trefoil* (fig. 118) a digitately *ternate* [1] leaf, all the leaflets springing from the top of the petiole.

The leaf is *bipinnate* (*f. bipinnatum*), when the secondary petioles, instead of each ending in a leaflet, form so many pinnate leaves (*Gleditschia triacanthos*, fig. 119); *tripinnate* (*f. tripinnatum*), when the secondary petioles bear as many bipinnate leaves (*Actæa spicata*, fig. 120); *tri-ternate*, when the common petiole bears three secondary petioles, which each bear three tertiary petioles, each of which again bears as many digitately tri-foliolate leaves (*Actæa racemosa*, fig. 121). A pinnate leaf with all its leaflets in lateral pairs is termed *pari-pinnate* (*f. pari-pinnatum*); when in addition it is terminated by a solitary leaflet, the leaf is *impari-pinnate* (*f. impari-pinnatum, Robinia*, fig. 114).

122. Caucalis. Decompound leaf.   123. Potato.   124. Agrimony.   125. Orobus.
Interruptedly pinnate leaf.   Pinnatisect leaf.   Pinnate leaf with unequal leaflet changed into a very short filament.

A leaf is *laciniate* or *decompound* (*f. laciniatum, decompositum*), when, without being really compound, it is cut into an indefinite number of unequal laciniæ (*Caucalis Anthriscus*, fig. 122; *Water Crowfoot*, fig. 71), as in most umbelliferous plants (*Parsley, Chervil, Hemlock, Carrot, Angelica*, &c.).

A leaf is *interruptedly-pinnate* or -*pinnatisect* (*f. interrupti-pinnatum, -pinnati-sectum*), when the leaflets or divisions are alternately large and small (*Potato*, fig. 123; *Agrimony*, fig. 124).

Tendrils.—*Tendrils* (*cirri*) are thread-like, more or less irregularly spiral organs, which usually coil round neighbouring bodies, and thus support the plant. The leaf is *cirrhose* (*f. cirrosum*), when one or more of its leaflets is reduced to its median nerve, and becomes a tendril. In the *Bitter Vetch* (fig. 125), the tendril is simple and very short, because it is only the terminal leaflet which is thus transformed. In the *Pea* (fig. 126), and in *Vetches* (fig. 127), the three terminal leaflets are changed into tendrils. In another Vetch (*Lathyrus Aphaca*, fig. 128) all the leaflets are suppressed, and the whole leaf is reduced to a filament without parenchyma (v); in

---

[1] More correctly a digitately tri-foliate leaf.—ED.

compensation, the stipules (s, s) are very much developed, and perform the office of leaves. In *Smilax* (fig. 129), the petiole bears two lateral tendrils below the single

126. Aphaca. Petiolar tendrils.

126. Pea. Leaf with tendrils formed from leaflets.

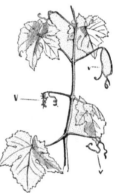

130. Vine. Tendrils formed from peduncles.

129. Smilax.
Stipulary tendrils.

cordate blade, which may be regarded as the lateral leaflets of a compound leaf, reduced to their median nerves. The lateral position of the solitary tendril in the *Melon* (fig. 61) and other *Cucurbitaceæ*, together with the fact that two tendrils occasionally occur, one on each side of the leaf, has led most botanists to regard it as a stipule, of which the corresponding one is suppressed; but the occurrence of two tendrils is very rare, and these are never precisely on the same level. On the other hand, unlike a true stipule, the

127. Vetch.
Pinnate leaf with foliolar tendrils and winged petiole.

tendril of *Cucurbitaceæ* is derived from a vascular bundle remote from that which produces the leaf, and is separated from the petiole of the latter by buds. Upon the whole, the simplest explanation of the tendril of the Melon is, that, unlike that of the Pea and other Leguminosæ, it represents a leaf, reduced to one or more of its nerves: thus, when simple, it represents the petiole and mid-rib; when branched, it represents the principal nerves of the leaf, which are themselves palmately divided. In the *Vine* (fig. 130) the tendril is leaf-opposed, and formed of a branching peduncle (v, v), of which the pedicels are suppressed, but which sometimes bears imperfect flowers.

INFLORESCENCE.

This term (*inflorescentia*) is used in two senses, signifying both the arrangement of the flowers upon a plant, and a collection of flowers not separated by leaves properly so called; the latter being the more special meaning of the term.

The organs of inflorescence are, (1) the supports of the flowers, *peduncle, pedicel, receptacle*; (2) the *bracts* (*bracteæ*), or altered leaves, from the axils of which the floral axes spring, and which are altered in colour and form, as they approach the flower; these are sometimes absent (*Stock* and other *Cruciferæ*).

The *peduncle* (*pedunculus*) is a branch directly terminated by a flower; and its extremity forms the *receptacle* (*receptaculum*). This name is also given to a more or less branched flowering axis, differing in appearance from the rest of the

131. Lime. Bract joined to the peduncle.

123. Currant. Simple raceme.

stem, bearing bracts, and of which the ultimate divisions are called *pedicels* (*pedicelli*).

Bracts vary in shape; they are usually small (*Currant*, fig. 132), and may be thin, transparent, and *membranous* (*br. membranaceæ*); or thin, dry, stiff, coloured, and *scarious* (*br. scariosæ, Geranium*); or *coloured*, like petals (*Bugloss*); they are very large in the *Lime* (fig. 131), which is peculiar in having the peduncle adnate to the midrib of the bract, and, though really axillary to it, appearing to rise from its centre.

The *primary axis* of the inflorescence is the common peduncle, whence spring the *secondary, tertiary*, &c. axes, according to their order of development.

The inflorescence is *axillary* (*inf. axillaris*), when the primary axis, instead of terminating in a flower, is indefinitely elongated, and the flowers are borne upon secondary axes, springing from the axils of its leaves (*Pimpernel*, fig. 30); it is *terminal* (*i. terminalis*), when the primary and secondary axes both terminate in a flower (*Poppy, Columbine*, fig. 31).

D

In every inflorescence the flowers are *solitary* (*fl. solitarii*), when each peduncle is undivided, and springs directly from the stem, and is isolated from the others by normal leaves (*Pimpernel*, fig. 30). Inflorescence, in its restricted sense, consists of a group of pedi-celled flowers, bracteate or not, all springing from a common peduncle which bears no true leaves.

Indefinite inflorescences are—the *raceme, corymb, umbel, spike,* and *head.*

1. The *raceme* (*race-mus*) is an inflorescence of which the nearly equal secondary axes rise along the primary axis; it is *simple,* when the pedicels spring directly from the primary axis, and terminate in a flower (*Lily, Lily of the Valley, Snapdragon; Cur-rant,* fig. 132; *Mignonette,* fig. 133); it is *compound,*

133. Mignonette. Simple panicle.

134. Yucca Gloriosa. Branch of compound panicle.

and called a *panicle* (*pani-cula*), when the secondary axes branch once or oftener before flowering (*Yucca Gloriosa,* fig. 134). A *thyrsus* (*thyrsus*) is a panicle of an ovoid shape, the central pedicels of which are longer than the outer ones.

135. Cerasus Mahaleb.
Indefinite corymb.

136. Cherry. Simple umbel.

137. Fennel.
Umbel and umbellule without involucre.

2. The *corymb* (*corymbus*) resembles the raceme, but the lower pedicels are so much longer than the upper, that the flowers are nearly on a level (*Cerasus Mahaleb,*

fig. 135). In the *Stock* and many allied plants, the inflorescence is at first a corymb, but changes to a raceme as the primary axis lengthens.

3. In the *umbel* (*umbella*) the secondary axes are equal in length, and starting from the same point, flower at the same height, diverging like the rays of a parasol; it is a raceme of which the primary axis is reduced almost to a point. The umbel is *simple* (*sertulum*), when the secondary axes flower (*Cherry*, fig. 136) ; it is *compound*, when these bear umbellately arranged tertiary axes, called *partial umbels* (*umbellulæ*, *Fennel*, fig. 137 ; *Carrot*, fig. 138 ; *Fool's Parsley*, fig. 139).

138. Carrot. Umbels with involucre ; umbellulæ with involucels.

139. Æthusa. Umbel without involucre ; umbellulæ with involucels.

The bracts, which in most racemes spring, like the pedicels, from different heights, in many umbelliferous plants rise on a level, like the secondary and tertiary axes, and form a whorl. The name *involucre* (*involucrum*) is given to the bracts at the base of the umbel

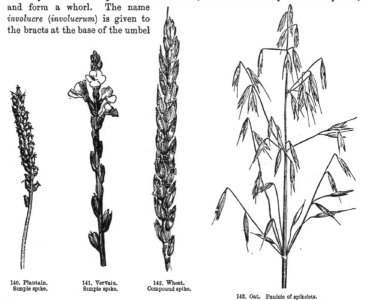

140. Plantain. Simple spike.

141. Vervain. Simple spike.

142. Wheat. Compound spike.

143. Oat. Panicle of spikelets.

(*Carrot*, fig. 138), and that of *involucel* or *partial involucre* (*involucellum*) to the bracts at the base of the partial umbel (*Fool's Parsley*, fig. 139).   Both involucre and involucel may be absent (*Fennel*, fig. 137).

4. In the *spike* (*spica*), the flowers are sessile or subsessile on the primary axis (*Plantain*, fig. 140; *Vervain*, fig. 141). In the *compound spike* (*s. composita*) the secondary axes each bear a small

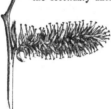

144. Willow. Staminiferous catkin.   145. Willow. Pistilliferous catkin.

distichous spike (*spikelet*, *Wheat*, fig. 142).   In many grasses, the spikelets are borne on long branching pedicels, forming a panicle (*Oat*, fig. 143).

The *catkin* (*amentum*) is a spike, the flowers of which are incomplete (*i.e.* they want either stamens or pistil), and which is deciduous when mature (*Mulberry; Willow*, figs. 144, 145; *Oak*, fig. 146).

The *cone* (*strobilus*) is a catkin with large thick scales, principally found on certain evergreen trees, hence named *Conifers* (*Pine*, fig. 147).   The spike of the *Hop* (fig. 148) is a cone with large membranous bracts.

146. Oak.
Staminiferous catkin.

149. Arum.
Spadix exposed by the removal of part of the spathe.

The *spadix* (*spadix*) is a spike of *incomplete* flowers, which, when young, is enveloped in a large bract or *spathe* (*spatha*).   The axis of the spadix sometimes flowers throughout its length, sometimes the upper portion is flowerless (*Arum*, fig. 149). The branched spadix of Palms is called a 'régime' (in French).

147. Pine. Cone.   148. Hop. Cone.

5. In the *head* (*capitulum*) the flowers are collected into a head or depressed spike, of which the primary axis is vertically contracted, thus gaining in thickness what it has lost in length (*Scabious*, fig. 150; *Trefoil*, fig. 151), and the

depressed axis is called the common receptacle (*clinanthium*). As in the umbel, the head is usually bracteate, each flower springing from the axil of a bract. There

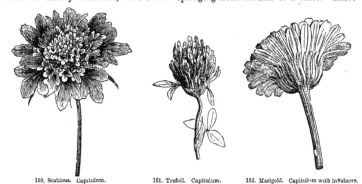

150. Scabious. Capitulum.   151. Trefoil. Capitulum.   152. Marigold. Capitulum with involucre.

should hence be as many bracts as flowers, but, owing to the crowding of the flowers, some of the bracts are usually suppressed. The outer bracts, or those below the outer flower, form the involucre (*involucrum, periclinium, Marigold,* fig. 152). The bracts of the centre flowers are usually reduced to *scales, bristles,*

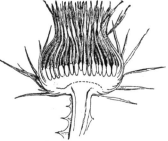

153. Camomile.
Paleate receptacle, cut
vertically.

154. Cornflower.
Bristly receptacle, cut
vertically.

155. Onopordon. Alveolate receptacle, cut vertically.

or *hairs.* The receptacle is *paleate* (*r. paleatum*), when covered with scale-like bracts separating the flowers (*Camomile,* fig. 153) ;—*setose* (*r. setosum*), when these are bristly ; such are often cut into fine hairs (*Cornflower,* fig. 154) ;—*pitted* (*r. alveolatum*), when the flowers are seated in depressions, separated by variously shaped membranes, which represent the bracts (*Onopordon,* fig. 155). When these inner bracts are absent the receptacle is described as *naked* (*r. nudum, Dandelion,* fig. 156). Sometimes the base of the head is naked, or only protected by some normal leaves (*Trefoil,* fig. 151), but each flower may still be accompanied by a bract.

The inflorescences of *Dorstenia Contrayerva,* and of the *Fig* (*hypanthodium*), are also heads. In *Dorstenia* (fig. 157) the receptacle is much depressed or slightly

concave, bearing incomplete flowers inserted in pits with ragged edges; in the *Fig*
(fig. 158) the inflorescence is similar, but the receptacle is still more concave, inso-
much that the male flowers, which are at the top of
the fig, answer in position to the lowest flowers of the
primary axis, and the small scales (bracts) at the
mouth represent an involucre, which in the normal
state would gird
the base of the
common recepta-
cle, as in an or-
dinary head.

It is obvious
that every indefi-
nite inflorescence
must be a modifi-
cation of the ra-
ceme; thus the
corymb is a ra-
ceme with unequal

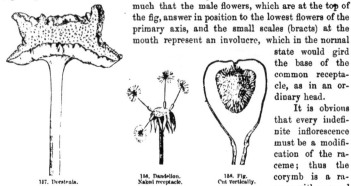

157. Dorstenia.   156. Dandelion.   158. Fig.
                    Naked receptacle.   Cut vertically.

secondary axes, reaching the same level; an umbel is a raceme whose primary
axis is undeveloped; the spike is a raceme whose secondary axes are undeveloped; the capitulum is a spike with the primary axis vertically thickened and
dilated.

The difference between the raceme, corymb, umbel, spike, and head being simply
due to the amount of development of the primary and secondary axes, these terms
cannot be precisely limited, and intermediate terms are therefore frequently resorted to;
as spiked racemes and panicles, when the pedicels are very short; a globose spike
approaches the head; and an ovoid or spiked head approaches the spike. Amongst
Trefoils, capitulate, spiked, and umbelled flowers all occur.

In the raceme, panicle, corymb, and spike, the pedicels flower from below upwards, i.e. the lowest flowers open first. In simple and compound umbels, the outer
flowers open first; whence we may conclude that the umbel is a depressed raceme.
In the head, as in the depressed spike, the flowers really open from below upwards, but as the surface of the inflorescence in both these cases is nearly
horizontal, they appear to open from the circumference to the centre, and are
called centripetal, a term which is applied to every indefinite inflorescence, whether
the flowers open from below upwards, or from without inwards.

Definite Inflorescences.—These are all included under the general term *cyme*
(*cyma*), however much they may be branched; they are, the *definite-* or *cymose-raceme;*
true corymb; umbellate-cyme; spicate-cyme, scorpioid cyme; and *contracted cyme,* which
comprises the *fascicle* and the *glomerule.*

1. In the *definite-* or *cymose-raceme* (*Campanula,* fig. 159), the flowering pedicels
are of nearly equal length, as in the raceme; from which it differs in the primary
axis (A, A, A), terminating in a flower, which is necessarily the first to expand;

whilst of the secondary axes (B, B, B), the lowest, being the oldest, flowers first; and the tertiaries (C, C, C), although often lower than the axis whence they spring, flower last. The result is, that of the expanded flowers some are above, some below the buds, according to the order of the succession of their axes. When examining such inflorescences, the student must look for the axis terminated by a flower, for the lateral leaf or bract which it bears, and for the shoot or secondary axis which springs between this axis and itself.

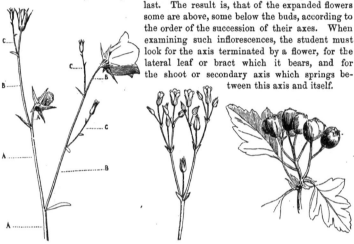

159. Campanula. Cymose raceme.   160. Cerastium. Dichotomous cyme.   161. Hawthorn. Definite corymb.

The racemose cyme is called a panicle or thyrsus when much branched (*Privet*); but in reality the difference between the definite raceme and panicle is not analogous to the difference between the indefinite raceme and panicle, for the indefinite raceme consists of a primary and many secondary axes; while the indefinite panicle consists of primary, secondary, tertiary, quaternary axes; just as is the case both in the definite raceme and definite panicle; the only difference between these two, then, is in appearance.

The definite raceme becomes a *dichotomous cyme*, when the primary axis terminates in a flower between two opposite leaves or bracts, from the axils of which spring two secondary axes, each again terminated by a flower between two bracts, from the axils of which spring two tertiary axes, and so on (*Cerastium*, fig. 160); this evolution of subordinate axes, each terminating between two opposite axes, is continued till the last axis fails, from deficient nutrition, to repeat the process. When, instead of two opposite leaves or bracts, there are three in a whorl below each successive central flower, with again three in their axils, the cyme becomes trichotomous.

2. In the *definite* (or true) *corymb*, the different flowering axes, although of unequal length, attain pretty much the same level (*Hawthorn*, fig. 161).

In the definite raceme and the corymb, the central flowers are first developed; in other words, the flowers open from within outwards, or centrifugally.

3. In the *definite umbel* or *umbellate cyme* the pedicels appear to start from the same point as in the indefinite umbel, but the central flowers open first, and the outer pedicels are evidently the youngest and shortest: being a definite umbel, flowering centrifugally, it is truly a cyme (*Chelidonium*, fig. 162).

4. The *definite spike* or *spicate-cyme* (*Sedum*, fig. 163)

162. Chelidonium.
Definite umbel.

163. Sedum. Spicate cyme.

164. Myosotis. Scorpioid cyme.

is composed of a succession of independent axes, alternating to the right and left, each terminating in an apparently sessile flower.

5. The *scorpioid cyme* (*Myosotis*, fig. 164) is a raceme which rolls up in a crozier shape, like the tail

165. Theoretical figure of the scorpioid cyme.

of a scorpion; it is composed of a succession of independent axes, which do not alternate right and left, but form an interrupted line, which tends to turn back upon itself; in this inflorescence, the bracts are usually suppressed (fig. 165).

6. In the *con-tracted cyme* the

166. Box. Glomerule.

167. Lamium. Fascicles on an indefinite stem.

flowers are crowded, owing to the extreme shortness of the axes; it is called *fascicled*, when the axes are somewhat lengthened, and are regularly distributed

(*Sweet William*);—*glomerate*, when the axes are almost suppressed, and extremely irregular (*Box*, fig. 166).

Mixed inflorescences are those in which the definite and indefinite both appear. In the *Labiatæ* (*Lamium*, fig. 167) the general inflorescence is indefinite, while

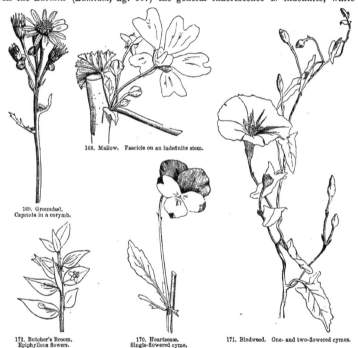

168. Mallow. Fascicle on an indefinite stem.

169. Groundsel. Capitula in a corymb.

172. Butcher's Broom. Epiphyllous flowers.

170. Heartsease. Single-flowered cyme.

171. Bindweed. One- and two-flowered cymes.

the separate heads are true axillary cymes or fascicles. In the *Mallows* the same arrangement occurs (fig. 168). In *Compositæ* (*Groundsel*, fig. 169) the general inflorescence is a definite corymb, and the separate portions are heads. The definite inflorescence is sometimes reduced to a single flower, and resembles the one-flowered pedicels of an indefinite inflorescence (*Heartsease*, fig. 170); but a little below the flower two small bracts (*bracteoles*) will be found, in the axils of which are two obvious or suppressed shoots, which sometimes flower (*Bindweed*, fig. 171). The two bracteoles of a one-flowered pedicel are therefore the evidences of a two- or three-flowered cyme, of which the primary axis only is developed.

The inflorescence of certain plants has been called *epiphyllous*, from the flowers appearing to spring from leaves or bracts. In the *Lime* (fig. 131), the peduncle is joined to the bracts. In *Xylophylla* the floral branch, dilated and flattened like a leaf, bears flowers along its edges. In the *Butcher's Broom* (fig. 172), as in *Xylophylla*, the peduncles, enlarged into green leaves, rise in the axils of small scales which are the true leaves, and bear on their centre one or more shortly pedicelled flowers, forming a cyme.

THE FLOWER.

The *flower*, in phænogamic plants, is a collection of several whorls (usually four), formed of variously modified leaves arranged one above another in rings or stages, so close that their internodes are not distinguishable.

The leaves which form each floral whorl are not always precisely on the same level, but often form a close spiral, and consequently not a true whorl; the term whorl is, however, always applied to the calyx, corolla, andrœcium, and pistil.

The flower may be regarded as a true shoot, terminating the peduncle or pedicel, and therefore *terminal* as regards the branch from which it springs: limiting the growth of that branch. Its terminal position may be theoretically explained by supposing that the floral whorls exhaust the supply of nutriment provided by the axis, and with this the vegetative force necessary to prolong it. In the normal condition of the flower, the reproductive and nutritive forces are in equilibrium; but there are cases in which this equilibrium is disturbed, and in which the axis lengthens beyond the floral whorls, and reproduces the plant by branch-buds; in which cases the seed-buds are usually suppressed: this is seen in many plants, and especially in proliferous roses (fig. 173), of which the peduncle is prolonged into a supplementary axis, ending usually in an imperfect flower (fig. 174) of sepals (s) and petals (P), in the middle of which are a few imperfect stamens and carpels.

174. The same rose cut vertically, and deprived of calyx and lower petals, to show the position of all the parts along the axis.

173. Proliferous rose. c, c, calyx transformed into leaves; P, petals multiplied at the expense of the stamens; A. prolonged axis bearing an imperfect flower; F. coloured blades representing abortive carpels.

The variously transformed leaves composing the floral whorls, though modified in tissue, colour, and texture, to form the calyx, corolla, andrœcium, and pistil, sometimes reveal their origin by resuming the aspect of normal leaves. The term *anomaly* or *monstrosity* is given to casual departures from the normal structure occurring in animals and plants, which anomalies are most frequently induced by cultivation.

The first whorl or *calyx*, being the exterior, and therefore the nearest to the leaves, resembles these most.

The second whorl or *corolla* is more altered; the tissue of its petals is more delicate, and their colour more brilliant, but their *claw, limb,* and *nerves,* and their usually flat shape, all reveal their foliar nature.

The third whorl, or *andrœcium,* bears much analogy to the second; the relative position of the stamens and petals is always the same, and these sometimes present an insensible transition from one to the other; as in semi-double flowers, where some of the stamens are changed into petals; in partially double flowers, where all the stamens are so changed; and in full double flowers, where the carpels also have become petaloid (*Ranunculus, Columbine, Rose*). In *Rosa centifolia* (fig. 175), par-

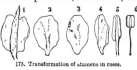

175. Transformation of stamens in roses.

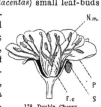

176. Hooded Columbine, showing one series of anthers transformed and connected together.

ticularly, the successive steps by which a stamen becomes a petal are obvious; sometimes the anther enlarges, and one cell reddens (6); or both cells lengthen (5); or the connective reddens and dilates, and bears on one side a yellow scale, which recalls an anther-cell (4, 3); oftenest the stamen expands at once into a complete petal (2); sometimes (1) the proximity of the calyx seems to influence this petal; a green midrib traverses its coloured blade, and it becomes sepaline in the middle, petaline on the sides. In the *double Columbine* (fig. 176), the anther swells, and forms a hooded petal; and sometimes, but more seldom, the filament dilates into a flat petal.

The fourth whorl or *pistil* is the central; its position and the pressure of the surrounding organs influence its form in many ways, and hence disguise its origin; but when the carpellary leaves are free (*Columbine,* fig. 12), or solitary (*Pea,* fig. 14), their foliaceous nature is obvious, and especially in anomalous cases, as the following.

Anomalies.—In the *Columbine* (fig. 177) the five carpellary leaves (F.c) instead

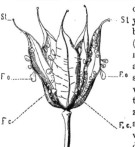

177. Monstrous Columbine.

of being folded to form a protecting cavity for the young seed, have been found to remain flat, and bear along their edges (or *placentas*) small leaf-buds (F.o); these buds, which normally would have contained an embryo, were mostly open; some few, though empty, were curved, and suggestive of their normal function; fertilization had not taken place, and the unfertilized stigma was reduced to a small glandular head (St), terminating the midrib of the carpellary

178. Double Cherry. Flower cut Vertically; s, sepals; p, petals; F.c, carpellary leaves; N.m, median nerve or style.

leaf. In the double *Cherry* (fig. 178), the free edges of the two carpels (F.c) bear no buds, and their blade or ovary, which altogether resembles an ordinary leaf, folded along its midrib (N.m), is lengthened into a style-like neck, terminated by a spongy tubercle representing the stigma.

The *Alpine Strawberry* (fig. 179) presents a curious metamorphosis of the floral whorls. The calyx (s) is normal, the five outer leaves are bifid, and accurately represent the stipules of the leaves. The petals (P) appear as green, strongly veined, nearly sessile leaves

179. Alpine Strawberry.   180. Alpine Strawberry.   181. Alpine Strawberry.   182. Alpine Strawberry.   183. Alpine Strawberry.
                          Green petal (mag.).   Green stamens.   Carpel (mag.).   Carpel without the
                                                                                                        ovary (mag.).

with five acute ciliate lobes (fig. 180). The twenty stamens (fig. 179, E) are arranged in four whorls, and are also expanded into green petioled simple or three-lobed leaves (fig. 181); most of them bear on each side of the base of the blade a yellow boss (A, A), representing a suppressed anther. The carpels (fig. 179, c), which have also reverted to leaves, are arranged spirally on a receptacle, which becomes succulent as the green flower grows. The carpellary leaf (fig. 182 F.c), the integument of the seed (F.o), called the ovulary leaf, and the embryo are transformed through excessive development into overlapping leaves. Of these, the outer leaf, often bifid (F.c), represents the ovary; its base sheaths the inner leaf (fig. 183, F.o), which should have formed the outer integument of the ovule. At the inner base of this ovulary leaf (183, F.o) is a pointed shoot (P); this is the embryo, of which a vertical section (fig. 184) shows rudimentary leaves or cotyledons (co) and a plumule (G).

In this curious flower, an excessive supply of nutrition has deranged the reproductive organs, and the whorls, which should have been modified in

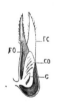

184. Alpine Strawberry.   185. Rumex.   186. Lily.   187. Narcissus.
     Carpel   Flower with a double   Flower with a double   Flower with a double petaloid perianth,
     cut vertically.   calycoid perianth.   petaloid perianth.   furnished with a cup simulating a corolla.

subservience to the function of reproduction, have preserved their original form of green leaves. Such a metamorphosis of all the floral organs into ordinary leaves is not uncommon throughout the Vegetable Kingdom; it is called *chloranthy*.

An *incomplete flower* (*fl. incompletus*) is one in which calyx, or corolla, or andrœcium, or pistil is absent. The single or double whorl which surrounds the andrœcium and pistil (or essential organs of the flower) is called a *perianth* (*perianthium, perigonium*).

A *dichlamydeous* flower (*fl. dichlamydeus*) is one with a double perianth, i.e. with two whorls, calyx, and corolla (*Wallflower*, fig. 7); which are similar in form or colour or not. When both whorls are green and calyx-like (*Rumex*, fig. 185), the perianth is called calycoid, calycine, or foliaceous (*p. foliaceum*), and when both are coloured or corolla-like (*Lily*, fig. 186), it is called *petaloid* (*p. petaloideum*). In *Narcissus* (fig. 187) there is a fringed cup within the petaloid perianth, which is greatly developed in the common species here figured, but is much less so in the *Narcissus poeticus*, and other species. In these latter it is cut into six lobes, alternating with those of the double perianth, whence some botanists have concluded that it represents two confluent whorls analogous to the outer ones. Others regard this cup of *Narcissus* as formed by lateral expansion of the confluent filaments. In *Orchis* (fig. 188) the petaloid

188. Orchis.
Flower with a double
irregular petaloid
perianth.

189. Chenopodium.
Monoperianthed flower.

191. Ash.
Naked flower.

190. Aristolochia.
Monoperianthed
flower with irregular
perianth.

192. Carex.
♂ flower.

perianth has six unequal, spreading lobes, of which the upper are erect and form the *hood* (*galea*); the lowest is dilated, variable in shape, and called the *lip* (*labellum*); it is sometimes produced into a *sac*, or *spur* (*calcar*).

A *monochlamydeous perianth* (*p. simplex*) is usually considered as a calyx, and the flower is said to be *apetalous* (*fl. apetalus*). It may be *foliaceous* (*Chenopodium*, fig. 189), or *petaloid* (*Anemone*, fig. 230), or *irregular* (*Aristolochia*, fig. 190).

An *achlamydeous* flower (*fl. achlamydeus*) has neither calyx nor corolla; it may be protected by one or more bracts (*Carex*, figs. 192, 193), or altogether unprotected (*fl. nudus, Ash*, fig. 191).

A *hermaphrodite* flower (*fl. hermaphroditus*, ☿) possesses both andrœcium and pistil (*Wallflower*, fig. 7);—the flower is *male* (*fl. masculus*, ♂) when it has andrœcium without pistil (*Carex*, fig. 192);—*female* (*fl. fœmineus*, ♀), when it has pistil without andrœcium (*Carex*, fig. 193);—and *neuter* or *sterile* (*fl. sterilis, neuter*), when it has neither andrœcium nor pistil (outer flowers of the *Cornflower*, fig. 194);—

193. Carex. ♀ flower.

monœcious (*fl. monoici*), when the male and female flowers are on the same plant (*Carex*, figs. 192, 193; *Oak*, fig. 146; *Hazel-nut*, fig. 195, 195 *bis*, 195 *ter*; *Arum*, figs. 196,

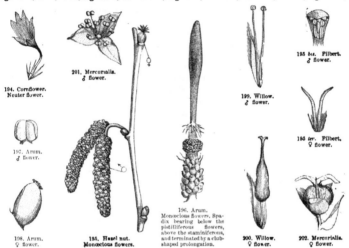

194. Cornflower. Neuter flower.

201. Mercurialis. ♂ flower.

195 *bis*. Filbert. ♂ flower.

199. Willow. ♂ flower.

195 *ter*. Filbert. ♀ flower.

197. Arum. ♂ flower.

198. Arum. ♀ flower.

195. Hazel nut. Monœcious flowers.

196. Arum. Monœcious flowers. Spadix bearing below the pistilliferous flowers, above the staminiferous, and terminated by a club-shaped prolongation.

200. Willow. ♀ flower.

202. Mercurialis. ♀ flower.

197, 198);—*diœcious* (*fl. dioici*), when on separate plants (*Willow*, figs. 199, 200; *Dog's Mercury*, figs. 201, 202);—*polygamous* (*fl. polygami*), when hermaphrodite flowers occur amongst the male or female (*Pellitory*). The general term *diclinous* (*diclinus*) is applied to monœcious, diœcious, and polygamous flowers.

THE CALYX.

The *calyx* (*calyx*) is the whorl placed outside of the corolla and androecium. It is usually *simple* (*Wallflower*), sometimes *double* (*Magnolia*, *Trollius*); its component leaves are termed *sepals* (*sepala*). It is *polysepalous* (*c. polysepalus*), when its sepals are wholly separate (*Wallflower*, fig. 8; *Columbine*, fig. 31); *gamo-* or *mono-sepalous* (*c. gamo-* or *mono-sepalus*), when its sepals cohere more or less.

A monosepalous calyx is *partite* (*c. partitus*), when the sepals are united at the base only; and it may be *bi- tri- multi-partite* (*Pimpernel*,

203. Pimpernel. Five-partite calyx and pistil.

204. Erythræa. Five-fid calyx.

205. Lychnis. Five-toothed calyx.

fig. 203);—it is *bi- tri- multi-fid*, when the sepals cohere about half-way up (*Comfrey*,

Erythrœa, fig. 204) ;—it is *bi- tri- multi-dentate* or *-toothed* (*c. dentatus*), when the sepals are united nearly to the top (*Lychnis*, fig. 205).

In the monosepalous calyx, the connected portion of the sepals is the *tube* (*tubus*), the free portion the *limb* (*limbus*), and the point of union of these the *throat* (*faux*).

Sepals are sometimes prolonged into *appendages* at the base, as in *Myosurus* (fig. 206) and *Heartsease* (fig. 500), where the five sepals are attached to the receptacle by their

  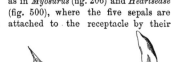

206. Myosurus. Flower with appendiculate calyx.

207. Campanula. Appendiculate calyx.

208. Lamium. Irregular calyx.

209. Larkspur. Calyx prolonged into a hollow horn.

centres ; in some *Campanulas* (fig. 207) the appendage is formed by the union of two lobes belonging to two contiguous sepals, between which it is placed.

The calyx is *regular* (*c. regularis, æqualis*), when its sepals, whether equal or unequal, form a symmetrical whorl (*Wallflower*, fig. 8; *Pimpernel*, fig. 203 ; *Erythrœa*, fig. 204; *Lychnis*, fig. 205) ;—it is *irregular* (*c. irregularis, inæqualis*), when the whorl is unsymmetrical (*Lamium*, fig. 208). In the *Aconite* the upper sepal forms a hood ; in *Larkspur* (fig. 209) it is prolonged into a hollow horn or *spur*. In the *Tropæolum* (fig. 210), the spur is formed by the united and lengthened three upper sepals. In *Pelargonium* the upper sepal is produced downwards, and forms a tube adherent to the pedicel. In *Scutellaria* the five sepals form two lips ; of which the upper

211. Scutellaria. Young calyx.

210. Tropæolum. Flower with calyx prolonged into a hollow horn or spur.

212. Scutellaria. Ripe calyx.

214. Henbane. Urceolate calyx.

213. Winter Cherry. Vesicular calyx.

protuberant one (fig. 211), after flowering, forms a shield to the ovaries, arching over them so as completely to envelop them, and meet the lower lip (fig. 212).

The tube of the monosepalous calyx may be *cylindric* (*cylindricus, Pink*, fig. 226) ; —*cup-shaped* (*cupuliformis, Orange*) ;—*club-shaped* (*clavata, claviformis, Silene, Armeria*) ; —*bladdery* (*vesiculosus*), when swollen like a bladder (*Winter Cherry*, fig. 213) ;—*tur-binate* (*turbinatus*), when it resembles a top or pear (*Black Alder*) ;—*bell-shaped* (*campanulatus*, *Kidney-bean*);—*urceolate* (*urceolatus*), when it resembles a small pitcher (*Henbane*, fig. 214).

The calyx is *connivent* (*s. conniventia*), when the sepals bend towards each other (*Ceanothus*);—*closed* (*s. clausus*), when their edges touch without joining (*Wallflower*, fig. 8);—*erect* (*s. erectus*), when the sepals are vertical (*Rocket*, fig.

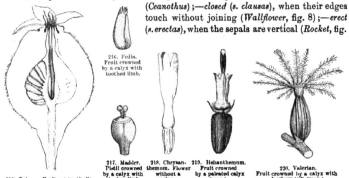

216. Fedia. Fruit crowned by a calyx with toothed limb.

215. Quince. Fruit cut vertically.

217. Madder. Pistil crowned by a calyx with obsolete limb.

218. Chrysan- themum. Flower without a calyx.

219. Helianthemum. Fruit crowned by a paleated calyx (mag.).

220. Valerian. Fruit crowned by a calyx with feathery tuft (mag.).

250);—*patent* (*s. patentia*), when they spread horizontally (*Mustard*);—*reflexed* (*s. reflexus*), when turned back so as to expose their inner surface (*Bulbous Crowfoot*).

The calyx-limb may be *petaloid* (*Iris*);—*foliaceous* (*Quince*, fig. 215);—*toothed* (*Fedia*, fig. 216);—reduced to a small membranous crown (*Field Camomile*)—or ring (*c. margo obsoletus, Madder*, fig. 217);—or altogether suppressed (*Chrysanthemum*, fig. 218); in the latter case the calyx is said to be *entire* (*c. integer*), because its tube is considered to be confluent with the ovary, and undivided.

222. Dandelion. Fruit crowned by a calyx with a limb in a simple tuft.

223. Scabious. Fruit open (mag.). Calyx with a stipitate taft.

221. Salsify. Fruit crowned by a calyx with a feathery tuft.

The calyx-limb may be reduced to *scales* (*squamæ* or *paleæ*, *Helianthemum*, fig. 219); or to radiating bristles or hairs, called a *pappus* (*pappus*). Such a pappus may be *plumose* (*p. plumosus*) when each of its hairs is covered with long secondary hairs or barbs visible to the naked eye (*Valerian*, fig. 220; *Salsify*, fig. 221); —*simple* (*p. simplex*) when the hairs or bristles are smooth and silky (*Dandelion*, fig. 222).

The pappus, whether simple or plumose, is sessile (*p. sessilis*), when the hairs are inserted directly on the top of the ovary (*Cornflower, Valerian*, fig. 220); *stipitate* (*p. stipitatus*), when the calyx-tube is prolonged into a slender neck above the ovary (*Dandelion*, fig. 222; *Salsify*, fig. 221; *Scabious*, fig. 223).

The calyx is *deciduous* (*c. deciduus*), when it falls with the corolla after flowering (*Wallflower*, fig. 8) ;—*caducous* or *fugacious* (*c. caducus*), when it falls as soon as

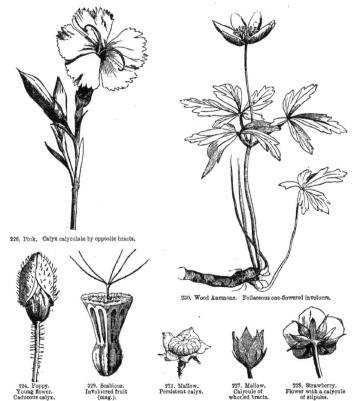

226. Pink. Calyx calyculate by opposite bracts.

230. Wood Anemone. Foliaceous one-flowered involucre.

224. Poppy. Young flower. Caducous calyx.

229. Scabious. Involucred fruit (mag.).

225. Mallow. Persistent calyx.

227. Mallow. Calycule of whorled bracts.

228. Strawberry. Flower with a calycule of stipules.

the flower begins to expand (*Poppy*, fig. 224) ;—*persistent* (*c. persistens*), when it remains after flowering (*Pimpernel*, fig. 203) ;—marcescent (*marcescens*), when it withers and dries up, and remains attached to the fruit (*Mallow*) ;—accrescent (*c. accrescens*), when it continues to grow after flowering (*Winter Cherry*, fig. 213).

 ' **Calycules** ' **and Calyciform Involucres.**—The calyx is sometimes accompanied by whorled or opposite bracts, simulating an accessory calyx ; to these have been given the name of calycule or outer calyx (*calyculus*). The Pink (fig. 226) has a ' calycule '

E

of four bracts in opposite pairs. The Mallow (figs. 225, 227) has, outside the **five-fid** calyx, a calycule of three bracts, and the Marsh Mallow one of six to nine bracts.

The five green bodies beneath and adherent to the calyx of the Strawberry (fig. 228), and which alternate with the five sepals, are not a calycule of bracts but of pairs of stipules belonging to the sepals. The pitted cup with fringed margins which encloses each flower of the *Scabious* (figs. 223, 229) may be considered a calycule.

Calycules are true one-flowered involucres, analogous to the many-flowered involucres of heads and umbels.

The following are also one-flowered involucres: the three foliaceous cut bracts of *Anemone* (fig. 230), placed far below the calyx;—the three entire bracts of *Hepatica* (fig. 231), also placed just below the calyx;—the many foliaceous bracts of the *Winter Hellebore* (fig. 231 *bis*), placed

231 *bis.* Winter Hellebore.   231. Hepatica.   232. Oak.
Calyciform involucre near the flower.   Calyciform involucre near the flower.   Fruit with a scaly cup.

235. Euphorbia.
Calyciform
many-flowered cup.

233. Filbert. Fruits with foliaceous cup.   234. Chestnut. Prickly involucre, containing three fruits.

almost in contact with the calyx;—the *cup* (*cupula*) of the acorn (fig. 232), which is composed of small imbricated scales;—the foliaceous cup, with cut margins, of the *Filbert* (fig. 233). The prickly cup of the *Chestnut* (fig. 234), and the calyciform cup of *Euphorbia* (fig. 235), only differ from the preceding in being many-flowered.

THE COROLLA.

The *corolla* (*corolla*) is the whorl next within the calyx; it is usually *simple* (*Rose*), sometimes *double*, i.e. composed of several whorls (*Magnolia, Nymphœa*) ; its leaves are *petals* (*petala*).

Petals are usually *coloured*, that is, not green like the (usually foliaceous) sepals; some plants, however (*Buckthorn, Vine, Narcissus viridiflorus*), have green petals, while others (*Helleborus, Aconite, Larkspur, Columbine, Fennel*) have coloured or petaloid sepals.

In the *polypetalous corolla* (*c. polypetala, dialypetala*) the petals are entirely separate from each other (*Wallflower, Strawberry, Columbine*) ;—in the *monopetalous* or *gamopetalous* corolla (*c. mono- gamo-petala*) the leaves cohere more or less, so as to form a corolla of a single piece.

The corolla is *regular* (*c. regularis*), when its petals, whether free or united, are equal, and form a symmetrical whorl; *irregular* (*c. irregularis*), when the reverse. A corolla may be formed of unequal divisions, and yet be regular; this is when the petals are alternately large and small, the small being all alike and the large all alike; or when its divisions are oblique, but all alike, the whole corolla being still symmetrical (*Periwinkle*, fig. 274).

Polypetalous Corollas.—The petals are clawed (*p. unguiculata*), when the broad part, or *limb* (*lamina*, fig. 9, L), is narrowed at the base into a petiole called the *claw*

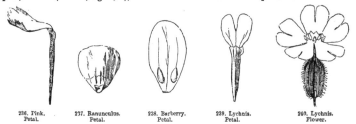

236. Pink.   237. Ranunculus.   238. Barberry.   239. Lychnis.   240. Lychnis.
Petal.   Petal.   Petal.   Petal.   Flower.

(*unguis, Wallflower*, fig. 9 ; *Pink*, fig. 236); the petals of the *Rose* and *Ranunculus* (fig. 237) are shortly clawed ; those of the *Philadelphus* and *Orange* are *sessile*.

The claw of the petal is *nectariferous* (*u. nectarifer*), when it bears a honey-secreting gland (*Ranunculus*, fig. 237) ; this gland may be protected by a scale (fig. 237), or naked (*Barberry*, fig. 238) ; and the claw itself is *naked* (*u. nudus*), when it bears neither gland nor scale (*Wallflower*, fig. 9 ; *Pink*, fig. 236); the claw is *winged* (*u. alatus*), when it bears a longitudinal membrane on its inner surface (*Rose Campion*). Little pits (*fornices*) are often found at the point of junction of the claw and limb, or forming small swellings inside the tube (*Lychnis Chalcedonica*) ; they also sometimes occur in monopetalous corollas. Small scales, placed within and on

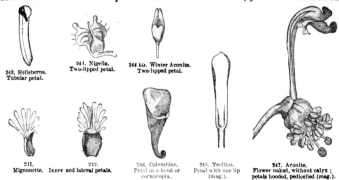

243. Helleborus. Tubular petal.

241. Nigella. Two-lipped petal.

244 *bis*. Winter Aconite. Two-lipped petal.

241. Mignonette. Inner and lateral petals.

242.

246. Columbine. Petal in a hood or cornucopia.

245. Trollius. Petal with one lip (mag.).

247. Aconite. Flower naked, without calyx ; petals hooded, pedicelled (mag.).

the top of the claw, forming a sort of crown around the androecium and pistil, are collectively called a *corona* (*coronula, Lychnis dioica*, figs. 239, 240; *Mignonette*, figs. 241, 242).

The limb of the petal may be *entire* (*Wallflower*, fig. 9), or *toothed* or *fringed* (*Mignonette*, figs. 241, 242).

Petals are generally *flat* (*p. plana*), like the leaves; but may be *concave* (*p. concava, Barberry*, fig. 238);—*tubular* with entire margins (*p. tubulosa, Helleborus fœtidus*, fig. 243);—*bilabiate* (*p. bilabiata*), or tubular with the mouth two-lipped (*Nigella*, fig. 244; *Winter Aconite*, fig. 244 *bis*);—*labiate* (*p. labiata*), when the tube terminates in a single lip (*Trollius*, fig. 245);—*hooded* (*p. cuculliformia, Columbine*, fig. 246; *Aconite*, fig. 247);—*spurred* (*p. calcariformia*), i.e. forming a spur or horn (*Heartsease*, fig. 248;

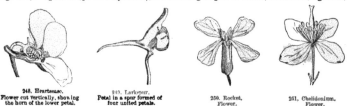

248. Heartsease. Flower cut vertically, showing the horn of the lower petal.

249. Larkspur. Petal in a spur formed of four united petals.

250. Rocket. Flower.

251. Chelidonium. Flower.

Larkspur, fig. 249). Hollow petals, of whatever form, usually enclose at the base a gland which is nectariferous when the flower expands, and the anthers open to shed their pollen.

The regular polypetalous corolla is *cruciform* (*c. cruciformis*), when it consists of four petals placed crosswise (*Rocket*, fig. 250; *Chelidonium*, fig. 251);—*rosaceous* (*c. rosacea*),˙when of five spreading, shortly-clawed, or sessile petals (*Rose, Strawberry*, fig. 252);—*caryophyllaceous* (*c. caryophyllea*), when of five clawed petals (*Lychnis*, figs. 239, 240).

The irregular polypetalous corolla is *papilionaceous* (*c. papilionacea*, *Cytisus*, figs. 253, 254), when composed of five petals, of which the upper or *standard*

252. Strawberry.   253. Cytisus.   254. Cytisus.   255. Cytisus.   257. Cytisus.
Flower.   Flower in profile.   Front view of flower.   Standard.   Petals forming the keel.

256. Cytisus. Left wing.

(*vexillum*, fig. 255) is placed next to the axis, and encloses the four others in bud; of these the two lateral *wings* (*alæ*, fig. 256) cover the two lower, which are contiguous, and often adhere by their lower margins, and together form the *keel* (*carina*, fig. 257).

Other irregular corollas are called *anomalous* (*c. anomala*, *Aconite*, *Pelargonium*, *Heartsease*, fig. 170).

Monopetalous Corollas.—In these, the *tube* consists of the united portions of the petals, the *limb* is the upper or free portion, the *throat* (*faux*) is the top of the tube, and is usually reduced to a circular opening, but is sometimes lengthened or dilated (*Comfrey*, fig. 268). It must be borne in mind that the term *limb*, as

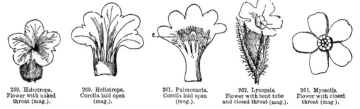

259. Heliotrope.   260. Heliotrope.   261. Pulmonaria.   262. Lycopsis.   263. Myosotis.
Flower with naked   Corolla laid open   Corolla laid open   Flower with bent tube   Flower with closed
throat (mag.).   (mag.).   (mag.).   and closed throat (mag.).   throat (mag.).

applied to the corolla, has two meanings; being used both to designate the blade of the leaf or petal, and the free upper portion above the tube of a gamopetalous corolla.

The throat is *appendiculate* (*f. appendiculata*) when furnished inside with, and often closed by, variously formed appendages, which often answer to external pits;—it is *naked* (*f. nuda*) when these are absent (*Heliotrope*, ·figs. 259, 260);—it is furnished with, but not closed by, long pencils of hairs in *Pulmonaria* (fig. 261);—closed by six swellings, each tipped with a pencil of hairs, and answering

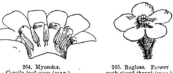

264. Myosotis.   265. Bugloss. Flower
Corolla laid open (mag.).   with closed throat (mag.).

to so many external pits, in *Bugloss* (figs. 265, 266);—closed by five swellings, answering to pits, in *Myosotis* (figs. 263, 264) and *Lycopsis* (fig. 262);—closed by

266. **Bugloss.**
Flower cut vertically.
(mag).

267. Borage.
Flower.

268. Comfrey.
Flower showing the pits between the sepals.

269. Comfrey.
Corolla laid open, showing the five scales between the five stamens.

five scales, conniving, and forming a conical roof over the tube, and answering to five external pits, in *Comfrey* (figs. 268, 269);—furnished with five emarginate scales, in *Borage* (fig. 267); and bearing a crown of long, narrow, cut scales, in *Oleander*.

The monopetalous limb is *bi- multi-partite*, when the petals cohere at their bases only (*Pimpernel*, fig. 277; *Borage*, fig. 267);

270. Campanula.
Flower.

271. Cerinthe.
Flower.

272. Chrysanthemum.
Flower with tubular corolla.

274. Periwinkle.
Flower.

273. Bindweed.
Flower.

—*bi- multi-fid*, when they cohere about half-way up, and the sinuses, as well as the segments, are acute (*Tobacco, Campanula,* fig. 270);—*bi- multi-lobate*, when the segments are obtuse or rounded (*Myosotis,* figs. 263, 264; *Heliotrope,* figs. 259, 260; *Bugloss,* fig. 265; *Comfrey,* fig. 268);—*toothed,* when the segments are very short (*Heath,* fig. 276).

The regular monopetalous corolla is *tubular* (*c. tubulosa*), when the tube is long and the limb erect and continuous with it (*Cerinthe,* fig. 271). The central flowers, called *florets* (*flosculi*), of *Chrysanthemum* (fig. 272) and allied plants with involucrate heads, have small tubular corollas. Such heads are called *flosculose.* They are *infundibuliform* (*c. infundibuliformis*), when the tube insensibly widens upwards like a funnel (*Bindweed,* fig. 273);—*hypocrateriform* (*c. hypocrateri-formis* or *-morpha*), when the straight and long tube abruptly terminates in a flat spreading limb, like an antique patera (*Lilac, Jessamine, Periwinkle,* fig.

275. Campanula.
Flower.

276. Heath.
Flower.

274; *Bugloss*, fig. 265);—*campanulate* (*c. campanulata*), when bell-shaped (*Campanula*, fig. 275);—*urceolate* (*c. urceolata*), when the tube is swollen in the middle, and the mouth contracted, like a small pitcher (*Heath*, fig. 276);

  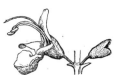

		281. Rosemary.	279. Lamium.	280. Galeobdolon.
277. Pimpernel.	278. Lamium.	Labiate corolla with upper	Front view of	Front view of
Flower.	Flower in profile.	lip upright.	flower.	flower.

—*rotate* (*c. rotata*), when the tube is suppressed, and the segments horizontal, and divergent like the spokes of a wheel (*Pimpernel*, fig. 277; *Borage*, fig. 267);— *stellate* (*c. stellata*), when rotate, with the segments very acute (*Galium*).

The irregular monopetalous corolla is *bilabiate* (*c. labiata*, *bilabiata*), when the limb is cut into two principal superimposed divisions (*lips*), and the throat is open; the upper lip consisting of two petals, and the lower of three. The upper lip may be entire, by the confluence of the two petals (*Lamium*, figs. 278, 279; *Galeobdolon*, fig. 280); or slightly split (*Sage*, *Rosemary*, fig. 281); or so deeply divided (*Germander*, figs. 282, 283) that the two petals stand widely apart, and are confluent with the lower lip rather than with one another. In this case the corolla appears to consist of one five-lobed lower lip. Lastly, the upper lip is sometimes wholly suppressed, or distinguishable from the tube only by a notch (*Bugle*, fig. 284). The mid-lobe of the lower lip may be *entire* (*Rosemary*, fig. 281);—*bifid* (*Lamium*, fig. 279; *Bugle*, fig. 284);—*trifid* (*Galeobdolon*, fig. 280).

The *personate* corolla (*c. personata*) is a form of the labiate, with the throat closed by a projection of the lower lip, called the *palate* (*palatum*); in many personate corollas the tube is tumid at the

282. Germander.	283. Germander.	284. Bugle.	285. Snapdragon.
Back of flower.	Flower in profile.	Labiate corolla with upper lip almost obsolete.	Flower.

base in the direction of the lower lip, and called *gibbous* (*c. gibbosa*, Snapdragon, fig. 285), or even *spurred* (*c. calcarata*, *Linaria*, fig. 286).

Two-lipped corollas are often described as *ringent* (*c. ringens*), but this term being equally applied to both the labiate and personate corollas, it is superfluous.

The *ligulate* corolla (*c. ligulata*) consists of five confluent petals, of which the two upper join at their base only, but unite almost throughout their length with the three others, as do these with each other, so that the corolla has a very short tube, and a limb entirely formed of a finely-toothed

291. Centranthus. Flower (mag.).

286 Linaria. Flower.

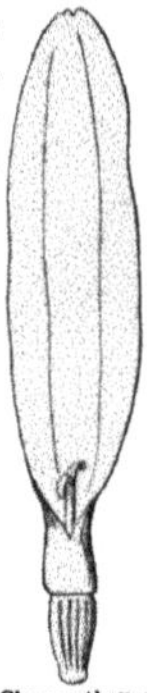

287. Chrysanthemum. Flower with ligulate corolla.

288. Foxglove. Flower with anomalous corolla.

289. Cornflower. Sterile floret (mag).

ligule (*Chrysanthemum*, fig. 287). Ligulate flowers are usually collected in an involucrate head, and are called *semi-florets* (*semi-flosculi*). A head (*capitulum*) composed of semi-florets is called *semi-flosculose* (*Dandelion*); one with tubular florets in the centre, and ligulate ones in the circumference, is *rayed* (*c. radiatum*, Chrysanthemum, Marigold).

290. Scabious. Ray floret.

All other irregular monopetalous corollas are considered to be *anomalous* (*c. anomala*). Of these the corolla of the *Foxglove* (fig. 288) resembles a thimble; the flowers on the circumference of the *Cornflower* (fig. 289) are large, irregular, and neuter; those of the *Scabious* (fig. 290) are also very irregular and almost labiate; and *Centranthus* (fig. 291) has an irregularly hypocrateriform corolla, with an inferior spur to the tube.

THE ANDRŒCIUM.

The *andrœcium* (*andrœcium*) is the simple or double whorl, placed within or above the corolla; the leaves composing it are called *stamens* (*stamina*).

A complete stamen (fig. 292) consists of a petiole or *filament* (*filamentum*, F) and a limb or *anther* (*anthera*, A); the anther is halved vertically by a median nerve, the *connective* (*connectivum*, C); each half consists of a *cell* (*loculus*, L) formed of two valves, the junction of which is marked externally by a furrow or *suture*. The back of the anther faces the corolla, and its face is opposite the pistil.

The cellular tissue of the anther-cells is originally soft, pulpy, and continuous; but when the anther is mature, this tissue becomes dry and powdery; the two valves then separate along the suture; the cell opens, and the parenchyma cells, now called *pollen*, are ready to be conveyed to the stigma. The anther is rarely sessile, i.e. without filament (*Arum*, fig. 293).

When the corolla is monopetalous, the stamens almost invariably adhere to it

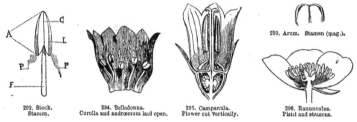

293. Arum. Stamen (mag.).

292. Stock.
Stamen.

294. Belladonna.
Corolla and andrœcium laid open.

295. Campanula.
Flower cut vertically.

296. Ranunculus.
Pistil and stamens.

(*Belladonna*, fig. 294);—amongst the few exceptions are *Heaths* and *Campanulas* (fig. 295).

Insertion of the Stamens.—This term relates to the position on the floral axis which the stamens occupy relative to the other whorls. The insertion of the corolla always coinciding with that of the stamens, in the staminiferous monopetalous corolla the insertion of the stamens may be inferred from that of the corolla. Thus the stamens, like the corolla, are *hypogynous* (*st. hypogyna*), when they do not adhere to the pistil or calyx, but spring from the receptacle below the base of the pistil (*Ranunculus*, fig. 296; *Primrose*, fig. 297);—*perigynous* (*st. perigyna*), when inserted on the calyx, rather above the base of the pistil, to which they are relatively lateral (*Apricot*, fig. 298; *Campanula*, fig. 295);—*epigynous* (*st. epigyna*) when inserted on the pistil itself (*Coriander*, fig. 299; *Madder*, fig. 300).

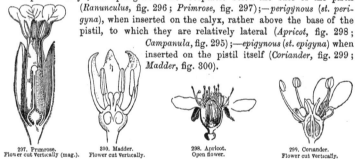

297. Primrose.
Flower cut vertically (mag.).

300. Madder.
Flower cut vertically.

298. Apricot.
Open flower.

299. Coriander.
Flower cut vertically.

The perigynous and epigynous insertions being easily confounded, the term *calycifloral* (*pl. calyciflorœ*) has been given to all plants whose corolla (whether mono- or poly-petalous) and stamens are inserted on the calyx, and this whether the calyx be below the ovary (*Apricot*, fig. 298), or above it (*Campanula*, fig. 295; *Coriander*, fig. 299; *Madder*, fig. 300). The term *thalamifloral* (*pl. thalamiflorœ*) has been given to plants whose polypetalous corolla and stamens are inserted below the pistil, or hypogynous; and *corollifloral* to plants with a monopetalous staminiferous corolla inserted below the pistil, or hypogynous (*Primrose*, fig. 297).

Number of the Stamens.--The flower is *isostemonous* (*fl. isostemoneus*), when the stamens equal the free or united petals in number (*Coriander*, fig. 299; *Primrose*, fig. 297);—*anisostemonous* (*fl. anisostemoneus*), when they are fewer than the petals (*Valerian*, fig. 301; *Centranthus*, fig. 291; *Snapdragon*, fig. 305), or more numerous than the petals (*Sedum*, fig. 302; *Horse-chestnut*, fig. 303; *Ranunculus*, fig. 296);—

diplostemonous (*fl. diplostemoneus*), when more than double the petals (*Ranunculus*, fig. 296; *Myrtle*, fig. 304). The flower, according to the number of stamens, from one to ten, is said to be

301. Valerian.
Flower (mag.).

302. Sedum.
Flower.

303. Horse-chestnut.
Flower.

304. Myrtle.
Flowering branch.

mon-, di-, tri-, tetr-, pent-, hex-, hept-, oct-, enne-, dec-androus; when above ten, the stamens are called *indefinite* (*st. plurima*), and the flower *polyandrous* (*fl. polyandrus*).

Proportions of the Stamens.—Stamens are not always equal: they are *didynamous* (*st. didynama*, Snapdragon, fig. 305), when four, of which two are the longest; this occurs in irregular monopetalous normally pentandrous flowers, in which four stamens alternate with four of the five lobes of the corolla, and the fifth stamen is suppressed. Stamens are said to be *tetradynamous* (*st. tetradynama*) when six, of which two are small and opposite, and four large, and placed in opposite pairs (*Wallflower*, fig. 306); these pairs being in juxtaposition, their fila-

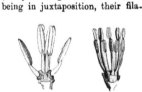

305. Snapdragon.
Androecium and half
of corolla.

306. Wallflower.
Androecium.

307. Stellaria.
Androecium.

308. Meconopsis.
Flower cut vertically.

ments sometimes cohere, so that each pair has been supposed to represent a double stamen. In polystemonous or diplostemonous flowers, the whorls of stamens are often unequal (*Stellaria*, fig. 307), but there is no special term for this modification.

Cohesion of the Stamens.—Stamens are *free* (*st. distincta, libera*), when completely independent of each other (*Meconopsis*, fig. 308) ;—*monadelphous* (*st. monadelpha*), when the filaments are more or less united in a single tube (*Oxalis*, fig. 309 ; *Mallow*,

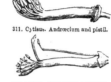

311. Cytisus. Andrœcium and pistil.

309. Oxalis.
Andrœcium and pistil.

310. Mallow.
Andrœcium (mag.).

312. Lotus. Andrœcium and pistil.

313. St. John's Wort.
Flower cut vertically.

fig. 310; *Cytisus*, fig. 311) ;—*diadelphous* (*st. diadelpha*), when united into two columns (*Lotus*, fig. 312) ;—*triadelphus* (*st. triadelpha*), when in three bundles (*St. John's Wort*, fig. 313) ;—*polyadelphus* (*st. polyadelpha*), when in several simple or branched bundles (*Orange*, fig. 314 ; *Castor-oil*, fig. 315) ;—*syngenesious* (*st. syngenesa*), when the anthers cohere (*Thistle*, fig. 316). Sometimes the cohesion extends to the filaments also (*Lobelia, Melon*, fig. 317). The stamens are said to be *gynandrous* (*st. gynandra*),

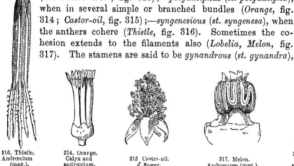

316. Thistle.
Andrœcium
(mag.).

314. Orange.
Calyx and
andrœcium.

315 Castor-oil.
♂ flower.

317. Melon.
Andrœcium (mag.).

318. Aristolochia.
Andrœcium
and pistil (mag.).

when they are united throughout their length to the pistil (*Orchis*, fig. 188; *Aristolochia*, fig. 318); in this case they are necessarily epigynous.

The *filament* may be cylindric or *filiform* (*Rose*), or capillary (*Wheat*, fig. 335), or *subulate* or *awl-shaped* (*Tulip*, fig. 345), or flat and dilated at its base (*Campanula*, fig. 319). It is said to be *bi- tricuspidate*, when forked at the top, or three-toothed, with the mid-tooth antheriferous (*Garlic-onion*, fig. 320; *Crambe*, fig. 321) ;—*appendiculate*, when it bears an appendage; such

319. Campanula.
Pistil and stamen.

320. Onion.
Stamen (mag.).

321. Crambe.
Andrœcium and pistil.

322. Borage.
Stamen (mag.).

appendages are of various shapes and sizes, and may be produced before or behind the anther. In *Borage* (fig. 322), the anther is *horned*; in *Self-heal* (fig. 323), it is forked, &c. In the *Mountain Alyssum*, the filaments of the long stamens bear a

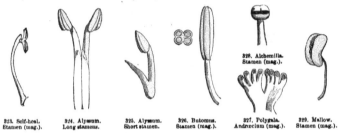

328. Alchemilla.
Stamen (mag.).

323. Self-heal.   324. Alyssum.   325. Alyssum.   326. Butomus.   327. Polygala.   329. Mallow.
Stamen (mag.).   Long stamens.   Short stamen.   Stamen (mag.).   Andrœcium (mag.).   Stamen (mag.).

toothed wing on their inner face (fig. 324), and those of the short stamens have an oblong appendage at the base in front (fig. 325).

Anther.—The anther is *two-celled* (a. *bilocularis*), when the two cells are separated by a connective (*Wallflower*, fig. 11) ; each cell being originally divided in two by a partition or plate springing from the connective, of which no trace remains at maturity ;—*four-celled* (a. *quadrilocularis*), when this partition remains (*Butomus*, fig. 326) ;—*one-celled* (a. *unilocularis*), when it presents only one cavity (*Polygala*, fig. 327 ; *Alchemilla*, fig. 328) ; this often happens, either by suppression of one cell (*Mallow*, fig. 329), when the filament is lateral ; or by fission of the stamen (*Hornbeam*, fig. 330). Sometimes the anther is seated on a flat-lobed connective, when it contains as many cells as there are lobes of the connective (*Yew*, fig. 331).

The anther is *adnate* (a. *adnata*), when its cells are confluent with the connective throughout their length (*Hepatica*, fig. 332). The connective is sometimes very short, connecting the anthers by a mere point. The anther is *didymous* (a. *didyma*), when the point of union of the cells is above their middle (*Euphorbia*, fig. 333) ; — *two-horned* (a. *bicornis*), when, the point of the union being at the base of the cells, the latter are

330. Hornbeam.   331. Yew.   332. Hepatica.   333. Euphorbia.   334. Heath.   335. Wheat.
Stamen (mag.).   Flower (mag.).   Stamen (mag.).   ♂ Flower (mag.).   Stamen (mag.).   Spikelet (mag.).

erect and slightly diverge (*Heath*, fig. 334) ;—*cruciate*, when the point of union of the cells is precisely in the middle, and their extremities are free (*Wheat*, fig. 335) ;—

sagittate (a. sagittata), when the upper portions only of the cells are united by the connective, and the lower portions slightly diverge (*Wallflower*, fig. 11; *Oleander*, fig. 340). The anther is usually *ovoid*, but may be *oblong, elliptic, globose, square*, &c.; it is *acute* in the *Borage* (fig. 322), and *sinuous* in the *Melon* (fig. 317).

The connective is sometimes developed

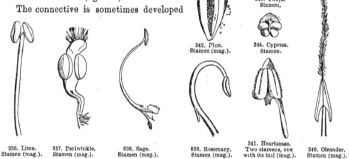

342. Pine.
Stamen (mag.).

343. Thuja.
Stamen.

344. Cypress.
Stamen.

336. Lime.
Stamen (mag.).

337. Periwinkle.
Stamen (mag.).

338. Sage.
Stamen (mag.).

339. Rosemary.
Stamen (mag.).

341. Heartsease.
Two stamens, one
with its tail (mag.).

340. Oleander.
Stamen (mag.).

transversely, when the two cells are placed wide apart; in the *Lime* (fig. 336) the filament appears to bear two unilocular anthers; in the *Periwinkle* (fig. 337), the cells are separate and tipped by a very thick connective; in the *Sage* (fig. 338) the connective is greatly produced, forming a bent arm, longer than the filament, and bearing a cell at either extremity; of these cells one alone contains pollen, the other usually enlarges into a petaloid scale; in the *Rosemary* (fig. 339) the second cell completely disappears.

The anther is often *appendiculate*. In the *Heath* (fig. 334), the appendages

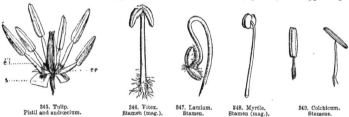

345. Tulip.
Pistil and andrœcium.

346. Vitex.
Stamen (mag.).

347. Lamium.
Stamen.

348. Myrtle.
Stamen (mag.).

349. Colchicum.
Stamens.

appear at the base of the cells as two small petaloid scales. In the *Oleander* (fig. 340) the connective is lengthened into a long feathery bristle. In the *Periwinkle* (fig. 337), the prolongation of the connective is large and hairy at the tip. In the *Heartsease* (fig. 341), the connective of two of the stamens lengthens above into a yellow, flat, triangular scale, and below into a glandular spur, which is lodged in the hollow spur of the petal. In the *Pine* (fig. 342), the anther is tipped by a bract-like connective. In *Thuja* (fig. 343), the filament bears a lateral three-celled anther, above which it dilates into a peltate disk. In the *Cypress* (fig. 344), the arrangement is the same, but the anther is four-celled. The anther is *basifixed* (*a. basifixa*), when attached to the filament by its base (*Wallflower*, fig. 11; *Tulip*, fig. 345) ;—*suspended* (*a. apicifixa*), when attached by its top (*Vitex*, fig. 346 ; *Lamium*, fig. 347); in this case the cells often diverge, their tops touch, and it becomes difficult to decide whether they are two-celled ;—*dorsifixed* (*a. dorsifixa*), when attached by the back (*Myrtle*, fig. 348) ;—*versatile* (*a. versatilis*), when it rocks upon its filament, which in this case is not contiuent with the connective, but attached to it by a finely pointed end (*Lily*, *Colchicum*, fig. 349).

The anther is *introrse* (*a. introrsa*), when the sutures are turned towards the centre of the flower (*Campanula*, fig. 319 ; *Thistle*, fig. 316 ; *Heartsease*, fig. 341) ;—*extrorse* (*a. extrorsa*), when the sutures are turned towards the circumference of the flower (*Iris, Ranunculus, Hepatica*, fig. 332) ; in these two cases the valves of each cell are unequal. The sutures are *lateral* when the valves are equal (*Myrtle*, fig. 348).

Dehiscence. — The *dehiscence*, or separation of the valves of each cell, may be *vertical* or *longitudinal* (*a. longitudinalis*), and either from top to bottom, or the reverse (*Wallflower*, fig. 11 ; *Campanula*, fig. 319) ;—or *transverse* (*a. transversa*), when it is horizontal, which principally occurs in unilocular anthers (*Alchemilla*, fig. 328) ;—or *apical* by pores or slits (*a. apice dehiscens*), in *Nightshade* (fig. 350), when the sutures open above only ;—or *valvate* (*a. valvula dehiscens*), when one valve of a cell comes away in one piece; in *Berberis* (fig. 351) the posterior valve dehisces near the connective, and ascends elastically like a trap; in *Laurel* the anterior valve does this ; in some Laurels with a four-celled anther, the dehiscence is by four such valves.

350. Nightshade. Stamen (mag.).

351. Berberis. Stamen (mag.).

Pollen.—Pollen varies in different plants, but is always alike in the same species ;[1] its grains are commonly ellipsoid (fig. 357) or spheroid (fig. 352), but sometimes *polyhedral* or *triangular* (*Œnothera*, fig. 353); their surface is *smooth, rugged, spinous* (*Rose-mallow*, fig. 352), or *reticulate*, &c.

The ripe pollen-grain generally consists of two membranes, the inner lining the outer, and containing a thick granular liquid, often mixed with minute oil-globules; this liquid, called the *fovilla*, is the essential part of the pollen.

The structure of the pollen-grains may be easily observed when they are moistened, which causes them to burst, from the inner membrane expanding more than the outer, and rupturing the latter. At certain points of its surface the outer membrane is thinner than elsewhere, and there folded inwards, or it presents dots which are regarded as *pores*. In most cases the membrane swells ; at these points the fold disappears, the dots or pores enlarge, and the outer membrane bursts at the thin part; the inner membrane, thus set free, emerges from

352. Rose-mallow. Adult pollen.

353. Œnothera. Ripe pollen.

---

[1] To this there are many exceptions.—ED.

,the openings in the shape of a small tubular bladder called the *pollen-tube* (fig. 353) ; this again soon swells, bursts in its turn, and allows the fovilla to escape in an irregular jet (fig. 354). Sometimes the thin portions are circular, and surround a sort of cap or covering (*operculum*), which is pushed off by the inner membrane (*Melon*, fig. 355).

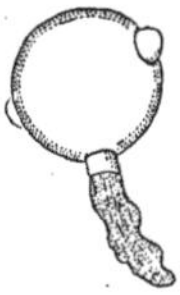

| 354. Cherry. Ripe pollen, ejecting the fovilla (mag.). | 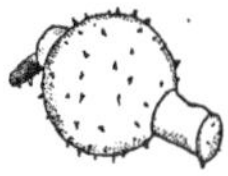 355. Melon. Ripe pollen (mag.). | 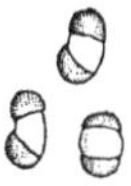 356. Pine. Ripe pollen (mag.). | 357. Polygala. Pollen, seen lengthwise (mag.). | 358. Polygala. Pollen, seen from above (mag.). |  359. Orchis. Pollen-masses, separated from the style, with their retinacula (mag.). |

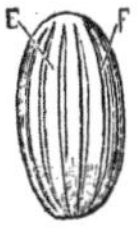

The pollen of the *Cherry* (fig. 354) and *Œnothera* (fig. 353) opens by three pores, giving passage to three pollen-tubes ; that of the *Melon* (fig. 355) by pushing off six discoid caps, which open like doors, or are completely removed by the pollen-tube. In *Pine* pollen (fig. 356) the outer membrane splits into halves by the distension of the inner. The pollen of *Polygala* (figs. 357, 358) resembles a little barrel, of which the staves, formed by the outer membrane (E), open by longitudinal clefts to allow of the passage of the inner membrane (F). The pollen of *Orchis* (fig. 359), instead of being powdery as in the previous cases, is composed of two waxy *masses* (*massœ pollinis*) supported on two small elastic stalks, named caudicles (*caudiculi*), and resting

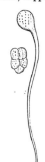

on a flat glandular base, called the *retinaculum* ; these masses present a series of small angular corpuscles (*massulœ*) joined by an elastic network, continuous with the caudicle ; each corpuscle again is formed of four pollen-grains, and each pollen-grain consists of a single membrane, which lengthens into a long tube containing the fovilla (fig. 360). The *retinaculum* is a portion of the anterior face of the style ; it secretes a viscous fluid, which agglutinates the originally free pollen-grains ; this viscous fluid is infiltrated between the grains, and adheres to them, then hardens, and forms the network which unites the grains together, and to the small stalk which bears the network (fig. 359).

361. Asclepias. Pistil and pollen masses adhering to the stigma.

360. Orchis. Pollen mass and tube.

The pollen of *Asclepias* (fig. 361) is very analogous to that of *Orchis* ; the five bilocular anthers are introrse, and rest against the sides of the stigma, which has five rounded angles ; each cell contains a compact mass of pollen, the grains of which are provided with a single membrane, and are closely united. At each angle of the stigma, between each pair of stamens, are two small viscous bodies (*retinacula*), from each of which a furrow descends towards and abuts on the contiguous cells of two adjacent anthers. These furrows contain a soft viscid fluid, secreted by the rectinacula; this fluid extends from the retinacula to the pollen-masses; soon the two retinacula unite and solidify, and the viscid fluid in the furrows solidifying at the same time, forms a double filament. This filament in hardening unites the two pollen-masses contained in the ' contiguous cells of two adjacent anthers, which thus form one body with the retinaculum, and remain suspended to it, much as the scales of a balance are suspended to the beam.

THE PISTIL.

The *pistil* or *gynœcium* (*pistillum, gynœcium*) is the whorl which crowns the receptacle and occupies the centre of the flower, of which it terminates the growth, just as the whole flower terminates the flowering branch.

In most cases, the pistil is inserted directly on the receptacle; but in some cases the internode from which it springs lengthens, when it is called a *gynophore* (*gynophorum*), and the pistil is said to be *stipitate* (*Fraxinella*, fig. 362; *Rue*, fig. 363).

The leaves composing the pistils are the *carpels* (*carpella, carpidia*); their number varies; they may form a single whorl (*Sedum, Columbine, Thalictrum*, fig. 364); or several (*Trollius*, fig. 365), or be solitary, by the suppression of one

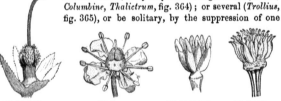

362. Fraxinella. Pistil and calyx. 363. Rue. Flower. 364. Thalictrum. Pistil. 365. Trollius. Pistil. 366. Bladder Senna. Pistil.

or more (*Bladder Senna*, figs. 366, 367; *Peach*, fig. 368). Under certain circumstances, the suppressed carpels may be developed, and complete the whorl (*Cherry*, fig. 369), which has then two carpels; or, as in some *Mimosas*, which have three to five, &c. The pistil is said to be *mono- bi- poly-carpellary*, according as there are one, two, or many carpels.

In the very young pistil, each carpel makes its appearance as a small round or pointed, more or less spreading scale, the edges of which gradually approach, and finally unite

367. Bladder Senna. Pistil cut vertically. 368. Peach. Pistil, portion of calyx and androecium. 369. Cherry. Pistil with two carpels.

and form a closed cavity; or, instead of uniting together, they may adhere to the edges of the contiguous carpels. The edges of the carpellary leaf (or sometimes its inner surface) present one or more small round bodies, attached to it directly or by a cord; these are the *ovules*, and will eventually become the seeds; the edges or surfaces bearing the ovules are the *placentæ*; the cord uniting the ovule to the placenta is the *funicle*; the limb of the carpellary leaf is the *ovary*; the upper portion of this limb, when it forms a slender prolongation,

370. Hellebore. Pistil.   371. Fennel. Pistil.   372. Flax. Pistil (mag.).   373. Stellaria. Pistil.   376. Primrose. Pistil (mag.).   374. Cactus. Pistil.   375. Lily. Pistil.

becomes the *style*; the extremity or top, which is variable in form, and always formed of a different tissue, is the *stigma*.

In the polycarpellary pistil the carpels are:—1, entirely separate (*c. distincta*, *Columbine*, fig. 12; *Thalictrum*, fig. 364; *Hellebore*, fig. 370); 2, coherent by their ovaries at the base only, or half-way up (*Fennel*, fig. 371), or to the top (*Flax*, fig. 372; *Stellaria*, fig. 373); 3, coherent by their ovaries and styles (*Cactus*, fig. 374; *Lily*, fig. 375); 4, coherent by their ovaries, styles and stigmas, so as to simulate a solitary carpel (*Primrose*, fig. 376; *Heartsease*, fig. 377); 5, coherent by their styles and stigmas only, their ovaries being free (*Periwinkle*, fig. 454; *Asclepias*, fig. 361).

Modern botanists, in deference to old usage, have continued to give the name of ovary to the union of several ovaries, which thus form a *compound ovary*; they have similarly retained the names of *style*, *stigma*, *placenta*, for the confluent styles, stigmas and placentas of several carpels.

377. Hearts-ease. Pistil.   378. Sedum. Carpel open (mag.).   379. Pine. Ovuliferous scale representing a carpel spread out, with neither style nor stigma.

When the ovaries are free, their edges, being folded inwards and united towards the centre of the flower, form an apparently single, but really double placenta, which, when the fruit ripens, often splits into two partially seed-bearing placentas (*Columbine*, fig. 13; *Sedum*, fig. 378). In some very rare cases (*Pine*, fig. 379; *Fir, Cypress, Thuja*) the carpels remain long spread open and quite free; later they approach and their surfaces unite, but without consolidating, and they thus form closed cavities in which the seeds are sheltered.

The ovary, whether simple or compound, is *superior* or *free* (*ov. superum, liberum*), when it adheres to none of the neighbouring organs (*Lychnis*, fig. 380; *Primrose*,

F

fig. 297). It is *inferior* (*ov. inferum*) when, instead of being placed above the level of the andrœcium, corolla and calyx, it is (apparently) below them, although still

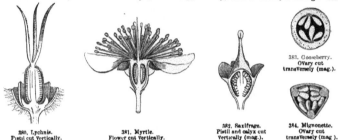

380. Lychnis.
Pistil cut Vertically.

381. Myrtle.
Flower cut Vertically.

382. Saxifraga.
Pistil and calyx cut
Vertically (mag.).

383. Gooseberry.
Ovary cut
transversely (mag.).

384. Mignonette.
Ovary cut
transVersely (mag).

retaining its central position (*Myrtle*, fig. 381). Most modern botanists explain this latter arrangement by assuming that the ovary is consolidated with the calyx-tube; —a theory which prevailed during the first half of the present century, and the expressions 'ovary adhering to the calyx' and 'calyx adhering to the ovary' have been employed in all Floras and descriptive works. But a closer study of the development of organs has shown that the so-called adherent calyx-tube is in reality a cup-shaped expansion of the receptacle, which has enveloped the ovary, and that the calyx only commences at the same point as the stamens and petals. Hence, what has hitherto been called an adherent calyx-tube, ought to be called a receptacular tube or cup. We shall return to this question when speaking of the Torus.

The ovary is said to be *half-inferior* (*ov. semi-inferum*) or *half-adherent* (*ov. semi-adhærens*), when it does not wholly adhere to the receptacular tube (*Saxifrage*, fig. 382).

In the compound ovary (whether free or inferior) the partial ovaries may be variously united:—1, the edges touch (*Gooseberry*, fig. 383; *Mignonette*, fig. 384;

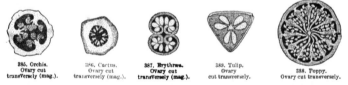

385. Orchis.
Ovary cut
transVersely (mag.).

386. Cactus.
Ovary cut
transversely (mag.).

387. Brythræa.
Ovary cut
transversely (mag.).

389. Tulip.
Ovary
cut transversely.

388. Poppy.
Ovary cut transversely.

Orchis, fig. 385; *Cactus*, fig. 386), when their union is marked by two contiguous placentas belonging to two different carpels; the placentas are then said to be *parietal* (*pl. parietales*), and the compound ovary is one-celled (*ov. uniloculare*); 2, they are folded inwards so as to form vertical partitions, each composed of two confluent plates called *septa* (*septa, dissepimenta*), belonging to different carpels; ·these septa are *incomplete* if they do not reach the axis of the flower, so as to unite;

the placentas are then *parietal*, and the ovary one-celled (*Erythræa*, fig. 387; *Poppy*, fig. 388); the septa are *complete* if their edges meet in the axis of the flower; a prolongation of the receptacle sometimes traverses this axis, which then forms a *column* (*columella: Mallow, Tulip*, fig. 389; *Campanula*, fig. 390); through this column, whether in its origin it be receptacular, or (as is more usual) through the placentas, the nourishment of the ovules is conveyed, as well as through the carpels. When the septa are complete, there are as many cells as carpels, and the compound ovary is two- or more celled (*ov. duo- pluri-loculare*); and the placentas, united in pairs (two to each carpel), are central.

390. Campanula.
Ovary cut transversely.

The septa are usually formed from the endocarp of the carpels, with an interposed expansion of the mesocarp. *Spurious dissepiments* (*d. spuria*) are vertical or horizontal septa, which are not formed by the union of the inflexed faces of two contiguous carpels; thus, in *Astragalus* (fig. 391), the solitary carpel is almost two-celled by an intruded vertical plate formed by a fold of the dorsal face; in *Flax* (fig. 392), where there are ten septa, five project from the midribs of the carpels towards the axis, which they do not always reach. In *Datura* (fig. 393), the three carpellary ovary is four-celled from the inflexed contiguous faces of the carpels, after uniting in the axis, being reflexed inwards, and meeting a prolongation from the midrib of the carpel: the placentas are thus borne on a septum composed partly of

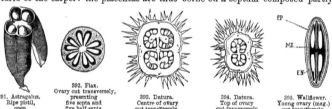

391. Astragalus.
Ripe pistil,
open.

392. Flax.
Ovary cut transversely,
presenting
five septa and
five half-septa.

393. Datura.
Centre of ovary
cut transversely.

394. Datura.
Top of ovary
cut transversely.

395. Wallflower.
Young ovary (mag.)
cut transversely.

the inflexed and then reflexed carpellary faces, and partly of a prolongation from the midrib. In the upper part of the ovary, the accessory septa (formed from the midrib) disappear, and two cells only are seen (fig. 394). In the *Wallflower*, and allied plants (fig. 395), the two carpels are pressed together; along each of their two edges runs a double seed-bearing fibro-vascular bundle; these are the four placentas arranged in pairs; the pistil is two-celled, by a delicate and almost transparent false septum, to which the placentas form a sort of frame. This septum is supposed to be formed by the placentas; for, when young, it is seen to be composed of four plates, which spring in pairs from each pair of placentas, and advance inwards till they join together; later, this false septum appears formed of a single membrane, but it retains in the centre the trace of its double origin, in a vertical median line, along which it is easily divided without tearing.

In *Coronilla* and *Cassia* (fig. 502), the young carpel is one-celled, but at a later period is divided into superimposed cells by septa formed of the parenchyma of the ovary, which is intruded horizontally between the seeds.

Spurious cells (*loculi spurii*) are cavities in the ovary which do not contain seeds. The young ovary of *Nigella* présents five cells, each containing two piles of ovules; later (fig. 396) there appear ten cells, of which five in the centre of the fruit contain seeds attached to their interior angle; the other five are exterior to these, and are empty, and due to the inflation of the epicarp (EP), which in swelling has dragged with it the mesocarp (M), whilst the endocarp (EN) has re- mained in its place.

396, Nigella.
Ripe ovary
cut transversely.

398, Lychnis.
Young ovary (mag.)
cut transversely.

397, Cyclamen.
Pistil
cut vertically.

Central placentas are said to be *free* (*p. centrales, liberæ*), when they are not united by septa to the walls of the ovary, and appear com- pletely independent of the carpels; this placentation is characteristic of *Primulaceæ* (*Pimpernel, Primrose, Cyclamen*, fig. 397). To explain this isolation of the placentas, it is assumed that the edges of the carpellary leaves join throughout their length, and constitute a one-celled ovary, but that their basal edges dilate, and ascend in the middle of the cell to form a central mass of placentas. The placentas of *Primulaceæ* are thus confined to the bases of the carpels. The reverse is the case in the one-celled ovary of *Combretaceæ*, where the ovules spring from the top of the cell.

In most *Caryophylleæ* (*Pink, Lychnis*), the placentas appear to be free, but this arises from the early evanescence of the septa, which can only be well seen in the very young flower (fig. 398).

Some German and French botanists regard the carpellary leaf as a protective organ merely; denying that it has the power of producing buds, and limiting this power to the floral axis. According to these, the axis alone produces ovules, and the carpellary leaves protect them. In the case of many-celled ovaries, they regard the edges of the carpellary leaves as folded inwards till they reach and cohere with the axile placentas (which in no wise belong to them), the fibro-vascular bundles of the placentas losing themselves in the tissue of the styles, which are continuations of the midribs of the carpels. In unilocular compound ovaries they consider that the placentiferous axis branches like the spokes of a half-opened parasol, and that the branches run along the contiguous edges of the carpellary leaves (*Heartsease, Mignon- ette*, fig. 384; *Orchis*, fig. 385).

This modification of the carpellary theory of placentation rests on the isola- tion of the placentas in *Primulaceæ* (fig. 397); on the enormous disproportion of the placentas relatively to the carpellary leaves in various plants (*Lychnis*, fig. 398; *Campanula*, fig. 390); and on the arrangement of the nerves in certain ovaries (*Pea*, fig. 14; *Columbine*, fig. 13), wherein two systems of fibro-vascular bundles are distinctly visible; the one coming from the median nerve, the others rising from the placentas, and communicating with the first; which seems to indicate a union between the axis and carpels.

The flower is *isogynous* (*fl. isogynus*), when the carpels of which the pistil is composed equal the sepals in number (*Sedum*);—*anisogynous* (*fl. anisogynus*), when the carpels are fewer in number than the sepals (*Saxifrage, Snapdragon, Comfrey*);—*polygynous* (*fl. polygynus*), when the carpels are more numerous than the sepals (*Ranunculus, Poppy*).

In pistils formed of consolidated carpels, the number of the latter is determined, either by the number of styles, when these are free, or by the number of septa, or by the number of placentas, which are usually in pairs, and form vertical series, or fleshy protuberances. In pistils with parietal ovules (*Butomus, Poppy, Gentian*) the number of stigmas or styles or septa must be examined.

The two- or more ovuled ovary (whether simple or compound, free or adherent) is always called *many-ovuled* (*ov. pluriovulatum*). All ovaries are supposed to be normally many-ovuled, for each carpel having two placentas, and each placenta being normally one- or more ovuled, it follows that no ovary should have fewer than two ovules. A one-ovuled ovary (*ov. uniovulatum*) is hence regarded as resulting from the suppression of one

399. Peach.
Young ovary (mag.)
cut transversely.

400. Oak.
Young ovary (mag.)
cut transversely.

or more ovules. The young ovary often contains two or more ovules, of which all but one are subsequently suppressed, as in the *Peach* (fig. 399), which is always two-ovuled when young; and in the *Horse-chestnut* and *Oak*, which have six ovules (fig. 400). The compound ovary is usually globose or ovoid; it is *lobed* (*ov. lobatum*), when the dorsal faces of the carpels are very convex, and separated by deep furrows (usually indicating the lines of junction, fig. 225), and according to the number, it is *bilobed, trilobed*, &c.

The carpels are not always whorled; but are sometimes arranged in a spiral, when they form a head or spike; the receptacle at the same time lengthening into a hemispheric, conical, or cylindric axis (*Strawberry*, fig. 401; *Raspberry*, fig. 402; *Ficaria*, fig. 403; *Adonis*, fig. 404). *Roses* (fig. 405) present a precisely reverse arrangement; the carpels (ov), instead of rising

   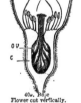

401. Strawberry.
Flower cut vertically.

402. Raspberry.
Ripe pistil, cut vertically.

403. Ficaria.
Carpels arranged
in a head.

404. Adonis.
Pistil (mag.).

405. Rose.
Flower cut vertically.

from a plane or convex surface, spring from the walls of a cavity (c); which will be described under the *torus*. In this (exceptional) case, the carpels are said to be *parietal* (*ov. parietalia*).

The compound style is improperly said to be *simple* (*st. simplex*), when wholly undivided; it is *bi- tri-fid*, &c., when the component styles cohere beyond the middle; *bi- partite*, &c., when they do not cohere to the middle. The styles of each carpel rarely bifurcate once or.twice; when they do, they are double or quadruple in number to the carpels (*Euphorbia*, fig. 406).

The style is *terminal* (*st. terminalis*), when it springs from the top of the ovary (*Apricot*, fig. 411);—*lateral*, when it springs more or less from the side of the carpel, the top of which appears bent downwards (*Strawberry*, fig. 407);—*basilar* (*st. basilaris*), when the top of the ovary is

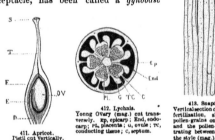

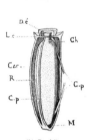

| 408. Alchemilla. Carpel (mag.). | 406. Euphorbia. Pistil. | 407. Strawberry. Carpel (mag.). | 409. Comfrey. Pistil and calyx cut vertically. | 410. Sage. Lower portion of flower, cut vertically. |

bent down to a level with its base (*Alchemilla*, fig. 408). When there are many ovaries, with confluent *basilar* styles, the style is said to be *gynobasic* (*st. gynobasicus*, *Comfrey*, fig. 409), and the dilated base of this composite style, extending below the ovaries and surface of the receptacle, has been called a *gynobase*

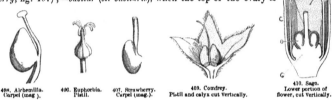

| 411. Apricot. Pistil cut vertically. | 412. Lychnis. Young Ovary (mag.) cut transversely. Ep, epicarp; End, endocarp; Pl, placenta; o, ovule; Tc, conducting tissue; c, septum. | 413. Snapdragon. Vertical section of style during fertilisation, showing two pollen-grains on the stigma, and the pollen-tubes penetrating between the cells of the style (mag.). | 414. Dandelion. Young pistil (mag.), open to show the two cords, c.p, of the conducting tissue, of which one is broken. Car, ovary; L.c, calyx; D.é, epigynous disk; R, raphe; Ch, chalaza; M, micropyle. |

(*gynobasis*). The gynobase is sometimes prolonged into a *gynophore* (*Sage*, fig. 410, G); but a gynophore proper must not be confounded with the gynobase; the gynobase belongs to the styles, that is, to the carpels; the gynophore proper belongs to the axis itself, of which it is the termination.[1]

The style is a portion of the carpellary leaf, contracted into a sort of longitudinal tube, filled with a moist and loose parenchyma, named *conducting tissue*

[1] Except under the view that the placentæ are productions of the axis.—Ed.

(fig. 411, T); it is this tissue, which, spreading over the top or sides of the style, forms the spongy surface called the stigma (s). The same tissue descends from the style into the cavity of the ovary (fig. 412, TC), passes along the placentas (PL), and covers with its loose cells the micropyle of each ovule (G); and it is between these cells (fig. 413) that the pollen-tube, leaving the pollen-grain on the stigma, effects a passage to and fertilizes the ovule.

In *Compositæ*, the conducting tissue consists of two threads (fig. 414, C.p, C.p), which descend from the base of the style upon the sides of the ovule, without adhering to it; at its base they join and enter the base of the funicle, near the micropyle.

In *Statice* (fig. 415), according to Mirbel, the conducting tissue (tis. c) resembles a pestle; it enters the cavity of the ovary, immediately above the gaping micropyle of the ovule (ov.), which is suspended from a basal cord (cor.). This conducting tissue rests on the micropyle like the stopper of a decanter, and is visible after fertilization (fig. 416).

417. Daphne.
Pistil.

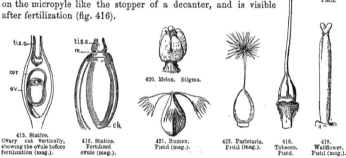

415. Statice.
Ovary cut vertically, showing the ovule before fertilization (mag.).

416. Statice.
Fertilized ovule (mag.).

420. Melon. Stigma.

421. Rumex.
Pistil (mag.).

422. Parietaria.
Pistil (mag.).

418.
Tobacco.
Pistil.

419.
Wallflower.
Pistil (mag.).

The stigma (figs. 413 and 411 s) is nothing but the conducting tissue spread out; the stigmatic surface has no epidermis, and is usually spongy, damp, and papillose, and thus suited to retain the pollen.

The stigma (whether simple or compound) is *complete* (*st. completum*) when it is continuous with the style, and clearly distinguishable. The complete stigma may be *globular* (*Daphne*, fig. 417), *hemispheric* (*Primrose*, fig. 376), *round* (*Tobacco*, fig. 418), *forked* (*Wallflower*, fig. 419), *bi-lamellate* (*Datura*), *lobed* (*Lily*, fig. 375; *Melon*, fig. 420), *laciniate* or *fringed* (*Saffron*, *Rumex*, fig. 421), *penicillate* (*Parietaria*, fig. 422), *plumose* (*Wheat*, fig. 423), *discoid*, *conical*, *cylindric*, *club-shaped*, *awl-shaped*, &c. It is *superficial* (*st. superficiale*) when confined to the surface of a part of the style or ovary, and only

424. Vetch.
Pistil.

425. Ranunculus.
Carpel (mag.).

423. Wheat.
Flower (mag.).

distinguishable by its papillæ.   The superficial stigma is *terminal* in *Fraxinella* (fig. 362), *Strawberry* (fig. 407), *Sweet Vetch* (fig. 424); *lateral* in *Ranunculus*, where it is hooked (fig. 425); and in *Hearts-ease* (fig. 377), where it forms a hollow ball with a two-lipped

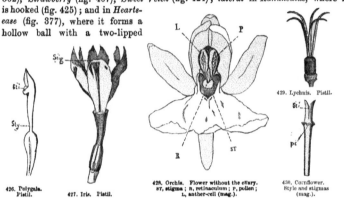

426. Polygala. Pistil.

427. Iris. Pistil.

428. Orchis. Flower without the ovary. ST, stigma; R, retinaculum; P, pollen; L, anther-cell (mag.).

429. Lychnis. Pistil.

430. Cornflower. Style and stigmas (mag.).

orifice; and in *Polygala* (fig. 426), where it forms a small very short lip (Sti.) on the sides of a style (Sty.) hollowed into a funnel, and spoon-shaped at the end;— in *Iris* (fig. 427), in which the composite style divides into three petaloid plates with two unequal lips, the interior of which is bifurcate, the stigmatic surface (Stig.) occupies a small transverse fissure between these lips;— in *Orchis* (fig. 428), where it forms a shining and viscous cup (ST) situated below the retinaculum (R);—in *Lychnis* (fig. 429) where it is papillose and transparent, clothing

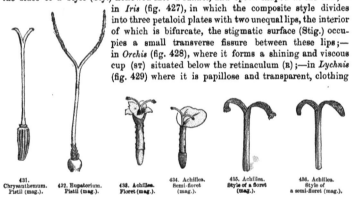

431. Chrysanthemum. Pistil (mag.).

432. Eupatorium. Pistil (mag.).

433. Achillea. Floret (mag.).

434. Achillea. Semi-floret (mag.).

435. Achillea. Style of a floret (mag.).

436. Achillea. Style of a semi-floret (mag.).

the furrowed inner faces of the styles;—in *Plantain*, where its papillæ form two velvety lines along the style.

The stigma must not be confounded with certain peculiar hairs which sometimes garnish the style, and are almost always directed obliquely upwards, and intended to catch the pollen; they are most frequent in flowers with contiguous introrse anthers. In these plants the young style is much shorter than the stamens; it grows rapidly as the flower expands, and traverses the tube formed by the stamens, where its hairs, rubbing the anther-cells, open them, and sweep out the pollen which adheres to them; they are hence called *collecting hairs* or *brushes* (*pili collectores*). In the *Cornflower* (fig. 430) the stigmas (Sti.) are lateral and superficial, as in the Lychnis, and below them is a small swelling clothed with a tuft of very small collecting hairs (pc). In the *Chrysanthemum* (fig. 431), the two style-branches are papillose on their inner faces, and tipped by a little tuft of collecting hairs. In *Eupatorium* (fig. 432) the two style-branches are cylindric and bristle with collecting hairs; and the stigmatic surfaces form a little band which extends from the fork half-way up the branches. In *Achillea* (figs. 433, 434), the heads of which are rayed, the central florets are tubular and hermaphrodite (fig. 433), and the circumferential are female semi-florets (fig. 434). Here the style-arms of the central florets (hermaphrodite) are papillose on the inner face, and tipped with a brush of collecting hairs; the semi-florets again, being female only, their style-arms (fig. 436) have no collecting hairs (fig. 435), but, as the pollen of the centre florets may reach them, their style-arms are papillose, so as to retain the pollen and secure fertilization. In *Campanula* the five style-branches (fig. 437) are papillose on the inner face, and subtended by five rows of collecting hairs, each row being double, and answering to the two halves of each anther. Before expansion, the style grows rapidly, the anthers open, and their pollen-grains, which bristle with hooks, adhere firmly to the hairs which have swept them; this accomplished, the collecting

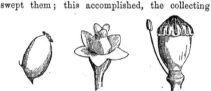

437. Campanula.   439. Vine.   438. Arum.   440. Elder.   441. Poppy.
Pistil.   Androecium and pistil (mag.).   Pistil (mag.).   Pistil and calyx (mag.).   Pistil.

hairs are retracted within themselves, like the horns of a snail; the pollen then disappears, and the style becomes clean, its surface being merely a little rough.

The stigma is *sessile* (*st. sessile*) when, there being no style, it is seated directly on the ovary. In the *Tulip* (fig. 345) it forms three bi-lobed crests;—in the *Nettle*, a pencil;—in *Arum* (fig. 438), a little papillose tuft;—in the *Vine* (fig. 439), a flattened head;—in the *Elder* (fig. 440), three rounded lobes; in the *Poppy* (fig. 441), velvety radiating double crests, clothing the depressed styles, which together resemble a shield or cap with scalloped edges.

The stigma is sometimes absent, and then the ovary remains open; this is the case with *Pine* (fig. 379), *Cypress*, and *Thuja*, the female flowers of which are arranged in a spike; each is furnished with an outer bract, which soon withers and disappears; each is formed of a scale representing an open carpel, without style or stigma, bearing at its base two ovules with a gaping micropyle; after fertilization, these carpels thicken, harden, and become appressed, and form closed cavities which protect the seeds during their ripening.

TORUS, DISK, NECTARIES.

The *torus* is the part of the receptacle situated between the calyx and the pistil on which the corolla and andrœcium are inserted. It is merely the periphery of the receptacle, and not a special organ; but for convenience of description it is so considered.

The torus produces, besides stamens and petals, nectariferous glands and sundry

442. Columbine.
Pistil surrounded by scales. Torus showing the scars left by the stamens (mag.).

443. Tree Peony.
Flower without the corolla and most of the stamens.

444. White Water-Lily.
Pistil and cup bearing the petals and stamens.

445. Orange.
Vertical section of pistil and receptacle.
T, torus ; c, calyx.

expansions analogous to petals or stamens. Thus, in the *Columbine* (fig. 442), between the andrœcium and the pistil, are ten membranous silvery white scales, with folded edges, larger at the base than at the top, which may be considered as filaments, and which sometimes bear an anther at their extremity. In the *Tree Peony* (fig. 443), the thick swollen torus elongates into a membranous cup surrounding the carpels, without adhering to them, and open at the top to afford a passage to the stigmas : it appears to form a part of the fruit, from which it is nevertheless very distinct. This petaloid involucre sometimes bears anthers. In the *White Water Lily* (fig. 444) the stamens and petals cohere with the torus, which envelops the ovary, so that they appear to adhere to the ovary; they die after flowering, leaving the torus marked with their scars. In the *Yellow Water Lily*, the thick cup, externally green and

446. Mignonette.
Flower without corolla (mag.).

flaccid, which some botanists have considered as a torus enveloping the ovary, is nothing but the epicarp of the ovary; at maturity it bursts irregularly, and comes away, leaving the seeds retained by the endocarp, when they fall to the bottom of the water and germinate.

The torus often forms, below the ovary, a projecting ring or swelling, from which spring the stamens and petals (*Orange*, fig. 445 T ; *Mignonette*, fig. 446); but more often this ring, reduced to its most simple form, only appears as a circular line on the receptacle, between the pistil and the calyx (*Chelidonium*, fig. 447). In every case the andrœcium and corolla, being inserted on this ring and below the pistil, are hypogynous, and the plant thalamifloral if the petals are free, corollifloral if they are coherent.

In many plants the receptacle dilates into a cup, which represents a calycinal tube, over which the torus is spread, and the stamens and pistils spring from its outer margin (*Strawberry*, fig. 401; *Apricot*, fig. 449). In others it rises upon the carpels, envelops them closely, and forms with

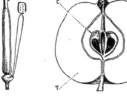

448. Apple.
Young fruit cut transversely.

447. Chelidonium.
Pistil (mag.).

449. Apple.
Fruit cut vertically.

them but one body, upon the circumference of which the stamens, petals and calyx are inserted at a higher level than the ovary (*Myrtle*, fig. 381; *Saxifrage*, fig. 382). This cup, enveloping the carpels and formed by the growth of the receptacle, is the *calyx-tube* of modern Floras, which it would be better to call a *receptacular* tube or cup.

This hypertrophy of the receptacle is particularly striking in orchard fruits. If we halve an unripe pear or apple (fig. 448), we find five carpels, forming five two-ovuled cells, surrounded by a fleshy mass, the so-called *calycine-tube* (better called *receptacular cup*), which has closely enveloped them, and agglutinated them by their lateral faces, but left their inner edges free. A vertical section of a ripe apple (fig. 449) exhibits a fibro-vascular bundle, extending from the peduncle, with which it is continuous, to the carpels (ε); it is the parenchyma of the receptacle, which has here enormously increased in bulk to envelop the ovaries (τ); at the summit of this mass, that is to say, at the top of the fruit, the remains of the sepals and stamens may be seen carried up by the expansion of the receptacle.

The receptacular theory of the calycine-tube completely explains the arrangement of the carpels of a *Rose* (fig. 405). In this, the position of the carpels on the internal wall of a calycine-tube was difficult to admit; the whorls of the flower being lateral expansions of the axis, it was impossible, in defiance of the law of the evolution of floral whorls, to attribute to the calyx the power of producing carpels. The position of the coloured ring from which the petals and stamens rise is the key to the apparently abnormal position of the carpels; this ring surmounts the ovoid body enclosing the carpels; the torus has therefore reached that point before emitting laterally the petals and stamens; and since the torus is nothing but the circumference of the receptacle, it is evident that it must be the latter organ which constitutes the hollow body enclosing the carpels. In fact, the receptacle, instead of forming, as in the *Strawberry* (fig. 401), a hemisphere, has swollen, risen much above its ordinary level, and formed a sort of cup; thus resembling the finger of a glove turned inside out, the normally outer or convex surface becoming the inner, or concave, one. Were the convex receptacle of the Strawberry reduced to a thin membrane, and turned inside out, the sepals would then form a ring round the mouth of a sort of bottle, represented by the inverted receptacle, whose throat would be occupied by the stamens and petals, and its inner surface by the ovaries; and the strawberry would be thus changed into a rose. The last evidence of the hollow body of the rose being a cup-shaped expansion of the axis rests on the cases in which the receptacle forms, instead of a cup, a central convex projection, which bears carpels; the rose thus being converted into a strawberry.

In all these cases the plant is calycifloral; the stamens and petals are not hypogynous, as in the *Lychnis* (fig. 380) and *Primrose* (fig. 297), but are inserted above the base of the pistil, at the distal end of the torus (*Sumach*, fig. 450), or on the outer circumference of a ring or cup formed by the torus (*Circæa*, fig. 450 *bis*; *Alchemilla*, fig. 451); they are thus either *perigynous* or *epigynous*, accord-

451. Alchemilla.
Flower cut vertically (mag.).

450. Sumach.
Flower cut vertically (mag.).

452. Nasturtium.
Flower cut vertically.

ing to their insertion *around* (fig. 450) or *above* the ovary (fig. 450 *bis*). When the torus both spreads over the base of the calyx and around that of the ovary, the andrœcium may be *hypogynous*, and the corolla *perigynous*; this is very rare, but occurs in *Tropæolum* (fig. 452).

The term *disk* has been reserved for the tumid ring which, in hypogynous flowers, surrounds the base

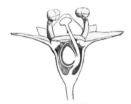 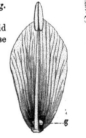

449 *bis* Circæa.
Flower cut vertically.

453. Radish.
Pistil and nectaries.

455. Sedum.
Pistil and nectaries (mag.).

457. Parnassia.
Petal and nectaries.

456. Fritillary.
Stamen, petal and nectary.

454. Periwinkle.
Pistil and nectaries.

of the ovary (*Orange*, fig. 445); and for the thickening which crowns the inferior ovary, enclosing the base of the style (*Circæa*, fig. 450 *bis*). These thickenings of the torus are glandular, and usually secrete honeyed fluids, whence they have been classed with nectaries, of which we are about to speak.

Nectaries or *nectariferous glands* are usually developed from the torus, and placed upon it or the organs developed from it.

The receptacle of the *Radish* (fig. 453), *Wallflower* (fig. 10), and other *Cruciferæ*, bears four or six glands;—the *Periwinkle* (fig. 454) two;—*Sedum* (fig. 455) five;—most *Gesneriaceæ* also five; but in this Order all intermediates between five free glands and a large hypogynous or epigynous disk are to be found. In the *Straw-berry* (fig. 401), *Peach* (fig. 368), and other *Rosaceæ*, the orange-yellow layer of the torus, which is spread over the calyx, secretes superficially a honeyed liquor; but often for so short a time that it is difficult to observe it. In *Ranunculus* (fig. 237) a small nectary occurs, protected by a scale, at the base of the claw of each petal. In *Berberis* (fig. 238), each petal bears, a little above the base, two naked ovoid necta-ries. In the *Fritillary* (fig. 456), the six petaloid perianth-segments each bear a nectary a little above the base, which, instead of projecting, forms a furrow. In the *Lily*, a double nectariferous furrow extends along the face of the midrib of each petal. In *Parnassia* (fig. 457), opposite each petal there is a petaloid scale which ramifies into three, five, seven, nine, or fifteen branches, each tipped by a globular nectary.

Nectaries are sometimes on the tip or base of the connective of the stamens, as in *Adenanthera, Prosopis*, &c. In *Heartsease* (fig. 458), two nectaries pro-ceed from two of the stamens, and, projecting from the connective at the base of the anther, form two recurved tails, sheathed in the hollow horn of the lower petal, at the base of which they secrete a sweet liquor from their tips.

458. Heartsease. Pistil and androecium.

It has already been remarked that hollow petals contain a nectary in their cavity (*Columbine*, fig. 246; *Aconite*, fig. 247; *Nigella*, fig. 244; *Helle-bore*, fig. 243; *Winter Aconite*, fig. 244 *bis*).

In monopetalous corollas the nectaries may be superficial (*Honeysuckle, Lilac*), or occupy a cavity which externally forms a boss or spur (*Linaria*, fig. 286; *Snap-dragon*, fig. 285; *Centranthus*, fig. 291); in the latter case the corolla is irregular, and the stamens are often imperfect; but it is difficult to say whether the necta-ries are the cause or effect of this irregularity.

Nectaries are not confined to the torus; they are found on the external surface of the calyx in *Malpighiaceæ*; and a glandular secreting layer occurs in the thick-ness of the septa of the ovary of *Liliaceæ*, named by Brongniart 'glandes septales.' In unisexual flowers, it often happens that the absent organs are replaced by necta-ries (*Melon*, and many other diclinous plants).

ARRANGEMENT OF APPENDICULAR ORGANS AROUND THE AXIS.

Appendages or appendicular organs are lateral developments from the vegetable axis:—the leaves, bracts, sepals, petals, stamens, and carpels.

It has been .stated (*Introd.* p. 2), that leaves are either opposite, whorled, or alternate; as also (p. 42) that the floral organs (calyx, corolla, andrœcium and pistil) are normally whorled; but we have warned the reader that very frequently the leaves of each series, instead of forming a true whorl, are arranged in successive flattened spirals, though still retaining the name of whorls.

We will now advert somewhat in detail: 1. To the arrangement of *leaves* properly so called, *carpellary* leaves, and *bracts* (this branch of Botany is called *Phyllotaxy*) ; 2. To the arrangement of the petals and sepals, an arrangement termed *Vernation*, because it can only be satisfactorily studied before the flower expands.

PHYLLOTAXY.

When leaves are clearly whorled, either in twos (*opposite*), threes, fours, fives, &c., they are generally separated by equal intervals, and consequently the arc comprehended between the bases of two contiguous leaves is equal to the circumference of the stem, divided by the number of leaves in the whorl. This arc will therefore embrace half the circumference if the whorl consists of two leaves ; one-third of the circumference if it consists of three ; one-fourth, one-fifth, one-sixth, if it consists of four, five, or six leaves.

It has also been observed that the leaves of a whorl are not placed directly above those of the whorl immediately above or below them, but opposite the intervals which separate the leaves, and either exactly opposite, or to one or the other side of the interval. When the leaves are opposite, and each pair crosses the upper and lower pair at right angles, the leaves occupy four rectilinear lines, and, seen from above, form a cross ; such leaves are *decussate* (*f. decussata*). Whorls of three or four leaves will in like manner occupy six or eight longitudinal lines. Whorled leaves are relatively few ; many more plants have opposite leaves, and by far the largest number have alternate leaves ; and it is by the latter that the arrangement of leaves on the stem must especially be studied.

We have seen (p. 3) that the *Oak* presents five leaves (1, 2, 3, 4, 5), spirally arranged around the stem, so that the one (6) which succeeds the fifth is placed vertically above the first. In a longer branch, the seventh would be placed above the second, the eighth above the third, &c. This spiral arrangement prevails in many woody and herbaceous plants, as the *Peach, Plum, Cherry, Rose, Raspberry, Hawthorn, Spiræa, Cytisus, Poplar, Willow, Sumach, Wallflower, Mignonette, Heartsease, Groundsel, Poppy*, &c.

The naturalist Ch. Bonnet, who was the first to observe this arrangement of alternate leaves, remarked that their points of insertion were separated from each other by equal intervals, and discovered some more complicated arrangements, as that, instead of the sixth leaf, it is often the ninth or even the fourteenth which is placed vertically above the first, indicating a series of eight or of thirteen leaves. Modern botanists have followed up this subject, and have formulated as *laws* the facts which Ch. Bonnet had not generalized.

To begin with the simplest example of alternation of leaves, that in which the leaves alternate on opposite sides of the stem (*Lime, Ivy, Elm, Hazel*, &c.) : if a thread be carried round the stem so as to touch the insertions of these leaves, it will describe a regular spiral. If one of these leaves be taken as a starting point, and if they be counted from below upwards, it will be perceived that 3 is above 1, 4 above 2, &c.; and all are arranged on two equidistant vertical lines, being separated by half the circumference of the stem. Leaves thus placed are called *distichous* (fig. 69).

If three leaves complete one turn of the spiral, the fourth will be vertically above the first, the fifth above the second, &c., and all will be arranged on three equidistant vertical lines, and separated from each other by a third of the circumference of the stem. Such leaves are termed *tristichous* (*Galingale, Carex*, and many monocotyledons).

In the *Oak, Poplar, Plum*, &c., where the leaves are arranged in fives, and occupy five vertical equidistant lines on a branch, these lines divide the circumference of the branch into five equal portions, and are separated by an arc equal to one-fifth of the circumference of the stem. But here it is important to remark, that if, taking one of these leaves as the starting-point, we examine the successive leaves of the spiral, the leaf which follows or precedes number one is not situated on the nearest vertical to that to which number one belongs, but on that which comes after number two, and that this vertical is at two-fifths the circumference from the first. Here the spiral is not completed in one turn by two or three leaves, as in the two preceding cases; for the intervals between the five leaves are such that, before arriving at the sixth, which is immediately above the first, the spiral passing through their points of insertion would make two complete turns round the stem; the distance between the leaves will therefore be two-fifths of the circumference. This arrangement is called the *quincunx*.

The name *cycle* is given to a system of leaves in which, after one or more turns of the spiral, a leaf is found immediately above the one from which we started, and beginning a new series. To obtain a complete idea of the cycle, we must therefore consider, besides the number of leaves which compose it, the number of spiral turns they occupy.

The angle of divergence of two consecutive leaves is measured by the arc between them. Thus the fraction $\frac{1}{3}$ expresses the angle of divergence of *tristichous* leaves, and the fraction $\frac{2}{5}$ the angle of divergence of *quincunx* leaves. As to *distichous* leaves, the term angle cannot apply to their divergence, being half a circumference, but it is expressed by the fraction $\frac{1}{2}$. These fractions have for their numerator the number of the spiral turns of which the cycle is composed, and for denominator the number of leaves in the cycle, or, to speak more exactly, the number of spaces separating the points of insertion of these leaves. A cycle may therefore be designated by the fraction expressing the angle of divergence, since the denominator of this fraction indicates the number of leaves, and its numerator the number of turns.

Besides the three cycles mentioned above, designated by the fractions $\frac{1}{2}$, $\frac{1}{3}$, $\frac{2}{5}$, we find cycles of eight leaves in three turns, i.e. $\frac{3}{8}$; thirteen leaves in five turns, $\frac{5}{13}$; twenty-one leaves in eight turns, $\frac{8}{21}$; thirty-four leaves in thirteen turns, $\frac{13}{34}$;

fifty-five leaves in twenty-one turns, $\frac{21}{55}$; eighty-nine leaves in thirty-four turns, $\frac{34}{89}$; one hundred and forty-four leaves in fifty-five turns, $\frac{55}{144}$, &c.

Now, if we arrange this series of fractions progressively,

$$\tfrac{1}{2}, \tfrac{1}{3}, \tfrac{2}{5}, \tfrac{3}{8}, \tfrac{5}{13}, \tfrac{8}{21}, \tfrac{13}{34}, \tfrac{21}{55}, \tfrac{34}{89}, \tfrac{55}{144}, \text{&c.,}$$

several curious analogies will appear, of which the most striking is, that each fraction has for its numerator the sum of the numerators of the two preceding fractions, and for denominator the sum of the two preceding denominators. In like manner any one of these fractions may be obtained by taking the two fractions which immediately follow it, and finding the quotient of their numerators and denominators. It is easy to obtain these fractions when the leaves are neither too distant nor too crowded on the stem, as often happens. The spiral which takes in all the leaves is called a *primitive* spiral. But if the internodes are long, the leaves consequently remote, and the cycle composed of a considerable number of leaves, it becomes difficult to ascertain by inspection which leaf is vertical to the first, and hence to estimate the angle of divergence between two consecutive leaves. This becomes still more difficult when the leaves are crowded, as in the rosettes of the *Houseleek*, in *Plantains* and other so-called *stemless* plants, in the bracts of heads (*Artichoke*); or in the scales or open carpels which compose the cones of *Pines, Firs, Larches,* &c.

In the case of crowded leaves, we can, however, by a very simple calculation, ascertain the angle of divergence, and thus determine the

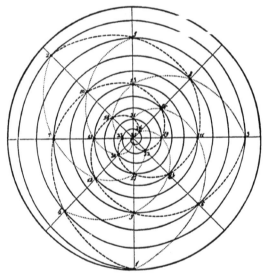

450 a. Primitive spiral from right to left, and bearing three cycles, each of eight leaves, shown by the numbered points, and inserted on three turns of the spiral. The secondary spirals, formed to the right by the numbers in fives, are indicated by the finely dotted lines; the secondary spirals, formed to the left by the numbers in threes, are indicated by the lines - - - - - -.

primitive spiral. Take, for example, a stem bearing a series of cycles of eight leaves moderately separated on three turns of the spiral; the cycle will be easily recognized, and the expression of the angle of divergence will be ⅜. This arrangement obtains in many succulent plants, and especially in *Sedum Telephium*. Suppose the stem to be shortened, so that the leaves become crowded into a rosette, it follows that the spiral will become a very close one, comparable to a watch-spring of which the coils contract in approaching the axis (fig. 459 c). Let us suppose, further, that the inner end of this spring represents the top of the spiral, and its outer extremity the base; it is obvious that on this depressed spiral the leaves nearest the centre would have been the nearest to the top of the more open spiral, and those nearest the circumference would have been the lowest. Now, knowing the angle of divergence of the leaves of *Sedum* in a normal state, it remains to find it for the same leaves gathered into a rosette; for this it suffices to represent or plan three or four cycles, of three leaves each, according to the fraction ⅜, that is, each cycle to contain eight leaves, that shall occupy three turns of a right-to-left spiral, and be separated by an arc equal to ⅜ of the circumference (fig. 459 a). A circle must then be drawn around this spiral, of which the radius shall join the two extremities of the spiral; it is by means of this circle that we must be guided in laying down the angular divergence of the leaves, which being ⅜, it follows that the circle must be divided into eight equal portions by as many radii, when three of these portions will represent ⅜ of the circumference, or in other words the angle of divergence. This done, we place a number (1) on the position of the first leaf, which is where the spiral touches the circumference; then follow the coils of the spiral, and after clearing the three first arcs (⅜ of the circumference) indicate the position of the next leaf (2), which will be at the intersection of the spiral and radius which bounds the third arc; and so on, a leaf position being marked at the intersection of every third radius with the spiral; till the centre of the spiral being reached, the plan will represent the entire series of leaves, numbered in order.

Let us now examine the relative positions of the leaves, as indicated by their numbers. If we examine the radius bearing leaf No. 1, we shall see above it on the same radius, Nos. 9 and 17, the difference between which is eight, and it is obvious that this horizontal radius would represent a vertical line on the *Sedum* stem, along which the leaves 1, 9, and 17 are inserted, each marking the commencement of a cycle; as also that these leaves are separated by three turns of the spiral. Commencing at any other radius (say Nos. 2, 10, 18, &c.), the result is the same, the fraction ⅜ being clearly expressed.

There are other relations between these leaves, which this plan clearly demonstrates. Thus, between Nos. 1 and 4, situated on the next radius to the left, there is a difference of three; the same between 4 and 7, &c.; and starting from leaf No. 2 or 3, we shall find the same numerical relations as in the first instance; the number expressing the difference (3) being the same as that of the series. If we now draw a line through the positions of all the leaves of each series, we shall see that each line is a portion of a spiral, and that these three partial spirals take the same direction, and include within their course the points of insertion of all the leaves.

If, on the other hand, starting from No. 1, we examine its relations with No. 6, on the radius next to the right, we find between them a difference of five; and similarly with 6 and 11, 11 and 16, &c.; and between Nos. 2 and 7, and Nos. 12, 17 and 22, and along the series commencing with 4 and 5. Here again, from left to right, the number expressing the difference corresponds to that of the series. Each of these series may be shown more clearly by means of a curved line uniting all the leaves which compose it, and we shall then have five segments of a spiral turning symmetrically from left to right, and passing through the insertions of all the leaves. These segments of the spiral have been termed *secondary spirals*, to distinguish them from the primitive spiral, also termed *generating* spiral. Now it will be remarked that the secondary spirals proceeding from right to left are three

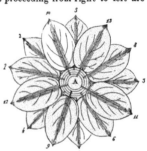

459 b. Rosette forming two cycles of eight leaves, of which the angle of divergence is ⅜.

459 c. Rosette forming a cycle of thirteen leaves, of which the angle of divergence is ⁵⁄₁₃; the axis A, where they are inserted, bears five turns of the spiral, showing the point of insertion of each leaf.

in number, which number is the numerator of the fraction ⅜; and that the sum of these three, and of the five going from left to right, is eight, or the denominator of the fraction. If therefore it is possible to count the secondary spirals to left and right, of rosettes, involucral bracts, or scales of Pine cones, in all of which the primitive spiral is obscured by the closeness of the parts, we may assume that the smaller number represents the numerator, and the sum of the two numbers the denominator of the desired fraction; which again gives the angle of divergence, the number of leaves in the cycle, and the number of turns of the spiral which they occupy.

This crowding of the leaves, which we have illustrated by *Sedum*, is frequent amongst plants with radical leaves, in many of which the cycle of the leaves is indicated by the fraction ⅜ (*Common Plantain*, fig. 459 b).

The number of secondary spirals to right and left being known, it is easy to number each leaf in the primitive spiral. Take, for example, the rosette (fig. 459 c), which represents a *Houseleek*, or the cone of the *Maritime Pine* (fig. 459 d). Their angle of divergence is $\frac{5}{13}$, which is easily found by counting the very obvious secondary spirals to right and left. We have only alluded to the most obvious secondary spirals; but it will readily be understood that there are many others, some more, some less oblique than these, and that every numerical series having the same relative differences between them would be a spiral. The secondary spirals are especially visible in *Pine* cones, the axis of which is much longer than that of the *Houseleek*, and in which they form very distinctly marked parallel series.[1]

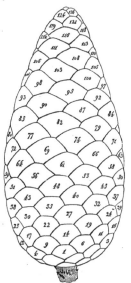

Begin by numbering as 1 one of the outer leaves of the rosette, or of the lower scales of the cone, and regard it as the first of a secondary spiral turning from left to right. To find No. 2 on it, remember that the numbers of a secondary spiral must be separated by a space equal to the number of the secondary spirals of which this forms a part; and as there are five parallel left-to-right spirals, the second leaf or scale must be numbered 6, the third 11, and so on to the top of the cone, or centre of the rosette. Having thus numbered all the scales or leaves of one of the five parallel secondary left-to-right spirals, these numbers may serve as starting-points from which to number all the other scales or leaves of the cone or rosette. We know that each of the numbered scales or leaves of the secondary left-to-right spiral equally forms one in the series of the right-to-left spirals, and we may number all the leaves or scales from any

459 d. Cone of Maritime Pine, with the scales numbered according to their relative heights. The most obvious secondary spirals are formed to the right by the series of numbers in fives; to the left by the series of numbers in eights.

starting-point, by adding 5 when turning to the right, and 8 when turning to the left.

Let us take, for example, No. 32; this number in the left-to-right spiral would (adding 5) lead us to No. 37, 37 leads to 42, and so on; but since No. 32 also enters into one of the eight secondary right-to-left spirals, the leaf or scale succeeding it in this spiral should be numbered 32 + 8, i.e. 40; and following this spiral, by additions of 8, we should have 40, 48, 56, 64, 72, &c.

To obtain in the same spiral the numbers below 32, we must deduct the number 8, which we had before added, and we shall have successively 32, 24, 16, 8.

If, in starting from the same No. 32, we descend the secondary spiral which turns from left to right, we must take 5 from 32, when we shall have successively 27, 22, 17, 12, 7, 2, &c.

All the leaves or scales of the rosette or cone being numbered, their succession

[1] Nothing is easier than to observe this, by numbering the scales of a ripe cone of the Maritime Pine.

indicates the generating spiral. But the direction of this generating spiral from left to right, or right to left, depends on the angle of divergence; if the fraction be $\frac{2}{5}$, or $\frac{5}{13}$, or $\frac{13}{34}$, and so on, the primitive or generating spiral will follow the most numerous secondary spirals; but if the fraction be $\frac{1}{8}$, or $\frac{3}{21}$, or $\frac{11}{55}$, &c., the generating spiral will follow the least numerous secondary spirals.

Take, for example, the fraction $\frac{3}{8}$ (fig. 459 a), and let us examine the relation between the genera-ting and secondary spirals. Whatever may be the direc-tion of the genera-ting spiral, the least numerous secondary spirals must follow the same, and *vice versâ*. Suppose the spiral to be a right-to-left one, as in 459 a, it follows that, placing No. 1 where the radius touches the outer end of the spiral, and successively numbering the leaves from $\frac{3}{8}$ to $\frac{3}{8}$, the nearest radius to the *left* will be occupied by a leaf before the nearest

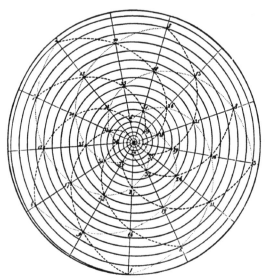

459 *a.* Primitive spiral from right to left, bearing five cycles, each of thirteen leaves, indicated by the numbered points, and inserted on five turns of the spiral. The secondary spirals, formed to the right by the numbers in fives, are indicated by the finely dotted lines; the secondary spirals, formed to the left by the numbers in eights, are indicated by the lines ------.

radius to the *right*. The first leaf on the left radius will

evidently be No. 4; for it will occur after traversing three $\frac{3}{8}$ $(\frac{9}{8})$; that is, after one entire revolution, plus $\frac{1}{8}$, and consequently on the left-hand radius nearest the one from which we started. The leaf which will be found on the right radius will evidently be No. 6, for it will occur after five times $\frac{3}{8}$ $(\frac{15}{8})$, that is, after one re-volution minus $\frac{1}{8}$, and consequently on the nearest right-hand radius. Now we know that the number of secondary spirals is equal to the difference between the numbers of two consecutive leaves on one of these spirals; therefore, if we suppose the fraction to be $\frac{3}{8}$, the number of the secondary spirals from right to left, that is, of the secondary spirals which follow the direction of the generating spiral, will be less than the number of the secondary spirals which follow an opposite direction. The same result can be obtained from the succeeding fractions.

On the contrary (fig. 459 e), with the fractions $\frac{3}{8}$, $\frac{5}{13}$, $\frac{14}{34}$, and so on, we find that the right-hand radius is occupied by a leaf sooner than the left-hand one, and that in consequence the number of the first leaf on the right-hand radius is less than the number of the first leaf on the left-hand radius. Therefore the number of secondary spirals which can be followed from left to right is less than those from right to left, or, in other words, the most numerous secondary spirals turn in the same direction as the generating spiral, and knowing the direction of the one, we know the direction of the other.

The direction of the generating spiral varies not merely in the individuals of a species, but sometimes in the same individual. Thus, in cones from the same specimen of *Maritime Pine*, right-to-left secondary spirals will be more frequent in some, and left-to-right in others; but in all cases the relative direction of the generating spiral follows the law just enunciated.

The angle of divergence itself is constant only in the fractions $\frac{1}{2}$, $\frac{1}{3}$, $\frac{2}{5}$, and when these cycles are more numerous, the one is often substituted for the other, which is owing to the distance between them being extremely small, and to the fact that the angles expressed by the fractions $\frac{5}{13}$, $\frac{8}{21}$, $\frac{14}{34}$, $\frac{22}{55}$, $\frac{34}{89}$, &c., if reduced to degrees and minutes, differ by a few minutes only; so that the angles of divergence actually oscillate between 137° and 138°. A slight twist of the stem or axis is sufficient to account for so small a variation, and may well occur in rosettes of leaves, in involucral bracts, and in cones, and cast a doubt on the value of the angle of divergence. Thus, in *Pines* (fig. 459 d), the rectilinear series indicating the successive cycles may deviate more or less to right or left, so that the secondary spirals, which were the most obvious at the base of the cone, become less so in ascending, and render it difficult to determine such fractions as $\frac{3}{8}$, $\frac{5}{13}$, $\frac{8}{21}$. A change in the shape of the stem will also lead to the substitution of one cycle for another, as in certain *Cacti* with ribbed or angular stems bearing tufts of prickles, and whose ribs double as they ascend, and offer cycles of a higher number.

Lastly, there are exceptional cases which perplex the student of Phyllotaxy; the above-named fractions are not the only ones which may be observed; $\frac{1}{4}$, $\frac{1}{5}$, $\frac{2}{7}$, $\frac{3}{11}$, &c., do occur, though very rarely; but when they do, they preserve among themselves the same relations as the preceding, i.e. that each successive fraction may be obtained by the addition of the numerators and denominators of the two preceding. We have seen that whorled leaves present a succession of circular groups; but here also, as in alternate leaves, the spiral arrangement is discernible. In a branch of *Oleander*, for instance, where the leaves are whorled in threes, a relation exists between any three vertically superimposed leaves of successive whorls; and a line successively passing through their insertions will describe a regular spiral; and if we examine the relations between the other leaves of these whorls, we shall perceive that the number of whorls represents as many parallel spirals as there are leaves in each of them.

Æstivation (*præfloratio, æstivatio*) is the arrangement of the floral organs in the bud, and is of especial importance in respect of the calyx and corolla.

The leaves of each floral whorl may be inserted exactly at the same level (forming a true whorl), or at unequal heights, when they form a depressed spiral, the lowest leaf of which is necessarily the outermost. The true whorl presents two modes of æstivation—the *valvate* and the *contorted*.

1. Æstivation is *valvate* (*æ. valvaris*) when the contiguous edges of the parts touch throughout their length, like the two leaves of a door (460 *a*); and it is then nearly always regular. It is *induplicative* (*æ. induplicativa*) when the contiguous parts cohere by a part of their back; *reduplicative* (*æ. reduplicativa*) when by a part of their faces

460 *a.* Valvate   460 *b* Valvate   461. Valvate   462. Contorted   463. Imbricate
æstivation.   induplicative æstivation.   reduplicative æstivation.   æstivation.   æstivation.

(fig. 461). 2. Æstivation is *twisted* or *contorted*[1] (*contorta*) when the leaves are so placed that each leaf partially covers one of the two between which it is placed, and is similarly covered by the other, as if each were twisted on its axis (fig. 462); in this case the whorl is always regular.

The depressed spiral presents two modes of æstivation: the *imbricate, properly so called*, and the *quincunzial*. These two are often indifferently termed *imbricate*. 1. In the true *imbricate* æstivation (*æ. imbricativa*, fig. 463) the parts (usually five) successively overlap, from the first, which is wholly exterior, to the last, which is wholly interior, and placed against the first; they thus complete one turn of a spiral. In *quincunzial* æstivation (*æ. quincuncialis*) two of the five pieces are exterior, two interior, and one intermediate, one side of the latter being covered by one of the outer, and on the other covering one of the inner (fig. 464). This arrangement corresponds to that of leaves expressed by ⅖. To explain this æstivation, which is nothing but a depressed spiral with two coils, we must consider the axis of the flower as a truncated cone, and draw a spiral line twice round it, from bottom to top; then mark off on this line five equidistant points, so that a sixth point at the top of the cone will be immediately above the first; it is clear that the interspaces will equal ⅖ the circumference of the cone, and the five spaces between the six points will constitute ¹⁰⁄₅, i.e. twice the circumference; which equals the two turns of the spiral traced on the conical axis of the flower. Now substitute for the five points five sepals or petals which shall be large enough to overlap; then depress the cone to a plane, and we shall have two exterior leaves (1, 2), a third, at once half

---

[1] Also called *convolute* by various botanists.—ED.

interior and half exterior (3), and two wholly interior (4, 5), which are both nearest to the top of the cone, and the most central. The *Rose* calyx (fig. 465) confirms this view; its outer sepals being next the axis of the flower, and consequently most vigorously developed, present small lateral leaflets, and often a terminal true leaflet, thus reducing the sepal to an unequally pinnate leaf like ordinary rose-leaves. As the sepals rise in the *quincunx*, the growth becomes weaker, the third bears small leaflets only on one side, and the upper or interior sepal terminates

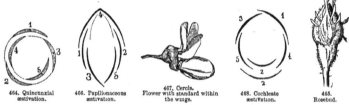

464. Quincunxial æstivation.

466. Papilionaceous æstivation.

467. Cercis. Flower with standard within the wings.

468. Cochleate æstivation.

465. Rosebud.

in a simple filament. The quincunx æstivation may be disturbed by unequal development of the leaves of the whorl, and this especially occurs in the corolla, owing to the relatively slow or rapid growth of some of the petals. Thus, in the *papilionaceous* corolla (fig. 466), the *standard*, which represents No. 4 of the quincunx, and ought to be internal, is wholly exterior, because, having developed more rapidly than the other petals, it covers the two wings representing Nos. 1 and 2; this æstivation is said to be *papilionaceous* (*æ. vexillaris*). In the St. John's Bread (*Cercis*), the standard retains its normal position, and the quincunx is properly formed (fig. 467). In the *Snapdragon* (fig. 468) and other *personate* plants, the second petal is interior instead of being exterior, either because it has developed before the others, or because the latter have grown the most rapidly; this mode of æstivation is called *cochleate* (*æ. cochlearis*). The calyx has a similar arrangement.

Amongst the varieties of imbricate æstivation is that termed *convolute* [1] (*æ. convolutiva*): it occurs when the sepals or petals overlap, so that each completely envelops all the others; as in the calyx of *Magnolia*, and the corolla of *Poppies* (fig. 470). Æstivation is *alternate* (*æ. alternativa*), when the leaves of the calyx or corolla form two whorls, of which the exterior encloses the interior whilst alternating with it, as in the calyx of the *Wallflower*, and corolla of *Fumitory* (fig. 472).

[Æstivation is *straight* (*æ. recta*), or open (*æ. aperta*), when the parts are so little developed or so distant that they do not meet.—ED.]

SYMMETRY OF THE FLOWER.

The term *symmetry* has been differently applied; according to De Candolle, it implies non-geometrical regularity in plants and animals; other botanists distinguish (often obscurely) symmetry from regularity: this we do not admit, but regard

---

[1] The term *convolute* is often used synonymously with *contorted* or *twisted* æstivation.

symmetry and regularity [1] as synonymous, and as implying a similarity between the leaves of a floral whorl; this relation including:—1, the *form*; 2, the *number*; 3, the *independence*; 4, the *relative position* of the parts of flowers: we have thus the symmetry of form, of number, of disjunction, and of position. Symmetry of form is regularity taken in its usual sense; as when portions of a whorl are alike, or when, being different, the one sort alternates with the other, so as to present a symmetrical whole around a common centre; this regularity might be termed *rayed symmetry* (calyx and corolla of *Columbine*, fig. 31; *Wallflower*, fig. 7; and *Buttercup*). A whorl that is not thus symmetrical is said to be irregular; though its two sides (or halves) may resemble each other, thus being analogous to the *longitudinal* [2] *symmetry* of animals, which is opposed to the *rayed symmetry* of *Zoophytes*. The corollas of the *Heartsease* (fig. 170), *Cytisus* (figs. 253, 254), *Tropæolum* (fig. 210), are irregular, but longitudinally symmetrical. The whorl is called regular, even though it forms a depressed spiral; but if the floral axis lengthens sensibly, the rayed symmetry disappears, and, to describe the symmetry, recourse is had to the comparative length of the spiral; thus the symmetry of the carpels is *hemispheric* in the *Strawberry* (fig. 401), *conical* in the *Raspberry* (fig. 402), *spiked* in *Adonis* (fig. 404).

Perfect numerical symmetry occurs when all the whorls consist of the same number of parts, as in *Crassula*, which has five sepals, petals, stamens and carpels.

Disjunctive symmetry occurs when the pieces of each whorl are entirely separated, and each whorl is entirely free (*Columbine*, *Hellebore*). *Symmetry of position* occurs when the pieces of each whorl *alternate* with those of the preceding and succeeding; and the normal position of the whorls (calyx, corolla, andrœcium, pistil) is undisturbed (*Crassula rubens*). Many botanists, regarding regularity as the normal feature in plants, assume it to be the *primitive type* adopted by Nature; they therefore look upon a combination of the above-named symmetries as indicating the

normal condition of the flower; which should thus consist of four whorls, each composed of the same number of leaves, all equal, free, alternating successively, and arranged in the order of calyx, corolla, andrœcium and pistil. Further, such a primitive type, whether real or imaginary, may be more or less completely and permanently modified by various single or combined causes, of which the principal are, — *inequality* of development, *cohesion* or *symphysis*, *multiplication*, *doubling*, *suppression* and *abortion*. This hypothesis has contributed largely to the progress

469 *bis*. Linaria.
Monstrous flower.

469. Diagram of an ideal
perfectly
symmetrical flower.

of organography, by stimulating investigations into the comparative anatomy of floral organs.

To ascertain the amount of symmetry a flower displays, its bud must be cut

---

[1] *Symmetry* in English and American works implies that the parts of successive whorls are isometric or equal in number; *regularity*, that the parts of a whorl are equal and similar.—Ed.

[2] Better called *bilateral*.—Ed.

through horizontally, when all the whorls will appear projected on the same plane; and the relative positions of the organs thus displayed is termed a diagram (fig. 469). *Inequality of development* necessarily interferes with symmetry of form (corolla of *Heartsease*, fig. 170; *Cytisus*, figs. 253, 254; *Tropæolum*, fig. 210); this inequality is frequently caused by the cohesion of parts, as in the bilabiate monosepalous calyx of *Lamium* (fig. 208), in the bilabiate corolla of *Snapdragon* (fig. 285), of *Linaria* (fig. 286), *Lamium* (figs. 278, 279), in the monadelphous andrœcium of *Mallow* (fig. 310), diadelphous of *Lotus* (fig. 312), didynamous of *Snapdragon* (fig. 305), tetradynamous of *Wallflower* (fig. 306); in the ovary of *Snapdragon*, the pistil of *Orchis*, &c.—irregularities which are usually accompanied with nectariferous glands (*Heartsease, Wallflower, Centranthus, Honeysuckle, Snapdragon, Linaria*, &c.). In *Linaria* (fig. 286) the calyx is monosepalous with five unequal divisions, the corolla is monopetalous with two unequal lips, of which the upper represents two petals, and the lower three, of which the centre one is prolonged below into a subulate spur; there are four stamens, of which the two longest are situated between the central and the two lateral petals of the lower lip; the two others, which are shorter, are opposite the fissures which separate the two lips; at the base of the upper lip a filament represents the fifth stamen. In certain circumstances all the petals of *Linaria* are developed like the centre one of the lower lip; the whorl is then perfectly regular, and presents a corolla with five lobes, and five equal spurs between them (fig. 469 *bis*). At the same time, the filament at the base of the upper lip develops into a stamen like the four others, which latter, usually unequal, become precisely alike, so that the flower is furnished with five symmetrical stamens: to this metamorphosis the name of *Peloria* has been given, which, according to the theory adverted to, would be regarded as a reversion to the normal state of the plant. *Violets* are also sometimes regular; sometimes presenting two opposite spurred petals, or three, or even five such; when the symmetry of form is established in the three first whorls.

Cohesion or *symphysis*, whether congenital or the result of growth, destroys the symmetry of disjunction by effecting either the cohesion of the leaves of the same whorl, or the cohesion [1] of one whorl with another; as in monosepalous calyces, monopetalous corollas, monadelphous, diadelphous and polyadelphous stamens, and compound ovaries; also in flowers with inferior ovaries (*Myrtle*, fig. 381; *Saxifrage*, fig. 382), and with monopetalous staminiferous corollas (*Belladonna*, fig. 294); in calycifloral (*Peach*, fig. 368) and gynandrous flowers (*Orchis*, fig. 188; *Aristolochia*, fig. 318).

Cohesion also masks numerical symmetry, by causing a compound organ to appear simple, as in the monosepalous calyx, the monopetalous corolla, the compound ovary, &c.; and it destroys the symmetry of position, as when the carpels are enclosed in the receptacular tube (*Quince*, fig. 215), or in causing the andrœcium to appear above the level of the pistil (*Orchis*, fig. 188; *Aristolochia*, fig. 318).

Multiplication consists in the repetition of the same whorl; thus, *Berberis* has

---

[1] In English works, the term *cohesion* is confined to the union of two or more organs of the same whorl; *adhesion*, to the union of the organs of different whorls.—ED.

three whorls of three sepals, two whorls of three petals, and two whorls of three stamens. The *Poppy* (fig. 470) has two whorls of two petals, and many whorls, each

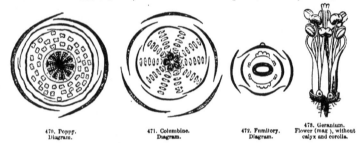

470. Poppy. Diagram.

471. Columbine. Diagram.

472. Fumitory. Diagram.

473. Geranium. Flower (mag), without calyx and corolla.

composed of two stamens. The *Columbine* (fig. 471) has ten whorls of five stamens and two whorls of five scales. The *Fumitory* (fig. 472) has two whorls of two petals, and two whorls of two stamens, of which the outer are normal two-celled stamens, and the inner stamens are divided into four, each one-celled (equal to two complete stamens). *Lythrum* has two whorls of six sepals, coherent and adherent. *Datura fastuosa* has two or three monopetalous corollas, one inside the other.

Deduplication or *chorisis* occurs when two or more organs take the place of one. This affects not only numerical symmetry, but symmetry of position; in which respect it differs from *multiplication*, when the whorls preserve their relative positions. Deduplication is *parallel*, when the organ is doubled from without inwards, and when the supernumerary piece is opposite to that from which it proceeds; it is *collateral*, when the supernumerary piece occurs by the side of the organ from which it proceeds, maintaining the same relative position on the receptacle; a parallel deduplication may double or treble the whorl, a collateral deduplication can only increase the number of parts in that whorl, which still continues simple. In the case of parallel deduplication, the supernumerary pieces are usually altered, and rather resemble those of the whorl which normally succeeds them, than those of the whorl to which they belong. In *Lychnis* (figs. 239, 240) and other *Caryophylleæ*, the petals give off a fringed petaloid layer, which coheres with the claw, and is free only where the claw meets the limb; in *Sedum* (fig. 476) the five petals produce a whorl of five stamens shorter than the five which alternate with them, and the normal and supernumerary androœcia are so close that their bases cohere. In *Geranium* (fig. 473), the five petals produce by deduplication five stamens shorter than and outside the others, but the five larger bear at their outer bases five nectaries, which re-establish the alternation disturbed by the five supernumerary stamens (fig. 474); in *Erodium* (fig. 475) the same arrangement exists, except that the extra stamens have no anthers; in *Sedum* (fig. 476) the stamens opposite to the petals are a deduplication of the latter; in *Flax* (fig. 477), the supernumerary stamens are reduced to sessile membranous teeth; in *Mignonette* (fig. 478), the petals with a fringed top bear within a small concave plate, which is a deduplication of the petal. The petals of *Ranunculus* (fig. 237) bear at their inner base a small scale, parallel to the claw,

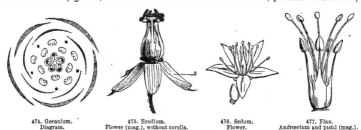

474. Geranium.
Diagram.

475. Erodium.
Flower (mag.), without corolla.

476. Sedum.
Flower.

477. Flax.
Andrœcium and pistil (mag.).

and forming with it a nectariferous cavity; the bilabiate petals of *Helleborus* are formed of two nearly equal plates, and may be regarded as originating by deduplication in a parallel direction. The petaloid laminæ of these plants must not be confounded with the different protuberances on the corolla of *Comfrey* (fig. 269), and other *Boragineæ*, nor with the sort of hairy palate on the lower lip of *Snapdragon* (fig. 285) and *Linaria* (fig. 286); which are not the result of deduplication, but are derived from the substance of the petal. Deduplications are chiefly confined to the corolla and andrœcium; they rarely occur in the pistil; in *Sedum* (fig. 455) there is externally at the base of each carpel a little green glandular scale, parallel to the carpel, and which might be looked upon as a deduplication of this.

Deduplications are not always a proof of superfluous vital action; they may arise from a misdirection of vegetative force; in fact, when one whorl is doubled, the succeeding one is either weakened, modified, or suppressed, as in the *Primrose, Pimpernel* (fig. 479), and other *Primulaceæ*, which have only five stamens, and these opposite to the petals, thus not forming a normally whorled andrœcium, but being referable to a parallel deduplication of the petals; they thus replace the normal andrœcium, which however sometimes appears, not as stamens, but as scales, alternating with the petals (*Samolus*, fig. 480). In the *Vine*

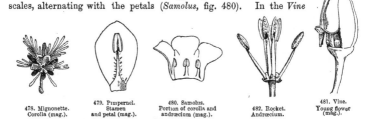

478. Mignonette.
Corolla (mag.).

479. Pimpernel.
Stamen
and petal (mag.).

480. Samolus.
Portion of corolla and
andrœcium (mag.).

482. Rocket.
Andrœcium.

481. Vine.
Young flower
(mag.).

(fig. 481), the five normal stamens are replaced by five nectaries, but fertilization is secured by five stamens opposite to the petals.

Collateral deduplication is less frequent than *parallel*; in the *Rocket* (fig. 482) and other *Cruciferæ*, the four stamens arranged in pairs alongside the pistil represent two doubled; the filaments of each pair are indeed often connected half-way up, or throughout their length. In the *Orange* (fig. 483), the andrœcium consists of a single whorl of thirty stamens, whose filaments cohere in bundles of four, five, or six; in *St. John's Wort* (fig. 484), the stamens form three or five bundles, of which each may be considered as a doubled stamen; and so in *Castor-oil* (fig. 315), the stamens of which form branched bundles. Each filament of the *Laurel* (fig. 485) bears on each side of its base a shortly stipitate gland, which firmly coheres to it, and is sometimes developed into a true stamen. This shows that the stamen of the *Laurel* with its two glands represents a stamen multiplied into three, of which the two lateral are rudimentary. In many *Garlics* (fig. 320) the filaments are dilated, and terminated by three teeth, of which the central only bears an anther; in *Pancratium* this dilatation is enormous; the lateral lobes of each filament cohere with the neighbouring filaments, and form with them a fringed tube; in *Narcissus* (fig. 486) this tube is still more remarkable, and assignable to the same origin.

Many plants present the case both of multiplication and deduplication; the flower of *Butomus* (fig. 487) has three sepals, three petals, six stamens in pairs opposite to the

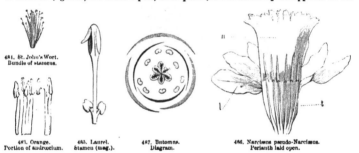

484. St. John's Wort.
Bundle of stamens.

483. Orange.
Portion of andrœcium.

485. Laurel.
Stamen (mag.).

487. Butomus.
Diagram.

486. Narcissus pseudo-Narcissus.
Perianth laid open.

sepals, three other stamens within the six preceding, also opposite the petals, and six carpels in two series: here we have a multiplication of the andrœcium and pistil, and besides this a collateral deduplication of the first whorl of the andrœcium. When the stamens are twice and thrice as many as the petals, and by their extreme closeness seem to form but a single whorl, it may be difficult to decide whether this is a case of collateral deduplication of the andrœcium, or of multiplication, or of a deduplication of the corolla added to the normal andrœcium. This difficulty is increased when the stamens all cohere. If the stamens are placed exactly on a level, they may be formed by a collateral deduplication (*Orange*, fig. 483); if some are a little within or without the others, which is easily distinguishable, in spite of coherence, then it is a case either of *multiplication* or of *parallel* deduplication. It is a case of multiplication when the outer stamens alternate with the petals (*Berberis*), but of parallel deduplication when they are opposite to the petals (*Geranium*, fig. 473).

Arrests and *suppressions* are due to failures of development, and affect more than all other causes the symmetry of the flower. *Arrest* is the condition of an organ the growth of which has stopped, so that it is reduced to a sort of stump, sometimes glandular; *suppression* implies that an organ has never even been developed. The outer whorls are more seldom arrested or suppressed than the andrœcium and especially the pistil, which occupies but a narrow area of the receptacle. The suppression or arrest of one or more pieces of a whorl affects the symmetry of number, position and form. For example, *Berberis*, whose calyx, corolla and andrœcium are in threes or multiples of three, has for pistil a single carpel; the *Pink* (fig. 488), whose other whorls are quinary, has but two carpels; the *Heartsease* three (fig. 489); in the *Bitter Vetch* (fig. 490) and other *Papilion-aceæ*, the two first whorls are quinary, the third decennary,

488. Pink.   489. Heartsease.   490. Vetch.   492. Scrophularia.   491. Snapdragon.
Diagram.   Diagram.   Diagram.   Diagram.   Diagram.

whilst the pistil is mono-carpellary; it is the same with the pistil of the *Plum* and *Peach*. The *Snapdragon* (fig. 491), of which the calyx and corolla are quinary, has (owing to arrest) four stamens, and two carpels due to suppression.

In *Scrophularia*, with the same arrangement, the fifth stamen is represented by a petaloid scale (fig. 492). The *Periwinkle* and other *Apocyneæ*, as well as many monopetalous families, have five sepals, five petals, five stamens, and two carpels; *Polygala* (fig. 493) has five sepals, three petals (sometimes five, alternating with the sepals), eight half anthers (equivalent to four stamens), and two carpels. *Umbelliferæ* (fig. 494) have five sepals, five petals, five stamens and two carpels. The *Cornflower*, *Dandelion*, *Chrysanthemum* and other *Compositæ* have quinary corollas and andrœcia and a single carpel; in most, the calyx degenerates into a pappus, though in some (*Asteriscus*, *Hymenoxys*) it presents five scales. In most *Cucurbitaceæ* (*Melon*, *Pumpkin*, *Cucumber*) the calyx and corolla are quinary and the stamens are reduced to two and a half.

In apetalous, monœcious, and diœcious flowers, an entire whorl is suppressed or arrested (*Lychnis*, *Sagina*, *Chenopodium*, fig. 189); sometimes several whorls are absent, as in the *Nettle* and *Mulberry* (fig. 495), which

493. Polygala.   494. Coriander.   495. Mulberry.
Diagram.   Diagram.   ♂ flower (mag.).

present only a calyx with an andrœcium, or a pistil. Sometimes several whorls are suppressed, together with one or more pieces of the remaining whorl; the male flower of *Euphorbia* (fig. 333) consists of one whorl, reduced to one stamen; and the female flower (fig. 406) of one whorl of three carpels; the flowers of *Arum* (figs. 196, 197, 198) consist of a solitary stamen or carpel.   Seeds, like the floral whorls, are subject to suppression and arrest; in *Geranium* (fig. 474) the five carpels are two-ovuled, and but single-seeded; the *Oak* (fig. 400) has three carpels forming three two-ovuled cells; the septa become speedily absorbed through the rapid growth of one of the ovules, and the ripe fruit is one-celled and one-seeded.   The *Horse-chestnut* presents

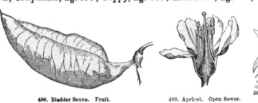

a similar arrest.   In the *Cornflower* and other *Compositæ*, in *Wheat* and other *Gramineæ*, the ovule is solitary from the first; at least, a second has never been discovered; thus offering a case of suppression and not arrest.

The causes which disguise or disturb symmetry in any one flower are not always isolated.   In *Larkspur* we have *unequal development* and *symphysis* in the calyx and corolla, *multiplication* in the andrœcium, and *suppression* in the pistil; in *Asclepias* (fig. 496) *symphysis* in all its whorls, *multiplication* in its corolla, *deduplication* in the second whorl of the corolla, and *suppression* in the pistil.   *Mignonette* is an example of *unequal development* in its calyx, corolla and andrœcium; of *symphysis* in its pistil, of *parallel deduplication* in its corolla, of *collateral deduplication* in its andrœcium, and of *suppression* in its pistil.

406. Asclepias. Flower (mag.).

THE FRUIT.

The *fruit* (*fructus*) is the fertilized and ripe pistil, that is, a pistil enclosing seeds capable of reproducing the plant.   It may be accompanied by accessory organs, which are considered as forming an integral part of it, and to which we shall return.

The fruit is *apocarpous*—1, when its carpels are separate from each other (*Columbine*, fig. 497; *Ranunculus*, fig. 524; *Bramble*, fig. 521; *Rose*, fig. 525), when each carpel is considered to be a fruit; 2, when the pistil is formed of a single carpel (*Pea, Bladder Senna*, fig. 498; *Apricot*, fig. 499; *Wheat*). It is *syncarpous*, when its carpels are consolidated into a single body (*Tulip*, fig. 389; *Iris, Campanula*, fig. 390; *Poppy*, fig. 388; *Heartsease*, fig. 500).

497. Columbine. Fruit.   498. Bladder Senna. Fruit.   499. Apricot. Open flower.   500. Heartsease. Ripe pistil.

According as each free carpel, or each cell of a syncarpous fruit, or each unilocular composite ovary contains one, few, or many seeds, this carpel, cell, or ovary is said to be *monospermous* (*monosperma*), *oligospermous* (*oligosperma*), or *many-seeded* (*polysperma*). The ripe ovary is called a *pericarp* (*pericarpium*); we have already described the three layers of which it is composed (figs. 15, 16), *epicarp*, *endocarp*, and *mesocarp* or *sarcocarp*.

Changes caused by Maturation.—In ripening, the fruit undergoes changes, some of which have been already mentioned: it may be *dry*, and then, according to its consistency, it is said to be *membranous, corky, coriaceous, woody, bony*; the latter quality is found in the *Filbert* (fig. 233); sometimes it becomes *fleshy* through the abundant pulp of the seed;[1] in *Belladonna* (fig. 567) the mesocarp is succulent; in the *Orange* (fig. 568) the pulp consists of long

501. Gooseberry.
Fruit
cut vertically.

503. Cneorum.
Fruit cut
vertically (mag.).

504. Tribulus.
Fruit cut
vertically (mag.).

502. Cassia.
Portion of open
fruit.

505. Radish.
Flower cut vertically.

spindle-shaped cells, fixed to the endocarp by one of their extremities, and free at the other; in the *Tomato* it is the placenta, in the *Gooseberry* (fig. 501) and the *Pomegranate* it is the testa itself of the seed which is pulpy.

In fruits with a succulent mesocarp, as *Plum, Cherry, Peach, Apricot, Walnut,* &c., the endocarp thickens at the expense of a portion of the mesocarp (figs. 16, 520), becomes bony, and forms the *stone* (*putamen*). The septa sometimes disappear in the pericarp; as in *Lychnis* (fig. 398) and other *Caryophyllaceæ*, where the rapid growth of the walls of the ovary breaks and effaces them; in the *Oak* (fig. 400), where one ovule stifles the other five, and destroys the three septa; in the *Ash* (fig. 561), where one of the two cells contains a seed, while the other is reduced to an almost imperceptible cavity by the destruction of the septum. Sometimes transverse septa are developed in the ripening ovary; these are horizontal expansions of the endocarp and mesocarp, which sometimes become woody (*Cassia*, fig. 502). In *Cneorum* (fig. 503) and *Tribulus* (fig. 504), the endocarp and mesocarp are gradually intruded from the inner wall of the ovary, so as to form oblique septa, which at maturity divide the cavity into small superimposed cells. The membranous transverse septa of the cells of the *Radish* pod (fig. 505), *Raphanistrum*, and some other *Cruciferæ*, are longitudinal septa which the growth of the seeds has driven to right and left by the resistance of the endocarp; in this case, the fruit dehisces transversely, each segment containing one seed.

Suture.—The *ventral suture* (*sutura ventralis*) is the line indicated by the

---

[1] The pulp rarely contributes to the formation of the seed; it aids in the dispersion of fruits by tempting birds, &c., and it is often an aid to the germination of the seed.—Ed.

cohering edges of a carpellary leaf, and which faces the axis of the flower; what is (somewhat improperly) called the *dorsal suture* (*s. dorsalis*) is nothing but the median nerve of the carpel, which consequently faces the periphery of the flower.￪ This nerve may be masked by the parenchyma developed from the carpel, as in the *Peach*; it is usually indicated either by a rib (*Columbine*) or furrow (*Astragalus*). The ventral suture may also be indicated by a rib (*Pea*) or furrow (*Peach*). In a many-celled ovary, the ventral sutures, occupying the axis of the flower, cannot be seen externally, and each cell is indicated by a dorsal line or rib; besides which, we generally see, on the walls of the compound ovary and between its dorsal furrows, other sutures, named *parietal* (*suturæ parietales*), which indicate the union of two septa, or of two parietal placentas (*Mallow*, fig. 225). In inferior ovaries, those are not sutures which we perceive on the walls of the fruit, but fibro-vascular bundles, which belong to the calyx-tube according to some, to the receptacular tube according to others (*Currant*). In this case, the calyx-limb often crowns the fruit, in the form of *teeth* (*Fedia*, fig. 216), or *bristles* (*Scabious*, fig. 229), or a *pappus* (*Dandelion*, fig. 222), or a *crown* (*Pomegranate, Medlar*).

Accessory Organs.—The style sometimes remains upon the ovary, and grows with the pericarp as it matures; it forms a flattened beak in the *Radish* and *Rocket* (fig. 506), a feathery tail in *Pulsatilla* and *Clematis*. The receptacle, which in some cases adheres to a part of the fruit; such is the receptacular tube which encloses the carpels in *Apples, Pears, Quinces, Medlars, White-beam, Azarole, Haws*, &c.; such is also the receptacle of the *Strawberry* (fig. 507), which, though almost dry at first, gradually enlarges, becomes fleshy, and encloses the ovaries in its crimson parenchyma; it is not then the pistil alone, but the enlarged receptacle which is prized in the strawberry, and which is usually regarded as the fruit; the carpels of the strawberry are insipid, and crack under the teeth, and the little black styles appear as dry deciduous threads. In the *Fig* (fig. 158), a fleshy receptacle encloses innumerable minute flowers, the lower female, the upper male.

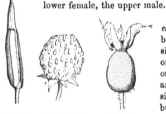

Exuviæ.—The name *exuviæ* (*induviæ*) has been given to the persistent withered remains of the calyx or corolla, or sometimes of the andrœcium, which persist around the fruit but do not adhere to it; in *Campanula* (fig. 544) the corolla withers

506. Rocket. 507. Strawberry. 509. Rose.
Fruit. Fruit. Fruit.

508. Winter Cherry.
Fruit shown by the removal
of half the calyx.

and persists on the calyx; in the *Marvel of Peru* the base of the petaloid perianth envelops the ovary, and resembles one of the integuments of the seed; in the *Winter Cherry* (fig. 508) the whole calyx persists, enlarging enormously, and enclosing the ovary in an inflated coloured bladder. In the *Rose* (fig. 509), the calyx-limb dries and decays, but the receptacular tube persists and becomes fleshy. In the ripe *Mulberry* (fig. 571), the female flowers of which form a dense spike, the four sepals are succulent, and enclose the pistil; they may thus themselves be regarded as belonging to the fruit. Involucres, which we have described in the paragraph on bracts, usually persist around the fruit and grow with it; such is the case with the involucres of *Compositæ*, the cups of the *Acorn* (fig. 232), of the *Nut* (fig. 233), and of the *Chestnut* (fig. 234).

Dehiscence.—Dehiscence is the act by which the ripe pericarp opens to let the seeds escape. Fruits which thus burst spontaneously are called *dehiscent* (*dehiscens* : *Tulip, Iris,* fig. 531); the term *indehiscent* (*indehiscens*) is applied to—1, fleshy fruits which do not open, but decay, and thus free the seeds (*Apple,* figs. 448, 449; *Peach,* fig. 519; *Melon, Pumpkin*); 2, dry fruits, whose pericarp is pierced by the embryo in germination (*Wheat, Buckwheat, Oat,* fig. 526; *Anemone,* fig. 523).

Valves (*valvæ, valvulæ*) are the pieces into which the pistil separates when ripe, to allow the seeds to escape; according to the number of these, the fruit is said to be *univalved, bivalved, &c.* (*univalvis, bivalvis, &c.*); sometimes the separation is incomplete, the valves only opening to a half or a quarter of their length, or at the top only. Apocarpous fruits dehisce by the ventral suture (*Columbine,* fig. 497; *Larkspur,* fig. 512; *Caltha,* fig. 511), or by the dorsal nerve (*Magnolia*), or by both at once (*Pea,* fig. 516, and other *Leguminosæ*); in the latter case, there are two valves to one carpel.

The dehiscence of *plurilocular syncarpous* fruits is *septicidal* (*d. septicida*) when the septa split into two parallel plates, and the united carpels separate (*St. John's Wort,* fig. 527; *Colchicum,* fig. 529; *Mullein, Scrophularia,* fig. 528); each valve then represents a carpel. The placentas may fall away with the valves, or form a solid central column (*Salicaria,* fig. 530). In all cases, the edges of the valves are said to be inflexed. The dehiscence of *plurilocular syncarpous* fruits is *loculicidal* (*d. loculicida*) when it takes place by the dorsal suture; this results from the septa being more firmly united than the median fibro-vascular bundles of the carpels; each valve then represents the halves of two carpels, and the valves are described as *septiferous in the middle* (*v. medio-septiferæ*). Sometimes the placentas are continued along the septa (*Lily, Iris,* fig. 531), at others they remain consolidated into a central column; sometimes, again, the placentas may retain a portion or the whole of each septum, and the central column then presents as many wings or plates as there were septa in the ovary before its dehiscence (*Rhododendron, Datura,* fig. 532); this variety of loculicidal dehiscence is called *septifragal.*

The same fruit may be both septicidal and loculicidal; thus, in *Foxglove,* which is two-carpellary, the septa first separate, then the dorsal nerve of each carpel splits, and each of the four resulting valves represents half a carpel.

Syncarpous fruits with parietal placentas usually dehisce by placental sutures, when each valve represents a carpel, and has *placentiferous margins* (*val. marginibus placentiferæ, Gentian,* fig. 533),—or by the dorsal sutures, when each valve represents the halves of two contiguous carpels, and is *placentiferous in the middle* (*v. medio-placentiferæ, Heartsease,* fig. 534; *Willow,* fig. 535),—or by the separation of the valves,

H

which leave the placentas in their places (*Wallflower*, fig. 547; *Chelidonium*, fig. 546).

In some *syncarpous* fruits, the dehiscence is by valvules or teeth, variously placed, which, by diverging or ascending, form openings for the seeds to escape (*Primrose, Lychnis*, fig. 542; *Snapdragon*, fig. 545; *Harebell*, fig. 544; *Poppy*, fig. 543). Dehiscence is *transverse* (*d. transversalis*) when a compound ovary is halved transversely (*Pimpernel*, fig. 537; *Henbane*, fig. 539; *Purslane*, fig. 538; *Plantain*);— as also when apocarpous fruits break up transversely into one-seeded segments (*Coronilla, Sainfoin*, fig. 518). Dehiscence is *irregular* (*d. ruptilis*) in fruits with resisting septa and dorsal sutures,

510. Linaria. Fruit.

but uniformly thin walls; thus, the pericarp of some *Linarias* (fig. 510) splits into longitudinal ribbons; the fruit of *Momordica, Wild Cucumber*, &c., rupture thus elastically.

Classification of Fruits.—Many authors have attempted this; but their efforts, though resulting in many valuable scientific observations, have sometimes given rise to a very obscure botanical terminology. Linnæus admitted five sorts of fruit; Gaertner, thirteen; Mirbel, twenty-one; Desvaux, forty-five; Richard, twenty-four; Dumortier, thirty-three; Lindley, thirty-six. The following classification, adapted from these several authors, appears to us the simplest and easiest of application; it includes most of the modifications of form observable in the fruits of phænogamous plants.

Apocarpous Fruits.—1. The *follicle* (*folliculus*) is dry, dehiscent, many-seeded,

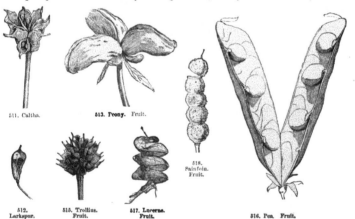

511. Caltha.

513. Peony. Fruit.

518. Sainfoin. Fruit.

512. Larkspur.

515. Trollius. Fruit.

517. Lucerne. Fruit.

516. Pea. Fruit.

and opens by its ventral suture (*Caltha*, fig. 511; *Larkspur*, fig. 512; *Peony*, fig. 513), or very rarely by the dorsal only (*Magnolia*). Follicles are rarely solitary, but almost always form a whorl (*Columbine*, fig. 497; *Peony*, fig. 513; *Caltha*, fig. 511), or head (*Trollius*, fig. 515).—2. The *legume* (*legumen*) is a follicle opening into two valves by

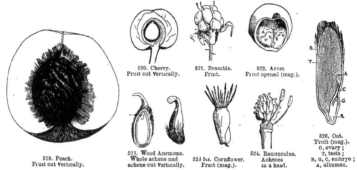

520. Cherry.
Fruit cut vertically.

521. Bramble.
Fruit.

522. Arum
Fruit opened (mag.).

526. Oat.
Fruit (mag.).
T, testa ;
R, G, C, embryo ;
A, albumen.

519. Peach.
Fruit cut vertically.

523. Wood Anemone.
Whole achene and
achene cut vertically.

523 *bis*. Cornflower.
Fruit (mag.).

524. Ranunculus.
Achenes
in a head.

its dorsal and ventral sutures (*Pea*, fig. 516). Some *Leguminosæ* have spirally twisted fruits (*Lucerne*, fig. 517); of others the fruit is indehiscent and one-seeded, hence a true achene (*Trefoil*); of others it is a *lomentum*, i.e. the legume is contracted at intervals into many cells by transverse septa; when ripe, the fruit separates through the septa of the cells into one-seeded joints (*Coronilla*, *Sainfoin*, fig. 518); other legumes are vertically more or less perfectly two-celled, by the inflexion of the dorsal (*Astragalus*, fig. 391), or ventral suture (*Oxytropis*). —3. The *drupe* (*drupa*) is indehiscent, usually one-seeded, with a fleshy mesocarp, and stony or bony endocarp (*Peach*, fig. 519; *Cherry*, fig. 520; *Apricot*, *Plum*, *Almond*, *Walnut*). *Acini* are the small drupes forming the fruit of the *Raspberry* and *Bramble*, &c. (fig. 521).—4. The *simple berry* only differs from the compound berry by originating in a solitary carpel (*Berberis*, *Arum*, fig. 522).—5. The *achene* (*achenium*) is dry, indehiscent, with a single free seed (not adhering to the pericarp); it is solitary in the *Cornflower* (fig. 523 *bis*) and *Dandelion*; agglomerated in the *Ranunculus* (fig. 524), *Anemone* (fig. 523), *Rose*, (fig. 525), and *Strawberry* (fig. 401). The *utricle* (*utriculus*) is an achene with a very thin and almost membranous pericarp (*Scabious*, *Amaranth*, *Statice*).—6. The *caryopsis* (*caryopsis*) is dry, indehiscent, with a single seed adhering to the pericarp (*Wheat*, *Maize*, *Oat*, fig. 526).

525. Rose.
Fruit
cut vertically.

Syncarpous Fruits.—7. The *capsule* (*capsula*) is dry, one- or many-celled, and dehiscent; it is *plurilocular* and *septicidal* in *St. John's Wort* (fig. 527), *Scrophularia* (fig. 528), *Mullein*, *Colchicum* (fig. 529), *Salicaria* (fig. 530); *loculicidal* in *Lilac*, *Lily*, *Iris* (fig. 531); septifragal in *Datura* (fig. 532), *septicidal* and *loculicidal* in *Digitalis* and *Linum catharticum*. The valves of the *unilocular* capsule are placentiferous at the edges in *Gentian* (fig. 533); placentiferous at the middle in *Heartsease* (fig. 534) and *Willow* (fig. 535). The capsule of *Orchis* (fig. 536) opens into three valves placentiferous at the middle, and the median nerves of the three carpels, united by their bases and tops, and crowned by the dry floral envelopes, persist after the valves fall away. In the *circumsciss* capsule (*pyxidium, c. circumscissa*), the dehiscence is transverse (*Plantain, Pimpernel,* fig. 537; *Purs-*

527.
St. John's Wort.   528. Scrophularia. Fruit.   530. Salicaria. Fruit (mag.).   531. Iris. Fruit.   536. Orchis. Fruit.

lane, fig. 538; *Henbane,* fig. 539). In *Mignonette* (fig. 540), the capsule opens by the separation of the three sessile connivent stigmatiferous lobes, without dividing into

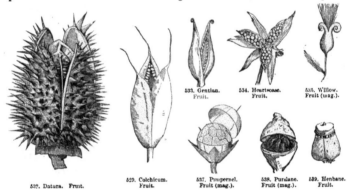

532. Datura. Fruit.   529. Colchicum. Fruit.   533. Gentian. Fruit.   534. Heartsease. Fruit.   535. Willow. Fruit (mag.).   537. Pimpernel. Fruit (mag.).   538. Purslane. Fruit (mag.).   539. Henbane. Fruit.

teeth or valves, and leaves an opening between them. In the *Primrose,* the capsule is five-valved at the top, by the fission of the dorsal nerves of the carpels. In the *Pink* (fig. 541), both the dorsal nerves and placental sutures split. In *Lychnis* (fig. 542), the capsule is similarly incompletely ten-valved. In the *Poppy* (fig. 543), the capsule opens by small tooth-like valves between the septa, below the disk formed by the style and stigmatic rays. In the *Harebell* (fig. 544), the capsule opens by five small valves at the base of the receptacular tube; these openings are formed by the lower portions of the septa separating from the central axis, and carrying up with them a portion of the pericarp, in the shape of a little open door. In other species of *Campanula* the opening occurs at the upper part of the receptacular tube, where the edge of the septum is thickened and forms a border with the concavity outside; the bottom of this border rolls over the concavity, and ruptures the wall of the ovary, forming between each sepal a little round protuberance, and the seeds escape by pores which are on a level with their placentas. In the *Snapdragon* (fig. 545),

540. Mignonette. Fruit.   542. Lychnis. Fruit.   545. Snapdragon. Fruit.   541. Pink. Fruit.   543. Poppy. Fruit.   544. Harebell. Fruit.

the upper carpel, that next the axis, opens near the persistent style by small free valves; the lower carpel, which is gibbous below, opens by two similar collateral valves, also near the style. The entire fruit, when seen in front, resembles a monkey's face, the style being the nose, the hole of the upper carpel the mouth, the two other holes the eyes, and the persistent calyx a head-dress.

The *siliqua* (*siliqua*) is a capsule with two carpels; it is properly one-celled (*Chelidonium*, fig. 546), but usually

546. Chelidonium Fruit.   547. Wallflower. Fruit.   548. Whitlow-grass. Fruit (mag.).   549. Cochlearia. Fruit.   550. Thlaspi. Fruit (mag.).   551. Bunias. Fruit.   552. Bunias. Fruit, open.

two-celled by a spurious membranous septum, and opens from bottom to top by two valves, the seed-bearing parietal placentas persisting (*Wallflower*, fig. 547). The *silicule* (*silicula*) is a siliqua of which the length does not much exceed the breadth (*Whitlow-grass*, fig. 548; *Cochlearia*, fig. 549; *Thlaspi*, fig. 550). In some cases the siliqua is *lomentaceous*, separating transversely into one-seeded joints (*Radish*). In the *Bunias* (figs. 551, 552), each of the two cells of the silicule is two-seeded and two-celled, by a longitudinal septum. In *Crambe* (fig. 553), the silicule is compressed, and originally consists of two unequal one-seeded cells, but

553. Crambe. Fruit, open.   554. Myagrum. Fruit, open.

whilst the seed of the upper cell becomes developed, that of the lower cell is arrested, its funicle being strangled in the septum; and the result is a one-seeded indehiscent fruit. In *Myagrum perfoliatum* (fig. 554), the silicule contains only one seed, which occupies its lower half, and pushes up the septum; the two upper cells are empty. In plurilocular capsules, the name *cocci* (*cocci*) has been given to one- or two-seeded carpels, which separate (often elastically), and carry the seeds with them, but

555. Geranium. Fruit (mag.).   556. Fraxinella. Fruit.   557. Æthusa. Fruit.   558. Bugle. Fruit (mag.).   559. Cerinthe. Fruit.

usually leave the placentas attached to a central column (*Cneorum, Fraxinella, Euphorbia, Geranium, Mallow*). In *Euphorbia* this central column consists of the placentas, and three double plates, which are portions of the septa, of which the other portions were carried away by the dehiscence of the valves.

In *Geranium* (fig. 555) the five carpels separate elastically upwards, and roll over upon themselves; the central column consists of the placentas and the edges of the carpellary leaves. In *Mallows*, the septa of the ten to fifteen carpels split, but the carpels do not wholly separate from the column; a considerable portion of the septa adhering to it. In *Fraxinella* (fig. 556) the five carpels separate completely, and leave no column.

The fruit of *Angelica, Æthusa* (fig. 557), and other *Umbelliferæ* is a capsule with two one-seeded cells, divided by a narrow septum; its two carpels separate like *cocci*, and remain suspended at the top to the filiform axis or prolongation of the receptacle. Most botanists consider this fruit to be composed of two achenes; but achenes are apocarpous fruits, and this fruit, being syncarpous, constitutes a true two-celled septicidal capsule, of which the only opening to the carpels is a narrow cleft, previously occupied by a filiform axis. The fruit of the *Bugle* (fig. 558) is composed of four one-seeded lobes, which separate when ripe, often called *achenes* and nucules; but the fruit of *Borragineæ* and *Labiatæ* is now considered to be formed

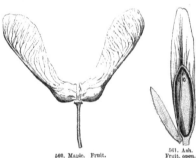

560. Maple. Fruit.   561. Ash. Fruit, open.

of two carpels, each distinctly two-lobed, and containing two seeds; this is obvious in *Cerinthe* (fig. 559). It has been demonstrated that in the very young buds of the

Sage and other *Labiatæ*, there really are only two carpellary leaves, opposite to the two lips of the corolla. Such fruits are not collections of achenes, but syncarpous, the carpels being united below by the dilated style-base (fig. 409); it is a true capsule of two carpels, each of which becomes two-celled, and hence it simulates four distinct carpels. The *samara* (*samara*) is a dry, one- to two-seeded fruit, of which the pericarp forms a membranous wing above or round the cell (*Maple, Ash, Elm*, &c.); these, which are often placed among apocarpous fruits, are evidently composed of two united carpels. In the *Maple* (fig. 560), the two cells are distinct, and the fruit separates, as in *Umbelliferæ*, into two *cocci* hanging at the top of a filiform axis; it is therefore a true septicidal capsule, the only opening of the carpels being the narrow slit previously occupied by the axis. In the Ash (fig. 561), the septum is perpendicular to the faces of the ovary, and consequently the two sharp edges answer to the backs of the carpel; after flowering, all the ovules but one are arrested; the septum is pushed back, one of the cells almost completely disappears, and the other is filled with the seed. ·

The fruit of the *Elm* (fig. 562) is similar; one of the cells is one-ovuled, the other is empty from the first. The *nucule* (*nucula*) is an indehiscent capsule, with a bony or coriaceous pericarp, plurilocular when young, but one-celled and one-seeded by arrest (*Oak*, fig. 232; *Filbert*, fig. 233; *Hornbeam, Beech, Chestnut*, fig. 563; *Lime*, fig. 564). To the same category belong also the fruits of *Fedia* (figs. 565, 566) and other *Valerianeæ*, sometimes for convenience, but not accurately, called achenes.— 8. The *berry* (*bacca*) (whether compound or simple) is succulent, indehiscent, and has no stone; it differs from the capsule only in its fleshy consistence, which frequently induces the suppression of the septa, and arrest of some of the seeds (*Vine*). There are some fruits

| 562. Elm. Fruit. | 563 Chestnut. Fruit. | 564. Lime. Fruit. | 565. Fedia. Fruit cut transversely (mag.). | 566. Fedia. Fruit cut vertically (mag.). | 567. Belladonna. Fruit. |

which may equally be termed a berry or a capsule (*Capsicum, Winter Cherry*). Among species of the same genus, some are provided with a capsule, others with a berry (*Galium, Asperula, Campion, Hypericum*). The *Privet, Nightshade, Belladonna* (fig. 567), *Vine*, have a two-celled berry; *Asparagus* and *Lily of the Valley*, a three-celled berry; *Herb Paris*, a four- to five-celled berry. Among plants with an inferior ovary, the berry of *Sambucus* is three-celled, that of the *Myrtle* four- to five-celled; *Ivy*, five-celled; *Coffee*, two-celled; *Gooseberry*, one-celled, with parietal placentas (fig. 501). The *hesperidium* (*hesperidium*) is a plurilocular berry, with an aromatic glandular epicarp, a dry and spongy mesocarp, an endocarp covered with small watery cells which spring from the walls of the cavities, and extend to the seeds (*Orange*, fig. 568). The *gourd* (*pepo*) is a berry composed of three to five (rarely one) carpels, united to the receptacular tube, and forming a single cell with very fleshy seed-bearing parietal placentas (*Melon, Pumpkin, Sechium, Briony*). The *pome* (*pomum, melonida*, figs. 569, 570), is a berry composed of many (usually five) cartilaginous carpels (E), forming five cells, and united to the receptacular tube (T) (*Apple,*

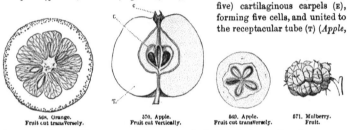

568. Orange.
Fruit cut transversely.

570. Apple.
Fruit cut vertically.

569. Apple.
Fruit cut transversely.

571. Mulberry.
Fruit.

Pear, Quince).—9. The *compound drupe* (*nuculanium*) is fleshy, and encloses many stones, which are sometimes connate (*Dogwood*), sometimes free (*Medlar, Beam, Sapotilla*).

Aggregate fruits is the name given to fruits that result from the union of several flowers; these component fruits are included amongst the above-described varieties. In the *Honeysuckle*, the fruit is formed of two connate, but originally free berries. In the *Mulberry* (fig. 571), the true fruit consists of a spike or head of small drupes, each enveloped in a succulent calyx. The *Fig* (fig. 158) is a pyriform body, fleshy, hollow, bracteate at the base, the mouth furnished with little scales, and serving as a common receptacle to the flowers enclosed in its cavity, the males above, the females below. In the *Pine-apple* (fig. 572), the flowers are spiked and pressed round an axis

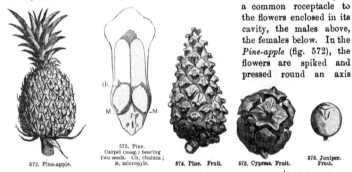

573. Pine.
Carpel (mag.) bearing
two seeds. Ch, chalaza;
M, micropyle.

572. Pine-apple.

574. Pine. Fruit.

575. Cypress. Fruit.

576. Juniper.
Fruit.

terminating in a tuft of leaves; the ovaries form so many berries, but the calyces, the bracts, and the axis itself become fleshy. The *Pine-cone* (*conus, strobilus*) is an aggregate fruit, which has nothing in common with the preceding; the carpels, represented by scales (fig. 573), have neither style nor stigma, and do not close to shelter the seeds, but protect them by closely overlapping each other until ripe. They are sometimes woody, when they form either a conical spike (*Pine*, fig. 574), or a globular head (*Cypress*, fig. 575); when fleshy and connate, they simulate a drupe (*Juniper*, fig. 576).

SEED.

The *seed* (*semen*) of phænogams is the ovule when fertilized, ripe, and ready for germination; it contains the *embryo* (*embryo, plantula, corculum*), which is destined to reproduce the mother-plant. Let us recapitulate the structure of the embryo in the *Pea* (fig. 577). It is composed of a *caulicle* (*cauliculus*, T), a *radicle* (*radicula*, R), two *cotyledons* (*cotyledones*, C), and a *plumule* (*gemmula, plumula*); it is enveloped by a double integument, of which the outer (I), or *testa* (*testa*), is attached to the *hilum* (*hilus, umbilicus*) by the *funicle* (*funiculus*, F), which rises from the *placenta* (*placenta*, P); and the inner (E), or *endopleure*,[1] (*endopleura*) provides a passage for the nourishing juices by the *chalaza* (*chalaza*, H), which communicates with the hilum by means of a cord (A), the *raphe* (*raphe*). Near the hilum is a small opening (M), the *micropyle* (*micropyle*), by which the ovule is fertilized by the pollen. As a general rule, the radicular end of the embryo answers to the micropyle, and the cotyledonary end to the chalaza; the exceptions to this rule, which are rare, and do not invalidate it, will be specified.

577. Pea seed (mag.), deprived of half its integument and one of its cotyledons.

Relative Positions of Seed and Embryo.—It is important to observe that, in the early condition of the ovule, the hilum and chalaza are united; consequently the raphe does not exist, and the micropyle occupies the opposite, or free end of the ovule; also that 1, the *base* of the fruit (ovary [2]), is the point by which this is attached to the receptacle, and its top is the point from which the style springs; 2, the *base* of the *seed* is the point by which it is attached to the funicle or placenta, and which is indicated by the hilum; the *top* of the seed is the extremity of an imaginary straight or curved line drawn through the axis of the seed. The axis of the ovary is defined in the same manner. The embryo has also its *axis*; its base is its radicular, and its top its cotyledonary extremity.

578. Nettle.
Achene cut vertically (mag.), showing one of the large sepals; the ovary terminated by a sessile stigma, St; the seed, fixed without a funicle to the placenta, PL; the testa, T, distinct from the ovary, and radicle, R, facing the top of the ovary.

The top of the seed is obvious whenever the hilum occupies either extremity of the long axis of the seed, as is usually the case (*Nettle*, fig. 578; *Sage*, fig. 579; *Chicory*, 580);

580. Chicory. Seed cut vertically (mag.).

579. Sage. Achene cut vertically (mag.). ov, ovary; Gr, seed.

but sometimes the hilum is placed at the middle of the long axis of the seed

---

[1] Sometimes called *tegmen*.—Ed.

[2] Throughout this section the authors speak of the seed in relation to the *ovary*, where we should say *carpel* or *fruit*.—Ed.

(*Lychnis*, fig. 587); it is then *ventral* (*h. ventralis*), and when the seed is flattened (*Madder*) it is described as *depressed* (*s. depressum*), or *peltate* (*s. peltatum*) if it is convex on one side and concave on the other (*Lychnis, Stellaria*). In these latter cases it is difficult and superfluous to determine the top of the seed, but it is easy and

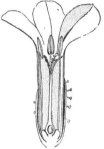

581. Sedum.
Flower cut vertically (mag.).

582. Valerian.
Flower cut vertically (mag.).

583. Plumbago.
Flower cut vertically (mag.).

important to distinguish the *ventral face*, i.e. that facing the placenta, and the *dorsal* or opposite face. The seed is *erect* (*s. erectum*) when it is fixed to the bottom of the cavity of the fruit (*Nettle*, fig. 578; *Sage*, fig. 579). It is *ascending* (*s. ascendens*), when, being fixed to a central or parietal placenta, its top is turned towards that of the fruit (*Sedum*, fig. 581; *Apple*, fig. 570). The seed is *reversed* (*s. inversum*) when its base corresponds to the top of the fruit, whether the placenta is immediately under the style (*Valerian*, fig. 582), or at the bottom of the ovary, in which case the seed is suspended from an ascending basal funicle (*Plumbago*, fig. 583). The seed is *suspended, pendent* (*s. pendulum*), when it is fixed to a central or parietal placenta, with its top turned towards the base of the fruit (*Apricot, Almond*, fig. 583 *bis*). The distinction between *reversed* and *pendulous* seeds is often very slight, and these terms are often used indifferently to describe a seed of which the free end faces the bottom of the fruit. The seed is *horizontal* when fixed to a central or parietal placenta, with its axis at right angles to that of the fruit (*Aristolochia, Lily*, fig. 584).

583 *bis*. Almond.

584. Lily.
Ovary cut
transversely (mag.).

585. Horse-chestnut.
Pistil cut
vertically (mag.).

In certain two-ovuled ovaries one ovule may be *pendulous* and the other *ascending* (*Horse-chestnut*, fig. 585); in others with many seeds or ovules, some are *ascending*, others *pendulous*, and those in the centre *horizontal* (*Columbine*). All the terms indicating the position of the *seed* are equally applicable to that of the *ovule*.

The radicle is *superior* (*r. supera*) when it points to the top of the ovary; it is inferior (*r. infera*) when it faces the bottom; thus corresponding to the erect and ascending seeds. . Thus, the *Nettle* (fig. 578) has an *erect* seed and a *superior* radicle; the axis of the seed is straight, the radicular end being furthest from the cotyledonary, which answers to the hilum. In the *Sage* (fig. 579) and *Chicory* (fig. 580), the seed is erect, with an inferior radicle; here the embryo seems to have twisted half round upon itself; the cotyledonary end, which ought to answer to the hilum, being at the opposite extremity, and the radicle nearly occupying its place; this movement has taken place in the cavity of the ovule before fertilization, as we shall presently explain; the result is a long raphe, which runs along one side of the seed, and the chalaza is consequently diametrically opposite to the hilum. The radicle is *centripetal* (r. centripeta) when it faces the central axis of the fruit (*Lily*, fig. 584); *centrifugal* (r. centrifuga), when it faces the circumference (*Mignonette*, fig. 384). The

586. Wallflower. Seed cut vertically (mag.). 587. Lychnis. Seed cut vertically (mag.). 588. Datura. Seed cut vertically (mag.). 589. Marvel of Peru. Fruit cut vertically. 591. Plantain. Ventral surface of seed (mag.). 592. Plantain. Seed cut vertically (mag.).

embryo is *antitropal* (e. antitropus), when, its axis being straight, the micropyle (and radicle) is furthest from the hilum (*Nettle*, fig. 578; *Rumex*, fig. 644); it is *homotropal* (e. homotropus) when, its axis being straight, the micropyle (and radicle) is next the hilum, while the chalaza (and cotyledonary end) is distant from the hilum, and only connected with it by a raphe; then the base of the seed (hilum) and of the embryo (radicle) correspond (whence the term homotropal, *Sage*, fig. 579; *Chicory*, fig. 580; *Pear, Apricot, Rose, Strawberry, Scabious, Centranthus, Campanula, Heartsease, Iris*, &c.). The embryo is *amphitropal* (e. amphitropus) when, its axis being bent, the micropyle and chalaza are both close to the hilum (*Wallflower*, fig. 586; *Lychnis*,

fig. 587; *Datura*, fig. 588; *Marvel of Peru*, fig. 589; *Mulberry*, fig. 590). The embryo is *heterotropal* (e. heterotropus) when, from the unequal growth of the coats, neither extremity of the embryo corresponds to the hilum, and the radicle does not correspond

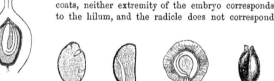

596. Pine. Seed. 590 Mulberry. Ovary cut vertically (mag.). 593. Palm. Seed cut vertically. 594. Asparagus. Seed vertically (mag.). 595. Spergularia. Seed (mag.). 597. Columbine. Seed (mag.).

to the micropyle; in this case, the axis of the embryo is sometimes *parallel* to the plane of the hilum (*Pimpernel, Plantain*, figs. 591, 592), sometimes *oblique* to it (*Wheat, Chamærops*, fig. 593; *Asparagus*, fig. 594); the radicle is then said to be excentric (r..vaga, excentrica).

Form and Surface of the Seeds.—According to their form, seeds are *globular, ovoid, reniform, oblong, cylindric, turbinate, flattened, lenticular, angular*, &c.; some are irregular, flat, and rather like grains of sawdust, and are said to be *scobiform* (*s. scobiformia*); flattened seeds with thick and projecting edges are said to be *margined* (*s. marginata*) (*Spergularia*, fig. 595), or winged if these margins become broad and membranous (*Bignonia, Pine*, fig. 596). The surface may be *smooth* (*s. læve, Columbine*, fig. 597); *wrinkled* (*s. rugosum, Fennel*, fig. 598); *striate* (*s. striatum, Tobacco*, fig. 599); *ribbed* or *furrowed* (*s. costatum, Larkspur*, fig. 600); *reticulate* (resembling a sort of network (*s. reticulatum, Cress*, fig. 601); *punctate* (*s. punctatum*), i.e. marked with little dots; *alveolate* (*s. alveolatum*), i.e. covered with little pits resembling

598. Fennel. 599. Tobacco. 600. Larkspur. 601. Cress. 602. Poppy. 603. Stellaria.
Seed (mag.). Seed (mag.). Seed (mag.). Seed (mag.). Seed (mag.). Seed (mag.).

honeycomb (*Poppy*, fig. 602); *tubercled* (*s. tuberculatum*), i.e. furnished with small rounded projections (*Stellaria*, fig. 603); *aculeate* (*s. aculeatum*), i.e. bristling with small points (*Snapdragon*, fig. 604); *glabrous* (*Flax*); *hairy* (*Cotton*). Some seeds have a pulpy testa (*Gooseberry*, fig. 605; *Pomegranate*); others are covered with oily glands, often arranged in bands (*Angelica*, fig. 606); sometimes placed in furrows (*Juniper*, fig. 607).

The *hilum*, or point by which the seed is attached to the funicle or placenta, forms a depressed or prominent scar; in the middle or towards one side of this scar, is the *umbilicus*, a very small simple or compound orifice, indicating the passage of the nourishing vessels of the funicle into the seed. The *chalaza*, or internal hilum,

604. Snapdragon. 605. Gooseberry. 606. Angelica. 607. Juniper. 608. Orange. 609. Orange.
Seed (mag.). Seed cut Seed (mag.). Seed (mag.). Seed, open. Whole seed.
        vertically (mag.).

forms sometimes a more or less distinct protuberance, sometimes a sort of knob, sometimes a simple blotch (*Orange*, fig. 608, *Almond*). The *raphe*, which maintains the communication between the hilum and chalaza when these are separated during the development of the ovule, appears like a band along one side of the seed; often it branches out in the thickness of the testa (*Almond, Orange*, fig. 609). The *micropyle*, which in the ovule formed a large gaping opening, remains visible on some seeds (*Bean, Kidney-bean, Pea*); it disappears in most, but its position is usually indicated by that of the tip of the radicle.

The Proper and Accessory Coats of the Seeds.—Seeds do not always possess a distinguishable *testa* and *endopleure*; often, when ripe, all the coats merge into one, or one splits up into several layers, and the seed presents three or four coats. The origins of these envelopes will be described under the development of the ovule.

Arils are accessory organs, which mostly develop after fertilization, and cover the seed more or less completely, without adhering to the testa; some are expansions of the funicle, and are specially designated as *aril* (*arillus*) (*Nymphœa, Passiflora, Opuntia, Willow, Yew*); others arise from the dilatation of the edges of the micropyle, and are called by some authors *arillodes* or *false arils* (*arillodes*).

In the *White Water Lily* (fig. 610) a swelling (A, A), rising from the funicle (F), gradually spreads over and caps the ovule, and ends by closely enveloping the seed, without adhering to it, leaving scarcely a trace of an opening over the chalaza (Ch.). In *Passion-flowers*

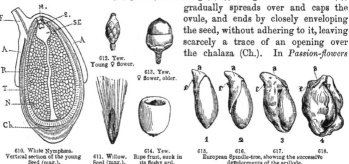

610. White Nymphæa.
Vertical section of the young Seed (mag.).

611. Willow.
Seed (mag.).

612. Yew.
Young ♀ flower.

613. Yew.
♀ flower, older.

614. Yew.
Ripe fruit, sunk in its fleshy aril.

615. 616. 617. 618.
European Spindle-tree, showing the successive developments of the arillode.

an annular swelling, with a free membranous torn margin, forms at the shortened end of the funicle, round the hilum; this gradually expands, and ends by enclosing the seed in a loose fleshy bag, with a large opening towards the chalaza. In *Willows* (fig. 611), the very short thick funicle expands into an erect pencil of hairs, which envelops the seed. In *Cactus Opuntia*, two concave boat-shaped expansions spring laterally from the funicle, into which the ovule is pushed, and within which it is developed; this accessory envelope thickens, hardens, and forms a sort of stone, covered with pulp. In the *Yew*, the female flower (fig. 612) consists of a single ovule, which is at first protected only by the scales of the bud from which it issued, and after fertilization disengages itself from these, when it is completely *naked*, with a gaping micropyle at its summit. Soon (fig. 613), between the ovule and the scales at its base, a small cup is developed, which gradually swells, becomes red and succulent, and ends by almost entirely covering the seed (fig. 614); this cup is nothing but an enormous development of the funicle, which thus furnishes an envelope to the fruit, which had not even the protecting scale of the *Pines* and *Firs* (fig. 379). In the *Spindle-tree* (figs. 615, 616, 617, 618), the successive stages of development of the *arillode* (*a*) are easily followed (1, 2, 3, 4); it does not spring from the funicle (*f*), but from the micropyle, the edges of which dilate by degrees so as to form around the seed a succulent, loose, folded bag, open towards the chalaza.

It must be observed that this arillode, starting from the micropyle, which is very near the hilum, unites at a very early stage with the funicle, of which it appears to be an appendage, but its origin may be recognized in very young ovules. In the *Nutmeg*, the fleshy and honeycombed envelope of the seed, which forms the aromatic substance called *mace*, may be looked upon as an expansion of the micropyle. In *Euphorbia* (fig. 619), the circumference of the micropyle, which formed at first a little swelling, thickens enormously after fertilization, and forms a small fleshy disk, of which the central canal, at first filled with the conducting tissue, becomes by degrees stopped up. In *Polygala* (fig. 620), the little three-lobed body at the base of the seed has the same origin as the disk of

621. Asclepias. Seed (mag.) with hairy arillode.　619 Euphorbia. Seed crowned by a fleshy arillode (mag.).　620. Polygala. Seal capped by a cartilaginous arillode (mag.).　622. Heartsease. Seed (mag.).　623. Chelidonium. Seed cut vertically (mag.).　624 Asarum. Seed (mag.).

Euphorbia, and the micropyle is visible long after fertilization. In *Asclepias* (fig. 621) the tuft of hairs which crowns the seed is also an arillode proceeding from the micropyle.

The name of *strophioles* (*strophiolæ, carunculæ*) has been given to excrescences on the testa which are independent of the funicle or micropyle, as the glandular crest which in *Heartsease* (fig. 622) and *Chelidonium* (fig. 623) marks the passage of the raphe; the cellular mass which in *Asarum* (fig. 624) extends from the hilum to beyond the chalaza, and the tuft of hairs at the chalaza in *Epilobium* (fig. 625). The name *aril* having been indifferently applied to *true arils, arillodes, strophioles*, &c., it would be advisable to keep this term as a general name for excrescences of various sorts which appear upon seeds, and to limit the meaning by an adjective indicating their origin. We should thus have a *funicular aril* (*Willow, Nymphæa, Yew*); a *micropylar aril* (*Spindle-tree, Euphorbia, Polygala, Asclepias*); a *raphean aril* (*Chelidonium, Asarum*); a *chalazian aril* (*Epilobium*), &c. A second adjective might denote its *membranous, fleshy, hairy*, &c. texture.

625. Epilobium (mag.). Chalazian tuft of hairs.　626. Pine. Embryo Vertically (mag.).　627. Berberis. Seed cut Vertically (mag.).　628. Walnut. Seed cut vertically.　629. Lime. Embryo spread out (mag.).　630. Cuscuta. Embryo coiled round its albumen (mag.).

Embryo or Young Plant.—In most phænogamic plants the embryo is dicotyledonous, whence the name *Dicotyledons.* Some species (*Pines,* fig. 626), possess six, nine, and even fifteen whorled cotyledons.

Other phænogamic plants have only one cotyledon; whence the name *Mono-cotyledons.* The colour of the embryo varies; it is *white* in most plants, *yellow* in some *Cruciferæ,* blue in *Salpiglossis,* green in the *Spindle-tree* and *Maples,* and *pink* in *Thalia.* The cotyledons are generally fleshy, their parenchyma is oily in the *Walnut* and the *Almond,* and mealy in the *Kidney-bean;* they have sometimes distinct nerves (*Berberis,* fig. 627); they are sessile or petioled, or reduced to a petiole without a limb; this is especially the case in monocotyledons. They are usually entire and equal, but may be lobed (*Geranium, Walnut,* fig. 628), or palmate (*Lime,* fig. 629), or very unequal, with the smaller so minute that the plant might be mistaken for a monocotyledon (*Trapa*). Those of the *Nasturtium* and *Horse-chestnut* unite as they grow old into a compact mass. In some parasites they entirely disappear, and the embryo is reduced to its axis; as in *Cuscuta* (fig. 630), whose thread-like stem

631. Mallow. Embryo (mag.).   632. Bindweed. Embryo spread out (mag.).   633. Wallflower. Transverse section of seed (mag.).   634. Rocket. Transverse section of seed (mag.).   635. Orange. Seed without its testa.   637. Almond. Doub'e embryo.   638. Almond. Embryos separated.

is attached to the plants it preys on, by suckers (p. 16, fig. 48); and living on their juices, it needs no leaves to elaborate sap; and the adult plant, like the embryo, possesses no leaves. Cotyledons are sometimes folded in halves, along their median line; or convolute (*Mallow,* fig. 631); or spiral (*Hop*); or crumpled (*Bindweed,* fig. 632); the embryo itself is straight, or curved, or zigzag, or annular, or spiral, or rolled into a ball, &c. Often the radicle is turned up on the cotyledons; if it is then placed against their commissure it is said to be *lateral,* and the cotyledons are *accumbent* (*c. accumbentes, Wallflower,* fig. 633); if it is on the back of one of the cotyledons, it is said to be *dorsal,* and the cotyledons are *incumbent* (*c. incumbentes, Rocket,* fig. 634).

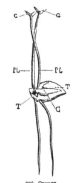

636. Orange. Germinating seed. T, testa; c, included cotyledons; PL, stems; G, plantules.

Some seeds contain several embryos; the *Orange* (fig. 635) has often two, three, or four unequal, irregular, and convolute, the cotyledonary ends of all facing the chalaza, and their radicles facing the micropyle; they all leave the seed at the period of germination (fig. 636). The seed of the *Almond* frequently presents two superimposed embryos, one of which appears to proceed from the first, like successive internodes (fig. 637); they may be easily separated (fig. 638), when their respective radicles and two cotyledons can be plainly seen.

The *monocotyledonous* embryo is usually cylindric or ovoid; to distinguish the

parts of which it is composed, it must be cut vertically, when it usually discloses an
elongated axis with a small protuberance marked with an oblique or vertical fissure:
this protuberance represents the plumule; the fissure through which the two first
leaves will appear marks the separation between the caulicle and cotyledon.
Owing to the small size of the parts, it is sometimes difficult to distinguish the
cotyledonary from the radicular end; but the latter, which answers to the micropyle,
is usually nearer to the integument than the former; this is evident in the seed of
Arum (fig. 639). In *Oats* and other *Gramineæ* (fig. 640), the
seed, if halved longitudinally along the furrow on its inner
face, discloses a very abundant farinaceous parenchymà (A),
of which we shall presently speak; from the base of the seed
along its dorsal face rises the embryo (R, G, C), of a yellow, semi-
transparent colour; within this is a fleshy leaf (C), which
extends one-third of the length of the seed; this leaf encloses
several others, successively smaller (G), which enfold each
other, and are placed between the largest leaf (C)
and the dorsal face of the ovary (O); all rise from

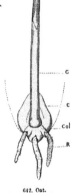

639. Arum.
Seed cut
vertically (mag.).

641. Oat.
Isolated embryo seen on
its outer face (mag.).

643. Aconite.
Seed cut
vertically (mag.).

640. Oat.
Vertical section of
fruit (mag.).

642. Oat.
Germinating embryo.
(mag.).

an enlarged neck which narrows towards the base into an obtuse cone; the interior
leaf (C) is the *cotyledon*, the others (G) form the *plumule*, the conical disk is the
caulicle, terminated by the radicular end (R).

If we extract the entire embryo (fig. 641), we perceive the cotyledon, which is
large, and hollowed into a sort of spoon-shape, in the middle of which lies the
plumule, forming a closed bag; in the middle of this bag is a very small longitudinal
slit, which enlarges later into a sheath, to open a passage for the contained leaves;
below is the caulicle, bearing the cotyledon on its side, and the plumule in its axis;
its free end is terminated by rounded protuberances, in which holes will form,
whence radicular fibres will emerge at the period of germination, as from so many
sheaths (fig. 642, Col.).

Albumen.—Many seeds contain, besides the embryo, a disconnected accessory
mass of parenchyma, named *albumen* (*albumen, perispermum*), the formation of
which will be explained in the chapter on the Ovule. It is destined to nourish the
embryo, and exists at an early period in all seeds; if only a portion of it is absorbed
by the embryo, the rest hardens, up to the period of germination, and the embryo is
said to be *albuminous* (*e. albuminosus*); if it be absorbed, the embryo is *exalbuminous*

(*e. exalbuminosus*). The albumen may be very copious (*Aconite*, fig. 643), or extremely thin and almost membranous; in general it is largest when the embryo is smallest, and *vice versa*. It is said to be *farinaceous* (*a. farinaceum*), when its cells are filled with starch (*Buckwheat, Barley, Oats*, fig. 640; *Rumex*, fig. 644); *fleshy* (*a. carnosum*), when its parenchyma, without being farinaceous, is thick and soft (*Berberis*, fig. 627; *Heartsease*, fig. 645; *Nightshade*); *mucilaginous* (*a. mucilaginosum*), when it is succulent and almost liquid; it is then rapidly absorbed, and may almost entirely dis-

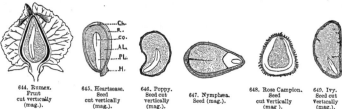

| 644. Rumex. Fruit cut vertically (mag.). | 645. Heartsease. Seed cut vertically (mag.). | 646. Poppy. Seed cut vertically (mag.). | 647. Nymphæa. Seed (mag.). | 648. Rose Campion. Seed cut vertically (mag.). | 649. Ivy. Seed cut vertically (mag.). |

appear (*Bindweed*); *oleaginous* (*a. oleaginosum*), when its parenchyma contains a fixed oil (*Poppy*, fig. 646); *horny* (*a. corneum*), when its parenchyma thickens and hardens (*Galium, Coffee, Iris*); *like ivory* (*a. eburneum*), when it has the consistency and polish of ivory (*Phytelephas*). In *Pepper* and *Nymphæa* (fig. 647), &c., the seed contains two sorts of albumen; which will be noticed when treating of the ovule.

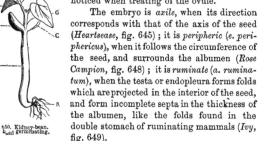

The embryo is *axile*, when its direction corresponds with that of the axis of the seed (*Heartsease*, fig. 645); it is *peripheric* (*e. periphericus*), when it follows the circumference of the seed, and surrounds the albumen (*Rose Campion*, fig. 648); it is *ruminate* (*a. ruminatum*), when the testa or endopleura forms folds which are projected in the interior of the seed, and form incomplete septa in the thickness of the albumen, like the folds found in the double stomach of ruminating mammals (*Ivy*, fig. 649).

650. Kidney-bean. Seed germinating.

651. Orange seed germinating. c, cotyledons enclosed in the testa, T.

Germination.—Germination is the action by which the embryo grows and throws off its coats, finally supporting itself by drawing its nourishment from without.

The free end of the caulicle (fig. 650, T), terminated by the radicle, usually enlarges the orifice of the micropyle, and emerges; soon the entire caulicle throws off its envelopes, with the cotyledons (c) and the plumule (G); the latter lengthens in its turn, and its little leaves expand as it rises; at the same time the radicle develops, and descends into the earth. If the caulicle, which is the first internode of the plant, lengthens during germination, the cotyledons are raised, and appear above ground; they are then said to be *epigeal* (*c. epigæi*, *Kidney-bean*, fig. 650; *Radish*, *Lime*). When the caulicle is very short, and the plumule (which forms the second internode) rapidly lengthens, the cotyledons remain in the ground, often even within the seed-coats; they are then said to be *hypogeal* (*c. hypogæi*, *Spanish Kidney-bean*, *Oak*, *Gramineæ*, *Orange*, fig. 651).

In monocotyledons, the evolution of the radicle presents a remarkable peculiarity: it is provided at the base (fig. 642) with a sort of sheath, named the *coleorhiza*; this is nothing but an outer cellular layer which, having been unable to accompany the development of the radicle (R), has been pierced by it.

ANATOMY.

In Organography we have described the *fundamental organs* which provide for the growth and reproduction of plants; namely, the root, the stem, the leaves, the floral whorls, and the seed; but these are themselves composed of parts which cannot be studied without the aid of the microscope. These parts, the structure of which varies but little in different plants, and which are elementary vegetable tissues, are named *elementary organs*; and the science which treats of them is called *Vegetable Histology*, or *Vegetable Anatomy*.

ELEMENTARY ORGANS.

If we examine microscopically the thinnest possible slice of a stem, root, leaf, or floral organ, it will present many different cavities, some entirely enclosed in walls, others having no proper walls, but being interspaces between the first; taken together, they present the appearance of a fabric or tissue; whence the name *vegetable tissue*.

The closed cavities present three principal modifications:—1. *Cells.*—Their diameter is [originally] nearly equal every way. 2. *Fibres.*—These are longer than broad, and their two ends are spindle-shaped. 3. *Vessels,* or lengthened sacs, the two ends of which cannot be seen at once under the microscope.

Cells are very variable in shape, depending on the manner in which they are

652. Elder.
Punctate cells.

arranged. If they are not crowded, they retain their primitive form of spheroids or ovoids (fig. 652); but if the contiguous faces become pressed together in the course of their growth, they become polyhedral, and may be dodecahedrons, or four-sided prisms, either lengthened into columns, or tabular, or cubical. A transverse section of prismatic cells presents equal squares, a vertical section of dodecahedral cells presents hexagons (fig. 653) like a

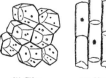

653. Elder.
Cellular tissue of the
central pith.

654. Lily.
Elongated
cells.

honeycomb; whence the name of *cellular tissue* given to these cells collectively. Lastly, the cells may be placed end to end, like superimposed cylinders or barrels (fig. 654).

When the cellular tissue (*parenchyma*) is very compact, there are no interstices

I 2

between the surfaces of the cells; but if the tissue is loose, the cells retain their
rounded form, and leave larger or smaller *intercellular canals* (fig. 655). These
spaces occur between polyhedral cells when an interposed liquid or gas displaces
them; and it may happen (fig. 655) that if a regular pressure is exerted in neighbouring spaces, each of which is circumscribed by a small number of cells, the latter
may be disjointed, and a portion of their walls pressed inwards; but where there are
two contiguous spaces, the pressures from without will counteract each other, and
the cells remain coherent; they then take the shape of stars, the contiguous rays
forming isthmuses which separate the spaces.

Sometimes the intercellular space is circumscribed by a great many cells; it is
then called a *lacuna*. These *lacunæ* do not always result from the displacement of the
surrounding cells, but from the destruction of several of them, or the rapid growth
of the plant.

In their earliest condition cells are sacs surrounded by a thin homogeneous
membrane, which is soft and moist at first, but dries by degrees. Sometimes this
membrane constitutes the sole wall of the cell, sometimes it is lined by a second; but
the latter does not form a continuous sac; it is wanting here and there, and only
partially lines the outer membrane; the result is that there are thin areas where

655. Bean.   656. Elder.   657 Mistleto.   658. Mistleto.   659. Orchid.
Starred cells.   Rayed cells.   Rayed and reticulate cell. Annular cell.   Spiral cell.

there is but one membrane, and thick areas where there are two. When the inner
membrane is deficient only in small spots, these appear as *punctures* (fig. 652) or
short *lines* (fig. 656); when it is absent over considerable irregular areas, the thin
places form an irregular *network* (fig. 657), of which the open parts answer to those
where the inner membrane is wanting, and the threads to the parts where it lines
the outer membrane. Lastly, when the solution of continuity of the inner membrane
is extremely regular, the open spaces are separated by parallel 'thickened rings
(fig. 658), or a thickened spiral which passes from one end of the cell to the other
(fig. 659). Cells may either be *homogeneous*, or *punctate*, or *rayed*, or *reticulate*, or
spiral, or *annular*; and in many cases the same cell passes successively through
more than one of these forms. It frequently happens that a third, fourth, or fifth
membrane is developed within the second, by which the wall of the cell is corre-
spondingly thickened. It has been observed that these successive membranes usually
mould themselves upon the second, so that the thin and thick portions of the cell
correspond throughout.

Fibres.—The length of these varies, but most have a very thick wall, formed at
first of a single membrane, lined by a succession of others developed within it; and
as the cavity of the fibre diminishes more and more with age, the fibre finally appears

nearly filled up. The canal which forms its axis is cylindric; but its outer walls, which are pressed against those of the neighbouring fibres, are flattened and prismatic, as may be seen by a transverse section of fibrous tissue (fig. 660).

The fibres, being spindle-shaped at their extremities, cannot be in juxtaposition throughout their length, but the extremities of other fibres are inserted between their free portions, and hermetically close the conical interspaces above and below them (fig. 661). When the successive inner layers completely line the outer layer, as frequently happens, the cavity of the fibre remains smooth; if the second layer does not completely line the first, spiral or reticulate thickenings are the result (*spiral* or *reticulate fibre*); and *dotted* or *punctate* fibre (fig. 661), the most common form of all, is the result of the failure of the inner layer over minute areas.

Vessels are much elongated tubes, the walls of which are never smooth, but present either slender *spots* or *lines*, or a close *network*, or *rings*, or *spiral lines*; they are cylindric, and constricted at intervals (fig. 668). The contractions are circular and horizontal and close set, or oblique and distant. If the vessel be boiled in dilute nitric acid, it breaks

660. China regia.
Fibres cut transversely.

661. Clematis.
Punctate fibre.

up at the striæ. Where the constrictions occur, membranous folds often project as rings or perforated diaphragms into the interior; whence it has been concluded

that the vessel is formed partly of cells, partly of fibres joined end to end, of which the ends, which at first formed septa, have gradually become obliterated or perforated. The vessels, like the cells and fibres, are named, according to the appearance of their walls, *punctate, striate, reticulate, annular, spiral*. The spiral vessels, or *tracheæ* (fig. 662), are membranous tubes, uninterruptedly traversed within by a pearly white spiral thread; this thread is neither tubular nor channelled, but cylindric, flattened (fig. 663), or a four-sided prism. The tracheæ being spindle-shaped at each end (fig. 662), are regarded as elongated fibres. Nothing is easier than to examine these tracheæ: if young shoots of *Rose* or *Elder* be gently broken, there will be seen by the naked eye between the ruptured surfaces a spiral thread, lengthening and shortening like a piece of elastic. The outer membrane is not so obvious,

662. Melon.
Tracheæ.

except when the coils of the spiral thread are very remote. In most cases the spiral thread is single, but it may be double, and sometimes as many as twenty form a ribbon (*Banana*) and can be unrolled together. Finally, a spiral thread, which was originally single, may become folded and broken up into finer threads (*Beet-root*).

663. Mamillaria.
Trachea.

Annular vessels (fig. 664) are membranous tubes girt within by rings, which may be incomplete, or spirally twisted (fig. 665), whence they have been mistaken for old tracheæ; they, however, differ from tracheæ in that they never present in their earliest condition a regular and continuous spiral, and that many intermediate forms between the ring and the spiral occur in every such vessel; as, however, they terminate in tapering cones, they have evidently the same origin as the tracheæ. *Reticulate* vessels are a modification of the annular; if rings are so placed as to touch at intervals, they resemble a network, and the same vessel may be both annular and reticulate (fig. 666). *Striate* vessels are membranous tubes, cylindric or prismatic, the inner membrane of which resembles a web, whose interstices form thin more or less regular striæ. In *prismatic* vessels (fig. 667) the striæ extend to

the angles, and the interstices resemble the rungs of a ladder, whence their name of *scalariform* vessels. Striate vessels originate as a series of superimposed cells; others as *fibres*, as shown by their spindle-shaped ends. *Dotted* vessels (fig. 668) are membranous tubes of which the

| 664. Melon. Annular vessel. | 665. Melon. Spiral and annular vessel. | 666. Melon. Reticulate and annular vessel. | 667. Brake. Rayed prismatic vessels. | 668. Melon. Punctate moniliform vessel. |

inner membrane is perforated by small holes forming parallel series of oblique or horizontal dots; the vessel presents equidistant constrictions corresponding to

circular folds in the interior, clearly indicating that the punctate vessel is formed by superimposed cells of which the connecting surfaces have been absorbed. *Punctate* vessels with deep constrictions resemble chaplets of beads, whence their name of *moniliform* or *beaded* vessels.

Laticiferous Vessels.—We have seen that *proper* vessels present inequalities resulting from the modifications of the inner membrane; there are others with smooth transparent and homogeneous walls, which contain a peculiar juice named the *latex* (fig. 669); these anastomose, and form a complicated network, of which the tubes meet at right or acute angles; these tubes are usually cylindric, and swollen here and there (fig. 670), from the accumulation of latex in certain places;

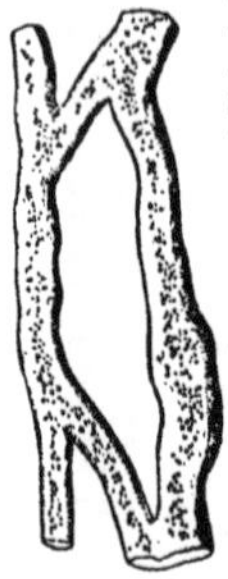

669. Chelidonium. Laticiferous Vessels.

670. Dandelion. Laticiferous vessels.

below these swellings the vessel is gradually constricted, and **the communication** between the constricted and swollen portions is interrupted. The laticiferous vessels are thus distinguished from proper vessels by their transparent walls and by branching.

Union of the Elementary Organs.—Botanists are divided in opinion as to the forces which cause the walls of the elementary organs to cohere; some think that the walls of the cells are originally semi-fluid, and hence become agglutinated, and remain so even after the plant has ceased to live; others consider that an intercellular secretion cements the adjacent cell-walls. A third opinion is that vegetable tissue originates as a homogeneous plasma, which gradually thickens, and ·ends by forming vacuoles, which afterwards become the cavities of the cells; a common septum therefore separates the neighbouring cells; but soon each cell becomes individualized, the septum doubles more or less completely, and the cohesion between the cells is due to an *interposed cellular tissue.* This theory differs from the second, inasmuch that in the latter the cells are cemented by a subsequently secreted matter, while in the former the cells are united by an unorganized tissue, developed cotemporaneously with themselves; this unorganized tissue then itself becomes cellular, and finally separates the previously individualized cells which it originally united. Communication is established between elementary organs in various ways; it has been stated that it takes place by means of the destruction of the contiguous surfaces of cells and fibres placed end to end, from which there results a vessel; communication can also be established through the walls of cells, either by the disappearance of the outer membrane, or by slits or holes at different points of its wall, or simply by pores rendering these membranes permeable.

Contents of the Elementary Organs.—The contents of these are very various: gaseous, liquid, or solid. Cell-contents appear as scattered or agglomerated granules, which in very young cells usually assume a lenticular form, and rest against the wall, or are even buried in its thickness (fig. 654); this body (the *nucleus, cytoblast,* or *phacocyst* of the cell) is regarded by botanists as a germ which, by its development, will produce a new cell. In most cases the nucleus becomes less distinct as the cell develops. According to the recent labours of M. Hartig, the nucleus is principally formed of small particles of matter analogous to albumen, a certain number of which are transformed into vesicles, which again give origin to *cellulose, fecula, chlorophyll* and *aleurone.* Cellulose is an insoluble substance forming the cell-walls, fibres, and vessels, the composition of which is identical in all plants. *Woody tissue* or *lignine* is nothing but the thickened and condensed cellulose; to its density wood owes its hardness; the stony particles in the flesh of pears and the stones of fruits are also formed of it.

Fecula or *starch* may be recognized by its blue-violet tinge when acted on by iodine, by its insolubility in cold water and its coagulation in hot water; its chemical composition is that of cellulose. Starch-grains are generally spheroidal or irregularly ovoid (fig. 671); their surface presents concentric circles around a point which usually occupies one of the ends of the granule. These circles indicate so many layers, superimposed around a small nucleus; thus the starch-grain is developed from within outwards, that is, in the reverse way to the cell

671. Pea.
Starch cells.

which contains it. Starch-grains may be easily examined by moistening a slice of cellular tissue containing them; a drop of iodine will then colour the starch-grains blue-violet, and bring out clearly the distinction between the cell and its contents. If there be grains of albumen accompanying the starch-grains, the iodine will colour them brown or yellow.

Chlorophyll or *chromule* is a green substance, which forms flakes of a gelatinous consistence floating in the colourless liquid of the cells; these flakes have a tendency to gather around or collect on the inner cell-walls, or on the contained starch or *aleurone* grains. Chlorophyll constitutes the green colour of plants; it is dissolved by alcohol, whence it has been supposed to be of a resinous nature.

The yellow colouring matter of cells is similar in consistence and properties to chlorophyll; but red, violet, or blue colouring matters are always liquid.

Aleurone abounds in ripe seeds, and is always found either in the embryo or albumen. Hartig considers an aleurone grain to be a vesicle with a double membrane, containing a colourless waxy mass, which is coloured yellow by iodine, and is ordinarily soluble in water.

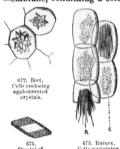

672. Beet.
Cells enclosing
agglomerated
crystals.

675.
Crystal of
aleurone.

673. Rumex.
Cells containing
raphides.

In certain plants it assumes a well-defined crystalline form (figs. 674, 675); in others, the nucleus of the aleurone mass has crystallized, while the surrounding layers remain amorphous, and thus the grain presents a round or ovoid form. Aleurone is essentially formed of substances which are collectively termed *protein* (for which see the section on Vegetable Physiology).

674. Lathræa.
Cell containing crystals of
aleurone, in the midst
of cells containing chlo-
rophyll.

According to Hartig, the particles of the *nucleus* undergo the following transformations : 1, the nucleus is transformed directly into chlorophyll, fecula, or aleurone; 2, it is transformed into starch and the starch into aleurone; 3, it is transformed into chlorophyll, and that into starch, which again passes into aleurone.

The laticiferous vessels contain a large quantity of powdery granules, which float in the *latex*, some of which are very large and colourless, and partake of the nature of starch.

As to the sap which fills these cells, and rises in the vessels, it is a colourless liquid, holding in solution the materials for cell-formation and cell-contents. The other liquids, either contained in the cells, or in the intercellular spaces, are fixed or volatile oils, turpentines, sugar or gum, dissolved in water. Finally, we find gases occupying the intercellular spaces, sometimes at considerable depths.

Besides the solid organic substances above described as occurring in the cellular tissue, special cells occur, containing mineral substances, the elements of which, either compound or simple, have been carried up by the sap, and have crystallized in the cells. Those of which the elements were originally in combination would crystallize at once ; but for the others, it is necessary that the elements which have a reciprocal affinity should be united in proper proportions.   In all cases, it is only during life that this crystallization is carried on, for the crystals are found in special cellular tissues, the forms of which determine theirs ; the same salt being, in fact, found to crystallize very differently according to the tissues in which it is formed.

The crystals contained in the cells are either solitary or clustered ; in the latter case they are grouped into radiating masses (fig. 672), or bundles of parallel needles (fig. 673) named *raphides* (R) ; and they may often be seen escaping from the cells (c) when the tissue containing them is dissected under the microscope.   Finally, the cells and even the intercellular spaces often contain silex, one of the most abundant of minerals, which constitutes sand and flint; this silex even encrusts the tissues of certain plants, and notably the straw of *Gramineæ*.   Certain mineral concretions are observable in the leaves of some *Urticeæ* ; if the leaf of a *Nettle* be viewed with a lens, transparent spots may be distinguished ; this is due to the presence of calcareous particles deposited in the outer cells, to which Weddell has given the name of *cystoliths*.   These cystoliths differ from the crystals represented in fig. 672, in being deposited in calcareous layers around a nucleus formed at the expense of the cell-wall, which has been pulled aside by the accumulation of mineral matter, and has lengthened into a very delicate pedicel from which the cystolith is suspended. This formation may be compared with that of stalactites.

Epidermis.—Before treating of the anatomy of the *fundamental organs*, we shall describe the *epidermis*, which covers the surface of the vegetable.   If the leaf of a

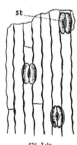

676. Lily.
Epidermis and stomata.

677. Balsam.
Epidermis and stomata.

Lily or Iris be torn, a shred of transparent colourless membrane is detached from one of the fragments, together with some cellular tissue, filled with green chlorophyll; a simple lens shows on this membrane several parallel (fig. 676) or reticulate lines (fig. 677) and small, more opaque spots. Under a microscope, it is seen to be composed of large flat cells, which may be hexagonal or quadrilateral, or irregularly waved, and which contain a colourless liquid; their lateral walls are closely united, whence the solidity of the epidermis; their lower surface slightly adheres to the subjacent cellular tissue; their exposed walls are usually thicker than the others, and may be flat or raised in the centre, according as the surface of the epidermis is smooth or rough.

In most cases, the epidermis is composed of a single layer of cells ; when there is a second, it is usually formed of much smaller cells.   The lateral walls of all the epidermal cells are not contiguous ; many of them present interspaces, occupied by little bodies resembling a button-hole with a double rim or border (figs. 676, 677), formed of two curved cells whose concavities face each other. These two small lip-like cells are termed *stomata*. *Stomata*, though epidermal orgeâs, differ from the epidermis in that their cells are much smaller, and nearly always situated below those of the epidermis ; they further present different contents, and especially granules of chlorophyll ; whence they may be regarded as intermediate between the epidermis and the subjacent parenchyma.

Stomata are variously distributed over the surface of the leaves : usually solitary,

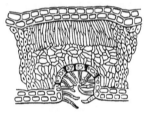

often arranged in series, sometimes crowded in the base of a cavity (as in some *Proteaceœ*, figs. 678, 679). Their number varies : the *Iris* contains 12,000 in a square inch ; the *Pink*, 40,000 ; the *Lilac*, 120,000. When moistened, their lips swell and become more curved, and hence gape ; when dry, they shorten and close.

678. Vertical section of part of a Banksia leaf (mag.).[1]

679. Part of a Banksia leaf, presenting three sections parallel to the lower surface, and at different depths (mag.).[2]

Stomata always correspond to intercellular passages, and are found on the ordinary leaves of Phœnogams, principally on their lower surface, on stipules, on

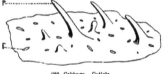

680. Cabbage. Cuticle.

herbaceous bark, calyces, and ovaries ; they are wanting on roots, rhizomes, non-foliaceous petioles, most petals, and seeds ; acotyledons, and submerged aquatic plants, which have no epidermis, equally want stomata.

If a fragment of a stem or leaf be macerated, the cellular tissue beneath the epidermis is rapidly destroyed, and the latter divides into two layers, an external epidermis proper, and a very thin membrane (fig. 680), moulded on the epidermis and extending even over its hairs, which are sheathed in it like fingers in a glove (P) ; it presents openings (F) corresponding to the stomata. Brongniart has called this membrane the *cuticle* (*little skin*) ; it is not cellular, like the epidermis which it covers.

[1] Fig. 678 is a section perpendicular to the thickness of the leaf, showing : 1, on the upper and lower faces two layers of epidermal cells ; 2, fibro-vascular bundles to the right and left, cut perpendicularly to their length ; 3, on the lower face, a depression, clothed with hairs, and pierced by stomata which communicate with the interstices of a very loose cellular tissue. Above this tissue, the upper half is a mass of elongated and erect cells, perpendicular to the epidermis.

[2] Fig. 679. Three sections parallel to the plane of the leaf, carried through three such depressions, each circum-scribed by the fibro-vascular bundles of the nerves. In the cavity at the bottom of the figure, the section has carried away the hairs clothing the walls of the depression, leaving the stomata and epidermal cells visible ; in the cavity on the right the loose cellular tissue which underlies the stomata of the epidermis is seen through the latter ; in the third depression the section has removed all but this subjacent tissue with its intercellular spaces.

The cuticle is more constantly present than the epidermis; submerged plants and acotyledons are clothed in it; and some botanists have considered that it should be regarded as the true epidermis. Its formation is attributed to the overflow of that intercellular secretive tissue which we have already spoken of as spreading itself upon all the organs, and which deposits a sort of varnish or continuous layer over their outer surface. Recent experiments of Frémy seem to show that the chemical composition of the cuticle is analogous to that of india-rubber, which makes it a suitable protection for the underlying tissues. Frémy has also discovered that woody fibre is sometimes clothed with a cuticle similar to that which clothes the epidermis.

FUNDAMENTAL ORGANS.

We shall now describe the anatomy of the fundamental organs in succession; i.e. the vegetable axis (stem and root), and its lateral expansions (leaves, sepals, petals, stamens, carpels, ovules). We have described the embryo as a diminutive plant, from which all the parts enumerated above will be developed; we must therefore first describe its structure, and then trace its stages of development from its birth till it becomes a plant similar to its parent.

The embryo invariably commences as a cell with granular contents. In cotyledonous plants, this cell does not retain its form and structure; from spherical it becomes oval; then at one of the extremities, if the plant is *monocotyledonous*, a rounded lobe (*cotyledon*) appears, obliquely and laterally to the axis; if *dicotyledonous*, two lateral lobes (*cotyledons*) appear, crowning the axis; the elongated summit of the axis becomes the *plumule*; from the opposite end the *radicle* will be developed, and the body of the cellular mass will form the *caulicle*. Following the growth of these fundamental organs, we begin with the stem, which differs remarkably, according to whether the embryo is mono- or di-cotyledonous.

Stem of Dicotyledonous Plants.—Take the *Melon* as a type. In the caulicle, which before germination is entirely cellular, some *cells* elongate into *fibres*; certain of these fibres, together with other superimposed cells, break the transverse walls which separated them, and become *vessels*. This change takes place in definite positions, and a horizontal section of the stem (fig. 681) will show in the centre a disk (M) of large, loose, nearly transparent polyhedral or spheroidal cells; at the circumference, a ring of dark green, more closely packed cells; communication being established between this ring and the disk by radiating bands of cells (RM), extending from the centre to the circumference, and dilating in the same direction; the whole resembling a wheel, of which the tire is the circle, the axle the central disk,

681. Melon.
Horizontal slice of the stem (mag.).

and the spokes the radiating bands. Between the disk and the circle, and separated by the bands, are wedge-shaped plates, which together form a circular group, and consist of fibro-vascular tissue, and vessels which have been formed in the middle of the cellular mass, and become united into bundles. The gaping mouths of these vessels and fibres are very evident, as is the relative thickness of their walls: we shall return to this immediately. The cellular tissue of the ring, disk, and bands, constitutes the medullary system. The medullary system of the disk (M) is called the *pith*; that of the ring is the *cortical pith*; and the radiating cellular bands (RM) are the *medullary rays*. The wedges of fibres and vessels, separated by the rays, are the *fibro-vascular system*. If we now dissect one of the bundles in a well-formed stem of Melon, whose duration is annual (fig. 682), it is found to be tolerably stout, and completely surrounded by the *cellular tissue*, the *pith* (M), *bark* (PC), and *medullary rays* (RM). Beginning from the interior, we find, 1, spiral vessels (T), and opaque white fibres with thick walls; 2, fibres (F) with thinner walls, and consequently larger cavities, arranged in series, and altogether occupying half of the wedge; together with *annular, rayed,* and *dotted* vessels (V P), recognizable, especially the latter, by the size of their walls;

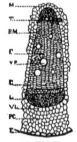

3, a greenish cellular tissue (C); 4, thick-walled fibres (L) like those next the pith, but more abundant; 5, some branching (*laticiferous*) vessels (V L) with soft walls; 6, the cortical parenchyma (PC), covered by a membrane (E) consisting of the epidermis and cuticle. In a horizontal section of the stem (fig. 681), the tracheæ (T) and fibres next the pith form with the neighbouring vessels a ring (interrupted by the medullary rays), which has received the collective name of medullary sheath; the fibres outside this sheath are the *woody fibres*; the outer fibres, separated from the former by a cellular zone, and resembling those of the medullary sheath, are the *woody fibres* of the bark; finally, the cellular zone which separates the cortical from the woody fibres is called the *cambium* layer. In the *Melon,* this zone dies each year, together with the fibro-vascular bundle, which it divides into two

682. Melon.
Horizontal slice of one of the fibro-vascular bundles of the stem (mag.).

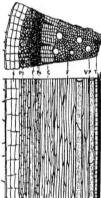

683. Maple.
Fibro-vascular bundle of the stem at the beginning of the second year. Transverse and vertical sections (mag.).

unequal parts; but in a woody-stemmed, and hence perennial plant (*Oak, Elder*), fresh layers are annually formed in the thickness of this zone, by which the thickness of the stem increases. Young branches, therefore, one or two years old or more, must be examined, to trace the further development of the wood and bark. A fibro-vascular bundle in a one-year-old branch of *Oak, Elder,* or *Maple* (fig. 683), coincides in structure with that of the *Melon* stem; but in the cortical system (P C) there will be found, between the epidermis and central layer of cells, a layer of close-set cubical or tabular cells (S); these contain no chlorophyll, are white or brown, and are readily distinguishable from the subjacent cortical cells, which are polyhedral, coloured by green granules, and separated by numerous interstices. This layer is the *suber*, which in certain trees attains a considerable development, and forms *cork*. Take now a vertical section of the same branch, and the disposition, &c. of the fibres and vessels will appear as in fig. 683.

The *cambium*, which does not become organized in annual or. herbaceous stems like the Melon, in perennial stems becomes highly organized (fig. 684). During the

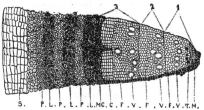

second year, this gelatinous tissue undergoes the following changes : outside the woody

684. Maple.

Horizontal slice, showing the development of a woody bundle in a three-year-old branch. c, cambium layer, separating the wood from the bark. 1. Pith (M), tracheæ (T), punctate Vessels and fibres of the first year (V).—2. Punctate Vessels (v) and fibres (F) of the second year.—3. Vessels (v) and fibres of the third year. Within the bark (B) is seen the cortical layer of the first year (P. L), then that of the second year (P. L), and of the third year (P. L.), separated by the cambium (C) from the contemporaneous woody layer (mag.).

685. Oak.

Horizontal slice of a twenty-five years' old trunk.

fibres interspersed with large vessels (1. v), is formed one fresh cambium layer (2. F, v) ; within the fibres of the liber and of the cortical system another is formed; these layers become moulded upon the older ones, and the zone of cambium which is transformed to produce them presents a cellular organization at those points, which corresponds to the cells of the medullary rays, so that these continue without interruption from the pith to the cortical layers.

Each ring of vascular bundles was hence from its earliest condition enclosed in two cambium layers, of which one belongs to the wood, the other to the bark; each of these vascular bundles, again, is in its turn separated by a cambium layer, which in the third year repeats the process, producing within ligneous fibres (3. F) and large vessels (3. v), and outside liber (L) and cortical parenchyma (P), and so on each year. Now, each wood bundle being composed of two elements, and the large-sized vessels being usually towards the interior of the bundle, we can, by counting their number (which is easily ascertained by the gaping mouths of the large vessels), reckon the number of annual layers, or, in a word, the age of the stem or branch (fig. 685).

It must be remarked that the secondary ligneous bundles differ from the primary in the total absence of tracheæ; these vessels being confined to the medullary sheath.

We have said that the medullary rays are not interrupted by the formation of new vascular bundles, because the cambium zone remains cellular at the points corresponding to these rays. If each newly formed bundle was undivided, like that in juxtaposition with it, the number of medullary rays would be always the same; but this is not the case ; at the circumference of the primitive bundle one or more longitudinal series of cells is developed, which reach to the circumference, and divide the new bundle into two or three parts (fig. 686). These cellular rays (2, 3, 4), which are termed *secondary medullary rays*, to distinguish them from the primary

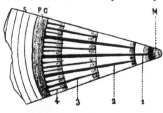

686. Cork Oak.
Horizontal slice showing the development of two woody bundles in a four-year-old branch (mag.).

686 *bis.*
Rhizophora.
Cortical fibre.

(1), which start from the pith (M), are thus doubled in each annual ring, and, like the large rays between the fibro-vascular vessels, form a sort of vertical septa or radiating walls, composed of elongated and superimposed cells; whence the name of *muriform tissue* for the medullary rays.

Hence, in its totality the stem presents two very distinct systems, the woody (*wood*), and the cortical (*bark*). 1. The woody system is formed of the central pith and zones of fibro-vascular bundles, separated by medullary rays. The innermost of these is the medullary sheath, formed of tracheæ and fibres analogous to the liber, and outwardly composed of woody fibres and rayed, annular, and dotted vessels. The other zones are similarly organized, except that they never possess tracheæ. 2. The bark system is formed of the epidermis, the cork, the endophleum, and the bast fibres (*liber*), external to and amongst which the laticiferous vessels ramify. With age the cells of the pith lose colour, dry, separate, and finally die; the woody fibres thicken, and usually darken; of these the *heart-wood* (*duramen*) differs from the more recently formed or *sap-wood*, which is more watery, softer, and brighter coloured. The liber fibres (fig. 686 *bis*) are more slender, longer, and more tenacious than the woody fibres; and are of great use in the manufacture of thread, cord, and textiles. Their bundles descend vertically and rectilinearly in thin concentric plates, whence their name *liber* (book); but in some plants, as the *Oak* and *Lime*, they form a network, the interstices of which are occupied by the medullary rays.

From the mode of development of the wood and bark systems, it is obvious that the wood must harden, and the bark decay; for in all the bark tissues, the later formed are constantly pushing towards the periphery, within which they have been developed; this produces the exfoliation of the several elements of the cortical system; the epidermis first, then the cork-cells, the endophleum, and sometimes the liber.

It is not necessary to describe any of those anomalous dicotyledonous stems which present peculiar tissues or hyper-development of certain elements, or the absence of others; except that of *Conifers* (*Pine, Fir, Larch, Yew*, &c.), the wood of which, with the exception of a few tracheæ in the medullary sheath, is entirely composed of regularly dotted fibres. The walls of these wood-fibres (fig. 687) are hollowed into small cups, like watch-glasses, which are arranged in two straight lines, occupying the opposite sides of each fibre. These cups are so placed in contiguity that their concavities correspond (fig. 688), leaving an interposed lens-like space. The dot is placed in the centre of each cup (and corresponds to a thinned portion resulting from the absence of the inner membranes) ; from this thin portion there proceeds, on the convexity of each cup, a short canal, with only one opening, which leads into the interior of the fibre. The lens-shaped cavity arising from the contact of two fibres is usually filled with resin (*turpentine*), which infiltrates into the cavity of the fibres and destroys them by degrees; the result is those resinous deposits which are often found occupying large cavities in the wood of conifers (fig. 689, *la*).

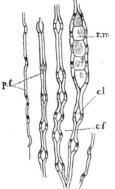
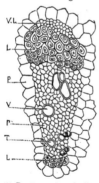

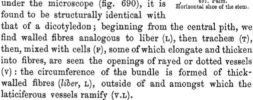

688. Pine.
Vertical section of the stem (mag.). p. f, fibre wall ; c. l, lenticular cavity ; r. m, medullary ray ; c. f, cavity in a fibre.

687. Pine.
Punctate fibre (mag.).

689. Pine.
Horizontal slice showing the development of two woody bundles in a three-year-old branch (mag.).

Stems of Monocotyledons.—When the monocotyledonous embryo, which is entirely cellular before germination, begins to elongate, fibro-vascular bundles form in its stem. These are at first arranged in a ring as in young dicotyledons; but soon, as the leaves develop, the bundles multiply without any apparent order in the cellular tissue, becoming more numerous and close as they approach the circumference of the stem. If a fully developed bundle be examined under the microscope (fig. 690), it is found to be structurally identical with that of a dicotyledon; beginning from the central pith, we find walled fibres analogous to liber (L), then tracheæ (T), then, mixed with cells (P), some of which elongate and thicken into fibres, are seen the openings of rayed or dotted vessels (v) : the circumference of the bundle is formed of thick-walled fibres (*liber*, L), outside of and amongst which the laticiferous vessels ramify (v.L).

691. Palm.
Horizontal slice of the stem.

690. Transverse section of a fibro-vascular bundle of a monocotyledonous stem (mag.). (The part answering to the centre of the stem is lowest.)

But, though individual bundles resemble those of a first year's dicotyledonous stem, when taken all together they present a very important difference (fig. 691), in not being grouped in concentric zones but

(F) remaining isolated and scattered through the medullary system (M) without any
medullary rays of muriform tissue.   Here there is no symmetrical arrangement ; the
bundles are scattered throughout the pith, and may multiply without being impeded
by lateral pressure ; further, each remains simple ; at no period does it develop
between its bark and wood systems a layer of cambium destined to form new
bundles.   In dicotyledons, on the contrary, the bundles are pressed into zones
from the first year, and their wood and bark systems being concentric, they can
only multiply by fresh wood and bark bundles being formed between
them.   The consequence of this arrangement of the fibro-vascular
bundles is, that in dicotyledons the stem is hardest towards the centre,
whilst in monocotyledons the stem is hardest towards the circumference ; as is very apparent in the woody (fig. 692), and even in
the herbaceous stems of monocotyledons.   In a longitudinal section
of a woody (fig. 693) or herbaceous (fig. 694) monocotyledonous stem,

those differences are still
more apparent; starting
from the insertion of a
leaf, each bundle descends
at first obliquely inwards,
then vertically, then again
obliquely outwards ; crossing in its path all the
bundles which have origi-
nated below it, and are
hence older than itself, and
ending by taking up a
position outside of them
all.   In dicotyledons also,

692. Palm.
Stem cut vertically.

693. Theoretical section of a
Palm stem.

694. Iris.
Stem cut
vertically.

the youngest bundles are the outermost, those of the same age follow nearly
parallel ; but whereas in their courses they unite so as to form a cylinder, in
monocotyledons they diverge below and converge above.   The composition of monocotyledonous bundles also differs in different parts of their course, the wood system
predominating over the cortical in the upper part, where it descends obliquely
inwards, the cortical system predominating in the lower part, where it descends
obliquely outwards, and finally the cortical system alone being developed where the
bundle reaches the periphery.   Here the bundle becomes more slender, and divides
into thread-like branches, which interlace with those of the neighbouring bundles,
and form together, within the cellular periphery, a layer of fibres comparable,
according to many botanists, with a liber zone.

It is obvious that these fibro-vascular bundles, being composed of different
elements at different heights, and becoming so slender towards the periphery, must
appear very dissimilar in a horizontal cut of the stem ; the scattered small bundles
with large vessels, which occupy the middle of the stem, are the upper portions of
bundles in which that which we have called the wood system (though it is rather

cellular and vascular than fibrous) predominates. The coloured and dense bundles, which form a more solid zone towards the periphery, are the lower portions of bundles in which fibres analogous to liber predominate; and, finally, the less compressed bundles which are usually seen outside of the coloured zone are these same fibres after having branched and spread out, and before being lost in the periphery, which is a cellular zone representing the bark.

A monocotyledonous stem usually retains about the same diameter throughout. This is because the fibro-vascular bundles, gradually attenuated towards their lower extremity, do not, as in dicotyledons, unite and descend to the bottom of the stem; hence, any two truncheons of a monocotyledonous stem, being equally rich in bundles, can differ but little in diameter.

Root.—In the embryo, the radicle is the simple cellular lower end of the caulicle, which elongates downwards as the latter ascends with its plumule and cotyledons. A monocotyledonous seed usually presents several radicles (fig. 642); these are not, however, naked like those of dicotyledons, but are originally enveloped in an outer layer (serving as bark), which they push forward and pierce, emerging from it as from a sheath; whence the name of *coleorhiza* for this organ (fig. 642).

Examples have been given of stems emitting accessory or adventitious roots from various parts of their surface; the structure of these is precisely the same as that of the radicle; and they may even be regarded as identical, the radicle being considered as a production of the caulicle, and all roots, whether primary or secondary, as adventitious.

In its earliest stage the root presents an axis of densely packed cells; the central of these elongate and form vessels which interlace with those of the stem (fig. 695). The root may be simple or branched, but its branches do not start from the axil of a leaf, and are not regularly arranged, like the shoots of the ascending axis. They terminate in fibrils, together called root-fibres, which decay, and are replaced by fresh ones which usually spring from near the base of the youngest branch. Like the stem, the root-branches and fibres are clothed with an epidermis or cuticle, except at the tips, which some botanists call spongioles (SP). The root elongates at the tips of its branches, but not of its root-fibres, which are caducous; and as the fresh cells of the root-branches are at first deprived of epidermis, it is supposed that roots absorb moisture from the soil by these, as well as by their root-fibres.

The fibrous and vascular tissues of roots are the same as those of stems, but no trachea are ever found in them; the cells are distended with juice or filled with fecula (*Orchis,* fig. 695).

695. Orchis. Vertical section of a rootlet, much enlarged. The cells (c,c) become gradually organized into punctate fibres (F P) and vessels; those at the bottom, more recently formed, constitute the spongiole (S P).

In dicotyledons, the root is distinguished from the stem by the absence of pith and medullary sheath, and by its axis being occupied by woody fibres; there is scarcely an exception to this. Its diameter increases, like that of the stem, by the annual formation of two concentric zones between the wood and bark; it elongates at its extremity only, while the stem and its branches elongate throughout their length; this may easily be proved by marking off an inch of a root and an inch of a stem. Monocotyledons, instead of having a *tap-root* (i.e. one main axis which branches), usually emit compound roots, i.e. composed of simple or slightly branched bundles, rising from the neck. Their anatomical structure is exactly similar to that of stems.

Leaves.—The anatomical structure of leaves is the same as that of the stem; they consist of a fibro-vascular bundle and parenchyma; this bundle, which is wholly formed before leaving the stem, spreads into a blade as it emerges (*sessile* leaf), or remains undivided for a certain distance before expanding (*petiolate* leaf); the nerves of the blade are formed of fibres and vessels; both it and the petiole are covered with a layer of epidermis bearing stomata on every part except the nerves and petiole. The petiole, before expanding, often forms a sheath or stipules; the sheath exists when the partial bundles of which it is composed separate from each other, but without diverging; the stipules are the result of the divergence of the lateral bundles of the petiole.[1]

Where the fibro-vascular bundle (fig. 696, FV) leaves the stem to form the petiole (F), the fibres composing it are shortened, and narrowed at each end, whence their surfaces of contact are contracted; they are hence not solidly united at the point of emergence; and it is this defective cohesion which causes the fall of most leaves. The stem presents a little swelling at the base of the petiole, called the *cushion* (c), which is visible after the disconnection of the petiole (fig. 54), together with the scar (F) left by the petiole. The relative position of the elements of the fibro-vascular bundle which passes from the stem into the leaf, shows clearly that the leaf-blade may be compared to a flattened stem, the fibres and vessels of which have been spread out, and thus allowed plenty of room for the development of parenchyma between their ramifications. As in the stem the fibro-vascular bundle consists of tracheæ in the centre, then rayed or dotted vessels and woody fibres, and on the outside laticiferous vessels and thick-walled liber-fibres, so in the leaf-blade each nerve (which is a partial bundle) presents tracheæ on its upper surface, rayed or dotted vessels with woody fibre on its lower surface, and laticiferous

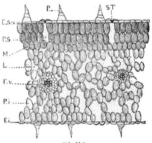

696. Branch cut vertically, showing the petiole springing from the stem (mag.).

697. Melon.
Section perpendicular to the surface of a leaf (mag.). P, hair, ST, stoma; F.V, fibro-vascular bundle; KK, upper epidermis; EI, lower epidermis.

---

[1] This theory of the origin and development of stipules requires considerable modification.—ED.

vessels and liber-fibres. The lower surface of the leaf, which corresponds to the cortical system, is generally more hairy and presents more stomata than the upper, which corresponds to the wood system. The parenchyma of the leaf, filled with green chlorophyll, usually presents ' (fig. 697), in flat leaves, two well-marked divisions; the upper, belonging to the woody system, consists of one or more series of oblong cells (P.s), arranged perpendicularly side by side beneath the epidermis (E.s), leaving very small interspaces (M); the lower division, belonging to the cortical system, consists of irregular cells (P.i), with interspaces (L) corresponding with the stomata. The parenchyma of fleshy leaves (as *Sedum*) consists of cells with few interspaces, which cells become poorer in chlorophyll towards the centre of the leaf. Submerged leaves (fig. 698) have no epidermis, stomata, fibres, or vessels; their parenchyma is reduced to elongated cells, arranged in few series, and is consequently very permeable by water.

The leaf originates as a small cellular tumour, which afterwards dilates into a blade, the cells on the median line of which elongate and form fibres, then, as in the stem, first tracheæ, and lastly other vessels.

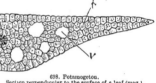

698. Potamogeton.
Section perpendicular to the surface of a leaf (mag.). P, parenchyma without epidermis; V, interstices.

In his treatise ' On the Formation of Leaves,' Trécul admits four principal types according to which these organs are formed: the *basifugal, basipetal, mixed,* and *parallel.* In the *basifugal,* the leaf is developed from below upwards, i.e. the oldest parts are those at the base of the leaf, and the tip is the last part formed; the stipules appear before the leaflets and secondary nerves of the leaf. In the *basipetal* type, the rachis or axis of the leaf appears first, and on its sides the lobes and leaflets spring from above downwards; the tip is hence developed before the base. The stipules are developed before the lowest leaflets, and sometimes even before the upper. In this type, not only the leaflets, but their secondary nerves and teeth, appear in succession downwards. In the *mixed* arrangement, both these types are followed. In the *parallel* type the nerves are all formed in parallel lines, but the sheath appears first. The elongation of the leaf takes place at the base of the blade, or base of the petiole. The sheath, although the first formed, does not increase till the leaf has developed to a certain extent.

The nerves of leaves are arranged very differently in monocotyledons and dicotyledons. In the former (fig. 33), they are usually simple, or, if branched, the branches do not inosculate. In dicotyledons, on the contrary (fig. 6), the nerves branch into *veins* and *venules,* which inosculate with those of the neighbouring nerves, and form a fibro-vascular network of which the interstices are filled with parenchyma. Nevertheless, in some monocotyledons, the basal nerves are not all parallel and simple; but secondary nerves spring from one or more of the principal nerves, and diverge in other directions; but these secondary nerves are parallel, and the convexity of the arc which they describe is turned towards the principal nerve (this nervation is rare among dicotyledons); lastly, the nerves in monocotyledons may anastomose into a network, and the blade, instead of being *entire*, as is usual in this class, may be lobed (*Arum*). On the other hand, some dicotyledons occur with parallel and simple nerves; but these exceptions do not invalidate the general rule indicated above. In all cases of determining the class of a plant, the examination of the nerves must be supplemented by that of the fibro-vascular bundles of the stem, which are symmetrically arranged in dicotyledons (fig. 685); and dispersed without order, though more closely packed towards the circumference, in monocotyledons (fig. 691).

Buds.—The bud (fig. 696, B) appears at first under the bark as a cellular point continuous with the extremity of a medullary ray; it soon pushes through the bark, and forms a tumour on the stem, when its cellular tissue becomes organized into fibres and vessels communicating with those of the stem; the medullary sheath, however, of the young branch is closed at first, and does not communicate with the medullary ray of the axis from which it emanates.

Sepals.—The anatomical structure of these organs completes the analogy between them and leaves. The nerves of the sepals are bundles of trachea and fibres, parenchyma is spread out between them, and their surfaces are covered by an epidermis, of which the upper presents more stomata than the lower. As with the leaves, the nerves of the sepals are usually parallel and simple in monocotyledons, branched and anastomosing in dicotyledons. The sepals first appear as small cellular papillæ, connected at the base by an annular disk referable to the receptacle; their tips are free in both the monosepalous and polysepalous calyx; it is only later that the calycinal tube appears. Vascular bundles are gradually formed in the sepals as in the leaves.

Petals.—The corolline leaves have often, like ordinary leaves, a petiole, which is called the claw. When this is present, the fibro-vascular bundles traverse its entire length, and only separate to form the nerves of the blade; these nerves, usually dichotomous, are composed of trachea and elongated cells; the parenchyma which fills their interstices is formed of a few layers of cells, covered by an epidermis presenting very few stomata on the upper surface only, or none at all.

Very young petals, like sepals, appear as cellular papillæ; but in petals these soon dilate, and form dark or light-green disks, which at a later period always change colour. Although the petals are placed below the stamens on the floral axis, they generally expand later, as if they had been developed later, which is not the case.

In a monopetalous corolla, the *torus* is raised above its ordinary level so as to form a little circular cushion which connects the leaves to which it gave birth, and the segments of the corolla appear as projections upon this cushion.

Finally, whether the corolla be monopetalous or polypetalous, its petals are developed like ordinary leaves; the tip and base are first formed, and the development takes place towards the central veins from below upwards, from above downwards, and laterally.

Stamens.—The complete stamen consists of *filament, connective, anther,* and *pollen* :

let us examine their structure in the adult, and their mode of development in the young stamen.

The *filament* consists of a central bundle of tracheæ which traverses its length, of a layer of cells enveloping this bundle, and of a thin superficial epidermis. The *connective*, which is the continuation of the filament, is formed of cells of the consistence of glandular tissue, in which the bundle of tracheæ terminates.

The *anther* is usually divided into two cavities, separated by the connective, and containing the pollen. The walls of these cells consist of an outer or epidermal layer of cells (fig. 699, CE) with many stomata, and of an inner simple or multiple layer of fibrous (E), annular, spiral, or reticulated cells; this layer becomes thinner as it approaches the line of dehiscence of the anther, where it ends. At the period of dehiscence the outer membrane of these cells is destroyed, and the little netted, ringed, or spiral bands which lined it alone enclose the pollen,

699. Melon.
Remains of the fibrous cells lining the epidermis of the anther (mag.).

the emission of which they assist when they dry up, contract, and separate the valves of the anther. The young stamen appears as a cellular green papilla, which usually turns yellow. The anther is the first formed; it presents a median furrow (the connective), and two lateral ones (the future lines of dehiscence); the filament appears next, at first wholly cellular, then traversed by a bundle of tracheæ. The tissue of the anther is at first a uniform cellular mass (fig. 700), in the middle of which a certain number of cells are absorbed and leave usually four spaces, which gradually enlarge and form as many cavities, nearly equidistant from the centre and the periphery. Each pair of these small cavities eventually represents an entire cell (fig. 701). All these four

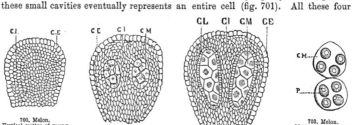

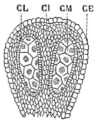

700. Melon.
Vertical section of young anther (mag.), showing the epidermal cells (CE), and the inner cells (CI), all alike and homogeneous, in the middle of which spaces will form.

701. Melon.
Vertical section of an anther-cell (mag.) with two cellules. CE, epidermal cells; CI, inner cells; CM, mother-cells contained in the cellules.

702. Melon.
Vertical section of an anther-cell where the cellules are filled with mother-cells. CL, walls of the cellules (mag.).

703. Melon.
Mother-cells (CM), originally hexagonal, of which the septa are destroyed, and containing each four pollen-grains (P) (mag.).

cavities gradually fill with mucilage, out of which are elaborated two forms of cellular tissue, one of small cells (fig. 702, CL) that line the cavity, the other of large cells (CM) that fill the cavity, and within which the pollen is developed. The latter, called *mother-cells* (CM), soon become filled with a fluid full of granules; the granules again aggregate, and form four nuclei floating in the liquid, which thickens by degrees from without inwards, and finally forms four septa dividing the mother-cells into as many cellules. Each nucleus then becomes coated with a membrane (fig.

703), after which both the septa and the walls of the mother-cells disappear, and the four nuclei (P) which filled them are set free as *pollen-grains* (fig. 704). As they grow (figs. 705 and 706), the cellular tissue of the anther, in the middle of which the cavities had been formed, are absorbed; a layer of cells which formed the walls of the cavities now lines the membrane of the epidermis (fig. 699, CE), and rapidly

changes into a layer of fibrous cells (E); the tissue which separated the small cavities becomes gradually thinner, and forms a septum which projects from the connective towards the line of

704. Melon.
Young pollen-grains,
free (mag.).

705. Melon.
Nearly adult pollen
(mag.).

706. Melon.
Ripe pollen
(mag.).

dehiscence; this septum is soon destroyed, and the two cavities form but one (*anther-cell*). In some plants this septum is persistent, and the anther remains quadrilocular (*Butomus*, fig. 326). In many plants the remains of the mother-cells only partially disappear, and the rest connect the pollen-grains, as in *Orchis* (figs. 359 and 360), where an elastic network causes them to cohere in small masses.

Carpels.—The anatomy of the carpellary leaves is analogous to that of ordinary

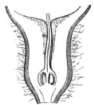

707. Pear.
Very young flower, cut
vertically to show the
petals, stamens and carpellary mammæ free on the
receptacle (mag.).

708. Pear.
Young carpels, seen from
within, at first concave,
and the edges afterwards
approaching to form
the style and placentas
(mag.).

709. Pear.
Young flower, cut vertically
to show the growth of the
receptacle, the arrangement
of the carpels, and the insertion of the petals and stamens
(mag.).

711. Pear.
Flower cut vertically, with the
stamens and petals removed, showing the carpels enveloped by the
receptacular cup (mag.).

leaves; a cellular tissue (sometimes very succulent, as in berries and drupes), traversed by fibro-vascular bundles, is covered with an epidermis, the outer surface only of which bears stomata; the bundles ascend from the ovary into the style, occupying its circumference, its centre being hollowed into a canal. The inner walls of this canal, which is formed by the convolution of the upper end of the carpellary leaf, is covered with projecting cells, and its axis is occupied by soft cellular filaments, named *conducting*

710. Pear.
Young flower, from
which calyx, petals and
stamens have been re-
moved, to show the
five carpels, sunk in
the receptacular cup
(mag.).

tissue; it is this tissue which constitutes, on the top or sides of the style, the true stigmatic tissue. The *placenta*, which transmits nourishment to the seed, and the *funicle*, which is a prolongation of the placenta, consist of a bundle of tracheæ surrounded by elongated cells.

Inferior ovaries have the carpels encased in a receptacular cup, which is sometimes enormously hypertrophied, especially in *Rosaceæ* and *Pomaceæ* (figs. 707 to 711), and bears on the top the stamens, petals, and calyx.

Ovule.—Botanists often apply this name to the undeveloped seed; but, to be precise, they ought to confine it to the unfertilized seed.

To trace the development of the ovule, it must be examined long before the bud opens : it then appears as a papilla on the placenta, called the nucleus (fig. 712); around the base of the nucleus (fig. 713) a circular ring is formed (s), which at first grows at the same rate as itself, but, rising on its surface, it eventually overtops and finally almost entirely envelops the nucleus ; but before this takes place, a second circular ring is developed (fig. 714, P) outside the first (s), which follows it in its growth, and ends by reaching and overtopping it ; the nucleus (N) is hence enclosed in two sacs, whose mouths are contracted, and on a level with its top, thus forming a little cylindrical or cup-shaped cavity, consisting of two superimposed rings touching at all points of their circumference. The upper opening, belonging to the outer coat, is named *exostome* (Ex) ; the lower, belonging to the inner coat, is named *endostome* (End). The union of the endostome and exostome constitutes the *micropyle*, which always corresponds to the top of the nucleus. The outer

712. Mistleto. Ovule (mag).

713. Walnut. Ovule (mag.).

714. Polygonum. Ovule (mag.).

715. Polygonum. Ovule cut vertically (mag.). P, primine; s, secundine; N, nucleus; s.e, embryonic sac.

coat is called the *primine* (P), the inner *secundine* (s) ; the nucleus (N) has also been called the *tercine*; terms which refer to their order of superposition from without inwards, not that of development. The funicle (F) is inserted on the primine, and its contained bundle of tracheæ, after traversing the primine and secundine, expands at the base of the nucleus into a swollen coloured cellular tissue, termed *chalaza*, opposite to which there is almost always a corresponding swelling on the primine. As the ovule, which is wholly composed of cellular tissue, grows, a cavity is formed near the centre (fig. 715) of the nucleus, by the dilatation of one of its cells; this cavity, which extends through the length of the nucleus, and adheres by its two ends to the neighbouring cells, is the embryonic sac (s.E), or quintine. Its walls shortly become lined with a mucilaginous cellular tissue, developed from the circumference towards the centre, which fills the cavity of the sac ; this tissue, together with that of the nucleus, constitutes the alimentary deposit destined for the embryo, and is called *albumen* (*perispermum*). The ovule, thus organized before fertilization, undergoes one of the three following changes :—most frequently the embryonic sac pushes away the nucleus on all sides, and its own parenchyma alone is developed, when the albumen is more or less fleshy.; sometimes, on the contrary, the nucleus presses upon the embryonic sac, and reduces it to a narrow tube, when the albumen is farinaceous ; sometimes, again, the action is reciprocal, and two kinds of *albumen* result,—the white *Nymphæa* (figs. 610, 647) affords a remarkable instance of this. For this reason Gaertner, comparing the ovule with a bird's egg, limited the term *albumen* (white of egg) to the tissue developed within the nucleus (fig. 610, N), and gave that of *vitellus* (yolk of egg) to the tissue developed in the embryonic sac (S.E).

Fertilization is announced by the appearance of a body (fig. 716) destined to form the embryo, suspended from or near the top of the embryonic sac (s.e). At first it consists of a vesicle (v.e), named *embryonic vesicle*, filled with a granular matter, in which is formed first one cell, then others, each of which bears a *cytoblast* on its wall. The upper and slender portion of this vesicle (fig. 717) is the *suspensor*; in the lower and swollen portion the embryo is developed; the vesicle and its suspensor soon disappear, when the embryo develops, according to whether it is monocotyledonous or dicotyledonous, as we have already shown, and increases within the cavity of the ovule, which it invades by absorbing

716. Polyxenum. Fertilized ovule, cut vertically (mag.).

717. Dicotyledonous embryo, in different stages of development in the ovule (mag.).

the albumen. If the albumen has solidified before the growth of the embryo, the latter remains small and takes up less room; and the absorption of the albumen is then delayed till the period of germination. The ovule is not always provided with two coats; sometimes the inner coat alone (secundine) is developed (*Walnut*, fig. 713); in others the nucleus remains naked (*Santalaceæ, Mistleto*, fig. 712). It is important to understand the changes the ovule may undergo before fertilization; changes due to unequal development altering the relative positions of its different parts. In theory, the hilum and chalaza correspond, and occupy the base of the ovule, the micropyle being at the top or opposite end. If the ovule develops uniformly, the arrangement is not disturbed, and the ovule is *straight* or *orthotropous* (*ov. orthotropum*, fig. 716), and the embryo will also be *straight*. In this case the position of the radicle answers to that of the micropyle, i.e. opposite the hilum and chalaza, and the embryo is said to be *antitropous* (*ov. antitropus*, Nettle, fig. 578).

When the ovule develops unequally, one of two things may happen: 1, the chalaza (Ch, fig. 718) may be removed from the hilum towards the position occupied by the top of the ovule; which top, by a reverse movement, may be turned towards the hilum; the axis of the ovule thus making a half turn upon itself, like a compass-needle turning from the north to the south pole. In this case the hilum not having been displaced, the vascular bundle which connects

718. Dandelion. Anatropous ovule cut vertically (mag.).

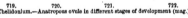

719.   720.   721.   722.
Chelidonium.—Anatropous ovule in different stages of development (mag.).

723. Vertical section of fig. 722.

it with the chalaza is forced to follow the latter in its revolution, and form a more or less projecting cord (R) in the thickness of the primine, named the *raphe*; the ovule is then *reversed* or *anatropous* (*ov. anatropum*, figs. 719 to 723). Here the embryo will be *straight*, as in the *Nettle*, but the .chalaza will be the 'antipodes of the hilum, the micropyle nearly touching the latter, and the radicle corresponding to the base of the ovule; such an embryo is called *homotropous* (*ov. homotropus*). There are many examples of this (*Sage*, fig. 579 ; *Chicory*, fig. 580).

2. When the hilum and chalaza (figs. 724 ch, 725) are inseparable, and one

724. Wallflower.
Campylotropous ovule
(mag.).

side of the primine (p) is more developed than the opposite side, the one lengthens while the other remains stationary; the resistance of the stationary side causes the lengthening side to turn around the centre of resistance ; the ovule (n) thus bent back upon itself is said to be *campylotropous* (*ov. campylotropum*). Here the embryo will follow the curvature of the ovule, and the micropyle and chalaza (ch)

725. Wallflower.
Campylotropous ovule, cut
vertically (mag.).

being both close to the hilum, the radicle and cotyledonary ends will be only separated by the hilum, and the embryo is called *amphitropous*. The *Wallflower* (figs. 724, 725) and the *Mallow* (figs. 726–730) are well-marked instances of the curved ovule and amphitropous embryo.

To these three types (*orthotropous, anatropous,* and *campylotropous*) all ovules properly belong; but there are many cases of intermediate types, which it is necessary to take into account. In one case which, although very rare, runs through the whole family of *Primulaceæ,* and occurs in *Vinca,* one side of the ovule develops

726.   727.   728.   729.   730. Vertical section
Mallow.—Campylotropous ovule in various stages of development (mag.).   of fig. 729.

enormously, while the other gradually atrophies; this action continues after fertilization, and the micropyle, approaching the hilum more and more, ceases to correspond to the radicle, which may hence be variable in direction; most commonly the axis of the embryo becomes *parallel* to the hilum, and the embryo is called *heterotropous* (*ov. heterotropus, Plantain,* fig. 592 ; *Asparagus,* fig. 594).

When the seed is mature, it becomes difficult to distinguish in its coats (*testa* and *endopleura*) the primine, secundine, tercine (nucleus), and quintine (embryonic sac), which all enter into its composition. The testa evidently represents the primine ; and, as the raphe has pursued its course between it and the secundine, this latter must be represented by the endopleura ; but the nucleus and embryonic sac are either pushed back by the embryo, and reduced to membranes lining the inner wall of the secundine, or they completely disappear; the secundine itself may indeed disappear, and the embryonic sac alone remain with or without the nucleus. Lastly, these membranes may be united and confounded, so as to become indistinct. The primine therefore cannot be identified with the testa, except in cases when the latter can be cleanly removed, exposing the raphe between it and the endopleura ; and then the endopleura is obviously formed by the secundine, with or without the tercine and quintine, as may be easily seen in the *Orange.*

The three typical modifications in the positions of the parts of the ovule being known, we will indicate the corresponding portions of the embryo in the seed :—

FIRST TYPE.—Ovule straight (*orthotropous*), and consequently embryo *antitropous*; —the seed may be : 1, *erect* (radicle superior) ; 2, *pendulous* (radicle inferior) ; 3, *horizontal-parietal* (radicle centrifugal) ; 4, *horizontal-axile* (radicle centrifugal).

SECOND TYPE.—Ovule *reversed* (anatropous), and embryo *homotropous* ;—the seed may be : 1, *erect* (radicle inferior) ; 2, *pendulous* (radicle superior); 3, *horizontal-parietal* (radicle centrifugal) ; 4, *horizontal-axile* (radicle centripetal).

THIRD TYPE.—Ovule *curved* (*campylotropous*), and embryo *amphitropous*; if the embryo is not much curved, the radicle is inferior, superior, centripetal, or centrifugal, according to the position of the micropyle ; if neither extremity of the embryo is turned towards the hilum, owing to the unequal growth of the coats, it is said to be *heterotropous*; it may then be either straight, curved, or flexuous, and the radicle is inferior, superior, centripetal, centrifugal, or vague.

ACCESSORY ORGANS.

To complete the anatomy of the *elementary* and *fundamental* organs, we must describe that of certain modifications of the cellular tissue: these are *prickles, hairs, glands,* and *lenticels.*

Prickles.—These are composed of a cellular tissue analogous to that of the *bark;* they must not be confounded with *spines,* which are fibro-vascular, and are merely transformed organs, whose nature is indicated by their position ; i.e. they are aborted branches (*Blackthorn,* fig. 51), hardened stipules (*Robinia,* fig. 114), petioles of pinnate leaves become spiny after the fall of the leaflets (*Astragalus Tragacantha*), leaves of which the nerves have lengthened into spiny points, to the destruction of the parenchyma (*Berberis,* fig. 94) ; *cushions,* which elongate greatly, and become pungent (*Gooseberry,* fig. 95). *Prickles,* on the contrary, are dispersed without order on the stem and leaves, and even on the corolla, and are thickened, hardened, and pungent hairs. When young, they exactly resemble *hairs,* of which we are about to speak, and it is only when older that they thicken, lengthen, and harden ; they occur on the *Rose* (fig. 50) in every stage of its growth.

731. Cabbage. Simple one-celled hair (mag.).

Hairs.—Cellular organs, which principally occur on branches, petioles, and the nerves and under surface of leaves, especially young ones ; they are lengthened epidermal cells, covered by cuticle, like those cells which do not lengthen. Hairs are *unicellular,* when formed of one elongated, vertical, oblique, or horizontal cell, which may remain simple (fig. 731), or branch in a fork (fig. 732), trident, star (fig. 733), &c. Some branch in stages, and resemble superimposed whorls (fig. 734). *Chambered, septate,* or *jointed* hairs are composed of cells joined end to end, and forming simple beads (figs. 735, 736) or branches; sometimes a bundle of hairs radiates horizontally from a common centre, and, being united

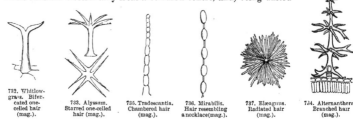

732. Whitlow-grass. Bifur-cated one-celled hair (mag.).

733. Alyssum. Starred one-celled hair (mag.).

735. Tradescantia. Chambered hair (mag.).

736. Mirabilis. Hair resembling a necklace(mag.).

737. Elæagnus. Radiated hair (mag.).

734. Alternanthera. Branched hair (mag.).

by the cuticle, resembles the rays of the sun (fig. 737). The small brown scales observable on *ferns* are considered as *scarious* hairs.

Glands.—These are organs of *secretion,* i.e. they extract a peculiar liquid from the materials with which they come in contact; they are entirely cellular; the cells of some glands project, and are called glandular hairs, which only differ from ordinary hairs by the liquid they contain; some are swollen at the tip; most are unicellular, as those on the calyx of the *Sage* (fig. 738), and on the velvety palate of the *Snapdragon* (fig. 739).

The stinging hairs of the *Nettle* (fig. 740) are formed of a single conical cell, of which the base is swollen into a bulb, and surrounded by a group of epidermal cells; the top is lightly bent, and it is the fragile tip of this hair which, breaking in the skin which it has penetrated, intro-duces the venomous juice contained in the cell. The sting-ing hairs of the *Wigandia* have a lanceolate tip (fig. 741). Glandular hairs may be chambered, when the terminal

738. Sage. Glandular one-celled hair (mag.).

739. Snapdragon. Glandular one-celled hairs (mag.).

740. Nettle. Stinging one-celled hair, bent at the top (mag.).

742. Snapdragon. Glandular chambered hair (mag.).

741. Wigandia. Stinging hair with lanceolate point (mag.).

cell alone is glandular, as in the calyx of the *Snapdragon* (fig. 742); or there may be several superimposed cells; but it is invariably the upper ones alone which secrete. Peltate hairs are composed of one cell lying horizontally on the leaf, and adhering by its centre to the epidermis, by means of a gland which forms its base (*Malpighia*).

True glands differ from glandular hairs only in projecting slightly or not at all above the epidermis; they pass insensibly into each other, as in glandular roses. The superficial glands covering the bracts and flowers of the *Hop* (fig. 743) are simple vesicles (fig. 744) containing a liquid, and a resinous principle called by chemists *lupuline*; these vesicles burst and soon disappear, when the resinous principle remains in the form of powder. Sometimes the glands are sunk in the thickness of the bark, but they are always near the epidermis; such are the glands called *vesicular* of the leaves of *St. John's Wort* and *Myrtle*, and of the bark of the *Orange*, which contain a volatile oil (fig. 745).

We have already described the *nectariferous glands* or *nectaries*, which secrete a sweet liquid (p. 74).

The cavities called *reservoirs of proper juice*, in which gums, resins, &c., are elaborated and accumulated, are lined with peculiar cells; they are analogous to the vesicular glands, but more deeply immersed in the tissue.

Lenticels, formerly called *lenticular glands*, are not glandular; they

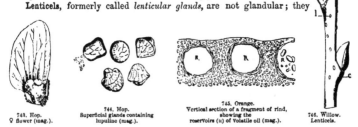

743. Hop.
♀ flower (mag.).

744. Hop.
Superficial glands containing
lupuline (mag.).

745. Orange.
Vertical section of a fragment of rind,
showing the
reservoirs (R) of volatile oil (mag.).

746. Willow.
Lenticels.

are prominences on the surface of the stem (fig. 746, 1), produced by excrescences of the endopleura which have pierced the bark. Adventitious roots often spring from lenticels; but they also spring from many other points; which invalidates the opinion of De Candolle, who regarded lenticels as the buds of aërial roots.

ANATOMY OF ACOTYLEDONS.

Stem.—The stems of *Ferns* more nearly resemble those of cotyledonous plants than do those of any other acotyledonous order. A transverse section of a *Tree-fern*

747. Cyathea.
Transverse section of the stem.

stem (fig. 747) shows fibro-vascular bundles (f, v) of various forms, disposed in a more or less irregular circle, which surrounds a yellowish central disk (m), and is itself surrounded by a zone of the same colour (p); this disk and zone are cellular, and communicate by larger or smaller passages between the bundles. The outermost blackish zone is an envelope formed subsequently to the epidermis, of the bases of the *fronds*. A transverse section of the bases of these fronds displays a structure analogous to that of the stem, on which their bases, when detached, leave remarkable scars. The same structure and scars characterize the stems of the herbaceous ferns of Europe (figs. 748, 749). The fibro-vascular bundles of ferns, whether exotic or indigenous, consist of a pale portion (fig. 747, v), formed of annular and radiating prismatic (scalariform) vessels, surrounded by a very narrow black zone (f), formed of woody fibres. Tracheæ are invariably wanting.

A few other acotyledonous families contain fibro-vascular bundles in their stem; in *Mosses* and *Hepaticæ* the stem is composed of elongated cells, which sometimes become fibres; the tissues of *Lichens, Fungi, Algæ*, &c., are entirely cellular.

Root.—The roots of the higher acotyledons, such as ferns, present the same structure as the stems; i.e. bundles of fibres and vessels, surrounded by cellular tissue; these roots are always *adventitious* and often *aërial.* In the lower acotyledons they are formed of cells which reach the ground, and then lengthen and bury themselves.

748. Male Fern.
Transverse section of the rhizome.

749. Male Fern.
Rhizome showing the scars (c) of the old fronds.

Leaves.—The leaves of acotyledons present the same structure as their stems; in ferns, we find radiating prismatic vessels and black fibres; in *Marsileaceæ*, the nerves are numerous; in *Lycopodiaceæ*, the leaf is a cellular plate traversed by a single fibro-vascular bundle; in *Mosses* and *Hepaticæ*, the nerves are represented by elongated cells; in the lower acotyledons, the leaves and stem are represented by a *frond* entirely composed of cellular tissue.

Reproductive Organs.—*Antheridia* are little sacs, at first perfectly closed, then opening at a certain period at one point of their surface, and emitting by this opening a mass of corpuscules, usually cohering by means of a mucilaginous liquid; these organs are considered analogous to anthers; we shall explain their nature in the description of the Orders.

Spores are little membranous sacs, full of liquid, which germinate by lengthening at some undetermined point of their circumference, and develop into a little plant similar to that which produced them. Spores are formed in particular cavities, called *sporangia*; they are the analogues of seeds, with regard to their functions, but they possess neither coats, caulicle, radicle, plumule, nor cotyledons; they are developed freely in the sporangium, and never adhere to its walls, as the seeds of cotyledons adhere to their placenta. The sporangium, which fulfils the functions of a carpel, has neither style, stigma, nor ovarian cavity; it is filled with a continuous cellular mass, in the midst of which are certain isolated cells, destined to reproduce the plant. We shall explain the spores and sporangia when describing the characters of the Orders.

ELEMENTS OF VEGETABLE PHYSIOLOGY.

The food necessary for the development of the plant is drawn from the soil by the root, and is absorbed by means of the *spongioles* which terminate the *root-fibres*, and which are composed of a renewable cellular tissue having no epidermis.

The substances drawn from the soil are, *carbonic acid, ammonia,* and alkaline and earthy salts dissolved in water. Carbonic acid comes: 1, from the rain, which has dissolved it in passing through the atmosphere; 2, from the slow decomposition of *humus* or *mould,* the carbon of which combines with the oxygen of the air, which the water holds in solution. The ammonia comes: 1, from rain during storms, when, by the influence of electricity, it is formed from the nitrate of ammonia; 2, from the putrefaction of vegetable or animal matter, at the commencement of which azote and hydrogen combine. This decomposition is aided by adding chalk to cultivated soil; for chalk, as Boussingault has proved, attacks insoluble azotized matters, and favours the formation of ammonia. The alkaline and earthy salts, and notably the sulphates, and phosphate of lime are derived from the soil; the sulphates are decomposed by the ammonia, which substitutes itself as their base, and forms a sulphate of ammonia, which, being soluble in water, and containing azote, hydrogen, sulphur, and oxygen, is eminently adapted for the nourishment of plants. Phosphate of lime, which is insoluble in pure water, is soluble in water containing either an ammoniacal salt or carbonic acid only, as is the case with rain. The water which holds in solution these different inorganic substances is a colourless liquid, which rises by the vessels into the root, stem, and leaves, fills the cells and their interstices, in which, during life, are formed the organic matters which are to be deposited in the tissue of the vegetable, or to assist in its growth.

The above-mentioned inorganic substances are all binary compounds, which sometimes remain isolated, sometimes enter into combination with one another. But the organized substances which are found in the plant are the results of more complicated combinations; we have already spoken of *cellulose* and *starch,* allied to which is a third substance named *dextrine,* which does not turn blue with iodine, and which is soluble in water; its chemical composition is exactly the same as that of cellulose and starch, which are *ternary* bodies, composed of carbon, together with hydrogen and oxygen in the same proportions as water. These three bodies, formed from the same elements in similar proportions, are called isomerous bodies; the difference between them consists entirely in the manner in which their molecules are grouped; it is therefore simply necessary to derange these molecules to convert dextrine, cellulose, and starch into each other.

The *sugar* yielded by the *Sugar-cane, Beetroot,* and many other vegetables is also a ternary compound very similar to the preceding ones, containing one molecule more of water than starch, dextrine, and cellulose contain.

Glucose or *grape-sugar* only differs from cane-sugar in containing three molecules more of water. Thus starch or dextrine, with an additional molecule of water, becomes cane-sugar; and grape-sugar from which three molecules of water are abstracted becomes cane-sugar.

Organic *acids,* such as *acetic* acid, which is found in the sap of plants, and forms in sour wine, *pectic* acid in the gooseberry, *tartaric* acid in grapes, *malic* acid in apples, *citric* acid in the lemon and other fruits, *gallic* acid in oak-galls and bark, &c., are ternary compounds which contain carbon and the elements of water (oxygen and hydrogen), plus a certain quantity of oxygen. *Oils, essences, resins, chromule* or *chlorophyll,* are ternary compounds, formed by the combination of carbon with the elements of water, plus a certain quantity of hydrogen.

Besides these, vegetables contain, especially in their bark, quaternary compounds of carbon, hydrogen, oxygen, and nitrogen; these crystallize, and are always found in union with an organic acid which forms a *salt* with them, whence their name of *vegetable alkalies.*

The *Poppy* contains *morphine, narcotine,* &c.; the *Nux vomica, strychnine;* the genus *Cinchona, quinine, cinchonine,* and *cusconine.* Experience has proved that the poisonous and medicinal properties of vegetables reside in the organic alkalies.

Other organic substances frequently found in vegetables are still more complicated; for, besides oxygen, hydrogen, carbon, and nitrogen, plants contain sulphur and phosphorus: these are *albumen, fibrine,* and *casein;* the proportions of their elements are similar, although their physical properties are different; whence the name of *protein* by which chemists designate the essential principle of all those substances that are collectively designated *albuminous.* Protein has been alluded to under the *nucleus;* it constitutes the nutritious element of vegetables, for without it no blood can be formed, and it is always found in this liquid. *Fibrine* is a compound substance, insoluble in water, like *cellulose;* it may be looked upon as the origin of all the parts of a plant; it always exists in them, and especially in the seeds of cereals. *Albumen* coagulates with heat like starch; it constitutes nearly all the serum of blood and the white of eggs, and abounds in the juices of plants. *Casein,* which forms with starch the nutritive part of beans, lentils, and peas, constitutes essentially, in the milk of animals, the nutriment that the young receives from its mother. *Gluten,* which forms the base of leaven or yeast, exists in most seeds, and is composed of the same elements (less the sulphur and phosphorus) as albumen, fibrine, and casein.

The elements of carbonic acid (*oxygen* and *carbon*), of ammonia (*hydrogen* and *nitrogen*), of water (*oxygen* and *hydrogen*), and the *sulphur* of soluble sulphates, supply most of the materials of vegetables.   The carbon of carbonic acid by uniting with the elements of water forms *cellulose, sugar, gum, starch*, &c.; an excess of oxygen produces vegetable *acids* (malic, citric, acetic, gallic, &c.); an excess of hydrogen produces chlorophyll, oils, resins; the azote of ammonia, added to the elements of water and of carbonic acid, gives rise to vegetable *alkalies* (*quinine, morphine*, &c.); finally, sulphur and phosphorus, combined with azote, oxygen, hydrogen, and carbon, form three organic substances of similar composition, namely, *fibrine, albumen*, and *casein*; these supply the animal kingdom with essentially nutritious elements; as stated above, they are always found in the blood, united with other substances, and notably with a certain quantity of *phosphate of lime*, a salt which constitutes the solid part of bones.

Humus or *mould* is the name given to the black carbonaceous matter which results from the decomposition of organic substances; vegetable mould is nothing but cellulose, which burns slowly under the influence of the oxygen of the atmosphere, and changes into carbonic acid, which, dissolving in the water of the soil, passes into the substance of the vegetable.   The decomposition of the mould is assisted by mineral alkalies (*potash, soda, chalk, magnesia*), which induce the formation of carbonic acid, and form with it soluble carbonates, absorbed by the roots; then, under the influence of these same alkalies, the water and carbonic acid decompose, and vegetable acids are formed, more or less oxygenized, with which they combine; finally, these acids change, and become *sugar, starch*, or *cellulose*.

Thus, vegetable acids are indispensable to the existence of plants, and their formation depends: 1, on the water and carbonic acid which combine to form them; 2, on the mineral alkalies which induce this combination.   Now these alkaline bases, which play so important a part in vegetation, reside in hard or soft rocks, named *feldspar, mica, granite, gneiss, basalt*, the elements of which are *silica, alumina, potash, magnesia, lime*, &c.; these bases are liberated by the disintegration or decomposition of the rocks, of which the débris, more or less changed, constitute *arable soil*.   The rocks are disintegrated by the water which, having penetrated them, expands in passing to the state of ice, and thus overcomes the cohesion of their elements. These elements are then dissolved by water, either pure, or containing oxygen, or loaded with carbonic acid; it is thus that the aluminous and alkaline silicates are disintegrated and dissolved, previous to forming argillaceous soils.

Alkalies, and especially potash, when mixed with soil, are rendered soluble by the addition of sulphate of lime, as Dehérain has proved.   Since the sulphate of lime changes the salts of potash into sulphate of potash, it has been supposed that the greater solubility of potash after being thus treated is attributable to this transformation; this hypothesis has not yet been practically proved, and we do not know whether the sulphate acts chemically on the potash, or whether its effects are purely physical, the object being to liquefy the soluble salts, to preserve them from the absorbent action of the earth, and to facilitate their absorption by the roots of the plant. But, whatever be the explanation, this property of sulphate of lime proves the advantage of adding it to the soil in which leguminous fodders are cultivated (*Trefoil, Lucerne,*

Sainfoin), of which the ashes are rich in potash; while, on the contrary, the addition of carbonate of lime, which induces the formation of ammonia, is very usefully employed in the cultivation of cereals, for which azotized manures are necessary.

Silica is useful, because, being powdery and insoluble, it admits air and moisture, *alumina*, because it retains moisture, around the roots; *lime*, because, under the influence of water acidified by carbonic acid, it replaces the alkaline bases of the silicates; hence the importance of marl, which is a mixture of clay and lime. If the soil is composed of pure *silica* or of pure *chalk*, it is absolutely sterile; if it is wholly of *clay*, the roots cannot penetrate it. The best soil is that in which clay is mixed with *carbonate of lime* and *sand* (*silica*), in such proportions that air and moisture readily permeate it.

Tillage improves the soil by breaking it up, and multiplying the surfaces which ought to be in contact with carbonic acid, the ammonia of the rain, and the oxygen of the air, so that the débris of the rocks may be rendered soluble, and form *arable land*. The period of fallow is that during which the soil is left to atmospheric influences. While the land is thus left fallow as a preparation for certain crops, it may be occupied by some other plant which does not rob the soil of the materials required for such crops; this explains the theory of the *rotation of crops*.

NUTRITION OF VEGETABLES.

Absorption.—The roots are the principal organs of absorption; they pump up the liquid into which they are plunged, by means of their permeable cells. The upward movement of the sap is explained by a recent discovery in physics:—if a tube closed below by a porous membrane, and filled with a dense liquid, is plunged into a less dense coloured liquid, there is soon a tendency to establish an equilibrium of density, and the dense liquid in the tube becomes coloured by the addition of the less dense liquid outside it, and the two liquids stand at different heights; that in the plunged tube rises above its level, and only stops rising when its density is no longer greater than that in the outer tube. But to produce this equilibrium, the exterior liquid must receive a certain quantity of that within; thus there is a double current established through the porous membrane; the one from without inwards, called *endosmose*; the other, less in degree, from within outwards, called *exosmose*. This action accompanies the absorption of fluid by the roots; the damp soil contains water laden with ammonia, carbonic acid, and different salts; the roots, as well as the stem, are composed of a series of superimposed cells, some of which are filled with a dense juice, and others with vessels in which the liquid can easily rise by capillary action; the spongioles which terminate the root-fibres having no epidermis, are very permeable, the water of the soil penetrates them, the juice which they contain is diluted by this water, and to establish equilibrium the sap rises from cell to cell to the top of the plant.

Circulation.—When the water of the soil, laden with the carbonic acid, ammonia, and mineral matters dissolved in it, has penetrated the plant, it takes the name of *ascending sap*; this sap thickens as it ascends, in proportion as it dilutes and dissolves the materials in the cells; but to the motive force of the endosmose and capillary action is added another not less powerful: this is the attraction exerted from above by the buds, which draw up the food necessary to their development, and by the already formed leaves, from the surface of which copious evaporation is carried on. The empty spaces resulting from this evaporation and from the substance assimilated by the buds, are filled by the sap in the parts immediately below; these repair their losses in their turn, and this action is continued from above down to the roots, for which the soil is the reservoir.

The buds are the first organs of the vegetable which awake in spring from their winter torpor; when they begin to swell, the resulting movement of the sap stimulates the roots, which recommence their functions; from this time the ascending current, assisted by the endosmose, is established through the swollen tissues of the thickened materials deposited the preceding year. Nevertheless, although it is the buds which give the roots the signal to recommence their work, the work of the roots is carried on independently of the influence of the buds; for these remain closed long after the sap has begun to rise with remarkable force and abundance. If at the period of the *spring sap* an incision is made in a stem, a stream of sap flows from it, and the proof that neither the buds nor the leaves are the cause of this phenomenon is that it occurs just the same on a stem deprived of buds and leaves. An example of this is seen in the tears of the vine, which flow from the stem when the plant is pruned, and even when it is cut almost to the ground; but as the buds lengthen, and as the branches resulting from their elongation become covered with leaves, the suction of the young branch and the evaporation from the surface of the leaves become active forces, which join those of the endosmose and capillary action to assist the ascension of the sap.

When the branches are developed and consolidated, the movement of the sap slackens, but without ceasing; its only object now is to provide for the daily requirements of the plant, and to prepare materials for the vegetation of the following year. When the spring rise of sap has taken place early, these materials are prepared before autumn, and then the *August sap* is produced, which represents a second spring.

In the autumn, the tissues, more and more solidified, dry up; the leaves, of which the canals become obstructed by a continual efflux of materials, cease to vegetate, and fall; evaporation is thenceforth arrested, and with it the movement of the sap; and finally, life is suspended for several months.

The ascent of the sap does not always take place in the same manner; in spring it rises across all the woody tissues; in old branches, across the sap-wood only. Later, most of the vessels are empty except of gases; it is then by the cellular tissue that the sap rises to support the vegetation.

When the sap, laden with the materials that it has dissolved in its ascending and diverging march, has reached the young branches, it penetrates their cortical pith and the parenchyma of the leaves; there it finds itself in contact with the air which has penetrated by the stomata into the intercellular spaces; then it undergoes important modifications, and loses a large portion of its water, which evaporates on the outside. The cells of the green parts of the bark and leaves fill with chloro-phyll. The latex of the laticiferous vessels becomes charged with coloured granules, and the sap, thickened and enriched with new principles, descends from the leaves along the inner surface of the bark towards the roots. This descending movement is easily proved; it is sufficient to prune the bark of a young branch to see the sap, if it is coloured, ooze from the upper lip of the incision and not from the lower. If the stem be tightly corded, after some time the bark swells, and forms a cushion above the ligament, while the stem below will preserve its original diameter. For this reason the *elaborated sap* is also called *descending sap*.

The elaborated sap furnishes the *cambium,* a gelatinous fluid which permeates the cellular zone, and in which are formed the elementary organs which combine to produce growth in the vegetable.

In dicotyledonous stems, the cambium is principally deposited between the woody and cortical systems, within the layer of laticiferous vessels and the fibres of the liber, in contact with which the descending sap flows. The young buds springing from the axil of a leaf are placed in the direction of the flow of latex from that leaf, and which, accumulating at the base of the petiole, elaborates there the elements of cambium.

In monocotyledonous stems, the fibres analogous to the liber and the vessels of the latex, which each fibro-vascular bundle contains, furnish an elaborated sap, which deposits cambium in heaps dispersed through the stem; so that their terminal bud profits by the sap elaborated by the leaves of the preceding bud.

Finally, rain, containing the materials for the food of the vegetable, is absorbed by the tips of the roots, rises in the stem, crosses the wood system, reaches the parenchyma of the leaves and the cellular tissue of the bark, where it undergoes the action of the air, becomes elaborated sap, descends through the bark, deposits a zone of cambium between the liber and alburnum, and arrives at the tips of the roots, whence it started; thus establishing a true circulation.

Cyclosis is a peculiar circulation which Schultz has discovered in the laticiferous vessels; he observed that the coloured granules flow in sinuous tracks, being carried by the latex currents in various directions along the courses of the anastomosing laticiferous vessels.

Physiologists have proposed different theories to account for the propelling force which puts the latex in motion; but Mohl has shown that this motion is not a vital phenomenon, but that it always arises, either from a rent in the tissue, whence the latex necessarily escapes, or from a mechanical pressure on the tissue, which sets the latex in motion; as also that this motion soon ceases.

But if *cyclosis* is an obscure and doubtful phenomenon, this is not the case with the intercellular circulation (*rotation*), which can be observed in the septate hairs of certain plants (*Tradescantia*), and especially in the cells of certain aquatics (as *Chara*). *Chara* is a leafless, frondless acotyledon; its internodes, whether isolated or in bundles, consist of cylindrical cells placed end to end; each internode pro-duces at its top a whorl of cells similar to itself, which speedily become similarly septate. If one of these cells be placed under a microscope, and cleared from the calcareous crust which often envelops it like a bark, numerous granules are seen floating in a transparent liquid within the cell, and forming a current which rises along one of the lateral walls, then flows horizontally along the upper wall, then descends along the other lateral wall, and becomes again horizontal along the lower wall of the cell. It is this intracellular motion which has been called *rotation*, a very inappropriate term, for which it would be better to substitute that of *cyclosis* (abolished by Hugo Mohl), which expresses much more exactly the circular movement of the sap in the cell.

Respiration.—The carbon of plants is derived from the carbonic acid contained in the air; the roots absorb it with the water of the soil which holds it in solution; whilst the carbon of the air enters the leaves through their stomata. Many experiments prove that the leaves and green parts exclusively possess the power of decomposing carbonic acid, thus separating the oxygen, and restoring it to the atmosphere; they also decompose water and retain the hydrogen; this power is only exercised under the influence of sunlight. Now animals are constantly burning carbon by means of the oxygen of the air, and exhaling carbonic acid, in which operation they consume an enormous quantity of oxygen; but plants, by their respiration, restore the balance, for they provide an inexhaustible store of pure oxygen, and incessantly repair the loss which the atmosphere has sustained through the respiration of animals.

The power possessed by leaves of decomposing carbonic acid ceases at night or in darkness; then the carbonic acid, absorbed by the roots with the water of the soil, enters the stem, and remains dissolved in the sap with which the plant is impregnated; soon this water evaporates through the leaves, and carries off the carbonic acid which it held in solution.

The green parts of plants absorb oxygen during the night by a chemical process, which tends to produce a change in the materials contained in their tissues. To blanch plants, they must be placed under the same conditions as the green parts of vegetables are during the night, namely, in continuous darkness; the carbonic acid is then not assimilated, the green chlorophyll is not formed, and their tissues contain an excess of water; and the horticulturist is thus enabled to expel the bitter principle from stems or leaves.

This exclusive property of the green parts is perhaps due to their having absorbed the chemical rays of the solar light, which rays may aid in the decomposition of carbonic acid in the chlorophyll.

Respiration, which is the reciprocal action of the sap upon the air, and of the air upon the sap, is carried on in the intercellular spaces (*lacunæ*) beneath the stomata, where the air comes into contact with the parenchyma. Submerged plants, which have no epidermis, and whose parenchyma is hence exposed to the fluid, decompose the carbonic acid which the water always contains, under the influence of light transmitted through the water; they fix the carbon and reject the oxygen, which remains in solution, and supports the life of aquatic animals. Here, as in the air, the Animal and Vegetable Kingdoms reciprocate only under the stimulus of light; and if the water be too deep, the plant becomes pale and etiolated.

Besides the elaboration of sap by the green tissues, other truly respiratory processes are carried on in the plant for the purposes of assimilation : thus, when the seed germinates, it absorbs oxygen and liberates carbonic acid ; a process analogous to the respiratory in animals, and which is continued until the first leaves of the embryo are developed. Similar respiratory processes accompany flowering ; the petals and stamens absorb, by day as well as by night, much oxygen, and emit much carbonic acid ; hence the noxious quality of the air in a room full of plants ; which is greatly increased by the exhalation of carburetted hydrogen, contributed by the volatile oils to which the perfume of the corolla is due.

Evaporation is a phenomenon analogous to the pulmonary perspiration of animals, and should be treated of after respiration. Evaporation is one of the most active agents in the ascent of the sap; it goes on through all the pores on the surface of the green parts, but especially through the stomata; increasing or diminishing as the surrounding air is drier or moister.

Leaves possess in a slight degree only the power of absorbing the watery vapour in the air ; and though certain uprooted plants remain fresh for some time, this is due to their losing little by evaporation. So, too, leaves floating with their lower surface on the water do not wither, not because they absorb water, but because their stomata being stopped up, evaporation is arrested.

Excretions.—A plant, after being nourished by the materials of the elaborated sap, rejects by its leaves, glands, bark, and especially by its root, all useless or noxious matters. Thus, to express in a few words the nutritive functions of its life, a vegetable may be said to *absorb, breathe, assimilate, perspire,* and *excrete.*

Direction of the Axis.—The stem tends always to ascend, and the root as uniformly to descend, and even in underground stems the tip of the rhizome always turns upwards. In the *Mistleto*, a parasite, the seed, fixed to the branch of a tree, germinates on the bark, and always directs its radicle towards the centre of the branch, and its plumule in the opposite direction ; here the tree takes the place of the soil, and the root obeys a centripetal, the stem a centrifugal force.

Attempts have been made to elude this general law of the direction of axes, by reversing the seeds of young plants, when the root bends round to the earth, and the stem turns upwards. A box of damp earth has been so suspended that seeds could be planted on the lower surface of the earth, the soil being above, air and light below; still the stems rose into the earth, the roots descended into the air.

Movements of Leaves and Flowers.—Leaves constantly direct their inner surface towards the sky, and their outer towards the earth; if this direction is reversed by twisting the base of the petiole, the leaf constantly tends to turn round in spite of all obstacles, and if these obstacles be insuperable, it gradually dies; if the branch be reversed artificially, the petiole twists ; if the reversion is natural, as in *weeping* trees, the torsion of the petiole is spontaneous, and the inner surface turns towards the sky ; if, finally, a leaf be so suspended that its blade is horizontal and its inner surface is turned downwards, the blade speedily turns round, and resumes its normal position. This *instinct* of the leaf depends neither on air nor on light, for it is displayed in water and in darkness. But with many species, the state of the atmosphere, whether gloomy or bright, dry or moist, hot or cold, gives rise to singular movements in leaves and flowers.   Thus, during the night the leaflets of the *Bean* and of *Trefoils* rise; those of the *Liquorice* and of *Robinias* hang vertically. This phenomenon has been called the *sleep of plants*; and to prove that this sleeping and waking depends on the absence and presence of light, plants have been caused to sleep at mid-day, by placing them in the dark; whilst others have been wakened at night by a strong artificial light.

There are various exotic plants, which, waking by day and sleeping by night in their native country, retain in our houses the habits of their climate, which being the reverse of ours, they sleep during our day, and wake when the sun has sunk below our horizon.   Tropical plants wake and sleep with us as if we had a perpetual equinox.[1]   Certain plants exhibit movements induced by accidental external stimuli; such is the *Sensitive Plant* (*Mimosa pudica*).   Its periods for sleeping and waking do not precisely coincide with our night and day, its waking periods being subject to vicissitudes depending on the slightest causes : a gentle shake, a breath of wind, the passage of a storm-cloud, the falling of a shadow, offensive vapours, the most delicate touch, cause the leaflets to droop suddenly, and closely overlap each other along the petiole, which then droops also; but soon after, if the cause be removed, the plant recovers from this sort of faint, all its parts revive and resume their first position.

Venus' Fly-trap (*Dionœa muscipula*) is a small North American herb, whose excitability is fatal to the insects which approach it; its leaves terminate in two rounded plates, joined by a hinge like the boards of a book, and fringed with marginal bristles; on their upper surface are two or three little glands which distil a liquid attractive to insects; when a fly touches these, the two plates close sharply and seize the insect, whose efforts to escape increase the irritation of the plant, which finally crushes it; when the insect is dead and all movement has ceased, the plates expand again, and await a fresh victim.   These phenomena, which are the effect of excitement, are not so exceptional as might be supposed; many plants of our climate offer analogous though much less remarkable examples.

The opening of some flowers is due to the stimulus of light: most open by day, though some by night, as the *Marvel of Peru* (*Mirabilis longiflora* and *Jalapa*); others open and close at various hours, and the hour of the day may be ascertained by watching their habits.   Linnæus arranged his *floral clock* in accordance with these periodical changes; but such a clock, in our variable climate, is often too slow or too fast; it can only be correct in the torrid zone, where there are but few atmospheric changes.

The heat and moisture of the atmosphere also influence the daily motions of flowers: certain species foretell rain by closing in the middle of the day, or by remaining open in the evening, or by not opening in the morning.   Attempts have been made to construct a *floral barometer* from these observations, but its performances are far more irregular than those of the floral clock.

[1] These statements are opposed to all the established phenomena of plant-life, as known to English observers.—ED.

PHENOMENA OF REPRODUCTION.

Fertilization.—Under Organography, the fertilizing action of the pollen on the ovules was alluded to, but not explained; we shall now analyse some details of this wonderful process, the most important of all departments of Vegetable Physiology.

The ancients had confused ideas as to the nature of the stamens; the botanists who wrote after the Renaissance hazarded some vague conjectures on this subject; and it was only towards the end of the seventeenth century that their true functions were assigned with precision to the pistil and stamen. Tournefort rejected the fact of fertilization, and persisted in considering the stamens as organs of excretion. After his death, the most devoted of his disciples, Sebastian Vaillant, in a discourse delivered in 1716 at the King's garden, explained the functions of the stamens, and demonstrated incontrovertibly the phenomena of fertilization in plants. Thanks to this discovery, the date of which is known, France claims the honour of the most important discovery which had hitherto been made in Botany. Eight years later, Linnæus popularized the doctrine of fertilization by his writings, which were no less remarkable for their learning than for their logical accuracy and poetic charm.

A few examples will suffice to prove the necessity of the pollen to fertilize the ovule. The *Date* is a diœcious tree, whose fruit is the principal food of certain eastern nations. From time immemorial these have habitually suspended panicles of male flowers on the female plants, when fertilization invariably ensues. These nations, when at war, destroy their enemies' male Date-trees, and so starve their owners by rendering the female plants sterile.

When the rainfall is excessive at the flowering season of the Vine, the growers say that the vine *runs*, i.e. that the pistils are abortive; which is owing to the pollen having been washed away, and fertilization having consequently not been effected. In newly-discovered Pacific islands, diœcious *Cucurbitaceæ* introduced for the first time have produced female flowers; but there being no males, fertilization has never taken place. Botanists can prevent or produce fertilization by cutting away all or some only of the stigmas of a pistil; in the latter case the ovaries corresponding to these stigmas do not produce seed. A pistilliferous Palm cultivated in a hothouse at Berlin had been sterile for eighty years, when some pollen from a staminiferous plant of the same species was sent by post from Carslruhe, by which the Berlin tree was fertilized; it was then left sterile for eighteen years, after which time it was again artificially fertilized, and the operation succeeded as at first.

Experimenters have employed other means to demonstrate the physiological action of the stamen; they have placed the pollen of one species on the stigma of a different species, belonging to the same genus, when individuals have been produced partaking of the nature of both species. Plants thus produced by cross fertilization are termed hybrids; their organs of vegetation are pretty well developed, but those of reproduction are imperfect, and their seeds are unfertile after one or two generations.

Connected with this interesting subject of fertilization is the history of *Cœlebogyne ilicifolia*, an Australian Euphorbiaceous shrub, which cannot be omitted here. Its flowers are diœcious, and for many years female individuals have been cultivated in English Botanic Gardens, which, without the co-operation of stamens (for there is not a single male plant in Europe), have produced seeds which have germinated, and produced in their turn individuals perfectly resembling the mother plant. Here the production of fertile seeds without the intervention of pollen is incontestable. But we do not think that this exceptional phenomenon (which has, however, been almost authentically paralleled by *Hemp* and *Mercurialis*, both indigenous diœcious plants) will overturn the admitted doctrine of the fertilization of the ovule by the pollen; and we find no difficulty in admitting that Nature has given to the seeds of certain diœcious plants a power of multiplied reproduction, which may extend to several generations, such as is proved to exist in the case of *Aphides*. Besides, the force of the anomaly presented by *Cœlebogyne* cannot be estimated at its true value until time shall have shown whether this power is limited or indefinite.

The period of fertilization is that at which the flower exhales its perfume and appears in its full beauty; the stamens and pistil then exhibit spontaneous motions, which in some species are very remarkable. Thus, in the *Berberis*, the filaments of the stamens are at first pressed between the two glands of each petal, which as they spread force the filaments to spread also; these soon free themselves under the stimulus of the sun, aided by a slight evaporation which has contracted these and the glands which retained them; when they quickly resume their original bent position and approach the pistil, on which the anthers shed their pollen. This action, which is effected by the solar rays, may be artificially induced, either by gently irritating the filaments, or by shaking the flower; for the least shake or slightest touch releases the stamen. The same irritability is observable in *Parietaria* and in *Nettles*, the filaments of which lie curved back within the calyx, but instantly spring up, if lightly touched; when the anther, which was previously pressed down at the bottom of the flower, is carried up, and sheds a little cloud of pollen. *Rue* sheds its pollen with less force but with better aim; it has four or five petals and eight or ten stamens; on most flowers there is one stamen which, instead of spreading horizontally over one or between two petals, bends over the pistil, against which the filament presses. If patiently watched, the anther will be found to open and emit the pollen; when the stamen, having fulfilled its function, falls back, and another rises to take its place, and so on in succession till all the anthers have in turn shed their pollen on the pistil. The elasticity of the anthers is not always sufficient to discharge the pollen on the stigma. The conditions under which the pollen is discharged are very various; in many cases the flower is fertilized before expansion; in many others, the anthers are placed above the pistil, and the pollen is brought directly into contact with the stigma; but it frequently happens that the position of the stamens is unfavourable to their pollen reaching the stigma, when its transmission is effected by the wind, and especially by insects. Butter-flies, flies, moths, bees, and often very small Coleoptera may be seen at the bottom of flowers, eagerly seeking the honey, and thus becoming useful auxiliaries in the fertilization of the pistil, either by dispersing the pollen with their wings, or by carrying the pollen of one plant on the hairs of their bodies to another plant of the same species. Here we must notice a very interesting series of coincidences : when the anthers open to shed their pollen, the stigma becomes viscous to retain it; nectar is distilled by the glands, and nectar-feeding insects make their appear-ance; lastly, at the same—often very brief—period, the corolla expands, whose colour and scent must affect the powerful sight of insects and their subtle sense of smell.

Mr. Darwin has recently published, on the fertilization of certain plants, experi-ments which throw a new light on Natural Science, and plainly reveal the marvellous precautions taken by Nature to prevent the degeneration of species. He has en-deavoured to give the rationale of the differences observable in the flowers of *Primula*. In this genus the same species presents two very remarkable forms : a long-styled, in which the stigma is globular and wrinkled, and exactly reaches the mouth of the corolla-tube, far overtopping the anthers, which only reach half-way up the tube. In the other form the style is not half the length of the corolla, and the stigma is de-pressed and soft, but the anthers occupy the upper part of the tube, their pollen is larger, and the capsule contains more seeds than that of long-styled individuals. This *dimorphism* between *longistyled* and *brevistyled* primroses is constant; the two forms are never met with on the same individual, and the individuals of each form are about equal in number. Mr. Darwin covered with netting plants of both the long-styled and short-styled forms, most of which flowered ; but as neither produced seed, he concluded that insects are necessary to their fertilization. But as, in spite of his utmost vigilance, he never saw any insects approach uncovered primroses during the day, he supposes that they are visited by moths, which find abundant nectar in them. He endeavoured to imitate the action of insects, which, while extracting honey from flowers, are the agents of their fertilization, and his experiments led him to very interesting conclusions. If we introduce into the corolla of a *short-styled* primrose the trunk of a moth, the pollen of the anthers placed at the mouth of the tube adheres to the base of the trunk, and it may be concluded that this pollen will necessarily be deposited on the stigma of the *long-styled* primrose when the insect visits it. But in this fresh visit, made to the long-styled primrose, the trunk, descending to the bottom of the corolla, finds the pollen of the anthers which are situated there; this pollen adheres to the end of the trunk, and if the insect visits a third flower, which is *short-styled*, the end of its trunk will touch the stigma placed at the base of the tube, and will deposit the pollen on it. Besides this it may be admitted as very probable that in its visit to the long-styled flower, the insect, in drawing back its trunk, may leave on the stigma a portion of the pollen from the anthers placed lower down, and the flower would be thus fertilized by itself. It is besides nearly certain that the insect, when plunging its trunk into a short-styled corolla, will have rubbed the anthers inserted at the top of the tube, and deposited on the stigma of the flower a portion of its own pollen. Finally, the corolla of primroses contains many tiny hemipterous insects, of the genus *Thrips*, which, moving about the flower in all directions, transport from the anthers to the stigma the pollen which adheres to them; by which means again the plant will be self-fertilized.

In the fertilization of *dimorphic* species four operations are thus possible:—1, self-fertilization of the *long-styled* flower; 2, self-fertilization of the *short-styled* flower; 3, fertilization of the *short-* by the *long-styled*; and 4, that of the *long-* by the *short-styled*: the two first Mr. Darwin calls *homomorphic*; the two others, *heteromorphic*.

Mr. Darwin has artificially fertilized flowers in these different ways, by protecting them from insects, and he has found in the wild primrose (*Primula veris*) and in the Chinese primrose (*Primula Sinensis*) that *heteromorphic* unions produce considerably more capsules and good seeds than *homomorphic* unions. Thus primroses present two sets of individuals, which, although belonging to the same species, and both possessing stamens and pistils, are mutually dependent on each other for perfect fertilization. Mr. Darwin concludes that Nature, in establishing *dimorphism* in primroses, and in distributing the two forms in equal numbers of individuals, has evidently had in view the crossing of distinct individuals; the relative heights of the anthers and stigmas obliging insects to deposit the pollen of one set on the stigma of the other. Nevertheless it is impossible not to admit that the stigma of the visited flower may receive its own pollen. Now it is a well-known fact, that if the pollen of several varieties fall on the stigma of one individual, that of one of the varieties is prepotent, and its pollen takes effect to the exclusion of that of all the others. Mr. Darwin thinks that it may be inferred from this that in primroses the *heteromorphic* pollen, which is known to be the most potent, will overcome the action of the *homomorphic* pollen whenever the two come into collision; thus, he adds, indicating the efficacy of *dimorphism* in producing crosses between individuals of the two forms. These two forms, although both bearing stamens and pistils, are in this case truly dioecious; each of them is fertile, though the pollen of each is less potent on its own stigma than on that of the other form.

Mr. Darwin has studied dimorphism in the different species of *Linum*, and he has instituted on *L. grandiflorum* and *perenne* a series of experiments which confirm the preceding conclusions.

The scarlet-flowered *L. grandiflorum* has also the two types of long and short-styled flowers; in the short-styled form the five stigmas diverge, project between the filaments, and rest against the tube formed by the petals. In the long-styled form, on the contrary, the stigmas are erect, and alternate with the anthers. Mr. Darwin selected twelve flowers of two long-styled individuals which he fertilized heteromorphically, i.e. with pollen from the short-styled form; most produced good capsules and seeds; those which were not touched remained absolutely sterile, although their stigmas were covered with a thick layer of their own pollen. He next sought to ascertain the probable cause of this sterility, by placing the pollen of a short-styled flower on the five stigmas of a long-styled flower, and after thirteen hours he found the latter discoloured, withered, and deeply penetrated by a number of pollen-tubes; he then made the reverse experiment on a long-styled flower, and this heteromorphic fertilization had the same result as the first.   Lastly, he placed the pollen of a long-styled flower on the stigmas of a similar flower, but belonging to another plant; but at the end of three days not a single pollen-grain had emitted a tube.   In another experiment, Mr. Darwin placed on three of the stigmas of a long-styled flower pollen belonging to the same type, and on the two others pollen from a short-styled flower.   At the end of twenty-two hours these two stigmas were discoloured and penetrated by numerous pollen-tubes; the three other stigmas covered with pollen of their own type remained fresh, and the pollen-grains scarcely adhered to them.   In *Linum perenne*, dimorphism is even more obvious that in *L. grandiflorum*; the pistil of the one form is much longer, and the stamens much shorter than in the other.   Mr. Darwin has ascertained, by numerous experiments on each of the two forms, that the stigmas of one can be impregnated only by pollen from the stamens of the other.

It is hence absolutely necessary that insects should carry the pollen from the flowers of one form of *Linum* to those of the other; and to these they are attracted by five minute drops of nectar secreted on the exterior of the base of the stamens: to reach these drops, the insect is obliged to insert its trunk between the staminal whorl and the petals.   Now, in the short-styled form, if the stigmas, which were originally vertical and faced the floral axis, had preserved this position, their backs only would have been presented to the insect, and the flower could never be thus fertilized; but the styles having diverged, and protruding between the filaments, the stigmatic surfaces are turned upwards, and rubbed by every insect which enters the flower, thus receiving the pollen which fertilizes them.

In the long-styled form of *L. grandiflorum* the styles diverge very slightly, and the stigmas project a little above the corolla-tube, so as directly to overhang the passage leading to the drops of nectar; consequently, after an insect has visited the flowers of either form, it withdraws its trunk well covered with pollen.   If it then plunges its trunk into a long-styled flower, it necessarily leaves some of this pollen on the papillæ of the stigmas; if it plunders a short-styled flower, it still deposits pollen on its stigmas, the papillæ of which are here turned upwards.   Thus the stigmas of the two forms receive indifferently the pollen of both, though fertilization of each can only be effected by the pollen of the opposite form.

In the long-styled type of *L. perenne* the styles do not sensibly diverge, but they twist so as to reverse the position of the stigmas, whence the inner surfaces are turned outwards; thus an insect seeking nectar in the flower brushes against the stigmatic surfaces, and leaves on them the pollen collected from another flower.

The facts here recorded demonstrate both the object of dimorphism, and the important part which insects play in the fertilization of plants.   Mr. Darwin complains that certain botanists attribute the transport of pollen to the wind and insects indifferently, as if there were no important difference between the action of these two agents.   Diœcious plants, or even hermaphrodite ones, in the fertilization of which the wind is a necessary auxiliary, present peculiarities of structure fitted for this mode of transport: in some the pollen is powdery and abundant, as in *Pines*, *Spinach*, &c.; the pendent anthers of others at the least breath scatter the pollen; in others the perianth is wanting, or the stigmas project beyond the flower at the moment of fertilization; in some the flowers appear before the leaves; and some have feathery stigmas, as *Gramineæ*, *Mercurialis*, &c. Wind-fertilized flowers do not secrete nectar; the pollen is too dry to adhere to insects, and the corolla is either absent, or possesses neither the colour, scent, nor nectar which attract them.

We shall conclude these remarks by mentioning the curious phenomena respecting the fertilization of *Vallisneria spiralis*, which grows submerged in stagnant waters in the south of France. It is diœcious, but the male plants always grow near the female; the female flower, protected by a spathe, is borne on a long peduncle which rises from a tuft of radical leaves; and the ovary bears three forked stigmas. The male flowers are borne on a very short peduncle, and are sessile on a conical axis enveloped in a spathe. At the flowering period the female peduncle gradually lengthens, so that the flower finally floats on the surface of the water, and opens its perianth of six very minute segments. Then the male flowers, which have hitherto remained submerged, detach themselves spontaneously from their peduncle, and rise to the surface, where numbers of them may be seen floating around the female flower, on which the anthers elastically project an abundance of pollen. After fertilization, the peduncle of the female flower contracts spirally, and the ovary descends to the bottom of the water to ripen its seeds.

In describing the anther, we spoke of the fibrous cells which, after the maturing of the pollen, form a layer upon the inner wall; which layer gets thinner as it approaches the line of dehiscence, where it disappears. At the moment when the pollen is ready to be discharged, the moisture of the anther evaporates, its hygrometrical tissue, pulled different ways by the variations of the atmosphere, produces a strain along the line where the fibrous cells are interrupted, and these by their contraction favour the emission of the pollen. At the same time the cells of the stigma become viscous, so as to retain the pollen projected on to them from the anther, or carried thither by the wind or by insects. Thereupon the pollen swells, through the action of endosmose; the inner membrane ruptures the outer at one of the points which touch the stigma; the *pollen-tube* (fig. 413) lengthens, traverses the interstices of the stigmatic cells, and reaches the conducting tissue which fills the canal of the style, and which is charged, like the stigma, with a thick fluid. Still lengthening, the pollen-tube finally enters the cavity of the ovary, traverses the conducting tissue which lines the placentas, and at last reaches the ovule (fig. 750), when it enters the micropyle and comes in contact with the cell of the nucleus (*embryonic sac*), its tip resting on the membrane of the sac, and partly adhering to it. Soon after this contact of the pollen-tube, one, or oftener two vesicles (*embryonic vesicles*, fig. 750) usually appear within the embryonic sac, below the tip of the pollen-tube. These vesicles elongate; the upper and thinner end adhering to the membrane of the sac. While one of the two shrinks and disappears, the other develops, and fills more or less completely with its free end the cavity of the embryonic sac. The embryonic vesicle, which will be developed into the embryo, is at first filled with a transparent fluid, but soon presents transverse septa at the upper contracted part which forms the *suspensor*; then a longitudinal septum is formed in the swollen part, which answers to the free end; on which free end is afterwards developed either one lobe, or two opposite lobes (*cotyledons*), and the opposite end becomes the caulicle.

All physiologists concur in the above, but there are different opinions as to the part the pollen plays. Schleiden contended that the tip of the pollen-tube forms the embryo by forcing inwards the membrane of the embryonic sac, folding it around itself, and occupying its cavity, where it speedily develops into the embryo. Thus, according to Schleiden, the ovule is merely a receptacle, destined to receive the embryo, to protect and nourish it, the true reproductive organ residing in the anther. But a closer examination instituted by the most skilful anatomists of the French school has repeatedly disproved the existence of the embryonic vesicle before the arrival of the pollen-tube. Nevertheless, it is certain that the pollen ma-

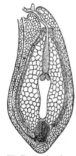

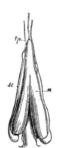

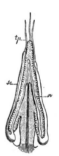

750. Œnothera longiflora. Anatropous ovule, cut vertically at the moment of fertilization, to show the pollen-tube, the end of which is in contact with the embryonic sac; within, at the top of this sac, are two vesicles, one of which will wither, and the other form the embryo (mag.).

751. Santalum. Placenta bearing three nuclei, whence issue three embryonic sacs, which receive three pollen-tubes (mag.).

752. Vertical section of fig. 751 (mag.), showing two of the embryonic sacs within and outside the nucleus.

753. Santalum. Portion of nut, cut vertically, to show the embryonic sac which has burst the nucleus at the bottom, and has ascended to the pollen-tube, the free end of which it sheaths (mag.).

terially assists in the formation of the embryo by means of its *fovilla*, which passes by endosmose from the pollen-tube into the interior of the ovule.

The fertilization of the ovule in *Santalaceæ* presents a quite exceptional phenomenon, which deserves to be mentioned (figs. 751, 752, 753). The ovary is unilocular, and the free central placenta bears several suspended ovules; each is a *naked* nucleus (without primine or secundine). At the period of fertilization, the nucleus *n* bursts at the lower part, the embryonic sac *se* emerges by this opening and ascends along the whole length of the outer surface of the nucleus, to meet the pollen-tube *tp* a little below the top of the nucleus. The latter soon withers, and the embryonic sac, which alone grows, forms the integument of the seed.

After fertilization flowers rapidly lose their freshness; the corolla and stamens wither and fall; the style dries up, together with the conducting tissue which filled it, and that portion of this tissue which abutted on the ovule disappears. Soon the ovary, receiving the nourishment which was previously distributed to other parts of the flower, increases, as do the ovules; many of these become arrested by the overwhelming development of the others, and the number arrested is often constant; sometimes also the septa disappear. Finally, the fertilized pistil becomes more or less modified in form, volume, and consistency.

Maturation.—This marks the period when the changes which take place in the fruit, from fertilization till the dispersion of the seed, are completed. Those fruits which remain foliaceous continue, like the leaves, to decompose carbonic acid and disengage oxygen by day, whilst by night they absorb oxygen and disengage carbonic acid. At maturity their tissue dries, their colour changes, their fibrovascular bundles separate, and dehiscence takes place.

Those fruits which lose their foliaceous consistency and become fleshy respire like the preceding until maturity; then the parenchyma is fully developed, its watery contents are decomposed, and fixed in new combinations; the *cellulose* loses some of its carbon and hydrogen, and becomes *starch*; and the latter, by the addition of water, is changed into *sugar*. Vegetable acids are the equivalents of starch and oxygen; to change these acids into sugar, all that is necessary is either that the carbon assimilated by the plant shall take up their oxygen, or that water shall be formed at the expense of the latter. In most fruits these acids are not entirely converted into sugar, but enter into combination with alkaline bases, thus modifying their acid flavour. The proportions of acid and sugar vary according to the nature of the fruit.

When maturation is complete, the fruit disengages carbonic acid formed at the expense of the sugar, and the latter gradually disappears; but the decomposing fruit, by disengaging carbonic acid around it, materially contributes to the nourishment of the young seed. At complete maturity the fruit breaks up, and the seed thereupon commences an independent existence.

Dissemination is the act by which the ripe seeds are scattered over the surface of the earth. In capsular fruits the seeds are freed by the dehiscence of the carpels; in fleshy fruits they are retained longer.

Nature has infinitely varied the methods tending to disseminate seeds: wind, water, and frugivorous animals are the principal agents; and man himself assists, often unwittingly, by his labours or voyages, in the transport and multiplication of seeds.

Germination.—The agents in germination are water, air, heat, and darkness. Seeds buried for many centuries[1] in dry soil, preserved from the air and from variations of atmospheric temperature, have been known to germinate and reproduce their species, when placed under favourable conditions.

Water softens the integuments, penetrates the tissue of the seed, and is

---

[1] The statements supposed to prove this are not generally trustworthy.—ED.

decomposed; its hydrogen is absorbed; its oxygen, like that of the air, combines with the carbon of the seed to form carbonic acid, which is set free. Heat is indispensable to germination; and in the series of phenomena which accompany this process, it acts alternately as cause and effect, for a seed is the theatre of chemical combinations. Light retards germination, by causing the decomposition of carbonic acid, and hence preventing the formation of this gas. Under a combination of favourable circumstances, the seed absorbs water, together with the oxygen of the air; the albumen, under the chemical action of these agents, loses a portion of its carbon, and at the same time combines with the elements of water; it soon changes into a saccharine, milky, soluble matter, fit to be absorbed by the embryo; if the albumen has been absorbed previous to germination, the cotyledons enlarge and nourish the plumule. When the latter has emerged from the ground and become green, the phenomena are reversed; the young plant, instead of absorbing oxygen to combine with its carbon, and disengaging carbonic acid, absorbs carbonic acid, separates the carbon, and assimilates it.

TAXONOMY.

TAXONOMY is the part of Botany which treats of *classification*; i.e. the methodical distribution of plants in groups, named *Classes, Families* (or *Orders*), *Genera*, and *Species*. All the individuals or separate beings of the Vegetable Kingdom which resemble each other as much as they resemble their parents and their posterity, form collectively a *species*.

All the species which resemble each other, although differing in certain characters which become the distinctive sign of each, form collectively a *genus*, which takes the name of the principal species.[1] Thus, the *Cabbage*, the *Turnip*, the *Colza*, the *Radish*, are species of the same genus, which has received the name of *Cabbage*. As a necessary consequence, each plant belonging to a genus and to a species has received two names, that of the genus and that of the species, i.e. the *generic* and *specific* name, and we say the *Drum-head Cabbage*, the *Turnip Cabbage*, the *Colza Cabbage*, the *Radish Cabbage*.

All the genera which resemble each other form collectively a *family* (or *order*); thus, the genus *Cabbage*, the genus *Stock*, the genus *Thlaspi*, the genus *Cochlearia*, belong to the same family, namely that of *Cruciferæ*.

Those families which are allied are united into *classes*; and thus all the species composing the Vegetable Kingdom are classified.

But the species itself may be subdivided: individuals of the same species may be placed under different conditions; one may vegetate on a barren rock, another in a swamp; this will be shaded, that torn by the wind; man himself may intentionally create such differences, and combine them according to his wants. The vegetable under these diverse influences will finally undergo changes in its sensible qualities, such as size of root; the size, consistency, and duration of stem; the form, colour, and scent of its floral whorls, the taste of its fruit, &c. But these changes, however considerable, will not destroy the primitive character of the species, which will always be discoverable throughout its modifications. A collection of individuals of the same species which have undergone such modifications bears the name of *variety*. The characters of a variety, depending on accidental causes, are never constant; as soon as the influencing cause ceases, the change ceases, and the primitive species reappears in its original form. The *cultivated Cabbage* is an example, of which six

---

[1] This holds only in a limited sense.—ED.

varieties are known in France :—1. the *Wild Cabbage*, which is the primitive type of the species ; 2. the *Common Kale*, with a long stem and spreading leaves ; 3. the *Scotch Kale*, of which the leaves are almost in a head when young, then spreading and wrinkled ; 4. the *Drumhead Cabbage*, of which the stem is short, the leaves green or red, concave, and gathered into a head before flowering ; 5. the *Kohlrabi*, the stem of which is swollen and globular below the insertion of the leaves ; 6. the *Cauliflower*, of which the floral branches are gathered closely together before flowering ; the sap enters this inflorescence exclusively, and transforms it into a thick, succulent, and granular mass, which furnishes an excellent food. Such are modifications induced by cultivation ; they are wholly due to the excessive development of the parenchyma, which accumulates, sometimes in the leaves (*Drumhead Cabbage*), sometimes only at the edge of these leaves (*Scotch Kale*), sometimes at the base of the stem (*Kohl-rabi*), and sometimes in the peduncles or floral branches (*Cauliflower*).

The seed does not preserve the variety ; it always tends to reproduce the primitive type. Nevertheless there are plants of which the varieties are propagated by seed, provided that the conditions which have modified the species be faithfully repeated ; such are the *Cereals*, which form, not varieties, but *races*, the original type of which is lost.

The older classifiers arranged plants according to their *properties* or *habitats* ; others on characters drawn from the *stem, roots, leaves*, or *hairs*. It was at last perceived that the *flower*, containing the seed which was to perpetuate the species, and composed of leaves of which the form, colour, number, and connection notably differ in each genus and species, is the part of the plant which ought to furnish the best characters for classification. Hence the flower furnishes the basis of the systems of *Tournefort* and *Linnæus*, the method of *A. L. de Jussieu*, and that of *A. P. de Candolle*, which is a slightly modified arrangement of De Jussieu's. Tournefort established his system on the consistency of the stem, on the presence or absence of a corolla (and he considered every floral envelope which is not green as a corolla), on the isolation or the contrary of the flowers, and on the shape of the petals. This method, which appeared in 1693, and comprised 10,000 species, being based on the most prominent part of the plant, was intelligible and easy of application, and was once universally accepted ; but as the knowledge of species increased, many were found that would not fall into any of its classes, and it was hence abandoned.

The system of Linnæus, which appeared forty years after that of Tournefort, was received with an enthusiasm which still exists, especially in Germany. He took as the base of his twenty-four *classes* the characters furnished by the stamens in their relations to each other and to the pistil.

TABLE OF THE ARTIFICIAL METHOD OF TOURNEFORT.

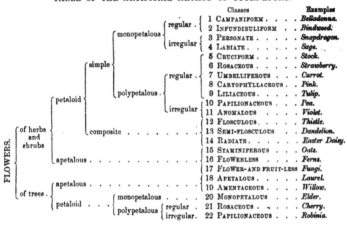

	Classes	Examples
	1 CAMPANIFORM	*Belladonna.*
	2 INFUNDIBULIFORM . .	*Bindweed.*
	3 PERSONATE	*Snapdragon.*
	4 LABIATE	*Sage.*
	5 CRUCIFORM	*Stock.*
	6 ROSACEOUS	*Strawberry.*
	7 UMBELLIFEROUS . . .	*Carrot.*
	8 CARYOPHYLLACEOUS .	*Pink.*
	9 LILIACEOUS	*Tulip.*
	10 PAPILIONACEOUS . . .	*Pea.*
	11 ANOMALOUS	*Violet.*
	12 FLOSCULOUS	*Thistle.*
	13 SEMI-FLOSCULOUS . .	*Dandelion.*
	14 RADIATE	*Easter Daisy.*
	15 STAMINIFEROUS . . .	*Oats.*
	16 FLOWERLESS	*Ferns.*
	17 FLOWER-AND FRUIT-LESS	*Fungi.*
	18 APETALOUS	*Laurel.*
	19 AMENTACEOUS	*Willow.*
	20 MONOPETALOUS . . .	*Elder.*
	21 ROSACEOUS	*Cherry.*
	22 PAPILIONACEOUS . . .	*Robinia.*

FLOWERS. — of herbs and shrubs: petaloid (simple — monopetalous: regular, irregular; polypetalous: regular, irregular — composite), apetalous; of trees: apetalous, petaloid (monopetalous — polypetalous: regular, irregular).

KEY TO THE LINNÆAN SYSTEM.

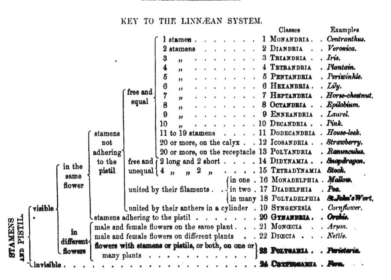

	Classes	Examples
1 stamen	1 MONANDRIA .	*Centranthus.*
2 stamens	2 DIANDRIA .	*Veronica.*
3 „	3 TRIANDRIA .	*Iris.*
4 „	4 TETRANDRIA .	*Plantain.*
5 „	5 PENTANDRIA .	*Periwinkle.*
6 „	6 HEXANDRIA .	*Lily.*
7 „	7 HEPTANDRIA .	*Horse-chestnut.*
8 „	8 OCTANDRIA .	*Epilobium.*
9 „	9 ENNEANDRIA .	*Laurel.*
10 „	10 DECANDRIA .	*Pink.*
11 to 19 stamens . . .	11 DODECANDRIA .	*House-leek.*
20 or more, on the calyx . .	12 ICOSANDRIA .	*Strawberry.*
20 or more, on the receptacle	13 POLYANDRIA .	*Ranunculus.*
2 long and 2 short	14 DIDYNAMIA .	*Snapdragon.*
4 „ „ 2 „	15 TETRADYNAMIA .	*Stock.*
united by their filaments . . in one	16 MONADELPHIA .	*Mallow.*
in two	17 DIADELPHIA .	*Pea.*
in many	18 POLYADELPHIA	*St.John'sWort,*
united by their anthers in a cylinder .	19 SYNGENESIA .	*Cornflower.*
stamens adhering to the pistil	20 GYNANDRIA .	*Orchis.*
male and female flowers on the same plant .	21 MONOECIA .	*Arum.*
male and female flowers on different plants .	22 DIOECIA .	*Nettle.*
flowers with stamens or pistils, or both, on one or many plants	23 POLYGAMIA .	*Parietaria.*
invisible	24 CRYPTOGAMIA .	*Fern.*

STAMENS AND PISTIL. — visible: in the same flower (stamens not adhering to the pistil: free and equal, free and unequal), in different flowers; invisible.

The first thirteen of the Linnæan *classes* are divided into *orders* founded on the number of ovaries or free styles composing the pistil. In *monogynia* the pistil is formed of a single carpel, or of several carpels united into one by their ovaries and styles; in *digynia* there are two distinct ovaries or styles; in *trigynia* three; in *tetragynia* four; in *pentagynia* five; in *hexagynia* six; in *polygynia* any number above ten. The 14th class contains two orders: *gymnospermia*, in which the pistil is composed of four lobes simulating naked seeds; *angiospermia*, in which the seeds are enclosed in a capsule. The 15th class is said to be *siliquose* or *siliculose*, according as the fruit is or is not three times longer than broad. The 16th, 17th, 18th, 20th, 21st, and 22nd classes have their orders founded on the number and connection of the stamens and styles (*triandria, pentandria, polyandria, monogynia, polygynia, monadelphia*, &c.). The 19th class is divided into *polygamia æquales*, in which all the centre flowers of the capitulum have stamens and pistils, and those of the circumference have pistils and are fertile; *polygamia frustranea*, where the flowers of the circumference are female and sterile; *polygamia necessaria*, where the flowers of the centre are male, and those of the circumference female and fertile, &c. The 23rd class is divided into *monœcious, diœcious, triœcious*. The 24th class is divided into *Ferns, Mosses, Algæ*, and *Fungi*.

A complete classification ought to satisfy two conditions: the first that of enabling one quickly to ascertain the name given by botanists to a plant, and to separate it from the rest of the Vegetable Kingdom by differential characters, as salient as possible. This object ought to be fulfilled by the *system*, which should be a true alphabetical dictionary, facilitating research; and its divisions ought, therefore, to be established on the most apparent characters, however bizarre and dissimilar they may be. From this point of view the Linnæan classification is a *chef d'œuvre* which will perhaps never be surpassed, in spite of the inconveniences resulting from the not very numerous difficulties to be overcome in applying it. *Dichotomous keys* are systems which consist in placing before the student a series of questions wherein the choice lies between two contradictory propositions, in such a manner that, the one being granted, the other must be necessarily rejected.

The second condition is that of placing each species and genus amongst those with which it agrees in the most essential points of resemblance: if this object be fulfilled, the *method* becomes a true science, its divisions being founded on the most important organs, without regard to their number, or to the difficulty of observing them. *System* enables us to discover the name of an individual from its description; *method* enables us to ascertain its position in the Vegetable Kingdom;—method is hence the complement of system.

The *affinities* which should form the basis of every natural method were first established by A. L. de Jussieu. Before him, *Magnol*, of Montpellier, had introduced into Botany *families* of which the arrangement was founded on the structure of the calyx and corolla; *Rivin* had published a classification based on the form of the corolla, on the number of the seeds, on the form, consistency, and cells of the fruit; *Ray* had classed upwards of 18,000 species, which he divided according to the number of cotyledons, the separation or aggregation of the flowers, the presence or absence of the corolla, the consistency of the fruit, and the adhesion or not of the ovary with the receptacular tube. The problem of a classification by *natural affinities* had thus been long propounded; it was solved by A. L. de Jussieu, who discovered the grand principle of the *relative value of characters*. In a Memoir on *Ranunculaceæ*, he enunciated and developed *the relative and subordinate importance of the different organs of a plant*; this was followed by his great work on the *Families* and *Genera* of the Vegetable Kingdom; and the clear principle of the subordination of characters, which had guided him in his labours, thereupon threw great light on all other branches of Natural History.

TABLE OF THE NATURAL METHOD OF A. L. DE JUSSIEU.

				Classes	Examples
ACOTYLEDONS	. . . .			1 ACOTYLEDONIA .	*Fungi.*
	Stamens inserted on the receptacle		.	2 MONO-HYPOGYNIA	*Oats.*
MONOCOTYLEDONS	,, ,, calyx	.	.	3 MONO-PERIGYNIA	*Iris.*
	,, ,, ovary	.	.	4 MONO-EPIGYNIA .	*Orchis.*
	,, ,, ovary	.	.	5 EPISTAMINIA .	*Aristolochia*
Apetalous flower.	,, ,, calyx	.	.	6 PERISTAMINIA .	*Rumex.*
	,, ,, receptacle	.	.	7 HYPOSTAMINIA .	*Amaranth.*
	Staminiferous corolla inserted on the receptacle			8 HYPOCOROLLIA .	*Belladonna.*
	,, ,, ,, calyx	.	.	9 PERICOROLLIA .	*Campanula.*
Monopetalous flower.				10 { EPICOROLLIA SYNANTHERIA	} *Cornflower.*
	,, ,, ,, ovary	.		11 { EPICOROLLIA CORISANTHERIA	} *Elder.*
	Stamens inserted on the ovary	.	.	12 EPIPETALIA .	*Carrot.*
Polypetalous flower.	,, ,, receptacle	.	.	13 HYPOPETALIA .	*Ranunculus.*
	,, ,, calyx	.	.	14 PERIPETALIA .	*Strawberry.*
Male and female flowers on different plants		.	.	15 DICLINIA .	*Nettle.*

(left margin, vertically: DICOTYLEDONS)

The successors of A. L. de Jussieu have followed in his path, but have differed as to the relative value of his characters; and it has further been shown that single characters of great importance may in certain cases be equalled, or even surpassed, by several characters of secondary importance: here quality is replaced by quantity, much as twenty sous are equal to one franc.

It may, however, be considered as proved, that the most constant characters should rank the highest: now this constancy especially prevails in the reproductive organs, and in accordance with the importance of their functions; therefore the floral organs have been rightly chosen to group species into genera, genera into families (or orders), and these into classes. As regards constancy of characters, the reproductive organs observe the following order:—the number of cotyledons, the cohesion or separation of the petals, the insertion of the stamens, the presence or absence of albumen and its nature, the direction of the radicle, the æstivation, the degree of symmetry in the position, number, and form of the floral whorls, &c.

In addition to the preceding synoptical tables, it is well to give the *Arrangement* of A. P. de Candolle, as followed in his 'Prodromus of the Vegetable Kingdom;' the *Classification* of Ad. Brongniart, according to which the Botanical

School of the Jardin des Plantes in Paris is arranged; and, finally, the *Succession of the Families* established by A. de Jussieu, which we have adopted for our ' Flora of Gardens and Fields,' and which we shall follow in the present work.

[In this, the English edition, the system of De Candolle is followed; of which a synopsis will be given at the end of the work.—ED.]

ARRANGEMENT OF A. P. DE CANDOLLE.

			Classes	Examples
VASCULAR PLANTS, or COTYLEDONS.	Exogens.[2]	Polypetalous corolla, and stamens inserted on the receptacle	1 THALAMIFLORAL .	*Ranunculus.*
		Polypetalous or monopetalous corolla, and stamens inserted on the calyx . . .	2 CALYCIFLORAL .	*Strawberry.*
		Monopetalous staminiferous corolla, inserted on the receptacle	3 COROLLIFLORAL .	*Belladonna.*
		A single floral envelope, or similar calyx and corolla	4 MONOCHLAMYDEOUS	*Nettle.*
	Endogens.[3]	Visible and regular fructification . .	5 PHANEROGAMIC .	*Iris.*
		Invisible or irregular „ . . .	6 CRYPTOGAMIC .	*Ferns.*
CELLULAR[4] PLANTS, or ACOTYLEDONS.		Foliaceous expansions	7 FOLIACEOUS . .	*Mosses.*
		No foliaceous expansions	8 APHYLLOUS . .	*Fungi.*

CLASSIFICATION OF M. AD. BRONGNIART.

(The Families being enumerated in the description of the Classification of A. de Jussieu, we here confine ourselves to the enumeration of the Classes.)

CRYPTOGAMS.—Vegetables deprived of stamens, pistil, and even of ovules. Embryo simple, homogeneous, without distinct organs, usually formed of a single vesicle.

AMPHIGENS.—No axis or appendicular organs evident; growth peripheric; reproduction by naked spores. *Algæ, Fungi, Lichens.*

ACROGENS.—Axis and appendicular organs evident; stems growing at the extremity only, without the addition of fresh portions at the base. Reproduction by spores covered by an integument, but not adhering by a funicle to the walls of the capsules which contain them. *Muscineæ, Filicineæ.*

PHANEROGAMS.—Reproductive organs evident, formed of stamens, and of ovules, which are either naked or enclosed in an ovary. Embryo compound, cellular, heterogeneous or formed of many distinct parts. Old parts of the living stem increasing by the addition of new tissues.

MONOCOTYLEDONS.—Embryo with a single cotyledon. Stem composed of fibro-vascular bundles scattered through the mass of the cellular tissue, not forming a regular circle; the living stem not increasing by distinct concentric zones of wood and bark.

---

[1] Provided with cells and vessels.
[2] The fibro-vascular bundles arranged in concentric layers, with the youngest outside.
[3] The fibro-vascular bundles arranged without order, the youngest in the centre of the stem.
[4] Deprived of vessels, and composed only of cells.

Albuminous.—Embryo accompanied by albumen.

Perianth none, or sepals not resembling petals. Albumen farinaceous. *Glumaceæ, Junceæ, Aroideæ.*

Perianth absent or double, with sepals or petals. Albumen not farinaceous. *Pandaneæ, Phœniceæ, Lirioideæ.*

Perianth double, the inner or both petal-like. Albumen farinaceous. *Bromeliaceæ, Scitamineæ.*

Exalbuminous.—Albumen wanting. *Orchideæ, Fluviales.*

DICOTYLEDONS.—Embryo with two opposite, or more (and then whorled) cotyledons. Stem with fibro-vascular bundles forming a cylinder around a central pith, separable into an inner woody zone and an outer bark zone, and increasing by concentric layers.

ANGIOSPERMS—Ovules contained in a closed ovary, and fertilized through the medium of a stigma.

GAMOPETALOUS.—Petals united.

PERIGYNOUS.—Stamens and corolla inserted on the calyx. Ovary inferior. *Campanulaceæ, Asteroideæ, Lonicerineæ, Coffeineæ (Rubiaceæ).*

HYPOGYNOUS.—Stamens and corolla inserted below the ovary.

ANISOGYNOUS.—Pistil composed of a less number of carpels than there are sepals.

Isostemonous.—Number of stamens equal to the divisions of the corolla, and alternating with them. *Asclepiadeæ, Convolvulaceæ, Asperifoliæ, Solaneæ.*

Anisostemonous.—Stamens partly abortive, four didynamous, or two. *Personatæ, Selagineæ, Verbenaceæ.*

ISOGYNOUS.—Pistil usually composed of a number of carpels equal to the sepals. *Primulaceæ, Ericoideæ, Diospyros.*

DIALYPETALOUS.—Petals free or absent.

HYPOGYNOUS.—Stamens and petals independent of the calyx, inserted below the ovary.

PERFECT FLOWERS, having petals, in most of the genera of each class.

Calyx usually persistent after flowering.

Polystemonous.—Stamens usually indefinite in number. *Guttiferæ, Malvaceæ.*

Oligostemonous.—Stamens usually definite in number. *Crotonineæ, Polygaleæ, Geraniaceæ, Terebinthaceæ, Hesperideæ, Æsculineæ, Celastrineæ, Violaceæ.*

Calyx falling, during or after flowering.

Albumen absent, or very thin. *Cruciferæ.*

Albumen thick, fleshy or horny. *Papaveraceæ, Berberideæ, Magnoliaceæ, Ranunculaceæ.*

Albumen double, the outer farinaceous. *Nymphæaceæ.*

IMPERFECT FLOWERS.—Corolla always absent. *Piperaceæ, Urticeæ, Polygoneæ.*

PERIGYNOUS.—Stamens and petals inserted on the calyx.

Cyclospermous.—Embryo bent around a farinaceous albumen. *Caryophyllaceæ, Cactaceæ.*

Perispermous.—Embryo straight, in the axis of a fleshy or horny albumen. *Crassulaceæ, Saxifrageæ, Passifloreæ, Hamamelideæ, Umbelliferæ, Santalaceæ, Asarineæ.*

Aperispermous.—Albumen wanting or scanty. *Cucurbitaceæ, Œnotheræ, Daphnaceæ, Proteoceæ, Rhamnaceæ, Myrtaceæ, Rosaceæ, Leguminosæ, Amentaceæ.*

GYMNOSPERMS.—Ovules naked, that is to say, not contained in a closed ovary and surmounted by a stigma, but directly receiving the influence of the pollen. *Coniferæ, Cycadeæ.*

SUCCESSION OF FAMILIES.

ACCORDING TO THE CLASSIFICATION OF A. DE JUSSIEU.

CRYPTOGAMS or ACOTYLEDONS.

CELLULAR.

ANGIOSPORES.

(Spores enclosed in the mother-cell, which persists, under the name of *theca*). *Algæ, Characeæ, Fungi, Lichens.*

GYMNOSPORES.

(Spores become, by the absorption of the mother-cell, free in a common cavity). *Hepaticæ, Mosses.*

VASCULAR.

Lycopodiaceæ, Isoetæ, Equisetaceæ, Ferns, Salviniaceæ, Marsileaceæ.

MONOCOTYLEDONOUS PHANEROGAMS.

AQUATIC AND EXALBUMINOUS.

Naiadeæ, Potameæ, Zosteraceæ, Juncagineæ, Alismaceæ, Butomaceæ, Hydrocharideæ.

ALBUMINOUS.

SPADICIFLORAL. (Flowers in a spadix). *Lemnaceæ, Aroideæ, Typhaceæ, Palmaceæ.*

GLUMACEOUS. (Perianth absent, replaced by bracts). *Gramineæ, Cyperaceæ.*

ENANTIOBLASTEÆ. (Radicle antipodal to the hilum). *Eriocauloneæ, Commelyneæ.*

HOMOBLASTEÆ. (Radicle facing the hilum),

Superovarian. (Ovary free). *Junceæ, Pontederiaceæ, Aphyllantheæ, Liliaceæ, Asparagineæ, Melanthaceæ.*

Inferovarian. (Ovary adherent). *Dioscoreæ, Irideæ, Amaryllideæ, Hypoxideæ, Hemadoraceæ, Bromeliaceæ, Musaceæ, Canneæ.*

ASCHIDOBLASTEÆ. (Embryo undivided). *Orchideæ.*

DICOTYLEDONOUS PHANEROGAMS.

GYMNOSPERMS.

Cycadeæ, Abietineæ, Cupressineæ, Taxineæ, Gnetaceæ.

ANGIOSPERMS.

DICLINOUS.

PENEANTHEÆ. (Incomplete flowers, that is to say, with one or no perianth). *Casuarineæ, Myriceæ, Betulineæ, Cupuliferæ, Juglandeæ, Salicineæ, Balsamiferæ, Plataneæ, Moreæ, Celtideæ, Ulmaceæ, Urticeæ, Cannabineæ, Cynocrambeæ, Ceratophylleæ, Saurureæ.*

PLOUSIANTHEÆ. (Flowers with a double perianth).

Ovules 1-2, axile. *Euphorbiaceæ, Empetraceæ.*

Ovules numerous, parietal. *Datisceæ, Begoniaceæ, Cucurbitaceæ, Nepentheæ.*

RHIZANTHEÆ. (Parasites on the roots of other plants). *Balanophoreæ, Cytineæ.*

☿ APETALOUS.

(Flowers with stamens and pistils, and 1 perianth.)

GYNANDROUS. (Stamens united to the pistil.) *Asarineæ.*

PERIGYNOUS. *Santalaceæ, Loranthaceæ, Proteaceæ, Eleagneæ, Thymeleæ, Laurineæ.*

CYCLOSPERMEÆ. (Embryo annular). *Polygoneæ, Phytolacceæ, Nyctagineæ, Amarantaceæ, Chenopodiaceæ, Tetragoniaceæ.*

☿ POLYPETALOUS.

CYCLOSPERMEÆ. *Portulaceæ, Paronychieæ, Sileneæ, Alsineæ, Elatineæ.*

Hypogynous.

PLEUROSPERMEÆ. (Placentation parietal). *Frankeniaceæ, Tamariscineæ, Violarieæ, Cistineæ, Resedaceæ, Capparideæ, Cruciferæ, Fumariaceæ, Papaveraceæ, Sarracenieæ, Droseraceæ, Parnassieæ.*

CHLAMYDOBLASTEÆ. (Embryo enveloped in the embryonic sac, which is thickened into an internal albumen). *Nymphæaceæ, Nelumbieæ.*

AXOSPERMOUS. (Placentation axile). *Dilleniaceæ, Magnoliaceæ, Anonaceæ, Schizandreæ, Berberideæ, Lardizabaleæ, Menispermaceæ, Coriarieæ, Xanthoxyleæ, Diosmeæ, Rutaceæ, Zygophylleæ, Oxalideæ, Lineæ, Limnantheæ, Tropæoleæ, Ranunculaceæ, Balsamineæ, Geraniaceæ, Malvaceæ, Sterculiaceæ, Byttneriaceæ, Tiliaceæ, Camelliaceæ, Hypericineæ, Polygalæ, Sapindaceæ, Hippocastaneæ, Acerineæ, Malpighiaceæ, Meliaceæ, Hesperideæ.*

Perigynous.

EXALBUMINOUS AXOSPERMS. (Seeds axile, without albumen). *Terebinthaceæ, Papilionaceæ, Cæsalpineæ, Mimosæ, Amydaleæ, Spireaceæ, Dryadeæ, Sanguisorbæ, Rosaceæ, Pomaceæ, Calycantheæ, Granateæ, Myrtaceæ, Lythrarieæ, Melastomaceæ, Hippurideæ, Callitrichineæ, Trapeæ, Halorageæ, Onagrarieæ.*

PLEUROSPERMEÆ. *Loaseæ, Passifloræ, Ribesieæ, Cactaceæ, Mesembryantheæ.*

ALBUMINOUS AXOSPERMS. (Seeds axile, provided with albumen). *Crassulaceæ, Francoaceæ, Saxifrageæ, Hydrangeæ, Cunoniaceæ, Escalloniceæ, Brexiaceæ, Philadelphieæ, Hamamelideæ, Corneæ, Garryaceæ, Gunneraceæ, Araliaceæ, Umbelliferæ.*

Peri-hypogynous.

(Insertion perigynous or hypogynous, often ambiguous.)
Rhamneæ, Ampelideæ, Celastrineæ, Staphylcaceæ, Pittosporeæ.

☿ MONOPETALOUS.

SEMI-MONOPETALOUS.

(Petals free in some).

Ericaceæ, Rhodoraceæ, Vaccineæ, Diapensieæ, Epacrideæ, Pyrolaceæ, Monotropeæ, Styraceæ, Jasmineæ, Oleineæ, Ilicineæ, Ebenaceæ, Myrsineæ, Primulaceæ, Plumbagineæ, Plantagineæ.

EU-MONOPETALOUS.

(Corolla always clearly monopetalous and staminiferous).

Hypogynous.

ANISANDROUS. (Stamens 4, dissimilar, or 2 by abortion). *Utriculariæ, Globulariæ, Selagineæ, Myoporineæ, Stilbineæ, Verbenaceæ, Labiatæ, Acanthaceæ, Sesameæ, Bignoniaceæ, Cyrtandraceæ, Gesneriaceæ, Orobancheæ, Personatæ.*

ISANDROUS. (Stamens similar, in number equal to the divisions of the corolla). *Solaneæ, Cestrineæ, Nolaneæ, Borragineæ, Cordiaceæ, Hydrophylleæ, Hydroleaceæ, Polemoniaceæ, Dichondreæ, Cuscutæ, Convolvulaceæ, Gentianeæ, Asclepiadeæ, Apocyneæ, Desfontaineæ, Loganiaceæ.*

Perigynous.

Rubiaceæ, Caprifoliaceæ, Valerianeæ, Dipsaceæ, Campanulaceæ, Lobeliaceæ, Goodeniaceæ Brunoniaceæ, Stylidieæ, Compositæ.

ATLAS OF BOTANY.

SECOND PART.

ILLUSTRATIONS AND DESCRIPTIONS OF FAMILIES.

I. RANUNCULACEÆ.

(Ranunculaceæ, *Jussieu.*—Pæoniaceæ and Ranunculaceæ, *Bartling.*)

Calyx *polysepalous.* Corolla *polypetalous, hypogynous, regular or irregular, imbricate, sometimes* 0. Stamens *numerous.* Anthers *adnate.* Carpels *usually distinct.* Fruit *an achene or follicle, rarely a capsule or berry.* Seeds *erect, pendulous, or horizontal.* Embryo *dicotyledonous, minute, at the base of a usually · horny albumen.*

Herbs, rarely shrubs (*Pæonia Moutan*), or woody climbers (*Clematis*). Leaves radical or alternate, rarely opposite (*Clematis*), simple or compound, petiole often dilated, amplexicaul, or rarely furnished with stipuliform appendages (*Thalictrum, Ranunculus*). Flowers usually terminal, solitary, racemed, or panicled, usually regular, sometimes irregular (*Delphinium, Aconitum*), ☿, or rarely diœcious by suppression (*Clematis*). Sepals 3—∞, usually 5, free, rarely persistent (*Helleborus, Pæonia*), often petaloid, usually imbricate, rarely valvate (*Clematis*). Petals equal and alternate with the sepals, or more (*Ficaria, Oxygraphis,* &c.), hypogynous, free, clawed, imbricate, deciduous, equal or unequal, various in form, often 0. Stamens usually many, many-seriate, hypogynous; *filaments* filiform, free; *anthers* terminal, 2-celled, cells adnate, extrorse or lateral. Carpels few or many, rarely solitary (*Actæa*), free, rarely coherent (*Nigella*); *style* simple; *stigma* on the inner surface of the top of the style, or sessile; *ovules* anatropous, sometimes solitary, ascending with a ventral raphe, or pendulous with a dorsal raphe; sometimes numerous, attached to the ventral suture, 2-seriate and horizontal. Fruit of pointed or feathery achenes, or of follicles, which are rarely united into a capsule (*Nigella*), or of 1-few-seeded berries (*Actæa*). Seeds erect, pendulous, or horizontal; *testa* coriaceous in the achenes, and raphe little prominent; *testa* crustaceous and fleshy-fungoid in the follicles, with the raphe very prominent and almost carunculate. Embryo minute at the base of a horny, or rarely fleshy albumen (*Pæonia*).

We have illustrated the family of *Ranunculaceæ* in greater detail than the others, in commemoration of the fine work of A. L. de Jussieu, read before the Académie des Sciences, in 1773, which may be regarded as the date of the birth of the *Natural System.* B.·de Jussieu, the uncle of Antoine Laurent, had long studied the relationships existing between the different large groups of the Vegetable Kingdom ; but he did not attempt to estimate the relative value of the characters they presented, contenting himself with arranging, in accordance with his views the flower-beds of the garden of the Trianon, which he formed for the instruction of Louis XV. Thirty years later, having become old and infirm, his nephew, Antoine Laurent, was charged with the completion of this garden, and made a special study of *Ranunculaceæ*, which at once revealed to him the scientific basis of his uncle's classification. He could not have chosen a more instructive family in a philosophical point of view, because of the numerous anomalies it presents in the form and structure of the calyx and corolla of such genera as *Columbine*, *Aconite*, *Larkspur*, *Hellebore*, *Ranunculus*, *Anemone*, *Clematis*, *Actæa*, *Thalictrum*, &c., which, nevertheless, all agree in the separation of their sepals and petals, the insertion of their numerous stamens, the direction of their anthers, the form of their ovaries, and especially in the structure of their seed; all of which, no doubt, led to Antoine Laurent's discovery of the grand principle of the *relative value of characters*. From these he was at once able to reason out and to formulate the pregnant axiom which his uncle had foreshadowed : that it is not by the number of characters, but by their value and importance, that the problem of the *Natural Method* can be solved. In his paper on *Ranunculaceæ* he enumerated and developed those views of the relative and subordinate importance of the organs of plants which other botanists, Linnæus included, had failed to perceive ; together with those principles of a natural classification which decided the Academy of Sciences to elect him a member. 'Antoine Laurent,' says his son, in the ' Dictionnaire Universelle d'Histoire Naturelle ' (art. TAXONOMIE), 'supplemented this discovery in the following year (1774) by a second paper, in which he extended his examination of a single family to all others. It hence became necessary to reconstruct the "Ecole Botanique " of the King's Garden, which was rapidly increasing under the powerful influence of Buffon ; and where the method of Tournefort, hitherto adopted, no longer kept pace with the progress and wants of science. Nor could the system of Linnæus, though it prevailed almost throughout the rest of Europe, be accepted in the Paris garden, which was controlled by Buffon under the directorship of Bernard de Jussieu. The latter was now old and nearly blind, and as he did not insist on his nephew's following his own arrangement of the Trianon garden, it would appear that he was not himself fully satisfied with it.' Antoine Laurent hereupon adopted the new classification which he had proposed in 1774 to the Academy, and thus became, as his son expressed it, both the lawgiver and administrator of the law— *legis simul lator et minister*.[1]

From this memorable epoch dates the commencement of Jussieu's preparations for his great work on the Families and Genera of the Vegetable Kingdom, upon which he worked unceasingly and single-handed for fifteen years, analysing all the genera, investigating the germination of their seeds, and finally embodying his materials in the 'Genera Plantarum,' which was published in 1789. In the Introduction, A. L. de Jussieu enunciates the lucid principles which guided him, and illustrates their application ; whilst in his co-ordination of the families and genera, he supplements by profoundly judicious notes the artificial character inherent in every linear series. He further indicates the manifold relationships that exist between the various groups of the Vegetable Kingdom ; and the very doubts which he expresses betray that fine instinct for affinities with which he was gifted.

Science has advanced since 1789; new types have been added to the family of *Ranunculaceæ*, without disturbing any of the characters assigned to it by Jussieu. The most recent work on the subject is the ' Genera Plantarum ' of the eminent botanists Bentham and Hooker fil. ; from which we shall borrow the description of all the known genera of *Ranunculaceæ*.

Tribe I. CLEMATIDEÆ, *D.C.*—Sepals valvate, petaloid. Petals 0, or narrow, flat, shorter than the sepals and staminoid. Carpels 1-ovuled ; ovule pendulous, raphe

---

[1] The precious manuscript, which is wholly in the author's handwriting, and which describes the arrangement of this garden, still exists. It is headed by an observation of its previous possessor, André Thouin, Professor of Horticulture in the Royal Garden, to the effect that this catalogue was the first drawn up in Paris in which both the binary nomenclature of Linnæus and the natural families established by Jussieu are adopted. Thouin adds that the laying out of the Botanic Garden was begun in the autumn of 1773, and was completed in the following spring, that is, during the vacation, so as not to interfere with the Botanical course.

dorsal. Achenes numerous, often plumose.—Stem herbaceous or woody, climbing. Leaves opposite. Flowers cymose, sometimes diœcious.

1. *Clematis.—Petals 0, or represented by the outer stamens becoming petaloid. Carpels numerous. Achenes capitate, sessile or sub-stipitate, tipped by the persistent naked or bearded or feathery style.—Stem woody, climbing, or sub-woody, or herbaceous. Leaves several- (rarely 1-) foliolate, petiole often twining, but not becoming a tendril. Flowers solitary or panicled, often polygamo-diœcious. *Nearly cosmopolitan.*

Section I. *Flammula.*—Involucre 0. Petals 0. Achenes with a feathery tail.
Section II. *Viticella.*—Involucre 0. Petals 0. Achenes with a short tail, not feathery.
Section III. *Cheiropsis.*—Involucre calyciform, of 2 connate bracts, situated under the flower. Petals 0. Achenes with a feathery tail.
Section IV. *Atragene.*—Involucre 0. Outer filaments dilated, and passing into petaloid staminodes. Achenes with feathery tails.

2. **Naravelia.**—Petals linear or clavate, quite distinct from the stamens. Carpels numerous. Achenes stipitate on a thick hollow receptacle, terminated by the persistent bearded style.—Stem woody, climbing. Leaves 2-foliolate, petiole cirrhiform. Flowers panicled. *Tropical Asia.*

Tribe II. ANEMONEÆ, *D.C.*—Sepals imbricate, usually petaloid, sometimes spurred (*Myosurus*). Corolla 0, or petals plane, claw not nectariferous (*Adonis*) or

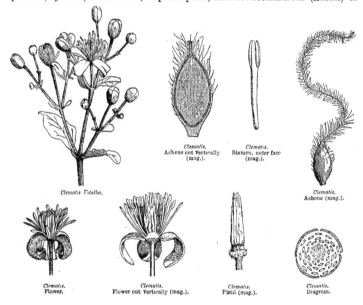

Clematis Vitalba.

Clematis.
Achene cut Vertically
(mag.).

Clematis.
Stamen, outer face
(mag.).

Clematis.
Achene (mag.).

Clematis.
Flower.

Clematis.
Flower cut Vertically (mag.).

Clematis.
Pistil (mag.).

Clematis.
Diagram.

nectariferous (*Callianthemum, Myosurus*).   Carpels 1-ovuled ; ovule pendulous, raphe dorsal.   Achenes dry, rarely fleshy (*Knowltonia*)₁—Stem herbaceous, erect.   Leaves all radical, or the cauline alternate.   Flowers often involucrate.

3. **Thalictrum.**—Involucre 0.   Sepals 4–5, petaloid.   Petals 0.   Carpels more or less numerous, on a narrow receptacle ; style short, deciduous, or 0.   Achenes often stipitate, ribbed, nerved, or winged.—Herbs with perennial rootstock.   Leaves 2–3-pinnatisect. Flowers often polygamous, panicled or racemose, dull green, yellow, purple, or whitish, usually small.   *Europe, Asia, America.*

4. **Anemone.**—Involucre usually distant from the flower, of three whorled leaflets.   Sepals petaloid, 4–20.   Petals 0, or represented by the outer stamens changed into stipitate glands. Carpels numerous.   Achenes in a head, terminated by the persistent naked or bearded style. —Scapigerous herbs with perennial rootstock.   Leaves radical, lobed, or dissected.   Flowers rarely yellowish.   Stamens shorter than the sepals.   *Extra-tropical regions.*

Section I. *Pulsatilla.*—Outer stamens without anthers, gland-like.   Achenes terminated by a bearded feathery tail.
Section II. *Anemone.*—Stamens all fertile.   Achenes terminated by a short point, not feathery.
Section III. *Hepatica.*—Involucre close to the flower, and simulating a calyx.   Achenes terminated by a short style, not bearded.

5. **Knowltonia.**—Sepals 5, herbaceous, deciduous.   Petals 5–16, without a nectariferous hollow.   Carpels numerous.   Achenes capitate, fleshy or pulpy ; style deciduous.—Herbs with perennial rootstock, and the habit of *Umbelliferæ.*   Radical leaves stiff, ternately decom-

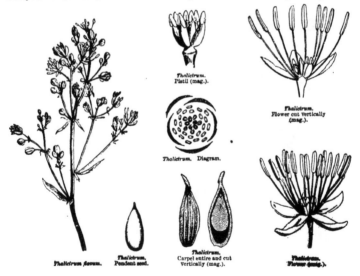

Thalictrum.
Pistil (mag.).

Thalictrum.
Flower cut vertically
(mag.).

Thalictrum.   Diagram.

Thalictrum flavum.

Thalictrum.
Pendent seed.

Thalictrum.
Carpel entire and cut
vertically (mag.).

Thalictrum.
Flower (mag.).

pound, cauline small or bract-like, or 0. Flowers greenish or yellowish, peduncles often irregularly umbellate. *Sou·h Africa.*

6. ***Adonis.**—Sepals 5–8, coloured, deciduous. Petals 5–16, often spotted at base, but with no nectariferous hollow. Carpels numerous. Achenes in a head or spike; style short, persistent, straight or hooked.—Annual or perennial herbs. Leaves pinnatipartite, multifid, segments narrow. Flowers solitary, yellow or red. *Europe, Asia.*

7. ***Callianthemum.**—Sepals 5, herbaceous, deciduous. Petals 5–15, with a basal nectariferous pit. Carpels numerous. Achenes capitate; style short, persistent. Alpine, low

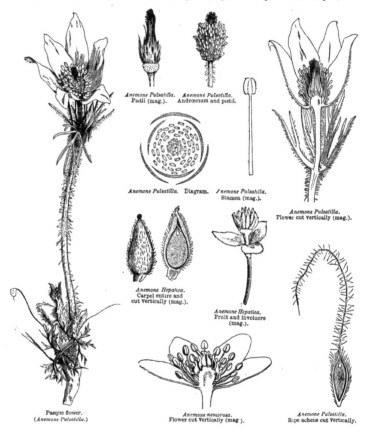

Anemone Pulsatilla.
Pistil (mag.).

Anemone Pulsatilla.
Andrœcium and pistil.

Anemone Pulsatilla. Diagram.

Anemone Pulsatilla.
Stamen (mag.).

Anemone Pulsatilla.
Flower cut vertically (mag.).

Anemone Hepatica.
Carpel entire and cut vertically (mag.).

Anemone Hepatica.
Fruit and involucre (mag.).

Pasque flower.
(*Anemone Pulsatilla.*)

Anemone nemorosa.
Flower cut vertically (mag.).

Anemone Pulsatilla.
Ripe achene cut vertically.

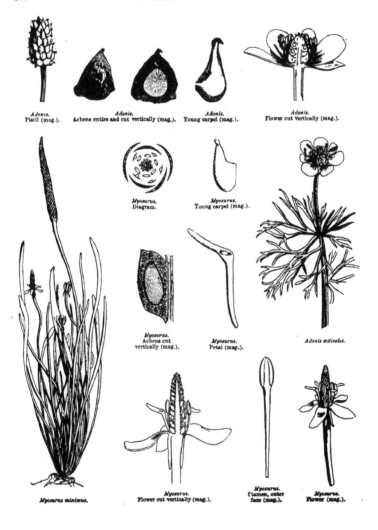

Adonis.
Pistil (mag.).

Adonis.
Achene entire and cut vertically (mag.).

Adonis,
Young carpel (mag.).

Adonis.
Flower cut vertically (mag.).

Myosurus.
Diagram.

Myosurus.
Young carpel (mag.).

Myosurus.
Achene cut
vertically (mag.).

Myosurus.
Petal (mag.).

Adonis æstivalis.

Myosurus minimus.

Myosurus.
Flower cut vertically (mag.).

Myosurus.
Stamen, outer
face (mag.).

Myosurus.
Flower (mag.).

herbs; rootstock perennial. Radical leaves decompound; cauline few or 0. Flowers white. *Europe, Asia.*

8. **Myosurus.**—Sepals 5 (rarely 6–7), prolonged into a spur below their insertion. Petals as many as the sepals, narrow, spathulate; claw filiform, tubular, nectariferous at the top. Carpels numerous. Achenes minute, in a long spike, laterally attached; style short, persistent.—Annual, small herbs. Leaves entire, all radical. Scapes 1-flowered, naked. Flowers minute. *Europe, Asia, Africa, Australia, New Zealand.*

Tribe III. RANUNCULEÆ, *D.C.*—Sepals imbricate. Petals with a nectariferous claw, rarely 0 (*Trautvetteria*). Carpels 1-ovuled; ovule ascending, raphe ventral. Achenes dry.—Herbs. Leaves radical or alternate.

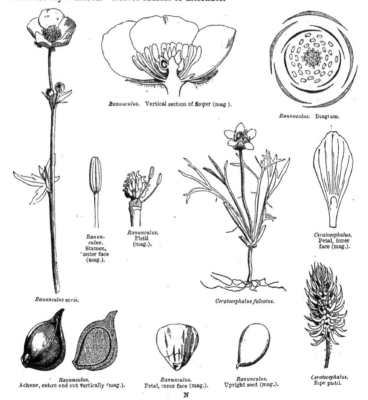

Ranunculus. Vertical section of flower (mag.).

Ranunculus. Diagram.

Ranunculus. Stamen, outer face (mag.).

Ranunculus. Pistil (mag.).

Ceratocephalus. Petal, inner face (mag.).

Ranunculus acris.

Ceratocephalus falcatus.

Ranunculus. Achene, entire and cut vertically (mag.).

Ranunculus. Petal, inner face (mag.).

Ranunculus. Upright seed (mag.).

Ceratocephalus. Ripe pistil.

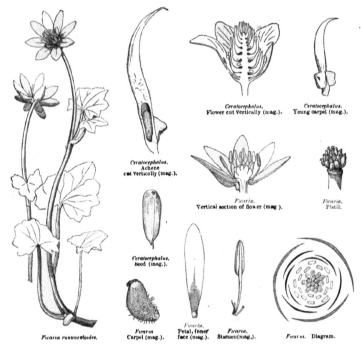

Ceratocephalus.
Flower cut Vertically (mag.).

Ceratocephalus.
Young carpel (mag.).

Ceratocephalus.
Achene
cut Vertically (mag.).

Ficaria.
Vertical section of flower (mag.).

Ficaria.
Pistil.

Ceratocephalus.
Seed (mag.).

Ficaria ranunculoides.

Ficaria.
Carpel (mag.).

Ficaria.
Petal, inner
face (mag.).

Ficaria.
Stamen (mag.).

Ficaria. Diagram.

9. **Trautvetteria.**—Sepals 3–5, concave. Petals 0. Carpels numerous. Achenes capitate, membranous; style very short. Embryo rather large.—Herbs with perennial rootstock. Leaves palmatilobed, cauline few. Flowers in a corymbose panicle. *North America and Japan.*

10. *****Ranunculus.**—Sepals 3–5, caducous. Petals as many, or more numerous, with a basal nectariferous pit or scale. Carpels numerous. Achenes in a head or spike, beaked by the short style.—Annual, or oftener perennial herbs. Leaves entire or cut. Flowers white, yellow or red, solitary or panicled. *Almost cosmopolitan.*—The aquatic species have been made into a separate genus (*Batrachium*) by several modern botanists, on account of their transversely wrinkled achenes, and habitat. *Ficaria* has been separated, from having three sepals, 6–9 petals, and obtuse carpels ; and *Ceratocephalus*, because the base of the carpels presents two external gibbosities, and internally two empty cells, and the carpels are further produced into a horn five to six times as long as the seed.

11. **Hamadryas.**—Flowers diœcious by suppression. Sepals 5–6, caducous or subpersistent. Petals 10–12, with a basal scale. Carpels numerous. Achenes capitate, tipped by the short

style.—Low herbs, with perennial rootstock, only differing from *Ranunculus* in the diœcious flowers. *Antarctic America.*

12. **Oxygraphis.**—Sepals 5, persistent. Petals 10–15, with a basal nectariferous pit. Carpels numerous. Achenes capitate, beaked by the persistent style.—Low herbs, rootstock perennial. Leaves radical, entire. Scapes naked. Flowers solitary, golden-yellow. *Mountains of extra-tropical Asia.*

Tribe IV. HELLEBOREÆ, *D.C.*—Flowers regular or irregular (*Aconitum, Delphinium*). Sepals imbricate, petaloid. Petals small, or irregular and nectariferous, or 0 (*Caltha, Hydrastis*). Carpels several-ovuled, dehiscent when ripe, rarely berry-like (*Actæa, Hydrastis*), follicular, free, rarely connate into a several-celled capsule (*Nigella*). Herbs. Leaves all radical, or the cauline alternate.

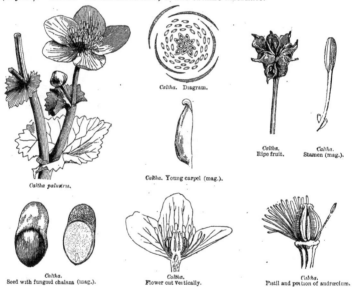

Caltha. Diagram.

Caltha. Ripe fruit.

Caltha. Stamen (mag.).

Caltha paluitris.

Caltha. Young carpel (mag.).

Caltha. Seed with fungoid chalaza (mag.).

Caltha. Flower cut vertically.

Caltha. Pistil and portion of androecium.

13. ***Caltha.**—Sepals 5–∞, equal, coloured, deciduous. Petals 0. Carpels few or many, sessile; ovules many, 2-seriate, follicular when ripe. Seeds obovoid; testa crustaceous, smooth, raphe prominent.—Glabrous perennial herbs, tufted, or with a perennial rootstock, Leaves radical, palminerved, entire or crenulate, cordate or auricled, cauline few or 0. Flowers yellow or white, one or few. Stamens and carpels numerous or few. *Europe, Asia, America, Australia, New Zealand.*

14. **Calathodes.**—Sepals 5, regular, coloured, deciduous. Petals 0. Carpels numerous, sessile, distinct; ovules 8–10, 2-seriate near the base of the suture.—A perennial erect herb,

habit of *Trollius*. Leaves cauline, palmatilobed or dissected. Flowers yellow, solitary. *Eastern Himalaya.*

15. **Glaucidium.**—Sepals 4, regular, deciduous. Petals 0. Carpels 1 or few, sessile, slightly coherent at the base; ovules numerous, many-seriate along the ventral suture. Follicles square, with dorsal dehiscence. Seeds numerous, oblong, depressed; testa finely crustaceous; raphe very prominent, almost winged.—A perennial upright herb. Leaves palmatilobed. Flowers solitary, ample, lilac or pink. *Japan.*

16. **Hydrastis.**—Sepals 3, regular, petaloid, caducous. Petals 0. Carpels numerous, sessile, distinct, 2-ovuled, fleshy when ripe, and forming a head, as in *Rubus*.—A perennial erect herb. Leaves palmatilobed, or dissected. Flowers solitary, small, white. Stamens a little longer than the sepals. *North America.*

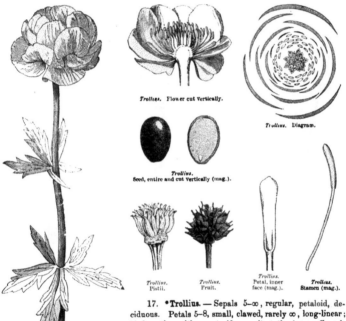

Trollius. Flower cut vertically.

Trollius. Diagram.

Trollius. Seed, entire and cut vertically (mag.).

Trollius. Pistil.

Trollius. Fruit.

Trollius. Petal, inner face (mag.).

Trollius. Stamen (mag.).

Trollius europæus.

17. ***Trollius.**—Sepals 5-∞, regular, petaloid, deciduous. Petals 5-8, small, clawed, rarely ∞, long-linear; blade entire, with a nectariferous pit at the base. Carpels many, free, sessile, many-ovuled, follicular when ripe. Seeds oblong, usually angular; testa crustaceous, rather smooth.—Erect herbs; rootstock perennial. Leaves palmati-lobed or -sect. Flowers solitary or few, large, yellow or lilac. *Europe, Asia, North America.*

18. *Helleborus.—Sepals 5, regular, petaloid or sub-herbaceous, usually persistent. Petals small, clawed, nectariform ; blade furnished at the base with an inner lip, or a scale. Carpels many, sessile or subsessile, distinct or coherent at the base, many-ovuled, dehiscing inwards at the top when ripe. Seeds 2-seriate ; testa crustaceous, shining.—Erect herbs ; rootstock perennial. Leaves palmati-sect or -lobed, or digitate, cauline few, the upper sometimes involucriform or all bracteiform. Flowers large, white, greenish, yellowish or livid, solitary or panicled. Sepals large. Follicles coriaceous or membranous. *Europe and Western Asia.*

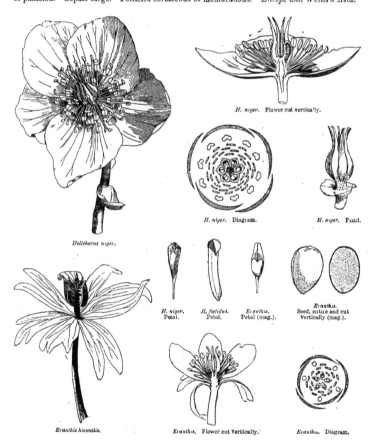

Helleborus niger.

H. niger. Flower cut vertically.

H. niger. Diagram.

H. niger. Pistil.

Eranthis hiemalis.

H. niger. Petal.

H. fœtidus. Petal.

Eranthis. Petal (mag.).

Eranthis. Seed, entire and cut vertically (mag.).

Eranthis. Flower cut vertically.

Eranthis. Diagram.

19. *Eranthis.—Sepals 5–8, regular, petaloid, deciduous. Petals small, nectariform, clawed; blade furnished at the base with an inner scale-like lip. Carpels many, distinct, stipitate, many ovuled, follicular when ripe. Seeds ovoid or sub-globose; testa crustaceous, smooth.—Low herbs; rootstock perennial, tuberous. Leaves radical, palmatisect, cauline solitary, amplexicaul beneath the flower or peduncle, segments simulating the whorled leaflets of an involucre. Flower solitary, yellow; sepals narrow. *Europe, and Mountains of Asia.*

20. Coptis.—Sepals 5–6, regular, petaloid, deciduous. Petals 5–6, small, cucullate or linear. Carpels many, stipitate, distinct, many-ovuled, follicular when ripe. Seeds with crustaceous shining testa.—Low herbs; rootstock perennial. Leaves radical, ternately dissected. Scapes naked, 1–3-flowered. Flowers white. *Europe, Asia, North America.*

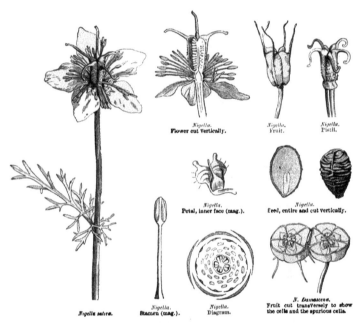

Nigella.
Flower cut vertically.

Nigella.
Fruit.

Nigella.
Pistil.

Nigella.
Petal, inner face (mag.).

Nigella.
Seed, entire and cut vertically.

Nigella sativa.

Nigella.
Stamen (mag.).

Nigella.
Diagram.

N. Damascena.
Fruit cut transversely to show the cells and the spurious cells.

21. *Isopyrum.—Sepals 5–6, regular, petaloid, deciduous. Petals 6, very short, nectariform or 0. Carpels 2–20, sessile, distinct, 3–∞ -ovuled, follicular when ripe.—Slender low herbs; rootstock perennial. Leaves ternately decompound; cauline alternate or subopposite, or 0. Flowers solitary or loosely panicled, white. Petals variable in form. Stamens sometimes reduced to about 10. *Europe, Asia, North America.*

22. *Nigella.—Sepals 5, regular, petaloid, deciduous. Petals 5, clawed; blade small, 2-fid. Carpels 3–10, sessile, more or less coherent, many-ovuled, opening when ripe at the top

of the ventral suture. Seeds angular; testa crustaceous or sub-fleshy, often granular.—Erect glabrous herbs. Cauline leaves pinnatisect, segments very narrow. Flowers white, blue, or yellowish, sometimes involucrate with one floral leaf. *Europe, Western Asia.*

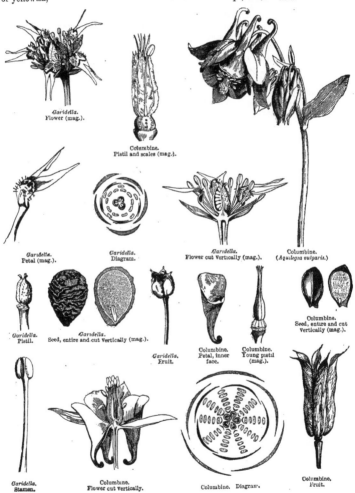

Garidella.
Flower (mag.).

Columbine.
Pistil and scales (mag.).

Garidella.
Petal (mag.).

Garidella.
Diagram.

Garidella.
Flower cut vertically (mag.).

Columbine.
(*Aquilegia vulgaris.*)

Garidella.
Pistil.

Garidella.
Seed, entire and cut vertically (mag.).

Garidella.
Fruit.

Columbine.
Petal, inner face.

Columbine.
Young pistil (mag.).

Columbine.
Seed, entire and cut vertically (mag.).

Garidella.
Stamen.

Columbine.
Flower cut vertically.

Columbine. Diagram.

Columbine.
Fruit.

22b. **Garidella.**—Sepals 5, petaloid, caducous. Petals 5, 2-labiate. Follicles 2–3, sessile, coherent at the base, and opening at the top; style very short. Seeds 2-seriate.—Slender herbs. Leaves finely multifid. Flowers small, white. *Mediterranean Region.*

23. ***Aquilegia.**—Sepals 5, regular, petaloid, deciduous. Petals 5, like a cornucopia or hood, attached by the margin of the limb, and nectariferous at the base of the cavity. Lower stamens reduced to scale-like staminodes. Carpels 5, sessile, distinct, many-ovuled, follicular when ripe. Seeds with crustaceous, smooth or granular testa.—Erect herbs; rootstock perennial. Leaves ternately decompound. Flowers conspicuous, blue, yellow, scarlet, or parti-coloured, solitary or panicled. *Europe, Asia, North America.*

24. ***Delphinium.**—Sepals 5, petaloid, unequal, subcoherent at the base, the posterior turned up in a horn or spur. Petals 2 or 4, small, all sometimes united, the two upper pro-longed into a pointed spur included in that of the calyx; the two lateral not spurred, or 0. Carpels 1–5, sessile, distinct, many-ovuled, follicular when ripe. Seeds subfleshy.—Annual herbs, or with perennial rootstock, erect, branched. Leaves palmatilobed or dissected.

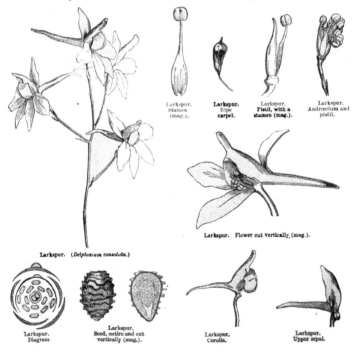

Larkspur. Stamen (mag.).

Larkspur. Ripe carpel.

Larkspur. Pistil, with a stamen (mag.).

Larkspur. Androecium and pistil.

Larkspur. Flower cut vertically (mag.).

Larkspur. (*Delphinium consolida.*)

Larkspur. Diagram.

Larkspur. Seed, entire and cut vertically (mag.).

Larkspur. Corolla.

Larkspur. Upper sepal.

Flowers rather large, in a loose raceme or panicle, blue, purplish, pink or white, rarely yellow. Filaments sometimes dilated at the base. *Europe, Asia, North America.*

25. *Aconitum.—Sepals 5, petaloid, unequal; posterior large, helmet-shaped, covering the corolla; 2 lateral larger than the 2 anterior. Petals 2–8, small, very unequal, the two upper with long claws, cucullate at the top, hidden under the helmet; the lower minute, filiform, often 0. Carpels 3–5, sessile, distinct, many-ovuled, follicular when ripe. Seeds with spongy testa, deeply wrinkled.—Erect herbs; rootstock perennial. Leaves palmati-lobed or -sect. Flowers racemed or panicled, blue, purplish, yellow or white; pedicels bracteolate. Filaments usually dilated at the base. *Europe, Asia.*

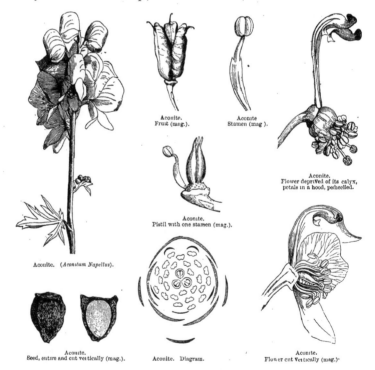

Aconite.
Fruit (mag.).

Aconite.
Stamen (mag.).

Aconite.
Flower deprived of its calyx,
petals in a hood, pedicelled.

Aconite.
Pistil with one stamen (mag.).

Aconite. (*Aconitum Napellus*).

Aconite.
Seed, entire and cut vertically (mag.).

Aconite. Diagram.

Aconite.
Flower cut vertically (mag.).

26. *Actæa.—Sepals 3–5, subequal, petaloid, deciduous. Petals 4–10, small, clawed, spathulate, flat. Carpel solitary, many-ovuled, berried when ripe. Seeds depressed; testa crustaceous, smooth.—Herbs; rootstock perennial, fusiform; stem erect. Leaves ternately

decompound. Flowers small, in short racemes that lengthen after flowering. Stamens longer than the sepals. Stigma sessile, dilated. *Europe, Asia, North America.*

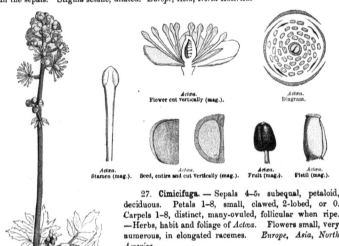

Actœa.
Flower cut vertically (mag.).

Actœa.
Diagram.

Actœa.
Stamen (mag.).

Actœa.
Seed, entire and cut vertically (mag.).

Actœa.
Fruit (mag.).

Actœa.
Pistil (mag.).

Actœa spicata.

27. **Cimicifuga.** — Sepals 4–5, subequal, petaloid, deciduous. Petals 1–8, small, clawed, 2-lobed, or 0. Carpels 1–8, distinct, many-ovuled, follicular when ripe. —Herbs, habit and foliage of *Actœa.* Flowers small, very numerous, in elongated racemes. *Europe, Asia, North America.*

28. ***Botrophis.**—Sepals 4–5, petaloid, equal. Petals 0. Outer stamens dilated, terminated by an imperfect anther. Carpel solitary, 1-celled; ovules 2-seriate. Follicle substipitate. — Herbs, leaves 2–3-sect, segments incised, toothed. Flowers racemose, white. *North America.*

29. **Xanthorhiza.**—Sepals 5, subequal, petaloid, deciduous. Petals 5, small, clawed, gland-like, dilated at the top. Stamens 5, alternate with the petals, or 10. Carpels 5–10, distinct, sessile, 2-ovuled on the middle of the inner suture, opening in follicles when ripe, and one-seeded by suppression. Seed pendulous.—Shrubs or under-shrubs, dwarf; stem yellow within. Leaves pinnatisect, proceeding in early spring from a scaly bud. Racemes compound, pendulous. Flowers small, blackish-purple, often polygamous. Stamens short. *North America.*

Tribe V. PÆONIEÆ, *D.C.*

30. ***Pæonia.**—Sepals 5, imbricate, herbaceous, persistent. Petals 5–10, conspicuous, broad, without a nectariferous pit. Carpels 2–5, many-ovuled, girt at the base with a fleshy disk, which is often spread over the base of the calyx, or forms an irregular cup more or less enveloping the ovaries. Fruit of coriaceous follicles. Seeds large, albumen fleshy.—Herbs with perennial fusiform rootstock, or with branching, more or less woody stems. Leaves

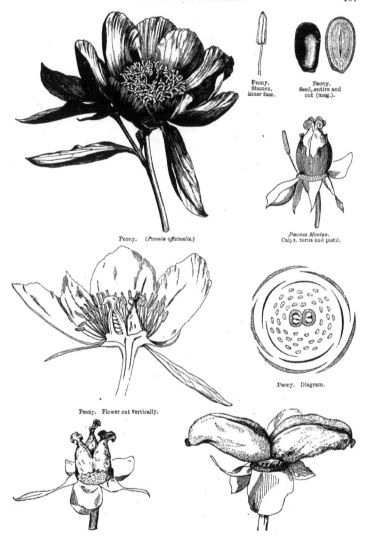

Peony. Stamen, inner face.

Peony. Seed, entire and cut (mag.).

Peony. (*Pæonia officinalis.*)

Pæonia Moutan.
Calyx. torus and pistil.

Peony. Diagram.

Peony. Flower cut vertically.

ample, pinnatisect or decompound. Flowers conspicuous, purplish, white, or red. *Europe, Asia.*

Ranunculaceæ approach *Dilleniaceæ* in the distinct imbricate sepals, polypetalism, hypogyny, polyandry, adnate anthers, distinct carpels, anatropous ovules, capsular or follicular fruit, erect albuminous seed, minute basilar embryo, and terminal inflorescence. *Dilleniaceæ* only differ in habit, persistent sepals, and especially in having arillate seeds. *Magnoliaceæ* offer the same analogies and differences; and are also distinguished by their habit and their many-seriate petals. *Berberideæ* have, like *Ranunculaceæ*, distinct sepals and petals, often nectariferous, adnate anthers, one or more free carpels, and albuminous seed; but their flower is iso- or diplo-stemonous, their anthers open by valves, and their embryo is axile and not basilar. *Papaveraceæ* differ in their syncarpous pistil, 2-merous flower and milky juice. Similar relations exist with *Nymphæaceæ*, which further differ in their habit, 1-flowered scape, many-seriate petals, largely dilated filaments, rayed stigma, and arillate seeds. Finally, some affinity has been discovered with *Sarraceniæ*, which are distinguished by their peltate and petaloid stigma, their radical leaves with tubular petiole, undeveloped blade, and 1-flowered scape. *Ranunculaceæ* are universally distributed, but most inhabit temperate and cold regions of the northern hemisphere; as throughout Europe, from the sea-shore to the limit of perpetual snow. They are rarer in North America and temperate Asia. *Clematis* alone is tropical, and is distinguished from all the other genera by its sarmentose habit and opposite leaves. Few *Ranunculi* inhabit the high mountains of the equator. *Ranunculus, Caltha,* and *Clematis* occur nearly everywhere. *Adonis, Ceratocephalus, Eranthis, Helleborus, Garidella, Nigella, Pæonia,* &c., belong exclusively to the Old World; *Cyrtorhyncha, Hydrastis, Trautvetteria, Botrophis,* and *Xanthorhiza* are their New World representatives. *Knowltonia* inhabits South Africa, *Hamadryas* extra-tropical South America, and *Naravelia* tropical Asia; the other genera are dispersed over the northern hemisphere.

Most *Ranunculaceæ* are acrid, and more or less poisonous; but these properties are volatile, and driven off by cooking and drying; except in some cases, where they are alkaline, and consequently more fixed and powerful. Their roots, when perennial, contain, besides the acrid, a bitter extractive principle, contained in various proportions, with a volatile oil, which renders them drastic and emetic. Their seeds are acrid; some contain both a fixed and a volatile oil, and are aromatic. *Clematis erecta, Vitalba* and *Flammula,* are very acrid and vesicant. The juice of the leaves of *C. Vitalba* is used by beggars to produce superficial sores and thus excite pity. *C. cirrhosa* from the Mediterranean region, *C. crispa* from North America, and *C. mauritiana* from Madagascar, replace cantharides in those countries. The numerous *Ranunculi* are often popularly used as vesicants; the most acrid are *R. Thora,* an alpine plant, and *R. sceleratus,* named by the Romans *Sardonia,* because it excites convulsive sardonic laughter: slow cooking dissipates its poisonous properties, and renders it eatable as a potherb. So it is with *Clematis Flammula,* one of the most acrid species, the young shoots of which may be eaten without danger. *Ranunculus Ficaria,* a common plant in damp hedges and woods, is very acrid before flowering, but the mucilage and starch which are developed later render it eatable. *R. alpestris* is a vesicant and strong purgative; yet the Alpine hunters chew its leaves to keep off giddiness and to strengthen them.

Anemones are equally vesicant. *A. nemorosa* is used as such in some parts of Europe, and *A. helleborifolia* replaces cantharides among the Peruvians; as does *Knowltonia,* of South Africa. The Italians prepare a rubefacient water with *A. apennina,* which the ladies are said to use to heighten their complexion. *A. ranunculoides,* a common northern species, is so acrid that the Kamtschatkans poison their arrows with it. *A. Pulsatilla* is the richest in medical properties: though nearly inodorous, yet if bruised, it emits a vapour that violently irritates the mucous membrane of the eyes, nose, and back of the mouth, owing to the presence of a volatile acid, an alkali named *anemonine,* and a volatile oil. In a fresh state it is used in paralysis, especially of the retina, in rheumatism, and in obstinate cutaneous diseases.

Thalictrum flavum, 'rhubarbe des pauvres,' is administered in jaundice and intermittent fevers. *T. Cornuti* is regarded in North America as a powerful alexipharmic. *Delphinium Consolida,* Larkspur, is aperient, diuretic and vermifuge; the seeds of *D. Staphisagria* are drastic, emetic, and employed externally in a powder to destroy lice, and in skin diseases. The seeds of the *Nigellæ* are slightly acrid and aromatic; they are used in the South of Europe and in the East to flavour bread. *Coptis trifoliata* is a sub-arctic plant of both worlds, renowned for its stomachic properties; it yields a yellow colouring principle. The root of *C. Teeta* is much celebrated in India and China as a powerful stimulant of the digestive organs. *Hydrastis canadensis* yields both a dye and a tonic medicine.

Helleborus niger, fœtidus, viridis, and *orientalis* contain a bitter substance, united to a resinous principle, which is a drastic purgative, and poisonous in large doses. The Aconites are narcotic acrid herbs, containing an alkaloid called *aconitine,* combined with a peculiar acid, and resinous or volatile principles; the leaves and seeds of *Aconitum Napellus* and *A. paniculatum* are of use in small doses for exciting the glandular and lymphatic systems, but are very poisonous in large doses. *A. ferox,*[1] a native of Nepal, is reputed to be the most poisonous of all. *Actæa spicata* was formerly given internally for asthma and scrofula, and externally for skin complaints. *Cimicifuga serpentaria,* of a nauseous smell and bitter taste, is in North America reputed to be a specific against the bite of the rattlesnake. *C. fœtida,* a widely diffused plant of cool northern regions, was formerly used in dropsy as a purgative: its name is derived from its supposed property of driving away bugs. The root and wood of *Xanthorhiza apiifolia,* a North American undershrub, contain a bitter resin and yellow dye, and are renowned as tonics. *Pæonia officinalis* was formerly famous in sorcery; its fresh seeds were used as emetics in epilepsy; and in some countries, necklaces made of them are still used to ward off convulsions from children. The Siberian *P. anomala* has a bitter root without acridity, smelling of violets, which is very useful in intermittent fevers.

II. *DILLENIACEÆ.*

(DILLENEÆ, *Salisbury.*—DILLENIACEÆ, *D.C.*)

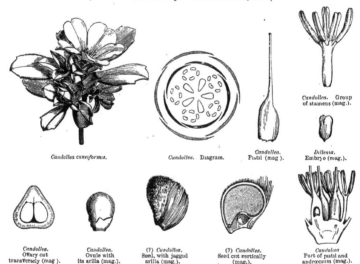

Candollea cuneiformis.

Candollea. Diagram.

Candollea. Pistil (mag.).

Dillenia. Embryo (mag.).

Candollea. Group of stamens (mag.).

Candollea. Ovary cut transversely (mag.).

Candollea. Ovule with its arilla (mag.).

(?) Candollea. Seed, with jagged arilla (mag.).

(?) Candollea. Seed cut vertically (mag.).

Candollea. Part of pistil and androecium (mag.).

SEPALS *usually 5, imbricate, persistent.* PETALS *usually 5, hypogynous, imbricate,* *deciduous.* STAMENS ∞, *hypogynous.* OVARIES *usually distinct, 1-celled, 1-several-*

---

[1] Referred to a var. of *A. Napellus* by Hooker fil. and Thomson (Fl. Ind. i. 57).

ovuled. OVULES *anatropous.* CARPELS *follicular, or berried.* SEEDS *erect or ascending,* *usually arillate, albuminous.* EMBRYO *dicotyledonous, minute, straight, axile.*

STEM arborescent or frutescent, sometimes climbing, rarely sub-woody or herbaceous (*Acrotrema, Hibbertia*). LEAVES alternate, very rarely opposite (*Hibbertia*), entire or toothed, rarely pinnatifid or 3-fid; *stipules* 0, or adnate to the petiole, and caducous. FLOWERS ⚥, or polygamous, rarely diœcious, solitary, or racemose, or panicled, usually yellow. SEPALS 5, rarely fewer (*Tetracera,* &c.), or ∞ (*Empedoclea*), imbricate, persistent. PETALS 5, or fewer (*Davilla,* &c.), alternate with the sepals, hypogynous, imbricate, deciduous. STAMENS ∞, rarely definite (*Hibbertia,* &c.), hypogynous, sometimes unilateral (*Hibbertia*), usually free, rarely mon–poly-adelphous (*Hibbertia, Candollea*); *anthers* introrse or extrorse, cells linear or sub-globose, adnate, often separate and overtopped by the connective, opening vertically or by an apical pore. OVARIES several, distinct or coherent, sometimes solitary (*Empedoclea, Doliocarpus, Delima,* &c.); *styles* terminal or sub-dorsal, divergent; *stigmas* simple or sub-capitate; *ovules* 2 or several, 2-seriate, ascending, raphe ventral, rarely solitary and erect (*Schumacheria*), anatropous or half-anatropous. CARPELS sometimes dehiscing by the ventral or dorsal suture or indehiscent, crustaceous or berried. SEEDS solitary or few, ovoid, arillate (except *Dillenia*), testa crustaceous, aril pulpy or membranous, cup-shaped, laciniate; *albumen* fleshy. EMBRYO minute, straight, basilar; *radicle* near the hilum, inferior.

PRINCIPAL GENERA.

| Candollea. | Dillenia. | Acrotrema. | Delima. |
| Hibbertia. | Wormia. | Tetracera. | Davilla. |

Dilleniaceæ are more or less closely allied to *Magnoliaceæ, Anonaceæ,* and *Ranunculaceæ.* (See these families.)

Dilleniaceæ are chiefly natives of the southern hemisphere. Tropical America and Asia possess about an equal number of species; they are rare in Africa. *Dillenia* is confined to tropical Asia; *Hibbertia* and *Candollea* are specially extra-tropical Australian. Hitherto none have been found in South Africa or temperate South America.

Dilleniaceæ are astringent and some are so used medicinally. The fruits of a very few are acidulous; others are reputed tonic stimulants. The leaves of *Davilla elliptica,* a Brazilian shrub, are vulnerary; those of *Curatella Cambaiba,* applied to ulcers, are detergent. *Tetracera Tigarea,* of Guiana and the Antilles, is a sudorific and diuretic; a decoction of it is given for syphilis; and a vinous infusion of its seeds is said to be efficacious in intermittent fevers, chlorosis, and scurvy. The astringent bark of *Dillenia serrata* is employed in Asia for ulcerated sores. The acid but uneatable fruit of *D. speciosa* serves to season dishes; and a syrup of the juice of the unripe fruit allays coughs, assists expectoration, and cures angina and aphthæ; its bruised bark is applied as a cataplasm in arthritis, and, like that of other species, is used for tanning.

III. *CALYCANTHEÆ, Lindl.*

COROLLA 0. STAMENS *numerous, inserted on the calyx.* CARPELS *numerous, free, inserted within the receptacular tube.* EMBRYO *dicotyledonous, exalbuminous.* STEM *woody.* LEAVES *opposite, exstipulate.*

SHRUBS with 4-angled stems. LEAVES opposite, petiolate, entire, exstipulate.

FLOWERS ☿, regular, appearing with the leaves or earlier, terminal or axillary, often sweet-scented or aromatic. CALYX coloured, segments numerous, many-seriate, im-

Calycanthus lævigatus.

Chimonanthus. Flower.

Chimonanthus.
Carpel, entire and cut vertically (mag.).

Chimonanthus. Flower cut vertically.

Chimonanthus. Diagram.

Chimonanthus.
Flower-bud (mag).

bricate, all alike, or the outer bracteiform and the inner petaloid, rising from a receptacular cup (calyx-tube of old botanists), short, urceolate. COROLLA 0. STAMENS numerous, inserted on a fleshy ring lining the calyx-throat, outer fertile, inner sterile, persistent or deciduous, free, or coherent at the base; *filaments* short, subulate or filiform; *anthers* extrorse, 2-celled, ovoid or oblong, adnate, dehiscence longitudinal. OVARIES numerous, inserted on the inner wall of the receptacular cup, free, 1-celled, 1-ovuled; *styles* as many as ovaries, terminal, simple, filiform or compressed, subulate; *stigmas* undivided, obtuse, terminal; *ovules* solitary, or rarely two, of which one is smaller, superimposed, ascending from the bottom of the cell, anatropous, raphe ventral. ACHENES numerous, included in the receptacular tube, accrescent, herbaceous, sub-fleshy, ovoid or oblong. SEED solitary, upright; *testa* membranous. EMBRYO exalbuminous; *cotyledons* foliaceous, convolute; *radicle* superior.

PRINCIPAL GENERA.

Chimonanthus.   Calycanthus.

The affinity of *Calycantheæ* with *Myrtaceæ* will be pointed out in the description of the latter. They also approach *Granateæ* in their coloured calyx, the number and insertion of the stamens, the carpels enclosed in the receptacular tube, exalbuminous embryo, convolute cotyledons, woody stem, generally opposite leaves and terminal flowers; they are distinguished by their apetalous flowers, extrorse anthers, free and one-ovuled ovaries, and dry fruit. They have also some affinity with *Monimieæ*, from their apetalous flowers, two-seriate calyx, numerous stamens inserted on the calyx-throat, numerous free ovaries inserted on the inner wall of the receptacular cup, one-celled and one-ovuled anatropous ovules, simple styles, woody stem and opposite leaves; but in *Monimieæ* the flowers are diclinous, the perianth is calycoid, the ovule is pendulous, the fruit is a drupe and the embryo small in a copious albumen. Finally, *Calycanthus* has been compared with *Rosa*; but its four-angled stem, opposite exstipulate leaves, sterile stamens and extrorse anthers readily distinguish it.

Calycanthus, of which two species are known, inhabits North America; *Chimonanthus* grows in Japan. *Calycantheæ* are aromatic, and the bark of *Calycanthus floridus* is used in America as a stimulating tonic.

IV. *MAGNOLIACEÆ.*

(MAGNOLIÆ, *Jussieu.*—MAGNOLIACEÆ, *D.C.*—MAGNOLIEÆ ET WINTEREÆ, *Br.*)

FLOWERS ☿. SEPALS *usually* 3. PETALS 6-∞, *free, hypogynous.* STAMENS ∞, *hypogynous; anthers adnate.* CARPELS *usually* ∞, *distinct or coherent,* 1-*celled,* 1-2-∞-*ovuled.* OVULES *anatropous.* ALBUMEN *copious, not ruminate.* EMBRYO *dicotyledonous, straight, minute, basilar.* STEM *woody.* LEAVES *alternate.*

TREES or SHRUBS. LEAVES alternate, simple, coriaceous, entire or rarely lobed (*Liriodendron*), penninerved, reticulate, sometimes minutely pellucidly dotted; *stipules* membranous, convolute in bud, or opposite, rarely 0 (*Drimys, Illicium*). FLOWERS ☿, or very rarely incomplete (*Tasmannia*), usually large, terminal or axillary, solitary, rarely racemose or fascicled. SEPALS 3, rarely 6, or 2-4, usually petaloid, free, imbricate, deciduous. PETALS 6-∞, inserted at the base of a stipitiform torus, 1-2-∞-seriate, imbricate, deciduous. STAMENS ∞, several-seriate, inserted with the petals; *filaments* free; *anthers* 2-celled, adnate, extrorse (*Liriodendron, Drimys, Illicium*), or bursting laterally, or introrse (*Magnolia, Talauma, Michelia,* &c.), dehiscence longitudinal or transverse (*Tasmannia*). OVARIES ∞ or few, sometimes many-seriate in a head or spike, free or rarely coherent (*Manglietia*), sometimes whorled at the top of the receptacle (*Illicium*), always 1-celled; *styles* continuous with the ovary, stigmatiferous within and near the top; *ovules* on the ventral suture, either 2, collateral or superimposed (*Magnolia, Liriodendron*), or more and 2-seriate (*Michelia, Manglietia*); pendulous, rarely erect at the base of the cell, and solitary (*Illicium*), anatropous. FRUIT various: *carpels* subpedicelled, free or coherent, either 2-valved and capsular, with dorsal or ventral dehiscence (*Mag-*

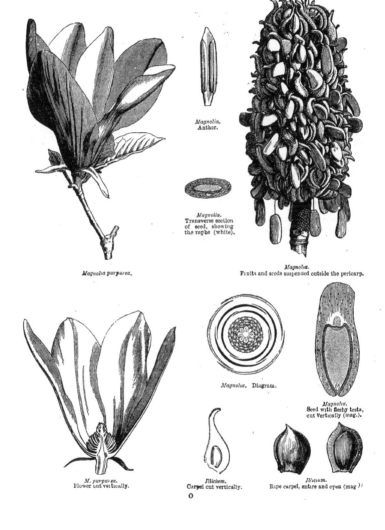

Magnolia.
Anther.

Magnolia.
Transverse section
of seed, showing
the raphe (white).

Magnolia purpurea.

Magnolia.
Fruits and seeds suspended outside the pericarp.

Magnolia. Diagram.

Magnolia.
Seed with fleshy testa,
cut vertically (mag.).

M. purpurea.
Flower cut vertically.

Illicium.
Carpel cut vertically.

Illicium.
Ripe carpel, entire and open (mag)

O

Illicium.
Fruit.

Illicium.
Seed with crustaceous testa,
cut vertically (mag.).

Tasmannia.
Flower (mag.).

Tasmannia.
Calyx and pistil cut
vertically.

nolia, Michelia, Manglietia, Illicium), or indehiscent and fleshy (*Drimys*), or woody and breaking transversely at the base (*Talauma*), or a samara (*Liriodendron*). SEEDS sessile or funicled, often suspended outside the pericarp (*Magnolia*); testa fleshy (*Magnolia*) or crustaceous (*Illicium*). EMBRYO minute, straight, at the base of a fleshy copious albumen; *radicle* and *cotyledons* very short.

Tribe I. MAGNOLIEÆ, *D.C.*—Flowers ☿. Carpels imbricate, many-seriate, in a head or spike.—Stipules enveloping the leaves.

PRINCIPAL GENERA.

Talauma.   Magnolia.   Liriodendron.   Michelia.

Tribe II. ILLICIEÆ, *D.C.*—Flowers ☿ or polygamous. Carpels whorled and 1-seriate, or solitary.—Leaves minutely pellucidly dotted, exstipulate.

PRINCIPAL GENERA.

Drimys.   Illicium.   Tasmannia.

Tribe III. TROCHODENDREÆ,[1] *Benth. et Hook. fil.*—Sepals and petals 0. Flowers polygamo-diœcious. Carpels whorled, 1-seriate.

GENERA.

Trochodendron.   Euptelea.   Cercidiphyllum.

Magnoliaceæ, which are very near *Schizandreæ*, *Anonaceæ* and *Myristiceæ* (see these families), are equally connected with *Dilleniaceæ*, by their hypogynism, æstivation, polyandry, adnate anthers, free ovaries, anatropous ovules, capsular fruit, albuminous seed, straight minute basilar embryo, woody stem and alternate leaves. *Dilleniaceæ* differ only in the quinary flowers, the often unilateral and polyadelphous stamens, the erect or ascending ovules, and the arillate seed. *Magnoliaceæ* also approach *Ranunculaceæ*, through *Dilleniaceæ*; but are easily distinguished by their habit. *Magnolieæ* are chiefly North American; they are also numerous in subtropical Asia, Japan and India. *Illicieæ* are scattered over America, Eastern Asia, Australia, New Zealand, and the Moluccas.

The properties of *Magnoliaceæ* resemble those of *Anonaceæ*, but their leaves and bark are more intensely bitter, owing to extractive resinous principles. The pericarps and seeds contain a fixed oil, with an often acrid aroma. The fruits are rarely eatable, but many are tonic and stimulant, and are sometimes used as condiments. *Michelia Champaca* is cultivated throughout tropical Asia, on account of its sweet-scented

[1] This tribe, which embraces three species of very anomalous structure, has been added to *Magnoliaceæ* by Bentham and Hooker fil. (Gen. Pl. i. 954). *Euptelea* contains two species, one Assamese, the other Japanese Trochodendron one, and *Cercidiphyllum* two, all from Japan.—ED.

flowers, which, however, become fœtid as they wither; all parts of the tree are aromatic, bitter, and acrid; an infusion of its powdered bark is a powerful emmenagogue; its young buds are administered in urethritis, and its powdered leaves are recommended for gout, and are applied as a lotion in rheumatic and arthritic pains; lastly, its seeds, which contain a very acrid substance, are rubbed, with ginger and galenga, over the region of the heart, to cure infantile intermittents. The bark of *Talauma* (*Aromadendron*) *elegans* is a renowned stomachic in Java; its slightly bitter leaves are antispasmodic and antihysteric. Of the Asiatic deciduous-leaved species, *Magnolia Yulan* has been cultivated from time immemorial in China, and its very bitter seeds used as a febrifuge; and the Himalayan *M. Campbellii*, a lofty tree with large red flowers, is one of the most splendid plants hitherto discovered. The principal American species are *M. grandiflora*, *auriculata*, and *macrophylla*, of which the bitter and slightly aromatic bark is a tonic. The fruit and seeds of *M. glauca* and *acuminata* are stimulants. The bitter, pleasantly aromatic bark of the Tulip-tree (*Liriodendron tulipifera*), which attains a height of 100 feet in North America, is regarded as an excellent substitute for cascarilla and quinine.

In *Illicieæ*, which, from their punctate pellucid leaves and 1-seriate whorled carpels, rather form a distinct order than a tribe of *Magnoliaceæ*, the aroma of the volatile oil and resin supersedes the bitterness, and gives them stimulating virtues; as in *Drimys Winteri* of Antarctic America, *D. Granatensis* of New Granada, *D. axillaris* of New Zealand, the *Tasmannias* of Australia, and especially the 'Badiane' (*Illicium anisatum*), a Chinese shrub, the fruit of which, called Star Aniseed from its smell and whorled carpels, is a powerful stimulant, which enters into the composition of Dutch aniseed cordial. *I. religiosum*, transported from China to Japan, and perhaps only a variety of the latter, possesses the same properties, but in an inferior degree.

V. *SCHIZANDREÆ*.

(SCHIZANDREÆ, *Blume.*—MAGNOLIACEARUM *tribe* III., *Benth. et Hook. fil.*)

Sarmentose glabrous SHRUBS, with mucous juice. LEAVES alternate, simple, penninerved, entire or toothed, sub-coriaceous, often pellucidly dotted (*Schizandra*), exstipulate. FLOWERS diclinous, axillary, solitary, small, usually scented. PERIANTH ternary, multiple; sepals and petals hypogynous, 9-12-15, 3-∞-seriate; passing gradually from the small outer to the petaloid inner. FLOWERS ♂: STAMENS ∞, or 5-15, distinct, or united into a globular mass; *filaments* very short, thick, free or coherent; *anthers* adnate, cells short, rounded, more or less separated by the connective. FLOWERS ♀: CARPELS ∞, in a head (*Kadsura*) or spike (*Schizandra*); *stigmas* sessile, decurrent on the inner edge of the ovary; *ovules* 2-3, superimposed, pendulous, anatropous. BERRIES indehiscent. SEEDS sunk in pulp; *albumen* oily, copious. EMBRYO minute, straight, basilar; *cotyledons* divaricate; *radicle* near the hilum, oblong, superior.

PRINCIPAL GENERA.

Schizandra.   Kadsura.

This little family, annexed by Bentham and Hooker fil. to *Magnoliaceæ*, is in fact only distinguishable from them by its climbing stem, exstipulate leaves, diclinous flowers, and fleshy 2-3-seeded carpels. It also approaches *Menispermeæ*, *Lardizabaleæ*, and *Anonaceæ* (see these families). *Schizandreæ* inhabit Eastern temperate and tropical Asia; one species grows in the warm regions of North America. The mucilaginous berries of some are eatable, but tasteless.

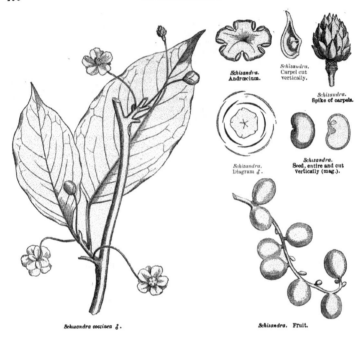

Schizandra
Androecium.

Schizandra.
Carpel cut
vertically.

Schizandra.
Spike of carpels.

Schizandra.
Diagram ♂.

Schizandra.
**Seed, entire and cut
vertically (mag.).**

Schizandra coccinea ♂.

Schizandra. Fruit.

VI. *ANONACEÆ.*

(ANONÆ, *Jussieu.*—GLYPTOSPERMÆ, *Ventenat.*—ANONACEÆ, *Dunal.*)

FLOWERS ☿. SEPALS 3. PETALS *usually 6, 2-seriate, hypogynous, most often valvate.* STAMENS ∞, *rarely definite, hypogynous* [on a large torus—ED.],. *many-seriate; anthers adnate.* OVARIES *usually many, distinct, 1-2-∞-ovuled; ovules erect or ascending.* FRUIT *a capsule or berry* [very rarely capsular—ED.]. ALBUMEN *ruminate.* EMBRYO *minute, basilar.*

TREES OR SHRUBS, sometimes climbing, generally aromatic, with acrid juice. LEAVES alternate, distichous, plaited in vernation, simple, entire, penninerved, pubescent when young; petiole usually jointed or swollen at the base; *stipules* 0. FLOWERS ♀, rarely diclinous, solitary, or fascicled on axillary peduncles, rarely lateral or leaf-opposed; corolla membranous, coriaceous or fleshy, greenish purple or

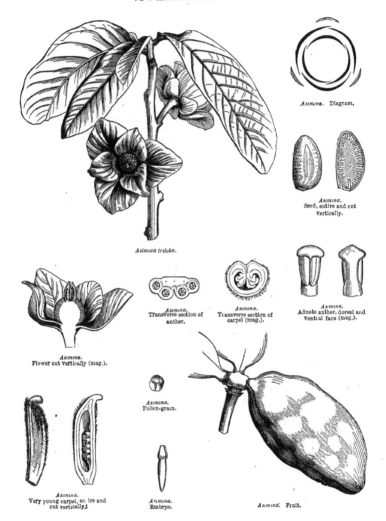

Asimina. Diagram.

Asimina.
Seed, entire and cut
Vertically.

Asimina triloba.

Asimina.
Flower cut vertically (mag.).

Asimina.
Transverse section of
anther.

Asimina.
Transverse section of
carpel (mag.).

Asimina.
Adnate anther, dorsal and
Ventral face (mag.).

Asimina.
Pollen-grain.

Asimina.
Very young carpel, entire and
cut vertically.t

Asimina.
Embryo.

Asimina. Fruit.

yellowish. SEPALS 3, rarely 2 (*Disepalum*), distinct or united at the base, or coherent in a 3-lobed or -toothed calyx (*Cyathocalyx*), valvate or imbricate in bud. PETALS usually 6 2-seriate, rarely 4 2-seriate (*Disepalum*), or only 3 1-seriate (*Unona*), rarely coherent (*Hexalobus*), hypogynous, valvate, or rarely imbricate in bud. STAMENS indefinite, multi-seriate on a thick torus; *anthers* adnate, 2-celled, cells dorsal or lateral, opening by a longitudinal slit, contiguous or separate, usually concealed by the overlapping dilated tops of the connectives; rarely definite, with the anthers not concealed by the connective, hardly or not at all dilated (*Miliusa, Orophea, Bocagea,* &c.). CARPELS ∞, rarely definite (*Asimina, Xylopia, Bocagea,* &c.), or solitary (*Cyathocalyx*), distinct, or rarely coherent (*Anona, Monodora*), sessile on the top of the torus; *style* short, thick, or 0; *stigma* thick, capitate, or oblong, sometimes furrowed or 2-lobed or radiate (*Monodora*) ; *ovules* 1–2, erect, basilar, or 1–∞, fixed either to the suture, or very rarely all over the walls of the ovary (*Monodora*), anatropous, raphe ventral, micropyle inferior. Ripe CARPELS sessile or stipitate, distinct, or united into a ∞-celled fruit (*Anona*), or 1-celled (*Monodora*), dry, fleshy or pulpy, indehiscent or 2-valved. SEEDS with copious ruminate albumen. EMBRYO minute, basilar ; *radicle* near the hilum, inferior.

PRINCIPAL GENERA.

*Uvaria.	Ellipeia.	Guatteria.	Duguetia.	Artabotrys.
Hexalobus.	Unona.	*Asimina.	Polyalthia.	Anaxagorea.
Popowia.	Oxymitra.	Goniothalamus.	Mitrephora.	Monodora.
Rollinia.	*Anona.	Melodorum.	Xylopia.	Miliusa.
Orophea.	Alphonsea.	Bocagea.	*Eupomatia.	

The genus *Eupomatia*, which is closely allied to *Anonaceæ*, presents remarkable anomalies of structure. *Sepals* and *petals* united into a conical mass, inserted on the upper edge of a turbinate torus, from which it is transversely separated like an operculum. *Stamens* numerous, perigynous; inner many-seriate, sterile, petaloid; outer few-seriate, linear-lanceolate ; connective longer than the anther-cells, acuminate. *Ovary* inferior, buried in the torus, and composed of several thick carpels; ovules numerous, inserted on the ventral suture; styles connate, terminated by a plane stigma, hollowed into as many areolæ as there are carpels. *Fruit* a truncate berry, crowned by the margin of the torus. *Seeds* angular.—The sterile inner stamens are connivent, and very closely imbricate over the stigma, which are thus shut off from communication with the outer fertile stamens, rendering fertilization impossible; but, as R. Brown observed, they are gnawed by insects, whose introduction thus assists the transport of the pollen to the stigma.

Anonaceæ are near *Myristiceæ* (see this family). They approach *Schizandreæ, Lardizabaleæ* and *Menispermeæ* in the ternary arrangement of the calyx and corolla, hypopetalism, extrorse anthers, berried fruit, copious albumen, basilar embryo (at least in *Lardizabaleæ* and *Schizandreæ*), alternate leaves and axillary flowers; but in *Schizandreæ* the flowers are diclinous, the æstivation imbricate, the ovules pendulous, the radicle superior, and the albumen is not ruminate. The affinity with *Magnoliaceæ* rests on the same features, and the diagnosis on the same differences; besides which, in the latter, the leaves are stipulate, and the seeds have generally a fleshy testa (*Magnolia*). *Anonaceæ* are also near *Dilleniaceæ* in hypopetalism, polyandry, adnate anthers, polygyny, erect anatropous ovules, copious fleshy albumen, basilar embryo, woody stem and alternate leaves; but in *Dilleniaceæ* the leaves are sometimes stipulate, the flowers are terminal and quinary, the æstivation is imbricate, and the albumen is not ruminate.

Anonaceæ are nearly all tropical. Some (*Asimina*) reach 33° N. latitude in America. Asia and America possess about the same number of species; somewhat fewer are met with in Africa. *Anona* and *Rollinia* have not yet been observed in Asia. Several *Anonas* inhabit Africa.

- The bark of *Anonaceæ* is usually more or less aromatic and stimulating; in some species the taste is acrid and almost nauseous; the leaves possess similar but less powerful properties, the fruits are aromatic and hot (*Xylopia*), or nearly inodorous, and these alone are eatable. The Malayans use the bark of several *Anonaceæ*, reduced to pulp, for bruises and rheumatic pains, and the fruit of others as a stomachic. With the flowers of *Uvaria odorata*, and with other aromatics and *Curcuma* root, they prepare an ointment with which they anoint themselves, to ward off fever in the rainy season. European women in India, it is said, macerate these scented flowers in cocoa-nut oil, as a hair oil. The root of *Polyalthia macrophylla* is strongly aromatic, and the Javanese mountaineers use an infusion of it in eruptive fevers; they also use the fruits of *P. subcordata* to allay nervous colics. *Artabotrys suaveolens* grows in nearly all the islands of the Malay Archipelago; from its infused leaves is prepared an aromatic medicine, which is very efficacious in inducing reaction during the cold stage of cholera. The aromatic fruit of *Xylopia grandiflora* furnishes the Brazilians with a condiment and a stimulating drug; that of *X. frutescens*, a shrub found throughout tropical America, is used as pepper by the negros; that of *X. longifolia*, which grows on the shores of the Orinoco, is reckoned one of the best substitutes for quinine. *X. æthiopica* furnished the ancients with Ethiopian pepper, before black pepper was introduced from India. The *Asiminas* of North America are remarkable for their nauseous odour; the leaves of *A. triloba* are used to hasten the ripening of abscesses; its berries are eatable, but its seeds are emetic. Many species of *Anona* produce agreeable fruits, much esteemed in the tropics, as the Peruvian Cherimoya (*Anona Cherimolia*), the Sweet Sop (*A. squamosa*), and the Custard-Apple (*A. muricata*). The West Indian *A. reticulata* has a mucilaginous, astringent, disagreeably tasted fruit, and is employed as an anti-dysenteric and vermifuge. All these are natives of America, whence they have been transported by man to the Old World.

VII. *MENISPERMEÆ.*

(MENISPERMA, *Jussieu.*—MENISPERMOIDEÆ, *Ventenat.*—MENISPERMEÆ, *D.C.*)

FLOWERS *diœcious.* SEPALS *usually 6, free, 2-seriate, imbricate.* PETALS *hypogynous, usually 6, imbricate, 2-seriate.* STAMENS *inserted on the receptacle, equal and opposite to the petals, rarely more or fewer, sterile or 0 in the ♀ flowers.* CARPELS *usually 3, rarely ∞, distinct, 1-ovuled, rudimentary or 0 in the ♂ flowers.* DRUPES *with the stylary scar often basal.* SEEDS *albuminous or not.* EMBRYO *usually curved; radicle facing the stylary scar.*—STEM *usually woody, climbing or twining.* LEAVES *alternate, exstipulate.*

STEM climbing; branchlets finely striate, sometimes twining, woody, or suffruticose, rarely herbaceous and springing from a woody rhizome (*Cissampelos*). LEAVES alternate, exstipulate, usually palminerved, entire or palmilobed or peltate, rarely compound (*Burasaia*), petiole spuriously jointed at the base, and sometimes at the top. FLOWERS diœcious, small, in a panicle, raceme or cyme, rarely solitary, sometimes accompanied by cordate bracts (*Cissampelos*). SEPALS usually 6, 2-seriate, sometimes 9 3-seriate or 12 4-seriate, rarely 4 (*Cissampelos*), sometimes 4 or 8 (*Menispermum*), very rarely 5 (*Sarcopetalum*), usually distinct, very rarely coherent (*Synclisia, Cyclea*). PETALS usually '6' 2-seriate, imbricate, but equal and simulating a single series, smaller than the inner sepals, rarely 4 or 8 (*Cyclea*), very rarely 1, 3, or 5 (*Stephania*), or 2 (*Cissampelos*), or 0 (*Anamirta, Abuta,* &c.), very rarely united (*Cissampelos*). STAMENS as many as petals, usually 6, opposite to the petals, very rarely 3 (*Triclisia,* &c.), or 4-8 (*Cyclea*), or 9 (*Limacia,* &c.), or ∞ (*Menispermum*);

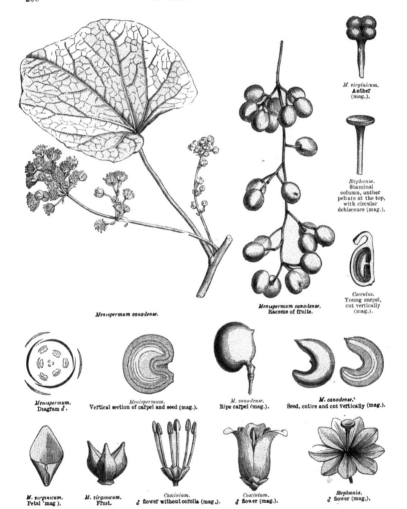

M. virginicum.
Anther
(mag.).

Stephania.
Staminal
column, anther
peltate at the top,
with circular
dehiscence (mag.).

Cocculus.
Young carpel,
cut vertically
(mag.).

Menispermum canadense.

Menispermum canadense.
Raceme of fruits.

Menispermum.
Diagram ♂.

Menispermum.
Vertical section of carpel and seed (mag.).

M. canadense.
Ripe carpel (mag.).

M. canadense.
Seed, entire and cut vertically (mag.).

M. virginicum.
Petal (mag.).

M. virginicum.
Fruit.

Coscinium.
♂ flower without corolla (mag.).

Coscinium.
♂ flower (mag.).

Stephania.
♂ flower (mag.).

filaments more or less free, or united in a monadelphous column; *anthers* various, free or united, usually extrorse, 1–2-celled, dehiscence longitudinal, transverse, or circular (*Stephania*). CARPELS usually 3, rarely 6 (*Coscinium, Sarcopetalum, Fibraurea*), or 9–12 (*Tiliacora, Sciadotenia*), or 2 4 (*Menispermum*), or 1 (*Cissampelos, Cyclea, Stephania*); *styles* terminal, simple or lobed, often becoming basilar from the curvature of the ovary; *ovules* solitary [1] in the carpels, half-anatropous, peltately attached to the ventral suture of the carpel, or very rarely anatropous; *micropyle* superior, *chalaza* facing the base of the ovary. Ripe CARPELS drupaceous, sessile, or stipitate; scar of style subterminal, excentric, or more often sub-basal; endocarp straight, or often curved like a horse-shoe, with its ventral surface intruded as a hemispherical, peltate, or flattened projection, to which the seed is ventrally attached. SEED of the form of the cavity, concave or furrowed on its ventral surface; *testa* thin, membranous; *albumen* more or less copious, sometimes ruminate (*Anamirta, Abuta*, &c.), or 0 (*Pachygone, Botryopsis, Triclisia*, &c.). EMBRYO usually curved, rarely straight (*Anomospermum*); *radicle* facing the scar of the style; *cotyledons* linear and contiguous, or large and thick, or foliaceous and divaricate.

PRINCIPAL GENERA.

Aspidocarya.	Parabœna.	Tinospora.	Jateorhiza.	Anamirta.
Coscinium.	Tiliacora.	Limacia.	* Cocculus.	Menispermum.
Stephania.	* Cissampelos.	Cyclea.	Pachygone.	Hyperbœna.
Triclisia.	Fibraurea.	Centitaxis.	Antitaxis.	Burasaia.

Menispermeæ are closely allied to *Anonaceæ, Lardizabaleæ, Berberideæ,* and *Schizandreæ.* They are connected with *Lardizabaleæ* by their hypopetalism, two-seriate sepals and petals, usually monadelphous stamens, extrorse anthers, distinct and fleshy carpels, woody twining stem, alternate exstipulate and sometimes compound leaves (*Burasaia*), and diclinous racemed flowers; but in *Lardizabaleæ* the carpels nearly always contain several ovules scattered over the walls of the ovary, and the embryo is minute, at the base of a very abundant horny albumen. *Berberideæ* are similarly related, but differ in their ꝗ flowers, free filaments, the dehiscence of their anthers, their solitary carpel, erect stem, penninerved leaves, &c. *Anonaceæ* are connected with *Menispermeæ* by their woody stem, alternate and exstipulate leaves, often diclinous flowers, hypogynous 2-seriate petals, extrorse anthers, free often 1-ovuled and fleshy carpels; they are distinguished from them by their habit, inflorescence, aromatic odour, rarely climbing stem, penninerved leaves, ruminate albumen, &c. *Schizandreæ* approach *Menispermeæ* in their woody climbing stem, alternate exstipulate leaves, diclinous flowers, several-seriate sepals and petals, extrorse anthers, and free carpels, fleshy when ripe; they are distinguished by the 2-ovuled ovaries, the straight minute embryo at the base of an abundant albumen, and the penninerved leaves.

Menispermeæ principally inhabit the intertropical regions of both worlds. Few are met with in North America, Western Asia, South Africa and extra-tropical Australia; none in Europe.

This family has several species used in medicine; some possess a bitter principle in their root, which stimulates the digestive organs; others are acrid and diuretic. Many contain in their herbaceous organs an abundant emollient mucilage. The pericarp of some is narcotic, acrid, and very poisonous. *Cocculus palmatus* is a tropical African and Madagascar perennial, whose turnip-shaped root (the Calumba-root of commerce) is one of the most efficacious of tonics, and prescribed for obstinate colic, dysentery and sickness; it is adulterated with other roots from India and Barbary, which bear its name without having its properties. *C. peltatus*, of Malabar, and *C. flavescens*, of the Moluccas, are the best substitutes for it. Among the Brazilian *Menispermeæ* with a bitter tonic root, are *Cocculus platyphyllus, cinerascens,* and *Cissampelos*

---

[1] Two in *Fibraurea*, of which one alone is further developed.—ED.

oralifolia: C. *Pareira* is a southern Brazilian shrub, the root of which, called 'Pareira Brava,' is woody, inodorous, of a taste at first mild, then bitter and somewhat acrid; it was formerly renowned as a lithotriptic, and is still used in Martinique against the bite of the Trigonocephalus. *Cissampelos Caapeba,* from the Antilles, and *C. mauritiana,* from the Mascarine islands, are used as substitutes for the Pareira Brava. The root of *Coscinium fenestratum,* of Ceylon, is a stomachic. The negros of Senegambia employ that of *Cocculus Bakis* as a diuretic and febrifuge. The roots of *Cissampelos glaberrimus* and *ebracteatus* are administered in Brazil in cases of snake-bites. *Cocculus crispus,* of the Moluccas, contains a glutinous and bitter juice, commonly used by the Indians in intermittent fevers, jaundice, and intestinal worms.

The bark of several species is extremely bitter; some yield a yellow dye. *Anamirta Cocculus* is a tropical Asiatic shrub, whose extremely poisonous fruits are used in India to intoxicate and poison fish, which are thus obtained in abundance, but are sometimes dangerous to eat, the narcotic principle contained in the seed *(pierotoxrine)* being scarcely less deleterious than strychnine. In England beer is sometimes adulterated with *Cocculus indicus.*

VIII. *BERBERIDEÆ.*

(BERBERIDES, *Jussieu.*—BERBERIDEÆ, *Ventenat.*—BERBERACEÆ, *Lindl.*)

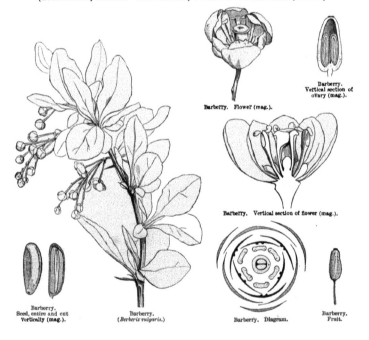

Barberry. Flower (mag.).

Barberry.
Vertical section of
ovary (mag.).

Barberry. Vertical section of flower (mag.).

Barberry.
Seed, entire and cut
Vertically (mag.).

Barberry.
(*Berberis vulgaris.*)

Barberry. Diagram.

Barberry.
Fruit.

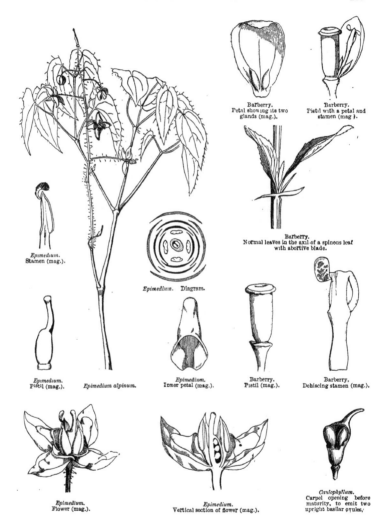

Barberry.
Petal showing its two
glands (mag.).

Barberry.
Pistil with a petal and
stamen (mag.).

Barberry.
Normal leaves in the axil of a spinous leaf
with abortive blade.

Epimedium.
Stamen (mag.).

Epimedium. Diagram.

Epimedium.
Pistil (mag.).

Epimedium alpinum.

Epimedium.
Inner petal (mag.).

Barberry.
Pistil (mag.).

Barberry.
Dehiscing stamen (mag.).

Epimedium.
Flower (mag.).

Epimedium.
Vertical section of flower (mag.).

Caulophyllum.
Carpel opening before
maturity, to emit two
upright basilar ovules.

Caulophyllum.
Fruit, leaving naked two seeds, one of which is abortive, the other globular and drupe-like.

Caulophyllum.
Fertile drupe-like seed, cut vertically, borne on a well-developed funicle.

Berberidopsis.
Flower (mag.).

Berberidopsis,
Andrœcium and pistil (mag.).

Berberidopsis.
Pistil (mag.).

SEPALS 3–4–9, 1–3-*seriate.* PETALS *hypogynous,* 4–6–8–9– ∞, 1–2–3-*seriate.* STAMENS *as many as the petals, hypogynous.* ANTHERS *extrorse, usually opening by valves, bursting from below upwards.* CARPEL *solitary, 1-celled, several-ovuled.* FRUIT *a berry or capsule.* SEEDS *albuminous.* EMBRYO *dicotyledonous, small, axile.*

HERBS or SHRUBS with cylindric stem and branches, juice watery. LEAVES sometimes simple (1-foliolate), or pinnate, with usually spiny teeth, occasionally reduced to simple or branching spines (*Berberis*), sometimes 2–3-pinnate (*Epimedium, Nandina, Leontice,* &c.), sometimes palmilobed (*Diphylleia, Jeffersonia*); *stipules* 2, petiolar, minute, caducous. FLOWERS ♀, regular, very rarely without perianth (*Achlys*), axillary, solitary (*Jeffersonia, Podophyllum*), or in racemes (*Berberis, Epimedium, Leontice*), panicles (*Bongardia*), cymes (*Diphylleia*), or spikes (*Achlys*). CALYX of 3, 4, or 9 1–3-seriate sepals, often petaloid, quite distinct, æstivation imbricate. PETALS inserted on the receptacle, as many as the sepals in the several-seriate calyces, double in the 1-seriate calyces, biglandular at the base (*Berberis*), or with a nectariferous pore at the claw (*Bongardia*), or nectariform (*Leontice, Caulophyllum*), or hooded or lengthened into a spur (*Epimedium*), or like the sepals (*Aceranthus, Diphylleia*). STAMENS inserted on the receptacle, usually as many as the petals, very rarely twice as many (*Podophyllum*); *filaments* free, short, flattened, often irritable; *anthers* 2-celled, extrorse, opening by two valves raised from below upwards, or sometimes by two longitudinal slits (*Nandina, Podophyllum*). CARPEL solitary,[1] 1-celled; *style* terminal, very short, often 0; *stigma* usually large, peltate, umbilicate; *ovules* numerous, ascending from a parietal placenta, or few, erect and basal or sub-basal, anatropous. BERRY or CAPSULE fleshy or membranous, indehiscent, sometimes dehiscent (*Epimedium, Vancouveria, Jeffersonia*); sometimes the ovary breaks up after fertilization, and disappears when ripe, leaving the drupe-like seeds exposed (*Caulophyllum*). SEEDS ovoid or globose, erect or horizontal; *testa* crustaceous, membranous or fleshy; *hilum* sublateral, near the base, sometimes carunculate; *albumen* fleshy, or sub-horny. EMBRYO straight, axile; *cotyledons* flat, elliptic; *radicle* longer than the cotyledons.

[1] The very anomalous Chilian genus *Berberidopsis* is a climber with three carpels connate into a 1-celled ovary, with many almost orthotropous ovules on three parietal placentas. It is referred to *Bixaceæ* by Baillon.—ED.

Berberideæ are closely allied to *Lardizabaleæ* (see that order); the genus *Berberidopsis*, of which the 1-celled ovary, with three parietal placentas, is an exception to the normal *Berberideæ*, establishes the passage of these to *Lardizabaleæ*, through the genus *Decaisnea*, of which the carpels with two series of ovules gape when ripe. *Berberideæ* also approach *Magnoliaceæ* in their hypopetalism, the distinct several-seriate sepals, the adnate anthers, albuminous seed, straight embryo, woody stem, and alternate leaves with caducous stipules; but *Magnoliaceæ* differ in their habit, scented wood, polyandry, and the number and mode of dehiscence of their carpels. *Berberideæ* resemble *Ranunculaceæ* (see this order), and slightly *Papaveraceæ*, which differ in habit, milky juice, terminal inflorescence, polyandry, introrse anthers, and the structure of the fruit and seed. *Berberideæ* have also some analogies with *Anonaceæ*, founded on the ternary arrangement of the flower, the free sepals, hypogynous 2-seriate petals, isostemony (*Bocagea*), adnate anthers, thick capitate and often sessile stigma, the erect anatropous ovules, and berried fruit; but *Anonaceæ* are distinguished by their habit, their distichous entire often climbing stem, terminal usually polyandrous flowers, the ternary carpels, and ruminate albumen.

Berberideæ grow in the temperate regions of the northern hemisphere, the Andes, and extra-tropical South America. They are absent from tropical and South Africa, Australia and New Zealand.

The berries and herbaceous parts of *Berberideæ* contain free malic acid ; the root and bark of several species yield a yellow bitter extractive (*berberine*) possessing properties analogous to those of rhubarb. This is especially the case with the common Barberry (*Berberis vulgaris*), a shrub which grows on calcareous soils in Europe and Northern Asia; the tincture yielded by the bark of its root and stem is purgative, and its yellow colour induced doctors formerly to administer it for jaundice; its acid berries form a very pleasant preserve, and the young acid astringent leaves are used to strengthen the gums. *B. fascicularis* (*Mahonia*) is esteemed in California for its mildly acid berries. The wood of the Indian and South American *Berberis* is used as a dye. [*B. Lycium*, the *Lycium indicum* of Dioscorides, is used to this day in India for ophthalmia.—ED.] The root of *Caulophyllum thalictroides* is valued as a sudorific in North America, and its seeds form a substitute for coffee. *Bongardia Chrysogonum* grows in Greece and the East; its leaves are eaten like those of sorrel; its root was formerly considered an alexifer. The powdered root of *Leontice Leontopetalum*, a plant of Asia Minor, is used at Aleppo as a soap for cleaning stuffs; the Turks also use it as a corrective for an overdose of opium. *Podophyllum peltatum*, from North America, connects *Berberideæ* with *Papaveraceæ* ; its herbaceous parts are narcotic and poisonous; its root contains a bitter gum-resin, which purges as effectively as jalap; its berries are very acid, but may be eaten with impunity, as may those of *P. himalayense* [which are insipid.—ED.].

IX. *LARDIZABALEÆ, Br.*

(MENISPERMACEARUM *tribus, D.C.*—BERBERIDEARUM *tribus, Benth. et Hook. fil.*)

Twining, or rarely erect SHRUBS (*Decaisnea*), branches striate. LEAVES alternate, compound, 2-3-ternate (*Lardizabala*), or 3-5-foliolate (*Boquila, Parvatia*), or pinnate (*Decaisnea*), or digitate (*Akebia, Holbœllia*) ; *leaflets* toothed or sinuate, petioles and petiolules swollen at the base and top, exstipulate. FLOWERS diclinous or polygamous, in solitary or fascicled axillary racemes, naked or bracteolate. FLOWERS ♂ : CALYX coloured, 6-phyllous, 2-seriate, or rarely 3-phyllous (*Akebia*), æstivation imbricate, the outer sepals often valvate. PETALS 6, inserted on the receptacle, much smaller than the sepals, or 0. STAMENS 6, inserted on the receptacle, opposite to

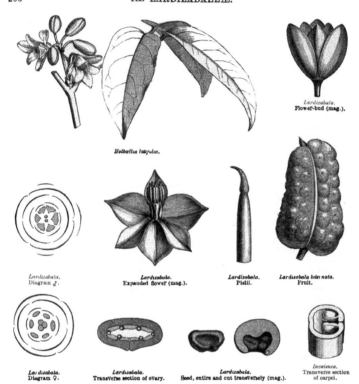

Holbœllia latifolia.

Lardizabala.
Flower-bud (mag.).

Lardizabala.
Diagram ♂.

Lardizabala.
Expanded flower (mag.).

Lardizabala.
Pistil.

Lardizabala bitei nata.
Fruit.

Lardizabala.
Diagram ♀.

Lardizabala.
Transverse section of ovary.

Lardizabala.
Seed, entire and cut transversely (mag.).

Decaisnea.
Transverse section
of carpel.

the petals; *filaments* cylindric, monadelphous or rarely free (*Akebia, Holbœllia*); *anthers* extrorse, apiculate, very rarely muticous (*Akebia*). OVARIES rudimentary, 2–3 or more, fleshy. FLOWERS ♀, a little larger, calyx and corolla as in the ♂. STAMENS 6; *anthers* imperfect or 0. OVARIES 3, rarely 6–9 (*Akebia*), distinct, sessile, 1-celled; *styles* short or obsolete; *stigmas* terminal, papillose, peltate, obtuse or conical; *ovules* many, rarely few (*Boquila*), inserted all over the wall in separate alveolæ, or 2-seriate along the ventral suture (*Decaisnea*), anatropous or campylotropous. Ripe CARPELS baccate, indehiscent or dehiscent (*Decaisnea*). SEEDS buried in the pulp, *testa* thin. EMBRYO usually minute, at the base of a copious fleshy-horny albumen; *cotyledons* flat.

GENERA.

| Lardizabala. | Buquila. | Parvatia. | Decaisnea. | Stauntonia. |
| Holbœllia. | Akebia. | | | |

Lardizabaleæ approach *Berberideæ* in their hypopetalism, petaloid sepals and 2-seriate petals, isostemonism, extrorse anthers, berried fruit, albuminous seed, alternate compound leaves, and racemed flowers; they differ in their diclinism, united filaments, anthers dehiscing longitudinally, and several carpels.

For the affinity with *Menispermeæ*, see that order. *Lardizabaleæ* also approach *Magnoliaceæ* in their ternary perianth, hypopetalism, extrorse anthers, free carpels, albuminous seeds, and basilar embryo; but in *Magnoliaceæ* the stem is not twining, the leaves are always simple and stipulate, the flowers are usually ☿ and polyandrous, the ovules are constantly attached to the ventral suture, and the ripe carpels are follicles or samaras. The affinity is more marked with *Schizandreæ*, which are sarmentose shrubs with alternate exstipulate leaves, diclinous flowers, ternary free hypogynous perianth-segments in 2-several series, extrorse anthers, free berried carpels, seeds sunk in the pulp, copious albumen, and basilar embryo; but in *Schizandreæ* the leaves are simple, often pellucidly dotted, the stamens are numerous, the carpels 2-ovuled, and the ovules attached to the ventral suture.

The majority of *Lardizabaleæ* inhabit India, China, and Japan; *Lardizabala* and *Boquila* are Chilian.

Lardizabaleæ contain neither a bitter nor an aromatic principle; their berries are mucilaginous and eatable; the flowers of many are scented. The Nepalese eat the fruits of *Holbœllia latifolia*, and the Chilians those of *Lardizabala* and *Boquila*; [those of *Decaisnea* are sweet and fleshy, and very grateful.—ED.]. The shoots of *Lardizabala*, passed through fire and macerated in water, form cords of great strength.

X. *NYMPHÆINEÆ*, Brongniart.

SEPALS 3–5. PETALS 3–∞, *hypogynous or perigynous, i.e. inserted at different heights on a torus enveloping the ovaries and more or less united with them.* STAMENS *usually numerous, inserted with the petals.* OVARIES *several, free or coherent.* FRUIT *a berry bursting irregularly, or an indehiscent nut.* SEEDS *furnished with a double albumen, or rarely exalbuminous. Aquatic* HERBS.

Aquatic HERBS; *rootstock* perennial, submerged, tuberous, creeping, with sometimes milky juice, emitting leaves and 1-flowered scapes, rarely floating branches (*Brasenia*). LEAVES alternate or opposite, long-petioled, blade floating, rarely emerged (*Nelumbium*), usually cordate-peltate, sometimes oblong or linear (*Barclaya*), often dissected and capillary in the submerged leaves (*Cabomba*). FLOWERS large ☿, regular, usually floating, rarely emerged (*Nelumbium*); *peduncles* axillary, 1-flowered. SEPALS usually 4–5–6, rarely 3 (*Cabomba, Brasenia*). PETALS imbricate in æstivation, usually numerous, rarely 3 (*Cabomba*); sometimes all free and hypogynous (*Nuphar*); sometimes all, or the inner only, inserted at different heights, on a torus enclosing the carpels. STAMENS ∞, or rarely 6 (*Cabomba*), inserted as the petals; *anthers* erect, connective continuous with the filament, cells adnate, opening by a longitudinal slit, introrse, or rarely extrorse (*Cabomba*). CARPELS ∞, or 8–10, rarely 3–4 (*Cabomba*), sometimes distinct (*Cabomba, Brasenia*), sometimes cohering in a whorl, and forming a several-celled ovary, either free and superior (*Nuphar*), or adherent to the torus, and then inferior or half-inferior (*Nymphæa, Victoria, Euryale,* &c.), sometimes sunk without order in the alveolæ of an obconic torus (*Nelumbium*); *stigmas* either distinct (*Nelumbium, Cabomba*), or united into a radiating or annular disk (*Nymphæa, Euryale, Victoria*); *ovules* anatropous, attached to the ovarian walls, or pendulous either from the ventral suture (*Cabomba, Brasenia*), or from the top of the cell (*Nelumbium*). Ripe CARPELS indehiscent, distinct, or united into a fleshy, pulpy, or spongy fruit. SEEDS arillate or naked; *albumen* copious, floury or fleshy, or very rarely 0 (*Nelumbium*). EMBRYO straight, enclosed in the embryonic sac, sunk in the albumen, and thus forming a second albumen; *cotyledons* thick; *radicle* very short.

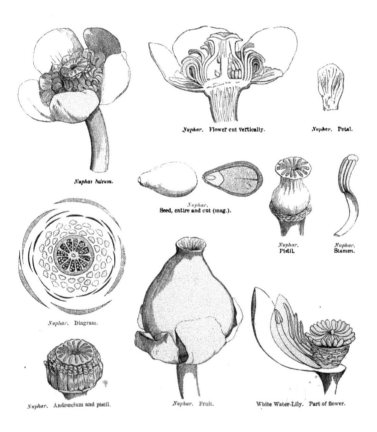

Nuphar. Flower cut vertically.   *Nuphar.* Petal.

Nuphar luteum.

Nuphar. Seed, entire and cut (mag.).

Nuphar. Pistil.   *Nuphar.* Stamen.

Nuphar. Diagram.

Nuphar. Androecium and pistil.   *Nuphar.* Fruit.   White Water-Lily. Part of flower.

SUB-ORDER I. *NYMPHÆACEÆ.*

(HYDROCHARIDUM *genera, Jussieu.*—NYMPHÆACEÆ, *Salisbury.*)

SEPALS 4–6. PETALS ∞, pluri-seriate, hypogynous (*Nuphar*) or perigynous, i.e. inserted at different heights on the torus enveloping the carpels (*Nymphæa*), or epigynous, i.e. inserted on the top of the torus (*Euryale, Victoria, Barclaya*), the inner narrower than the outer, and passing gradually, in some genera, into stamens (*Nymphæa, Victoria*). (In *Barclaya* the coherent bases of the petals are inserted on the top of the torus which envelops the carpels, the calyx remaining free.) *Filaments* often flattened (especially the outer) and sub-petaloid, usually prolonged above the anthers. CARPELS whorled, coherent, and forming a several-celled superior ovary (*Nuphar*), or half-inferior (*Nymphæa*), or inferior (*Victoria, Euryale, Barclaya*); *styles* connate into a peltate radiating stigma, longer than the torus, sessile or stipitate, more or less

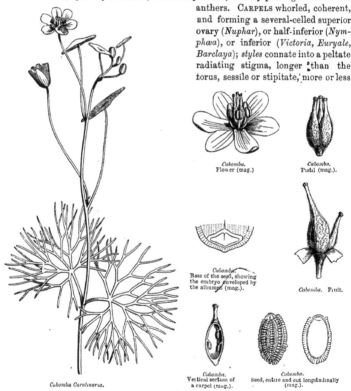

Cabomba.
Flower (mag.)

Cabomba.
Pistil (mag.).

Cabomba.
Base of the seed, showing the embryo enveloped by the albumen (mag.).

Cabomba. Fruit.

Cabomba Caroliniana.

Cabomba.
Vertical section of a carpel (mag.).

Cabomba.
Seed, entire and cut longitudinally (mag.).

depressed in the middle, with sometimes a central umbilicate gland (*Nymphœa*); *ovules* usually numerous, inserted on the walls of the septa. FRUIT a several-celled berry, bursting irregularly when ripe, rarely separating into distinct carpels. SEEDS often furnished with a saccate pulpy aril (*Nymphœa, Euryale*); *albumen* farinaceous.

GENERA.

Nuphar. *Nymphæa. Barclaya. *Euryale. *Victoria.

SUB-ORDER II. *CABOMBEÆ.*

(CABOMBEÆ, *Richard.*—HYDROPELTIDEÆ, *D.C.*—CABOMBACEÆ, *Asa Gray.*)

SEPALS 3-4. PETALS 3-4, hypogynous, persistent. STAMENS 6, 12, or 18; *filaments* subulate; *anthers* extrorse or lateral. OVARIES 3-2-4, or 6-18, free, whorled, inserted on a narrow torus, narrowed into styles, stigmatiferous at the top (*Cabomba*), or throughout their length (*Brasenia*); *ovules* 2-3, pendulous. Ripe CARPELS enclosed in the persistent calyx and corolla, often solitary by arrest, follicular, indehiscent. SEEDS with a fleshy copious albumen.

GENERA.

Cabomba.   Brasenia.

SUB-ORDER III. *NELUMBONEÆ.*

(NELUMBONEÆ, *Bartling.*—NELUMBIACEÆ, *Lindl.*)

SEPALS 4-5. PETALS and STAMENS ∞, hypogynous, several-seriate at the base of the torus; *filaments* filiform, dilated above; *anthers* introrse, connective prolonged beyond the cells as a flat or clubbed appendage. OVARIES several, sunk separately in the pits of a fleshy, obconic, flat-topped torus; *style* short; *stigma* terminal, sub-dilated; *ovules* 1-2 in each ovary, pendulous from a basilar funicle, which ascends along its wall, and is free above; raphe dorsal. NUCULES sub-globose, indehiscent. SEEDS exalbuminous, testa thin. EMBRYO floury, plumule foliaceous.

GENUS.

Nelumbium.

Nymphæineæ approach the polypetalous hypogynous apocarpous families, although the principal genera are syncarpous, and the ovary is adherent to the torus. The sub-order of *Nymphæaceæ* is allied to *Papaveraceæ* in the many-ovuled ovary with placentas on the septa, radiating stigmas, polyandry, and truly milky juice; but is distinguished by its frequent perigynism or epigynism, aquatic habitat, and especially the embryo with its enveloping sac being immersed in a superficial cavity of the amylaceous albumen; the latter is almost the only character which distinguishes *Cabombeæ* from *Ranunculaceæ*. *Nymphæaceæ* have also a real affinity with *Sarraceniaceæ* (see that order).

The species of *Nymphæa* are dispersed over nearly all regions; *Nuphar* is confined to the extra-tropical northern hemisphere; *Barclaya* and *Euryale* inhabit tropical Asia; *Victoria*, equatorial America. The three or four species of *Cabombeæ* are American; *Brasenia* is also found in India and Australia. [*Nelumbium* inhabits the southern United States, tropical Asia, Africa, and Australia.—ED.]

Some species of this family were venerated by the ancients, not only for the magnificence of their flowers and leaves, carpeting the surface of the tranquil waters, but also on account of their utility. Their young rootstocks contain abundance of starchy, mucilaginous and sugary matters, which render them

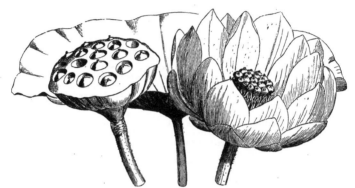

Nelumbium luteum. Leaf, flower, and carpels sunk in an alveolate receptacle.

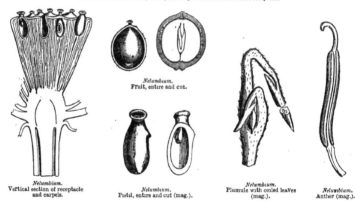

Nelumbium.
Vertical section of receptacle
and carpels.

Nelumbium.
Fruit, entire and cut.

Nelumbium.
Pistil, entire and cut (mag.).

Nelumbium.
Plumule with coiled leaves
(mag.).

Nelumbium.
Anther (mag.).

nutritive; in the adult rootstock these principles are replaced by gallic acid. The flowers, which have a peculiar scent, are narcotic. The seeds, filled with a floury albumen, are edible; the negros of Nubia use them as millet, and the Egyptians still eat the seeds and rootstock of *Nymphæa Lotus* and *cærulea. N. alba* is the greatest ornament of still waters in the northern hemisphere; its mucilaginous and somewhat acrid rootstock is administered in some countries for dysentery, and its flowers are reputed to be anti-aphrodisiac. *Nuphar luteum* is indigenous, like the preceding; its flower exhales an alcoholic odour, and is said to have the same soothing properties as *Nymphæa*; its leaves are astringent, and are given in Germany in cases

Nelumbium.
Embryo with spreading
lobes (mag.).

Nelumbium.
Embryo, one lobe re-
moved to show the
plumule (mag.).

of hæmorrhage; it is also used by nurses to reduce the secretion of milk, and its rootstock is eaten in Russia and Finland. The same is the case with the rootstock and seeds of *Euryale ferox*, a native of India, and cultivated in China under the name of Kiteou. The *Maruru*, dedicated to the Queen of England (*Victoria regia*), is the most beautiful of the *Nymphæaceæ*; it inhabits the tranquil waters of the lagunes formed by the overflow of the large rivers of South America. Its leaves are floating and peltate, their circular blade is 12–15 feet in girth, and its edge 2½–6 inches high; the upper surface is of a brilliant dark green, the under of a red brown, furnished with large reticulated prominent cellular ribs, full of air, and bristling, like the petiole and peduncle, with elastic prickles. The flowers, which rise a few inches above the water, are more than 30 inches in circumference: at first of a pure white, in twenty-four hours they successively pass through pale pink to bright red; they exhale an agreeable scent during the first day of their blossoming; at the end of the third day the flower withers and sinks into the water to ripen its seeds. The fruit, which is inferior, attains when ripe the size of a large depressed apple, covered with prickles. The seeds, which are known in the province of Corrientes as *water maize*, are rich in starch, and are roasted by the natives, who consider them excellent food. *Brasenia peltata* is used in North America as a mild astringent. *Nelumbium specionum* was the Lotus of the Egyptians; its leaves, peltate and saucer-shaped, are represented on their monuments and the statues of their gods; its pink flowers resemble enormous tulips, and its fruit-bearing peduncles served as a model for the columns of their buildings. This species grows in several parts of Asia, as far as the mouths of the Volga, but is no longer met with in Egypt; its seeds, formerly called *Egyptian beans*, still serve as food to the Indians and Chinese, who also use its petals as an astringent. *N. luteum* inhabits the large rivers of Louisiana and Carolina.

XI. *SARRACENIACEÆ, Endlicher.*

Perennial HERBS, inhabiting the turfy spongy bogs of North America and Guiana. ROOT fibrous. LEAVES all radical, with a tubular or amphora-shaped petiole; blade small, rounded, usually lying on the orifice of the petiole. SCAPES naked, or furnished with a few bracts, 1-flowered (*Sarracenia, Darlingtonia*), or terminated by a few-flowered raceme (*Heliamphora*). FLOWERS large, nodding. SEPALS 4–5, free, very much imbricated at the base, sub-petaloid, persistent. PETALS 5, free, hypogynous, imbricate, deciduous, rarely 0 (*Heliamphora*). STAMENS ∞, hypogynous, free; *filaments* filiform; *anthers* 2-celled, versatile, opening by 2 longitudinal slits. OVARY free, 3–5-celled, placentas prominent at the inner angle of the cells; *style* terminal, short, sometimes dilated at the top, as a 5-angled or -lobed petaloid parasol with 5 radiating nerves (*Sarracenia*), or 5-fid, lobes narrow, spreading, reflexed, stigmatiferous (*Darlingtonia*), or obtuse and terminated by an obscurely 3-lobed stigma (*Heliamphora*); *ovules* numerous, many-seriate, sub-horizontal, anatropous, raphe lateral. CAPSULE 3–5-celled, loculicidally 3–5-valved. SEEDS ∞, small; testa crustaceous, sometimes loosely reticulate (*Darlingtonia*), or membranous and winged (*Heliamphora*); *albumen* copious, fleshy. EMBRYO minute, near the hilum.

GENERA.

Sarracenia.　　　　　Darlingtonia.　　　　　Heliamphora.

This little family approaches *Papaveraceæ* in hypopetalism, polyandry, numerous ovules, capsular fruit, fleshy copious albumen, and minute basilar embryo; but *Papaveraceæ* differ much in habit, proper juice, caducous dimerous calyx, and one-celled ovary with parietal placentation. *Sarraceniaceæ* are connected with *Nymphæaceæ* by the same characters, and also by the always radical leaves, one-flowered

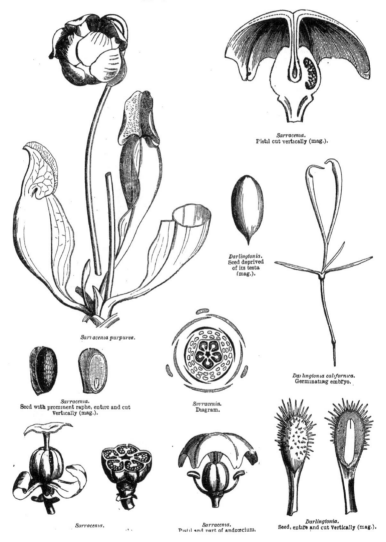

Sarracenia.
Pistil cut vertically (mag.).

Darlingtonia.
Seed deprived
of its testa
(mag.).

Sarracenia purpurea.

Darlingtonia californica.
Germinating embryo.

Sarracenia.
Seed with prominent raphe, entire and cut
vertically (mag.).

Sarracenia.
Diagram.

Sarracenia.

Sarracenia.
Pistil and part of androecium.

Darlingtonia.
Seed, entire and cut vertically (mag.).

scape, and aquatic habitat; but *Nymphæaceæ* differ in their numerous several-seriate petals, placentation, sessile stigma, and double albumen. Certain affinities or analogies have also been indicated, which link *Sarraceniaceæ* with *Droseraceæ, Pyrolaceæ, Nepentheæ,* and *Cephaloteæ.*

[All are natives of America, and chiefly of the United States. *Darlingtonia* inhabits the Rocky Mountains, and *Heliamphora* the Roruma Mountains of Venezuela.

Of the properties of *Sarraceniaceæ* little is known. *Sarracenia rubra* has been vaunted in Canada as a specific against small-pox, but has not proved such. The pitcher-shaped leaves are effective insect traps: a sugary secretion exudes at the mouth of the pitcher, and attracts the insects, which descend lower in the tube, where they meet with a belt of reflexed hairs, which facilitate their descent into a watery fluid that fills the bottom of the cavity, and at the same time prevent their egress.—ED.]

XII. *PAPAVERACEÆ, Jussieu.*

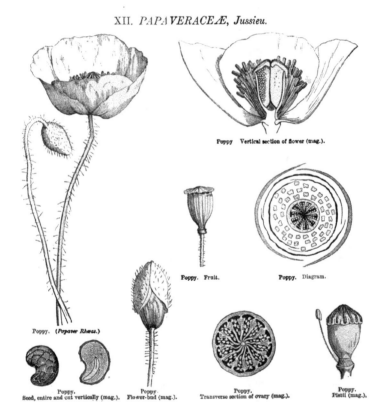

Poppy　Vertical section of flower (mag.).

Poppy. Fruit.

Poppy. Diagram.

Poppy. (*Papaver Rhœas.*)

Poppy.
Seed, entire and cut vertically (mag.).

Poppy.
Flower-bud (mag.).

Poppy.
Transverse section of ovary (mag.).

Poppy.
Pistil (mag.).

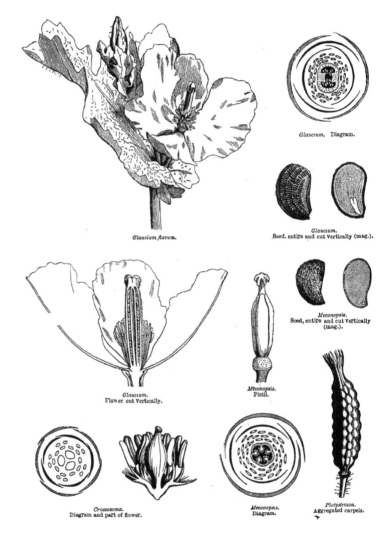

Glaucium flavum.

Glaucium. Diagram.

Glaucium.
Seed. entire and cut vertically (mag.).

Meconpsis.
Seed, entire and cut vertically
(mag.).

Glaucium.
Flower cut vertically.

Meconpsis.
Pistil.

Crossosoma.
Diagram and part of flower.

Meconpsis.
Diagram.

Platystemon.
Aggregated carpels.

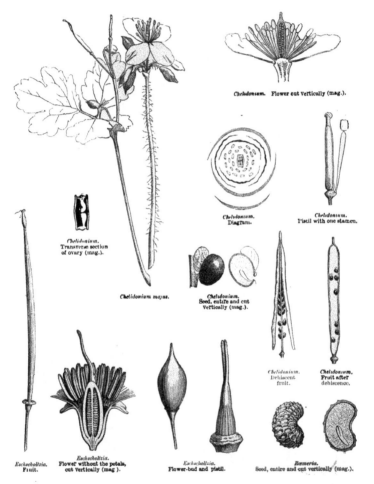

Chelidonium. Flower cut vertically (mag.).

Chelidonium. Diagram.

Chelidonium. Pistil with one stamen.

Chelidonium. Transverse section of ovary (mag.).

Chelidonium majus.

Chelidonium. Seed, entire and cut vertically (mag.).

Chelidonium. Dehiscent fruit.

Chelidonium. Fruit after dehiscence.

Eschscholtzia. Fruit.

Eschscholtzia. Flower without the petals, cut vertically (mag.).

Eschscholtzia. Flower-bud and pistil.

Rœmeria. Seed, entire and cut vertically (mag.).

SEPALS 2, *rarely* 3. PETALS *double or multiple the number of the sepals, free, regular, usually hypogynous.* STAMENS ∞, *hypogynous, free.* OVARY *1-celled, with parietal placentas, many-ovuled.* FRUIT *capsular or siliquose.* SEEDS *albuminous.* EMBRYO *minute, basilar.*—STEM *herbaceous.* LEAVES *alternate.*

Annual or perennial HERBS, rarely suffrutescent (*Bocconia, Dendromecon*); juice milky, yellow, white, or red, rarely watery (*Eschscholtzia, Hunnemannia, Platystemon,* &c.). LEAVES alternate, simple, penninerved, toothed or pennilobed. INFLORES-CENCE terminal; *peduncles* usually 1-flowered, rarely bearing an umbellate cyme (*Chelidonium*) or panicle (*Bocconia, Macleya*). FLOWERS ☿, regular, yellow or red, very rarely blue (*Meconopsis Wallichii,* &c.). SEPALS 2 (rarely 3), free, or very rarely coherent into a cap (*Eschscholtzia*), lateral, overlapping each other, caducous. PETALS hypogynous, very rarely perigynous (*Eschscholtzia*), equal, free, usually double the number of the sepals, rarely 8 or 12, 2–3-seriate (*Sanguinaria*), rarely 0 (*Bocconia, Macleya*), often crumpled before expansion, the outer equitant on the inner. STAMENS hypogynous, very rarely perigynous (*Eschscholtzia*), free, usually ∞ many-seriate, rarely 4–6 1-seriate (*Platystigma,* section *Meconella*); *filaments* filiform ; *anthers* 2-celled, basifixed, dehiscence longitudinal. CARPELS connate into an ovoid or oblong 1-celled ovary; placentas 2-∞, parietal, sometimes prolonged into vertical incomplete septa (*Papaver*), sometimes marginal and filiform (*Chelidonium, Argemone, Rœmeria,* &c.); *style* short or obsolete ; *stigmas* as many as placentas, persistent, more or less connate, sub-sessile, or arranged in rays on the surface of an orbicular disk formed by the styles, and which crowns the ovary (*Papaver*) ; *ovules* anatropous, ascending or horizontal, micropyle inferior, raphe superior or lateral. Ripe CARPELS very rarely distinct (*Platystemon*), generally connate into a capsule, or one-celled siliqua, rarely 2-celled owing to a cellular development of the placentas (*Glaucium*); opening either by valves between the placentas (*Papaver*), or in two or four valves separating from below upwards (*Chelidonium*), or from above downwards, leaving the placentas exposed (*Glaucium, Stylophorum*) ; rarely fleshy when young (*Bocconia, Sanguinaria*). SEEDS usually numerous, rarely definite (*Macleya*), or solitary (*Bocconia*), globose or ovoid, sub-reniform (*Papaver*), or crested along the raphe (*Chelidonium,* &c.) or not; *albumen* copious, oily. EMBRYO minute, basilar; *radicle* near the hilum and centrifugal.

PRINCIPAL GENERA.

* Platystemon.	Platystigma.	* Papaver.	* Argemone.	* Meconopsis.
Stylophorum.	* Sanguinaria.	* Bocconia.	* Glaucium.	Rœmeria.
* Chelidonium.	Hunnemannia.	* Eschscholtzia.	Dendromecon.	

Papaveraceæ are closely allied to *Fumariaceæ*, which are only distinguished by their irregular petals, their definite, usually diadelphous stamens, and their non-oleaginous albumen. They approach *Cruciferæ* in their flower formed on the binary type, in hypopetalism, parietal placentation, capsular siliquose fruit, polyandry (*Megacarpæa*), and oily seed ; but *Cruciferæ* are usually tetradynamous, their ovary is two-celled, their ovules are campylotropous, and their seed exalbuminous. *Papaveraceæ* are also near *Ranunculaceæ, Berberideæ,* and *Nymphæaceæ* (see these families). One monotypal Californian genus, *Crossosoma,* placed among *Ranunculaceæ,* approaches *Papaveraceæ* in its monosepalous calyx, polyandry, perigynous petals and stamens (as in *Eschscholtzia*), and in the separation of the carpels (as in *Platystemon*) ; it differs in its isomerous calyx and corolla, and multifid axil enveloping the seeds.

Papaveraceæ inhabit the temperate and subtropical regions of the northern hemisphere; but few are met with in the tropics or southern hemisphere. Some species are now dispersed over cultivated ground throughout the world. The most important of the *Papaveraceæ* with milky juice is the *Papaver somniferum,* an annual herb, a native of Asia. Its juice, obtained by a superficial incision of its capsule, and thickened by exposure, is *opium*, a substance containing several principles, and notably an alkaloid (*morphine*), whose powerful properties render this one of the most valuable of drugs. Taken in a large dose, it is a deadly poison; but habit rapidly weakens its action, and by degrees considerable quantities can be swallowed with impunity. Orientals, and especially the Chinese, drink, chew, or smoke opium to procure intoxication, the daily renewal of which becomes a want, which they satisfy at any price; when they soon fall into a state of physical and moral degradation, from which nothing can rescue them.

In the north of France a variety of *P. somniferum* is largely cultivated, the seeds of which are blackish when ripe, and yield by expression a bland oil, used like olive oil, and known as *white oil* and *oleolum*. The petals of the Field-Poppy (*P. Rhœas*) are mucilaginous, emollient, and slightly narcotic. *Chelidonium majus* is a perennial herb, found in cultivated spots. The yellow and acrid milky juice, which fills all parts of the plant, is used in Europe to destroy warts and to remove specks on the cornea; it is regarded in Brazil,[1] either rightly or wrongly, as efficacious against the bite of venomous serpents. The juice of *Argemone mexicana* possesses, it is said, the same virtues. The root of *Sanguinaria canadensis*, which contains a red juice, is acrid and bitter, and colours the saliva of a bright red; sedative properties similar to those of *Digitalis* are attributed to it, and its narcotic seeds are considered equally powerful as those of *Datura Stramonium* (Thorn-Apple).

XIII. *FUMARIACEÆ, D.C.*

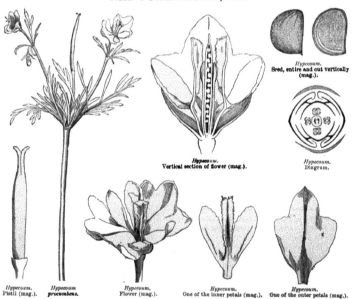

Hypecoum.
Seed, entire and cut vertically (mag.).

Hypecoum.
Vertical section of flower (mag.).

Hypecoum.
Diagram.

Hypecoum.
Pistil (mag.).

Hypecoum procumbens.

Hypecoum.
Flower (mag.).

Hypecoum.
One of the inner petals (mag.).

Hypecoum.
One of the outer petals (mag.).

[1] There is probably some error here, *Chelidonium* not being a native of Brazil.—ED.

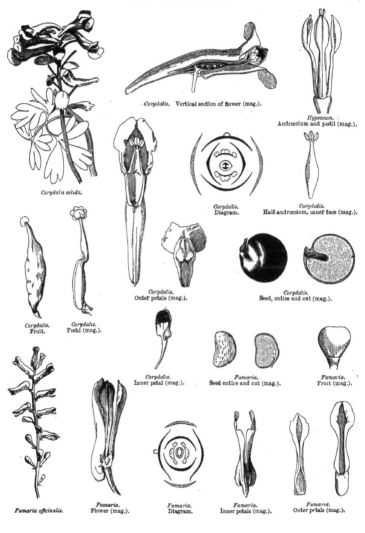

Corydalis, Vertical section of flower (mag.).

Hypecoum.
Andrœcium and pistil (mag.).

Corydalis solida.

Corydalis.
Diagram.

Corydalis.
Half-andrœcium, inner face (mag.).

Corydalis.
Outer petals (mag.).

Corydalis.
Seed, entire and cut (mag.).

Corydalis.
Fruit.

Corydalis.
Pistil (mag.).

Corydalis.
Inner petal (mag.).

Fumaria.
Seed entire and cut (mag.).

Fumaria.
Fruit (mag.).

Fumaria officinalis.

Fumaria.
Flower (mag.).

Fumaria.
Diagram.

Fumaria.
Inner petals (mag.).

Fumaria.
Outer petals (mag.).

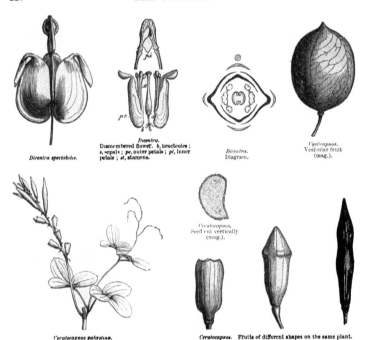

Dicentra spectabilis.

Dicentra.
Dismembered flower. *b*, bracteoles; *s*, sepals; *pe*, outer petals; *pi*, inner petals; *st*, stamens.

Dicentra.
Diagram.

Cysticapnos.
Vesicular fruit
(mag.).

Ceratocapnos.
Seed cut vertically
(mag.).

Ceratocapnos palæstina.

Ceratocapnos. Fruits of different shapes on the same plant.

Herbaceous, annual or perennial, usually glaucous plants, with watery juice. STEM sometimes tuberous, rarely sarmentose. LEAVES alternate, cut. FLOWERS ☿, irregular, terminal, in a raceme or spike, or sometimes solitary. SEPALS 2, antero-posterior, free, petaloid or scale-like, imbricate in bud, caducous. PETALS hypogynous, 4, free or connate at the base, 2-seriate, the 2 outer lateral, alternate with the sepals, differing from the inner and equitant upon them, equal (*Hypecoum, Dicentra, Adlumia*) or unequal, one being spurred or gibbous, the other flat, 2 inner petals placed crosswise to the outer, oblong-linear, sub-callous, and coherent at the tips, which enclose the anthers and stigmas. STAMENS rarely 4, free, with 2-celled anthers (*Hypecoum*), usually 6, united by their filaments in two bundles opposite to the outer petals, and each composed of 3 anthers, of which the 2 lateral are 1-celled, and the median is 2-celled (*Fumaria, Sarcocapnos, Corydalis, Adlumia, Dicentra*); anthers extrorse, dehiscence sometimes lateral (*Hypecoum*). OVARY free; *style* simple, sometimes 2-fid (*Hypecoum*); *stigma* usually forming two crenulated lobes ; *ovules* half-anatropous, one or several on parietal placentas. FRUIT siliquose, many-seeded, 2-valved (*Corydalis, Adlumia, Dicentra*), or vesicular (*Cysticapnos*), or 1–2-seeded and indehiscent (*Fumaria, Sarcocapnos*), or jointed and divided by transverse septa into 1-seeded indehiscent cellules (*Ceratocapnos, Hypecoum*). SEEDS horizontal, hilum usually naked, sometimes strophiolate (*Diçentra, Corydalis*) ; *albumen* fleshy. EMBRYO usually minute, nearly straight, basilar, often only visible at the moment of germination, and having apparently only one oval cotyledon (*Corydalis*).

GENERA.

Hypecoum.	*Dicentra.	Pteridophyllum.	*Adlumia.	*Corydalis.
Sarcocapnos.	*Fumaria.			

Fumariaceæ are so closely allied to *Papaveraceæ* (see that family), that many modern botanists have united them, for they differ chiefly in the dissimilarity between their inner and outer petals, and in their definite stamens. Like *Papaveraceæ*, they approach *Cruciferæ* in their corolla, hypopetalism, parietal placentation, curved ovule, and the structure of the fruit, but differ in their irregular flowers with two sepals, diadelphous stamens, albuminous seed, and minute and basilar embryo. They inhabit the temperate northern hemisphere, and especially the Mediterranean region and North America. Some (*Cysticapnos, Phacocapnos*) inhabit South Africa ; none have been observed in the hot regions of the tropics.

Most *Fumariaceæ* contain in their herbaceous parts mucilage, saline substances, and a peculiar acid or acrid juice, so combined that they are classed among tonic and alterative medicines. The Common Fumitory (*Fumaria officinalis*) occurs everywhere in corn-fields and on rubbish heaps ; its juice is bitter, stomachic, and depurative. The roots of *Corydalis bulbosa* and *fabacea* are sub-aromatic, very bitter and slightly astringent, and employed as emmenagogues and vermifuges. The rather bitter and very acrid foliage of *C. capnoïdes* is a reputed stimulant.

XIV. *CRUCIFERÆ.*

(TETRAPETALÆ, *Ray.*—SILIQUOSÆ, *Magnol.*—CRUCIFORMES, *Tournefort.*—TETRADY-NAMÆ, *L.*—ANTISCORBUTICÆ, *Crantz.*—CRUCIATÆ, *Haller.*—CRUCIFERÆ, *Adanson.*—BRASSICACEÆ, *Lindl.*)

SEPALS 4. PETALS 4, *hypogynous.* STAMENS 6, *tetradynamous.* OVARY *sessile,* 2- (*rarely* 1-) *celled, with* 2 *parietal placentas.* FRUIT *a siliqua or silicula, or nut or lomentum.* SEEDS *exalbuminous.* EMBRYO *oily, bent, rarely straight.*

Usually HERBS, rarely suffruticose, with watery juice, often rather acrid ; hairs when present simple or stellate, or fixed by the middle, very rarely glandular. STEM cylindric or angular, sometimes spinescent. LEAVES simple, alternate, rarely opposite, entire, lobed or dissected, the radical often runcinate, and the cauline often auricled at the base ; *stipules* generally 0. FLOWERS ⚥, in a raceme, rarely solitary on a scape ; racemes usually terminal, corymbose when young, rarely bracteate. COROLLA white, yellow, or purplish, rarely blue or pink. SEPALS 4, free, 2-seriate, the 2 outer opposite, antero-posterior, answering to the placentas, the 2 inner lateral, often larger, and gibbous at the base, imbricate in bud, very rarely valvate (*Ricotia, Savignya,* &c.). PETALS 4, hypogynous, rarely 0 (*Armoracia, Lepidium* (some), *Cardamine* (some), &c.), arranged crosswise, usually entire, equal, or the outer

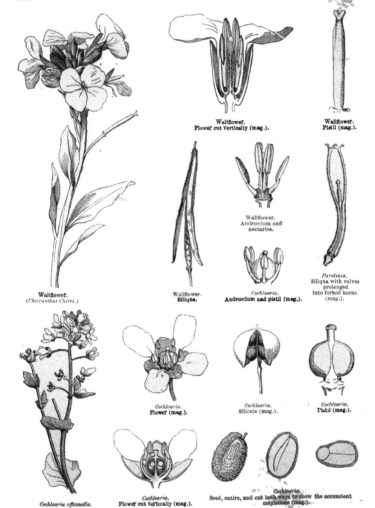

Wallflower.
Flower cut vertically (mag.).

Wallflower.
Pistil (mag.).

Wallflower.
Androecium and
nectaries.

Wallflower.
(*Cheiranthus Cheiri.*)

Wallflower.
Siliqua.

Cochlearia.
Androecium and pistil (mag.).

Parolinia.
Siliqua with valves
prolonged
into forked horns
(mag.).

Cochlearia.
Flower (mag.).

Cochlearia.
Silicule (mag.).

Cochlearia.
Pistil (mag.).

Cochlearia officinalis.

Cochlearia.
Flower cut vertically (mag.).

Cochlearia.
Seed, entire, and cut both ways to show the accumbent
cotyledons (mag.).

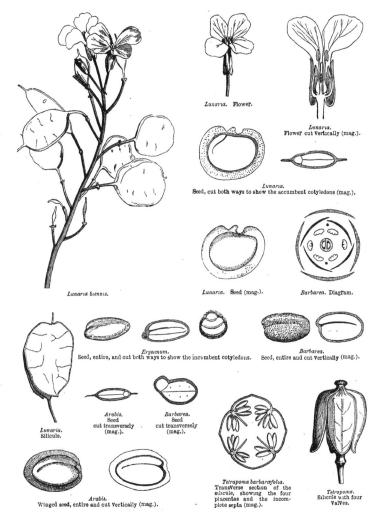

Lunaria. Flower.

Lunaria.
Flower cut Vertically (mag.).

Lunaria.
Seed, cut both ways to show the accumbent cotyledons (mag.).

Lunaria biennis.

Lunaria. Seed (mag.).

Barbarea. Diagram.

Erysimum.
Seed, entire, and cut both ways to show the incumbent cotyledons.

Barbarea.
Seed, entire and cut Vertically (mag.).

Arabis.
Seed
cut transversely
(mag.).

Barbarea.
Seed
cut transversely
(mag.).

Lunaria.
Silicule.

Arabis.
Winged seed, entire and cut Vertically (mag.).

Tetrapoma barbaræfolia.
Transverse section of the
silicule, showing the four
placentas and the incom-
plete septa (mag.).

Tetrapoma.
Silicule with four
Valves.

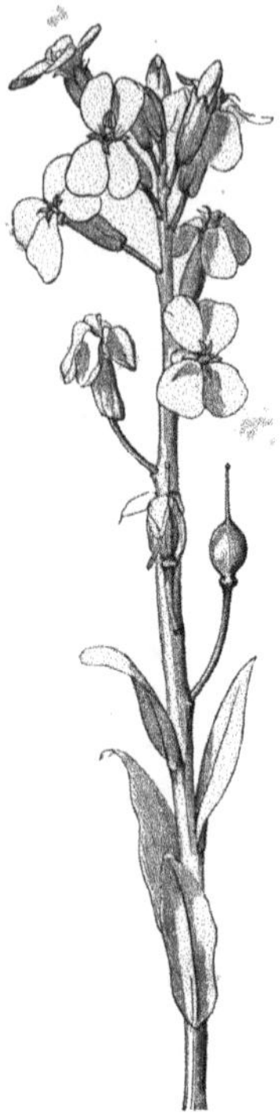

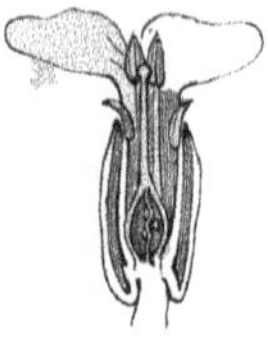

Vesicaria.
Flower cut vertically
(mag.).

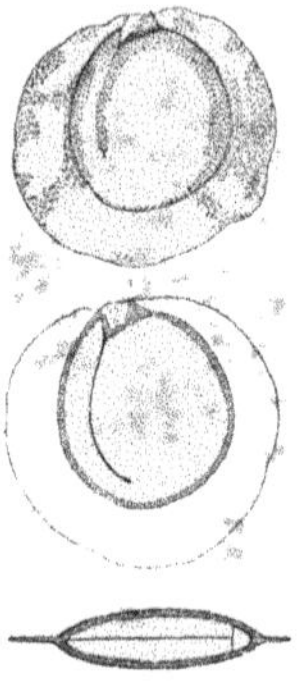

Vesicaria.
Seed, entire, and cut both
ways to show the accumbent
cotyledons.

Vesicaria utriculata.

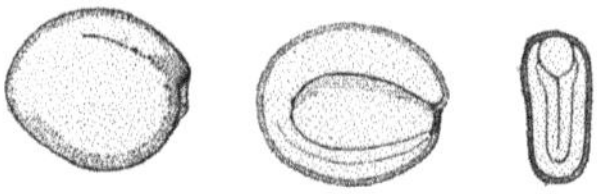

Eruca.
Seed, entire, and cut both ways to show the cotyledons
folded lengthwise, and accumbent.

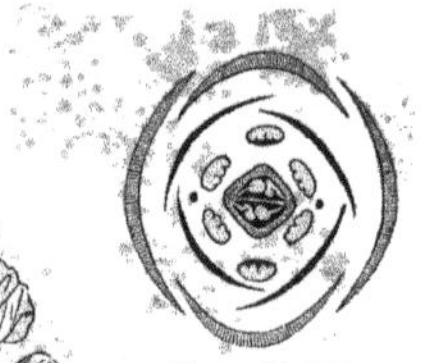

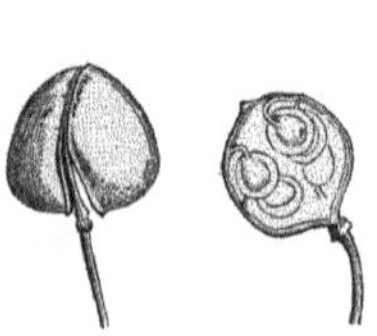

Eruca. Diagram.

Eruca.
Siliqua
terminated
by the style
enlarged
into a beak.

Eruca.
Pistil with
an ensiform
style (mag.).

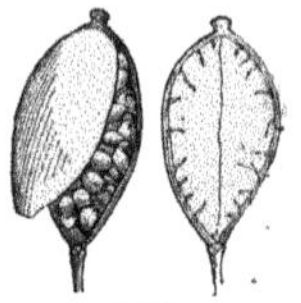

Eruca sativa.

Erophila.
Silicule   Placentas and
(mag.).   septum (mag.).

Vesicaria.
Silicule, entire and without its valves
(mag.).

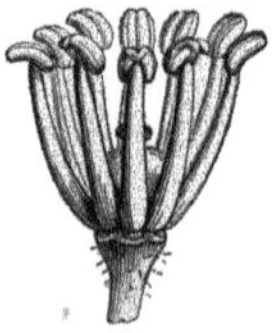

Megacarpœa polyandra.
Andrœcium.

Megacarpœa.
Pistil (mag.).

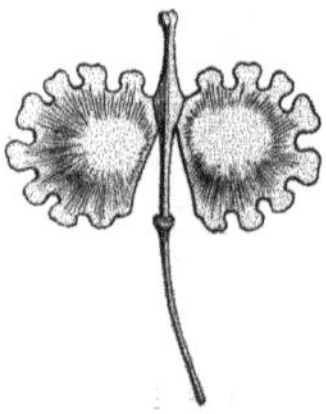

Cremolobus sinuatus.
Silicule (mag.).

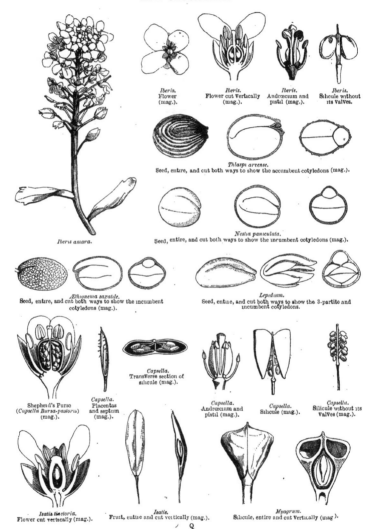

Iberis.
Flower
(mag.).

Iberis.
Flower cut Vertically
(mag.).

Iberis.
Androecium and
pistil (mag.).

Iberis.
Silicule without
its Valves.

Thlaspi arvense.
Seed, entire, and cut both ways to show the accumbent cotyledons (mag.).

Nestia paniculata.
Seed, entire, and cut both ways to show the incumbent cotyledons (mag.).

Iberis amara.

Æthionema saxatile.
Seed, entire, and cut both ways to show the incumbent
cotyledons (mag.).

Lepidium.
Seed, entire, and cut both ways to show the 3-partite and
incumbent cotyledons.

Shepherd's Purse
(*Capsella Bursa-pastoris*)
(mag.).

Capsella.
Placentas
and septum
(mag.).

Capsella.
Transverse section of
silicule (mag.).

Capsella.
Androecium and
pistil (mag.).

Capsella.
Silicule (mag.).

Capsella.
Silicule without its
Valves (mag.).

Isatis tinctoria.
Flower cut vertically (mag.).

Isatis.
Fruit, entire and cut vertically (mag.).

Myagrum.
Silicule, entire and cut Vertically (mag.).

Q

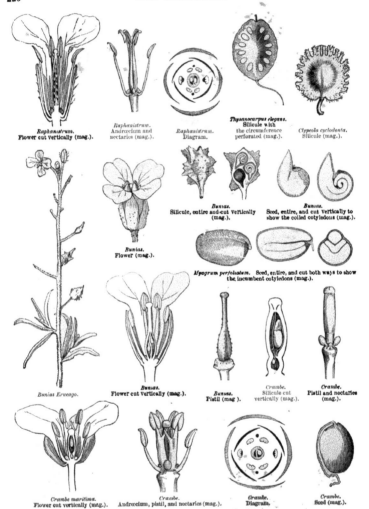

Raphanistrum.
Flower cut vertically (mag.).

Raphanistrum.
Andrœcium and
nectaries (mag.).

Raphanistrum.
Diagram.

Thysanocarpus elegans.
Silicule with
the circumference
perforated (mag.).

Clypeola cyclodonta.
Silicule (mag.).

Bunias.
Silicule, entire and cut vertically
(mag.).

Bunias.
Seed, entire, and cut vertically to
show the coiled cotyledons (mag.).

Bunias.
Flower (mag.).

Myagrum perfoliatum. Seed, entire, and cut both ways to show
the incumbent cotyledons (mag.).

Bunias Erucago.

Bunias.
Flower cut vertically (mag.).

Bunias.
Pistil (mag).

Crambe.
Silicule cut
vertically (mag.).

Crambe.
Pistil and nectaries
(mag.).

Crambe maritima.
Flower cut vertically (mag.).

Crambe.
Andrœcium, pistil, and nectaries (mag.).

Crambe.
Diagram.

Crambe.
Seed (mag.).

larger, variously imbricate in bud. *Glands* sessile at the base or on the circumference of the torus, usually 4, opposite to the sepals, or 2 or 6, or forming a continuous variously lobed ring, sometimes 0. STAMENS hypogynous, 6, of which 2 are short, and opposite the lateral sepals, and 4 longer facing the placentary sepals, and close together in pairs, or coherent; sometimes reduced to 4 or 2 (*Lepidium* (some), *Capsella* (some), *Senebiera* (some)), rarely ∞ (*Megacarpæa polyandra*); *filaments* subulate, the longest sometimes 1-toothed, rarely dilated or appendaged (*Lepidostemon*); *anthers* 2- (very rarely 1-) celled (*Atelanthera*), introrse, dehiscence longitudinal, basifixed, cordate or sagittate, sometimes linear (*Parrya*) or twisted (*Stanleya*). OVARY of 2 connate carpels (very rarely 3-4, *Tetrapoma*), placed right and left of the floral axis, sessile, rarely stipitate (*Warea*, &c.), placentation parietal, usually 2-celled by cellular plates springing from the placentas, and dilated into a false vertical septum; sometimes 1-celled, with parietal, basilar or apical placentation; sometimes divided into several superimposed cellules by spongy transverse septa (*Raphanus*); *style* simple, or dilated, or appendaged, below the stigmas; *stigmas* 2, opposite to the placentas, erect or divergent, or united into one, sometimes decurrent on the style; *ovules* ∞, or few or solitary, pendulous or horizontal, very rarely solitary and basilar in the 1-celled ovaries (*Clypeola*, *Dipterygium*), or apical (*Isatis*, *Tauscheria*, *Euclidium*), campylotropous or half-anatropous, raphe ventral, and micro-. pyle superior. FRUIT elongated (*siliqua*), or short (*silicula*), usually 2-celled, or 1-celled from arrest of the septum (*Isatis*, *Clypeola*, *Calepina*, *Myagrum*, &c.), usually with 2 valves separating from the placentas, rarely with 3-4 valves (*Tetrapoma*), sometimes indehiscent from the cohesion of the valves (*Raphanus*), rarely transversely divided into 2 one- or more-seeded joints, of which the upper (*Erucaria*, *Morisia*, &c.), or the lower (*Crambe*, *Rapistrum*, *Cakile*, *Enarthrocarpus*) is indehiscent. SEEDS subglobose or margined or winged; *testa* cellular, usually becoming mucilaginous when moistened. EMBRYO oily, curved, very rarely straight (*Leavenworthia*), exalbuminous or very rarely enveloped in a layer of fleshy albumen (*Isatis*, some); *cotyledons* subaerial, usually plano-convex, accumbent (*Pleurorhizeæ*) or incumbent (*Notorhizeæ*) relatively to the radicle (which is usually ascendent), rarely oblique, sometimes folded in two along their length and embracing the radicle (*Orthoploceæ*), rarely linear, and folded twice transversely (*Diplecolobeæ*), very rarely linear and coiled transversely upon themselves (*Spirolobeæ*).

TRIBE I. *ORTHOPLOCEÆ, D.C.*

Cotyledons longitudinally conduplicate, embracing the dorsal radicle.

PRINCIPAL GENERA.

* Sinapis.	Eruca.	* Brassica.	Hirschfeldia.	Erucastrum.
Diplotaxis.	Vella.	Moricandia.	Calepina.	* Crambe.
Morisia.	Rapistrum.	Enarthrocarpus.	* Raphanus.	Raphanistrum.

TRIBE II. *PLATYLOBEÆ.*

(PLEURORHIZEÆ ET NOTORHIZEÆ, *D.C.*)

Cotyledons plane. Radicle lateral or dorsal.

PRINCIPAL GENERA.

(Siliquose Platylobeæ.)

* Hesperis.	* Malcolmia.	* Cheiranthus.	* Matthiola.	* Erysimum.
* Barbarea.	Sisymbrium.	Alliaria.	* Nasturtium.	* Arabis.
* Cardamine.	Dentaria.			

(Siliculose Platylobeæ.)

* Lunaria.	* Farsetia.	* Aubrietia.	* Vesicaria.	* Alyssum.
Clypeola.	Peltaria.	Draba.	Erophila.	* Armoracia.
* Cochlearia.	Tetrapoma.	Neslia.	* Myagrum.	* Camelina.
Biscutella.	Megacarpæa.	* Lepidium.	Hutchinsia.	* Iberis.
Teesdalia.	* Æthionema.	Thlaspi.	Capsella.	Cakile.
Isatis.	Anastatica.			

TRIBE III. *SPIROLOBEÆ, D.C.*

Cotyledons linear, coiled transversely upon themselves. Radicle dorsal.

PRINCIPAL GENERA.

* Bunias.   * Schizopetalum.

TRIBE IV. *DIPLECOLOBEÆ, D.C.*

Cotyledons linear, folded twice transversely upon themselves. Radicle dorsal.

PRINCIPAL GENERA.

Coronopus.   Subularia.   * Heliophila.

CLASSIFICATION OF CRUCIFERÆ.

BY BENTHAM AND HOOKER FIL.

Series A.—Siliqua long or short, dehiscent throughout its length. Valves continuous within, rarely septiferous, plane or concave, not compressed, in a perpendicular direction to the plane of the septum. Septum the same breadth as the valves.

TRIBE I. ARABIDEÆ.—Siliqua narrow, long, seeds often 1-seriate. Cotyledons accumbent. *Matthiola, Cheiranthus, Atelanthera, Nasturtium, Barbarea, Arabis, Cardamine, Lonchophora, Anastatica,* &c.

TRIBE II. ALYSSINEÆ.—Siliqua often short, large, seeds 2-seriate. **Cotyledons** accumbent. *Lunaria, Farsetia, Aubrietia, Vesicaria, Alyssum, Draba, Erophila, Cochlearia,* &c.

TRIBE III. SISYMBRIEÆ.—Siliqua narrow, long, seeds often 1-seriate. Cotyledons incumbent, straight or coiled, or transversely folded. *Schizopetalum, Hesperis, Malcolmia, Streptoloma, Sisymbrium, Conringia, Erysimum, Heliophila,* &c.

TRIBE IV. CAMELINEÆ.—Siliqua short or long, oblong, ovoid or globular. Seeds 2-seriate. Cotyledons incumbent. *Stenopetalum, Braya, Camelina, Tetrapoma, Subularia,* &c.

TRIBE V. BRASSICEÆ.—Siliqua short or long, dehiscent throughout its length, or at the top only. Cotyledons folded longitudinally. *Brassica, Sinapis, Erucastrum, Hirschfeldia, Diplotaxis, Eruca, Moricandia, Vella, Carrichtera, Succovia,* &c.

Series B.—Siliqua short, dehiscent throughout its length. Valves continuous within, very concave, compressed in a direction perpendicular to the plane of the septum. Septum usually very narrow.

TRIBE VI. LEPIDINEÆ.—Cotyledons incumbent, straight or bent, or conduplicate longitudinally, or coiled upon themselves. *Capsella, Senebiera, Lepidium, Æthionema, Campyloptera,* &c.

TRIBE VII. THLASPIDEÆ.—Cotyledons accumbent, straight. *Cremolobus, Biscutella, Megacarpœa, Thlaspi, Iberis, Teesdalia, Hutchinsia, Iberidella,* &c·

Series C. - Siliqua short (rarely long), indehiscent, not jointed, often crustaceous or bony, 1-celled, 1- (rarely 2-) seeded, or 2–4-celled with parallel 1-seeded cells. Pedicels often very slender, drooping in fruit. Seed often furnished with a thin albumen; testa not mucilaginous.

TRIBE VIII. ISATIDEÆ.—Characters of the series. *Peltaria, Clypeola, Isatis, Tauscheria, Neslia, Calepina, Myagrum, Euclidium, Bunias, Zilla,* &c.

Series D.—Siliqua transversely 2-jointed, short or long; lower joint indehiscent, empty or longitudinally 2-celled, 2-∞-seeded; upper joint indehiscent, 1-celled, 1-seeded, or 2-∞-celled, with parallel or superimposed cellules.—Siliqua always upright or nearly so, pedicel straight.

TRIBE IX. CAKILINEÆ.—Characters of the series. *Crambe, Muricaria, Rapistrum, Cakile, Enarthrocarpus, Erucaria, Morisia,* &c.

Series E.—Siliqua long, not jointed, indehiscent, cylindric or moniliform, 1-celled, many-seeded, or with several 1–2-seriate, 1-seeded cellules, separating when ripe.

TRIBE X. RAPHANEÆ.—Characters of the series. *Raphanus, Raffenaldia, Anchonium, Parlatoria,* &c.

A Cruciferous flower is not strictly symmetrical in relation to the floral axis. The arrangement of the calyx and corolla at first appears to follow the quaternary type, four sepals alternating with four petals; but the slightest examination shows that the two antero-posterior sepals are inserted lower than the two lateral; the petals, however, evidently form a single whorl. The exceptional structure of the androecium has given rise to many contradictory theories. The two lateral stamens are shorter and lower than the other four, which are in pairs, and alternate with the two lateral. It is these two pairs of long stamens which have especially exercised the sagacity of botanists. De Candolle, and after him Seringe, Saint-Hilaire, Moquin-Tandon, and Webb, admit the quaternary type for the calyx and corolla, and extend it equally to the androecium, where, according to them, each pair of long stamens represents a double stamen. This theory does not account for the lower position of the two antero-posterior stamens relatively to the lateral sepals, nor for the situation of the short stamens opposite to the carpels, which would be contrary to the laws of alternation.

Later botanists (Lestiboudois, Kunth, Lindley, and then J. Gay, Schimper, Wydler, Krause, Duchartre, Chatin, Godron) advocate an entirely different theory. They do not admit the doubling of the long stamens; they affirm, contrary to the organogenic observations of Payer, that in the very young flower each group of twin stamens springs from two distinct protuberances, separate from each other and exactly opposite to the petals. They consider the andrœcium to be composed of two quaternary whorls: 1st, the lower whorl, represented by the two lateral stamens only, and which is an imperfect one, from the constant arrest of two stamens which should be developed in front of the antero-posterior sepals; 2ndly, the upper whorl, composed of the four large stamens which were originally developed opposite to the petals, but which approach each other afterwards, so as to form two pairs. As to the pistil they consider it to be normally formed of four carpels opposite to the four sepals, an arrangement which is observable in the genus *Tetrapoma*. Thus the original plan of the flower may, according to them, be formulated as follows: four sepals, four petals, four outer stamens, of which two are never developed, four inner stamens, and four carpels, of which the two antero-posterior are developed in *Tetrapoma* only; all these whorls exactly alternating with each other.

More recently, A. G. Eichler has published (in 1865), in the 'Flora of Brazil,' the result of his researches. He affirms, with De Candolle, that each pair of long stamens results from the splitting up, or *chorisis*, of a single stamen, because, according to his organogenic researches, the protuberance from which each pair of stamens springs is originally simple, and only divides afterwards. With regard to the two-celled anthers of the twin stamens, which, according to the partisans of the non-development theory, ought to be one-celled, Eichler avers that this objection is valueless; that the question here is not of a doubling, which divides an entire organ in two halves; but of a chorisis, which results in a sort of multiplication of the organs; and that, further, in the genus *Atelanthera* the long stamens are constantly one-celled.

With regard to the polyandry observable in some species of *Megacarpæa*, in which the andrœcium is composed of 8–16 stamens, we may, according to Eichler, admit that it results from an unusual multiplication of the long stamens, and that the chorisis has been extended to the lateral stamens. We must, besides, remember that this tendency to multiplication (which is really exceptional, and not found in all the species of *Megacarpæa*) also appears in *Cleomeæ*, a tribe of *Capparideæ*, a family closely allied to *Cruciferæ*; their andrœcium, which is normally hexandrous, and arranged like that of *Cruciferæ*, presents in some species of *Cleome* four stamens, and in *Polanisia* eight or ∞, collected in antero-posterior bundles, the two lateral remaining solitary, or very rarely being represented by stamens. According to Eichler's organogenic observations, the anterior sepal appears first, and then the posterior, after which the two lateral sepals appear together; the four petals then appear simultaneously, and occupy four points, diagonally crossing the lateral sepals. The andrœcium first appears as two large obtuse gibbosities, opposite to the lateral sepals, which remain simple, and become the short stamens. Soon after their appearance the two similar antero-posterior gibbosities appear, inserted higher than the preceding, larger and more obtuse; and these, enlarging more and more, gradually divide into two protuberances, which finally become two long stamens. Never, says Eichler (who energetically maintains his opinion against that of Duchartre, Chatin, and Krause), never are these stamens, when young, exactly opposite to the petals; they are, on the contrary, then nearer to the median line, an arrangement which is still more obvious in some hexandrous *Capparideæ*. He affirms, contrary to the observations of Chatin, that in *Cruciferæ* with fewer than six stamens (*Lepidium ruderale, latifolium, virginianum,* &c.) the lateral stamens are inserted lower than the two antero-posterior, which proves that the latter do not belong to a lower whorl, as the incomplete-development theory demands. Eichler consequently regards the theory of chorisis as true, but his application of it differs from that of De Candolle, in that the latter makes three tetramerous whorls (andrœcium, corolla, and calyx), while Eichler only admits one, i.e. corolla, and assigns a binary type to the andrœcium and calyx; and his view of the composition of the flower would be:—two antero-posterior sepals, two lateral sepals, four petals diagonally crossing the lateral sepals, two short lateral stamens, two antero-posterior stamens (each doubled), and two lateral carpels valvately juxtaposed.

The family of *Cruciferæ* is closely allied to *Capparideæ, Papaveraceæ,* and *Fumariaceæ* (see these families). It also approaches *Resedaceæ* in habit, æstivation, hypopetalism, parietal placentation, curved ovule, and exalbuminous seed.

Cruciferæ are dispersed over the world; reaching, in the polar regions and on the highest mountains, the limits of phænogamic vegetation. Most of the genera and species inhabit the South of Europe and Asia Minor; they are rarer in the tropics, in extra-tropical and temperate North America.

The name *Antiscorbutics*, given by Crantz to the plants of this family, designates their most important property. They contain, besides oxygen, hydrogen, and carbon, a notable quantity of sulphur and azote. These elementary bodies form by their various combinations mucilage, starch, sugar, a fixed oil, albumine, and especially the elements of a peculiar volatile and very acrid oil, to which Crucifers owe their stimulating virtue. When dead, these ternary and quaternary products rapidly decompose to form binary compounds, and especially hydro-sulphuric acid and ammonia, the fœtid odour of which is insupportable.

The principal edible species is the Cabbage (*Brassica oleracea*), which has been cultivated from the most ancient times, and which yields *varieties* or *races* known under the names of Colza, Kail, Cabbage, Savoy, Cauliflower, Broccoli, &c. The Rape (*B. Rapa*) and the Turnip (*B. Napus*) have a fleshy root, rich in sugar and albumine; and their seeds contain a fixed oil, used for burning. The Radish (*Raphanus*), of which two species are cultivated, the one with a root black outside and white within; the other (Small Radish), with a white, pink, or violet root, is used as a condiment.

At the head of the antiscorbutic Crucifers must be placed the *Cochlearia officinalis*, a biennial herb which inhabits the shores of the seas and salt lakes of the North of Europe; its congeners of the European Alps, the Mediterranean region, Asia, and North America, possess similar properties, but in a less degree. The Garden Cress (*Lepidium sativum*) and Water Cress (*Nasturtium officinale*) are also used as condiments. *Lepidium oleraceum*, which grows on the shores of New Zealand, is an excellent antiscorbutic, and also an agreeable vegetable, which has proved invaluable to seamen; *Cardamine hirsuta*, *amara* and *pratensis*, indigenous species, which rival Water Cress, have an acrid and slightly bitter taste; *C. asarifolia* replaces Cochlearia in Piedmont; *C. nasturtioides* is eaten in Chili as cress is in France; *C. maritima*, which grows on the shores of the Atlantic and Mediterranean, has fallen into disuse; but *C. americana* has in North America and the Antilles a great reputation as an antiscorbutic. *Barbarea vulgaris* (Winter Cress), an indigenous plant, of an acrid and piquant taste, has been unjustly abandoned. *Sisymbrium officinale*, another common indigenous species, was formerly used as a cough medicine. *S. Alliaria*, whose bruised leaves exhale a strong odour of garlic, was long employed as a vermifuge, diuretic and depurative.

Rose of Jericho
(*Anastatica hierochuntica.*)

Sea-kale (*Crambe maritima*), which grows on the shores of the Atlantic and British Channel, is now cultivated; its spring shoots are blanched, and when cooked have so.uewhat the taste of Cauliflowers. *C. tatarica* inhabits the sandy plains of Hungary and Moravia; its large root, commonly called *Tartar bread*, is eaten, cooked or raw, seasoned with oil, vinegar, and salt.

Black Mustard (*Sinapis nigra*) grows in fields throughout Europe. Its powdered seeds are used as a condiment and rubefacient; it contains a fixed and very acrid volatile oil, to which latter its pungent quality is due. But this volatile oil does not exist there ready formed; it is produced by the action of a peculiar albumine (*myrosine*) on the *myronic* acid contained in the seed; it is this acid which becomes the volatile oil; and to effect this change the albumine must be soaked in cold water, which, by dissolving it, renders it fit to change the acid into a volatile oil. White Mustard (*S. alba*) contains principles analogous to those of the preceding species, the mucilaginous testa of the seed being superadded to an active principle, which stimulates the digestive organs. *S. chinensis* is valued in India as much as *S. nigra.*

Horseradish (*Cochlearia rusticana* or *Armoracia*) is cultivated in all gardens of central Europe; its root contains much sugar, starch, fatty oil, and albumine, and is eaten as a condiment. The acrid principle which it contains, and which is developed by the action of water, like that of *Sinapis*, gives it antiscorbutic properties.

The seeds of the Wild Radish (*Raphanistrum arvense*), of *Eruca sativa*, of Mithridate Mustard (*Thlaspi arvense*), and of Honesty (*Lunaria rediviva*), indigenous plants, have fallen into disuse, in spite of their stimulating acridity. Those of *Camelina sativa* contain a fixed oil, used for burning.

The leaves of Woad (*Isatis tinctoria*), a herb common throughout France, yield a blue dye, similar to indigo, but inferior, with which the Picts and Celts used to paint themselves; and from these early times blue has remained the national colour for our royal robes.

Anastatica hierochuntica is a small annual which grows in sandy places in Arabia, Egypt, and Syria. Its stem branches from the base, and bears sessile flowers, which give place to rounded pods; as these ripen, the leaves fall, the branches harden, dry, and curve inwards, and the plant contracts into a rounded cushion, which the autumn winds soon uproot, and carry even to the sea shore. Thence it is brought to Europe, where it fetches a high price, on account of its hygrometric properties; if the tip of its root be placed in water, or even if the plant be exposed to damp, the pods open and the branches uncurl, to close afresh when dry. This peculiarity, together with its native country (whence its name, Rose of Jericho), has given rise to the popular superstition that the flower expands yearly on the day and hour of Christ's birth. Women sometimes place the plant in water at the commencement of labour, hoping that its expansion may be the signal for their deliverance. Many other plants possess a similar hygrometric property;[1] as certain *Compositæ* of the genus *Asteriscus*, *Plantago cretica*, *Selaginella circinalis*, &c.

XV. *CAPPARIDEÆ.*

(CAPPARIDES, *Jussieu.*—CAPPARIDEÆ, *Ventenat.*—CAPPARIDACEÆ, *Lindl.*)

SEPALS 4–8, *free or coherent.* PETALS *hypogynous or perigynous*, 4–8 *or* 0. STAMENS *usually* 6, *or* ∞, *hypogynous or perigynous.* OVARY *usually stipitate and* 1-*celled, with parietal placentas.* OVULES *curved.* FRUIT *a siliquose capsule, or berry.* SEEDS *exalbuminous.* EMBRYO *arched or folded.*

Herbaceous annuals, or rarely perennials, often shrubby, sometimes arborescent (*Morisonia, Cratæva*, &c.), with watery juice. STEM and branches terete, glabrous, glandular, cottony, or rarely scaly (*Atamisquea, Capparis*). LEAVES alternate, or very rarely opposite (*Atamisquea*), petioled, simple or digitate, leaflets entire, very rarely toothed (*Cleome*), or lobed (*Thylachium*); *stipules* usually 0 or inconspicuous, setaceous

[1] In England, Mesembryanthemum capsules are sold as the Rose of Jericho; as is the Mexican *Selaginella lepidophylla.*—ED.

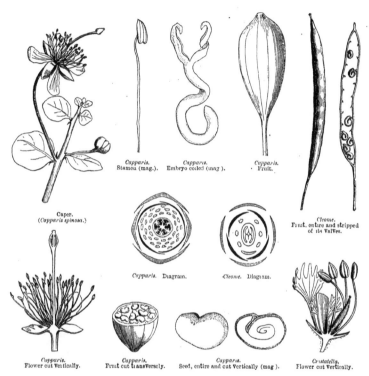

Capparis.
Stamen (mag.).

Capparis.
Embryo coiled (mag.).

Capparis.
Fruit.

Caper.
(Capparis spinosa.)

Cleome.
Fruit, entire and stripped
of its valves.

Capparis. Diagram.

Cleome. Diagram.

Capparis.
Flower cut vertically.

Capparis.
Fruit cut transversely.

Capparis.
Seed, entire and cut vertically (mag.).

Cristatella.
Flower cut vertically.

or spinescent (*Capparis*). FLOWERS ⚥, very rarely diœcious (*Apophyllum*), regular or sometimes sub-irregular, axillary, fascicled, solitary, or in a terminal raceme or corymb. SEPALS 4–8, sometimes free, 1–2-seriate, subequal or unequal; sometimes variously connate in a tubular calyx, sometimes closed and opening irregularly (*Cleome, Thylachium, Steriphoma*), æstivation imbricate, or rarely valvate. PETALS usually 4, rarely 0 (*Thylachium, Boscia, Niebuhria, &c.*), very rarely 2 (*Cadaba, Apophyllum*) or 8 (*Tovaria*), sessile or clawed, æstivation imbricate or twisted, very rarely valvate (*Ritchiea*), inserted on the edge of the torus. TORUS short or long, symmetrical or unsymmetrical, or discoid, or prolonged behind into an appendage, or depressed, or narrowed into a pedicel, or lining the bottom of the calyx, edge glandular or fringed. STAMENS inserted at the base or top of the torus, usually 6, rarely 4–8 (*Polanisia*,

Cadaba), often in multiples of 6 or 8, all fertile, or some sterile (*Dactylæna*, *Cleome*, *Polanisia*, &c.) ; *filaments* filiform, sometimes thickened at the top (*Cleome*), free or united to the torus, or connate at the base (*Gynandropsis*, *Cadaba*, *Boscia*, &c.) ; *anthers* introrse, 2-celled, oblong or ovoid, basi-dorsally fixed, dehiscence longitudinal. OVARY usually stipitate, rarely sessile, 1-celled, or sometimes 2-8-celled by false septa springing from the placentas (*Morisonia*, *Capparis*, *Tovaria*, &c.) ; *style* usually short or 0, simple (3, and hooked in *Roydsia*) ; *stigma* usually orbicular, sessile ; *ovules* numerous, fixed to parietal placentas, campylotropous or semi-anatropous, rarely solitary (*Apophyllum*). FRUIT a capsule, siliquose and 2-valved, or a berry, very rarely a drupe (*Roydsia*). SEEDS reniform or angular, often sunk in the pulp of the fleshy fruits, exalbuminous, or very rarely albuminous (*Tovaria*) ; *testa* smooth, coriaceous or crustaceous. EMBRYO curved or arched ; *cotyledons* incumbent or accumbent, folded, coiled or induplicate, rarely flat.

TRIBE I. *CLEOMEÆ*.

Fruit a 1-celled capsule, usually siliquose. Mostly annual herbs.

PRINCIPAL GENERA.

* Cleome.   Isomeris.   * Polanisia.   * Gynandropsis.

TRIBE II. *CAPPAREÆ*.

Fruit a berry or drupe. Shrubs or trees.

PRINCIPAL GENERA.

| * Morisonia. | Mærua. | Boscia. | Roydsia. | Ritchiea. |
| Niebuhria. | Cadaba. | * Capparis. | Tovaria.[1] | |

Capparideæ approach *Cruciferæ* in the number of sepals, petals and stamens, the æstivation, the ovary with parietal placentation, with or without a false septum, the campylotropous ovules, siliquose fruit, exalbuminous seed, curved embryo, and acrid volatile principles. They scarcely differ, except in the sometimes perigynous insertion, the never tetradynamous stamens, the usually stipitate ovary, and the often fleshy fruit. They are equally closely allied to *Moringeæ* (which see). They also resemble *Tropæoleæ* in their habit, exalbuminous seed, and acrid principle. *Resedaceæ* are separated only by their habit and the structure of their fruit. *Capparideæ* are distributed nearly equally over the tropical and subtropical regions of both hemispheres ; the frutescent species are mostly American.[2]

The herbaceous capsular *Capparideæ* rival *Cruciferæ* in their stimulating properties, which depend on an acrid volatile principle. The species with fleshy fruit, which are mostly woody, possess this acridity in their roots, leaves and herbaceous parts ; their bark is bitter, and some have a pleasant fruit. *Cleome gigantea* is used as a rubefacient in tropical America. The herbage of *Gynandropsis pentaphylla*, a native of the tropics in both worlds, has the qualities of *Cochlearia* and *Lepidium*, and its oily seed is as acrid as that of *Sinapis*. *Polanisia felina* and *icosandra*, natives of India, are epispastics and vermifuges ; the fresh juice is used as a condiment. *Cleome heptaphylla* and *polygama*, American plants, have a balsamic odour, whence they have been reputed vulneraries and stomachics. *Polanisia graveolens*, a native of North America, and a very fœtid plant, possesses the same qualities as *Chenopodium anthelminthicum*.

Among the *Capparideæ* with fleshy fruit, *Capparis spinosa* must rank first. It is a shrub of the Mediterranean region, the bitter, acrid and astringent bark of whose root has been esteemed from the most ancient times for its aperient and diuretic qualities. The flower-buds, preserved in salt and vinegar,

---

[1] *Tovaria* has been transferred to *Papaveraceæ* by Eichler.—ED.

[2] As many are natives of the old world as of the new.—ED.

are known as Capers, and much used as a condiment. Other species of *Capparis* from Greece, Barbary and Egypt, are similarly used. *Capparis sodada* is a native of tropical Africa; the negresses eat its acidulous and stimulating fruit, which they believe will make them prolific. The bitter and astringent bark of *Cratæva Tapia* and *gynandra*, American trees, is reputed a febrifuge. Their fruit, which has an oily odour, is eatable. *C. Nurvala*, of tropical Asia, produces succulent and vinous berries; its acidulous leaves are diuretic.

XVI. *MORINGEÆ, Endlicher.*

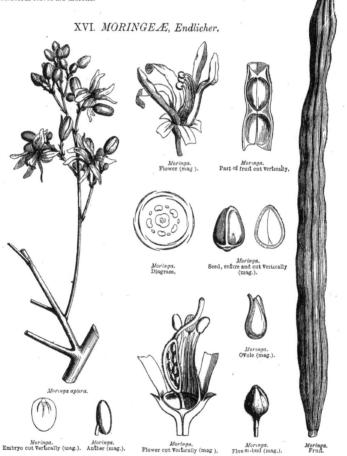

Moringa.
Flower (mag).

Moringa.
Part of fruit cut vertically.

Moringa.
Diagram.

Moringa.
Seed, entire and cut vertically (mag.).

Moringa.
Ovule (mag.).

Moringa aptera.

Moringa.
Embryo cut vertically (mag.).

Moringa.
Anther (mag.).

Moringa.
Flower cut vertically (mag).

Moringa.
Flower-bud (mag.).

Moringa.
Fruit.

TREES. LEAVES 2-3-imparipinnate; *leaflets* very caducous; *stipules* deciduous. FLOWERS ⚥, irregular, in panicled racemes. CALYX 5-partite, with oblong subequal segments, imbricate in bud. PETALS 5, inserted on the calyx, linear-oblong, the two posterior rather the longest, ascending, imbricate in bud. STAMENS 8-10, inserted on a cup-shaped disk lining the base of the calyx; *filaments* flattened at the base, connivent in a tube which is split behind, united above the middle, free above and below, unequal, the posterior longest, all fertile, or those opposite the calyx-segments shorter and imperfect; *anthers* introrse, 1-celled, ovoid-oblong, dorsally fixed, dehiscence longitudinal. OVARY pedicelled, 1-celled, with three parietal slender placentas; *style* terminal, simple, thickened [tubular, open at the truncate top]; *ovules* numerous [bi-seriate], pendulous, anatropous, [raphe ventral]. CAPSULE siliquiform, 3-many-angled, torulose, 3-valved, valves with the placentas on the middle. SEEDS 1-seriate, separated by spongy septa, ovoid-trigonous, angles apterous or winged; *chalaza* apical, corky. EMBRYO straight, exalbuminous; *cotyledons* plano-convex, fleshy [plumule many-leaved]; *radicle* very short, superior.

ONLY GENUS.
Moringa.

The genus *Moringa* has been by some botanists placed in *Papilionaceæ* on account of a slight resemblance in the flower, which, however, indicates no true affinity. Hooker [following Lindley] compared it with *Violarieæ*, which resemble it in their irregular flower with unequal dorsal petal, in their perigynous insertion, tubular style, one-celled ovary with three parietal nerviform placentas and anatropous ovules; but *Moringeæ* are widely separated by habit, one-celled anthers, and exalbuminous seeds. It is amongst *Capparideæ* that we must search for the real affinities of *Moringeæ*, through their polypetalous imbricate corolla, perigynism, stamens more numerous than petals, stipitate one-celled ovary, parietal placentation, siliquose capsule, exalbuminous embryo, alternate leaves and caducous stipules; to which must be added the acrid root, leaves, and bark, which are common to both families, recalling the smell and taste of the Horse-radish, and associating *Moringa* also with *Cruciferæ*, themselves so closely allied to *Capparideæ*.

Moringeæ are tropical Asiatic, Arabian and Madagascan trees. The best known species is *Moringa aptera*, of which the seed, called Ben nut, yields a fixed oil, of much repute in the East, because it does not become rancid [and used by perfumers and machinists from its not freezing. The root of *M. pterygosperma* is used as a stimulant in paralysis and intermittent fevers; and a colloid gum, like tragacanth, exudes in great quantities from its bark.—ED.].

XVII. *RESEDACEÆ, D.C.*

CALYX 4-8-*partite*. PETALS *generally hypogynous,* 4-8 (*rarely* 2 *or* 0). STAMENS 3-40, *inserted within a fleshy disk.* CARPELS *usually united into a* 1-*celled ovary.* FRUIT *a capsule or berry.* SEEDS *exalbuminous.* EMBRYO *curved.*

Annual or perennial HERBS, sometimes UNDERSHRUBS, rarely SHRUBS (*Ochradenus*), juice watery, stem and branches terete. LEAVES scattered, simple, entire, 3-fid or pinnatipartite; *stipules* minute, gland-like. FLOWERS ⚥, rarely diclinous, more or less irregular, in a raceme or spike, bracteate. CALYX persistent, 4-8-partite, more

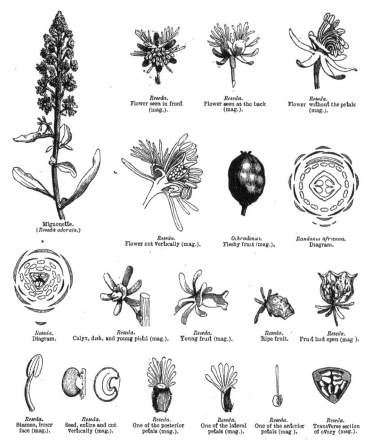

Reseda.
Flower seen in front (mag.).

Reseda.
Flower seen at the back (mag.).

Reseda.
Flower without the petals (mag.).

Mignonette.
(Reseda odorata.)

Reseda.
Flower cut vertically (mag.).

Ochradenus.
Fleshy fruit (mag.),

Randonia africana.
Diagram.

Reseda.
Diagram.

Reseda.
Calyx, disk, and young pistil (mag.).

Reseda.
Young fruit (mag.).

Reseda.
Ripe fruit.

Reseda.
Fruit laid open (mag.).

Reseda.
Stamen, inner face (mag.).

Reseda.
Seed, entire and cut vertically (mag.).

Reseda.
One of the posterior petals (mag.).

Reseda.
One of the lateral petals (mag.).

Reseda.
One of the anterior petals (mag.).

Reseda.
Transverse section of ovary (mag.).

sub-coherent (*Oligomeris*), equal or unequal, open in æstivation. DISK hypogynous, sessile or stipitate, more or less concave, fleshy, often prolonged behind, rarely 0 (*Oligomeris*). STAMENS 3–40, inserted within the disk, rarely perigynous (*Randonia*), not covered by the petals in æstivation; *filaments* equal or unequal, often pendulous, free or rarely connate at the base (*Oligomeris*); *anthers* introrse, 2-celled, dehiscence longitudinal. OVARY sessile or stipitate, of 2–6 carpels, sometimes coherent into a

1-celled ovary, closed or gaping at the top, with many-ovuled parietal placentas, more rarely distinct, or sub-coherent at the base, many-ovuled and with basilar placentation, gaping (*Caylusea*), or 1–2-ovuled and closed (*Astrocarpus*); *stigmas* sessile, terminating the 2-lobed top of the carpels; *ovules* campylotropous or half-anatropous. FRUIT usually a capsule, indehiscent, closed or gaping at the top, rarely a berry (*Oehradenus*), sometimes follicular (*Astrocarpus*). SEEDS reniform, exalbuminous, epidermis membranous, adhering to the testa, or detaching when ripe; *testa* crustaceous. EMBRYO curved or folded; *cotyledons* incumbent; *radicle* near the hilum.

GENERA.

Astrocarpus.   Randonia.   Caylusea.   *Reseda.   Ochradenus.   Oligomeris.

The small family of *Resedaceæ* is allied to *Cruciferæ* and *Capparideæ* (see these families). It also approaches *Moringeæ* in its irregular polypetalous flowers, fleshy disk, stamens more numerous than the petals, parietal placentation, capsular fruit, exalbuminous embryo, alternate stipulate leaves, and finally in the acrid principle found in the root of several species; but *Moringeæ* are separated by their habit, arborescent stem, two-three-pinnate leaves, straight embryo, filaments united into a tube above the middle, and one-celled anthers.

Most *Resedaceæ* grow in southern Europe, northern Africa, Syria, Asia Minor and Persia. Some reach the Indian frontier: a few inhabit central and northern Europe. Three species belong to the Cape of Good Hope.

Resedaceæ, so named because sedative qualities were formerly attributed to them, are no longer used in medicine, in spite of the acridity of their root, which contributes, with other characters, to bring them near *Cruciferæ* and *Capparideæ*; the root of *Reseda lutea* in particular has the odour of the Radish, and was long reckoned an aperient, sudorific and diuretic. Dyer's Weed (*R. luteola*) has intensely bitter leaves, and all parts yield a yellow dye much in demand. Mignonette (*R. odorata*), a plant whose origin was long considered unknown, but which Griffith asserts to be a native of Affghanistan, is extensively cultivated for its sweet scent.

XVIII. *CISTINEÆ.*

(CISTI, *Jussieu.*—CISTOIDEÆ, *Ventenat.*—CISTINEÆ, *D.C.*—CISTACEÆ, *Lindl.*)

PETALS 5–3, *hypogynous.* STAMENS ∞, *hypogynous.* OVARY 1-*celled, with* 3–5 *parietal placentas.* OVULES *orthotropous.* STYLE *simple.* CAPSULE *with the placentas on the centre of the valves.* SEEDS *albuminous.* EMBRYO *bent, coiled or folded.*

HERBS, UNDERSHRUBS, or SHRUBS; stem and branches terete or sub-tetragonous, often glandular, pubescent or tomentose, with simple or sometimes stellate hairs. LEAVES simple, opposite, rarely alternate, sometimes whorled, entire, sessile or petioled; *stipules* foliaceous, free at the contracted base of the petiole, or 0 when the petiole is amplexicaul. FLOWERS ☿, regular, terminal, solitary, or in cymes or unilateral racemes, peduncle outside of the axil of the bracts. SEPALS 3, twisted in bud, often furnished with 2 usually smaller calyciform bracts. PETALS hypogynous, 5, very rarely 3, or 0 (*Lechea*), twisted in æstivation in an opposite direction to the sepals, scarcely clawed, spreading, very fugacious. STAMENS ∞, hypogynous; *filaments* free, filiform; *anthers* 2-celled, introrse, ovoid or lanceolate, dehiscence longitudinal. OVARY free, sessile, 1-celled, or with 3–5 imperfect cells formed by

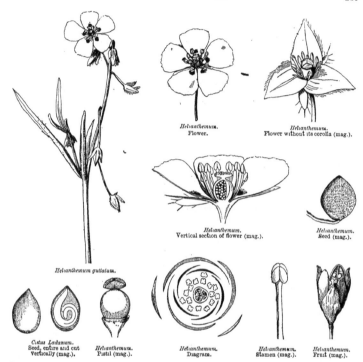

Helianthemum.
Flower.

Helianthemum.
Flower without its corolla (mag.).

Helianthemum.
Vertical section of flower (mag.).

Helianthemum.
Seed (mag.).

Helianthemum guttatum.

Cistus Ladanum.
Seed, entire and cut
vertically (mag.).

Helianthemum.
Pistil (mag.).

Helianthemum.
Diagram.

Helianthemum.
Stamen (mag.).

Helianthemum.
Fruit (mag.).

placentiferous septa only united at the bottom of the ovary; *placentas* 3–5, parietal, or fixed to the half-septa, 2– ∞-ovuled; *style* simple; *stigmas* 3–5, free, or united in a head; *ovules* with short or longer funicles, usually ascending, orthotropous, or half-anatropous. Capsule membranous or coriaceous, opening to the base, or above only, into 3–5 half-placentiferous valves. Seeds with crustaceous testa, and floury or sub-horny albumen. Embryo excentric or sub-central, bent, coiled, or folded, rarely nearly straight (*Lechea*); *hilum* and *chalaza* contiguous, diametrically opposite to the radicle, except in some species (*Lechea*), in which the funicle is adnate to the seed.

GENERA.

* Cistus.   * Helianthemum.   Hudsonia.   Lechea.

Cistineæ are near *Droseraceæ, Violarieæ* and *Bixineæ* in polypetalism, placentation, capsular fruit with seminiferous valves and albuminous seeds; they are polyandrous like *Bixineæ* and *Dionæa*, and the albumen of these latter is floury. But, besides the different habit, *Droseraceæ* have extrorse anthers, anatropous ovules, and a straight embryo; *Violarieæ* proper have irregular imbricate isostemonous flowers, anatropous ovules, straight embryo, and fleshy albumen; *Birineæ* scarcely differ save in the anatropous ovules. There is a decided affinity between *Cistineæ* and *Hypericineæ* (which see). They have also been compared with *Capparideæ*, from which, however, they differ in habit, fugacious petals, albuminous seed, etc.

Cistineæ mostly inhabit the Mediterranean region; some grow in North America; a very few in central Europe and eastern Asia, and still fewer in South America.

The herbage of *Cistineæ* is slightly astringent; some *Cisti* yield a balsamic resin, named *ladanum*, which is used in perfumery. *Helianthemum vulgare*, a species of central Europe, is sometimes administered as a vulnerary.

XIX. *VIOLARIEÆ*.

(*Genera* CISTIS *affinia, Jussieu.*—IONIDIA, *Ventenat.*—VIOLARIEÆ, *D.C.*—VIOLACEÆ, *Lindl.*—VIOLEÆ, *Br.*)

PETALS 5, *more or less unequal, hypogynous or slightly perigynous, imbricate.* STAMENS 5, *inserted like the petals.* OVARY 1-*celled, placentation parietal.* STYLE *simple.* FRUIT *a capsule with the placentas on the centre of the valves, or rarely an indehiscent berry.* SEEDS *albuminous.* EMBRYO *straight.*

HERBS, UNDERSHRUBS, or SHRUBS, rarely sarmentose (*Agation*). LEAVES alternate, rarely opposite (*Ionidium, Alsodeia*), simple, petioled, usually involute in bud, sometimes arranged in radical rosettes, and spotted with brown below (*Viola cotyledon* and *rosulata*); *stipules* free, foliaceous, or small, usually deciduous in the woody species. FLOWERS ⚥, often dimorphous and apetalous, irregular or sub-regular, pentamerous, or very rarely tetramerous (*Tetrathylacium*), axillary, solitary or in a cyme, panicle or raceme; pedicels usually 2-bracteolate. SEPALS 5, distinct, or connate at the base, usually persistent, equal or unequal, æstivation imbricate. PETALS 5, hypogynous or slightly perigynous, alternate with the sepals, æstivation imbricate and convolute, sometimes equal or subequal, clawed, connivent, or cohering in a tube at the base (*Paypayrola, Tetrathylacium, Gloiospermum, Sauvagesia*); sometimes very unequal, the two upper exterior, the two lateral within the others, and not clawed, the inner (lowest by the reversal of the flower) larger, clawed, and prolonged into a hollow spur below its insertion. STAMENS 5, inserted on the receptacle or bottom of the calyx; *filaments* very short, dilated, free, or sometimes connate at the base (*Leonia, Gloiospermum, Alsodeia*, &c.); *anthers* introrse, 2-celled, connivent, or coherent round the ovary, cells adnate by the back to the inner surface of the connective, and opening by a longitudinal slit; connective prolonged above the cells in a membranous appendage, those of the 2 or 4 lower stamens (in the irregular flowers) gibbous and glandular on their dorsal face, or prolonged into filiform spurs, which are included in that of the lower petal. OVARY free, sessile, often girt with a basal annulus, 1-celled; *placentas* parietal, slender, generally 3, rarely 2 (*Hymenanthera*), or 5 (*Melicytus*), or 4 (*Tetrathylacium*); *style* simple, sometimes thickened at the top, or bent with a dorsal stigmatic cavity, or of various form, sometimes

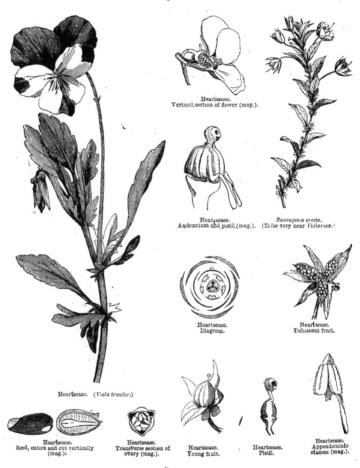

Heartsease.
Vertical section of flower (mag.).

Heartsease.
Androecium and pistil (mag.).

Sauvagesia erecta.
(Tribe very near Violorieæ.)

Heartsease.
Diagram.

Heartsease.
Dehiscent fruit.

Heartsease. (Viola tricolor.)

Heartsease.
Seed, entire and cut vertically
(mag.).

Heartsease.
Transverse section of
ovary (mag.).

Heartsease.
Young fruit.

Heartsease.
Pistil.

Heartsease.
Appendiculate
stamen (mag.).

subulate with a terminal stigma, rarely 3–5-fid, or style 0 with 3–5 free stigmas (*Melicytus*); *ovules* anatropous, usually many, very rarely 1–2 (*Isodendrion, Hymenanthera, Scyphellandra*). FRUIT a capsule, often opening elastically by as many seminiferous valves as there are placentas; or an indehiscent berry (*Leonia, Tetra-*

R

thylacium, Melicytus, Hymenanthera). SEEDS ovoid or subglobose; *testa* crustaceous or membranous, raphe sometimes thickened and separating when ripe; *albumen* fleshy, copious. EMBRYO axile, straight; *cotyledons* flat, broad or narrow; *radicle* cylindric, near the hilum.

TRIBE I. *VIOLEÆ.*

Corolla irregular, lower petal dissimilar. Fruit a capsule.

PRINCIPAL GENERA.

* Viola.   Ionidium.   Agation.

TRIBE II. *PAYPAYROLEÆ.*

Petals subequal, claws contiguous, and sub-coherent in a tube.   Fruit a capsule.

PRINCIPAL GENERA.

Isodendrion.   Paypayrola.   Amphirrhox.

TRIBE III. *ALSODINEÆ.*

Petals equal or subequal, very shortly clawed. Fruit a berry or capsule.

PRINCIPAL GENERA.

Alsodeia.   Leonia.   Hymenanthera.   Melicytus.

Sauvagesieæ, of which we have given a figure near *Violarieæ*, are so closely allied to them that several botanists have united them. They are distinguished only by the presence of five-∞ staminodes placed outside the stamens, and by the three valves of the capsule being seminiferous on their edges. *Violarieæ* also approach *Droseraceæ* in isostemony, the one-celled ovary with parietal placentation, the capsule with placentiferous valves, and the albuminous seed; but in *Droseraceæ* the anthers are extrorse, the styles are distinct, the embryo is minute and basilar. They have the same affinities with *Frankeniaceæ*, which have also a simple style and an axile embryo, but their calyx is tubular and elongate, their anthers are extrorse, their ovules ascending, their leaves usually opposite and exstipulate. They are also connected with *Cistineæ* (which see).

The herbaceous species of the tribe *Violeæ* principally inhabit the northern hemisphere; they are rare in the temperate regions of the southern hemisphere and in the tropics; the woody species of this tribe are chiefly natives of equatorial America. The other tribes inhabit the intertropical region of both worlds, and especially of America. *Hymenanthera* inhabits Australia and New Zealand.

The active principle of *Violeæ* (*violine*) is a substance analogous in properties to *emetine* (see *Cephælis*), the emetic and laxative properties of which it shares. Violine is principally found in the root and rootstock, from which, as well as from the leaves, has also been extracted a peculiar acid; and the scented petals contain a volatile oil. The root of the European violets, and especially of the Sweet Violet (*Viola odorata*), is slightly bitter and acrid, recalling the taste of *Ipecacuanha*; the flowers, which are sweet-scented but nauseous in taste, are used in syrup and infused as emollients and cough-mixtures. The stem and leaves of the Wild Pansy (*V. tricolor*) are frequently administered as a depurative tisane in cutaneous disorders. The American Violets (*V. pedata* and *palmata*) are similarly employed. *V. ovata* is a reputed specific against rattlesnake bites. Some species of *Ionidium*, in South America, are used as substitutes for Ipecacuanha; the root of *I. Ipecacuanha* especially, the White Ipecacuanha of commerce, is a powerful emetic, peculiarly suited to lymphatic temperaments. The root of *Anchietea salutaris*, a small Brazilian tree, is purgative, and useful, like our wild Pansy, in skin affections. *Ionidium microphyllum*, a species growing at the foot of Chimborazo, is supposed to yield the root called Cuichunchulli, prescribed by the Americans for tubercular elephantiasis.

The medical properties of *Alsodineæ* are very obscure, and entirely differ from those of *Violeæ*. The leaves and bark of *Alsodeia Cuspa*, which grows in New Granada, are bitter and astringent. The leaves of *A. castaneæfolia* and *Lobolobo*, Brazilian species. are mucilaginous, and are cooked and eaten by the negros.

XX. *CANELLACEÆ.*[1]

Glabrous aromatic trees. LEAVES alternate, quite entire, penninerved, pelluciddotted; *stipules* 0. FLOWERS ⚥, regular, in terminal, lateral, or axillary cymes; *bracteoles (sepals of some authors)* 3, orbicular, close under the calyx, much imbricate, persistent; *sepals (petals of some)* 4–5, free, thick, deciduous, much imbricate, the inner narrowest. PETALS *(petaloid scales of some)* as many as the sepals, thin, imbricate or 0. STAMENS hypogynous; *filaments* connate into a tube; *anthers* 20 or fewer, linear, adnate to the outer surface of the tube, longitudinally 2-valved. DISK 0. OVARY free, 1-celled; *placentas* 2–5, parietal, 2- or more-ovuled; *style* short, thick; *stigmas* 2–5; *ovules* horizontal or ascending, almost anatropous. BERRY indehiscent, 2–many-seeded. SEEDS with a shining crustaceous testa; *albumen* oily and fleshy. EMBRYO straight or curved?, radicle next the hilum; *cotyledons* oblong.

GENERA.

Canella.   Cinnamodendron.   Cinnamosma.

A very small order, placed by Martius near *Guttiferæ*, included by Lindley under *Pittosporeæ*, and placed by Miers near *Magnoliaceæ*; but according to Bentham and Hooker fil. it has less affinity with any of the above orders than with *Violarieæ* and *Bixineæ*, differing from the first of these chiefly in the absence of stipules, aromatic properties, and more numerous anthers, which are extrorse, and adnate to the staminal column.

Canellaceæ, of which only five species are known, are natives of tropical America, with one Madagascan species; all are highly aromatic. The Canella bark of commerce is the Wild Cinnamon of the West Indies, and is a well-known carminative and stomachic; it is exported from the Bahamas as ' White-wood bark,' on account of the white appearance of the trees when stripped of the bark; the inner layers alone are used, and yield by distillation a warm aromatic oil. The bark of a Brazilian species is used as a tonic and antiscorbutic; it is prescribed in low fevers, and made into a gargle is useful in cases of relaxation of the tonsils.

XXI. *BIXINEÆ.*

(BIXINEÆ, *Kunth,*—BIXACEÆ ET COCHLOSPERMEÆ, *Endlicher.*—FLACOURTIACEÆ ET PANGIACEÆ, *Lindl.*—FLACOURTIANEÆ, *L. C. Richard, D. Clos.*)

SEPALS distinct or connate, usually imbricate. COROLLA polypetalous, hypogynous, or 0. STAMENS usually ∞, hypogynous or sub-perigynous. OVARY free, usually 1-celled, placentation parietal. STYLE simple, or divided to its base. BERRY or CAPSULE with half-seminiferous valves. SEEDS albuminous. EMBRYO usually straight, axile.

---

[1] This order is omitted in the original.—ED.

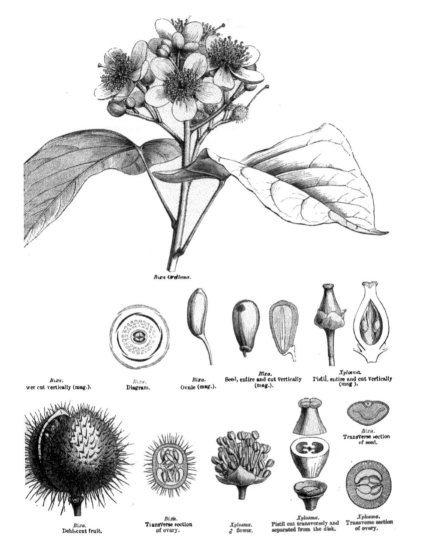

Bixa Orellana.

Bixa.
wer cut vertically (mag.).

Bixa.
Diagram.

Bixa.
Ovule (mag.).

Bixa.
Seed, entire and cut vertically
(mag.).

Xylosma.
Pistil, entire and cut vertically
(mag.).

Bixa.
Dehiscent fruit.

Bixa.
Transverse section
of ovary.

Xylosma.
♂ flower.

Xylosma.
Pistil cut transversely and
separated from the disk.

Bixa.
Transverse section
of seed.

Xylosma.
Transverse section
of ovary.

TREES or SHRUBS. LEAVES alternate, simple, toothed, rarely entire, sometimes palmilobed or compound (*Cochlospermum, Amoreuxia*), sometimes pellucid-dotted; *stipules* minute, caducous or 0. FLOWERS ⚥ or unisexual, regular, axillary or terminal, solitary or more often fascicled, or corymbose, racemose or panicled. SEPALS 4–5, or 2–6, free or connate, æstivation imbricate, rarely sub-valvate (*Azara*, &c.), or united into 2 more or less regular valves (*Pangium*, &c.). PETALS hypogynous, as many as sepals, or ∞, æstivation imbricate and twisted, deciduous, or 0. STAMENS hypogynous, or obscurely perigynous, indefinite, or rarely definite (*Azara, Erythrospermum*, &c.) ; *anthers* opening by slits, or rarely by an apical pore (*Bixa, Cochlospermum*, &c.). TORUS often glandular, thick, or dilated (*Xylosma*), sometimes adnate to the calyx base, rarely annular and adnate to the ovary (*Peridiscus*). OVARY free, usually 1-celled, with 2–∞ parietal placentas, sometimes several-celled (*Flacourtia, Amoreuxia*, &c.) ; *styles* as many as placentas, united, or more or less free; *ovules* 2–∞ on each placenta, anatropous or half-anatropous. FRUIT fleshy or dry, indehiscent, or opening by seed-bearing valves. SEEDS usually ovoid or pisiform, rarely reniform, or cochlear and velvety (*Cochlospermum*), smooth, or pulpy on the outside (*Bixa, Dendrostylis*); *albumen* fleshy, more or less copious. EMBRYO axile, straight or curved ; *radicle* near the hilum ; *cotyledons* large, usually cordate.

TRIBE I. *BIXEÆ.*

Flowers ⚥, or rarely polygamous. Petals large, without a scale, twisted in bud. Anthers linear or oblong, opening by 2 terminal pores, or short valves. Capsule dehiscent; endocarp membranous, separating from the valves.

PRINCIPAL GENERA.

Cochlospermum.   * Bixa.

TRIBE II. *ONCOBEÆ.*

Flowers diœcious or polygamous. Sepals and petals imbricate, the latter most numerous and without a scale. Anthers linear, opening by slits.

PRINCIPAL GENERA.

Oncoba.   Dendrostylis.

TRIBE III. *FLACOURTIEÆ.*

Flowers ⚥ or diœcious. Petals.0, or equal to the sepals, imbricate, without a scale. Disk surrounding the stamens or the ovary. Anthers short, linear, opening by slits.

PRINCIPAL GENERA.

| Lætia. | *Azara. | Erythrospermum. | Xylosma. |
| Ludia. | Scolopia. | Flacourtia. | Aberia. |

Tribe IV. *PANGIEÆ*.

Flowers diœcious.　Petals with a scale at the base.

PRINCIPAL GENERA

* Kiggelaria.　　　Pangium.　　　Hydnocarpus.　　　Gynocardia.

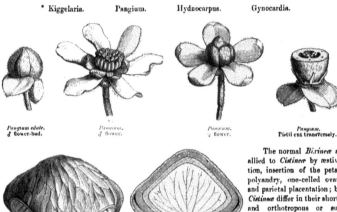

Pangium edule.
♂ flower-bud.

Pangium.
♂ flower.

Pangium.
♀ flower.

Pangium.
Pistil cut transversely.

Pangium edule. Seed, entire and cut vertically (mag.).

The normal *Bixineæ* are allied to *Cistineæ* by æstivation, insertion of the petals, polyandry, one-celled ovary, and parietal placentation; but *Cistineæ* differ in their shorter and orthotropous or sub-orthotropous embryo, and their usually floury albumen. *Bixineæ* bear some relation to *Capparideæ*, but are separated by their albuminous seeds. They differ from *Tiliaceæ* in the one-celled ovary and placentation. The oligandrous *Bixineæ* approach *Violarieæ*, which are separated by their irregular corolla and connivent anthers. They also approach *Papayaceæ*, through the tribe of *Pangieæ*. *Bixineæ* inhabit the tropical regions of both worlds.

The most important of the tribe *Bixieæ* is the Arnotto (*Bixa Orellana*), a tropical American tree, cultivated throughout the tropics; the reddish pulp of its seeds smells of violets, and is bitter and astringent. A refreshing decoction is prepared from it, which is considered antifebrile, and is also used in cases of hæmorrhage, diarrhœa and gravel. The aromatic bitter seeds and root are reputed stomachic. The seeds, steeped in hot water and allowed to ferment, furnish a red dye, which by evaporation becomes a solid paste, the arnotto of commerce, used largely by painters, and especially by dyers, as also to colour butter and wax; the Caribbeans formerly tattooed themselves with it to prevent mosquito bites. The soft wood of *Bixa* serves as tinder to Indians, who obtain fire by rubbing together two pieces of wood of different species. *Cochlospermum insigne*, which grows in Brazil, is supposed to cure abscesses in the viscera. The root of *C. tinctorium*, which contains a yellow dye, is useful in amenorrhœa. The gum of the East Indian *C. Gossypium*, called Cuteera, is used as a substitute for tragacanth. The fruit of *Oncoba*, which inhabits tropical Africa from Nubia to the Cape de Verd, yields a sweet and eatable pulp. *Lætia apetala*, from tropical America, secretes a balsamic resin similar to sandarac. The more or less acid juicy berries of *Flacourtia cataphracta*, *sapiaria*, *sapida* and *inermis*, Asiatic species, and of *F. Ramontchi*, are eatable. The bitter shoots of *F. cataphracta* taste like rhubarb, and are used as a tonic. The Cingalese use the fruits of *Hydnocarpus inebrians* to intoxicate fish.

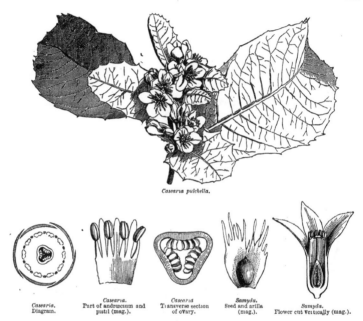

Casearia pulchella.

Casearia. Diagram.	*Casearia.* Part of andrœcium and pistil (mag.).	*Casearia* Transverse section of ovary.	*Samyda.* Seed and arilla (mag.).	*Samyda.* Flower cut vertically (mag.).

SAMYDEÆ form a small group of trees and shrubs inhabiting the tropics, especially in America; they are connected with *Bixineæ* by most characters, and are only separated by their apetalous flower, strongly perigynous sub-monadelphous stamens, and apical embryo. They also approach *Homalineæ* and *Passifloreæ* in apetalism, perigyny, one-celled ovary, parietal placentation, albuminous seed, alternate stipulate leaves, &c.

XXII. *PITTOSPOREÆ.*

(PITTOSPOREÆ, *Br.*)

COROLLA *polypetalous, hypogynous, isostemonous, œstivation imbricate.* STAMENS *5, alternate with the petals.* OVARY *of 5 more or less perfect many-ovuled cells.* OVULES *anatropous.* FRUIT *dry or fleshy.* EMBRYO *alluminous.* STEM *woody.* LEAVES *alternate.*

TREES or erect SHRUBS, sometimes climbing (*Sollya*). LEAVES alternate, petioled, simple, sub-coriaceous, exstipulate. FLOWERS ☿, regular, axillary or terminal, racemose, corymbose, or cymose. CALYX 5-partite or -phyllous, æstivation

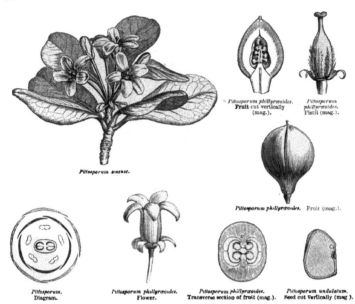

Pittosporum sinense.

Pittosporum phillyræoides.
Fruit cut vertically
(mag.).

*Pittosporum
phillyræoides.*
Pistil (mag.).

Pittosporum phillyræoides. Fruit (mag.).

Pittosporum.
Diagram.

Pittosporum phillyræoides.
Flower.

Pittosporum phillyræoides.
Transverse section of fruit (mag.).

Pittosporum undulatum.
Seed cut vertically (mag).

imbricate, deciduous.   PETALS 5, inserted on the receptacle, usually erect, claws
connivent or sometimes coherent, æstivation imbricate, deciduous.   STAMENS 5,
alternating with the petals; *filaments* filiform or subulate; *anthers* introrse, cells
opening by short or long longitudinal slits.   OVARY free, sessile or stipitate, of 2
perfect cells, or incompletely 2–5-celled; *style* terminal, simple; *stigma* obtuse or
capitate; *ovules* 2-seriate, horizontal or sub-ascending, anatropous.   FRUIT a capsule
with 2–5 half-septiferous valves, or a more or less fleshy indehiscent berry.   SEEDS
often few from arrest, often immersed in a pulp or viscous juice; *testa* loose, raphe
short, thick.   EMBRYO minute, at the base of a fleshy dense copious albumen;
cotyledons indistinct.

PRINCIPAL GENERA.

| * Pittosporum. | * Bursaria. | * Sollya. | * Billardiera. |

Pittosporeæ are connected with *Celastrineæ* by the polypetalous isostemonous corolla, imbricate
æstivation, ascending anatropous ovules, dry or fleshy fruit, albuminous embryo, woody stem, and alternate
leaves.   But in *Celastrineæ* the stamens and petals are inserted outside a fleshy disk lining the bottom of
the calyx; the cells of the ovary are perfect; the seeds are enveloped in a pulpy aril, and the embryo is
axile in the albumen.   There is also a real affinity between *Pittosporeæ* and the polypetalous pentandrous
Ericineæ (*Ledum*), founded on the insertion, the æstivation and isostemony of the corolla, the many-
celled ovary, the simple style, the anatropous ovules, the structure of the fruit, the albuminous embryo,

the texture of the stem, and the alternate leaves; besides which, in many *Pittosporeæ* (*Sollya, Cheiranthera*) the anther-cells open near the top by little slits.

Pittosporeæ principally inhabit extra-tropical Australia, but many are Indian and Malayan, and some African and Oceanic. Some are cultivated in Europe for ornament. All contain resinous aromatic and bitter principles, which give their berries a tart, disagreeable taste; but the natives of Australia, who to appease their hunger are reduced to filling their stomachs with clay mixed with organic detritus, eagerly devour the fleshy fruits of this family.

XXIII. *POLYGALEÆ* AND *TREMANDREÆ.*

(POLYGALEÆ, *Jussieu.*—POLYGALACEÆ ET KRAMERIACEÆ, *Lindl.*)

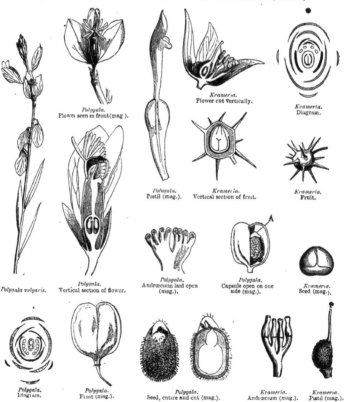

Polygala.
Flower seen in front(mag.).

Krameria.
Flower cut vertically.

Krameria.
Diagram.

Polygala.
Pistil (mag.).

Krameria.
Vertical section of fruit.

Krameria.
Fruit.

Polygala vulgaris.

Polygala.
Vertical section of flower.

Polygala.
Andrœcium laid open (mag.).

Polygala.
Capsule open on one side (mag.).

Krameria.
Seed (mag.).

Polygala.
Diagram.

Polygala.
Fruit (mag.).

Polygala.
Seed, entire and cut (mag.).

Krameria.
Andrœcium (mag.).

Krameria.
Pistil (mag.).

FLOWERS *irregular.* PETALS *hypogynous, unequal.* STAMENS *usually double the number of the petals.* ANTHERS *1- (rarely 2-) celled, opening at the top by 1–2 pores.* OVARY *2-celled.* OVULES *pendulous, anatropous.* FRUIT *a fleshy capsule, rarely indehiscent.* EMBRYO *albuminous or exalbuminous.*

HERBS or UNDERSHRUBS, sometimes twining, or erect, sometimes climbing SHRUBS, rarely arborescent, glabrous, cottony or velvety, hairs not stellate. LEAVES alternate, rarely opposite, simple, entire. FLOWERS ⚥, irregular, solitary, spiked or racemed, or rarely panicled, axillary or terminal; *pedicels* usually jointed at the base, bracteate and 2-bracteolate. SEPALS 5, free, imbricate, 2 inner largest, often winged and petaloid. PETALS 3 or 5, hypogynous, the 2 lateral free, or united at their base with the lower, concave or galeate (*keel*), in the gamopetalous corolla split behind, rarely 0; upper 2 sometimes equal to the lateral, enveloping the keel in æstivation, sometimes small, scale-like, or 0 (*Securidaca*). STAMENS 8, rarely 5–4 (*Salomonia*), inserted on the receptacle; *filaments* rarely free (*Xanthophyllum*), usually monadelphous, forming a sheath split on its upper edge, and more or less united outside with the petals; *anthers* erect, basifixed, 1- (rarely 2-) celled (*Xanthophyllum, Securidaca*), opening at the top by a pore (rarely 2), more or less oblique; *pollen of Polygala* ovoid, external membrane splitting in longitudinal bands, allowing the inner membrane to protrude, and resembling the staves of a barrel. DISK small, often 0, or rarely expanded into an imperfect unilateral ring. OVARY free, with 2 antero-posterior cells, rarely 1-celled by arrest (*Securidaca*), very rarely 3–5-celled (*Trigoniastrum, Montabea*); *style* terminal, curved, dilated at the top, undivided or 2–4-lobed; *stigma* terminal, or situated between the lobes of the style; *ovules* pendulous, usually solitary in each cell, or rarely twin, collateral (*Krameria*), or very rarely 2–6, scattered (*Xanthophyllum*), anatropous, raphe ventral. FRUIT usually a loculicidal or indehiscent capsule, a drupe (*Carpolobia, Mundtia*), or samara (*Securidaca, Trigoniastrum*). SEEDS pendulous; *testa* crustaceous, often velvety (*Comespermum*); *hilum* often strophiolate (*Polygala*); *albumen* sometimes copious, fleshy or mucilaginous, sometimes scanty or 0. EMBRYO axile, straight; *cotyledons* planoconvex, fleshy and thick in the exalbuminous seeds; *radicle* short, superior.

PRINCIPAL GENERA.

* Polygala.	Comesperma.	Bredemeyera.	Securidaca.	Carpolobia.
Montabea.	Xanthophyllum.	Krameria.	Salomonia.	Muraltia.

The affinities of *Polygaleæ* are obscure. They were formerly placed near *Rhinantheæ* on account of the irregular hypogynous apparently monopetalous corolla, the two-celled ovary, and compressed capsule; but their other characters are all opposed to this affinity. They have since been compared with *Papilionaceæ*; but in these, besides the perigynous insertion and a host of other differences, the odd petal is next the axis, whilst it is opposite it in *Polygaleæ*. The affinity with *Sapindaceæ* is also very distant, and almost confined to the hypogynous imbricate and often irregular corolla, the 1–2-ovuled ovarian cells, simple style, capsular or samaroid fruit, and often arillate or strophiolate seeds. There is a much closer affinity with *Tremandreæ*: similar habit, ovary with two one-ovuled cells, pendulous ovules, compressed capsule, strophiolate seeds, one-celled anthers opening by pores, pollen-granules opening by longitudinal slits; but in *Tremandreæ* the flower is regular, the æstivation of the calyx is valvate, the stamens are in pairs opposite to the petals, the filaments are free, the anthers extrorse, the hairs stellate and glandular; but, nevertheless, *Tremandreæ* may be considered as regular-flowered *Polygaleæ*, &c.

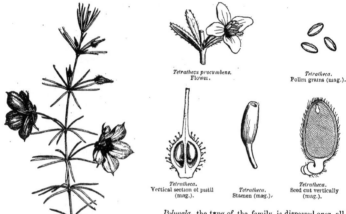

Tetratheca procumbens.
Flower.

Tetratheca.
Pollen grains (mag.).

Tetratheca.
Vertical section of pistil (mag.).

Tetratheca.
Stamen (mag.).

Tetratheca.
Seed cut vertically (mag.).

Tetratheca verticillata.

Polygala, the type of the family, is dispersed over all the globe, though least frequent in extra-tropical South America. The other genera are distributed over the tropical and warm southern temperate zones.

Polygaleæ contain a bitter principle which gives them tonic and astringent properties; this is often accompanied by an acrid principle, named *senegine,* which renders some species emetic. The root of *P. Senega* is used in Europe on account of its stimulating action on the pulmonary mucous membrane; the natives of Virginia use it as an antidote to snake-bites, as do the South Africans the *P. Serpentaria.* The European Polygalas are still prescribed for lung diseases. *Badiera diversifolia,* a shrub of the Antilles, is a sudorific analogous to guaiacum. The bark of the root of *Monnina polystachya* is employed in Peru as an astringent and antidysenteric; the ladies of that country also use it in smoothing their hair. The drupe of the South African *Mundtia spinosa* is eatable. The root of *Krameria triandra* possesses astringent and tonic properties, due to its containing much tannin.

XXIV. *VOCHYSIACEÆ.*[1]

TREES, often gigantic, with copious resinous juice, rarely erect or sarmentose (*Trigonia*), or climbing SHRUBS. BRANCHES usually opposite or whorled. LEAVES opposite or whorled (alternate in *Lightia*), shortly petioled, coriaceous, quite entire; *stipules* small or 0, or reduced to glands. INFLORESCENCE various, often racemed or panicled. FLOWERS irregular, ⚥, often large, pedicels jointed and bracteate. SEPALS 5, free or connate at the base, or rarely adnate to the ovary, 2 outer often

---

[1] This order is omitted in the original.—ED.

smaller, 2 anterior larger, posterior often largest, spurred or gibbous at the base.
PETALS hypogynous, or inserted on the top of the calyx-tube, 1, 3, or·rarely 5,
when one is protruded between the anterior sepals, clawed, blade obcordate, æstivation
convolute.   STAMENS inserted with the petals, usually 1 fertile, the rest imperfect
(except *Lightia* and *Trigonia*) ; *filaments* usually thick, excrescent, subulate ; *anthers*
oblong-linear or linear-cordate, connective thickened, cells sub-distant, including the
style.   OVARY free, rarely adnate to the sepals, often oblique and inserted by a
broad base ; *style* simple, subulate, filiform, or gradually dilated upwards ; *stigma*
capitate, truncate or oblique, entire or obscurely lobed ; *ovules* twin, collateral,
or ∞ 2-seriate, usually inserted in the axis, ascending or pendulous, micropyle
superior, raphe ventral.   FRUIT usually capsular (a winged samara in *Erisma*),
oblong, terete or trigonous, coriaceous, loculicidally or septicidally 3-valved ; valves
coriaceous after parting from the seed-bearing axis, endocarp often parting from the
epicarp.   SEEDS 1, few or many, sometimes imbricate in 2 series, often winged; *testa*
membranous or coriaceous, often hairy or cottony ; *albumen* 0, fleshy in *Trigonia*.
EMBRYO straight ; *cotyledons* flat, wrinkled, or membranous and convolute ; *radicle*
short or long, superior.

PRINCIPAL GENERA.

Callisthene.   Qualia.   Erisma.   Vochysia.   Trigonia.   Lightia.

Vochysiaceæ were placed by De Candolle amongst *Calyciflorœ* next *Onagrarieœ*, but by Lindley near
Polygaleœ with more reason.   *Lightia* presents various points of analogy with *Chrysobalaneœ* ; and *Erisma*
with *Dipterocarpeœ*, in its fruit, convolute petals, often contorted or folded cotyledons, and resinous juice.
The order is wholly tropical American ; of its properties nothing is known.

XXV. *FRANKENIACEÆ, Saint-Hilaire.*

CALYX *tubular, 4-5-fid.*   PETALS *4-5, hypogynous, equal, long-clawed.*   STAMENS
usually 6, hypogynous.   OVARY *free, with 3-4-2 parietal placentas.*   STYLE *3-4-2-
partite at the top.*   CAPSULE *of 3-4 valves, bearing at the base seeds with floury albumen.*
EMBRYO *straight, axile.*

STEM herbaceous or suffruticose.   BRANCHES many, terete, jointed at the nodes.
LEAVES opposite, small, entire, subsessile or petioled, often fascicled when young,
exstipulate.   FLOWERS ☿, regular, pink or violet, solitary in the forks of the
branches, sessile, or in a terminal dense leafy cyme.   CALYX monosepalous, tubular,
persistent, 4-6-lobed, æstivation induplicate-valvate.   PETALS 4-6, inserted on the
receptacle, long-clawed, free, imbricate in æstivation, claw with an adnate scale in
front, limb spreading.   STAMENS usually 6, sometimes 4-5-∞, hypogynous, free, or
connate at the base into a very short ring ; *filaments* filiform or flattened ; *anthers*
extrorse, versatile, didymous or ovoid, cells parallel, opening longitudinally.   OVARY
free, sessile, 3-4-gonous, 1-celled, with 3 or sometimes 4 parietal slender placentas ;
style filiform, with as many branches as placentas, branches stigmatiferous inside at
the ˙top ; *ovules* ∞, 2-seriate, semi-anatropous, micropyle inferior, funicle long,

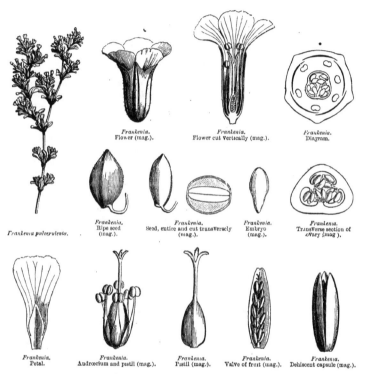

Frankenia.
Flower (mag.).

Frankenia.
Flower cut vertically (mag.).

Frankenia.
Diagram.

Frankenia pulverulenta.

Frankenia.
Ripe seed
(mag.).

Frankenia.
Seed, entire and cut transversely
(mag.).

Frankenia.
Embryo
(mag.).

Frankenia.
Transverse section of
ovary (mag.).

Frankenia.
Petal.

Frankenia.
Andrœcium and pistil (mag.).

Frankenia.
Pistil (mag.).

Frankenia.
Valve of fruit (mag.).

Frankenia.
Dehiscent capsule (mag.).

ascending. CAPSULE included in the calyx-tube, valves 3–4, placentiferous in their lower half. SEEDS ascending, ovoid; *testa* crustaceous; *raphe* linear; *chalaza* apical. EMBRYO straight, in the axis of a floury albumen; *cotyledons* ovoid-oblong; *radicle* very short, inferior.

ONLY GENUS.
Frankenia.

This small family is closely allied to the tribe *Sileneæ* of *Caryophylleæ*, but is distinguished by the extrorse anthers, parietal placentation, seed with sub-terminal hilum, and straight embryo. It also approaches *Tamariscineæ* in hypopetalism, one-celled ovary, parietal placentation, ascending anatropous ovules, capsule-valves seminiferous towards the base, and straight embryo; but *Tamariscineæ* differ in their nearly free imbricate sepals, introrse anthers, exalbuminous seed, alternate leaves, and spiked inflorescence.

Frankeniaceæ inhabit extra-tropical maritime shores, and principally the Mediterranean and Atlantic, extending into central Asia and north-western India; they are very rare in the tropics and southern latitudes.

Frankeniæ are mucilaginous and slightly aromatic. *F. portulacifolia,* which grows on maritime rocks in St. Helena, was formerly used by the colonists as tea.

XXVI. *CARYOPHYLLEÆ, Jussieu.*

SEPALS *free or united.* PETALS 4–5, *hypogynous or sub-perigynous, sometimes* 0. STAMENS *usually twice as many as the petals, and inserted with them.* OVARY 1-*celled, or with 2–5 imperfect cells.* OVULES *ventrally attached, placentation central or basilar.* SEEDS *smooth or granular, albumen usually floury.* EMBRYO *more or less curved.—* LEAVES *opposite.*

Annual or perennial HERBS, rarely shrubby. STEM and branches often thickened at the nodes, and sometimes jointed. LEAVES opposite, entire, usually 1–3-nerved, sometimes without nerves, often united at the base, exstipulate, or furnished with small scarious stipules. FLOWERS regular, ☿, or rarely unisexual. INFLORESCENCE centrifugal, sometimes many-flowered, in a simple or dichotomous loose or dense cyme, rarely in a thyrsoid or panicled raceme; sometimes few-flowered, simply forked, or reduced to a single flower; *bracts* opposite, at the forks, upper often scarious. SEPALS 4–5, persistent, free or united into a 4–5-toothed calyx, æstivation imbricate. PETALS inserted on a hypogynous or sub-perigynous disk, entire, 2-fid or laciniate, claw naked or appendiculate within, æstivation imbricate or twisted; sometimes minute, scale-like, or 0. STAMENS 8–10, inserted with the petals, sometimes equalling and alternate with them, very rarely alternate with the sepals (*Colobanthus*), sometimes fewer than the petals; *filaments* filiform; *anthers* introrse, dorsally fixed, cells opening longitudinally. TORUS usually small, sometimes (in some *Sileneæ*) elongated into a gynophore, and bearing the stamens on its summit beneath the ovary; sometimes (in many *Alsineæ*) forming a staminiferous annular disk, slightly adnate to the base of the calyx, or swelling into short glands between the stamens, or bearing, outside the stamens, staminodes opposite to the sepals. OVARY of 5 or 4 united carpels, or of 3 of which 2 are anterior, or of 2 which are times forming a thin layer on its dorsal surface, rarely 0 (*Velezia*, sp., *Dianthus*). EMBRYO more or less curved, peripheric or annular (*Drypis*), or nearly straight in the scutiform seeds ; *cotyledons* narrow, plano-convex or half-cylindric, incumbent or very rarely accumbent ; *radicle* cylindric, inferior or superior.

TRIBE I. *SILENEÆ, D.C.*

Sepals united into a 5-toothed or -lobed calyx. Petals and stamens hypogynous, inserted on an erect gynophore, rarely sessile. Petals with scales at the top of the claw, or naked. Styles completely distinct.—Leaves exstipulate.

1. **Lychnideæ.**—Corolla twisted or imbricate in æstivation. Calyx with commissural nerves. Petals usually furnished at the base of the limb with scales forming a coronet, very rarely with small winged bands at the claw (*Agrostemma*). Fruit 3–5-merous. Embryo arched, circular or coiled (*Drypis*).

PRINCIPAL GENERA.

Petrocoptis.	* Agrostemma.	* Lychnis.	* Viscaria.
Melandrium.	* Silene.	Cucubalus.	* Drypis.

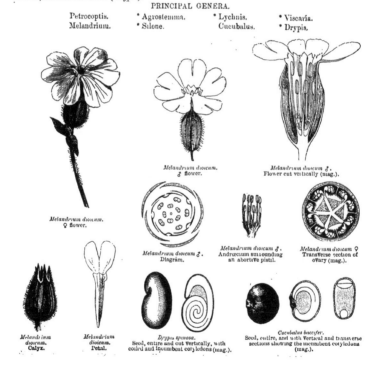

Melandrium dioicum.
♀ flower.

Melandrium dioicum.
♂ flower.

Melandrium dioicum ♂.
Flower cut vertically (mag.).

Melandrium dioicum ♂.
Diagram.

Melandrium dioicum ♂.
Andrœcium surrounding an abortive pistil.

Melandrium dioicum ♀
Transverse section of ovary (mag.).

Melandrium dioicum.
Calyx.

Melandrium dioicum.
Petal.

Drypis spinosa.
Seed, entire and cut vertically, with coiled and incumbent cotyledons (mag.).

Cucubalus baccifer.
Seed, entire, and with vertical and transverse sections showing the incumbent cotyledons (mag.).

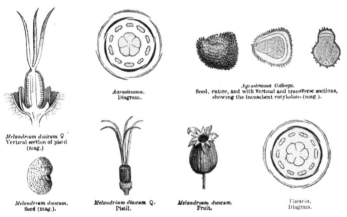

Melandrium dioicum ♀
Vertical section of pistil
(mag.)

Agrostemma.
Diagram.

Agrostemma Githago.
Seed, entire, and with vertical and transverse sections,
showing the incumbent cotyledons (mag.).

Melandrium dioicum.
Seed (mag.).

Melandrium dioicum ♀.
Pistil.

Melandrium dioicum.
Fruit.

Viscaria.
Diagram.

2. **Diantheæ.**—Corolla always twisted to the right in bud. Calyx with no commissural nerves. Petals usually furnished with small winged bands at the claw, or with a coronet of scales at the base of the limb (*Saponaria*, *Velezia*). Fruit 2-merous. Embryo peripheric, or rarely straight, and then albumen scanty or 0.

　　　* Saponaria.　　　　* Gypsophila.　　　　* Dianthus.　　　　Velezia.

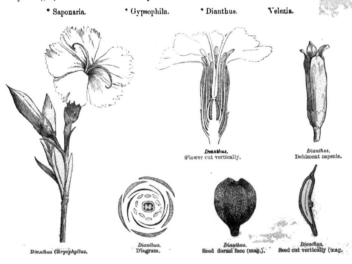

Dianthus.
Flower cut vertically.

Dianthus.
Dehiscent capsule.

Dianthus Caryophyllus.

Dianthus.
Diagram.

Dianthus.
Seed dorsal face (mag.).

Dianthus.
Seed cut vertically (mag.).

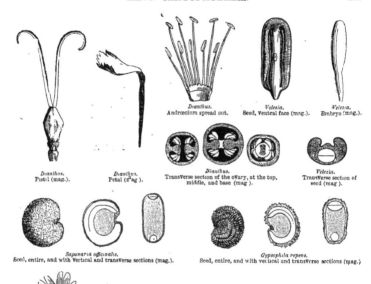

Dianthus.
Androecium spread out.

Velezia.
Seed, Ventral face (mag.).

Velezia.
Embryo (mag.).

Dianthus.
Pistil (mag.).

Dianthus.
Petal (mag.).

Dianthus.
Transverse section of the ovary, at the top,
middle, and base (mag.).

Velezia.
Transverse section of
seed (mag).

Saponaria officinalis.
Seed, entire, and with vertical and transverse sections (mag.).

Gypsophila repens.
Seed, entire, and with vertical and transverse sections (mag.)

TRIBE II. *ALSINEÆ, D.C.*

Sepals free, or united at their base by the disk.
Petals and stamens hypogynous on a slightly developed
disk, or shortly perigynous. Petals with a short or
obtuse base, without claw or scales. Styles quite
distinct. Leaves exstipulate, or sometimes with small
scarious stipules.

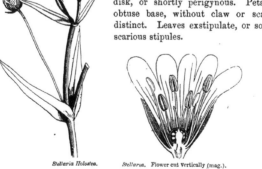

Stellaria Holostea.

Stellaria. Flower cut vertically (mag.).

Stellaria. Diagram.

Stellaria.
Pistil
and andrœcium
(mag.).

Stellaria.
Compressed seed, entire, and with vertical
and transverse sections,
with incumbent cotyledons (mag.).

Stellaria.
Dehiscent fruit.

Colobanthus.
Apetalous flower, stamens
alternate with
the sepals (mag.).

Buffonia macrosperma.
Seed, entire, and with vertical and transverse sections showing the
accumbent cotyledons (mag.).

Spergularia marginata.
Winged seed, entire, and with vertical and
transverse sections,
with incumbent cotyledons (mag.).

Holosteum.
Depressed seed, ventral
keeled face (mag.).

Holosteum umbellatum.
Seed, cut vertically and transversely, with
incumbent cotyledons (mag.).

Cerastium arvense.
Seed, entire, and with vertical and transverse
sections, with incumbent cotyledons (mag.).

PRINCIPAL GENERA.

Holosteum.	*Cerastium.	Stellaria.	*Arenaria.	Buffonia.
Sagina.	Colobanthus.	Queria.	*Spergula.	Spergularia.

Tribe III. *POLYCARPEÆ, D.C.*

Sepals free, or united at the base by the disk. Petals as in *Alsineæ*, usually small, hypogynous, inserted with the stamens on a slightly developed torus, or shortly perigynous. Style simple at the base, 3-2-fid above. Stamens 5 or fewer. Leaves usually furnished with scarious stipules.

PRINCIPAL GENERA.

Drymaria.	Polycarpon.	Ortegia.
Lœflingia.	Polycarpæa.	Stipulicida.

Caryophylleæ, with *Paronychieæ, Portulaceæ, Amarantaceæ, Baselleæ, Chenopodieæ, Phytolacceæ, Nyctagineæ*, and even *Polygoneæ*, form a group of plants of which the common character is a curved embryo surrounding a floury albumen (see these families). Those *Caryophylleæ* which have petals, definite stamens, a one-celled and many-ovuled ovary, and opposite leaves, are easily distinguished from all these families; but the apetalous and few-ovuled genera approach several of them. Notwithstanding their parietal placentation, we may unite to this group *Mesembryanthemeæ*, which have a curved embryo surrounding a floury albumen, and *Cacteæ*, which have a curved but usually exalbuminous embryo.

Caryophylleæ mostly inhabit the extra-tropical regions of the northern hemisphere, extending to the Arctic regions and to the tops of the highest Alps. They are rarer in the southern hemisphere, and still more so in the tropics, where they are almost confined to the mountains.

Some *Caryophylleæ* possess refreshing and slightly demulcent properties, but they have fallen into disuse. Such are *Holosteum umbellatum*, *Cerastium arvense*, *Stellaria Holostea* and *media*; the latter, which grows everywhere, is the common Chickweed whose seeds form the food for many cage-birds. The seeds of *Spergula* were formerly recommended for consumption. The root of *Saponaria officinalis*, an indigenous species, contains a gum, a resin, and a peculiar matter which froths in water like soap, whence it has been placed among demulcent and depurative medicines; some doctors even substitute it for Sarsaparilla in cases of syphilis. The White Lychnis (*Melandrium dioicum*) and *Lychnis chalcedonica* are also used as demulcents. *Silene Otites*, a bitter and astringent herb, is prescribed for hydrophobia. The root of *Silene virginica* is used as an anthelminthic in North America. Pinks, and especially *Dianthus Caryophyllus*, have sweet-scented petals, with which chemists prepare a syrup and a distilled water. The Rose Campion (*Lychnis Githago*) is common amongst corn: its seeds are acrid, and render bread poisonous when mixed with the flour in too great quantities.

XXVII. *PORTULACEÆ, Jussieu.*

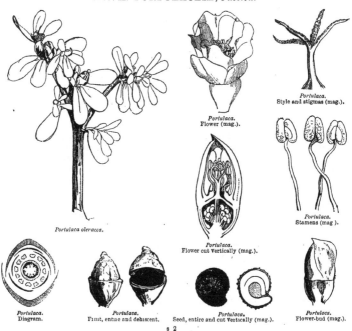

Portulaca.
Style and stigmas (mag.).

Portulaca.
Flower (mag.).

Portulaca oleracea.

Portulaca.
Flower cut vertically (mag.).

Portulaca.
Stamens (mag).

Portulaca.
Diagram.

Portulaca.
Fruit, entire and dehiscent.

Portulaca.
Seed, entire and cut vertically (mag.).

Portulaca.
Flower-bud (mag.).

s 2

FLOWERS ⚥. COROLLA 0, *or petals sometimes coherent at the base, very fugacious.* STAMENS *hypogynous or perigynous, equal and alternate with the calyx-lobes, or double, triple, or multiple in number.* OVARY *usually free, rarely inferior, 1–8-celled.* FRUIT *indehiscent, or a pyxidium, or a loculicidal capsule.* EMBRYO *peripheric, arched or annular, surrounding a floury albumen.*

Herbaceous annuals or perennials, often suffruticose or fruticose; stem and branches diffuse, glabrous or with simple rarely stellate or hooked hairs. LEAVES alternate or opposite, very various in form, entire, sessile or sub-sessile, often fleshy, with a single nerve, or nerveless, sometimes stipulate. FLOWERS ⚥, usually regular and axillary, solitary or variously disposed, æstivation imbricate. CALYX diphyllous, or monosepalous with 2, 3, 4, 5 divisions. PETALS 5, 4, 3, hypogynous, or rarely sub-epigynous (*Portulaca*), distinct, or connate at the base, very tender and fugacious, often 0. STAMENS 1–∞, inserted on the receptacle or on the calyx, free or in bundles; *filaments* filiform or subulate; *anthers* introrse, dehiscence longitudinal. DISK hypogynous, girding the base of the ovary, very often inconspicuous. OVARY sessile, usually free, sometimes half-inferior (*Portulaca*), 1–5-celled, cells 1–few–many-ovuled; *style* terminal, with 2–8 branches, stigmatiferous on their inner surface; *ovules* semi-anatropous, rarely solitary in the 1-celled ovaries (*Portulacaria*), usually numerous, inserted by separate funicles on a central free placenta, or pendulous to funicles ascending from the bottom of the cell; in the many-celled ovaries 1 or few or many in each cell, attached to the central angle throughout its length, or to its centre or top. FRUIT a dehiscent capsule, or rarely indehiscent (*Portulacaria*). SEEDS reniform, ovoid, globose or lenticular; *albumen* floury. EMBRYO peripheric, curved or annular, surrounding the albumen; *cotyledons* incumbent; *radicle* facing the hilum.

TRIBE I. *CALANDRINIEÆ.*

Calyx diphyllous, or 2-partite or 2–3-fid. Petals 5, 4, 3, hypogynous, distinct, sometimes more or less connate into a tube (*Montia*). Stamens fewer or more than the sepals, or indefinite, hypogynous, inserted alone or in bundles at the base of the petals; filaments free, or united at the base. Ovary 1-celled, few–several–many-ovuled, placentation basilar or free, central; style filiform, 2–5-fid. Capsule 2–5-valved.—Herbaceous or frutescent plants. Leaves alternate or opposite, often fleshy, sometimes furnished with intrafoliar stipules cut into hairs or laciniæ. Flowers solitary, or collected into racemes, or into axillary or terminal cymes.

PRINCIPAL GENERA.

Portulacaria.	Anacampseros.	Talinum.'
'Calandrinia.	'Claytonia.	Montia.

TRIBE II. *SESUVIEÆ.*

Calyx 5-tid, rarely 2-fid or -partite. Petals 0, or rarely 4–6, and epigynous (*Portulaca*). Stamens 5–10–∞, inserted singly or in pairs, or in bundles, at the base or throat of the calyx, and between its segments. Ovary free, rarely inferior

(*Portulaca*), 1–5-celled, many-ovuled; ovules ascending, fixed to a basilar placenta, or pendulous to the central angle of the cells; stigmas 2–5. Capsule opening transversely by circular dehiscence.—Fleshy glabrous herbs with opposite or alternate leaves, often stipulate, or bearing stipuliform hairs in the leaf-axils. Flowers axillary, sessile, solitary, or glomerate, in spiked or umbelled cymes.

PRINCIPAL GENERA.

| *Portulaca. | Sesuvium.[1] | Trianthema. |

TRIBE III. *AIZOIDEÆ.*

Calyx 4–5-fid or -partite. Corolla 0. Stamens 5–15, inserted singly or in pairs, or in bundles, on the calyx, between its segments. Ovary free, with 2–5 1–2–∞ -ovuled cells; ovules pendulous to the central angle of the cells; stigmas 5–2. Capsule loculicidal.—Herbaceous or frutescent plants, covered with simple or bi-acuminate hairs. Leaves alternate or opposite. Flowers axillary, sessile.

PRINCIPAL GENERA.

| Aizoon. | Galenia. | Plinthus. |

ALLIED TRIBE. *MOLLUGINEÆ.*

Calyx 5–4-partite, or 5-fid, persistent. Petals 0, or very numerous, ligulate, sub-perigynous. Stamens hypogynous or perigynous, equal and alternate with the sepals, or fewer, or more, or indefinite, distinct, or aggregated in bundles, the outer alternate with the calyx-segments. Ovary free, many-ovuled, 2–3–5-celled; ovules fixed to the inner angle of the cells by separate funicles, or rarely solitary and basilar (*Acrosanthes*). Capsule usually angular or compressed, loculicidal. Seeds as in *Portulaceæ*.—Herbaceous or sub-woody plants, glabrous, or covered with stellate hairs. Leaves opposite or alternate, or fascicled and pseudo-whorled, often stipulate. Flowers crowded in racemes or cymes, or in axillary or leaf-opposed umbellules, rarely solitary (*Acrosanthes*).

PRINCIPAL GENERA.

| Orygia. | Gleinus. | Mollugo. | Pharnaceum. | Acrosanthes. |
| Psamnotrophe. | Adenogramme. | Giesekia. | Limeum. | |

Portulaceæ approach *Tetragonieæ, Mesembryanthemeæ,* and *Paronychieæ* (see these families). The tribe of *Mollugineæ* is also connected with *Portulaceæ* by habit, the entire fleshy leaves, inflorescence, perigynous corolla, often 0, the isostemonous or indefinite stamens, distinct or aggregated into bundles alternate with the sepals, and especially by the structure of the ovule and the nature of the albumen.

---

[1] In Bentham and Hooker's ' Genera Plantarum,' *Portulaca*, from its disepalous calyx and 1-celled ovary, is regarded as a very close ally of *Montia* and the other disepalous *Portulaceæ*, included in this work under the tribe *Calandrineæ*; whilst *Sesuvium* (and *Trianthema*) with its 5-merous isomerous perianth and several-celled ovary, is placed with the other *Aizoideæ* under *Ficoideæ*, and referred to the Calycifloral sub-order of Dicotyledons. No doubt the *Ficoideæ* and *Portulaceæ* are members of one great group (which should also include *Tetragonieæ*), but the exigencies of a linear classification render it convenient to keep them apart.—ED.

Portulaceæ are not absolutely absent from any climate, although more rare in the temperate regions of Europe and central Asia than in North America. Most inhabit the subtropical regions of the southern hemisphere. *Aizoideæ* abound in South Africa, and occur in Arabia Petræa, and in very small numbers in the Mediterranean region. *Sesuvieæ* are much more widely dispersed; none, however, have been met with in America north of the tropic, and very few are found in temperate Asia and Europe. *Calandrinieæ* are nearly cosmopolitan: they penetrate into the cold regions of the North, and abound beyond the tropics, and rather in the northern than in the southern hemisphere. *Mollugineæ* are most frequent in tropical and subtropical regions.

Most of the species are mucilaginous; some are slightly bitter, astringent, and have been classed amongst mild tonics and diuretics. The herbage of *Portulaca oleracea* has long enjoyed a reputation as refreshing, sedative and antiscorbutic. It is also eaten as a salad; its seed, steeped in wine, acts as an emmenagogue. Several American and Asiatic *Calandrinieæ* are also used as potherbs, as are *Sesuvium Portulacastrum* and *repens*, which grow in tropical Asia. The root of *Claytonia tuberosa*, a native of eastern Siberia, is eatable. *Talinum* and *Pharnaceum* are bitter and astringent, and are popular remedies in Asia and America. Soda is obtained in abundance from *Aizoon canariense* and *hispanicum*, by calcining.

XXVIII. *TAMARISCINEÆ.*

(PORTULACEARUM *genus, Jussieu.*—TAMARISCINEÆ, *Desvaux.*—TAMARICACEÆ, *Lindl.*)

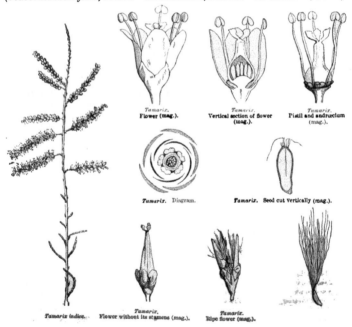

Tamarix.
Flower (mag.).

Tamarix.
Vertical section of flower (mag.).

Tamarix.
Pistil and andrœcium (mag.).

Tamarix. Diagram.

Tamarix. Seed cut vertically (mag.).

Tamarix indica.

Tamarix.
Flower without its stamens (mag.).

Tamarix.
Ripe flower (mag.).

Sepals 5–4. Petals 5, *hypogynous, imbricate, marcescent.* Stamens 5 *or* 10. Ovary 1-*celled, placentas parietal or basilar, usually* 3, *many-ovuled.* Seeds *ascending*; *chalaza apical, bearded.* Embryo *straight, exalbuminous.* Leaves *alternate, rather thick.*

Undershrubs, shrubs, or small trees, with both persistent and annual caducous branchlets. Leaves alternate, sessile, small, sub-imbricate, rather fleshy, sometimes amplexicaul, entire, often dotted, usually glaucous, exstipulate. Flowers perfect, regular, white or pink, bracteolate, in terminal racemed spikes. Calyx free, persistent, of 5 (rarely 4) sepals, imbricate, 2-seriate, sometimes connate at the base. Petals 5, inserted on the receptacle, imbricate in bud, marcescent. Stamens equal to and alternate with the petals, or double in number, inserted on the edge of a hypogynous disk; *filaments* free, united at their base into a ring, cup, or tube; *anthers* introrse, dorsifixed, dehiscence longitudinal. Ovary free, sessile, usually 3-gonous, 1-celled, with 3–4 (rarely 2–5) parietal or basilar placentas; *styles* equal in number to the placentas; *stigmas* obtuse or truncate, dilated, sometimes sessile; *ovules* numerous, ascending, anatropous. Capsule 1-celled, or incompletely several-celled by the development of the placentas, 2–5-valved, valves placentiferous at the base. Seeds numerous, ascending, with membranous testa, furnished at their apical chalaza with a dense beard, or beaked and furnished with spreading plumose hairs. Embryo exalbuminous, straight; *cotyledons* oblong, obtuse, plano-convex; *radicle* short, conical, inferior.

PRINCIPAL GENERA.

*Myricaria. *Tamarix.

Bentham and Hooker fil. have combined with the small family of *Tamariscineæ*, *Reaumuriaceæ* and the genus *Fouquiera*; which approach it, in fact, in their somewhat fleshy leaves, æstivation, hypogyny, often isostemonous or diplostemonous corolla, one-celled ovary with parietal placentation, capsular fruit erect and hairy seeds; but differ in the seeds being albuminous, and hairy over their entire surface. *Reaumuria* is distinguished by the solitary flowers and floury albumen; *Fouquiera* has a monopetalous five-fid corolla with a long tube, 10–8 hypogynous stamens of unequal length; the seeds are surrounded by a membranous wing, or by transparent hairs which simulate a wing; the albumen is fleshy, and the flowers are spicate or in thyrsoid panicles.

[The including of these genera in *Tamariscineæ* requires that the ordinal characters should be modified, and the order itself be broken up into the following tribes, as proposed in the 'Genera Plantarum.'

Tribe I. *TAMARISCEÆ.*

Petals free or nearly so. Seeds exalbuminous, hairy. Flowers racemed or spiked. *Tamarix, Myricaria.*

Tribe II. *REAUMURIEÆ.*

Petals free. Seeds hairy all over; albumen between fleshy and floury. Flowers solitary, axillary, and terminal. *Hololachne, Reaumuria.*

Tribe III. *FOUQUIERIEÆ.*

Petals connate into a long tube. Seeds winged or furnished with long hairs. Flowers large, panicled. *Fouquieria.* Ed.]

Tamariscineæ are near *Caryophylleæ, Portulaceæ,* and *Frankeniaceæ,* which are separated principally by the structure of their ovules and their floury albumen; they further differ from *Caryophylleæ* and *Frankeniaceæ* in their alternate and fleshy leaves, from *Portulaceæ* in habit, insertion, &c. They have also some affinity with *Crassulaceæ.*

Tamariscineæ (proper) are confined to the Old World, where they extend from 9° to 55° of north latitude. They prefer sea-shores, the margins of brackish lakes, the banks of rivers and torrents, in sandy or clayey soils. [*Reaumurieæ* extend from the Levant to Central Asia; *Fouquiera* is a Mexican shrub.—ED.]

Tamariscineæ contain tannin, resin and a volatile oil, which render them bitter and astringent. The bark of *Myricaria germanica* is employed in Germany for jaundice; that of *Tamarix gallica* is aperient. *T. mannifera,* which grows on Mount Sinai and elsewhere in Arabia, secretes, as the result of the puncture of a Cynips, a saccharine matter, supposed by some to be the manna which fed the Hebrews in the desert. The galls of other species (also produced by the puncture of an insect) are valued for their strongly astringent properties.

XXIX. ELATINEÆ.

(ELATINEÆ, *Cambessèdes.*—ELATINACEÆ, *Lindl.*)

SEPALS 2–5. PETALS 2–5, *hypogynous, imbricate.* STAMENS *equal or double the number of the petals, hypogynous.* OVARY *3–5-celled.* OVULES *anatropous.* FRUIT *a capsule.* SEEDS *exalbuminous.*—LEAVES *opposite or fascicled, stipulate.*

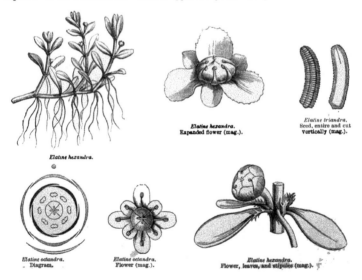

Elatine hexandra.

Elatine hexandra.
Expanded flower (mag.).

Elatine triandra.
Seed, entire and cut
vertically (mag.).

Elatine octandra.
Diagram.

Elatine octandra.
Flower (mag.).

Elatine hexandra.
Flower, leaves, and stipules (mag.).

| *Elatine Hydropiper.*
Flower cut vertically (mag.). | *Elatine.*
Ovule (mag.). | *Merimea.*
Flowers, leaves, and stipules (mag.). | *Merimea.*
Diagram. |

Dwarf HERBS, or marsh UNDERSHRUBS; stems creeping or spreading. LEAVES opposite, rarely whorled, sessile or sub-sessile, entire or toothed, stipulate. FLOWERS ☿, small, regular, axillary, solitary or cymose. SEPALS 2–5, distinct, æstivation imbricate. PETALS 2-5, hypogynous, æstivation imbricate. STAMENS equal or double the number of the petals, hypogynous; *filaments* filiform-subulate, free; *anthers* introrse, dorsifixed, versatile, dehiscence longitudinal. OVARY free, cells as many as sepals; *styles* as many as cells, distinct; *stigmas* capitate; *ovules* ∞, fixed to the central angle of the cells, horizontal or sub-ascending, anatropous, raphe lateral or superior. CAPSULE septicidal, valves flat or inflexed, leaving the placentiferous central column free. SEEDS numerous, cylindric, straight or curved, strongly striate transversely, rarely smooth (*Merimea*), hilum basilar, exalbuminous. EMBRYO straight or curved; *cotyledons* short, obtuse; *radicle* cylindric, long, near the hilum.

GENERA.

Elatine.   Bergia.   Merimea.

Elatineæ, formerly placed in *Caryophylleæ*, tribe *Alsineæ*, are distinguished by the capitate stigmas, dehiscent capsule, exalbuminous seed, and straighter embryo. They approach *Hypericineæ* in hypopetalism, the 3–6-celled ovary with many ovules in each cell, the free styles, capsular fruit, straight or curved exalbuminous seeds, and opposite or whorled leaves; but in *Hypericineæ* the petals are twisted, the stamens usually numerous and polyadelphous, and the leaves exstipulate. *Elatineæ* approach some *Lythrarieæ*, which have also isostemonous or diplostemonous flowers, an ovary with two or several many-ovuled cells, anatropous ovules, a septicidal capsule, exalbuminous seeds, and opposite leaves; but they differ in the tubular calyx, perigynism, simple style and exstipulate leaves.

The genus *Tetradiclis* (or *Anatropa*) appears much nearer *Elatineæ* than *Zygophylleæ*, in which Bentham and Hooker fil.[1] place it; it differs from *Zygophylleæ* in the number of the parts of the flower, the dehiscence of the capsule, the nature of the seeds, &c., and is only separated from *Elatineæ* by its exstipulate and laciniate leaves.

Elatineæ are widely dispersed, especially in the Old World, inhabiting ditches and the submerged shores of ponds and rivers. They are of no use to man.

---

[1] Bentham and Hooker place it in *Rutaceæ*, tribe *Ruteæ*, and not in *Zygophylleæ*.—ED.

XXX. *HYPERICINEÆ.*

(HYPERICA, *Jussieu.*—HYPERICINEÆ, *D.C.*—HYPERICAÇEÆ, *Lindl.*)

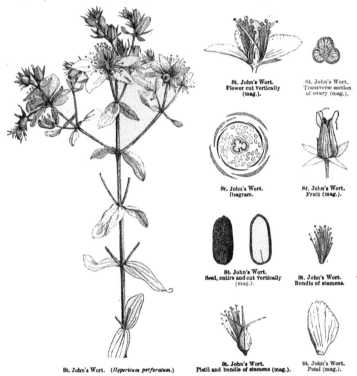

St. John's Wort.
Flower cut Vertically
(mag.).

St. John's Wort.
Transverse section
of ovary (mag.).

St. John's Wort.
Diagram.

St. John's Wort.
Fruit (mag.).

St. John's Wort.
Seed, entire and cut Vertically
(mag.).

St. John's Wort.
Bundle of stamens.

St. John's Wort. (*Hypericum perforatum.*)

St. John's Wort.
Pistil and bundle of stamens (mag.).

St. John's Wort.
Petal (mag.).

PETALS *hypogynous, claw naked, or furnished with a scale-like pit.* STAMENS *numerous, hypogynous, monadelphous or polyadelphous.* OVARY *5–3-celled, or 1-celled by imperfection of the septa.* OVULES *numerous, anatropous.* FRUIT *a capsule, rarely fleshy.* EMBRYO *exalbuminous.*—LEAVES *opposite, entire, usually dotted with pellucid glands.*

STEM woody or herbaceous, perennial, rarely annual, juice resinous or limpid, branches opposite or rarely whorled, generally 4-gonous, sometimes compressed or cylindric, sometimes heath-like. LEAVES opposite or rarely whorled, simple, penni-nerved, entire or with glandular teeth, usually sprinkled with pellucid glands sunk in the parenchyma, and edged with vesicular black glands ; *stipules* 0. FLOWERS ☿, regular, usually terminal, panicled, or in dichotomous cymes. CALYX persistent, of 4–5 more or less connate sepals, 2-seriate, the two outer often the smallest, rarely 4 decussate, the 2 outer largest, and covering the inner. PETALS inserted on the receptacle, as many as the sepals, sessile or clawed, equal, more or less inequilateral, veins radiating, æstivation contorted or imbricate ; *claw* naked (*Hypericum*), or fur-nished within, above the base, with a fleshy scale, or furrowed. STAMENS inserted on the receptacle, usually indefinite, rarely definite, always more numerous than the petals ; *filaments* in 3 or 5 bundles, sometimes alternating with glands or hypo-gynous scales, or irregularly polyadelphous, or united into a tube, or quite free ; *anthers* small, subglobose, introrse, sub-didymous, often tipped by a gland, cells parallel, opening longitudinally. OVARY of 3-5 carpels, or of 1 (*Endodesmia*), 3-5-celled, or with as many imperfect cells; *styles* as many as carpels, filiform; *stigmas* terminal, capitate, peltate or clavate; *ovules* numerous in each cell, 2-seriate, rarely few or solitary (*Endodesmia*), usually horizontal, rarely ascending (*Haronga, Psoro-spermum*), anatropous, very rarely pendulous (*Endodesmia*). FRUIT a capsule, usually septicidal, rarely loculicidal (*Cratoxylon, Eliæa*), or an indehiscent berry. SEEDS straight, rarely curved, hilum basilar, funicle sub-lateral ; *testa* crustaceous or mem-branous, dotted or smooth, sometimes loosely cellular, arilliform; *chalaza* diametrically opposite to the hilum, often dilated into a membranous wing (*Eliæa, Cratoxylon*). EMBRYO straight or curved, exalbuminous ; *cotyledons* flat, half-cylindric or rarely coiled ; *radicle* cylindric, obtuse, usually longer than the cotyledons, and near the hilum.

PRINCIPAL GENERA.

Hypericum.　　Vismia.　　Cratoxylon.　　Ascyrium.　　Psorospermum.

Hypericineæ are closely connected with *Guttiferæ* and *Camelliaceæ*; they are allied to *Guttiferæ* in their resinous juice, their tetragonous branches, opposite entire leaves, free or nearly free decussate unequal sepals, contorted or imbricate petals, indefinite stamens, filaments usually in several bundles, or monadelphous, one-pluri-celled ovary, horizontal or ascending anatropous ovules, capsular or fleshy fruit, and exalbuminous embryo; the diagnosis almost wholly rests on the usually herbaceous stem of *Hyperi-cineæ*, their not jointed branches, less coriaceous leaves, always perfect flowers, and filiform styles. They approach *Camelliaceæ* in their free sepals, imbricate or contorted petals, indefinite monadelphous or polyadel-phous stamens, and connective often glandular at the top, their capsular or fleshy fruit, and exalbuminous seed; they are principally separated by their resinous juice, opposite leaves and inflorescence. They have also a close affinity with *Cistineæ* in their two-seriate sepals, hypogynous contorted petals, numerous stamens, one-celled or sub-pluricelled ovary, capsule with septicidal valves with placentiferous margins; but in *Cistineæ* the stamens are completely free, the style is simple, the embryo is much curved or coiled, the albumen is floury, the leaves stipulate and usually alternate. Finally, more than one analogy has been noticed between *Hypericineæ* and *Myrtaceæ* (see this family).

Hypericineæ are spread over the temperate and hot regions of the globe, and especially in the northern hemisphere. They are not rare in tropical America; but become so in equinoctial Asia and Africa.

Hypericineæ, like *Guttiferæ*, possess balsamic resinous juices which flow abundantly from the woody species, and which in the herbaceous ones are secreted by black or pellucid glands sunk in the paren-

chyma of the leaves. With these juices is present a certain quantity of volatile oil and a bitter extractive in the bark, which give different properties to *Hypericineæ*. The indigenous species of St. John's
Wort, formerly recommended as astringent, are no longer used, except *Hypericum perforatum*, of which
the tips infused in olive oil are rubbed in for gouty pains. The Tutsan (*H. Androsæmum*), formerly
used as a vulnerary, has fallen into disuse without good reason. *Cratoxylon Hornschuchii*, a small Javanese
tree, is employed in that country as an astringent and diuretic.

XXXI. *GUTTIFERÆ.*

(GUTTIFERÆ, *Jussieu.*—GARCINIEÆ, *Bartling.*—CLUSIACEÆ, *Lindl.*)

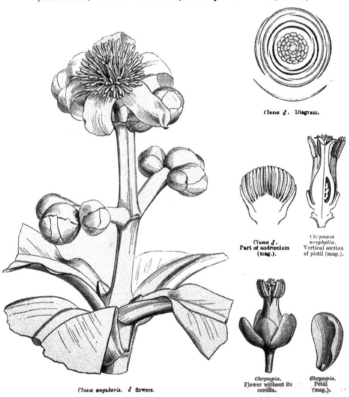

Clusia ♂. Diagram.

Clusia ♂.
Part of andrœcium
(mag.).

*Chrysopia
urophylla.*
Vertical section
of pistil (mag.).

Chrysopia.
Flower without its
corolla.

Chrysopia.
Petal
(mag.).

Clusia angularis. ♂ flowers.

Garcinia Mangostana. Fruit.

Garcinia Mangostana.
Berry with thick bark, the upper portion removed to show the cells.

FLOWERS *polygamo-diœcious, rarely* ☿ . CALYX *4–6-poly-phyllous.* PETALS *hypogynous, equal with the sepals, rarely more numerous.* STAMENS *indefinite, rarely definite, free, or monadelphous or polyadelphous.* OVARY *2–∞ -celled, rarely 1-celled.* OVULES *1–∞ in the cells, ascending or erect, anatropous.* FRUIT *a capsule, drupe, or berry.* EMBRYO *exalbuminous, straight.*—STEM *woody.* LEAVES *opposite.*

Pilosperma caudatum.
Vertical
section of ovary
(mag.).

Pilosperma.
Embryo
(mag.).

Pilosperma.
Seed,
entire, with its
arillode.

Pilosperma.
Seed cut vertically.
h, hilum ;
m, micropyle.

TREES or SHRUBS, sometimes climbing or epiphytal, with resinous usually yellow or green juice, branches opposite, generally tetragonous, jointed. LEAVES opposite, usually decussate, rarely whorled. coriaceous, mostly shining, penninerved, secondary nerves transverse, rarely pellucid-punctate ; *petiole* jointed at its base to the branch, entire and exstipulate, or very rarely pinnatisect and stipulate (*Quiina*). FLOWERS white, yellow or red, regular, polygamo-diœcious, or ☿ , terminal or axillary, solitary, or in fascicles or few-flowered cymes, trichotomous panicles, or racemes. SEPALS 2–6, rarely more, imbricate, or decussate in pairs, sometimes furnished with pairs of decussate bracts. PETALS 2–6, rarely more, hypogynous, imbricate or contorted, rarely decussate in pairs, very rarely 4, sub-valvate.—FLOWERS ♂ : STAMENS inserted on the receptacle, numerous, or rarely definite and equal or double the number of the petals ; *filaments* often thick or short, free or variously connate, sometimes united into a fleshy mass, or in bundles equalling in number and opposite to the petals, sometimes long and filiform ; *anthers* 2- (rarely 1-) celled, cells usually linear, adnate or terminal, extrorse or rarely introrse, sometimes sessile, or plunged in the mass of the filaments, opening longitudinally, or by an apical pore. OVARY rudimentary, or more or less developed.—FLOWERS ♀ and ☿ : STAMINODES or STAMENS surrounding the ovary, often definite, fewer and less coherent than those of the ♂ flower. OVARY seated on a flat receptacle or a fleshy disk, 2–many-celled, rarely 1-celled ; *stigmas* as many as cells, sessile or sub-sessile, radiating, or coherent and peltate, or radiating at the top of a single elongated style, sometimes distinct on as many styles; *ovules* 1–∞ in the cells, fixed to the central angle, or erect and basal, anatropous. · FRUIT usually between fleshy and coriaceous, sometimes indehiscent, berried or drupaceous, sometimes with as many septicidal valves as cells. SEEDS large, often arillate or strophiolate ; *testa* thin, coriaceous, or rarely spongy. EMBRYO straight, exalbuminous, filling the seed, sometimes with a voluminous radicle and minute or scaly cotyledons, sometimes divided into 2 cotyledons, which are connate, or separable with difficulty ; *radicle* very short, inferior.

PRINCIPAL GENERA.

*Clusia.	Garcinia.	Calophyllum.	Havetiopsis.	Chrysochlamys.
Tovomita.	Rheedia.	Mesua.	Mammea.	Quiina.

Guttiferæ are very near *Hypericineæ* and *Marcgraviaceæ* (see these families). They are equally close to *Camelliaceæ*, in their more or less distinct sepals, the æstivation of their petals and connection of filaments ; they are distinguished by their opposite leaves, their usually diclinous four-merous flowers with decussate sepals and petals, their straight embryo, and often inconspicuous cotyledons. All the *Guttiferæ* are intertropical, except a few natives of the warm regions of North America ; they are more numerous in America than in Asia, and are rather rare in Africa.

Guttiferæ owe their name to the yellow or greenish juice which flows on the incision of their stem, and which contains an acrid resin held in solution by a volatile oil, sometimes mixed with a gummy principle. The acidulous-sugary berries of several species are eatable. The seeds of others contain a fixed oil, and the wood of all is durable, and hence valuable. The inspissated juice of *Hebradendron cambogioides*, a Ceylon tree, is the saffron-red colored, opaque, smooth, shining substance called *gamboge*, which is a rich golden-yellow pigment and a powerful purgative. The same is the case with *Clusia rosea*, a West Indian tree, whose blackish bitter juice, thickening in the air, is frequently used instead of scammony. That of *C. flava*, which is also cultivated in European hot-houses, is praised in Jamaica as a vulnerary. The berries of *Calophyllum* are sweet, acidulous, and agreeable. *C. inophyllum*, an Indian plant, affords a purgative and emetic resin, and its root is considered diuretic. That of *C. turiferum*, a native of Peru, emits a balsamic odour when burned, and is used for incense. *C. Calaba*, of the Antilles, yields a juice (*aceite de Maria*) which rivals copal. *Mesua speciosa* and *ferrea*, of India, have very hard and excellent woods ; their aromatic and bitter root and bark are powerful sudorifics. The fruit of the *Mangosteen* (*Garcinia mangostana*), a native of the Moluccas (now introduced into the Antilles), possesses a bitter and astringent rind, but a delicious pulp which is refreshing and antibilious. The fruit of the *Mammea* is also eatable, the water distilled from its flowers (*eau de Créole*) is eminently digestive, and the juice of its young shoots yields a very agreeable vinous liquor.

XXXII. *CAMELLIACEÆ.*

(Ternstrœmieæ, *Mirbel.*—Ternstrœmiaceæ, *D.C.*—Theaceæ, *Mirbel.*—Camellieæ, *D.C.*—Camelliaceæ, *Bartling.*)

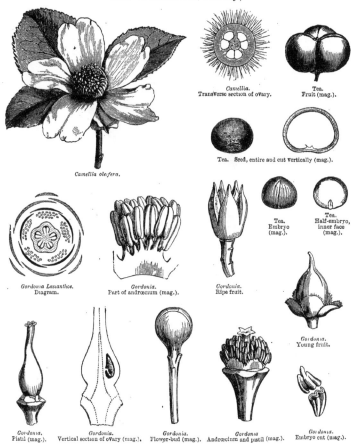

Camellia oleifera.

Camellia.
Transverse section of ovary.

Tea.
Fruit (mag.).

Tea. Seed, entire and cut vertically (mag.).

Tea.
Embryo
(mag.).

Tea.
Half-embryo,
inner face
(mag.).

Gordonia Lasianthos.
Diagram.

Gordonia.
Part of andrœcium (mag.).

Gordonia.
Ripe fruit.

Gordonia.
Young fruit.

Gordonia.
Pistil (mag.).

Gordonia.
Vertical section of ovary (mag.).

Gordonia.
Flower-bud (mag.).

Gordonia.
Andrœcium and pistil (mag.).

Gordonia.
Embryo cut (mag.).

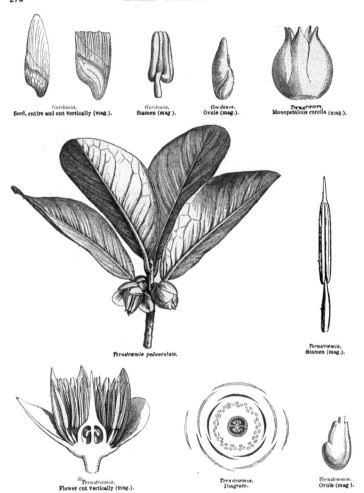

Gordonia.
Seed, entire and cut vertically (mag.).

Gordonia.
Stamen (mag.).

Gordonia.
Ovule (mag.).

Ternstræmia.
Monopetalous corolla (mag.).

Ternstræmia pedunculata.

Ternstræmia.
Stamen (mag.).

Ternstræmia.
Flower cut vertically (mag.).

Ternstræmia.
Diagram.

Ternstræmia.
Ovule (mag.).

PETALS *hypogynous, usually 5, free or nearly so, imbricate or contorted.* STAMENS *usually indefinite, hypogynous.* OVARY *usually 3–5-celled.* OVULES *pendulous or ascending.* FRUIT *indehiscent or capsular.* EMBRYO *exalbuminous or albuminous.—* STEM *woody.* LEAVES *generally alternate.*

Large or small TREES with watery juice and cylindric branches. LEAVES alternate, often fascicled at the top of the branches, very rarely opposite (*Caryocar, Haploclathra,* &c.), usually simple, rarely digitate (*Caryocar, Anthodiscus*), coriaceous or membranous, penninerved, entire or toothed ; *stipules* 0, or very rarely 2, minute, caducous. FLOWERS ☿, rarely diclinous (*Actinidia, Omphalocarpum,* &c.), regular, sometimes axillary, solitary or fascicled, sometimes in a terminal raceme or panicle; *peduncle* jointed at its base, naked or bracteate. SEPALS 5, rarely 4–6–7, free, or slightly connate at the base, imbricate. PETALS 5, rarely 4–6–9, hypogynous, free, or oftener coherent at the base into a ring or short tube, æstivation imbricate or contorted. STAMENS usually indefinite, rarely equal with the petals (*Pentaphylax, Pelliciera*), or double (*Stachyurus*), hypogynous, free or variously coherent at the base, or adherent to the base of the corolla; *anthers* basifixed and erect, or dorsifixed and versatile, cells parallel, opening by a slit, or sometimes by an apical pore (*Saurauja, Pentaphylax*). OVARY free, sometimes more or less buried in the torus (*Anneslea, Visnea*), base large and sessile, 3–5- (rarely 2-) celled (*Pelliciera*), or many-celled (*Anthodiscus, Omphalocarpum,* &c.); *styles* as many as cells, free or more or less connate; *stigmas* pointed or obtuse; *ovules* 2–∞ in each cell, rarely solitary, erect, or horizontal and anatropous, or pendulous and anatropous, or campylotropous, sometimes fixed laterally and semi-anatropous. FRUIT fleshy or coriaceous and indehiscent, or a loculicidal or septicidal capsule. SEEDS numerous or few, fixed to the inner angle of the cells on projecting fleshy or spongy placentas; *albumen* often scanty or 0, rarely copious (*Actinidia, Saurauja, Stachyurus*). EMBRYO straight, curved or coiled; *cotyledons* sometimes semi-cylindric, continuous with the radicle, but shorter, sometimes larger, flat, crumpled, folded lengthwise, or thick and fleshy.

TRIBE I.[1] *RHIZOBOLEÆ.*

Petals imbricate, or united in a cap. Anthers dorsifixed, sub-versatile. Fruit indehiscent. Seeds solitary in the cells; albumen 0 or very scanty; radicle superior, very large, bent at top, or coiled; cotyledons minute.—Leaves digitate. Racemes terminal.

GENERA.

Caryocar. Anthodiscus.

TRIBE II. *TERNSTRŒMIEÆ.*

Petals imbricate. Anthers basifixed. Fruit rarely dehiscent. Seeds generally few ; albumen fleshy, usually scanty ; embryo inflexed or arched ; cotyledons shorter and not broader than the radicle.—Trees or shrubs. Peduncles one-flowered.

PRINCIPAL GENERA.

Visnea.   *Ternstrœmia.   Pentaphylax.   Adinandra.   Cleyera.   Freziera.   Eurya.

[1] These tribes are taken from the 'Genera Plantarum,' omitting *Marcgravieæ,* for which see p. 275.—ED.

T

Tribe III. *SAURAUJEÆ.*

Petals imbricate. Anthers versatile. Fruit very rarely sub-dehiscent, usually pulpy. Seeds numerous, small; albumen copious; embryo straight, or slightly bent, radicle usually longer than the cotyledons.—Trees or upright or twining shrubs. Peduncles many-flowered.

PRINCIPAL GENERA.

* Saurauja.   Actinidia.   Stachyurus.

Tribe IV. *GORDONIEÆ.*

Petals imbricate. Anthers versatile. Fruit loculicidal (*Camellia, Thea, Stuartia*), or indehiscent (*Pelliciera, Omphalocarpum*). Albumen usually 0, or scanty; cotyledons thick, flat, or crumpled or folded; radicle short, straight or inflexed.—Trees or erect shrubs. Peduncles 1-flowered.

PRINCIPAL GENERA.

* Stuartia.   * Gordonia.   * Camellia.   * Thea.   Schima.   Pyrenaria.   Laplacea.

Tribe V. *BONNETIEÆ.*

Petals contorted. Anthers versatile or sub-basifixed. Capsule septicidal. Albumen 0, or very scanty; embryo straight; cotyledons large; radicle short. —Erect trees. Flowers in terminal panicles or in axillary racemes.

PRINCIPAL GENERA.

Bonnetia.   Mahurea.   Caraipa.   Marila.   Kielmeyera.

Camelliaceæ have many affinities: 1. With polypetalous polyandrous hypogynous families with a plurilocular ovary (see *Hypericineæ* and *Guttiferæ*). They scarcely differ from *Bixineæ*, to which they are united by *Cochlospermeæ*, except in their ovary with perfect cells, and their exstipulate leaves. They approach *Dipterocarpeæ* in the polysepalous calyx, the polypetalous hypogynous corolla with imbricate æstivation, the polyandry, the several-celled ovary which is broadly sessile or slightly buried in the torus, the exalbuminous seed, the woody stem and the alternate leaves; but *Dipterocarpeæ* are separated by their persistent and usually accrescent calyx, their one-celled and one-seeded fruit, their habit, and especially by their resinous juice. They have also some affinities with *Tiliaceæ*, which principally differ in the valvate calyx. 2. With monopetalous families (see *Ericineæ, Styraceæ* and *Ebenaceæ*). They are further connected with *Sapoteæ* through *Eurya* and *Ternstræmia*, whose corolla is monopetalous, imbricate, diplo- or triplo-stemonous, the fruit a berry, the stem woody, and the leaves alternate and coriaceous; but *Sapoteæ* have extrorse anthers.

Camelliaceæ principally inhabit tropical America and eastern Asia; very few are met with in North America, and one species only (*Visnea Mocanera*) in the Canaries.

Some species of *Saurauja* and *Kielmeyera* are mucilaginous and emollient. *Gordonia* contains an astringent principle, and is used in tanning leather. The seeds of *Camellia japonica*, introduced into Europe in 1739, are valued in eastern Asia on account of the fixed oil which they contain. Its leaves have a slight tea-like scent. The most important species of this family is the Tea (*Thea chinensis*), which some authors place in the genus *Camellia*. Two centuries have not elapsed since Tea was first used in Europe, and the annual importation now exceeds twenty-two millions of pounds. The stimulating property of Tea is due to an astringent principle, an azotized substance called *théine*, and especially a small proportion of slightly narcotic volatile oil; the leaves also contain a considerable quantity of *casein*, a very nutritious substance, which is not soluble in water; whence the Tibetans, after drinking the infusion, eat the boiled leaves mixed with fat, which forms a substantial food. The two principal Teas of commerce, *green* and *black*, belong to the same species; their difference is simply due to a peculiar preparation of the leaf before it is dried. Many varieties of green and black Teas are distinguished; that called Pekoe is a green tea much prized for its scent, which is given to it by the flowers of *Olea fragrans*. The Chinese perfume other teas with various scented flowers, such as the *Jasminum Sambac* and *Camellia Sesanqua*. Many attempts have been made to cultivate tea in Brazil and in Europe, but the produce cannot be compared with that from China.

XXXIII. *MARCGRAVIACEÆ.*

(MARCGRAVIACEÆ, *Jussieu.*—MARCGRAVIEÆ, *Planchon.*—TERNSTRŒMIACEARUM *tribus,*
Benth. and *Hook. fil.*)

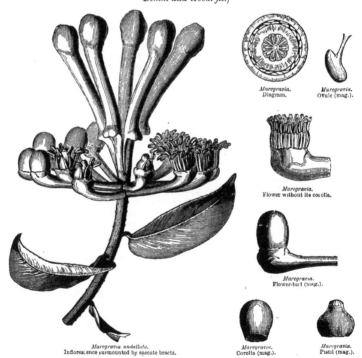

Marcgravia.
Diagram.

Marcgravia.
Ovule (mag.).

Marcgravia.
Flower without its corolla.

Marcgravia.
Flower-bud (mag.).

Marcgravia umbellata.
Inflorescence surmounted by spacate bracts.

Marcgravia.
Corolla (mag.).

Marcgravia.
Pistil (mag.).

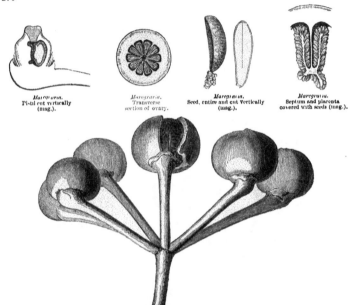

Marcgravia.
Pistil cut vertically
(mag.).

Marcgravia.
Transverse
section of ovary.

Marcgravia.
Seed, entire and cut vertically
(mag.).

Marcgravia.
Septum and placenta
covered with seeds (mag.).

Marcgravia umbellata. Umbel of fruits.

TREES or unarmed SHRUBS, erect, climbing, or epiphytal. LEAVES alternate,
simple, penninerved, entire, glabrous, shining, jointed to the branches, exstipulate.
FLOWERS ☿, regular, in umbels, racemes or terminal spikes; *peduncles* jointed at the
base, usually furnished with bracts, which are sometimes saccate or hooded and
petioled; *bracteoles* at the base of the calyx, minute and resembling an outer calyx,
or 0. CALYX of 2–3–5–6 subequal sepals, distinct or slightly connate at the base,
imbricate, coriaceous, usually coloured, deciduous. PETALS imbricate, inserted on
the receptacle, free or connate at the base, equal and alternate with the sepals; or
more numerous, united into a cap which circumscisses at the base. STAMENS
inserted either below the ovary, or on the edge of a flat disk girding the base of the
ovary, very rarely equal and opposite to the sepals (*Ruyschia*), generally more
numerous (*Marcgravia*); *filaments* free or connate at the base, sometimes adnate to
the base of the petals; *anthers* introrse, ovoid, linear or oblong, basifixed, cells
opposite, contiguous, opening longitudinally. OVARY sessile, free, sometimes girt at
its base by the staminiferous disk, 3–5–many-celled; *stigma* sessile or subsessile,
radiating; *ovules* numerous, attached to the fleshy and projecting lobes of the

ascending or horizontal placentas. FRUIT indehiscent, or opening gradually at its base, loculicidal, valves semi-septiferous (*Ruyschia*). SEEDS few, sunk in the fleshy placentas, ascending, oblong; *testa* areolate, *hilum* lateral, *endopleura* membranous. EMBRYO exalbuminous, subclavate, straight or slightly arched; *cotyledons* obtuse; *radicle* long, conical, acute, near the hilum, inferior.

GENERA.

Ruyschia.   Marcgravia.   *Norantea.

This little group is closely allied to the various tribes of *Ternstræmiaceæ*, and is considered one of them by Bentham and Hooker fil. Of these tribes, however, 1, *Rhizoboleæ* differ in their versatile anthers, superior radicle, and opposite and digitate leaves; 2, *Ternstræmieæ* differ in their one-flowered peduncle; 3, *Saurauijeæ* in their versatile anthers and copious albumen; and others in their contorted petals and septicidally dehiscing capsule,[1] &c. *Marcgraviaceæ* are also distinguished by their sessile radiating stigma, and especially by the singular conformation of their bracts, saccate in *Marcgravia*, and hooded in *Norantea*. They are also very near *Guttiferæ*, from which they are only separated by their basifixed anthers, their alternate leaves, and their saccate bracts.

Marcgraviaceæ inhabit tropical America. The root, stem, and leaves of *M. umbellata* are renowned West Indian diuretics and antisyphilitics.

XXXIV. *DIPTEROCARPEÆ*,[2] *Blume.*

[TREES, rarely SHRUBS, often gigantic, exuding a resinous juice, rarely climbing (*Ancistrocladus*). LEAVES alternate, penninerved, quite entire, rarely crenate; *stipules* small or large, caducous or persistent, sometimes sheathing and leaving an annular scar on the branch. FLOWERS regular, ☿, often odorous, in axillary panicles, ebracteate or with minute deciduous rarely large persistent bracts. CALYX in flower free and campanulate, rarely short or adnate to the torus or base of the ovary; segments 5, imbricate when young, sometimes sub-valvate in age; *fruiting calyx* enlarged, segments unaltered, or 2 or all foliaceous or variously expanded. PETALS 5, strongly contorted, free or connate at the base. STAMENS either ∞ ∞ - seriate, or 15 2-seriate, or 10 in pairs, 5 exterior and 5 interior, or 5 or 10 1-seriate, inserted on a hypogynous or sub-perigynous torus; *filaments* short, often dilated at the base, free or connate at the base, or cohering with the petals; *anthers* erect, 2-celled, dehiscence introrse or lateral, cells equal or one smaller, connective sometimes acuminate or aristate. OVARY inserted by a broad base or sub-immersed, 3- (rarely 1- or 2-)celled; *style* subulate or thickened; *stigma* simple or 3-lobed; *ovules* in pairs, pendulous or laterally attached, anatropous with superior micropyle and ventral raphe, or 1 or more erect in the 1-celled ovary. FRUIT free or adnate to the calyx, 1- (rarely 2-) seeded, indehiscent or at length 3-valved. SEED usually inverted, rarely erect, testa thin, albumen 0; *cotyledons* either thick and equal or unequal, straight or lobed and plaited, or thin and corrugated; *radicle* next the hilum, either short and exserted, or long and included in a fold of the cotyledons.

[1] This is an error: *Gordonieæ* are loculicidal, and have imbricate petals like *Marcgravieæ*.—ED.
[2] This order is omitted in the original.—ED.

| Dryobalanops. | Dipterocarpus. | Anisoptera. | Vatica. | Lophira. | Shorea. |
| Hopea. | Dovua. | Vateria. | Monoporandra. | | ? Ancistrocladus. |

Dipterocarpeæ are allied to *Tiliaceæ* and *Ternstrœmiaceæ*, differing from both in their resinous juice, from the former in their imbricate calyx, and from the latter in their enlarged fruiting calyx and solitary exalbuminous seed. The remarkable Indian and African genus *Ancistrocladus*, which is a climber, has little affinity with the rest of the order, and has been referred to *Terebinthaceæ* by Thwaites, and doubtfully placed near *Gynocarpeæ* by Oliver.

All are natives of the hot damp forests of India and the Malayan islands, except a few African species. Many *Dipterocarpeæ* are valued for their magnificent timber; as the Sal (*Shorea robusta*), which also yields the Dammar resin, called Ral or Dhooma in India. The famous Borneo camphor is the produce of *Dryobalanops Camphora*; it is found in the form of yellow rectangular prisms in fissures of the wood, and is chiefly exported to China, where it is employed as a tonic and aphrodisiac; in Borneo itself it is used as a diuretic and in nephritic affections, and as a popular remedy for rheumatism. The tree yielding this drug is the noblest in the Bornean forests, attaining 130 feet in height, with gigantic buttresses. The wood is dense and hard, and preferred to all others for boat planks; it is reddish and fragrant when first cut; the flowers are deliciously fragrant, and the leaves give a blue tinge to water.

The resin of *Dipterocarpus trinervis* is made into plasters, as also into a tincture with alcohol, and into an emulsion with eggs, useful in diseases of the mucous membrane. The Javanese smear the resin on plantain leaves, and thus make torches which yield a white light and have no unpleasant smell. Dammar resin is also yielded by *Vatica baccifera* and *Tumbugaia*. *Vateria indica* yields the Indian Copal, Piney varnish, or white Dammar, sometimes called Indian Animi, which is also used as a medicine and made into candles. Wood-oil is the produce of various species of *Dipterocarpus*, as *lævis, angustifolius, zeylanicus, hispidus*. *Dovua zeylanica* exudes a colourless gum-resin, much used in Ceylon as a varnish.—ED.]

XXXV. *CHLÆNACEÆ*,[1] *Thouars.*

[SHRUBS or TREES. LEAVES alternate, quite entire, coriaceous, penninerved, folded in bud; *stipules* 0 or very caducous. FLOWERS ☿, regular, in dichotomous cymes or panicles, bracteolate or involucellate. SEPALS 3, free, imbricate. PETALS 5-6, contorted in bud, free, hypogynous. STAMENS 10-∞, inserted within an entire or toothed cup; *filaments* filiform; *anthers* versatile, dehiscing longitudinally, connective often produced. OVARY 3-celled; *style* long, simple; *stigma* 3-lobed; *ovules* 2 pendulous, or ∞ horizontal in each cell, anatropous. CAPSULE loculicidally 3-valved, or by arrest 1-celled and 1-seeded. SEEDS pendulous or sub-horizontal; *testa* coriaceous; *albumen* fleshy or horny. EMBRYO straight; *cotyledons* leafy, flat or plaited; *radicle* superior.

GENERA.

| Sarcolæna. | Leptolæna. | Schizolæna. | Rhodolæna. |

A small and little known Madagascan order, allied to *Tiliaceæ*, but with imbricate sepals; also allied to *Ternstrœmiaceæ*, but distinguished by the stipules, inflorescence, staminal cup, and uniformly contorted petals. From *Dipterocarpeæ* it differs in having albuminous seeds.

Nothing is known of the uses of this order.—ED.]

---

[1] This order is omitted in the original, and is supplied here from the 'Genera Plantarum.'—ED.

XXXVI. *MALVACEÆ.*

(MALVACEÆ, *Jussieu, Br., Kunth, Bartling, Lindl.*)

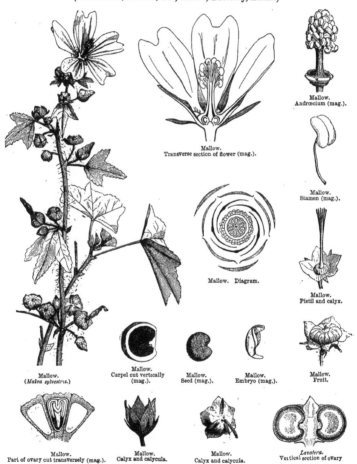

Mallow.
Transverse section of flower (mag.).

Mallow.
Androecium (mag.).

Mallow.
Stamen (mag.).

Mallow. Diagram.

Mallow.
Pistil and calyx.

Mallow.
(*Malva sylvestris.*)

Mallow.
Carpel cut vertically (mag.).

Mallow.
Seed (mag.).

Mallow.
Embryo (mag.).

Mallow.
Fruit.

Mallow.
Part of ovary cut transversely (mag.).

Mallow.
Calyx and calycula.

Mallow.
Calyx and calycula.

Lavatera.
Vertical section of ovary.

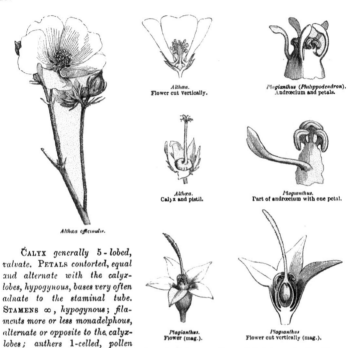

Althæa.
Flower cut vertically.

Plagianthus (Philippodendron).
Andrœcium and petals.

Althæa.
Calyx and pistil.

Plagianthus.
Part of andrœcium with one petal.

Althæa officinalis.

Plagianthus.
Flower (mag.).

Plagianthus.
Flower cut vertically (mag.).

CALYX *generally 5-lobed, valvate.* PETALS *contorted, equal and alternate with the calyx-lobes, hypogynous, bases very often adnate to the staminal tube.* STAMENS ∞, *hypogynous; filaments more or less monadelphous, alternate or opposite to the calyx-lobes; anthers 1-celled, pollen echinulate.* OVARY *of many carpels, whorled or agglomerated into a head.* FRUIT *usually dry, rarely a berry.* SEEDS *reniform, ascending, horizontal or pendulous; albumen scanty.* EMBRYO *curved; cotyledons folded on each other.*

HERBS, SHRUBS, or TREES, with light and soft wood. LEAVES alternate, simple, usually palminerved, entire or palmilobed, hairs usually stellate; *stipules* 2, lateral, persistent or deciduous. FLOWERS ☿, regular, axillary, solitary or agglomerated, sometimes in a raceme, corymb, or panicle. CALYX with an involucel of whorled bracts, rarely naked (*Sida, Abutilon*), 5-fid or -partite, rarely 3–4-fid, valvate in æstivation, persistent or rarely deciduous. PETALS equal and alternate with the calyx-segments, inserted on the receptacle, claw very often adnate to the staminal tube, limb usually inequilateral, æstivation contorted. STAMENS connate in a tube or column enclosing the ovary with its dilated base, sometimes divided at the top into segments alternate or opposite to the calyx-lobes, and separating into numerous antheriferous filaments, sometimes emitting shortly stipitate or sessile anthers from its outer surface; *anthers* reniform, simple, 1-celled, opening in 2 valves by a semi-circular slit; *pollen* echinulate. OVARY sessile, composed of 5 or more carpels, rarely 3–4, sometimes whorled around a more or less developed central axis, sometimes dilated at the top, sometimes attenuated into a column, free or connate, sometimes agglomerated into a head; *styles* terminal, united below, stigmatiferous at the top (*Abutilon, Hibiscus,* &c.), or throughout their length (*Malva, Lavatera, Malope,* &c.); *ovules* one or more on the ventral angle of each carpel, campylotropous or semi-anatropous, sometimes ascending or horizontal, with a ventral or superior raphe (*Callirhoe*), sometimes pendulous with a dorsal raphe (*Sida*). FRUIT of several free cocci, or septicidally splitting into cocci which are indehiscent or ventrally dehiscent, sometimes a loculicidal capsule, with 5, 3, or several septiferous valves; very rarely fleshy (*Malvaviscus*). SEEDS reniform; *testa* crustaceous, usually wrinkled, sometimes hairy (*Gossypium, Fugosia, Hibiscus*), rarely pulpy; *albumen* mucilaginous, scanty or 0. EMBRYO curved; *cotyledons* foliaceous, plaited, or variously contorted; *radicle* next the hilum, inferior in the ascending seeds, bent upwards in the pendulous ones.

TRIBE I. *MALOPEÆ.*

Calyx involucelled, or rarely naked. Carpels numerous, 1-celled, 1-ovuled, joined into a capitulum, separating from the axis when ripe.

PRINCIPAL GENERA.

* Malope.   * Kitaibelia.   * Palava.

TRIBE II. *MALVEÆ.*

Calyx involucelled. Carpels 5–∞, whorled, separating from the axis when ripe, or united into a capsule with several cocci.

PRINCIPAL GENERA.

* Althæa.   * Lavatera.   * Malva.   * Sphæralcea.   * Pavonia.   * Goethea.

TRIBE III. *HIBISCEÆ.*

Calyx involucelled. Carpels 3–5–10, united into a loculicidal capsule, rarely inde-hiscent (*Thespesia*), or berried (*Malvaviscus*).

PRINCIPAL GENERA.

* Hibiscus.   * Malvaviscus.   * Lagunaria.   * Gossypium.

TRIBE IV. *SIDEÆ.*

Calyx naked. Carpels 5–∞, rarely 1–2 (*Plagianthus*), whorled, united into a loculicidal capsule of many cocci.

PRINCIPAL GENERA.

* Sida.   * Abutilon.   * Plagianthus.

[As the above enumeration omits *Bombaceæ*, I have given the following description of the tribes and genera of this extensive family, which is that adopted in the 'Genera Plantarum.'

TRIBE 1. MALVEÆ.—Staminal column antheriferous to the top. Styles as many as ovarian cells. Carpels separating from the axis (except *Bastardia* and *Howittia*). Cotyledons foliaceous, folded or contorted.

Sub-tribe 1. MALOPEÆ.—Carpels ∞, densely congested; ovule 1, ascending. *Malope, Kitaibelia, Palava.*

Sub-tribe 2. EUMALVEÆ.—Carpels in one whorl'; ovule 1, ascending. *Althæa, Lavatera, Malva, Calirrhoe, Sidalica, Malvastrum.*

Sub-tribe 3. SIDEÆ.—Carpels in one whorl; ovule 1, pendulous. *Plagianthus, Hoheria, Anoda, Cristaria, Gaya, Sida, Bastardia.*

Sub-tribe 4.' ABUTILEÆ.—Carpels in one whorl; ovules 2–∞ (except one species of *Wissadula*). *Howittia, Kydia, Wissadula, Abutilon, Sphærrhea, Modiola.*

TRIBE II. URENEÆ.—Staminal column truncate or toothed, anthers on its outer surface. Style-branches 10. Carpels 5, separating from the axis. Cotyledons of Malveæ. *Malachva, Urena, Pavonia, Goethea, Malvaviscus.*

TRIBE III. HIBISCEÆ.—Staminal column of Ureneæ. Styles as many as the ovarian cells. Carpels loculicidal, persistent. Cotyledons of Malveæ. *Kostelclykya, Desaschista, Hibiscus, Lagunaria, Fugosia, Thespesia, Gossypium,* &c.

TRIBE IV. BOMBACEÆ.—Staminal column 5–8-cleft at the top, or rarely to the base, rarely entire. Anthers free, reniform, or cells adnate, globose, linear, oblong, or contorted. Style entire or with as many branches as ovarian cells. Capsule loculicidal or indehiscent, carpels usually persistent. Cotyledons variable.—Trees.

Sub-tribe 1. ADANSONIEÆ.—Leaves digitate. Bracteoles distinct or 0. Cotyledons crumpled or convolute. *Adansonia, Pachira, Bombax, Eriodendron, Chorisia.*

Sub-tribe 2. MATISCIEÆ.—Leaves simple, 3–5-nerved. Bracteoles distinct or 0. Petals 5. Cotyledons twisted or convolute. *Hampia, Cavanillesia, Matisia, Ochroma.*

Sub-tribe 3. FREMONTIEÆ.—Leaves of Matisieæ. Petals 0. Anthers 5 2-celled, or 10 1-celled, adnate in pairs to the branches of the staminal column. Cotyledons flat. (Probably a tribe of *Sterculiaceæ.*)

Sub-tribe 4. DURIONEÆ.—Leaves simple, penninerved, quite entire, lepidote beneath. Flowers involucellate. Fruit muricate. Cotyledons various. *Cullenia, Durio, Boschia, Neesia,* &c.—ED.]

Malvaceæ are closely allied to *Sterculiaceæ* and *Tiliaceæ* (see these families). They are so near *Bombaceæ* that Bentham and Hooker fil. have united them; having separated them from *Sterculiaceæ* on account of their one-celled anthers, which are only apparently two-celled in certain genera in which they are united in pairs; whereas *Sterculiaceæ*, whose anthers are apparently one-celled, are so through the confluence of the cells. In *Bombaceæ* the staminal column is more or less deeply divided into 5–8 branches, each bearing 2–∞ anthers, sometimes free and reniform (*Adansonia*), sometimes adnate, globose (*Cælostegia*), or linear (*Matisia*), or sinuous (*Ochroma*); the capsule is loculicidal or indehiscent; the cotyledons are coiled (*Ochroma*), or folded and contorted (*Adansonia, Bombax,* &c.), or flat (*Cheirostemon*). *Malvaceæ* have also some vegetative characters in common with *Urticeæ.*

To the normal monogynous or digynous species of *Plagianthus* have been joined *Philippodendron*, of Poiteau, a plant remarkable for the tenacity of its liber fibres.[1] *Malvaceæ* are essentially tropical, diminishing rapidly as they recede from the equator, and they are more numerous in the northern tropics and in America than in the Old World.

An emollient mucilage abounds in most of the species; some contain free acids, and are employed as refreshing drinks; others are classed among stimulants, on account of a contained hydrocarbon, which acts on the mucilage. The seeds contain a fixed oil, and their testa is often woolly; the bark of many is very tenacious. The leaves and flowers of Mallows (*Malva sylvestris* and *rotundifolia*), the root and flowers of Guimauve (*Althæa officinalis*) and of *Althæa rosea* are emollients. Those *Malvaceæ* which have acid juice are principally the white and red Ketmies, *Oseilles de. Guinée* (*Hibiscus Sabdariffa* and *digitatus*), natives of tropical Africa, but now cultivated throughout the tropics on account of the free oxalic acid which exists in their mucilage. *H. esculentus* is a widely diffused tropical annual; its green capsule is largely consumed, either by dissolving its mucilage in boiling water, to thicken soups, or else cooked and seasoned. The root of the Indian *Pavonia odorata* is aromatic and a febrifuge. That of *Sida lanceolata* is praised by the Indians as a stomachic. *Hibiscus Abelmoschus* is an annual herb, a native of India and Egypt, which has been introduced into the Antilles; its seeds (*graina d'Ambretta*) have a powerful musky principle, used by perfumers. *Hibiscus Rosa-Sinensis* contains a colouring principle in its flowers which the Chinese make use of to blacken their shoes and eyebrows. *Althæa Cannabina*, a native of South Europe, has tenacious fibres like hemp, as which it is used. The genus *Gossypium* consists of herbs or shrubs, whose capsule contains numerous ovoid seeds with a spongy testa covered with woolly hairs called Cotton, easily spun, and the source of an immense commerce between the two worlds. These plants are indigenous throughout the tropics, but their cultivation has been gradually extended into temperate latitudes. The principal species, *G. herbaceum*, from Upper Egypt; *G. arboreum* and *religiosum*, from India; *G. peruvianum* and *hirsutum*, from the New World, &c., are as yet imperfectly defined. Cotton was known in Egypt in the earliest times. Cotton seeds further yield by expression a fixed oil, which is used for burning, cattle food, and for the manufacture of soap. Among *Bombaceæ*, the Durian produces a large fœtid fruit, the flavour of which is pronounced to be unequalled, after habit has overcome its disgusting smell. The Baobab has an oblong fruit, the size of a melon, filled with acidulous white pulp, much sought by the negros as a preservative from dysentery; its bark is a febrifuge.

Bombaceæ are all arborescent, and principally tropical, and include some of the largest trees in the Vegetable Kingdom; as *Bombax*, *Adansonia*, *Pachira*, *Durio*, *Neesia*, &c. The most remarkable is the Baobab (*Adansonia digitata*), a tree of tropical Africa, introduced into Asia and America; the thickness of its trunk is enormous, sometimes attaining 100 feet in circumference. Adanson observed in the Cape de Verd Islands Baobabs which had been measured by travellers three centuries before, and from the little growth they had made during that period he calculated that their age must be more than 6,000 years. [Such estimates are altogether fallacious; the Baobab is now well known to be a very fast-growing and short-lived tree.—ED.]

XXXVII. *STERCULIACEÆ.*

(STERCULIACEÆ, *Ventenat.*—STERCULIACEÆ ET BUTTNERIACEÆ, *Endlicher.*)

CALYX 5–4–3*-merous, valvate.* COROLLA 0, *or* PETALS *as many as the calyx lobes, hypogynous.* STAMENS *equal and opposite to the petals, or multiple, often mixed with staminodes opposite to the calyx-lobes; filaments variously coherent; anthers extrorse.* CARPELS *distinct or more or less united.* OVULES *ascending or horizontal, anatropous or orthotropous.* FRUIT *usually a capsule.* EMBRYO *straight or arched, albuminous or exalbuminous.*

---

[1] No more so than the other species of *Plagianthus.*—ED.

Büttneria gracilipes.

Theobroma Cacao.
Fruit, one-third natural size.

Hermannia.
Diagram, showing the stamens enveloped
by the base of the petals.

Hermannia.
Petal
(mag).

Hermannia.
Styles
joined at the top.

Theobroma.
Seed, entire and cut vertically
(mag).

Hermannia.
Diagram of corolla twisted to
the right.

Hermannia.
Diagram of corolla twisted to
the left.

Hermannia.
Ovule
(mag.).

Hermannia.
Stamen, outer
face (mag.).

Hermannia.
Stamen, inner
face (mag.).

TREES or SHRUBS with soft wood, erect, sometimes climbing perennial or annual (*Ayenia*) HERBS, pubescent with starred or forked hairs, often mixed with simple hairs, rarely scaly. LEAVES alternate, sometimes simple, penninerved or palminerved; sometimes digitate with 3–9 leaflets. *Stipules* free, deciduous, rarely foliaceous and persistent, very rarely 0 (*Lasiopetalum*). FLOWERS regular, ⚥, or

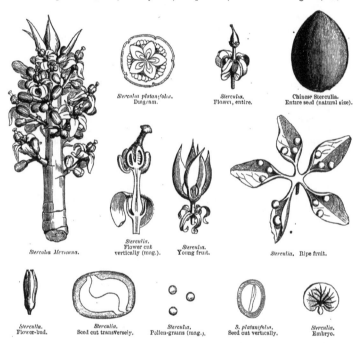

Sterculia platanifolia.
Diagram.

Sterculia.
Flower, entire.

Chinese Sterculia.
Entire seed (natural size).

Sterculia Mexicana.

Sterculia.
Flower cut
vertically (mag.).

Sterculia.
Young fruit.

Sterculia. Ripe fruit.

Sterculia.
Flower-bud.

Sterculia.
Seed cut transversely.

Sterculia.
Pollen-grains (mag.).

S. platanifolia.
Seed cut vertically.

Sterculia.
Embryo.

unisexual: *inflorescence* very various, usually axillary. CALYX usually persistent, 5–4–3-fid, or of 5 free sepals, æstivation valvate. COROLLA 0, or PETALS hypogynous, free, or adnate by their base to the staminal tube, æstivation imbricate, convolute or contorted. STAMENS very various; *filaments* more or less connate into a tubular or urceolate column: (1) staminal tube divided at the top into 5 teeth or tongues (*staminodes*) alternate with the petals, and bearing in the intervals between the staminodes 1–2–5–∞ anthers opposite to the petals, stipitate or subsessile (*Buttneria*); (2) staminodes 0, anthers numerous, many-seriate, inserted on the column from the middle to the top (*Eriolæna*), or 1-seriate at the top of the cup

(*Astiria*); (3) anthers adnate to the top of the column, and arranged in a ring or without order (*Sterculia*); (4) fertile stamens 5, free or nearly so, opposite to the petals, without staminodes, or alternating with 5 staminodes opposite to the sepals (*Seringia*); *anthers* extrorse, of 2 parallel or diverging cells, very rarely confluent at the top (*Helicteres*), sometimes dehiscing at the top by 2 pores or small slits (*Lasiopetalum, Guichenotia*). OVARY free, sessile or substipitate, 4–5- (or rarely 10–12-celled, sometimes of 4–5 (rarely more or fewer) connate or distinct carpels (*Seringia*); *styles* as many as the cells, distinct or more or less connate; *ovules* 2–∞ (rarely 1) in each cell, fixed to the inner angle, ascending or horizontal, anatropous or semi-anatropous, raphe ventral or lateral, micropyle inferior, rarely orthotropous (*Sterculia Balanghas*). FRUIT dry, or rarely fleshy (*Theobroma*), carpels sometimes united into a loculicidal or woody indehiscent capsule, sometimes dividing into follicles or 2-valved cocci. SEEDS globose or ovoid, sometimes compressed and prolonged above into a membranous wing, shortly strophiolate, or more often naked; *testa* coriaceous or crustaceous, sometimes covered with a succulent epidermis (*Sterculia*); *albumen* fleshy, often thin or 0. EMBRYO straight or arched, sometimes dividing the albumen into 2 parts (*Sterculia*); *cotyledons* usually foliaceous, flat, or folded and crumpled, or convolute, rarely fleshy; *radicle* short, inferior, pointing to the hilum, or not.

<h3 style="text-align:center">TRIBE I.[1] LASIOPETALEÆ.</h3>

Flowers ☿. Calyx petaloid. Petals 0, or scale-like, flat, shorter than the sepals. Stamens slightly monadelphous at the base, 5 fertile, alternate with the sepals; staminodes 5 or fewer, opposite to the sepals, sometimes 0; anthers incumbent, with parallel cells, or opening at the top by 2 pores. Carpels free, or united into a 3–5-celled ovary; ovules 2 or 8 in each cell, ascending. Seeds strophiolate; albumen fleshy. Embryo straight or slightly bent, axile; cotyledons foliaceous, flat.

PRINCIPAL GENERA.

Seringia.	Guichenotia.	* Thomasia.	* Lasiopetalum.	Keraudrenia.

<h3 style="text-align:center">TRIBE II. BUTTNERIEÆ.</h3>

Flowers ☿. Petals sessile or clawed, concave or hooded, often produced and tongue-shaped. Staminal tube lobed, some lobes 1–3-antheriferous, opposite to the petals, the others sterile (staminodes), opposite to the sepals. Ovary of 5–2- (or more) ovuled cells. Fruit usually a loculicidal or septicidal capsule. Seeds straight or arched, naked or strophiolate. Embryo albuminous or exalbuminous, straight or curved; cotyledons sometimes flat and foliaceous, or coiled or crumpled.

PRINCIPAL GENERA.

Commersonia.	* Buttneria.	* Theobroma.	Guazuma.
Abroma.	Herranea.	Ayenia.	**Rulingia.**

---

[1] These tribes are taken, in inverse order, from the 'Genera Plantarum.'—ED.

XXXVII. STERCULIACEÆ.

TRIBE III. *HERMANNIEÆ.*

Flowers ☿. Petals flat, marcescent, linear, and sometimes convolute (*Visenia*). Stamens more or less monadelphous, equal and opposite to the petals; staminodes 0, or rarely toothed. Ovary of 1 or several 1-∞ -ovuled cells. Capsule loculicidal. Seeds obovoid or reniform; albumen fleshy. Embryo axile, straight or arched; cotyledons foliaceous, flat.

PRINCIPAL GENERA.

*Hermannia.   *Mahernia.   *Melochia.   Waltheria.

TRIBE IV. *DOMBEYEÆ.*

Flowers ☿. Petals flat, often marcescent. Anthers 10–20 with parallel cells, inserted at or near the top of a shortly urceolate or rarely elongated column; staminodes 5 or 0. Ovary sessile, with 5 or more 2-many-ovuled cells. Capsule loculicidal or septicidal. Albumen fleshy, scanty. Cotyledons foliaceous, often 2-fid, or folded and contorted, rarely flat.

PRINCIPAL GENERA.

*Pentapetes.   *Dombeya.   *Astrapæa.   Trochetia.   Astiria.   Melhania.

TRIBE V. *ERIOLÆNEÆ.*

Flowers ☿. Petals flat, deciduous. Anthers numerous, multiseriate, stipitate, inserted on the column from the middle to the top; staminodes 0. Ovary of 5–10 many-ovuled cells. Capsule loculicidal. Albumen fleshy. Embryo straight, axile.

ONLY GENUS.

Eriolæna.

TRIBE VI. *HELICTEREÆ.*

Flowers ☿. Petals 5, deciduous. Anthers 5–15, sessile, or stipitate on the top of an elongated column, alternating in five groups with as many staminodes, or shorter or longer teeth of the column. Albumen fleshy. Embryo straight or arched.

PRINCIPAL GENERA.

Helicteres.   Reevesia.   Kleinhovia.   Petrospermum.   Myrodia.

TRIBE VII. *STERCULIEÆ.*

Flowers diclinous or polygamous. Calyx often coloured. Corolla 0. Anthers sometimes 5–15, adnate on the top of a short or elongated column, sometimes shortly polyadelphous or 1-seriate in a ring; pollen smooth. Ripe carpels free, sessile, or shortly stipitate. Seeds albuminous or exalbuminous.

PRINCIPAL GENERA.

* Sterculia.   Heritiera.   Tarrietia.   Cola.

The above-named tribes, united into one family by Ventenat, and then divided into two, have again been united by Bentham and Hooker fil. They are closely allied on the one hand to *Malvaceæ*, on the other to *Tiliaceæ* ; they are distinguished from the first by their two-celled anthers and their generally smooth pollen, and from the second by their extrorse anthers, alternate with the epals when definite, or monadelphous when indefinite. *Tremandreæ*, which we have annexed to *Polygaleæ* (see p. 240), and which approach the tribe of *Lasiopetaleæ* in the æstivation of the calyx, the apical dehiscence of the extrorse anthers, the two ovuled ovary cells, the anatropous ovules, the fleshy albumen, the straight axile embryo, are separated by the free filaments, the simple stigma, the pendulous ovule with superior micropyle, the exstipulate leaves, &c.

Sterculiaceæ belong to tropical and subtropical regions. The tribe of *Lasiopetaleæ* inhabits Australia and Madagascar. Some *Buttneriæ* are common to the tropics of both continents (*Buttneria, Guazuma*), others are peculiar to America (*Theobroma*), or Asia (*Abroma*), or Australia and Madagascar (*Rulingia*). *Commersonia* is Australian and tropical American ; *Hermannieæ* principally South African ; *Dombeyeæ* inhabit the hot regions of Asia and Africa. *Eriolæna* is exclusively an Asiatic genus, as are most *Helictereæ* ; although *Helicteres* itself belongs to both continents, *Ungeria* to Norfolk Island, and *Myrodia* to America. The tribe of *Sterculieæ* is dispersed over the tropical zone ; *Sterculia* is almost wholly Asiatic, there being but few African or American species. The known species of *Cola* are all from Africa. *Heritiera* is tropical Asiatic and Australian.

Sterculiaceæ, like *Malvaceæ*, contain an abundant mucilage, combined, in the old bark of the woody species, with a bitter astringent extractable matter. and are stimulants and emetics. The seeds are oily. The fleshy envelope of the seed of *Sterculia* is eatable ; their seeds, oily and slightly acrid, are used for seasoning food ; their bark is strongly astringent, and some species produce a gum analogous to tragacanth.

The most important species of *Buttneriæ* is the *Theobroma Cacao*, an American tree, cultivated in Asia and Africa. The seeds, which are enveloped in pulp, contain a fixed and solid oil (*cocoa butter*), a red colouring matter, a substance analogous to tannin, a gum, and a crystallizable azotized principle, called *theobromine*. The dried and split cotyledons of its seeds are called cocoa-nibs ; and when ground and made into a paste, chocolate, which is rendered more digestible by being flavoured with vanilla or cinnamon. The mucilaginous astringent fruit of *Guazuma* is used in America for skin diseases. Its sugary and eatable pulp is fermented, and furnishes a kind of beer. Many species of *Buttneria, Waltheria* and *Pterospermum* are used in America and Asia as emollients. The bitter and fœtid root of *Helicteres Sacarrotha* is a reputed stomachic in Brazil, and its bark is frequently used in syphilitic affections.

XXXVIII. *TILIACEÆ*.

(TILIACEÆ, *Jussieu.*—ELÆOCARPEÆ, *Jussieu.*—ELÆOCARPACEÆ, *Lindl.*)

CALYX *valvate, deciduous.* PETALS 4–5, *hypogynous, æstivation contorted, imbricate or valvate.* STAMENS *double or multiple the number of the petals, all fertile, or the outer sterile, free, or in bundles; anthers 2-celled.* FRUIT *dry or fleshy.* EMBRYO *generally albuminous.*—STEM *woody, or very rarely herbaceous.* LEAVES *stipulate, usually alternate.*

TREES or SHRUBS, rarely HERBS. LEAVES alternate, rarely opposite or sub-opposite (*Plagiopteron*) simple, penninerved or palminerved, entire or palmilobed, crenulate or dentate, very often coriaceous, reticulate beneath ; *stipules* 2, free, deciduous or rarely persistent. FLOWERS ☿, very rarely imperfect, regular, axillary or terminal,

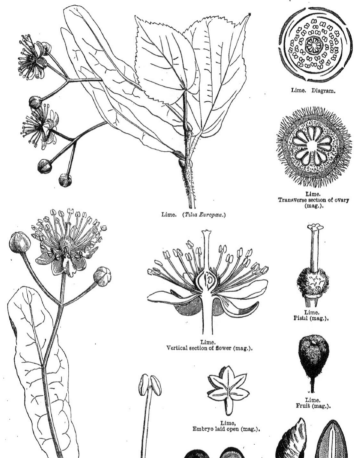

Lime. (*Tilia Europæa.*)

Lime. Diagram.

Lime.
Transverse section of ovary
(mag.).

Lime.
Vertical section of flower (mag.).

Lime.
Pistil (mag.).

Lime.
Fruit (mag.).

Lime,
Embryo laid open (mag.).

Lime.
Flowering and bracteate peduncle.

Lime.
Stamen, outer
face (mag.).

Lime.
Seed, entire and cut
vertically (mag.).

Vallea.
Seed, entire and cut vertically
(mag.).

U

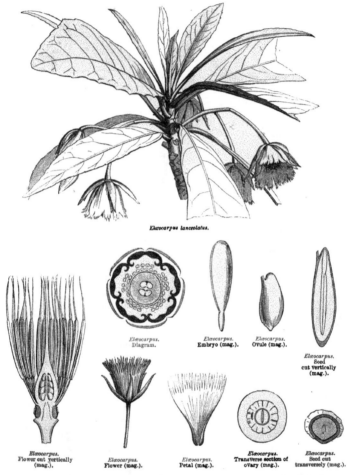

Elæocarpus lanceolatus.

Elæocarpus.
Diagram.

Elæocarpus.
Embryo (mag.).

Elæocarpus.
Ovule (mag.).

Elæocarpus.
Seed
cut vertically
(mag.).

Elæocarpus.
Flower cut vertically
(mag.).

Elæocarpus.
Flower (mag.).

Elæocarpus.
Petal (mag.).

Elæocarpus.
Transverse section of
ovary (mag.).

Elæocarpus.
Seed cut
transversely (mag.).

solitary or in small cymes, corymbs, or panicles. SEPALS 5, rarely 4–3, free or connate, usually valvate, very rarely imbricate (*Ropalocarpus, Echinocarpus*). PETALS as many as sepals, inserted around the base of the torus, entire or cut, æstivation contorted or variously imbricate, induplicate or valvate, very rarely gamopetalous (*Antholoma*). STAMENS usually indefinite, rarely double the number of the sepals (*Triumfetta, Corchorus*), sometimes 10-seriate on the top of a stipitiform torus and distinct from the petals, sometimes covering the whole surface of a discoid torus, sometimes inserted around the edge of the torus next the petals or enveloped by them; *filaments* free, or connate into a ring or 5-10 bundles, filiform, all antheriferous, or some not (*Sparmannia, Luhea, Diplodiscus*, &c.), sometimes irritable (*Sparmannia*); *anther-cells* parallel, contiguous, dehiscing longitudinally, or at the top by a pore or transverse slit (*Elæocarpus, Sloanea, Vallea, Aristotelia*, &c.), rarely divergent and confluent at the top (*Brownlowia, Diplodiscus*). OVARY free, sessile, 2-10-celled; *style* simple; *stigmas* as many as cells, free. or connate, sometimes sessile (*Carpodiptera, Muntingia*); *ovules* attached to the inner angle of the cells, solitary or geminate, pendulous from the top of each cell, or ascending from the base; sometimes few, inserted at the middle of the cell, sometimes numerous, 2-many-seriate, anatropous or sub-anatropous, raphe ventral or lateral. FRUIT of 2-10 cells, or 1-celled by suppression, or many-celled by false septa, sometimes indehiscent, nutlike (*Tilia*), or a drupe (*Grewia, Elæocarpus*), rarely a berry (*Aristotelia, Muntingia*); sometimes separating into cocci (*Colombia*); sometimes loculicidally dehiscent, rarely septicidally (*Dubouzetia*). SEEDS solitary or numerous in each cell, ascending or pendulous or horizontal, exarillate, ovoid or angular; *testa* usually coriaceous or crustaceous, often velvety; *endopleura* sometimes hardened at the chalaza; *albumen* fleshy, copious or thin, rarely 0 (*Brownlowia*); *cotyledons* foliaceous, flat, entire or lobed; *radicle* inferior or superior or centripetal.

SECTION I. *TILIEÆ.*

Petals entire, or very rarely emarginate, æstivation imbricate, or more often contorted.

PRINCIPAL GENERA.

Brownlowia.	*Grewia.	*Triumfetta.	*Sparmannia.	*Corchorus.
Luhea.	*Tilia.	Apeiba.	Vallea.	

SECTION II. *ELÆOCARPEÆ.*

Petals often cut, sometimes entire (*Dubouzetia*) or 0, usually pubescent, æstivation valvate or induplicate, never contorted. Stamens some in groups opposite to the petals, the others solitary and alternate.

PRINCIPAL GENERA.

Prockia.	Hasseltia.	Sloanea.	*Aristotelia.	*Elæocarpus.	*Monocera.

[The following tribes of this extensive family have been proposed in the 'Genera Plantarum':—

Series A. Holopetaleæ.—Petals usually glabrous, membranous, contracted at the base or clawed, rarely notched, usually contorted.

TRIBE I. BROWNLOWIEÆ.—Sepals connate into a 3–5-fid cup. Anthers usually globose or didymous, cells at length confluent at the tips. *Brownlowia, Benga, Christiana,* &c.

TRIBE II. GREWIEÆ.—Sepals distinct. Petals pitted at the base. Stamens inserted at the top of the torus; anthers short; cells parallel, distinct. *Grewia, Colombia, Triumfetta, Heliocarpus.*'

TRIBE III. TILIEÆ.—Sepals distinct. Petals not pitted at the base. Stamens inserted close to the petals. *Entelea, Sparmannia, Corchorus, Luhea, Mollia, Muntingia, Tilia,* &c.

TRIBE IV. APEIBEÆ.—Sepals distinct. Petals not pitted at the base. Stamens inserted close to the petals; anthers erect, linear, tipped by a membrane, cells parallel. *Glyphæa, Apeiba.*

Series B. Heteropetaleæ.—Petals 0 or sepaloid, or incised, often pubescent, not clawed, valvate, rarely imbricate or contorted.

TRIBE V. PROCKIEÆ.—Anthers digynous, subglobose, cells dehiscing longitudinally: *Prockia, Hasseltia.*

TRIBE VI. SLOANEÆ.—Anthers linear, cells dehiscing at the top. Torus flat or turgid; sepals and petals inserted close to the stamens. *Vallea, Sloanea, Echinocarpus, Antholoma.*

TRIBE VII. ELÆOCARPEÆ.—Anthers linear, cells dehiscing at the top. Petals inserted around the base of a glandular torus with stamens at its top. *Aristotelia, Elæocarpus, Dubouzetia,* &c.—ED.]

Tiliaceæ and *Elæocarpeæ*, which formerly formed two families, have been united by Endlicher and Bentham and Hooker fil.; the latter have joined *Prockia* to them, which was formerly placed in *Bixineæ*, but from which they differ in their axile placentation. *Tiliaceæ* are connected with the tribe *Buttnerieæ* of *Sterculiaceæ* by their valvate calyx, their petals which are hypogynous or 0, numerous stamens, fleshy albumen, woody stem, alternate stipulate leaves, and stellate pubescence; besides which, in *Elæocarpeæ* the induplicate base of each petal embraces a group of stamens as in *Buttnerieæ*, though in *Elæocarpeæ* there is further one isolated stamen left opposite each sepal. They are similarly related to *Malvaceæ*, which also differ in their one-celled anthers. They also approach *Camelliaceæ* in polypetalism, hypogyny, polyandry, the connate filaments, apical dehiscence of the anthers (observable in the genera *Saurauja* and *Pentaphyla*), the several-celled ovary, &c.; but the valvate æstivation of the calyx separates them. Finally, they have more than one point of analogy with *Chlænaceæ*; but these are distinguished by their imbricate calyx, and especially by the urceolate filaments.

Most *Tiliaceæ* are tropical; a few inhabit the temperate northern hemisphere, and some are found beyond the tropic of Capricorn. *Brownlowia* and the allied genera are tropical Asiatic and African; *Grewia* and *Corchorus* are natives of the hot parts of the Old World; *Sparmannia,* of tropical and South Africa; *Luhea,* of tropical and subtropical America; *Tilia,* of Europe, temperate Asia, and North America; *Prockia, Hasseltia, Vallea, Sloanea,* of tropical America; *Aristotelia,* of Chili and New Zealand; *Elæocarpus,* of tropical Asia and Australasia; *Antholoma* and *Dubouzetia,* of New Caledonia, &c.

The useful species of *Tiliaceæ* are rather numerous. The inner bark of *Tilia parvifolia* and *grandifolia* (the European Limes) contains an astringent mucilage, which is used in Germany as a vulnerary, and its tenacious fibres are used for making cord; the sugary sap of their trunk is fermentable and yields an agreeable vinous liquor; their wood is easily worked and affords excellent charcoal; the flowers, of a balsamic odour, are much used in infusions, and are antispasmodic and diaphoretic, and become astringent when used with the large bract which accompanies them. *Triumfetta* and *Sparmannia africana* are mucilaginous plants, and used as emollients. Throughout the tropics, the young fruits and leaves of *Corchorus olitorius,* when cooked and seasoned, are used as potherbs; its seeds are purgative. *C. tridens, acutangulus,* and *depressus* are also eatable, and the Arabs employ their tenacious cortical fibres for making cords or coarse mats. [Jute, the fibre of *C. capsularis,* has of late become the rival of hemp in the English market; many thousands of tons being annually imported from India.—ED.] *Grewia orientalis* and *microcos* are valuable for their aromatic bitter bark and astringent leaves; the wood of *G. elastica* is much valued on account of its flexibility, which fits it for bow-making. [The fruits of *G. microcos* and *asiatica* are grateful, and extensively used for sherbet in North-western India. Various other species yield cordage, and from *G. oppositifolia* a paper is made.—ED.] The bitter and resinous bark of *Elæocarpus* is renowned as a tonic. Their acidulous sugary fruit is eatable, and a stomachic. The kernels of several species, which are elegantly marked, are made by the Indians into necklaces and bracelets. [*E. Hinau*, of New Zealand, yields an excellent dye.—ED.]

XXXIX. LINEÆ.

(LINEÆ, *D.C.*—LINACEÆ, *Lindl.*)

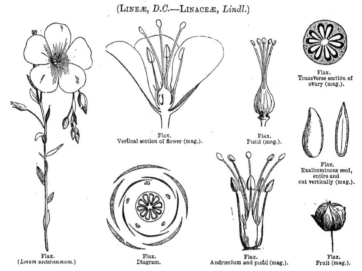

Flax.
Transverse section of ovary (mag.).

Flax.
Vertical section of flower (mag.).

Flax.
Pistil (mag.).

Flax.
Exalbuminous seed, entire and cut vertically (mag.).

Flax.
(*Linum usitatissimum.*)

Flax.
Diagram.

Flax.
Andrœcium and pistil (mag.).

Flax.
Fruit (mag.).

CALYX 5–4-*merous, imbricate.* PETALS 5–4, *hypogynous, contorted.* STAMENS *fertile, as many as the petals, with usually as many staminodes.* OVARY *of 5–4 2-ovuled cells.* OVULES *pendulous, anatropous.* STYLES 3–5, *free.* CAPSULE *globose, septicidal.* SEEDS *more or less albuminous, rarely exalbuminous.* EMBRYO *straight.*

STEM sometimes herbaceous, annual or perennial; sometimes sub-woody or woody. LEAVES alternate or opposite, rarely whorled, simple, sessile, entire, 1–3 nerved, sometimes biglandular at the base, exstipulate (*Linum, Radiola*) or with 2 minute lateral stipules, caducous (*Reinwardtia*) or intra-axillary (*Anisadenia*). FLOWERS ☿, regular, usually terminal, in racemes, panicles, corymbs, heads, fascicles or spikes. CALYX persistent, imbricate in æstivation, of 5 entire sepals, or rarely 4-partite with 3-fid lobes (*Radiola*). PETALS clawed, usually 5, rarely 4 (*Radiola*), claw naked, or furnished with a sort of crest (*Anisadenia, Reinwardtia*), inserted on the torus, contorted in æstivation, caducous. STAMENS equal and alternate with the petals, sometimes accompanied by as many toothed staminodes opposite to the petals; *filaments* flattened-subulate, usually connate into a short cup furnished on the outside with 5 little glands, sometimes 0 (*Radiola*); *anthers* introrse, 2-celled, linear or oblong, basi- or dorsi-fixed, dehiscence longitudinal. OVARY usually 5-celled, rarely 4-celled (*Radiola*) or 3-celled (*Anisadenia, Reinwardtia*), cells subdivided by a more or less perfect dorsal septum; *styles* 5, rarely 4–3, filiform, free; *stigmas* simple, linear or sub-capitate; *ovules* 2–4 ?, pendulous, anatropous. CAPSULE globose, enveloped by the persistent calyx and andrœcium, tipped by the style-base, septicidally dehiscing into as many cocci as cells, or into twice as many by the splitting of each through the dorsal septum. SEEDS pendulous, compressed; *testa* coriaceous, shining when dry, and developing in water an abundant mucilage; *albumen* copious, scanty, or 0. EMBRYO straight; *cotyledons* flat; *radicle* contiguous and parallel to the hilum, superior.

PRINCIPAL GENERA.

*Linum.　　　　　Radiola.

[*Lineæ*, including *Erythroxyleæ*, are thus divided in the 'Genera Plantarum' :—

TRIBE I. EULINEÆ.—Petals contorted, fugacious. Perfect stamens as many as petals. Capsule septicidal, rarely indehiscent and 1-seeded. *Radiola, Linum, Reinwardtia, Anisadenia.*

TRIBE II. HUGONIEÆ.—Petals contorted, fugacious. Stamens all antheriferous, twice or thrice the number of petals. *Hugonia, Roucheria.*

TRIBE III. ERYTHROXYLEÆ.—Petals at length deciduous, imbricate, rarely contorted, with a villous line or ridge on the inner face. Stamens all antheriferous, twice as many as petals. Drupe indehiscent. *Erythroxylon, &c.*

TRIBE IV. TRIONANTHEÆ.—Petals persistent, contorted. Stamens twice or more than twice as many as the petals. Capsule septicidal. *Durandea, Sarcotheca, Phyllocosmus, Xoranthus,* &c.—ED.]

Lineæ are closely related to *Erythroxyleæ*, and more or less to *Geraniaceæ* and *Oxalideæ* (see those families). They approach the latter in polypetalism, hypogynism, contorted corolla, clawed caducous petals, false diplostemony (*Arerrhoa*), filaments connate at the base, 5-celled ovary with pendulous anatropous ovules, free styles, capitate stigmas, and capsular fruit; but *Oxalideæ* are separated by their compound leaves, copious albumen, and usually arillate seed.

The species of *Linum* are found in all temperate regions; *Radiola* is European and Asiatic; *Reinwardtia*, tropical Asiatic; *Anisadenia*, temperate Himalayan. The common Flax (*Linum usitatissimum*) is one of the most useful of vegetables; the tenacity of its cortical fibres places it at the head of textile plants. It is indigenous in the South of Europe and in the East, and its cultivation, which has been carried on from the oldest times, extends to 54° N. lat. The testa contains an abundant mucilage, and the embryo a fixed emollient oil, which is very drying (and hence extensively used by painters). *Linum catharticum*, an abundant indigenous species, of a slightly bitter and salt taste, was formerly used as a purgative. *L. selaginoides* is considered by the Peruvians as a bitter aperient. *L. aquilinum*, a herb of Chili, is there looked upon as refreshing and antifebrile. Many species with red, yellow, blue, and white flowers are ornamental garden plants.

XL. *ERYTHROXYLEÆ, Kunth.*

UNDERSHRUBS, SHRUBS or TREES; branches usually flattened or compressed at the tip when young. LEAVES alternate or rarely opposite, simple, entire, generally glabrous, penninerved, folded lengthwise in bud, and preserving two impressions parallel to the midrib; *stipules* intra-axillary, concave, scarious, scaly, bracteiform

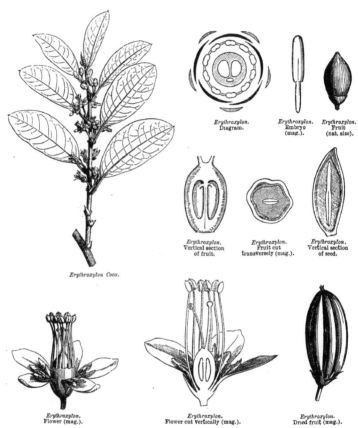

Erythroxylon.
Diagram.

Erythroxylon.
Embryo
(mag.).

Erythroxylon.
Fruit
(nat. size).

Erythroxylon.
Vertical section
of fruit.

Erythroxylon.
Fruit cut
transversely (mag.).

Erythroxylon.
Vertical section
of seed.

Erythroxylon Coca.

Erythroxylon.
Flower (mag.).

Erythroxylon.
Flower cut vertically (mag.).

Erythroxylon.
Dried fruit (mag.).

on the aphyllous peduncles. FLOWERS ⚥, regular, solitary or geminate, or fasci-cled in the axis of the leaves or stipules; *peduncles* 5-angled, gradually thickened at the top. CALYX persistent, 5-partite, or rarely 5-fid, imbricate. PETALS 5, hypogy-nous, equal, appendiculate above their base within by a double ligule, lamella, or rib, æstivation usually imbricate, rarely contorted. STAMENS 10, inserted on the receptacle; *filaments* flattened at the base, and connate into a short tube, filiform and free above; *anthers* introrse, 2-celled, ovoid-subglobose, dorsi-fixed, mobile, dehiscence longitudinal. OVARY free, 2-3-celled ; *styles* 3, distinct or connate ; *stigmas* 3, capitate ; *ovule* solitary in the cells, pendulous from the top of the inner angle, anatropous, often wanting in 1 or 2 cells. DRUPE ovoid, angular, 1-celled and 1-seeded by suppression. SEED inverted, testa coriaceous. EMBRYO straight, in the axis of a cartilaginous scanty albumen ; *cotyledons* elliptic or linear, flat, foliaceous ; *radicle* short, cylindric, superior.

PRINCIPAL GENUS.

Erythroxylon.

Erythroxyleæ are closely related to *Lineæ*, differing only in their always appendiculate and diplostemonous petals, their drupaceous fruit and woody stem. They are near *Malpighiaceæ* in hypogynism, diplostemony, connate filaments, ovary cells with one pendulous ovule, distinct styles, woody stem, and stipulate leaves ; but in *Malpighiaceæ* many of the stamens are often suppressed, and the petals have long claws and no appendages, and the stipules are at the base of the petiole. The same affinity exists with *Sapindaceæ*, in which also the claw of the petals is glandular or velvety, and which scarcely differ save in their capsular or samaroid fruit and their exalbuminous seed. *Erythroxyleæ* also approach *Geraniaceæ* in their persistent calyx, hypogynism and diplostemonous petals, more or less distinct styles, and stipulate leaves ; but in *Geraniaceæ* the carpels are nearly free, the fruit is capsular, the embryo is curved and exalbuminous.

Erythroxyleæ inhabit the intertropical regions of the Old and New Worlds. The wood of several species contains a red dye. The young shoots of *Erythroxylon areolatum* are refreshing ; its bark is tonic, and the juice of its leaves is used externally against herpetic affections. The leaves of *E. Coca* contain a very volatile stimulating principle, producing in those who chew them an excitement of the nervous system, eagerly sought by the Peruvians, and which they cannot do without when they have acquired the habit.

XLI. OXALIDEÆ.

(OXALIDEÆ, D.C.—OXALIDACEÆ, Lindl.)

CALYX 5-*merous, imbricate.* PETALS 5, *hypogynous, contorted.* STAMENS *double the number of the petals.* OVARY 5-*celled.* OVULES 1-*seriate, pendulous, anatropous.* STYLES *free.* FRUIT *a capsule, or rarely a berry.* SEEDS *with abundant albumen.* EMBRYO *axile.*—LEAVES *alternate.*

Herbaceous annual or perennial plants, stemless or caulescent, rootstock creeping, bulbous or tuberous, rarely sub-frutescent (*Connaropsis*), very rarely arborescent (*Averrhoa*). LEAVES alternate, petiolate, digitate, rarely pinnate, sometimes appearing simple by suppression of the lateral leaflets ; *leaflets* spirally coiled when young, sessile or rarely petiolulate, entire, often obcordate, usually closing at night ; *stipules* 0. FLOWERS ☿, regular, sometimes dimorphous, some perfect, others minute, apetalous ; *peduncles* axillary or radical, 1-flowered, or branched in an umbel, raceme, panicle or cyme. CALYX 5-fid or -partite or -phyllous, æstivation imbricate. PETALS 5, equal, inserted on the receptacle, longer than the sepals, obtuse, shortly clawed, free, or shortly connate at the base, contorted in æstivation, deciduous. STAMENS 10, inserted on the receptacle, connate at the base, the 5 opposite the petals shortest, fertile or antherless (*Averrhoa*) ; *filaments* filiform or subulate, flattened ; *anthers* introrse, 2-celled, ovoid or elliptic, dorsifixed, dehiscence longitudinal.

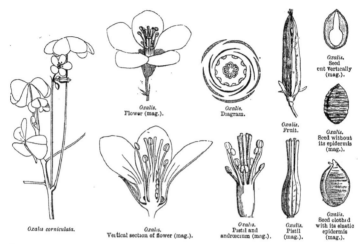

Oxalis corniculata.

Oxalis.
Flower (mag.).

Oxalis.
Diagram.

Oxalis.
Fruit.

Oxalis.
Seed
cut vertically
(mag.).

Oxalis.
Seed without
its epidermis
(mag.).

Oxalis.
Vertical section of flower (mag.).

Oxalis.
Pistil and
andrœcium (mag.).

Oxalis.
Pistil
(mag.).

Oxalis.
Seed clothed
with its elastic
epidermis
(mag.).

OVARY 5-lobed, cells 5, opposite to the petals; *styles* 5, filiform, free, or shortly connate, persistent; *stigmas* capitate, sometimes 2-fid or laciniate; *ovules* pendulous from the inner angle of the cells, solitary or numerous, 1-seriate, anatropous. FRUIT generally capsular, cylindric, ovoid or subglobose, 5-lobed, cells opening longitudinally at the back, valves not separating from the placentiferous column, rarely an oblong 5-furrowed and indehiscent berry (*Averrhoa*). SEEDS pendulous, usually clothed with a fleshy arilliform epidermis which separates elastically; *testa* crustaceous; *albumen* fleshy, abundant. EMBRYO axile, straight or sub-arched; *cotyledons* often elliptic; *radicle* short, superior.

PRINCIPAL GENERA.

| *Oxalis. | Biophytum. | Connaropsis. | Averrhoa. |

Oxalideæ are near *Geraniaceæ*, to which Bentham and Hooker fil. have united them (see this family). They are equally near *Balsamineæ* and *Lineæ* (see these families). They likewise approach *Zygophylleæ* in their hypopetalism, diplostemony, plurilocular ovary, pendulous superimposed and anatropous ovules, capsular or berried fruit, herbaceous or woody stem, compound leaves, and axillary inflorescence; but in *Zygophylleæ* the petals are often imbricate, the style is simple, the seeds sometimes without albumen, and the leaves are opposite and stipulate. *Oxalideæ* have also some affinity with *Rutaceæ*: in both families the calyx is 5-partite and imbricate; the corolla is polypetalous, hypogynous, and diplostemonous, the ovary is lobed and several-celled, the ovules are pendulous and anatropous, the fruit is a capsule, the seed furnished with a fleshy albumen, the leaves alternate and exstipulate; but in *Rutaceæ* the petals are imbricate, the base of the ovary is furnished with a glandular well-developed disk, the embryo is curved, and the plant is scented and usually glandular-dotted. *Oxalideæ* are also connected with *Connaraceæ* through *Averrhoa* and *Cnestis*, both of which are polypetalous, hypogynous, and diplostemonous, with distinct styles, capitate stigmas, albuminous seed, woody stem, alternate exstipulate imparipinnate leaves, and axillary inflorescence; but *Cnestis* differs in its polygamo-diœcious flowers, valvate calyx, free filaments, distinct carpels, and non-arillate seeds. *Oxalideæ* are distantly related to *Droseraceæ* in their spirally coiled young leaves, their hypogynism, and diplostemonous flowers (at least in *Dionæa* and *Drosophyllum*), pendulous anatropous ovules, capsular fruit, albuminous seed, and axile embryo. Finally, a certain analogy has been noticed between *Oxalis* and *Mimoseæ* (see this family).

The species of *Oxalis* inhabit both worlds, especially abounding in South Africa and tropical and subtropical America; they are rare in temperate regions, and are entirely absent from very cold countries. *Averrhoa* and *Connaropsis* are tropical Asiatic.

Oxalideæ contain in their herbaceous parts and fruit, when it is fleshy, an acid salt, tempered by a sufficient quantity of mucilage, which gives them refreshing, antibilious, and antiseptic properties. The tubers of the stemless species are farinaceous and eatable. Sorrel (*Oxalis acetosella*), a European and North American herb, together with *Rumex acetosa* and *acetosella*, yields binoxalate of potash, commonly called *oxalic acid*, a deadly poison, but used for the removal of ink-stains, &c. Many American species of *Oxalis* possess in their underground stem a feculent, wholesome, light, and very nourishing substance, the slightly acid taste of which almost entirely disappears in cooking; these occur as tubers like potatos (*O. crenata*), or bulbs (*O. esculenta*), or as swollen and fleshy roots (*O. Deppei*). The slightly bitter leaves of *O. sensitiva* are tonic and stimulating; its root is recommended for calculous diseases and scorpion-bites. The berries of *Averrhoa Carambola*, an Indian tree, are very acid in a wild state, but cultivation renders them sugary-acidulous and eatable. Those of *A. Bilimbi* are more tart than the preceding, and are only eatable when cooked and mixed with other condiments. The leaves of both these species are used to hasten the ripening of tumours.

XLII. *HUMIRIACEÆ*,[1] *Jussieu.*

[TREES or SHRUBS, mostly with balsamic juice, glabrous or puberulous. LEAVES alternate, simple, coriaceous, entire or crenulate; *stipules* 0. FLOWERS regular, ☿, white, in corymbiform, axillary, terminal, or lateral cymes. SEPALS 5, small, subconnate at the base, imbricate in bud. PETALS 5, hypogynous, deciduous, slightly contorted. STAMENS 10–∞, hypogynous, more or less connate at the base; *filaments* flattened or filiform, tips slender; *anthers* versatile, cells 2, adnate to the base of a thick fleshy connective. DISK annular, truncate, toothed or of separate scales, closely girding the base of the ovary. OVARY free, sessile, 5- rarely 6-7-celled; *style* simple, filiform; *stigma* entire or obscurely toothed; *ovules* solitary in the cells, or 2-3 suspended by unequal funicles from the top of the inner angle of the cell, anatropous, raphe ventral. FRUIT a drupe; endocarp bony or woody, cells usually in part suppressed. SEEDS solitary, or geminate and then separated by a transverse septum, oblong, pendulous; *testa* membranous; *albumen* copious, fleshy. EMBRYO axile; *cotyledons* short, obtuse; *radicle* usually elongate, superior.

PRINCIPAL GENERA.

Vantanea.   Humiria.   Sacoglottis.   Aubrya.

A small order, formerly associated with *Ebenaceæ* or *Olæineæ*; but differing from the latter in æstivation, ovary, ovules, &c., and from *Ebenaceæ* in polypetalism, hypogyny, &c. They appear to be more closely allied to the tribe *Ixionantheæ* of *Lineæ*, from which their habit and anthers distinguish them. From the 1-2-ovuled genera of *Meliaceæ* they are distinguished by their stamens, simple leaves, &c.

[1] This order is omitted in the original.—ED.

Humiriaceæ are natives of Guiana and Brazil, with the exception of one tropical African species. The above four genera comprise about twenty known species.

The Balsam of Unieri, which possesses the properties of the Peruvian and of Copaiva, is the produce of *Humiria floribunda*, a native of Brazil. A preparation of the juice both of this and *H. balsamifera* has the odour of Storax, and is made into an ointment used for pains in the joints, and used internally as a remedy for tænia and hæmorrhage.—ED.]

XLIII. *MALPIGHIACEÆ, Jussieu.*

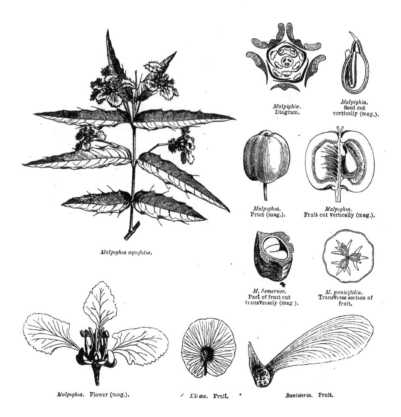

Malpighia.
Diagram.

Malpighia.
Seed cut
vertically (mag.).

Malpighia.
Fruit (mag.).

Malpighia.
Fruit cut vertically (mag.).

Malpighia aquifolia.

M. Semeruco.
Part of fruit cut
transversely (mag.).

M. punicifolia.
Transverse section of
fruit.

Malpighia. Flower (mag.).

Eliœa. Fruit.

Banisteria. Fruit.

CALYX 5-*merous, persistent, segments usually biglandular.* PETALS 5, *usually unguiculate, isostemonous or diplostemonous, inserted either on the receptacle or on a hypogynous or perigynous disk.* STAMENS *inserted with the petals, usually monadelphous, when several are antherless.* OVARY *composed of 3 or 2 carpels, connate, or distinct at the top, of 3 or 2 1-ovuled cells.* OVULE *nearly orthotropous.* FRUIT *a drupe, or of 3-2 cocci.* EMBRYO *exalbuminous.* STEM *woody.*

TREES or SHRUBS, rarely UNDERSHRUBS, often climbing; branches usually pubescent; hairs fixed by the centre and stinging, or silky, of metallic lustre, closepressed, and not stinging. LEAVES generally opposite, petiole jointed to the stem, entire, flat (rarely alternate, or whorled, sessile, sinuate-toothed or -lobed, margins recurved); petiole or under surface or margin of the leaf often glandular; *stipules* usually geminate at the base of the petiole, below (rarely above) the joint, generally rudimentary or suppressed, rarely large, sometimes the 2 belonging to the same leaf coherent in 1 and axillary, sometimes the 4 of 2 opposite leaves joining in pairs so as to form 2 interpetiolar stipules. FLOWERS ☿, or polygamous by suppression, sometimes dimorphous (*Aspicarpa, Janusia*), axillary or terminal, in a corymb, umbel, raceme or panicle; *peduncle* bracteate at its base; *pedicels* jointed, 2-bracteolate below the joint. CALYX 5-partite, segments imbricate, or very rarely valvate, all or 4 or 3 biglandular outside. PETALS 5, inserted on the receptacle, or on a hypogynous disk, or on a disk lining the base of the calyx, generally equal, fringed or toothed, claw slender, æstivation imbricate. DISK inconspicuous. STAMENS usually 10, hypogynous or subperigynous, all fertile, or some anantherous, or one or more or all of those opposite the petals wanting; *filaments* filiform or subulate, usually connate at the base; *anthers* short, introrse, cells sometimes winged, connective often thickened, sometimes prolonged at the top into a glandular appendage. OVARY free, composed of 3 (rarely 2) carpels, coherent, or distinct at the top, of 3-2 1-ovuled cells; *styles* 3, inserted between the lobes of the ovary, distinct or connate; *stigma* simple, inconspicuous; *ovule* nearly orthotropous, pendulous by a short funicle to the inner angle of the cell or middle of the septum, ascending, erect or curved, raphe ventral, micropyle superior. Ripe CARPELS 3 or fewer, 1-seeded, sometimes connate into a fleshy or woody drupe, sometimes distinct, and separating into usually winged samaras, rarely 2-valved. SEEDS obliquely pendulous below the top of the cell, exalbuminous; *testa* double, usually membranous. EMBRYO straight, curved, or hooked, very rarely annular; *cotyledons* flat or thick, often unequal; *radicle* short, above the hilum, superior.

PRINCIPAL GENERA.

*Malpighia.　　*Stigmaphyllon.　　*Banisteria.　　Hiptage.　　Hiræa.　　Aspicarpa.

[The following is the disposition of the tribes, &c., of this large family, proposed by the authors of the ' Genera Plantarum ' :—

TRIBE I. MALPIGHIEÆ.—Stamens 10, usually all perfect; anthers mostly appendiculate; styles usually 3, free. Carpels on a flat torus, never winged, free or connate into a fleshy or woody drupe.—Usually erect, with opposite leaves and connate stipules. *Byrsonima, Malpighia, Bunchosia, Thryallis, Galphinia, Spachea,* &c.

TRIBE II. BANISTERIEÆ.—Stamens 10, usually all perfect; anthers appendaged. Styles normally 3, free. Fruit of 1-3 dorsally (not laterally) winged nuts or samaras on a conical torus (rarely wingless or feathery).—Usually climbers, with often alternate exstipulate leaves. *Heteropterys, Acridocarpus, Brachypteris, Stigmaphyllon, Ryssopteris, Banisteria, Peixotoa*, &c.

TRIBE III. HIREÆ.—Stamens usually all perfect. Styles normally 3, free. Samaras 1-3, on a pyramidal torus, 1-7-winged, lateral wings broad.—Usually scandent, with often alternate exstipulate leaves. *Tristellateia, Hiptage, Aspidopterys, Triopterys, Tetrapterys, Hiræa*, &c.

TRIBE IV. GAUDICHAUDIEÆ.—Stamens 8 or fewer, some or all of those opposite the petals, and sometimes others, anantherous. Style 1. Carpels winged or not, usually pendulous from a raphe-like thread.—Erect or climbing; leaves often alternate. Flowers often dimorphic. Calyx 8-10-glandular. *Gaudichaudia, Aspicarpa, Camarea, Janusia, Schuannia*.—ED.]

Malpighiaceæ are near *Erythroxyleæ* (which see), *Acerineæ* and *Sapindaceæ*; they are separated from *Acerineæ* by their glandular calyx, long clawed petals, monadelphous stamens, usually 3-merous fruit, 1-ovuled ovarian cells, curved ovules, and superior radicle. The affinity is still more close with *Sapindaceæ*, from which they only differ in the inconspicuous disk and solitary abnormally-formed ovules. *Malpighiaceæ* mostly inhabit the plains and virgin forests of the southern tropic of America; they are less numerous beyond this region, rarer still in equatorial Asia, and very rare in Africa and Australia.

Many *Malpighiaceæ* owe to the colouring matter and tannin contained in their bark astringent properties, which are useful in various disorders, and especially in dysentery and intermittent fevers; as the different species of the American genus *Byrsonima*. The acidulous-sugary fruits of *Malpighia urens* and *glabra* are recommended as refreshing and antiseptic.

XLIV. *CORIARIEÆ, Endlicher.*

PETALS 5, *hypogynous, small, fleshy.* STAMENS 10, *hypogynous.* OVARY 5-*lobed, cells alternate with the petals,* 1-*ovuled.* STYLES 5, *stigmatiferous throughout their length.* FRUIT *of* 5 *carpels.* EMBRYO *scarcely or not albuminous.*—STEM *woody.* LEAVES *opposite.*

Unarmed SHRUBS, branches angular, the lower opposite or ternately whorled, the upper opposite, often sarmentose; buds scaly. LEAVES opposite, rarely ternately whorled, ovate or cordate or lanceolate, 1-5-nerved, entire, glabrous, exstipulate. FLOWERS ⚥ or polygamous, in terminal [or axillary] racemes; pedicels opposite or the upper alternate, bracteate at the base, and often 2-bracteolate. SEPALS 5, oval-triangular, imbricate in bud, persistent, spreading, margins membranous. PETALS 5, hypogynous, shorter than the sepals and alternate, triangular, fleshy, keeled within, accrescent, persistent. TORUS conical, fleshy, lobed. STAMENS 10, hypogynous, free, or the inner adnate to the keel of the petals; *filaments* short, filiform; *anthers* large, introrse, basifixed, dehiscence longitudinal; *pollen* very fine, subglobose. CARPELS 5-10, free, oblong, whorled on the torus, conical, alternate with the petals, 1-ovuled; *styles* as many as carpels, free, thick, elongated, distant, entirely covered with stigmatic papillæ; *ovules* pendulous from the top of the cells, anatropous, raphe dorsal. FRUIT of 5-8 indehiscent cocci, embraced by the accrescent and fleshy petals, compressed, oblong; *pericarp* crustaceous, keeled on the back and sides,

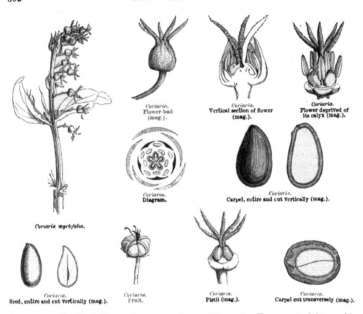

Coriaria.
Flower-bud
(mag.).

Coriaria.
Vertical section of flower
(mag.).

Coriaria.
Flower deprived of
its calyx (mag.).

Coriaria.
Diagram.

Coriaria.
Carpel, entire and cut vertically (mag.).

Coriaria myrtifolia.

Coriaria.
Seed, entire and cut vertically (mag.).

Coriaria.
Fruit.

Coriaria.
Pistil (mag.).

Coriaria.
Carpel cut transversely (mag.).

SEEDS compressed; *testa* membranous; *albumen* thin or 0. EMBRYO straight, ovoid, compressed; *cotyledons* plano-convex; *radicle* very short, obtuse, superior.

ONLY GENUS.

*Coriaria.

The affinities of *Coriaria* are very obscure. It has been placed near *Malpighiaceæ* (see this family). It somewhat distantly recalls *Rutaceæ* and *Zanthoxyleæ* in its hypopetalism, diplostemony, free carpels, &c., but is separated from them by its pendulous ovule with dorsal raphe and superior micropyle. Compared with *Sapindaceæ* and *Terebinthaceæ*, it differs in habit, the inner stamens usually adherent to the petals, and the styles stigmatiferous throughout their length. It has some analogies with *Phytolaceæ*, founded on the distinct and whorled carpels, papillose styles and fruit [but differs in the petals, pendulous ovules, fleshy albumen (when present), and straight thick embryo].

Coriaria comprises a few species, dispersed over the Mediterranean region, Nepal, Japan, New Zealand, and western South America. *C. myrtifolia* (Myrtle-leaved Sumach), which grows especially in the West of the Mediterranean region, abounds in tannin, utilized by curriers; its leaves and fruits contain a narcotic-acrid crystallizable principle (*coriariine*) which renders them poisonous. These leaves are used to adulterate senna, which is thus often a fatal medicine. The fruit of *C. sarmentosa*, a New Zealand shrub, is full of a vinous juice, which the natives and colonists drink with impunity, taking care not to swallow the seeds, which are eminently poisonous. The same may be said of *C. nepalensis*. *C. ruscifolia*, of China, yields a black colour, used by shoemakers.

XLV. *ZYGOPHYLLEÆ.*

(RUTACEARUM *sectio, Jussieu.*—ZYGOPHYLLEÆ, *Br.*—ZYGOPHYLLACEÆ, *Lindl.*)

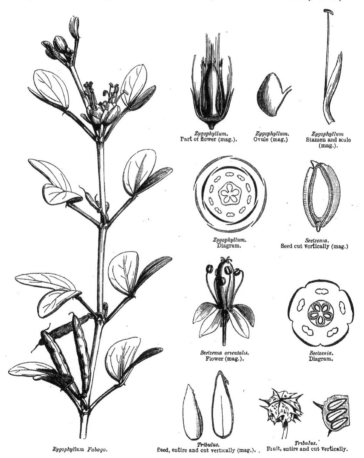

Zygophyllum.
Part of flower (mag.).

Zygophyllum.
Ovule (mag.)

Zygophyllum.
Stamen and scale (mag.).

Zygophyllum.
Diagram.

Seetzenia.
Seed cut vertically (mag.)

Seetzenia orientalis.
Flower (mag.).

Seetzenia.
Diagram.

Zygophyllum Fabago.

Tribulus.
Seed, entire and cut vertically (mag.).

Tribulus.
Fruit, entire and cut vertically.

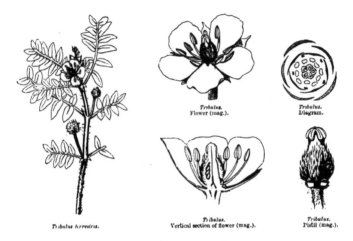

Tribulus.
Flower (mag.).

Tribulus.
Diagram.

Tribulus terrestris.

Tribulus.
Vertical section of flower (mag.).

Tribulus.
Pistil (mag.).

CALYX 4–5-*merous, generally imbricate.* PETALS *hypogynous, usually imbricate.*
STAMENS *usually double the number of the petals, hypogynous; filaments usually with a
scale inside.* OVARY *several-celled.* FRUIT *a loculicidal capsule, septicidally dividing
into cocci.* EMBRYO *exalbuminous, or enclosed in a cartilaginous albumen.—Scentless
plants.* LEAVES *opposite, pinnate, stipulate.*

HERBS, SHRUBS or TREES; branches often divaricate and jointed at the nodes.
LEAVES opposite, or alternate from the suppression of one, stipulate, compound,
sometimes pinnate or imparipinnate, sometimes 2- (rarely 1-) foliolate (*Zygophyllum*);
petiole sometimes flattened and winged; *leaflets* sessile, entire, not punctate, often
inequilateral, flat, or fleshy, or terete; *stipules* geminate at the base of the petioles,
persistent, sometimes spinescent. FLOWERS ☿, regular or irregular, white, red or
yellow, rarely blue; *peduncles* usually 1–2, springing from the axil of the stipules,
1-flowered, ebracteate. SEPALS 5–4, usually persistent, free, rarely connate at the
base, æstivation imbricate or very rarely valvate (*Seetzenia*). PETALS 5–4, very rarely
0 (*Seetzenia*), hypogynous, free, æstivation usually imbricate, sometimes contorted
(*Zygophyllum*); *disk* hypogynous, convex or depressed, rarely annular (*Tribulus*),
sometimes inconspicuous (*Fagonia, Guaiacum,* &c.), or 0 (*Seetzenia*). STAMENS
usually double the number of the petals, rarely equal (*Seetzenia*), inserted on the receptacle, 2-seriate, the outer opposite to the sepals; *filaments* filiform, usually with
a small scale on the base within, or on their centre; *anthers* introrse, dorsifixed
above their base, versatile, dehiscence 'longitudinal. OVARY free, sessile, or rarely
borne on a short gynophore (*Larrea, Guaiacum*), furrowed, angular, or winged, 4–5-
(rarely 10–12-) celled (*Tribulus, Augea*), or 2–3-celled (*Zygophyllum*), cells sometimes

divided into several by transverse septa (*Tribulus*) ; *style* simple, terminal, angular or furrowed, sometimes short or 0 (*Tribulus*) ; *stigma* simple ; *ovules* 2, superimposed in each cell, or several 2-seriate, pendulous or ascending, raphe ventral, and micropyle superior. FRUIT coriaceous or crustaceous, sometimes septicidal, and dividing into 2–10 dehiscent or indehiscent cocci, which are connate, or separable from the placental column ; sometimes a loculicidal capsule. SEEDS usually solitary in each cell, rarely 2 or several, pendulous ; *testa* membranous or crustaceous, or thick and mucilaginous ; *albumen* cartilaginous, rarely 0 (*Tribulus*, &c.). EMBRYO green, straight or slightly curved ; *cotyledons* foliaceous ; *radicle* short, straight, superior.

PRINCIPAL GENERA.

Tribulus. *Zygophyllum. Larrea. Porliera. *Fagonia. Guaiacum. ?Nitraria.

Zygophylleæ are near *Geraniaceæ* and *Oxalideæ* (see these families). They are connected with *Rutaceæ* by their hypopetalism, diplostemony, several-celled ovary, hypogynous disk, loculicidal or septicidal capsule, albuminous seeds, straight or arched embryo, and superior radicle ; but *Rutaceæ* differ in habit, alternate and glandular-dotted exstipulate leaves, eglandular filaments, and basilar style. The genus *Nitraria*, doubtfully placed here by Bentham and Hooker fil., certainly approaches them in its stipulate leaves, oblong anthers, and the structure of its ovary ; but it is separated by its habit, simple leaves, valvate induplicate corolla, naked filaments, one-ovuled cells of the ovary, and drupaceous fruit. *Zygophylleæ* scarcely differ from *Simarubeæ*, except in their ovary attenuated into a simple terminal style. *Batideæ* also display some analogy with *Tribulus*.

Zygophylleæ principally inhabit the extra-tropical and hot regions of both hemispheres, especially abounding from the north-west of Africa, through the Mediterranean region, to the northern limit of India ; they are rarer in South Africa, Australia, and South America. Except *Fagonia*, which is spread over the Mediterranean region and Central Asia, and *Zygophyllum*, which grows throughout Africa and Asia, and *Tribulus*, which is dispersed over tropical and sub-tropical countries, each genus has a special habitat. *Seetzenia* inhabits tropical Africa and Eastern Asia ; the other genera are exclusively American. *Nitraria* grows in saline ground in the North of Africa, West Asia, and Australia.

The most useful plant of this order is *Guaiacum officinale*, a West Indian tree with very hard faintly aromatic wood, much heavier than water, and with an acrid and bitter taste. Cabinet-makers use it for balls, castors, pulleys, and other objects exposed to weight and friction. The raspings of guaiacum form a valuable medicine, acting powerfully on the functions of the skin and the secretions of the kidneys, and recommended as a depurative in syphilis. These properties are due to a resinous substance (*guaiacine*) contained in the wood, and which exudes from it when cut ; it is also obtained by steeping the guaiacum in alcohol, and evaporating the tincture. The Holy Guaiacum, another American species, possesses the same virtues, but is only used in the New World. *Zygophyllum Fabago* is a reputed antisyphilitic and vermifuge ; its flower-buds are used instead of capers. The *Z. simplex*, a species very common in the most arid deserts, is used by the Arabs to remove freckles ; this plant, as well as its congeners, exhales so foetid a smell, that all herbivorous animals, including even the camel, reject it. *Tribulus terrestris*, formerly praised as an astringent, has fallen into disuse.

XLVI. *BATIDEÆ*, Lindl.

Littoral saline plants of a grey colour. STEMS branching, diffuse, fragile. LEAVES opposite, oblong-linear or obovate-oblong, sessile or subsessile, plane above, convex below, fleshy, exstipulate. FLOWERS in four rows, in conical oblong spikes, opposite, sessile, green, diœcious. FL. ♂ : FLOWERS distinct ; *bracts* cochleariform, obtuse or very shortly acuminate, concave, entire, persistent, close together. CALYX

x

membranous, campanulate, or forming a compressed truncate sub-bilabiate cup. PETALS 4, claws united at the base, limb rhomboid. STAMENS 4, alternate with the petals, exsertèd; *filaments* subulate, glabrous; *anthers* 2-celled, versatile, oblong, didymous, incumbent, introrse, dehiscence longitudinal. OVARY rudimentary or 0. FL. ♀: FLOWERS in a fleshy spike; *bracts* as in the ♂, deciduous, the two lower connate. CALYX and COROLLA 0. OVARIES 8-12, coherent, adherent to the base of the bracts, 4-celled; *style* 0; *stigma* capitate, sub-bilobed; *ovules* solitary, erect, anatropous. RIPE CARPELS 4-celled, connate, and forming a fleshy ovoid-conical fruit; endocarp coriaceous. SEEDS erect, oblong, straight; *testa* membranous; *albumen* 0. EMBRYO conformable to the seed; *cotyledons* fleshy, oblong, compressed; *radicle* short, near the hilum.

ONLY GENUS.
Batis.

The affinities of *Batis* are very obscure; in habit it resembles *Chenopodieæ*, but the structure of its flower appears to indicate a closer affinity with *Reaumuriaceæ, Tamariscineæ*, and with some *Zygophylleæ*, through *Tribulus*. It inhabits the seashores of tropical America.

XLVII. *GERANIACEÆ.*

(GERANIA, *Jussieu.*—GERANIOIDEÆ, *Ventenat.*—GERANIACEÆ, *D.C.*)

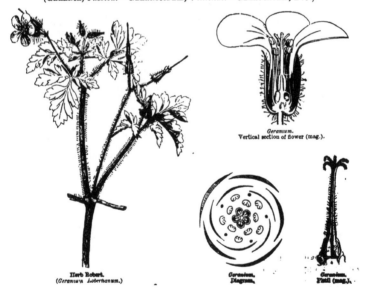

Geranium.
Vertical section of flower (mag.).

Herb Robert.
(*Geranium Robertianum.*)

Geranium.
Diagram.

Geranium.
Pistil (mag.).

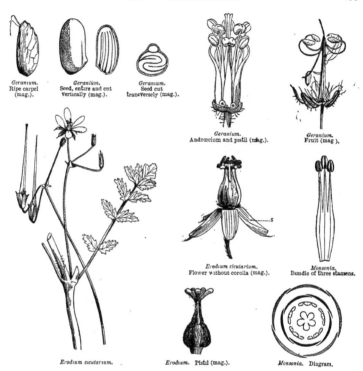

Geranium.
Ripe carpel
(mag.).

Geranium.
Seed, entire and cut
vertically (mag.).

Geranium.
Seed cut
transversely (mag.).

Geranium.
Andrœcium and pistil (mag.).

Geranium.
Fruit (mag).

Erodium cicutarium.
Flower without corolla (mag.).

Monsonia.
Bundle of three stamens.

Erodium cicutarium.

Erodium. Pistil (mag.).

Monsonia. Diagram.

CALYX 5-*merous, imbricate.* PETALS *hypogynous, æstivation contorted.* STAMENS *hypogynous, usually twice the number of the petals.* CARPELS 5, *cohering by their inner edges into a 5-celled ovary prolonged into a beak; cells 2-ovuled.* FRUIT *a capsule, opening elastically from below upwards.* SEEDS *exalbuminous.* EMBRYO *curved; cotyledons folded or coiled.*—LEAVES *stipulate.*

HERBS or UNDERSHRUBS, sometimes fleshy (*Pelargonium, Sarcocaulon*). Lower LEAVES opposite ; upper alternate or opposite, petiolate, simple, usually palminerved and with palmate divisions, rarely pinnatisect, sometimes entire or crenulate; *stipules* twin at the base of the petioles, foliaceous or scarious. FLOWERS ⚥ , regular or irregular (*Pelargonium*), rarely solitary, usually in pairs or involucrate umbels ; *peduncles* opposite to the alternate leaves, or axillary in one of a pair of opposite leaves, or in the fork of a branch, sometimes radical. CALYX free, persistent, 5-

phyllous or partite, æstivation imbricate ; sepals equal or unequal, with the posterior
then prolonged into a spur longitudinally adnate to the peduncle (*Pelargonium*).
PETALS inserted on the receptacle, equal in number with the sepals, or fewer by suppression, and sub-perigynous (*Pelargonium*), clawed, equal or unequal (*Pelargonium*),
contorted in æstivation, caducous.   STAMENS inserted with the petals, and usually
twice as many, 2-seriate; the inner fertile, the outer shorter, and opposite to the
petals, all fertile (*Geranium*), or sometimes all (*Erodium*) or some (*Pelargonium*)
antherless; rarely triple the number of the petals, and disposed in 5 triandrous
bundles, opposite to the petals (*Monsonia, Sarcocaulon*) ; *filaments* membranous, flattened, and more or less monadelphous below, tips subulate, the lower with external
basal glands that alternate with the petals, rarely eglandular (*Pelargonium*) ; *anthers*
introrse, oblong, dorsifixed, versatile, dehiscence longitudinal.   PISTIL of 5
carpels, cohering into a 5-lobed 5-celled ovary, prolonged above into a beak terminated by the styles; *styles* continuous with the ovaries, connate below, free
above, stigmatiferous along their inner edge; *ovules* geminate on the ventral suture
of each cell, more or less superimposed, semi-anatropous.   FRUIT a capsule, opening
elastically from below upwards, by septifragal dehiscence, into five 1-seeded cocci
with spirally twisted beaks, and detaching themselves from the placentiferous
column.   SEED 3-gonous ; *testa* crustaceous ; *hilum* ventral, a little below the base.
EMBRYO exalbuminous, curved ; *cotyledons* foliaceous, flexuous ; *radicle* conical, incumbent on the inner cotyledon, near the hilum, and, together with the tips of the
cotyledons, pointing towards the base of the fruit.

PRINCIPAL GENERA.

*Monsonia.	*Geranium.	*Erodium.	*Pelargonium.

[Disposition of the *Geraniaceæ* proposed in the ' Genera Plantarum ' :—

TRIBE I. GERANIEÆ.—Flowers regular or nearly so. Sepals imbricate. Petals alternating
with glands.   Antheriferous stamens as many as the petals, or twice or thrice as many.
Biebersteinia, Monsonia, Sarcocaulon, Geranium, Erodium.

TRIBE II. PELARGONIEÆ.—Flowers irregular. Posticous sepal spurred. Petals perigynous,
two upper external. Glands 0. Stamens declinate. *Pelargonium, Tropæolum.*

TRIBE III. LIMNANTHEÆ (see order LIMNANTHEÆ, p. 313).—Flowers regular.   Sepals
valvate. Petals alternating with glands. Capsule indehiscent, not beaked.   Ovary 1 ; ovule
with an inferior micropyle. *Floerkia, Limnanthes.*

TRIBE IV. VIVIANIEÆ.—Flowers regular. Calyx with valvate lobes. Petals alternating
with glands. Stamens twice as many as the petals. Capsule loculicidal ; ovules geminate.
Viviania.

TRIBE V. WENDTIEÆ.—Flowers regular. Sepals imbricate. Glands 0. Stigmas ligulate.
Ovules 2–∞ . Leaves small, often opposite, entire or 2–5-lobed. *Rhynchotheca, Wendtia,
Ledocarpon.*

TRIBE VI. OXALIDEÆ (see order OXALIDEÆ, p. 296).—Flowers regular. Sepals imbricate.
Glands 0. Stigmas capitate. Ovules 2–∞ . Leaves compound.

TRIBE VII. BALSAMINEÆ (see order BALSAMINEÆ, p. 309).—**Flowers irregular.** Posticous

sepal spurred. Petals hypogynous. Stamens 5, short; anthers coherent. Stigmas sessile.—
Ed.]

Geraniaceæ are so closely allied to Limnantheæ, Vivianieæ, Ledocarpeæ, Oxalideæ, and Balsamineæ, that Bentham and Hooker fil. have united them into one family, connecting with them also Tropæoleæ, which are near Pelargonium in their irregular anisostemonous eglandular flower, spurred posterior sepal, and perigynous petals with the two upper exterior; but which are separated by their free fertile stamens, 1-ovuled indehiscent carpels without beak, and exstipulate leaves. Limnanthes is distinguished from Geraniaceæ by its valvate calyx, free stamens, gynobased style, 1-ovuled beakless indehiscent carpels, straight embryo, and exstipulate leaves. Vivianieæ differ in their valvate calyx, 3-celled capsule with loculicidal dehiscence, seed with copious albumen, and exstipulate leaves. Ledocarpeæ differ in the absence of glands, the ligulate stigmas, loculicidal capsule, and exstipulate leaves. Oxalideæ are separated by their compound exstipulate leaves, eglandular receptacle, capitate stigmas, capsular or berried fruit, albuminous seed, and straight or scarcely curved embryo. Balsamineæ have, like Pelargonium, irregular flowers, a spurred posterior sepal, and an elastically opening fruit; but they differ in their exstipulate leaves, eglandular disk, pentandrous androcium with connate or connivent anthers, sessile stigma, loculicidal capsule, and straight embryo with superior radicle. Geraniaceæ approach Zygophylleæ in their jointed stem, æstivation, diplo- or triplo-stemonous flower, capsular fruit dividing into cocci, opposite and stipulate leaves; but in Zygophylleæ the filaments do not cohere and are filiform, the style is simple, the seeds are often albuminous, and the embryo is straight or hardly curved. Lineæ (especially the genus Linum) also approach Geraniaceæ in their corolla and in their androcium composed of 5 fertile stamens and 5 staminodes opposite to the petals, their dilated filaments monadelphous at the base, their entire ovary with 2-ovuled cells, pendulous anatropous ovules, terminal capitate stigmas, straight embryo, and usually exstipulate leaves. Finally, Geraniaceæ offer some affinity with Malvaceæ in their stipulate palmilobed leaves, monadelphous stamens, exalbuminous seeds, and coiled embryo.

Geraniaceæ principally inhabit warm, temperate, and tropical countries of both worlds. Geranium and Erodium especially belong to the northern hemisphere; Monsonia to South Africa and tropical and West Asia. Pelargonieæ are met with in south temperate latitudes, and especially at the Cape of Good Hope; they are rarer in Australia and in the South Pacific Islands. One species (P. Endlicherianum) advances into Asia Minor. [Tropæolum is wholly South American, and chiefly western; Wendtieæ and Vivianieæ are also Peruvian and Chilian; Limnantheæ North American. About 600 species are known, belonging to the 13 genera enumerated above.]

Geraniaceæ contain tannin and gallic acid, and thus possess astringent properties. Several contain resins, and a volatile oil, mixed with an abundant mucilage; others contain free acids. Herb Robert (Geranium Robertianum) and G. sanguineum, indigenous species, formerly used as astringents and slight stimulants, have fallen into disuse; G. pratense is still used as a vulnerary. The root of G. maculatum, a native of North America, is there administered for dysentery, that of G. nodosum and striatum replaces in Italy the Tormentilla. Erodium moschatum, which exhales a strong odour of musk, is a reputed stimulant and diaphoretic. Some Pelargoniums are as remarkable for their scent as for the brilliancy of their colours which causes them to be cultivated as ornamental plants; such are P. zonale and inquinans, the bruised leaves of which cause rusty spots. The pale yellow flowers of P. triste are spotted with brown, and are sweet-scented by night. From some species a very sweet volatile oil is obtained by distillation, which is used to adulterate essence of roses. The tubers of P. antidysentericum, triste, &c., are employed by the Namaquois in diarrhœa. The leaves of P. acetosum and pellatum have a very agreeable sharpish taste. The resinous balsamic stem of Monsonia spinosa burns with a flame, and the natives of South Africa make torches of it.

XLVIII. BALSAMINEÆ.

(Balsamineæ, A. Richard.—Balsaminaceæ, Lindl.—Hydrocereæ, Blume.)

Calyx irregular, 3-5-phyllous, imbricate, caducous. Petals 3-5, hypogynous, unequal, imbricate. Stamens 5, hypógynous, connate above, and covering the ovary.

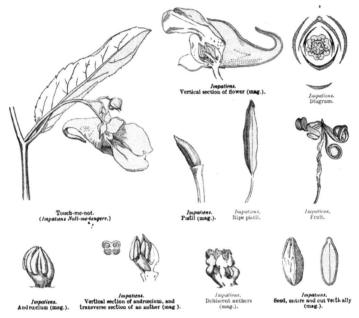

Impatiens.
Vertical section of flower (mag.).

Impatiens.
Diagram.

Touch-me-not.
(*Impatiens Noli-me-tangere.*)

Impatiens.
Pistil (mag.).

Impatiens.
Ripe pistil.

Impatiens.
Fruit.

Impatiens.
Andrœcium (mag.).

Impatiens.
Vertical section of andrœcium, and
transverse section of an anther (mag.).

Impatiens.
Dehiscent anthers
(mag.).

Impatiens.
Seed, entire and cut vertically
(mag.).

OVARY *of* 5 *many–few-ovuled cells*; *stigma sessile.* FRUIT *a capsule, dehiscing elasti-cally, or an indehiscent drupe.* EMBRYO *straight, exalbuminous.*

Soft succulent HERBS, usually annual, sometimes suffruticose, erect, full of watery juice; *root* fibrous or sometimes tuberous. LEAVES sometimes all radical, long-petioled, cordate or reniform; cauline opposite, alternate or ternately whorled, penninerved, crenate or dentate; *stipules* 0, but petioles sometimes furnished at the base with substipitate cupular glands; *peduncles* axillary, solitary or aggregated, 1–∞ -flowered, bracteate; *pedicels* naked or bracteolate. FLOWERS ☿, irregular, spurred, often resupinate from the weakness of the pedicel and weight of the spur, which hence appears posterior. SEPALS 3–5, irregular, coloured, caducous, imbricate in æstivation, the 2 outer lateral, opposite, small or minute, incumbent on the 2 anterior, which are sometimes wanting; the posterior large, concave, prolonged at the base into a boss or spur, and enveloping the ovary in bud. COROLLA inserted on the receptacle, of 5 petals alternate with the sepals, all free (*Hydrocera*), or appearing reduced to 3 by the confluence of the lateral with the 2 posterior petals (*Impatiens*); anterior petal concave, much larger than the others, and enveloping them; posterior petals smaller, enveloping the 2 lateral. STAMENS 5, inserted on the receptacle, as long as the ovary, and covering it like an operculum; *filaments* short, flattened, cohering at the top; *anthers* introrse, 2-celled, connivent or coherent. OVARY free, sessile, cylindric-oblong, or obtusely angled, 5-celled; *stigma* sessile, entire or 5-partite; *ovules* numerous in each cell (*Impatiens*), or 2–3 (*Hydrocera*), 1-seriate, pendulous, anatropous. FRUIT a loculicidal 5-valved capsule, elastically coiling inwards from the top downwards, or outwards from the bottom upwards; sometimes an indehiscent 5-celled drupe with a bony 5-lobed endocarp. SEEDS pendulous, slightly compressed; *testa* membranous, dotted or tubercular, glabrous or velvety. EMBRYO exalbuminous, straight; *cotyledons* plano-convex, fleshy; *radicle* very short, superior.

GENERA.

*Impatiens.   Hydrocera.

Balsamineæ are closely allied to *Geraniaceæ* (see that family [1]). They are equally near *Oxalideæ* in hypopetalism, 5-celled ovary with pendulous anatropous ovules, usually loculicidal fruit, and axillary flowers; but *Oxalideæ* differ in their regular diplostemonous flower, free styles, the valves of the capsule remaining attached to the placentiferous column,[2] the usually arillate albuminous seed, and the compound leaves. *Lineæ* have some analogies with *Balsamineæ*, founded on the 5-phyllous calyx, the polypetalous hypogynous isostemonous corolla with pendulous anatropous ovules, the capsular fruit, and the seed with little or no albumen; but they are separated by their regular flower, contorted corolla, long styles, terminal inflorescence, and entire leaves. The affinity with *Tropæoleæ* is more obvious: irregular coloured calyx with one sepal prolonged into a spur, several-celled ovary, pendulous anatropous ovules, capsular fruit, exalbuminous straight embryo, axillary flowers, and herbaceous stem; but in *Tropæoleæ* the calyx is persistent, and it is the posterior sepal which is prolonged into a spur, the corolla is anisostemonous, the ovary has three 1-ovuled cells, and the fruit divides into three indehiscent lobes.

Balsamineæ mostly grow in the mountainous regions of India and Africa; a few inhabit South Africa, North America, Europe, and Central Asia. The European is the *Impatiens Noli-me-tangere*, or Touch-me-not, so called from the elasticity of its capsules, which open from below upwards, and scatter their seeds at the slightest touch: it was formerly classed among diuretic medicines. *I. Balsamina* is an East Indian annual, everywhere cultivated on account of the abundance and variety of the colours of its flowers, which have been so doubled by florists as to have become almost regular. *I. parviflora* is almost naturalized in shady places near Paris. *Hydrocera* belongs to the marshes of tropical Asia.

XLIX. *TROPÆOLEÆ.*

(TROPÆOLEÆ, *Jussieu.*—BALSAMINACEARUM *subordo, Lindl.*)

FLOWERS *irregular*. SEPALS *unequal, the posterior spurred*. PETALS *perigynous, unequal, equal and alternate with the sepals, or fewer.* STAMENS 8, *hypogynous.* OVARY *free, of three* 1-*ovuled cells.* SEEDS *exalbuminous.* EMBRYO *straight, with thick cotyledons.*—STEM *herbaceous.* LEAVES *alternate.*

---

[1] A glance at the seven tribes of *Geraniaceæ* proposed in the 'Genera Plantarum,' and here introduced under that order (p. 308), will show how the characters used to separate *Geraniaceæ, Oxalideæ, Balsamineæ,* and

Tropæoleæ are so intermixed that it appears impossible to keep these groups distinct as orders.—ED.

[2] Not in *Biophytum.*—ED.

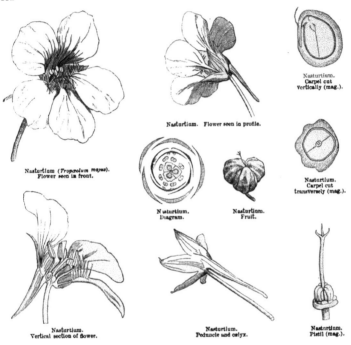

Nasturtium (*Tropæolum majus*).
Flower seen in front.

Nasturtium. Flower seen in profile.

Nasturtium.
Carpel cut
vertically (mag.).

Nasturtium.
Diagram.

Nasturtium.
Fruit.

Nasturtium.
Carpel cut
transversely (mag.).

Nasturtium.
Vertical section of flower.

Nasturtium.
Peduncle and calyx.

Nasturtium.
Pistil (mag.).

Succulent, prostrate, glabrous or twining HERBS with watery juice; root usually
tuberous. Primordial LEAVES opposite, 2-stipulate; the others alternate, exstipulate,
petioled, simple, peltate, entire or lobed, or deeply palmipartite with entire or cut
lobes, sometimes pellucid-dotted. FLOWERS ☿, irregular, axillary, long-peduncled.
CALYX coloured, persistent, 5-fid, 2-lipped, the upper lip 2-fid, the lower 3-fid and prolonged at the base into a hollow spur, æstivation imbricate or subvalvate. PETALS
inserted at the bottom of the calyx, equal and alternate with the sepals, or fewer,
imbricate in æstivation, the 2 upper inserted on the throat of the spur, and exterior,
spreading, and different from the 3 lower, which are usually smaller and sometimes
0. STAMENS 8, unequal, inserted on the receptacle, surrounding the ovary; *filaments*
subulate, free; *anthers* introrse, basifixed, erect, dehiscence longitudinal. OVARY
free, sessile, 3-lobed, 3-celled; *style* central, filiform, divided at the top into three
branches terminated by a minute stigma; *ovules* solitary, pendulous from the top of
the inner angle of each cell, anatropous. FRUIT of 3 indehiscent dry or spongy-

fleshy carpels, rugose, separating from a short persistent column. SEED inverted; *testa* cartilaginous, often confluent with the endocarp. EMBRYO exalbuminous, straight; *cotyledons* thick, hemispheric when young, afterwards often cohering, auricled at the base, auricles appressed but distinct, concealing the short radicle, which is near the hilum and superior.

GENUS.

*Tropæolum.

Tropæoleæ, placed beside *Pelargonium* by Bentham and Hooker fil.,[1] approach *Limnantheæ* (which these authors have also annexed to *Geraniaceæ*) in their persistent calyx, their imbricate perigynous or sub-perigynous anisostemonous or diplostemonous petals, their whorled carpels adherent around a central column into an ovary with 1-ovuled cells separating from the column, their exalbuminous seeds, herbaceous stem, alternate leaves, and axillary 1-flowered peduncles; *Limnantheæ* being hardly separated, except by the regularity of their flowers, the staminiferous ring lining the base of the calyx, their deeply 3- or 5-lobed ovary and ascending ovule. For the affinity with *Balsamineæ*, see this family. Some analogies have been noticed between *Tropæoleæ* and *Lineæ*, but the latter differ in their hypogynous regular petals, the 2-ovuled cells of their ovary, the free styles, and terminal inflorescence. *Tropæoleæ* may with more reason be compared with *Capparideæ* (see this family).

Tropæoleæ are all natives of South America, and chiefly of the western slopes of the Andes. They contain an acrid antiscorbutic principle, analogous to that of Cress. The large and small Nasturtium (*T. majus* and *minus*) are cultivated in Europe; their buds and young fruits are employed as a condiment instead of capers. The farinaceous tubers of *T. tuberosum*, made into jelly with treacle, furnish a sort of food to the Peruvians.

L. *LIMNANTHEÆ*.

(LIMNANTHEÆ, *Br.*—LIMNANTHACEÆ, *Lindl.*)

SEPALS 3–5, *valvate, or slightly imbricate.* PETALS *equal and alternate with the sepals, contorted in æstivation, inserted on a glandular ring lining the bottom of the calyx.* STAMENS *double the number of the petals, sub-perigynous.* OVARY *deeply 5-lobed, style gynobasic, cells 1-ovuled.* OVULE *erect, anatropous.* Ripe CARPELS *free, indehiscent.* EMBRYO *exalbuminous.*—STEM *herbaceous.* LEAVES *alternate.*

Annual succulent glabrous diffuse marsh plants. LEAVES alternate, long-petioled, pinnatifid or bipinnatifid, with narrow lanceolate or oval divisions; *stipules* 0. FLOWERS regular, ☿; *peduncles* axillary, long, 1-flowered, solitary, ebracteate and thickened at the top. SEPALS 5, valvate in æstivation (*Limnanthes*), or 3, slightly imbricate (*Flœrkea*). PETALS 5 or 3, sub-perigynous, contorted in æstivation, marcescent. STAMENS 10 (*Limnanthes*), or 6 (*Flœrkea*), sub-perigynous; *filaments* plano-subulate or subulate-filiform, marcescent, some opposite to the petals, others opposite to the sepals and furnished with a gland at the base; *anthers* introrse, subglobular-didymous, dehiscence longitudinal. CARPELS 5 (*Limnanthes*), or 3 (*Flœrkea*), opposite to the sepals, nearly free, connate at the base into a deeply 5–3-lobed ovary, 5–3-celled; *style* gynobasic, with 5–3 short stigmatiferous branches; *ovules* solitary in each cell, ascending, anatropous, micropyle inferior, raphe dorsal·

[1] See *Geraniaceæ*, p. 308.

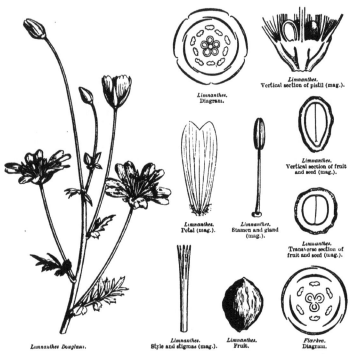

Limnanthes.
Diagram.

Limnanthes.
Vertical section of pistil (mag.).

Limnanthes.
Vertical section of fruit
and seed (mag.).

Limnanthes.
Petal (mag.).

Limnanthes.
Stamen and gland
(mag.).

Limnanthes.
Transverse section of
fruit and seed (mag.).

Limnanthes Douglasii.

Limnanthes.
Style and stigmas (mag.).

Limnanthes.
Fruit.

Flœrkea.
Diagram.

Ripe CARPELS free, indehiscent (achenes), at first fleshy, then coriaceous, rugose.
SEED erect, testa membranous. EMBRYO exalbuminous, straight; *cotyledons* fleshy,
thick, green, cordate at the base, and enclosing the very short, inferior radicle.

PRINCIPAL GENERA.
*Limnanthes.   Flœrkea.

Limnantheæ are closely allied to *Tropæoleæ* (see this family). By Bentham and Hooker fil. they
have been annexed to *Geraniaceæ* (see this family).

Limnantheæ inhabit the temperate regions of North America. *Limnanthes Douglasii*, a Californian
herb, is cultivated in Europe as an ornamental plant; it has a subacrid taste, which confirms its affinity
with *Tropæolum.* ·

[*RUTAL FAMILY OR ALLIANCE.*

The following is the arrangement of the great family of *Rutaceæ* proposed by the
Editor and Bentham; most of the tribes of which are treated as orders in this work :—

Series A.—Ovary deeply 2–5-lobed; styles basilar or ventral, free or connate by their stigmas only. Fruit capsular or of 3–5 cocci, endocarp usually separating.

TRIBE I. CUSPARIEÆ (see p. 322).—Flowers usually irregular. Corolla usually tubular, gamopetalous. Stamens 5, some often anantherous, free or adherent to the corolla. Ovules 2. Albumen scanty or 0. Cotyledons convolute.—Leaves usually broad and compound, leaflets 3–5. Tropical America. *Almeidea, Erythrochiton, Galipea, Ticorea, Monneira*, &c.

TRIBE II. RUTEÆ (see RUTACEÆ, p. 315).—Flowers regular (except *Dictamnus*), ♂ ☿ ?, often 4-merous. Petals and stamens free, spreading. Disk free, thick. Ovules 3 or more. Albumen fleshy. Embryo often curved.—Herbs, often shrubby below. Leaves often pinnatisect. Chiefly natives of the north temperate regions. *Ruta, Bœnninghausenia, Peganum, Dictamnus.*

TRIBE III. DIOSMEÆ (see p. 321).—Flowers regular, usually ☿. Petals free, usually erect. Stamens 4–5, inserted beneath the free edge of a disk that is adnate to the calyx-tube. Ovary usually deeply 4-lobed. Ovules usually geminate in each cell. Carpels often beaked. Albumen 0. Embryo straight; cotyledons fleshy.—Usually heath-like shrubs. All South African. *Calodendron, Euchætis, Macrostylis, Diosma, Coleonema, Acmadenia, Adenandra, Barosma, Agathosma*, &c.

TRIBE IV. BORONIEÆ (see p. 322).—Flowers regular, ☿. Petals and stamens free. Disk free, cupular or annular, or 0. Ovules 2, superimposed in each cell, rarely geminate. Carpels 2–5. Albumen fleshy, embryo terete.—Shrubs. All Australasian. *Zieria, Cyanothamnus, Phebalium, Philotheca, Chorilæna, Boronia, Eriostemon, Crowea, Correa, Diplolæna.*

TRIBE V. ZANTHOXYLEÆ (see p. 323).—Flowers regular, usually polygamo-diœcious. Petals and stamens free, spreading. Disk free, annular or tumid, or 0. Ovules 2, geminate or superimposed. Carpels 2–5. Embryo straight or curved; cotyledons usually flat.—Tropical trees or shrubs with compound leaves. *Melicope, Zanthoxylum, Pilocarpus, Evodia, Geijera, Esenbeckia.*

Series B.—Ovary entire or slightly lobed; style terminal. Fruit drupaceous or coriaceous, rarely dehiscent.

TRIBE VI. TODDALIEÆ (see ZANTHOXYLEÆ, p. 323).—Flowers regular, often polygamo-diœcious. Petals and stamens free. Disk free. Albumen usually fleshy. Cotyledons flat.— Shrubs or trees, usually tropical. *Toddalia, Hostia, Acronychia, Casimiroa, Phellodendron, Ptelea, Skimmia.*

TRIBE VII. AURANTIEÆ (see order AURANTIACEÆ, p. 318).—Flowers regular, ☿. Petals and stamens free or connate. Ovules 1–2 or more. Berry pulpy, usually with a thick rind. Seeds exalbuminous. Leaves 1–3-foliolate or pinnate.—Tropical trees and shrubs. *Glycosmis, Triphasia, Clausena, Atalantia, Feronia, Micromelum, Murraya, Luvunga, Citrus, Ægle.*

As thus extended, *Rutaceæ* embrace 83 genera and 650 known species.—ED.]

LI. *RUTACEÆ.*

(RUTACEARUM *pars, Jussieu.*—RUTEÆ, *Adr. Jussieu.*—RUTACEÆ, *Bartling.*)

PETALS 4–5, *hypogynous.* STAMENS *generally double the number of the petals, inserted at the base of a thick disk.* OVARY 3–5-*lobed, cells* 2–4–∞ *-ovuled.* FRUIT *a capsule.* EMBRYO *albuminous, more or less curved.*—*Scented* PLANTS. LEAVES *alternate, usually pellucid-dotted, exstipulate.*

LI. RUTACEÆ.

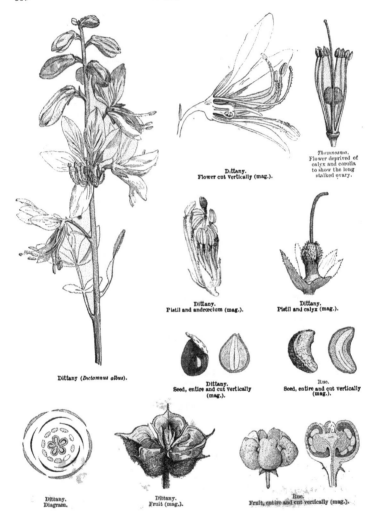

Dittany.
Flower cut vertically (mag.).

Thamnosma,
Flower deprived of
calyx and corolla
to show the long
stalked ovary.

Dittany.
Pistil and andrœcium (mag.).

Dittany.
Pistil and calyx (mag.).

Dittany (*Dictamnus albus*).

Dittany.
Seed, entire and cut vertically
(mag.).

Rue.
Seed, entire and cut vertically
(mag.).

Dittany.
Diagram.

Dittany.
Fruit (mag.).

Rue.
Fruit, entire and cut vertically (mag.).

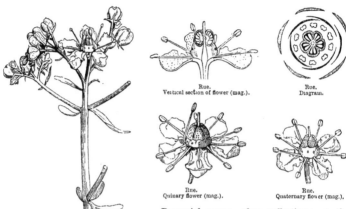

Rue.
Vertical section of flower (mag.).

Rue.
Diagram.

Rue.
Quinary flower (mag.).

Rue.
Quaternary flower (mag.).

Rue (*Ruta graveolens*).

Perennial HERBS, often suffruticose at the base. LEAVES alternate, simple, variously cut, rarely quite entire, usually pellucid-dotted or tubercled; *stipules* 0, or replaced by 2 bristle-like teeth at the base of the leaves (*Peganum*). FLOWERS ☿, regular, terminal, in racemes or a corymb, yellow, or sometimes white. CALYX persistent, 4-5-partite, æstivation imbricate. PETALS 4-5, inserted at the base of a shorter or longer gynophore, æstivation imbricate. STAMENS inserted with the petals, generally double, sometimes triple (*Peganum*), rarely equal (*Thamnosma*) in number; *filaments* filiform, free, or sometimes shortly monadelphous, base often dilated; *anthers* introrse, connective sometimes glandular at the tip (*Haplophyllum*), dehiscence longitudinal. OVARY deeply 2-3-5-lobed, 2-3-5-celled, seated on a gynophore usually dilated at the base into a glandular disk; *styles* central, sometimes distinct at the base and top (*Bœnninghausenia*), usually connate, stigmatiferous at the top, or on the angles; *ovules* 3-4-∞ in each cell, inserted on a projecting placenta at the inner angle of the cell, 2-seriate, anatropous or semi-anatropous. FRUIT a capsule, sometimes opening in 3-4 loculicidal valves (*Peganum*), sometimes in 4-5 lobes opening at the top (*Ruta*), sometimes fleshy and indehiscent (*Ruteria*), sometimes separating into cocci (*Dictamnus, Bœnninghausenia*), endocarp sometimes crustaceous or cartilaginous, and separating from the epicarp (*Dictamnus*). SEEDS pendulous, or ventrally fixed; *testa* crustaceous or spongy, pitted and granular; *albumen* fleshy. EMBRYO axile, curved or rarely straight (*Dictamnus*); *radicle* superior.

PRINCIPAL GENERA.

| *Ruta. | Peganum. | *Dictamnus. | *Bœnninghausenia. |

Rutaceæ are very closely allied to *Diosmeæ*, which can only be distinguished from them by the woody stem, the 2-ovuled ovarian cells, and the usually straight embryo. They also approach *Zygophylleæ*,

Oxalideæ, and *Simarubeæ* (see these families). Bentham and Hooker fil. have united with them *Diosmeæ*, *Aurantieæ* and *Zanthoxyleæ* (see these families). *Rutaceæ* all belong to the Old World; they especially abound in the north temperate hemisphere, the shores of the Mediterranean, and South Siberia; and they become very rare towards the poles and equator. *Bœnninghausenia* inhabits the Himalayas and Japan.

Rutaceæ owe their stimulating properties to a bitter substance, a resinous acrid principle, and especially to a volatile oil, secreted by the glands of the leaves and flowers. The Rue (*Ruta graveolens*), a native of the Mediterranean region, and cultivated in all gardens, is remarkable for its strong smell and acrid taste, and its essence, obtained by distillation, is employed as a sudorific, vermifuge, and emmenagogue. Vinegar of Rue was regarded during many centuries as a certain remedy against the plague. The Romans used Rue as a condiment, as do the Germans still. *Ruta montana*, which grows in Spain, is so extremely acrid that it produces erysipelas and ulcerous pustules on the skin of those who gather it. *Haplophyllum tuberculatum* is so much less acrid that the Egyptian women bruise its leaves in water, and use it as a hair-wash. The peduncles and flowers of the European Dittany (*Dictamnus albus*) are laden with pedicelled glands which secrete an abundant volatile oil so copiously that the plant ignites at the approach of a candle; its resinous scented and bitter root is tonic and stimulating. *Peganum Harmala* grows in sandy soil in the Mediterranean region; its smell is repulsive and its taste acrid and bitter; the Turks use its seeds as a condiment, and obtain a red dye from them.

LII. *AURANTIACEÆ.*

(AURANTIORUM *sectio*, *Jussieu.*—HESPERIDEARUM *sectio*, *Ventenat.*—AURANTIACEÆ, *Correa.*)

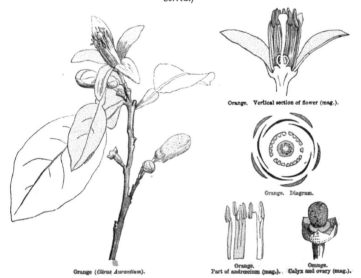

Orange. Vertical section of flower (mag.).

Orange. Diagram.

Orange (*Citrus Aurantium*).

Orange.
Part of andrœcium (mag.).

Orange.
Calyx and ovary (mag.).

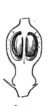

| Orange.
Transverse section of
fruit. | Orange.
Calyx and
andrœcium (mag.). | Orange.
Calyx and
pistil (mag.). | *Bergera.*
Flower without
corolla (mag.). | *Bergera.*
Vertical section of
ovary (mag.). | *Bergera.*
Pistil
(mag.). |

Orange. Seed, entire and cut (mag.).

PETALS *hypogynous, œstivation imbricate.* STAMENS *hypogynous, double or multiple the number of the petals, free, monadelphous or polyadelphous.* OVARY *several-celled.* OVULES *1–2 or more in each cell, pendulous or horizontal, anatropous.* FRUIT *a 1–many-seeded berry.* EMBRYO *exalbuminous.*—STEM *woody.* LEAVES *alternate, imparipinnate or 1-foliolate.*

TREES or SHRUBS, usually glabrous; bark, leaves, calyx, corolla, filaments and epicarp covered with vesicles containing a volatile oil. LEAVES persistent, alternate, compound, often 1-foliolate by suppression, with the leaflet jointed, or the top an often winged petiole; *stipules* 0; *buds* axillary, the outer often spinescent. FLOWERS generally ☿, regular, terminal, solitary, or corymbose or racemed. CALYX short, urceolate or campanulate, 4–5-toothed or -fid, rarely 3-fid (*Triphasia*), sometimes nearly entire, imbricate in æstivation, marcescent. PETALS inserted below the ovary at the base of a stipitiform, annular or cupuliform disk, free or slightly connate at the base, imbricate in æstivation, deciduous. STAMENS inserted on the receptacle, double or multiple the number of the petals; *filaments* free, or monadelphous at their base or to the middle, or polyadelphous, linear-subulate, usually dilated, thinner above, equal or the alternate shorter; *anthers* introrse, dorsi- or basifixed, incumbent, dehiscence longitudinal. OVARY free, sometimes girt by the cupuliform disk, 5–many-celled; *style* terminal, simple, stout; *stigma* capitate, undivided or lobed; *ovules* inserted at the inner angle of the cells, solitary or 2, collateral or superimposed, or numerous and 2-seriate, pendulous or rarely horizontal, anatropous. BERRY fleshy or dry, bark thick, indehiscent, of 2 or several usually 1-seeded cells filled with mucilage or vesiculose cellules. SEEDS inverted or sub-horizontal, testa membranous, raphe branching, chalaza coloured; *albumen* 0. EMBRYOS (sometimes several in each cell) straight; *cotyledons* either fleshy, amygdaloid, plano-convex, often unequal, or thick, green, lobulate, auricled; *radicle* short, near the hilum, superior.

PRINCIPAL GENERA.

*Triphasia.	*Limonia.	*Murraya.	*Cookia.	*Citrus.

Aurantiaceæ have been united by Lindley and Hooker and Bentham to *Rutaceæ*, *Diosmeæ*, and *Zanthoxyleæ*, with which they are, in fact, closely allied, these latter being scarcely separated, except by their more or less distinct carpels with a basilar or ventral style, by their capsular fruit and albuminous seed ; and even these differences disappear in many *Diosmeæ* which are exalbuminous, and in some *Zanthoxyleæ* which have a fleshy fruit ; a genus of the latter family, *Skimmia*, has in fact been classed among *Aurantiaceæ*, and placed in the genus *Limonia*. *Aurantiaceæ* approach *Meliaceæ* in the petals inserted at the base of a hypogynous disk, in the connate filaments, several-celled ovary, simple style, fleshy fruit, woody stem and alternate leaves ; but in *Meliaceæ*, besides the difference in habit, the leaves are eglandular, the sepals are more or less distinct, and the seed is sometimes albuminous. *Cedrelaceæ* are connected with *Aurantiaceæ* through the genus *Flindersia*, the leaves of which are dotted, the stamens inserted at the base of the disk, and the embryo albuminous ; the principal difference is the capsular fruit of *Flindersia*. *Humiriaceæ* are also connected with *Aurantiaceæ* by most of the characters, and are scarcely separated save by their anthers, drupaceous fruit, and abundant albumen. *Burseraceæ* are also related to them, especially the genus *Amyris*, and are distinguished by their drupaceous fruit. Finally, an affinity has been noticed by Planchon between *Aurantiaceæ* and *Hypericineæ*, founded on the glandular leaves and flowers, hypogynous petals, polyadelphous stamens, sometimes fleshy fruit (*Vismia*), and exalbuminous embryo. The same observation holds good for *Myrtaceæ*, and especially for the genera with free ovaries (*Fremya*).

Aurantiaceæ are almost wholly tropical Asiatic, but various species are now cultivated in the warm regions of both continents.

The wide-spread celebrity of the genus *Citrus* from the earliest times is firstly due to the free acids (citric and malic) contained in the cells of its parenchyma, which fill the fruit ; and secondly, to the sweet and pungent volatile oil secreted by the glands which abound in nearly all parts of the plant. The acids are used in domestic economy and in medicine, as refreshing, laxative, and antiseptic. A small quantity of the aromatic principle dissolved in water, by infusion of the leaves or distillation of the flowers, affords a stimulating antispasmodic. The volatile oil, obtained by distillation of the flowers and epicarp, is employed in perfumery, either mixed with fat as a pomade, or dissolved in alcohol as the cosmetic known as *eau de Cologne.*

The Orange (*Citrus Aurantium*) is universally sought for its acid-sweet fruit. The Seville Orange (*C. communis*) has a bitter fruit, but it is not less useful ; its leaves are used in infusions ; its flowers yield the distilled water so much used in medicine, and the volatile oil called essence of Neroli. From its young fruits, gathered soon after flowering, is distilled a Neroli oil called essence 'de petit grain.' The epicarp of the Seville or bitter Orange is used in the preparation of a dye ; it is also made into a syrup and a marmalade, and is one of the principal ingredients of the highly esteemed liqueur Curacoa. The 'Cedrat'[1] (*C. medica*[1]) bears large oblong fruits with a rough surface ; the rind yields a perfume by expression or distillation ; the inner rind is thick and fleshy, and a pleasant preserve is prepared from it. The Lemon (*C. Limonum*) has an ovoid fruit, mamillate at the top ; its rind adheres strongly to the very acid pulp. The variety of this species known under the name of Citron[2] yields the medicinal preparation called syrup of lemons.

The Lime (*C. Limetta*) bears a globular berry with mild and insipid juice ;[3] another species, perhaps only a variety of the preceding, *C. Bergamota*, produces small pyriform fruits, the pulp of which is acid and bitter ; but their rind is thin, of a golden yellow, and filled with a sweet essence ; formerly sweetmeats called *bergamottes* were made of it ; now it is only used for the expression of essence of Bergamot. The fruits of *C. myrtifolia* and *C. deliciosa*, preserved in brandy, are called 'Chinois.' The Tangerine Orange is well known for its peculiar scent, insipid juice, and flaccid loose rind. The berries of some other genera from China and Japan are eatable ; as *Glycosmis citrifolia*, *Triphasia trifoliata*, *Ægle marmelos*, *Cookia punctata*, &c.

---

[1] This is best known in England as the Citron.—ED.

[2] The Citron of the English markets is *C. medica*.

[3] The juice of the Lime is the tartest of the genus, and forms a well-known ingredient of Glasgow punch.—ED.

LIII. *DIOSMEÆ.*

(RUTACEARUM *genera, Jussieu.*—DIOSMEARUM *genera, Br.*—DIOSMEÆ, *Adr. Jussieu.*)

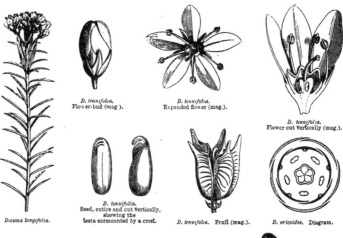

D. *tenuifolia.*
Flower-bud (mag).

D. *tenuifolia.*
Expanded flower (mag.).

D. *tenuifolia.*
Flower cut vertically (mag.).

D. tenuifolia.
Seed, entire and cut vertically,
showing the
testa surmounted by a crest.

Diosma longifolia.

D. tenuifolia. Fruit (mag.).

D. ericoides. Diagram.

PETALS 5-4, *inserted on a hypo-gynous disk, imbricate.* STAMENS *equal and alternate with the petals, or double the number.* OVARIES *distinct or co-herent, 2-ovuled.* FRUIT *a capsule, divid-ing into cocci.* SEEDS *albuminous or exalbuminous.* EMBRYO *usually straight.* —*Scented PLANTS.* STEM *woody.* LEAVES *usually glandular-dotted, exstipulate.*

D. tenuifolia.
Seed deprived of its testa,
showing the inner chalaza
(mag.).

D. alba.
Seed with lateral
hilum (mag.).

Diosma.
Exalbuminous
embryo (mag.).

SMALL TREES OR SHRUBS. LEAVES opposite or alternate, coriaceous, usually simple, sometimes 3-foliolate (*Spiranthera, Zieria,* &c.), rarely pinnate (*Boronia*), usually glandular-dotted ; *stipules* 0, or replaced by glands at the base of the petioles. FLOWERS ☿, or very rarely imperfect by suppression (*Empleurum*), regular, axillary or terminal, solitary, or in an umbel, corymb or panicle, rarely in an involu-crate capitulum (*Diplolœna*). CALYX 4–5-fid or -partite, imbricate in æstivation. PETALS 4–5, inserted under a free or rarely sub-perigynous disk, imbricate in æstiva-tion, usually free, rarely connate, or conniving by their dilated bases into a cylin-drical tube, and then valvate in æstivation (*Correa, Nematolepis,* &c.), or 0 (*Em-pleurum*). STAMENS inserted with the petals, generally equal, rarely double in

Y

number (when those opposite the petals are imperfect, or shorter than the others) ; *filaments* subulate, usually free, rarely monadelphous (*Erythrochiton*), or adherent to the petals .(*Galipea*, &c.) ; *anthers* introrse, 2-celled, dorsifixed near their base ; cells parallel, opening longitudinally ; connective often prolonged into a glandular appendage (*Crowea, Eriostemon, Philotheca*, &c.). CARPELS 5–3, rarely 1 (*Empleurum*), sessile or borne on a gynophore, girt with a disk at their base, or buried in this disk, united into a deeply-lobed ovary, lobes distinct, cohering only by the styles ; *styles* as many as ovaries, springing from their ventral edge, distinct at the base, connate above ; *stigmas* united in a head, lobed or 3–5-furrowed ; ovules 2 in each cell, inserted at the middle of the ventral suture, collateral or superimposed. CAPSULES of 3–5 cocci, distinct or connate at the base, 1-seeded by suppression ; *epicarp* dry, sub-coriaceous, glandular-dotted or muricate ; *endocarp* smooth, cartilaginous, often elastic and 2-valved. SEEDS oblong or sub-reniform ; *testa* cartilaginous, smooth ; *albumen* 0, or fleshy. EMBRYO usually straight, rarely curved (*Almeida, Spiranthera*, &c.) ; *cotyledons* flat or crumpled, enveloping each other, foliaceous in germination ; *radicle* usually superior, straight or inflexed.

TRIBE I. *EUDIOSMEÆ.*

Fertile stamens equal in number with the petals, often alternating with as many staminodes, inserted below the free edge of a disk which lines the calyx-tube.   Carpels 2-ovuled.   Testa coriaceous or sub-crustaceous.   Embryo exalbuminous, straight.—Heath-like shrubs with alternate or opposite simple coriaceous small and imbricate leaves, rarely trees with large leaves (*Calodendron*).

PRINCIPAL GENERA.

*Calodendron.	*Coleonema.	*Acmadenia.	*Adenandra.
*Diosma.	*Agathosma.	*Barosma.	

TRIBE II. *BORONIEÆ.*

Stamens hypogynous, double in number to the petals, and all fertile, rarely equal and perigynous (*Zieria*). Disk free, cupuliform or annular, sometimes inconspicuous. Testa crustaceous. Albumen fleshy. Embryo straight, cylindric.—Shrubs, rarely trees. Leaves simple or 3-foliolate or pinnate.

PRINCIPAL GENERA.

*Correa.	*Diplolæna.	*Phebalium.	*Crowea.
*Eriostemon.	*Boronia.	*Zieria.	

TRIBE III. *CUSPARIEÆ.*

Flowers often irregular, corolla often tubular. Stamens 5, some often imperfect, sometimes hypogynous, sometimes connate, or adnate to the corolla. Disk usually cupular (*Almeida, Naudinia, Ticorea*, &c.), or urceolate (*Erythrochiton*), sometimes columnar (*Spiranthera*), rarely depressed (*Galipea*), or squamiform, unilateral (*Monnieria*). Carpels 2-ovuled; *testa* coriaceous; *embryo* exalbuminous, curved; *cotyledons* crumpled, convolute.—Leaves usually alternate, 1–3-foliolate.

PRINCIPAL GENERA.

| Monnieria. | Galipea. | *Erythrochiton. | Almeidea. | Spiranthera. |

Diosmeæ cannot be separated from *Rutaceæ* (see this family); the genus *Dictamnus* unites them by its irregular flowers, straight embryo, 4-ovuled ovarian cells, albuminous seed, herbaceous stem, and imparipinnate leaves. *Zanthoxyleæ* are equally near, in their regular flowers, polypetalous hypogynous imbricate isostemonous or diplostemonous corolla, free or nearly free carpels with 2-ovuled cells, often elastic endocarp, straight (rarely curved) embryo, woody stem, and usually pellucid-dotted simple or compound alternate and opposite exstipulate leaves; they scarcely differ save in the diclinous flowers and fruit. *Diosmeæ* have also some analogy with *Simarubeæ* (see this family).

Eudiosmeæ all belong to South Africa; *Boronieæ* to Australia, and *Cusparieæ* to tropical America. The volatile oil and the aromatic resin of *Eudiosmeæ* are stimulating and antispasmodic, and many species are thus used by the natives and colonists of the Cape. The leaves of *Barosma crenata* also contain a principle (*diosmine*), owing to which they are now admitted into European medicine as diuretics and diaphoretics, in affections of the kidneys and bladder, in rheumatism, and even cholera. The properties of *Boronieæ* are little known; in Australia the leaves of *Correa* are made use of as tea. In the bark of some *Cusparieæ* there is a bitter alkaloid (*angusturine* or *cusparine*), united with a soft resin and a little volatile oil, which places them, after quinine, amongst the most efficacious tonics and febrifuges. According to some this bark is yielded by the *Galipea Cusparia*, a large tree forming vast forests on the banks of the Orinoco; according to others, by *G. officinalis*, a shrub of the same country. The bark of *Ticorea febrifuga*, a Brazilian and Guiana tree, is also recommended as a substitute for quinine. The aromatic and acrid root of *Monnieria trifolia*, a plant distinguished from its woody congeners by its herbaceous stem, is praised in America as a diaphoretic, diuretic and alexipharmic.

LIV. *ZANTHOXYLEÆ*, *Adr. Jussieu*.

(DIOSMEARUM *genera et* PTELEACEÆ, *Kunth.*—XANTHOXYLACEÆ, *Lindl.*)

FLOWERS *regular, very often polygamo-diœcious*. PETALS 5–4–3, *imbricate or valvate, inserted at the base of a free disk*. STAMENS *inserted with the petals, equal or double them in number*. CARPELS *distinct or connate, 2-ovuled*. FRUIT *a drupe or samara, or of separate dehiscent cocci*. ALBUMEN *fleshy, rarely 0*. EMBRYO *axile, straight or curved.*—STEM *woody*. LEAVES *exstipulate, usually glandular-dotted.*

Large or small TREES or SHRUBS, unarmed, thorny, or aculeate. LEAVES alternate or opposite, rarely whorled (*Pitavia, Pilocarpus*), generally pinnate, or imparipinnate, often 1-foliolate by suppression of the lateral leaflets (*Zanthoxylum, Evodia*, &c.), usually pellucid-dotted, rarely simple (*Skimmia*), petiole sometimes margined or winged (*Zanthoxylum*); *stipules* 0. FLOWERS usually imperfect, regular, axillary or terminal, mostly arranged in axillary cymes, panicles or corymbs, rarely in racemes or spikes (*Pilocarpus, Esenbeckia*), very rarely solitary (*Astrophyllum*). CALYX persistent or deciduous, 4-5- (rarely 3-) partite (*Zanthoxylum*), æstivation imbricate, or rarely valvate (*Melanococca*). PETALS inserted at the base of a free disk, in a ring or cushion, sometimes inconspicuous, imbricate or valvate in

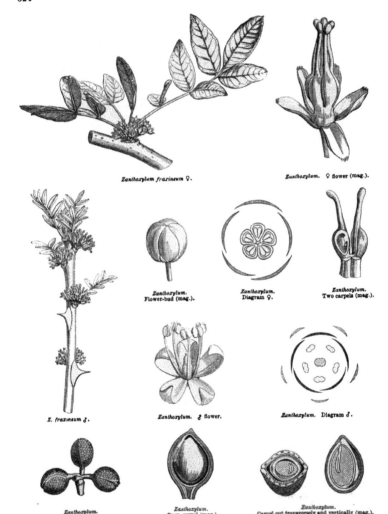

Zanthoxylum fraxineum ♀.

Zanthoxylum. ♀ flower (mag.).

Zanthoxylum.
Flower-bud (mag.).

Zanthoxylum.
Diagram ♀.

Zanthoxylum.
Two carpels (mag.).

Z. fraxineum ♂.

Zanthoxylum. ♂ flower.

Zanthoxylum. Diagram ♂.

Zanthoxylum.
Fruit.

Zanthoxylum.
Open carpel (mag.).

Zanthoxylum.
Carpel cut transversely and vertically (mag.).

æstivation, deciduous, rarely 0 (*Zanthoxylum*). FLOWERS ♂ : STAMENS inserted with the petals, equal and alternate, or double in number; *filaments* filiform or subulate, free; *anthers* dehiscing longitudinally. OVARY rudimentary, stipitate, sometimes 0. FLOWERS ♀ : STAMENS 0 or rudimentary, inserted at the base of the disk, shorter than the ovary. CARPELS equal with the petals or fewer, distinct, or connate at the base, or throughout into a several-celled ovary; *ovules* 2 in each cell, superimposed or collateral, very rarely solitary (*Skimmia*), usually anatropous or semi-anatropous. FRUIT sometimes simple, 2–5-celled, fleshy (*Toddalia, Acronychia, Skimmia*), or rarely a samara (*Ptelea*), usually a capsule opening in dehiscent shells by their inner edge, sometimes compound, formed of several drupes (*Melanococca,* &c.) or capsules (*Zanthoxylum, Boymia*); *endocarp* sometimes separating elastically. SEEDS pendulous; *testa* coriaceous or crustaceous, usually smooth, shining; *albumen* fleshy, rarely 0 (*Pilocarpus, Esenbeckia, Casimiroa*). EMBRYO axile, straight or slightly curved; *cotyledons* oval or oblong, flattened; *radicle* shorter than the cotyledons, superior.

PRINCIPAL GENERA.

*Skimmia.	*Zanthoxylum.	Pitavia.	Toddalia.	*Ptelea.	Acronychia.

Zanthoxyleæ are allied to *Rutaceæ, Diosmeæ* and *Simarubeæ* (see these families). They also approach *Burseraceæ* in the woody stem, pellucid-dotted compound exstipulate leaves, often polygamo-diœcious flowers, æstivation of the calyx and corolla, annular or cupuliform disk, diplostemony, 2-ovuled ovarian cells, and drupaceous fruit. *Zanthoxyleæ* also present more than one point of analogy with *Anacardiaceæ*; but they have most affinity with *Aurantiaceæ*, so much so that a *Skimmia* has been described as a species of *Limonia*.

Zanthoxyleæ inhabit the tropical regions of Asia, and especially of America; they are less numerous in extra-tropical America, South Africa, and Australia. *Zanthoxylum* belongs to the tropical zone of both worlds; *Skimmia* to Japan and the Himalayas; *Toddalia* to tropical Asia and Africa; *Ptelea* to North America. Australia possesses the genera *Acronychia, Pentaceras, Medicosma,* &c.

Some species of this family are medicinal: the bark of *Zanthoxylum,* and especially that of the root, contains a bitter crystallizable principle (*zanthopicrite*), an acrid resin, and a yellow colouring matter. The aromatic root of *Z. nitidum* is classed in China amongst sudorifics, emmenagogues, and febrifuges; the leaves contain a little volatile oil, whence they are used as a condiment. *Z. Budrunga* in India is similarly employed. *Z. Rethsa* grows on the Indian mountains; its young fruits have the taste of orange-rind, and its seeds that of black pepper. The capsules of *Z. piperitum,* all parts of which have an acrid aroma, afford the Japan pepper of commerce. The bark of *Z. fraxineum,* a native of North America, is a reputed diuretic and sudorific; it is also chewed to excite salivation and to alleviate toothache. *Z. ternatum* and *Clava-Herculis* are similarly used; they are West Indian shrubs which yield a yellow dye; their bark is regarded as antisyphilitic and their bitter astringent leaves as a vulnerary. Finally, the seeds of some species are employed to poison fish. *Ptelea trifoliata,* commonly called Three-leaved or Samarian Elm, is a small Carolina tree cultivated in Europe; its leaves are considered in North America as a vermifuge, and detergent for ulcers. Its bitter aromatic capsules are a substitute for hops in brewing; but this substitution is not without its inconveniences. *Toddalia aculenta,* a shrub of tropical Asia, all parts of which contain an aromatic bitter acrid principle, is used by the natives of the Indian Archipelago as a stomachic, a febrifuge, and as seasoning for food.

LV. *SIMARUBEÆ*, D.C.

(Simarubaceæ, *Richard.*—Simarubeæ, *D.C., Planchon, Benth. et Hook. fil.*)

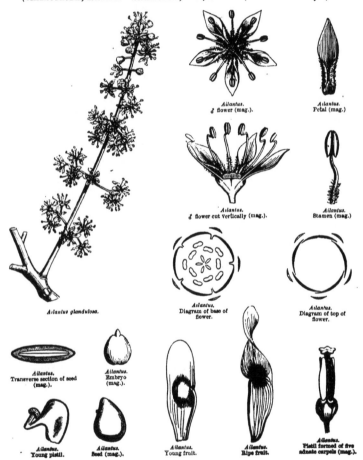

Ailantus.
♂ flower (mag.).

Ailantus.
Petal (mag.)

Ailantus.
♂ flower cut vertically (mag.).

Ailantus.
Stamen (mag.)

Ailantus glandulosa.

Ailantus.
Diagram of base of flower.

Ailantus.
Diagram of top of flower.

Ailantus.
Transverse section of seed (mag.).

Ailantus.
Embryo (mag.).

Ailantus.
Young pistil.

Ailantus.
Seed (mag.).

Ailantus.
Young fruit.

Ailantus.
Ripe fruit.

Ailantus.
Pistil formed of five adnate carpels (mag.).

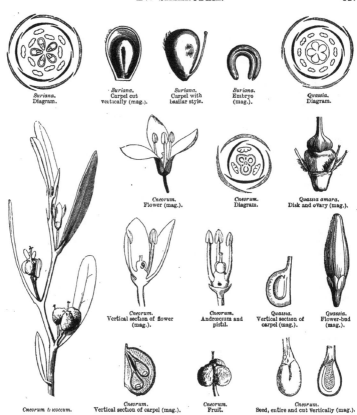

Suriana.
Diagram.

Suriana.
Carpel cut
vertically (mag.).

Suriana.
Carpel with
basilar style.

Suriana.
Embryo
(mag.).

Quassia.
Diagram.

Cneorum.
Flower (mag.).

Cneorum.
Diagram.

Quassia amara.
Disk and ovary (mag.).

Cneorum.
Vertical section of flower
(mag.).

Cneorum.
Androecium and
pistil.

Quassia.
Vertical section of
carpel (mag.).

Quassia.
Flower-bud
(mag.).

Cneorum tricoccum.

Cneorum.
Vertical section of carpel (mag.).

Cneorum.
Fruit.

Cneorum.
Seed, entire and cut vertically (mag.).

FLOWERS *diclinous or polygamous.* PETALS *3–5, rarely 0, hypogynous, imbricate or valvate.* STAMENS *inserted at the base of a hypogynous disk, as many or twice as many as the petals, rarely* ∞. CARPELS *2–5, free, or connate into a deeply-lobed 1–5-celled ovary.* OVULES *usually solitary in each cell.* FRUIT *a drupe, capsule or samara.* SEEDS *pendulous, albuminous or not.* EMBRYO *straight or bent.*—STEM *woody.* LEAVES *usually alternate and pinnate, not dotted.*

SHRUBS or scentless TREES; bark often bitter, sometimes extremely so. LEAVES alternate, or rarely opposite (*Brunellia, Cneoridium*), pinnate, rarely 1–3-foliolate

(*Harrisonia, Brunellia*), or 2-foliolate (*Balanites*), or simple (*Cneorum, Castela, Soulamea,* &c.), not dotted, very rarely stipulate (*Brunellia, Irvingia, Cadellia*). Flowers diclinous or polygamous, regular, usually axillary, panicled or racemed, or rarely the ♂ in a spike, the ♀ solitary (*Picrodendron*). CALYX 3–5-lobed or -partite, regular, very rarely sub-bilabiate (*Hannoa*), æstivation imbricate or valvate. PETALS 3–5, very rarely 0 (*Brunellia, Amaroria*), free, or very rarely connivent into a tube (*Quassia*), hypogynous ; æstivation imbricate, valvate or contorted. DISK annular, cupuliform or tumid, entire or lobed, sometimes elongated into a column (*Quassia, Cneorum*), rarely inconspicuous (*Suriana, Picrolemma*), or 0 (*Spathelia, Eurycoma, Cadellia*). STAMENS inserted at the base of the disk, double the number of the petals, or equal and alternate, very rarely opposite to the petals (*Picrolemma, Picramnia*), very rarely more than 10 (*Mannia*) ; *filaments* free, naked or more often hairy, or furnished with a scale at the base ; *anthers* oblong, usually introrse, 2-celled, dehiscence longitudinal. CARPELS 2–5, rarely solitary (*Cneoridium, Amaroria*), completely free (*Brunellia, Suriana*), or connate at the base only, or cohering by the styles only, or completely united into a 2–5-celled ovary ; *styles* 2–5, free at the base and top ; *stigmas* free or cohering in a head ; *ovules* usually solitary in each cell, sometimes geminate, very rarely 4–5 (*Dictyoloma*), or ∞ (*Kœberlinia*), fixed to the inner angle of the cell, anatropous, raphe ventral and micropyle superior, very rarely ascending with a dorsal raphe and inferior micropyle (*Cneoridium*). FRUIT usually of fleshy or dry drupes, rarely of 2-valved capsules (*Dictyoloma, Brunellia*), or indehiscent (*Soulamea*), very rarely of samaras (*Ailantus*). SEEDS pendulous, usually solitary ; *testa* membranous ; *albumen* usually 0 or scanty, rarely copious (*Cneorum, Brucea, Brunellia, Spathelia*). EMBRYO straight or rarely curved (*Cneorum, Suriana, Dictyoloma*) ; *cotyledons* plano-convex, or flat, rarely coiled or folded (*Harrisonia, Cadellia*) ; *radicle* superior.

TRIBE I. *EUSIMARUBEÆ.*

Carpels free or nearly so.

PRINCIPAL GENERA.

*Quassia.	Simaba.	Hannoa.	Simaruba.	*Ailantus.
Samadera.	Castela.	Cneorum.	Brucea.	Prinera.
Euryloma.	Dictyoloma.	Cadellia.	Suriana.	Brunellia.

TRIBE II. *PICRAMNIEÆ.*

Carpels united into a non-lobed ovary, 2-5-1-celled.

PRINCIPAL GENERA.

Irvingia.	Harrisonia.	Balanites.	Spathelia.	Picramnia.	Picrodendron.

The family of *Simarubeæ*, as reconstituted by Bentham and Hooker fil. and Planchon, only differs from *Rutaceæ, Diosmeæ, Zanthoxyleæ,* and *Hesperideæ* in its eglandular leaves, often bitter bark, and filaments usually furnished with a scale ; characters which, although not of a high intrinsic value;

naturally connect the genera of *Simarubeæ*, and clearly distinguish them from the above-named families. The affinity is less close with *Zygophylleæ* (see this family). They appear somewhat nearer to *Ochnaceæ* (which see).

Simarubeæ mostly grow in the torrid zone. *Quassia, Simaba, Simaruba, Castela, Picramnia,* &c., belong to tropical America ; *Hannoa, Samadera, Brucea, Balanites,* to tropical Africa ; the three latter also inhabit Asia, as well as *Picrasma* and *Ailantus. Suriana* is found on sea-coasts throughout the intertropical zone. *Soulamea, Eurycoma,* and *Harrisonia,* are natives of the Malayan Archipelago and Pacific Islands ; the latter also grows in Australia, as does *Cadellia. Cneorum* inhabits the Mediterranean region and the Canaries.

Many plants belonging to the genera of the first tribe (*Quassia, Simaba, Simaruba*) contain a peculiar principle, extremely bitter, in combination with salts, a resinous matter, and a small quantity of volatile oil, which gives them tonic properties, and renders them very digestive. *Quassia amara* occupies the first rank amongst bitter medicines. The bark of the root and trunk of *Simaruba guianensis* and *amara* yields the Simaruba of the druggist, the virtues of which rival those of Quassia. *S. versicolor* is much renowned among the Brazilians, who apply a decoction of its bark and leaves to snake-bites and syphilitic exanthema. The *Simabas* of Guiana and Brazil, and the *Samaderas* of India, are also extremely bitter, and possess similar properties. The inner bark of *Brucea antidysenterica,* an Abyssinian shrub, is regarded as an admirable medicine in cases of dysentery and obstinate intermittent fevers. *B. sumatrana,* which grows in the Moluccas and India, possesses the same properties. *Ailantus glandulosa,* a native of China, and naturalized in the temperate parts of Europe, is commonly called Japan varnish, a name which perpetuates an error (see *Terebinthaceæ,* p. 363). [*Balanites ægyptiaca,* a spinous shrub of Syria, Arabia, and North Africa, bears drupes which are acrid, bitter and purgative when young, but edible when old. Its seeds yield a fatty oil, the *zachun* of the Arabs.—ED.]

LVI. *OCHNACEÆ, D.C.*

(OCHNACEÆ, *D.C.—Planchon, Benth. et Hook. fil.*)

SEPALS 4--5. PETALS *as many or twice as many.* STAMENS *double or multiple the number of the petals ; anthers dehiscing at the top.* CARPELS 4–5 *or more, united at the base by the gynobasic style,* 1-*ovuled.* FRUIT *fleshy.* ALBUMEN *scanty or* 0.—STEM *woody.* LEAVES *alternate, stipulate.*

SHRUBS or TREES with watery juice. LEAVES alternate, stipulate, glabrous, simple, or very rarely pinnate (*Godoya*), coriaceous, shining, often toothed, margin sometimes thickened, midrib strong, the lateral nerves close, parallel. FLOWERS ☿, usually panicled, rarely axillary and solitary, or fascicled. SEPALS 4–5, free, imbricate, very often scarious, concave and striate. PETALS hypogynous, 5, rarely 3–4 or 10 (*Ochna*), free, longer than the calyx, deciduous, spreading, æstivation imbricate or contorted. DISK elongating after flowering, never annular nor glandular, often inconspicuous or 0 ; *staminodes* 1–3-seriate, accompanying the stamens in some genera (*Wallacea, Pœcilandra, Blastemanthus*). STAMENS inserted at the base or top of the torus, 4–5 or 8 or 10 or ∞, erect, equal or unequal, unilateral or declinate ; *filaments* free, short, persistent ; *anthers* linear-elongate, basifixed, cells straight or flexuous, usually opening by terminal pores. OVARY central or excentric, short and deeply 2–10-lobed, or elongated and 2–10-celled, rarely 1-celled, with 3 parietal placentas (*Wallacea*) ; *style* central, gynobasic, simple, subulate, sharp, straight or curved, rarely divided in as many branches as carpels (*Ochna*) ; *stigma*

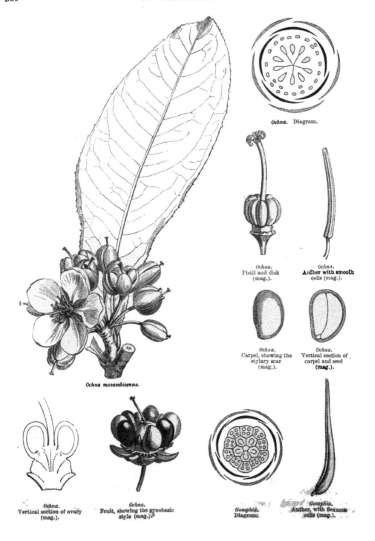

Ochna. Diagram.

Ochna.
Pistil and disk
(mag.).

Ochna.
Anther with smooth
cells (mag.).

Ochna.
Carpel, showing the
stylary scar
(mag.).

Ochna.
Vertical section of
carpel and seed
(mag.).

Ochna mozambicensis.

Ochna.
Vertical section of ovary
(mag.).

Ochna.
Fruit, showing the gynobasic
style (mag.).

Gomphia.
Diagram.

Gomphia.
Anther, with flexuous
cells (mag.).

simple, terminal; *ovules* solitary in each cell (*Ochna, Gomphia,* &c.), or geminate (*Euthemis*), or numerous (*Luxemburgia, Godoya,* &c.), ascending, rarely pendulous, raphe ventral, and micropyle superior. FRUIT of 3–10 1-seeded drupes, whorled on the enlarged gynophore (*Ochna, Gomphia,* &c.), or 2–4-lobed, 1–4-seeded, coriaceous, indehiscent (*Elvasia*), or fleshy with 5 nucules (*Euthemis*), or a 1-celled capsule, coriaceous (*Luxemburgia*), or woody, 2–5-celled, and septicidal (*Godoya, Pœcilandra,* &c.). SEEDS with fleshy albumen (*Luxemburgia, Pœcilandra, Cespedesia, Euthemis,* &c.), or exalbuminous (*Ochna, Gomphia, Elvasia,* &c.); *testa* usually membranous, sometimes winged or margined (*Luxemburgia, Pœcilandra*). EMBRYO large, subcylindric, straight, or very rarely curved (*Brackenridgea*); *cotyledons* plano-convex (*Ochna, Gomphia,* &c.) or linear (*Luxemburgia, Pœcilandra,* &c.); *radicle* inferior or superior.

PRINCIPAL GENERA.

Ochna.	Gomphia.	Euthemis.	Luxemburgia.
Godoya.	Blastemanthus.	Wallacea.	Pœcilandra.

[From Bentham and Hooker's ' Genera Plantarum ' :—

TRIBE I. OCHNEÆ.—Ovary 2–10-celled, cells 1-ovuled. Seeds exalbuminous. *Ochna, Gomphia, Brackenridgea, Elvasia, Tetramerista.*

TRIBE II. EUTHEMIDEÆ.—Ovary incompletely 5-celled, cells 2-ovuled. Berry with 5 pyrenes. Seeds albuminous. *Euthemis.*

TRIBE III. LUXEMBURGIEÆ.—Ovary excentric, 2–5- or 1-celled, cells ∞ -ovuled. Capsule many-seeded. Seeds albuminous. *Luxemburgia, Blastemanthus, Godoya, Cespedesia, Wallacea, Pœcilandra.*—ED.]

Ochnaceæ, which are near *Rutaceæ, Diosmeæ,* and *Zanthoxyleæ,* are separated from them by their stipulate not dotted leaves, neither annular nor glandular disk, acute gynobasic style, and never free carpels. They differ from *Simarubeæ* in their disk, filaments without scales, anthers opening by terminal pores, and style undivided at the base.

Ochnaceæ are dispersed over all tropical regions; the capsular-fruited genera are American, the drupaceous Asiatic and South African. *Ochnaceæ* are bitter, like *Simarubeæ,* but their bitterness is tempered by an astringent principle. The aromatic root and the leaves of *Gomphia angustifolia,* an Indian tree, are employed as tonics and stomachics. The bark of *G. hexasperma,* a Brazilian shrub, is astringent, and very useful for the cure of ulcers caused by the stings of flies. The berries of *G. jabotapita,* a tree of the Antilles and Brazil, are edible, like those of the Bilberry; its seeds are oily.

LVII. *BURSERACEÆ,*[1] *Kunth.*

[TREES or SHRUBS, often lofty, abounding in resinous or oily secretions. LEAVES alternate, very rarely opposite, exstipulate, 3- (rarely 1-) foliolate or imparipinnate ; *leaflets* rarely pellucid-dotted, the lowest pair sometimes stipuliform. FLOWERS ☿, often small, racemed or panicled, regular. CALYX 3–5-fid or -partite, imbricate or valvate in bud. PETALS 3 to 5, erect or spreading, free or rarely connate, deciduous, imbricate or valvate in bud. DISK annular or cupular, rarely obsolete, free or adnate to

---

[1] This order is not described in the original, but is mentioned as an ally of *Terebinthaceæ* ; the characters and disposition of the genera are taken from Bentham and Hooker's ' Genera Plantarum.'—ED.

the calyx-tube. STAMENS usually twice as many as the petals, rarely as many, inserted at the margin or base of the disk, equal or unequal, or the alternate longer ; *filaments* free, naked, subulate, staminodes 0; *anthers* oblong or subglobose, often versatile. OVARY free, trigonous, ovoid or globose, 2-5-celled, usually contracted into the short style ; *stigma* undivided or 2-5-lobed; *ovules* 2 (rarely 1) in each cell, usually collateral and pendulous from the top of the cell, rarely ascending ; micropyle superior and raphe ventral. FRUIT a drupe, indehiscent, with 2-5 pyrenes, or with a bony or thin endocarp, or a capsule with the epicarp dehiscing and exposing bony cocci, which are connate, and separate from a central column. SEEDS pendulous, testa membranous, albumen 0; *cotyledons* usually membranous, contortuplicate, rarely plano-convex or thick ; *radicle* superior.

TRIBE I. BURSEREÆ.—Ovary 2-5-celled.
a. Calyx free. Petals free. Drupe with a valvate epicarp and 3 separable pyrenes. *Boswellia, Trionema.*
b. Calyx free. Petals free. Drupe with a valvate epicarp and 3 connate pyrenes.

* Stamens 6–10.		** Stamens 5.
Garuga.	*Balsamodendron.*	*Filicium.*
Protium.	*Bursera.*	*Ganophyllum.*
Canarium.	*Santeria, &c.*	*Nothoprotium.*

c. Calyx free. Corolla gamopetalous. *Trattinichia, Hedwigia.*
d. Calyx adnate to the ovary. *Darryodes.*

TRIBE II. AMYRIDEÆ.--Ovary 1-celled. *Amyris, Hemprichia.*—ED.]

Burseraceæ, whose close affinity to *Terebinthaceæ* we have noticed, and to which has been annexed the genus *Amyris* (which only differs in its one-celled ovary and generally opposite leaves), yield spontaneously, or by incision of their stems, balsamic resinous substances, employed in medicine. The incense called Olibanum is a resin of balsamic odour and stimulating properties, obtained from *Boswellia thurifera*, an Indian tree ; the Arabian incense is the product of one or more allied species. The resin Elemi of Ceylon, which is yellow, and of a penetrating odour, is furnished by *Canarium commune*, and the Javanese Elemi by *Bursera gummifera*. The Mexican Elemi comes from *Elaphrium elemiferum*. The Balm of Mecca or Gilead is a sweet smelling turpentine obtained by incision from two species of the genus *Balsamodendron*, natives of Arabia. Bdellium, a gum-resin of a sweet smell and bitter taste, used externally medicinally, comes from *B. africanum* (*Heudelotia africana*). Guggur is furnished by *B. Mukul*, a tree of the province of Scinde in India. The Kafal (*B. Kafal* and *B. Opobalsamum*) produces gum and red aromatic wood, which are the objects of considerable commerce in Arabia. Myrrh, a gum-resin, whose use as an aromatic and medicament goes back to the highest antiquity, is furnished by *B. Myrrha*, a tree of Arabia and Abyssinia. [*B. Roxburghii* yields the Gogul balsam of Bengal.] *Icica guianensis*, a Guiana tree, commonly called Incense-wood, yields a resin used similarly to Olibanum. *I. altissima* gives the Carana gum, which takes the place of Balm of Gilead in America. The resin Chihou or Cachibou comes from the American Gum-tree (*Bursera gummifera*), which grows from Guiana to Mexico. [*Amyris balsamifera* yields the Jamaica Lignum Rhodium.—ED.]

Hedwigia balsamifera is a tree of the Antilles, yielding in abundance a resin called 'Baume à cochon,' because the wild pigs, when wounded by hunters, pierce, it is said, the bark with their tusks, to rub their wounds with the balsamic juice which exudes. [Tacamahac is the resin of *Elaphrium tomentosum*. *Bursera altissima* is a very lofty American tree, of whose aromatic wood canoes forty-two feet long have been made. Many species of *Canarium* are very lofty Indian forest-trees, abounding in resinous balsams; that of *C. Zeylanica* is used for torches in Ceylon. The oil expresed from the nuts of *C. commune* is used for lamps and as food, but if eaten fresh brings on diarrhœa ; its bark is said to yield a pungent terebinthaceous oil, which has the properties of Copaivi. *C. strictum* is the Black Dammar tree of Malabar, which yields a transparent resin.—ED.]

LVIII. *MELIACEÆ.*

(MELIACEÆ ET CEDRELACEÆ, *Adr. Jussieu.*)

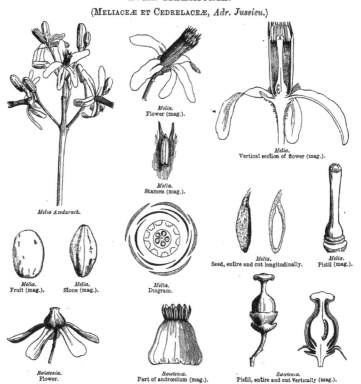

Melia.
Flower (mag.).

Melia.
Stamen (mag.).

Melia.
Vertical section of flower (mag.).

Melia Azedarach.

Melia.
Seed, entire and cut longitudinally.

Melia.
Pistil (mag.).

Melia.
Fruit (mag.).

Melia.
Stone (mag.).

Melia.
Diagram.

Swietenia.
Flower.

Swietenia.
Part of andrœcium (mag.).

Swietenia.
Pistil, entire and cut vertically (mag.).

PETALS *hypogynous,* 4-5 *or* 3-7, *distinct or coherent, or adnate to the staminal tube, æstivation contorted, imbricate or valvate.* STAMENS *usually double the number of the petals, inserted with them; filaments joined into a tube.* OVARY *free, girt or sheathed at the base by a more or less developed disk of 2–several 1-2- or several-ovuled cells.* OVULES *ascending or pendulous, hilum usually ventral, micropyle superior.* FRUIT *dry or fleshy.* EMBRYO *albuminous or exalbuminous.*—STEM *woody.* LEAVES *alternate.*

Large or small TREES, or rarely UNDERSHRUBS; wood often hard, coloured, sometimes odoriferous. LEAVES alternate, exstipulate, very rarely dotted (*Flindersia*), pinnate or rarely simple, entire. FLOWERS ⚥, or rarely polygamo-diœcious, regular, terminal or axillary, panicled. CALYX generally small, 4–5-fid or -partite, æstivation usually imbricate. PETALS hypogynous, 4–5, rarely 3–7, sometimes free and contorted or imbricate, sometimes connate, or adnate to the staminal tube, and valvate. STAMENS generally 8 or 10, rarely 5, very rarely 16–20, inserted with the petals outside of the base of a hypogynous disk; *filaments* united by their margins into a more or less complete tube, entire or toothed or variously laciniate, very rarely free (*Cedrela*); *anthers* introrse, sessile or sub-sessile on the staminal tube, included or exserted, dehiscence longitudinal, connective sometimes lengthened. DISK various, usually annular or tubular and sheathing, free, or adnate to the ovary (*Trichilia*) or staminal tube (*Mallea*). OVARY free, usually 3–5-celled; *style* simple; *stigma* disciform or pyramidal; *ovules* usually 2 in each cell, collateral or superimposed, very rarely solitary, sometimes 6 or more (*Cedrela, Swietenia*), ascending or pendulous, raphe ventral, micropyle superior. FRUIT various: sometimes a drupe (*Melia, Mallea*) or berry (*Vavæa, Sandoricum*), sometimes a capsule, loculicidal (*Trichilia, &c.*) or septifragal (*Cedrela, Swietenia, &c.*). SEEDS exalbuminous, or with fleshy albumen, winged (*Swietenia, Cedrela, &c.*) or not (*Melia, Trichilia, &c.*). EMBRYO flat, hilum usually ventral; *cotyledons* fleshy; *radicle* usually sunk between the cotyledons and superior, sometimes vague.

TRIBE I. *MELIEÆ.*[1]

Stamens united into a tube. Ovary with 2-ovuled cells. Seeds not winged; albumen thin, fleshy. Fruit a capsule (*Quivisia, Turræa, &c.*) or drupe (*Melia, Mallea*) or berry (*Vavæa*). Cotyledons plano-convex or foliaceous.—Leaves simple, 3-foliolate or pinnate, or decompound.

GENERA.

Quivisia.	Turræa.	Vavæa.	Munronia.	Melia.	Mallea.
Azadirachta.		Naregamia.		Calodryum.	

TRIBE II. *TRICHILIEÆ.*

Stamens united in a tube. Ovary with 1–2-ovuled cells. SEEDS not winged, exalbuminous. Fruit a capsule (*Carapa, Trichilia, Guarea, &c.*) or berry (*Sandoricum, Milnea, Dasycoleum, Lansium, &c.*). Cotyledons thick.—Leaves pinnate.

PRINCIPAL GENERA.

Dysoxylum.	Chisocheton.	Epicharis.	Cabralea.	Sandoricum.	Aglaia
Milnea.	Lansium.	Amoora.	Guarea.	Walsura.	Ekebergia.
Odontandra.	Heynea.	Moschoxylon.	Trichilia.	Carapa.	Owenia.

[1] These tribes are those proposed by Bentham and Hooker fil. in the 'Genera Plantarum.'—ED.

Tribe III. *SWIETENIEÆ.*

Stamens united in a tube. Ovary with many-ovuled cells. Capsule septifragal at the top by 3–5 bilamellate valves detached from the axis. Seeds numerous, albuminous or not, usually winged, hilum lateral or apical, raphe along the wing.— Leaves pinnate.

GENERA.

Khaya. Soymida. Swietenia. Chickrassia. Elutheria.

Tribe IV. *CEDRELEÆ.*

Stamens free. Ovary with many-ovuled cells. Capsule septifragal or loculicidal at the top by 3–5 valves detached from the axis. Seeds numerous, compressed, winged, albuminous or not.—Leaves usually pinnate.

GENERA.

Cedrela. Chloroxylon. Flindersia.

Melieæ and *Cedreleæ* are near *Aurantiaceæ* and *Rutaceæ* (which see). They approach *Sapindaceæ* in hypopetalism, diplostemony, union of filaments, hypogynous disk, 1–2-ovuled ovarian cells, simple style, woody stem, and alternate leaves; but in *Sapindaceæ* the filaments, when connate, are only so at the base, the stamens are inserted within the disk, and the radicle is inferior. Between *Meliaceæ* and *Humiriaceæ* there are also some points of analogy, founded on the insertion of the petals, the number of the stamens, and the connection of their filaments, simple style, 1–2-ovuled ovarian cells, simple style, lobed stigma, berry, drupe or capsule, woody stem, and alternate and often dotted leaves. The same analogy exists with *Burseraceæ*, which are distinguished by their exalbuminous seed and folded contorted cotyledons.

Melieæ inhabit the tropics of Africa and Asia. *Trichilieæ* are more common, especially in Asia and America. *Swietenieæ* and *Cedreleæ* inhabit the tropics of both continents, and some grow in the Moluccas and Australia.

This family is useful to man both in medicine and manufactures. The acrid, bitter, astringent, and aromatic principles which it possesses in various proportions are tonic, stimulating, purgative, or emetic. Some species have agreeable sugary and refreshing fruits.

Melia Azedarach is a small Asiatic tree naturalized in all warm climates, all parts of which are bitter, purgative, vermifuge, but poisonous in large doses; the seeds contain a fixed oil, useful for burning. The fruit of *M. sempervirens*, or Indian Lilac, is poisonous. The bark of *M. Azadirachta* is bitter and powerfully tonic, and the oil in its seeds is a reputed remedy for headache from sunstroke. The aromatic root of *Sandoricum indicum* is employed for heartburn. *Trichilia* and *Guarea*, American species, are very energetic purges and emetics. Several species of *Dysoxylum* have a strong smell of garlic. In Asia the bark of *Walsura piscidia* is used to stupefy fish; that of *Carapa guianensis* is considered in America to be a febrifuge, the fatty oil of its seeds (Crab-oil) is a reputed anthelminthic, and is largely used in machinery. The *Xylocarpi* of Asia are praised as stomachics. The pulp which surrounds the seed of *Milnea edulis*, an Asiatic species, is delicious. The pericarp of *Lansium* is acidulous and sugary. *Soymida febrifuga*, celebrated in India for the virtues of its bitter, astringent, and aromatic bark, is admitted by European doctors among the substitutes for quinine, as is also *Cedrela febrifuga*, a native of Java. The *Khaya* used in Senegal, and the *Chickrassia* in tropical Asia, possess the same properties. The bitter and styptic bark of *Swietenia Mahogoni*, a native of tropical America. is employed, mixed with quinine, in intermittent fevers. The wood of most of the species of this family, often called Cedar, is esteemed, not only on account of its sweet scent, but especially for its density and fine colour. The most celebrated species is the *Swietenia*

Mahogoni, which yields Mahogany, a fine wood, close in texture, of a reddish colour shaded with brown, becoming darker when exposed to the air, and much used because it is easy to work, and takes a fine polish.

LIX. *CHAILLETIACEÆ.*[1]

[SHRUBS or small TREES. LEAVES alternate, petioled, quite entire, penninerved, coriaceous; *stipules* petiolar or close to the axils, deciduous. INFLORESCENCE of dense sometimes almost capitate corymbose cymes; *peduncles* axillary, often adnate to the petiole, the flowers thus appearing to be placed at the base of the leaf-blade. FLOWERS small, ⚥ or unisexual. SEPALS 5, free or connate, sometimes unequal, coriaceous, imbricate in bud. PETALS 5, inserted on the calyx, rather longer than the sepals, imbricate or open in bud, free and equal, or connate and unequal, broadly clawed, often narrow, 2-fid or -lobed, with a terminal inflexed ligule, the edges of which are adnate to those of the fissure of the petal. DISK various, cup-shaped, entire or lobed or broken up into hypogynous glands or 5 free scales. STAMENS 5, inserted with the petals, alternate with the scales or lobes of the disk, free or adnate to the gamopetalous corolla-tube; *anthers* shortly oblong, dehiscing longitudinally, connective usually thickened at the back. OVARY free, usually depressed, globose, pubescent or villous, 2–3-celled; *styles* 2–3, terminal, short or long, free or connate; *stigmas* sub-terminal, simple or capitate; *ovules* 2, geminate, pendulous from the top of each cell, anatropous, raphe ventral, micropyle superior. DRUPE oblong or compressed, pubescent, dry; *epicarp* coriaceous, entire or dehiscing and disclosing the 1–2-celled, sometimes 2-valved bony or crustaceous endocarp, cells 1-seeded. SEEDS pendulous, adnate by a broad hilum to the top of the cell; *testa* membranous; *albumen* 0. EMBRYO large; *cotyledons* amygdaloid; *radicle* small, superior.

GENERA.

| Chailletia. | Stephanopodium. | Tapura. |

A small order of three genera and about forty species, allied to *Celastrineæ* and *Rhamneæ*, but differing in the disk, pendulous ovules, position of the raphe, and amygdaloid cotyledons. Muller (Argan) placed *Moacurra,* Roxb., which is a true *Chailletia,* in *Euphorbiaceæ,* but he was ignorant of the structure of the fruit.

Chailletiaceæ are natives of the tropics of Asia, Africa, and America, with one South African species. Of their properties little is known; the fruit of *Ch. toxicaria* is said to be poisonous, and is called Ratsbane by the Sierra Leone colonists.—ED.]

LX. *OLACINEÆ, Endl.*

TREES or SHRUBS, erect, climbing or twining, rarely suffruticose. LEAVES usually alternate, exstipulate. FLOWERS ⚥ or unisexual, regular, axillary, in a raceme, corymb or spike, very rarely terminal and paniculate; *peduncles* jointed at the base; *receptacular cup (calycode)* toothed or lobed, often accrescent when ripe.

---

[1] This order is omitted in the original.—ED.

PERIANTH [1] single. SEPALS 4–5 (rarely 6), distinct, or coherent in a campanulate or tubular calyx, valvate in æstivation. STAMENS 4–10 (rarely 12), often adnate to the sepals, all antheriferous, or rarely some sterile (*Olax, Liriosma*); *filaments* free, or very rarely monadelphous (*Aptandra*); *anthers* erect, versatile or rarely adnate (*Cathedra, Lasianthera*, &c.). DISK very various, free, or adnate to the calycule or ovary, sometimes 0 or inconspicuous. OVARY free, or the base partially sunk in the disk, 1-celled, or falsely 3–5-celled by imperfect septa, or very rarely truly 3-celled (*Emmotum*); *ovules* anatropous, without coats and reduced to the nucleus, sometimes 2–3 (rarely 4–5), collaterally pendulous, either from the top of a central placenta, or from the top of the ovarian cavity, and excentric (in consequence of the lateral union of the placenta with the wall), rarely solitary and similarly pendulous (*Opilia, Pennantia*), very rarely nearly erect and basilar (perhaps owing to the suppression of the placenta) (*Cansjera, Agonandra*). FRUIT mostly a drupe, 1-celled, 1-seeded, superior, or becoming inferior by the accrescence and adherence of the calycule. SEED with fleshy copious albumen, which is entire or sometimes ruminate, lobed, or partite (*Aptandra, Gomphandra*, &c.). EMBRYO small, at the top of the albumen, or nearly as long as the albumen, and straight; *radicle* cylindric, superior; *cotyledons* small, or large and foliaceous, entire or cut.

TRIBE I. OLACEÆ.

Stamens unequal in number with the sepals (*Olax*) or double (*Ximenia, Heisteria*, &c.), or equal and opposite (*Erythropalum, Anacolosa, Strombosia*, &c.). Ovary 1-celled (*Erythropalum, Olax, Ptychopetalum*, &c.), or with 3–5 incomplete 1-ovuled cells (*Ximenia Heisteria, Liriosma, Schœpfia, Anacolosa, Aptandra*). Ovules pendulous from the top of a central placenta.

PRINCIPAL GENERA.

| Aptandra. | Heisteria. | Ximenia. | Olax. | Liriosma. |
| Erythropalum. | Strombosia. | Cathedra. | Anacolosa. | Schœpfia. |

TRIBE II. OPILIEÆ.

Flowers isostemonous. Stamens opposite to the sepals. Ovary 1-celled, 1-ovuled. Ovule basal, nearly erect.

PRINCIPAL GENERA.

| Cansjera. | Agonandra. | Lepionurus. | Opilia. |

TRIBE III. ICACINEÆ.

Flowers isostemonous. Stamens alternate with the sepals. Ovary 1-celled, with 1–2 ovules pendulous from the top of the cell, very rarely with 3 perfect 1–2 ovuled cells.

PRINCIPAL GENERA.

| Lasianthera. | Gomphandra. | Desmostachys. | Apodytes. | Mappia. | Emmotum. |
| Pennantia. | Poraqueiba. | Icacina. | Platea. | Villaresia. | |

[1] Most authors regard the receptacular cup here alluded to as a calyx, and the calyx as a corolla.—ED.

z

[The above tribes are adopted from the 'Genera Plantarum' of Bentham and Hooker, in which work a fourth is added, often regarded as an order, but which is omitted in the original of this book, i.e.—

TRIBE IV. PHYTOCRENEÆ.—Flowers and fruit of *Icacineæ*. Embryo more developed; cotyledons broad, foliaceous or thick and fleshy.—Stem climbing. Flowers diœcious. *Phyto-crene, Miquelia, Sarcostigma, Natsiatium, Iodes.*

To this order the above authors have further appended the singular Indian and Javanese genus *Cardiopteris*, a slender annual climbing glabrous herb with milky juice, imbricate sepals, and gamopetalous five-lobed imbricate corolla. Stamens and ovary of *Icacineæ*, but two short styles with capitate stigmas, one (?) of which grows remarkably after impregnation. The one-celled superior ovary is succeeded by a broadly two-winged coriaceous white fruit with a narrow central longitudinal cell containing one pendulous seed with very minute embryo next the hilum, in a dense granular albumen. Of the two collateral pendulous ovules one only is impregnated; it consists of a naked nucleus, the embryo-sac of which is exserted as a very slender long tube.

About 170 species of *Olacineæ* are known, included in 36 genera; they are dispersed over the tropical and sub-tropical regions of the whole globe, but are rare in South Africa and Australia; *Pennantia* inhabits New Zealand and extra-tropical Australia. Various species of *Villaresia* advance into Chili. *Phytocreneæ* are tropical Asiatic and African.

Olacineæ are so closely allied to *Santalaceæ* and *Cornaceæ* that it is impossible to separate them by any natural characters; and these, together with *Loranthaceæ*, form one great family. *Ilicineæ* are separated from *Olacineæ* solely by the complete cells of their ovary, *Villaresia* being in this respect quite intermediate. *Cornaceæ* differ in their completely inferior ovary.

Little is known of the properties of *Olacineæ*. The drupes of *Ximenia* are eaten in Senegal, and said to be sweet and aromatic, but rough to the palate. *Olax zeylanica* has a fœtid wood with a saline taste, and is employed in putrid fevers; its leaves are used as salad. The stem of *Phytocrene* is very curious, being white and very porous, and discharging when cut a stream of limpid potable water.—ED.]

LXI. *ILICINEÆ.*

(RHAMNORUM *genera, Jussieu.*—AQUIFOLIACEÆ, *D.C.*—ILICINEÆ, *Brongniart.*)

COROLLA *sub-polypetalous or polypetalous, hypogynous, isostemonous, æstivation imbricate.* STAMENS *inserted at the base of the petals, or on the receptacle.* OVARY *of many 1-ovuled cells.* OVULES *pendulous from the central angle of the cells.* FRUIT *a drupe.* EMBRYO *minute, albuminous.* RADICLE *superior.*—STEM *woody.*

TREES or SHRUBS with persistent or caducous leaves. LEAVES alternate or opposite, simple, coriaceous, glabrous, shining, exstipulate. FLOWERS ☿, or rarely unisexual, small, solitary, or fascicled in the axils of the leaves, on simple peduncles, sometimes branching into dichotomous cymes. CALYX 4–6-fid or -partite, persistent, with obtuse segments. COROLLA inserted on the receptacle, of 5–4–3 free or nearly free petals, æstivation imbricate. STAMENS 5–4–3, alternating with the petals, and cohering to their base, or on the receptacle; *filaments* filiform or subulate, shorter than the petals; *anthers* introrse, dorsally adnate, dehiscence longitudinal. OVARY free, fleshy, sub-globose, 2–6–8-celled; *stigma* subsessile, lobed; *ovules* solitary in each cell, pendulous from the top of the central angle, anatropous. DRUPE fleshy,

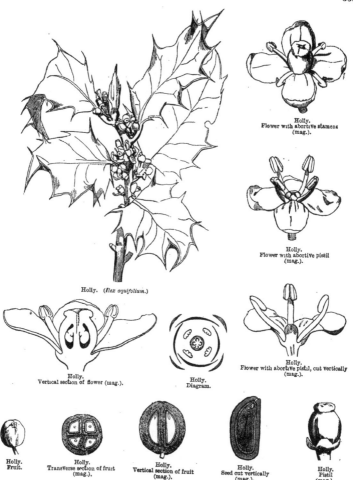

Holly. (*Ilex aquifolium.*)

Holly.
Flower with abortive stamens
(mag.).

Holly.
Flower with abortive pistil
(mag.).

Holly.
Vertical section of flower (mag.).

Holly.
Diagram.

Holly.
Flower with abortive pistil, cut vertically
(mag.).

Holly.
Fruit.

Holly.
Transverse section of fruit
(mag.).

Holly.
Vertical section of fruit
(mag.).

Holly.
Seed cut vertically
(mag.).

Holly.
Pistil
(mag.).

of 2–8–∞ woody or bony one-seeded indehiscent pyrenes. SEED inverted; *testa* membranous, raphe dorsal; *hilum* turned towards the top of the cell, naked, or capped by the funicle dilated into a cupule; *albumen* fleshy, copious. EMBRYO straight, minute, at the top of the albumen, sub-cylindric or globular; *radicle* near the hilum, superior.

GENERA.

Cassine.　　Ilex.　　Prinos.　　Byronia.　　Nemopanthes.

Ilicineæ were for a long time placed in the family of *Celastrineæ*, which approach them in their persistent calyx, hypogynism, isostemony, æstivation of their corolla, their many-celled ovary, anatropous ovule, sessile or subsessile stigma, sometimes drupaceous fruit (*Elæodendron*), straight albuminous embryo, woody stem, alternate leaves, and axillary small and greenish flowers; but *Celastrineæ* differ in the fleshy disk which lines the bottom of the calyx and often the base of the ovary, in the erect or ascending ovule, and finally in their corolla, which is clearly polypetalous. We shall indicate the affinity of *Ilicineæ* with *Ebenaceæ* under that family. They have also a connection with *Olacineæ*, founded on the hypogynous corolla, pendulous and anatropous ovule, fleshy fruit, albuminous seed, and woody stem; but *Olacineæ* have an anisostemonous corolla with valvate æstivation, and their embryo is axile, and not apical.[1]

Ilicineæ are rare in Europe; they are more numerous in Asia, South and Central America, and at the Cape of Good Hope.

Ilicineæ contain a bitter principle, the *ilicine* of chemists, combined in various proportions with an aromatic resin and a glutinous matter, to which some species of Holly owe medicinal properties. *Ilicine* has been proposed as a substitute for quinine. An infusion of the leaves of *Ilex vomitoria* is diuretic and diaphoretic; in large doses it produces vomiting, and is the usual emetic of the savages of South America. *Ilex Paraguayensis* yields *maté*, which takes the place of tea in South America. The bark of *Prinos verticillata* is astringent, and is used in the United States as a tonic and antiseptic. Many kinds of Holly are cultivated in Europe as ornamental plants (*I. Dahoon, balearica, maderensis, latifolia*, &c.); but the most interesting species is the common Holly (*Ilex aquifolium*), which grows in the hilly forests of western Europe, and the spiny and persistent leaves of which were formerly used as a febrifuge. The berries are of a brilliant red, and with the shining green leaves greatly contribute to the beauty of our winter bouquets. The bark yields bird-lime, used by birdcatchers, and formerly employed topically to reduce tumours. Holly-wood is close and hard, and much esteemed for cabinet-work.

LXII. *EMPETREÆ*, Nuttall.

Low heath-like dry erect or prostate much-branched shrubs, branches cylindric. LEAVES alternate, sometimes sub-whorled, coriaceous, entire, exstipulate. FLOWERS small, regular, usually diœcious, rarely polygamous, sessile in the axils of the upper leaves, solitary (*Empetrum*), or few together (*Ceratiola*), rarely crowded at the top of the branches (*Corema*), naked, or furnished with scaly imbricate bracts. CALYX 3–2-phyllous; *leaflets* imbricate in æstivation, coriaceous or membranous, and like the bracteoles. PETALS hypogynous, shortly clawed, persistent, marcescent. STAMENS (rudimentary or 0 in the ♀) inserted with the petals, equal and alternate with them; *filaments* filiform, free, persisting after the fall of the anthers; *anthers* extrorse, sub-globose, didymous or oblong, dehiscence longitudinal. OVARY (rudimentary in the ♂) seated on a fleshy sub-globose disk, 2–3–6–9-celled, cells 1-ovuled; *style* short, angular, or obsolete; *stigma* lobed, radiating, lobes truncate, laciniate or

---

[1] Various *Olacineæ* have isostemonous imbricate corollas and apical embryos, notably *Villaresia*, which has further the habit of *Ilex*, and stamens cohering to the base of the connate petals.—ED.

incised; *ovules* ascending from the base of the central angle, anatropous. DRUPE fleshy, spherical, sub-depressed, umbilicate, of 2-3-6-9 connate or distinct bony 1-

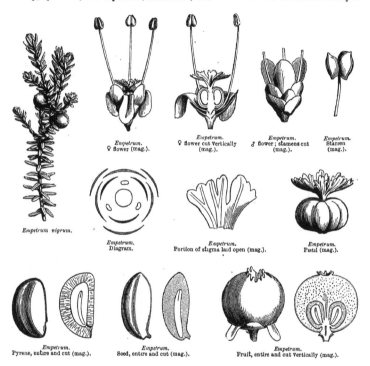

Empetrum.
♀ flower (mag.).

Empetrum.
♀ flower cut Vertically (mag.).

Empetrum.
♂ flower; stamens cut (mag.).

Empetrum.
Stamen (mag.).

Empetrum nigrum.

Empetrum.
Diagram.

Empetrum.
Portion of stigma laid open (mag.).

Empetrum.
Pistil (mag.).

Empetrum.
Pyrene, entire and cut (mag.).

Empetrum.
Seed, entire and cut (mag.).

Empetrum.
Fruit, entire and cut Vertically (mag.).

seeded pyrenes. SEEDS triangular, erect; *testa* membranous; *albumen* fleshy, dense. EMBRYO straight, axile, cylindric; *cotyledons* short, obtuse; *radicle* near the hilum, inferior.

GENERA.

Empetrum.   Corema.   Ceratiola.

The little family of *Empetreæ* approaches *Celastrineæ, Ilicineæ*, and especially *Ericineæ* properly so called. It has the habit of the latter, their marcescent hypogynous corolla, several-celled ovary with anatropous ovules, albuminous seed and straight embryo, and it also recalls the tribe of *Rhodoraceæ* by the structure of the stigma; but *Ericineæ* are gamopetalous and gamosepalous, their flower is perfect, and their fruit is a capsule or berry. The affinity with *Ilicineæ* is not doubtful; in both families we find diclinism, hypocorollism, isostemony, imbricate æstivation, several-celled and 1-ovuled ovary, anatropous ovules, very short style, drupaceous fruit, fleshy albumen, woody stem, alternate leaves, and axillary flowers. *Celastrineæ* have, like *Empetreæ*, small axillary flowers, polypetalous and isostemonous imbricate corollas, a fleshy disk, an ovary with several 1-ovuled cells, ascending and anatropous ovules, a sub-sessile lobed stigma, drupaceous fruit, albuminous seed, and straight and axile embryo. *Celastrineæ* chiefly differ in habit, stipulate leaves, perigynism, introrse anthers, often 2-ovuled ovarian cells, and the fleshy aril of the seed. The few species of this family are dispersed over the Iberian peninsula, Central Alpine and North Europe, North America, and the Magellanic region.

The leaves and drupes are acidulous; the fruits of *Empetrum nigrum* are eaten in the North of Europe for their antiscorbutic and diuretic properties. The Greenlanders ferment them, and obtain a spirituous liquor. From the drupes of *Corema* an acid drink is prepared in Portugal, and employed as a febrifuge in popular medicine.

LXIII. *CELASTRINEÆ.*

(RHAMNORUM *sectio, Jussieu.*—CELASTRINEÆ, *Br.*—CELASTRACEÆ, *Lindl.*)

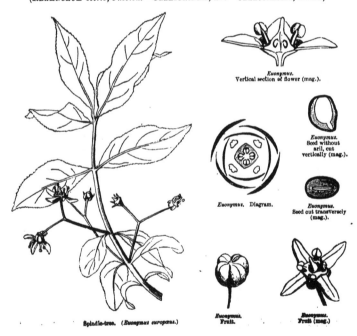

Euonymus.
Vertical section of flower (mag.).

Euonymus.
Seed without aril, cut vertically (mag.).

Euonymus. Diagram.

Euonymus.
Seed cut transversely (mag.).

Spindle-tree. (*Euonymus europæus.*)

Euonymus.
Fruit.

Euonymus.
Fruit (mag.)

Euonymus.
Seed with its aril (mag).

E. latifolius.
Pendulous ovules (mag.).

E. latifolius.
Upright ovules (mag.).

COROLLA *polypetalous, perigynous, isostemonous, æstivation imbricate.* PETALS *4–5, inserted on a fleshy disk, surrounding the ovary and occupying the bottom of the calyx.* STAMENS *4–5, inserted like the petals.* OVARY *2–3–5-celled, cells 1–2-ovuled.* OVULES *ascending or erect, anatropous.* FRUIT *dry or fleshy.* SEEDS *usually arillate, albuminous.*—STEM *woody.* LEAVES *simple, stipulate.*

Small TREES or SHRUBS, often climbing. LEAVES alternate, or rarely opposite, simple, entire or toothed, often coriaceous; *stipules* small, very caducous. FLOWERS ☿, or unisexual, regular, axillary, cymose, small, greenish or reddish. CALYX 4–5-fid or -partite, segments equal, imbricate in æstivation, persistent. DISK fleshy, annular or orbicular, lining the bottom of the calyx, and sometimes adnate to the ovary. PETALS 4–5, alternate with the sepals, inserted under the edge of the disk by a broad base, sessile, imbricate in æstivation, deciduous. STAMENS 4–5, inserted under, on, or within the edge of the disk; *filaments* short; *anthers* introrse, erect, fixed by the base or back, connective often dilated, dehiscence longitudinal. OVARY sessile, more or less buried in the disk, and sometimes adhering to it by its base, of 2–3–5 1–2–many-ovuled cells; *style* short, thick; *stigma* of 2–3–5 lobes; *ovules* usually 2, collateral, erect or ascending, with a ventral raphe, sometimes pendulous, and then with a dorsal raphe (rarely many–2-seriate). FRUIT 2–5-celled, sometimes an indehiscent drupe or samara with 1-seeded cells, sometimes a loculicidal capsule with semi-septiferous valves. SEEDS erect or ascending, usually furnished with a pulpy coloured sometimes very much developed and cupular aril; *testa* crustaceous or membranous, traversed by a longitudinal raphe. EMBRYO straight, occupying the axis of a fleshy copious albumen; *cotyledons* foliaceous, flat; *radicle* cylindric, inferior.

[Disposition of the tribes and genera of *Celastrineæ* in the ' Genera Plantarum,' including *Hippocrateaceæ*, which are omitted in the original of this work :—

TRIBE I. CELASTREÆ.—Stamens 4–5, very rarely 10, inserted (except *Schæfferia*) on or beneath the margin of a conspicuous disk. Filaments subulate, often incurved. Seeds albuminous (exalbuminous in *Hartogia, Kokoona*, and some species of *Maytenus*). *Euonymus, Microtropis, Maytenus, Elæodendron, Catha, Lophopetalum, Gymnosporia, Perrottetia, Kokoona, Hartogia, Kurrimia, Cassine, Celastrus, Myginda.*

TRIBE II. HIPPOCRATIEÆ (order *Hippocrateaceæ*, Endl., Lindl.).—Stamens 3, rarely 2, 4, or 8, inserted on the face of the disk; filaments flattened, recurved, sometimes adnate to the disk; anthers extrorse from the reflexion of the filaments. Seeds exalbuminous.—Leaves usually opposite. *Hippocratia, Salacia.*—ED.]

For the affinities of *Celastrineæ* with *Jlicineæ, Pittosporeæ,* and *Staphyleaceæ,* see these families. Their affinity with *Rhamneæ* is very close, and Jussieu placed them in the same family; it is founded on the woody stem, stipulate leaves, axillary small and greenish flowers, fleshy disk lining the calyx and often adhering to the ovary, isostemonous and perigynous corolla, 1-2-ovuled cells of the ovary, upright and anatropous ovules, fleshy or capsular fruit, the often arillate seeds, and the albuminous embryo; but in *Rhamneæ* the æstivation is valvate, the stamens are opposite to the petals, and the fruit, if a capsule, usually divides into cocci. The two families inhabit the same country. *Euonymus* inhabits the temperate regions of the northern hemisphere; *Celastrus* and the other genera are, with few exceptions, tropical and sub-tropical, and dispersed over Asia, America, the Pacific islands, Australia, and South Africa. *Hippocratieæ* are also pretty equally distributed through Asia, Africa, and America. *Celastrineæ* usually possess purgative and emetic properties, but are not used in European practice; the bark of *Celastrus* is used as an emetic in South America. The root and leaves of *Myginda* are esteemed as diuretics in tropical America. *Catha edulis* is an East African shrub, called *Khat,* cultivated by the Arabs, with whom it is an article of commerce; the bruised leaves produce an agreeable excitement, analogous, it is said, to that induced in Peru by the use of *coca;* it is also landed by them as a sovereign remedy for the plague.

[In India the bark of *Euonymus tingens* is used to dye a yellow colour, with which the Hindoos make the sacred mark on the forehead; it is also used in eye complaints. The bark of *E. Roxburghii* is an astringent, used to reduce swellings in India. The seeds of the European *Euonymi* are nauseous and purgative, and said to poison sheep; an ointment made of them was formerly used to kill lice in the head. The spines of *Celastrus veneratus* are said to inflict a poisoned wound. The drupes of *Elæodendron Kuba* are eaten in South Africa by the colonists. The seeds of *Celastrus nutans* and *paniculatus* are acrid and stimulant, and used as a medicine in India.

Of the *Hippocrateaceæ* the fruit of *Salacia pyriformis,* a native of West Africa, is eatable, as are the nuts of *Hippocratia comosa,* the 'Amandier du Bois' of the French West Indies. The fruits of others are mucilaginous and edible.—ED.]

LXIV. *STAPHYLEACEÆ.*[1]

(CELASTRINEARUM *tribus, D.C.*—STAPHYLEACEÆ, *Bartling.*)

COROLLA *polypetalous, sub-hypogynous, isostemonous, æstivation imbricate.* PETALS 5, *inserted on a hypogynous disk.* STAMENS 5, *inserted with the petals.* OVARY 2-3-*lobed.* OVULES *anatropous.* FRUIT *dry or fleshy.* EMBRYO *albuminous.*—STEM *woody.* LEAVES *compound, bistipulate.*

TREES or erect SHRUBS. LEAVES generally opposite, 3-foliolate, or imparipinnate; *leaflets* opposite, petiolulate; *stipules* twin, at the base of the petioles, deciduous. FLOWERS ☿ or imperfect, regular, racemed or panicled, pedicels bracteate at the base. CALYX free, coloured, 5-partite, æstivation imbricate. PETALS inserted on or beneath a' hypogynous disk, crenulated, æstivation imbricate, deciduous. STAMENS 5, inserted like the petals; *filaments* subulate, free, equal; *anthers* introrse, opening longitudinally. CARPELS 2-3, united at the base, or throughout their length, into a 2-3-celled and -lobed ovary; *styles* equal in number to the lobes of the ovary, distinct or cohering, finally free; *stigma* undivided; *ovules* many, inserted along the ventral suture, 1-2-seriate, horizontal or ascending, anatropous. FRUIT a membranous turgid capsule, its lobes opening at the top by the ventral suture; or a berry, 3-celled, or 2-celled by suppression. SEEDS few or solitary

[1] See *Sapindaceæ,* Sub-order V., p. 353.

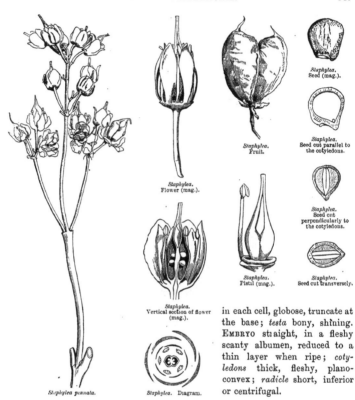

Staphylea.
Flower (mag.).

Staphylea.
Fruit.

Staphylea.
Seed (mag.).

Staphylea.
Seed cut parallel to
the cotyledons.

Staphylea.
Seed cut
perpendicularly to
the cotyledons.

Staphylea.
Pistil (mag.).

Staphylea.
Seed cut transversely.

Staphylea.
Vertical section of flower
(mag.).

Staphylea pinnata.

Staphylea. Diagram.

in each cell, globose, truncate at the base; *testa* bony, shîning. EMBRYO straight, in a fleshy scanty albumen, reduced to a thin layer when ripe; *cotyledons* thick, fleshy, planoconvex; *radicle* short, inferior or centrifugal.

PRINCIPAL GENERA.

Euscaphis.   Staphylea.   Turpinia.

Staphyleaceæ, joined by De Candolle to *Celastrineæ*, are connected with them by the polypetalous isostemonous corolla, imbricate æstivation, fleshy disk on which the petals and stamens are inserted, ascending and anatropous ovules, woody stem and stipulate leaves; but *Celastrineæ* have simple and alternate leaves, usually arillate seeds, and a copious albumen. A still more legitimate affinity links *Staphyleaceæ* with *Sapindaceæ* and *Acerineæ*; for in the latter the petals are imbricate, and inserted, like the stamens, on a fleshy hypogynous disk, the ovary is two-lobed and of two carpels, the fruit is a capsule, the seeds are ascending, and scarcely or not albuminous, the stem is woody, and leaves opposite. There is the same relationship with *Sapindaceæ*, in many genera of which the leaves are stipulate, though not opposite.

These two families are scarcely separated from *Staphyleaceæ* except by the diplostemonous corolla and curved embryo; and they are united in the 'Genera Plantarum' as tribes of *Sapindaceæ*. It is the same in *Hippocastaneæ* (see this family).

The few species of this little family are scattered over temperate Europe, North America, the Antilles, Mexico, Japan, and tropical Asia. Their useful properties are little known. The root of a Japan shrub (*Euscaphis*) is employed as an astringent in dysentery. [The seeds of *Staphylea* are oily, austere, and slightly purgative.—Ed.]

LXV. *STACKHOUSIEÆ*.[1]

[Small HERBS with watery juice, usually woody, simple or branched, rootstocks giving off many erect simple or sparingly divided slender leafy branches., LEAVES scattered, alternate, rather fleshy or coriaceous, linear or spathulate, quite entire; *stipules* 0 or very minute. FLOWERS ☿, regular, in terminal spikes or racemes at the ends of the branches, or fascicled, 3-bracteate, white or yellow. CALYX small, hemispheric, 5-lobed or -partite, lobes rather unequal, imbricate in bud. PETALS 5, perigynous, inserted on the throat of the calyx, linear or spathulate; claws long, free or connate; limb reflexed, imbricate in bud. DISK thin, clothing the base of the calyx-tube. STAMENS 5, inserted on the edge of the disk, erect, included; *filaments* slender, the alternate shorter; *anthers* oblong, dehiscing longitudinally; *pollen* obscurely 4-lobed, rough. OVARY sessile, free, sub-globose, 2-5-lobed or -partite, 2-5-celled; *styles* 2-5, free or connate; *stigma* 5-lobed or stigmas 5, capitate; *ovule* solitary in each cell, erect from its base, anatropous, raphe ventral. FRUIT of 2-5 globose or angular smooth or reticulate or winged indehiscent 1-seeded cocci, which separate from a central persistent column. SEED erect; *testa* membranous; *albumen* fleshy. EMBRYO axile, straight, as long as the albumen; *cotyledons* short, obtuse; *radicle* inferior.

ONLY GENUS.

Stackhousia.

A small and geographically limited order, embracing some twenty species, common in extra-tropical Australia, with a solitary representative in New Zealand, and another that wanders north to the Philippine Islands. It appears to agree most nearly with *Celastrineæ* in technical characters, but its affinities are quite unknown. In the disk and fruit it approaches *Rhamneæ*. Robert Brown indicated an affinity with *Euphorbiaceæ*, but on what grounds is not stated, nor have these been apparent to succeeding botanists. Nothing is known of its uses.—Ed.]

LXVI. *RHAMNEÆ*.

(RHAMNORUM *genera, Jussieu.*— RHAMNEÆ, *Br.*—FRANGULACEÆ, *D.C.*—RHAMNACEÆ, *Lindl.*)

COROLLA polypetalous, perigynous, isostemonous, æstivation valvate. PETALS 4-5, inserted on a perigynous disk, lining the calyx, and sometimes the ovary also. STAMENS 4-5, opposite to and inserted with the petals. OVARY *free, or adnate to the disk, of*

---

[1] This order is omitted in the original.—Ed.

2–3–4 1–2-*ovuled cells.* OVULES *erect, anatropous.* FRUIT *a drupe or capsule.* EMBRYO *large, albumen scanty.*—STEM *woody.* LEAVES *simple, 2-stipulate.*

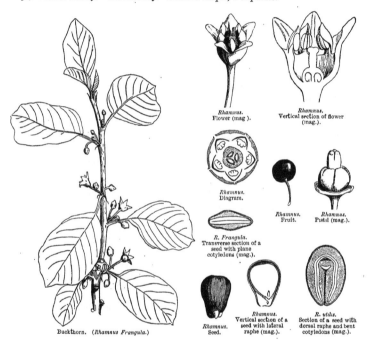

Rhamnus.
Flower (mag).

Rhamnus.
Vertical section of flower
(mag.).

Rhamnus.
Diagram.

Rhamnus.
Fruit.

Rhamnus.
Pistil (mag.).

R. Frangula.
Transverse section of a
seed with plane
cotyledons (mag.).

Rhamnus.
Seed.

Rhamnus.
Vertical section of a
seed with lateral
raphe (mag.).

R. utilis.
Section of a seed with
dorsal raphe and bent
cotyledons (mag.).

Buckthorn. (*Rhamnus Frangula.*)

TREES, SHRUBS, or UNDERSHRUBS, branches sometimes spinescent, sometimes climbing by their extremity, which is bare of leaves (*Gouania*). LEAVES simple, usually alternate, rarely sub-opposite or opposite, entire or toothed, petioled, sometimes minute or suppressed (*Colletia*) ; *stipules* small or 0, sometimes transformed into thorns. FLOWERS ☿ or unisexual, regular, small, greenish, usually axillary, solitary, or variously fascicled. CALYX 5-fid or -partite, æstivation valvate. DISK adnate to the calyx, and lining it with a single or double layer, of various form. PETALS 4–5, usually inserted at the edge of the disk, æstivation induplicate-valvate, rarely 0 (*Colletia, Pomaderris*). STAMENS 4–5, opposite to and inserted with the petals ; *filaments* sometimes adnate to the base of the petals, but not connate ; *anthers* introrse, dorsifixed, versatile, sometimes ovoid, with longitudinal dehiscence, sometimes reniform and 1-celled by confluence of the cells at the top, and opening into 2 valves by an arched slit. OVARY free, or buried in the disk, or more or less adnate to the calyx-tube, of 3-2-4 1-2-ovuled cells; *styles* equal in number to the cells, more or less connate; *stigmas* simple, distinct or connate; *ovules* usually solitary in each cell, erect, sessile or funicled, anatropous. FRUIT superior or inferior, rarely 1-celled by suppression, sometimes an indehiscent fleshy spongy or membranous drupe, sometimes winged, with a hard, fibrous or woody 2-3-celled endocarp, sometimes a capsule with 2-3 crustaceous cocci separating at the top, pendulous from the axis, and opening when ripe at the base of their inner edge. SEEDS erect; *testa* loose, raphe lateral or dorsal, chalaza thick; *albumen* fleshy, scanty. EMBRYO large, straight, yellow or green; *cotyledons* flat, fleshy; *radicle* short, inferior.

PRINCIPAL GENERA.

Paliurus.   Zizyphus.   Hovenia.   Rhamnus.   Ceanothus.   Phylica.   Pomaderris.

[Conspectus of the tribes and genera, from the ' Genera Plantarum ' of Bentham and Hooker fil. :—

TRIBE I. VENTILAGINEÆ.—Ovary superior or semi-superior. Disk filling the calyx-tube. Fruit dry, 1-celled, 1-seeded, girt at the base or up to the middle by the calyx-tube. Seeds exalbuminous.—Unarmed climbing shrubs. Leaves alternate. *Ventilago, Smythea.*

TRIBE II. ZIZYPHEÆ.—Ovary superior or semi-superior. Disk filling the calyx-tube. Drupe dry or fleshy, girt at the base or up to the middle by the calyx-tube; endocarp 1-3-celled. *Paliurus, Condalea, Beschemia, Zizyphus, Sarcomphalus, Karwinskia, Microrhamnus.*

TRIBE III. RHAMNEÆ.—Ovary inferior or superior. Disk various or 0. Fruit dry or drupaceous, containing 3-4 cocci or pyrenes. *Rhamnus, Ceanothus, Colubrina, Trymalium, Hovenia, Scutia, Phylica, Spyridium, Sageretia, Alphitonia, Cryptandra, Pomaderris.*

TRIBE IV. COLLETIEÆ.—Ovary free or semi-superior. Calyx-tube deep, produced much beyond the disk. Stamens inserted on its mouth. Fruit coriaceous, of 2-3 cocci or a 1-3-celled drupe.—Trees or shrubs, often spinescent. Leaves opposite or small or 0. *Colletia, Discaria, Retanilla, Trevoa, Talguenea.*

TRIBE V. GOUANIEÆ.—Ovary inferior. Disk various. Fruit coriaceous, containing 3 cocci, usually 3-gonous or 3-alate.—Shrubs. Leaves alternate, often broad. *Crumenaria, Gouania, Helinus, Reissekia.*—ED.]

The affinities of *Rhamneæ* with *Celastrineæ* and *Ampelideæ* will be found under these families. Those with *Araliaceæ* are the same as with *Ampelideæ*. They also approach *Elæagneæ* in the valvate calyx, stamens inserted alternately with the sepals (at least in the isostemonous flowers) on a perigynous disk, the erect anatropous ovule, albuminous straight axile embryo, woody stem, usually alternate leaves, and axillary flowers. But *Elæagneæ* are apetalous (which is also the case with some *Rhamneæ*), their ovary is one-celled and -ovuled, their leaves are covered with scales, and are exstipulate. The same analogies and differences are observable between *Rhamneæ* and *Proteaceæ*, and the latter also differ in the entire absence of albumen.

Rhamneæ inhabit the moderately hot regions of all countries of both hemispheres; they are not rare in the torrid zone, but are never met with in glacial regions. [*Colletieæ*, so remarkable for their leafless branches with cruciate spines, are wholly South American, New Zealand, and Australian.

Of all the genera in this family, the most useful to man are *Rhamnus* and *Zizyphus*. *Rhamnus catharticus* bears berries which contain a bitter principle, much used in the form of a purgative syrup. The fruits of many allied species (especially *R. infectorius*) yield a yellow or green colour, and as dyes are the objects of a considerable commerce. *R. utilis* and *chlorophorus* produce Chinese green. The bark of *R. catharticus* is also used for dyeing yellow, like that of *R. Frangula,* a common shrub throughout temperate Europe, the tender and porous wood of which yields a very light charcoal, which is used, like that of *Euonymus,* in the manufacture of gunpowder. The *Zyziphi* contain in every part astringent and bitter principles; but in the fruit this bitterness is corrected by a quantity of sugar and mucilage which render them edible. *Z. vulgaris,* a native of Syria, was imported into Italy towards the first century of our era, and has long been naturalized in the south of France; its drupe is used as an emollient and laxative. The *Z. Lotus,* the Nabk of the Arabs, is very abundant along the African shore of the Mediterranean; its pulpy and agreeable fruit was very celebrated among the ancients, and is still eaten. [The succulent peduncles of *Hovenia elatus* are much eaten in China as a fruit. The *Quina* of Brazil is the acrid root of *Discaria febrifuga.* The bitter bark of *Colubrina* is said to bring on violent fermentation. The Chinese employ the leaves of *Sageretia theezans* as a kind of tea.—ED.]

LXVII. *AMPELIDEÆ.*

(VITES, *Jussieu.*—SARMENTACEÆ, *Ventenat.*—AMPELIDEÆ, *Kunth.*—VITACEÆ, *Lindl.*)

COROLLA *polypetalous or sub-polypetalous, isostemonous, æstivation valvate.* PETALS 4–5, *inserted outside a disk lining the calyx, and surrounding the base of the ovary.* STAMENS

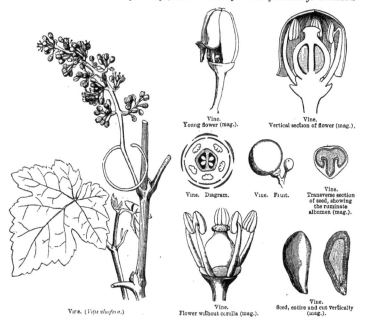

Vine. Young flower (mag.).

Vine. Vertical section of flower (mag.).

Vine. Diagram.

Vine. Fruit.

Vine. Transverse section of seed, showing the ruminate albumen (mag.).

Vine. (*Vitis vinifera.*)

Vine. Flower without corolla (mag.).

Vine. Seed, entire and cut vertically (mag.).

4–5, opposite to and inserted with the petals. OVARY *free, of 2–3–6 1–2-ovuled cells.* OVULES *erect, anatropous.* BERRIES *2–3–6-celled.* EMBRYO *albuminous.* RADICLE *inferior.*—STEM *woody.*

TREES or sarmentose SHRUBS, usually climbing, stem and branches nodose. LEAVES petioled, simple, palmate, digitate or imparipinnate, the lower opposite, the upper alternate, opposite to the peduncles which are often changed into branching tendrils ; *stipules* petiolar, sometimes 0. FLOWERS ⚥ or unisexual, usually small, greenish, in racemes, panicles or thyrsi. CALYX small, 4–5-toothed or entire, clothed with a disk. PETALS 4–5, inserted on the outer base of the disk, coherent at the top, sometimes connate at the base (*Leea*), æstivation valvate. STAMENS 4–5, opposite to the petals, and inserted with them, or fixed to the dorsal face of a sub-globose 5-lobed cup, adnate to the base of the corolla (*Leea*) ; *filaments* short, distinct or sub-monadelphous at the base (*Leea*) ; *anthers* introrse, dehiscence longitudinal. OVARY free, of 2 2-ovuled cells, or of 3–6 1-ovuled cells (*Leea*) ; *style* short or 0 ; *stigma* capitate or peltate ; *ovules* anatropous when solitary, erect when 2, collateral and ascending. BERRY 2–3–6-celled. SEEDS erect, testa bony, endopleura often rugose, or folded within. EMBRYO short, at the base of a cartilaginous often ruminate albumen ; *radicle* inferior.

GENERA.

| Cissus. | Ampelopsis. | Vitis. | Leea. | Pterisanthes. |

The genus *Pterisanthes*, which inhabits the Indian Archipelago, has a peculiar inflorescence, which deserves to be mentioned : the flowers are unisexual, and inserted on a large flattened membranous receptacle : the ♂ are marginal and pedicelled ; the ♀ sessile on the disk.

The affinities of *Ampelideæ* are rather obscure. They approach *Araliaceæ*, and especially the Ivy, by their climbing stem, palmately lobed leaves, valvate æstivation of the petals, dorsifixed incumbent anthers, berried fruit, and small embryo with often ruminate albumen. The most important difference is in the position of the stamens, which in *Araliaceæ* are alternate with the petals ; the epigyny and the inverted ovule, which also distinguish these latter from *Ampelideæ*, are perhaps of less importance. *Rhamneæ* also are connected with *Ampelideæ* by the woody stem, often climbing by tendrils, the alternate or opposite stipulate leaves, the valvate isostemonous petals inserted on a perigynous disk, the stamens opposite to the petals, the ovary often buried in the disk, its 1–2-ovuled cells, and the erect ovules. They scarcely differ except in their penninerved leaves and in the albumen being scanty or 0. Finally, some distant relations have been observed between *Ampelideæ* and *Meliaceæ*, almost entirely founded on the monadelphous stamens of the genera *Melia* and *Leea*.

Ampelideæ abound in the tropics, but are much rarer in the temperate zones. None are found wild in Europe. The *Vitis vinifera* is apparently a native of Georgia and Mingrelia ; it is now cultivated in all countries of which the mean summer temperature is not below 66° Fahr.; where the temperature is lower, the saccharine principle does not develop, and the grapes remain sour. The Vine, if cultivated in the tropics, grows rapidly, but the grapes wither before ripening. The *V. vinifera* is almost the only species of the family useful to man ; the berries of the allied species which grow in the North American forests are acid and little sought.[1]

[1] The *Vitis Labrusca* of the United States is much used in making wine, and is the origin of the Isabella Grape. *V. æstivalis*, of the same country, is also pleasant and edible, as are the berries of various Indian species. The Sultana Raisin is the seedless fruit of a variety of the common Grape cultivated in the Levant ; and the Currant or Corinth Grape of commerce is the fruit of another, cultivated in the Ionian Islands. All parts of *Cissus cordata* and *setosa* are acrid, and applied in India to indolent tumours. The leaves and fruit of *C. tinctoria* abound in a green colouring matter, which soon turns blue, and is used to dye cotton fabrics in Brazil. *V. latifolia* and others are famous in India for their real or supposed properties in indolent ulcers and toothache, and as detergents and purifiers of the blood. —ED.

Cissus grows in the tropics ; its berries are refreshing, and the young leaves of some species, when cooked, serve for food.

LXVIII. *SAPINDACEÆ, Jussieu.*

COROLLA *either 0, or composed of 5-4 petals, imbricate, inserted outside a glandular or annular disk.* STAMENS *usually inserted within the disk, double, equal, or fewer than the sepals, rarely more.* OVARY *central or excentric, usually 3-celled; cells usually 1-*

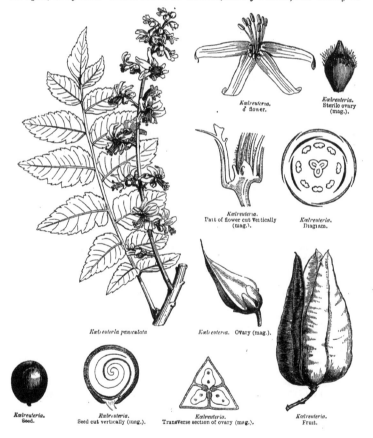

Kœlreuteria.
♂ flower.

Kœlreuteria.
Sterile ovary
(mag.).

Kœlreuteria.
Part of flower cut vertically
(mag.).

Kœlreuteria.
Diagram.

Kœlreuteria paniculata

Kœlreuteria. Ovary (mag.).

Kœlreuteria.
Seed.

Kœlreuteria.
Seed cut vertically (mag.).

Kœlreuteria.
Transverse section of ovary (mag.).

Kœlreuteria.
Fruit.

Stadmannia.   *Alectryon.*   *Alectryon.*   *Alectryon.*
Fruit.   Ovary.   Seed cut vertically (mag.).   Fruit.

(*sometimes 2–several-*) *ovuled.* FRUIT *a capsule, samara, drupe, or berry.* EMBRYO *curved, or rolled into a spiral.*

TREES, SHRUBS or UNDERSHRUBS, rarely HERBS. STEM with watery juice, erect or climbing, often furnished with tendrils. LEAVES alternate, or very rarely opposite, generally compound, sometimes appearing simple by the suppression of the lateral leaflets; *petiole* sometimes winged; *stipules* caducous, or often 0. FLOWERS ☿, or polygamo-diœcious, racemed or panicled, pedicels bracteolate at their base, the lower often changed into tendrils. CALYX of 5 usually unequal sepals, the 2 posterior often united in one, all more or less connate, æstivation imbricate. DISK fleshy, free, or lining the base of the calyx, sometimes regular, forming an entire ring, or lobed between the petals and stamens, sometimes unilateral, shorter, or wanting at the posterior part of the flower, prolonged anteriorly into a lamina, sometimes double, or divided into glands opposite to the petals. PETALS 0, or inserted outside the disk, alternate with the sepals, the posterior often wanting, the others equal or unequal, imbricate or rarely sub-valvate; claw velvety, or glandular within, furnished in all the petals, or the anterior only, with a hooded or crest-like scale which often terminates in an inflexed appendage. STAMENS usually 8, sometimes 10, rarely 5 (very rarely 2, 4, 12 or ∞), usually inserted within the disk, rarely on its edge or around its base, often excentric or unilateral; *filaments* filiform-subulate, free or united at the base, equal or unequal; *anthers* introrse, dorsifixed, dehiscence longitudinal. OVARY free, central or excentric, a little oblique, 3- (rarely 2–4-) celled; *style* terminal, simple; *stigmas* as many as cells; *ovules* anatropous or campylotropous, inserted at the angle of the cells, usually solitary, sometimes geminate, usually ascending, with a ventral raphe and inferior micropyle, very rarely numerous and horizontal, or inverted, funicle often swollen. FRUIT 2-3-4-celled, or 1-celled by suppression, rarely 5–6-celled (*Dodonæa*); sometimes a woody, coriaceous or membranous loculicidal, septicidal, or circumsciss capsule; sometimes samaras winged on their back, base or top, usually indehiscent; sometimes a drupe or berry. SEEDS globose or compressed; *testa* crustaceous or membranous, sometimes winged, often arillate or largely umbilicate. EMBRYO exalbuminous, rarely straight, often bent, or rolled into a crozier; *cotyledons* incumbent, sometimes transversely folded, frequently confluent into a fleshy mass; *radicle* facing the hilum, usually inferior, very rarely superior.

[Sub-orders, tribes and genera of *Sapindaceæ* according to the authors of the 'Genera Plantarum':—

Sub-order I. Sapindeæ.—Stamens inserted within the disk, around the base of the ovary, or unilateral. Seeds exalbuminous. Leaves rarely opposite.

A. Flowers usually irregular and 4-petalous. Disk unilateral or very oblique.

* Ovules solitary.—*Urvillea, Serjania, Toulicia, Cardiospermum, Paullinia, Hemigyrosa, Erioglossum, Schmidelia,* &c.

** Ovules 2 or more in the cells.—*Kœlreuteria, Cossignia, Æsculus, Ungnadia, Stocksia, Diplopeltis, Magonia,* &c.

B. Flowers regular or nearly so. Disk complete.

* Ovule usually solitary. Fruit capsular.—*Cupania, Thouinia, Ratonia, Atalaya,* &c.

** Ovule solitary. Fruit indehiscent, not lobed.—*Talisia, Hippobromus, Melicocca, Schleichera.*

*** Ovule solitary. Fruit indehiscent, deeply lobed, or of 1–3 cocci.—*Sapindus, Euphoria, Capura, Pappea, Deinbollia, Pometia, Nephelium, Heterodendron.*

**** Ovules 2 or more.—*Harpullia,* ? *Hypelate, Xanthoceras, Llagunoa.*

Sub-order II. Acerineæ.—Flowers regular. Sepals and petals (when present) isomerous. Stamens variously inserted. Ovarian cells 1–2-ovuled. Fruit with indehiscent lobes. Seeds exarillate, albuminous. Leaves opposite. See order *Acerineæ*, p. 354.

Sub-order III. Dodoneæ.—Flowers regular. Sepals and petals (when present) isomerous. Disk 0 or complete. Stamens inserted outside the disk (when present), or between its lobes. Ovarian cells 1–2-ovuled. Fruit various. Seeds albuminous. Leaves very rarely opposite. *Dodonæa, Alectryon, Distichostemon, Pteroxylon, Alvaradoa, Aitonia.*

Sub-order IV. Meliantheæ.—Flowers irregular. Stamens inserted within the disk, hypogynous. Seeds albuminous. Embryo straight. Leaves opposite. See order *Meliantheæ*, p. 358.

Sub-order V. Staphyleæ.—Flowers ☿, regular. Stamens inserted outside the disk at its base. Seeds albuminous. Embryo straight. Leaves opposite. See order *Staphyleaceæ*, p. 344.—Ed.]

This family is very closely allied to *Acerineæ* and *Malpighiaceæ*, as well as to *Hippocastaneæ* and *Staphyleaceæ* (see these families). It has equally an affinity with *Meliantheæ*, which only differ in their albuminous seed. Through *Staphyleaceæ* it is also connected with *Celastrineæ*; but is distinguished from them by its generally compound leaves, often irregular flowers, rarely isostemonous petals, stamens inserted within the disk,[1] calyx with free sepals, and generally curved embryo.

Sapindaceæ abound in tropical regions, especially in America; they are rare beyond the tropic of Capricorn, and have not yet been observed north of the tropics, except in the North of China and in India (*Xanthoceras*[2]); *Dodonæa* abounds in Australia.

Sapindaceæ possess very various properties. Many contain astringent and bitter principles, sometimes

[1] They are inserted outside the disk in *Dodonæa* and other undoubted *Sapindaceæ*.—Ed.

[2] *Kœlreuteria, Stocksia, Ungnadia, Cardiospermum,* and *Sapindus* offer other exceptions.

A A

combined with a resinous matter and a certain quantity of volatile oil. The berries of several species and the aril of others have an agreeable taste, owing to the mucilage, sugar, and free acids which they contain; while others possess narcotic principles which render them eminently poisonous. The seeds of most yield a fixed oil by expression. The bark, root, and pulp of the fruit of the Soap Tree (*Sapindus Saponaria*) are regarded as tonics; besides which, this pulp, like that of its Asiatic congeners, froths with hot water like soap, and is used in washing, being said to cleanse more linen than sixty times its weight of soap. The berries of *S. senegalensis* are sought by the negros for their sugary and vinous taste. The succulent and well-tasted aril of *Melicocca* serves as food in Asia and America, as well as that of the Akee (*Blighia* or *Cupania sapida*), an African plant cultivated over the tropics of both worlds. The fruit, cooked with sugar and cinnamon, is taken for dysentery, and when roasted is applied to indolent tumours. The species of the genus *Nephelium* rank high among Asiatic fruits; *N. Litchi* (Litchi), *N. Longanum* (Longan), *N. lappaceum* (Rambutan), and their congeners, are cultivated for their excellent fruits, which are used in inflammatory and bilious fevers. The *Serjania* and the *Paullinia*, American genera, are poisonous; the Brazilians use their juice to stupefy fish; and it is from the flower of *Serjania lethalis* that the Lecheuquana bee collects a narcotic-acrid honey, of which a small quantity produces raving madness, and even death. The juice of *Paullinia Cururu* is used by the savages of Guiana to poison their arrows; the negro slaves prepare a poison with the root and seeds of *P. pinnata*; the expressed juice of its leaves furnishes the Brazilian Indians with a powerful vulnerary. The seed of *P. sorbilis* is bitter and astringent; the Brazilians powder it, and make it into a paste called *guarana* bread, which they roll into little balls or cylinders; on their journeys they dilute this dried paste with sugared water, when it forms a refreshing and febrifuge drink. [It owes its properties to a principle, *guaranine*, identical with theine.—ED.] *Cardiospermum Halicacabum*, a herb growing throughout the tropics, produces a mucous nauseous root to which aperient and lithontriptic virtues are attributed, [but its leaves are cooked as a vegetable in the Moluccas. —ED.]. *Dodonæa* owes its scent to a resinous principle which exudes from its leaves and capsules; the leaves of *D. viscosa* are used for baths and fomentations; its seeds are edible. [The *Rambeh* and *Choopa* of Malacca, species of *Pierardia*, and the *Tampui* (*Hedycarpus malayanus*) yield esculent drupes; these genera, hitherto regarded as Euphorbiaceous, have recently been referred to *Sapindaceæ*. The seeds of the wild Prune of South Africa (*Pappea*) abound in oil. Many produce most valuable timber, and the structure of the wood of many climbing species is most remarkable. The African Teak (*Oldfieldia africana*), doubtfully placed in *Euphorbiaceæ*, is referred to *Sapindaceæ* by T. Mueller. The wood of *Pteroxylon utile*, of South Africa, is hard, and as handsome as Mahogany; its sawdust causes sneezing, whence the Cape names of Neishout and Sneezewood. *Hippodromus aluta*, of the same country, also yields a valuable timber.—ED.]

LXIX. *ACERINEÆ*.[1]

(ACERA, *Jussieu.*—ACERINEÆ, *D.C.*—ACERACEÆ, *Lindl.*)

PETALS 4-5, *hypogynous, imbricate, sometimes* 0. STAMENS *equal or more in number than the petals.* OVARY *2-lobed, of two 2-ovuled cells; style central.* OVULES *pendulous, curved.* FRUIT *a samara.* EMBRYO *exalbuminous; cotyledons folded or convolute; radicle descending.*—STEM *woody.* LEAVES *opposite.*

TREES with sugary, usually limpid, but sometimes milky juice; *buds* scaly. LEAVES opposite, petiolate, usually simple, palmi-nerved and -lobed, rarely entire or imparipinnate, leaflets petiolulate; *stipules* 0. FLOWERS ☿, or often polygamodiœcious, regular, in a simple or compound raceme or corymb, axillary or terminal; *pedicels* with a caducous bract. CALYX 4-5- (rarely 6-8-) partite, segments often

---

[1] See *Sapindaceæ*, Suborder III., p. 353.—ED.

coloured, imbricate, deciduous. PETALS 4–5 or 0, often sepaloid, inserted on the
edge of a free disk surrounding the base of the ovary, shortly clawed, æstivation
imbricate. STAMENS inserted with the petals, equal or more numerous, 4–12, oftener

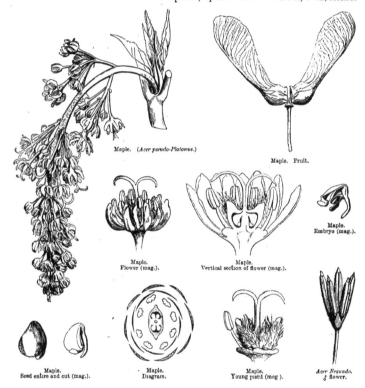

Maple. (*Acer pseudo-Platanus.*)

Maple. Fruit.

Maple.
Embryo (mag.).

Maple.
Flower (mag.).

Maple.
Vertical section of flower (mag.).

Maple.
Seed entire and cut (mag.).

Maple.
Diagram.

Maple.
Young pistil (mag).

Acer Negundo.
♂ flower.

8; *filaments* filiform, free, sometimes very short; *anthers* 2-celled, introrse, oblong,
basifixed or versatile, dehiscence longitudinal. OVARY free, sessile, 2-celled, 2-lobed,
compressed perpendicularly to the septum; *style* central, sub-basilar; *stigma* bifid;
ovules 2 in the central angle of each cell, superimposed or collateral, pendulous,
campylotropous. FRUIT of 2 samaras or 1- (rarely 2-) seeded cocci, prolonged into a
dorsal coriaceous or membranous wing, reticulate, and remaining suspended to a
carpophore, as in *Umbelliferæ*. SEEDS ascending; *testa* membranous; *endopleura*

fleshy. EMBRYO exalbuminous; *cotyledons* foliaceous, green, accumbent, irregularly folded or convolute ; *radicle* descending, facing the hilum.

GENERA.

Acer.   Negundo.   Dobinea.

Acerineæ, regarded by Bentham and Hooker fil. as a sub-order of *Sapindaceæ* (see p. 353), only differ from these in their always opposite leaves and non-appendiculate petals, and in their occasionally trinerous fruit (*Acer pseudo-Platanus*), as in most *Sapindaceæ*. *Hippocastaneæ* only differ from *Acerineæ* in their capsule with semi-septiferous valves ; and are further allied, since in both orders the buds bear both leaves and flowers. For the affinity with *Malpighiaceæ* see p. 301.

Acerineæ inhabit northern temperate regions, and especially Japan, the Himalayas, and eastern North America. The curious genus *Dobinea*, which is placed in *Acerineæ*, is Himalayan ; [it has apetalous ♀ flowers without a disk, and a one-celled ovary, which ripens into a small broadly-winged achene.—ED.].

Acerineæ contain a sugary sap, milky in some, limpid in others, which is obtained by incision of the trunk, and is either evaporated for sugar (the Maple Sugar of Canada), or allowed to ferment and thus form spirituous or acid liquors. Their bark is astringent, and yields reddish or yellow colouring principles. [The wood of various species is of great value, especially the Bird's-eye Maple of America.—ED.]

LXX. *HIPPOCASTANEÆ*,[1] *Endlicher.*

Large or small TREES with scaly buds. LEAVES opposite [alternate in *Ungnadia.*—ED.], generally digitate, rarely imparipinnate, with toothed or crenate leaflets ; *stipules* 0. FLOWERS ☿ or polygamous, in racemes or terminal thyrsi. CALYX campanulate or tubular, 5-fid, lobes unequal, imbricate. PETALS 4-5, inserted on the receptacle, unequal, clawed, not appendiculate, imbricate. DISK hypogynous, entire, annular or unilateral. STAMENS 5-8, usually 7, inserted within the disk, and free ; *filaments* filiform, exserted, ascending ; *anthers* 2-celled, dehiscence longitudinal. OVARY sessile, oblong or lanceolate, of 3 2-ovuled cells ; *style* conical or filiform ; *stigma* pointed ; *ovules* curved, fixed to the central angle of the cell, superimposed or horizontal, or one ascending, the other pendulous. CAPSULE coriaceous, naked or spinous, 3-celled or 2-1-celled by suppression, loculicidal, valves semi-septiferous. SEEDS usually solitary in each cell ; *testa* coriaceous, shining ; *hilum* basilar, large. EMBRYO exalbuminous, curved ; *cotyledons* large, thick, fleshy, often more or less confluent ; *radicle* short, near the hilum.

PRINCIPAL GENERA.

Æsculus.   Pavia.   Ungnadia.

The little group of *Hippocastaneæ*, which evidently belongs to the family of *Sapindaceæ*, is only distinguished from the latter by its opposite digitate leaves and the 2-ovuled ovarian cells ; and it must further be observed that the genus *Ungnadia*, placed by botanists near *Æsculus*, has alternate and imparipinnate leaves, which bring it still nearer to *Sapindaceæ*. For the affinity with *Acerineæ*, see this family.

Hippocastaneæ are chiefly North American, except *Castanella* [which is not referable to *Hippocastaneæ*,

---

[1] Included in the ' Genera Plantarum ' in *Sapindaceæ*, where it does not even form a tribe ; the leaves of *Ung-* *nadia* being alternate invalidates the only character distinguishing it from *Sapindaceæ* proper.—ED.

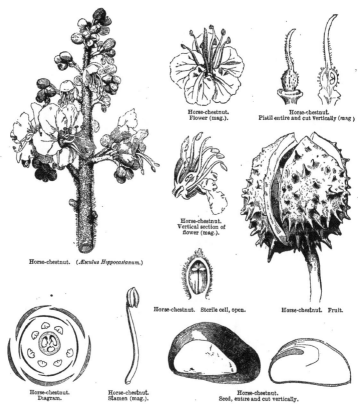

Horse-chestnut.
Flower (mag.).

Horse-chestnut.
Pistil entire and cut vertically (mag.)

Horse-chestnut.
Vertical section of
flower (mag.).

Horse-chestnut. (*Æsculus Hippocastanum.*)

Horse-chestnut. Sterile cell, open.

Horse-chestnut. Fruit.

Horse-chestnut.
Diagram.

Horse-chestnut.
Stamen (mag.).

Horse-chestnut.
Seed, entire and cut vertically.

in its restricted sense], which belongs to New Grenada; one species, the Horse-chestnut (*Æ. Hippocastanum*) is cultivated in Asia and Europe [but its origin is wholly unknown. There are also two Indian species, one Himalayan, and the other tropical, found in Silhet and Assam.—ED.]. The bark of the Horse-chestnut contains gallic acid and a bitter principle, which make it rival, as a tonic, that of the Willow; its seeds, the taste of which is at once mild and bitter, are rich in starch, and are given in Turkey to broken-winded horses; reduced to powder they serve as soap; roasted they are used as coffee; and fermented they yield a spirituous liquor, which yields alcohol by distillation; the young aromatic buds have been employed in place of the Hop in the manufacture of beer. [The fruit and leaves of the American *Æ. Ohioensis* (Buckeye) are considered to be deadly poison.—ED.]

LXXI. *MELIANTHEÆ,*[1] *Endlicher*

Melianthus.
Flower (mag.).

Melianthus.
Fruit.

Melianthus minor.

Melianthus.
Seed (mag.).

Melianthus.
Seed, entire and cut vertically
(mag.).

Melianthus.
Flower deprived of part of its
calyx and petals (mag.).

Bersema abyssinica.
Diagram. (From
Planchon's ' Mémoire.')

SHRUBS, glabrous, glaucous or whitish. LEAVES alternate, stipulate, imparipinnate, leaflets unequilateral, toothed, decurrent; *stipules* 2, free, or united into
one, very large, intrapetiolar. FLOWERS ⚥, in axillary and terminal racemes,
shortly pedicelled, bracteate, the lower sometimes apetalous, with 2 fertile and 2
sterile stamens. CALYX compressed, of 5 unequal segments, the inner very short,
distant, gibbous below, hooded at the top; the others lanceolate, flat; the two upper
larger, covering the lateral. PETALS 5, excentric (the fifth minute or 0), subperigynous, narrow, long-clawed, cottony in the middle. DISK thickened, unilateral,
lining the gibbous bottom of the calyx, and distilling an abundant nectar. STAMENS

[1] A sub-order of *Sapindaceæ*; see p. 353.—ED.

4, hypogynous, inserted within the disk, nearly central, didynamous and a little inclined; *anthers* introrse, 2-celled, ovoid-oblong, dehiscence longitudinal. OVARY oblong, 4-lobed, 4-celled; *style* central, arched, fistular, furrowed; *stigma* 4-toothed; *ovules* 2–4 in each cell, 2-seriate on the inner angle above the middle, ascending or horizontal, anatropous. CAPSULE papery, deeply 4-lobed, with 4 1-seeded cells opening ventrally at the top. SEEDS sub-globose, without aril; *testa* crustaceous, shining; *hilum* conical, foveolate; *albumen* copious, fleshy or horny. EMBRYO small, green; *cotyledons* linear-oblong; *radicle* thickened at the tip.

GENERA.

Melianthus.   Bersama.   Greyia.

The genus *Melianthus*, which was formerly placed in *Zygophylleæ*, is separated from it by its irregular and racemed flowers, perigynous and isostemonous petals, and ascending ovules. It is only distinguished from *Sapindaceæ* by its albuminous[1] seed; and it has therefore been annexed to this family by Planchon and Bentham and Hooker, together with the genus *Bersama*, which differs in its often polygamous flowers, its stamens all (or two only) united at the base, its one-ovuled ovarian cells, its capsule with four semi-septiferous valves, and its arillate seeds. [The remarkable and beautiful Natal genus *Greyia* is referred here by Bentham and Hooker fil. (Gen. Pl. p. 1000), and though exstipulate, is regarded as a member of *Meliantheæ* with partially consolidated carpels, which, being united by the margins only, enclose one cell with parietal placentas. The fruit breaks up into five follicles.—ED.]

Melianthus inhabits South Africa; one species has been introduced into Nepal [no doubt in gardens only. The whole plant has a remarkably heavy smell.—ED.].

The sugary-vinous nectar secreted by the disk of *M. major* is much sought by colonists and natives of the Cape. That of *M. minor* is thicker and less esteemed.

LXXII. SABIACEÆ, Blume.[2]

[SHRUBS or TREES, erect or scandent or sarmentose, with watery juice, glabrous or with simple hairs. LEAVES alternate, exstipulate, simple or pinnate, entire or serrate, penninerved. INFLORESCENCE various, usually panicled. FLOWERS ⚥ or polygamo-diœcious, small or minute. CALYX 4–5-partite, imbricate in bud. PETALS 4–5, equal or unequal, inserted on the receptacle, alternate with or opposite to the sepals, imbricate in bud. DISK small, annular, lobed, rarely tumid. STAMENS 4–5, inserted at the base or on the top of the disk opposite the petals, free or adnate to the petals, rarely all equal and fertile with thick filaments, often 2 opposite the inner smaller petals, perfect with clavate or obcuneate filaments, and 3 antherless scale-like; *anthers* didymous, separated by a thick connective, or dehiscing transversely or by a deciduous calyptra; *pollen* minute, globose. OVARY sessile, 2–3-celled, compressed or 2–3-lobed; *styles* connate or cohering, with stigmatiferous tips, or *stigmas* simple, sessile on the lobes of the ovary; *ovules* 1–2 in each cell, collateral or superimposed, horizontal or pendulous, raphe ventral, micropyle inferior, remote from the funicle. FRUIT of 1 or 2 dry or drupaceous indehiscent carpels, often sub-globose,

---

[1] *Staphylea* having albuminous seeds completely unites these families.—ED.

[2] This order is omitted in the original.—ED.

top usually deflexed, compressed and reniform in *Sabia*; *endocarp* crustaceous or bony, 1-seeded. SEEDS compressed or sub-globose, adnate by a broad hilum to the base of the cell; *testa* membranous or coriaceous; *albumen* 0, or a thin layer adnate to the testa. EMBRYO with thick rugose or membranous contorted *cotyledons*, and an inferior curved *radicle* pointing upwards to the hilum.

GENERA.

Sabia.   Meliosma.   Phoxanthus.   Ophiocaryon.

A small but well-defined order of four genera and about thirty-two species, differing from its allies in the isomerous stamens opposite the petals. It is related to *Terebinthaceæ* and *Sapindaceæ*, but differs from both not only in the above character, but from *Terebinthaceæ* in the always few stamens, anthers, filaments, two-ovuled ovarian cells, and deflexed carpels; and from *Sapindaceæ* in the short stamens, never eight in number, which are not declinate. *Sabia* is very remarkable for the opposition of its bracts, sepals, petals, stamens, and ovarian carpels, which is perhaps unique in the Vegetable Kingdom. Its fruit and certain other characters have caused authors to assign it to *Menispermaceæ*, with which it has nothing whatever in common.

Sabiaceæ are for the most part tropical Indian, but *Sabia* is Himalayan also. *Meliosma* is common to Asia and America; *Phoxanthus* and *Ophiocaryon*, monotypic genera, are from North Brazil and Guiana.

The wood of the Indian *Meliosma* is of excellent quality, and in considerable demand for house-building. The singular embryo of *Ophiocaryon*, resembling a snake coiled up inside the nut, gives the name of Snake-nut to the fruit.—ED.]

LXXIII. *TEREBINTHACEÆ*.

(ANACARDIEÆ, Br.—TEREBINTHACEÆ, Kunth.—ANACARDIACEÆ, Lindl.)

FLOWERS *very often diclinous.* PETALS *inserted on an annular disk, equal in number to the calyx-lobes, sometimes 0.* STAMENS *equal or double the number of the petals.* OVARY *generally solitary, 1-celled, 1-ovuled.* OVULE *suspended from a basilar or lateral funicle.* FRUIT *usually a drupe.* EMBRYO *exalbuminous.*—STEM *woody.* LEAVES *exstipulate.*

TREES, large or small, with gummy or milky-resinous juice, often poisonous. LEAVES alternate, or very rarely opposite (*Bouea*), simple, 3-foliolate or impari-pinnate, exstipulate. FLOWERS ☿, or polygamo-diœcious, or monœcious, regular, small, axillary or terminal, fascicled, spiked or panicled. CALYX 3–5-fid or -partite, often persistent, sometimes accrescent (*Loxostylis*). PETALS equal in number to the calyx-lobes, inserted at the base or top of an annular disk, æstivation usually imbricate, sometimes accrescent (*Melanorrhœa*), or 0 (*Pistacia*). STAMENS inserted with the petals, or double in number, very rarely more (*Melanorrhœa*), and then some imperfect; *filaments* subulate or filiform; *anthers* very often versatile, introrse, dehiscence longitudinal. OVARY 1-celled (*Anacardieæ*), or 2–5-celled (*Spondieæ*), or very rarely of 5–6 distinct carpels, of which all but one are sterile, or reduced to the style (*Buchanania*); *style* simple, terminal or sublateral, sometimes several by the sup-

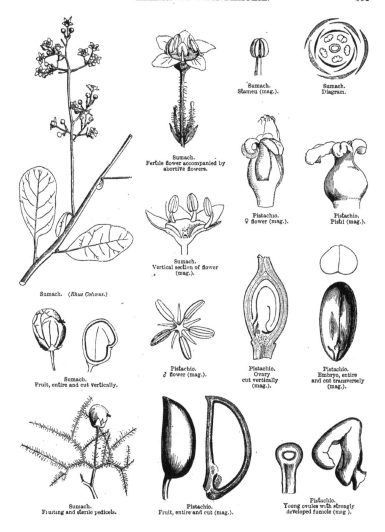

Sumach.
Stamen (mag.).

Sumach.
Diagram.

Sumach.
Fertile flower accompanied by
abortive flowers.

Pistachio.
♀ flower (mag.).

Pistachio.
Pistil (mag.).

Sumach.
Vertical section of flower
(mag.).

Sumach. (*Rhus Cotinus*.)

Sumach.
Fruit, entire and cut vertically.

Pistachio.
♂ flower (mag.).

Pistachio.
Ovary
cut vertically
(mag.).

Pistachio.
Embryo, entire
and cut transversely
(mag.).

Sumach.
Fruiting and sterile pedicels.

Pistachio.
Fruit, entire and cut (mag.).

Pistachio.
Young ovules with strongly
developed funicle (mag.).

pression of the ovaries, which have become confluent with the fertile one; *ovules* solitary, pendulous, or broadly adnate to the wall of the cell, or suspended to a basal ascending funicle, micropyle superior and raphe dorsal, rarely erect with the micropyle inferior and raphe ventral (*Anacardium, Mangifera*, &c.). FRUIT usually superior, rarely inferior (*Holigarna*), free, or girt by a receptacular cúp, sometimes seated on a broadened pyriform fleshy receptacle (*Anacardium*), usually a drupe, indehiscent, or with a dehiscent endocarp, rarely nut-like (*Anacardium*). SEED erect, horizontal or inverted; *testa* membranous, sometimes confluent with the endocarp; *hilum* usually ventral. EMBRYO exalbuminous; *cotyledons* plano-convex; *radicle* more or less curved, superior or inferior.

[The following is the new arrangement of *Anacardiaceæ* (*Terebinthaceæ*) in the ' Genera Plantarum ' : —

TRIBE I. ANACARDIEÆ.—Ovary 1-celled.

A. Ovule usually suspended from a basilar funicle.

* Sepals and petals not accrescent.—*Rhus, Pistacia, Sorindeia, Buchanania, Comocladia, Mangifera, Anacardium, Bonea, Gluta.*

* * Sepals or petals accrescent.—*Melanorrhœa, Swintonia, Loxostylis.*

B. Ovule suspended from above the middle of the cell.

* Leaves compound. Calyx not accrescent.—*Schinus, Solenocarpus, Smodingium, Odina.*

* * Leaves compound. Calyx accrescent.—*Astronium, Parishia.*

* * * Leaves simple (or pinnate in *Mauria*).—*Semecarpus, Corynocarpus, Drimycarpus, Mauria, Holigarna, Duvana.*

TRIBE II. SPONDIEÆ.— Ovary 2–5-celled.   Ovules pendulous.   *Spondias, Dracontomelum, Sclerocarya, Harpephyllum,* &c.—ED.]

Terebinthaceæ approach *Rosaceæ*, tribe *Amygdaleæ*, in their habit, woody stem, alternate leaves, perigynous insertion of their polypetalous corolla and (sometimes polyandrous) andrœcium; in the solitary carpel, usually drupaceous fruit, and exalbuminous seed. They approach some *Leguminosæ* in the same points, and also in the frequently monadelphous stamens, and in the more or less curved embryo. *Terebinthaceæ* are allied to *Juglandeæ*, which, like them, have diclinous flowers, a one-celled one-ovuled ovary, drupaceous fruit, exalbuminous embryo, woody stem, and alternate usually pinnate leaves. They are also closely allied to *Connaraceæ* and *Zanthoxyleæ*, and have therefore been placed in the same class. *Burseraceæ* scarely differ save in the two-ovuled ovarian cells and the ovules with superior micropyle and ventral raphe. *Connaraceæ* differ in their distinct carpels with two collateral and erect ovules, and their capsular fruit. Finally, *Zanthoxyleæ* chiefly differ in their seed, which is furnished with a more or less copious albumen.

Terebinthaceæ are frequent in the intertropical zone of both continents; they diminish rapidly beyond this zone, so that they are rare in the Mediterranean region,[1] in South Africa, North America, and Australia, where however, five genera occur.

Terebinthaceæ yield medicinal substances, edible fruits, [fine varnishes, ED.], and many woods useful to dyers and cabinet-makers. The principal species are the following :—

---

[1] This refers to genera and species, and not to individuals , for *Pistacia* of three species abounds, as do several of *Rhus*, in some districts of the Mediterranean.—ED.

Pistacia vera (Pistachio). A tree of Persia and Syria,[1] now cultivated throughout the Mediterranean region. Fruit with an oily green seed of an agreeable taste, used by confectioners and pharmacists.—*P. Lentiscus.* A small tree, cultivated in the Greek Archipelago, and especially at Scio, yielding by incision of its trunk an aromatic resin called *mastic*, softening in the mouth, slightly tonic and astringent, and much used in the East to perfume the breath and strengthen the gums [as also to flavour wines and confectionery. In England it is used for varnishing pictures and in dentistry.—ED.].—*P. atlantica*, from Mauritania, also yields a mastic, employed in the same way.—*P. Terebinthus*, a Mediterranean tree, yields by incision the Chio or Cyprus turpentine, formerly used in medicine, but now unjustly depreciated; [it further produces curious horn-shaped galls, used for tanning leather in the East.—ED.].

Schinus Molle (False Pepper) is a small tree of tropical America with a sugary edible drupe and a mastic with the odour of pepper, slightly purgative. [Fragments of the leaf floated on water move about by jerks, owing to the discharge of a volatile oil from the tissues.—ED.].

Duvaua dependens is a small Chilian tree with fermentable seeds which yield an intoxicating drink.

Rhus Coriaria (Varnish-tree) is a small Mediterranean tree, the dried and pulverized leaves of which furnish a tan much used in the preparation of leather; its acid fruit is used in Turkey as a condiment.— The flowers and fruits of *R. typhina*, a North American shrub, are there used to sharpen vinegar, whence its name of Vinegar-tree.—*R. Cotinus*, a South European shrub, yields Venetian Sumach, or Young Fustic, a valuable orange-yellow dye. Its bark is aromatic and astringent, and is used as a febrifuge.—*R. Toxicodendron* (the Poison Sumach or Poison Ivy), of North America, has a milky volatile very acrid juice, the touch of which, or even an exhalation from it, brings on violent erysipelas [in many persons, whilst others are wholly unaffected by it.—ED.]. An extract is prepared from the leaves, and used in some cutaneous disorders.—*R. Vernix*, a Japanese shrub with milky juice, of which is composed Japanese varnish. Other trees of the same family, natives of China and India, yield also by incision a very deleterious resinous juice, employed in the composition of Chinese lacquer. The juice of *R. venenata*, of North America, is not less deleterious, and is similarly employed. *R. Succedanea* yields the Vegetable Wax of Japan [which is found as a thick white coating of the seed within the capsule.—ED.].

Melanorrhœa usitatissima yields the celebrated black varnish of Burmah and Martaban. [A similar varnish is yielded in India by the fruits of *Holigarna longifolia*.—ED.]

Mangifera indica, an East Indian tree introduced into the Antilles, yields the Mango, a large drupe, variable in colour and size, of a perfumed and sugary-acid taste, becoming purgative when eaten to excess; [but which is one of the best of tropical fruits. Its bitter aromatic root is used medicinally.—ED.].

Anacardium occidentale, an American tree now naturalized throughout the tropics, yields the Cashew-nut, which contains in its pericarp a caustic oil [and black varnish], and in its seed a sweet oil. Its greatly enlarged and pear-shaped receptacle, called the Cashew-apple, is juicy, fleshy, sugary and acid in taste, but a little acrid. The true Cashew wood is furnished by a tree of the Antilles belonging to the family of *Cedrelaceæ*. The juice of the pericarp of *Semecarpus* gives an indelible black dye, used for marking linen.

Spondias purpurea (Spanish Plum) is a West Indian tree with an acidulous sugary drupe. *S. dulcis* is cultivated in the Friendly and Society Islands for its wholesome and refreshing fruit. [*S. lutea, Mombra*, and *tuberosa*, yield the Hog Plum of the West Indies.—ED.] *S. birrea*, a native of Senegambia, has a fermentable fruit from which the negros make a spirituous liquor.

[The gum of *Odina Odier* is used for plasters in India.—ED.]

---

[1] The native country of the cultivated Pistachio nut (*P. vera*) is unknown, and it is doubtful whether it may not be a cultivated form of *P. atlantica*, or some other species. The plant is omitted in De Candolle's 'Géographie Botanique.' I did not find it in Syria, and was informed that it was known there in cultivation only told.—ED.

LXXIV. *CONNARACEÆ.*[1]

[Erect or climbing TREES or SHRUBS with watery juice. LEAVES alternate, exstipulate, 1–3-foliolate or imparipinnate; *leaflets* coriaceous, always quite entire. FLOWERS rather small, usually ☿, regular or nearly so, in racemes or panicles. CALYX 5-fid or 4–5-partite, usually persistent and embracing the base of the fruit, imbricate or valvate in bud. DISK 0, or membranous, or annular and embracing the bases of the stamens, sometimes unilateral. STAMENS 5 or 10, perigynous or hypogynous, sometimes declinate, if 10 those opposite the petals usually smaller and often imperfect; *filaments* filiform, usually monadelphous; *anthers* short, didymous, bursting longitudinally, introrse or extrorse after flowering. PETALS 5, linearoblong, free, or cohering slightly in the middle, imbricate or very rarely valvate in bud. CARPELS 5, rarely 1–3, globose, free, hirsute, 1-celled; *styles* subulate or filiform; *stigmas* capitellate, simple or 2-lobed; *ovules* 2, collateral, ascending from the inner angle of the cell, orthotropous. FRUIT of usually a solitary sessile or stipitate follicle, dehiscing by the ventral (rarely by the dorsal) suture, often hairy within, 1-(rarely 2-) seeded. SEED erect, arillate or not; *testa* thick, sometimes fleshy, coloured, and aril-like around the lower half of the seed; *aril* fleshy, coloured, entire, dimidiate or cupular. EMBRYO either exalbuminous with amygdaloid cotyledons, or with fleshy albumen and foliaceous cotyledons; *radicle* superior, very rarely ventral.

PRINCIPAL GENERA.

Byrsocarpus.	Agelæa.	Rourea.	Connarus.
Manotes.	Cnestis.	Tricholobus.	Ellipanthus.

A considerable order, of a dozen genera and 140 species, of very complex affinities, perhaps most closely allied to *Terebinthaceæ*, but differing in the hermaphrodite flowers, geminate orthotropous ovules, and constantly superior radicle. The exstipulate leaves, usually albuminous seed, orthotropous collateral ovules and superior radicle distinguish it from *Leguminosæ*. Planchon has pointed out its many points of analogy with *Oxalideæ*, especially through the genus *Connaropsis*; but their corolla is not contorted, their carpels are free, and they differ essentially in the ovules, seed, and embryo. *Byrsocarpus* presents the remarkable structure of lacunose cotyledons, &c.; the petals are valvate in *Tricholobus*, and (according to Blume) its ovules are anatropous, which character may remove it from the order.

Connaraceæ are altogether tropical, and chiefly Asiatic and Malayan; but many are American, African, and several Australian and Pacific. They are chiefly valuable in an economic point of view for their woods: of these one is the beautiful Zebra wood (*Connarus Lambertii*), a native of Guiana. The aril of some species is edible. The seeds of *C. speciosus*, a large Rangoon tree, yield an abundance of oil, and its wood is very useful, as is that of *C. paniculatus.*—ED.]

LXXV. *LEGUMINOSÆ.*

(PAPILIONACEÆ *et* LOMENTACEÆ, *L.*—LEGUMINOSÆ, *Jussieu.*)

HERBS, SHRUBS, or TREES. LEAVES *alternate, usually compound, stipulate.* FLOWERS *irregular or regular,* ☿*, or sometimes diclinous.* COROLLA *perigynous or hypogynous, regular and valvate in æstivation, or irregular and imbricate, rarely 0.*

---

[1] This order is not described in the original.—ED.

STAMENS *inserted with the corolla, double in number to the petals, or indefinite ; anthers 2-celled.* PISTIL *generally 1-carpelled, becoming a pod or an indehiscent fruit, often jointed.* EMBRYO *generally exalbuminous.*

SUB-ORDER I. *MIMOSEÆ,* Br.

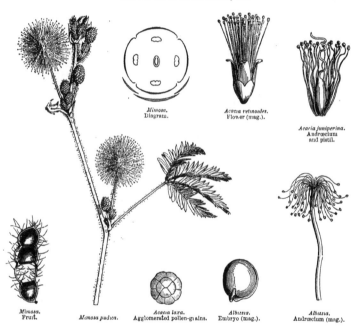

Mimosa.
Diagram.

Acacia retinoides.
Flower (mag.).

Acacia juniperina.
Andrœcium
and pistil.

Mimosa.
Fruit.

Mimosa pudica.

Acacia laxa.
Agglomerated pollen-grains.

Albizzia.
Embryo (mag.).

Albizzia.
Andrœcium (mag.).

STEM woody, rarely herbaceous, unarmed or thorny, straight or sarmentose, sometimes aquatic and floating (*Neptunia*). LEAVES simple (*phyllodes*) or 2–3-pinnate, sometimes irritable ; *petiole* inserted on a small cushion, usually furnished with petiolar or spinal glands; *stipules* free, caducous, sometimes persistent and spinescent. FLOWERS ☿, often polygamous, regular, in a spike or head, rarely in a panicle or corymb. CALYX 4–5-fid or -partite, æstivation generally valvate. PETALS as many as the sepals, inserted on the base of the calyx, or distinctly hypogynous, free (*Parkia, Prosopis*), or more or less coherent into a tube (*Acacia, Mimosa, Inga,* &c.), æstivation generally valvate. STAMENS usually double or multiple the petals, very rarely equal (*Desmanthus*); *filaments* free (*Adenanthera, Desmanthus, Entada,*

Gagnebina, &c.), or monadelphous (*Albizzia, Parkia, Prosopis, Inga*, &c.) ; *anthers* small, rounded, dorsifixed; pollen-granules often agglomerated in fours or sixes. Ripe CARPELS solitary, or very rarely many, free (*Affonsea*) ; *ovary* 1-celled ; *ovules* anatropous. POD 1-celled and 2-valved, or pluricelled owing to transverse or lomentaceous septa, which sometimes attain enormous proportions (*Entada*). SEEDS often marked with an areola. EMBRYO straight, usually exalbuminous, very rarely albuminous (*Fillœa*).

PRINCIPAL GENERA.

Parkia.	Entada.	Adenanthera.	*Calliandra.
Gagnebina.	Neptunia.	Desmanthus.	*Albizzia.
*Mimosa.	*Acacia.	*Inga.	

[The Sub-order *Mimoseæ* is subdivided into five tribes by Bentham :—

TRIBE I. PARKIEÆ.—Calyx-teeth short, broad, imbricate. Corolla 5-fid. Stamens (perfect) 5-10 ; anthers crowned with a deciduous gland. *Parkia, Pentaclethra.*

TRIBE II. ADENANTHEREÆ.—Calyx valvate. Stamens free, usually twice as many as the petals ; anthers usually crowned with a stalked gland ; pollen-grains many, distinct. *Entada, Piptadenia, Adenanthera, Prosopis, Dichrostachys, Neptunia*, &c.

TRIBE III. EUMIMOSEÆ.—Calyx valvate, pappose, or 0. Stamens free ; anthers eglandular ; pollen-grains many, distinct. *Desmanthus, Mimosa, Schranckia, Leucœna, Xylia*, &c.

TRIBE IV. ACACIEÆ.—Calyx valvate, rarely 0. Stamens indefinite, usually many, free or connate at the base ; pollen-grains aggregated in 2-6 masses. *Acacia.*

TRIBE V. INGEÆ.—Calyx valvate. Stamens indefinite, rarely 10-15, connate at the base, rarely beyond the middle ; anthers small ; pollen-grains usually aggregated in 2-6 masses. *Calliandra, Albizzia, Pithecolobium, Inga, Affonsea*, &c.—ED.]

SUB-ORDER II. SWARTZIEÆ, D.C.[1]

Unarmed TREES. LEAVES unequally imparipinnate or simple, stipulate. FLOWERS ☿, sub-irregularly racemed. CALYX valvate in æstivation, 4-5-lobed, or rarely splitting along one side (*Zollernia*). PETALS more or less unequal, imbricate in æstivation, 5-3-1, sometimes 0, generally hypogynous, rarely perigynous (*Aldina*). CARPEL solitary. OVARY 1-celled, stipitate. POD 1-celled, 2-valved, few-seeded, rarely an indehiscent drupe (*Detarium*). EMBRYO exalbuminous.

PRINCIPAL GENERA.

Swartzia.	Aldina.	Detarium.[2]

[1] Regarded as the last tribe of Sub-order *Papilionaceæ* by Bentham.
[2] Referred by Bentham to Tribe VI. of *Cæsalpinieæ.*—ED.

Sub-order III. *CÆSALPINIEÆ, Br.*

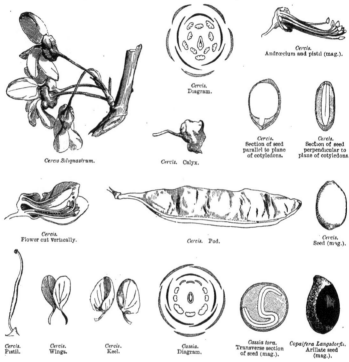

Cercis.
Andrœcium and pistil (mag.).

Cercis.
Diagram.

Cercis Siliquastrum.

Cercis. Calyx.

Cercis.
Section of seed
parallel to plane
of cotyledons.

Cercis.
Section of seed
perpendicular to
plane of cotyledons.

Cercis.
Flower cut vertically.

Cercis. Pod.

Cercis.
Seed (mag.).

Cercis.
Pistil.

Cercis.
Wings.

Cercis.
Keel.

Cassia.
Diagram.

Cassia tora.
Transverse section
of seed (mag.).

Copaifera Langsdorfii.
Arillate seed
(mag.).

STEM woody, straight or climbing, sometimes flexuous, flattened, ribbon-like (*Bauhinia*). LEAVES generally compound, stipulate. FLOWERS ☿, rarely diœcious (*Ceratonia*), imbricate in æstivation, nearly regular or sub-papilionaceous, in racemes or spikes. CALYX usually 5-merous. PETALS inserted on the calyx, usually 5, alternate with the sepals, rarely 3, 2, or 1, sometimes 0 (*Copaifera, Ceratonia*). STAMENS 10 or fewer, inserted with the petals; *filaments* generally free, rarely coherent (*Leptolobium*), more or less unequal. CARPEL solitary; *ovules* anatropous. POD dehiscent or often indehiscent, sometimes pluricelled by transverse septa (*Cassia, Gleditschia*). SEEDS often marked with an areola. EMBRYO straight, exalbuminous, or often albuminous.

[The Sub-order *Cæsalpinieæ* is subdivided into seven tribes by Bentham :—

TRIBE I. SCLEROLOBIEÆ.—Leaves usually unequally pinnate. Sepals free, imbricate. Petals usually 5, subequal. Ovary-stalk free in the bottom of the calyx-tube ; ovules 3–∞ . *Sclerolobium, Pœppigia, Cenostigma,* &c.

TRIBE II. EUCÆSALPINIEÆ.—Leaves 2-pinnate. Sepals free. Petals usually 5, subequal. Ovary-stalk free in the bottom of tho calyx-tube ; ovules 3–∞, rarely 1. *Pellophorum, Mezoneurum, Cæsalpinia, Hoffmanseggia, Hæmatoxylon, Pterolobium, Gleditschia, Poinciana, Parkinsonia.*

TRIBE III. CASSIEÆ.—Leaves equally or unequally pinnate. Sepals 5, rarely 3–4, free, imbricate, rarely subvalvate. Petals 5–0. Anthers bursting by slits or pores. Ovary free in tho base of the calyx-tube. Seeds albuminous. *Cassia, Labichea, Dialium, Ceratonia,* &c.

TRIBE IV. BAUHINIEÆ.—Leaves entire, 2-lobed or -foliolate.· Calyx gamosepalous. Anthers versatile. Ovary-stalk free or adnate to tho calyx-tube. Seeds albuminous. *Bauhinia, Cercis,* &c.

TRIBE V. AMHERSTIEÆ.—Leaves usually pinnate. Sepals free, rarely valvate. Anthers versatile. Ovary-stalk laterally adnate to the calyx-tube ; ovules 3–∞ . *Brownea, Amherstia, Heterostemon, Humboldtia, Macrolobium, Afzelia, Tamarindus, Schotia, Hymenæa, Crudia.*

TRIBE VI. CYNOMETREÆ.—Leaves equally pinnate, or 2-8-foliolate. Flowers small. Sepals free, imbricate or valvate. Petals 5–0. Anthers versatile. Ovary 1-2-ovuled. *Detarium, Copaifera, Hardwickia, Cynometra.*

TRIBE VII. DIMORPHANDREÆ.—Leaves 1-2-pinnate. Flowers small, spicate. Calyx 5-toothed or -lobed. Petals 5, imbricate. Anthers versatile. Ovules ∞ . *Backia, Dimorphandra, Erythrophlœum.*—ED.]

SUB-ORDER IV. *PAPILIONACEÆ, Br.*

STEM woody or herbaceous. ROOTLETS often covered with small tuberous excrescences. LEAVES stipulate, with no petiolar glands, often terminating in tendrils, sometimes 0, and replaced either by stipules (*Lathyrus Aphaca*) or by herbaceous and membranous wings edging the stem (*Crotalaria Vespertilio*) ; sometimes opposite when young (*Phaseolus*). FLOWERS perfect, very rarely polygamous (*Arachis*), inflorescence axillary, in a raceme, spike, head, or umbel, or solitary, imbricate in æstivation. CALYX more or less irregular, 5-toothed, -fid, or -partite, or 2-labiate, the two posterior divisions forming the upper lip, the two lateral and the anterior forming the lower lip. PETALS usually 5, alternate with the sepals, sometimes 4, 3, 2, or 1 (*Amorpha*), inserted on a disk lining the bottom of the calyx, usually free, or rarely coherent (*Trifolium*), unequal, the posterior petal (*standard*) embracing the others, the two lateral (*wings*) alike, pressed upon the two anterior, which also are alike, often conniving, and simulating a single petal (*keel* or *boat*). STAMENS 10, or less by abortion ; *filaments* either monadelphous, or diadelphous by the separation of the stamen opposite to the standard ; or completely free (*Sophora, Cladrastis, Anagyris,* &c.). OVARY solitary, opposite to the anterior sepal, sessile or pedicelled, usually pluriovuled ; *ovules* situated along the suture facing the standard, campylotropous ;

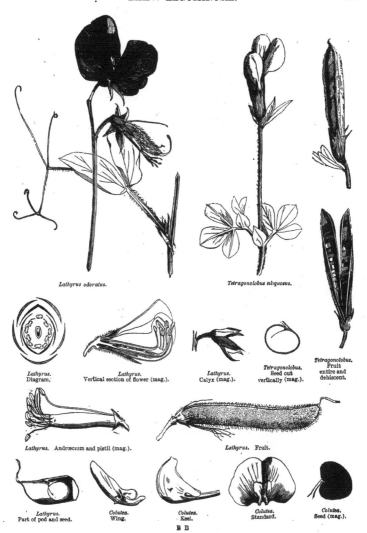

Lathyrus odoratus.

Tetragonolobus siliquosus.

Lathyrus.
Diagram.

Lathyrus.
Vertical section of flower (mag.).

Lathyrus.
Calyx (mag.).

Tetragonolobus.
Seed cut
vertically (mag.).

Tetragonolobus.
Fruit
entire and
dehiscent.

Lathyrus. Andrœcium and pistil (mag.).

Lathyrus. Fruit.

Lathyrus.
Part of pod and seed.

Colutea.
Wing.

Colutea.
Keel.

Colutea.
Standard.

Colutea.
Seed (mag.).

B B

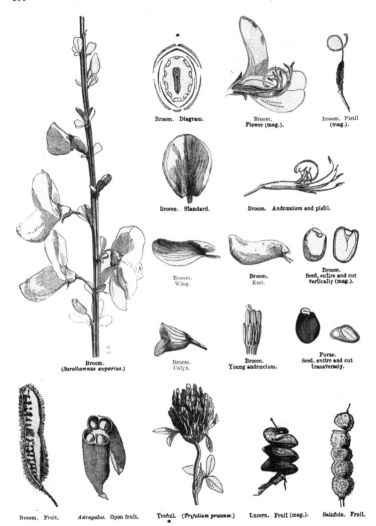

Broom. Diagram.

Broom.
Flower (mag.).

Broom. Pistil
(mag.).

Broom. Standard.

Broom. Andrœcium and pistil.

Broom.
Wing.

Broom.
Keel.

Broom.
Seed, entire and cut
Vertically (mag.).

Broom.
(Sarothamnus scoparius.)

Broom.
Calyx.

Broom.
Young andrœcium.

Furze.
Seed, entire and cut
transversely.

Broom. Fruit.

Astragalus. Open fruit.

Trefoil. (Trifolium pratense.)

Lucern. Fruit (mag.).

Sainfoin. Fruit.

style filiform; *stigma* terminal, or situated laterally on the inner side, below the extremity of the style. Pod either 1-celled, opening in 2 valves, which sometimes separate from the placentas (*Carmichælia*); or divided into 2 cells by a longitudinal septum (*Astragalus, Oxytropis*); or divided by transverse septa into superimposed cellules; or divided by strangulation (*isthmus*) into 1-seeded joints, which separate when ripe. Seed with a soft, generally carunculate testa. Embryo exalbuminous or albuminous; *radicle* bent.

[Divided by Bentham in the 'Genera Plantarum' into the following eleven tribes.

Tribe I. Podalyrieæ.—Shrubs, rarely herbs. Leaves simple or digitately compound. Stamens 10, free. *Anagyris, Thermopsis, Baptisia, Podalyria, Brachysema, Oxylobium, Chorizema, Mirbelia,* and numerous other (chiefly Australian) genera.

Tribe II. Genisteæ.—Shrubs or herbs. Leaves simple or digitately compound, leaflets quite entire. Flowers solitary, fascicled, or in terminal or leaf-opposed racemes. Stamens 10, usually monadelphous. *Priestleya, Bossiæa, Hovea, Goodia, Rafnia, Lotononis, Lebeckia, Aspalathus, Crotalaria, Lupinus, Laburnum, Genista, Ulex, Cytisus,* &c.

Tribe III. Trifolieæ.—Herbs, rarely shrubby. Leaves 3-foliolate, usually toothed by the excurrent nerves. Stamens 10, monadelphous or diadelphous. *Ononis, Trigonella, Medicago, Melilotus, Trifolium, Parochetus.*

Tribe IV. Loteæ.—Herbs or shrubs. Leaves pinnately 3-foliolate, leaflets quite entire. Flowers usually umbelled or capitate. Stamens 10, monadelphous or diadelphous; alternate filaments often dilated. *Anthyllis, Lotus, Hosackia,* &c.

Tribe V. Galegeæ.—Herbs (not climbing), trees, or erect or climbing shrubs. Leaves pinnately 5-∞ - (rarely 3-1-) foliolate; leaflets usually quite entire. Pod 2-valved or small, 1-2-seeded, and vesicular. *Psoralea, Amorpha, Dalea, Indigofera, Galega, Tephrosea, Milletia, Wistaria, Robinia, Sesbania, Carmichælia, Clianthus, Lessertia, Swainsonia, Colutea, Caragana, Astragalus, Oxytropis, Glycirrhiza.*

Tribe VI. Hedysareæ.—Habit of IV., V., and VIII., but pod inarticulate. *Scorpiurus, Ornithopus, Coronilla, Hippocrepis, Hedysarum, Onobrychis, Ormocarpium, Æschynomyne, Smithia, Adesmia, Arachis, Desmodium, Uvaria, Alysicarpus, Lespedeza.*

Tribe VII. Vicieæ.—Herbs. Leaves abruptly pinnate, petiole ending in a tendril or point, leaflets usually toothed at the tip. Stamen and pod of VIII. *Cicer, Vicia, Lens, Lathyrus, Pisum, Abrus,* &c.

Tribe VIII. Phaseoleæ.—Herbs, usually twining, rarely erect, or shrubs or trees. Leaves usually pinnately 3- (rarely 1-7-) foliolate; leaflets entire or lobed. Pod 2-valved. *Centrosema, Clitoria, Glycine, Hardenbergia, Kennedya, Erythrina, Mucuna, Butea, Galactia, Camptosema, Dioclea, Pueraria, Canavalia, Phaseolus, Vigna, Dolichos, Rhynchosia,* &c.

Tribe IX. Dalbergieæ.—Erect or climbing trees or shrubs. Leaves pinnately 1-3-5-∞ - foliolate. Stamens monadelphous or diadelphous. Pod indehiscent. *Dalbergia, Pterocarpus, Lonchocarpus, Derris, Dipteryx, Inocarpus,* &c.

Tribe X. Sophoreæ.—Trees or shrubs, erect or climbing, rarely small or subherbaceous. Leaves pinnately 5-∞ -foliolate, or broad and 1-3-foliolate. Stamens 10, free. *Dalhousiea, Baphia, Calpurnea, Sophora, Ormosia, Myroxylon,* &c.

Tribe XI. Swartzieæ. See sub-order Swartzieæ, p. 366.—Ed.]

The extensive family of *Leguminosæ* is closely allied to *Amygdaleæ* (see this family). *Mimoseæ* obviously, as M. Planchon has demonstrated, approach *Oxalideæ*, in which, as in many *Mimoseæ*, the corolla is diplostemonous, the stamens monadelphous, the ovules anatropous, the embryo albuminous and straight, the seeds arillate, and the leaves alternate, compound and irritable; but the calyx of *Oxalideæ* is imbricate, the ovary 5-celled, the leaves are exstipulate, and the stem is generally herbaceous (except in the genus *Averrhoa*). *Papilionaceæ* are also near *Terebinthaceæ*, which resemble them in habit, their alternate often compound leaves, perigynous stamens, often solitary ovary, campylotropous ovule, and exalbuminous embryo; but which differ in their regular flower, free stamens, usually fleshy fruit, and exstipulate leaves; but here again the affinity is re-established by some *Cæsalpineæ* (*Ceratonia*), which are apetalous and diœcious, like many *Terebinthaceæ*, and the flower of which is sub-regular, and the stamens nearly free.

Mimoseæ abound in the tropical zone; they are rare in the sub-tropical regions of the northern hemisphere, and are especially numerous in Africa and Australia. Tropical America produces a great number of species, belonging to the group of *Inga*.

Swartzieæ inhabit intertropical America and Africa; none have yet been met with in Asia. *Papilionaceæ* are found in all climates, but they mostly grow between and near the tropics, and in the Old World more than in the New. Some *Astragali* ascend the highest mountains. *Cæsalpineæ* are numerous in tropical regions; they scarcely pass the tropic in the Old World, and are rather rare in North America.

The order *Leguminosæ* is extremely serviceable to mankind; the farinaceous seeds of *Papilionaceæ* are very nutritious, whilst their herbage forms an excellent fodder. *Leguminosæ* also yield more substances used in medicine and the arts than any other order in the Vegetable Kingdom. We will enumerate the most important species, in their botanical order, and notice, in passing, certain noxious ones.

Albizzia anthelminthica, an Abyssinian tree, the bark of which is employed against tænia.

Acacia vera and *arabica*, North-east African, Arabian, and Indian trees, produce gum arabic. *A. Verek*, *Segal*, and *Adansonii*, Senegambian trees, produce gum senegal, used similarly to gum arabic. A decoction of the wood of the Indian *A. Catechu* yields Catechu, a thick juice, soluble in water, which is an astringent tonic.

Adenanthera pavonina, an Indian tree, of whose hard red seeds, named Kuara, necklaces and bracelets are made.

Detarium Senegalense, a Senegambian tree, yields an edible drupe.

Swartzia tomentosa is a tropical American tree with a resinous sudorific bark.

Ceratonia Siliqua (Carob). A tree of the Mediterranean region, whose lomentaceous fruit contains a russet insipid edible pulp, serving as forage in Spain [and extensively imported into England for cattle food].

Copaifera officinalis, *coriacea*, *cordifolia*, &c. Trees of tropical America, yielding by incision of the trunk a turpentine called Balsam of Copaiba, used in catarrhal affections.

Hymenæa verrucosa. A Madagascar tree, yielding a yellow resin named *copal*, which is insoluble in alcohol, but soluble after fusion in linseed oil, and then in essence of turpentine, and is much used as a varnish.

Aloexylon Agallochum. A tree of Cochin China, whose veined resinous aromatic wood, called Aloe-wood, burns with a fragrant flame.

Cassia obovata, *acutifolia*, *lanceolata*, &c., the Sennas, are plants of Upper Egypt, Syria, Arabia, India, and Senegal, the leaves of which contain an active purgative principle, much used in medicine; their flattened pods are much weaker purgatives. *C. fistula*, an Indian tree, bears a woody indehiscent septate pod, named Cassia, the cells of which contain a sugary laxative black pulp. The seeds of *C. Absus* are used in Egypt to cure chronic ophthalmia.

Tamarindus indica. A tree of India, West Asia, and Egypt, whose pulpy, acid, and sugary pods, called Tamarinds, are used in medicine.

Hæmatoxylon campechianum, a Central American and West Indian tree, affords Logwood, which contains a colouring principle (*hematine*), much used in dyeing black or dark red.

- *Cæsalpinia echinata*, a Brazilian tree, yields Brazil-wood, containing a red colouring principle (*brasiline*).

C. coriacea, of tropical Asia, yields the very astringent pods [called Divi Divi], used in tanning leather. [*C. Sappan* yields the red Sappan-wood of Eastern India and Ceylon.]

Castanospermum australe, an Australian tree, yields edible seeds called Australian chestnuts.

Sophora tomentosa. A tree whose roots and seeds are used in India to arrest choleraic vomiting. The flowers of *Styphnolobium japonicum* are used in China as a yellow dye.

Myroxylon peruiferum, a Peruvian [Central American] tree, yields a sweet-smelling liquid balsam, composed of a resin, an oil, and a peculiar acid (*cinnamic*). *M. toluiferum,* a Columbian tree, [is supposed to] produce the Balsam of Tolu, a similar substance, used in chronic pulmonary catarrh.

Coumarouna [Dipteryx] odorata, a Guiana tree with very hard and heavy wood, yields Tonquin Beans, which contain a very odoriferous crystallizable principle (*coumarine*), and are employed to perfume snuff.

Andira surinamensis, inermis, racemosa, &c., tropical American trees, contain narcotic-acrid principles, which are emetic, narcotic and vermifuge.

Geoffroya vermifuga and *spinulosa,* Brazilian trees, of which the seeds possess an acrid and volatile principle, and are used as anthelminthics.

Dalbergia latifolia [and other species], of Brazil, India, and Africa, afford Rosewood, as do many species of *Machærium.*

Pterocarpus Draco, a West Indian tree, yields by incision of its bark Dragon's Blood, a red astringent resin. [*P. erinaceus* yields the African Rosewood and Kino, and *P. santalinus,* the red Sanders-wood, used to dye red-brown.]

Butea frondosa [and *B. superba*], trees of tropical Asia, yield by incision an astringent juice, named Eastern Kino; [and the flowers of *B. frondosa* afford an orange-yellow dye].

Drepanocarpus senegalensis, an African tree, produces the true or Gambia Kino.

Abrus precatorius is a tropical African and Asiatic climber, introduced into America, whose root yields a liquorice, and its red shining seeds, with a black bilum, are used for chaplets and necklaces [and as weights, called *Retti* (the origin of the word *carat*) ; each seed weighs one grain very exactly].

Dolichos Lablab [and other species] Indian herbs with farinaceous edible seeds. Some neighbouring genera (as *Pachyrrhizus*) have tuberous rhizomes and edible seeds.

Phaseolus vulgaris is an Indian or American climber, or dwarf herb, the young sugary pods of which are mucilaginous, and the seeds (*haricots*) farinaceous and edible.

[*Cicer arietinum,* the Chick Pea or Gram of India, is extensively cultivated in South Europe and the East for its edible seeds ; its herbage yields so strong an acid (oxalic ?) that shoes are spoiled by walking through a field of it.—ED.]

Faba vulgaris (Bean), *Pisum sativum* (Pea), *Ervum Lens* (Lentil), are annual herbs with farinaceous edible seeds ; those of *Ervum Ervilia* are poisonous, [as are those of some *Phaseoli* and *Lathyri*].

Apios tuberosa, Psoralea esculenta and *hypogæa,* are North American herbs with tuberous, starchy, edible rhizomes.

Alhagi Maurorum is a West Asiatic and tropical and subtropical African shrub, from which exudes Persian Manna, a substance analogous to the Manna of the Ash, and possessing the same properties.

Mucuna pruriens is an Indian annual, the pod of which is covered with stiff stinging hairs [called Cowitch, and used as an anthelminthic]. The seed is called Donkey's Eye, from the large, pupil-like areola on the testa.

Onobrychis sativa (Sainfoin) is a perennial cultivated herb which furnishes an excellent fodder.

[*Æschynomene aspera,* a marsh shrub of India, has a very soft light wood, extensively used for making hats, under the name of *Shola.*—ED.]

Arachis hypogæa is an annual Brazilian herb which buries its fruit to ripen the seeds. Its oily and starchy seeds (Earth-nuts) are both used as food and much valued by manufacturers on account of their bland oil ; three and a half millions of pounds of these are annually imported into France alone.

Voandzeia subterranea, a Madagascar herb with hypogæous pods like those of *Arachis,* also yields edible seeds.

Lathyrus tuberosus is a perennial climbing herb with a feculent sugary rhizome, much cultivated before the introduction of the potato.

Vicia sativa (Vetch) is an annual climbing herb, cultivated for forage.

Astragalus creticus, verus and *aristatus* are Oriental shrubs, from the trunk of which exudes Gum Tragacanth, a gelatinous juice [colloid], swelling in water, much used in pharmacy and manufactures.

Coluten arborescens (Bladder-Senna). A shrub indigenous to Southern Europe. Leaves purgative, and seeds emetic.

Herminiera elaphroxylon. A shrub of Senegambia with a very light wood, used instead of corks for floating fishing nets.

Glycirrhiza glabra, echinata, and *glandulifera* are perennial herbs of North-western Europe, with sugary rhizomes, used in medicine as emollients. By decoction and evaporation their roots yield the dry extract called Liquorice [or Spanish Juice].

Indigofera tinctoria, Anil and *argentea* are tropical Asiatic undershrubs, containing in their leaves Indigo, a colouring principle which is extracted by fermentation in water.

Melilotus officinalis (Melilot). An indigenous herb becoming odoriferous when dried, and sweetening hay for cattle. An infusion of the flowers is used as an antiophthalmic.

Trigonella Fœnum-græcum (Fenugreek) is a herb with aromatic and bitter seeds, used as a resolvent poultice, and mixed as a stimulant with oats for horses.

Medicago sativa, lupulina, &c. (Lucern) ; *Trifolium pratense, repens,* &c. (Clover), are indigenous excellent fodder-herbs.

Genista tinctoria (Dyer's Weed) is an indigenous dye-plant with diuretic flowers and purgative and emetic seeds formerly prescribed for hydrophobia.

Sarothamnus scoparius (Broom) is an indigenous shrub with slender flexible branches [extensively used for basket-work and as a diuretic]. The flowers are infused in milk, and used as a lotion for skin diseases ; the flower-buds, preserved in vinegar, are used like capers.

Ulex europæus (Furze, Gorse, Whin), an indigenous shrub, much used for firewood [in France], affords winter cattle-food.

Lupinus albus, varius, luteus, &c. Annual herbs with feculent eatable seeds ; stem and leaves used as green fodder.

Crotalaria juncea is a Bengal shrub [whose fibre is extensively used as Sunn-hemp].

Anagyris fœtida is a Mediterranean shrub, called Stinking-wood, with purgative stimulating leaves and very poisonous seeds.

Physostigma venenosum yields seeds called Calabar Beans ; [they contain one of the most virulent of poisons, which possesses the curious property of causing contraction of the pupil.—Ed.].

LXXVI. *ROSACEÆ, Jussieu.*

Stem herbaceous or woody. Leaves alternate [opposite in Rhodotypus], *stipulate,* or very rarely exstipulate (Spiræa, Aruncus, &c.). Inflorescence various. Flowers ☿, sometimes diclinous [regular, asymmetrical in Tribe Chrysobalaneæ]. Calyx 5–4-merous, imbricate or valvate in æstivation. Petals as many as sepals, *free,* inserted on the calyx, imbricate in æstivation, sometimes 0. Stamens usually indefinite, many-seriate, inserted like the petals ; anthers 2-celled, introrse, dorsifixed. Pistil very various. Ovules anatropous. Embryo straight, exalbuminous, or very rarely albuminous (Neviusia).

Tribe I. *POMACEÆ, Jussieu.*

Stem woody. Leaves with free caducous stipules. Flowers ☿, terminal, in a corymb, cyme, raceme or umbel. Receptacular cupule (calyx-tube of many botanists) enveloping the ovaries, and adnate to them, terminated by a 5-lobed calyx.

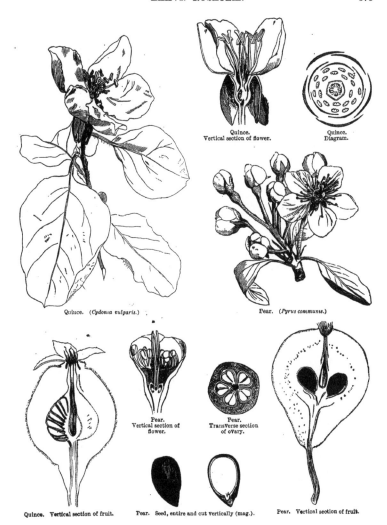

Quince.
Vertical section of flower.

Quince.
Diagram.

Quince. (*Cydonia vulgaris.*)

Pear. (*Pyrus communis.*)

Pear.
Vertical section of
flower.

Pear.
Transverse section
of ovary.

Quince. Vertical section of fruit.

Pear. Seed, entire and cut vertically (mag.).

Pear. Vertical section of fruit.

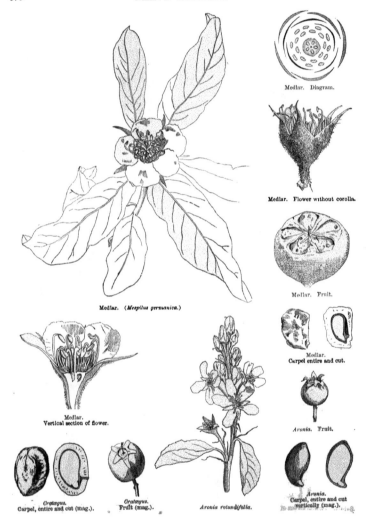

Medlar. Diagram.

Medlar. Flower without corolla.

Medlar. Fruit.

Medlar. (*Mespilus germanica.*)

Medlar.
Carpel entire and cut.

Medlar.
Vertical section of flower.

Aronia. Fruit.

Cratægus.
Carpel, entire and cut (mag.).

Cratægus.
Fruit (mag.).

Aronia rotundifolia.

Aronia.
Carpel, entire and cut
vertically (mag.).

Sorbus.
Diagram.

Cotoneaster.
Carpel, ripe and cut Vertically (mag.).

Cotoneaster.
Fruit.

Cotoneaster.
Fruit cut Vertically (mag).

Petals 5. Stamens numerous. Ovaries 5, sometimes 3, 2, 1, adnate to the receptacular cupule, 1-celled, 2–pluri-ovuled; ovules ascending; styles as many as ovaries, free, or coherent by their base. Fruit formed by the carpels and the succulent receptacular cupule, crowned by the calyx-limb or its scar, with 5 cells or fewer, enclosing 1, 2, or several seeds; pericarp bony, indehiscent, perforated at the base; or cartilaginous or membranous, partially opening on the side of the axis. Seed ascending; radicle inferior.

PRINCIPAL GENERA.

*Pyrus.	*Malus.	*Sorbus.	*Mespilus.	*Cydonia.
*Cotoneaster.	*Eriobotrya.	*Photinia.	*Raphiolepis.	*Stranvæsia.
*Cratægus.	*Aronia.	Osteomeles.		

TRIBE II. *ROSEÆ, D.C.*

Rose. Flower cut Vertically (mag.).

Sweetbriar. (*Rosa rubiginosa.*)

Rose.
Diagram.

Rose.
Young carpel (mag.).

Rosebud.   Rose. Fruit (mag.).   Rose. Ripe carpel, entire and cut vertically (mag.).

Stem woody, usually thorny, erect or sarmentose. Leaves unequally pinnate, with stipules adnate to the petiole, rarely simple, sometimes 0, and replaced by the stipules. Flowers ☿, in a terminal corymb, white, pink or yellow. Receptacular cupule (calyx-tube) usually ovoid, and contracted beneath the calyx, sometimes cyathiform. Sepals foliaceous, imbricate. Petals 5, readily multiplying under cultivation. Stamens numerous. Carpels numerous, inserted at the bottom or on the inner wall of the receptacular cupule, which becomes fleshy when ripe. Seed pendulous; radicle superior.

GENERA.

*Rosa.   *Hulthemia.

Rosa berberifolia, Pallas, is a small aphyllous shrub of Central Asia, in which the leaf is replaced by two connate stipules simulating a simple reticulate wedge-shaped toothed leaf with an entire or bifid tip. A rudimentary normal leaf is sometimes developed in the notch or bifurcation of the stipule.

TRIBE III. *SANGUISORBEÆ*, *A. Gray.*

(POTERIEÆ, *Benth. et Hook. fil.*)

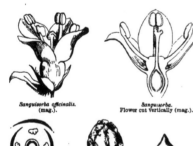
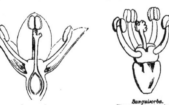

Sanguisorba officinalis. (mag.).   Sanguisorba. Flower cut vertically (mag.).   Sanguisorba. Flower without calyx and corolla (mag.).   Sanguisorba. Pistil (mag.).

Sanguisorba. Diagram.   Sanguisorba. Fruit (mag.).   Sanguisorba. Seed (mag.).   Sanguisorba. Ripe carpel cut vertically (mag.).   Agrimony. Ripe carpel, entire and cut vertically (mag.).

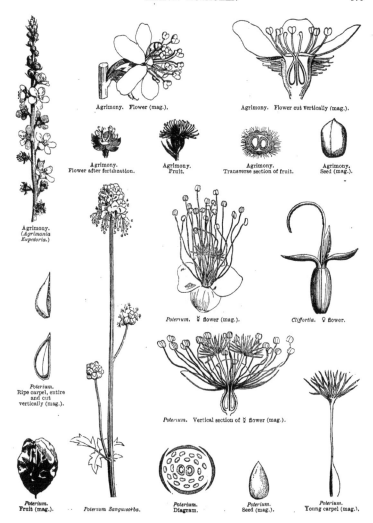

Agrimony. Flower (mag.).

Agrimony. Flower cut vertically (mag.).

Agrimony.
Flower after fertilization.

Agrimony.
Fruit.

Agrimony.
Transverse section of fruit.

Agrimony.
Seed (mag.).

Agrimony.
(*Agrimonia
Eupatoria.*)

Poterium. ☿ flower (mag.).

Cliffortia. ♀ flower.

Poterium.
Ripe carpel, entire
and cut
vertically (mag.).

Poterium. Vertical section of ☿ flower (mag.).

Poterium.
Fruit (mag.).

Poterium Sanguisorba.

Poterium.
Diagram.

Poterium.
Seed (mag.).

Poterium.
Young carpel (mag.).

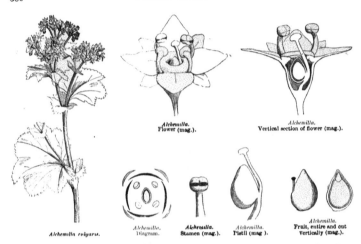

Alchemilla.
Flower (mag.).

Alchemilla.
Vertical section of flower (mag.).

Alchemilla vulgaris.

Alchemilla.
Diagram.

Alchemilla.
Stamen (mag.).

Alchemilla.
Pistil (mag.).

Alchemilla.
Fruit, entire and cut
Vertically (mag.).

Stem herbaceous, rarely woody.   Leaves generally compound; stipules adnate
to the petiole.   Flowers terminal, small, ☿ or diclinous; receptacular cupule of the
☿ and ♀ flowers urceolate, contracted at the top, and bearing a 4–5–3-fid calycinal
limb; calyx of the ♂ flowers 4-phyllous (*Poterium*) or 3-phyllous (*Cliffortia*); petals
usually 0, rarely 4.   Stamens equal in number to the calyx-lobes (*Sanguisorba*), or
fewer (*Margyricarpus, Tetraglochin*, &c.), or double, treble, or multiple (*Agrimonia,
Aremonia, Cliffortia, Poterium*, &c.).   Carpels 1–4, free, enclosed in the receptacular
cupule, ripening into achenes; styles sub-basilar, lateral or terminal; stigmas capitate or penicillate.   Seed pendulous; radicle superior.

PRINCIPAL GENERA.

Alchemilla.	Brayera.	* Agrimonia.	Margyricarpus.	Acœna.
, * Poterium.	* Sanguisorba.	Cliffortia.	Braya.	* Aremonia.

TRIBE IV. *DRYADEÆ, Ventenat.*

(RUBEÆ et POTENTILLEÆ, *Benth. et Hook. fil.*)

Herbs or shrubs.   Leaves usually compound; stipules adnate to the petiole.
Flowers ☿.   Calyx 5-4-partite, persistent, naked or calyculate, æstivation valvate.
Petals 5-4.   Carpels free, usually numerous, sometimes 5-10, arranged in a head
on a convex receptacle; style ascending from the ventral margin; ovule solitary
[or 2], pendulous, rarely ascending (*Geum, Dryas*); achenes naked (*Potentilla,*

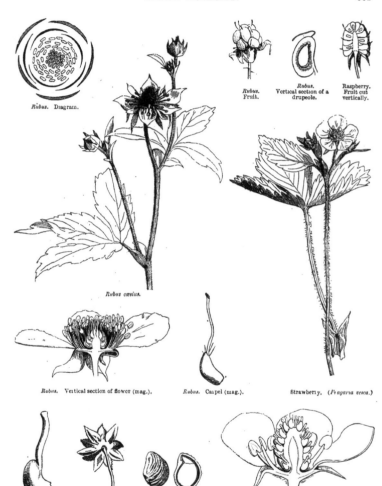

Rubus. Diagram.

Rubus.
Fruit.

Rubus.
Vertical section of a
drupeole.

Raspberry.
Fruit cut
vertically.

Rubus cæsius.

Rubus. Vertical section of flower (mag.).

Rubus. Carpel (mag.).

Strawberry. (*Fragaria vesca.*)

Potentilla.
Young carpel (mag.).

Potentilla.
Calyx and calyculus.

Potentilla.
Ripe carpel, entire and
cut vertically.

Strawberry.
Vertical section of flower (mag.).

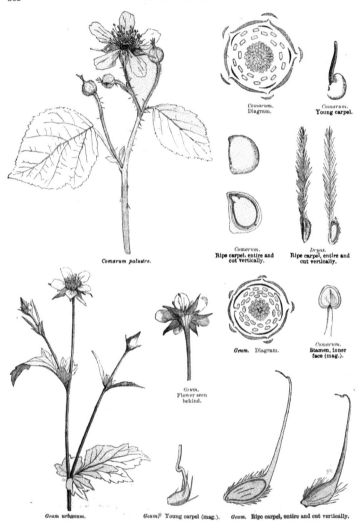

Comarum.
Diagram.

Comarum.
Young carpel.

Comarum palustre.

Comarum.
Ripe carpel, entire and
cut vertically.

Dryas.
Ripe carpel, entire and
cut vertically.

Geum. Diagram.

Comarum.
Stamen, inner
face (mag.).

Geum.
Flower seen
behind.

Geum urbanum.

Geum. Young carpel (mag.).

Geum. Ripe carpel, entire and cut vertically.

Comarum, &c.), or terminated by a feathery style (*Dryas, Cercocarpus*), or drupeoles on a usually dry receptacle (*Rubus*), sometimes fleshy (Strawberry). Seed pendulous, rarely ascending (*Geum, Dryas*); radicle superior, or rarely inferior.

PRINCIPAL GENERA.

*Rubus.	*Fragaria.	Comarum.	*Potentilla.
Sibbaldia.	*Geum.	Dryas.	Waldsteinia.

[Divided by Bentham and Hooker fil. into two tribes; of which one, RUBEÆ, with an ebracteolate calyx and 2 pendulous ovules, contains but one genus, *Rubus*; the other, POTENTILLEÆ, has usually a bracteolate calyx and always one ovule.—ED.]

TRIBE V. *SPIRÆACEÆ, D.C.*

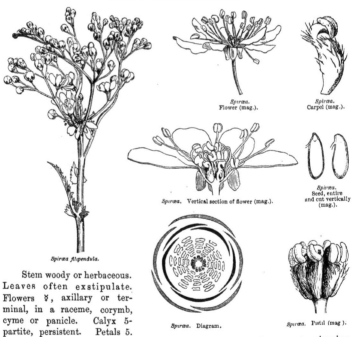

Spiræa.
Flower (mag.).

Spiræa.
Carpel (mag.).

Spiræa. Vertical section of flower (mag.).

Spiræa.
Seed, entire
and cut vertically
(mag.).

Spiræa filipendula.

Spiræa. Diagram.

Spiræa. Pistil (mag.).

Stem woody or herbaceous. Leaves often exstipulate. Flowers ☿, axillary or terminal, in a raceme, corymb, cyme or panicle. Calyx 5-partite, persistent. Petals 5. Stamens numerous. Carpels usually 5, whorled, free, rarely connate, ripening into follicles; style short; stigma thick; ovules 2–12, pendulous. Seed pendulous; radicle superior.

PRINCIPAL GENERA.

*Kerria.   *Spiræa.   *Gillenia.   Neviusia.   Neillia.

TRIBE VI. *NEURADEÆ, D.C.*

Herbs. Leaves sinuate-pinnatifid, stipulate. Receptacular tube accrescent. Petals 5. Stamens 10. Ovaries 10, dorsally adnate to the receptacular tube, and ventrally free; [styles subulate; stigma capitate; ovule solitary]. Fruit a capsule with 1-seeded cells [orbicular, formed of 10 1-seeded follicles sunk in the hardened receptacle]. Seeds pendulous.

GENERA.

Neurada.   Grielum.

TRIBE VII. *AMYGDALEÆ, Jussieu.*

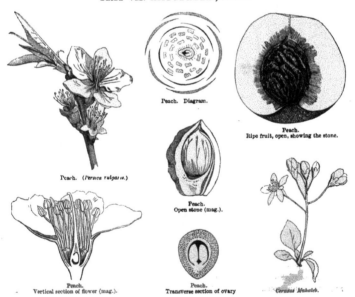

Peach. Diagram.

Peach.
Ripe fruit, open, showing the stone.

Peach. (*Persica vulgaris.*)

Peach.
Open stone (mag.).

Peach.
Vertical section of flower (mag.).

Peach.
Transverse section of ovary

Cerasus Mahaleb.

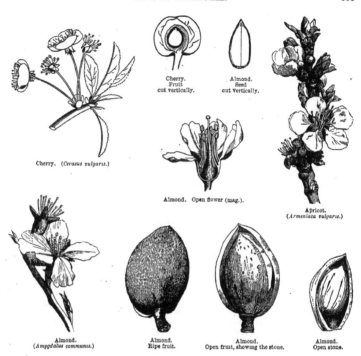

Cherry.
Fruit
cut vertically.

Almond.
Seed
cut vertically.

Cherry. (*Cerasus vulgaris.*)

Almond. Open flower (mag.).

Apricot.
(*Armeniaca vulgaris.*)

Almond.
(*Amygdalus communis.*)

Almond.
Ripe fruit.

Almond.
Open fruit, showing the stone.

Almond.
Open stone.

Stem woody, yielding gum, branches sometimes spinescent.   Leaves simple, entire or toothed, glandular; stipules free, caducous.   Flowers ⚥, axillary, solitary or geminate, or in a raceme, corymb, or umbel. Calyx deciduous. Petals 5. Stamens numerous.   Carpel solitary, very rarely several, ripening into a drupe; [style sub-terminal; stigma capitellate;] ovules 2, pendulous.   Seed usually solitary by arrest, pendulous; radicle superior.

PRINCIPAL GENERA.

*Amygdalus.	*Persica.	*Prunus.	*Armeniaca.
*Cerasus.	Pygeum.	Nuttallia.	Prinsepia.

[The two following tribes are omitted in the original; the first of which is usually ranked as a distinct order.

c c

CHRYSOBALANEÆ, Benth. et Hook. fil.

Trees or shrubs. Leaves simple, quite entire. Flowers asymmetrical. Calyx ebracteolate; lobes usually deciduous. Stamens unilateral, or in a complete whorl. Carpel 1; style basilar; ovules 2, ascending. Fruit coriaceous or drupaceous, not included in the calyx-tube (receptacular cup). Radicle inferior.

PRINCIPAL GENERA.

Chrysobalanus.	Licania.	Moquilea.	Parinarium.
Parastemon.	Couepia.	Lecostemon.	Stylobasium.

QUILLAJEÆ, Benth. et Hook. fil.

Trees or shrubs. Leaves simple, coriaceous, rarely pinnate. Calyx ebracteolate; lobes usually persistent. Stamens 5–10–20. Carpels free or connate; ovules one or more, ascending or pendulous. Fruit of 5 cocci or follicles, or a capsule, not included in the calyx-tube (receptacular cup), inferior and indehiscent in *Pterostemon*. Seeds usually broadly winged.

PRINCIPAL GENERA.

Quillaja.	Kagenackia.	Lindleya.	Eucryphia.	ED.]

Each of the tribes composing the entire group of *Rosaceæ* may be considered as a separate family. *Amygdaleæ* approach *Chrysobalaneæ* [1] in their 5-merous calyx and corolla, the insertion and number of the stamens, the drupaceous fruit, the exalbuminous embryo, the woody stem, and the alternate simple stipulate leaves; *Chrysobalaneæ* differ in their eglandular petiole, the [usually] inequilateral calyx, unequal stamens smaller and often sterile on the side on which the calyx is least developed, erect seed, and the absence of hydrocyanic acid. *Rosaceæ*, as a whole, strongly resemble *Leguminosæ* in habit, in the alternate stipulate often pinnate leaves, sometimes precisely like those of some *Papilionaceæ* (*Osteomeles, Horkelia*). *Amygdaleæ* especially are closely allied to *Leguminosæ* by their axillary inflorescence, 5-merous calyx, imbricate corolla, solitary carpel, and glandular petioles resembling those of *Mimoseæ*, like which order they further secrete gum. The leguminous *Papilionaceæ* scarcely differ save in the irregular corolla, definite stamens, and [often] connate filaments; but in some genera of *Leguminosæ* the flowers are regular or sub-regular (*Hæmatoxylon, Labichea, Bauhinia*, &c.), the filaments are free (*Cæsalpinia, Cassia, Gymnocladus, Gleditschia, Hymenæa*, &c.), and, as an additional affinity, the fruit is sometimes a drupe (*Detarium*). Hence the only absolute distinctive character between these orders is in the calyx, of which the odd sepal is anterior in *Leguminosæ*, and next the axis in *Rosaceæ* (see the diagrams).

Rosaceæ are connected with *Saxifrageæ* by *Spiræaceæ*, and particularly by the genus *Neviusia*, the seed of which is albuminous [as in various other genera, as *Neillia, Eucryphia*, &c.]. They approach *Cephaloteæ* through *Dryadeæ* (see these families). Ad. Brongniart has noticed an incontrovertible affinity between *Pomaceæ* and *Cupuliferæ*—woody stem rich in tannin, alternate stipulate leaves, inferior ovary with many two-ovuled cells, anatropous ovules, exalbuminous embryo; the diagnosis almost entirely rests on the absence of petals and on the ovules, which in *Cupuliferæ* are pendulous. Finally, an analogy has been indicated by R. Brown between *Amygdaleæ* and *Thymeleæ*, founded on the insertion of the stamens, the monocarpellary pistil, oblique 1-celled one-ovuled ovary, pendulous ovule, subterminal style, drupaceous fruit, and exalbuminous embryo with fleshy cotyledons; but *Thymeleæ* differ in the

---

[1] Omitted in the original. Regarded as a tribe of *Rosaceæ* by Bentham and Hooker fil., and as such inserted above.—ED.

nature of the elementary organs of their stem, in their acrid and blistering principles, exstipulate leaves, and the definite number of their stamens. [The points of similarity between *Rosaceæ* and *Cupuliferæ* and *Thymeleæ* are more probably analogical resemblances.—ED.] *Pomaceæ* all belong to the northern hemisphere; they inhabit Europe, Asia, and North America; they are common on the mountains of India, and rare in Mexico, Madeira, North Africa, and the Sandwich Islands. *Pyrus* [in its limited sense] is confined to the Old World; the Mountain Ash often accompanies the Birch to the highest northern latitudes; *Rosa* is only found north of the tropics; *Fragaria* inhabits all northern temperate regions, as also extra-tropical South America and the Moluccas; [an Indian yellow-flowered species is sub-tropical]. Brambles abound especially in the northern temperate regions of both worlds, [but there are comparatively few species in America]; they are rare in the tropics, and some are found in the southern hemisphere as far as New Zealand. *Potentilla, Geum, Dryas, Agrimonia, Sanguisorba, Poterium,* and *Aichemilla* mostly inhabit the temperate and cool parts of the northern hemisphere; [*Alchemilla* is essentially Andean, and *Agrimonia* is found in South Africa, South America, and Australia]. Some *Sanguisorbeæ* belong to tropical and sub-tropical America. The true *Spiræaceæ* live north of the tropics; the others in Peru and Chili. *Neuradeæ* are confined to the south and north of Africa [and West Asia]. *Amygdaleæ* for the most part inhabit the north temperate zone; a few only occur in tropical America, and some in the Canaries, and Azores, and Sandwich Islands; none have yet been met with beyond the tropic of Capricorn. [*Prunus* itself is not uncommon in tropical America, and *Pygeum* is essentially tropical Asiatic, and is also found in Africa and Australia. *Chrysobalaneæ* are chiefly tropical American, and many of them Brazilian; a few occur in tropical Asia and Africa; *Stylobasium* is Western Australian. *Quillajeæ* are for the most part Western American; but *Eucryphia* is also found in Australia.—ED.]

The fruits of *Pomaceæ* contain mucilage, sugar, and malic acid, in proportions so well developed and modified by cultivation, that this family has become one of the most useful in the Vegetable Kingdom.

Cydonia vulgaris (Quince). Fruit astringent, edible with sugar in jelly and syrup. Seeds with an emollient mucilage.—*Pyrus communis* (Pear). Fruit obconic or sub-globular; flesh saccharine, savoury and melting, containing near the heart stony concretions of cells; it contains the same principles as that of Apples, and its juice is fermented to make perry. Its wood, which is close-grained, and much sought by joiners, was formerly used by engravers.—*Pyrus Malus* (Apple). Fruit usually globose, always umbilicate at the base, and not narrowed into the peduncle; flesh firm, brittle and acidulous, never stony, containing, besides sugar and malic acid, gum, pectine, and albumine. Apples are used for preserves, syrup, and jelly; and their fermented juice yields cider and vinegar.—*Pyrus (Sorbus) domestica* (Service). Fruit at first harsh, becoming pulpy and sweet after gathering, edible, fermentable. Wood very finely grained, and takes a good polish.—*Pyrus (Sorbus) Aucuparia* (Mountain Ash, or Rowan-tree). Fruit pulpy, containing malic acid, of a nauseous taste, but fermentable, and yielding a spirituous liquor when distilled.—*Pyrus (Sorbus) Aria* (Beam-tree). Fruit with a sugary pulp, scarcely acid. Wood of a very fine texture, and more valuable than that of the Pear.—*Pyrus (Sorbus) torminalis* (Service). Fruit harsh, then acid. Bark formerly employed as an astringent in dysentery.

Cratægus Azarolus. Fruit pulpy, edible.—[*C. Oxyacantha* is the Hawthorn, so useful for hedges.] —*Mespilus germanica* (Medlar). Fruit harsh, becoming pulpy and sweet after it is gathered, edible, astringent.—[*Eriobotrya japonica,* the Loquat, a dessert fruit of China and Japan, is now cultivated in all warm countries.]

Rosa canina (Dog-Rose). Fruit pulpy, making an astringent antiseptic preserve. The achenes are a vermifuge; young leaves infused as tea. Root formerly praised as a specific against hydrophobia (whence its name). Stem frequently presenting a mossy excrescence (Robin's Pincushion), caused by the puncture of an insect, and formerly used, under the name of *bedeguar,* as a diuretic, anthelminthic, and antiscorbutic.—*R. Gallica (Rose de Provence).* Petals astringent, and afford conserve of Roses and honey of Roses.—*R. centifolia, Kalendarum, moschata,* [*damascena*], &c. The petals yield Rose-water by distillation (employed in making an astringent eyewater), and, by maceration in oil of sesamum, the attar of Roses used in perfumery.—*Agrimonia Eupatoria* and *A. odorata* (Agrimony). Leaves astringent, employed against angina, nephritis, pulmonary catarrhs, &c.—*Alchemilla vulgaris* (Lady's Mantle). Leaves astringent, vulnerary.—*Sanguisorba officinalis* (Great Burnet). Plant astringent.—*Poterium Sanguisorba* (Salad Burnet). Plant used for forage, and as a condiment, astringent.

Rubus fruticosus (Bramble or Blackberry).   Fruits edible, astringent [sometimes not when ripe], as are the buds.—*R. Idæus* (Raspberry). Fruit perfumed, acidulous and sugary, employed in the preparation of jelly and Raspberry vinegar.—*Fragaria vesca* (Strawberry). Plant edible and medicinal.   Fruit succulent, perfumed.   Root astringent and diuretic.—*Tormentilla erecta, Potentilla reptans, P. anserina.* Roots and leaves astringent —*Geum urbanum* and *ritale* (Avens).   Root aromatic, bitter, tonic and stimulating.—*Dryas octopetala.* Plant astringent, tonic.—The flowers of *Brayera anthelminthica*, an Abyssinian tree, are, with the bark of the Pomegranate root, the most efficacious known remedy for tænia. The roots of *Spiræaceæ*, like those of *Dryadeæ*, are astringent, and contain resinous and aromatic principles, which render them bitter, tonic, and stimulating; such are *Spiræa Filipendula* (Drop-wort), *Aruncus*, and *Ulmaria* (Meadowsweet); the flowers of the latter are used to give a bouquet to wine, and their watery infusion is sudorific and cordial.

The tribe of *Amygdaleæ* is not less useful than that of *Pomaceæ*, and the excellence of its fruits is due to cultivation, the sugary matter in them overpowering the acid, without entirely disguising it, and giving a delicious taste to the drupe.   Many species, and especially the Bitter Almond, further contain in their seed, and even in their leaves, the elements of hydrocyanic acid, joined to a peculiar volatile oil, which is only developed in contact with water, and which gives them narcotic qualities.   Their wood, like that of *Pomaceæ*, is much used for joiners' work.   The most useful species is *Amygdalus communis* (Almond), a tree of the Mediterranean region.   The drupe, contrary to that of [many of] the other *Amygdaleæ*, is fibrous, coriaceous and dry.   The seed yields by expression a mild alimentary medicinal fixed oil, rendered soluble in water by the gum, sugar, and albumine which accompany it, and forming a milky emulsion, with which 'loochs' and the 'sirop d'orgeat' are prepared.

Persica vulgaris (Peach).   A tree, originally from China, with edible fruit.   The seed contains the elements of hydrocyanic acid; the fruit and crushed kernel are used in the composition of *noyau*.[1]   The flowers are used in a purgative syrup.—*P. lævis* (Nectarine).   Fruit with a soft epicarp, edible.   Origin unknown [probably a variety of the Peach].

Armeniaca vulgaris (Apricot).   Tree originally from northern Asia (China?).   Fruit edible, flesh succulent, perfumed, [much used by travellers, dried and preserved in the form of flat cakes, throughout Central and Western Asia].

Prunus spinosa (Sloe).   An indigenous tree.   Flowers purgative.   Fruits very harsh, only becoming edible when the frost has softened the pulp.   Bark astringent, bitter, and febrifuge.—*P. domestica* (Wild Plum) and its wild congener the Bullace (*P. insititia*), are spread over the temperate regions of the world.   Fruit edible and medicinal.

Cerasus Avium (Wild Cherry).   A European species, drupe yielding by fermentation and distillation *kirschwasser* and Cherry wine.   Wood reddish yellow, valued by cabinet-makers.   *C. duracina* (Bigarou). A species near the former.   Drupe edible, and flesh adhering to the stone.   Native country unknown.— *C. Juliana* (Gean).   Drupe edible, flesh easily separating from the stone.   Native country unknown. —*C. caproniana* (Griotte).   A tree, originally from Asia, brought, it is said, from Cerasonte by Lucullus, after his victories over Mithridates.   Fruit edible, much changed by cultivation, flesh acidulous and refreshing.—*C. Mahaleb.* Wood sought after by cabinet-makers under the name of 'wood of St. Lucia. Seed of a mild taste and sweet smell, renowned amongst the Arabs against calculus in the bladder, yielding by expression a fixed oil, employed in perfumery.—*C. Padus* (Bird Cherry).   Bark bitter and astringent, proposed as a substitute for quinine.—*C. Lauro-Cerasus.* A tree of Asia Minor, with aromatic leaves yielding by distillation a volatile oil and a considerable amount of hydrocyanic acid.   Distilled water medicinal and narcotic, even in small doses.

Chrysobalaneæ produce edible drupes, among which is the Cocoa Plum of the West Indian *Chrysobalanus Icaco*, and various other species, as also some of the genera *Moquilia* and *Parinarium*.   The latter genus inhabits both the New World and tropical West Africa.   *P. excelsum* is the Rough Skin or Grey Plum of Sierra Leone; and *P. macrophyllum* the Gingerbread Plum of the same colony.   The leaves of the Polynesian *P. laurinum* are used for thatching, its rough wood for spars, and its seeds for a perfume.

---

[1] Perhaps a French liqueur is here alluded to; the true *noyau* is a West Indian liqueur flavoured with the seeds of *Cerasus occidentalis.*—ED.

[The bark of *Moquilia utilis*, the Pottery Tree of the Amazons, abounds in silica to such an extent that when pulverized and mixed with clay it is used in making pottery by the natives of Para. *Quillaja Saponaria* and *brasiliensis* yield bark, used as soap, and containing a sternutatory allied to saponine.—ED.]

LXXVII. SAXIFRAGEÆ.

(SAXIFRAGÆ, *Jussieu.*— SAXIFRAGEÆ, *Ventenat.*—SAXIFRAGACEÆ, *D.C.*)

COROLLA *polypetalous, perigynous or epigynous, isostemonous or diplostemonous, æstivation imbricate.* STAMENS *inserted with the petals.* CARPELS *usually 2, distinct, or cohering into an ovary with more or less complete cells.* OVULES *anatropous.* FRUIT *dry.* EMBRYO *albuminous, axile.*

STEM herbaceous or sub-woody, sometimes woody, variable in appearance. LEAVES alternate or opposite, sometimes whorled; *stipules* 0 in the herbaceous species, interpetiolar in the woody, deciduous. FLOWERS ☿, regular or rarely irregular, variously arranged. CALYX usually pentamerous; *sepals* distinct or connate. PETALS 5, rarely fewer, inserted on a disk lining the calyx-tube, and alternate with the sepals, generally imbricate in æstivation, very rarely 0 (*Chrysosplenium*). STAMENS inserted alternately with the petals, or double them in number, very rarely indefinite (*Bauera*); *filaments* filiform, subulate; *anthers* introrse, 2-celled, ovoid, dehiscing longitudinally. CARPELS usually 2, sometimes 1 (*Neillia*), rarely 3 or 5, free, or united with the receptacular cup into an ovary of more or less perfect cells; *styles* and *stigmas* terminal, simple, sometimes cohering (*Polyosma*); *ovules* usually numerous, fixed either throughout their length, or to the bottom or top of the placentas, horizontal, ascending or pendulous, anatropous; *stigmas* terminal, simple, sometimes cohering (*Polyosma*). FRUIT capsular, rarely indehiscent, more rarely fleshy (*Polyosma*); *carpels* separating when ripe at their inner edge, either from the top downwards, or the reverse. SEEDS usually numerous, very rarely solitary, or definite, small; *testa* smooth or punctate, sometimes winged. EMBRYO straight, in the axis of a fleshy abundant albumen, and nearly equalling it in length; *cotyledons* short, semi-cylindric; *radicle* near the hilum, direction various.

SUB-ORDER I. *SAXIFRAGEÆ, D.C.*[1]

Stem herbaceous. Leaves alternate or rarely opposite, exstipulate, or with petioles of which the dilated bases resemble stipules. Flowers all fertile, racemed or panicled, rarely solitary. Petals 5, regular or dimorphous, sometimes 0. Stamens 5 or 10. Ovary free or inferior.

PRINCIPAL GENERA.

* Saxifraga.	Chrysosplenium.	* Hoteia.	* Astilbe.

[1] For the most recent classification of *Saxifrageæ*, see p. 392.—ED.

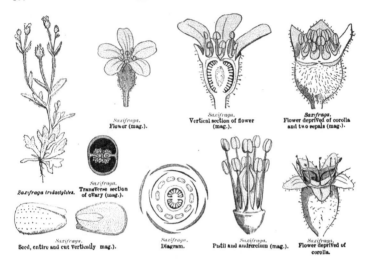

Saxifraga.
Flower (mag.).

Saxifraga.
Vertical section of flower (mag.).

Saxifraga.
Flower deprived of corolla and two sepals (mag.).

Saxifraga tridactylites.

Saxifraga.
Transverse section of ovary (mag.).

Saxifraga.
Seed, entire and cut vertically (mag.).

Saxifraga.
Diagram.

Saxifraga.
Pistil and androecium (mag.).

Saxifraga.
Flower deprived of corolla.

SUB-ORDER II. *CUNONIEÆ, D.C.*

Cunonia capensis.

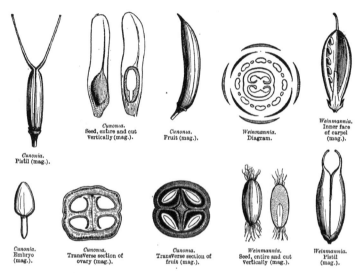

Cunonia.
Pistil (mag.).

Cunonia.
Seed, entire and cut
vertically (mag.).

Cunonia.
Fruit (mag.).

Weinmannia.
Diagram.

Weinmannia.
Inner face
of carpel
(mag.).

Cunonia.
Embryo
(mag.).

Cunonia.
Transverse section of
ovary (mag.).

Cunonia.
Transverse section of
fruit (mag.).

Weinmannia.
Seed, entire and cut
vertically (mag.).

Weinmannia.
Pistil
(mag.).

Shrubs or trees, chiefly from the southern parts of both continents. Leaves opposite, simple or compound; stipules interpetiolar. Calyx imbricate or valvate in æstivation. Petals 4, 5 or 0. Stamens 4-5, or 8-10, or 12-14, or ∞. Ovary free or rarely adherent, usually of 2-3 carpels, sometimes of 5 free carpels with coherent styles (*Spiræanthemum*).

PRINCIPAL GENERA.

* Callicoma. * Cunonia. * Bauera. Weinmannia. Curtisia. Crypteronia.[1]

SUB-ORDER III. *POLYOSMEÆ.*

Stem woody. Leaves opposite, exstipulate. Petals 4, valvate in æstivation. Stamens 4. Ovary inferior, 1-celled; placentas 2, parietal; style elongated; stigma simple. Berry 1-seeded.—Shrubs of tropical Asia and Australia, near *Marlea* (*Alangieæ*).

GENUS.

Polyosma.

[1] Referred to *Lythrarieæ* in the 'Genera Plantarum.'—ED.

SUB-ORDER IV. *HYDRANGEÆ, D.C.*

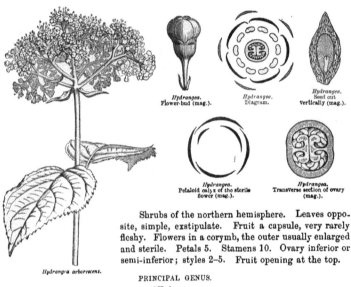

Hydrangea.
Flower-bud (mag.).

Hydrangea.
Diagram.

Hydrangea.
Seed cut
vertically (mag.).

Hydrangea.
Petaloid calyx of the sterile
flower (mag.).

Hydrangea.
Transverse section of ovary
(mag.).

Hydrangea arborescens.

Shrubs of the northern hemisphere. Leaves opposite, simple, exstipulate. Fruit a capsule, very rarely fleshy. Flowers in a corymb, the outer usually enlarged and sterile. Petals 5. Stamens 10. Ovary inferior or semi-inferior; styles 2–5. Fruit opening at the top.

PRINCIPAL GENUS.
* Hydrangea.

SUB-ORDER V. *ESCALLONIEÆ, D.C.*

Shrubs or trees. Leaves alternate, exstipulate. Petals 5–6. Stamens 5–6. Ovary free or inferior; styles 2. Fruit opening at the base.

PRINCIPAL GENERA.
* Escallonia.   * Itea.

[The following is the most recent elaboration of the great family of *Saxifrageæ*, as undertaken for the ' Genera Plantarum ' :—

TRIBE I. SAXIFRAGEÆ.—Herbs, usually scapigerous. Leaves usually alternate, exstipulate. Flowers usually 5-merous. Ovary 1–3-celled. *Donatia, Astilbe, Saxifraga, Vahlia, Tiarella, Heuchera, Chrysosplenium, Parnassia,*[1] &c.

TRIBE II. FRANCOEÆ. (See order FRANCOACEÆ, p. 401).—Scapigerous herbs. Flowers 4.

---

[1] See order *Parnassieæ*, p. 402.

merous. Stamens 4 or 8, alternating with scales. Ovary 4- (rarely 2-) celled. (Intermediate between SAXIFRAGEÆ and CRASSULACEÆ.) *Francoa, Tetilla.*

TRIBE III. HYDRANGEÆ.—Shrubs or trees. Leaves opposite, simple, exstipulate. Petals often valvate. Stamens usually epigynous. Ovary usually 3–5-celled. *Hydrangea, Dichroa, Deutzia,[1] Decumaria,[1] Philadelphus,[1] Jamesia, Fendlera,* &c.

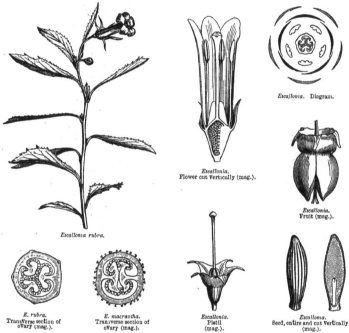

Escallonia. Diagram.

Escallonia.
Flower cut vertically (mag.).

Escallonia.
Fruit (mag.).

Escallonia rubra.

E. rubra.
Transverse section of
ovary (mag.).

E. macrantha.
Transverse section of
ovary (mag.).

Escallonia.
Pistil
(mag.).

Escallonia.
Seed, entire and cut vertically
(mag.).

TRIBE IV. ESCALLONIEÆ.—Trees or shrubs. Leaves alternate, simple, exstipulate, often coriaceous, and gland-serrate. Stamens usually as many as the petals. *Escallonia, Quintinia, Brexia,[2] Carpodetus, Itea, Polyosma, Anopterus, Argophyllum,* &c.

TRIBE V. CUNONIEÆ.—Trees or shrubs. Leaves opposite, rarely whorled, simple or compound, stipulate. Petals never valvate. *Codia, Callicoma, Spiræanthemum, Ceratopetalum, Acrophyllum, Ackama, Weinmannia, Cunonia,* &c.

TRIBE VI. RIBESIEÆ. (See order RIBESIACEÆ, p. 398).—Shrubs. Leaves alternate, simple; stipules 0, or adnate to the petiole. Flowers usually racemose. Ovary 1-celled, 2-carpellary. Seeds immersed in pulp; raphe free. *Ribes.*

ANOMALOUS GENERA. *Bauera, Cephalotus.*[3]—ED.]

[1] See order *Philadelpheæ,* p. 394. [2] See order *Brexiaceæ,* p. 396. [3] See order *Cephaloteæ,* p. 399.

Saxifrageæ, divided here into five sub-orders, are connected more or less closely with a good many families. The true *Saxifrageæ* approach *Crassulaceæ* in æstivation, diplostemony and insertion of the corolla, capsular fruit, herbaceous stem and cymose flowers. They resemble *Lythrarieæ* in their perigynous petals which are imbricate in æstivation and isostemonous or diplostemonous, and in the capsular fruit; but in *Lythrarieæ* the embryo is exalbuminous. There is also an evident analogy between some genera (*Hoteia, Lutkea, Astilbe*) and *Spiræa Aruncus* belonging to *Rosaceæ*. Besides the resemblance in habit, the corolla is polypetalous, imbricate, perigynous, polyandrous or diplostemonous; the carpels are distinct (at least in *Lutkea*), and open by the inner edge, the leaves are alternate, and in *Hoteia* clearly stipulate. We have indicated the affinities of *Hydrangeæ* with *Philadelpheæ* (see this family). They also recall, by habit and inflorescence, the genus *Viburnum*, belonging to *Caprifoliaceæ*; but in *Hydrangea* the sepals become petaloid, and in *Viburnum* the corolla is enlarged. For a comparison between *Escallonieæ* and *Cunoniaceæ* on the one hand, and *Hamamelideæ* on the other, see the latter family. The *Saxifrageæ* have also some points of resemblance with *Parnassieæ* (see this family). Finally, we must notice a real relation between *Escallonieæ* and *Grossularieæ*; in both the petals are isostemonous and imbricate in æstivation, the ovary is inferior and one-celled; there are two styles, and the embryo is albuminous, the stem woody, and the leaves alternate. But in *Grossularieæ* the placentation is more clearly parietal, the fruit a berry, the testa of the seed gelatinous, the embryo minute, and the leaves palminerved.

Near *Saxifrageæ* should be placed the little group of *Diamorpheæ* (consisting of *Diamorpha* and *Penthorum*), placed by most authors in *Crassulaceæ*, from which it differs in its many-celled ovary, and especially in its habit; it may be allied on the one hand to *Saxifrageæ*, and on the other perhaps to *Pongatium* (*Sphenoclea*, of Gaertner), which Jussieu placed near *Portulaceæ* [a monopetalous genus or order placed near *Campanulaceæ*].

The different tribes of this large family occupy different countries. The true *Saxifrageæ* mostly inhabit the high mountains of the northern hemisphere, and are most fully represented in America; they are very rare in the tropics and antarctic regions. *Cunonieæ* are frequent in the south temperate zone; they are less common in tropical America, and have never been found north of the tropic. *Hydrangeæ* are not rare in upper India, Japan and South America, but become so in Peru and Java. *Escallonieæ* all belong to America, and are for the most part trans-tropical. [Various genera are natives of Australia, New Zealand, tropical and temperate Asia, South Africa, and the islands of Mauritius and Madagascar.]

The useful properties of *Saxifrageæ* are unimportant. The mucilaginous acidulous leaves and the root-bulbs of *Saxifraga granulata* were formerly praised as powerful lithontriptics. *S. tridactylitis* was employed in diseases of the liver, and *Chrysosplenium* was a reputed tonic. The resinous buds and aromatic leaves of the *Escallonieæ* are similarly employed in Peru and Chili. [The leaves of various Hydrangeas make a highly esteemed tea in Japan. *Weinmannia* yields an astringent bark, used both as a medicine and for tanning purposes.]

LXXVIII. *PHILADELPHEÆ.*

(MYRTI, *partim, Jussieu.*—PHILADELPHEÆ, *Don.*)

COROLLA *polypetalous, epigynous, valvate or contorted in æstivation.* STAMENS *double or a multiple of the number of the petals.* OVARY *inferior, many-celled, with many-ovuled central placentas.* OVULES *pendulous or ascending, imbricate.* FRUIT *a capsule.* SEEDS *with a membranous loose testa.* EMBRYO *albuminous, axile.*—STEM *woody.* LEAVES *opposite.*

Erect SHRUBS. LEAVES opposite, simple, petioled, quite entire or toothed, deciduous, exstipulate. FLOWERS ☿, regular, white, [often] sweet-scented, in a terminal

Syringa.
Diagram.

Syringa.
One of the cells of the
ovary (mag.).

Syiinga. (*Philadelphus coronarius.*)

Syringa. Vertical section of flower (mag.).

Syrliga.
Pistil and calyx.

Syringa.
Seed, entire and cut (mag.).

Syringa.
Transverse section of ovary (mag.).

Syringa.
Fruit (mag.).

Decumaria.
Expanded flower
(mag.).

Decumaria.
Closed flower
(mag.).

Decumaria.
Vertical section of pistil
(mag.).

Decumaria.
Transverse section of ovary
(mag.).

cyme. CALYX superior, 4–10-partite, valvate in æstivation, persistent. PETALS 4–5–7–10, inserted under an annular disk crowning the ovary and lining the calyx, alternate with its segments, æstivation induplicate or contorted. STAMENS double or a multiple of the number of the petals, inserted with them, 1-2-seriate; *filaments* filiform or compressed; *anthers* introrse, 2-celled, ovoid or sub-globose, didymous, dehiscence longitudinal. OVARY inferior or semi-inferior, of 3-4–10 cells; *styles* as many as the cells, distinct or more or less coherent; *stigmas* free or connate; *ovules* numerous, ascending or pendulous, imbricate on projecting central placentas. CAPSULE 3–10-celled, dehiscing at the top loculicidally or septicidally, or rupturing longitudinally along the sides of the receptacular tube. SEEDS with a membranous reticulate loose and ample testa. EMBRYO straight, axile, as long as the copious fleshy albumen; *cotyledons* short, semi-cylindric or oval; *radicle* long, next the hilum, superior or inferior.

PRINCIPAL GENERA.

* Philadelphus.   * Decumaria.   * Deutzia.

Philadelpheæ approach *Saxifrageæ*, tribe *Hydrangeæ*, in the epigyny, æstivation, and diplostemony of the corolla, the many-ovuled cells of the ovary, distinct styles, capsular fruit, straight albuminous axile embryo, woody stem and opposite leaves. They have some affinities with *Onagrarieæ*, founded on the insertion and æstivation of the petals, the numerous pendulous or ascending ovules, and the loculicidal or septicidal capsule; but *Onagrarieæ* differ in the structure of the testa, and in being exalbuminous.

Philadelpheæ inhabit South Europe, Upper India and Japan, but are nowhere numerous.

The very strongly scented flowers of the Syringa (*Philadelphus coronarius*), formerly employed as a tonic, have fallen into disuse. The rough leaves of *Deutzia scabra* are used in Japan to polish wood.

LXXIX. *BREXIACEÆ*, Endlicher.

TREES or SHRUBS. LEAVES alternate, sub-coriaceous, entire or spinous-toothed, exstipulate. FLOWERS in axillary and terminal umbels. CALYX 5-fid, persistent, æstivation imbricate. PETALS 5, shortly clawed, inserted on the edge of an annular disk, perigynous, æstivation contorted (*Brexia*) or imbricate (*Ixerba*). STAMENS alternate, accompanied by palmate scales which are opposite to the petals, and connect the bases of the subulate filaments; *anthers* introrse, 2-celled, dehiscence longitudinal. OVARY free, of 5 many-ovuled cells; *style* short; *stigma* 5-lobed; *ovules* 2-seriate, horizontal, anatropous. FRUIT a 5-sided drupe with papillose epicarp and bony endocarp, or a loculicidal capsule (*Ixerba*). SEEDS horizontal, shortly funicled, ovoid-angular, shining; testa membranous. EMBRYO [almost] exalbuminous, straight; *cotyledons* obtuse; *radicle* cylindric.

This little group is composed of the genera *Brexia*, *Ixerba* and *Argophyllum*; [1] *Brexia* inhabits Madagascar, *Ixerba* Australia, and *Argophyllum* New Caledonia. Endlicher places *Brexiaceæ* after *Saxifrageæ*, as being near the tribe *Escallonieæ*; in both, in fact, the stem is woody, the leaves alternate,

----

[1] *Argophyllum* differs notably in the valvate calyx, copious albumen and minute embryo.—ED.

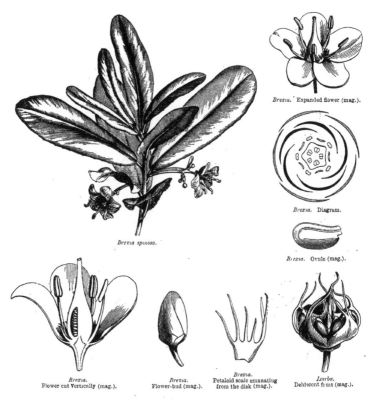

Brexia. Expanded flower (mag.).

Brexia. Diagram.

Brexia. Ovule (mag.).

Brexia spinosa.

Brexia.
Flower cut vertically (mag.).

Brexia.
Flower-bud (mag.).

Brexia.
Petaloid scale emanating
from the disk (mag.).

Ixerba.
Dehiscent fruit (mag.).

and the corolla polypetalous and isostemonous; the ovary is free (at least in *Itea*), with many-ovuled cells; the ovules are anatropous with central placentation, and style simple; but in *Brexia* the embryo has no albumen. A. Brongniart places them doubtfully in the class *Ericoideæ*.

LXXX. *RIBESIACEÆ.*

(Cactorum *genus, Jussieu.*—Grossularieæ, *D.C.*—Grossulaceæ, *Lindl.*)

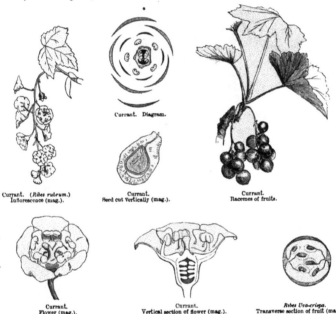

Currant. Diagram.

Currant. (*Ribes rubrum.*)
Inflorescence (mag.).

Currant.
Seed cut vertically (mag.).

Currant.
Racemes of fruits.

Currant.
Flower (mag.).

Currant.
Vertical section of flower (mag.).

Ribes Uva-crispa.
Transverse section of fruit (mag.).

Corolla *polypetalous, epigynous, isostemonous, æstivation imbricate.* Stamens *5, inserted alternately with the petals.* Ovary *inferior, 1-celled; placentas parietal.* Ovules *horizontal, anatropous.* Fruit *a berry.* Embryo *albuminous.*—Stem *woody.* Leaves *scattered or fascicled.*

Shrubs, unarmed, or armed with infra-axillary or scattered spines; *branches* cylindric or angular. Leaves scattered or fascicled, simple, petioled, palminerved, often glandular, folded or convolute in bud; *petiole* channelled, dilated at its base. Flowers ☿, or often imperfect through arrest, regular, usually racemed, and terminating in very short leafy or sometimes leafless branchlets or buds; *pedicels* 2-bracteolate, jointed near the top. Calyx coloured, marcescent, superior, cylindric, campanulate or cupular, 5–4-fid. Petals inserted on the throat of the calyx, alternate with its segments, æstivation imbricate, marcescent. STAMENS inserted alternately with the petals; *filaments* filiform; *anthers* introrse, 2-celled, ovoid or oblong, tip emarginate or pointed or glandular, dehiscence longitudinal. OVARY inferior, 1-celled, crowned by a thin disk; *placentas* 2, rarely 3 or 4, nerviform, parietal or edging semi-septa; *styles* as many as placentas, distinct or more or less coherent; *stigmas* short, distinct, obtuse; *ovules* usually numerous, pluriseriate, horizontal, shortly funicled, anatropous. BERRY crowned by the calyx and the withered petals, 1-celled, pulpy. SEEDS angular; *testa* gelatinous; *endopleura* crustaceous, adhering to the albumen. EMBRYO very small, straight, at the base of a horny albumen.

PRINCIPAL GENUS.

*Ribes.

Ribesiaceæ have many analogies with *Cacteæ* (which see). They are near *Saxifrageæ*, tribe *Escalloniæ*, in their woody stem, alternate leaves, racemed flowers, polypetalous isostemonous epigynous corolla, inferior generally 2-carpellary ovary, and albuminous embryo; they are separated by their habit, fleshy fruit, pulpy seeds, and minute embryo.   [*R. Grossularia* is indigenous on the Morocco Atlas.—ED.]

Ribesiaceæ inhabit the temperate and cold regions of the northern hemisphere, especially North America; they are rare in South America, and absent from Africa.

USEFUL SPECIES.—*Ribes rubrum* (Currant). Berries red or white, containing a sugary mucilage combined with citric and malic acids; much used for dessert, and in the preparation of a syrup and a jelly. —*R. Uva-crispa* (Gooseberry). Fruit with a sugary taste, sourish and slightly aromatic, juice fermentable, and used in England in the preparation of a spirituous liquor [probably Gooseberry wine is here alluded to.—ED.].—*R. nigrum* (Black Currant). Berries containing a resinous aromatic principle, formerly employed medicinally, now forming the base of the popular drink called ' cassis.'

LXXXI. *CEPHALOTEÆ, Endlicher.*

Perennial HERBS with short subterranean rhizomes. LEAVES in a rosette at the top of the rhizome, of two forms: some flat, elliptic, entire, nerveless, with a subcylindric petiole dilated at its base; others (*ascidia*), scattered amongst the first, are composed of a petiole which is dilated at the top into·two lips, of which the lower is large, hollowed into a cup, and opens by a circular orifice; the upper is smaller, flat, and serves as a lid to the cup. SCAPE simple, with scattered alternate bracts, terminating in a spike composed of 4–5-flowered partial spikes furnished below with linear bracts. FLOWERS in a corymb, small, white, ebracteate. CALYX coloured, 6-fid; *segments* ovate-lanceolate, valvate in æstivation, with a small tooth at the top within, clothed at the thickened base with capitate hairs. COROLLA 0. STAMENS 12, inserted at the top of the calycine tube, shorter than its limb, the six which alternate with the sepals longer and forwarder than the others; *filaments* subulate; *anthers* rounded, didymous, with opposite cells opening longitudinally; *connective* sub-globose, spongy. OVARIES 6, crowded, sessile, whorled on a flat receptacle around a central bundle of hairs, alternate with the sepals, sub-compressed, 1-celled; *styles* terminal, cylindric; *stigmas* simple; *ovule* solitary, erect, sub-basilar, anatro-

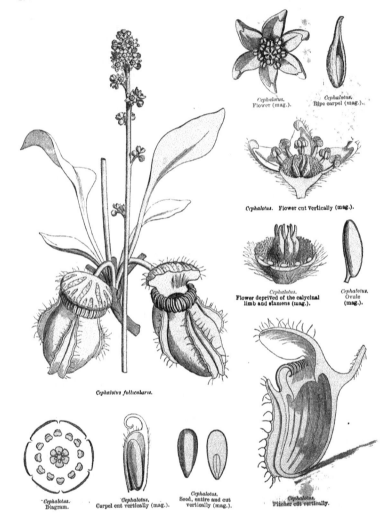

Cephalotus.
Flower (mag.).

Cephalotus.
Ripe carpel (mag.).

Cephalotus. Flower cut vertically (mag.).

Cephalotus.
Flower deprived of the calycinal
limb and stamens (mag.).

Cephalotus.
Ovule
(mag.).

Cephalotus follicularis.

Cephalotus.
Diagram.

Cephalotus.
Carpel cut vertically (mag.).

Cephalotus,
Seed, entire and cut
vertically (mag.).

Cephalotus.
Pitcher cut vertically.

pous, raphe dorsal. ACHENES [follicles] membranous, surrounded by the accrescent calyx and stamens, detaching circularly near their persistent base, which is composed of a simple membrane; the upper portion composed of a double membrane, thickly hairy externally, terminated by the style, and opening longitudinally. SEED with a membranous loose testa, a lateral slender raphe, and an apical chalaza. EMBRYO straight, very short, occupying the base of a fleshy oily albumen; *cotyledons* plano-convex; *radicle* cylindric, inferior.

ONLY GENUS.

* Cephalotus.

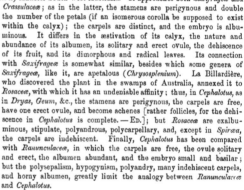

The only species (*C. follicularis*) inhabits South-west Australia. It approaches both *Saxifrageæ* and *Crassulaceæ*; as in the latter, the stamens are perigynous and double the number of the petals (if an isomerous corolla be supposed to exist within the calyx); the carpels are distinct, and the embryo is albuminous. It differs in the æstivation of its calyx, the nature and abundance of its albumen, its solitary and erect ovule, the dehiscence of its fruit, and its dimorphous and radical leaves. Its connection with *Saxifrageæ* is somewhat similar, besides which some genera of *Saxifrageæ*, like it, are apetalous (*Chrysosplenium*). La Billardière, who discovered the plant in the swamps of Australia, annexed it to *Rosaceæ*, with which it has an undeniable affinity; thus, in *Cephalotus*, as in *Dryas*, *Geum*, &c., the stamens are perigynous, the carpels are free, have one erect ovule, and become achenes [rather follicles, for the dehiscence in *Cephalotus* is complete. — ED.]; but *Rosaceæ* are exalbuminous, stipulate, polyandrous, polycarpellary, and, except in *Spiræa*, the carpels are indehiscent. Finally, *Cephalotus* has been compared with *Ranunculaceæ*, in which the carpels are free, the ovule solitary and erect, the albumen abundant, and the embryo small and basilar; but the polysepalism, hypogynism, polyandry, many indehiscent carpels, and horny albumen, greatly limit the analogy between *Ranunculaceæ* and *Cephalotus*.

LXXXII. *FRANCOACEÆ*, *Endlicher*.

Perennial HERBS. LEAVES subradical, lyrate-pinnatifid or palminerved, sinuous-toothed. FLOWERS bracteate, in

Francoa appendiculata.   *Francoa.* Half of the flower (mag.).   *Francoa.* Diagram.

D D

Francoa.   *Francoa.*   *Francoa.*   *Francoa.*
Pistil (mag.). Transverse section of ovary (mag.). Seed, entire and cut vertically (mag.). Ovule (mag.).

terminal racemes. CALYX 4-partite. PETALS 4, rarely 5, inserted at the base of
the calyx, clawed. STAMENS inserted with the petals, 8 or 10 fertile alternating
with as many sterile; *filaments* distinct, subulate; *anthers* introrse, 2-celled,
dehiscence longitudinal. OVARY free, tetragonous, oblong, 4-lobed at the top,
4-celled; *stigma* sessile, with four lobes alternating with the cells; *ovules* numerous,
2-seriate at the central angle of the cells, horizontal, anatropous. CAPSULE loculicidal. SEEDS numerous, tuberculate and striate. EMBRYO straight, cylindric, in
the axis of a fleshy albumen.

GENERA.

Francoa.   Tetilla.

Francoaceæ are Chilian herbs, placed near *Crassulaceæ* and *Saxifrageæ*. They approach the latter
in the polypetalous perigynous diplostemonous corolla, many-ovuled ovarian cells, anatropous ovules,
capsular fruit, straight albuminous embryo, herbaceous stem, alternate leaves, and racemed flower; they
are scarcely separated but by the contorted æstivation of the corolla, sterile stamens, sessile 4-lobed
stigma, and dehiscence of the capsule. They approach *Crassulaceæ* in the perigynous and diplostemonous
petals, the horizontal anatropous ovules, capsular fruit, herbaceous stem, and alternate leaves. They
have also some connection with *Lythrarieæ* in the corolla, stamens and ovary; but the embryo in *Lythrarieæ* is exalbuminous, and the leaves are opposite.

LXXXIII. *PARNASSIEÆ*, *Endlicher*.

Perennial glabrous HERBS. STEMS scape-like, simple, 1-flowered. RADICAL
LEAVES long-petioled, cordate [oblong] or reniform, cauline sessile. FLOWERS ☿,
regular, white or yellowish. CALYX 5-partite, imbricate, persistent. PETALS 5,
perigynous, alternate with the sepals, imbricate in æstivation, deciduous. STAMENS
5, fertile, [hypogynous or perigynous,] inserted alternately with the petals; *filaments*
subulate; *anthers* extrorse, 2-celled, ovoid or sub-globose, dehiscence longitudinal;
petaloid scales 5, opposite to the petals, perhaps representing bundles of sterile
stamens, and dividing into 3, 5, 7, 9, or 15 branches, each terminated by a globose
nectariferous gland. OVARY superior (*P. palustris*), or semi-inferior (*P. himalayensis*,
&c.), 1-celled, with 3–4 parietal placentas; *stigma* sessile, 3–4-partite; *ovules*

SEEDS very small; *testa* membranous, reticulate, loose, extending beyond the endopleura. EMBRYO exalbuminous [or albumen scanty], straight, oblong-cylindric.

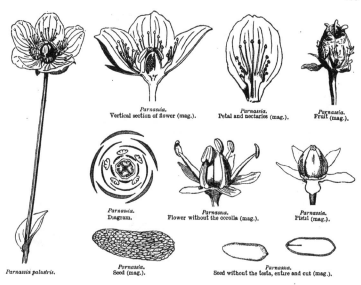

Parnassia.
Vertical section of flower (mag.).

Parnassia.
Petal and nectaries (mag.).

Parnassia.
Fruit (mag.).

Parnassia.
Diagram.

Parnassia.
Flower without the corolla (mag.).

Parnassia.
Pistil (mag.).

Parnassia palustris.

Parnassia.
Seed (mag.).

Parnassia.
Seed without the testa, entire and cut (mag.).

ONLY GENUS.
Parnassia.

The genus *Parnassia*, long annexed to *Droseraceæ*, only approaches them in its parietal placenta (see this family). Its bundles of staminodes and its exalbuminous seeds bring it near *Hypericineæ*, but other characters separate it from them, and especially the extrorse anthers. It has been more correctly compared with *Saxifrageæ*. The few species of this genus inhabit the temperate and cool regions of the northern hemisphere, especially in America; they are rare [common] on the high mountains of tropical Asia. The Grass of Parnassus (*P. palustris*), an indigenous plant, is a bitter and astringent herb, formerly used as a diuretic and anti-ophthalmic; a decoction of it is in Sweden added to beer, and stomachic virtues are attributed to it.

LXXXIV. *CRASSULACEÆ.*

(SEMPERVIVÆ, *Jussieu.*—SUCCULENTÆ, *Ventenat.*—CRASSULÆ, *Jussieu.*—CRASSULACEÆ, *D.C.*)

COROLLA *generally polypetalous, perigynous, diplo- (rarely iso-) stemonous, œstivation imbricate.* STAMENS *inserted with the petals at the bottom of the calyx.* OVARIES *equalling the petals in number, generally distinct, furnished with a scale at their outer*

base, and pluri-ovuled; OVULES *anatropous.* FRUIT *follicular.* EMBRYO *exalbuminous.*
—*Succulent plants.*

HERBS or UNDERSHRUBS with cylindric more or less fleshy stem and branches.
LEAVES usually scattered, fleshy, sometimes cylindric or subulate, simple, entire, or
very rarely pinnately lobed (*Bryophyllum, Kalanchoe*), exstipulate. FLOWERS ☿, or
imperfect by arrest, regular, in unilateral cymes, or in terminal often dichotomous

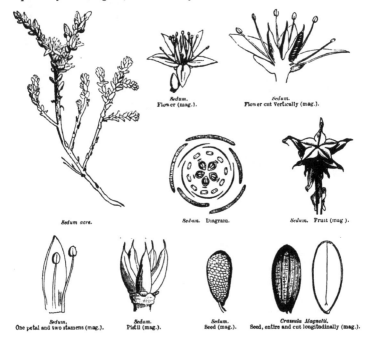

Sedum.
Flower (mag.).

Sedum.
Flower cut vertically (mag.).

Sedum acre.

Sedum. Diagram.

Sedum. Fruit (mag.).

Sedum,
One petal and two stamens (mag.).

Sedum.
Pistil (mag.).

Sedum.
Seed (mag.).

Crassula Magnolii.
Seed, entire and cut longitudinally (mag.).

corymbs, rarely spiked, sometimes axillary and solitary. CALYX usually 5-fid or
-partite, rarely 3–20-partite, æstivation imbricate, persistent. COROLLA inserted at
the base of the calyx; *petals* free or connate into a tube, alternate with the sepals,
æstivation imbricate or valvate. STAMENS adnate to the monopetalous corolla, or
inserted alternately with the petals, sometimes double them in number; *filaments*
distinct, subulate; *anthers* introrse, basifixed, with 2 opposite cells, dehiscence
longitudinal. SCALES hypogynous, as many as carpels, at their outer base. CARPELS
opposite the petals, and of equal number [except *Triactina*], whorled, 1-celled, many-

ovuled, distinct [connate in *Diamorpha, Penthorum, Triactina*]; *styles* continuous with the back of the ovaries; *stigma* subterminal, on the ventral face; *ovules* 2-seriate along the ventral suture, horizontal or pendulous, anatropous. FOLLICLES free, dehiscence ventral. SEEDS very small, testa membranous. EMBRYO straight, exalbuminous, according to Brongniart, [albuminous according to others]; *cotyledons* very short; *radicle* next the hilum.

PRINCIPAL GENERA.

| Tillæa. | * Crassula. | * Rochea. | * Bryophyllum. | * Kalanchoe. |
| * Cotyledon. | * Sedum. | * Sempervivum. | * Echeveria. | |

To *Crassulaceæ* belongs the *Bryophyllum calycinum*, an undershrub of the tropical regions of the Old World, quoted in the Introduction (p. 7). *Crassulaceæ* are connected with *Saxifrageæ, Francoaceæ* and *Cephaloteæ* (see those families). They inhabit warm temperate regions of the Old World, and, owing to the fewness of their stomata, and consequent slight transpiration, they remain green in the most arid countries. They especially abound beyond the tropic of Capricorn; half of the known species live in South Africa, the sixth part in Europe and the Mediterranean region, and an equal number in Central Asia and the Canaries, and in sub-tropical America, Southern Asia and Australia.

The watery juice of *Crassulaceæ* contains, besides an abundance of albumine, astringent principles and malic acid, free or combined with lime. The useful species are the following :—*Sempervivum Tectorum* (House Leek). Juice taken as a refreshing drink, and united to some fatty body applied externally for burns and bleeding. The leaves are also used to remove corns; a property also possessed by *Crassula Cotyledon* and *arborescens.—Sedum album* (White Stonecrop). Juice astringent, refreshing.—*S. Telephium* (Orpine). Formerly cultivated as a pot-herb. Juice employed to remove corns and to heal wounds.—*S. acre* (Biting Stonecrop). Purgative and emetic if taken internally, rubefacient outwardly, and recommended for bad ulcers.—*S. reflexum.* Refreshing, diuretic, vulnerary.—*Crassula rubens.* Leaves employed as a vulnerary.—*Umbilicus pendulinus* (Nàvelwort). An emollient, employed outwardly for hard nipples.

LXXXV. *DROSERACEÆ.*

(DROSEREÆ, *Salisbury.*—DROSERACEÆ, *D.C.*)

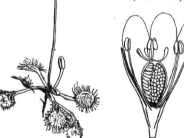

Sundew. (*Drosera rotundifolia.*)
Vertical section of flower (mag.).
Drosera. Diagram.
Drosera. Fruit (mag.).
Drosera. Seed (mag.).

Drosera. Vertical section of flower (mag.).

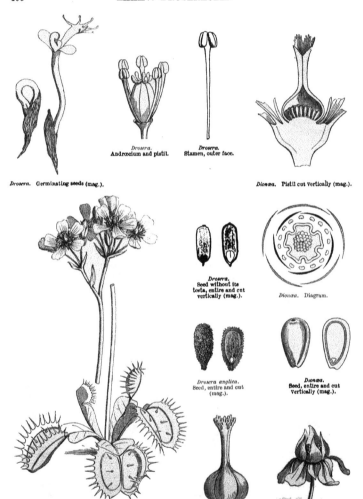

Drosera.
Androecium and pistil.

Drosera.
Stamen, outer face.

Dionæa. Pistil cut vertically (mag.).

Drosera. Germinating seeds (mag.).

Drosera.
Seed without its
testa, entire and cut
vertically (mag.).

Dionæa. Diagram.

Drosera anglica.
Seed, entire and cut
(mag.).

Dionæa.
Seed, entire and cut
vertically (mag.).

Dionæa muscipula.

Dionæa. Pistil (mag.).

Dionæa. Fruit (mag.).

PETALS 5, *hypogynous, imbricate.* STAMENS 5, *or rarely more;* ANTHERS *extrorse.* OVARY *usually 1-celled, and with parietal placentation.* CAPSULE *with semi-placentiferous valves.* EMBRYO *albuminous.*

HERBS, sometimes UNDERSHRUBS, stemless or caulescent, sprinkled and ciliated with glandular tracheiferous hairs. LEAVES alternate, usually in a rosette, simple, entire, or rarely cut, coiled from top to bottom before expansion [outwards or backwards in *Drosophyllum,* inwards or forwards in the other genera]; blade contracted into a petiole, sometimes jointed, with median nerve irritable, and the two halves of the blade closing quickly at the least touch (*Dionæa,* p. 150); *stipules* 0, represented by hairs edging the dilated base of the petiole [or scarious]. FLOWERS ☿, regular, solitary, or in unilateral circinnate cymes [or subcorymbose]. CALYX of 5 sepals, free or almost free, imbricate. PETALS 5, hypogynous, or united at the base with the sepals, shortly clawed, imbricate in bud, marcescent. STAMENS hypogynous [or perigynous, rarely epipetalous], sometimes equal and alternate with the petals; sometimes double (*Dionæa, Drosophyllum*); sometimes triple or quadruple (*Dionæa*), and then some opposite to the sepals singly, and the others opposite to the petals in twos or threes; *filaments* filiform, linear, free; *anthers* extrorse, 2-celled, erect (*Dionæa, Drosera, Roridula*), or incumbent and versatile (*Drosophyllum, Byblis*), dehiscence longitudinal, or rarely apical (*Byblis, Roridula*). OVARY free, sessile, 1-celled, with 3–5 parietal placentas, sometimes united in one, basilar (*Dionæa, Drosophyllum*), rarely 2-celled (*Byblis*) or 3-celled (*Roridula*); *styles* as many as the placentas, undivided or bifid or laciniate (*Drosera*), or coherent in a simple style (*Dionæa, Roridula,* &c.); *stigmas* capitate, lobed or fringed; *ovules* anatropous, usually upright or ascending, rarely pendulous from the top of the cells in the 3-celled ovaries (*Roridula*). CAPSULE sometimes 1-celled, opening throughout its length into 3–5 semi-placentiferous valves, or at the top only, with a free basilar placenta (*Drosera, Aldrovanda, Drosophyllum, Dionæa*); or 2-celled with 2 loculicidal valves, bearing on the middle a seminiferous semi-septum (*Byblis*); or 3-celled with 3 loculicidal semi-septiferous valves separated from the persistent seminiferous columella (*Roridula*). SEEDS with a crustaceous granular or striate rarely loose and cellular testa; *albumen* fleshy. EMBRYO straight, axile or basilar; *cotyledons* truncate; *radicle* very short, inferior or superior.

GENERA.

| Drosera. | Drosophyllum. | Aldrovanda. | * Dionæa. | Roridula. | Byblis. |

Droseraceæ, which are near *Violarieæ* in the usually parietal placentation, æstivation, insertion, structure of the fruit and presence of albumen, are separated from them by habit, vernation, the absence of stipules and extrorse anthers. With *Frankeniaceæ* they have somewhat analogous relations, together with that of the extrorse anthers; but in *Frankeniaceæ* the calyx is tubular, and the leaves are opposite or quaternary. *Nepentheæ* and *Sarraceniæ* have also some connection with *Droseraceæ,* founded on the loculicidal capsule, the nature of the seeds, the albuminous embryo, and the exceptional structure of the leaves; but *Nepentheæ* are diœcious (see these families). *Parnassieæ,* which contain but a single genus, have been annexed to *Droseraceæ* by many botanists, but they differ in habit, the petaloid glanduliferous scales opposite to the petals, the sessile stigma and exalbuminous seed.

Droseraceæ inhabit nearly all climates. *Drosera* are widely scattered, being especially frequent in Australia, equatorial America and South Africa. They are met with in the turfy prairies of Europe and

North America. The other genera are monotypic. *Aldrovanda* floats on stagnant waters in the South of France and North Italy [and Bengal], *Drosophyllum* in the Spanish peninsula [and Morocco], *Dionæa* in the savannahs of South Carolina, *Roridula* in South Africa, *Byblis* in Australia.

The properties of *Droseraceæ* are imperfectly known. The indigenous *Droseræ* are acidulous-acrid, bitter, vesicant, and very hurtful to sheep. They have been found useful in dropsy and intermittent fevers. Their name of Sundew is derived from the tiny drops secreted by the glandular hairs of the leaves. [The glandular hairs on the leaves of various species are irritable, curving round insects that get entangled by their viscid tips.—ED.]

LXXXVI. *HAMAMELIDEÆ.*

(HAMAMELIDEÆ, *Br.*—HAMAMELACEÆ, *Lindl.*)

COROLLA 0 *or polypetalous.* PETALS 4–5, *perigynous, æstivation valvate.* APETALOUS FLOWERS *polyandrous;* PETALOID FLOWERS *diplostemonous.* STAMENS *some fertile, opposite to the petals; others sterile, squamiform, alternate.* OVARY *semi-inferior, 2-celled;* OVULES *pendulous, anatropous.* FRUIT *a capsule.* EMBRYO *albuminous, axile;* RADICLE *superior.*—STEM *woody.* LEAVES *alternate, stipulate.*

SHRUBS, or small or large TREES, with cylindric branches glabrous or stellately hairy. LEAVES alternate, petioled, simple, penninerved; *stipules* geminate at the base of the petioles, deciduous. FLOWERS ☿, or ♂ ♀ by arrest, sub-sessile, in a fascicle, head or spike, usually bracteate. CALYX superior, sometimes of 4–5 lobes, imbricate in bud, deciduous; sometimes truncate, with 5–7 sinuous teeth, callous. COROLLA 0 (*Fothergilla*) or polypetalous. PETALS perigynous, alternate with the calyx-lobes, valvate in bud, deciduous. STAMENS of the petaloid flowers inserted with the petals, and double them in number, those opposite to the petals fertile, the others sterile, squamiform; those of the apetalous flowers indefinite; *filaments* free, short

Hamamelis virginica.

Hamamelis. Flower cut vertically (mag.).

Hamamelis. Portion of flower (mag.).

Hamamelis.
Flower seen from
above (mag.).

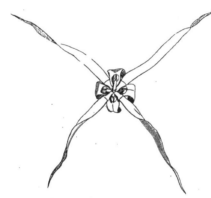

Hamamelis. Expanded flower (mag.).

Hamamelis.
Capsular woody fruit, opening in two bifid
valves, with horned endocarp, separating into
two shells (mag.).

Hamamelis.
Seed cut vertically
(mag.).

Hamamelis.
Embryo
(mag.).

Rhodoleia Championi.
Flower (mag.).

Hamamelis.
Seed (mag.).

Hamamelis.
Anther (mag.).

Hamamelis.
Transverse section of
fruit (mag.).

and dilated at the base in the petaloid flowers, elongated and sub-clavate in the apetalous; *anthers* introrse, 2-celled, sometimes ovoid-quadrangular, cells opposite, adnate to a connective which is usually prolonged into a point, divided by a longitudinal median septum, and opening by valves; sometimes (*Fothergilla*) horse-shoe shaped, opening by a semicircular slit. OVARY semi-inferior, 2-celled; *styles* 2, distinct; *stigmas* simple; *ovules* solitary, or very rarely many, of which only the lower is fertile (*Bucklandia*), suspended to the top of the septum, anatropous or semianatropous. CAPSULE semi- or quite superior, 2-valved, valves 2-fid at the top. SEEDS pendulous. EMBRYO straight, occupying the axis of a fleshy or cartilaginous albumen, and nearly equalling it in length; *cotyledons* foliaceous, flat or curled at the edges; *radicle* cylindric, superior.

PRINCIPAL GENERA.

*Hamamelis.   *Fothergilla.   *Rhodoleia.

[This and the following order have been recently studied carefully for the ' Genera Plantarum,' and combined; with the following arrangement of the genera, to which a few new discoveries are added.

A. Ovarian cells 1-ovuled.

PETALS 0.—*Parrotia, Davidia, Fothergilla, Disanthus, Distylium, Sycopsis.*

PETALOID (petals sometimes reduced to scales). — *Corylopsis, Dicoryphe, Maingaya, Hamamelis, Trichocladus, Loropetalum, Tetrathyrium, Eustigma.*

B. Ovarian cells 2–∞-ovuled.

*Rhodoleia, Ostrearia, Bucklandia, Altingia, Liquidambar, Disanthus.—*ED.]

Hamamelideæ approach *Corneæ* in insertion, number and æstivation of the petals, the pendulous anatropous ovules, albuminous axile embryo, woody stem and capitulate inflorescence. *Corneæ* are separated by their complete epigynism, the alternation of the stamens, simple style, fleshy fruit, [usually] opposite and exstipulate leaves. The same relations exist between *Araliaceæ* and *Hamamelideæ*. There is also a resemblance between them and *Saxifrageæ*, tribe *Cunoniaceæ*, both being perigynous, diplostemonous, digynous, and having anatropous ovules, capsular fruit, albuminous axile embryo, and woody stem; *Cunoniaceæ* differ principally in their interpetiolar stipules, opposite leaves, and imbricate æstivation of the petals. The same applies to *Escalloniæ*, excepting that these have alternate and exstipulate leaves. Finally, an affinity exists between the diclinous apetalous *Hamamelideæ* and *Plataneæ*, consisting in polyandry, inferior ovary, pendulous ovules, dry fruit, albuminous and axile embryo, woody stem, alternate stipulate leaves. For the diagnosis, *Plataneæ* differ in the flowers having no perianth, and being arranged in catkins, and their fruit being a nucule. The affinity is still more close with *Liquidambar*, which Mr. Bentham unites to *Hamamelideæ*. They also approach *Grubbiaceæ* in the general structure of their flowers and the valvular anthers. (See also *Garryaceæ*.) This family, which contains but few species, is dispersed over both hemispheres; it inhabits South America, Japan, China, India, Persia, Madagascar, and the Cape of Good Hope. With regard to the useful species, the seed of the Virginian Hamamelis is oily, the leaves and bark are medicinal. The wood of *Parrotia* is extremely hard, and in Persia is called Iron-wood.

LXXXVII. *BALSAMIFLUÆ, Blume.*

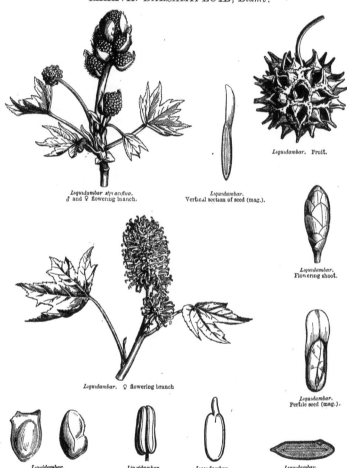

Liquidambar styraciflua.
♂ and ♀ flowering branch.

Liquidambar.
Vertical section of seed (mag.).

Liquidambar. Fruit.

Liquidambar.
Flowering shoot.

Liquidambar. ♀ flowering branch

Liquidambar.
Fertile seed (mag.).

Liquidambar.
Ovules, perfect and abortive (mag.).

Liquidambar.
Stamen (mag.).

Liquidambar.
Embryo (mag.).

Liquidambar.
Transverse section of seed (mag.).

Liquidambar.
Transverse section of
♂ catkin

Liquidambar.
Vertical section of a portion of
♀ catkin (mag.).

Liquidambar.
Stigmas and abortive
anthers (mag.).

Liquidambar.
Vertical section of
ovary (mag.).

FLOWERS *monœcious, in catkins or capitula.* PERIANTH *single or* 0. STAMENS
numerous. ♀ FLOWERS *with a single accrescent perianth.* OVARIES *connate,* 2-*celled;*
STYLES 2; OVULES *numerous, sub-anatropous.* FRUIT *compound, of several* 2-*valved
capsules.* SEEDS (*fertile*), *elliptic, peltate, albuminous.* EMBRYO *axile; radicle
superior.*—STEM *woody.* LEAVES *alternate.* STIPULES *caducous.* JUICE *resinous.*

TREES with alternate branches, balsamic juices exuding from their bark.
LEAVES alternate, petioled, entire or lobed, with glandular teeth, edges of the lobes
folded inwards before expansion; *stipules* fugacious; floral *buds* terminal, scaly,
preceding the leaves. FLOWERS monœcious, in catkins or unisexual capitula; *bracts*
4, caducous. ♂ FLOWERS achlamydeous, composed of stamens agglomerated between
the bracts of the capitula. *Anthers* pyramidal-linear, 4-angular, with 2 opposite
cells; *filaments* short or 0. ♀ FLOWERS: CALYX infundibuliform, entire or glandular-
lobed. PETALS 0. STAMENS sterile, often 4–9, inserted around the top of the calyx.
OVARY semi-inferior, with 2 antero-posterior cells, many-ovuled; *styles* 2, linear,
pointed, recurved, papillose on their inner face; *ovules* sub-anatropous, inserted in 2
rows at the inner angle of each cell. CAPSULES connate by their edges, septicidal
above the middle. SEEDS few, or solitary by arrest, the arrested ones numerous,
deformed; the fertile sub-peltate, elliptic, membranous, or shortly winged towards the
top; *albumen* thin. EMBRYO axile; *cotyledons* flat; *radicle* short, superior.

ONLY GENUS.

* Liquidambar.

Balsamifluæ are connected with *Plataneæ* (see this family) and *Hamamelideæ*, to which they are
joined by Bentham; they differ in their inflorescence and aggregate fruit. *Liquidambar* also approaches *Salicineæ*, and especially the Poplars, in inflorescence, diclinous achlamydeous and polyandrous
flowers, many-ovuled ovary, capsular fruit, woody stem, and stipulate leaves; but *Salicineæ* are diœcious,
the ovary is one-celled, with parietal placentation, the ovules anatropous, and the funicle hairy. But
four species are at present known of *Liquidambar:*—*L. Altingia,* a gigantic tree, forms vast forests in Java,
Asia, New Guinea, &c., under the names of *Rosa-mallos, Rassa-mala,* &c. *L. orientale,* a small tree
resembling a Maple, inhabits the isle of Cyprus and Asia Minor. *L. macrophylla* and *styraciflua* grow
in North America [another is Chinese]. *L. styraciflua* yields [the North American] Liquidambar
Balsam, obtained by incisions in the trunk. This balsam contains a tolerable quantity of benzoic acid;
it is of the consistence of a thick oil or of soft pitch. Liquid *styrax,* a sweet balsam, much used by the
Orientals as a perfume, and entering into the composition of several medicaments, is the produce of
L. Altingia and perhaps also *L. orientale.* [The bark of all is a hot, bitter stomachic.]

LXXXVIII. *BRUNIACEÆ*.

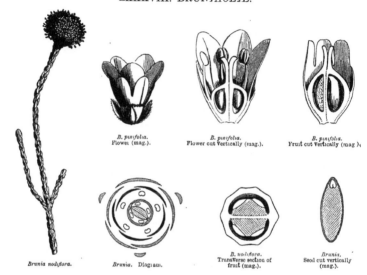

B. *pinifolia*.
Flower (mag.).

B. *pinifolia*.
Flower cut vertically (mag.).

B. *pinifolia*.
Fruit cut vertically (mag.).

Brunia *nodiflora*.

Brunia. Diagram.

B. *nodiflora*.
Transverse section of fruit (mag.).

Brunia.
Seed cut vertically (mag.).

SHRUBS or UNDERSHRUBS from the Cape of Good Hope, Heath-like in habit. LEAVES small, acerose, sub-trigonous, entire, usually imbricated in 5 rows, exstipu-late. FLOWERS ☿, small, regular, generally in a spike or head, sessile, 5-bracteate. CALYX 5–4-partite, persistent or deciduous, imbricate in bud. RECEPTACULAR CUP enveloping the ovary, very rarely spread into an epigynous disk (*Thamnea*). PETALS 5–4, inserted on the edge of the cup, alternate with the calyx-lobes, usually free, sometimes connate with the stamens into a tube at the base, imbricate in bud. STAMENS inserted with the petals, equal in number and alternate with them; *fila-ments* free, or sometimes adnate to the claws of the petals; *anthers* introrse, 2-celled, cells opposite, parallel or diverging at their base, and opening longitudinally. OVARY semi-inferior or inferior, very rarely free (*Raspailia*), 1–2–3-celled; *styles* 2–3, terminal, more or less cohering; *stigmas* minute, papillose; *ovules* anatropous, solitary or 2, collateral in the cells of the 2–3-celled ovary, and pendulous from the central angle or near the top of the septum, solitary in the 1-celled ovaries (in *Thamnea* the ovary is 1-celled, with 10 pendulous 1-seriate ovules). FRUIT crowned by the calyx, and sometimes by the persistent corolla and androecium, dry, indehis-cent or capsular, often with two 1–2-seeded cocci, dehiscence internal, longitudinal.

SEEDS inverted, testa crustaceous, hilum naked, or covered with a fleshy cupule. EMBRYO minute, straight, at the top of a copious fleshy albumen; *cotyledons* short; *radicle* conical, superior.

PRINCIPAL GENERA.

| Berzelia. | *Brunia. | Raspailia. | Berardia. | Staavia. | Linconia. |

Bruniaceæ approach *Hamamelideæ, Corneæ, Araliaceæ,* and *Umbelliferæ,* in the polypetalous and isostemonous corolla, epigyny, solitary or geminate pendulous anatropous ovules, and albuminous embryo; but in all these families, independently of other differences, the æstivation of the petals is [usually] valvate. *Bruniaceæ* have besides, in the genus *Raspailia,* a quite exceptional character in their ovary, superior to the calyx, and inferior to the petals. [The properties of *Bruniaceæ* are quite unknown.]

LXXXIX. *HALORAGEÆ.*

(ÓNAGRARUM *genera, Jussieu.*—HALORAGEÆ, *Br.*—CERCODIACEÆ, *Jussieu.*—
HYGROBIÆ, *Richard.*)

CALYX *superior.* PETALS *inserted on the calyx* [or epigynous] *alternately with its lobes, sometimes* 0. STAMENS *inserted with the petals, equal or double in number to the calyx-lobes, sometimes reduced to one.* OVARY *inferior, of one or several* 1-*ovuled cells;* OVULES *pendulous.* EMBRYO *straight, in the axis of a fleshy albumen.*

Aquatic HERBS, or terrestrial [HERBS or] UNDERSHRUBS. LEAVES usually opposite or whorled (*Myriophyllum, Hippuris*), simple, entire or toothed, the submerged usually pectinate, rarely entire (*Myriophyllum*); *stipules* 0. FLOWERS ☿ (*Haloragis, Hippuris*), or monœcious through arrest (*Myriophyllum, Hippuris*), regular, inconspicuous, sessile in the axil of the leaves, solitary or fascicled, often 2-bracteolate at the base, sometimes whorled in a spike, rarely pedicelled, sometimes panicled [or racemed]. CALYX superior, usually 4-fid or -partite, sometimes truncate or almost wanting. COROLLA 0, or petals inserted on the calyx, alternate with its segments, usually longer, sub-concave, valvate or imbricate in æstivation, spreading after flowering, and deciduous. STAMENS inserted with the petals, usually equal and opposite to the calyx-lobes, or double in number, sometimes reduced to one (*Hippuris*); *filaments* filiform; *anthers* introrse [or dehiscence lateral], 2-celled, oblong or ovoid, basifixed (*Myriophyllum, Haloragis*) or dorsifixed (*Hippuris*), dehiscence longitudinal. OVARY inferior, of 2-3-4 1-ovuled cells, rarely 1-celled (*Hippuris*); *styles* as many as ovules, often short or nearly 0; *stigmas* tomentose or penicillate; *ovules* pendulous from the top of the cell, anatropous. FRUIT nutlike, often crowned by the calyx-limb, 2-3-4-celled, or 1-celled normally or by arrest. SEEDS inverted, *testa* membranous. EMBRYO straight, in the axis of a more or less fleshy albumen; *cotyledons* short, obtuse; *radicle* longer, next the hilum, superior.

PRINCIPAL GENERA.[1]

| Hippuris. | Myriophyllum. | Haloragis. | Serpicula. | Proserpinaca. | Meionectes. |

[1] The two anomalous genera, *Gunnera* and *Callitriche,* which have been included in this order by Bentham and Hooker fil, are treated as distinct natural orders in this work.—ED.

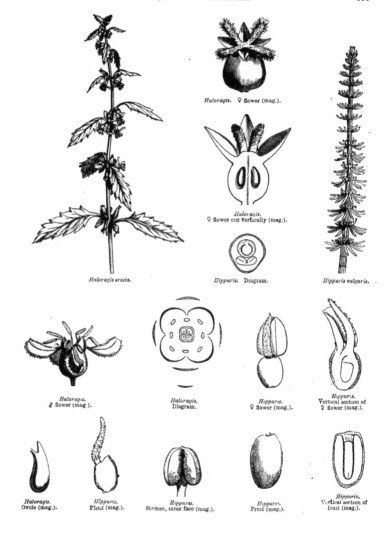

Haloragis.   ♀ flower (mag.).

Haloragis.
♀ flower cut vertically (mag.).

Hippuris.   Diagram.

Haloragis erecta.

Hippuris vulgaris.

Haloragis.
♂ flower (mag).

Haloragis.
Diagram.

Hippuris.
♀ flower (mag.).

Hippuris.
Vertical section of
♀ flower (mag.).

Haloragis.
Ovule (mag.).

Hippuris.
Pistil (mag.).

Hippuris.
Stamen, inner face (mag.).

Hippuris.
Fruit (mag.).

Hippuris.
Vertical section of
fruit (mag.).

The affinities of *Halorageæ* with *Onagrarieæ* will be indicated in the description of the latter.[1] They are near *Trapeæ*, which were formerly placed in the same family, and which only differ in their hemispherical stigma and exalbuminous embryo. They also approach *Combretaceæ*, which are separated by their ovary (which is always 1-celled and 2-4-5-ovuled), their simple style, drupaceous fruit, and exalbuminous embryo.

Halorageæ are rare in the tropics; they are found most abundantly in temperate and cold countries (*Hippuris*), especially beyond the tropic of Capricorn; *Haloragis* is only found in Australia and the neighbouring islands. They are of no use to man.

XC. *GUNNERACEÆ*,[2] *Endlicher.*

Gunnera. ♂ flower (mag.).   Gunnera. Vertical section of ♀ flower.

Gunnera.   Gunnera.
Seed cut vertically (mag.).   Embryo (mag.).   Gunnera.
                                                    Fruit (mag.).

Gunnera. Fruit opened (mag.).

Gunnera scabra.   Gunnera. Fruit cut vertically (mag.).

[1] The affinity between *Halorageæ* and *Onagrarieæ* is extremely slight, and was based chiefly on an erroneous view of the relations of *Trapa.*—ED

[2] See under *Halorageæ*, p. 414.—ED.

STEM herbaceous. LEAVES all radical, petiole long, blade usually reniform, crenulate, hairy. FLOWERS ☿, or imperfect, monœcious or diœcious, ebracteate; scape bearing a very close spike, composed of several spikelets, each furnished with a bract. PERIANTH 4-partite, 2 segments small, tooth-shaped, alternating with 2 larger, petaloid, caducous, sometimes obsolete, reduced to scales in the ♂ flowers. [Perianth otherwise described as, CALYX-LOBES 2–3, equal or unequal, or 0. PETALS 0 or 2, hooded.] STAMENS 2, opposite to the petaloid segments; *filaments* short [or long]; *anthers* 2-celled, [basifixed], dehiscence longitudinal [lateral]. OVARY inferior, 1-celled; *styles* 2, covered with stigmatic papillæ; *ovule* solitary, pendulous from the top of the cell. FRUIT a drupe. EMBRYO minute, at the top of a fleshy albumen; *radicle* superior.

GENUS.

* Gunnera.

A. de Jussieu considered that, in many cases, apetalous and diclinous structures were to be regarded as arrested conditions of perfect types, which. masked affinities without annulling them; and this explains the position he gives to *Gunneraceæ*, between *Araliaceæ* and *Corneæ*, which they approach in their flower, which is hermaphrodite in some species, in their epigyny, the single pendulous anatropous ovule in each carpel, the drupaceous fruit, and the minute embryo at the top of a fleshy albumen. The same considerations establish the affinity of *Gunneraceæ* with *Haloregeæ*; in both these families may be observed, on the one hand, a perfect organization; on the other, the absence of petals, and the abortion of the reproductive organs; and the analogy is increased by the stigmatic papillæ along the styles.

The few species of this little group inhabit tropical Southern Africa and America, the high mountains of tropical America, the Sandwich and Society Islands, Java, [Australia] and New Zealand.

The fruit of *Gunnera·macrophylla* is used in Java as a stimulant. The roots of *G. scabra*, called *Panqué* in Chili, and cultivated in Europe for the beauty of its leaves, which are sometimes more than six feet in diameter, contain astringent principles, and are used in Chili for tanning skins, and as an anti-dysenteric.

XCI. *CALLITRICHINEÆ, Léveillé.*[1]

Floating flaccid annual HERBS, simple or branched, stem cylindric. LEAVES opposite, sessile, the lower [submerged], often linear, the upper oval, 1–3-nerved, entire, emerged, often rosulate; *stipules* 0. FLOWERS ☿ or monœcious-diœcious by arrest, solitary and sessile in the axil of the leaves; *involucre* diphyllous [or 0], of 2 lateral opposite curved somewhat fleshy and coloured [white] persistent or deciduous leaflets. PERIANTH 0. STAMEN posterior, rarely 2 antero-posterior, inserted below the ovary in the ☿ flowers; *filament* filiform, elongated; *anther* reniform, basifixed, 1-celled, opening at the top by an arched slit. OVARY free, at first sessile, then stipitate, formed of 2 bilobed carpels with two 2-ovuled cells;[2] *styles* 2, distant; *stigmas* acute, papillose over the whole surface; *ovules* curved, fixed to the central angle near the top of the cell; *micropyle* lateral and internal, placed -

---

[1] See under *Haloregeæ*, p. 414.—ED.
[2] The ovary is 4-lobed and 4-celled with 1-ovuled cells, but considered as originally 2-celled, each cell being divided into two by the inflection of its walls.—ED.

below the top. FRUIT fleshy-membranous, indehiscent, 4-celled and -lobed, sides rounded or sharp. SEEDS with a finely membranous testa. EMBRYO a little arched,

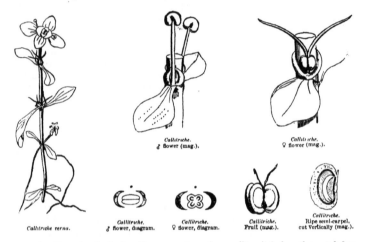

Callitriche vernn.

Callitriche.
♂ flower (mag.).

Callitriche.
♀ flower (mag.).

Callitriche.
♂ flower, diagram.

Callitriche.
♀ flower, diagram.

Callitriche.
Fruit (mag.).

Callitriche.
Ripe semi-carpel,
cut vertically (mag.).

occupying the axis of a fleshy albumen, and nearly equalling it in length ; *cotyledons* very short ; *radicle* superior.

ONLY GENUS

Callitriche.

Callitricheæ were formerly included in *Halorageæ*, which they approach in the aquatic herbaceous stem, the opposite leaves, axillary flowers, 1-ovuled cells, distinct [papillose] styles, [pendulous ovules,] and albuminous embryo ; being separated by the achlamydeous flowers, and fruit of four cocci. They present some remarkable analogies with *Euphorbiaceæ*, in their diclinous achlamydeous involucred flowers, stamens inserted on the receptacle, ovary with 1-ovuled cells, pendulous anatropous ovules, distinct stigmas, fruit of cocci, and albuminous axile embryo ; analogies which have led some modern botanists to consider *Callitriche* as aquatic *Euphorbiaceæ* : they, however, differ in their 4-lobed ovary, and the structure of their seeds.

Callitrichineæ inhabit stagnant water in Europe and North America [and most other temperate parts of the globe]. They are useless.

XCII. *RHIZOPHOREÆ.*[1]

[TREES or SHRUBS, usually quite glabrous; branches terete, swollen at the nodes. LEAVES opposite and stipulate, rarely alternate and exstipulate (*Combretocarpum, Anisophyllea*), petioled, thickly coriaceous, usually quite entire; *stipules* inter-

---

[1] This order is omitted in the original.—ED.

petiolar, very caducous. FLOWERS ☿, in axillary cymes, racemes, panicles or spikes.
CALYX-TUBE more or less adnate to the ovary, rarely free; *limb* 3–14-lobed or entire,
persistent, lobes valvate. PETALS as many as the calyx-lobes, usually small, concave
or involute, and embracing the stamens, notched, 2-fid or lacerate, rarely entire,
convolute or inflexed in bud. STAMENS 2–3–4 times as many as the petals, rarely
equal in number, often inserted in pairs opposite the petals, on the edge or at the
base of a perigynous disk; *filaments* long or short; *anthers* basi- or dorsi-fixed,
2-celled or multi-locellate. OVARY usually inferior, 2–5- (rarely 3–6-) celled, or 1-celled
by the suppression of the septa; *style* simple (several in *Anisophyllea*); *stigma* simple
or lobed; *ovules* usually 2, collaterally suspended to the inner angle of each cell
above its middle, rarely 4 or more, pendulous, anatropous, raphe ventral. FRUIT
usually coriaceous, crowned with the calyx-limb, indehiscent or rarely septicidal,
1-celled and 1-seeded, or 2–5-celled with 1-seeded cells. SEEDS pendulous, arillate
or not, testa coriaceous or membranous, albumen fleshy or 0. EMBRYO inverted in
the albuminous seeds, usually small and axile, with terete radicle and semi-terete
cotyledons; in the exalbuminous seed the embryo is usually elongate, with small or
inconspicuous often connate cotyledons.

TRIBE I. RHIZOPHOREÆ.—Ovary inferior; style 1. Embryo exalbuminous, radicle very
large, protruded from the fruit when still on the tree. Leaves opposite, glabrous, quite entire,
stipulate. *Rhizophora, Ceriops, Kandelia, Bruguiera.*

TRIBE II. CASSIPOUREÆ.—Ovary inferior, semi-inferior, or superior; style 1. Embryo in
the axis of a fleshy albumen. Leaves opposite, usually glabrous and quite entire, stipulate.
Carallia, Gynotroches, Weihea, Cassipourea, &c.

TRIBE III. ANISOPHYLLEÆ.—Ovary inferior; styles 3–5. Embryo exalbuminous. Leaves
alternate. *Anisophyllea, Combretocarpus.*

Rhizophoreæ are nearly allied to *Myrtaceæ* in habit, opposite sometimes pellucid dotted leaves, and
in the exalbuminous seeds of the tribe *Rhizophoreæ*; but they differ in their valvate calyx-lobes, stipules, and
usually definite stamens. The same characters distinguish them from *Melastomaceæ*. *Lythrarieæ*, which
agree with them in the valvate calyx and usually definite stamens, have many ovules and large usually
crumpled petals. *Combretaceæ*, which are still more closely allied, differ in their monocarpellary ovary,
folded or convolute cotyledons, and exstipulate leaves.

This order is almost wholly tropical, and to a great extent littoral; the species of the tribe *Rhizophoreæ* forming, with *Avicennia* (see *Verbenaceæ*), the Mangrove forests of tidal rivers in both the Old and
New Worlds. Most of these root from the branches into the mud below, and thus form dense thickets,
the roots becoming stems after a time. The tribe *Cassipoureæ* is, with the exception of *Cassipourea* itself,
confined to the Old World. *Anisophyllea* is remarkable for its distichous alternate leaves, of which every
second one is reduced to a minute stipuliform body.

Of the properties of *Rhizophoreæ* little is known. The wood of several is hard and durable; the
bark of *Bruguiera* yields a black dye; and the fruit of *Rhizophora Mangle* is described as edible, and when
fermented produces a light wine, drunk by the natives of the Indian Peninsula.—ED.]

XCIII. *COMBRETACEÆ*,[1] Br.

(TERMINALIACEÆ, *St. Hilaire.*—MYROBALANEÆ, *Jussieu.*)

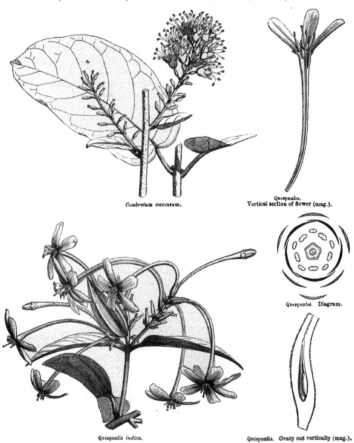

Combretum coccineum.

Quisqualis.
Vertical section of flower (mag.).

Quisqualis. Diagram.

Quisqualis indica.

Quisqualis. Ovary cut vertically (mag.).

[1] The tribe *Gyrocarpeæ*, which is referred to *Laurineæ* in this work, is included in the 'Genera Plantarum' under *Combretaceæ*, from which it differs chiefly in the structure of the stamens, and the short funicles of the seed.—ED.

Quisqualis. Transverse section of ovary (mag.).	*Quisqualis.* Transverse section of the floral tube, with the style joined to one of the sides.	*Quisqualis.* Seed, entire and cut Vertically (mag.).	*Quisqualis.* Ovule (mag.).

COROLLA *polypetalous, epigynous, isostemonous, or 2-3-stemonous, contorted or valvate in bud, sometimes 0.* STAMENS *inserted with the petals.* OVARY *inferior, 1-celled;* OVULES *pendulous from the top of the cell.* EMBRYO *exalbuminous.*

TREES or SHRUBS, erect or climbing. LEAVES alternate or opposite, simple, penninerved, entire or toothed, coriaceous; *petiole* often biglandular at the top, exstipulate. FLOWERS regular, ⚥, or imperfect by arrest, in spikes, racemes or heads, naked or involucrate, axillary or terminal, each flower furnished with a bract and two lateral opposite bracteoles. CALYX superior, 4–5-fid, lobes valvate in bud, deciduous, or persisting with the fruit. COROLLA 0, or petals inserted on the calyx, alternate with its lobes, valvate in bud. STAMENS inserted with the petals, sometimes alternate with them, sometimes double in number, of which the alternate five are inserted higher and opposite to the petals, rarely triple the number; *filaments* free, filiform or subulate; *anthers* introrse, 2-celled, dehiscing longitudinally. OVARY inferior, 1-celled, usually crowned with a sometimes rayed or indented disk; *style* terminal, simple; *stigma* undivided; *ovules* 2 or 4, rarely 5, pendulous from the top of the cell, with long funicles, anatropous. FRUIT a drupe, often longitudinally winged. SEED usually solitary by arrest; *testa* membranous, thin; *endopleura* swollen, intruded between the folds of the cotyledons. EMBRYO exalbuminous, straight; *cotyledons* foliaceous, rolled into a spiral, or thick, folded lengthwise, or crumpled; *radicle* near the hilum, superior.

PRINCIPAL GENERA.

Conocarpus.	Anogeissus.	Laguncularia.	Lumnitzera.
*Quisqualis.	*Combretum.	Terminalia.	

Combretaceæ are near *Onagrarieæ, Haloragex, Napoleoneæ,* &c. (see these families) The diagnosis principally rests·on the 1-celled ovary, the ovules pendulous from the top of the cell, the 1-seeded fruit and the structure of the seed. All are inter-tropical [or nearly so].

The trees of this family are useful from the hardness and closeness of their wood; their bark contains astringent principles, which render it fit for tanning and dyeing. The seed [of *Terminalia Chebula*], known medicinally as Myrobalans, which is oily, and much eaten in India, was formerly used as a laxative. [The astringent nuts of *T. Chebula* make a capital ink with sulphate of iron; its leaves yield an excellent yellow dye with alum. The seeds of *T. angustifolia, Bellerica,* and *Catappa,* are likewise much eaten, and yield an excellent bland oil. *T. angustifolia* produces by incisions in the bark a white benzoin of agreeable scent, much used as a cosmetic, and burnt as an incense in Mauritius. The wood of *Conocarpus latifolius* is reckoned one of the best in India.—ED.]

XCIV. *MYRTACEÆ.*

(MYRTI, MYRTEÆ, *Jussieu.*—MYRTOIDEÆ, *Ventenat.*—MYRTINEÆ, *D.C.*—
MYRTACEÆ, *Br.*)

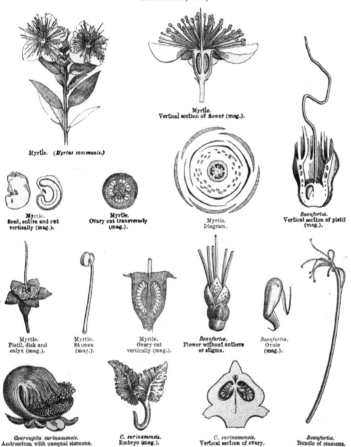

Myrtle. (*Myrtus communis.*)

Myrtle.
Vertical section of flower (mag.).

Myrtle.
Seed, entire and cut
vertically (mag.).

Myrtle.
Ovary cut transversely
(mag.).

Myrtle.
Diagram.

Beaufortia.
Vertical section of pistil
(mag.).

Myrtle.
Pistil, disk and
calyx (mag.).

Myrtle.
Stamen
(mag.).

Myrtle.
Ovary cut
vertically (mag.).

Beaufortia.
Flower without anthers
or stigma.

Beaufortia.
Ovule
(mag.).

Couroupita surinamensis.
Androecium with unequal stamens.

C. surinamensis.
Embryo (mag.).

C. surinamensis.
Vertical section of ovary.

Beaufortia.
Bundle of stamens.

Fabricia læ.igata.

Lecythis urnigera. Fruit, ½ nat. size.

Bertholletia excelsa, Seed, nat. size

Fabricia.
Fruit.

Bertholletia excelsa.
Transverse section of seed.

Eucalyptus Globulus.
Section of lower part of ovary (mag.).

Eucalyptus.
Lower ovule.

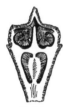

Eucalyptus.
Flower-bud cut
vertically.

Eucalyptus Globulus.
Section of upper part of ovary
(mag.).

Eucalyptus.
Upper
ovule.

Calycothrix.
Ovules fixed to a basilar filiform
placenta (mag.).

Caryophyllus.
Flower-bud
(mag.).

CALYX *valvate [or imbricate] in bud.* COROLLA *polypetalous, inserted on the calyx, imbricate in bud.* STAMENS *numerous, inserted with the petals.* OVARY *inferior or semi-inferior, sometimes free, 1–several-celled.* FRUIT *a capsule or berry.* EMBRYO *exalbuminous.*

STEM arborescent or suffrutescent, very rarely herbaceous (*Careya*). LEAVES opposite, rarely alternate or whorled, simple, entire or very rarely denticulate, flat or cylindric or semi-cylindric, 3-nerved or penninerved, nerves often marginal, usually coriaceous, very often dotted from pellucid glands sunk in the parenchyma, narrowed into a petiole at the base [or sessile]; *stipules* generally 0, rarely geminate at the base of the petioles, minute, caducous. FLOWERS ☿, regular, very rarely sub-irregular from the unequal length of the stamens, axillary and solitary, or in a spike, cyme, corymb or panicle, sometimes even capitate, naked or involucrate; often 2-bracteolate, white, pink, purplish or yellow, never blue. CALYX superior or semi-superior; *limb* 4–5–multi-fid or -partite, persistent or deciduous, valvate in bud; sometimes entire, closed in bud and falling away like an operculum at the expansion of the flower. PETALS inserted on a disk edging the calyx-throat, and usually expanding into a plate or cushion which crowns the top of the ovary, equal in number and alternate with the calyx-lobes, very rarely 0, æstivation imbricate or convolute. STAMENS numerous, inserted with the petals, very rarely equal in number and alternate, often double or treble, and then some without anthers, most often indefinite, pluri-seriate, and then usually all fertile; *filaments* filiform or linear, free, or more or less monadelphous at the base, or united in bundles opposite to the petals, rarely united into a cup which is abbreviated on one side, and on the other prolonged into a concave petaloid blade which is bent down upon the style and antheriferous within; *anthers* small, rounded, introrse, of 2 contiguous or separated cells, opening longitudinally or transversely. OVARY inferior or semi-inferior, covered by a fleshy disk, sometimes free (*Fremya*), either 1-celled with numerous erect anatropous ovules on a basilar placenta, or 2-pluri-celled with numerous anatropous ovules inserted at the inner angle of the cells, or ovules rarely solitary and fixed to the inner angle by their ventral surface; *style* terminal or rarely lateral (in the 1-celled ovary), simple, naked or barbed at the top; *stigma* undivided. FRUIT generally crowned by the calyx-limb, sometimes 1-celled and 1-seeded by arrest, dry, indehiscent, or quite 2-valved at the top, sometimes 2–many-celled, and then either a capsule opening at the top loculicidally or septicidally or by the removal of the epigynous disk, or an indehiscent berry with many- (or by arrest 1-) seeded cells. SEEDS straight, angular, cylindric or compressed, sometimes dimorphous in each cell, some being turgid and fertile, the others linear and sterile (*Eucalyptus, Fabricia*); *testa* crustaceous or membranous, winged or furnished with membranous scales at the base [arrested ovules?] (*Spermolepis*). EMBRYO exalbuminous, straight or curved or rolled spirally; *cotyledons* usually short, obtuse, sometimes confluent with one another and with the radicle, very rarely foliaceous; *radicle* very often thick, next the hilum.

TRIBE I. *CHAMÆLAUCIEÆ.*

Stamens often definite, some usually sterile. Ovary 1-celled, with one or several basilar ovules. Fruit 1-seeded, indehiscent, or imperfectly 2-valved at the top.—Shrubs of Australia, often resembling Heaths, and especially *Blaeria.* Leaves opposite, or rarely alternate, punctate; stipules 0 or rarely 2 (*Calycothrix*).

PRINCIPAL GENERA.

*Calycothrix. *Verticordia. *Chamælaucium. Darwinia. Lhotskya. Thryptomene.

Tribe II. *LEPTOSPERMEÆ.*

Stamens very often indefinite, free, or polyadelphous, rarely monadelphous. Ovary 2–many-celled ; ovules numerous, rarely solitary in each cell. Capsule loculicidal or septicidal, rarely indehiscent. Seeds sometimes dimorphous (*Eucalyptus, Fabricia,* &c.).—Shrubs or trees, abundant in Australia, less numerous in tropical Asia. Leaves opposite or alternate, exstipulate, entire or rarely denticulate, punctate, sometimes presenting the appearance of phyllodes.

PRINCIPAL GENERA.

Scholtzia.	*Bœckia.	Hypocalymna.	Agonis.	*Leptospermum.
*Callistemon.	*Melaleuca.	*Beaufortia.	*Calothamnus	*Eucalyptus.
*Tristania.	*Metrosideros.	Backhousia.	Xanthostemon.	*Angophora.
*Billiottia.	*Fabricia.			

Tribe III. *MYRTEÆ.*

Stamens indefinite, free. Ovary 2- or more- celled, ovules numerous. Berry 2- or more- celled, cells often 1-seeded.—Tropical and sub-tropical trees or shrubs. (The common Myrtle is found as far as 43° N. lat.) Leaves opposite, entire, punctate, exstipulate.

PRINCIPAL GENERA.

Campomanesia.	Psidium.	Rhodomyrtus.	*Myrtus.	Rhodamnia.	Nelitris.
Myrcia.	Marliera.	*Calyptranthes.	Pimenta.	*Eugenia.	Caryophyllus.
*Acmenas.	*Jambosa.				

Tribe IV. *BARRINGTONIEÆ.*

Stamens numerous, often monadelphous. Ovary inferior, 2- or more- celled ; ovules definite or numerous. Berry corticate, 1- or more- celled, 1-few-seeded.— Trees of tropical Asia and America. Leaves alternate, rarely opposite or whorled, exstipulate, eglandular, entire or dentate.

GENERA.

Barringtonia.	Careya.	Grias.	Planchonia.	Gustavia.

Tribe V. *LECYTHIDEÆ.*

Stamens numerous, filaments united into a cup, shortened on one side, prolonged on the other into a concave petaloid tongue, sterile or antheriferous within. Ovary several-celled ; ovules numerous. Fruit dry or fleshy, indehiscent or opening by the raising of the disk.—Trees of tropical America. Leaves alternate, not punctate, entire or rarely toothed ; stipules 0 or caducous.

PRINCIPAL GENERA.

Courateri.	Couroupita.	Bertholletia.	Lecythis.

[Tribe NAPOLEONEÆ : see p. 426.]

Myrtaceæ are related to *Melastomaceæ*, and through these to *Onagrarieæ* and *Lythrarieæ* (which see). They approach *Granateæ* in the valvate calyx, imbricate corolla, polyandry, inferior ovary with many-ovuled cells, simple style and stigmas, fleshy fruit, exalbuminous embryo, woody stem, and generally opposite leaves; they differ principally in the ovary, which in *Granateæ* presents two super-imposed rows of cells. *Myrtaceæ* are also near *Olineæ*, as shown by the inferior several-celled ovary, subulate style, fleshy fruit, exalbuminous embryo, woody stem, and opposite coriaceous leaves; but *Olineæ* are isostemonous, and the ovarian cells are only 3-ovular (which is also the case in some *Myrtaceæ*). Points of resemblance have also been traced between certain *Myrtaceæ* and the little family of *Calycan-theæ*; in the numerous stamens, of which some are antherless, inserted on a fleshy ring crowning the calyx-throat, the exalbuminous embryo, woody stem, and generally opposite leaves; but *Calycantheæ* are apetalous, their anthers are extrorse, their ovaries are free, and the fruit is composed of achenes. Finally, in the gland-punctate leaves, polypetalous corolla, monadelphous or polyadelphous stamens, and exal-buminous seed of *Aurantieæ* and *Hypericineæ* we find an analogy between these families and *Myrtaceæ*.

Myrtaceæ contain tannin, fixed and volatile oils, free acids, mucilage and sugar; these principles, associated in different proportions, give to some species properties which are useful to man. The Myrtle (*Myrtus communis*), a shrub of the Mediterranean region, has astringent berries, and its leaves were formerly renowned for their tonic and stimulating virtues. *Eugenia* (*Caryophyllus*) *aromatica*, a native of the Moluccas, yields cloves, which are the flower-buds, and which contain a very aromatic volatile oil, and are hence universally used as a condiment, in medicine, and as a perfume. The fruit of *E. pimenta*, a tree of the Antilles, possesses an aroma and taste which combine the qualities of the nutmeg, the cinnamon and clove, whence its name of Allspice. The berries of the Guava (*Psidium*), the Jambosa, and many other species, are much esteemed for their aromatic taste, and are made into preserves. The capsular *Myrtaceæ* also contain a volatile oil in their leaves and fruit; the principal species is the *Melaleuca Cajaputi*, a shrub of the Moluccas, which yields by distillation a green oil of a mild and pene-trating scent, recalling at once camphor, rose, mint and turpentine, and much esteemed as an anti-spasmodic. The *Eucalypti* are gigantic trees of Australia, of which the wood is very useful for building purposes (*E. robusta*, *Globulus*, &c.). The *Couroupita* is a large tree, called in Guiana Cannon-ball, on account of the shape and size of its fruit, which contains an acid-sweet pulp, very agreeable and refreshing.

The seeds of *Bertholletia excelsa* are edible, and are sold in Europe as Brazil-nuts. We may also mention the Sapucaya (*Lecythis ollaria*), a Brazilian tree the capsule of which is woody, very large and thick, and opens circularly by the raising of its epigynous cap-shaped disk. Vases and pots are made of this cap-sule, whence the name of Monkey-pot given to the fruit.

[Humboldt describes the fruits of *Gustavia speciosa* as eaten by children, and causing them to turn temporarily yellow. *G. brasiliana* is an emetic. Rose-apples are the fruits of *Eugenia Malaccensis*. The fruits of the Chilian *E. Ugni* are very agreeable. Myrtle buds and berries are still eaten as spice in Italy, and made into wine; the flowers yield *eau d'ange*. Various species of *Eucalyptus* yield a red astringent gum like Kino, and a manna; one Tasmanian species (*E. Gunnii*) is tapped for its juice, which is fermented and drunk. Various *Leptosperma* have been used as substitutes for tea by the early settlers in Australia. Lastly, some species of *Eucalyptus* are the most gigantic of timber trees, attaining nearly 500 feet in height and 110 feet in girth.—ED.]

XCV. *NAPOLEONEÆ*,[1] *Endlicher.*

SHRUBS of tropical Africa. LEAVES alternate, exstipulate, entire or unequally 2-3-dentate at the top. FLOWERS ☿, regular, solitary on axillary peduncles (*Asteranthos*), or scattered on the branches and sessile (*Napoleona*). CALYX superior, 5-partite (*Napoleona*) or many-toothed (*Asteranthos*). COROLLA epigynous, simple,

[1] Regarded as a tribe of *Myrtaceæ* in the ' Genera Plantarum.'—ED.

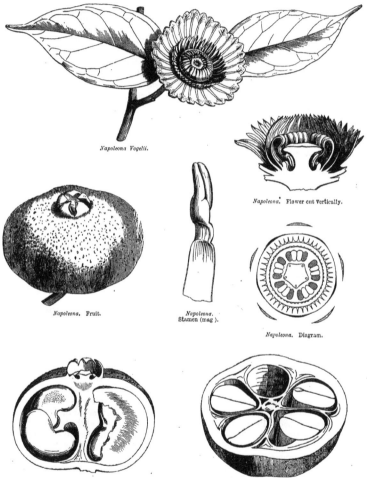

Napoleona Vogelii.

Napoleona. Flower cut vertically.

Napoleona. Fruit.

Napoleona.
Stamen (mag).

Napoleona. Diagram.

Napoleona. Fruit cut vertically.

Napoleona. Fruit cut transversely.

rotate; *limb* shortly multifid (*Asteranthos*) or double, the outer sub-rotate, plaited, entire, the inner radiating, multifid (*Napoleona*). STAMENS inserted at the base of the corolla, 5 petaloid, with 2 anthers [filaments connate in a petaloid cup, bearing many anthers] (*Napoleona*), or [filaments] indefinite, filiform (*Asteranthos*); *anthers* 2-celled, dehiscence longitudinal (*Asteranthos*) [or 1-celled in *Napoleona*]. OVARY inferior [or semi-inferior, 5–6-celled]; *style* short; *stigma* a depressed head, lobed (*Asteranthos*) or peltate, angular (*Napoleona*); *ovules* several in each cell, pendulous from the inner angle. BERRY crowned by the calyx-limb. SEEDS numerous, sunk in the pulp [in *Napoleona*; few, large, angled or reniform, with a membranous testa and large thick cotyledons, between which the radicle is retracted;—unknown in *Asteranthos*.—ED.].

GENERA.

Napoleona.   Asteranthos.

[A very singular family or tribe of *Myrtaceæ*, evidently closely allied to *Lecythis* and *Bertholletia*, with the valvate calyx and alternate leaves of *Lythrarieæ*, to which they approach in the crumpled corolla.

Napoleona is a native of west tropical Africa, and *Asteranthos* of North Brazil. Of their properties nothing is known.—ED.]

XCVI. MELASTOMACEÆ.

(MELASTOMA, Jussieu.—MELASTOMACEÆ, Br.)

COROLLA *polypetalous, usually diplostemonous, inserted on the calyx, contorted in bud.* STAMENS *3–12, inserted with the petals.* OVARY *free or adherent, of 1–20 many-ovuled cells.* FRUIT *a capsule or berry.* EMBRYO *straight or curved, exalbuminous.*

STEM arborescent or frutescent or suffrutescent, rarely herbaceous, erect, climbing or epiphytal, with cylindric or tetragonal branches swollen at the nodes. LEAVES opposite or whorled, simple, equal or unequal, entire, rarely toothed, usually narrowed into a sometimes swollen petiole; lateral nerves 2, 4, 6, or 8, nearly as prominent as the median, and directed like it from the base to the top of the leaf, preserving a uniform thickness, and united by finer transverse nerves. FLOWERS ☿, regular, usually in panicled or contracted cymes, rarely solitary, naked, or diversely bracteolate, sometimes furnished with a sort of coloured involucre (*Blakea*). RECEP-TACULAR CUP campanulate, urceolate or oblong-tubular, quite free, or adhering to the ovary by longitudinal septa. CALYX-LIMB 5–6–3-partite, sometimes entire, imbricate or contorted in bud. PETALS free, or sometimes slightly united at the base, inserted on the calyx-throat, on a fleshy annular layer, alternate with the calyx-segments, shortly clawed, contorted in bud. STAMENS inserted with the petals, usually double (or sometimes multiple) their number, sometimes equal (*Sonerila, Poter-anthera*), sometimes all equal and fertile, sometimes those opposite to the petals smallest or sterile, rarely rudimentary or 0; *filaments* free, inflexed in bud; *anthers*

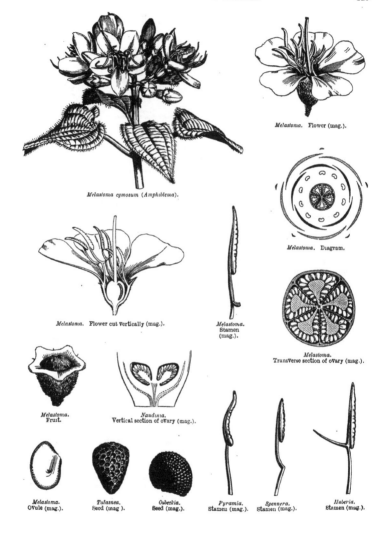

Melastoma cymosum (Amphiblema).

Melastoma. Flower (mag.).

Melastoma. Diagram.

Melastoma. Flower cut vertically (mag.).

Melastoma. Stamen (mag.).

Melastoma. Transverse section of ovary (mag.).

Melastoma. Fruit.

Naudinia. Vertical section of ovary (mag.).

Melastoma. Ovule (mag.).

Tulasnea. Seed (mag.).

Osbeckia. Seed (mag.).

Pyramia. Stamen (mag.).

Spennera. Stamen (mag.).

Huberia. Stamen (mag.).

| *Kibessia.* Flower-bud raising its calyx. | *Kibessia.* Flower. | *Kibessia.* Vertical section of andrœcium and pistil. | *Kibessia.* Transverse section of ovary (mag.). |

terminal, 2-celled, pendulous before flowering, and sunk in the spaces separating the ovary from the receptacular cupule, globose, ovoid or elongate; *cells* parallel, generally opening at the top (often prolonged into a beak) by a common pore, or by 2 distinct pores, rarely by longitudinal slits (*Kibessia*); *connective* polymorphous, not prolonged below the anther and without appendages (*Pyramia*), prolonged below the anther and without appendages (*Spennera*), prolonged below the anther with anterior appendages (*Melastoma*), not prolonged below the anther, but furnished with posterior appendages (*Huberia*). OVARY quite free, adhering to the nerves of the receptacular cup, or completely adherent; *cells* 4–5, rarely 6–20, or one only (*Memecylon*); *style* simple; *stigma* undivided; *ovules* numerous [rarely few], anatropous or semi-anatropous. FRUIT with usually axile placentation, or parietal (*Kibessia*), or basilar (*Naudinia*), of 1–20 cells; sometimes a berry from the development of the receptacular cupule, indehiscent or ruptile (*Astronia*); sometimes a drupe (*Mouriria*), sometimes dry and indehiscent, sometimes a loculicidal capsule with semi-septiferous valves, the placentas frequently remaining united into a central column (*Melastoma*). SEEDS numerous, testa crustaceous, dotted or areolate, sometimes reniform or cochleate; sometimes ovoid, oblong, angular, pyramidal or scobiform (*Huberia*), rarely margined (*Castanella*); *hilum* basilar. EMBRYO exalbuminous, straight or curved; *cotyledons* equal, or the outer much the larger in the irregularly formed seeds; *radicle* next the hilum.

PRINCIPAL GENERA.

Blakea.	* Medinilla.	* Osbeckia.	* Melastoma.	* Pleroma.	Monochætum.
* Rhexia.	* Sonerila.	* Bertolonia.	* Rhynchanthera.	* Centradenia.	

[The following classification of this most extensive order is that of the ' Genera Plantarum.'

SUB-ORDER I. *MELASTOMEÆ.*

Ovary 2–∞-celled; ovules numerous, inserted on placentas which project from the inner angle of the cells. Fruit polyspermous. Seeds minute; embryo very minute.

Series I.—Fruit usually capsular. Stamens unequal.

TRIBE I. · MICROLICIEÆ.—Connective produced at the base, appendaged in front. Seeds oblong or ovoid (American). *Pyramia, Cambessedesia, Chætostoma, Microlicia, Lavoisiera, Rhynchanthera, Centradenia,* &c.

Tribe II. Osbeckieæ.—Connective produced at the base. Seeds cochleate (Old and New Worlds). *Acisanthera, Comolia, Pterolepis, Macacria, Pleroma, Brachyota, Chætolepis, Aciotis, Osbeckia, Tristemna, Melastoma, Dissotis*, &c.

Tribe III. Rhexieæ.—Connective hardly produced at the base, spurred posteriorly, not appendaged in front. Seeds cochleate (American). *Monochætum, Rhexia*.

Tribe IV. Merianieæ.—Connective hardly produced at the base, appendaged or spurred in front, appendage usually erect and inflated. Seeds cuneate, angled or fusiform (American). *Axinæa, Meriania, Centronia, Graffenriedia*, &c.

Tribe V. Oxysporeæ.—Connective hardly produced at the base, acute or spurred behind, not appendaged in front. Seeds angled or oblong, raphe usually thickened and produced (Old World). *Oxyspora, Allomorpha, Ochthocharis, Bredia*, &c.

Tribe VI. Sonerileæ.—Ovary and capsule depressed at the top. Connective simple or appendaged before and behind. Seeds straight or nearly so (Old and New Worlds). *Sonerila, Amphiblemma, Bertolonia, Monolena*, &c.

Series II.—Fruit a berry. Stamens usually equal.

Tribe VII. Medinilleæ.—Anthers usually recurved; connective appendaged behind, or before and behind. Seeds not cochleate (Old World). *Marumia, Dissochæta, Anplectrum, Medinilla, Pachycentria*, &c.

Tribe VIII. Miconieæ.—Anthers incurved or recurved, opening by 1-2 pores or slits; connective rarely produced or appendaged. Seeds various (American). *Oxymeris, Calycogonium, Tetrazygia, Conostegia, Miconia; Tococa, Heterochitum, Clidemia, Sagræa, Bellucia, Loreya, Henriettea, Ossæa*, &c.

Tribe IX. Blakeæ.—Anthers large, 1-2-pored; connective simple or spurred behind. Flowers involucrate. Leaves with strong veins between the ribs (American). *Blakea, Topobea*.

SUB-ORDER II. *ASTRONIEÆ.*

Ovary 2-∞-celled; ovules numerous, placentas at the base or on the outer walls of the cells near their base. Seeds minute (Old World). *Astronia, Pternandra, Kibessia*.

SUB-ORDER III. *MEMECYLEÆ.*

Ovary either 1-celled with the ovules whorled round a central placenta, or several-celled with 2-3 collateral ovules in the inner angles of the cells. Fruit 1-few-seeded. Seeds large (Old and New Worlds). *Mourisia, Memecylon*.

Melastomaceæ approach *Lythrarieæ* in the valvate calyx, insertion of the petals, diplostemony, several many-ovuled ovarian cells, exalbuminous embryo, opposite or whorled leaves, and especially in the singular structure of the stamens. There is also a real affinity between them and *Myrtaceæ*; in both the petals are inserted on the calyx, the ovary has several many-ovuled cells, the style is simple, the embryo is exalbuminous, the leaves opposite, sometimes even 3-nerved (*Rhodomyrtus*), and the stem woody; but *Myrtaceæ* are generally sweet-scented, and have punctate leaves; their anthers are short, rounded, deprived of appendages, and the petals imbricate in bud.

Melastomaceæ mostly grow in tropical America; a few advance into North America to 40° (*Rhexia*). None have yet been found in Chili, a few inhabit Asia and Africa.

The leaves of *Melastomaceæ* are astringent, and of many species slightly acid. The berries also contain free acids united to a certain quantity of sugar, whence result various medicinal properties; some possess a small quantity of volatile oil or balsamic resin, which renders them stimulating. The bark, the fruit, and especially the leaves of some others contain colouring principles.

[The berries of *Melastoma* are edible, but dye the mouth black, whence the name (*Melastoma*); others yield yellow, black and red dyes.—ED.]

XCVII. *LYTHRARIEÆ.*

(SALICARIÆ, LYTHRARIEÆ, *Jussieu.*—CALYCANTHEMÆ, *Ventenat.*—LYTHRACEÆ, *Lindl.*)

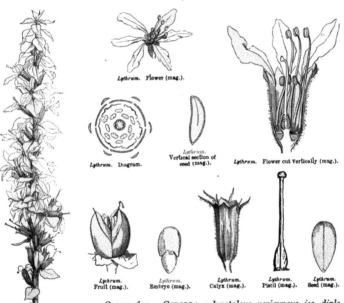

Lythrum. Flower (mag.).

Lythrum. Diagram.

Lythrum. Vertical section of seed (mag.).

Lythrum. Flower cut vertically (mag.).

Lythrum. Fruit (mag.).

Lythrum. Embryo (mag.).

Lythrum. Calyx (mag.).

Lythrum. Pistil (mag.).

Lythrum. Seed (mag.).

Lythrum Salicaria.

CALYX *free.* COROLLA *polypetalous, perigynous, iso- diplo- or triplo-stemonous, imbricate in bud, rarely* 0. STAMENS *inserted on the calyx-tube.* OVARY *with 2 or several many-ovuled cells.* EMBRYO *exalbuminous.*

HERBS, SHRUBS or TREES. LEAVES opposite or whorled, rarely opposite and alternate on the same plant, simple, penninerved, entire, petioled or sessile, sometimes punctate-glandular, always exstipulate. FLOWERS ⚥, regular, or rarely irregular (*Cuphea*), solitary in the axil of the leaves, or fascicled or cymose, sometimes spiked or racemed, and accompanied by floral bracteiform leaves, rarely panicled, peduncle and pedicels 2-bracteolate at the base, middle or top. CALYX free, persistent, tubular or campanulate, rarely urceolate, 3–many-toothed; tube usually ribbed with nerves, straight or rarely oblique, or gibbous or sub-spurred at the base (*Cuphea*); limb more or less deeply toothed, teeth sometimes 1-seriate, equal, valvate in bud, sometimes 2-seriate; the outer alternate, narrower, incumbent on the commissure of the valvate inner ones. PETALS (rarely 0, *Peplis*, *Abatia*, &c.) inserted on the top of the calyx-throat, equal in number and alternate with its inner teeth, sessile or clawed, obovate or ovate or oblong, equal, or the dorsal very rarely larger, imbricate in bud, often folded, waved on the edges (*Lagerstrœmia*), spreading after flowering, deciduous or fugacious (*Suffrenia*, *Peplis*). STAMENS inserted on the calyx-tube above its base, continuous with its nerves, equal in number and alternate with the petals, very rarely less numerous (*Suffrenia*), very often double or triple (*Dodecas*, *Antherylium*, *Lagerstrœmia*), 1–several-seriate, included or exserted, equal or unequal, all fertile, or rarely some sterile; *filaments* filiform, free; *anthers* introrse, 2-celled, orbicular or ovoid or oblong, erect or incumbent, dorsifixed, debiscence longitudinal. OVARY free, sessile or shortly stipitate, rarely girt at the base with a fleshy ring, or accompanied by a unilateral gland (*Cuphea*), 2–6-celled, sometimes sub-1-celled by failure of the septa (*Diplusodon*), or completely 1-celled with parietal placentation (*Cryptotheca*); *placentas* attached to the middle of the septum, or the inner angle of the cells, or uniting the base of the semi-septa; *style* terminal, simple, more or less long; *stigma* simple, obtuse or capitate, rarely emarginate or bilobed; *ovules* usually numerous, ascending or horizontal, anatropous. CAPSULE membranous, or rarely coriaceous, woody, crowned by the persistent or accrescent calyx, 2-several- (rarely 1-) celled, bursting irregularly or by circumsciss, or by regular semi-septiferous valves, the placentas remaining united in a free column. SEEDS usually numerous, ovoid-angular, cuneiform, plano-compressed, or edged with a membrane (*Lagerstrœmia*); *testa* coriaceous; *hilum* marginal or basilar. EMBRYO exalbuminous, straight; *cotyledons* sub-orbicular, plano-convex, 2-auricled at the base, rarely semi-cylindric or convolute; *radicle* short, near the hilum.

[Classification of *Lythrarieæ*, from the ' Genera Plantarum ' :—

TRIBE I. AMMANNIEÆ.—Herbs, often aquatic. Calyx membranous, not ribbed or striate. Flowers small or minute. Petals flat or 0. *Ammannia*, (*Suffrenia*), *Peplis*, &c.

TRIBE II. LYTHREÆ.—Shrubs or trees, more rarely herbs. Calyx coriaceous or herbaceous, usually costate or striate. Flowers rarely small. Petals usually crumpled. *Adenaria*, *Grislea*, *Woodfordia*, *Cuphea*, *Lythrum*, *Nesœa*, *Pemphis*, *Diplusodon*, *Lafœnsia*, *Lawsonia*, *Crypteronia*, *Lagerstrœmia*, *Sonneratia*.

ANOMALOUS GENERA.—*Punica* (see Order *Granateæ*, p. 435), *Olinia* (see Order *Olinieæ*, p. 434), *Axinandra*, *Heteropyxis*.—ED.]

F F

Lythrarieæ are near *Onagrarieæ*, *Melastomaceæ* and *Saxifrageæ* (see these families). They also approach *Rhizophoreæ* in their persistent valvate calyx, stamens more numerous than the petals, many-celled ovary, and straight exalbuminous embryo; but *Rhizophoreæ* have an inferior or semi-inferior ovary, the cells of which only contain two pendulous ovules [often more, see this order, p. 419].

Lythrarieæ are chiefly tropical American; they are much more rare in the temperate regions of both hemispheres.

The species of this family possess different properties; some (*Lythrum Salicaria*) contain tannin, which renders them astringent; others (*Heimia*, *Cuphea*) secrete resinous and acrid principles, which caused them to be used as purgatives, emetics or diuretics. *Lawsonia alba* (Henna), an Egyptian shrub, is renowned throughout the East for the perfume of its flowers, and for the orange-red colouring matter contained in its leaves, with which the women dye their nails and hair; its root (Alkanna) is astringent and yields a red dye. [The *Ammannia vesicatoria*, an Indian weed, is a strong vesicant. *Pemphis acidula* is used as a pot-herb in tropical Asia. The flowers of *Woodfordia tomentosa*, the Dhak of India, are much used as a red dye. The Lagerstrœmias produce excellent timber, called Jarul in India. The wood of *Sonneratia acida* is considered the best substitute for coal on the Indus steamers.—ED.]

XCVIII. *OLINIEÆ*, Arnott.

Olinia. Flower (mag.).   *Olinia.* Diagram.   *Olinia.* Fruit (mag.).   *Olinia.* Ovary cut vertically (mag.).   *Olinia.* Pendent anatropous ovule (mag.).

SHRUBS of South Africa, Brazil,[1] and Australia.[2] LEAVES opposite, coriaceous, penninerved, entire, not dotted, exstipulate. FLOWERS axillary or terminal, in small cymes, or solitary, bibracteolate at the base. CALYX 5-4-toothed, or 4-fid or 5-partite. PETALS 5-4, inserted on the calyx, alternate with its segments, oblong or obovate, obtuse, sometimes (*Olinia*) with 5-4 scales alternate with the petals, scales pubescent on the back and connivent. STAMENS 5-4-∞, inserted with the petals; *filaments* flexuous in bud; *anthers* 2-celled, globose-didymous or oblong, dehiscence longitudinal. OVARY inferior, 2-4-5-celled, cells 2-3-∞-ovuled; *style* subulate or flexuous; *stigma* simple; *ovules* pendulous, anatropous. FRUIT a berry (*Myrrhinium*, *Fenzlia*) or drupe (*Olinia*) crowned by the calyx-limb or by its scar; kernel woody, 3-4-celled. SEEDS oval. EMBRYO exalbuminous, rolled in a spiral, or arched; *cotyledons* indistinct.

[1] The Brazilian plant here alluded to, *Myrrhinium*, is unquestionably Myrtaceous, and closely allied to *Pimenta*.—ED.

[2] The Australian *Fenzlia* is also Myrtaceous, and near *Myrtus* itself.—ED.

The little group of *Oliniceæ*, composed of the genera *Olinia, Myrrhinium,* and *Fenzlia,* is placed between *Myrtaceæ* and *Melastomaceæ.* The berries of *Myrrhinium atropurpureum,* [a Brazilian shrub,] are edible.

XCIX. *GRANATEÆ*,[1] *Endlicher.*

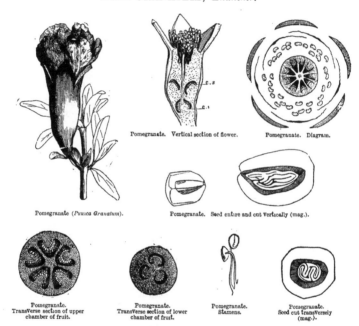

Pomegranate. Vertical section of flower.

Pomegranate. Diagram.

Pomegranate (*Punica Granatum*).

Pomegranate. Seed entire and cut vertically (mag.).

Pomegranate. Transverse section of upper chamber of fruit.

Pomegranate. Transverse section of lower chamber of fruit.

Pomegranate. Stamens.

Pomegranate. Seed cut transversely (mag.).

STEM woody, branches sometimes spinescent. LEAVES generally opposite, often fascicled, entire, not punctate, glabrous, exstipulate. FLOWERS ☿, terminal, solitary or aggregate. CALYX coloured; limb many-partite, many-seriate, valvate in bud. PETALS 5-7, inserted on the calyx-throat, alternate with the sepals, imbricate in bud. STAMENS numerous, many-seriate, inserted below the petals, and included; *filaments* filiform, free; *anthers* introrse, 2-celled, ovoid, dorsifixed, dehiscence

[1] See end of *Lythrarieæ*, p. 433.—ED.

longitudinal. OVARY adhering to the receptacular cup (calyx-tube), forming 2 superimposed rows, the lower 3-celled with central placentation, the upper 5–7-celled with parietal placentation; *style* filiform, simple; *stigma* capitate; *ovules* numerous, anatropous. BERRY spherical, crowned by the calyx-limb, cells separated by membranous septa. SEEDS numerous, integument full of an acid pellucid pulp. EMBRYO exalbuminous, straight; *cotyledons* foliaceous, convolute; *radicle* oblong, short, pointed.

ONLY GENUS.

Punica.

The fruit of the Pomegranate is probably monstrous, and analogous to certain fruits singularly modified by cultivation, as the Tomato (*Lycopersicum esculentum*), and that variety of Orange called 'Bizarrerie' or 'Mellarose.' [The Pomegranate is indigenous in Upper India, where its fruit altogether resembles that of the cultivated state.—ED.]

The genus *Punica* is very near *Myrtaceæ* (see that family). The single species of which it is composed (*P. Granatum*) is a native of Mauritania,[1] whence its name of *Punica*. It grows all along the shores of the Mediterranean, and extends thence throughout the temperate regions of the world; its fruit (Pomegranate) is covered with a coriaceous bark, called 'malicor,' very rich in tannin, and used by curriers; its pulpy seeds are refreshing. Its flowers, called *Balaustium*, were formerly administered as a vermifuge; but its anthelminthic properties principally exist in the bark of its root. This bark contains an astringent substance, with a mild and an acrid principle, which latter destroys tapeworm.

C. ONAGRARIEÆ.

(EPILOBIACEÆ, *Ventenat.*—ONAGRACEÆ, *Lindl.*—ONAGREÆ, *Spach.*—ŒNOTHEREÆ, *Endl.*)

COROLLA *polypetalous, epigynous, contorted in bud.* STAMENS *inserted with the petals, equal or double them in number, rarely fewer.* OVARY *inferior, many-celled, many- (rarely few-) ovuled.* EMBRYO *exalbuminous.*

Terrestrial or aquatic HERBS, or SHRUBS. LEAVES opposite or alternate, simple, penninerved, entire or toothed, exstipulate. FLOWERS ☿, usually regular, often fugacious, axillary and solitary, or racemed or spiked. CALYX herbaceous or coloured; limb 4- (rarely 3-2-) partite, persistent or deciduous, valvate in bud. PETALS (very rarely 0) inserted on the top of the calyx-throat, on an epigynous, flat or annular glandular disk, alternate with the calyx-segments, more or less distinctly clawed, sometimes emarginate or bifid, contorted in bud. STAMENS inserted with the petals, either equal in number and alternate, or double and 1-2-seriate, rarely fewer; *filaments* filiform or subulate, free; *anthers* 2-celled, introrse, dehiscence longitudinal; *pollen* of trigonous granules, often cohering by threads. OVARY inferior, often crowned by the glandular edge of the disk, usually 4- (rarely 2-) celled; *style* filiform; *stigmas* as many as cells, linear, papillose on their inner

<hr>

[1] A. de Candolle ('Géographie Botanique,' ii. 891) rightly points out that the Pomegranate exists in North Africa only in a cultivated state. It ranges in a wild state from Asia Minor to the Punjab Himalaya.—ED.

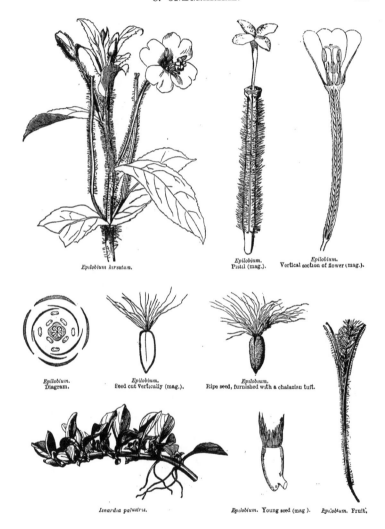

Epilobium hirsutum.

Epilobium.
Pistil (mag.).

Epilobium.
Vertical section of flower (mag.).

Epilobium.
Diagram.

Epilobium.
Seed cut vertically (mag.).

Epilobium.
Ripe seed, furnished with a chalazian tuft.

Isnardia palustris.

Epilobium. Young seed (mag.).

Epilobium. Fruit.

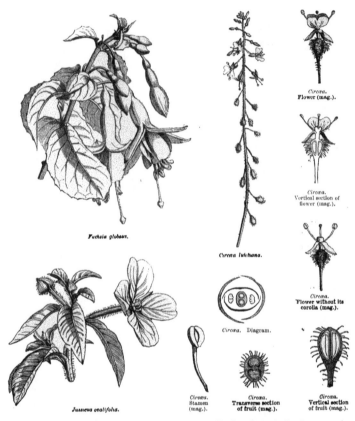

Fuchsia globosa.

Circæa lutetiana.

Circæa.
Flower (mag.).

Circæa.
Vertical section of
flower (mag.).

Circæa.
Flower without its
corolla (mag.).

Circæa. Diagram.

Jussieua ovalifolia.

Circæa.
Stamen
(mag.).

Circæa.
Transverse section
of fruit (mag.).

Circæa.
Vertical section
of fruit (mag.).

surface, rarely coherent; *ovules* numerous in the cells, inserted at the inner angle, rarely few, ascending or pendulous, anatropous. FRUIT generally a capsule, sometimes a berry (*Fuchsia*), rarely nut-like (*Gaura*); *capsule* 4–2-celled, sometimes 1-celled by obliteration of the septa, loculicidal (*Œnothera*), or septicidal (*Jussieua, Isnardia*), with semi-septiferous valves, or having the seeds on a free columella. SEEDS numerous, rarely few or solitary, ascending or pendulous; *testa* crustaceous or membranous, sometimes winged (*Montinia*), sometimes fringed (*Godetia, Clarkia*), or hairy at the chalaza (*Epilobium*). EMBRYO exalbuminous, straight; *cotyledons* foliaceous or a little fleshy, often auricled at the base; *radicle* conical, cylindric, near the hilum, superior or inferior, rarely centripetal.

PRINCIPAL GENERA.

* Epilobium.	* Jussieua.	Ludwigia.	* Clarkia.	* Œnothera.
* Fuchsia.	* Lopezia.	* Godetia.	* Eucharidium.	Isnardia.
* Zauschneria.	Circæa.	* Gaura.	[Trapa.¹]	

ANOMALOUS GENUS.

Montinia.

Onagrarieæ are connected with *Halorageæ*, *Trapeæ* and *Combretaceæ* by the valvate calyx, the iso- or diplo-stemonous corolla and the inferior ovary; but *Halorageæ* differ in their albuminous embryo, *Trapeæ* in the imbricate æstivation of their corolla, and *Combretaceæ* in their 1-celled ovary. *Lythrarieæ* have some affinity with *Onagrarieæ*; in both families the calyx is valvate, the corolla iso- or diplo-stemonous, the ovarian cells many-ovuled, the style simple, the fruit a capsule, and the embryo straight and exalbuminous; but *Lythrarieæ* have a free ovary and imbricate petals. *Onagrarieæ* are widely diffused; but chiefly in the extra-tropical temperate regions of the northern hemisphere, and especially of the New World. *Fuchsia* extends from Mexico to the Straits of Magellan and New Zealand. Many Epilobia are found in the southern hemisphere [they are characteristic of the herbaceous vegetation of New Zealand].

Onagrarieæ contain mucous and sometimes slightly astringent principles, and for this reason *Circæa lutetiana* and the narrow-leaved Epilobia are used in some countries, especially externally; in Sweden they eat the young shoots of the latter. Several Œnotheras, and especially *Œ. biennis*, have a sweet and edible root. *Fuchsia excorticata* has poisonous berries [?]. [Those of other species are sweet and edible. *Montinia*, a Cape genus, is said to be acrid.— ED.]

CI. *TRAPEÆ*,² *Endlicher.*

Lacustrine floating HERBS. LEAVES, some submerged, others emerged: the submerged opposite, pinnatisect (like rootlets), the upper alternate; the emerged in a rosette, petioled, rhomboid; *petiole* vesicular during flowering, exstipulate. FLOWERS axillary, solitary, shortly peduncled. CALYX 4-partite, valvate in bud, with spiny lobes. PETALS 4, inserted on an annular fleshy sinuous disk, crowning the top of the ovary, alternate with the calyx-segments, imbricate in bud, edges folded. STAMENS 4, inserted with the petals, the alternate shorter; *filaments* filiform-subulate; *anthers* introrse, 2-celled, dorsifixed, dehiscence longitudinal. OVARY semi-inferior, of two 1-ovuled cells; *style* cylindric, simple; *stigma* flattened, obtuse; *ovules* pendulous from the top of the septum, anatropous, raphe dorsal. FRUIT coriaceous [or woody], indehiscent, crowned by the spinescent calyx-limb, which resembles 2-4 horns, capped by the hardened disk, 1-celled and 1-seeded from arrest. SEED inverted; *testa* membranous, adherent, the upper part spongy. EMBRYO exalbuminous, straight; *cotyledons* very unequal, one very large, thick, and farinaceous, the other minute, squamiform, inserted a little lower; *radicle* slightly

¹ See order *Trapeæ*, p. 439.—ED.   ² See also under *Onagrarieæ*, p. 439.—ED.

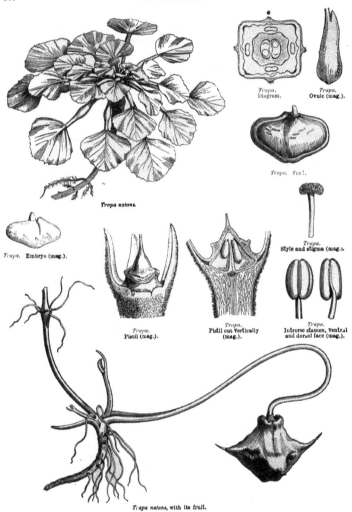

Trapa natans.

Trapa.
Diagram.

Trapa.
Ovule (mag.).

Trapa. Seed.

Trapa. Embryo (mag.).

Trapa.
Style and stigma (mag.).

Trapa.
Pistil (mag.).

Trapa.
Pistil cut vertically
(mag.).

Trapa.
Introrse stamen, ventral
and dorsal face (mag.).

Trapa natans, with its fruit.

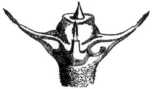

Trapa. Young fruit (mag.).   *Trapa.* Ripe fruit (mag.).   *Trapa.* Petal (mag.).

bent, superior, piercing the top of the fruit in germination; *plumule* very small, concealed within the small cotyledon.

ONLY GENUS.

Trapa.

Trapeæ are closely related to *Halorageæ* (see this family). *T. natans,* the Water Chestnut or Calthrop, inhabits stagnant water in Central and Southern Europe. Its seeds afford a farinaceous food, as do those of the Kashmirian *T. bispinosa,* and the Chinese *T. bicornis,* called by the natives Ling or Ki-chi.

CII. *SAMYDACEÆ.*[1]

(HOMALINEÆ, *Endl.*)

[TREES or SHRUBS, glabrous, or pubescent or tomentose. LEAVES petioled, simple, alternate, distichous, quite entire or serrate, teeth sometimes gland-tipped; *stipules* small, usually deciduous or 0. FLOWERS ☿, regular, inconspicuous, racemed, fascicled or panicled, pedicels articulated and 2-bracteolate. CALYX coriaceous, persistent, tube usually free; lobes 3–7, imbricate or valvate. PETALS as many as the calyx-lobes, and like them, rarely more or 0, perigynous, imbricate in bud. DISK cupular, annular or glandular. STAMENS definite or indefinite, 1–∞-seriate, usually alternating with staminodes, equidistant or collected in fascicles; *filaments* filiform or capillary, free or connate; *anthers* didymous or oblong, introrse or extrorse, dehiscence longitudinal. OVARY free, rarely adnate to the calyx-tube, sessile, 1-celled; *style* long or short, rarely 3; *stigma* entire or 3-fid; *ovules* few or many, inserted on 3–5 parietal placentas, sometimes confined to the upper part of the cell, pendulous or ascending, anatropous, raphe lateral or ventral. FRUIT indehiscent or capsular, 1-celled, 1–∞-seeded, 3–5-valved at the top or throughout its length, valves alternating with the placentas. SEEDS usually few by arrest, oblong or angled, ascending or pendulous, outer coat thin and fleshy or a torn aril; *testa* crustaceous or coriaceous, striate or rugose; *albumen* copious, fleshy. EMBRYO axile; *cotyledons* oblong or orbicular or cordate, often foliaceous; *radicle* short or long.

---

[1] This order is merely alluded to in the original work at the end of *Bixineæ* (p. 247), to which it is most closely allied, and of which it is perhaps best regarded as a tribe, connecting *Bixineæ* with *Passifloreæ.*—ED.

TRIBE I. CASEARIEÆ.—Leaves alternate. Calyx free, 4–5-merous. Petals 0. Stamens 6–30, inserted in one series within the calyx-tube, usually alternating with as many staminodes. *Casearia, Iannaria, Samyda,* &c.

TRIBE II. BANAREÆ.—Leaves alternate. Calyx free, 4–5-merous. Petals 4–5, or more. Stamens many, inserted in many series on a perigynous disk. *Banara, Kuhlia,* &c.

TRIBE III. ABATIEÆ.—Leaves opposite, sub-opposite, or whorled. Calyx 4-partite, valvate. Petals 0. Stamens 8 or indefinite, 1- or many-seriate, staminodes 0. *Abatia, Raleighia, Aphærema.*

TRIBE IV. HOMALIEÆ.—Leaves alternate, rarely sub-opposite or whorled. Calyx free or adnate to the ovary, 4–15-merous. Petals 4–15. Stamens as many as and opposite to the petals, or if more, in bundles opposite to the petals, and alternating with glands. *Homalium, Byrsanthus,* &c.

A small order intermediate between *Passifloreæ* and *Bixineæ,* differing from the former in habit and the want of a corona, and from the latter in perigynism: it is also near *Violarieæ* and *Canellaceæ.* The species are few, wholly tropical, and found in both the Eastern and Western hemispheres.

The general properties of the order are astringent, but none are of any known value; some are bitter, others mucilaginous, and others again are said to be acrid.—ED.]

[For illustrations, see p. 247.]

CIII. *LOASEÆ.*

(LOASEÆ, *Jussieu.*—LOASEÆ VERÆ, *Kunth.*—LOASACEÆ, *Lindl.*)

COROLLA *polypetalous, epigynous, anisostemonous, valvate or imbricate in bud.* STAMENS *more numerous than the petals, rarely all fertile, the outer usually fertile and united in bundles, the inner sterile.* OVARY *inferior, 1-celled, placentation parietal.* OVULES *pendulous, anatropous.* FRUIT *a capsule.* EMBRYO *albuminous.* RADICLE *superior.*

Erect or climbing HERBS, often dichotomous, usually covered with stiff often hooked stinging bristles. LEAVES opposite or alternate, simple, usually palmilobed, exstipulate. FLOWERS ☿, regular, solitary or aggregated, on 2-bracteolate peduncles, axillary or terminal, or leaf-opposed from the elongation of the axillary branch. CALYX superior, 4–5-partite, lobes usually 3-nerved, imbricate or valvate in bud. PETALS deciduous, inserted on the calyx, rarely equal to its lobes in number, and alternate with them, often double in number, of which the 4–5 outer are alternate with these same lobes, and induplicate-valvate or imbricate in bud, and as many inner opposite to the calyx-lobes, smaller than the outer, sometimes antheriferous, generally squamiform, dorsally naked, or awned below the top. STAMENS inserted with the petals, double in number or indefinite, rarely all fertile; the outer usually fertile, various in number, grouped in bundles before the largest petals, the inner sterile, differently shaped, in groups of fours, opposite to the smallest petals; *filaments* filiform or subulate, free or united in bundles; *anthers* introrse, 2-celled, dorsifixed, dehiscence longitudinal. OVARY inferior, 1-celled, with 3–5–4 parietal

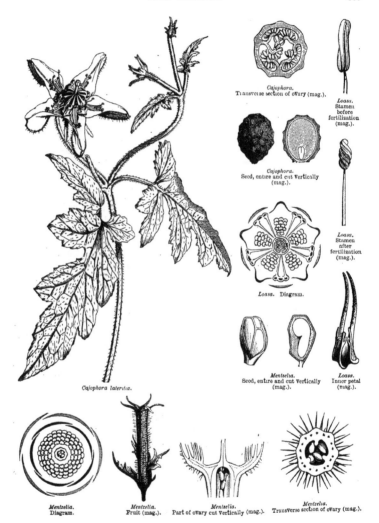

Cajophora.
Transverse section of ovary (mag.).

Loasa.
Stamen
before
fertilization
(mag.).

Cajophora.
Seed, entire and cut vertically
(mag.).

Loasa.
Stamen
after
fertilization
(mag.).

Loasa. Diagram.

Cajophora lateritia.

Mentzelia.
Seed, entire and cut vertically
(mag.).

Loasa.
Inner petal
(mag.).

Mentzelia.
Diagram.

Mentzelia.
Fruit (mag.).

Mentzelia.
Part of ovary cut vertically (mag.).

Mentzelia.
Transverse section of ovary (mag.).

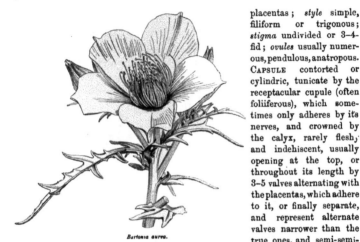

Bartonia aurea.

placentas; *style* simple, filiform or trigonous; *stigma* undivided or 3-4-fid; *ovules* usually numerous, pendulous, anatropous. CAPSULE contorted or cylindric, tunicate by the receptacular cupule (often foliiferous), which sometimes only adheres by its nerves, and crowned by the calyx, rarely fleshy and indehiscent, usually opening at the top, or throughout its length by 3-5 valves alternating with the placentas, which adhere to it, or finally separate, and represent alternate valves narrower than the true ones, and semi-seminiferous. SEEDS usually numerous, pendulous, funicles short; *testa* loose, reticulate; *endopleura* membranous. EMBRYO straight, in the axis of a fleshy albumen, and nearly equalling it in length; *cotyledons* flat, small; *radicle* cylindric, longer than the cotyledons, superior.

PRINCIPAL GENERA.

| * Mentzelia. | * Bartonia. | * Loasa. | Cajophora. | [Gronovia]. |

Loaseæ approach *Passifloreæ* (see this family). Like *Cucurbitaceæ*, they are generally climbers, with palmilobed leaves; their ovary is inferior and one-celled, with parietal placentation; the ovules are numerous and anatropous; but *Cucurbitaceæ* have definite stamens, extrorse and usually syngenesious anthers, diclinous flowers, tendrils, an exalbuminous embryo and corolla, imbricate in bud. The same affinity exists between *Loaseæ* and *Gronovieæ*,[1] which again are separated by their pentandrous androecium [*Cevallia* and other *Loaseæ* are pentandrous], the fleshy ring crowning the ovary, their dry fruit, which is a nucule, and their exalbuminous seed. *Loaseæ* have also an affinity with *Turneraceæ*, in the contorted æstivation, one-celled ovary, parietal placentation, numerous anatropous ovules, capsular fruit, and straight albuminous axile embryo; but in *Turneraceæ* the ovary is free, the stamens definite, the valves of the capsule are semi-placentiferous, and the stem is erect.

Loaseæ are all American, except the genus *Fissenia*, which is African. Most of them grow on the slopes of the Cordilleras facing the Pacific Ocean, beyond the equator, but not in cold regions. The species are little used, excepting *Mentzelia hispida*, which is a strong purgative, and employed by the Mexicans in syphilitic affections.

[1] *Gronovieæ*, alluded to in this work under *Cucurbitaceæ*, consists of one genus, which undoubtedly belongs to *Loaseæ*, with the habit of *Cucurbitaceæ*.—ED.

CIV. *TURNERACEÆ.*

(Loasearum *sectio, Kunth.*—Turneraceæ, *D.C.*)

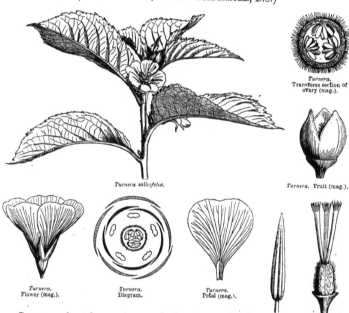

Turnera salicifolia.

Turnera,
Transverse section of
ovary (mag.).

Turnera. Fruit (mag.).

Turnera.
Flower (mag.).

Turnera.
Diagram.

Turnera.
Petal (mag.).

Turnera.
Stamen (mag.).

Turnera.
Pistil (mag.).

Turnera.
Seed, entire and cut vertically
(mag.).

Corolla *polypetalous, perigynous, isostemonous, contorted in bud.* Stamens 5, *sub-hypogynous.* Ovary *free,* 1-*celled, with* 3 *parietal placentas.* Capsule *with* 3 *semi-seminiferous valves.* Seeds *strophiolate.* Embryo *albuminous.*

Herbs, or undershrubs, or shrubs, chiefly of tropical America, with simple rarely stellate hairs. Leaves alternate, simple, petioled, entire or toothed, rarely pinnatifid, exstipulate, but often furnished at the base with 2 lateral glands. Flowers ☿, regular, axillary, sessile or peduncled; *peduncle* free or adnate to the petiole, simple, 2-bracteolate, or jointed below the middle and ebracteate, very rarely branching and many-flowered. Calyx coloured, deciduous, 5-fid, imbricate in bud. Petals 5, inserted near the base or in the throat of the calyx, alternate with its lobes, shortly clawed, equal, contorted in bud, deciduous. STAMENS 5, inserted at the bottom of the calyx-tube and opposite to its lobes; *filaments* free, plano-subulate; *anthers* introrse, 2-celled, erect, dehiscence longitudinal. OVARY free, 1-celled, with 3 nerviform placentas alternating with the sutures of the carpels; *styles* 3, terminal, opposite to the placentas; *stigmas* 3 or 6, fan-shaped; *ovules* numerous, ascending, anatropous. CAPSULE with 3 semi-placentiferous valves. SEEDS numerous, 2-seriate, ascending, cylindric, curved; *testa* crustaceous; *hilum* basilar; *raphe* filiform; *chalaza* projecting; *strophiolus* membranous, appressed to the base of the seed on the side of the raphe. EMBRYO straight, in the axis of a fleshy albumen; *cotyledons* sub-elliptic, plano-convex; *radicle* reaching to the hilum, inferior.

GENERA.

* Turnera.　　　　Erblichia.　　　　Wormskioldia.

This small family is near *Loaseæ* (which see). It has many affinities with *Malesherbiaceæ*; but these differ in their stamens, which are hypogynous as in *Passifloreæ*, their dorsal styles alternating with the placentas, their undivided stigmas, and their non-strophiolate and pendulous seeds. [*Turnera* is a large American genus with one Cape species; *Erblichia* is also American, and *Wormskioldia* Asiatic.]

Turneraceæ possess little-used tonic properties, due to astringent and mucous principles, with a small quantity of volatile oil.

CV. *PASSIFLOREÆ*, Jussieu.

PERIANTH *free, petaloid.* STAMENS *sometimes inserted at the base, or on the throat of the perianth, sometimes hypogynous, united with the gynophore.* OVARY *usually*

Passion-flower. (*Passiflora cærulea.*)

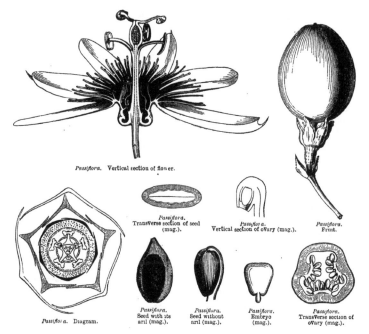

Passiflora. Vertical section of flower.

Passiflora. Transverse section of seed (mag.).

Passiflora. Vertical section of ovary (mag.).

Passiflora. Fruit.

Passiflora. Diagram.

Passiflora. Seed with its aril (mag.).

Passiflora. Seed without aril (mag.).

Passiflora. Embryo (mag.).

Passiflora. Transverse section of ovary (mag.).

stipitate, with 3–5 parietal placentas. FRUIT a berry or capsule. EMBRYO albuminous.—LEAVES alternate.

STEM herbaceous or woody, generally climbing, very rarely arborescent (Ryania, Smeathmannia). LEAVES alternate, sometimes simple, entire, lobed or palmate, sometimes compound, imparipinnate; stipules geminate at the base of the petioles, rarely 0; tendrils axillary, arising from sterile pedicels. FLOWERS ☿, or imperfect through arrest, regular; peduncles usually 1-flowered, jointed at the flower, and [usually] furnished at the joint with a 3-phyllous or -partite involucel. PERIANTH petaloid, monophyllous; tube urceolate or tubular, sometimes very short; limb 4–5-partite or 8–10-partite and 2-seriate, the outer segments sometimes herbaceous, equivalent to a calyx, the inner more coloured, equivalent to a corolla; throat usually crowned by one or many series of subulate filaments; gynophore cylindric, more or less elongating, supporting the pistil and stamens. STAMENS usually equal in number to the [inner] perianth-segments, and opposite to them, or very rarely alternate, sometimes double the number, inserted either at the bottom of the perianth, or at the base or top of the gynophore; filaments subulate or filiform, free or monadelphous and sheathing the gynophore; *anthers* introrse, 2-celled, usually versatile, dehiscence longitudinal. OVARY more or less stipitate, very rarely sessile, 1-celled; *styles* equal in number to the placentas, cohering at the base, distinct at the top,

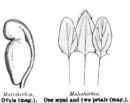

Malesherbia.
Flower (mag.).

Malesherbia.
Diagram.

Malesherbia.
OVule (mag.).

Malesherbia.
One sepal and two petals (mag.).

Malesherbia.
Andrœcium and pistil.

Malesherbia.
Seed, entire and cut vertically (mag.).

Malesherbia.
Transverse section of ovary (mag.).

spreading; *stigmas* clavate or peltate, sometimes sub-2-lobed; *ovules* numerous, anatropous, 1–2-seriate, attached to 3–5 parietal linear placentas by longer or shorter funicles, enlarged into a cupule at the umbilicus. FRUIT 1-celled, an indehiscent berry, or a capsule with 3–5 semi-placentiferous valves. SEEDS numerous; *funicle* dilated into a pulpy cupuliform or saccate aril; *testa* crustaceous, foveolate, easily separable from the membranous endopleura, which bears a longitudinal raphe. EMBRYO straight, occupying the axis of a fleshy dotted albumen; *cotyledons* foliaceous, flat; *radicle* cylindric, near the hilum, centrifugal.

[The following is the classification of *Passifloreæ* made for the 'Genera Plantarum':—

TRIBE I. MALESHERBIÆ.—Flowers ☿. Calyx-tube elongate, lobes triangular-subulate. Corona and corolla membranous. Stamens 5, adnate to the gynophore. Styles 3, remote at the base. Seeds oblong. (South America.) *Malesherbia, Gynopleura.*

TRIBE II. PASSIFLOREÆ.—Flowers ☿. Corona simple or double. Petals herbaceous. Styles 1 or 3–5, connate at the base. Seeds flattish. *Passiflora, *Tacsonia, Paropsia, Smeathmannia,* &c.

TRIBE III. MODECCEÆ.—Flowers ☿, or if unisexual perianths of the sexes alike. Corona small or 0. Petals usually included in the calyx-tube. Connective of the anthers usually produced. (Old World.) *Modecca, Ophiocaulon, Machudoa.*

TRIBE IV. ASCHARIEÆ.—Flowers unisexual, perianth of the sexes alike. Calyx very short. Corolla campanulate, 3–5-fid. Stamens inserted on the throat or base of the corolla. (South Africa.) *Ceratosicyos, Ascharia.*

TRIBE V. PAPAYACEÆ.[1]—Flowers unisexual, or ☿ and unisexual in the same inflorescence, perianths of the sexes dissimilar. Corona 0. Stamens 2-seriate, inserted on the corolla-tube. *Carica, Jacaratia.*]

[1] See end of *Cucurbitaceæ*, p. 452.—ED.

A. L. de Jussieu placed *Passifloreæ* in the family of *Cucurbitaceæ*, which they approach in their climbing stem furnished with tendrils, their alternate palminerved leaves, double perianth, and one-celled ovary with parietal placentation ; in addition to which some genera are diclinous (*Modecca*) ; but in *Cucurbitaceæ* the ovary is inferior, the anthers are extrorse, usually reduced to three, and syngenesious, the embryo is exalbuminous, the leaves exstipulate, and the tendrils are leaves arising from a branch joined to the stem, arrested near the point of departure, and merged in the petiole of the leaf which it bears. Some *Loaseæ* have, like *Passifloreæ*, a climbing stem, palminerved leaves, a one-celled ovary with parietal placentation, numerous pendulous and anatropous ovules, the fruit a capsule or berry, and a straight axile albuminous embryo ; but they want stipules and tendrils, and the placentas occupy the interspaces and not the middle of the valves of the fruit. *Passifloreæ* have an affinity with *Homalineæ*,[1] founded on the 2-seriate perianth, the 1-celled ovary, parietal placentation, styles equalling the placentas in number, the berry or capsule, albuminous seed, and the alternate stipulate leaves ; but in *Homalineæ* the ovary is usually inferior, and the stamens inserted high up the calyx-tube ; *Papayaceæ* also approach *Passifloreæ* in their palminerved leaves, their free usually 1-celled ovary with parietal placentation, their fleshy fruit, and their arillate seeds ; but they are separated by being diclinous [Tribes *Modecceæ* and *Achariæ* are both diclinous], by the insertion of the stamens, the radiating sub-sessile stigma, &c.

Passifloreæ mostly inhabit the tropical regions of the New World ; they are much rarer in Asia, Australia, and tropical Africa, where we find *Smeathmannia*, a shrub without tendrils. The pulpy aril of *Passifloreæ* and *Tacsonia* is used in America in the preparation of cooling drinks. The flowers and fruits of *Passiflora rubra* [called Dutchman's Laudanum] are prescribed in the Antilles for their narcotic properties. *P. quadrangularis* (the Grenadilla) is valued, like the allied species, for the refreshing pulp of its seeds, but its root is very poisonous ; if administered in a small dose it is a vermifuge, like many of the other species.

[The Papaw is the insipid berry of *Carica*, the juice of whose fruit is a powerful vermifuge and antiseptic, and contains fibrine, a substance otherwise supposed to be peculiar to the Animal Kingdom. The whole tree has the singular property of rendering tough meat tender by separating the muscular fibres ; its roots smell like decaying radishes, and its leaves are used as a soap by negros. The juice of the Brazilian *C. digitata* is a deadly poison.—Ed.]

CVI. *CUCURBITACEÆ*, Jussieu.

FLOWERS *monœcious, diœcious, or polygamous.* COROLLA *5-merous, imbricate.* STAMENS *5–3, of which one is usually 1-celled.* OVARY *inferior, 1-several-celled, 1-many-ovuled.* FRUIT *a berry.* SEEDS *exalbuminous.* EMBRYO *straight.*—STEM *furnished with tendrils.* LEAVES *alternate.*

Annual or perennial HERBS or UNDERSHRUBS with fibrous or often tuberous roots. STEM cylindric or angular, climbing, juice watery. LEAVES alternate, petioled, palminerved, often palmilobed, usually cordate ; *tendrils* simple or branching, springing singly in the same plane with the leaves. FLOWERS monœcious or diœcious, very rarely ☿, axillary, solitary fascicled racemed or panicled, white or yellow, rarely red. CALYX usually campanulate, limb 5-toothed or -lobed, imbricate in bud. COROLLA monopetalous, rotate or campanulate, 5-lobed, sometimes a little irregular (*Thladiantha*), lobes entire or fringed, imbricate in bud, inserted on the calycinal limb, and alternate with its divisions, distinct, or more often coherent, and then adnate to the calyx, and as if continuous with its limb. ANDRŒCIUM inserted at the bottom of the corolla, or of the calyx, composed of two 2-celled and

---

[1] *Samydaceæ*, tribe *Homalieæ* ; see p. 442.—Ed.

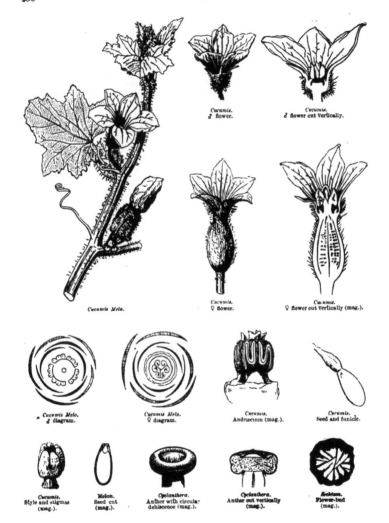

Cucumis.
♂ flower.

Cucumis.
♂ flower cut vertically.

Cucumis Melo.

Cucumis.
♀ flower.

Cucumis.
♀ flower cut vertically (mag.).

Cucumis Melo.
♂ diagram.

Cucumis Melo.
♀ diagram.

Cucumis.
Andræcium (mag.).

Cucumis.
Seed and funicle.

Cucumis.
Style and stigmas
(mag.).

Melon.
Seed cut
(mag.).

Cyclanthera.
Anther with circular
dehiscence (mag.).

Cyclanthera.
Anther cut vertically
(mag.).

Sicyium.
Flower-bud
(mag.).

| *Sechium.* Young ♀ flower. | *Sechium.* Ovary cut vertically, with pendulous ovule. | *Echinocystis.* Pistil cut vertically, with erect ovules. | *Echinocystis.* Ovary cut transversely (mag.). |

one 1-celled anther, or of 5 normal 2-celled stamens; *filaments* short, thick, free or monadelphous, connective sometimes prolonged beyond the anthers; *anthers* extrorse, cells usually sinuous, adnate to the connective, rarely straight or arched, and opening longitudinally or circularly. OVARY inferior, usually composed of 3–5 carpels (rarely 1), coherent, with parietal placentas reflexed towards the circumference; *style* terminal, short, 3-fid or -partite; *stigmas* thick, lamellate, lobed or fringed; *ovules* solitary or numerous, several-seriate, erect pendulous or horizontal, anatropous. BERRY fleshy, (rarely dry), usually indehiscent, sometimes opening elastically at the base by the separation of the peduncle (*Ecbalium*), or at the top by the raising of an operculum (*Luffa*), or by irregular rupture (*Momordica*), sometimes dehiscent in 3 valves, or circumsciss (*Actinostemma*). SEEDS numerous, horizontal, or erect at the bottom of the cell (*Echinocystis, Abobra*), rarely one single pendulous (*Sicyos, Sechium*) [or two, one pendulous and one erect (*Raphanocarpum*)], usually compressed, sessile or shortly funicled; *testa* membranous crustaceous or horny, often girt with a thick border, rarely linear and samaroid; *albumen* 0. EMBRYO straight; *cotyledons* foliaceous, veined; *radicle* short, reaching to the hilum, centrifugal; *plumule* with 2 distinct leaves.

[The following is the most recent classification of the *Cucurbitaceæ*, by Bentham and Hooker fil., in the ' Genera Plantarum ':—

TRIBE I. CUCUMERINEÆ, *Naud.*—Stamens usually 3, free or connate; anther-cells straight or curved or flexuous. Ovary with 3, rarely 2–5 placentas; ovules horizontal. *Hodgsonia, Telfairia,* *Trichosanthes, Trochomeria, *Lagenaria, *Luffa, *Benincasa, *Momordica, Thladiantha, *Cucumis, *Citrullus, Ecbalium, Cephalandra, *Cucurbita, Bryonia, Zehneria, Melothria, Rhynchocarpa, Anguria,* &c.

TRIBE II. ABOBREÆ. *Naud.*—Stamens 3, free; anther-cells flexuous. Ovary 3-4-celled; ovules 1–2 in each cell, erect or ascending from the base. *Abobra, Trianosperma,* &c.

TRIBE III. ELATERIEÆ, *Hook. fil.*—Stamens 1–3. Ovary usually oblique, 1-4-celled, or 2-∞-locellate. Berry bursting elastically, exposing a naked fleshy seed-bearing columella, rarely dehiscing by pores. Seeds erect, ascending, or horizontal. *Echinocystis, Elaterium, Cyclanthera,* &c.

TRIBE IV. SICYOIDEÆ, *Naud.*—Stamens 3–5; filaments usually connate. Ovary 1-celled; ovule solitary, pendulous. *Sicyos,* *Sechium, &c.

TRIBE V. GOMPHOGYNEÆ, *Hook. fil.*—Stamens 5, free; anthers 1-celled, dehiscing longitudinally. Ovary 1-celled; ovules 2–3, pendulous from the top of the cell. *Gomphogyne, Actinostemma.*

TRIBE VI. GYNOSTEMMEÆ, *Hook. fil.*—Stamens 3–5, free or connate; anthers 1-celled, dehiscence longitudinal. Ovary 3-celled; ovules 1–2, pendulous in each cell. *Gynostemma, Schizopepon.*

TRIBE VII. ZANONIEÆ, *Hook. fil.*—Stamens 5, free; anthers oblong, 1-celled; dehiscence longitudinal. Ovary with 3 thick placentas; ovules many. Fruit 1-celled, cylindric, truncate, opening at the depressed vertex by 3 valves. Seeds winged. *Zanonia, Alsomitra, Gerrardanthus.*

TRIBE VIII. FEVILLEÆ, *Hook. fil.*—Stamens 5, free; anthers 2-celled. Ovary 3-celled; styles 3, distant; ovules inserted on the axis of the ovary. Fruit large, indehiscent. Seeds large, orbicular, attached to a large central trigonous column. *Fevillea.*—ED.]

The tendrils of *Cucurbitaceæ* were long considered to consist of a solitary stipule, but Naudin's observations render it more probable that the upper portion represents the midrib or principal nerve of a leaf-blade, and the base a prolongation of the axis, *i.e.* a branch which is stunted near the point where it leaves the stem and merges into the petiole of the leaf which springs from it. *Cucurbitaceæ* are more or less closely connected with *Gronovieæ* [see *Loaseæ*], *Loaseæ, Begoniaceæ, Papayaceæ,* &c. *Gronovia* only differs in its ☿ flowers, didymous anthers, 1-ovuled ovary and nut. We have indicated the affinities existing with *Loaseæ* and *Passifloreæ* (which see). *Cucurbitaceæ* also somewhat resemble *Begoniaceæ* in the palminerved leaves, diclinism, extrorse anthers and inferior ovary; but the latter differ in polyandry, loculicidal capsule, very small seeds, stem without tendrils and clearly stipulate leaves. There is also some analogy between *Cucurbitaceæ* and *Papayaceæ* [see *Passifloreæ* Tribe V., p. 448], in diclinism, the berried fruit and palminerved leaves; but in *Papayaceæ* the flower is diplostemonous, the corolla is valvate [or contorted] and hypogynous, the anthers introrse, the seed albuminous, &c. Finally, *Cucurbitaceæ,* like *Aristolochieæ,* have an inferior ovary, extrorse anthers, climbing stem, alternate leaves and axillary flowers, but the resemblance goes no further.

Cucurbitaceæ are met with in the tropical and sub-tropical regions of both worlds; they are quite absent from cold countries, and rare in temperate, although several tropical species can be cultivated there in consequence of their short life, every phase of which can be passed through in a summer. In comparing the Melon with the Colocynth, it might be thought that their properties were very different; they are nevertheless identical in most species, and only vary in intensity according to the nature and development of the organs, and the presence of certain accessory principles, the chief of which is sugar. Many species, in fact, owe to bitter substances, extractable and sub-resinous, crystallizable or not, a drastic and emetic quality, violent in some, weak in others, usually concentrated in the root, sometimes in the pulpy fruit. The rind of many berries is bitter, while the flesh is rendered agreeable by the sugar, mucilage, salts, free acids and aromatic principles which it contains, besides which the quality varies with age and ripeness. The seeds are oily and rather bitter. *Bryonia alba* and *dioica* have a large root, containing a milky juice, acrid and bitter, with a nauseous smell, and strongly drastic; even externally, if applied fresh to the abdomen, it is purgative. The exotic Bryonias have the same property; the root of *B. abyssinica,* rich in starch like all its congeners, is used as food in Abyssinia, after having been cooked. The Colocynth (*Citrullus Colocynthis*) is an Eastern [and North African] plant, the fruits of which are more bitter than those of any other species; their spongy pulp, insipid and nauseous in smell, contains a fixed oil, a resin, and an extractable principle, to which are due the drastic properties known to the ancients. *Ecbalium agreste,* [the Squirting Cucumber,] commonly called Wild Cucumber, a plant common about ruins throughout the Mediterranean region, and formerly renowned for its bitterness and purgative properties, is now fallen into oblivion. The fruit of *Luffa* is eatable in India and Arabia before it is ripe, but when ripe it becomes strongly purgative. The same is the case with that of *Trichosanthes anguina,* which grows in China and India; and with the Momordicas of America. The berry of *Momordica Balsamina,* infused in olive oil, enjoys among the inhabitants of tropical Asia a great reputation as a vulnerary. The leaves of *M. Charantia* possess the same properties. Among edible *Cucurbitaceæ* we must place in the first rank the Pumpkin (*Cucurbita moschata, C. Pepo, C. Pepo* var., and *C. maxima*), Gourd (*Lagenaria vulgaris*), Cucumber (*Cucumis sativus*), [Vegetable Marrow (*Cucur-*

bita ovifera)], Water Melon (*Citrullus vulgaris*), and Melon (*Cucumis Melo*), all Asiatic or African species, cultivated in Europe from the highest antiquity. All these run into numerous varieties, with a reticulated or smooth bark, tubercular sides, and white, yellow or red flesh, &c. The Water Melon provides the natives of hot countries with a refreshing food. The juice of the Cucumber, mixed with calves' fat, is largely used as a cosmetic; the fruit of one of its varieties, gathered before it is ripe and preserved in vinegar, forms a condiment known by the name of Gherkin. [The true or West India-1 Gherkin is the unripe fruit of *Cucumis Anguria*.] *Cucumis Dudaim* is cultivated in Turkey for its fruit, which has a delicious scent, but insipid pulp. The seeds of these different species contain a fixed oil and mucilage, which lead to their employment as an emulsion; the seeds of the Cucumber, Melon, Pumpkin, and Gourd are called in pharmacy the 'four larger cold seeds.' *Telfairia pedata*, a shrub growing wild on the shores of East Africa, and cultivated in the Mascarine Islands, is renowned for the edible fixed oil which is contained abundantly in its cotyledons, [as *T. occidentalis*, found in West Africa]. All the cultivated *Cucurbitaceæ* are remarkable for polymorphism and the variety of their fruits. *Lagenaria* produces both the small gourds of pilgrims and enormous calabashes. *Trichosanthes colubrina*, from equatorial Asia, has slender cylindrical fruits, red, yellow and green, coiled like a serpent, and six feet long. We may also mention *Luffa*, the fruit of which, dried and reduced to its fibrous part, serves as a sponge or dishcloth in the Antilles, [and is sold as 'Egyptian Bath-sponge' in England. *Benincasa cerifera*, of tropical Asia, or White Gourd, yields a wax on the surface of its fruit; it is considered a type of fertility in India, and presented to newly-married couples. *Acanthosicyos*, a remarkable erect furze-like leafless spinescent plant of the desert regions of South Africa, produces a small gourd whose edible pulp is much sought by the natives. The oily seeds of *Fevillea* are intensely bitter, emetic, and purgative, and its oil is largely used for lamps, &c.—ED.].

CVII. *BEGONIACEÆ, Br.*

FLOWERS *monœcious*. STAMENS *numerous*. ANTHERS *extrorse*. OVARY *inferior*, 3-celled, many-ovuled. CAPSULE *with 3 cells winged on the back, and 3 loculicidal valves.* SEEDS *numerous; albumen scanty or 0.* EMBRYO *straight, axile.*

HERBS with a fleshy tuberous rhizome, or UNDERSHRUBS or SHRUBS with watery acidulous juice. STEM alternately branched, cylindric, swollen at the nodes, jointed. LEAVES alternate, sometimes distichous, rarely sub-whorled, petioled, simple, usually palmi- or peltate- or penni-nerved, sides usually unequal, cordate at the base, denticulate, teeth often mucronate, rarely entire and linear-lanceolate, sometimes variously cut, folded within the stipules before expansion, hairs usually simple, rarely stellate, scattered over the upper surface, and principally situated on the nerves of the lower surface; *stipules* free, often caducous. FLOWERS monœcious, on axillary peduncles, branching into cymes, ♂ in the middle, ♀ at the circumference, furnished beneath the inflorescence with membranous bracts. FLOWERS ♂: PERIANTH petaloid, with 2-seriate leaflets, which may be considered as calyx and corolla; outer leaflets 2, opposite, valvate in bud; inner usually 2, folded in bud, alternate with the outer, sometimes 3-7, or 0. STAMENS numerous, in the centre of the flower; *filaments* distinct, or variously monadelphous, continuous with the connective; *anthers* extrorse [or dehiscing laterally], 2-celled, cells adnate to the connective and separated by it, opening by 2 longitudinal slits, or rarely by 2 terminal pores. Rudimentary OVARY 0. FLOWERS ♀: lobes of *perianth* (sepals and petals) nearly alike in form and colour; sometimes 2, valvate in bud, and opposite; sometimes 3-4, of which 1-2 are inner and smaller; sometimes 5-6-8, imbricate in bud. OVARY inferior, usually divided into cells corresponding to the styles, and

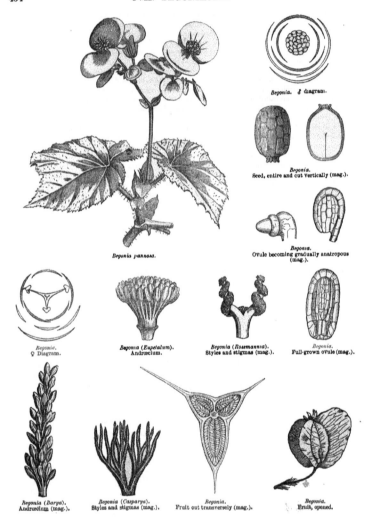

Begonia. ♂ diagram.

Begonia.
Seed, entire and cut vertically (mag.).

Begonia pannosa.

Begonia.
Ovule becoming gradually anatropous
(mag.).

Begonia.
♀ Diagram.

Begonia (Eupetalum).
Andrœcium.

Begonia (Rossmannia).
Styles and stigmas (mag.).

Begonia.
Full-grown ovule (mag.).

Begonia (Barya).
Andrœcium (mag.),

Begonia (Casparya).
Styles and stigmas (mag.).

Begonia.
Fruit cut transversely (mag.).

Begonia.
Fruit, opened.

winged on the back, rarely almost 1-celled (*Mezierea*); *placentas* occupying the inner angle of the cells, thick, single or 2-partite, rarely semi-parietal (*Mezierea*); *styles* usually 3, short, thick, 2-fid or pluripartite; *stigmas* usually arranged on the stylary branches in flexuous or spiral bands, united at the base outside; *ovules* very numerous, anatropous. FRUIT a capsule, rarely a berry (*Mezierea*). CAPSULE 3- (rarely 1-2-4-5-) celled, with loculicidal valves bearing on their centre membranous septa, separated from the seminiferous axis, and coherent below and above. SEEDS minute, rarely funicled, obovoid, globose, ellipsoid or sub-cylindric; *testa* reticulate-foveolate, crustaceous; *endopleura* sub-fleshy; *albumen* scanty or 0. EMBRYO straight, axile; *cotyledons* very short; *radicle* elongated, touching the hilum, centripetal.

PRINCIPAL GENERA.

Casparya.   * Begonia.   Mezierea.   [Begoniella.]   [Hillebrandia.]

The affinities of *Begoniaceæ* are very uncertain, and their place has therefore been repeatedly changed. Of the families with which they have some affinity it may be sufficient to mention *Cucurbitaceæ* and *Datisceæ*. The latter have, like *Begoniaceæ*, diclinous flowers, a polyandrous androecium with extrorse and adnate anthers, several styles opposite to the sepals when they are isomerous, and furnished with stigmatic papillæ on their inner face, an inferior ovary (with parietal placentation in *Mezierea*), numerous anatropous ovules, a capsular fruit, sub-exalbuminous seeds, a straight cylindric axile embryo with very short cotyledons; but in *Datisceæ*, besides the different habit, the leaves are usually imparipinnate and exstipulate; the branches are not jointed and knotty; the capsule is always 1-celled, gaping at the top, &c. *Begoniaceæ* inhabit almost exclusively intertropical regions; nevertheless one species from the north of China, *Begonia discolor*, stands our winter. They are more frequent in America than in Asia, much rarer in Africa [many have of late been discovered in tropical Africa], and hitherto unknown in Australia. *Begoniaceæ* contain oxalic acid, and have the properties of Sorrel, which name is given to them in the Antilles; in some species it is united with astringent and drastic substances. Several American and Asiatic species are ranked among refreshing, antibilious and antiscorbutic medicines. *Begonia malabarica* and *tuberosa* [and other Indian species] are edible. The bitter root of *B. tomentosa* and *grandiflora* is considered by the Peruvians to be a powerful astringent. Some Mexican Begonias have a drastic root, used in syphilitic and scrofulous diseases. Many Begonias are ornamental hothouse plants, as *B. incarnata, semperflorens, manicata, coccinea, Rex, fuchsioides, cinnabarina, heracleifolia, argyrostigma, zebrina, diversifolia*, &c. *B. discolor* is a species from China, the branches of which are tinted with red above each joint, and the leaves of which are green above and dark red below; its tubers resist the most severe winters. We have noticed (page 8) the vital energy of the Begonias.

[The discovery of *Hillebrandia*, a Begoniaceous genus, in the Sandwich Islands, has enabled Professor Oliver to settle the affinities of *Begoniaceæ*, which are undoubtedly close to *Saxifrageæ*. Another recently discovered Begoniaceous genus (*Begoniella*), a native of New Granada, has a simple gamophyllous campanulate perianth and definite stamens.]

CVIII. *DATISCEÆ.*

(DATISCEÆ, *Presl.*—DATISCACEÆ, *Lindl.*)

HERBS or TREES. LEAVES alternate, imparipinnate or palminerved and sub-inequilateral, exstipulate. FLOWERS usually diœcious, sometimes ☿ or polygamous, greenish, small, in a panicle or spicate raceme. FLOWERS ♂: CALYX 3-9-fid. COROLLA 0. STAMENS 3-15, inserted on the calyx; *anthers* 2-celled, dorsifixed, dehiscence longitudinal, extrorse. FLOWERS ♀ and ☿: CALYX with a superior 3-8-toothed limb. COROLLA 0. STAMENS in the ☿ equal and alternate to the calycinal

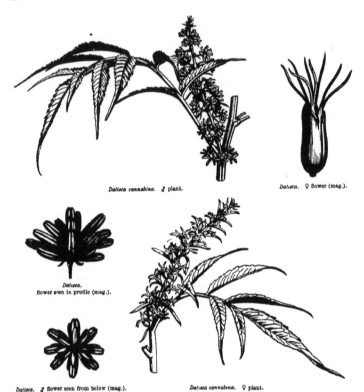

Datisca cannabina. ♂ plant.

Datisca. ♀ flower (mag.).

Datisca.
flower seen in profile (mag.).

Datisca. ♂ flower seen from below (mag.).

Datisca cannabina. ♀ plant.

Datisca.
Seed, whole and cut
vertically (mag.).

Datisca.
Transverse section of ovary,
showing the parietal many-
ovuled placentas (mag.).

teeth (*Tricerastes*); *anthers* extrorse, dehiscence longitudinal. OVARY inferior, 1-celled, usually gaping at the top, with parietal placentas alternating with the calyx-lobes; *styles* opposite to the calyx-lobes, simple, stigmatiferous within the top, or 2-fid with linear branches stigmatiferous along their inner face; *ovules* numerous, sub-horizontal, anatropous. CAPSULE membranous, crowned by the calyx-limb and styles. SEEDS numerous, small, oblong ; *testa* reticulate and foveolate ; *hilum* bearing a membranous cupuliform strophiole ; *albumen* scanty. EMBRYO cylindric ; *cotyledons* very short ; *radicle* long, near the hilum.

GENERA.

| * Datisca. | Tetrameles. | Octomeles. |

Datisceæ have a certain affinity with *Begoniaceæ* (which see). Some botanists have placed them near *Loaseæ* on account of their epigyny and the arrangement of the placentas ; others near *Resedaceæ*, on account of the one-celled ovary gaping at the top from the bud, and the parietal placentation, but there are no other points of affinity. The genera of this little order are singularly scattered over the globe; *Datisca* inhabits West Asia and Nepal; *Tricerastes* [which is *Datisca* itself] is met with in California, *Tetrameles* is a large tree of Java and India, and *Octomeles* is Malayan. The herbage of *Datisca cannabina* is nauseous, purgative and emetic ; it is frequently used in Italy for intermittent fevers and disorders of the stomach. The root contains a peculiar starch named by chemists *datiscine.* Nothing is known of the properties of *Tetrameles.*

CIX. *CACTEÆ.*

(CACTI, *Jussieu.*—CACTOIDEÆ, *Ventenat.*—NOPALEÆ, *D.C.*)

PETALS *numerous, several-seriate, epigynous, free or cohering below.* STAMENS *numerous, many-seriate, inserted at the base of the corolla.* OVARY *inferior, 1-celled,*

Echinocactus Decaisneanus.

Rhipsalis funalis.

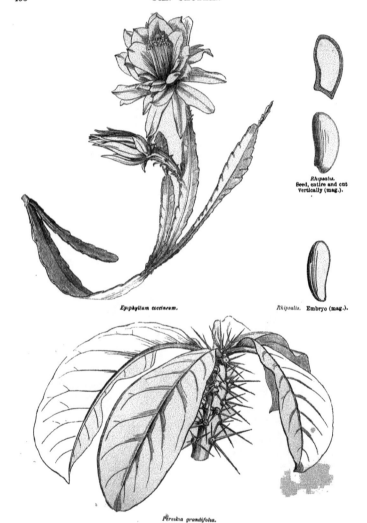

Epiphyllum coccineum.

Rhipsalis.
Seed, entire and cut
vertically (mag.).

Rhipsalis. Embryo (mag.).

Pereskia grandifolia.

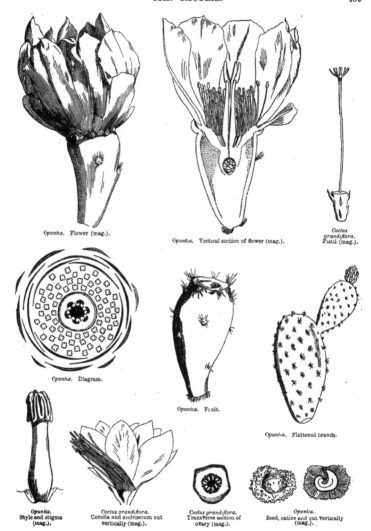

Opuntia. Flower (mag.).

Opuntia. Vertical section of flower (mag.).

Cactus
grandiflora.
Pistil (mag.).

Opuntia. Diagram.

Opuntia. Fruit.

Opuntia. Flattened branch.

Opuntia.
Style and stigma
(mag.).

Cactus grandiflora.
Corolla and andrœcium cut
vertically (mag.).

Cactus grandiflora.
Transverse section of
ovary (mag.).

Opuntia.
Seed, entire and cut vertically
(mag.).

with many-ovuled parietal placentas. BERRY *pulpy.* SEEDS *numerous.* EMBRYO *straight or curved, albumen 0 or scanty.—Fleshy* SHRUBS. LEAVES *generally 0 or rudimentary, rarely normal.*

SHRUBS with watery or milky juice. ROOT woody, simple or branching, bark soft. STEM branching, or simple from the arrest of the buds, covered with tubercles which represent the arrested buds, cylindric, angular, channelled, plane, winged, elongate or globose, fleshy, with thick usually green bark, loose cellular tissue, scanty and fragile, or rarely numerous and hard woody fibres, medullary sheath broad. LEAVES generally 0, indicated by a cushion under the bud, sometimes rudimentary, deciduous, rarely perfect, flat and petiolate (*Pereskia*). BUDS springing from the axil of the latent rudimentary or normal leaf, solitary or geminate and superimposed ; the lower arrested in its development, furnished with spines, naked or cottony ; the upper close to it, developing into a flower or branch ; *stipules* 0 ; *spines* springing from the abortive buds, fascicled, definite or indefinite, rarely absent. FLOWERS perfect, solitary, terminal or springing from the axil of an abortive branch, ebracteate. PERIANTH multiple, calyx hardly distinguishable from the corolla. CALYX generally petaloid, rarely foliaceous, superior. COROLLA epigynous; *petals* delicate, 2-several-seriate, the inner largest, distinct and rotate (*Opuntia, Rhipsalis*), or erect and cohering by their bases into an elongated tube (*Mamillaria, Melocactus, Echinocactus, Cereus, Epiphyllum*). STAMENS numerous, many-seriate, inserted at the base of the corolla, the inner generally the smallest; *filaments* filiform ; *anthers* introrse, 2-celled, dehiscence longitudinal. OVARY inferior, 1-celled ; *placentas* parietal, 3 or more, bilamellate; *style* simple, elongated, cylindric or pyramidal, hollow or solid ; *stigmas* equal in number to the placentas, linear or loriform, spreading or close ; *ovules* numerous, horizontal, anatropous. BERRY smooth or furnished with spines or bristles (in the axils of which often spring branches), umbilicate, 1-celled ; placentas parietal, pulpy. SEEDS numerous, buried in the pulp, globose or ampullaceous; *testa* nearly bony, black, shining, foveolate; *hilum* large, circular, pale ; *albumen* 0, or nearly so. EMBRYO sometimes straight, clavate or sub-globose, sometimes curved or semicircular; *cotyledons* free or united ; *radicle* facing the hilum.

[The following tribes are adopted in the ' Genera Plantarum ' :—

TRIBE I. ECHINOCACTEÆ.—Calyx-tube produced beyond the ovary. Stem covered with elongate tubercles, or ribs which are aculeate, rarely leafy. *Melocactus, *Mamillaria, Leuchtenbergia, *Echinocactus, *Cereus, *Phyllocactus, *Epiphyllum, &c.

TRIBE II. OPUNTIEÆ.—Calyx-tube not produced beyond the ovary. Stem branched, jointed. *Rhipsalis, Nopala, *Opuntia, *Pereskia.—ED.]

We have indicated the affinities of *Cacteæ* with *Mesembryanthemeæ* (which see). A. L. de Jussieu placed *Cactus* and *Ribes* in the same family ; and in fact there is a close affinity between them, founded on the polypetalous and epigynous corolla and its æstivation, the one-celled ovary with parietal placentation, the berried acid fruit, and the spiny cushion from the axil of which spring the leaves and flowers; but the habit of *Cacti*, the fleshy consistency of their stem, the indefinite number of their petals and stamens, and their scanty albumen, render the diagnosis easy.

Cacteæ are all American; a *Rhipsalis* has, however, recently been discovered on the west coast of Africa [and in Ceylon]. They are especially tropical, but many are found beyond the tropical zone, and as far north as 49°, and south to 30°; they abound in Texas, Mexico and California. It is in the Sonora, in the environs of Gila, that we find the most gigantic Cacti (*Cereus giganteus*), resembling candelabra of 50 to 60 feet high.

Opuntia vulgaris is now naturalized throughout the Mediterranean region, where its fruit is eaten under the name of Indian Fig; its taste recalls that of a pumpkin, and its pulp contains a gelatinous principle analogous to gum tragacanth. The berries of several *Cacteæ* are sub-acid, and hence refreshing, antibilious and antiscorbutic. The milky juice of some species is administered in America for intestinal worms. A decoction of the flowers of *Melocactus communis* is a reputed remedy for syphilis. The fruit of *Opuntia vulgaris* is diuretic, and colours urine of a deep red; its joints are applied as a topic to hasten the maturing of tumours. It is on this species and its congeners, known as Prickly Pear, and cultivated in Mexico and the Canaries, that the Cochineal insect lives—an hemipterous insect, much employed in the arts in the composition of carmine, crimson lake, and a dye called Cochineal red.

CX. *FICOIDEÆ.*[1]

[Annual or perennial HERBS, rarely SHRUBS, with often whorled and knotted branches. LEAVES opposite, alternate, or in false whorls, entire or with cartilaginous margins and teeth; *stipules* 0 or scarious. FLOWERS ☿, usually cymose, rarely unisexual, regular. CALYX free or adnate to the ovary, 4–5-lobed or -divided, persistent, imbricate in bud. PETALS 0 or small and white, large in *Mesembryanthemum*. STAMENS perigynous, rarely hypogynous, equal in number to and opposite the sepals, or more numerous and scattered, or combined in fascicles; *filaments* free or connate; *anthers* oblong, 2-celled, dehiscence longitudinal. DISK 0, or annular, or produced into staminodes. OVARY usually free, 2–∞-celled, rarely 1-celled; *styles* as many as the cells, free or connate, usually subulate and papillose on the inner surface; *ovules* solitary in the cells and basal, or numerous and inserted on the inner angles of the cells, amphitropous. FRUIT a membranous or hard capsule, or achene or drupe, or separating into utricles or cocci. SEEDS solitary or numerous, reniform, ovoid, globose or obovoid; *testa* membranous or crustaceous, hilum lateral or rarely facial; *albumen* scanty or copious, farinaceous, rarely fleshy. EMBRYO curved round the albumen, terete; *cotyledons* narrow, incumbent; *radicle* terete.

TRIBE I. MESEMBRYANTHEÆ.—Calyx adnate to the ovary. Leaves exstipulate. *Mesembryanthemum, Tetragonia.*

TRIBE II. AIZOIDEÆ.—Calyx-tube more or less elongate, free. Stamens perigynous, inserted on the calyx-tube, rarely sub-hypogynous. Fruit capsular. *Aizoon, Gatenia, Sesuvium, Trianthema,* &c.

TRIBE III. MOLLUGINEÆ.—Calyx 5-partite. Petals 3–5 or 0. Stamens hypogynous or sub-perigynous. Fruit capsular, or of 3–5 cocci. *Orygia, Telephium, Mollugo, Pharnaceum, Gisekia, Semonvillea, Limeum,* &c.

---

[1] [I have introduced this order here to show the distribution of the genera adopted in the 'Genera Plantarum,' but which are variously disposed by botanists; it includes most of the *Ficoideæ* of Jussieu, as also *Mesembryanthemeæ.* Endl. (p 462). *Portulaceæ,* Endl. in part (p. 259), *Tetragonieæ,* Lindl. (p. 464), and *Mollvgineæ,* Lindl. (p. 261).—ED.]

As thus constituted, *Ficoideæ* are intermediate between *Caryophylleæ, Portulaceæ* and *Paronychieæ*, and indeed form one aggregate order with them. It differs from *Caryophylleæ* in the 2-∞-celled ovary, usually alternate leaves, and absence of petals; from *Portulaceæ* in the never 2-sepalous calyx and the 2-∞-celled ovary; and from *Paronychieæ* in habit, 2-∞-celled ovary, simple stigma, and dehiscence of the capsule. *Ficoideæ* also rank near *Phytolacceæ* and *Polygoneæ*. As a whole they may be regarded as the perigynous many-celled representatives of those orders.

Most of the *Ficoideæ* are weeds of dry hot regions throughout the globe; their properties are unimportant.—Ed.]

CXI. *MESEMBRYANTHEMEÆ.*[1]

(FICOIDEARUM *genera, Jussieu.*—MESEMBRIANTHEMEÆ, *Fenzl.*)

CALYX *superior*. PETALS *and* STAMENS *indefinite, epigynous*. OVARY *several-celled, with linear parietal placentas, occupying the bottom of the cells.* CAPSULE *depressed, many-valved.* SEEDS *numerous.* EMBRYO *curved.* ALBUMEN *farinaceous.*

STEM sub-woody or rarely herbaceous or fleshy. LEAVES opposite or alternate, fleshy, plane or cylindric or trigonous, exstipulate. FLOWERS ☿, regular, axillary or terminal, inflorescence various, generally opening towards noon, sometimes in the evening, gold, saffron, purple, violet, pink or white. CALYX superior, 5- (rarely 2-8-) partite, segments herbaceous, leafy or semi-scarious, usually unequal, imbricate. PETALS numerous, inserted on the calyx, usually many-seriate, linear,

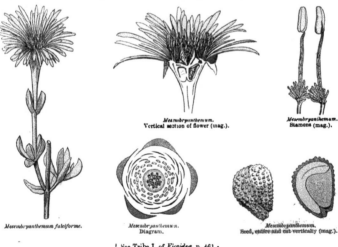

Mesembryanthemum.
Vertical section of flower (mag.).

Mesembryanthemum.
Stamens (mag.).

Mesembryanthemum falciforme.

Mesembryanthemum.
Diagram.

Mesembryanthemum.
Seed, entire and cut vertically (mag.).

[1] See Tribe I. of *Ficoideæ*, p. 461.

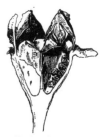

Mesembryanthemum.
Nearly ripe fruit (mag.).

Mesembryanthemum.
Ripe fruit (mag.).

Mesembryanthemum.
Ripe fruit cut vertically (mag.).

marcescent or deliquescent, imbricate in bud. STAMENS indefinite, many-seriate; *filaments* subulate or setaceous, unequal, free or united at the base; *anthers* introrse, 2-celled, ovoid, versatile, dehiscence longitudinal. CARPELS 4–20, cohering into an inferior ovary, 4–20-celled, ventral suture free, superior; *placentas* linear, parietal, occupying the bottom of each cell; *stigmas* 4–20, cristate, crowning the floral axis; *ovules* numerous, several-seriate, fixed by a ventral hilum to long funicles. CAPSULE at first fleshy, then woody and dry, top truncate, opening along the stigmatic crests by the centrifugal raising of the thick coriaceous epicarp as it separates from the endocarp, which persists under the form of geminate chartaceous triangular segments. SEEDS numerous; *testa* crustaceous, soft or granular; *albumen* farinaceous. EMBRYO peripheric, dorsal, curved or hooked, voluminous; *cotyledons* ovoid or oblong; *radicle* cylindric.

GENUS.

* Mesembryanthemum.

Mesembryanthemeæ approach *Cacteæ* in the polypetalous and epigynous corolla and its æstivation, in polyandry, parietal placentation and curved ovules; they are separated by their many-celled ovary, sessile stigmas, farinaceous albumen, and normal leaves. They have also some affinity with *Portulaceæ*, and especially with *Tetragonia*, in the more or less inferior ovary, polyandrous stamens, curved ovule, peripheric embryo, and farinaceous albumen; but in *Portulaca* the placentation is central and free, and in *Tetragonia*, which has a pluricelled ovary, the ovules are inserted at the top of the central angle of the cells.

Mesembryanthemeæ inhabit South Africa. A small number of species are met with in the Mediterranean region, America and Australia. The fruits of some (*M. edule*) contain sugar, and are edible. The leaves of *M. geniculiflorum* are used as a vegetable by the people on the borders of the great African desert, and the bruised seeds yield them flour. *M. crystallinum* (Ice-plant), naturalized in the Mediterranean region, is frequently cultivated on account of its singular appearance, its surface being covered with shining vesicles containing a gummy principle insoluble in water, and resembling in the sunlight a covering of hoar-frost. The inhabitants of the Canaries use the juice of many of these plants as a diuretic, and burn their leaves to obtain soda. The juice of *M. acinaciforme* is successfully employed at the Cape against dysentery. That of *M. tortuosum* is considered as a narcotic or sedative. [The leaves of *M. australe*, called Pig's Face, are eaten pickled in Australia. The seeds of the Shama are a most important article of food with the desert Arabs.]

CXII. *TETRAGONIEÆ*,[1] *Fenzl.*

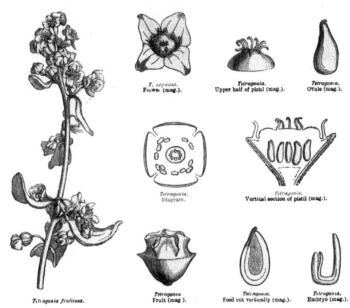

T. expansa.
Flower (mag.).

Tetragonia.
Upper half of pistil (mag.).

Tetragonia.
Ovule (mag.).

Tetragonia.
Diagram.

Tetragonia.
Vertical section of pistil (mag.).

Tetragonia fruticosa.

Tetragonia
Fruit (mag.).

Tetragonia.
Seed cut vertically (mag.).

Tetragonia.
Embryo (mag.).

Herbaceous, annual, or suffrutescent PLANTS, diffuse, succulent, glabrous or velvety. LEAVES alternate or sub-opposite, flat, fleshy, usually entire. FLOWERS ☿, regular, axillary or leaf-opposed, solitary or agglomerated, sometimes in a spike or raceme. CALYX superior, 3–5-lobed, fleshy, coloured within, induplicate-valvate in bud. COROLLA 0. STAMENS epigynous, 1–5–∞, solitary, or aggregated between the calycinal lobes; *filaments* filiform-subulate; *anthers* 2-celled, didymous, cells oblong, separated at the base and top, opening longitudinally. OVARY inferior, 3–5- (sometimes 8–9-) celled, or 1–2-celled by arrest; *styles* as many as the cells, short, stigmatiferous on their inner edge; *ovules* solitary in each cell, pendulous by a short funicle from the top of the inner angle, semi-anatropous, micropyle superior, raphe dorsal. DRUPE or angular NUT crowned by the accrescent calyx, which is often dilated into horns or longitudinal wings; crown of pericarp naked and marked with

[1] See Tribe I. of *Ficoideæ*, p. 461 —ED.

radiating furrows, cells 1-9. SEEDS pendulous, pyri-reniform; *testa* crustaceous, shining, brown, striate longitudinally; *hilum* naked. EMBRYO annular, surrounding a farinaceous albumen.

GENUS.

* Tetragonia.

This group, long united to *Mesembryanthemeæ*, is connected with it by the inferior ovary, peripheric embryo and farinaceous albumen; but it is separated by the apetalous flower, the plurality of the ovarian cells, and the placentation. *Tetragonieæ* are also very near *Portulaceæ*, but are distinguished by their always inferior ovary with 1-ovuled cells, the form and consistence of the fruit, &c. They also approach *Chenopodieæ* in the curved ovule and the nature of the albumen; but the latter are distinguished by their superior always one-celled ovary, their perigynous stamens opposite the sepals, &c. All the species of *Tetragonia* are dispersed over the isles and promontories of the southern hemisphere, beyond the tropic.

Tetragonia expansa is a plant of New-Zealand and the isles of the Southern Ocean, the properties of which were unknown to the natives before Captain Cook used it as food for his sailors and as a cure for scurvy. It was introduced into Europe by Sir J. Banks, and is now cultivated as Summer or New Zealand Spinach.

CXIII. *UMBELLIFERÆ.*

(UMBELLATÆ, *Tournefort.*—SCIADOPHYTUM, *Necker.*—UMBELLIFERÆ, *Jussieu.*— APIACEÆ, *Lindl.*)

COROLLA *polypetalous, epigynous, isostemonous, valvate in bud.* PETALS 5, *inserted on an epigynous disk.* STAMENS 5, *alternate with the petals.* OVARY *inferior, of 2 1-ovuled cells.* OVULES *pendulous, anatropous.* FRUIT *dry.* EMBRYO *albuminous, apical.* RADICLE *superior.*—LEAVES *alternate.*

Herbaceous or rarely woody PLANTS (*Myodocarpus*). STEM usually furrowed or channelled, knotty, fistular, or full of pith. LEAVES alternate, petiole dilated at the base, blade usually cut, rarely entire (*Bupleurum, Gingidium*). FLOWERS ☿, rarely diclinous by arrest, arranged in umbels and umbellules, sometimes in a head (*Eryngium*), sometimes in whorls (*Hydrocotyle*); *umbels* and *umbellules* each furnished with an involucre of bracts, or naked. PETALS 5, valvate or sub-imbricate in bud, inserted outside an epigynous disk, free, caducous, the point generally inflexed, sometimes 2-fid or -partite, the outer often largest. STAMENS 5, alternate with and inserted like the petals; *filaments* inflexed in bud; *anthers* 2-celled, sub-didymous, introrse. CARPELS 2, coherent into a 2-celled ovary, cells antero-posterior; *styles* 2, thickened at the base into [one or two] *stylopodes*, which crown the ovary; [*stigmas* minute, capitellate]; *ovules* originally geminate in each cell, afterwards usually reduced to one, pendulous, anatropous. FRUIT 2-celled, dividing into 2 mericarps, which [often] remain suspended at the top of a single or double filiform prolongation of the axis (*carpophore*). Surface of the fruit with 10 more or less prominent ridges (*juga*) named *primary*. The median dorsal ridge of each carpel is called the *carinal* or *dorsal*; the two to the right and left of

H H

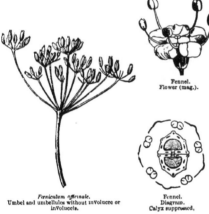

Fœniculum officinale.
Umbel and umbellules without involucre or involucels.

Fennel.
Flower (mag.).

Fennel. Flower cut vertically (mag.).

Fennel.
Diagram.
Calyx suppressed.

Fennel.
Fruit (mag.).
2-partite carpophore.

Fennel.
Transverse section of carpel (mag.).

Fennel.
Transverse section of fruit (mag.).
Carpels with five primary sides, furrows 1-vittate; commissural face 2-vittate.

Æthusa.
Vertical section of a semi-globose carpel, with thick wall, showing the minute embryo at the top of the albumen (mag.).

Æthusa.
Transverse section of fruit (mag.). Carpels with five raised and thickened sides; the lateral largest; furrows 1-vittate; commissural face 2-vittate.

Carrot. (*Daucus Carota.*)
Umbel and umbellules with involucre and involucels.

Æthusa Cynapium.
Naked umbel; umbellules with involucels.

Æthusa.
Flower (mag.).

Æthusa.
Flower cut vertically (mag.).

Æthusa.
Fruit.
Bifid carpophore (mag.).

Æthusa.
Umbellule with 3-phyllous involucel.

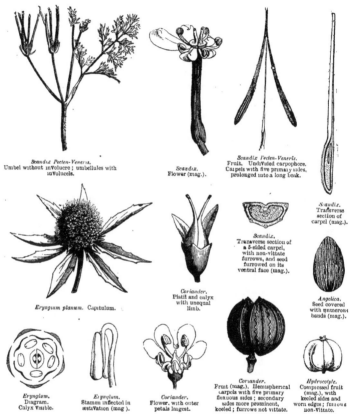

Scandix Pecten-Veneris.
Umbel without involucre; umbellules with involucels.

Scandix.
Flower (mag.).

Scandix Pecten-Veneris.
Fruit. Undivided carpophore.
Carpels with five primary sides,
prolonged into a long beak.

Scandix.
Transverse
section of
carpel (mag.).

Scandix.
Transverse section of
a 5-sided carpel,
with non-vittate
furrows, and seed
furrowed on its
ventral face (mag.).

Coriander.
Pistil and calyx
with unequal
limb.

Eryngium planum. Capitulum.

Angelica.
Seed covered
with numerous
bands (mag.).

Eryngium.
Diagram.
Calyx visible.

Eryngium.
Stamen inflected in
æstivation (mag).

Coriander.
Flower, with outer
petals largest.

Coriander.
Fruit (mag.). Hemispherical
carpels with five primary
flexuous sides; secondary
sides more prominent,
keeled; furrows not vittate.

Hydrocotyle.
Compressed fruit
(mag.), with
keeled sides and
worn edges; furrows
non-vittate.

it are called *intermediate* ridges; and the two on each side of the carpel are called *lateral* ridges; the intervals which separate the primary ridges are called *furrows*, and are sometimes occupied by other or *secondary* ridges. Longitudinal resiniferous canals, called *vittæ*, are developed in the thickness of the pericarp, and placed in the furrows, or on their commissural surface, or on the seed itself; they may be sometimes absent. SEED pendulous, free or adhering to the pericarp. EMBRYO straight, minute, at the top of a horny albumen; *radicle* superior.

The following is the Candollean division of the order :—

TRIBE I. UMBELLIFERÆ ORTHOSPERMÆ.—Seed flat or convex on its commissural face.

TRIBE II. UMBELLIFERÆ CAMPTLOSPERMÆ.—Seed channelled, furrowed, or concave on its commissural face, either from the incurved margins, or from the inflexion of the two ends.

[The following conspectus of the tribes of *Umbelliferœ* is that elaborated for the ' Genera Plantarum,' and is a sketch only ; the classification of the genera of this order being a most difficult and complicated task, and there being many exceptions to the characters given :—

Series I. HETEROSCIADIEÆ.—Umbels simple, or regularly (rarely irregularly) compound. Vittæ 0.

TRIBE I. HYDROCOTYLEÆ.—Fruit laterally compressed or constricted at the narrow commissure ; carpels dorsally acute or obtuse. *Hydrocotyle, Trachymena, Xanthosia, Azorella,* &c.

TRIBE II. MULINEÆ.—Fruit exceedingly narrow at the commissure, 4-angled, or of 2 discoid carpels placed face to face, which are dorsally flat or concave. *Bowlesia, Asteriscum, Mulinum, Hermas,* &c.

TRIBE III. SANICULEÆ.—Fruit sub-terete, commissure broad, dorsally compressed. *Eryngium, Arctopus, *Astrantia, Sanicula, Actinotus,* &c.

Series II. HAPLOZYGIEÆ.—Umbels compound. Primary ridges only of the fruit conspicuous ; vittæ rarely absent in the furrows.

TRIBE IV. ECHINOPHOREÆ.—Flowers ♀ solitary in the umbels, sessile ; fruit involucrate by the hardened pedicels of the ♂ flowers. One carpel perfect, sub-terete, the other arrested, slender or rudimentary. *Echinophora, Pycnocycla.*

TRIBE V. AMMINEÆ.—Fruit laterally compressed, or constricted on both sides towards the commissure, or grooved.

Sub-tribe 1. *Smyrnieœ.*— Fruit broadly ovate or didymous. Seed hollowed or furrowed in front. *Physospermum, Conium, Trachydium, *Arracacha, Smyrnium,* &c.

Sub-tribe 2. *Euamminecœ.*—Seed convex or flat in front. *Bupleurum, Lichtensteinia, Trinia, *Apium, Cicuta, Ammi, Carum, *Sium, Sison, Ægopodium, *Pimpinella.*

Sub-tribe 3. *Scandicineœ.*—Fruit ovate, oblong, or linear. Seed grooved in front. *Conopodium, *Myrrhis, Oreomyrrhis, Chœrophyllum, Scandix, *Anthriscus,* &c.

TRIBE VI. SESELINEÆ.—Fruit sub-terete or dorsally compressed, commissure broad ; lateral ridges distinct, thickened or slender, but not dilated.

Sub-tribe 1. *Euseselineœ.*—Fruit sub-terete or rarely dorsally sub-compressed ; primary ridges sub-equal, not winged. *Seseli, *Fœniculum.*

Sub-tribe 2. *Thecocarpeœ.*—Fruit hard, terete, equally 5-angled or 5-winged ; carpels connate ; vittæ obscure or scattered. *Thecocarpus,* &c.

Sub-tribe 3. *Cachrydeœ.*—Fruit sub-terete or dorsally compressed ; ridges obtuse, acute or winged. Vittæ indefinite, adhering to the seed, and separating from the corky exocarp. *Cachrys, Prangos, Crithmum,* &c.

Sub-tribe 4. *Œnantheœ.*—Fruit sub-terete or dorsally compressed, ridges wingless, the lateral coherent as a thick, almost corky, margin to the fruit. Vittæ solitary in the furrows. *Œnanthe, Æthusa, Siler, &c.*

Sub-tribe 5. *Schultzieæ.*—Fruit more or less dorsally compressed, ridges wingless, the lateral hardly thickened. Vittæ various or 0. *Schultzia, Silaus,* &c.

Sub-tribe 6. *Selineæ.*—Fruit dorsally compressed or sub-terete, primary ridges or the carinal only produced into thickened wings; wings equal, or the lateral broader. *Meum, Ligusticum, Aciphylla, Selinum, Cymopterus, Anesorhiza, Pleurospermum,* &c.

Sub-tribe 7. *Angeliceæ.* — Fruit dorsally compressed, dorsal and secondary ridges wingless or obscurely winged, lateral expanded into broad membranous or corky wings. *Levisticum, Angelica,* **Archangelica,* &c.

TRIBE VII. PEUCEDANEÆ.—Fruit dorsally much compressed, lateral ridges dilated into broad tumid or wing-like margins, those of the opposite carpels closely coherent, and appearing as one till dehiscence. **Ferula, Dorema, Peucedanum, Heracleum, Opopanax, Malabaila, Tordylium,* &c.

Series III. DIPLOZYGIEÆ.—Umbels compound. Furrows of the fruit thickened over the vittæ, or furnished with secondary ridges.

TRIBE VIII. CAUCALINEÆ.—Fruit sub-terete, ridges obtuse or prickly, or dilated into lobed or toothed wings. (Annuals or biennials.) **Coriandrum,* **Cuminum,* **Daucus, Caucalis,* &c.

TRIBE IX. LASERPITIEÆ.—Fruit sub-terete or dorsally compressed; secondary ridges all, or the lateral only, much raised or winged. (Perennials, rarely biennials.) *Laserpitium, Thapsia, Monizia, Elæoselinum,* &c.—ED.]

Umbelliferæ are allied to *Araliaceæ* by the inflorescence, alternate leaves, polypetalous, epigynous, and isostemonous corolla, valvate in bud, the inverted and anatropous ovule, and the minute embryo at the top of a copious albumen. *Araliaceæ* differ only in their fruit, which is usually fleshy [and often polycarpellary; and conspicuously in habit]. *Umbelliferæ* also approach *Corneæ* (which see).

Umbelliferæ principally belong to the northern hemisphere, where they inhabit temperate and cool countries, especially the Mediterranean region and Central Asia. But few are met with in the torrid zone, where they only grow on high mountains and near the sea, where the heat is moderate.

Umbelliferæ contain a great many species, some alimentary, others medicinal or poisonous. These very different properties are due to principles which exist in various proportions either in the leaves, root or fruit; the roots principally contain resins or gum-resins; the fruits possess a volatile oil in the vittæ of their pericarp or seed; the leaves of some species are aromatic and spicy, of others narcotic and acrid. Such *Umbelliferæ* are good for food as have a sufficient quantity of sugar and mucilage united to the hydro-carbon principles; when the volatile oil predominates, as in the fruit of many, they become a stimulating medicine, and an agreeable condiment. We will briefly mention the indigenous species most remarkable for their properties, of this numerous family:—

Cicutaria virosa (Water Hemlock). Root and stem with a yellowish very poisonous juice. Rarely used in medicine, as the Spotted Hemlock.

Apium graveolens. Root aromatic, bitter, acrid, aperient, as is the fruit. Celery is a cultivated variety, of which the root and blanched petioles are used for food, and possess exciting qualities.

Petroselinum sativum (Parsley). Herb and root used as a sauce. The expressed juice is recom-mended as an emollient and diuretic.

Ægopodium Podagraria (Gout-weed). A stimulant, diuretic and vulnerary.

Carum Carui (Caraway). A stimulating stomachic, employed in the North to flavour bread and cheese.

Bunium Bulbo-castanum (Earth-nut). A tuberous globose starchy edible rootstock.

Pimpinella Anisum (Anise or Aniseed). The fruit contains an aromatic volatile and a fixed oil; it is of a piquant and sweetish taste, and is much employed by confectioners and dealers in liqueurs; recommended as a carminative, diuretic, diaphoretic, and even expectorant.

Sium Sisarum and *S. Ninsi* (Water Parsnip, Skirrets). Natives of China and Japan, rarely cultivated in Europe. They have a sweet root with an agreeable aroma, considered to be an excitant.

Œnanthe crocata (Meadow Saffron). A plant growing by river-sides. Root composed of oblong fascicled tubercles, of a mild taste, containing a milky juice turning yellow when exposed to the air, and eminently poisonous.

Æthusa Cynapium (Lesser Hemlock, Fool's Parsley). A very poisonous plant with a nearly glaucous stem striped with reddish lines, with finely-cut dark green leaves with a disagreeable and suspicious scent when bruised. It grows in all cultivated places, where it is often mistaken for Parsley, which differs from it, besides the characters of the fruit, 1st, in its bright clear green foliage with rather large divisions, the teeth of which are terminated by a little white spot, and which have a fresh aromatic smell; 2nd, in the stem, which is neither glaucous nor marked below with reddish lines.

Phellandrium aquaticum. A poisonous plant; the aromatic fruit is employed in medicine as an antiphthisic and antidysenteric.

Fœniculum vulgare (Fennel). Fruit aromatic, stimulant, stomachic. Root and leaves aromatic, used in medicine, the one as nutritive, the other as stimulating.

Crithmum maritimum (Samphire). Juice a vermifuge; leaves aromatic, used as a condiment [and for pickling].

Levisticum officinale (Lovage, Mountain Hemlock). Roots and fruits with an agreeable smell, slightly stimulant and diuretic.

Angelica Archangelica (Angelica). Root a tonic. Fruit a stimulant and stomachic. Leaves vulnerary. Young stems preserved [in sugar] and eaten.

Imperatoria Ostruthium (Master-wort). Root bitter, aromatic and stimulating.

Peucedanum officinale (Sulphur-wort). Root containing a yellow fœtid juice, formerly employed against hysterics; an aperient and bechic.

Anethum graveolens (Bastard Fennel). Fruit exciting, tonic, carminative, employed in dyspepsia.

Pastinaca oleracea (Parsnip). An alimentary and stimulating root.

Heracleum Spondylium (Cow-Parsnip). Root acrid and bitter. Stem sugary, with a fermentable juice, which in the north yields a very intoxicating liquor.

Cuminum Cyminum (Cumin). An Egyptian and Asiatic plant. Fruit aromatic, of a bitter and hot taste, used as a stimulating medicine.

Thapsia villosa (Deadly Carrot). A Mediterranean plant. Root purgative.

Daucus Carota (Common Carrot). A sugary edible root; its juice is administered as an analeptic. Flowers very aromatic; infused in alcohol they produce the liqueur called Oil of Venus.

Myrrhis odorata (Sweet-Cicely). An aromatic plant, used for flavouring.

Conium maculatum (Hemlock). A poisonous plant, employed in cases of enlargement of the glands and viscera.

Anthriscus Cerefolium (Chervil). Cultivated in kitchen gardens, of an agreeable scent and perfumed taste, without acridity or bitterness.

Smyrnium Olusatrum (Alexanders) Formerly esteemed as a vegetable; leaves very aromatic; root diuretic.

Coriandrum sativum (Coriander). Fruit fœtid, with the odour of bugs, becoming aromatic when dry; used as a stimulating and stomachic medicine.

Hydrocotyle asiatica. Prescribed in India against leprosy.

Arracacha esculenta is an Umbellifer cultivated on the high table-lands of the Andes; its tubercled roots furnish an agreeable and digestible food.

The gum-resins of some exotic Umbellifers are used in medicine; the most important is the Asafœtida [Devil's Dung, *Narthex Asafatida*], which is procured from a Persian [West Tibetan] plant belonging to a genus near *Ferula*. This substance diffuses a very fœtid smell, and its taste is acrid and bitter. The Persians praise it as a delicious condiment; it is recommended by European doctors as the most powerful of anti-hysterical medicines, and is also administered in the treatment of asthma. The Sagapenum or Seraphic Gum is a strong-smelling substance, of an acrid and bitter taste, composed of a gum, a resin, and a volatile oil; it comes from Persia like the Asafœtida, and its properties are analogous, though less powerful; it probably belongs, like the latter, to a genus near *Ferula.* Galbanum has been employed for centuries as a stimulant of the nervous and vascular systems; it comes from Syria, but its origin is unknown; as is the case with the Laser [or Thapsia of the ancients], represented on some Phœnician medals or coins, and of which the juice was exported from Cyrenaica to Greece. Gum Ammoniac is procured from *Dorema Ammoniacum,* a native of Persia and Armenia. This resin is at first of a sweetish taste, then acrid and bitter; its qualities are the same as those of the Asafœtida, but it is less powerful in hysterical cases; it is also employed to stimulate the functions of the abdominal viscera and respiratory organs. [The Sumbal, a very fœtid musky drug, used as an antispasmodic, is the very large root of the *Euryangium Sumbal,* a native of Central Asia.—ED.]

CXIV. *ARALIACEÆ.*

(ARALIÆ, *Jussieu.*—ARALIACEÆ et HEDERACEÆ, *Bartling.*)

COROLLA *polypetalous, epigynous, usually isostemonous.* PETALS 5–10, *valvate in bud.* STAMENS *inserted alternately with the petals, rarely more.* OVARY *inferior, of 2-many 1-ovuled cells.* OVULES *pendulous, anatropous.* FRUIT *a berry.* EMBRYO *albuminous.* RADICLE *superior.*

STEM woody, rarely herbaceous, perennial, with cylindric, sometimes spiny branches, often climbing or attaching itself to other plants by fibrillæ, whence they appear parasitic. LEAVES alternate, very rarely opposite, simple, pinnate or digitate; *petioles* enlarged and thickened at the base; *stipules* 0. FLOWERS ⚥, or imper-

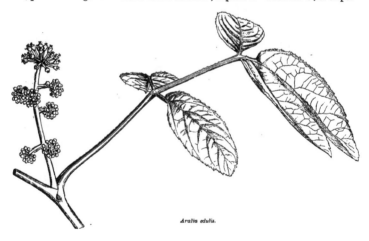

Aralia edulis.

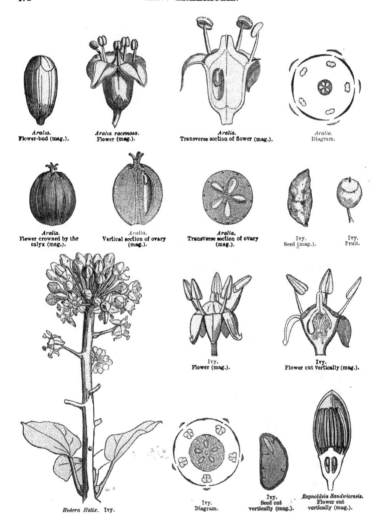

Aralia.
Flower-bud (mag.).

Aralia racemosa.
Flower (mag.).

Aralia.
Transverse section of flower (mag.).

Aralia.
Diagram.

Aralia.
Flower crowned by the calyx (mag.).

Aralia.
Vertical section of ovary (mag.).

Aralia.
Transverse section of ovary (mag.).

Ivy.
Seed (mag.).

Ivy.
Fruit.

Ivy.
Flower (mag.).

Ivy.
Flower cut vertically (mag.).

Hedera Helix. Ivy.

Ivy.
Diagram.

Ivy.
Seed cut vertically (mag.).

Reynoldsia Sandwicensis.
Flower cut vertically (mag.).

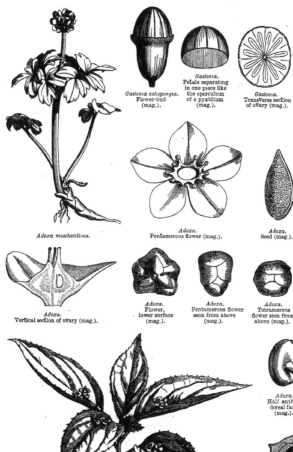

Adoxa moschatellina.

Gastonia cutispongia.
Flower-bud
(mag.).

Gastonia.
Petals separating
in one piece like
the operculum
of a pyxidium
(mag.).

Gastonia.
Transverse section
of ovary (mag.).

Gastonia.
Polygynous pistil
(mag.).

Adoxa.
Pentamerous flower (mag.).

Adoxa.
Seed (mag.).

Adoxa.
Seed cut
vertically (mag.).

Adoxa.
Vertical section of ovary (mag.).

Adoxa.
Flower,
lower surface
(mag.).

Adoxa.
Pentamerous flower
seen from above
(mag.).

Adoxa.
Tetramerous
flower seen from
above (mag.).

Adoxa.
Fruit crowned
by the calyx
(mag.).

Adoxa.
Half anther,
dorsal face
(mag.).

Adoxa.
Half anther,
inner face
(mag.).

Helwingia rusiflora.

Adoxa.
Vertical section of fruit (mag.).

fect through arrest, regular, in umbels, naked or involucrate capitula, racemes or panicles. CALYX superior, usually short, entire or toothed. PETALS 5, 10 or more, inserted on the edge of an epigynous disk, valvate or imbricate in bud, free, or cohering at the top and detaching like a cup. STAMENS inserted alternately with the petals, rarely double or treble in number (or indefinite); *filaments* short, distinct, very rarely 2-partite (*Adoxa*[1]); *anthers* ovoid or linear, introrse, incumbent, of 2 opposite cells opening longitudinally. OVARY inferior, crowned by the disk, of 2–15 1-ovuled cells; *styles* equalling the cells in number, sometimes cohering, often very short; *stigmas* simple; *ovules* suspended from the top of the cells, anatropous; *berry* fleshy or dry, crowned by the calyx. SEEDS inverted; *testa* crustaceous, sometimes margined. EMBRYO minute, straight, at the top of a fleshy copious albumen; *coty-ledons* short; *radicle* superior.

TRIBE I. *ARALIEÆ.*

Corolla quite polypetalous, æstivation valvate.   Stem usually woody.

PRINCIPAL GENERA.

* Aralia.	* Hedera.	* Panax.	* Oreopanax.	* Dendropanax.
* Paratropia.	* Sciadophyllum.	* Didymopanax.	* Gastonia.	

TRIBE II. *ADOXEÆ.*[1]

Corolla sub-polypetalous, æstivation imbricate.   Filaments 2-partite.   Stem herbaceous.

GENUS.

Adoxa.

[The genera of *Araliaceæ* have been re-examined and arranged as follows in the 'Genera Plantarum':—

TRIBE I. ARALIEÆ.—Petals more or less imbricate, attached by a broad base. *Stilbocarpa, Aralia, Pentapanax,* &c.

TRIBE II. MACKINLAYEÆ.—Petals shortly clawed, involute, valvate. *Mackinlaya.*

TRIBE III. PANACEÆ.—Petals valvate. Stamens as many as the petals. Albumen not ruminate. *Horsfieldia, Panax, Acanthopanax, Fatsia, Didymopanax, Helwingia, Meryta, Sciado-phyllum, Heptapleurum, Dendropanax,* &c.

TRIBE IV. HEDEREÆ.—Petals valvate. Stamens as many as the petals. Albumen ruminate. *Arthrophyllum, Cussonia, Oreopanax, Hedera,* &c.

TRIBE V. PLERANDREÆ.—Petals valvate or connate. Stamens very numerous. Styles 0, or cohering in a cone. *Plerandra, Tupidanthus,* &c.—ED.]

Araliaceæ approach *Umbelliferæ, Ampelideæ,* and *Caprifoliaceæ* (see these families). They are closely connected with *Corneæ*; in both the petals are epigynous and isostemonous and valvate, the

<hr>

[1] *Adoxa* is now referred to *Caprifoliaceæ.*—ED.

ovules are solitary in the cells, pendulous and anatropous; the fruit is fleshy and the embryo albuminous, the stem is generally woody, and the flowers are umbelled or capitulate. *Corneæ* only differ in their drupaceous fruit and opposite leaves.

We place near *Araliaceæ* the genus *Helwingia*, which is connected with them and with *Hamamelideæ* by the valvate æstivation, inferior ovary, pendulous and anatropous ovules, albuminous embryo, woody stem and alternate leaves.

Araliaceæ inhabit both hemispheres, but not beyond latitude 52°; they abound in America and particularly in the mountains of Mexico and New Grenada, and are rare in the parallel regions of Europe and Asia, although the genus *Paratropia* is numerously represented in the latter.

This family contains few species useful to man. The leaves of the Ivy (*Hedera Helix*) are aromatic, and their chlorophyll, dissolved in tallow or oil, serves as a dressing for ulcers; a decoction of them is also employed against vermin on the body. The root of *Panax Ginseng* is celebrated in Persia, China and India as a tonic and aphrodisiac. The Aralias of North America are esteemed there as sudorifics and depuratives, the rhizomes of *Aralia nudicaule*, the bark of the spiny Aralia and the mucilaginous aromatic root of the racemose Aralia are thus used. In Japan the young shoots of *Helwingia* are eaten. [The beautiful substance called rice paper is the pith of *Fatsia papyrifera.*]

CXV. *CORNEÆ.*

(CAPRIFOLIACEARUM *tribus, Kunth.*—CORNEÆ, *D.C.*—CORNACEÆ, *Lindl.*)

COROLLA *polypetalous, epigynous, isostemonous*; PETALS 4–5, *valvate.* STAMENS 4–5, *alternate with the petals.* OVARY *inferior, of 2–3 1-ovuled cells.* OVULES *pendulous, anatropous.* FRUIT *a drupe.* EMBRYO *albuminous, axile.* RADICLE *superior.*

STEM woody, sometimes subterranean and emitting herbaceous branches. LEAVES opposite or very rarely alternate (*Decostea*), penninerved, simple, entire or toothed, caducous or persistent, exstipulate. FLOWERS ☿, or diœcious by arrest (*Griselinia*), in a head or umbel with a usually coloured involucre, rarely in a corymb without an involucre. CALYX superior, 4-toothed. PETALS 4–5, inserted on the calyx and alternate with its teeth, valvate in bud, or sub-imbricate in the ♂ flowers (*Griselinia*), deciduous. STAMENS 4–5, alternate with the petals; *filaments* filiform, distinct; *anthers* introrse, dorsifixed, 2-celled, dehiscence longitudinal. OVARY inferior, 2- (sometimes 3-) celled, crowned by a disk, often scarcely visible; *style* simple; *stigma* capitate; *ovules* solitary in each cell, pendulous, anatropous. DRUPES distinct or cohering, stone bony, 2-3-celled, or 1-celled by arrest. SEEDS inverted, integument coriaceous. EMBRYO straight, in the axis of a fleshy albumen, and equalling it in length; *cotyledons* oblong, sub-foliaceous; *radicle* short, superior.

PRINCIPAL GENERA.

* Cornus   * Benthamia.   * Aucuba.   * Griselinia.

[*Corneæ*, as re-classified by Bentham and Hooker fil. for the 'Genera Plantarum,' contains many more exceptional genera than are included in former arrangements of the order. They are thus disposed:—

Flowers hermaphrodite.—*Alangium, Marlea, Curtisia, Corokia, Cornus, Mastixia.*
Flowers unisexual; leaves opposite.—*Aucuba, Garrya.*[1]
Flowers unisexual; leaves alternate.—*Griselinia, Kaliphora, Nyssa, Toricellia.*—ED.]

Corneæ were formerly included in *Caprifoliaceæ*; they are very near *Araliaceæ* (see these families); they also approach *Umbelliferæ* in the epigynous polypetalous and isostemonous corolla and its æstivation, in the pendulous anatropous ovule, albuminous embryo, and umbelled or capitulate inflorescence.

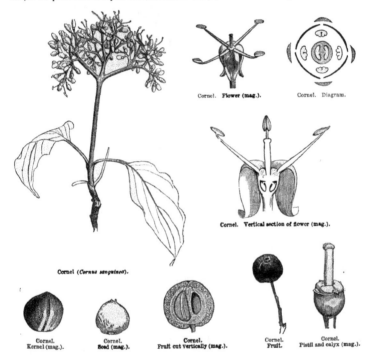

Cornel. Flower (mag.).   Cornel. Diagram.

Cornel. Vertical section of flower (mag.).

Cornel (*Cornus sanguinea*).

Cornel.
Kernel (mag.).   Cornel.
Seed (mag.).   Cornel.
Fruit cut vertically (mag.).   Cornel.
Fruit.   Cornel.
Pistil and calyx (mag.).

Umbelliferæ are separated from *Corneæ* by their two styles, dry fruit, minute embryo, and alternate cut leaves with dilated petiole. The connection of *Corneæ* with *Hamamelideæ* is noticed in the account of the latter family.

Corneæ almost exclusively belong to the northern hemisphere; they inhabit especially the Himalayas [and Japan], and the temperate and cool regions of America; they are very rare in tropical America. *Griselinia* and *Corokia* belong to New Zealand; [*Curtisia* is South African].

<hr />

[1] See order *Garryaceæ*, p. 479.—ED.

The wood of *Corneæ* is extremely hard; [that of *Cornus* being used in the manufacture of gun-powder]. The bark of Dogwood, especially of *Cornus florida*, is bitter and astringent, and yields a principle (*cornine*), which is administered in North America instead of quinine. The drupes of *C. mascula* have an acid-sweet taste, and possess astringent properties. [They are used in making sherbet in the East.] The seed of *C. sanguinea* contains a fixed oil, useful in the manufacture of soap. The *C. (Benthamia) fragifera* is a shrub of the Himalayas and Japan, the fruits of which resemble a strawberry, and have an agreeable taste. The *Aucuba* also comes from [the Himalayas and] Japan, and is extensively cultivated in Europe for its coriaceous variegated and persistent leaves.

CXVI. *GARRYACEÆ, Endlicher.*

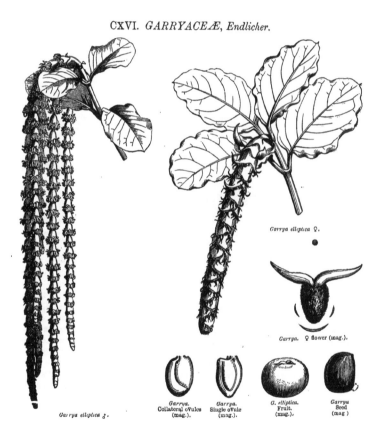

Garrya elliptica ♀.

Garrya. ♀ flower (mag.).

Garrya elliptica ♂.

Garrya. Collateral ovules (mag.).

Garrya. Single ovule (mag.).

G. elliptica. Fruit. (mag.).

Garrya Seed (mag.)

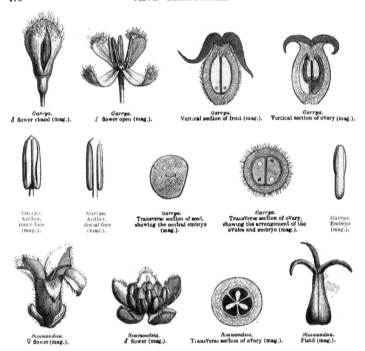

Garrya.
♂ flower closed (mag.).

Garrya.
♂ flower open (mag.).

Garrya.
Vertical section of fruit (mag.).

Garrya.
Vertical section of ovary (mag.).

Garrya.
Anther,
inner face
(mag.).

Garrya.
Anther,
dorsal face
(mag.).

Garrya.
Transverse section of seed,
showing the central embryo
(mag.).

Garrya.
Transverse section of ovary,
showing the arrangement of the
ovules and embryo (mag.).

Garrya.
Embryo
(mag.).

Simmondsia.
♀ flower (mag.).

Simmondsia.
♂ flower (mag.).

Simmondsia.
Transverse section of ovary (mag.).

Simmondsia.
Pistil (mag.).

STEM woody, branches 4-sided. LEAVES opposite, shortly petioled, entire, penninerved, evergreen; *petioles* united at the base; *stipules* 0. FLOWERS diœcious, arranged in little groups (*Simmondsia* [1]), or in long axillary catkins, ternate in the axils of decussate and coherent bracts (*Garrya*). ♂: PERIANTH calycinal, of 4 linear sub-membranous spreading (*Garrya*) or 5 (*Simmondsia*) sepals. STAMENS 4, alternate with the sepals (*Garrya*), or 10–12 (*Simmondsia*); *filaments* free, equal; *anthers* introrse, basifixed, of 2 opposite cells opening longitudinally. ♀: PERIANTH superior, of 2 setiform lobes, or without apparent lobes (*Garrya*), or replaced by involucrate bracts (*Simmondsia*). OVARY inferior, 1–3-celled; *styles* 2-3, alternate with the perianth-lobes, covered with stigmatic papillæ; *ovules* solitary or geminate, suspended by funicles from the top of the cell, anatropous. FRUIT a berry (*Garrya*) or capsule (*Simmondsia*), crowned by the persistent styles. SEEDS 2,

[1] *Simmondsia* is now placed in *Buxeæ.*—ED.

pendulous, oblong; *testa* thin, transversely rugose; *raphe* prominent and lateral; *albumen* copious, fleshy. EMBRYO minute, straight, axile; *cotyledons* hypogeous in germination; *radicle* superior.

PRINCIPAL GENERA.

*Garrya [see *Corneæ*, p. 476.]   Simmondsia [see *Buxeæ*].

The affinities of *Garrya* are obscure. A. de Jussieu makes the same observations as on *Gunneraceæ*, and places it between *Gunnera* and *Corneæ*. Like the latter, *Garrya* has epigynous stamens, suspended and anatropous ovules, a fleshy fruit, a minute embryo at the top of an abundant albumen, a woody stem, and opposite leaves. As in *Gunnera*, the styles are stigmatiferous throughout their length. *Garrya* approaches *Hamamelideæ* in the inferior ovary, the pendulous anatropous ovule, the two distinct styles, the albuminous and axile embryo, and the woody stem; but *Hamamelideæ* have frequently petals, and are polyandrous, the ovary is 2-celled, the fruit a septicidal capsule, the embryo large, and the leaves are alternate.

Garrya elliptica grows in Mexico and California [and *G. Fadyeni* in Jamaica and Cuba]. There is nothing to be noticed respecting their useful properties.

CXVII. *CAPRIFOLIACEÆ.*

(CAPRIFOLIA, *A. L. de Jussieu.*—CAPRIFOLIACEÆ, *De Candolle.*—CAPRIFOLIACEÆ and SAMBUCEÆ, *Kunth.*—CAPRIFOLIACEÆ and VIBURNEÆ, *Bartling.*—LONICEREÆ, *Endlicher.*)

COROLLA *monopetalous, epigynous, isostemonous, imbricate in bud.* STAMENS 5–4, *inserted on the corolla.* OVARY *with* 2–5 *one- or many-ovuled cells.* OVULES *pendulous, anatropous.* FRUIT *a berry.* EMBRYO *albuminous.*—LEAVES *opposite, exstipulate.*

Plants with a woody or partially woody stem, very rarely herbaceous perennials. LEAVES opposite; *stipules* absent, sometimes represented by filiform or glandular appendages, situated at the base of the petiole. FLOWERS perfect, regular or irregular; *inflorescence* various, generally definite. CALYX superior, 5-fid or -toothed. COROLLA superior, monopetalous, tubular or infundibuliform or rotate; *limb* 5-fid, regular or ringent, imbricate in bud. STAMENS inserted on the corolla-tube, alternate with its lobes; *filaments* filiform, equal or didynamous; *anthers* introrse, 2-celled, dehiscence longitudinal. OVARY inferior, 2–5-celled; *style* terminal, sometimes filiform with a capitate undivided or bilobed stigma, sometimes nearly or quite obsolete, with 3–5 stigmas; *ovules* sometimes solitary and pendulous near the top of the cell, sometimes many, 2-seriate at the central angle, anatropous. BERRY several-celled, rarely 1-celled by the disappearance of the septa. SEEDS inverted; *testa* bony or crustaceous, raphe dorsal or ventral. EMBRYO straight, occupying the axis of a fleshy albumen; *radicle* superior.

SUB-ORDER I. *LONICEREÆ.*

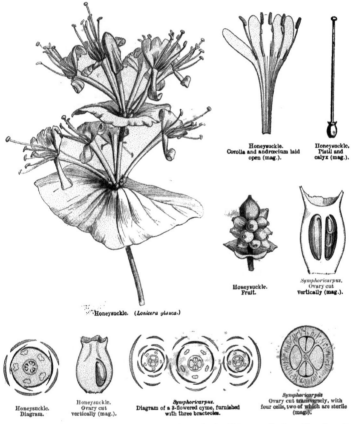

Honeysuckle.
Corolla and andrœcium laid
open (mag.).

Honeysuckle,
Pistil and
calyx (mag.).

Honeysuckle.
Fruit.

Symphoricarpus.
Ovary cut
vertically (mag.).

Honeysuckle. (*Lonicera glauca.*)

Honeysuckle.
Diagram.

Honeysuckle.
Ovary cut
vertically (mag.).

Symphoricarpus.
Diagram of a 3-flowered cyme, furnished
with three bracteoles.

Symphoricarpus.
Ovary cut transversely, with
four cells, two of which are sterile
(mag.).

Corolla tubular, limb regular or irregular.   Style filiform.   Seeds with a dorsal
raphe.

PRINCIPAL GENERA.

Triosteum.	* Symphoricarpus.	* Abelia.	Linnæa.	Lonicera.
* Leycesteria.	Diervilla.	Alseuosmia.	* Weigelia.	

Sub-order II. *SAMBUCEÆ.*

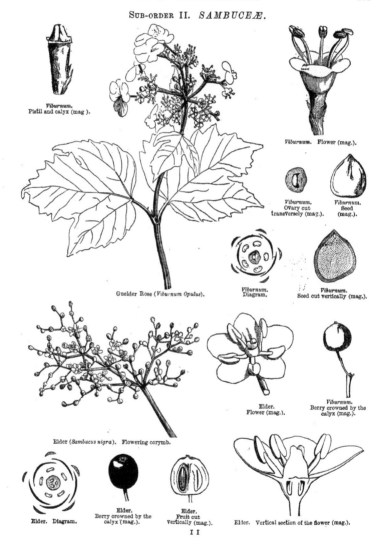

Viburnum.
Pistil and calyx (mag).

Viburnum. Flower (mag.).

Viburnum.
Ovary cut
transversely (mag.).

Viburnum.
Seed
(mag.).

Guelder Rose (*Viburnum Opulus*).

Viburnum.
Diagram.

Viburnum.
Seed cut vertically (mag.).

Elder.
Flower (mag.).

Viburnum.
Berry crowned by the
calyx (mag.).

Elder (*Sambucus nigra*). Flowering corymb.

Elder. Diagram.

Elder.
Berry crowned by the
calyx (mag.).

Elder.
Fruit cut
vertically (mag.).

Elder. Vertical section of the flower (mag.).

Elder. Pistil and calyx (mag.).

Elder (*Sambucus nigra*). Ripe fruit.

Elder. Seed entire and cut vertically (mag.).

Corolla regular, rotate.   Stigmas 3, sessile.   Seeds with a ventral raphe.

PRINCIPAL GENERA.

* Viburnum.   * Sambucus [Adoxa, see p. 474].

We shall hereafter indicate the affinities of *Caprifoliaceæ* with *Valerianeæ* and *Dipsaceæ*. The affinity with *Rubiaceæ* is much more obvious; it is founded on the epigynous isostemonous corolla, the several-celled ovary, the axile embryo in a thick fleshy albumen, the opposite leaves and knotty stem. Almost the only difference is in the imbricate corolla and the absence of stipules.   The sub-order of *Sambuccæ* is closely allied to *Corneæ*, which only differ in the many petals, valvate in bud.   The same affinity may be noticed with *Araliaceæ* and *Umbelliferæ*; but these, besides the polypetalous and valvate corolla, differ from *Caprifoliaceæ* in the alternate leaves and the umbelled or capitate inflorescence. An analogy has also been pointed out between the *Hydrangeæ* (of *Saxifrageæ*) and *Viburneæ*.

Caprifoliaceæ inhabit the temperate regions of the northern hemisphere, especially central Asia, the north of India, and of America.   A small number [many Viburnums] inhabit the intertropical zone, preferring the mountains, where the temperature is colder.   The Elder, a cosmopolitan genus, is represented by a very few species in the southern hemisphere.   [*Alseuosmia* is a New Zealand genus, remarkable for the intensely sweet odour of its flowers]. The flowers of most *Caprifoliaceæ* exhale a sweet odour, especially after sunset.   They contain an acrid, bitter and astringent principle, which has caused some to be placed among medicinal plants. The berries of the Honeysuckle (*Lonicera Caprifolium*) are diuretic; those of *L. Xylosteum* are laxative.   The stems of *Diervilla canadensis* are employed as a depurative in North America. The roots of the common *Symphoricarpos parviflora*, a Carolina shrub, are used by the Americans as a febrifuge. All these species are cultivated in European gardens. The common Elder (*Sambucus nigra*) produces numerous berries, which are cooked and eaten in Germany [and also extensively used in the manufacture of wine]. Pharmacists prepare from these berries, as also from those of *S. Ebulus*, an extract, or purgative *rob*. The dried flowers of the common Elder are an excellent sudorific, employed against snake-bites; they are also used to give to certain wines a Muscat flavour.

The *Linnæa borealis*, an elegant evergreen herb, abounds in the forests of Sweden, the country of Linnæus, to whom it has been dedicated. Swedish doctors recommend its stem and leaves as diuretics and sudorifics.

CXVIII.—*RUBIACEÆ.*

(RUBIACEÆ, *A. L. de Jussieu.*—LYGODYSODEACEÆ et RUBIACEÆ, *Bartling.*—CINCHO-NACEÆ, LYGODYSODEACEÆ et STELLATÆ, *Lindl.*)

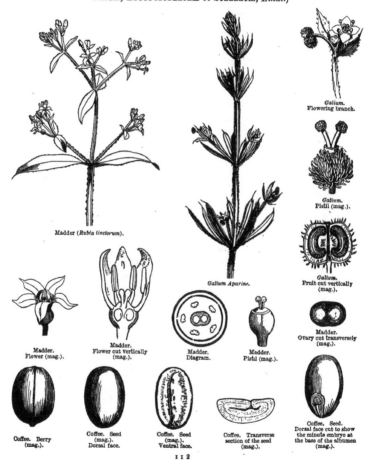

Galium.
Flowering branch.

Galium.
Pistil (mag.).

Galium.
Fruit cut vertically
(mag.).

Madder (*Rubia tinctorum*).

Galium Aparine.

Madder.
Ovary cut transversely
(mag.).

Madder.
Flower (mag.).

Madder.
Flower cut vertically
(mag.).

Madder.
Diagram.

Madder.
Pistil (mag.).

Coffee. Berry
(mag.).

Coffee. Seed
(mag.).
Dorsal face.

Coffee. Seed
(mag.).
Ventral face.

Coffee. Transverse
section of the seed
(mag.).

Coffee. Seed.
Dorsal face cut to show
the minute embryo at
the base of the albumen
(mag.).

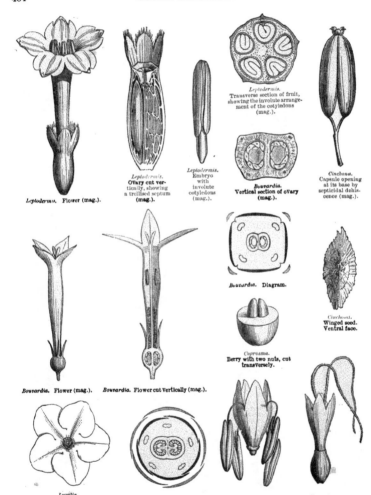

Leptodermis. Flower (mag.).

Leptodermis.
Ovary cut vertically, showing
a trellised septum
(mag.).

Leptodermis.
Embryo
with
involute
cotyledons
(mag.).

Leptodermis.
Transverse section of fruit,
showing the involute arrangement of the cotyledons
(mag.).

Bouvardia.
Vertical section of ovary
(mag.).

Cinchona.
Capsule opening
at its base by
septicidal dehis-
cence (mag.).

Bouvardia. Flower (mag.).   *Bouvardia.* Flower cut vertically (mag.).

Bouvardia. Diagram.

Coprosma.
Berry with two nuts, cut
transversely.

Cinchona.
Winged seed.
Ventral face.

Lucuĺia.
Flower seen from above, with convolute
æstivation (mag.).

Lucuĺia.
Diagram.

Coprosma.
♂ flower (mag.).

Coprosma.
♀ flower (mag.).

COROLLA *monopetalous, epigynous, isostemonous, æstivation valvate or* [*imbricate or*] *contorted.* STAMENS 4–6, *inserted on the corolla-tube.* OVARY *inferior, bi–pluri-locular;* OVULES *anatropous, or semi-campylotropous.* EMBRYO *almost always albuminous.*—LEAVES *opposite, stipulate.*

TREES, SHRUBS or HERBS, usually with tetragonous knotty jointed stems. LEAVES opposite [or whorled], simple, usually entire, stipulate; *stipules* various, sometimes free, sometimes united to the leaf or the neighbouring stipule, sometimes leaf-like, and appearing to form a whorl with the leaves, but distinguished by the absence of buds. FLOWERS usually ☿, very rarely unisexual, sometimes sub-irregular, generally cymose, panicled or capitate. CALYX superior or semi-superior, tubular, or deeply cut or 2–6-fid or -toothed or 0. COROLLA superior, monopetalous, infundibuliform or hypocrateriform or rotate [rarely 2-labiate, *Henriquezia, Dichilanthe,* &c.]; *limb* of 4–6 segments, usually equal, æstivation valvate, rarely contorted or imbricate. STAMENS 4–6 [2 in *Carlemannia* and *Sylvianthus*], inserted on the tube, very rarely coherent; *anthers* introrse, bilocular, dehiscence longitudinal, distinct, very rarely cohering in a tube. OVARY inferior, bi–pluri-locular, crowned by a more or less developed fleshy disk; *style* simple bifid or multifid, stigmatiferous at the top or inner surfaces or sides of the arms; *ovules* solitary or 2 or more in each cell, erect or pendulous, or ventrally attached to the central angle or septum of the cell, or to a prominent placenta, anatropous or semi-campylotropous. FRUIT a capsule, berry or drupe. SEEDS in various positions; *albumen* fleshy or cartilaginous, or almost horny, rarely scanty or 0, sometimes involute. EMBRYO straight or curved, in the base or axis of the albumen; *cotyledons* flat, rarely involute; *radicle* usually inferior.

[Sketch of the arrangement of *Rubiaceæ* into tribes, as recently constructed for the 'Genera Plantarum':—

SERIES A. Ovules indefinite.

Sub-series I. Fruit dry, capsular or indehiscent.

* *Flowers collected in a globose head.*[1]

TRIBE I. NAUCLEÆ.—Corolla narrow, infundibuliform, lobes never contorted. Stigma much exserted, entire. *Sarcocephalus, Cephalanthus, Adina, Nauclea, Uncaria,* &c.

* *Flowers not collected into spherical heads.*

TRIBE II. CINCHONEÆ.—Corolla-lobes valvate, imbricate or contorted. Capsule 2-celled, seeds winged. *Cinchona, Cascarilla, Ladenbergia, Bouvardia, Manettia, Hindsia, Hillia, Exostemma, Luculia,* &c.

TRIBE III. HENRIQUEZIEÆ.—Corolla 2-labiate. Ovules geminate. Seeds very broadly winged, exalbuminous. *Henriquezia, Platycarpum.*

TRIBE IV. CONDAMINIEÆ.—Corolla-lobes valvate. Capsule 2-celled. Seeds numerous, minute, not winged. Trees and shrubs. *Condaminea, Portlandia, Bikkia, Pinkneya,* &c.

TRIBE V. RONDELETIEÆ.—Corolla-lobes imbricate or contorted. Seeds very numerous, albuminous, not winged. *Rondeletia, Wendlandia, Augusta, Deppea, Sipanea,* &c.

TRIBE VI. HEDYOTIDEÆ.—Corolla-lobes valvate. Ovary 2–4-celled. Seeds numerous,

[1] See also under tribes *Gardenieæ* and *Morindeæ.*

angular, not winged. Herbs; rarely small shrubs. *Dentella, Argostemma, Pentas, Hedyotis, Oldenlandia, Houstonia, Kadua, Ophiorhiza, Carlemannia, &c.*

Sub-series II. Fruit fleshy or coriaceous, indehiscent.

TRIBE VII. MUSSÆNDEÆ.—Corolla-lobes valvate. Seeds very numerous, minute, usually angled. *Mussænda, Isertia, Gonzalea, Adenosacme, Urophyllum, Sabicea, Coccocypselum, &c.*

TRIBE VIII. HAMELIEÆ.—Corolla-lobes imbricate or contorted. Seeds very numerous, minute, angular. *Hamelia, Hoffmannia, Bertiera, Gouldia, &c.*

TRIBE IX. CATESBÆEÆ.—Corolla-lobes valvate. Seeds many, rather large, compressed. *Catesbæa, Pentagonia, &c.*

TRIBE X. GARDENIEÆ.—Corolla-lobes contorted. Seeds few or many, large and compressed, or smaller and angled. *Aliberta, Amajoua, Duvoia, Posoqueria, Tocoyena, Burchellia, Webera, Randia, Gardenia, Genipa, Pouchetia, Petunga, Diplospora, &c.*

SERIES B. Ovules geminate in each cell. (See also Tribe III. *Henriquezia*).

TRIBE XI. CRUCKSHANKIEÆ.—Corolla-lobes valvate. Ovary 2-celled, ovules numerous. Capsule 2-valved. *Cruckshankia, Oreopolus.*

TRIBE XII. RETINIPHYLLEÆ.—Corolla-lobes contorted. Ovary 5–7-celled. Drupe with 5–7 pyrenes. *Retiniphyllum, Kotchubœa.*

SERIES C. Ovules solitary in each cell.

Sub-series I. Radicle superior.

TRIBE XIII. GUETTARDEÆ.—Corolla-lobes imbricate or valvate. Stamens inserted on the corolla-throat. Seeds pendulous from the top of the cell, usually exalbuminous, with a thickened funicle. *Guettarda, Antirrhœa, Machaonia, Timonius, Chomelia, Malanea, Dichilanthe, &c.*

TRIBE XIV. KNOXIEÆ.—Corolla-lobes valvate. Stamens inserted on the throat of the corolla. Seeds compressed, albuminous. *Knoxia, Pentanisia.*

TRIBE XV. CHIOCOCCEÆ.—Corolla-lobes valvate or imbricate. Stamens inserted at the base of the corolla. Seeds albuminous. *Erithalis, Chiococca, Chione, &c.*

TRIBE XVI. ALBERTEÆ.—Corolla-lobes contorted. Stamens inserted on the throat of the corolla. Seeds albuminous. *Cremaspora, Alberta, &c.*

TRIBE XVII. VANGUERIEÆ.—Corolla-lobes valvate. Stamens inserted on the throat of the corolla. Seeds albuminous. *Plectronia, (Canthium), Vangueria, Cuviera, &c.*

Sub-series II. Radicle inferior.

* *Corolla contorted.*

TRIBE XVIII. IXOREÆ.—Ovules attached to the middle or about the middle of the cell, rarely basilar. *Ixora, Pavetta, Coffea, Myonima, Strumpfia, &c.*

* * *Corolla valvate.*

† *Ovules attached to the septum at or below the middle.*

TRIBE XIX. MORINDEÆ.—Flowers often united by the calyx-tube into heads. *Morinda, Damnacanthus, Prismatomeris, &c.*

† † *Ovules basilar, erect, anatropous.*

TRIBE XX. COUSSAREÆ.—Ovary 1-celled, or with an evanescent septum. Fruit 1-seeded. *Coussarea, Faramea, &c.*

TRIBE XXI. PSYCHOTRIEÆ.—Stamens inserted on the throat of the corolla. Stigma entire or style-arms short. Fruit indehiscent. *Psychotria, Palicourea, Rudgea, Declieuxia, Geophila, Cephaelis, Lasianthus, Suteria, Saprosma, Psathyra,* &c.

TRIBE XXII. PÆDERIEÆ.—Stamens inserted on the throat or base of the corolla. Ovary 2–5-celled; style-arms filiform. Fruit capsular or of 2 cocci. *Pæderia, Lygodisodea, Hamiltonia, Leptodermis,* &c.

TRIBE XXIII. ANTHOSPERMEÆ.—Flower usually unisexual. Stamens usually inserted at the base of the corolla. Ovary 1–4-celled. Style entire, or arms filiform. Fruit a berry, or indehiscent. *Putoria, Crocyllis, Mitchella, Serissa, Coprosma, Anthospermum, Phyllis, Opercularia, Pomax,* &c.

† † † *Ovules attached to the septum, amphitropous.*

TRIBE XXIV. SPERMACOCEÆ.—Herbs or small shrubs. Leaves usually opposite and stipules setose. *Triodon, Diodia, Gaillonia, Spermacoce, Emmeorhiza, Mitracarpum, Richardsonia,* &c.

TRIBE XXV. GALIEÆ.—Herbs. Leaves and stipules similar, forming a whorl. *Callipeltis, Vaillantia, Rubia, Galium, Asperula, Crucianella, Sherardia,* &c.—ED.]

We have indicated the affinity of *Rubiaceæ* with *Caprifoliaceæ* and *Dipsaceæ* (see these families). The *Rubiaceæ* with many-ovuled cells are allied to *Loganiaceæ* in all their characters, and are only distinguished by their epigyny. *Gentianeæ, Oleineæ,* and *Apocyneæ* also approach them, although hypogynous, in the opposite leaves, æstivation, isostemonous corolla and the presence of albumen. Some *Gesneraceæ* also approach the section *Coffeaceæ,* as shown by their whorled or opposite leaves, the development of their receptacular cupule, the varied nature of their fruit, and the presence of albumen; but they are widely separated by the didynamous stamens, unilocular ovary and parietal placentation.

Rubiaceæ mostly inhabit intertropical regions; [but *Galieæ* are almost exclusively temperate]. The principal medicinal species of this family are exotic; and of these the most important are Quinine and Ipecacuanha. The latter is the root of a little shrub, a species of *Cephaelis,* inhabiting the virgin forests of Brazil; the bark of this root has an acrid taste and a nauseous smell; it contains an alkaloid (*emetine*), but in practice the root is preferred to the alkaloid. This medicine is invaluable in dysentery, asthma, whooping-cough, and especially puerperal fever. Quinine is yielded by the bark of several species of *Cinchona*; they are evergreen trees or shrubs, inhabiting the valleys of the Andes of Peru, at heights varying from 4,000 to 11,000 feet above the level of the sea. The bark is bitter, and contains two organic alkalis (*quinine* and *cinchonine*), united to a special acid; it contains, besides colouring matter, a fatty matter, starch, gum, &c. The preparation of these vegetable alkalis is the most important service that chemistry has rendered to medicine since the beginning of the nineteenth century, for without exhausting the patient, enormous doses of quinine may be administered in a concentrated form, effecting the most difficult cures. Quinine is the most powerful specific in cases of intermittent fevers (of which marsh miasma is the most common cause); acting, not by neutralizing the miasma as a counter-poison would do, but by strengthening the system, and thus enabling it to resist the incessant attacks of the morbific cause. Besides its virtues as a febrifuge, quinine is a first-class tonic in hastening convalescence, and restoring the digestive functions. Lastly, Cinchona bark is used outwardly as an antiseptic to arrest the progress of gangrene; its antiseptic properties are however not due to its febrifugal principle, but to the astringent principles with which the bark abounds.

The American genus *Chiococca,* like *Cephaelis,* belongs to the uniovulate section, some species of which possess a root reputed valuable against snake-bites; this root, known as *cainça,* is used in Europe as a diuretic and purgative in cases of hydrophobia.

Of all the *Rubiaceæ* of the Old World the Coffee is the most noticeable; forming as it does, with cotton and sugar, the staple of the maritime commerce of Europe. The Coffee is an evergreen shrub, a native of Abyssinia, which was introduced three centuries ago into Arabia, towards the close of the seventeenth century into Java, and finally naturalized in 1720 in the Antilles. The seed of the Coffee yields, besides various oily, albuminous and gummy matters, a bitter principle containing an organic crystallizable alkali named *caffein,* associated with a peculiar acid. A slight roasting develops in this seed that agreeable aroma and taste of which advantage has been taken to prepare a drink which specially stimulates the functions of the brain. To those who do not habitually drink it, coffee may become a useful medicine; it succeeds in the treatment of intermittent fevers; it relieves asthma, and it is said gout also; and it counteracts the effects of wine or of opium. Its most prevalent use as a medicine is in curing headache.

Some indigenous *Rubiaceæ* were formerly used as medicines; thus an infusion of the flowering tops of the Yellow Galium was given to nurses to increase the secretion of milk, and as an antispasmodic. They are now employed in many countries, and especially in England, to give a yellow colour to cheese. The Squinancy (*Asperula cynanchica*) the leaves of which contain a bitter slightly astringent principle, was used in cases of angina. *Asperula odorata*, the perfume of which comes out when dried, was praised as a tonic and vulnerary; it is now only used to give a bouquet to Rhine wines, and gardeners cultivate it as an edging. Madder (*Rubia tinctorum*) grows wild in the Mediterranean region; it is cultivated at Avignon, in Alsace, and in Zealand, on account of the red colouring matter contained in the root, and which is largely used for dyeing fabrics. This dye, in a pure state, is called *alizarine*. It also exists, but in less quantity, in the root of the *Chaya-ver*, a Rubiaceous plant, which is cultivated on the Coromandel coast. [Probably the Bengal Madder (*Munjeet*) is here alluded to; it is cultivated throughout India.]

[The above notice of the useful *Rubiaceæ* must be largely extended; very many species contain bitter febrifuge principles, especially *Exostemma*, *Rondeletia*, and *Condaminea* in South America; as also *Pinkneya* in Carolina, *Hymenodictyon* in India, and the *Ophiorhiza Mungos* (Earth-gall) in the Malayan Islands. Gambir, one of the most important of astringents, is the produce of *Uncaria Gambir*. The American *Richardsonia scabra* and others yield one valuable false Ipecacuanha, and the *Psychotria emetica* another. The fruit of the Indian *Randia dumetorum* is a powerful emetic. Amongst edible fruits the Genipap is that of *Genipa americana*, the native Peach of Africa is the fruit of *Sarcocephalus esculentus*, and the Voavanga of Madagascar that of *Vangueria edulis*.—ED.]

CXIX. *VALERIANEÆ.*

(DIPSACEARUM *sectio, A. L. de Jussieu.*—VALERIANEÆ, D. C.—VALERIANACEÆ, *Lindl.*)

COROLLA *monopetalous, epigynous, æstivation imbricate.* STAMENS 5-4-3-1, *inserted on the corolla-tube.* OVARY *3-celled, two cells without ovules, the third 1-ovuled;* OVULE *pendulous, anatropous.* EMBRYO *exalbuminous.*

Annuals with slender inodorous roots, or perennials with a usually strong-scented rhizome. LEAVES: radical fascicled; cauline opposite, simple; petiole dilated, exstipulate. FLOWERS perfect, or unisexual by arrest, in a dichotomous cyme or close corymb, or solitary in the forks, and bracteate. CALYX superior, sometimes cut into 3-4 accrescent teeth, or reduced to a single tooth; sometimes of bristles which are involute before flowering, when they unfold into a plumose deciduous crown. COROLLA monopetalous, inserted on a disk crowning the top of the ovary, tubular-infundibuliform; *tube* regular, or produced at its base into a knob or hollow spur; *limb* with 5-4-3 equal or sub-labiate lobes, æstivation imbricate. STAMENS inserted above the middle of the corolla-tube, alternate with its divisions, rarely 5, usually 4 by suppression of the posterior stamen, or 3 by suppression of the posterior and a lateral stamen; sometimes the posterior only is developed; *filaments* distinct, exserted; *anthers* introrse, 2-celled, dehiscence longitudinal. OVARY inferior, 3-celled, two cells empty, the third fertile; *style* simple, filiform; *stigma* undivided or 2-3-fid; *ovule* solitary, pendulous from the top of the cell, anatropous. FRUIT dry, indehis-

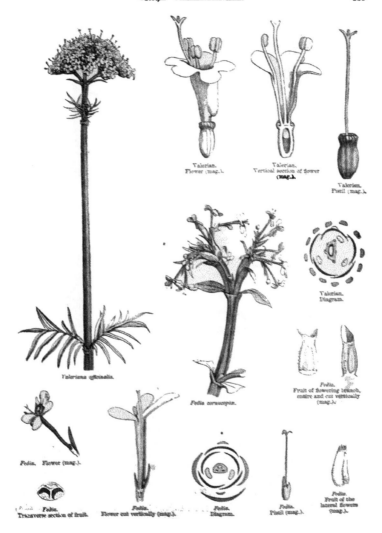

Valerian.
Flower (mag.).

Valerian.
Vertical section of flower
(mag.).

Valerian.
Pistil (mag.).

Valerian.
Diagram.

Fedia.
Fruit of flowering branch,
entire and cut vertically
(mag.).

Valeriana officinalis.

Fedia cornucopiæ.

Fedia. Flower (mag.).

Fedia.
Transverse section of fruit.

Fedia.
Flower cut vertically (mag.).

Fedia.
Diagram.

Fedia.
Pistil (mag.).

Fedia.
Fruit of the
lateral flowers
(mag.).

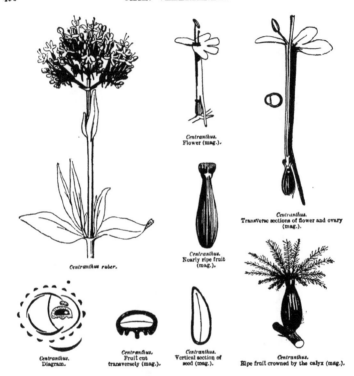

Centranthus.
Flower (mag.).

Centranthus.
Transverse sections of flower and ovary (mag.).

Centranthus.
Nearly ripe fruit (mag.).

Centranthus ruber.

Centranthus.
Diagram.

Centranthus.
Fruit cut transversely (mag.).

Centranthus.
Vertical section of seed (mag.).

Centranthus.
Ripe fruit crowned by the calyx (mag.).

cent, coriaceous or membranous, 3-celled or 1-celled by suppression of the empty cells, always 1-seeded. SEED inverted. EMBRYO straight, exalbuminous; *radicle* superior.

PRINCIPAL GENERA.

° Valeriana.   ° Centranthus.   ° Fedia.   ° Valerianella.   Nardostachys.   Patrinia.

The family of *Valerianeæ* is very near *Dipsaceæ*; as indicated by their opposite leaves, irregular terminal flowers, epigynous tubular staminiferous corolla, imbricate æstivation, solitary pendulous anatropous ovule, and simple style. The diagnosis consists in the corymbiform cyme, the three-celled ovary, and the exalbuminous seed. *Valerianeæ* approach *Compositæ* in their toothed or plumose calyx, epigynous staminiferous corolla, and solitary exalbuminous seed; they are separated from it by their inflorescence, æstivation, nervation of the corolla, free anthers, three-celled ovary and pendulous ovule. They offer some analogy with *Caprifoliaceæ* in the terminal inflorescence, opposite leaves, æstivation, epigynous corolla, several-celled ovary and pendulous ovule; but the latter differ in the woody stem, axile placentation, fleshy fruit and albuminous embryo.

Valerianeæ are mostly natives of the Old World, and principally of Central Europe, the Mediterranean and Caucasian regions, whence some species have advanced eastwards into Siberia, Nepal and Japan. They abound on the Cordilleras of South America, extending abundantly into Chili, Fuegia, and the Falkland Islands. They are very rare in North America.

Valerianeæ possess medicinal qualities known from an ancient period; but these properties are much more marked in the perennial than in the annual species, where they have not time to be elaborated. Their rhizomes contain a volatile oil, a peculiar acid, a bitter principle and starch; their taste is acrid and their odour penetrating. *Valerianeæ* now rank at the head of the vegetable antispasmodics; the principal species is the *Valeriana officinalis*, which grows in Europe in damp meadows. Celtic Nard is yielded by two Alpine species [*V. celtica* and *Saliunca*] which inhabit the limits of the eternal snows [in Styria and Carinthia], whence their roots are sent to Turkey, and largely used to scent baths and as a medicine. It also enters into the very complicated electuary called *thériaque*.

The Spikenard of the ancients, Indian Nard of the moderns, *Nardostachys Jatamansi*, is greatly esteemed in India on account of its aroma and stimulating properties. In the annual *Valerianeæ* the leaves are not bitter like those of the perennial species; this bitterness is replaced by a somewhat vapid mucilage, relieved by a slight quantity of volatile oil, which renders them edible; such are the Valerianellas [species of *Fedia*], Lamb's Lettuce, Corn Salad, of which the young leaves are used for salad.

CXX. *CALYCEREÆ*.

(CALYCEREÆ, *R. Br.*—BOOPIDEÆ, *Cassini*.)

FLOWERS *in an involucrate capitulum.* COROLLA *epigynous, monopetalous, isostemonous, æstivation valvate.* ANTHERS *syngenesious at the base.* OVARY *1-celled, 1-ovuled*; OVULE *pendulous, anatropous.* EMBRYO *albuminous.*

Annual or perennial HERBS. LEAVES alternate, sessile, without stipules. INFLORESCENCE in a capitulum, with an involucre of one or more series of bracts. FLOWERS sessile on a paleaceous or alveolate receptacle, sometimes all fertile, sometimes mixed with flowers of which the pistil is suppressed, the fertile sometimes cohering below. CALYX of 5 usually unequal segments, persistent. COROLLA inserted on an epigynous disk, monopetalous, regular; tube elongated, slender; *limb* 5-fid, segments with a dorsal and two sub-marginal nerves, æstivation valvate. STAMENS 5, inserted at the bottom of the corolla-tube, and alternate with its segments; *filaments* coherent to the corolla-tube throughout its length, free near the throat, and monadelphous or separate; *anthers* introrse, bilocular, cohering at the base, free at the top, dehiscence longitudinal. OVARY inferior, 1-celled, 1-ovuled, crowned by a conical disk uniting the base of the corolla to that of the style, lining the corolla-tube, and dilating near the throat into 5 glandular areolæ; *style* terminal, simple, exserted, tip clavate and glabrous; *stigma* terminal, globose; *ovule* pendulous from the top of the cell, anatropous. ACHENES usually crowned by the accrescent calyx and marcescent corolla, sometimes connate. SEED inverted, raphe longitudinal, chalaza apical. EMBRYO straight, in the axis of a fleshy albumen.

GENERA.

* Calycera.   * Boopis.   * Acicarpha.

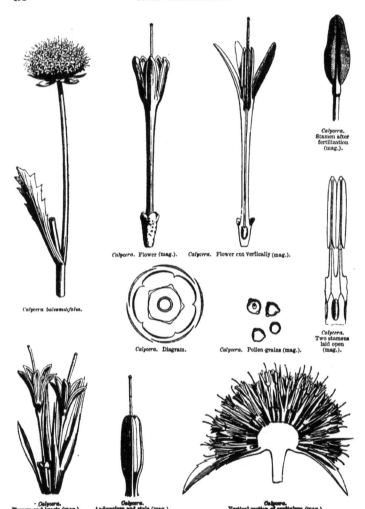

Calycera balsamifolia.

Calycera. Flower (mag.).   Calycera. Flower cut vertically (mag.).

Calycera.
Stamen after
fertilization
(mag.).

Calycera. Diagram.

Calycera. Pollen-grains (mag.).

Calycera.
Two stamens
laid open
(mag.).

Calycera.
Flowers and bracts (mag.).

Calycera.
Androecium and style (mag.).

Calycera.
Vertical section of capitulum (mag.).

| *Calycera.*
Ripe
fruit. | *Calycera,*
Fruit crowned by the
accrescent calyx. | *Calycera.*
Anatropous ovule
(mag.). | *Calycera.*
Cylindric embryo
(mag.). | *Calycera.*
Seed cut
vertically (mag.). |

Calycereæ is closely allied to *Compositæ*, but is distinguished by the nervation of the corolla, monadelphous filaments, position of the ovule, absence of collecting hairs, globose stigma, and albumen. It is equally near *Dipsaceæ*, from which it differs in the alternate leaves, valvate æstivation, and monadelphous and syngenesious stamens. The species are not numerous, and mostly inhabit temperate South America.

CXXI. *DIPSACEÆ.*

(DIPSACEÆ, *A. L. de Jussieu.*)

COROLLA *monopetalous, epigynous, æstivation imbricate.* STAMENS 4, *inserted on the tube of the corolla.* OVARY *1-celled, 1-ovuled, adnate to the receptacular tube throughout its length, or only at the top;* OVULE *pendulous, anatropous.* EMBRYO *albuminous.*

Annual or perennial HERBS. LEAVES opposite, rarely whorled, exstipulate. FLOWERS perfect, more or less irregular, gathered into a dense capitulum, involucrate, on a naked or paleaceous receptacle, very rarely in a whorl in the axil of the upper leaves, and each furnished with a calyciform obconic involucel, the tube of which is pitted at the top or furrowed lengthwise, and its limb scarious. CALYX superior, cup-shaped, or in setaceous segments forming a naked or bearded plumose crown. COROLLA superior, monopetalous, tubular, inserted at the top of the receptacular tube; *limb* 5–4-fid, usually irregular, sometimes labiate, æstivation imbricate. STAMENS 4, often unequal, rarely 2–3, alternate with the corolla-lobes, inserted at the bottom of the tube; *filaments* exserted, distinct, or rarely united in pairs; *anthers* introrse, 2-celled, dehiscence longitudinal. OVARY inferior, 1-celled, 1-ovuled, sometimes free in a receptacular tube which is closed at the top, sometimes adhering to this tube throughout its length, or at the top only; *style* terminal, filiform, simple, united at its base to the neck of the receptacular tube; *stigma* simple, clavate, or very shortly and unequally 2-lobed; *ovule* pendulous from the top of the cell, anatropous. UTRICLE enclosed in the receptacular tube and the involucel. SEED inverted, testa membranous, hardly separable from the pericarp. EMBRYO straight, in the axis of a fleshy scanty albumen; *radicle* superior.

GENERA.

* Dipsacus.   * Morina.   * Cephalaria.   Knautia.   * Scabiosa.   Pterocephalus.

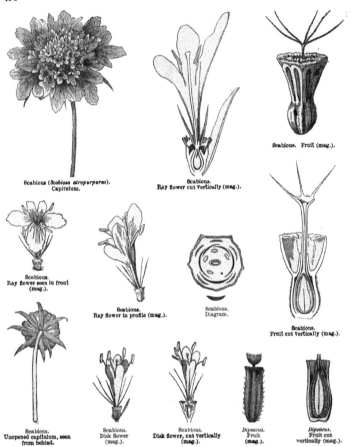

Scabious (*Scabiosa atropurpurea*).
Capitulum.

Scabious.
Ray flower cut vertically (mag.).

Scabious. Fruit (mag.).

Scabious.
Ray flower seen in front
(mag.).

Scabious.
Ray flower in profile (mag.).

Scabious.
Diagram.

Scabious.
Fruit cut vertically (mag.).

Scabious.
Unopened capitulum, seen
from behind.

Scabious.
Disk flower
(mag.).

Scabious.
Disk flower, cut vertically
(mag.).

Dipsacus.
Fruit
(mag.).

Dipsacus.
Fruit cut
vertically (mag.).

Dipsaceæ are so nearly allied to *Valerianeæ* that A. L. de Jussieu placed them in the same family (see *Valerianeæ*). They resemble *Compositæ* in their inflorescence, which in all the genera except *Morina* is an involucrate capitulum, in their usually paleaceous receptacle, epigynous staminiferous corolla, one-celled ovary crowned by a toothed or plumose calyx, and solitary anatropous ovule; they differ in the separate involucel of each flower, imbricate aestivation, nervation of the corolla, free anthers, pendulous ovule, simple style, terminal stigma, and albuminous embryo. They approach *Calycereæ* in their inflorescence, in the epigynous staminiferous tubular corolla, one-celled ovary, solitary pendulous ovule, and albuminous seed ; but the opposite leaves, imbricate æstivation, and free anthers render the identification easy.

Brongniart has placed *Dipsaceæ* and *Caprifoliaceæ* in the same class; their analogies are founded on their epigynous corolla, æstivation, pendulous and anatropous ovule, axile embryo in a fleshy albumen, and opposite leaves; but the inflorescence, 1-celled ovary, solitary ovule, and apical placentation form a prominent line of demarcation.

Dipsaceæ inhabit temperate and hot regions of the Old World and of Africa situated beyond the tropics. The rhizome and leaves of some *Dipsaceæ* are medicinal, containing a bitter-sweet slightly astringent principle. Scabious is administered as a depurative in cutaneous disorders. The roots of the Teasel (*Dipsacus sylvestris*) are diuretic and sudorific; its leaves and root were formerly considered to be a remedy for hydrophobia. The capitula of *Dipsacus fullonum*, a species of which the origin is unknown, are furnished with recurved hard and elastic bracts, which have led to their employment by clothiers for carding woollen and cotton fabrics ; hence its vulgar name of Fuller's Teasel.

CXXII. *COMPOSITÆ.*

(Compositæ, *Vaillant.*—Synantheræ, *L.-C. Richard.*)

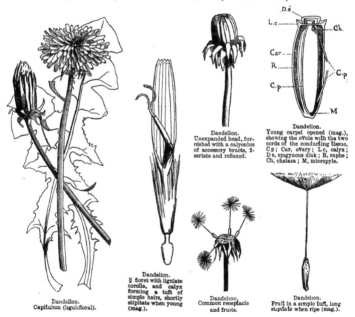

Dandelion.
Capitulum (ligulifloral).

Dandelion.
Unexpanded head, furnished with a calyculus of accessory bracts, 2-seriate and reflexed.

Dandelion.
Young carpel opened (mag.), showing the ovule with the two cords of the conducting tissue, C p ; Car, ovary ; L c, calyx ; D e, epigynous disk ; R, raphe ; Ch, chalaza ; M, micropyle.

Dandelion.
☿ floret with ligulate corolla, and calyx forming a tuft of simple hairs, shortly stipitate when young (mag.).

Dandelion,
Common receptacle and fruits.

Dandelion.
Fruit in a simple tuft, long stipitate when ripe (mag.).

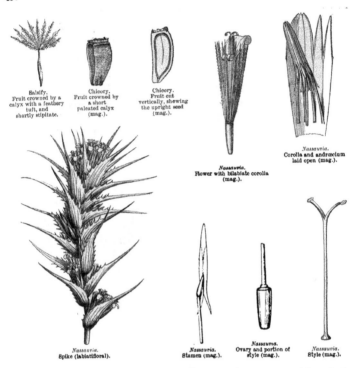

Salsify.
Fruit crowned by a
calyx with a feathery
tuft, and
shortly stipitate.

Chicory.
Fruit crowned by
a short
paleated calyx
(mag.).

Chicory.
Fruit cut
vertically, shewing
the upright seed
(mag.).

Nassauria.
Corolla and andrœcium
laid open (mag.).

Nassauria.
Flower with bilabiate corolla
(mag.).

Nassauria.
Spike (labiatifloral).

Nassauria.
Stamen (mag.).

Nassauria.
Ovary and portion of
style (mag.).

Nassauria.
Style (mag.).

FLOWERS *in an involucrate capitulum.* COROLLA *epigynous, monopetalous, isoste-monous, valvate in æstivation.* ANTHERS *syngenesious.* OVARY *1-celled, 1-ovuled;* OVULE *erect, anatropous.* EMBRYO *exalbuminous.*

PLANTS generally perennial, mostly herbaceous, sometimes woody below, rarely arborescent. LEAVES generally alternate, often very much cut, rarely compound, exstipulate, but sometimes furnished with stipuliform auricles. CAPITULA sometimes few-flowered, very rarely 1-flowered, generally many-flowered; *inflorescence* indefinite, but forming collectively a definite inflorescence, in a corymb, cyme or glomerule, and composed of flowers inserted on a common receptacle. RECEPTACLE sometimes furnished with bracteoles (*paleæ, scales, bristles, fimbrillæ*), sometimes naked and smooth, or with small pits (*foveolate*), or deeply pitted (*alveolate*), with entire or toothed margins, or cut into membranous segments, or covered with pentagonal areolæ, that enclose the bases of the flowers. INVOLUCRE (*peri-clinium*) composed of one or many series of bracts (*scales* or *leaflets*), sometimes furnished outside with accessory bracts (*calyculi*). FLOWERS ☿, or ♀ or ♂ or neuter, sometimes all ☿ in one capitulum; sometimes ♀, or neuter at the circumference, the inner ☿; sometimes ♂ at the centre, and ♀ at the circumference; capitula sometimes exclusively composed of ♀ or ♂ flowers, and then monœcious or diœcious. CALYX rarely foliaceous, generally scarious or membranous, sometimes cup-shaped, sometimes spread into a crown, entire toothed or laciniate; sometimes divided into paleæ, or teeth or scales or awns; sometimes reduced to capillary hairs or bristles, which are smooth or scabrid or ciliate or plumose, and forming a tuft, either sessile or stipitate; finally, sometimes reduced to a thin circular cushion, or even entirely wanting. COROLLA epigynous, monopetalous, sometimes regular, tubular, 5–4-fid or -toothed, æstivation valvate; sometimes irregular, either bilabiate or ligulate, each lobe furnished with two marginal nerves confluent in the tube. STAMENS 5–4, inserted on the corolla, and alternate with its divisions; *filaments* inserted at the base of the tube, free above, rarely monadelphous, articulated towards the top; *anthers* 2-celled, introrse, cohering into a tube which sheaths the style, very rarely free, usually prolonged into a terminal appendage, cells often terminating in a tail at the base. OVARY inferior, 1-celled, 1-ovuled, crowned with an annular disk which surrounds a concave nectary; *style* filiform, undivided in the ♂ flowers, bifid in the ♀ and ☿ flowers; branches of the style, commonly called *stigmas*, convex on the dorsal surface, flat on the inner, furnished toward their tops, or outside, with short stiff hairs (*collecting hairs*), and traversed on the inner edges by two narrow glandular (*stigmatic*) bands, constituting the true stigma; *style* much shorter than the stamens before the opening of the flower, but rapidly growing at the period of fertilization, traversing the hollow cylinder formed by the anthers, and gathering, by means of the collecting hairs, the pollen destined to fertilize the newly opened neighbouring flowers. ☿ flowers furnished with stigmatic glands and collecting hairs; the ♀ have stigmatic glands but no collecting hairs; the ♂ have collecting hairs and no stigmatic glands; *ovule* straight, anatropous. ACHENE articulated on to the common receptacle, generally sessile, provided with a basilar or lateral areola, indicating its point of insertion, often prolonged in a beak to the top. SEED erect. EMBRYO straight, exalbuminous; *cotyledons* plano-convex, very rarely convolute (*Robinsonia*); *radicle* inferior.

SUB-ORDER I. LIGULIFLORÆ.

Tribe I. CICHORACEÆ.—Capitula formed of flowers with a ligulate irregular corolla (*demi-florets*), all ☿. Style with filiform branches, pubescent; stigmatic bands separate, and not half as long as the branches of the style.—Milky plants. Leaves alternate.

PRINCIPAL GENERA.

Andryala.	Chondrilla.	Picridium.	Helminthia.
Geropogon.	Drepania.	Hyoseris.	Lapsana, &c.

[For others, see Tribe XIII. of new classification, p. 505.]

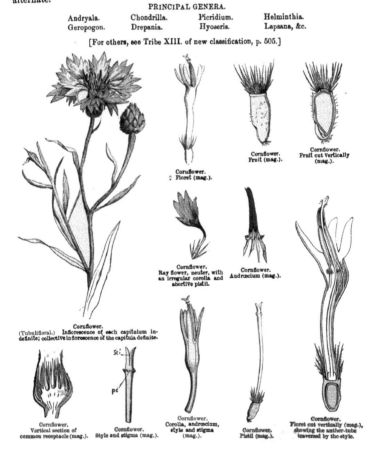

Cornflower.
♀ Floret (mag.).

Cornflower.
Fruit (mag.).

Cornflower.
Fruit cut vertically (mag.).

Cornflower.
Ray flower, neuter, with an irregular corolla and abortive pistil.

Cornflower.
Andrœcium (mag.).

Cornflower.
(Tubulifloral.) Inflorescence of each capitulum indefinite; collective inflorescence of the capitula definite.

Cornflower.
Vertical section of common receptacle (mag.).

Cornflower.
Style and stigma (mag.).

Cornflower.
Corolla, andrœcium, style and stigma (mag.).

Cornflower.
Pistil (mag.).

Cornflower.
Floret cut vertically (mag.), showing the anther-tube traversed by the style.

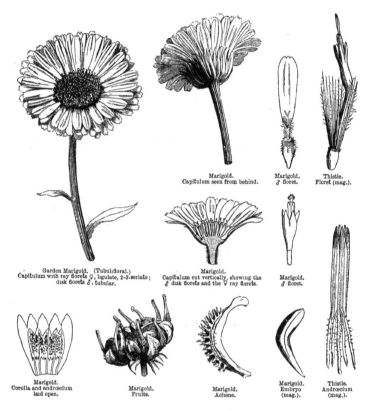

Marigold.
Capitulum seen from behind.

Marigold.
♂ floret.

Thistle.
Floret (mag.).

Garden Marigold. (Tubuliffloral.)
Capitulum with ray florets ♀, ligulate, 2-3-seriate;
disk florets ♂, tubular.

Marigold.
Capitulum cut vertically, showing the
♂ disk florets and the ♀ ray florets.

Marigold.
♂ floret.

Marigold.
Corolla and androecium
laid open.

Marigold.
Fruits.

Marigold.
Achene.

Marigold.
Embryo
(mag.).

Thistle.
Androecium
(mag.).

SUB-ORDER II. LABIATIFLORÆ.

Corolla of the ☿ flowers generally bilabiate; the ♂ and ♀ flowers ligulate or bilabiate.

Tribe II. MUTISIACEÆ.—Style of ☿ flowers cylindric, or almost nodose; stigmas obtuse, very convex and finely pubescent outside, equal, rarely 0.

PRINCIPAL GENERA.

* Mutisia.   * Chabræa [&c., see Tribe XII. (p. 504)].

K K 2

Tribe III. NASSAUVIACEÆ.[1]—Flowers all ☿. Style swollen at the base; stigmas truncate, with a pencil of hairs at the top, and within separate prominent stigmatic bands.

PRINCIPAL GENERA.

Nassauvia.   Moscharia.   Triptilion.

SUB-ORDER III. TUBULIFLORÆ.

Capitula sometimes formed of flowers with a regular tubular corolla (*florets*) all ☿, rarely irregular and sterile; sometimes *rayed*, i.e. composed of tubular flowers occupying the centre (*disk*), and of ligulate flowers (*demi-florets*) which are ♀ or neuter, occupying the circumference (*ray*).

Tribe IV. CYNAREÆ.—Capitula generally discoid. Style of the ☿ swollen above, nearly always furnished with a pencil; arms free or cohering, pubescent outside; stigmatic bands reaching the top of the stigma, and there uniting.—Leaves alternate.

PRINCIPAL GENERA.

Lappa.	Tyrimnus.	Galactites.	* Silybum.
Stœhelina.	Arctium.	* Xeranthemum.	* Gazania, X.[2]
* Venidium, X.	* Arctotis, X.	* Osteospermum, IX.	* Calendula, IX.

Tribe V. SENECIONIDEÆ.—Capitula generally rayed. Style cylindric at the top, bifid in the ☿ flowers; arms elongated, linear, truncate, or crowned with a pencil, above which they sometimes extend in a long appendage or a short cone; stigmatic bands prominent, prolonged but not reaching to the pencil.—Leaves alternate or opposite.

PRINCIPAL GENERA.

* Cacalia.	* Senecillis.	* Emilia.	Carpesium, IV.
* Phœnocoma, IV.	* Podolepis, IV.	* Rhodanthe, IV.	* Humea, IV.
* Cassinia, IV.	* Plagius, VII.	* Tanacetum, VII.	Artemisia, VII.
Athanasia, VII.	* Monolopia, VI.	* Madia, V.	* Spilanthes, V.
* Cosmos, V.	* Chrysanthemum, VII.	* Sphenogyne, X.	* Gamolepis, VIII.
* Sogalgina, V.	Bidens, V.	* Santolina, VII.	* Bæria, VI.
* Helianthus, V.	Diotis, VII.	* Helenium, VI.	* Coreopsis, V.
Ambrosia, V.	* Achillea, VII.	* Gaillardia, VI.	Xanthium, V.
* Calliopsis, V.	* Anthemis, VII.	* Porophyllum, VI.	* Chrysostemma, V.
* Silphium, V.	* Anacyclus, VII.	* Tagetes, VI.	* Rudbeckia, V.
Robinsonia.	* Oxyura, V.	* Œderia, V.	* Ximenesia, V.
* Zinnia, V.			

Included in Tribe XII. of the new arrangement, 504.—ED.   which the genera are referred under the new classification of the Order given at p. 503.—ED.

[2] The numerals indicate the number of the Tribe to

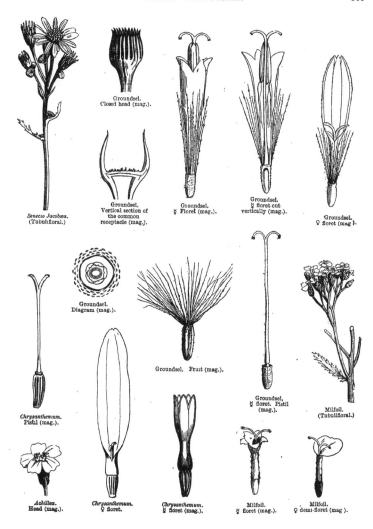

Senecio Jacobœa.
(Tubulifloral.)

Groundsel.
Closed head (mag.).

Groundsel.
Vertical section of
the common
receptacle (mag.).

Groundsel.
☿ Floret (mag.).

Groundsel.
☿ floret cut
vertically (mag.).

Groundsel.
♀ floret (mag.).

Groundsel.
Diagram (mag.).

Groundsel. Fruit (mag.).

Chrysanthemum.
Pistil (mag.).

Groundsel,
☿ floret. Pistil
(mag.).

Milfoil.
(Tubulifloral.)

Achillea.
Head (mag.).

Chrysanthemum.
♀ floret.

Chrysanthemum.
☿ floret (mag.).

Milfoil.
☿ floret (mag.).

Milfoil.
♀ demi-floret (mag.).

Astericus.
Fruit crowned by a calyx
with five scarious sepals
(mag.).

Heltanthemum.
☿ floret (mag.).

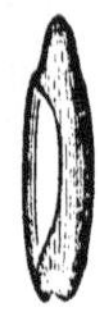

Robinsonia.
Embryo with coiled
cotyledons (mag.).

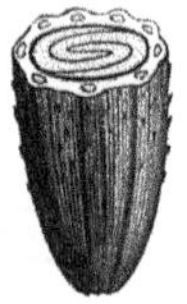

Robinsonia
Lower half of fruit.

Tagetes.
☿ floret, cut
vertically
(mag.).

Tagetes.
♀ floret cut vertically. Ovule
retaining a part of the
albumen (mag.).

Tribe VI. Asteroideæ.—Capitula generally rayed. Style of ☿ cylindric above; arms 2, a little flattened outside and puberulous ; stigmatic bands prominent, extending to the commencement of the external hairs.—Leaves alternate or opposite.

PRINCIPAL GENERA.

* Dahlia, V.	* Chrysocoma.	* Stenactis.	Linosyris.	Buphthalmum, IV.
* Vittadinia.	* Solidago.	* Schizogyne, IV.	* Inula, IV.	* Neja.
Micropus, IV.	Bellium.	Evax, IV.	* Boltonia.	* Brachylœna, IV.
* Charieis [&c.	See Tribe III., p. 503].			

Tribe VII. Eupatoriaceæ.—Capitula generally rayed. Style of ☿ flowers cylindric above, with long almost club-shaped arms, papillose externally ; stigmatic bands narrow, not prominent, usually stopping below the middle of the branches.—Leaves opposite or alternate.

PRINCIPAL GENERA.

* Tussilago, VII.	* Eupatorium.	* Ageratum.	* Nardosmia, VII.
* Liatris.	* Cœlestina.	Adenostyles.	* Stevia.

Tribe VIII. Vernoniaceæ.—Capitula usually discoid. Style of ☿ flowers cylindric ; arms long, hispid. Stigmatic bands prominent, narrow, stopping below the middle of the branches.—Leaves alternate or opposite.

PRINCIPAL GENUS.
* Vernonia.

[See also Tribe I. of the new classification, p. 503.]

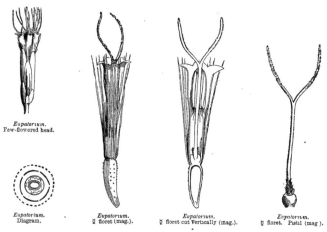

Eupatorium.
Few-flowered head.

Eupatorium.
Diagram.

Eupatorium.
♀ floret (mag.).

Eupatorium.
♀ floret cut vertically (mag.).

Eupatorium.
♀ floret. Pistil (mag).

[Tribes of *Compositæ*, as re-classified for the ' Genera Plantarum ' :—

TRIBE I. VERNONIACEÆ.—Heads homogamous ; flowers all tubular. Anthers sagittate at the base. Style-arms subulate, hirtellous.—Leaves usually alternate. Corolla never yellow. Pappus usually setose or paleaceous. *Ethulia, Vernonia, Piptocarpha, Stilpnopappus, Lychnophora, Eremanthis, Eliphantopus,* &c.

TRIBE II. EUPATORIACEÆ.—Heads homogamous ; flowers all tubular. Anthers nearly entire at the base. Style-arms sub-terete, obtuse, shortly papillose.—Leaves opposite or alternate. Corolla rarely pale ochreous, never truly yellow. Pappus often setose. *Adenostemma, Ageratum, Stevia, Trichogonia, Eupatorium, Mikania, Brickellia, Liatris,* &c.

TRIBE III. ASTEROIDEÆ.—Heads heterogamous, rayed or discoid, or with the ray suppressed and then homogamous. Anthers obtuse, almost quite entire at the base. Style-arms compressed. appendiculate (except in *Baccharideæ*).—Leaves usually alternate. Receptacle most often naked. Corollas of the disk usually yellow, of the ray the same, or blue or red or purple. *Solidago, Gutierrezia, Haplopappus, Pteronia, Lagenophora, Brachycome, Bellis, Amellus, Aster, Felicia, Olearia, Celmisia, Erigeron, Conyza, Psiadia, Baccharis,* &c.

TRIBE IV. INULOIDEÆ.—Heads heterogamous, discoid or rayed, or with the ray suppressed, and then homogamous. Anther-cells tailed or setose at the base. Style-arms linear, obtuse, inappendiculate, or styles of ♂ flowers undivided.—Leaves usually alternate. Corolla of the disk usually yellow, of the ray of the same (rarely of a different) colour. *Blumea, Pluchea, Epaltes, Evax, Filago, Anaphalis, Gnaphalium, Helipterum, Helichrysum, Cassinia, Angianthus, Stœbe, Metalasia, Relhania, Podolepis, Inula, Iphiona, Pulicaria, Buphthalmum,* &c.

TRIBE V. HELIANTHOIDEÆ.—Heads heterogamous, rayed, rarely discoid, or with the ray suppressed, and then homogamous. Receptacle paleaceous or rarely naked in the disk. Anther-cells not tailed, or mucronate only at the base. Style-arms truncate or appendiculate, or style of the ♂ flowers undivided. Achenes 3–4-angled, or terete, or often variously compressed, naked or crowned with 2–4 slender or sub-paleaceous awns, sometimes mixed with scales.—Leaves opposite, rarely alternate. Corolla of the disk usually yellow, of the ray usually of the same colour. *Clibadium, Polymnia, Espeletia, Melampodium, Silphium, Parthenium, Xanthium, Zinnia, Siegesbeckia, Eclipta, Rudbeckia, Wedelia, Viguiera, Helianthus, Encelia, Verbesina, Spilanthes, Coreopsis, Dahlia, Bidens, Calea, Hemizonia,* &c.

Tribe VI. Helenioideæ.—Heads heterogamous, radiate, rarely discoid, or with the ray suppressed and then homogamous. Receptacle naked. Anther-cells without tails. Style-arms in ☿ flowers truncate or appendiculate. Achenes narrow or turbinate, 4–5-angled or 8–∞-ribbed, rarely naked, usually crowned with paleæ or bristles.—Leaves opposite or alternate. Involucral bracts 1–2- (rarely 3–4-) seriate, herbaceous or membranous. Corollas of the ray usually yellow, of the disk the same, rarely differently coloured. *Bahia, Laphamia, Flaveria, Tagetes, Pectis, Gaillardia,* &c.

Tribe VII. Anthemideæ.—Heads heterogamous, rayed or discoid, or with the ray suppressed and then homogamous. Involucral bracts, 2–∞-seriate, dry or scarious at the tips. Receptacle paleaceous or naked. Anther-cells without tails. Style-arms truncate Pappus 0, or coroniform, rarely shortly paleaceous.—Leaves usually alternate. Corolla of the disk yellow, of the ray the same or of a different colour. *Eriocephalus, Athanasia, Achillea, Anthemis, Chrysanthemum, Matricaria, Cotula, Tanacetum, Artemisia,* &c.

Tribe VIII. Senecionideæ.—Heads heterogamous, rayed or discoid, or with the ray suppressed and then homogamous. Involucral bracts : inner 1-seriate, sub-equal ; outer small or 0, rarely ∞-seriate and imbricate. Receptacle usually naked. Anther-cells without a tail at the base or only mucronate. Style-arms of the ☿ flowers truncate or appendiculate. Pappus most often setose.—Leaves usually alternate. Corolla of the disk usually yellow, of the ray the same or differently coloured. *Liabum, Tussilago, Petasites, Arnica, Doronicum, Culcitium, Gynura, Cineraria, Senecio, Gynoxys, Werneria, Euryops, Othonna.*

Tribe IX. Calenduleæ.—Heads rayed. Involucral bracts 1–2-seriate, sub-equal, narrow. Receptacle naked. Anther-cells mucronate or shortly tailed at the base. Style-arms of ☿ flowers truncate, style of the ♂ undivided. Achenes usually of various forms or very thick, naked or crowned with wool.—Leaves usually alternate or radical. *Dimorphotheca, Calendula, Tripteris, Osteospermum.*

Tribe X. Arctotideæ.—Heads rayed, rarely homogamous. Involucral bracts ∞-seriate, usually scarious and spinescent at the tips. Receptacle naked, paleaceous or pitted. Anther-cells without tails. Style-arms in ☿ flowers usually short, rounded, obtuse, rarely truncate at the tip ; style of the ♂ flowers undivided. Achenes (usually thick) naked or with paleaceous or coroniform pappus.—Leaves radical or alternate. *Ursinia, Arctotis, Venidium, Gazania, Berkheya, Cullumia,* &c.

Tribe XI. Cynaroideæ.—Heads homogamous ; flowers all tubular. Involucral bracts many-seriate, imbricate, tips often scarious or spinescent or foliaceous. Receptacle often fleshy, densely setose or paleaceous, rarely shortly alveolate. Corolla-limb narrow, deeply 5-fid. Anther-cells without tails. Style sub-entire or with short erect arms, thickened and hairy at the base of the stigmatiferous portion. Achenes usually hard ; pappus setose or paleaceous, rarely 0.—Leaves alternate, usually spinous. *Echinops, Carlina, Atractylis, Cousinia, Carduus, Cnicus, Onopordon, Cynara, Saussurea, Jurinea, Serratula, Centaurea, Carthamus, Carduncellus,* &c.

Tribe XII. Mutisiaceæ.—Heads heterogamous or homogamous, rayed or not. Involucral bracts usually ∞-seriate, unarmed, rarely spinous. Receptacle rarely paleaceous. Corolla bilabiate or limb deeply 5–fid. Anther-cells usually without tails. Style-arms (very short or very long) rounded or truncate, inappendiculate. Achenes various; pappus setose, paleaceous, or 0.—Leaves radical or alternate, rarely opposite. *Barnadesia, Mutisia, Onoseris, Chuquirayua, Dicoma, Ainsliea, Chætanthera, Trichocline, Gerbera, Chaptalia, Leucœria, Perezia, Trixis, Jungia, Nassauvia*, &c.

TRIBE XIII. CICHORACEÆ.—Heads homogamous; flowers equal or sub-rayed. Involucre various. Receptacle with deciduous paleæ or 0. Corollas all ligulate, truncate and 5-toothed at the tip. Anther-cells with short tails or 0. Style-arms slender. Achenes various; pappus setose, paleaceous, or 0.—Juice usually milky. *Catananche, Cichorium, Microseris, Tolpis, Picris, Crepis, Hieracium, Hypochœris, Leontodon, Taraxacum, Lactuca, Prenanthes, Sonchus, Microrhynchus, Tragopogon, Scorzonera, Scolymus*, &c.—ED.]

Compositæ, of which about 10,000 species are known, form the tenth part of Cotyledonous plants, and ought perhaps rather to form a class than an order; nevertheless, the type which they present is so well characterized, that, in spite of their enormous numerical superiority over other natural groups, the term order has been retained for them. *Compositæ* are allied to *Calyceretæ, Dipsaceæ, Valerianeæ, Campanulaceæ, Brunoniaceæ* (see these families).

Compositæ chiefly inhabit temperate and hot regions. America produces the largest number of species; those with herbaceous stems grow in temperate and cold climates. The *Tubulifloræ* are most numerous in the tropics; the *Ligulifloræ* in the temperate regions of the northern hemisphere; the *Labiatifloræ* are chiefly natives of extra-tropical South America.

The rayed *Tubulifloræ* contain a bitter principle, which is usually combined with a resin or a volatile oil, according to the proportions of which certain species possess different medicinal properties; some being tonic, others excitant, or stimulant, or astringent. Many indigenous species of the large genus *Artemisia* (Wormwood, Southernwood, Tarragon, 'Génipi') owe to their aroma and bitterness decidedly stimulating properties. [From *Achillea moschata* the liqueur called Iva is made in the Engadin.] The common Tansy and the Balsamite (*Pyrethrum Tanacetum*) are also stimulants. The Camomiles contain an acrid or bitter volatile oil, which renders them antispasmodics and febrifuges. *Pyrethrum*, a Mediterranean *Anacyclus*, contains a resin and a very acrid volatile oil in its root, which lead to its employment in diseases of the teeth and gums; the 'Spilanthes,' or Para Cress, of tropical America, is an excellent cure for toothache. The flowers of *Arnica* and the root of Elecampane (*Inula Helenium*) are used as stimulants of the action of the skin. The Ayapana is an *Eupatorium* greatly esteemed in South America as a powerful sudorific and a sovereign remedy against snake-bites. Of all alexipharmics (snake-bite antidotes) the most celebrated are the *Guaco* and the *Herba-di-cobra*, tropical American species of a genus near *Eupatorium*. The *Tussilago* (Colt's-foot), and the *Gnaphalium dioicum* contain a gummy matter, united with a bitter and slightly astringent principle, which gives them sedative qualities; their capitula are therefore commonly employed as bechics under the name of Pectoral Flowers. The abundant tubercles of the Jerusalem Artichoke (*Helianthus tuberosus*), a perennial Brazilian plant cultivated throughout Europe, contain a principle analogous to starch (*inuline*), and a large proportion of uncrystallizable sugar. These tubercles furnish good food for cattle, and even for man, when cooked and seasoned. Some other rayed species are oleaginous and employed in commerce; the *Madia sativa* and *mellosa*, Chili plants, furnish an oil which many travellers declare to be preferable in taste to olive oil; it is distinguished from the latter, as well as from most fixed oils, by its solubility in alcohol. The seeds of the *Guizotia oleifera*, a plant cultivated in India and Abyssinia, yield an oil used for food and burning.

The Discoid *Tubulifloræ* (or *Carduaceæ*) contain a bitter principle, which is stimulating in some, diuretic and sudorific in others. As such are employed the Burdocks, the Milk Thistle (*Silybum Marianum*), and the Blessed Thistle, a species of *Centaurea*, to which genus also the Cornflower belongs, from which an eye-water was formerly distilled. Some *Carduaceæ* are edible when young; the flowers and leaves of some yield a dye; many have oleaginous seeds; none possess volatile oils. *Atractylis gummifera*, an exotic *Carduacea*, near *Centaurea*, contains a poisonous principle.

The genus *Cynara* contains many species, natives of the Mediterranean basin, the leaves of which are bitter and diuretic. The unexpanded capitula of the Common Artichoke (*C. Scolymus*) are eaten, as are the leaves of the Cardoon (*C. Cardunculus*) when blanched.

Amongst the *Carduaceæ* used in dyeing, the Safflower holds the first place; it is an Indian plant, now cultivated throughout the world, the flowers of which give a red dye (*carthamine*), employed for dyeing silk and cotton, and with which is prepared in Spain a much esteemed paint. The Dyer's Savory(*Serratula tinctoria*) contains a yellow colour of some value. Marigolds contain a bitter mucilage, various salts, and a little volatile oil; they were formerly celebrated as sudorifics and resolvents in cancerous obstructions.

The *Liguliflora* or *Cichoracea* possess a milky juice, which contains bitter, resinous, saline, and narcotic principles, the properties of which vary according to their relative proportions. Many of these, if gathered young, before the complete elaboration of the latex, are edible and have a pleasant taste. Their medicinal properties differ according to their development and that of their organs; thus the observations to be made on them must vary with the season. Amongst the medicinal *Cichoracea*, there are some in which the bitter, resinous, gummy and saline ingredients are united in such proportions that the result is highly nutritive. In the first rank must be placed the Dandelion, which is met with throughout Europe and the Mediterranean region. The Wild Chicory (*Cichorium Intybus*) possesses the same properties. The root of the cultivated Chicory is an important article of trade; it is employed, roasted, powdered and mixed with ground coffee, or used instead of the latter. The blanched leaves are edible. In Salsify (*Tragopogon polifolius*) and *Scorzonera hispanica*, the bitter of the root is corrected by the mucilage contained in the milky juice, and the root is edible.

The species of *Lactuca* have a bitter acrid juice with a poisonous smell; they contain wax, india-rubber, albumine, a resin, and a bitter crystallizable matter, with a peculiar volatile principle. It is to these different substances that they owe their medicinal properties. The thickened juice of the cultivated Lettuce, called *thridace*, is used as a narcotic, and preferred to opium in cases where there is reason to fear the stupifying action of the latter. The young leaves of the same species, which do not yet contain the milky juice, are much used as food.

CXXIII. *STYLIDIEÆ.*

(STYLIDEÆ, *R. Br.*—STYLIDIACEÆ, *Lindl.*)

COROLLA epigynous, monopetalous, anisostemonous, æstivation imbricate. STAMENS united to the style. OVARY with two many-ovuled cells; OVULES ascending, anatropous. EMBRYO albuminous.

Annual or perennial PLANTS, usually herbaceous, sometimes woody below. LEAVES simple, entire, exstipulate; cauline scattered, rarely whorled; radical in tufts. FLOWERS perfect, irregular, in a spike raceme or corymb, pedicels usually 3-bracteate. CALYX persistent, usually bilabiate, lower lip 2-fid or -toothed, upper lip 3-fid or -toothed. COROLLA monopetalous, irregular, tube short, limb 5-fid, 4 lobes large spreading, the fifth (*lip*) smaller, spreading or depressed, at first anterior, then becoming lateral through the torsion of the tube, to which it is sometimes attached by an irritable joint. STAMENS 2, parallel, inserted on a glandular disk crowning the ovary; *filaments* united into a column with the style, which is sometimes erect and continuous, sometimes with two bends, the lower of which is irritable; *anthers* forming 2 cells on the top of the column, and embracing the stigmas. OVARY inferior, more or less completely 2-celled, septum parallel to the calyx-lips; *stigma* obtuse, sometimes undivided, hidden between the anthers, sometimes divided into

Stylidium.
Flower (mag.).

Stylidium.
Corolla laid open (mag.).

Stylidium.
Seed cut
vertically (mag.).

Stylidium adnatum.

Stylidium.
Fruit (mag.).

two capillary branches terminated by a glandular head; *ovules* ascending, anatropous, on placentas fixed in the middle of the septum. CAPSULE 2-celled, or almost 1-celled by suppression of the septum, sometimes septifragally 2-valved, or with the anterior cell suppressed, the posticous fertile and loculicidal, sometimes indehiscent. SEEDS numerous, minute, sub-globose. EMBRYO minute, at the base of a fleshy oily albumen.

GENERA.

Forstera.   *Stylidium.   Levenhookia.

Stylidieæ approach *Campanulaceæ* in the epigynous corolla and stamens, introrse anthers, anatropous ovules, capsular fruit and fleshy albumen. But *Campanulaceæ* have an isostemonous corolla, free filaments, horizontal ovules, the style furnished with a series of collecting hairs, and a loculicidal capsule. *Stylidieæ* also resemble *Goodeniaceæ* in their irregular flower, epigynous corolla and stamens, 1-2-celled ovary, placentation, ascending and anatropous ovules, and fleshy albumen. But *Goodeniaceæ* differ in the induplicative æstivation, isostemonous corolla, indusiate stigma and axile embryo.

Stylidieæ belong to the southern hemisphere; most of its species inhabit extra-tropical Australia. [A few advance into Eastern India and South China. The species of *Forstera* inhabit the Alps of Australia, New Zealand and Fuegia.—ED.]

CXXIV. *GOODENIACEÆ*.

(GOODENOVIÆ, *R. Br.*—GOODENOVIEÆ, *Bartling.*—SCÆVOLACEÆ, *Lindl.*)

COROLLA *epigynous or perigynous, monopetalous, isostemonous, æstivation induplicate.* STAMENS *epigynous.* STIGMA *indusiate.* OVULES *erect, anatropous.* EMBRYO *albuminous.*

Usually herbaceous, sometimes woody below, erect or climbing. LEAVES scattered, sometimes all radical, simple, exstipulate. FLOWERS ☿, irregular, axillary or terminal. CALYX sometimes superior, absent or distinctly 5-fid; sometimes of 3-5 inferior sepals, coherent at the base. COROLLA inserted at the base or top of the

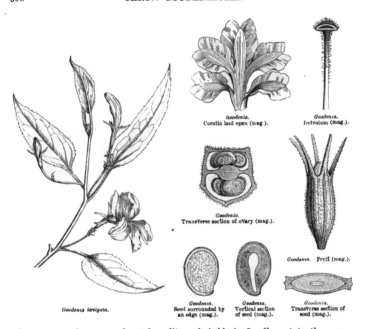

Goodenia.
Corolla laid open (mag.).

Goodenia.
Indusium (mag.).

Goodenia.
Transverse section of ovary (mag.).

Goodenia. Fruit (mag.).

Goodenia lævigata.

Goodenia.
Seed surrounded by
an edge (mag.).

Goodenia.
Vertical section
of seed (mag.).

Goodenia.
Transverse section of
seed (mag.).

calyx, monopetalous, irregular, tube split, or divisible in 5, adherent to the ovary; limb 5-partite, 1–2-labiate, æstivation induplicate; lobes with a lanceolate disk and dilated more membranous wing-like margins. STAMENS 5, inserted on the disk crowning the ovary, free, and alternate with the corolla lobes; *filaments* free, or cohering at the top; *anthers* free or cohering, erect, linear, 2-celled, introrse. OVARY inferior, or superior to the calyx and inferior to the corolla, 1-celled, or more or less completely 2- (rarely 4-) celled; cells sometimes 1–2-ovuled, with collateral erect ovules; sometimes with imbricated ascending ovules occupying both sides of the septum, anatropous; *style* usually simple; *stigma* fleshy, enveloped in a membranous cup-shaped sheath proceeding from a prolongation of the disk that is adnate to the style. FRUIT a drupe or nucule, or 2-celled capsule with 2 semi-septiferous valves, or 4-celled and 4-valved. SEEDS erect or ascending. EMBRYO straight, in the axis of a fleshy albumen; *radicle* inferior.

PRINCIPAL GENERA.

| * Leschenaultia. | * Goodenia. | Dampiera. | * Euthales. |
| Velleia. | Scævola. | Selliera. | |

We have indicated the affinity of *Goodeniaceæ* with *Brunoniaceæ* and *Stylidieæ* (see these families). It is also allied to *Lobeliaceæ* in its epigynous stamens, isostemonous corolla, æstivation, many-celled ovary, anatropous ovules, axile embryo and fleshy albumen. But *Lobeliaceæ* differs in the stigma having a ring of hairs, and not being sheathed by an indusium.

Goodeniaceæ are almost exclusively Australian, and especially extratropical. The species of *Scævola* have migrated to the Moluccas and the Indian continent, and thence to the South of Africa [and the Pacific Islands. *Selliera* inhabits the coasts of Australia, New Zealand, Chili and Fuegia]. We know little for certain respecting the properties of some Indian *Scævolæ*. The leaves and berry of the Mokal [*S. Taccada*] yield a bitter juice supposed to remove cataract, and its young leaves are eaten as a vegetable. The inhabitants of Amboina use the root to enable them to eat with safety poisonous crabs and fish. The pith is used in cases of exhaustion [and extensively in the construction of ornaments, models, &c.]. The leaves of *S. Bela-modogam* [probably identical with *S. Taccada*], a native of Malabar, are applied as a poultice on inflammatory tumours, and a decoction from them is diuretic. *Goodenia*, *Euthales*, and *Leschenaultia* are cultivated in European hothouses as ornamental plants.

CXXV. *BRUNONIACEÆ.*[1]

(BRUNONIACEÆ, *R. Br.*)

FLOWERS *fascicled, fascicles aggregated into an involucrate capitulum.* COROLLA *hypogynous, monopetalous, isostemonous, æstivation valvate.* STAMENS *hypogynous;* ANTHERS *syngenesious.* OVARY *free,* 1-*celled,* 1-*ovuled;* STIGMA *indusiate;* OVULE *erect, anatropous.* FRUIT *a utricle.* EMBRYO *exalbuminous.*

Perennial nearly stemless HERBS, resembling *Scabious*. LEAVES radical, close-set, spathulate, entire. FLOWERS ☿, sub-regular, each with 5 whorled bracteoles, agglomerated in fascicles united in an involucrate capitulum, and separated by bracts like those of the involucre. SCAPES many from the same root, simple, and terminating in a single capitulum. CALYX-TUBE short; limb divided into 5 subulate plumose segments. COROLLA hypogynous, monopetalous, infundibuliform, persistent, tube splitting after flowering, limb 5-fid, lobes spathulate, the 2 upper deeply divided. STAMENS 5, inserted on the neck of the ovary, included; *filaments* flat, articulate, free; *anthers* linear, 2-celled, introrse, coherent in a tube around the style. OVARY free, shortly stipitate, 1-celled; *style* terminal, simple, exserted, hairy above; *stigma* obconic, truncate, fleshy, in a sheath bifid at the tip; *ovule* solitary, basilar, anatropous. FRUIT an indehiscent utricle, enclosed in the enlarged and hardened calyx, and crowned by the plumose segments of the calyx-limb. SEED erect. EMBRYO straight, exalbuminous; *radicle* inferior.

ONLY GENUS.

* Brunonia.

Brunoniaceæ approach *Goodeniaceæ* by the indusiate stigma; *Campanulaceæ* and *Lobeliaceæ* by its inflorescence, isostemonism, æstivation of the corolla, free filaments, anatropous ovule and hairy style; it differs in its hypogynism, its solitary erect ovule, the absence of albumen, and especially in the indusiate stigma. The same analogies exist between *Brunoniaceæ* and *Compositæ*, and in addition, in both families the ovule is solitary, erect and exalbuminous, and the calyx expands into a pappus; the diagnosis, in fact,

[1] Considered as a genus of *Goodenieæ* by Bentham (Fl. Austral. v. 4. p. 38).—ED.

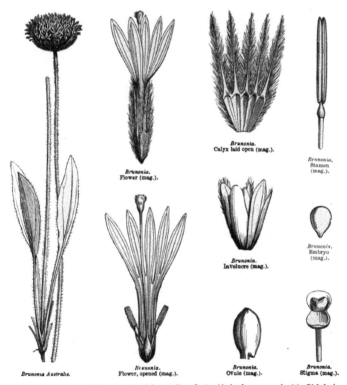

Brunonia.
Flower (mag.).

Brunonia.
Calyx laid open (mag.).

Brunonia,
Stamen
(mag.).

Brunonia.
Involucre (mag.).

Brunonia.
Embryo
(mag.).

Brunonia Australis.

Brunonia.
Flower, opened (mag.).

Brunonia.
Ovule (mag.).

Brunonia.
Stigma (mag.).

is the same, except as regards the nervation of the corolla. It should also be compared with *Globularieæ* and *Plumbagineæ* (see these families).

The only species inhabits extra-tropical Australia.

CXXVI. *CAMPANULACEÆ.*

(Campanulæ, *partim, Adanson.*—Campanulaceæ, *exclusis pluribus, Jussieu.*— Campanuleæ, *D.C.*—Campanulaceæ, *Bartling.*)

Corolla *epigynous, monopetalous, regular, isostemonous, æstivation valvate.* Stamens *epigynous.* Ovary *several-celled, many-ovuled;* stigma *without indusium;* ovules *anatropous.* Fruit *capsular.* Embryo *albuminous.*

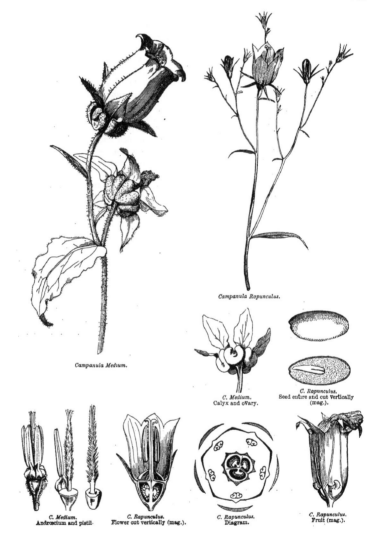

Campanula Rapunculus.

Campanula Medium.

C. Medium.
Calyx and ovary.

C. Rapunculus.
Seed entire and cut vertically
(mag.).

C. Medium.
Andrœcium and pistil.

C. Rapunculus.
Flower cut vertically (mag.).

C. Rapunculus,
Diagram.

C. Rapunculus,
Fruit (mag.).

Annual, biennial or perennial HERBS, rarely woody below, sometimes climbing, usually milky. LEAVES alternate, rarely opposite, simple, exstipulate. FLOWERS perfect, regular, in a raceme spike or glomerule, sometimes panicled, involucrate or not. CALYX superior or semi-superior, persistent, usually 5- (rarely 3–6–8-) partite, æstivation valvate. COROLLA monopetalous, marcescent, inserted on an epigynous ring, campanulate, infundibuliform or tubular, limb more or less deeply divided, æstivation valvate. STAMENS alternate with the corolla-lobes; *filaments* free or very rarely adhering to the base of the corolla, connivent, or sub-coherent by their usually dilated bases; *anthers* introrse, 2-celled, distinct or cohering into a tube around the style. OVARY inferior or semi-inferior, 2–8-celled; *style* simple, bristling with fugacious collecting hairs arranged in longitudinal series; *stigma* usually lobed, glabrous within, hairy on the back, very rarely undivided and capitate; *ovules* anatropous, numerous, horizontal at the inner angle of the cells, or attached to the surface of the septa. CAPSULE with many-seeded cells, sometimes loculicidal at the top, or opening by valves near the bottom or middle or beneath the calyx, or by as many pores as cells, very rarely by transverse slits. SEEDS numerous, minute, ovoid or angular. EMBRYO straight, in the axis of a fleshy albumen; *radicle* near the hilum.

PRINCIPAL GENERA.

Jasione.	* Roella.	* Specularia.	* Canarina.
* Phyteuma.	* Trachelium.	* Platycodon.	* Campanula.
* Adenophora.	* Wahlenbergia.	* Codonopsis.	* Michauxia.

We have noticed the affinities of *Campanulaceæ* with *Lobeliaceæ, Brunoniaceæ* and *Stylidieæ* (see these families). They approach *Compositæ* in the inflorescence of some genera, the synanthery of others, the epigyny, isostemonism and æstivation of the corolla, collecting hairs, and anatropous ovule; they are separated by the nervation of the corolla, plurality and horizontal direction of the ovules, collecting hairs in lines and not in a ring, capsular fruit and albuminous embryo. The *Campanulaceæ* with basal or lateral dehiscence inhabit the temperate regions of the Old World; those with apical dehiscence are most frequent in the southern temperate zone, and especially in South Africa, Australia and South America.

Campanulaceæ yield a milky juice, which differs from that of *Lobeliaceæ* in the acrid principles being neutralized by a sweet and very abundant mucilage, to which the fleshy roots of *Campanula Rapunculus* and its allies owe their alimentary properties; they are agreeable and easy of digestion, and being milky, they were recommended by the ancients in nursing. Many species were considered to cure hydrophobia in Russia. Two indigenous Campanulas (*C. Trachelium* and *cervicaria*) were formerly used in angina of the pharynx and trachea; whence their specific names.

CXXVII. *LOBELIACEÆ.*

(CAMPANULACEARUM pars, *R. Br.*—LOBELIACEÆ, *Jussieu, Bartling.*)

COROLLA *epigynous, isostemonous, irregular, æstivation valvate.* STAMENS *epigynous, cohering into a tube.* OVARY *1-2-3-celled;* STIGMA *not indusiate;* OVULES *numerous, generally horizontal, anatropous.* FRUIT *a capsule or berry.* EMBRYO *albuminous.*

Herbaceous annual or perennial PLANTS, often woody below, rarely shrubby, usually with milky juice. LEAVES alternate or radical, simple, exstipulate. FLOWERS perfect, very rarely diœcious, generally irregular; inflorescence axillary or terminal, usually racemed or spiked, rarely in a corymb or capitulum, sometimes solitary and axillary. CALYX superior or semi-superior, with 5 sub-regular or irregular segments. COROLLA inserted on the calyx, 1-2-labiate, of 5 very rarely free

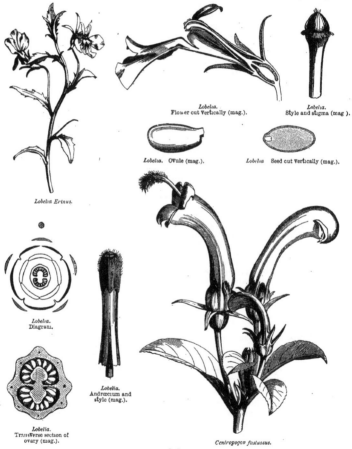

Lobelia.
Flower cut vertically (mag.).

Lobelia.
Style and stigma (mag).

Lobelia. Ovule (mag.).

Lobelia Seed cut vertically (mag.).

Lobelia Erinus.

Lobelia.
Diagram.

Lobelia.
Andrœcium and style (mag.).

Lobelia.
Transverse section of ovary (mag.).

Centropogon fastuosus.

and equal usually cohering and irregular petals, æstivation valvate. STAMENS 5, opposite to the calyx-lobes, inserted with the corolla on a ring which is often dilated into a disk crowning the top of the ovary; *filaments* usually free from the corolla-tube, distinct at the base and cohering above; *anthers* introrse, 2-celled, cohering into a usually curved cylinder. OVARY inferior or semi-superior, sometimes 2–3-celled by the inflection of the edges of the carpels, or sub-1-celled by the imperfection of the septa; or of 3 carpels joined by their edges, 2 of which are placentiferous on the median nerves, and the third is narrower and sterile; *style* simple; *stigma* usually emarginate, or of 2 lobes girt with a ring of hairs; *ovules* anatropous, numerous, generally horizontal, sessile, inserted on the inner angle of the cells, or on each side of the septum. FRUIT indehiscent and fleshy, or capsular, dehiscence loculicidal, longitudinal or apical, rarely transverse. SEEDS numerous, small; *hilum* marked by an orbicular pit; *raphe* indistinct. EMBRYO straight, in the axis of a fleshy albumen; *radicle* near the hilum.

PRINCIPAL GENERA.

| * Clintonia. | * Siphocampylus. | * Centropogon. | * Lobelia. |
| Laurentia. | * Tupa. | * Isotoma. | |

Lobeliaceæ are closely connected with *Campanulaceæ*, in which many botanists have placed them; they only differ in the irregular corolla, more complete cohesion of the stamens, and often fleshy fruit. They approach *Cichoraceæ*, a tribe of *Compositæ*, in their milky juice, epigynous irregular corolla, synanthery of the stamens, and stigmatic lobes furnished with collecting hairs: they are separated from it by the many-ovuled ovary, horizontal ovules and presence of albumen. We have indicated their affinities with *Goodeniaceæ* under that family.

Some *Lobeliaceæ* inhabit the north temperate zone; most are dispersed over tropical and southern regions, nearly in equal proportions in America and in the Old World, especially in temperate Australia and South Africa. They are very rare in the northern regions of Asia and Europe.

Lobeliaceæ contain in abundance a very acrid and narcotic bitter juice, which burns the skin, and taken internally produces mortal inflammation on the digestive canal; and they are hence amongst the most poisonous of plants. Some are employed in medicine by American practitioners, but with the greatest caution. *Lobelia inflata* (Indian Tobacco) is used as an expectorant and diaphoretic in the treatment of asthma; but from being incautiously administered has caused many deaths. The West Indian *Isotoma longiflora* [a violent cathartic], also employed as a therapeutic, is as dangerous. [*Lobelia syphilitica*, once in great repute, has fallen into disuse. *L. cardinalis* is an acrid anthelminthic, and the European *L. urens* is a vesicant.—ED.]

CXXVIII. *ERICINEÆ.*

(ERICÆ ET RHODODENDRA, *Jussieu.*—ERICACEÆ, *D.C.*—ERICINEÆ, *Desvaux.*)

COROLLA mono- or poly-petalous, hypogynous, usually diplostemonous. STAMENS hypogynous, or rarely inserted at the base of the petals; ANTHERS 2-celled, usually opening by 2 terminal pores. OVARY many-celled; OVULES anatropous. FRUIT dry or fleshy. EMBRYO albuminous, axile.—STEM woody.

SHRUBS or UNDERSHRUBS. LEAVES usually alternate, entire or toothed, exstipulate. FLOWERS ☿, axillary or terminal, solitary or aggregated. CALYX 4–5-fid or -partite, persistent. COROLLA hypogynous, 4–5-merous, usually monopetalous, inserted at the outer base of a hypogynous disk, æstivation contorted or imbricate. STAMENS usually double the petals in number, rarely equal, and then alternate with

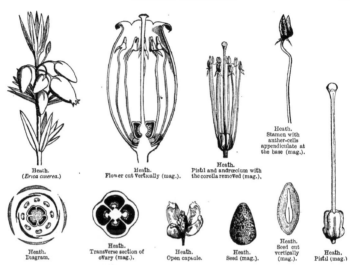

Heath.
(*Erica cinerea.*)

Heath.
Flower cut vertically (mag.).

Heath.
Pistil and androecium with the corolla removed (mag.).

Heath.
Stamen with anther-cells appendiculate at the base (mag.).

Heath.
Diagram.

Heath.
Transverse section of ovary (mag.).

Heath.
Open capsule.

Heath.
Seed (mag.).

Heath.
Seed cut vertically (mag.).

Heath.
Pistil (mag.)

them, not adhering to the corolla, and inserted like it on the disk, or scarcely adherent to its base; *filaments* free, sometimes more or less monadelphous (*Lagenocarpus, Philippia*); *anthers* dorsi- or basi-fixed, variously appendaged or not; cells 2, hard, dry, separate at the base or top, opening by terminal or lateral more or less oblique pores which are sometimes prolonged into 2 longitudinal slits (*Loiseleuria, Leiophyllum*). OVARY free, surrounded by a disk at its base, of several many-ovuled cells, rarely few-ovuled (*Calluna*) or 1-ovuled (*Arctostaphylos*); *style* simple; *stigma* capitate, peltate or cyathiform; *ovules* anatropous. FRUIT a capsule berry or drupe. SEEDS inserted on central placentas, small, numerous; *testa* very adherent and dotted, or loose reticulate and arilliform. EMBRYO straight, cylindric, in the axis of a fleshy albumen; *cotyledons* short; *radicle* opposite the hilum.

TRIBE I. ARBUTEÆ.—Corolla deciduous. Fruit a berry or drupe.—Evergreen shrubs.

PRINCIPAL GENERA.

Pernettia.   * Arbutus.   Arctostaphylos.

TRIBE II. ANDROMEDEÆ.—Corolla deciduous. Capsule loculicidal.—Shrubs with persistent or deciduous leaves. Buds generally scaly.

PRINCIPAL GENERA.

* Clethra.	* Zenobia.	* Oxydendrum.	* Cassandra.
* Epigæa.	* Pieris.	* Leucothoe.	* Cassiope.
* Gualtheria.	* Lyonia.	* Andromeda.	

TRIBE III. ERICEÆ.—Corolla persistent, usually 4-merous. Anthers often cohering before flowering. Capsule loculicidal (*Erica*) or septicidal (*Calluna*).— Evergreen shrubs. Buds not scaly.

PRINCIPAL GENERA.

* Erica.	* Calluna.	Philippia.	Blæria.
Grisebachia.	Lagenocarpus.	Salaxis.	

TRIBE IV. RHODORACEÆ.—Corolla deciduous, sometimes irregular (*Azalea*, *Rhodora*, *Rhododendron*). Disk hypogynous, glandular. Capsule septicidal.—Leaves flat. Flower-buds scaly, strobiliform.

PRINCIPAL GENERA.

Phyllodoce.	* Rhodora.	* Ledum.	* Daboœcia.	* Rhododendron.
* Bejaria.	Loiseleuria.	* Kalmia.	* Azalea.	* Leiophyllum.

Ericineæ are very closely allied to *Vaccinieæ*, *Pyrolaceæ*, *Monotropeæ* and *Epacrideæ* (see these families). They should also be compared with *Empetreæ* and *Diapensieæ* (which 'see). They evidently approach the exotic family of *Camelliaceæ* through the genera *Saurauja* and *Clethra*. In the latter genus, as in many *Rhodoraceæ*, the corolla is polypetalous, hypogynous and imbricate, the anther-cells diverge at the base and open by a pore, the ovary is 5-celled and surrounded at its base by a hypogynous disk,

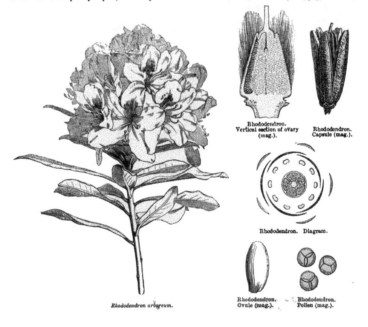

Rhododendron.
Vertical section of ovary
(mag.).

Rhododendron.
Capsule (mag.).

Rhododendron. Diagram.

Rhododendron.
Ovule (mag.).

Rhododendron.
Pollen (mag.).

Rhododendron arboreum.

| Rhododendron. Transverse section of ovary (mag.). | Rhododendron. Embryo (mag.). | Rhododendron. Style and stigma (mag.). | Rhododendron. Stamen with anther-cells perforated at the top (mag.). | Rhododendron. Seed, entire and cut longitudinally (mag.). |

the ovules are numerous and anatropous, the style is simple, the capsule loculicidal, the albumen fleshy, the embryo straight and axile, the stem woody, and the leaves alternate. The diagnosis rests on little beyond the polyandry of the one and the diplostemony of the others.

Ericineæ are scattered over the globe. A few species of Heath inhabit Central and Northern Europe, covering immense otherwise sterile tracts. The number of species increases in the Mediterranean region, and is very considerable at the Cape of Good Hope. There are no Ericas in America, Asia and Australia, in which [latter country] they are replaced by Epacrideæ. Arbutus and Andromeda, genera with a deciduous corolla, inhabit the north temperate zone; they are rare in Central Europe and the Mediterranean region, and abound in North America, where they descend towards the tropics and even cross the tropic of Capricorn. In tropical Asia they are sub-alpine; they are very rare in Australia, but several species occur in New Zealand. Rhodoraceæ chiefly inhabit the temperate and cool regions of the northern hemisphere, and especially of America. Some are found on the highest mountains of tropical America and Asia. [Their centre is in the Himalayas, whence they extend to the mountains of Borneo, where also they occur on the coast.—Ed.]

Ericineæ generally have a bitter and styptic taste, due to an extractive principle and tannin, to which is sometimes joined an aromatic resin; to this the diuretic properties of the leaves of Arctostaphylos Uva-ursi are due, and its use in cases of calculus in the bladder. Its berries are very tart; those of the Arbutus Unedo resemble a small strawberry, and have a mild taste; in some parts of Italy they are fermented, and yield a spirituous alcoholic liquor. The bark and leaves of the Arbutus contain a large quantity of gallic acid, and are used in the East to tan skins. The leaves of Gualtheria are much used in Canada under the name of Mountain Tea, and the fruit (Box-berry) is edible.

Rhodoraceæ possess, like the other Ericineæ, bitter and astringent properties, but they are also very narcotic, and must be used medicinally with great caution. The Rhododendron chrysanthos is given in North Asia for many internal and external maladies. The buds of R. ferrugineum are employed in Piedmont in the preparation of an antirheumatic liniment called Marmot Oil. The genera Rhododendron, Ledum, Kalmia and Azalea are narcotic; the honey extracted from their flowers is extremely poisonous; that which maddened Xenophon's soldiers during the retreat of the Ten Thousand was collected from either Azalea pontica or Rhododendron ponticum, which abound on the shores of the Euxine.

[The leaves of the Himalayan Andromeda ovalifolia poison goats, and sheep are killed by those of our native A. polifolia. A jelly is made in the Himalayas from the flowers of Rhododendron arboreum. Gualtheria procumbens yields a pungent volatile oil called Oil of Winter-green, used by druggists and perfumers.—Ed.]

CXXIX. MONOTROPEÆ.

(Monotropeæ, Nuttall.—Monotropaceæ, Lindl.)

Corolla hypogynous, persistent, mono- or ⚥ly-petalous, diplostemonous, æstivation imbricate. Stamens 8–10, hypogynous; anthers dehiscing variously. Ovary free,

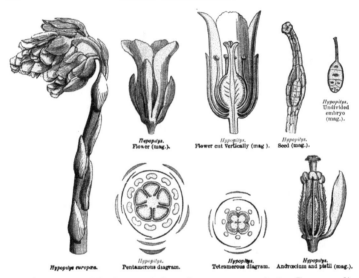

Hypopitys.
Undivided
embryo
(mag.).

Pæpopitys.
Flower (mag.).

Hypopitys.
Flower cut vertically (mag.).

Hypopitys.
Seed (mag.).

Hypopitys europæa.

Hypopitys.
Pentamerous diagram.

Hypopitys.
Tetramerous diagram.

Hypopitys.
Androecium and pistil (mag.).

5–4-celled, many-ovuled. FRUIT capsular. SEEDS numerous, minute. EMBRYO undivided, minute.—*Parasitic fleshy leafless scaly* HERBS.

Perennial HERBS resembling *Orobanche*, simple, fleshy, parasitic on the roots of trees, never green. LEAVES 0, replaced by alternating scales. FLOWERS ☿, subregular, sometimes solitary and 5-merous, sometimes in a raceme or spike, the terminal 5-merous, the others 4-merous. CALYX 5-partite, persistent, sometimes 0, or replaced by bracts. COROLLA hypogynous, white or pinkish, persistent, of 5–4 petals gibbous at the base, free or more or less cohering into a 5–4-fid corolla. STAMENS 10 or 8, inserted on the receptacle, sometimes accompanied by filiform appendages; *anthers* 1-celled, peltate, and opening by a transverse slit; or 2-celled, cells with a basal awn, opening by longitudinal slits, or awnless and opening by pores. OVARY free, ovoid or sub-globose, 4–5-celled, with 10 glands at the base; *style* simple, straight, hollow; *stigma* discoid, margined; *ovules* innumerable, on fleshy axile placentas filling the cells. CAPSULE 4–5-celled, with 4–5 loculicidal semi-septiferous valves; *placentas* fleshy. SEEDS numerous, minute, sub-spherical; *testa* loose, reticulate. EMBRYO undivided, minute.

PRINCIPAL GENERA.

Monotropa.   Hypopitys.   Schweinitzia.   Pterospora.

Monotropeæ approach *Ericineæ*, and especially *Pyrolaceæ*, in their polypetalous or sub-polypetalous diplostemonous corolla, their stamens distinct from the corolla, their anthers opening by pores or transverse slits, their many-celled and -ovuled ovary and loculicidal capsule. The diagnosis rests solely upon their parasitism and fleshy stem, provided with scales replacing the leaves.

Monotropeæ are European [Asiatic] and American parasites on the roots of trees, principally pines and beeches. Many species have the scent of violets or pinks. In some parts of Europe the shepherds give their sheep powdered *Hypopitys* to quiet their cough. The Canadian *Pterospora Andromedea* is employed by the Indians as a vermifuge and diaphoretic.

CXXX. *PYROLACEÆ*.

(ERICARUM *genera*, *Jussieu*.—PYROLACEÆ, *Lindl.*)

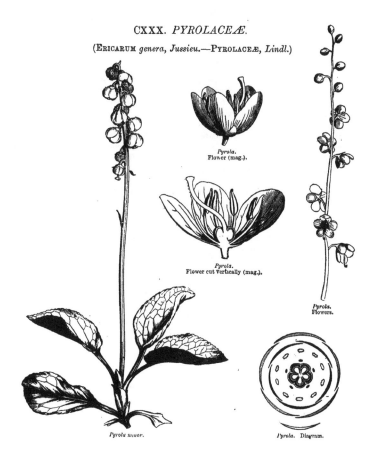

Pyrola.
Flower (mag.).

Pyrola.
Flower cut vertically (mag.).

Pyrola.
Flowers.

Pyrola minor.

Pyrola. Diagram.

Pyrola.	*Pyrola.*	*Pyrola.*	*Pyrola.*	*Pyrola.*
Transverse section of ovary (mag.).	Pistil (mag.).	Stamen (mag.).	Ovule (mag.).	Embryo taken out of the testa (mag.).

COROLLA *polypetalous or sub-polypetalous, hypogynous, diplostemonous, æstivation imbricate.* STAMENS 10, *not adhering to the petals ;* ANTHERS *usually 2-celled, opening by 2 pores, or by a transverse slit.* OVARY 3-5-*celled.* CAPSULE *loculicidal.* SEEDS *minute.* EMBRYO *undivided, minute.*

Perennial HERBS, sometimes sub-woody at the base, rarely woody. LEAVES scattered or sub-whorled, exstipulate. FLOWERS ☿, regular, in a raceme or umbel, or solitary, white or pinkish. CALYX 5-partite, persistent. COROLLA of 5 petals inserted on the receptacle, æstivation imbricate. STAMENS 10, hypogynous, all fertile and distinct, or monadelphous at the base, 5 fertile and 5 without anthers (*Galax*) ; *anthers* introrse, 2-celled, opening at the top by 2 pores or by an oblique slit ; or 1-celled, opening by 2 transverse valves. OVARY free, seated on a hypogynous disk; 3-5-celled ; *style* terminal ; *stigma* capitate, girt with a membranous ring ; [*ovules* very numerous, on placentas projecting from the inner angle of the cells, anatropous]. CAPSULE 3-5-celled, loculicidal, valves semi-septiferous, placentas fungoid. SEEDS numerous, minute ; *testa* loose, much larger than the nucleus. EMBRYO minute, undivided.

PRINCIPAL GENERA.

Pyrola.   Chimaphila.

The small family of *Pyrolaceæ* has been separated from *Ericineæ*, from which it scarcely differs, save in the structure of the seed. We have indicated the close affinity connecting it with *Monotropeæ* (see this family), which may be looked upon as parasitical *Pyrolaceæ*.

Pyrolaceæ inhabit the temperate and cool regions of the northern hemisphere. They owe their medicinal properties to bitter and resinous principles. *Chimaphila umbellata* is recommended in America as stimulating the functions of the kidneys and skin. *P. rotundifolia* was formerly employed in Europe as an astringent in dysenteric flux and hæmorrhage.

CXXXI. *VACCINIEÆ.*

(ERICARUM *genera, Jussieu.*—VACCINIEÆ, *D.C.*—VACCINIACEÆ, *Lindl.*)

COROLLA *monopetalous, epigynous, diplostemonous, æstivation imbricate.* STAMENS *epigynous ;* ANTHERS *of 2 bipartite cells, opening by 2 pores at the top.* OVARY *inferior, many-celled*; OVULES *anatropous.* FRUIT *fleshy.* EMBRYO *albuminous.*—STEM *woody.*

Branching SHRUBS. LEAVES scattered or alternate, simple, entire or toothed, exstipulate. FLOWERS solitary or racemed. CALYX 4–5–6-partite, deciduous or persistent. COROLLA monopetalous, epigynous, in 4–5–6 segments, deciduous, æsti-

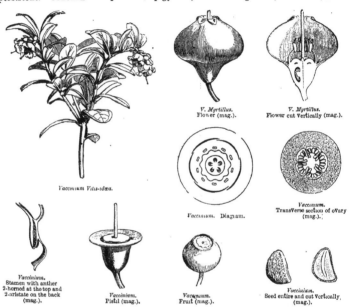

V. Myrtillus.
Flower (mag.).

V. Myrtillus.
Flower cut vertically (mag.).

Vaccinium Vitis-idæa.

Vaccinium. Diagram.

Vaccinium.
Transverse section of ovary
(mag.).

Vaccinium.
Stamen with anther
2-horned at the top and
2-aristate on the back
(mag.).

Vaccinium.
Pistil (mag.).

Vaccinium.
Fruit (mag.).

Vaccinium.
Seed entire and cut vertically,
(mag.).

vation imbricate. STAMENS twice as many as the corolla-segments, inserted on a disk crowning the ovary; *anthers* dorsifixed, vertical, 2-celled; cells parallel, often separate at the top, and terminating in a narrow tube, open at the tip. OVARY inferior, of 4–5–6–10–many 1-ovuled cells; *style* simple; *stigma* usually capitate; *ovules* [one, few or many, on placentas projecting from the inner angles of the cells,] anatropous. FRUIT a berry or drupe. SEEDS inserted on central placentas. EMBRYO straight, in the axis of a fleshy albumen.

PRINCIPAL GENERA.

* Thibaudia.   * Vaccinium.   Oxycoccos.   * Macleania.

Vaccinieæ only differ from *Ericineæ* in the epigyny; many botanists therefore persist in keeping them in the same family. To this structural affinity are added analogous properties. The berries of *Vaccinium* and *Oxycoccos* are acid, sweet, and slightly astringent; preserves are made of them, and in some countries they are used as antiscorbutics. The flowers of *Thibaudia melliflora* contain an abundance of honey, and the Peruvians of the Andes suck them with avidity.

Vacciniæ usually inhabit northern temperate regions, but many are South American and Indian, and they also occur on the high tropical mountains of Asia, Africa and Madagascar; [one genus, *Wittsteinia*, is Australian].

CXXXII. *EPACRIDEÆ.*

(ERICEARUM *sectio*, *Link*.—EPACRIDEÆ, *Br*.—EPACRIDACEÆ, *Lindl*.)

COROLLA *monopetalous, hypogynous, usually isostemonous, æstivation imbricate or valvate*. STAMENS *inserted on the receptacle or corolla*; ANTHERS 1-*celled*. OVARY *seated on a disk*, 2-*many-celled*; OVULES *pendulous, anatropous*. FRUIT *fleshy or dry*. EMBRYO *albuminous*.—STEM *woody*.

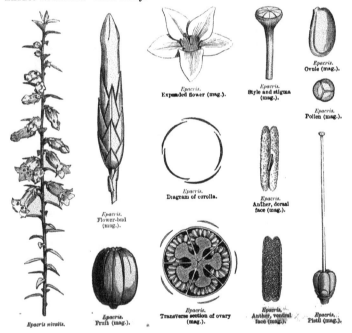

Epacris.
Ovule (mag.).

Epacris.
Expanded flower (mag.).

Epacris.
Style and stigma (mag.).

Epacris.
Pollen (mag.).

Epacris.
Diagram of corolla.

Epacris.
Anther, dorsal face (mag.).

Epacris.
Flower-bud (mag.).

Epacris.
Fruit (mag.).

Epacris.
Transverse section of ovary (mag.).

Epacris.
Anther, ventral face (mag.).

Epacris.
Pistil (mag.).

Epacris nivalis.

Archeria hirtella.
Vertical section of ovary,
showing the upright
ovules and basilar
style.

Leucopogon.
Fruit
(mag.).

Leucopogon.
Transverse section of
fruit (mag.).

Leucopogon.
Vertical section of fruit, showing
the pendent seeds and sub-
terminal style (mag.).

Leucopogon.
Seed cut verti-
cally (mag.).

SHRUBS or SMALL TREES, stems and branches inarticulate. LEAVES alternate, often close-set, rarely opposite, usually petioled, sometimes sheathed at the base, exstipulate. FLOWERS ⚥, rarely incomplete by suppression, in terminal spikes or racemes, or axillary and solitary, pedicels with 2 or more cylindrical bracts. CALYX 4–5-partite, persistent. COROLLA monopetalous, inserted on the receptacle, tubular, bell- funnel- or salver-shaped, base of tube naked within or with bundles of hairs or glands alternating with the stamens; *limb* 4–5-fid, regular, valvate or imbricate in bud; *lobes* rarely cohering, in which case the corolla dehisces transversely above the persistent base of the tube (*Richea, Cystanthe*). STAMENS 4–5, rarely fewer, hypogynous or epipetalous, alternate with the corolla-lobes; *anthers* basifixed, simple, with 2 longitudinal valves, and one complete polliniferous septum; *pollen* sub-globose or 3-globate. OVARY free, sessile on a disk, or surrounded at its base by free or connate hypogynous glands, 2–10- (rarely 1-) celled; *style* simple; *stigmas* undivided; *ovules* pendulous, rarely erect (*Archeria*), solitary from the top of the cell, or many on projecting placentas, anatropous. FRUIT 2 or more, rarely 1-celled (by suppression), either a drupe with many 1-seeded stones, or a septi-loculi-cidal many-seeded capsule with the placentas free or attached to the central column. EMBRYO straight, cylindric, in the axis of a fleshy albumen; *cotyledons* very short; *radicle* superior in the drupes, various in the capsules.

TRIBE I. STYPHELIEÆ.—Ovary with 1-ovuled cells. Fruit a drupe.

PRINCIPAL GENERA.

Pentachondra.	Acrotriche.	Melichrus.	* Styphelia.
* Leucopogon.	Lissanthe.	* Stenanthera.	Astroloma.
Monotoca.	Trochocarpa.	Conostephium.	Cyathodes.

TRIBE II. EPACREÆ.—Ovary-cells many-ovuled. Fruit capsular.

PRINCIPAL GENERA.

Lysinema.	Archeria.	* Epacris.	Prionotis.
Richea.	* Sprengelia.	Andersonia.	Dracophyllum.

Epacrideæ only differ from *Ericineæ* in the structure of their anthers. They are almost exclusively natives of Australia (where they represent the Heaths); a few only inhabiting the Moluccas, New Zealand and the Pacific Islands, and one South America. The *Epacrideæ* are mostly ornamental plants, cultivated in European greenhouses. Some, as *Lissanthe sapida*, have an edible drupe.

CXXXIII. *DIAPENSIACEÆ.*

UNDERSHRUBS [or HERBS] of Europe [North Asia] and North America. LEAVES alternate, imbricate, evergreen, nerveless. FLOWERS terminal, solitary. CALYX 3-bracteate, of 5 sepals, 2-seriate, unequal. COROLLA hypogynous, salver-shaped,

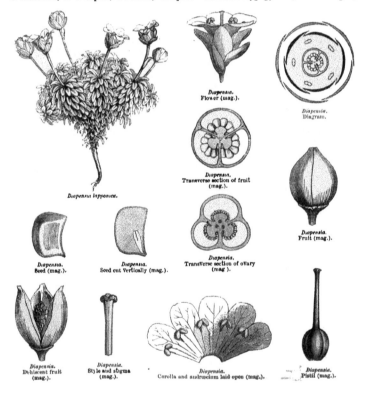

Diapensia.
Flower (mag.).

Diapensia.
Diagram.

Diapensia.
Transverse section of fruit
(mag.).

Diapensia lapponica.

Diapensia.
Seed (mag.).

Diapensia.
Seed cut vertically (mag.).

Diapensia.
Transverse section of ovary
(mag.).

Diapensia.
Fruit (mag.).

Diapensia.
Dehiscent fruit
(mag.).

Diapensia.
Style and stigma
(mag.).

Diapensia.
Corolla and andrœcium laid open (mag.).

Diapensia.
Pistil (mag.).

5-lobed, æstivation imbricate. STAMENS 5, inserted on the corolla, [often alternating with scales or staminodes]; *filaments* [usually] dilated; *anthers* 2-celled, opening by 2 transverse [rarely longitudinal] muticous valves, or one valve: sometimes aristate (*Pyxidanthera*); [*pollen* simple]. OVARY free, [not seated on a disk], 3- [rarely 4-] celled; *style* terminal, simple; *stigma* 3-lobed; [*ovules* very numerous, attached to the inner angles of the cells, anatropous or amphitropous]. CAPSULE thin, terminated by the persistent 3-celled style, opening at the top in 3 loculicidal semi-septiferous valves. SEEDS nearly cubical, fixed by a ventral hilum to central fungoid placentas; *testa* lax, membranous, areolate. EMBRYO filiform, [minute, terete], in the axis of a fleshy albumen; *cotyledons* very short; *radicle* long, parallel to the hilum.

[*Diapensiaceæ* have been recently studied by Professor A. Gray, who separates them from *Ericaceæ* on account of the absence of a disk and of a marginate stigma, and because they have peculiar anthers and simple pollen. The characters given above have been modified in accordance with his views. He associates *Galax* with them, and divides them into two tribes as follows:—

TRIBE I. DIAPENSIEÆ.—Filaments dilated, staminodes 0. Placentas thick, adnate to a persistent columella. Testa not lax and reticulate.—Tufted depressed evergreen leafy under-shrubs. Leaves small, sessile, nerveless, evergreen, quite entire. Flower solitary, terminal. *Pyxidanthera, Diapensia.*

TRIBE II. GALACINEÆ.—Filaments flattened, alternating with as many staminodes. Seeds with a lax reticulate testa.— Stemless herbs with a creeping rhizome. Leaves long-petioled, toothed, nerved, evergreen. Flowers on long scapes rising from the rhizome. *Shortia, Galax.*

Diapensia inhabits North Europe and the Himalayas; *Pyxidanthera* and *Galax* the United States; *Shortia* the United States and Japan.—ED.]

This little family, composed of the genera *Diapensia* and *Pyxidanthera*, naturally falls into place near *Ericineæ*, with which it is allied by the monopetalous hypogynous imbricate corolla, the two-celled anthers and their anomalous dehiscence, the many-celled and -ovuled ovary, central placentation, loculicidal capsule, fleshy albumen, axile embryo, woody stem and imbricating leaves. It scarcely differs from *Ericineæ*, except by the insertion of the stamens on the corolla-throat.

CXXXIV. *PLUMBAGINEÆ.*

(PLUMBAGINES, *Jussieu.*—PLUMBAGINEÆ, *Ventenat.*—PLUMBAGINACEÆ, *Lindl.*)

COROLLA *monopetalous or sub-polypetalous, hypogynous, isostemonous, æstivation contorted or imbricate.* STAMENS 5, *hypogynous, or inserted on the corolla, and opposite to its lobes.* OVARY 1-*celled;* OVULE *solitary, anatropous, pendulous from a funicle springing from the bottom of the cell.* FRUIT *dry.* EMBRYO *straight;* ALBUMEN *farina-ceous;* RADICLE *superior.*

Herbaceous or woody generally perennial PLANTS. LEAVES sometimes fascicled at the top of a rhizome, simple, entire, semi-amplexicaul, sometimes alternate on a branching stem with swollen nodes, sometimes shortened into a petiole dilated at its

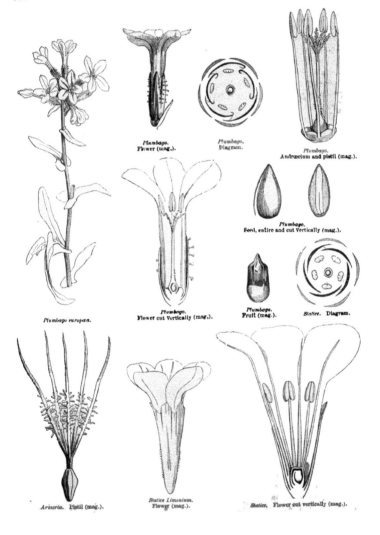

Plumbago.
Flower (mag.).

Plumbago.
Diagram.

Plumbago.
Andrœcium and pistil (mag.).

Plumbago.
Seed, entire and cut vertically (mag.).

Plumbago europæa.

Plumbago.
Flower cut vertically (mag.).

Plumbago.
Fruit (mag.).

Statice. Diagram.

Armeria. Pistil (mag.).

Statice Limonium.
Flower (mag.).

Statice. Flower cut vertically (mag.).

base and amplexicaul, exstipulate. FLOWERS ⚥, on simple or branched scapes, in unilateral-spikes, panicles, or scarious capitula; each flower 3-2-bracteate, usually scarious. CALYX persistent, tubular, scarious coriaceous or herbaceous, sometimes coloured, with 5 folds and teeth, rarely 5-phyllous. COROLLA inserted on the receptacle, sometimes monopetalous, hypocrateriform, with a narrow angular tube and 4-partite regular limb, imbricate in æstivation (*Plumbago*); sometimes of 5 petals cohering by their bases or quite free, contorted in æstivation (*Statice*). STAMENS 5, opposite to the petals or corolla-lobes, inserted on the receptacle when the flower is monopetalous, and on the claws of the petals when it is polypetalous; *filaments* filiform; *anthers* introrse, cells separate at the base, opening by longitudinal slits. OVARY free, 1-celled, with 5 prominences at the top; *styles* 5 (rarely 3–4), inserted on the prominences of the ovary, distinct, rarely connate; *stigmas* capillary, furnished on their inner surface with several lines of glands, rarely capitate; *ovule* solitary, anatropous, suspended from a funicle fixed to the bottom of the cell. FRUIT membranous, included in the calyx, sometimes a capsule and 5-valved at the top, sometimes a utricle breaking irregularly round its base and along its sides. SEED inverted, sometimes appearing erect by the union of the funicle with its integuments (*Plumbago*). EMBRYO straight, in a scanty farinaceous albumen; *cotyledons* flat; *radicle* short, superior.

TRIBE I. PLUMBAGINEÆ VERÆ.—Calyx herbaceous. Corolla monopetalous. Stamens inserted on the receptacle. Styles connate. Fruit a capsule.

PRINCIPAL GENERA

* Plumbago. Vogelia.

TRIBE II. STATICEÆ.—Calyx scarious or coriaceous. Corolla with 5 free or almost free petals. Stamens inserted at the base of the petals. Styles distinct. Fruit a utricle.

PRINCIPAL GENERA.

* Armeria. * Statice. Ægialitis. Acantholimon.

We shall indicate the affinities between *Plumbagineæ* and *Plantagineæ* under the latter family. The diagnosis of *Plantagineæ* rests on its two-celled and many-ovuled ovary, its peltate seeds, simple style, and non-farinaceous albumen. The affinity is closer between *Plumbagineæ* and *Primulaceæ*: in both the anthers are introrse and opposite to the petals, the æstivation contorted, at least in *Statice*, the ovary one-celled, and the central placentation free, the fruit opens circularly, or by more or less perfect valves, and the embryo is straight; the ovule is indeed solitary in *Plumbagineæ*, but, as Brongniart has observed, the ovary is symmetrical, with five nerves and five stigmas, which indicates a plurality of carpels. *Primulaceæ* are distinguished by their simple stigma, ovules with a ventral hilum and non-farinaceous albumen. Endlicher has noticed some relations between *Plumbagineæ*, *Brunoniaceæ* and *Globularieæ*, founded on the inflorescence, hypogyny, one-celled and one-ovuled ovary and anatropous ovule; but a full diagnosis weakens this affinity (see these families), which is truer with *Frankeniaceæ*. In this family, as in *Plumbagineæ*, we find a stem with swollen nodes, fascicled leaves, an hypogynous isostemonous corolla, contorted æstivation, a one-celled ovary, styles furnished with stigmatic papillæ on their inner edge, and a farinaceous albumen. *Plumbagineæ* and *Polygoneæ* may also be compared: in both the stamens are hypogynous, the ovary is one-celled and one-ovuled, the styles are distinct or coherent, and the albumen is farinaceous; but here again the differences exceed the resemblances.

Plumbagineæ are cosmopolitan plants. *Statice* inhabits maritime shores and salt lands of the temperate regions of both hemispheres. *Armeria* is dispersed over both continents; many species grow on mountains in the arctic and antarctic regions. *Plumbago europæa* is the only European species of the genus; the others are tropical and sub-tropical.

The leaves of *Armeria vulgaris* and the root of *Statice Limonium*, although possessing decided tonic and astringent properties, have fallen into disuse. The root of *S. latifolia*, a species near *Limonium*, which has been recently imported from Russia, contains a large quantity of gallic acid, which renders it useful in tanning and for black dyes. The Plumbagos contain a very caustic colouring matter; the root of the European species contains a fatty substance which gives a leaden colour to fingers and paper, and which was formerly used for toothache, cutaneous diseases and cancerous ulcers; surgeons have given up its use, but beggars make use of it to produce sores, and thus excite pity. Many American and Asiatic species (*Plumbago zeylanica, rosea, scandens*) are considered in India to be alexipharmics. Some others (*Pl. Larpentæ, cærulea*, &c.) are cultivated in Europe as ornamental plants.

CXXXV. *PRIMULACEÆ.*

(LYSIMACHIÆ, *Jussieu.*—PRIMULACEÆ, *Ventenat.*)

COROLLA *monopetalous, hypogynous (or rarely perigynous), isostemonous, æstivation contorted or imbricate, very seldom 0.* STAMENS *opposite to the corolla-lobes.* OVARY *free, or very rarely inferior, 1-celled; placenta central, globose, many-ovuled;* OVULES *fixed by their ventral face.* FRUIT *a capsule.* EMBRYO *albuminous.*—HERBS *with radical or opposite leaves.*

HERBS with a woody rhizome, sometimes tuberous, very rarely suffrutescent. STEM generally a subterranean rhizome. LEAVES gland-dotted, sometimes all radical, and tufted; sometimes cauline, opposite or whorled, very rarely alternate, exstipulate. FLOWERS ☿, regular, very rarely sub-irregular, solitary and radical, or in scapose umbels, or axillary, and then solitary or racemed in terminal spikes. CALYX tubular, 5-fid or -partite, rarely 4-6-7-fid. COROLLA monopetalous, rotate, campanulate, infundibuliform, or sub-2-labiate (*Coris*), very rarely 3-petaled (*Pelletiera*), or 0 (*Glaux*). STAMENS inserted on the corolla-tube or -throat, opposite to its divisions, often alternating with as many petaloid scales (*staminodes*); *filaments* filiform or subulate, usually very short; *anthers* introrse, 2-celled, sometimes shorter than the connective, dehiscence longitudinal. OVARY free, or rarely enclosed in the receptacular cupule (*Samolus*), 1-celled; *placenta* central or basilar, free, globose, sessile or stipitate, continuous with the conducting tissue of the style; *style* terminal, simple; *stigma* undivided; *ovules* numerous, peltate, semi-anatropous, or very rarely anatropous (*Hottonia, Samolus*). CAPSULE 1-celled, opening at the top, or throughout its length by valves or by entire or bifid teeth, or transversely. SEEDS sessile in the pits of the placenta, hilum ventral, rarely basilar. EMBRYO straight, parallel to the hilum, in the axis of a fleshy or sub-horny albumen; *cotyledons* semi-cylindric; *radicle* vague.

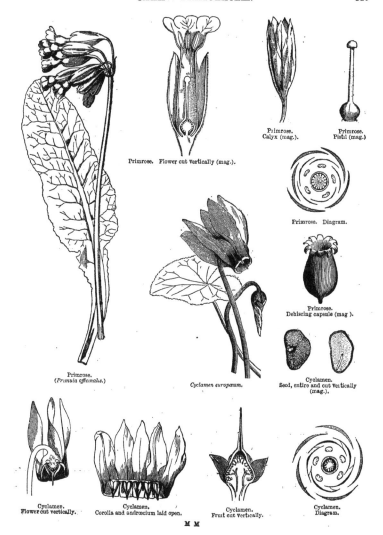

Primrose. Flower cut vertically (mag.).

Primrose.
Calyx (mag.).

Primrose.
Pistil (mag.).

Primrose. Diagram.

Primrose.
Dehiscing capsule (mag).

Cyclamen.
Seed, entire and cut vertically
(mag.).

Primrose.
(*Primula officinalis.*)

Cyclamen europæum.

Cyclamen.
Flower cut vertically.

Cyclamen.
Corolla and andrœcium laid open.

Cyclamen.
Fruit cut vertically.

Cyclamen.
Diagram.

M M

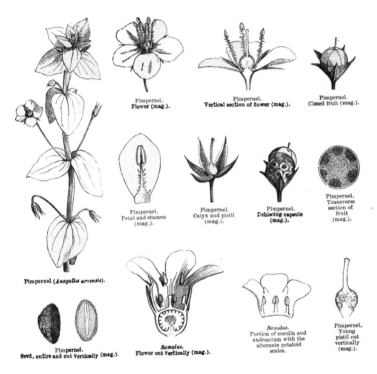

Pimpernel.
Flower (mag.).

Pimpernel.
Vertical section of flower (mag.).

Pimpernel.
Closed fruit (mag.).

Pimpernel.
Petal and stamen
(mag.).

Pimpernel.
Calyx and pistil
(mag.).

Pimpernel.
Dehiscing capsule
(mag.).

Pimpernel.
Transverse
section of
fruit
(mag.).

Pimpernel (*Anagallis arvensis*).

Pimpernel.
Seed, entire and cut vertically (mag.).

Samolus.
Flower cut vertically (mag.).

Samolus.
Portion of corolla and
andrœcium with the
alternate petaloid
scales.

Pimpernel.
Young
pistil cut
vertically
(mag.).

TRIBE I. PRIMULEÆ.—Ovary free. Capsule opening by valves or valvules. Seeds with ventral hilum.

PRINCIPAL GENERA.

Androsace.
* Soldanella.
Coris.

* Dodecatheon.
Trientalis.
* Cyclamen.

* Bernardina.
* Cortusa.
* Lysimachia.

* Primula.
Glaux.

TRIBE II. ANAGALLIDEÆ.—Ovary free. Capsule opening transversely. Seeds with ventral hilum.

PRINCIPAL GENERA.

Asterolinum.   Centunculus.   * Anagallis.   Euparea.

TRIBE III. HOTTONIEÆ.—Ovary free. Capsule opening by valves. Seeds with a basilar hilum.—Aquatic submerged plants.

GENUS.

Hottonia.

TRIBE IV. SAMOLEÆ.—Ovary semi-inferior. Capsule opening by valves. Seeds with a basilar hilum.

GENUS.

Samolus.

We have mentioned the affinities of *Primulaceæ* with *Plumbagineæ* and *Plantagineæ* (see these families). They are much more closely allied to *Myrsineæ*, by the hypogynous or perigynous corolla, stamens opposite to the corolla-lobes, 1-celled ovary, free central placentation, ventral hilum of the ovules, and albuminous embryo. *Myrsineæ* only differ in their woody stem and fleshy fruit.

Primulaceæ mostly inhabit the temperate regions of Europe and Asia; many species are alpine. Few are found in the southern hemisphere, except *Samolus*, the species of which are numerous in Australia. Some genera are met with on the mountains and shores of the tropical zone.

Primulaceæ are more remarkable for their beauty than for their utility. Many species contain an acrid and volatile substance in their roots, others an extractable bitter and resinous substance; the foliage of some is astringent; the flowers of most are sweet-scented. The rhizomes of the Primrose (*Primula veris*) were formerly employed for rheumatism in the joints and for diseases of the kidney and bladder; and an infusion of its flowers is still prescribed as a diaphoretic. Primrose wine is made from the flowers of *P. acaulis* and *veris*. The Auricula (*Primula Auricula*) is employed by the natives of the Alps against pulmonary consumption. The tuberous rhizome of the *Cyclamen europæum* is acrid, strongly purgative, and even emetic; it formerly entered into the composition of an ointment, which, applied to the stomach, purged or caused vomiting. In some countries the powdered rhizome is employed to stupefy fish; but, dried and roasted, it becomes edible on account of the starch which it contains, and pigs eat it with avidity, whence its common name of Sowbread. The Pimpernels were formerly prescribed in dropsy, epilepsy, and even in hydrophobia. The species of *Lysimachia*, and especially *L. nummularia*, were considered astringent, but have fallen into disuse, as has *Samolus*. The *Coris* of Montpellier is an undershrub, containing a bitter nauseous principle, the use of which has been suggested in syphilis.

CXXXVI. *MYRSINEÆ.*

(MYRSINEÆ, *Br.*—OPHIOSPERMEÆ, *Ventenat.*—MYRSINACEÆ, *Lindl.*—MYRSINEACEÆ, *A. D.C.*)

COROLLA monopetalous, regular, isostemonous, hypogynous or perigynous. STAMENS inserted on the corolla, and opposite to its lobes. OVARY 1-celled; PLACENTA central, free; OVULES campylotropous. FRUIT a drupe or berry. EMBRYO albuminous.—STEM woody.

TREES or SHRUBS. LEAVES generally alternate, simple, coriaceous, gland-

M M 2

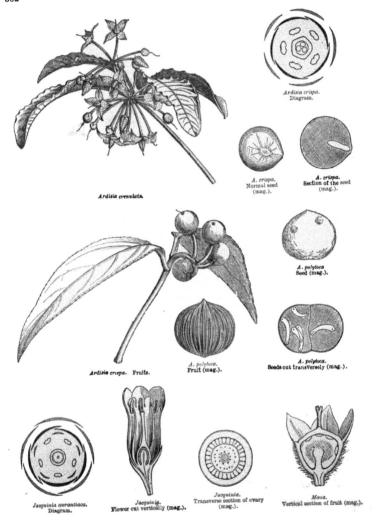

Ardisia crenulata.

Ardisia crispa.
Diagram.

A. crispa.
Normal seed
(mag.).

A. crispa.
Section of the seed
(mag.).

Ardisia crispa. Fruits.

A. polytoca.
Fruit (mag.).

A. polytoca
Seed (mag.).

A. polytoca.
Seeds cut transversely (mag.).

Jacquinia aurantiaca.
Diagram.

Jacquinia.
Flower cut vertically (mag.).

Jacquinia.
Transverse section of ovary
(mag.).

Mæsa.
Vertical section of fruit (mag.).

Monotheca.
Flower (mag.).

Monotheca.
Portion of corolla and andrœcium (mag.).

Monotheca.
Seed cut vertically (mag.).

Monotheca.
Ovule (mag.).

Monotheca.
Transverse section of ovary
(mag.).

Monotheca.
Fruit opened (mag.).

Ægiceras.   Flower cut vertically.

Monotheca.
Pistil (mag.).

Monotheca.
Pistil opened (mag.).

Monotheca.
Fruit (mag.).

Ægiceras.
Andrœcium detached from
the corolla-tube, and retaining
2 anthers.

Ægiceras.
Flower-bud, æsti-
vation imbricate,
contorted.

Ægiceras.
Follicular fruit, bent into
a horn, 1-seeded, splitting
at one side when ripe
and furnished with its
persistent calyx.

Ægiceras.
Seed suspended from the
placenta, and showing
the radicle, which has
pierced its integuments.

Ægiceras.
Embryo cut
longitudinally to
show the
cotyledons.

Ægiceras.
Embryo with coty-
ledons joined in a
cylindric tube
(mag.).

dotted, exstipulate. FLOWERS ☿, often imperfect through arrest, regular, usually
axillary, umbelled, corymbose, fascicled, racemed or panicled, often covered with
glands. CALYX 4–5-fid or -partite. COROLLA monopetalous, or sometimes polypetalous, campanulate or rotate, isostemonous. STAMENS inserted on the corolla-
tube or throat, and opposite to its lobes, sometimes alternating with as many petaloid
scales (*staminodes*) ; *filaments* short, free, or more or less cohering in a tube ; *anthers*
2-celled, sometimes connivent, dehiscence longitudinal or apical. OVARY free or
inferior, 1-celled ; *placenta* basilar or central, sessile or stipitate ; *style* short, simple ;
stigma usually undivided ; *ovules* fixed to the placenta by a ventral, linear, or punctiform hilum, exceptionally anatropous (*Monotheca*). FRUIT a drupe or berry, usually
few-seeded, or 1-seeded by arrest. SEEDS with a simple integument, often mucilaginous, sometimes with many embryos. EMBRYO cylindric, usually arched, parallel
to the hilum in the many-seeded fruits, and transverse in the single-seeded ; *albumen* fleshy or horny ; *cotyledons* semi-cylindric, or flat and sub-foliaceous ; *radicle*
terete, longer than the cotyledons, inferior or vague.

TRIBE I. ARDISIEÆ.—Æstivation contorted. Anthers introrse. Ovary free.
Fruit 1-seeded.

PRINCIPAL GENERA.

Myrsine.   * Ardisia.

TRIBE II. MÆSEÆ.—Æstivation induplicate-valvate. Anthers introrse. Ovary
inferior. Fruit many-seeded.

GENUS.

Mæsa.

TRIBE III. THEOPHRASTEÆ.—Æstivation imbricate. Staminodes 5. Anthers
extrorse. Fruit many-seeded. Placenta sometimes minute, and ovules anatropous
(*Monotheca*).

PRINCIPAL GENERA.

Theophrasta.   * Jacquinia.   Monotheca.

We have indicated the affinity between *Myrsineæ* and *Primulaceæ*, which is so close that they might
be united (see *Primulaceæ*).

Myrsineæ principally inhabit the tropical zone of Asia and America ; they are rare beyond the tropics,
at the Cape of Good Hope, in Australia, [New Zealand,] Japan and the Canaries. *Theophrasta* is an
American genus. *Mæsa* belongs to the Old World ; *Ardisia* to the hot regions of Asia, Africa and
America, and extends to the Canaries. The fruit of some species of *Ardisia* is edible. The leaves of
Jacquinia are used in America to stupefy fish, like the rhizomes of *Cyclamen*, and their fruit is poisonous.
The seeds of *J. armillaris* were strung like pearls by the Caribbeans to form bracelets. The crushed seeds
of *Theophrasta Jussieui*, called at St. Domingo *Petit Coco*, are used for making bread.

Near *Myrsineæ* is placed the genus *Ægiceras*, which comprises shrubs growing on the shores of tropical
Asia and Oceania, with alternate leaves and hermaphrodite flowers in an umbel. The corolla, stamens

and ovary present the same characters as *Myrsineæ* ; the fruit is a curved 1-seeded (by arrest) follicle. The seed is upright, and germinates in the pericarp ; its membranous integument tears during germination and caps the cotyledonary end.   The embryo, as in many aquatic plants, is exalbuminous ; the cotyledons form a cylindrical tube, and the radicle is inferior.

CXXXVII. *SAPOTEÆ.*

(SAPOTÆ, *Jussieu.*—SAPOTEÆ, *Br.*—SAPOTACEÆ, *Endl.*)

TREES or SHRUBS with milky juice.   LEAVES alternate, entire, coriaceous, exstipulate.   FLOWERS ☿, axillary.   CALYX 4–8-partite.   COROLLA monopetalous,

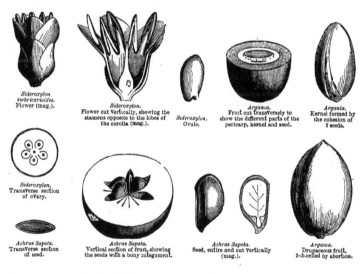

Siderozylon
imbricarioides.
Flower (mag.).

Siderozylon.
Flower cut vertically, showing the
stamens opposite to the lobes of
the corolla (mag.).

Siderozylon.
Ovule.

Argania.
Fruit cut transversely to
show the different parts of the
pericarp, kernel and seed.

Argania.
Kernel formed by
the cohesion of
2 seeds.

Siderozylon,
Transverse section
of ovary.

Achras Sapota.
Transverse section
of seed.

Achras Sapota.
Vertical section of fruit, showing
the seeds with a bony integument.

Achras Sapota.
Seed, entire and cut vertically
(mag.).

Argania.
Drupaceous fruit,
2–3-celled by abortion.

hypogynous, regular, [4–8- lobed, æstivation imbricate, sometimes in two series]. STAMENS inserted on the corolla, the fertile equalling in number the corolla-lobes, and opposite to them, or more numerous, 2- (or more) seriate, sometimes with alternating staminodes ; *anthers* usually extrorse, [dehiscence longitudinal].   OVARY many-celled ; *style* conical or cylindric ; *stigma* acute or capitellate ; *ovules* solitary in the cells, ascending from the base of the inner angle, anatropous. FRUIT a one- or many-celled berry [a 4-valved capsule in *Ponteria*]. SEEDS with a bony [or crustaceous, nearly shining] testa; [*hilum* often large and longitudinal]; *albumen* 0 or scanty, fleshy or oily. EMBRYO large; *cotyledons* broad, foliaceous; *radicle* inferior.

[PRINCIPAL GENERA.

Chrysophyllum.	Ponteria.	Lucuma.	Sapota.
Bassia.	Mimusops.	Argania.	Imbricaria.
Sideroxylon.	Isonandra.	Bumelia].	

Sapoteæ approach *Myrsineæ* in their hypogynous monopetalous corolla, stamens opposite to the corolla-lobes, usually extrorse anthers, straight albuminous embryo, woody stem, and alternate leaves; they are distinguished by their anisostemonous corolla, many-celled ovary and anatropous ovules. They have also an obvious affinity with *Ebenaceæ* in their arborescent stem, alternate entire leaves, axillary inflorescence, monopetalous hypogynous regular corolla, many-celled ovary, fleshy fruit, and albuminous embryo; but in *Ebenaceæ* the wood is very hard, and there is no milky juice, the flowers are often unisexual, the calyx and corolla always uniseriate, the anthers always introrse, the ovules geminate, pendulous and collateral.

This family inhabits tropical and sub-tropical regions, and includes several species useful to man. The fruits of *Lucuma mammosa* [the Marmalade of the West Indies] are a very agreeable food; as are those of *Achras Sapota* and [various species of] *Chrysophyllum*, which are much sought after in the Antilles; those of *Bassia* and *Imbricaria*, Asiatic genera, are also edible. From the seeds of *Bassia butyracea*, in India, and of *B. Parkii*, in Senegal, a fixed oil is expressed (Galam Butter), which quickly curdles, and is much used as food. Other *Sapoteæ*, both Asiatic and African (*Sideroxylon*, *Argania*), are employed for building purposes on account of the hardness of the wood, whence their name of Iron Wood. Finally, a Malayan tree (*Isonandra gutta*) furnishes Gutta Percha, a substance of a resinous nature, allied to india-rubber, which is so useful in various manufactures, from its plasticity.

[Other valuable tropical fruits are the Star-Apple (*Chrysophyllum Cainito*), the West Indian Medlar (*Mimusops Elengi*), the Bullet-tree of Guiana (*M. Balata*), the Abi or Abui of Peru (*Lucuma Caimito*), and those of two Mauritian species of *Imbricaria*. The genus *Bassia* contains *B. Parkii*, the Butter-tree of Park, which produces the Shea Butter of West Africa; *B. butyracea* and *B. longifolia*, Indian Butter-trees, which make an excellent soap; *B. latifolia*, the Mahoua of Bengal, from whose fleshy flowers an arrack is made, which is extensively drunk. The flowers of *Mimusops Elengi* yield a fragrant essence, and the seeds an oil much used by painters. The bark of different species of *Mimusops*, *Achras* and *Bumelia* is bitter, astringent and febrifuge. The seeds of *Achras* and *Sapota* are aperient and diuretic. The fruit of the Maroccan *Argania Sideroxylon* is greedily eaten by cattle and goats, and the seeds which they pass are afterwards collected and crushed for the bland oil which they contain, and which rivals olive oil as an article of food and illumination; its wood is intensely hard, as is that of the Guiana Bullet-tree, and of various species of *Mimusops* and *Sideroxylon*. Lastly, the Cow-tree of Para (*Massaranduba*) is probably a species of this family.—ED.]

CXXXVIII. *EBENACEÆ.*

(GUAJACANEÆ, *partim, Jussieu.*—EBENACEÆ, *Ventenat.*—DIOSPYREÆ, *Duby.*)

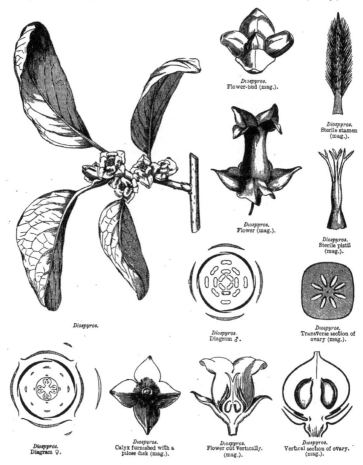

Diospyros.
Flower-bud (mag.).

Diospyros.
Sterile stamen
(mag.).

Diospyros.
Flower (mag.).

Diospyros.
Sterile pistil
(mag.).

Diospyros.

Diospyros.
Diagram ♂.

Diospyros.
Transverse section of
ovary (mag.).

Diospyros.
Diagram ♀.

Diospyros.
Calyx furnished with a
pilose disk (mag.).

Diospyros.
Flower cut vertically.
(mag.).

Diospyros.
Vertical section of ovary.
(mag.).

Diospyros.
Fruit with persistent calyx
(mag.).

Diospyros.
Vertical section of fruit
(mag.).

Diospyros.
Transverse section of fruit
(mag.).

Diospyros.
Vertical section of seed
(mag.).

COROLLA *monopetalous, hypogynous, regular, 3–7-lobed, æstivation imbricate.* STAMENS *inserted on the corolla or receptacle, equalling the corolla-lobes, or double or quadruple in number.* OVARY *free, of many 1–2-ovuled cells;* OVULES *pendulous, anatropous.* FRUIT *a berry.* EMBRYO *albuminous;* RADICLE *superior.*

TREES or SHRUBS with dense often very hard and black wood. LEAVES alternate, coriaceous, entire, exstipulate. FLOWERS rarely ☿, usually diœcious (ovary of ♂ rudimentary, stamens of ♀ imperfect or 0); ♂ in many-flowered cymes; ♀ one-flowered by arrest of the lateral flowers; *pedicels* jointed at the top. CALYX 3–6-fid, sub-equal, persistent. COROLLA inserted on the receptacle, monopetalous, deciduous, urceolate, coriaceous, usually pubescent outside, glabrous within; limb 3–6-fid, æstivation imbricate-convolute. STAMENS inserted at the bottom of the corolla, or sometimes on the receptacle, double the corolla-lobes in number, rarely quadruple, very rarely equal (*Maba*), and then alternate with them; *filaments* free, or united in pairs below; *anthers* introrse, 2-celled, basifixed, lanceolate, dehiscence longitudinal. OVARY sessile, 3–many-celled; *style* rarely simple; *stigmas* simple or 2-fid; *ovules* solitary in each cell, or geminate, pendulous from the top of the inner angle of the cell, anatropous, raphe external. BERRY globose or ovoid, more or less succulent, usually few-seeded from arrest. SEEDS inverted; *testa* membranous. EMBRYO axile, or oblique in a cartilaginous albumen, which is twice as long as the embryo; *cotyledons* foliaceous, oval, nearly equal in length to the superior radicle.

PRINCIPAL GENERA.

* Diospyros.	* Royena.	Euclea.	Maba.

Ebenaceæ were formerly united to *Styraceæ*: in both the corolla is 3–7-lobed, the stamens numerous and fascicled, the ovary many-celled, the fruit fleshy, the embryo albuminous and axile, the stem arborescent, the leaves alternate and flowers axillary; but *Styraceæ* differ in their racemed flowers, epigynous or perigynous corolla, semi-inferior or inferior ovary, more numerous ovules, and fleshy albumen. The affinity of *Ebenaceæ* with *Oleineæ* is founded on the hypogynous and regular corolla, many-celled ovary, geminate pendulous anatropous ovules, berried fruit, straight albuminous axile embryo, and woody stem. *Oleineæ* differ in the [more or less] valvate æstivation of the corolla, fleshy albumen, and opposite leaves. *Ebenaceæ* approach *Ilicineæ* in the hypogyny and æstivation of the corolla, many 1-ovuled ovarian cells, pendulous anatropous ovule, fleshy fruit, straight albuminous embryo, woody stem, and alternate leaves; but in *Ilicineæ* the corolla is nearly polypetalous and isostemonous; the fruit is a drupe, the embryo is minute at the top of the fleshy albumen, and the leaves are persistent. Planchon recognizes a certain relationship between *Ebenaceæ* and *Camelliaceæ*: they agree in the insertion and æstivation of the corolla, numerous stamens, coherent filaments, many-celled ovary, pendulous anatropous ovules, fleshy fruit, albuminous embryo (of many genera), woody stem, alternate leaves, and often unisexual axillary flowers; but in *Camelliaceæ* the corolla is polypetalous or sub-polypetalous, and the stamens are very numerous.

Ebenaceæ grow in the tropical [and sub-tropical] regions of Asia, South Africa, Australia and America; they are rare in the Mediterranean region. *Ebenaceæ* are less noticeable for the beauty of their flowers and the utility of their fruits or seeds than for the hardness and colour of their wood. Ebony is the product of *Diospyros Ebenum, melanoxylon, Ebenaster, tomentosa,* &c. [The best is the Mauritian, yielded by *D. reticulata*; the next best that from the *D. Ebenum.*] The heart-wood of these trees is usually perfectly black, though sometimes marked with fawn-coloured lines; and its grain is so fine that when it is polished no trace of woody fibre is perceptible. It is white when young, and darkens with age; the colour of the alburnum contrasting with that of the heart-wood. Some species of *Diospyros* have edible berries, as the *D. Lotus* of the Mediterranean region, *D. Virginiana* [the Persimon or Date Plum of the United States], and *D. Kaki* [of Japan and China], which are cultivated in the open air in European gardens, and the latter of which is much esteemed in China for its berries, which when mellow will bear comparison with our best apricots. [*D. quesita*, of Ceylon, yields the beautiful Calamander wood. The glutinous juice of the fruit of *D. Embryopteris* is extensively employed in caulking boats and coating fishing nets in India; it yields a powerful astringent used for tanning purposes. A spirituous liquor is distilled from the fruits of *D. Virginiana*, the bark of which is a bitter febrifuge.—ED.]

CXXXIX *CYRILLEÆ.*

South American SHRUBS. LEAVES alternate, membranous, entire, exstipulate. FLOWERS in terminal or axillary racemes. CALYX 5-fid or -partite. PETALS 5, slightly united at the base, and with the filaments inserted on the receptacle, æstivation contorted to left or right, sometimes convolute. STAMENS 5 or 10, inserted with the petals; *filaments* subulate, dilated below the middle; *anthers* introrse, 2-celled, dehiscence longitudinal. OVARY free, [not inserted on a disk], 2-4-celled; *style* short; *stigma* of 2 acute lobes, or sessile, peltate and obscurely 4-lobed; *ovules* 1 or more in each cell, pendulous. FRUIT either a fleshy 2-celled 2-valved 1-2-seeded capsule (*Cyrilla*), or a nearly dry drupe with 4 wings, 4 cells and 4 seeds (*Cliftonia*). SEEDS inverted. EMBRYO straight, cylindric, in the axis of a fleshy albumen; *radicle* superior.

GENERA.

Cyrilla.	Cliftonia.	Elliottia.

Cyrilleæ approach *Ericineæ* in their hypogynous isostemonous or diplostemonous corolla, their contorted æstivation, many-celled ovary with pendulous ovules, usually capsular fruit, albuminous embryo, woody stem and alternate leaves; the principal difference is in their anthers being normal in structure. The same characters connect them with *Ilicineæ*, which have besides, as in *Cyrilla*, the petals connected at the base by the stamens, normal anthers, and drupaceous fruit; but in *Cyrilla* the flowers are racemed, and the embryo more elongated.

Finally, *Cyrilleæ* may be compared with *Pittosporeæ*: both have five hypogynous and isostemonous petals, a many-celled ovary, a capsular or fleshy fruit, a woody stem and alternate leaves; but in *Pittosporeæ* the ovules are ascending, and the embryo is minute. [*Cyrilleæ* are all natives of the Southern States of North America, and have no known uses.—ED.]

Cyrilla.
Flower (mag.).

Cyrilla.
Vertical section of flower
(mag.).

Cyrilla.
Flower from which a petal has been
removed to show the ovary (mag.).

Cyrilla racemiflora.

Cyrilla.
Diagram.

Cyrilla.
Fruit (mag.).

Cyrilla.
Vertical section of fruit
(mag.).

Cyrilla.
Transverse section of ovary
(mag.).

CXL. *STYRACEÆ.*

(Guajacaneæ, *partim, Jussieu.*—Styraceæ, *Richard.*—Styracineæ, *Kunth.*—
Styracaceæ, *A. D.C.*)

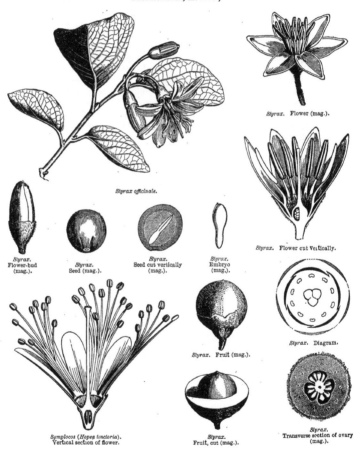

Styrax. Flower (mag.).

Styrax officinale.

Styrax.
Flower-bud
(mag.).

Styrax.
Seed (mag.).

Styrax.
Seed cut vertically
(mag.).

Styrax.
Embryo
(mag.).

Styrax. Flower cut vertically.

Styrax. Diagram.

Styrax. Fruit (mag.).

Symplocos (Hopea tinctoria).
Vertical section of flower.

Styrax.
Fruit, cut (mag.).

Styrax.
Transverse section of ovary
(mag.).

COROLLA *monopetalous or sub-polypetalous, perigynous or epigynous.* STAMENS *inserted at the base of the corolla, free, or filaments cohering, definite or indefinite, 1-many-seriate.* OVARY *inferior or semi-inferior, of 2-5 2-many-ovuled cells;* OVULES *anatropous.* FRUIT *usually fleshy.*—EMBRYO *albuminous, axile.*

SHRUBS or SMALL TREES. LEAVES alternate, simple, exstipulate. FLOWERS racemed or solitary, axillary, bracteate. CALYX 5-4-lobed. COROLLA usually 5- (rarely 4-6-7-) lobed, i.e. composed of 5-4-6-7 petals usually scarcely cohering at the base, sometimes increased by an inner whorl of petals, adhering to the outer and alternating with them, [æstivation various]. STAMENS inserted at the base of the corolla, free, or filaments cohering, 1-many-seriate, sometimes 8-10; sometimes numerous, pentadelphous or monadelphous the longest bundles or stamens alternating with the corolla-lobes; *anthers* 2-celled, dehiscence introrse or lateral. OVARY inferior or semi-inferior, 5-2-celled, cells opposite to the calyx-lobes when they equal them in number; *style* simple; *stigma* small, capitate, lobed; *ovules* geminate, or many in each cell, all pendulous, or the lower horizontal or ascending, and the upper pendulous, anatropous. FRUIT usually fleshy, nearly always 1-celled by arrest. SEEDS 5-1, usually solitary. EMBRYO straight, in the axis of a fleshy albumen; *cotyledons* flat; *radicle* usually superior.

TRIBE I. SYMPLOCEÆ. — Corolla sub-polypetalous, æstivation quincunxial. Stamens 1-many-seriate 15-∞, sometimes polyadelphous, sometimes 1-seriate, almost free, usually monadelphous. Anthers ovoid-globose. Ovules all pendulous.

ONLY GENUS.

Symplocos.

TRIBE II. STYRACEÆ PROPER.—Corolla 5-partite, æstivation convolute or subvalvate. Stamens 1-seriate, 7-12; anthers elongated, adnate. Ovules, the lower horizontal or ascending, the upper pendulous.

PRINCIPAL GENERA.

* Styrax.   * Halesia.

[To the above A. De Candolle adds a third tribe :—

TRIBE III. PAMPHILIEÆ.—Corolla 5-fid or -partite, valvate. Stamens 5 or 10, connate at the base and adnate to the corolla; anthers elongate, tubular above. Ovary free; ovules erect, anatropous. South American trees. *Pamphilia, Foveolaria.*—ED.]

Styraceæ are near *Ebenaceæ* (see that family). There is also an affinity between *Symploceæ* and *Camelliaceæ*; both have a woody stem, alternate leaves, sub-polypetalous or polypetalous corolla, and imbricate æstivation; the stamens are numerous and many-seriate, and the filaments polyadelphous; and in some genera of *Camelliaceæ* the style is simple, the ovary semi-inferior, and the embryo albuminous. *Styraceæ* proper present some analogy with *Philadelpheæ*, in the woody stem, axillary and terminal flowers, free or nearly free petals, numerous stamens, inferior and many-celled ovary, and albuminous and axile embryo; but in *Philadelpheæ* the leaves are opposite and the fruit capsular.

Styraceæ inhabit Asia and tropical America; there are a few in Japan, in the hotter parts of South America, and in the eastern Mediterranean region. [Many species of *Symploceæ* are Indian, and several temperate Himalayan.]

Storax and Benzoin are two balsams, composed of an aromatic resin in combination with a volatile oil and an acid which crystallizes into needles, named Benzoic acid. These balsams, formerly adminis-tēred internally as stimulants, are now only used externally. Storax flows spontaneously or from incisions in the stem of the *Styrax officinale*, a tree of the Mediterranean region; and Benzoin is derived from the *Styrax Benzoin*, which grows in the Moluccas. Some species of *Symplocos* yield dyes, and *S. Alstonia* is used as tea in Central America. [Several species are employed as tea, and for dyeing yellow in the Himalayas.]

CXLI. *JASMINEÆ.*

(JASMINEARUM *genera*, Jussieu.—JASMINEÆ, Br.—JASMINACEÆ, *Lindl.*)

COROLLA *monopetalous, 5–8-fid, hypogynous, anisostemonous, æstivation imbricate.* STAMENS 2, *inserted on the corolla.* OVARY *of two 1–2-ovuled cells;* OVULES *collateral, ascending, anatropous.* FRUIT *a berry or capsule.* ALBUMEN *disappearing when ripe.* RADICLE *inferior.*—STEM *woody.*

Small TREES or SHRUBS, often twining or climbing. LEAVES opposite or alternate, 1–3–5–7-foliolate, exstipulate. FLOWERS ☿, regular or sub-regular, in a corymb or panicle; *pedicels* trichotomous, many-flowered. CALYX 5–8-fid or -toothed, persistent. COROLLA hypogynous, monopetalous, hypocrateriform, 4–5–6-lobed, æstivation imbricate. STAMENS 2, inserted on the corolla-tube, included, opposite to the 2 outer petals in the 4-lobed corollas; in the 5-lobed corollas, when one of the outer petals is doubled, one of the stamens is inserted between these two petals, and the other remains opposite to the single outer petal; in the 6-lobed corollas the same change takes place in the 2 staminiferous petals; *filaments* very short, or 0; *anthers* 2-celled, introrse, basifixed, dehiscence longitudinal. OVARY free, 2-celled; *style* terminal, very short; *stigma* capitate or 2-lobed; *ovules* 1–2 in each cell, at first pendulous near the base of the septum, finally ascending, anatropous. BERRY didymous, often one-seeded by arrest (*Jasminum*), or a cordate 2-celled 2-partible capsule (*Nyctanthes*). SEEDS erect, sub-compressed; *testa* coriaceous, and *endopleura* thick; *albumen* at first copious, when ripe reduced to a thin membrane. EMBRYO straight; *cotyledons* plano-convex, fleshy; *radicle* short, inferior.

PRINCIPAL GENERA.

| Menodora. | Jasminum. | Nyctanthes. | Bolivaria. |

We have indicated the affinities of *Jasmineæ* with *Oleineæ* and *Verbenaceæ* (see these families); they approach *Apocyneæ* in their climbing or twining stem, usually opposite exstipulate leaves, hypogynous staminiferous corolla, carpels cohering into a 2-celled ovary (as in *Carissa*), and dry or fleshy fruit; but *Apocyneæ*, besides the separation of the ovaries in most genera, differ in their milky juice, the isoste-mony and æstivation of the corolla, and the persistence of the albumen. On the other hand, *Jasmineæ* approach *Ebenaceæ* in their woody stem, in the imbricate æstivation of the corolla, basifixed anthers, 1–2-ovuled ovarian cells, fleshy fruit and compressed seeds; differing in the diandrous flowers, erect ovules and absence of albumen.

Jasmineæ inhabit the hot regions [and cool mountains] of Asia. Some are African, Australian and

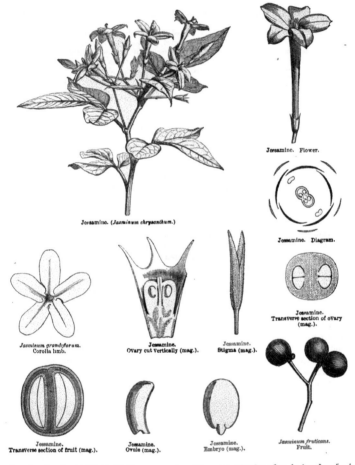

Jessamine. (*Jasminum chrysanthum.*)

Jessamine. Flower.

Jessamine. Diagram.

Jasminum grandiflorum.
Corolla limb.

Jessamine.
Ovary cut vertically (mag.).

Jessamine.
Stigma (mag.).

Jessamine.
Transverse section of ovary
(mag.).

Jessamine.
Transverse section of fruit (mag.).

Jessamine.
Ovule (mag.).

Jessamine.
Embryo (mag.).

Jasminum fruticans.
Fruit.

Oceanic, and a few inhabit the Mediterranean region. The genus *Menodora* alone is American [as is *Bolivaria*].

Jasmineæ are prized for the elegance of their foliage and their sweet-scented flowers; their perfume, due to a volatile oil, is fixed by means of Ben-oil [*Moringa pterygosperma*], which dissolves and preserves it, for the preparation of perfumes. The essence of Jessamine, so much used in perfumery, is prepared from the flowers of *Jasminum Sambac*, an Indian shrub, and of *J. grandiflorum*, or Spanish Jessamine. The *J. officinale*, cultivated in gardens, is a native of Asia; its flowers were formerly employed in medicine as a nervine, aperient and emollient. The *Nyctanthes Arbor-tristis* is an Indian shrub, the flowers of which open in the evening, and fall at daybreak; whence its name of Somnambulist. [The tube of the corolla affords a beautiful but fugacious yellow dye. The flowers of *J. Sambac* are sacred to Vishnu.]

CXLII. *OLEINEÆ.*

(JASMINEARUM *genera, Jussieu.*—OLEINEÆ ET FRAXINEÆ, *Martius.*—OLEINEÆ, *Link.*—OLEACEÆ, *Lindl.*)

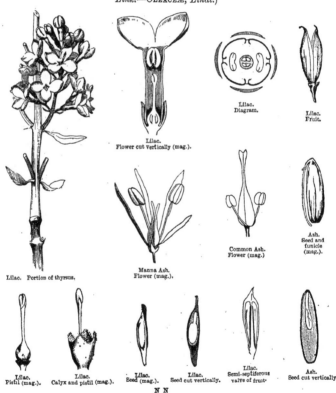

Lilac.
Flower cut vertically (mag.).

Lilac.
Diagram.

Lilac.
Fruit.

Manna Ash.
Flower (mag.).

Common Ash.
Flower (mag.)

Ash.
Seed and
funicle
(mag.).

Lilac. Portion of thyrsus.

Lilac.
Pistil (mag.).

Lilac.
Calyx and pistil (mag.).

Lilac.
Seed (mag.).

Lilac.
Seed cut vertically.

Lilac.
Semi-septiferous
valve of fruit·

Ash.
Seed cut vertically.

N N

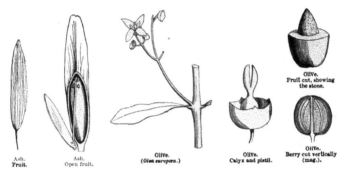

Ash. Fruit.   Ash. Open fruit.   Olive. (*Olea europæa*.)   Olive. Calyx and pistil.   Olive. Fruit cut, showing the stone.   Olive. Berry cut vertically (mag.).

COROLLA 4-*merous, monopetalous or sub-polypetalous, hypogynous, anisostemonous, æstivation valvate.* STAMENS 2, *inserted on the corolla.* OVARY 2-*celled*; OVULES *pendulous, anatropous.* FRUIT *a capsule, berry or drupe.* EMBRYO *albuminous, axile*; RADICLE *superior.*—STEM *woody.* LEAVES *opposite.*

TREES or SHRUBS. LEAVES opposite, petioled, simple or rarely imparipinnate, exstipulate. FLOWERS ☿, rarely diœcious and apetalous, in a raceme or trichotomous panicle, sometimes fascicled, pedicels opposite. CALYX monosepalous, 4-lobed or -toothed, sometimes obsolete. COROLLA hypogynous, of 4 petals united at the base in pairs by the filaments, or clearly gamopetalous, infundibuliform or sub-campanulate, æstivation valvate, very rarely 0 (*Ash, Olive*). STAMENS 2, inserted on the corolla and alternate with its lobes; *anthers* 2-celled, introrse, dorsifixed, dehiscence longitudinal. OVARY free, 2-celled, cells alternating with the stamens; *style* simple or 0; *stigma* undivided or 2-fid; *ovules* collateral, pendulous from the top of the septum, rarely 3 with the 2 lateral arrested (*Ash*), sometimes numerous, 2-seriate (*Forsythia*), anatropous, raphe dorsal. FRUIT various: either a drupe and often 1-celled and -seeded (*Olive*), or a 2-celled berry (*Privet*), or a loculicidal capsule (*Lilac*), or an indehiscent samara prolonged above into a foliaceous wing (*Ash*). SEEDS pendulous, generally more or less compressed. EMBRYO straight, in the axis of a thick fleshy or sub-horny albumen; *cotyledons* foliaceous; *radicle* cylindric, superior.

SUB-ORDER I. OLEINEÆ VERÆ.—Fruit a drupe or berry.

PRINCIPAL GENERA.

* Olea.   * Chionanthus.   Linociera.   Notelæa.   * Phillyrea.   * Ligustrum.

SUB-ORDER II. FRAXINEÆ.—Fruit an indehiscent samara or a 2-valved loculicidal capsule.

PRINCIPAL GENERA.

* Fraxinus.   * Syringa.   * Forsythia.   * Fontanesia.

Oleineæ were formerly united with *Jasmineæ*: in both the stem is woody, the leaves opposite, the flowers diandrous and racemed or panicled, the ovary has two 1-2-ovuled cells, the ovule is anatropous, and the fruit capsular or fleshy; but *Jasmineæ* differ in the imbricate corolla-lobes, the basifixed anthers, the ascending ovule, and the albumen being reduced to a thin membrane when ripe. A relation has also been observed between *Oleineæ* and *Apocyneæ*, both having a woody stem, opposite leaves, hypogynous staminiferous corolla, valvate æstivation in some, contorted in others, a 2-celled ovary (at least in *Carissa*), and an albuminous embryo. Finally, comparing *Oleineæ* with *Rubiaceæ*, we find in common the opposite leaves, staminiferous corolla, valvate æstivation, 2-celled ovary, pendulous ovules, albuminous embryo, and fleshy or dry fruit. *Rubiaceæ* principally differ in the epigynous and isostemonous corolla, and the stipulate or whorled leaves. The same observation applies to *Caprifoliaceæ*, which further differ in the imbricate æstivation of their corolla.

Oleineæ mostly inhabit the northern hemisphere. The *Oleineæ* proper prefer the northern temperate and warm regions; some are, however, tropical, and even extend beyond the tropic of Capricorn. [*Olea* occurs in New Zealand and South Africa, *Noteleæ* is Australian, and both *Chionanthus* and *Linociera* are American.] *Fraxineæ* all grow north of 23°, and Africa possesses none [except in Barbary and the Mediterranean region]. Most Ashes are American; some are scattered over Europe and temperate Asia. Lilacs are natives of the East.

The most useful species of this family is the Olive (*Olea europæa*), which has spread from the East throughout the Mediterranean region. The fixed oil expressed from the pericarp of its drupe holds the first place among alimentary oils. The unripe drupe macerated in brine is eaten, as are those of some exotic species (*O. americana, fragrans,* &c.). The bark and leaves of the Olive, Privet (*Ligustrum vulgare*) and *Phillyrea* were formerly used as bitter-astringent medicines. The bark of the Common Ash (*Fraxinus excelsior*) is bitter, and has been proposed as a substitute for quinine. Manna is a sugary concreted juice, gathered in Sicily on two species of Ash (*F. Ornus* and *rotundifolia*); it exudes spontaneously from the puncture of a Cigala (*Cicada Orni*), but its flow is induced by regular incisions made in the bark during summer. Manna is almost entirely composed of *mannite*, a proximate principle, which rapidly decomposes; whence Manna, which when fresh is simply nutritious, becomes nauseous, and is employed as a purgative, a quality which disappears when the manna has been boiled for a long time. [Olive wood is extremely hard and durable. The flowers of *O. fragrans* are used to scent teas in China. Lilac bark is a renowned febrifuge in certain malarious districts of France. Ash wood is well known as invaluable for its lightness, flexibility and strength.—ED.]

CXLIII. *SALVADORACEÆ, Lindl.*

Glabrous glaucous powdery SHRUBS; *branches* marked with transverse scars. LEAVES opposite, petioled, entire, coriaceous, obscurely veined, furnished with 2 minute stipules. FLOWERS inconspicuous, in spicate paniculate racemes. CALYX small, 4-toothed, æstivation imbricate. COROLLA hypogynous, monopetalous, membranous, æstivation imbricate. STAMENS 4, very short, inserted on the corolla, uniting its lobes and alternate with them; *anthers* 2-celled, introrse. DISK hypogynous, 4-lobed. OVARY free, 2-celled; *stigma* 2-lobed, sub-sessile; *ovules* geminate, ascending, anatropous. SEEDS 4–1, erect [pendulous, *Dobera* and *Monetia*]; *testa* pulpy, exalbuminous. EMBRYO with fleshy plano-convex cotyledons and inferior radicle.

ONLY GENUS.

Monetia.	Salvadora.	Dobera.

Planchon has grouped with *Salvadora* the genera *Monetia* and *Dobera*, both of which are mono-petalous, hypogynous, tetrandrous, and have a 2-celled ovary, a berry with exalbuminous seeds, woody stem and opposite leaves, and which scarcely differ from *Salvadora* except in the diœcious flowers and pendulous ovule. This affinity is confirmed by their geographical distribution, which extends over the tropical and sub-tropical regions of the Old World. In fact, *Monetia* is found from South Africa, through the Indian Peninsula and Ceylon, to Malacca; *Salvadora* from the coast of Benguela, through North

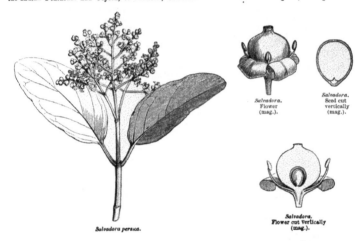

Salvadora.
Flower
(mag.).

Salvadora.
Seed cut
vertically
(mag.).

Salvadora.
Flower cut vertically
(mag.).

Salvadora persica.

Africa, to Palestine, Persia and India; and *Dobera* from Abyssinia and Arabia to the Indian Peninsula. As to the affinity, Gardner and Wight place *Salvadoraceæ* near *Oleineæ* and *Jasmineæ*, and Planchon is also disposed to adopt this arrangement.

The bark of the root of the *Salvadora persica* contains acrid and vesicant properties, and that of the stem is a tonic; its red berries are edible [aromatic and tasting like Cress], as are those of *S. indica* [in India they are not eaten], the leaves of which, like those of Senna, are purgative and vermifuge. This plant [according to Royle] is the Mustard-tree of the Jews, alluded to in the New Testament parables.

CXLIV. *APOCYNEÆ.*

(APOCYNEARUM pars, *A. L. de Jussieu.*—APOCYNEÆ, *R. Br.*—VINCEÆ, *D.C.*—APOCYNACEÆ, *Lindl.*)

COROLLA *monopetalous, hypogynous, regular, isostemonous, æstivation contorted or valvate.* STAMENS *inserted on the corolla;* POLLEN *granular.* CARPELS 2, *distinct or cohering;* STYLE *single.* FRUIT *various.* EMBRYO *albuminous, very rarely exalbuminous.*—JUICE *milky.* LEAVES *usually opposite or whorled.*

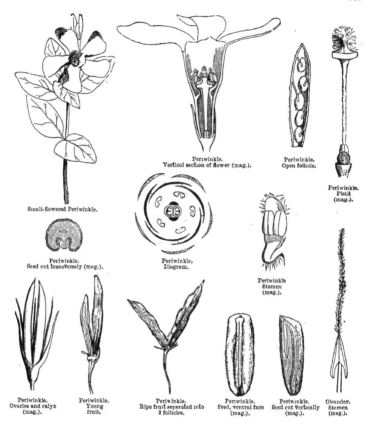

Periwinkle.
Vertical section of flower (mag.).

Periwinkle.
Open follicle.

Small-flowered Periwinkle.

Periwinkle.
Pistil
(mag.).

Periwinkle.
Seed cut transversely (mag.).

Periwinkle.
Diagram.

Periwinkle
Stamen
(mag.).

Periwinkle.
Ovaries and calyx
(mag.).

Periwinkle.
Young
fruit.

Periwinkle.
Ripe fruit separated into
2 follicles.

Periwinkle.
Seed, ventral face
(mag.).

Periwinkle.
Seed cut vertically
(mag.).

Oleander.
Stamen
(mag.).

TREES or SHRUBS, often climbing or perennial HERBS, generally with milky juice. LEAVES opposite or whorled, rarely alternate (*Plumiera, Rhazya, Lepinia*), simple, entire, exstipulate, or with rudimentary gland-like or ciliiform stipules. FLOWERS ☿, regular, terminal or axillary, in a corymboid cyme, rarely solitary. CALYX free, 5-fid or -partite, rarely 4-fid. COROLLA hypogynous, monopetalous, deciduous, infundibuliform or hypocrateriform, throat naked or furnished with scales, limb 5-4-fid or -partite, æstivation contorted or valvate. STAMENS inserted on the corolla-tube or throat, alternate with its segments ;. *anthers* introrse, 2-celled, ovoid, usually acuminate or mucronate, often sagittate, sometimes slightly coherent, dehiscence longitudinal; *pollen* granular, applied directly to the stigma. CARPELS 2, sometimes distinct, sometimes cohering into a 2-1-celled ovary, sometimes 3–4, at first undivided, separating after flowering into 3 or 4 long stipitate ovaries, united at their tips by the persistent base of the style (*Lepinia*); *style* single, uniting the ovaries, usually thickened towards the top, often dilated into a disk below the stigma; *stigma* generally bifid; *ovules* usually numerous, anatropous or semi-anatropous. FRUIT various. SEEDS usually compressed, often comose. EMBRYO straight; *albumen* cartilaginous or fleshy, sometimes scanty or 0; position and direction of the *radicle* various.

SUB-ORDER I. CARISSEÆ.—Ovary 2-celled; placentation on the septum. Fruit a berry.

PRINCIPAL GENERA.

Hancornia.	Vahea.	Melodinus.	Carissa.	Couma.
Ambelania.	Pacouria.	Collophora.	Carpodinus.	

SUB-ORDER II. ALLAMANDEÆ.—Ovary unilocular. Placentas 2, parietal. Capsule 2-valved.

GENUS.

* Allamanda.

SUB-ORDER III. OPHIOXYLEÆ.—Fruit fleshy. Drupes 2, I often aborted.

PRINCIPAL GENERA.

Tanghinia.	Ophioxylon.	* Cerbera.	Ochrosia.
Hunteria.	Alyxia.	Thevetia.	Rauwolfia.

SUB-ORDER IV. EUAPOCYNEÆ.—Fruit with 2 follicles, sometimes fleshy, pulpy, generally dry, often reduced to one by abortion, rarely united into a capsule.

PRINCIPAL GENERA.

* Tabernæmontana.	* Lochnera.	* Mandevillea.	Lepinia.
* Vinca.	* Beaumontia.	* Echites.	* Plumiera.
* Apocynum.	* Gelsemium.	Aspidosperma.	Rhazya.
* Nerium.	Malonetia.	Dipladenia.	* Amsonia.
* Wrightia.	Strophanthus.	Urceola.	

[The following genera do not fall into any of the above defined sub-orders.

Ovary 1-celled; placentas parietal. Fruit a berry. — *Willughbeia, Landolphia.*—ED.]

For the affinity of *Apocyneæ* with *Loganiaceæ* see this family. They are only distinguished from *Asclepindeæ* by the stamens, and from *Gentianeæ* by their milky juice, usually woody stem, and the distinct ovaries of many genera. They are connected with *Rubiaceæ* through *Loganiaceæ*. They are allied to *Oleineæ* by their woody stem, opposite leaves, æstivation and hypogynous corolla, and by their genera with a 2-celled ovary, single style, anatropous ovule, dry or fleshy fruit, and albuminous embryo ; but the *Oleineæ* have an anisostemonous corolla.

Apocyneæ principally inhabit the intertropical zone of the Old and New Worlds, especially Asia beyond the equator. They are [comparatively] rare in extra-tropical hot and temperate regions. Most of the species possess a milky juice, often rich in india-rubber (*Collophora utilis*) ; this juice is sometimes bitter and employed as a purgative or febrifuge, or depurative (*Allamanda cathartica, Carissa xylopicron, Plumiera alba*) ; sometimes acrid and very poisonous (*Tanghinia veneniflua, Cerbera Ahouai*) ; sometimes mild, scarcely bitter, and simply laxative (*Cerbera salutaris*) ; finally, sometimes acid-sweet or unctuous, and much sought as food (*Carissa Carandas, C. edulis, Carpodinus dulcis, Ambelania Pacouria, Couma, Tabernæmontana utilis*, &c.). [Other india-rubber yielding genera are *Willughbeia* in India, *Vahea* in Madagascar, *Hancornia* in Brazil, *Urceola* in the Malay Peninsula, and *Landolphia* in West Africa. *Tanghinia*, the Ordeal-tree of Madagascar, is the most poisonous of plants, a seed no larger than an almond suffices to kill twenty people. Oleander wood, flowers, and leaves are very poisonous ; death has followed using its wood as meat-skewers ; an infusion of its leaves is an active insecticide, and its bark a rat-poison ; that of *Wrightia antidysenterica* is a valuable Indian astringent and febrifuge. *W. tinctoria* leaves yield an indigo, and *W. tomentosa* a yellow dye. Edible fruits are produced by *Willughbeia edulis* and *Urceola elastica*. The wood of *Alstonia scholaris* is a bitter powerful tonic, much used in India.—ED.]

CXLV. *ASCLEPIADEÆ.*

(APOCYNEARUM *pars, A.-L. de Jussieu.*—ASCLEPIADEÆ, *Jacquin.*
—ASCLEPIADACEÆ, *Lindl.*)

COROLLA *hypogynous, regular, 5-fid, isostemonous, æstivation usually contorted.* STAMENS *inserted on the corolla, usually cohering in a tube ;* ANTHERS *introrse, 2–4-celled ;* POLLEN *agglutinated in as many masses as there are cells.* CARPELS 2 ; OVARIES *distinct ;* STYLES *juxtaposed, united by a common stigma ;* OVULES *pendulous, anatropous.* FRUIT *follicular.* EMBRYO *albuminous.*—LEAVES *opposite.* JUICE *milky.*

Woody, rarely herbaceous PLANTS, usually climbing and milky ; *stem and branches*

Asclepias.
Flower (mag.).

Asclepias.
Diagram, showing the relation of the appendages to the anthers, and that of the anthers to the stigmatic corpuscules.

Asclepias.
Pistil bearing pollen-masses (mag.).

Asclepias.
Stamen furnished with its appendages (mag.).

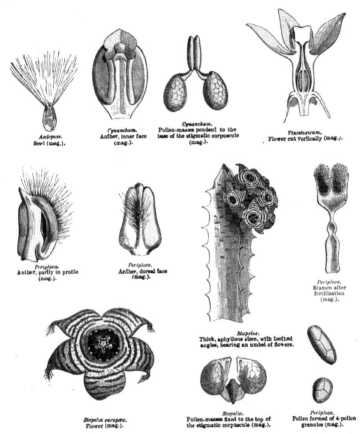

Asclepias.
Seed (mag.).

Cynanchum.
Anther, inner face
(mag.).

Cynanchum.
Pollen-masses pendent to the
base of the stigmatic corpuscule
(mag.).

Vincetoxicum.
Flower cut vertically (mag.).

Periploca.
Anther, partly in profile
(mag.).

Periploca.
Anther, dorsal face
(mag.).

Periploca.
Stamen after
fertilization
(mag.).

Stapelia.
Thick, aphyllous stem, with toothed
angles, bearing an umbel of flowers.

Stapelia europæa.
Flower (mag.).

Stapelia.
Pollen-masses fixed to the top of
the stigmatic corpuscule (mag.).

Periploca.
Pollen formed of 4-pollen
granules (mag.).

jointed, knotty, sometimes fleshy (*Stapelia*). LEAVES opposite, rarely whorled or
alternate, obsolete or rudimentary in the fleshy species, petioled, simple, entire, exstipulate, or with interpetiolar bristles. FLOWERS ☿, regular, very often in umbels
or panicles, rarely in cymes or racemes, very rarely solitary; peduncles axillary or
interpetiolar. CALYX free, 5-fid or -partite, æstivation imbricate. COROLLA hypogynous, monopetalous, deciduous, campanulate, urceolate, hypocrateriform, infundibu-

liform or terete, tube and throat furnished within with scales; *limb* contorted-imbricate or rarely valvate in æstivation. STAMENS 5, inserted at the bottom of the corolla, and alternate with its segments; *filaments* flattened, usually united in a tubular column surrounding the ovary, and furnished behind the anther with a crown with various appendages; *anthers* introrse or lateral, 2-celled, generally cohering in a tube; cells back to back, parallel, sometimes divided by a septum, opening by a longitudinal or apical slit, rarely transverse (*Gonolobus*); *pollen* agglutinated into a mass; *pollen-masses* (*pollinia*) pendulous (*Asclepias*) or horizontal (*Gonolobus*) or erect (*Stapelia*), either one for each cell, or united in pairs belonging to two contiguous cells, fusiform, enveloped in an oily matter, and adhering to the glandular prominences of the stigma. OVARIES 2, distinct, placentas nerviform, on ventral sutures; *styles* usually very short, closely appressed, and united by the common stigma; *stigma* with 5 rounded angles, their bases alternate with the anthers, and provided with cartilaginous corpuscules, or with a gland which retains the pollinia; *ovules* numerous, anatropous, pendulous, multiseriate. FOLLICLES 2, sometimes 1 by arrest, on a placenta which detaches itself when ripe. SEEDS numerous, compressed, imbricate, often comose. EMBRYO straight, in the axis of a fleshy albumen, rarely exalbuminous; *radicle* superior.

[The following arrangement is that of Decaisne, in De Candolle's "Prodromus."

Sub-order I. PERIPLOCEÆ.—Filaments more or less distinct; anthers with 20–10 pollinia, free, or applied to the top of the stigma; pollen of 3–4 grains. (Asiatic and African). *Periploca, Hemidesmus, Streptocaulon.*

Sub-order II. SECAMONÆ.—Filaments coherent; anthers 4-celled; pollinia 20, applied in fours to the top of the corpuscules of the stigma. (Asiatic and African). *Secamone, Toxocarpus.*

Sub-order III. EUASCLEPIADEÆ.—Filaments coherent; anthers 2-celled; pollinia 10, fixed in pairs to the prominences of the stigma, separated by a longitudinal furrow.

DIVISION I. ASTEPHANÆ.—Throat of the corolla without scales. Staminal corona 0 (African and American). *Astephanus*, &c.

DIVISION II. MICROLOMÆ.—Throat of the corolla furnished with fleshy scales. Staminal crown 0. (African and Arabian). *Microloma*, &c.

DIVISION III. HAPLOSTEMMÆ.—Staminal crown simple, of 5 segments; segments inserted at the base of the gynostegium, simple, entire or 2-fid. *Metastelma, Roulinia, Acerates, Vincetoxicum, Haplostemma*, &c.

DIVISION IV. CYNOCTONÆ.—Throat of the corolla naked. Staminal crown simple, cupshaped or tubular, mouth sub-entire or lobed. *Orthosia, *Cynoctonum, Holostemma, *Arauja*, &c.

DIVISION V. SARCOSTEMMÆ.—Throat of the corolla naked. Staminal crown usually double; outer short, innate-lobed; inner of 5 segments, which are fleshy or ligulate or more or less rounded and tumid. *Calotropis, Pentatropis, Sarcostemma, Dæmia*, &c.

DIVISION VI. EUSTEGIÆ.—Throat of the corolla naked. Staminal crown campanulate, double or triple; segments more or less connate below, opposite or alternate, produced into a linear appendage, surrounding the sessile or stipitate gynostegium. *Eustegia, Cynanchum*, &c.

DIVISION VII.. ASCLEPIADÆ.—Throat of the corolla naked. Staminal crown of 5 segments ; segments concave or hooded, inserted at the base, rarely at the top, of the gynostegium, with often a ligulate appendage on the inner face, or thickened in the middle and then toothed at the side. *Gomphocarpus, *Asclepias, &c.

DIVISION VIII. DITASSÆ.—Staminal column compound, of 2 opposite series, linear, equal or the outer ovate and minutely toothed. (America. Shrubs ; flowers small, hoary within). *Ditassa, Tassadia, &c.

DIVISION IX. OXYPETALÆ.—Staminal crown adnate to the corolla-tube, tubular or of 5 segments; segments simple or toothed internally. Pollinia fixed to a broad geniculate process. Stigmatic capsule linear, often horned or spurred at the base ; stigma long, often dilated, truncate or deeply 2-7-fid. (Perennial twining American plants). *Oxypetalum, &c.

Sub-order IV. GONOLOBÆ.—Filaments connate. Anthers 2-celled, dehiscence transverse. Pollinia 10, horizontal, fixed in pairs to a longitudinal bipartite furrow of the stigmatic process ; tips usually pellucid and hidden under the depressed stigma. (Perennial twining American herbs). *Gonolobus, Fischeria, &c.

Sub-order V. STAPELIÆ.—Filaments connate ; anthers usually terminated by a simple membrane. Pollinia 10, ascending or erect, fixed in pairs to the stigmatic process, opaque at both ends, or pellucid at the sides or above. (Twining plants, often fleshy herbs, of the Old World).

DIVISION I. PERGULARIÆ.—Pollinia opaque at both ends. *Tylophora, Marsdenia, Pergularia, *Stephanotis, Gymnema, Sarcolobus,* &c.

DIVISION II. CEROPEGIÆ.—Pollinia pellucid at the top or side. *Leptadenia, Dischidia,* *Hoya, *Ceropegia, Boucerosia, Huernia, *Stapelia,* &c.—ED.]

Asclepiadeæ were formerly placed in the same family as *Apocyneæ* ; the exceptional structure of the pollen and stigma, however, separates them, as do the usually coherent filaments. *Periploceæ,* however, by their nearly free filaments connect the two families. Their affinity with *Gentianeæ* is less than that of *Apocyneæ,* some genera of which have their carpels united into a 1-2-celled ovary.

Asclepiadeæ inhabit the same countries as *Apocyneæ* ; the fleshy species all belong to the Old World, and especially to South Africa. Their medicinal properties reside in their milky juice ; some are emetica (*Vincetoxicum officinale, Gomphocarpus crispus, Secamone emetica,* &c.) ; others are purgative (*Cynanchum monspeliense, Solenostemma Arghel*) ; some are sudorifics (*Hemidesmus indicus*) ; the acrid milky juice of others is used to poison arrows (*Gonolobus macrophyllus*), or wolves (*Periploca græca*), whence the names of Wolf's-bane and Dog's-bane, given to several species. In others, again, the milk has no acridity, and is alimentary (*Gymnema lactiferum,* the Cow-plant of Ceylon, and the Cape *Oxystelma esculentum*).

[The *Asclepias decumbens* of Virginia causes perspiration without increase of animal heat, and is used in pleurisy. *A. tuberosa* is a mild cathartic ; *A. curassavica* is the well-known American Wild Ipecacuanha, an emetic and purgative. *Tylophora asthmatica* is one of the most useful medicines in India as a cure for dysentery. *Sarcostemma glaucum* is the Ipecacuanha of Venezuela. *Cynanchum acutum* is the Montpellier Scammony. *Calotropis gigantea* yields Mudar, a celebrated Indian drug, a tonic, alterative and purgative ; the root of *Hemidesmus indicus* is in no less repute as a substitute for Sarsaparilla.

Some Indian species yield most tenacious fibres, as *Marsdenia tenacissima, Orthanthera viminea,* and *Calotropis gigantea* ; others dyes, as *Marsdenia tinctoria* ; and, lastly, others a good Caoutchouc.]

CXLVI. *LOGANIACEÆ.*

(LOGANIEÆ, *R. Br.*—POTALIEÆ, *Martius.*—STRYCHNEÆ, *D.C.*—STRYCHNACEÆ,
Blume.—LOGANIACEÆ, POTALIACEÆ ET APOCYNEARUM *pars, Lindl.*)

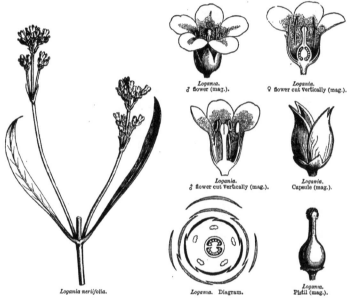

Logania.
♂ flower (mag.).

Logania.
♀ flower cut vertically (mag.).

Logania.
♂ flower cut vertically (mag.).

Logania.
Capsule (mag.).

Logania neriifolia.

Logania. Diagram.

Logania.
Pistil (mag.).

COROLLA *monopetalous, hypogynous, regular, generally isostemonous, æstivation
valvate, contorted or convolute.* STAMENS *inserted on the corolla.* OVARY *of* 2–4 1-
or many-ovuled cells; OVULES *anatropous or semi-anatropous.* EMBRYO *albuminous.*
—LEAVES *opposite.*

STEM woody, rarely herbaceous. LEAVES opposite, stipulate, or exstipulate
when the dilated and connate bases of the *petioles* embrace the stem, with a short sometimes obsolete border; *stipules* adnate on both sides to the petioles, or free and interpetiolar, or cohering in a sheath, or axillary, dorsally adnate to the base of the
petiole. FLOWERS ☿, regular, very rarely anisostemonous, axillary and solitary, or
racemose or corymbose; sometimes in a terminal corymb or panicle. CALYX monosepalous with valvate æstivation, or of 4–5 free imbricate sepals. COROLLA hypo-

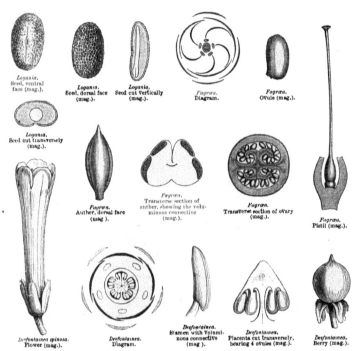

Logania.
Seed, ventral
face (mag.).

Logania.
Seed, dorsal face
(mag.).

Logania.
Seed cut vertically
(mag.).

Fagræa.
Diagram.

Fagræa.
Ovule (mag.).

Logania.
Seed cut transversely
(mag.).

Fagræa.
Anther, dorsal face
(mag).

Fagræa.
Transverse section of
anther, showing the volu-
minous connective
(mag.).

Fagræa.
Transverse section of ovary
(mag.).

Fagræa.
Pistil (mag.).

Desfontainea spinosa.
Flower (mag.).

Desfontainea.
Diagram.

Desfontainea.
Stamen with volumi-
nous connective
(mag).

Desfontainea.
Placenta cut transversely,
bearing 4 ovules (mag.).

Desfontainea.
Berry (mag.).

gynous, monopetalous, rotate, campanulate or infundibuliform, limb 5-4-10-fid, æstivation valvate contorted or convolute. STAMENS inserted on the corolla-tube or throat, alternate in the 4-5-fid corolla, opposite in the 10-fid corolla; *filaments* filiform or subulate; *anthers* introrse, 2-celled, dehiscence longitudinal. OVARY superior, 2-4-celled; *style* filiform, simple; *stigma* capitate or peltate or bilobed; *ovules* numerous, semi-anatropous, attached to the septum, or ascending from the base of the cell, rarely solitary, and peltate by their ventral face, very rarely erect at the base of the cells and anatropous (*Gærtnera*). FRUIT capsular, septicidally or septifragally 2-valved, or of 2 cocci with transverse dehiscence; sometimes a berry or drupe. SEEDS numerous or solitary, sometimes winged. EMBRYO straight, in the axis or base of a fleshy or cartilaginous albumen; *radicle* inferior or vague.

[The following is Mr. Bentham's classification of the *Loganiaceæ* (Journ. Linn. Soc. Lond. i. 88) :—

Tribe I. Antonieæ.—Ovules numerous in the ovarian cells. Seeds winged. *Antonia, Usteria*, &c.

Tribe II. Euloganieæ.—Ovules numerous in the ovarian cells. Fruit capsular. Seeds naked or hardly winged. *Spigelia, Mitreola, Mistrasacme, Logania, Nuxia, Buddleia.*

Tribe III. Fagræeæ.—Ovules numerous in the ovarian cells. Fruit a berry. *Desfontainea, Fagræa, Strychnos,* &c.

Tribe IV. Gærtnereæ.—Ovules solitary, rarely geminate in the ovarian cells. *Gardneria, Gærtnera,* &c.]

Loganiaceæ are very closely allied to *Rubiaceæ* (see this family). They approach *Gentianeæ* in the opposite and entire leaves, the insertion, æstivation and isostemony of the corolla, the capsular fruit, and the presence of albumen; but *Gentianeæ* differ in the 1- or incompletely 2-celled ovary, anatropous ovules and exstipulate leaves. The affinities and differences are the same in *Apocyneæ*, whose fruit, like that of *Loganiaceæ*, is a capsule, berry, or drupe; but they are distinguished by their milky juice, the always isostemonous corolla, and numerous genera with free carpels. The little group of *Desfontaineæ* also approaches *Loganiaceæ*, of which it has nearly all the characters; but its æstivation is contorted, the placentation is parietal, and the leaves are always exstipulate. *Loganiaceæ* are scattered over the tropical regions of Asia, Africa and America, and extra-tropical Australia.

Most *Loganiaceæ* have a very bitter juice. The species of *Strychnos* contain in the bark of their root and in their seeds two alkaloids (*strychnine* and *brucine*), combined with a peculiar acid (*igasuric acid*), principles which are extremely energetic; their action on the nervous system is most powerful, whether as invaluable medicines or as mortal poisons. A decoction of the root of *S. Tieuté* is the *tjettek*, with which the Javanese poison their arrows; and which, when taken internally, also acts as a poison, but less rapidly than when absorbed through the veins. The natives of South America also use two species of *Strychnos* to poison their arrows; this poison, called *curare*, is prepared by mixing the juice of the bark with pepper, the Indian berry, and other acrid plants, and is preserved in little vases of baked earth. It is supposed that the *curare* acts as a poison only through the blood, and that it may be swallowed without inconvenience; it is certain that chemists have found no alkaloid in it. The seeds of *S. Nux vomica* act as a powerful excitant of the spinal cord and nerves, and stimulate the functions of the organs of voluntary motion, in cases of paralysis which do not proceed from injury to the brain, for which the seed itself, or an extract, or its alkaloid, *strychnine*, are employed. *Spigelia anthelmintica*, an American plant, very poisonous in its fresh state, is innocuous when dry, and is a successful vermifuge. *S. marylandica* is a less active but also useful vermifuge. [*Strychnos pseudo-quina* is a reputed Brazilian febrifuge, and yields Copalche bark. *S. Ignatia* yields the Ignatius Bean of India, used as a remedy for cholera. *S. potatorum* yields the celebrated Clearing Nut of India, which clarifies foul water when this is put in a vessel of which the inside has been rubbed with it.]

CXLVII. *GENTIANEÆ*.

(GENTIANEÆ, *Jussieu.*—GENTIANACEÆ, *Lindl.*)

COROLLA *monopetalous, hypogynous, isostemonous, æstivation contorted or indupli-cate, 5-4-6-8-fid.* OVARY *1- or sub-2-celled;* OVULES *∞, anatropous, horizontal, placentation parietal.* CAPSULE *dehiscing along the margins of the carpels.* EMBRYO *albuminous.*

Annual or perennial HERBS, sometimes woody below, rarely throughout, sometimes climbing, usually glabrous, juice watery. LEAVES opposite, sometimes whorled, very

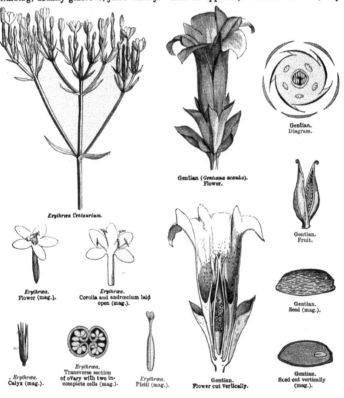

Gentian.
Diagram.

Gentian (*Gentiana acaulis*).
Flower.

Gentian.
Fruit.

Erythræa Centaurium.

Erythræa.
Flower (mag.).

Erythræa.
Corolla and andrœcium laid
open (mag.).

Gentian.
Seed (mag.).

. Erythræa.
Calyx (mag.).

Erythræa.
Transverse section
of ovary with two in-
complete cells (mag.).

Erythræa.
Pistil (mag.).

Gentian.
Flower cut vertically.

Gentian.
Seed cut vertically
(mag.).

rarely alternate or rosulate, nearly always simple and entire, exstipulate. Flowers ☿, generally regular, terminal or axillary, inflorescence various. Calyx persistent, of 5–4 sepals, rarely 6–8, distinct or more or less cohering, æstivation valvate or contorted. Corolla monopetalous, hypogynous, infundibuliform or hypocrateriform or sub-rotate; *throat* naked or furnished with a delicate fringed ring; *limb* naked or ciliate, or studded with glandular pits, æstivation valvate or induplicate. Stamens inserted on the corolla-tube or -throat, alternate with its lobes; *filaments* equal or nearly so, bases rarely dilated and united into a ring; *anthers* 2-celled, introrse, dehiscence usually longitudinal, sometimes apical. Carpels 2, connate into a 1- or more or less completely 2-celled ovary; *style* terminal, sometimes very short or wanting; *stigma* bifid or bilamellate; *ovules* numerous, many-seriate, anatropous. Capsule 2-valved, usually placentiferous at the edges of the valves. Seeds minute. Embryo minute, in the base of a fleshy copious albumen; *radicle* near the hilum, nearly always centrifugal.

[The following is Grisebach's arrangement of *Gentianeæ* :—

Tribe I. Eugentianeæ.—Corolla-lobes contorted. Albumen filling the cavity of the seed. —Leaves opposite,

Sub-tribe 1. Chironieæ.—Anthers erect, cells opposite, without a distinct connective, dehiscence lateral, often short and pore-like. *Chironia, Exacum,* &c.

Sub-tribe 2. Chloreæ.—Anthers with an obvious connective, often twisted. Style distinct, deciduous. *Sabbatia, Sebæa, Erythræa, Chlora,* &c.

Sub-tribe 3. Lisiantheæ.—Anthers with an obvious connective. Style persistent. *Lisianthus, Leianthus, Voyria,* &c.

Sub-tribe 4. Swertieæ.—Anthers with an obvious connective. Stigmas 2, persistent or confluent on the branches of a persistent style. *Gentiana, Crawfurdia, Pleurogyne, Ophelia, Halenia, Swertia,* &c.

Tribe II. Menyantheæ.—Corolla-lobes induplicate in æstivation. Albumen smaller than the cavity of the seed.—Leaves alternate. *Villarsia, Menyanthes, Limnanthemum,* &c.]

Gentianeæ are near *Loganiaceæ, Apocyneæ;* and *Asclepiadeæ* (see these families). They have also characters in common with *Gesneraceæ,* and especially with the genera with free ovaries, as the opposite leaves, anatropous ovules, parietal placentation, capsular fruit, and fleshy albumen; but *Gesneraceæ* have irregular anisostemonous corollas with imbricate æstivation, and an axile embryo, and are usually perigynous. *Orobancheæ* present the same affinity, and they have also, like *Gentianeæ,* a minute and basilar embryo, but they are parasites, and the scales which take the place of leaves are alternate. There is also some analogy between the true *Gentianeæ* and *Polemoniaceæ;* but the latter are separated by the many-celled ovary, axile placentation, loculicidal capsule, and alternate leaves.

Gentianeæ are scattered over the surface of the globe; they inhabit the mountains of the northern hemisphere; they especially abound on tropical [and temperate] mountains [whence their absence from the polar regions is very remarkable]. *Gentianeæ* supply tonic medicines, owing to their containing a bitter principle named *gentianin.* The chief indigenous species is the Yellow Gentian (*Gentianà lutea*), one of the earliest of known medicines. *G. cruciata* is also a febrifuge and vermifuge; its root was in repute among the ancients as an antidote to the plague and the bite of mad dogs. The Centaury (*Erythræa Centaurium*) is employed as a substitute for Gentian; its flowering tops contain, besides a bitter principle, an acrid substance which increases its tonic and febrifugal action. The Water-Trefoil, or Buckbean,

(*Menyanthes trifoliata*) has the same properties as the Centaury, and is also used as an antiscorbutic, as is *Villarsia nymphæoides*, both indigenous plants. [The rootstock of *Menyanthes* is intensely bitter, and an excellent tonic. Various species of *Ophelia* supply the celebrated Chirita of the Indian Pharmacopœia. American Calumba is the root of *Frazera Walteri.*—ED.]

CXLVIII. *HYDROPHYLLEÆ.*

(HYDROPHYLLEÆ, *R. Br.*—HYDROPHYLLACEÆ, *Lindl.*)

COROLLA *monopetalous, inserted on a hypogynous disk, isostemonous, æstivation imbricate.* STAMENS 5, *inserted at the bottom of the corolla-tube.* OVARY 1- *or incompletely 2-celled ;* OVULES *with a ventral hilum.* FRUIT *capsular, or almost fleshy.* EMBRYO *straight, albuminous.* RADICLE *vague.*

Annual or perennial HERBS with watery juice and angular stems. LEAVES

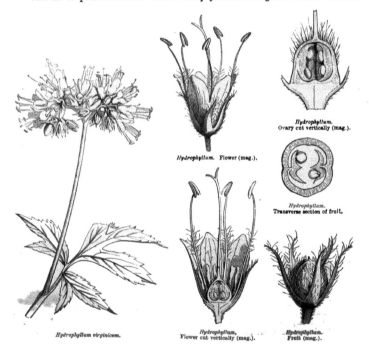

Hydrophyllum.
Ovary cut vertically (mag.).

Hydrophyllum. Flower (mag.).

Hydrophyllum.
Transverse section of fruit.

Hydrophyllum virginicum.

Hydrophyllum.
Flower cut vertically (mag.).

Hydrophyllum.
Fruit (mag.).

Phacelia. Flower-bud (mag.).   *Phacelia.* Diagram.   *Phacelia.* Pistil (mag.).   *Ellisia.* Seed cut vertically (mag.).   *Ellisia.* Cylindrical embryo (mag.).

alternate, the lower sometimes opposite, exstipulate. FLOWERS ☿, regular, in spikes or scorpioid racemes, very rarely solitary on axillary peduncles. CALYX 5-partite, persistent, æstivation imbricate. COROLLA monopetalous, inserted outside a hypogynous disk, campanulate or sub-rotate, very rarely infundibuliform, tube often furnished with scales within alternate with the stamens, limb 5-fid, æstivation imbricate. STAMENS 5, inserted at the bottom of the corolla-tube, and alternate with its lobes; *filaments* with various basal appendages; *anthers* introrse, 2-celled, dorsifixed, versatile, dehiscence longitudinal. OVARY 1- or incompletely 2-celled; *placentas* 2, linear or dilated, adnate by the base and apex to the ovary, but the dorsal face often free; *style* terminal, 2-fid at the top, lobes terminated by a papillose stigma; *ovules* 2 or more, attached by their ventral face, semi-anatropous. CAPSULE membranous or almost fleshy, 2-valved, placentas free or on the middle of the valves. SEEDS angular or sub-globose. EMBRYO straight, in a copious cartilaginous albumen, axile or excentric; *radicle* distant from the hilum, vague, or rarely superior.

PRINCIPAL GENERA.

* Hydrophyllum.	* Nemophila.	Ellisia.	* Eutoca.
* Phacelia.	* Cosmanthus.	* Whitlavia.	

Hydrophylleæ are near *Polemoniaceæ* (see this family). They also approach *Hydroleaceæ*, which only differ in the anatropous ovules and distinct styles. They were long confounded with *Borragineæ*, but their only resemblance is in the scorpioid inflorescence.

Hydrophylleæ abound in the temperate regions of North America; they are rare in extra-tropical South America, and still rarer in the tropics. One species only, *H. canadense*, is used in medicine, being considered in America a specific for snake-bites.

CXLIX. *HYDROLEACEÆ*.

(CONVOLVULORUM *genera,* Jussieu.—HYDROLEACEÆ, Br.)

COROLLA *monopetalous, hypogynous, isostemonous, æstivation imbricate.* STAMENS *inserted on the corolla-tube.* OVARY *more or less completely 2-celled;* OVULES *anatropous;* STYLES 2, *distinct.* CAPSULE *loculicidal or septifragal.* EMBRYO *albuminous.*

Annual herbaceous or sub-woody PLANTS with watery juice; *stem and branches*

o o

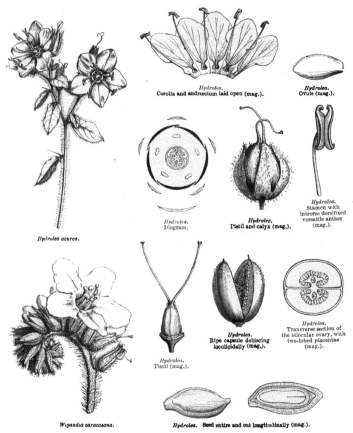

Hydrolea.
Corolla and andrœcium laid open (mag.).

Hydrolea.
Ovule (mag.).

Hydrolea.
Diagram.

Hydrolea.
Pistil and calyx (mag.).

Hydrolea.
Stamen with introrse dorsifixed versatile anther (mag.).

Hydrolea azurea.

Wigandia caracasana.

Hydrolea.
Pistil (mag.).

Hydrolea.
Ripe capsule dehiscing loculicidally (mag.).

Hydrolea.
Transverse section of the bilocular ovary, with two-lobed placentas (mag.).

Hydrolea. Seed entire and cut longitudinally (mag.).

often glandular-pubescent or clothed with stinging hairs, or sometimes with axillary spines. LEAVES alternate, exstipulate. FLOWERS ☿, regular, solitary or in corymbs or scorpioid spikes. CALYX herbaceous, 5-fid or -partite. COROLLA monopetalous, hypogynous, infundibuliform or sub-campanulate or sub-rotate, limb 5-fid, æstivation imbricate. STAMENS 5, inserted on the corolla-tube, alternate with its divisions; *filaments* filiform-subulate, sometimes dilated and arched at the base; *anthers*

2-celled, dehiscence longitudinal. OVARY with 2 more or less complete cells; *styles* 2, terminal, distinct; *stigmas* capitate; *ovules* numerous, anatropous, horizontal or pendent. CAPSULE with 2 valves, sometimes septiferous, leaving the placentiferous septum free, sometimes loculicidal and semi-septiferous. SEEDS numerous, minute, angular; *testa* membranous, loose, striate or areolate. EMBRYO straight, in the axis of a fleshy scanty albumen; *radicle* near the hilum, centripetal or superior.

[Arrangement of *Hydroleaceæ* by Choisy in De Candolle's 'Prodromus':—

TRIBE I. HYDROLEÆ.—Capsule 2-celled, septicidal; septum free, single, with a fungous placenta on each face. *Hydrolea.*

TRIBE II. NAMEÆ.—Capsule 1-2-celled, loculicidal; septum double, with lamellar placentas. *Nama, Wigandia, &c.*]

We have indicated the affinities of *Hydroleaceæ* with *Polemoniaceæ* and *Hydrophylleæ* (see these families). They are common in tropical and extra-tropical America; but *Hydrolea* itself is the only representative in the Old World, where it inhabits wet places in the tropics.

CL. *POLEMONIAOEÆ.*

(POLEMONIACEÆ, *Ventenat.*—POLEMONIDEÆ, *D.C.*—COBÆACEÆ, *Don.*)

COROLLA *monopetalous, hypogynous, regular, isostemonous, æstivation contorted.* STAMENS 5, *inserted at the middle or on the top of the corolla-tube.* OVARY 3-celled; OVULES *solitary and erect, or numerous and ascending.* CAPSULE 3-valved. EMBRYO *albuminous;* RADICLE *superior.*

Herbaceous PLANTS, rarely sub-woody or woody, with watery juice. LEAVES alternate, the lower sometimes opposite, exstipulate. FLOWERS ☿, rarely solitary, usually in a panicle or corymb or involucrate head. CALYX monosepalous, 5-fid. COROLLA monopetalous, hypogynous, infundibuliform or hypocrateriform, limb 5-partite, æstivation contorted. STAMENS 5, inserted on the corolla-throat or tube, and alternate with its lobes; *anthers* 2-celled, dehiscence longitudinal. OVARY seated on a more or less obvious glandular disk, 3–5-celled; *style* terminal, 3-fid, or 5-fid at the top, papillose on the inner surface; *ovules* either solitary and erect at the inner angle of the cell, anatropous, or numerous, biseriate, peltate by the ventral face, ascending and semi-anatropous. CAPSULE membranous or sub-woody, rarely fleshy, of 3–5 valves septiferous in the middle. SEEDS angular or compressed; *hilum* basilar or ventral; *testa* sometimes formed of mucilaginous cells with unrollable tracheæ. EMBRYO straight or nearly so, in the axis of a fleshy albumen; *cotyledons* foliaceous; *radicle* inferior.

[*Polemoniaceæ* have been divided by A. Gray into two groups:—

1. Stamens unequally inserted. **Phlox, *Collomia* (including **Leptosiphon*), *Nanarratia, Hugelia, *Gilia, Leptodactylon.*

2. Stamens equally inserted. **Gilia, *Polemonium, *Lœselia, *Cantua, *Cobœa.*—ED.]

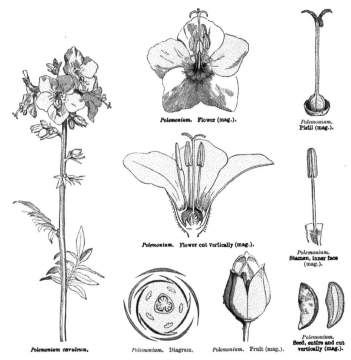

Polemonium. Flower (mag.).

Polemonium. Pistil (mag.).

Polemonium. Flower cut vertically (mag.).

Polemonium. Stamen, inner face (mag.).

Polemonium cæruleum.

Polemonium. Diagram.

Polemonium. Fruit (mag.).

Polemonium. Seed, entire and cut vertically (mag.).

Polemoniaceæ are closely allied to *Convolvulaceæ* (which see). They approach *Hydrophylleæ* in the alternate leaves, hypogynous and isostemonous corolla, loculicidal capsule, axile embryo and copious albumen ; but in *Hydrophylleæ* the æstivation of the corolla is imbricate, the ovary sub-2-celled with parietal placentation, the radicle superior, and the albumen cartilaginous. *Hydroleaceæ* agree and differ similarly ; besides which the styles are distinct and the ovules horizontal. *Polemoniaceæ* are also distantly allied to *Gentianeæ* (which see).

Polemoniaceæ mostly inhabit the west of extra-tropical America ; very few occur in the temperate and cold regions of the Old World. The Jacob's Ladder, or Greek Valerian (*Polemonium cæruleum*), is mucilaginous, nauseous in scent and bitter ; in some countries its leaves are applied to ulcers following contagious diseases, and the Russians give a decoction of it in cases of hydrophobia.

CLI. *CONVOLVULACEÆ.*

(CONVOLVULI, *Jussieu.*—CONVOLVULEÆ, *Ventenat.*—CONVOLVULACEÆ, *D.C.*)
[ERYCIBEÆ, *Endl.*]

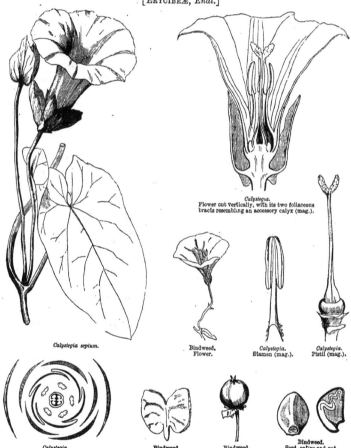

Calystegia.
Flower cut vertically, with its two foliaceous
bracts resembling an accessory calyx (mag.).

Calystegia sepium.

Bindweed.
Flower.

Calystegia.
Stamen (mag.).

Calystegia.
Pistil (mag.).

Calystegia.
Diagram of the flower and two foliaceous bracts.

Bindweed.
Embryo laid open (mag.).

Bindweed.
Fruit (mag.).

Bindweed.
Seed, entire and cut
vertically (mag.).

COROLLA *monopetalous, hypogynous, isostemonous, regular, æstivation contorted.* STAMENS 5, *inserted at the bottom of the corolla-tube.* OVARY *of* 4 1–2-*ovuled cells.* OVULES *collateral, erect, anatropous.* CAPSULE *with valves detaching from the septum, or a* BERRY. EMBRYO *curved;* ALBUMEN *mucilaginous;* COTYLEDONS *folded;* RADICLE *inferior.*

Herbaceous, sub-woody or woody PLANTS. STEM generally climbing, rarely erect, juice usually milky. LEAVES alternate, exstipulate. FLOWERS ☿, regular; *peduncles* axillary or terminal, simple or trichotomous, usually bibracteolate; *bracts* sometimes close together, enveloping the flower. CALYX of 5 sepals, usually free, persistent. COROLLA hypogynous, monopetalous, campanulate or infundibuliform or sometimes hypocrateriform, limb 5-fid or forming 5 folds, æstivation generally contorted. STAMENS 5, inserted at the bottom of the corolla-tube, alternate with its lobes; *filaments* usually dilated below, filiform above; *anthers* introrse, 2-celled, dehiscence longitudinal. OVARY sometimes girt by a disk, with 2–3–4 1–2-ovuled cells, or 1-celled and 1-ovuled by atrophy of the septum; *style* terminal, simple or 2-partite; *ovules* solitary or collateral, erect, anatropous. FRUIT either capsular, 1–4-celled, valves separating from the placentiferous column at its base; or fleshy, indehiscent. SEEDS erect; *testa* sometimes very villous; *albumen* mucilaginous, scanty. EMBRYO more or less curved; *cotyledons* foliaceous, folded or crumpled; *radicle* near the hilum, inferior.

[Arrangement of *Convolvulaceæ* :—

TRIBE I. ARGYREIEÆ.—Fruit indehiscent, coriaceous or sub-baccate. Carpels connate; style simple. Embryo with distinct cotyledons. *Rivea, Argyreia*, &c.

TRIBE II. CONVOLVULEÆ.—Fruit capsular. Carpels connate; style simple. Embryo with distinct cotyledons. *Quamoclit, Batatas, Pharbitis, Ipomœa, Jacquemontia, Convolvulus, Aniseia, Porana, Evolvulus*, &c.

TRIBE III. DICHONDREÆ.—Fruit of distinct dry carpels, each with one style. Embryo with distinct cotyledons. See DICHONDREÆ, p. 567.

TRIBE IV. CUSCUTEÆ.—Fruit capsular, often transversely dehiscing. Carpels connate; styles 2, rarely connate. Embryo spiral, without distinct cotyledons. See CUSCUTEÆ, p. 568.

TRIBE V. ERYCIBEÆ.—Fruit baccate. Carpels connate into a 1-celled ovary with a sub-sessile 5-lobed stigma. Embryo with distinct cotyledons. *Erycibe.*]

Convolvulaceæ are near *Cuscuteæ* and *Dichondreæ* (see these families). They approach *Polemoniaceæ* in the insertion, isostemony and æstivation of the corolla, structure of the ovary, anatropy and position of the ovules, capsular fruit, alternate leaves, and often climbing stem; but in *Polemoniaceæ* the ovary has three many-ovuled cells, the capsule has semi-septiferous valves, the embryo is straight and axile, and the fleshy albumen abundant. There is a certain analogy between *Convolvulaceæ* and *Cordiaceæ* in the form and æstivation of the corolla, the 2–4-celled ovary, bifid style and anatropous ovules; but in *Cordiaceæ* the radicle is superior, the straight embryo exalbuminous, and the cotyledons are folded longitudinally. The erect species approach *Solaneæ* in the insertion, isostemony, æstivation and form of the corolla, in the 2-celled ovary, capsular or berried fruit, curved embryo, inferior radicle, and alternate leaves; but in *Solaneæ* the ovule is campylotropous, the albumen copious, and the radicle is distant from the hilum. There is also a distant connection between *Convolvulaceæ* and *Hydrophylleæ* (which see).

Convolvulaceæ are chiefly tropical; they decrease northwards, and are very rare in our climate, and absolutely wanting in the arctic regions and on mountains. Many species possess a milky juice containing a highly purgative resin; this resin, which especially abounds in the rhizome, owes its properties solely to the presence of an aromatic principle; for rhizomes which have been pulverized and long exposed to the air lose it, although preserving the purely resinous principle. The species most in use are Jalap (*Convolvulus Jalapa* and *C. Schiedeanus*), from Mexico [the best is from *Exogonium Purga*]; the Turbith (*C. Turpethum*), a native of the East Indies; Scammony (*C. Scammonia* and *C. sayittæfolius*), from the Asiatic Mediterranean region [and the *Ipomœa pandurata* of the United States]. The rhizomes of our indigenous Bindweeds are also purgative, but the exotic species are much more active. The American genus *Batatas* comprises several species in the rhizomes of which (called Sweet Potato) the resinous principle is replaced by an abundant quantity of starch, and they are thus sought for as a food similar to potatos. [*Convolvulus dissectus* is said to abound in prussic acid, and to be used in the preparation of Noyau. Oil of Rhodium is the produce of the rootstock of *Rhodorhiza.*—ED.]

CLII. *DICHONDREÆ.*

(CONVOLVULACEARUM *genera, Endlicher.*)

This little family may be considered as a tribe of *Convolvulaceæ*, with which they agree in the insertion, regularity, and isostemony of the corolla, the number of carpels and of the erect anatropous ovules, the mucilaginous albumen, and the contortuplicate cotyledons; it has been separated on account of its

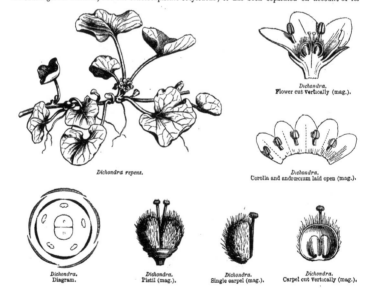

Dichondra.
Flower cut vertically (mag.).

Dichondra.
Corolla and androecium laid open (mag.).

Dichondra repens.

Dichondra.
Diagram.

Dichondra.
Pistil (mag.).

Dichondra.
Single carpel (mag.).

Dichondra.
Carpel cut vertically (mag.).

Dichondra.
Ovule (mag.).

Dichondra.
Embryo (mag.).

Dichondra.
Embryo (mag.).

Dichondra.
Seed (mag.).

Dichondra.
Seed cut transversely (mag.).

free carpels (of which there are 2–4, united in pairs), its basilar styles and the valvate æstivation of its corolla. The genus *Dichondra* comprises a few herbaceous, climbing, not milky species, which live in the hot region [chiefly] of the southern hemisphere and of America. The genus *Falkia* is founded on a South African shrub.

CLIII. *CUSCUTEÆ.*

(CONVOLVULORUM *pars, Jussieu.*—CUSCUTEÆ, *Presl.*—CUSCUTINÆ, *Link.*)

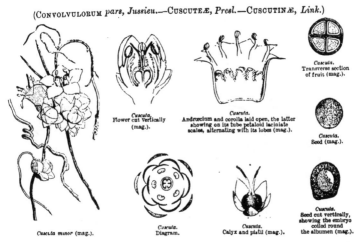

Cuscuta.
Transverse section
of fruit (mag.).

Cuscuta.
Flower cut vertically
(mag.).

Cuscuta.
Andrœcium and corolla laid open, the latter
showing on its tube petaloid laciniate
scales, alternating with its lobes (mag.).

Cuscuta.
Seed (mag.).

Cuscuta minor (mag.).

Cuscuta.
Diagram.

Cuscuta.
Calyx and pistil (mag.).

Cuscuta.
Seed cut vertically,
showing the embryo
coiled round
the albumen (mag.).

The genus *Cuscuta*, which constitutes this little family, is separated from *Convolvulaceæ* only by its filiform stems of a reddish or greenish-yellow colour, the absence of leaves, and its parasitism on other plants by means of suckers, by the [often] transverse dehiscence of its capsular or sometimes fleshy fruit, and by the acotyledonous embryo, which is coiled spirally round the albumen. The flowers are in a head or spike, and are usually bracteate.

Cuscuteæ inhabit all hot and temperate regions, as parasites on the stems of many herbaceous or even woody plants, which they exhaust by absorbing their elaborated sap. The Small Dodder (*C. minor*) lives on Field Clover, Lucerne, Thyme, Broom, Furze, Heath, &c.; the *C. densiflora* infests Flax fields; the Large Dodder (*C. major*) is parasitic on Nettles and Hops, and even invades the peduncles of the Vine, enclosing them in its thread-like branches, whence the name of Bearded Grapes, given to the clusters whose nutriment it has appropriated.

CLIV. *BORRAGINEÆ.*

(ASPERIFOLIÆ, *L.*—BORRAGINEÆ, *Jussieu.*—BORRAGINEÆ ET HELIOTROPICEÆ, *Schrader.*
—ARGUZIEÆ ET BORRAGINEÆ, *Link.*—EHRETIACEÆ ET BORRAGINACEÆ, *Lindl.*—
ASPERIFOLIÆ, *Endlicher.*)

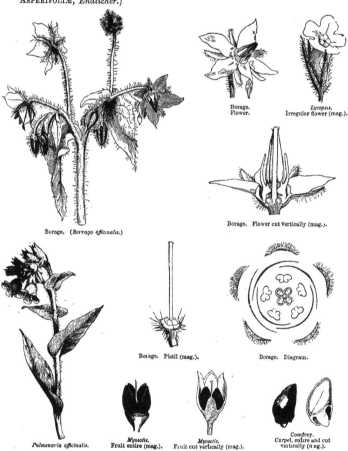

Borage. Flower.

Lycopsis.
Irregular flower (mag.).

Borage. Flower cut vertically (mag.).

Borage. (*Borrago officinalis.*)

Borage. Pistil (mag.).

Borage. Diagram.

Pulmonaria officinalis.

Myosotis.
Fruit entire (mag.).

Myosotis.
Fruit cut vertically (mag.).

Comfrey.
Carpel, entire and cut
vertically (mag.).

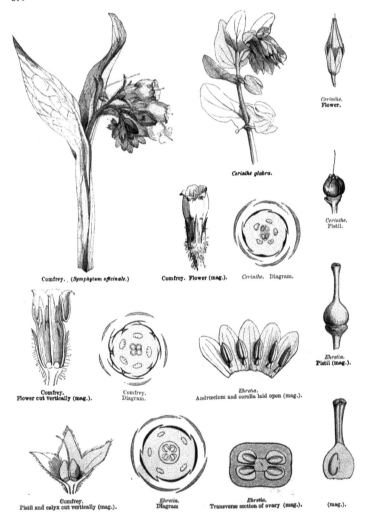

Comfrey. (*Symphytum officinale.*)

Cerinthe glabra.

Comfrey. Flower (mag.).

Cerinthe. Diagram.

Cerinthe.
Flower.

Cerinthe.
Pistil.

Comfrey.
Flower cut vertically (mag.).

Comfrey.
Diagram.

Ehretia.
Andrœcium and corolla laid open (mag.).

Ehretia.
Pistil (mag.).

Comfrey.
Pistil and calyx cut vertically (mag.).

Ehretia.
Diagram

Ehretia.
Transverse section of ovary (mag.).

(mag.).

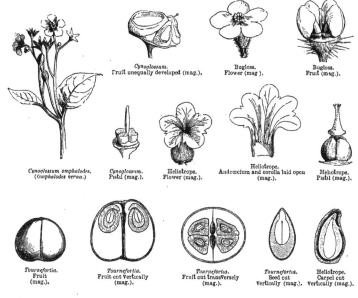

Cynoglossum.
Fruit unequally developed (mag.).

Bugloss.
Flower (mag).

Bugloss.
Fruit (mag.).

Cynoglossum omphalodes.
(Omphalodes verna.)

Cynoglossum.
Pistil (mag.).

Heliotrope.
Flower (mag.).

Heliotrope.
Andrœcium and corolla laid open
(mag.).

Heliotrope.
Pistil (mag.).

Tournefortia.
Fruit
(mag.).

Tournefortia.
Fruit cut Vertically
(mag.).

Tournefortia.
Fruit cut transversely
(mag.).

Tournefortia.
Seed cut
Vertically (mag.).

Heliotrope.
Carpel cut
vertically (mag.).

COROLLA *monopetalous, hypogynous, isostemonous, æstivation imbricate.* STAMENS *5, inserted on the corolla.* OVARY *with 2 bipartite carpels;* STYLE *gynobasic;* OVULES *4, appendiculate, anatropous or semi-anatropous.* EMBRYO *usually exalbuminous;* RADICLE *superior.*—INFLORESCENCE *a scorpioid raceme.* •

HERBS, SHRUBS or TREES, usually hispid. LEAVES generally alternate, simple, entire, exstipulate. FLOWERS ☿, rarely ♂ ♀, regular, sometimes irregular, solitary in the axils of the leaves, or in panicles, corymbs, or terminal scorpioid unilateral racemes. CALYX persistent, monosepalous, 4-5-partite. COROLLA hypogynous, monopetalous, deciduous, tubular-infundibuliform, campanulate or rotate; *throat* naked, or furnished with hairs, scales or protuberances; *limb* 5-fid, æstivation imbricate. STAMENS 5, inserted on the tube or throat of the corolla, alternate with its divisions; *anthers* introrse, 2-celled, with longitudinal dehiscence, usually free, sometimes slightly cohering at the base or top. CARPELS 2, antero-posterior, more or less distinct, with 2 more or less connected 1-ovuled cells, usually forming together a 4-lobed ovary, inserted on a central column (*gynobase*) formed by the thickened style-base at its union with the receptacle (*gynophore*); *style* either gynobasic or terminating the united carpels; *ovules* pendulous from the inner angle of the cell, anatropous or semi-anatropous. FRUIT composed of 4 distinct or geminate nucules, or a drupe with 2–4 kernels. SEEDS inverted, straight, or a little arched; *albumen* 0, or reduced to a fleshy layer. EMBRYO straight or a little curved; *radicle* superior.

[*Borragineæ* have been thus classified by De Candolle :—

TRIBE I. CORDIEÆ.—Ovary undivided ; style terminal (rarely 0), twice forked. Fruit indehiscent, usually fleshy, 4-seeded. Cotyledons longitudinally folded. Albumen 0. Shrubs or trees. See CORDIACEÆ, p. 573.

TRIBE II. EHRETIEÆ.—Ovary undivided ; style terminal, 2-lobed. Fruit indehiscent, 4-seeded. Cotyledons flat. Albumen scanty, fleshy. Shrubs or small trees. *Ehretia, Tourne-fortia,* &c.

TRIBE III. HELIOTROPIEÆ.—Ovary several-celled ; style terminal, simple. Fruit dry, entire, or separating into cocci. Albumen scanty or 0. Cotyledons flat. *Heliotropium, Heliophytum,* &c.

TRIBE IV. BORRAGEÆ.—Ovary of 2 2-celled or 2-partite carpels ; style ventral or basal. Fruit 2–4-partite. Seeds exalbuminous. Herbs, rarely shrubs.

Sub-tribe 1. CERINTHEÆ.—Corolla regular, throat naked. Carpels 2, 2-celled. Nucules with a flat areole, seated on a flat torus. *Cerinthe.*

Sub-tribe 2. ECHIEÆ.—Corolla irregular, throat naked. Carpels 4. Nucules distinct, with a flat imperforate areole, seated on a flat torus. *Lobostemon, Echium,* &c.

Sub-tribe 3. ANCHUSEÆ.—Corolla regular, with scales under the middle of the lobes. Nucules 4, dehiscing transversely at the base, which hence appears perforate or excavated. *Nonnea, Borrago, Symphytum, Anchusa, Lycopsis,* &c.

Sub-tribe 4. LITHOSPERMEÆ.—Corolla regular, with or without fornices. Nucules 4, distinct, 1-celled, with a minute flat imperforate base. *Onosma, Moltkia, Lithospermum, Mertensia, Pulmonaria, Alkanna, Myosotis,* &c.

Sub-tribe 5. CYNOGLOSSEÆ.—Corolla regular, with or without fornices. Nucules 4, usually echinate or winged, imperforate at the base, very obliquely inserted on the torus. *Eritrichium, Echinospermum, Cynoglossum, Omphalodes, Mattia, Trichodesma,* &c.

Sub-tribe 6. ROCHELIEÆ.—Corolla regular. Ovary of 2 1-celled 1-seeded carpels adnate to the style. *Rochelia.*]

Borragineæ approach *Labiatæ* and *Verbenaceæ* in the insertion and æstivation of the corolla, arrangement of the carpels and style, anatropous ovules, nature of the fruit, and usually the absence of albumen ; but in *Labiatæ* and *Verbenaceæ* the corolla is very irregular, the stamens are didynamous, the ovules erect or ascending, the stem square, and the leaves opposite. There is also an affinity between the tribe *Ehretieæ* and *Cordiaceæ,* founded on the insertion, regularity and isostemony of the corolla, the pendulous anatropous ovules, terminal bifid style, fleshy fruit, absent or scanty albumen, and alternate leaves ; the diagnosis principally rests on the contorted æstivation of *Cordiaceæ* and their longitudinally folded cotyledons. *Borragineæ* inhabit [chiefly] extra-tropical temperate regions, and especially the Mediterranean region and Central Asia. The tribe of *Ehretieæ* is chiefly tropical. Many species contain a mucilage, to which is often added a bitter astringent principle, to which they owe their medicinal qualities. The root of the Comfrey (*Symphytum officinale*) is employed in cases of hæmoptysis. The leaves of the Borage (*Borrago officinalis*) are filled with a viscous juice abounding in nitrates, whence their diuretic and sudorific properties. *Cynoglossum officinale,* the poisonous smelling root of which was a reputed narcotic, is now only administered with opium. The following are no longer used :—*Pulmonaria officinalis,* of which the white spotted leaves, like a tubercled lung, were employed in lung diseases ;

Viper's Bugloss (*Echium vulgare*), the flowering tops of which were recommended for the bite of the viper; Gromwell (*Lithospermum officinale*), commonly termed Pearlwort, on account of its hard and pearl-grey nucules, which were supposed to be of use in dissolving bladder-stones; and *Heliotropium europæum*, of which the bitter and salt leaves were applied to ulcers and warts. *Tournefortia umbellata* is still used in Mexico as a febrifuge. In tropical America and in India certain species of *Tiaridium* are used in herpetic affections. Some *Ehretieæ* have an edible fruit. Finally, the roots of several species of *Anchusa, Onosma, Lithospermum, Arnebia*, contain a red colouring matter, soluble in alcohol and fatty bodies, which is employed to colour certain unguents and other external applications.

CLV. *CORDIACEÆ.*

(CORDIACEÆ, *Br.*—CORDIEÆ, *Dumortier.*)

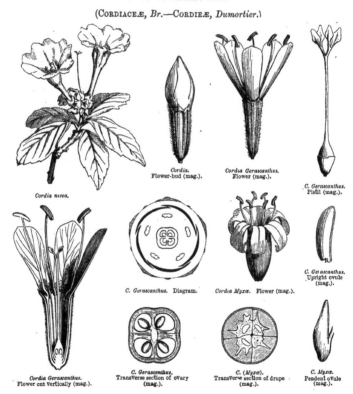

Cordia.
Flower-bud (mag.).

Cordia Gerascanthus.
Flower (mag.).

C. Gerascanthus.
Pistil (mag.).

Cordia nivea.

C. Gerascanthus. Diagram.

Cordia Myxæ. Flower (mag.).

C. Gerascanthus.
Upright ovule (mag.).

Cordia Gerascanthus.
Flower cut vertically (mag.).

C. Gerascanthus.
Transverse section of ovary (mag.).

C. (Myxæ).
Transverse section of drupe (mag.).

C. Myxæ.
Pendent ovule (mag.).

C. (Myxa).	*C. (Myxa).*	*C. (Myxa).*	*C. (Myxa).*	*C. (Myxa).*
Fruit enveloped in the calyx (mag.).	Fruit deprived of the calyx (mag.).	Kernel deprived of the pericarp (mag.).	Transverse section of the seed (mag.).	Embryo (mag.).

COROLLA *hypogynous, monopetalous, isostemonous, generally regular, æstivation contorted.* STAMENS 5, *inserted on the corolla.* OVARY 4–8-*celled ;* OVULES *appendiculate or erect, anatropous.* FRUIT *drupaceous.* EMBRYO *straight, exalbuminous ;* COTYLEDONS *longitudinally folded.*

TREES OR SHRUBS. LEAVES alternate, simple, coriaceous, scabrous, exstipulate. FLOWERS ☿, or ♂ ♀, terminal, in a panicle or corymb, sometimes in a more or less contracted spike, ebracteate. CALYX persistent or accrescent, 4-toothed or 4–5-partite. COROLLA monopetalous, hypogynous, infundibuliform or campanulate, limb usually 5-fid, æstivation convolute or contorted. STAMENS inserted on the corolla-tube, alternate with its lobes ; *filaments* filiform or subulate ; *anthers* 2-celled, dehiscence longitudinal. OVARY free, 4–8-celled ; *style* terminal, dichotomous, or twice dichotomous at the top ; *stigmas* 4 or 8 ; *ovules* solitary in each cell, appendiculate or erect, anatropous. DRUPE fleshy, with one bony 4–8-celled stone, or 1-celled by arrest. SEEDS with a membranous testa. EMBRYO exalbuminous, straight, with thick fleshy cotyledons, forming many contiguous longitudinal folds ; *radicle* short.

PRINCIPAL GENERA.

Cordia.   Varronia.

We have indicated the more or less real affinities between *Cordiaceæ* and *Borragineæ* and *Convolvulaceæ* (which see).

Cordiaceæ mostly inhabit the intertropical regions of the Old and New Worlds.

The drupe of *Cordia* is mucilaginous, of a pleasant and slightly astringent taste, acid in some species. The cotyledons contain a mild oil. *Cordia Myxa* is an Asiatic tree, which has been cultivated in Egypt from time immemorial. The ancients employed its fruit as an emollient in affections of the lungs, and its bark in astringent gargles. *C. Sebestena,* a tree of the Antilles, possesses the same properties. *C. Rumphii* produces a wood of a maroon brown, elegantly veined with black, and which smells like musk.

CLVI. *NOLANEÆ.*

(SOLANACEARUM *tribes, Dunal.*—NOLANEÆ, *G. Don.*—NOLANACEÆ, *Endlicher.*)

Herbaceous or sub-woody prostrate PLANTS. LEAVES alternate, geminate, entire. PEDUNCLE 1-flowered, extra-axillary. CALYX campanulate, 5-partite, persistent, [valvate]. COROLLA hypogynous, monopetalous, infundibuliform ; *limb* folded, 5–10-

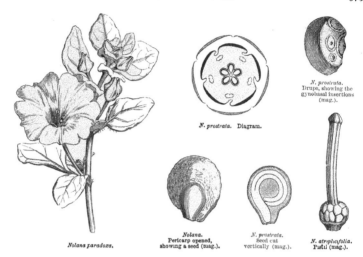

N. prostrata. Drupe, showing the gynobasal insertions (mag.).

N. prostrata. Diagram.

Nolana paradoxa.

Nolana. Pericarp opened, showing a seed (mag.).

N. prostrata. Seed cut vertically (mag.).

N. atriplicifolia. Pistil (mag.).

lobed. STAMENS 5, inserted on the corolla-tube, exserted. OVARIES numerous, inserted on a fleshy disk, hypogynous, distinct, 1–6-celled; *style* solitary, central, basilar, simple; *stigma* capitate; *ovules* solitary, erect in each cell. DRUPES distinct, fleshy, with a bony endocarp. SEEDS reniform, compressed; *albumen* fleshy. EMBRYO filiform, annular or spiral; *radicle* inferior.

This little order is formed by the genus *Nolana*, formerly placed by A. L. de Jussieu at the end of *Borragineæ* on account of its distinct carpels and its gynobasic style; it approaches nearer to *Solaneæ* and *Convolvulaceæ* in the insertion, regularity, isostemony and æstivation of the corolla, and the curved embryo; it is particularly close to *Solaneæ* in the alternate geminate leaves and the extra-axillary flowers; it is separated from it by its distinct carpels, its gynobasic style and its drupaceous fruit.

The species of the genus *Nolana* inhabit Chili and Peru.

[*Nolaneæ* are now reduced to a tribe of *Solaneæ* with five genera, of which the chief are *Nolana*, *Alibrexia*, and *Dolia.*—ED.]

CLVII. *SOLANEÆ.*

(LURIDÆ, *L.*—SOLANEÆ, *Jussieu.*—SOLANACEÆ, *Bartling.*)

COROLLA *monopetalous, hypogynous, isostemonous, æstivation induplicate or contorted.* STAMENS *inserted on the corolla.* OVARY *with 2 many-ovuled antero-posterior cells, placentation septate;* OVULES *campylotropous.* SEEDS *compressed.* EMBRYO *curved, albuminous.*

Herbaceous or woody PLANTS with watery juice. LEAVES alternate, the upper usually geminate, simple, exstipulate. FLOWERS ☿, often extra-axillary, with ebracteate pedicels. CALYX monosepalous, with 5 (rarely 6-4) segments, persistent. COROLLA hypogynous, monopetalous, more or less regular, rotate, campanulate, infundibuliform or hypocrateriform; *limb* with 5, rarely 4-6 segments; æstivation folded, contorted, induplicate or valvate. STAMENS inserted on the corolla-tube, alternate with its segments; *anthers* introrse, sometimes connivent, or even cohering at the top; cells opposite, parallel, dehiscence longitudinal or by an apical pore (*Nightshade*). CARPELS 2, one antero-posterior, cohering into a 2-celled ovary; *placentas* thick, attached on the middle of the septum by a broad or linear surface, sometimes bipartite, the lobes separated by a false septum, which subdivides each cell, except at the top; *style* terminal, simple; *stigma* undivided or lobed; *ovules* very numerous, campylotropous. FRUIT a septicidal (*Tobacco*), rarely loculicidal and septifragal (*Datura*) capsule, or a pyxidium (*Henbane*), or a pulpy (*Nightshade*) or dry (*Pimento*) berry. SEEDS numerous, compressed, hilum ventral; *albumen* fleshy, copious. EMBRYO curved or annular; *cotyledons* semi-cylindric; *radicle* next the hilum, or vague.

TRIBE I. NICOTIANEÆ.

Capsule 2-celled, septicidally 2-valved.

PRINCIPAL GENERA.

* Fabiana.　　* Nierembergia.　　* Petunia.　　* Nicotiana.

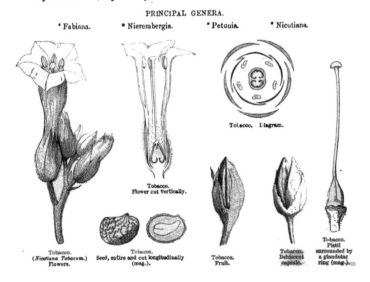

Tobacco. Diagram.

Tobacco.
Flower cut vertically.

Tobacco.
Pistil
surrounded by
a glandular
ring (mag.).

Tobacco.
(*Nicotiana Tabacum*.)
Flowers.

Tobacco.
Seed, entire and cut longitudinally
(mag.).

Tobacco.
Fruit.

Tobacco.
Dehiscent
capsule.

TRIBE II. *DATUREÆ.*

Capsule or berry incompletely 4-celled; primary septum bearing a placenta on each side, either on its centre, or near the parietal angle.

PRINCIPAL GENERA.

* Datura. * Solandra.

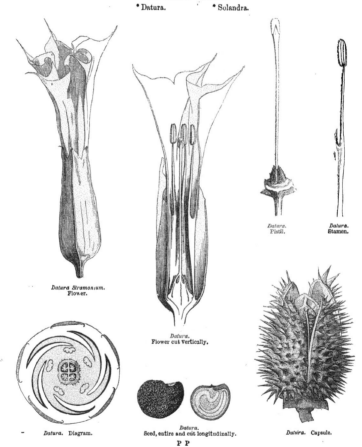

Datura Stramonium.
Flower.

Datura.
Flower cut vertically.

Datura.
Pistil.

Datura.
Stamen.

Datura. Diagram.

Datura.
Seed, entire and cut longitudinally.

Datúra. Capsule.

TRIBE III. *HYOSCYAMEÆ.*

Capsule 2-celled, dehiscence circumsciss.

PRINCIPAL GENERA.

Hyoscyamus.   Scopolia.

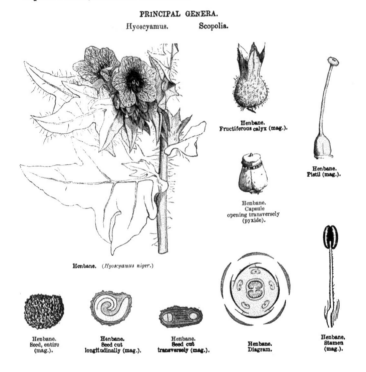

Henbane.
Fructiferous calyx (mag.).

Henbane.
Pistil (mag.).

Henbane.
Capsule
opening transversely
(pyxide).

Henbane. (*Hyoscyamus niger.*)

Henbane.
Seed, entire
(mag.).

Henbane.
Seed cut
longitudinally (mag.).

Henbane.
Seed cut
transversely (mag.).

Henbane.
Diagram.

Henbane,
Stamen
(mag.).

TRIBE IV. *SOLANEÆ VERÆ.*

Berry 2- or more-celled, placentation central; rarely a capsule without valves.

PRINCIPAL GENERA.

* Nicandra.	Physalis.	* Capsicum.	**Solanum.**
* Lycopersicum.	Atropa.	Mandragora.	* Iochroma.
* Lycium.	Withania.	Arnistus.	Cyphomandra.

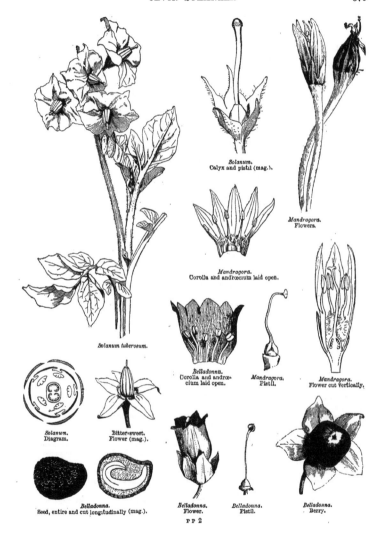

Solanum.
Calyx and pistil (mag.).

Mandragora.
Flowers.

Mandragora.
Corolla and androecium laid open.

Solanum tuberosum.

Belladonna.
Corolla and androecium laid open.

Mandragora.
Pistil.

Mandragora.
Flower cut vertically.

Solanum.
Diagram.

Bitter-sweet.
Flower (mag.).

Belladonna.
Seed, entire and cut longitudinally (mag.).

Belladonna.
Flower.

Belladonna.
Pistil.

Belladonna.
Berry.

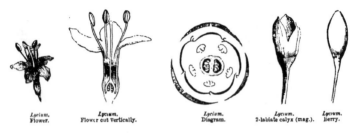

Lycium. Flower. *Lycium.* Flower cut vertically. *Lycium.* Diagram. *Lycium.* 2-labiate calyx (mag.). *Lycium.* Berry.

[To these tribes should be added :—

TRIBE V. NOLANEÆ.—See order NOLANEÆ, p. 574.

TRIBE VI. GRABOWSKIEÆ.—Carpels 2, 2–3-celled, united into a 2-partite or a 4-celled ovary; ovules solitary in each cell. *Grabowskia.*

TRIBE VII. TRIGUIEREÆ. Ovary 2–4-celled; ovules few in each cell. Fruit sub-glabrous, membranous, 2–4-celled, indehiscent; placentas central, connate. Embryo spiral. (Spain.) *Triguiera.*

TRIBE VIII. CESTRINEÆ.—See order CESTRINEÆ, p. 582.

TRIBE IX. RETZIEÆ.—Ovary 2-celled; fruit 2-celled, 2-valved. Seeds few; placentas on the middle of the septum. *Lenchostema, Retzia.*

TRIBE X. METTERNICHIEÆ.—Ovary 2-celled. Fruit a coriaceous 2-celled capsule, septifragal at the top, few-seeded. Seeds ascending, narrow, linear. (Tropical America.) *Metternichia, Sessea.*—ED.]

We have pointed out the affinities of *Solaneæ* with *Convolvulaceæ* (which see). They are near *Polemoniaceæ* in insertion, isostemony and imbrication of their corolla, capsular fruit and albuminous embryo; but *Polemoniaceæ* have a 3-celled ovary, axile placentation and straight embryo. The affinity is much closer between *Solaneæ* and *Scrophularineæ*; in both the ovary has two many-ovuled antero-posterior cells, the fruit is capsular or fleshy, the embryo is albuminous, and in some *Scrophularineæ* it is bent as in *Solaneæ*. The diagnosis rests on the irregularity, the æstivation and the anisostemony of the corolla in *Scrophularineæ*; and even this last difference disappears in some genera where there is a rudimentary fifth stamen. *Solaneæ* are mostly intertropical; they become rare in temperate regions, two species only (*Solanum nigrum* and *S. Dulcamara*) attaining high latitudes.

The medicinal properties of this family reside in narcotic alkaline substances combined with an acrid principle. The principal medicinal *Solaneæ* are Belladonna, Stramonium and Henbane; of these the roots and especially the leaves of *Atropa Belladonna* contain the alkaloid *atropine*, a most efficacious remedy for neuralgia and rheumatism. Belladonna further has a specific action on the muscular fibre, and is hence employed to dilate the pupil in diseases of the eyes, and to facilitate respiration in asthma and whooping-cough. The Mandragora, a genus allied to Belladonna, and possessing the same properties, was formerly used by sorcerers to produce hallucination in their dupes. The Henbane (*Hyoscyamus niger*) owes its narcotic virtues, which are, however, less energetic than those of Belladonna, to the alkaloid *hyoscyamine*. Stramonium seeds (*Datura Stramonium*) [and those of *D. Tatula* and *Metel*] contain the alkaloid *daturine*; these are highly narcotic, and were formerly employed by magicians to produce fantastic visions, and by thieves to stupefy their victims.

The American genus *Nicotiana* contains several species used for Tobacco; the chief of these,

N. Tabacum, was employed by the Caribbeans as a sedative, and called *tabaco* or *petun*, according as they smoked or snuffed it. It was introduced, about 1520, into Portugal and Spain by Doctor Hernandez of Toledo; into Italy by Tornabon and the Cardinal de Sainte-Croix; into England by Captain Drake, and into France by André Thevet, a gray friar. It was through John Nicot, ambassador at Lisbon, that Tobacco first acquired its popularity; he sent to Queen Catharine de Medicis, together with some Tobacco seeds, a little box full of powdered Tobacco; the queen acquired a taste for it, and the plant was thence called *Nicotian* and *Herbe à la Reine*. The Abbé Jacques Gohory, the author of the first book written in France on Tobacco, proposed to call it *Catherinaire* or *Médicée*, to record the name of Medicis and the medicinal virtues of the plant; but the name of Nicot superseded these, and botanists have perpetuated it in the genus *Nicotiana*. During the latter half of the sixteenth century the sovereigns of Europe, of Persia and of Turkey, vainly endeavoured, by more or less severe measures, to stem the increasing popularity of Tobacco; but in the following century, perceiving that its popularity might be made the means of raising a revenue, they tolerated its use and either heavily taxed it, or reserved to themselves the monopoly of it. It was in 1621 that the French government first put a duty on Tobacco of forty sols the quintal (less than five centimes the kilogramme). In 1674 the monopoly of Tobacco was granted to the farmer-general of taxes: he received in 1697, 250,000 livres of Tours; in 1718, four millions; in 1730, eight millions; in 1789, thirty-seven millions. The office was suppressed in 1791. From 1801 to 1804 this tax produced annually about 4,800,000 francs. The government monopoly was re-established in 1811, and from 1814 to 1844 Tobacco yielded a clear profit of 1,625,000,000 francs, an average of fifty-four millions yearly; and in 1840 seventy-five millions. From 1844 to 1864 the profit was two thousand millions; an increase so rapid that Tobacco, snuffed, smoked and chewed, will probably in a few years yield double the revenue it now does.

Tobacco is sometimes used medicinally, but only externally; its properties are those of other poisonous *Solaneæ*, and are due to a peculiar and extremely poisonous alkaloid, named *nicotine*. [Tobacco-oil is one of the most deadly poisons]. The *N. rustica*, also a native of America, is employed in the same way as *N. Tabacum*.

The Winter Cherry (*Physalis Alkekengi*) is a European plant of which the fruit, enclosed in a red accrescent calyx, is a diuretic. The Chili (*Capsicum annuum*), an Indian annual, bears a sub-succulent berry with an acrid [burning] principle, hence much used as a condiment in all countries. Cayenne Pepper, [the ground fruit of] a sub-woody *Capsicum*, is a much more powerful excitant. The Tomato, or Love Apple (*Lycopersicum esculentum*), now cultivated everywhere, is a bright red tropical American fruit, filled with an orange acid pulp, much used as a vegetable. The genus *Solanum* (Nightshade), which gives its name to the order, comprises nearly twice as many species as the other *Solaneæ*. The Bitter-sweet (*S. Dulcamara*), an indigenous shrub, the bitter stem of which leaves a mild taste in the mouth, is a depurative in cutaneous disorders. The species of *Solanum* all contain an emetic and narcotic alkaloid (*solanine*), which in the Black Nightshade (*S. nigrum*), a small herb with a poisonous smell, common near habitations, and in many exotic species (*S. guineense, S. pterocaulon*), is neutralized by an acid and diluted by a mucilage; owing to these the Nightshades, after being boiled to remove the poisonous odour, are employed like Spinach in tropical regions, under the name of *brèdes*. [In England both *S. Dulcamara* and *S. nigrum* are regarded as very dangerous plants].

The Brinjal, Aubergine or Egg-plant of Asia (*S. Melongena*), now cultivated in Europe, bears a large ovoid violet or yellowish fruit with a white flesh, which is edible when cooked; as is that of *S. oviferum*, which resembles a hen's egg.

Of all the *Solaneæ* the most useful to man is the Potato (*S. tuberosum*), a native of the Cordilleras of Peru and Chili, and now cultivated throughout the world. Besides the agreeable wholesome tubers, its starch yields a cheap sugar and alcohol. The tuber is the only edible part of the Potato plant; the leaves, fruit, and even the buds which spring from the eyes of the Potato contain *solanine*, and are narcotic. [The fruits of several are not only edible, but favourite articles of food; as those of *S. laciniatum*, the Kangaroo Apple of Australia; *S. quitoense*, the *Narangitas de Quito* (Quito Orange), and others.—ED.]

CLVIII. *CESTRINEÆ.*

(SOLANEARUM *genus, A.-L. de Jussieu.*—CESTRINEÆ, *Sendtn.*—CESTRACEÆ, *Lindl.*—SOLANACEARUM *tribes, Endlicher.*)

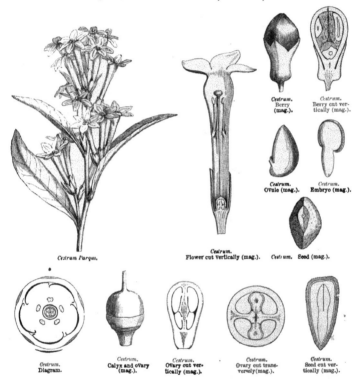

Cestrum. Berry (mag.).

Cestrum. Berry cut vertically (mag.).

Cestrum. Ovule (mag.).

Cestrum. Embryo (mag.).

Cestrum Parqui.

Cestrum. Flower cut vertically (mag.).

Cestrum. Seed (mag.).

Cestrum. Diagram.

Cestrum. Calyx and ovary (mag.).

Cestrum. Ovary cut vertically (mag.).

Cestrum. Ovary cut transversely (mag.).

Cestrum. Seed cut vertically (mag.).

COROLLA *monopetalous, hypogynous, isostemonous, æstivation induplicate.* STAMENS *inserted on the corolla.* OVARY *with* 2 *many-ovuled antero-posterior cells;* PLACENTÆ *on the septum.* OVULES *semi-anatropous.* FRUIT *a capsule or berry.* SEEDS *ovoid.* EMBRYO *straight, albuminous.*

PRINCIPAL GENERA.

Cestrum. * Habrothamnus. Vestia.

This little family, detached from *Solaneæ*, only differs in the straight embryo and foliaceous cotyledons. It is almost comprised in the tropical American genus *Cestrum*. Some are cultivated in Europe ; as *C. diurnum*, the flowers of which are odorous by day ; *C. vespertinum*, with a violet corolla, which exhales an odour of vanilla ; and *C. nocturnum*, the greenish flowers of which are odorous only at night. [Their properties are narcotic and diuretic.]

CLIX. *SCROPHULARINEÆ.*

(PEDICULARES ET SCROPHULARIÆ, *Jussieu.*—RHINANTHOIDEÆ ET PERSONATÆ, *Ventenat.*
—RHINANTHACEÆ ET PERSONATÆ, *Jussieu.*—SCROPHULARINEÆ, *Br.*—SCROPHULARIACEÆ, *Lindl.*)

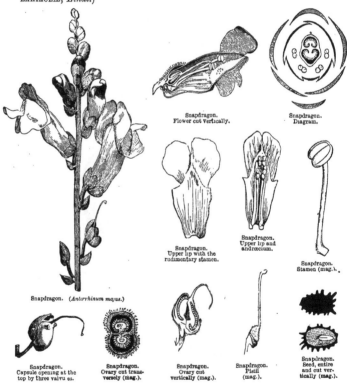

Snapdragon.
Flower cut vertically.

Snapdragon.
Diagram.

Snapdragon.
Upper lip with the
rudimentary stamen.

Snapdragon.
Upper lip and
androecium.

Snapdragon.
Stamen (mag.).

Snapdragon. (*Antirrhinum majus.*)

Snapdragon.
Capsule opening at the
top by three valvu es.

Snapdragon.
Ovary cut trans-
versely (mag.).

Snapdragon.
Ovary cut
vertically (mag.).

Snapdragon.
Pistil
(mag.).

Snapdragon.
Seed, entire
and cut ver-
tically (mag.).

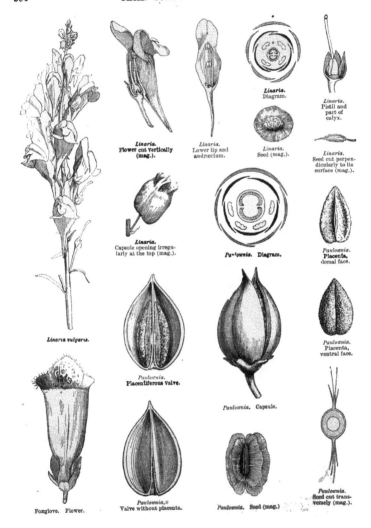

Linaria.
Diagram.

Linaria.
Pistil and
part of
calyx.

Linaria.
Flower cut vertically
(mag.).

Linaria.
Lower lip and
andrœcium.

Linaria.
Seed (mag.).

Linaria.
Seed cut perpen-
dicularly to its
surface (mag.).

Linaria.
Capsule opening irregu-
larly at the top (mag.).

Paulownia. Diagram.

Paulownia.
Placenta,
dorsal face.

Linaria vulgaris.

Paulownia.
Placentiferous valve.

Paulownia.
Placenta,
ventral face.

Paulownia. Capsule.

Foxglove. Flower.

Paulownia.
Valve without placenta.

Paulownia. Seed (mag.)

Paulownia.
Seed cut trans-
versely (mag.).

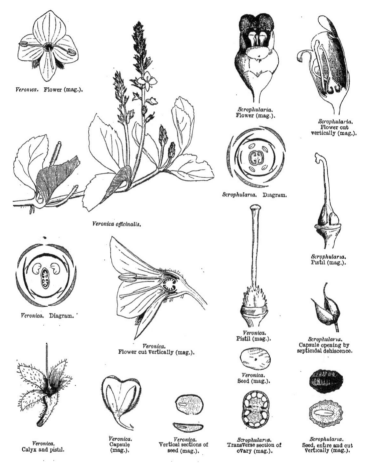

Veronica. Flower (mag.).

Scrophularia. Flower (mag.).

Scrophularia. Flower cut vertically (mag.).

Veronica officinalis.

Scrophularia. Diagram.

Scrophularia. Pistil (mag.).

Veronica. Diagram.

Veronica. Flower cut vertically (mag.).

Veronica. Pistil (mag.).

Scrophularia. Capsule opening by septicidal dehiscence.

Veronica. Seed (mag.).

Veronica. Calyx and pistil.

Veronica. Capsule (mag.).

Veronica. Vertical sections of seed (mag.).

Scrophularia. Transverse section of ovary (mag.).

Scrophularia. Seed, entire and cut vertically (mag.).

COROLLA *hypogynous, monopetalous, more or less irregular, anisostemonous, æstivation imbricate.* STAMENS 4, *didynamous, or 2 inserted on the corolla.* OVARY *with 2 antero-posterior çells, placenta on the septum;* OVULES *numerous and anatropous, or definite and semi-anatropous.* FRUIT *capsular, rarely fleshy.* EMBRYO *albuminous.*

HERBS or UNDERSHRUBS or SHRUBS. LEAVES alternate, opposite or whorled, simple, exstipulate. FLOWERS ☿, more or less irregular; inflorescence various. CALYX persistent, of 4-5 free or cohering sepals. COROLLA hypogynous, mono-petalous, tube sometimes gibbous or spurred at the base; *limb* irregular, rarely sub-regular, campanulate, rotate or bilabiate, upper lip bilobed, lower trilobed; æstiva-tion imbricate. STAMENS inserted on the corolla-tube, normally as many as and alternate with its lobes, but often fewer, the posterior being absent or rudi-mentary, the other 4 being didynamous; sometimes the 2 anterior are, like the posterior, sterile or wanting; *anthers* 2-celled, often 1-celled by the confluence of the sutures across the top of the connective. CARPELS 2, antero-posterior, cohering into a 2- (rarely 1-) celled ovary; *style* terminal, rarely bifid at the top; *stigma* often bilobed; *ovules* numerous, generally anatropous, rarely semi-anatropous. FRUIT generally a capsule, very rarely a berry. CAPSULE sometimes loculicidally, septici-dally or septifragally 2-valved; *valves* undivided, bifid or bipartite, sometimes opening at the top by 2–3 valvules, or operculate. SEEDS horizontal, ascending or pendu-lous; *hilum* basilar, rarely ventral. EMBRYO straight or a little curved, whitish or violet, in the axis of a fleshy or cartilaginous albumen.

SUB-ORDER I. SALPIGLOSSIDEÆ, *Bentham.*

Corolla folded or imbricate in æstivation, the two posterior lobes placed outside the others. Inflorescence definite from the first.

PRINCIPAL GENERA.

| *Anthocercis. | *Browallia. | Brunsfelsia. | *Salpiglossis. | *Schizanthus. |

SUB-ORDER II. ANTIRRHINIDEÆ, *Bentham.*

Corolla imbricate in æstivation, bilabiate, the posterior or upper lip placed out-side the lower. Inflorescence completely indefinite or mixed.

[The following tribes are established by Bentham :—

TRIBE I. CALCEOLARIEÆ.—Corolla 2-lobed; lobes entire, concave. Calyx valvate, 4-fid. Stamens declinate. Leaves opposite or whorled. Inflorescence composite. *Calceolaria.*

TRIBE II. VERBASCEÆ.—See order VERBASCEÆ, p. 588.

TRIBE III. HEMIMERIDEÆ. Corolla rotate, rarely tubular, 2-lipped, gibbous, saccate or spurred. Capsule 2-valved. Leaves (or the lower only) opposite. Inflorescence centripetal, uniform. *Alonsoa, Angelonia, Diascia, Hemimeris, *Nemesia, &c.

TRIBE IV. ANTIRRHINEÆ.—Corolla tubular, often saccate or spurred. Capsule dehiscing by pores. Leaves, lower or all opposite or whorled. Inflorescence centripetal, uniform. *Linaria, *Antirrhinum, *Maurandia, *Lophospermum, &c.

TRIBE V. CHELONEÆ.—Corolla tubular, not saccate or spurred. Capsule 2-4-valved, rarely an indehiscent berry. Inflorescence composite. Calyx imbricate. Phygelius, *Paulownia, *Scrophularia, *Chelone, *Pentstemon, Russelia, &c.

Tribe VI. Escobedieæ.—Corolla tubular, not saccate nor spurred. Capsule 2-valved. Leaves, all or the lower only opposite. Inflorescence centripetal, peduncles with opposite bracts. Calyx broad, lobes valvate. *Escobedia, Melasma, Alectra.*

Tribe VII. Gratioleæ.—Corolla tubular, not saccate nor spurred. Capsule 2-valved, rarely indehiscent. Inflorescence centripetal, uniform (composite in *Manulea*). Calyx-lobes imbricate. *Aptosimum, Nycterinia, *Polycarena, *Chænostoma, *Lyperia, *Manulea, *Mimulus, Mazus, Lindenbergia, Stemodia, Limnophila, Herpestis, Gratiola, Torenia, Vandellia,* &c.—Ed.]

Sub-order III. RHINANTHIDEÆ, *Benth.*

Corolla imbricate in æstivation, the two lateral lobes, or one of them, placed outside all the others, the posterior never. Inflorescence usually indefinite.

[Tribe I. Sibthorpieæ.—Leaves alternate or fascicled with the flowers at the nodes, rarely opposite, not connate; floral similar, or upper smaller. Flowers rarely cymose. *Limosella, Sibthorpia, Hemiphragma, Capraria, Scoparia,* &c.

Tribe II. Digitaleæ.—Leaves all alternate; lower crowded, petioled. Inflorescence centripetal, racemose. **Digitalis, Picrorhiza, *Wulfenia,* &c.

Tribe III. Veroniceæ.—Leaves, all or the lower only opposite. Inflorescence centripetal, racemose. Stamens distant. Anthers 2-celled or cells confluent. ** Veronica, Ourisia,* &c.

Tribe IV. Buchnereæ.—Leaves, all or the lower only opposite. Inflorescence centripetal, racemose. Stamens approximate in pairs; anthers dimidiate, 1-celled. *Buchnera, Striga, Cycnium, Hyobanche, Ramphicarpa.*

Tribe V. Gerardieæ.—Leaves, all or the lower only opposite. Inflorescence centripetal, racemose. Stamens approximate in pairs; anthers 2-celled, cells often spurred, equal or one empty. *Scymaria, Gerardia, Sopubia, Aulaya, Harveya, Centranthera,* &c.

Tribe VI. Euphrasieæ.—Inflorescence centripetal, racemose. Upper lip of corolla galeate or concave, erect. **Castilleja, Orthocarpus, Lamourouxia, Trixago, Bartsia, Odontites, Euphrasia, Rhinanthus, Pedicularis,* &c.—Ed.]

We have indicated the affinity between *Scrophularineæ* and *Solaneæ.* They have also many characters in common with other families having an irregular anisostemonous corolla, and especially with *Acanthaceæ* and *Bignoniaceæ. Acanthaceæ* differ in the æstivation of the corolla, the curved ovules, the processes of the placenta which support them, and the absence of albumen; *Bignoniaceæ* in the winged exalbuminous seeds, and the ovary girt with a fleshy ring. Many genera belonging to the sub-order *Rhinanthideæ* are root-parasites like *Orobancheæ* (*Rhinanthus, Melampyrum, Pedicularis, Odontites, Euphrasia, Bartsia, Castilleja*).

Scrophularineæ are found in all climates, but most abundantly in temperate regions, and very rarely towards the poles and equator. Their medicinal properties vary greatly, chemical analysis yielding bitter and acrid principles, combined with resinous and volatile substances. Some—as *Veronica officinalis,* called in France European Tea—are tonic, astringent and vulnerary; others are antiscorbutic (*V. Beccabunga*). *Scrophularia nodosa* and *aquatica,* fœtid and nauseous herbs, are resolvents and sudorifics.

The Large-flowered Snapdragon is acrid and bitter; it was formerly used as a diuretic. The common Toadflax is supposed to cure jaundice and skin diseases. Eyebright (*Euphrasia officinalis*) possesses a bitter principle; a water distilled from it is used in ophthalmia. The seeds of *Melampyrum pratense* are emollients if used externally, but mixed with wheat flour render the latter bitter and poisonous.

Gratiola officinalis (the *Gratia Dei,* or Poor Man's Herb) contains a resinous and acrid principle, and is hence a very energetic but sometimes dangerous purge. [It is the reputed basis of the *eau médicinale*, a famous gout medicine]. Of all the medicinal *Scrophularineæ* the most useful is the Foxglove; its very bitter and rather acrid leaves are poisonous in large quantities, but in small doses are diuretic, lower the pulse and subdue palpitations; its active principle, *digitaline*, is poisonous even in minute doses, so that many practitioners prefer administering the plant itself. [*Scoparia dulcis*, a common tropical weed, is a famed febrifuge in America. Some are intense bitters, as the Indian *Herpestes amara*, and the Himalayan Teeta (*Picrorhiza Teeta*), a renowned Indian ague medicine. *Brunsfelsia* (*Franciscea*) *uniflora* is the Mercurio Vegetal or Manaca of the Brazils, with a bitter nauseous bark that acts like mercury on the lymphatics, and is an active poison in overdoses.]

CLX. *VERBASCEÆ.*[1]

(SOLANEARUM *genus, Jussieu.*—SCROPHULARINEARUM *sectio, Endlicher.*—
VERBASCEÆ, *Bartling.*)

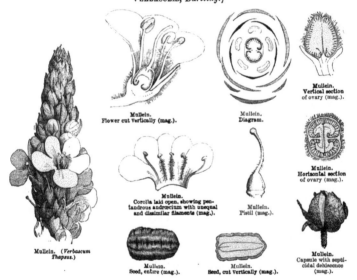

Mullein. (*Verbascum Thapsus.*)

Mullein.
Flower cut vertically (mag.).

Mullein.
Diagram.

Mullein.
Vertical section of ovary (mag.).

Mullein.
Corolla laid open, showing pentandrous androecium with unequal and dissimilar filaments (mag.).

Mullein.
Pistil (mag.).

Mullein.
Horizontal section of ovary (mag.).

Mullein.
Seed, entire (mag.).

Mullein.
Seed, cut vertically (mag.).

Mullein.
Capsule with septicidal dehiscence (mag.).

COROLLA *monopetalous, hypogynous, isostemonous, sub-irregular, æstivation imbricate.* STAMENS *inserted on the corolla-tube; filaments unequal;* ANTHERS *sub-1-celled.* OVARY *with 2 many-ovuled antero-posterior cells;* PLACENTÆ *on the septum.* OVULES

---

[1] See Tribe II. of *Scrophularineæ*, p. 586.—ED.

anatropous. CAPSULE 2-*valved, septicidal.* SEEDS *minute.* EMBRYO *straight, albuminous.*

Biennials, rarely perennials, generally cottony or woolly, with watery often mucilaginous juice. LEAVES alternate, often decurrent, exstipulate. FLOWERS ☿, a little irregular, fascicled, rarely solitary, in simple or branching spiciform racemes. CALYX monosepalous, 5-partite, persistent. COROLLA hypogynous, monopetalous, isostemonous, sub-rotate; *limb* 5-partite, caducous, æstivation imbricate. STAMENS 5, inserted on the corolla-tube, and alternate with its lobes; *filaments* unequal; *anthers* fixed by the middle, or throughout their length, of 2 confluent cells. OVARY with 2 antero-posterior cells, placentæ on the septum; *style* undivided, dilated at the top; *stigma* simple or bilobed; *ovules* numerous, anatropous. CAPSULE 2-celled, septicidally 2-valved, valves bifid at the top. SEEDS minute, rugose. EMBRYO straight, in the axis of a fleshy thick albumen.

PRINCIPAL GENERA.

Verbascum. Celsia, &c.

This little order, mainly composed of the genus *Verbascum* (Mullein), agrees with *Solaneæ* in its isostemonous corolla, and with *Scrophularineæ* in its straight embryo. The Mulleins inhabit the temperate regions of the Old World.

Some indigenous species (*V. Thapsus* and *phlomoides*) contain a bitter astringent principle in their leaves and a mucous principle in their flowers, combined with a little volatile oil of a sweet taste, whence their use as bechics and sedatives. [Their seeds are used to poison mice and to stupefy fish.]

CLXI. *UTRICULARIEÆ.*

(LENTIBULARIEÆ, *L.-C. Richard.*—UTRICULARINÆ, *Link.*—LENTIBULARIACEÆ, *Lindl.*)

COROLLA *monopetalous, hypogynous, irregular, anisostemonous.* STAMENS 2, *with* 1-*celled anthers, inserted on the corolla.* OVARY 1-*celled, placenta central, free;* OVULES *numerous, anatropous.* FRUIT *capsular.* SEEDS *minute.* EMBRYO *straight, exalbuminous.*

Aquatic or marsh HERBS. LEAVES all radical, sometimes rosulate, entire, somewhat fleshy; sometimes scattered or whorled, capillary and laden with vesicles, or peltate. SCAPES usually simple, naked or scaly, sometimes furnished with whorled vesicles, 1-flowered, or terminating in a spike or raceme. FLOWERS ☿, irregular, usually bracteate. CALYX persistent, of 2 or 5 sub-equal segments. COROLLA monopetalous, hypogynous, personate or 2-labiate; *tube* short, spurred at the base; *limb* with a 2-fid upper lip, the lower undivided or 3-fid; *palate* convex or depressed. STAMENS 2, inserted at the base of the corolla, under the upper lip, included; *filaments* cylindric, compressed, arched, convergent; *anthers* terminal, usually compressed in the middle, 1-celled, transversely 2-valved. OVARY free, of 2 carpels, 1-celled; *placenta* basilar, globose; *style* short, thick; *stigma* 2-lipped, upper lip shortest, or

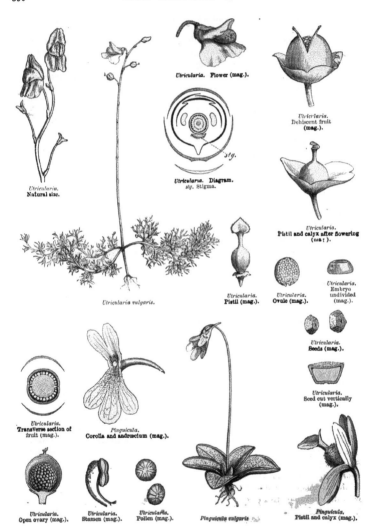

Utricularia. Flower (mag.).

Utricularia.
Natural size.

Utricularia. Diagram.
stg. Stigma.

Utricularia.
Dehiscent fruit
(mag.).

Utricularia.
Pistil and calyx after flowering
(mag.).

Utricularia vulgaris.

Utricularia.
Pistil (mag.).

Utricularia.
Ovule (mag.).

Utricularia.
Embryo
undivided
(mag.).

Utricularia.
Seeds (mag.).

Utricularia.
Seed cut vertically
(mag.).

Utricularia,
Transverse section of
fruit (mag.).

Pinguicula,
Corolla and androecium (mag.).

Utricularia.
Open ovary (mag.).

Utricularia.
Stamen (mag.).

Utricularia.
Pollen (mag.).

Pinguicula vulgaris

Pinguicula,
Pistil and calyx (mag.).

| *Pinguicula.*
Stamen (mag.). | *Pinguicula.*
Vertical section of
ovary (mag.). | *Pinguicula.*
Ovule (mag.). | *Pinguicula.*
Dehiscent fruit
(mag.). | *Pinguicula.*
Seed, entire and cut
longitudinally (mag.). |

0, lower dilated; *ovules* numerous, anatropous. CAPSULE bursting irregularly, or 2-valved. SEEDS numerous; *testa* rúgose; *hilum* basilar. EMBRYO exalbuminous, straight, undivided, or with very short cotyledons; *radicle* elongated, near the hilum.

PRINCIPAL GENERA.

<table>
<tr><td>Utricularia.</td><td>Pinguicula.</td></tr>
</table>

Utricularieæ derive their name from the bladders (*ascidia*) scattered over the submerged leaves of the principal genus. 'These bladders are rounded and furnished with a kind of moveable operculum; in the young plant they are filled with a mucus heavier than water, and the plant, submerged by this ballast, remains at the bottom. Towards the flowering season the leaves secrete a gas which enters the utricles, raises the operculum, and drives out the mucus, when the plant, now furnished with aerial bladders, rises slowly, and floats on the surface, and there flowers; this accomplished, the leaves again secrete mucus, which replaces the air in the utricles and the plant re-descends to the bottom, and ripens its seeds in the place where they are to be sown.'—De Candolle, *Vegetable Physiology.*

Utricularieæ approach *Scrophularineæ* in their corolla and andrœcium, and *Primulaceæ* in their central placentation and 1-celled ovary; they are distinguished by the exalbuminous embryo. They are cosmopolitan plants, but the greater number inhabit the tropical regions of both worlds and of Australia, where they vegetate in stagnant water, in swampy meadows, [on mossy tree trunks and rocks], and in places inundated during the rainy season.

The European Utricularias were formerly prescribed in cases of dysuria; they are now used as topics for wounds and burns. The leaves of *Pinguicula vulgaris* are reputed poisonous to sheep; in small quantities, when fresh, used by man, they purge gently, and are considered as a vulnerary. The Lapps use them to curdle reindeer's milk, and the Danish peasant girls employ their juice as a hair-pomade. [*Pinguicula* leaves, whether fresh or dry, are used by the Lapps to thicken fresh still warm milk, which neither curdles nor gives cream thereafter, but forms a delicious compact tenacious mass, a small portion of which will act similarly on another quantity of fresh milk.]

CLXII. *OROBANCHEÆ.*

(OROBANCHEÆ, *L.-C. Richard.*—OROBANCHACEÆ, *Lindl.*)

COROLLA *monopetalous, hypogynous, irregular, anisostemonous, persistent, æstivation imbricate.* STAMENS 4, *didynamous, inserted on the corolla.* OVARY *girt with a fleshy disk, 1-celled, placentation parietal;* OVULES *numerous, anatropous.* CAPSULE *with 2 semi-placentiferous valves.* SEEDS *minute;* ALBUMEN *copious.* EMBRYO *basilar.—Parasitic* HERBS, *leafless, stem scaly.*

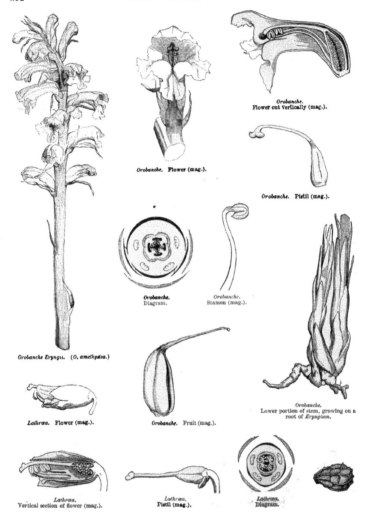

Orobanche.
Flower cut Vertically (mag.).

Orobanche. Flower (mag.).

Orobanche. Pistil (mag.).

Orobanche.
Diagram.

Orobanche.
Stamen (mag.).

Orobanche Eryngii. (O. amethystea.)

Orobanche.
Lower portion of stem, growing on a root of Eryngium.

Lathræa. Flower (mag.).

Orobanche. Fruit (mag.).

Lathræa.
Vertical section of flower (mag.).

Lathræa.
Pistil (mag.).

Lathræa.
Diagram.

HERBS, usually perennial, never green, parasitic on roots. STEM stout, fleshy. LEAVES replaced by coloured sessile scattered or imbricate scales. FLOWERS ☿, irregular, usually solitary in the axils of the upper scales, in a spike, rarely a raceme. CALYX persistent, tubular or campanulate, 4–5-fid, or of 4 sepals more or less completely united in lateral pairs. COROLLA monopetalous, hypogynous, tube circumsciss at the base, persistent and marcescent; *limb* 2-lipped, the upper usually hooded, entire or bifid, lower 3-fid or -toothed; *throat* usually with 2 gibbous oblique folds, æstivation imbricate. STAMENS inserted on the corolla-tube or throat, 4, didynamous; *filaments* with a dilated base; *anthers* 2- (very rarely 1-) celled, sometimes mucronate at the base, cells opening by a continuous or a basilar slit, connective sometimes spurred, curved at the top. OVARY superior, generally 1-celled; *carpels* 2, antero-posterior, coherent, usually with a basal unilateral fleshy disk; *placentas* parietal, 4, distinct or united in pairs; *style* terminal, simple, usually curved at the apex; *stigma* capitate, 2-lobed, undivided and sub-clavate; *ovules* usually numerous, anatropous. CAPSULE 1- (rarely 2-) celled, with 2 placentiferous valves separating at the top, or throughout their length, or more often in the middle only. SEEDS minute; *testa* thick, tubercled or punctulate. EMBRYO minute, sub-globose, at the base of a copious transparent albumen.

GENERA.

Orobanche.   Phelipæa.   Clandestina.   Lathræa.   Æginetia.   Hyobanche.

Orobancheæ approach *Scrophularineæ* in their regular corolla, didynamous stamens, capsular fruit and albuminous embryo; they differ in their leafless and scaly stem and parietal placentation. This placentation, their glandular disk, and the preceding characters, ally them to *Gesneraceæ*, from which they are separated by their scattered scales, parasitism, hypogynous corolla, and basilar embryo. We have noticed their connection with *Gentianeæ* (which see).

Orobancheæ mostly inhabit north temperate countries, and especially the Mediterranean region. Some species are pests of agriculture, from the damage they do to useful plants. The *Phelipæa ramosa* starves the Hemp, Maize and Tobacco; *Orobanche pruinosa*, the Beans; *O. cruenta*, the Sainfoin; *O. rubens*, the Lucern; *O. minor*, the Clover, &c. They are rare in tropical and South Africa, and appear to be absent from Australia and South America.

Orobancheæ are no longer used medicinally, though several species were formerly in much repute. They contain a bitter, acrid and astringent principle; some contain hydrocarbons, oils or resins [especially *Orobanche major*, which was formerly used as a detergent and astringent in diarrhœa]. The stock of the Thyme Orobanche was employed as a tonic, and its faintly scented flowers as an antispasmodic. *Lathræa* used to be given to epileptics. *Clandestina* was supposed by the ancients to confer fertility on women.

CLXIII. *COLUMELLIACEÆ.*

TREES or SHRUBS, evergreen, with compressed opposite branches. LEAVES opposite, exstipulate. FLOWERS terminal, yellow; *peduncles* short, 2-bracteolate. CALYX 5-partite. COROLLA monopetalous, epigynous, rotate, 5-fid, sub-irregular, æstivation imbricate. STAMENS 2, inserted on the corolla, between its posterior and lateral segments; *filaments* short, compressed, dilated into a 3-lobed connective; *anthers* with sinuous cells, confluent at the top. OVARY inferior, 2-celled; *placentas*

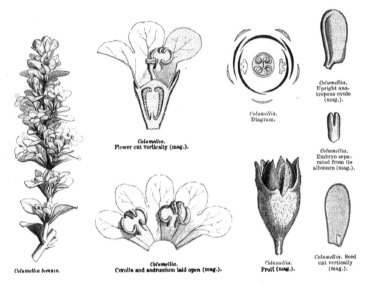

Columellia biennis.

Columellia.
Flower cut vertically (mag.).

Columellia.
Corolla and andrœcium laid open (mag.).

Columellia.
Diagram.

Columellia.
Upright ana-
tropous ovule
(mag.).

Columellia.
Embryo sepa-
rated from its
albumen (mag.).

Columellia.
Fruit (mag.).

Columellia. Seed
cut vertically
(mag.).

placed right and left of the floral axis; *style* short, thick, 2-sulcate; *stigma* 2-lobed; *ovules* numerous, ascending, anatropous. CAPSULE sub-woody, semi-superior by the growth of its top, septicidal; *valves* 2, bifid. SEEDS numerous, ascending, obovoid, compressed; *testa* coriaceous, soft; *hilum* basilar; *chalaza* apical; *raphe* almost wanting. EMBRYO straight; *albumen* fleshy; *cotyledons* ovoid, obtuse; *radicle* longer than the cotyledons, cylindric, inferior.

This little family, composed of the single genus *Columellia*, should normally be placed between *Rubiaceæ* and *Gesneraceæ*, as in the latter family the leaves are opposite, the corolla is monopetalous epigynous sub-irregular anisostemonous, the ovules are numerous and anatropous, the placentas are placed right and left of the floral axis, the embryo is straight and albuminous; but the sinuous anthers of *Columellia* and the septicidal dehiscence of the capsule render the diagnosis easy. They are also very near *Rubiaceæ* in the opposite leaves, epigynous corolla, septicidal capsule and albuminous embryo; they are principally separated from them by their æstivation and anisostemonous corolla [and exstipulate leaves].

Columelliæ are natives of Mexico and Peru. [Lindley describes the anthers as 6-celled, the cells arranged in three pairs on the 3-lobed fleshy connective. May not this genus be referable to *Logani-aceæ* ?—ED.]

CLXIV. *GESNERACEÆ.*

(GESNERIEÆ, *Richard.*—GESNEREÆ, *Martius.*—GESNERACEÆ, *Endl.*—CYRTANDRACEÆ, *Jack.*—DIDYMOCARPEÆ, *Don.*—GESNERACEÆ ET CYRTANDRACEÆ, *Lindley.*)

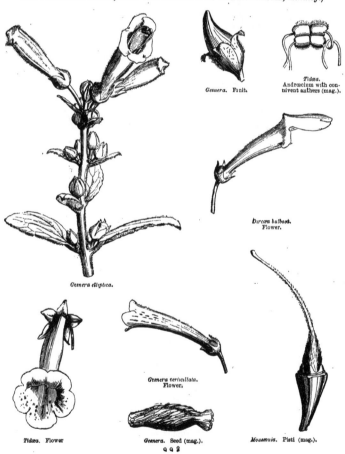

Tidæa.
Andrœcium with connivent anthers (mag.).

Gesnera. Fruit.

Dircæa bulbosa.
Flower.

Gesnera elliptica.

Gesnera verticillata.
Flower.

Tidæa. Flower

Gesnera. Seed (mag.).

Moussonia. Pisti (mag.).

Q Q 2

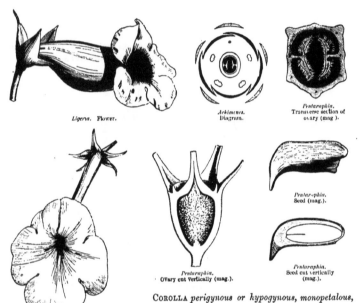

Ligeria. Flower.

Achimenes.
Diagram.

Pentaraphia.
Transverse section of
ovary (mag.).

Pentaraphia.
Seed (mag.).

Pentaraphia.
Ovary cut vertically (mag.).

Pentaraphia.
Seed cut vertically
(mag.).

Achimenes longiflora.
Flower.

COROLLA *perigynous or hypogynous, monopetalous, anisostemonous, irregular, 5-lobed, æstivation imbricate.* STAMENS *usually 4, didynamous, inserted on the corolla.* OVARY *inferior or semi-inferior or free, 1-celled, placentation parietal;* OVULES *anatropous.* EMBRYO *albuminous or not.*

Usually herbaceous, rarely sub-woody or woody. LEAVES usually opposite or whorled, simple, exstipulate. FLOWERS ☿, irregular, in a raceme or spike or cyme, sometimes fascicled; *peduncles* often 1- or 2-flowered. CALYX persistent, 5-partite, lobes unequal. COROLLA inserted on the receptacle, or on a fleshy disk between the ovary and the receptacular cup, monopetalous, tubular, infundibuliform or campanulate, more or less oblique, usually gibbous at the base; *limb* lengthened in front, 5-fid, 2-lipped, upper lip 2-lobed, lower 3-lobed, æstivation imbricate. STAMENS inserted on the corolla-tube, usually didynamous; the fifth (posterior) being arrested or absent, sometimes reduced to 2 by arrest of the 2 anterior or 2 lateral ones; *anthers* very often cohering, 2-celled (or 1-celled either by confluence of the cells or by the arrest of one), terminal, or lateral by the bifurcation of the connective, dehiscence longitudinal. OVARY 1-celled, free or semi-inferior, rarely inferior, girt or crowned with an annular or interrupted or unilateral disk; *placentas* parietal, 2, opposite, placed right and left of the floral axis; *style* filiform, simple; *stigma* capitate, concave or 2-lobed; *ovules* numerous, sessile or funicled, anatropous. FRUIT either fleshy with pulpy placentas, or capsular ovoid or subglobose, or siliquiform with 2 straight or twisted semi-placentiferous valves. SEEDS minute, oblong; *testa* loose or cellular. EMBRYO straight, exalbuminous, or in the axis of a fleshy albumen; *radicle* next the hilum.

TRIBE I. *GESNEREÆ.*

Seeds albuminous. Ovary semi-inferior or inferior. Fruit capsular.

PRINCIPAL GENERA.

| · Rytidophyllum. | Conradia. | * Gesnera. | Gloxinia. | * Achimenes. |
| Mitraria. | * Ligeria. | * Mandirola. | * Tidæa. | * Pentaraphia. |

TRIBE II. *BESLERIEÆ.*

Seeds albuminous. Fruit a berry or capsule. Ovary free.

PRINCIPAL GENERA.

| Besleria. | Hypocyrta. | * Columnea. | Nematanthus. |
| * Alloplectus. | Episcia. | * Mitraria. | Tapeinotes. |

TRIBE III. *CYRTANDREÆ.*

Seeds exalbuminous. Fruit a contorted capsule or berry.

PRINCIPAL GENERA.

[§ 1. *Fruit capsular.*]

| * Æschynanthus. | Lysionotus. | Didymocarpus. | * Chirita. |
| Streptocarpus. | Bœa. | Ramondia.[1] | Klugia. |

[§ 2. *Fruit fleshy.*]

| Cyrtandra. | Rhynchotechum. | Fieldia. |

Gesneraceæ are allied to *Rubiaceæ, Gentianeæ, Sesameæ, Orobancheæ* (see these families). They approach *Bignoniaceæ* in the generally opposite leaves, the imbricate æstivation (and *Beslerieæ* in the hypogynous corolla), the disk girding the base or middle of the ovary, and the anatropous ovules. They

Tapeinotes. Flower.    *Hypocyrta.* Flower.    *Alloplectus.* Diagram.

¹ See Order *Ramondieæ*, p. 599.

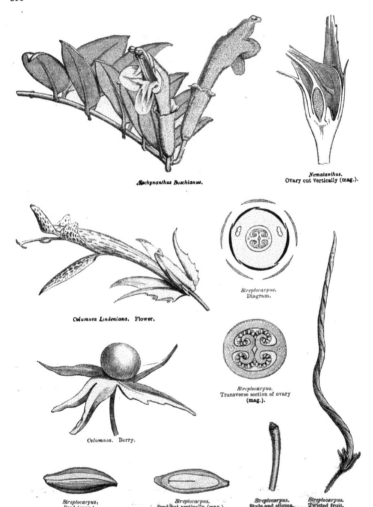

Æschynanthus Boschianus.

Nematanthus,
Ovary cut vertically (mag.).

Columnea Lindeniana. Flower.

Streptocarpus,
Diagram.

Streptocarpus,
Transverse section of ovary
(mag.).

Columnea. Berry.

Streptocarpus,
Seed (mag.).

Streptocarpus,
Seed cut vertically (mag.).

Streptocarpus,
Style and stigma.

Streptocarpus,
Twisted fruit.

are separated by their generally anisostemonous and perigynous corolla, 1-celled ovary, and wingless albuminous seeds.

The tribes *Beslerieæ* and *Gesnereæ* are abundant in the tropical regions of the New World; they are very rare beyond the tropics. *Cyrtandreæ* inhabit tropical Asia, especially the islands of the Pacific and the southern slopes of the [Eastern] Himalayas; they are rare in South Africa and [very rare in] Australia.

A family of little importance as regards properties; but largely cultivated as hothouse ornaments (*Ligeria, Achimenes, Gesnera*). The flowers of the climbing *Columnea* yield an abundant nectar from the glandular disk, whence its American name of 'Liane à Sirop.'

CLXV. *RAMONDIEÆ.*

(RAMONDIACEÆ, *Godron and Grenier.*)

STEM herbaceous. LEAVES collected at the base of the naked scape. FLOWERS 2–4, in a terminal corymb. CALYX 5-fid. COROLLA monopetalous, hypogynous, isostemonous, rotate, 5-partite; *lobes* obtuse, sub-equal, with a papillose gland on each side of the base, æstivation imbricate. STAMENS 5, inserted on the corolla-tube; *filaments* very short; *anthers* cordate, lobes parallel, introrse, opening by two

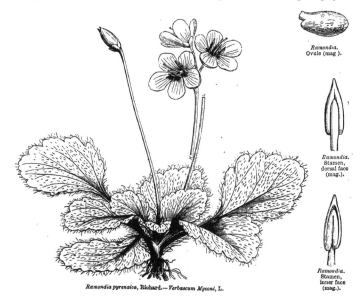

Ramondia.
Ovule (mag.).

Ramondia.
Stamen,
dorsal face
(mag.).

Ramondia.
Stamen,
inner face
(mag.).

Ramondia pyrenaica, Richard.— *Verbascum Myconi*, L.

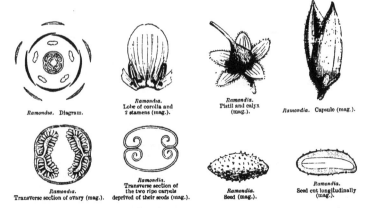

Ramondia. Diagram.

Ramondia.
Lobe of corolla and
2 stamens (mag.).

Ramondia.
Pistil and calyx
(mag.).

Ramondia. Capsule (mag.).

Ramondia.
Transverse section of ovary (mag.).

Ramondia.
Transverse section of
the two ripe carpels
deprived of their seeds (mag.).

Ramondia.
Seed (mag.).

Ramondia.
Seed cut longitudinally
(mag.).

longitudinal slits confluent at the top, so as to represent a pore. OVARY 1-celled, of
2 connate carpels, placentation parietal; *style* simple ; *stigma* obtuse ; *ovules* numerous,
anatropous, horizontal. CAPSULE with 2 valves placentiferous on their edges. SEEDS
hispidulous. EMBRYO straight ; *cotyledons* plano-convex.

The genus *Ramondia*, which constitutes this little family, is represented by a Pyrenean plant
formerly referred to *Verbascum* (*V. Myconi*). It is distinguished from *Solaneæ* by its one-celled fruit,
its anatropous ovules and straight embryo ; from *Scrophularineæ* by its regular pentandrous flower and its
1-celled ovary ; and from *Gesneraceæ* by its regular isostemonous corolla and the placentiferous dehiscence of the capsule.

CLXVI. *BIGNONIACEÆ*.

(BIGNONIEÆ, *A.-L. de Jussieu.*—BIGNONIACEÆ, *Br.*)

COROLLA *hypogynous, monopetalous, anisostemonous, more or less irregular, æstivation imbricate.* STAMENS *generally 4, didynamous, inserted on the corolla.* OVARY
1-2-celled, base girt by a glandular disk ; OVULES *usually horizontal, anatropous, inserted near the edge of the septum, or parietal.* CAPSULE *generally 2-valved.* SEEDS
transverse, compressed, winged. EMBRYO *straight, exalbuminous ;* RADICLE *usually
centrifugal.*

Woody PLANTS, often climbing or twining, very rarely herbaceous (*Incarvillea,
Tourretia*). LEAVES generally opposite, very often compound, sometimes terminating
in a tendril, exstipulate. FLOWERS ☿, in racemed or spiked cymes. CALYX monosepalous, 5-fid or -toothed, or 2-partite or -labiate, or with the limb nearly entire,
sometimes spathaceous (*Spathodea*), sometimes furnished externally with 5 teeth

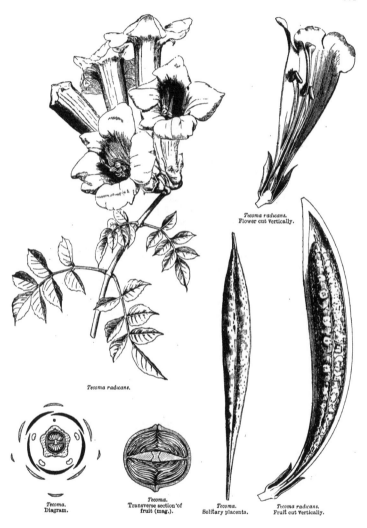

Tecoma radicans.
Flower cut Vertically.

Tecoma radicans.

Tecoma.
Diagram.

Tecoma.
Transverse section of
fruit (mag.).

Tecoma.
Solitary placenta.

Tecoma radicans.
Fruit cut Vertically.

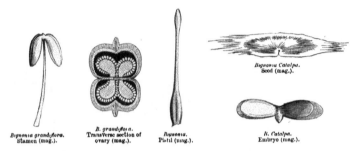

Bignonia grandiflora.   B. grandiflora.   Bignonia.   B. Catalpa.
Stamen (mag.).   Transverse section of   Pistil (mag.).   Embryo (mag.).
             ovary (mag.).

equalling its segments (*Incarvillea*). COROLLA monopetalous, hypogynous, deciduous; *tube* short; *throat* dilated; *limb* usually of 5 segments, bilabiate, rarely sub-regular (*Zeyhera*), æstivation usually imbricate. STAMENS 5, inserted on the corolla-tube, very rarely all fertile (*Calosanthes*) and sub-equal, generally 4, didynamous, the fifth with no anther or absent, sometimes the 2 posterior only fertile (*Catalpa*); *anthers* introrse, 2-celled, *cells* parallel and contiguous, sometimes united at the top only, and diverging, dehiscence longitudinal. DISK glandular, girding the base of the ovary. OVARY superior; *cells* 2, antero-posterior, sometimes 1-celled (*Eccremocarpus*); *style* simple, filiform; *stigma* 2-lamellate or -fid; *ovules* more or less numerous, usually horizontal, anatropous, inserted in vertical lines along the edges of the septum in the 2-celled ovaries, or parietal in the 1-celled. CAPSULE ovoid or siliquiform or compressed, 1–2-celled, sometimes pseudo-4-celled (*Tourretia*), generally 2-valved, rarely opening by a posterior longitudinal slit (*Incarvillea, Amphicome*); *valves* parallel or perpendicular to the septum, sometimes leaving the seminiferous septum free, sometimes septiferous or placentiferous. SEEDS generally transverse, imbricate, 1- or many-seriate, margin winged, wing sometimes fimbriate (*Amphicome, Catalpa*), very rarely apterous (*Argylea*). EMBRYO straight, exalbuminous; *radicle* near the hilum, usually centrifugal, with the raphe next the septum; sometimes centripetal, the seed being perpendicular to the septum (*Jacaranda*), sometimes superior from the seed becoming pendulous during development (*Incarvillea, Amphicome*); *cotyledons* plane, foliaceous, reniform or bilobed.

TRIBE I. *EUBIGNONIEÆ.*

Septum parallel to the valves; dehiscence marginal, that is, taking place along the edges of the septum.

PRINCIPAL GENERA.

* Bignonia.	Pachyptera.	Fredericia.	Calosanthes.
Arrabidœa.	* Lundia.	Cuspidaria.	Anemopaegma.
Pithecoctenium.	Adenocalymna.		

Tribe II. *TECOMEÆ.*

Septum perpendicular to the valves. Dehiscence loculicidal.

PRINCIPAL GENERA.

* Spathodea.	* Pandorea.	Argylea.	Tapebuja.
Zeyhera.	* Catalpa.	* Tourretia.	* Tecoma.
* Jacaranda.	Stereospermum.		

Tribe III. *INCARVILLEÆ.*

Capsule with 2 cells, the posterior only opening along its median line.

GENERA.

* Incarvillea.   * Amphicome.

Tribe IV. *ECCREMOCARPEÆ.*

Capsule 1-celled ; valves 2, placentiferous in the middle.

GENUS.

* Eccrémocarpus.

Bignoniaceæ are more or less nearly related to most monopetalous hypogynous anisostemonous regular-flowered families. We have indicated their affinities with *Scrophularineæ.* They similarly approach *Acanthaceæ,* as also in the 2-celled ovary, 2-valved capsule and exalbuminous embryo ; but *Acanthaceæ* differ in the contorted æstivation of the corolla [in some only], the campylotropous ovules, and the retinacula of the seed. Their winged seeds alone separate them from *Sesameæ,* and they further differ from *Pedalineæ* in being capsular. They are also allied to tribe *Cyrtandreæ* of *Gesneraceæ* (which see). Finally, they offer an evident resemblance to the genus *Cobæa,* in *Polemoniaceæ,* in the hypogynous corolla, the disk, many-celled ovary, the free seminiferous septum of the capsule, winged seeds, and compound leaves ending in a tendril ; but in *Cobæa* the leaves are alternate, the corolla is regular, isostemonous, and contorted in æstivation, and the embryo is albuminous.

The wood of some climbing *Bignoniaceæ* represents in section a sort of Maltese cross, resulting from the unequal development of the layers of liber (*Bignonia crucis* and *B. capreolata*). This family, remarkable for the beauty of its flowers, principally inhabits the tropics, especially of America. Some species are used in native practice, as Uruparaiba (*Bignonia Leucoxylon*), of tropical America, the bark of which is supposed to be an antidote to the Manchineel. The leaves of many species of *Jacaranda* contain an acrid and astringent principle, whence they are employed in Brazil as a prophylactic against glandular diseases.

[CLXVII. *CRESCENTIEÆ.*[1]

(Crescentineæ, *D.C.*—Crescentiaceæ, *Lindl.*)

Small TREES. LEAVES alternate or fascicled, simple, exstipulate. FLOWERS from the old wood. CALYX inferior, undivided, rupturing irregularly. COROLLA hypogynous, monopetalous, inferior, sub-2-lipped, lobes imbricate in æstivation. STAMENS 4, inserted on the corolla, didýnamous, fifth between the posterior pair and

[1] This Order is omitted in the original work.

rudimentary; *anthers* 2-celled, dehiscence longitudinal. OVARY seated on an annular disk, free, of 2 connate antero-posterior carpels, 1-celled; *placentas* 2 or 4, parietal, sometimes produced as false septa; *style* simple; *stigma* 2-lamellar; *ovules* very numerous, horizontal. FRUIT woody, indehiscent, many-seeded, pulpy within. SEEDS large, amygdaloid; *testa* loose, leathery. EMBRYO straight, exalbuminous; *cotyledons* plano-convex, fleshy; *radicle* short, thick, next the hilum.

TRIBE I. TANÆCIEÆ.—Fruit fleshy, elongate, 2- or more-celled. Leaves opposite, rarely whorled. *Colea, Phyllarthron, Parmentiera, Tanœcium,* &c.

TRIBE II. CRESCENTIEÆ.—Fruit corticate, ovoid or globose. Leaves alternate. *Crescentia, Kigelia,* &c.

Crescentieæ are clearly related to *Bignoniaceæ*, of which De Candolle, and subsequently Boreau, have regarded them as a sub-order, distinguished by their indehiscent fruit, pulpy within, and wingless seeds. They are all tropical and widely dispersed, abounding in Madagascar and the Mauritius.

The Calabash-tree (*Crescentia Cujete*) of America is the most important to man of all *Crescentieæ*; its sub-acid pulp is edible, and its dried pericarp, which is used as a substitute for bottles, &c., is so hard as to admit of water being repeatedly boiled in it. *Parmentiera cerifera*, the Candle-tree or Palo de Velas of Panama, bears a long candle-like fruit, greedily eaten by cattle, to the flesh of which it communicates its apple-like flavour. *P. edulis* is eaten by the Mexicans. Various species yield timber.—ED.]

CLXVIII. *ACANTHACEÆ*.

(ACANTHI, *A.-L. de Jussieu.*—ACANTHACEÆ, R. Brown.)

COROLLA *hypogynous, monopetalous, 5-cleft, usually irregular, anisostemonous, æstivation imbricate.* STAMENS *inserted on the corolla,* 4 *didynamous, or* 2. OVARY 2-*celled;* OVULES *campylotropous, seated on a prolongation of the placenta.* FRUIT *capsular.* EMBRYO *usually curved, exalbuminous.*—LEAVES *opposite or whorled.*

HERBS sub-woody at the base, or woody, stem and branches jointed, nodes tumid. LEAVES opposite, or in whorls of 3 or 4, exstipulate. FLOWERS ☿, irregular, axillary or terminal, spiked racemed or fascicled, rarely solitary, bracteate and 2-bracteolate; *bracteoles* minute, or very large when the calyx is small or obsolete. CALYX of 5 segments, equal or unequal, distinct or variously coherent or 4-fid or -partite, sometimes obsolete or reduced to a truncate entire or toothed ring. COROLLA monopetalous, tubular, hypogynous; *limb* usually bilabiate, upper lip bifid, sometimes obsolete, lower 3-lobed, æstivation imbricate. STAMENS inserted on the corolla-tube, usually 4 didynamous, the fifth or posterior rudimentary or obsolete, sometimes 2 by arrest of the 2 anterior; *filaments* filiform or subulate; *anthers* sometimes 2-celled with opposite parallel cells, often appearing 1-celled from the contiguity of the cells; sometimes 1-celled from the unequal insertion or obliquity or super-position or divergence of the cells, of which one is rudimentary or obsolete. OVARY superior, cells 2, antero-posterior, septum double, 2-3-4-many-ovuled; *style* terminal, simple, filiform; *stigma* usually 2-fid, rarely undivided; *ovules* campylotropous or semi-anatropous, 2-seriate along the middle of the septum, usually seated on a process of the placenta. CAPSULE membranous coriaceous or cartilaginous, sessile, or contracted into a pedicel, obtuse or acute, 2-celled, sometimes opening elastically into 2 boat-shaped semi - septiferous entire or 2-partite valves, sometimes indehiscent by arrest of one of the cells. SEEDS rounded or compressed, generally supported by subulate or hooked processes (retinacula) arising from the septum, which is sometimes reduced to a mere cupule; *testa* soft, or covered with mucilaginous hairs. EMBRYO exalbuminous, usually curved; *cotyledons* large, orbicular, plano-convex, sometimes crumpled; *radicle* cylindric, descending and centripetal.

[The Asiatic and African genera of *Acanthaceæ* have been grouped as follows by Dr. Thomas Anderson, who considers that the American and other genera will fall under the same tribes :—

SUB-ORDER I. THUNBERGIDEÆ.

Calyx reduced to a ring. Corolla-lobes contorted. Seeds with a cupuliform funicle. Stem usually twining. *Thunbergia* (*Hexacentris*).

SUB-ORDER II. RUELLIDÆ.

Calyx herbaceous, 5- (rarely 4-) partite. Corolla-lobes contorted. Seeds inserted on a hooked reti. naculum or on a papilla. Stem not twining.

TRIBE I. NELSONIEÆ. — Calyx small, herbaceous. Seeds minute, globose, inserted on a small papilla. *Elytraria, Ebermayera, Nelsonia, Adenosma.*

Acanthus u ollis.

TRIBE II. RUELLIEÆ.—Calyx small, herbaceous. Seeds large, compressed, retinaculum hooked. *Nomaphila, Hygrophila, Calophanes, Ruellia, Stenosiphonium, Strobilanthes, Goldfussia, Æchmanthera, Brillantaisia, Whitfieldia,* &c.

SUB-ORDER III. ACANTHIDEÆ.

Calyx herbaceous, 5- (rarely 4-) partite. Corolla-lobes imbricate. Seeds with a hooked retinaculum.

TRIBE III. BARLERIEÆ.—Corolla hypocrateriform or funnel-shaped (2-labiate in *Lepidagathis*). *Barleria, Neuracanthus, Crossandra, Lepidagathis, Blepharis, Acanthus, Geissomeria, Aphelandra, Andrographis, Gymnostachyum, Lankesteria,* &c.

TRIBE IV. PHLOGACANTHEÆ.—Corolla tubular; limb 2-labiate. Stamens 2; anthers 2-celled, cells parallel, not spurred. Capsule sub-terete, many-seeded. Inflorescence terminal, spicate. *Phlogacanthus.*

TRIBE V. JUSTICIEÆ.—Corolla 2-labiate; lower lip 3-fid; mid-lobe the largest; upper 2-toothed. Stamens 2. *Justicia, Adhatoda, Gendarussa, Beloperone, Anisacanthus, Rungia, Dicliptera, Peristrophe, Hypoestes, Rhinacanthus, Graptophyllum, Duvernoia,* &c.

TRIBE VI. ASYSTASIEÆ. — Corolla funnel-shaped or campanulate, rarely hypocrateriform, 2-lipped in æstivation. Stamens 4, 2 usually imperfect. *Eranthemum, Asystasia, Mackaya,* &c.—ED.]

Adhatoda.
Diagram.

Adhatoda. Anther with unequal cells (mag.).

Adhatoda betonica.

Adhatoda cupreata.
Corolla (mag.).

Adhatoda.
Style and stigma (mag.).

Ruellia patula.
Seed (mag.).

Ruellia,
Seed cut vertically (mag.).

Adhatoda.
Pistil and disk
(mag.).

Ruellia.
Capsular fruit with
2 cells, dehiscing
loculicidally
(mag).

Ruellia.
Valve of fruit, semi-
septiferous, showing
the process supporting
the seeds (mag.).

Ruellia.
Transverse section of
fruit (mag.).

Ruellia.
Transverse section of
seed (mag.).

Acanthaceæ approach *Labiatæ* and *Verbenaceæ* in the irregular anisostemonous corolla and its æstivation, and exalbuminous embryo with descending radicle and opposite leaves; they are separated by the curved ovules, capsular fruit with compressed valves, and retinacula. We have indicated their affinity with *Scrophularineæ* and *Bignoniaceæ* (which see).

This family almost exclusively inhabits the tropics.

Acanthaceæ furnish no species to European medicine. They contain, however, an abundant mucilage, sometimes combined with a bitter principle; others are somewhat acrid, others contain a stimulating volatile oil. The mucilaginous *Acanthaceæ* are employed in India as emollients and bechics; the bitter species are reported tonics and febrifuges; the acrid are considered to excite the functions of the skin and of the mucous membrane. Some are dyes [especially a Bengal *Ruellia,* which produces the blue Room dye of India. *Gendarussa vulgaris* is in India a famed cure for rheumatism, it is also a febrifuge, and its dried leaves preserve clothes from insects. The popular French tonic, *drogue amère,* is the tincture of *Justicia paniculata.*—ED.].

CLXIX. *SESAMEÆ.*

(PEDALINEÆ, *R. Br.*—SESAMEÆ, *Kunth.*—MARTYNIACEÆ, *Link.*)

COROLLA *hypogynous, monopetalous, irregular, usually anisostemonous, æstivation imbricate.* STAMENS *generally 4, didynamous, inserted on the corolla; anther-cells 2, shorter than the connective, glandular at the top.* OVARY *2- 4- or 1-celled, girt at the base by a glandular disk;* OVULES *anatropous.* FRUIT *a capsule, drupe or nucule.* EMBRYO *straight, exalbuminous or sub-exalbuminous.*

HERBS with vesicular glands. LEAVES opposite or alternate, simple, exstipulate. FLOWERS ☿, irregular, axillary, solitary, or racemed or spiked, usually 2-bracteolate. CALYX 5-partite or -fid, almost equal, sometimes split on one side and spathaceous (*Craniolaria*). COROLLA monopetalous, hypogynous; *tube* cylindric or gibbous; *throat* swollen; *limb* usually bilabiate, 5-lobed, æstivation imbricate or sub-valvate. STAMENS 5, inserted on the corolla-tube, the upper sterile, the other 4 fertile, didynamous, sometimes the 2 shortest sterile, and the fifth rudimentary (*Martynia*); *anther-cells* 2, equal, parallel or divergent, connective jointed on to the filament and prolonged upwards as a glandular appendage. OVARY superior, base girt with a glandular disk, 2- 4- or 1-celled by arrest of the septa; *style* terminal, simple; *stigma* bilamellate; *ovules* anatropous. CAPSULE or DRUPE often with an angular and coriaceous epicarp. SEEDS generally pendulous; *albumen* absent, or nearly so. EMBRYO straight; *cotyledons* plane or plano-convex; *radicle* superior, inferior, or centripetal.

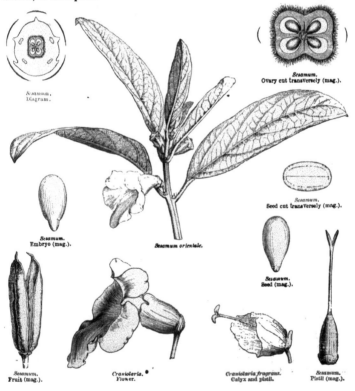

Sesamum.
Diagram.

Sesamum.
Ovary cut transversely (mag.).

Sesamum.
Seed cut transversely (mag.).

Sesamum.
Embryo (mag.).

Sesamum orientale.

Sesamum.
Seed (mag.).

Sesamum.
Fruit (mag.).

Craniolaria.
Flower.

Craniolaria fragrans.
Calyx and pistil.

Sesamum.
Pistil (mag.).

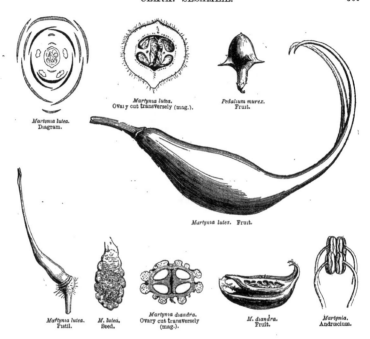

Martynia lutea.
Diagram.

Martynia lutea.
Ovary cut transversely (mag.).

Pedalium murex.
Fruit.

Martynia lutea. Fruit.

Martynia lutea.
Pistil.

M. lutea.
Seed.

Martynia diandra.
Ovary cut transversely
(mag.).

M. diandra.
Fruit.

Martynia.
Andrœcium.

TRIBE I. *EUSESAMEÆ.*

Capsule 4-celled, 2-valved, seminiferous septum free; seeds numerous, 1-seriate, fixed at the central angle of the cells, ascending or horizontal; albumen very scanty. —Stigma irritable.

PRINCIPAL GENUS.

* Sesamum.

TRIBE II. *PEDALINEÆ.*

Fruit 4- or pseudo-4-celled, sub-capsular or drupaceous, indehiscent or obscurely dehiscent at the top. Seeds usually few, pendulous or horizontal, rarely erect and solitary (*Josephinia*), completely exalbuminous.

PRINCIPAL GENERA.

* Craniolaria.	Pedalium.	Josephinia.	* Harpagophytum.
* Martynia.	Uncaria.	Pterodiscus.	Pretrea.

R R

Sesameæ are very near *Bignoniaceæ* (which see); they are connected with *Gesneraceæ* through *Craniolaria* and *Martynia*, and are further connected with *Verbenaceæ* and *Myoporineæ* (which see).

This family inhabits the tropics of both worlds and South Africa. Few of the species are useful. The seeds of *Sesamum orientale* and *S. indicum* yield a bland oil, used by Orientals as food, medicine, and as a cosmetic, called Sesamum or Gingilie oil. The cultivation of these plants, which was spread for ages over Asia and Africa, now extends to the New World. The importation of Sesamum seeds into France amounted in 1855 to sixty millions of kilogrammes (58,040 tons); the oil extracted from them is principally used in the manufacture of soap. *Pedalium Murex* exhales a strong musky odour, and the thick juice contained in its vesicular glands is employed in India to give a mucilaginous consistency to water, and thus render it emollient. The Creoles of America eat the raw root of *Craniolaria annua* with sugar; it is fleshy and mild-tasted, and when dried is employed in preparing a bitter and cooling drink. [The curious 2-horned fruit of *Martynia proboscidea* is the Testa di Quaglia of the Italians, notorious for its cleaving to clothes, &c.; *Uncaria procumbens* is the famous Grapple-plant of South Africa, the fruit of which is dispersed by animals to whose fur its hooked horns enable it to cling.]

CLXX. *MYOPORINEÆ.*

(MYOPORINÆ, *Br.*—MYOPORINEÆ, *Jussieu.*—MYOPORACEÆ, *Lindl.*)

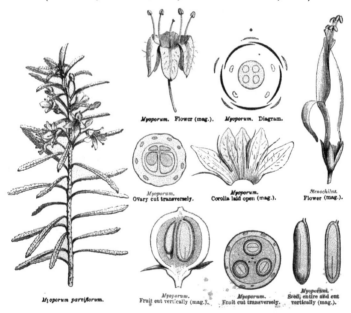

Myoporum. Flower (mag.).   *Myoporum.* Diagram.

Myoporum. Ovary cut transversely.   *Myoporum.* Corolla laid open (mag.).   *Stenochilus.* Flower (mag.).

Myoporum parviflorum.   *Myoporum.* Fruit cut vertically (mag.).   *Myoporum.* Fruit cut transversely.   *Myoporum.* Seed, entire and cut vertically (mag.).

COROLLA *monopetalous, hypogynous, irregular or sub-regular, anisostemonous.* STAMENS 4, *didynamous, inserted on the corolla.* OVARY 2–4-*celled;* OVULES *pendulous, anatropous.* FRUIT *a drupe, cells* 1–4-*seeded.* EMBRYO *straight, albuminous;* RADICLE *superior.*

SHRUBS or UNDERSHRUBS. LEAVES alternate, rarely opposite, simple, entire or toothed, usually studded with resinous glands, exstipulate. FLOWERS ☿, axillary; *pedicels* .1-flowered, rarely branched into a cyme, ebracteate. CALYX 5-partite or -fid, persistent, scarious. COROLLA monopetalous, 5-lobed, sub-regular or ringent, æstivation imbricate. STAMENS 4, inserted at the base of the corolla, alternate with its lobes; *filaments* filiform; *anthers* introrse, versatile, cells confluent. OVARY 2-celled, cells antero-posterior, sometimes more or less perfectly subdivided by a secondary septum from the axis; *style* terminal, simple; *stigma* emarginate, rarely 2-fid; *ovules* 2 collateral in each carpel, rarely 4 imbricate in pairs, pendulous, anatropous. DRUPE succulent or nearly dry, 2-celled or more or less completely 4-celled. SEEDS inverted. EMBRYO cylindric, in the axis of a scanty fleshy albumen; *cotyledons* semi-cylindric; *radicle* near the hilum, superior.

GENERA.

* Myoporum (Stenochilus).   Pholidia.   . Eremophila.   Bontia.

Myoporineæ are connected with *Verbenaceæ*, as we have already shown. They approach *Selagineæ* in hypogynism, the anisostemonous imbricate corolla, didynamous stamens, 1-celled anthers, 2 carpels, pendulous and anatropous ovules, albuminous embryo and alternate leaves; but *Selagineæ* differ in their terminal spiked inflorescence, 1-ovuled cells, and fruit composed of 2 achenes. The affinities are the same between *Myoporineæ* and *Globularieæ*, but these differ in their terminal capitulum, their 1-celled and -ovuled ovary, and their fruit being a caryopsis.

Myoporineæ are mostly natives of Australia and some of the Pacific Islands. One genus (*Bontia*) is found in the Antilles. They are generally studded with resinous glands, and in some species the resin exudes in transparent drops. They are of no use to man; some (*Myoporum parviflorum*, &c.) are cultivated in Europe as ornamental plants.

CLXXI. *SELAGINEÆ.*[1]

(SELAGINEÆ, *Jussieu.*—SELAGINACEÆ, *Lindl.*)

COROLLA *monopetalous, hypogynous, sub-regular, anisostemonous, æstivation imbricate.* STAMENS 4 *nearly equal, or* 2 *inserted on the corolla.* OVARY *of* 2 1-*ovuled cells;* OVULES *pendulous, anatropous.* FRUIT *of* 2 *achenes.* EMBRYO *albuminous;* RADICLE *superior.*

HERBS or branching UNDERSHRUBS. LEAVES alternate or fascicled, sometimes sub-opposite, simple, usually linear, exstipulate. FLOWERS ☿, generally irregular, bracteate, spiked, solitary or panicled or corymbose. CALYX persistent, mono-

---

[1] Reduced by Harvey to a sub-order of *Verbenaceæ.*—ED.

CLXXL. SELAGINEÆ.

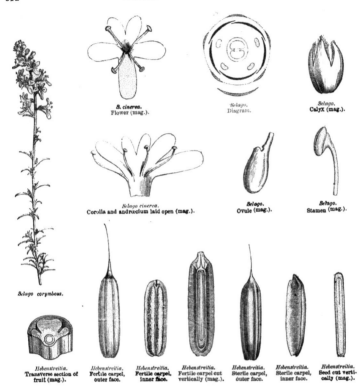

S. cinerea.
Flower (mag.).

Selago.
Diagram.

Selago.
Calyx (mag.).

Selago cinerea.
Corolla and andrœcium laid open (mag.).

Selago.
Ovule (mag.).

Selago.
Stamen (mag.).

Selago corymbosa.

Hebenstreitia.
Transverse section of
fruit (mag.).

Hebenstreitia.
Fertile carpel,
outer face.

Hebenstreitia.
Fertile carpel,
inner face.

Hebenstreitia.
Fertile carpel cut
vertically (mag.).

Hebenstreitia.
Sterile carpel,
outer face.

Hebenstreitia.
Sterile carpel,
inner face.

Hebenstreitia.
Seed cut verti-
cally (mag.).

sepalous, spathiform or tubular, 5–3-toothed or -partite, rarely of 2 free sepals.
COROLLA hypogynous, monopetalous, deciduous; *tube* entire or split lengthwise; *limb*
4–5-lobed, 1–2-labiate or sub-regular, spreading, œstivation imbricate. STAMENS
inserted on the corolla-tube, alternate with its segments, sometimes 4 sub-didyna-
mous, or equal with a rudimentary fifth, sometimes 2 only; *filaments* filiform; *anthers*
1-celled, dehiscence longitudinal. OVARY free, of 2 antero-posterior cells; *style*
terminal, simple; *stigma* undivided, sub-capitate; *ovules* 1 in each cell, pendulous
from the top cell, anatropous. FRUIT of 2 achenes, free when ripe, often unequal,
one sterile or obsolete; *pericarp* membranous, adherent to the seed, rarely spongy or
furrowed with cellules. SEEDS inverted; *testa* coriaceous. EMBRYO straight, cylin-

dric, in the axis of a fleshy albumen, which it equals in length; *cotyledons* semi-cylindric; *radicle* near the hilum, superior.

PRINCIPAL GENERA.

* Selago.   * Hebenstreitia.   * Polycenia.   Microdon.   Dischisma.

We have indicated the affinities of *Selagineæ* with *Verbenaceæ*, *Stilbineæ*, and *Myoporineæ* (see these families). They are closely allied to *Globulariæ* by the hypogynous 2-labiate corolla and imbricate æstivation, didynamous stamens and 1-celled anthers, pendulous and anatropous ovules, dry fruit, straight albuminous axile embryo and alternate leaves; but in *Globulariæ* the ovary is 1-celled, the fruit is a caryopsis, and the flowers are in a capitulum. All the *Selagineæ* inhabit South Africa. Some are cultivated under glass in Europe; the flowers of the *Hebenstreitia dentata* have no scent in the morning, but a strong and disagreeable one at noon, and are very sweet in the evening.

CLXXII. *STILBINEÆ.*[1]

(STILBINEÆ, *Kunth.*—STILBACEÆ, *Lindl.*)

COROLLA *monopetalous, hypogynous, sub-regular, anisostemonous, æstivation imbricate.* STAMENS 4, *fertile, equal, inserted on the corolla.* OVARY 2-*celled;* OVULES·

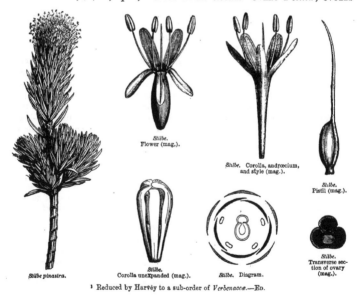

Stilbe pinastra.

Stilbe.
Flower (mag.).

Stilbe. Corolla, androecium, and style (mag.).

Stilbe.
Pistil (mag.).

Stilbe.
Corolla unexpanded (mag.).

Stilbe. Diagram.

Stilbe.
Transverse section of ovary (mag.).

[1] Reduced by Harvey to a sub-order of *Verbenaceæ.*—ED.

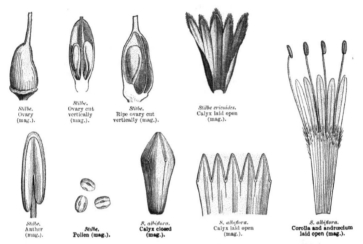

Stilbe.
Ovary
(mag.).

Stilbe.
Ovary cut
vertically
(mag.).

Stilbe.
Ripe ovary cut
vertically (mag.).

Stilbe ericoides.
Calyx laid open
(mag.).

Stilbe.
Anther
(mag.).

Stilbe.
Pollen (mag.).

S. albiflora.
Calyx closed
(mag.).

S. albiflora.
Calyx laid open
(mag.).

S. albiflora.
Corolla and androecium
laid open (mag.).

erect, anatropous. FRUIT *a capsule or utricle.* EMBRYO *straight, albuminous;* RADICLE *inferior.*—STEM *woody.* LEAVES *alternate.*

Heath-like SHRUBS. LEAVES whorled, close-set, jointed on to the stem, exstipulate. FLOWERS ☿, in dense spikes terminating the branches, each with one foliaceous bract and 2 lateral bracteoles. CALYX coriaceous, persistent, tubular-campanulate, 5-fid or -partite, the 2 lower segments more distinct than the 3 upper, or equal, æstivation valvate. COROLLA monopetalous, hypogynous; *tube* infundibuliform; *throat* bristling with close hairs; *limb* 5-partite, spreading, sub-bilabiate, æstivation imbricate. STAMENS exserted, 4 fertile, inserted on the corolla-throat and alternate with its lobes, the fifth upper one sterile or obsolete; *filaments* filiform, sub-equal; *anthers* introrse, dorsifixed, cells often apart at the base, and slits longitudinal, confluent at the top. OVARY free, of 2 antero-posterior unequal lobes, cells 1-ovuled or one cell only fertile; *style* filiform; *stigma* simple; *ovules* erect, anatropous. FRUIT a 2-celled capsule, loculicidally 4-valved at the top, or an indehiscent 1-seeded utricle. SEEDS erect. EMBRYO sub-cylindric, in the axis of a fleshy albumen, which is twice its length; *cotyledons* indistinct; *radicle* inferior.

PRINCIPAL GENERA.

Stilbe.   Campylostachys.

This small family approaches *Verbenaceæ* in the hypogynous irregular anisostemonous corolla, the 2-carpelled ovary, the erect and anatropous ovules, the spiked inflorescence, and the non-alternation of its leaves; but *Verbenaceæ* differ in the didynamous stamens, fleshy fruit, and exalbuminous embryo.

Selagineæ, Myoporineæ, and *Globularieæ* are similarly allied with *Verbenaceæ,* and in addition their embryo is albuminous, and the anther-cells are confluent after the opening of the flower. But in these three families the æstivation of the calyx is imbricate, the andrœcium is didynamous, the ovules are pendulous, and the leaves are alternate; besides which, the fruit of *Globulariæ* is a caryopsis, that of *Selagineæ* consists of 2 achenes, and that of *Myoporineæ* is a drupe with 2–4-cells.

Stilbineæ inhabit southern Africa. They are shrubs which possess no useful property.

CLXXIII. *VERBENACEÆ.*

(Vitices, *Jussieu* [1789].—Verbenaceæ, *Jussieu* [1806].)

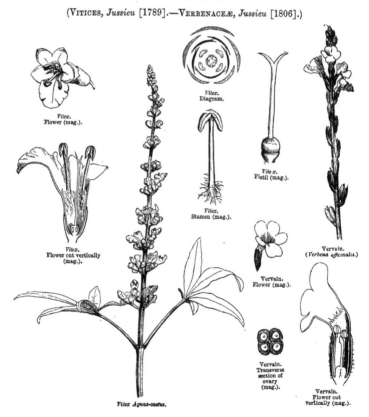

Vitex.
Flower (mag.).

Vitex.
Diagram.

Vitex.
Flower cut vertically
(mag.).

Vitex.
Stamen (mag.).

Vitex.
Pistil (mag.).

Vervain.
Flower (mag.).

Vervain.
(*Verbena officinalis*.)

Vervain.
Transverse
section of
ovary
(mag.).

Vervain.
Flower cut
vertically (mag.).

Vitex Agnus-castus.

Vervain.
Corolla laid open
(mag.).

Vervain.
Diagram.

Vervain.
Pistil (mag.).

Vervain.
Solitary carpel
(mag.).

Vervain.
Carpel cut
vertically (mag.).

COROLLA *hypogynous, monopetalous, irregular, anisostemonous, æstivation imbri-*
cate. STAMENS *inserted on the corolla, usually 4 didynamous, or 2.* OVARY 2–4–8-
celled, cells 1–2-ovuled; OVULES *erect or ascending;* STYLE *terminal.* FRUIT *fleshy.*
EMBRYO *scarcely or not albuminous;* RADICLE *inferior.*

Herbaceous or woody PLANTS, stems and branches usually 4-gonous. LEAVES
generally opposite, sometimes whorled, very rarely alternate (*Dipyrena, Amazonia*),
simple or compound, exstipulate. FLOWERS ⚥, irregular, rarely sub-regular, in a
spike, raceme, head, or cyme, rarely solitary, usually bracteate. CALYX monosepalous,
persistent, tubular; *limb* partite or toothed. COROLLA hypogynous, monopetalous,
tubular; *limb* 4–5-fid, usually unequal and labiate, rarely regular (*Tectona, Callicarpa, Ægiphila*), æstivation imbricate. STAMENS inserted on the corolla-tube or
throat, generally 4, didynamous by arrest of the fifth, sometimes 2 by arrest of the 3
upper, very rarely 5 fertile (*Tectona*); *anthers* 2-celled, sometimes diverging, dehiscence longitudinal. OVARY free, composed of 2–4 carpels, 2–4–8-celled; *style*
terminal, simple; *stigma* usually undivided; *ovules* solitary or geminate in each cell,
sometimes erect and anatropous, or ascending and semi-anatropous, rarely reversed
(*Holmskioldia*). FRUIT a drupe or berry; *drupe* with 2–3–4 1–2-celled pyrenes,
usually separating from the epicarp when ripe; *berries* 2–4-celled, sometimes 1-celled
by arrest. SEEDS solitary in each cell, erect or ascending. EMBRYO exalbuminous
or sub-exalbuminous, straight; *cotyledons* foliaceous; *radicle* inferior.

[The following subdivisions of the order are those of Schauer in De Candolle's
'Prodromus,' and are followed by Bentham as being more natural than any subsequently proposed :—

TRIBE I. VERBENEÆ.—Inflorescence indefinite. Ovules erect, anatropous,

Ovules 2 in each cell.
Sub-tribe 1. SPIELMANNIEÆ.—(Cape of Good Hope.) *Spielmannia,*

Ovules 1 in each cell.
Sub-tribe 2. MONOCHILEÆ.—Racemes lax. Calyx sub-2-lipped. Corolla 1-lipped. Drupe.
(Brazil.) *Monochilus.*

Sub-tribe 3. CASSELLIEÆ.—Racemes axillary, few-fid. Calyx tubular. Corolla infundibuliform. Drupe. (Tropical America.) *Cassellia, Tamonia.*

Sub-tribe 4. VERBENEÆ.—Flowers capitate, racemed or spiked. Calyx campanulate or tubular. Corolla-limb oblique or 2-lipped. Capsule 2-coccous. *Verbena, Bouchea, *Stachy. tarpheta, *Lippia.

Sub-tribe 5. LANTANEÆ.—Flowers capitate or spiked. Calyx short, membranous. Corolla. limb oblique. Drupe of 2 1-celled pyrenes. *Lantana.

Sub-tribe 6. DURANTEÆ.—Racemes lax. Calyx enlarged in fruit. Corolla campanulate or hypocrateriform. Drupe with 2–4 2-celled pyrenes. (America.) Citharexylon, *Duranta.

Sub-tribe 7. PETRÆEÆ.—Racemes lax. Calyx cup-shaped, 5-toothed, with a large invo. lucriform epicalyx. Fruit coriaceous, indehiscent. (America.) *Petræa.

TRIBE II. VITEÆ. Inflorescence cymose, definite. Ovules pendulous, amphitropous, or sub-anatropous.

Sub-tribe 8. SYMPHOREMEÆ.—Cymes few-flowered, contracted, involucrate. Fruit coria. ceous, indehiscent. (Indian twiners.) Symphorema, Congea, &c.

Sub-tribe 9. CARYOPTERIDEÆ. — Cymes not involucrate. Capsule at length 4-valved. (Asia.) *Caryopteris, &c.

Sub-tribe 10. VITICEÆ.—Cymes not involucrate. Drupe fleshy, rarely dry. Tectona, Premna, Callicarpa, Ægiphila, *Volkameria, *Clerodendron, Gmelina, *Vitex, &c.

TRIBE III. AVICENNIEÆ.—Inflorescence capitate, spiked, or centripetal; flower with im. bricate bracts. Calyx 5-leaved. Corolla 4-fid. Ovules geminate, pendulous, amphitropous. Fruit indehiscent; embryo germinating in the pericarp. Avicennia.—ED.]

The affinities of Verbenaceæ with Borragineæ, Labiatæ, and Acanthaceæ have been given under those families; they are very close to Stilbineæ in the irregularity of the corolla, number of the stamens, 2-celled ovary, 1-2-ovuled cells, erect and anatropous ovules, spiked inflorescence and whorled leaves; but in Stilbineæ the corolla is valvate, the 4 stamens are equal, the fruit dry, and the embryo is axile in the fleshy albumen. Their affinity with Myoporineæ is indicated by the insertion, irregularity, anisostemony, and imbrication of the corolla, the didynamous stamens, 1-2-ovuled 2-celled ovary, and drupaceous or baccate fruit; but in Myoporineæ the ovules are pendulous, the embryo axile in the fleshy albumen, the leaves are generally alternate, the flowers axillary and usually solitary. They have the same affinity with Selagineæ, which besides have spiked flowers, and the diagnosis is the same; added to which, in Selagineæ the anthers are reniform and 1-celled, and the fruit is dry. A comparison with Globularieæ shows the same similarities and differences, and Globularieæ are further distinguished by their dry fruit, which is a caryopsis. A close relationship is also observable between Verbenaceæ and Jasmineæ; in both the corolla is hypogynous, sub-irregular, anisostemonous, and imbricate in æstivation, the ovary is 2-celled, the cells 1-2-ovuled, the ovules collateral, ascending and anatropous, the fruit is fleshy, the embryo exalbuminous or nearly so, and the leaves opposite. Verbenaceæ principally inhabit tropical regions, decreasing towards the poles; the woody species grow in the torrid zone, the herbaceous in temperate climates. They are rare in Europe, Asia, and North America.

Verbenaceæ contain a little volatile oil, but bitter and astringent principles predominate, and their medicinal properties are little esteemed in Europe. The Vervain (Verbena officinalis), celebrated among the ancient Romans and the Druids of Gaul, was used in religious ceremonies and in incantations; its slightly aromatic bitter gave it formerly a place among tonics, whence its name of officinalis. Lippia citriodora is an undershrub of South America, cultivated in Europe, the dried leaves of which are infused like tea, and also used for flavouring cream. Many species of Lantana are also used as tea in Brazil (Lantana pseudothea), and their drupes are edible (L. annua and L. trifolia), as are those of Premna. The Asiatic Callicarpæ have a bitter aromatic bark, and their leaves are diuretic; the American species of the same genus are of repute in the treatment of dropsy. Some other Verbenaceæ are alexipharmics (Ægiphila, Gmelina); Gmelina villosa is a febrifuge, G. arborea is recommended for rheumatism. Verbena erinoides is employed in Peru as a uterine stimulant. The Clerodendrons are trees remarkable for the sweet scent of 'their flowers; the bitter leaves and the aromatic root of many are prescribed for scrofulous and syphilitic diseases; others [as *Stachytarpha*] are used in the superstitious ceremonies of Indian sorcerers, as the Vervain was in Europe [its leaves are the Brazil Tea]. *Vitex Agnus-castus* is a shrub, indigenous to southern France, to which the ancients attributed cooling virtues, whence its name.

[*V. littoralis* is one of the best woods in India. *Gmelina parviflora* has the property of imparting mucilage to water. Teak, one of the most important timbers in the world, is the wood of *Tectona grandis*, a gigantic Asiatic forest tree. *Avicenniæ* are trees growing in tidal swamps throughout warm countries, called White Mangrove in Brazil, where the bark is extensively used for tanning; a preparation from the ashes of the bark is used in washing in India.]

CLXXIV. *GLOBULARIEÆ.*

(GLOBULARIÆ, *D.C.*—GLOBULARINEÆ, *Fndl.*—GLOBULARIACEÆ, *Lindl.*)

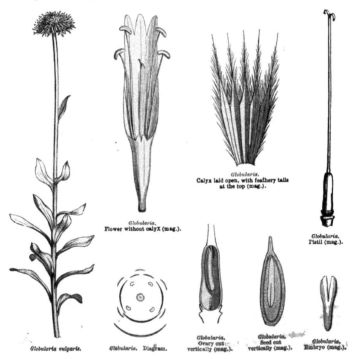

Globularia.
Calyx laid open, with feathery tails at the top (mag.).

Globularia.
Flower without calyx (mag.).

Globularia.
Pistil (mag.).

Globularia vulgaris.

Globularia. Diagram.

Globularia.
Ovary cut vertically (mag.).

Globularia.
Seed cut vertically (mag.).

Globularia.
Embryo (mag.).

COROLLA *monopetalous, hypogynous, 2-labiate, anisostemonous, æstivation imbricate.* STAMENS 4, *didynamous, inserted on the corolla.* OVARY 1-*celled;* OVULE *solitary, pendulous, anatropous.* FRUIT *a caryopsis.* EMBRYO *albuminous;* RADICLE *superior.* —FLOWERS *capitate.*

SHRUBS or UNDERSHRUBS, or evergreen HERBS. LEAVES alternate, simple, entire, aggregated at the base of the branches, the upper smallest, scattered, spathulate, contracted into a petiole, exstipulate, marcescent. FLOWERS ☿, irregular, capitate on a convex paleaceous receptacle, surrounded by an involucre of many series of bracts. CALYX herbaceous, monosepalous; *tube* tetragonal after flowering; *throat* usually closed by hairs; *limb* 5-fid, regular or rarely 2-lipped, the upper lip 3-fid, the lower 2-fid. COROLLA monopetalous, hypogynous; *tube* cylindric; *limb* 1-2-labiate, upper lip entire or 2-lobed, or very short or obsolete, lower longer, 3-partite or -fid or -toothed, æstivation imbricate. STAMENS 4, inserted at the top of the corolla-tube, alternate with its lobes, the fifth wanting between the lobes of the upper lip; *filaments* filiform, exserted, the upper a little the shortest; *anthers* reniform, 2-celled in bud, cells then confluent, opening at the top by a single slit. OVARY free, 1-celled, seated on a hypogynous minute disk, which is rarely reduced to an anterior gland, or absent; *style* terminal, simple; *stigma* undivided or shortly 2-lobed; *ovule* solitary, pendulous, anatropous. CARYOPSIS enveloped by the calyx, mucronate by the persistent style-base. SEED inverted. EMBRYO straight, in the axis of a fleshy albumen; *cotyledons* ovoid, obtuse; *radicle* next the hilum, superior.

GENUS.

Globularia.

We have indicated the affinities between *Globulariæ* and *Verbenaceæ, Stilbineæ, Myoporineæ,* and *Selagineæ,* affinities which are not disturbed by the 1-celled and 1-ovuled ovary of *Globulariæ,* since the base of the style is geniculate and a little furrowed on the back, which has led A. de Candolle to suspect that the pistil is 2-carpellary, and that the posterior carpel is arrested. Some botanists have noticed a close connection between *Globulariæ* and *Dipsaceæ,* the latter only differing in the epigynous corolla, and the opposite or whorled leaves. *Globulariæ* also present some analogy with *Brunoniaceæ,* founded on the capitate inflorescence, hypogynous corolla, 1-celled 1-ovuled ovary, and anatropous ovule; but in *Brunoniaceæ* the corolla is regular and isostemonous, and the æstivation is valvate, the stamens are hypogynous, the anthers 2-celled, the ovule is erect, and the embryo exalbuminous. There are also some points of resemblance between *Globulariæ* and *Calycereæ*; the inflorescence is the same, the ovary is 1-celled and 1-ovuled in both, the ovule is pendulous and anatropous, and the embryo albuminous; but *Calycereæ* differ in their epigyny, the regularity, isostemony and valvate æstivation of the corolla, and in the syngenesious stamens. *Globulariæ* principally inhabit the south-west countries of Europe, and are not met with farther north than 54°.

Some species were formerly used medicinally; the leaves of *Globularia communis* are reckoned among detergent and vulnerary medicines. The *G. Alypum* replaces Senna in the south of Europe, and is a very decided purge.

CLXXV. *LABIATÆ.*

(VERTICILLATÆ, *L.*—LABIATÆ, *A.-L. de Jussieu.*—LAMIACEÆ, *Lindl.*)

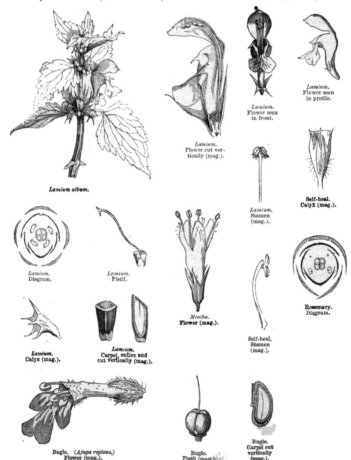

Lamium album.

Lamium.
Flower cut vertically (mag.).

Lamium.
Flower seen in front.

Lamium.
Flower seen in profile.

Lamium.
Stamen (mag.).

Self-heal.
Calyx (mag.).

Lamium.
Diagram.

Lamium.
Pistil.

Mentha.
Flower (mag.).

Self-heal.
Stamen (mag.).

Rosemary.
Diagram.

Lamium.
Calyx (mag.).

Lamium.
Carpel, entire and cut vertically (mag.).

Bugle. (*Ajuga reptans.*)
Flower (mag.).

Bugle.
Pistil (mag.).

Bugle.
Carpel cut vertically (mag.).

Germander.	Germander.	Germander.	Scutellaria.	Scutellaria.
Back view of flower.	Flower seen in profile.	Pistil cut vertically (mag.).	Carpel cut vertically (mag.).	Embryo (mag.).

COROLLA *hypogynous, monopetalous, irregular, anisostemonous, æstivation imbricate.* STAMENS 4 *didynamous, or 2 inserted on the corolla.* OVARY *4-lobed;* OVULES 4, *erect, anatropous;* STYLE *gynobasic.* FRUIT *separating into 4 achenes.* EMBRYO *exalbuminous;* RADICLE *inferior.*—STEM *4-gonous.* LEAVES *opposite or whorled.*

Whole plant often covered with vesicular glands containing an odoriferous volatile oil. STEM herbaceous or sub-woody, rarely woody, usually 4-gonous. LEAVES opposite or whorled, with pinnate reticulate nerves, exstipulate. FLOWERS ☿, irregular, very rarely sub-regular (*Mentha*), in the axils of leaves or bracts, solitary or geminate, or in clustered centrifugal cymes, which form false whorls by their union in pairs, and are scattered, or crowded into spikes. CALYX persistent, of 5 connate sepals, sometimes irregular, 2-lipped, upper lip 2-fid, lower 3-fid, sometimes sub-regular with 5 segments, or 4 by the arrest of the upper. COROLLA monopetalous, hypogynous; *tube* sometimes twisted (*Hyssopus lophantus, Ajuga orientalis*); *limb* 4–5-lobed, æstivation imbricate, sometimes 2-labiate, the upper lip entire or emarginate, the lower 3-lobed, sometimes 1-labiate from the upper lip being very short and deeply cleft (*Ajuga*), sometimes being bell- or funnel-shaped with 4 sub-equal lobes and sub-equal stamens (*Mentha*). STAMENS inserted on the corolla-tube, usually 4, didynamous, rarely 2 by arrest of the 2 upper (*Cunila, Lycopus, Salvia, Rosmarinus*); *anther-cells* often confluent at the top, sometimes separated by a well-developed filiform connective (*Salvia*). OVARY free, of 2 carpels, borne on a thick disk, with 4 lobes or cells, all free or cohering in pairs; *style* simple, rising from the base of the ovarian lobes, and dilating into a gynobase which lines the disk; *stigma* generally bifid; *ovules* solitary and erect in each cell, anatropous. FRUIT of 4 achene-like lobes or nucules, free or geminate, epicarp sometimes fleshy (*Prasium*). SEEDS erect. EMBRYO straight, very rarely curved (*Scutellaria*), exalbuminous, or with a thin fleshy albumen; *radicle* inferior.

[The following is Bentham's classification of this large order :—

TRIBE I. OCIMOIDEÆ.—Stamens declinate. *Ocimum, Mesona, Orthosiphon, *Plectranthus, Coleus, Hyptis, Eriope, *Lavandula, &c.

TRIBE II. SATUREIEÆ.—Stamens remote, straight, spreading or connivent under the upper lip, 4 or 2 with the anther-cells contiguous. Corolla-lobes flat. *Pogostemon, Elsholtzia, Perilla, *Mentha, Lycopus, Bistropogon, *Origanum, *Thymus, Micromeria, *Satureia, Calamintha, *Gardoquia, *Melissa, Hedeoma, Sphacele, &c.

TRIBE III. MONARDEÆ.—Stamens 2, straight or ascending. Anther-cells linear-oblong, solitary or separated by a long connective. *Meriandra*, *Salvia, *Rosmarinus, *Monarda, *Zizyphora, &c.

TRIBE IV. NEPETEÆ.—Stamens 4, the posticous (upper) pair always the longest. *Lophanthus, Nepeta,* *Dracocephalum, *Cedronella, &c.

TRIBE V. STACHYDEÆ.—Stamens 4, parallel and ascending under the upper lip. Nucules quite free, erect. *Prunella, *Scutellaria, *Melittis,* *Sideritis, Marrubium, Anisomeles, *Betonica, *Stachys, Leonurus, *Lamium, Ballota, Leucas, *Phlomis, &c.

TRIBE VI. PRASIEÆ.—Stamens of *Stachydeæ.* Nucules fleshy, sub-connate at the base. *Gomphostemma, Phyllostegium, Stenogyne, Prasium.*

TRIBE VII. PROSTANTHERÆ.—Nucules usually rugose, connate at the base, style persistent. Corolla-throat campanulate, lobes flat. (Australia.) *Prostanthera, Hemiandra, Microcoris, Westringia, &c.

TRIBE VIII. AJUGOIDEÆ.—Nucules rugose, sub-connate at the base. Stamens parallel, ascending. Upper lip of corolla minute or 2-fid with declinate lobes. *Teucrium, Ajuga, Trichostemma, &c.—ED.]

Labiatæ form one of the most natural groups of plants; the characters of its members are so uniform that it may be called *monotypic,* as if all the species could be comprehended in a single genus, and the discrimination of its genera is hence often very difficult. For the same reason the affinities of *Labiatæ* are but few. We have noticed their connection with *Scrophularineæ, Borragineæ,* and *Acanthaceæ.* They approach nearest to *Verbenaceæ,* which differ only in the coherence of the parts of the ovary, the terminal style, the berried or drupaceous fruit, the leaves not constantly opposite, and the absence of oleiferous vesicular glands. It is in the temperate regions of the Old World that the majority of *Labiatæ* are found; they are not numerous beyond 50° north latitude or in the tropics, and are less frequent in the southern hemisphere; from the arctic regions they are completely absent.

A volatile oil is contained in the vesicular glands of *Labiatæ* which in some species holds in solution a solid hydrocarbon (*stearoptene*) analogous to camphor; to the different proportions in which these substances are united to bitter and astringent principles the various properties of its members are due. The purely aromatic species are condiments, stimulants, [carminatives], or cosmetics; especially Peppermint [(*Mentha piperita*),Spearmint (*M. viridis*),and Pennyroyal (*M. Pulegium*)],Thyme (*Thymus vulgaris*), Savory (*Satureia hortensis* and *montana*), Balm (*Melissa officinalis*), Basil Thyme (*Calamintha Acinos*),Lemon Thyme (*Thymus citriodorus*), [Sweet Basil (*Ocymum Basilicum*), Bengal Sage (*Meriandra bengalensis*), Sage (*Salvia grandiflora* and *officinalis*), Marjoram (*Origanum Majorana, Onites,* &c.), Hyssop (*Hyssopus officinalis*)]. The powerful stimulating properties of Rosemary, utilized medicinally in Hungary water, are due to its volatile oil and stearoptene [it is also an ingredient in Eau de Cologne, and in the green pomades, having the power of encouraging the growth of hair]. When the aromatic principle is combined with the bitter one, they are stimulating and tonic (Marjoram, Lavender, &c.). The very strong-scented essence of *Lavandula Spica* (Oil of Lavender) is used as an embrocation in rheumatic affections, [and as Oil of Spike by painters]. The common Lavender (*L. vera*), cultivated in gardens, is used to preserve linen, woollen, and furs from insects; as is Patchouly, an Indian species of *Pogostemon.*

Teucrium, which contains gallic acid and a bitter principle, is a tonic. *Scutellaria galericulata* was formerly employed in tertian fevers. Ground Ivy (*Glechoma hederacea*) is bitter and slightly acrid; it is used as a bechic and antiscorbutic. *Marrubium,* in which the bitter overcomes the aroma, is recommended as a tonic. Finally, Sage (*Salvia officinalis*) combines all the medicinal properties of the other *Labiatæ,* whence its stimulating, tonic, and astringent virtues, and its trivial name. [Others are Horehound (*Marrubium vulgaris*), a popular and excellent remedy in coughs; and *Lycopus europæus,* which yields a black dye.]

CLXXVI. *PLANTAGINEÆ.*

(PLANTAGINES, *Jussieu.*—PLANTAGINEÆ, *Br.*—PLANTAGINACEÆ, *Lindl.*)

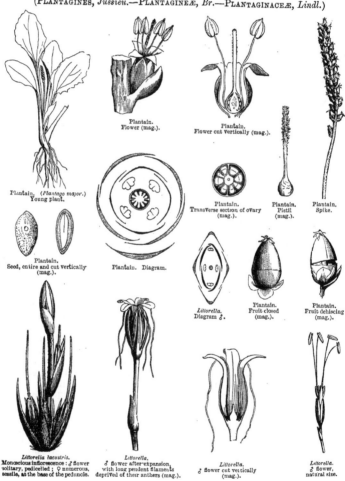

Plantain.
Flower (mag.).

Plantain.
Flower cut vertically (mag.).

Plantain, (*Plantago major.*)
Young plant.

Plantain.
Transverse section of ovary
(mag.).

Plantain.
Pistil
(mag.).

Plantain.
Spike.

Plantain.
Seed, entire and cut vertically
(mag.).

Plantain. Diagram.

Littorella.
Diagram ♂.

Plantain.
Fruit closed
(mag.).

Plantain.
Fruit dehiscing
(mag.).

Littorella lacustris.
Monœcious inflorescence : ♂ flower
solitary, pedicelled ; ♀ numerous,
sessile, at the base of the peduncle.

Littorella.
♂ flower after expansion,
with long pendent filaments
deprived of their anthers (mag.).

Littorella.
♂ flower cut vertically
(mag.).

Littorella.
♂ flower,
natural size.

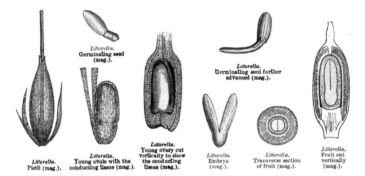

Littorella.
Germinating seed
(mag.).

Littorella.
Germinating seed further
advanced (mag.).

Littorella.
Pistil (mag.).

Littorella.
Young ovule with the
conducting tissue (mag.).

Littorella.
Young ovary cut
vertically to show
the conducting
tissue (mag.).

Littorella.
Embryo
(mag.).

Littorella.
Transverse section
of fruit (mag.).

Littorella.
Fruit cut
vertically
(mag.).

COROLLA *monopetalous, hypogynous, generally isostemonous, œstivation imbricate.* STAMENS 4 (*rarely* 1), *inserted on the corolla, or hypogynous.* OVARY 1–4-*celled;* OVULES *peltate.* FRUIT *a capsule or nucule.* SEEDS *fixed by a ventral hilum.* EMBRYO *parallel to the hilum, albuminous, straight or curved.*

Annual or perennial HERBS; *rhizome* subterranean, sometimes stoloniferous, giving off scapigerous peduncles, or leafy rarely woody stems. LEAVES all radical in most species, or rosulate, alternate, or opposite (*Psyllium*), simple, flat, nerved, entire, toothed, or pinnatifid (*Psyllium*), or semi-cylindric, sessile and fleshy, or contracted into a petiole dilated at its base, and accompanied by a woolly membrane. PEDUNCLES always springing from the axils of the lower leaves. FLOWERS usually ☿, spicate and spiked, bracteate, or rarely diclinous; the ♂ solitary, scapose; the ♀ crowded, sessile at the base of the scape.

1. FLOWERS ☿ (*Plantain*): CALYX ? herbaceous, 4-partite, persistent; anterior sepals distinct or cohering, imbricate, usually keel-shaped, edges membranous. COROLLA ? hypogynous, monopetalous, tubular, scarious, marcescent, with 4 imbricate lobes. STAMENS 4, inserted on the corolla-tube and alternate with its lobes, exserted or sometimes included and imperfect; *filaments* filiform, flaccid, inflexed before flowering; *anthers* versatile, apiculate, cells parallel, dehiscence longitudinal, introrse, deciduous. OVARY free, 2–4-celled; *style* filiform, exserted erect, or included, with 2 longitudinal lines of stigmatic papillæ; *ovules* 1–8 in each cell, peltate on the middle of the septum of the many-ovuled cells, or at the bottom of the 1-ovuled cells. CAPSULE circumsciss, sub-membranous, 1–4-celled, 1-many-seeded, edges of septum free, surfaces seminiferous. SEEDS peltate; *testa* mucilaginous. EMBRYO parallel to the hilum, straight, cylindric, in the axis of a fleshy albumen; *cotyledons* oblong or linear; *radicle* distant from the hilum, inferior, rarely centrifugal.

2. FLOWERS monœcious (*Littorella*) or polygamous (*Bougueria*).—♂ CALYX 4-partite, with membranous edges (*Littorella*), or with 4 sub-equal hairy sepals (*Bougueria*). COROLLA tubular, scarious, with 4 equal lobes (*Littorella*), or irregularly

3–4-lobed, with silky edges (*Bougueria*). STAMENS 4, hypogynous and alternate with the corolla-lobes (*Littorella*), or 1–2, inserted on the middle of the corolla-tube (*Bougueria*). OVARY rudimentary (*Littorella*), or obsolete · (*Bougueria*).— ♀ : CALYX with 3 unequal sepals (bracts ?), the anterior larger (*Littorella*), or of 4 sub-equal hairy sepals (*Bougueria*). COROLLA urceolate; *throat* short; *limb* 3–4-toothed (*Littorella*) or tubular, irregularly 3–4-lobed, with silky edges (*Bougueria*). STAMENS 0. OVARY 1-celled; *ovule* solitary, campylotropous. NUCULE bony. SEED peltate, *testa* membranous. EMBRYO straight, in the axis of a fleshy albumen (*Littorella*), or curved round the albumen (*Bougueria*).

GENERA.

| Plantago. | Littorella. | Bougueria. |

Plantagineæ, although very different in appearance, form a very homogeneous group; they approach *Plumbagineæ* in their inflorescence, hypogynous generally isostemonous corolla, and the stamens sometimes hypogynous, sometimes inserted on the corolla, as in *Statice* and *Plumbago*, and finally in their dry fruit and albuminous embryo. *Plumbagineæ* are separated from *Plantago* by their 1-celled and 1-ovuled ovary, from *Plantago* and *Littorella* by their many styles, anatropous ovule pendulous from a basal ascending funicle, and farinaceous albumen. *Plantagineæ* are allied rather closely with *Primulaceæ* by the direction of the ovules, the circumsciss capsule, the ventral hilum, and the embryo parallel to the hilum, as also with *Veronica*; another analogy results from the isolation of the placentiferous septum in *Plantago*, which, although only occurring at maturity, recalls the placentation of *Primulaceæ*.

As to the different position of the stamens, opposite to the corolla-lobes in *Plantagineæ* and alternate with them in *Primulaceæ*, this is of no account, if it be admitted (with certain authors) that in *Plantagineæ* the scarious and persistent corolla is a calyx, and the calyx an involucel; in this case the Plantains would be apetalous, as in *Glaux*, and the stamens would alternate with the sepals.

There is yet another affinity which deserves to be noticed; it is founded on the alternation of the stamens with the calyx, an alternation which exists, if we admit, with Grisebach, that both these families are apetalous, and regard the corolla of *Plumbagineæ* as a staminal crown, and that of *Plantagineæ* as a true calyx.

Plantagineæ inhabit the temperate regions of both hemispheres, but principally the northern, and especially Europe and America; they are more rare in the tropics, where they only grow on mountains. Many indigenous species of Plantain (*Plantago lanceolata, major, media*) are employed in medicine; their leaves are bitter and slightly astringent. An eye-water is distilled from the whole plant. The seeds of *Plantago Psyllium, arenaria* and *Bophula* contain an abundant mucilage in their testa, whence their use as emollients in inflammatory ophthalmia; and in Indian manufactures to stiffen muslins. *Plantago Coronopus* was considered by the ancients, on account of its toothed leaves, to be efficacious against hydrophobia, and was ranked amongst diuretics. It is cultivated for salad in some countries.

CLXXVII. *NYCTAGINEÆ, Jussieu.*

FLOWERS ☿ *or diclinous*. PERIANTH *petaloid or coloured*. STAMENS *hypogynous*. OVARY 1-*celled*, 1-*ovuled*; OVULE *campylotropous, erect*. ACHENE *included in the persistent base of the perianth*. ALBUMEN *farinaceous (rarely 0)*. EMBRYO *curved, rarely straight*; RADICLE *inferior*.

TREES, SHRUBS or HERBS. STEMS knotty, fragile, branches often spinescent. LEAVES usually opposite (that subtending a branch, peduncle or spine being smaller than its pair), rarely alternate, or scattered, petioled, entire. FLOWERS ☿, or rarely

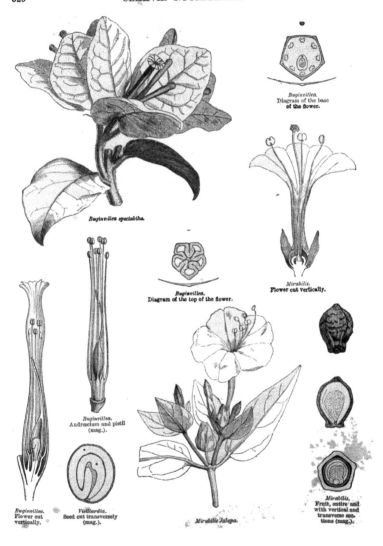

Buginvillea spectabilis.

Buginvillea.
Diagram of the base
of the flower.

Buginvillea.
Diagram of the top of the flower.

Mirabilis.
Flower cut vertically.

Buginvillea.
Andrœcium and pistil
(mag.).

Buginvillea.
Flower cut
vertically.

Vieillardia.
Seed cut transversely
(mag.).

Mirabilis Jalapa.

Mirabilis.
Fruit, entire and
with vertical and
transverse sec-
tions (mag.).

Mirabilis. Diagram. *Mirabilis.* Andrœcium. *Pisonia.* Fruit. *Pisonia aculeata.* Fruit cut transversely (mag.).

diclinous (*Pisonia*), axillary or terminal, solitary or aggregated, rarely in a simple spike, umbel, cyme or panicle; *floral bracts* 6, sometimes ovate or lanceolate, free or united, often forming a calyciform much dilated coloured involucre larger than the flowers; sometimes 1–3, minute; *involucre* calyciform, herbaceous, sometimes monophyllous, 3–5-toothed, 1–5-flowered, and often enlarged after flowering; sometimes polyphyllous, many-flowered. PERIANTH petaloid, tubular or tubular-campanulate or infundibuliform, variously coloured; the lower part of the tube hardest, sometimes striate, always persistent, enveloping the fruit and accrescent, the upper part resembling a corolla; *limb* membranous, folded in æstivation, rarely marcescent, usually falling after flowering. STAMENS hypogynous, fewer or more than the divisions of the perianth, 8–30, rarely equal in number, sometimes unilateral; *filaments* filiform, often unequal, included or exserted, inflexed in æstivation, free, or coherent at the base, sometimes even adnate below to the perianth-tube; *anthers* introrse, 2-celled, globose, dehiscence longitudinal; *pollen* of globose large granules. OVARY free, sessile or sub-stipitate, 1-carpelled, 1-celled; *style* terminal or sub-lateral, simple, involute in æstivation; *stigma* simple, pointed or globose, straight or coiled, sometimes branched or penicillate, multifid; *ovule* solitary, erect, sessile, micropyle inferior. ACHENE membranous, included in the hardened tube of the perianth. SEED erect, *testa* adherent to the endocarp. EMBRYO usually curved or folded, rarely straight (*Pisonia*); *cotyledons* foliaceous, enveloping a farinaceous albumen, or coiled upon themselves, and scarcely separated by a mucilaginous albumen (*Vieillardia*); *radicle* inferior.

[The following is Choisy's arrangement of the genera in De Candolle's 'Prodromus' :—

TRIBE I. MIRABILEÆ.—Involucre calyciform, 1- or many-leaved. *Mirabilis, Quamoclidion, *Oxybaphus, Abronia, Acleisanthes, Pentacrophys,* &c.

TRIBE II. BUGINVILLEÆ.—Involucre bract-like; bracts large, dilated. *Buginvillea (Bougainvillea), Tricycla,* ? *Cephalotomandra.*

TRIBE III. BOERHAAVIEÆ. *Salpianthus, Pisonia, Neea, Vieillardia, Boerhaavia.*—ED.]

Nyctagineæ are not closely allied to any other family; they have been placed near *Phytolacceæ*, *Chenopodieæ*, and *Polygoneæ* on account of the structure of their ovary, their curved seed and albumen; but they differ in their folded æstivation and exstipulate leaves. They have an apparent affinity with *Valerianeæ* (as observed by A.-L. de Jussieu), through *Boerhaavia*, several species of which have been often confounded with that family.

Nyctagineæ principally inhabit the tropical regions of the Old, and especially the New World. *Abronia* grows in North-west America, some *Boerhaaviæ* in extra-tropical Australia, and in South America. *Buginvillea* is limited to South America.

The roots of *Nyctagineæ* are purgative or emetic. That of the false Jalap, or Marvel of Peru (*Mirabilis Jalapa*), a tropical American plant, was long confounded with that of the true Jalap, of which it has the nauseous smell. It possesses similar qualities, but is much less efficacious, and is sometimes administered for dropsy, as are *M. dichotoma* and *longiflora*; all are cultivated in our gardens. *M. suaveolens* is recommended in Mexico for diarrhœa and rheumatic pains. The numerous species of *Boerhaavia* furnish the Americans with emetic and purgative roots. The juice of *B. hirsuta* is used in Brazil as a remedy for jaundice. The cooked root of *B. tuberosa* is eaten in Peru; an infusion of it is ranked as an antisyphilitic. A decoction of the herbage of *B. procumbens* is an Indian febrifuge. The properties of *Pisonia* are analogous to those of *Boerhaavia*.

CLXXVIII. *PHYTOLACCEÆ.*

(ATRIPLICUM *sectio, Jussieu.*—PHYTOLACCEÆ, *Br.*—RIVINEÆ ET PETIVEREÆ, *Agardh.* —PHYTOLACCACEÆ ET PETIVERIACEÆ, *Lindl.*—PHYTOLACCACEÆ, *Endl.*)

CALYX 4-5-*partite.* COROLLA *usually* 0. STAMENS *sub-hypogynous or hypogynous, as many as the sepals, or more numerous.* CARPELS *several, whorled, or one excentric, 1-ovuled;* STYLES *lateral and ventral, hooked.* FRUIT *fleshy or dry.* SEED *erect.* ALBUMEN *farinaceous, sometimes scanty or* 0. EMBRYO *annular or curved, rarely straight;* RADICLE *inferior.*

HERBS or UNDERSHRUBS [rarely TREES], usually glabrous. STEMS cylindric, or irregularly annular, rarely twining (*Ercilla*). LEAVES alternate or rarely sub-opposite, simple, entire, membranous or somewhat fleshy, sometimes pellucid-punctate; *stipules* 0, or geminate at the base of the petioles, free, deciduous, or changed into persistent thorns. FLOWERS ☿ or rarely diœcious (*Achatocarpus, Gyrostemon*), regular or sub-regular, in a spike, raceme, or glomerate cyme, axillary, terminal or leaf-opposed, pedicels naked or 1–3-bracteate. CALYX 4–3-partite; *lobes* herbaceous, often membranous at the edge, frequently coloured on their inner surface, equal or unequal, æstivation imbricate. COROLLA usually 0, rarely of 4–5 petals (*Semonvillea*, &c.) alternate with the sepals and inserted at their base, distinct, with narrow claws. STAMENS sub-hypogynous or hypogynous, inserted at the base of a disk lining the bottom of the calyx, or of a somewhat convex torus, or sometimes of a slender gynophore; either equal and alternate with the sepals, or more numerous, the outer alternate, the inner opposite; rarely united in alternate bundles; more rarely indefinite and arranged without order; *filaments* filiform or dilated at the base, distinct, or connate below; *anthers* introrse, 2-celled, basifixed or dorsifixed, erect or incumbent, dehiscence longitudinal. CARPELS several whorled, rarely

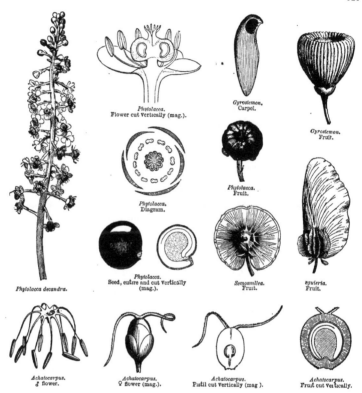

Phytolacca.
Flower cut vertically (mag.).

Gyrostemon.
Carpel.

Gyrostemon.
Fruit.

Phytolacca.
Diagram.

Phytolacca.
Fruit.

Phytolacca.
Seed, entire and cut vertically (mag.).

Phytolacca.
Fruit.

Semonvillea.
Fruit.

Seguieria.
Fruit.

Phytolacca decandra.

Achatocarpus.
♂ flower.

Achatocarpus.
♀ flower (mag.).

Achatocarpus.
Pistil cut vertically (mag).

Achatocarpus.
Fruit cut vertically.

solitary and sub-excentric, distinct or more or less coherent, seated on an inconspicuous gynophore, or fixed to a central column, 1-celled; *styles* ventral, distinct or rarely coherent at the base, tips recurved and stigmatiferous on their inner surface; *ovules* usually solitary, basifixed, campylotropous or rarely semi-anatropous. FRUIT a berry, utricle, coccus, nut or samara. SEED erect; *testa* membranous or crustaceous, usually shining and fragile. EMBRYO sometimes annular or arched, peripheric, surrounding a copious floury albumen with flat large or narrow and unequal cotyledons, the edges of the outer closing round the inner; sometimes straight, with foliaceous convolute cotyledons and little or no albumen; *radicle* inferior.

TRIBE I. *PETIVERIEÆ.*

Carpel solitary, becoming a samara or achene. Embryo curved (*Seguieria*) or straight (*Petiveria*), cotyledons convolute. Leaves stipulate.

PRINCIPAL GENERA.

Seguieria.　　　　Petiveria.　　　　Rivina.　　　　Mohlana.

TRIBE II. *PHYTOLACCEÆ.*

Fruit usually composed of 2 or more carpels, distinct or coherent, but without a column. Leaves exstipulate.

PRINCIPAL GENERA.

Microtea.　　　Limeum.　　　Gieseckia.　　　Phytolacca.　　　Semonvillea.
Anisomeria.　　Pircunia.　　Ercilla.　　　* Rivina.　　　Achatocarpus.

TRIBE III. *GYROSTEMONEÆ.*

Fruit compound, with a central column, resembling a 1-celled or 2–several-celled capsule. Cotyledons not rolled. Leaves exstipulate.

GENERA.

Didymotheca.　　　Gyrostemon.　　　Codonocarpus.　　　Tersonia.

Phytolacceæ, long confounded with *Chenopodieæ*, are connected with them by their alternate leaves, inflorescence, 1-ovuled carpels, farinaceous albumen and usually peripheric embryo; they are sufficiently distinguished by their frequently having petals, by the number and position of their stamens, their lateral style, plurality of carpels, and berried or coccus-like pericarp. They approach *Baselleæ* and *Amarantaceæ* in the coloured calyx and the structure of the seed; *Portulaceæ* in the alternate leaves, the stamens alternating with the sepals when they are the same in number, and the structure of the flower and seed. *Seguieria*, the cotyledons of which are coiled, and the albumen 0 or nearly so, and *Gyrostemon*, the carpels of which are whorled around a central column, establish a certain affinity between *Phytolacceæ* and *Malvaceæ*. *Phytolacceæ* inhabit the tropical and sub-tropical regions of the Old, and especially of the New World; they are much rarer in Asia than in Africa. *Petiverieæ* are all tropical American. *Phytolacceæ*, properly so called, mostly belong to the Old World. *Gyrostemoneæ* are all Australian.

Phytolacceæ owe their properties to acrid, vesicant, and drastic substances. *Phytolacca decandra* (Pokeweed or American Currant), a native of North America, has been naturalized in the Landes; its acrid leaves, its root and unripe berries are a strong purgative. Its ripe berries contain a purple juice which is by no means innocuous, and which is imprudently used to colour confectionery and wine, for which reason the Portuguese Government has forbidden its culture. Nevertheless the young leaves of this species and of its congeners (*P. esculenta*, &c.), are edible when cooked. *P. drastica* grows among rocks in Chili, and the natives chew its root as a purgative. Some *Phytolacceæ* blacken in drying (*Bosia, Achatocarpus*). *Petiverieæ*, remarkable for their alliaceous smell, are used in domestic medicine by Americans, as antifebrile, diaphoretic, diuretic, and vermifuge.

[The berries of *Phytolacca octandra* are used as soap in the West Indies. *Pircunia dioica*, a tree of La Plata, is now extensively cultivated in the south of Spain as Bella-sombra, and is a conspicuous feature in the gardens at Gibraltar, where its trunks, enormously swollen at the base, attract universal attention; the tree is of most rapid growth, but the wood is very spongy. The young shoots and leaves of *Phytolacca dioica* are recommended for cultivation as a potherb, being eaten, cooked like asparagus, in the United States, as are those of *P. acinosa* in the Himalayas. The turnip-shaped root of *P. drastica* is a violent drastic purge.—ED.]

CLXXIX. *POLYGONEÆ.*

(PERSICARIÆ, *Adanson.*—POLYGONEÆ, *Jussieu.*—POLYGONACEÆ, *Lindl.*)

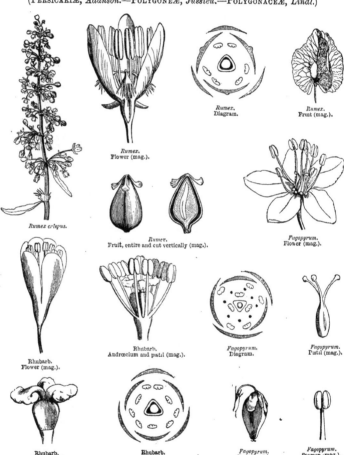

Rumex.
Diagram.

Rumex.
Fruit (mag.).

Rumex.
Flower (mag.).

Rumex crispus.

Rumex.
Fruit, entire and cut vertically (mag.).

Fagopyrum.
Flower (mag.).

Rhubarb.
Flower (mag.).

Rhubarb.
Andrœcium and pistil (mag.).

Fagopyrum.
Diagram.

Fagopyrum.
Pistil (mag.).

Rhubarb.
Pistil (mag.).

Rhubarb.
Diagram.

Fagopyrum.
Fruit (mag.).

Fagopyrum.
Stamen (mag.).

Eriogonum.   *Muhlenbeckia.*   *Kœnigia.*   *Kœnigia.*
Inflorescence.   Fruit and accrescent calyx (mag.).   Flower-bud (mag.).   Monandrous flower laid open (mag.).

FLOWERS ☿ *or diclinous.* PERIANTH *herbaceous or petaloid.* STAMENS *peri-gynous.* OVARY *1-celled, 1-ovuled;* OVULE *erect, orthotropous.* FRUIT *an achene.* SEED *erect;* ALBUMEN *farinaceous.* EMBRYO *straight, curved and lateral, or straight and axile;* RADICLE *superior.*—LEAVES *alternate, with an intrapetiolar stipule.*

Herbaceous or frutescent PLANTS, sometimes arborescent, erect [or prostrate] or twining, stem and branches jointed and knotty, leafy, rarely aphyllous and scapige-rous. LEAVES alternate, often collected at the base of the stem, very rarely opposite (*Pterostegia*), simple, entire, waved, crisped or crenulate, very rarely cut, usually penninerved, margins revolute when young, often glandular or pellucid-dotted; *petiole* dilated at the base and amplexicaul, or inserted on a close-sheathing intra-petiolar stipule (*ochrea*), sometimes inconspicuous. FLOWERS ☿ or diclinous, springing from the axil of the leaves or bracts (sometimes ochreiform), solitary or whorled, racemed spiked panicled or cymose, sometimes capitate, either naked, or enclosed singly or collectively in a tubular or cyathiform involucre; *pedicels* filiform (sometimes 0), usually jointed, often cernuous in fruit. PERIANTH calycine or petaline, of 3–4–5–6 sepals, distinct, or coherent at the base, rarely united in a tube, usually persistent and accrescent; *sepals* or segments sometimes 3, 1-seriate; sometimes 5, imbricate; sometimes 4 or 6, 2-seriate, imbricate or sub-valvate, equal or unequal, and in the latter the outer 2 or 3 herbaceous, rarely coloured, smaller or larger than the inner, usually concave or keeled, sometimes winged or spiny; the inner 2 or 3 petaloid (rarely herbaceous), plane, or concave and folded, entire, toothed, fringed, or spiny at the margins, usually becoming membranous-scarious when ripe, with netted veins, median nerve sometimes swollen and callous. STAMENS perigynous, 1–15, usually 6–8–9, very rarely ∞, inserted on a disk or glandular ring lining the base of the perianth, opposite or rarely alternate with the sepals, usually in twos or threes before the outer, solitary before the inner; *filaments* capillary or subulate, distinct, or very shortly coherent by their dilated bases; *anthers* 2-celled, dehiscence longitudinal, ovoid or oblong, dorsifixed and versatile, or rarely basifixed and erect, all introrse, or the 5 outer introrse and 3 inner extrorse, or all opening laterally. OVARY solitary, of 2–4 carpels valvately united, free or rarely adherent by the base, ovoid or elliptic, compressed or 3-gonous, 1-celled or rarely incompletely 3-celled by false septa; *styles* 2–4, answering to the angles of the ovary, distinct, or more or less united, very rarely adherent to the angles of the ovary; *stigmas* simple, capitate or discoid, sometimes feathery or penicillate; *ovule* solitary, basilar, orthotropous, erect, rarely pendulous from a basal funicle, with the micropyle downwards, but always erect when ripe. ACHENE or CARYOPSIS compressed-lenticular, 3–4-gonous, with projecting or obtuse angles, sometimes winged, entire, toothed, or spiny, rarely naked, usually covered by the accrescent and sometimes fleshy perianth. SEED conformable to the cell, erect, free or adnate to the endocarp; *testa* membranous; *hilum* basilar, large; *albumen* copious, farinaceous, rarely sub-fleshy and scanty. EMBRYO antitropous, curved and appressed to the side of the albumen, or straight and axile in the albumen; *cotyledons* linear or oval, incumbent or accumbent, sometimes largely foliaceous, flexuous; *radicle* superior.

TRIBE I. *ERIOGONEÆ.*

Flowers ☿, or rarely polygamous, within a 1- or more-flowered tubular involucre. Calyx 6-partite. Stamens 9. Ovary free; ovule basilar, erect. Embryo included in a scanty albumen.—Ochreas obsolete.

PRINCIPAL GENERA.

Eriogonum.	Oxytheca.	Nemacaulis.	Chorizantha.	Pterostegia.

TRIBE II. *POLYGONEÆ VERÆ.*

Flowers ☿ or polygamous; involucre 0. Stamens 1·9, usually 6 or 8, rarely 12–17. Ovary free or rarely adherent below; ovule basilar, erect.—Stipules ochreate.

PRINCIPAL GENERA.

Calligonum.	Emex.	Muhlenbeckia.	* Rheum.	* Rumex.
* Coccoloba.	Fagopyrum.	Oxyria.	* Atraphaxis.	Triplaris.
Oxygonum.	* Polygonum.	Ruprechtia.		

TRIBE III. *BRUNNICHIEÆ.*

Flowers ☿; involucre 0. Calyx 5-partite. Stamens 8. Ovary free, 3-gonous; ovule pendulous from a basal funicle, erect when ripe.—Ochreas 0 or obsolete. Stems usually woody, climbing, furnished with tendrils.

GENERA.

Antigonum.	Brunnichia.,

TRIBE IV. *SYMMERIEÆ.*

Flowers diœcious, polyandrous. Calyx of ♀ 6-partite. Ovary adherent.—Ochreas 0.

PRINCIPAL GENUS.

Symmeria.

Polygoneæ are clearly separated from all the *Cyclospermeæ* with farinaceous albumen by their orthotropous ovule and antitropous embryo. They are further distinguished from *Chenopodeæ* and *Amarantaceæ* by their ternary floral whorls, the often coloured perianth, and the prevalence of the ochrea. The tribe of *Eriogoneæ*, the flowers of which have an involucre, approach nearer to *Phytolacceæ*, but are separated by the superior radicle. *Polygoneæ* are also connected with *Caryophylleæ* through *Paronychieæ*, and have also some affinity with *Plumbagineæ* (which see).

Polygoneæ mostly inhabit the north temperate hemisphere; they are less frequent in the tropics. They are frutescent or arborescent in Central America, and become rare south of the tropic of Capricorn. The tribe of *Eriogoneæ* is principally North American and Chilian, *Brunnichia* inhabits Carolina, and *Antigonum* Mexico. The Rhubarbs (*Rheum*) grow on the mountains of Central Asia and North India. The woody genera *Calligonum* and *Atraphaxis* (*Tragopyrum*), grow in the plains of Central Asia. *Coccoloba*, *Triplaris*, &c., are large trees of tropical America;. *Kœnigia*, a minute herb [and *Oxyria*], are sub-arctic and arctic. The numerous species of *Polygonum* and *Rumex* are scattered everywhere, from the sea-shore to the snow-line. [*Muhlenbeckia* is Australian, *Symmeria* and *Ruprechtia* Brazilian.]

The properties of the genera of *Polygoneæ* sustain their affinities. Their herbage contains oxalic, citric, and malic acids, and is edible or medicinal. The seeds of some abound in starch, the root of most contains astringent matters, sometimes combined with a resinous principle to which they owe medicinal properties that have been recognized from the highest antiquity; the most important of these is Rhubarb, which is distinguished from all other purgatives by its restorative action on the functions of the stomach; it is further an antidysenteric and vermifuge. The botanical history of the Rhubarb [1] is obscure; in the tenth century the Arabs received it from the Chinese, and spread it through Europe, but the Chinese only vaguely indicated its habitat, and botanists have long doubted to what species it belongs. The root of *Rheum australe*, a Himalayan species cultivated at Calcutta, has the decided smell and bittertonic taste of the roots the Chinese sell to the Russians, and like it it grates under the teeth; added to which, the form of its leaves agrees with the description which the natives of Bokhara gave of the true Rhubarb to the illustrious naturalist Pallas. *Rheum Rhaponticum*, the species originally known to the ancients as *Rha*, grows wild in ancient Thrace and on the shores of the Euxine; it was called later *Rha-ponticum*, to distinguish it from the Scythian Rhubarb, which they called *Rha-barbarum*, whence our name of Rhubarb. In Europe, and especially in Germany and England, several varieties of *R. Rhaponticum* and *R. undulatum* are cultivated on account of the pleasant acid taste of their leaves, the petiole and principal nerves of which are used in tarts and preserves. [The root of the former furnishes the English medicinal Rhubarb, and is extensively grown, both as a substitute for the Oriental, and also to adulterate it.—Ed.]

Rumex is divided into two distinct groups: the one the Sorrels (*R. Acetosa, scutatus*, &c.), containing oxalate of potash in the stem and leaves, whence their acid taste and their use as food and their laxative properties; their root is red and scentless. The others, the Docks (*R. patientia, crispus, aquaticus*, &c.), have yellow and scented bitter roots containing sulphur, which are used as depuratives and antiscorbutics.

Some indigenous *Polygona* (*P. Bistorta, Hydropiper, Persicaria, aviculare, amphibium*), were formerly used medicinally, but all have fallen into disuse except the Bistort (*P. Bistorta*), the twisted root of which is an astringent tonic. *P. stypticum* is in great repute in Brazil on account of the astringent properties of the herbage and root. The natives of Colombia employ a decoction of *P. tamnifolium* for hæmorrhage. *P. perfoliatum* is outwardly applied in Cochin China for tumours and skin diseases. *P. cochinchinense*, administered as a topic and a drink, is considered to be an efficacious remedy for swellings of the knee, a common and obstinate disease in Cochin China. The tuberous root of *P. multiflorum* is a reputed cordial in Japan. *P. hæmorrhoidale* contains an acrid principle, and is used by the Brazilians as a condiment, and as a topical application, or in baths, for rheumatic pains.

Buckwheat (*Fagopyrum esculentum*), or Black Wheat, is valuable for the abundant and excellent

---

[1] Within the last two years the true Rhubarb plant has been introduced into France by the Chinese missionaries in East Tibet, and named *R. officinale*, Baillon.—Ed.

farina of its seed, which is a substitute for that of the cereals. It is a native of North Asia, grows in the poorest soil, requires little care in cultivation, ripens quickly, and is now extensively grown in the most sterile countries of Europe; it is also used for feeding fowls, and bees find a copious supply of honey in its flowers. Another species of Buckwheat (*P. tataricum*) is cultivated with the preceding; it is hardier, and succeeds on high mountains, but its farina is slightly bitter. The leaves of certain Polygonums yield a dye-stuff; as *P. tinctorium*, cultivated from time immemorial in China for the extraction of a blue dyeing substance identical with indigo; its cultivation was introduced into France in 1834. *Coccoloba uvifera*, the Seaside Grape, is a West Indian and South American littoral shrub, whose inspissated juice, called American Kino and False Rhatany, is a strong astringent. *Calligonum Pallasia* is a small leafless tree, growing in the sands of South Siberia, whose cooked root yields a gum and mucilage, which the Kalmucks eat to stay their hunger; they also appease their thirst with its young shoots and acidulous fruits. [Some Polygonums (as *P. Hydropiper*) are so acrid as to blister the skin. *Rumex alpinus*, or Monk's Rhubarb, a European species, was formerly in great repute. *R. scutatus* is still much cultivated as a Sorrel. The leaves of *Oxyria reniformis* are a most grateful acid.—ED.]

CLXXX. AMARANTACEÆ.

(AMARANTI, *Jussieu.*—AMARANTOIDEÆ, *Ventenat.*—AMARANTACEÆ, *Br.*)

Herbaceous or suffruticose PLANTS, sometimes frutescent, glabrous, pubescent or woolly. STEM and BRANCHES often diffuse, cylindric or sub-angular, continuous or jointed, erect or ascending, sometimes twining (*Hablitzia*). LEAVES opposite or alternate, simple, sessile or shortly petioled, membranous or a little fleshy, usually entire; *stipules* 0. FLOWERS small, regular or sub-regular, ☿ or diclinous, sessile, solitary or in glomerules heads or spikes, the lateral ones sometimes arrested or developed into crests awns or hooked hairs; *bracts* 3, rarely 2, usually contiguous, the lowest largest, usually persistent, rarely leafy, the lateral very often keeled, concave, never leafy, scarious, deciduous with the flower. CALYX of 3–5 sepals, or very rarely 1 (*Mengea*), distinct or sometimes more or less coherent at the base, equal or sub-equal, sub-scarious, glabrous or furnished with accrescent wool, petaloid or greenish, persistent, æstivation imbricate. COROLLA 0. STAMENS hypogynous, 5 fertile, opposite to the sepals (rarely 3 or fewer), with or without alternating staminodes, all free, or united below in a cup or tube; *filaments* filiform, subulate or dilated, sometimes 3-fid; *staminodes* entire or fringed, flat or rarely concave, sometimes very small and tooth-shaped, or lobulate; *anthers* introrse, 1–2-celled, erect, ovoid or linear, dorsifixed, dehiscence longitudinal. OVARY free, compressed, rarely depressed, 1-carpelled, 1-celled; *style* terminal, simple, various in length, sometimes obsolete; *stigma* capitate, emarginate, 2-lobed or 2–3-fid; *ovules* 1 or more, curved, basal, or suspended singly from separate erect funicles; *micropyle* inferior. FRUIT usually enveloped in the calyx, sometimes a membranous 2- or more-seeded utricle, or rupturing irregularly or circumsciss, or a caryopsis, rarely a berry. SEEDS usually somewhat compressed, reniform, vertical; *testa* crustaceous,

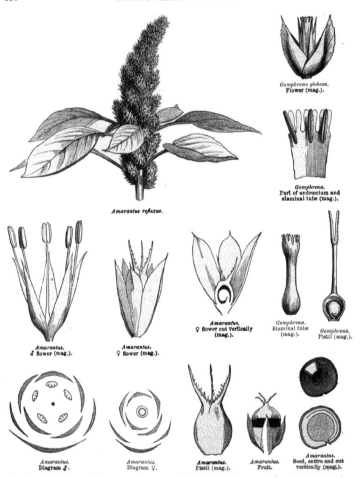

Gomphrena globosa.
Flower (mag.).

Gomphrena.
Part of andrœcium and
staminal tube (mag.).

Amarantus reflexus.

Amarantus.
♂ flower (mag.).

Amarantus.
♀ flower (mag.).

Amarantus.
♀ flower cut vertically
(mag.).

Gomphrena.
Staminal tube
(mag.).

Gomphrena.
Pistil (mag.).

Amarantus.
Diagram ♂.

Amarantus.
Diagram ♀.

Amarantus.
Pistil (mag.).

Amarantus.
Fruit.

Amarantus.
Seed, entire and cut
vertically (mag.).

black, shining; *endopleura* membranous; *hilum* naked, or rarely arillate; *albumen* abundant, central, farinaceous. EMBRYO peripheric, annular or curved; *cotyledons* incumbent; *radicle* near the hilum, inferior, sub-ascending.

TRIBE I. *CELOSIEÆ.*

Anthers 2-celled.   Ovary many-ovuled.

PRINCIPAL GENERA.

Deeringia.   * Celosia.   Hermbstædtia.

TRIBE II. *ACHYRANTHEÆ.*

Anthers 2-celled.   Ovary 1-ovuled.

PRINCIPAL GENERA.

Chamisson.	* Amarantus.	Euxolus.	Psilotrichum.
Digera.	Aerva.	Achyranthes.	Cyathula.
Trichinium.	Pupalia.	Polycnemum,	

TRIBE III. *GOMPHRENEÆ.*

Anthers 1-celled.   Ovary 1-ovuled.

PRINCIPAL GENERA.

* Iresine.   Alternanthera.   Telanthera.   * Gomphrena.   Frœlichia.

Amarantaceæ in their embryo and farinaceous albumen are near *Chenopodeæ, Baselleæ, Phytolacceæ,* and *Paronychieæ.* Their affinity with *Chenopodeæ* is so close (the latter only differing in their distinct styles and herbaceous calyx) that it is difficult to draw a clear diagnosis between them, although they are widely separated by habit. *Baselleæ* differ in habit, perigynous stamens, usually cubical pollen, &c.; *Phytolacceæ,* in their whorled ovaries borne on a gynophore; *Paronychieæ,* in their scale-like petals, perigynous stamens, scarious stipules, &c.

Amarantaceæ are mostly tropical, but are not rare in sub-tropical regions; very few are met with in the [north] temperate zone, and they are absolutely wanting in cold countries. [Many are Australian.] Various *Amarantaceæ* contain mucilage and sugar, and are hence alimentary and emollient; some are slightly astringent, other diaphoretic and diuretic, or tonic and stimulating. *Amarantus Blitum* is eaten as a spinach in the south of Europe, as are other species in China and India, where the natives abstain from animal food. *Gomphrena globosa, Celosia argentea* and *margaritacea, Aerva lanata,* &c., are used as resolvents. The flowers of *Celosia cristata* (Cockscomb) are astringent, and prescribed in Asia for diarrhœa, menorrhagia, vomiting of blood, &c. The tuberous roots of *Gomphrena officinalis* and *macrocephala,* from Brazil, are tonic and stimulating, and are hence regarded as a panacea, and under the name of *paratudo* they are a reputed remedy for weakness of the stomach and intestines, and are especially used as a febrifuge. *Amarantus frumentaceus* and *Anordana* are cultivated in the Himalayas on account of their edible seeds.

CLXXXI. *CHENOPODEÆ.*

(ATRIPLICES, *Jussieu.*—CHENOPODEÆ, *Br.*—CHENOPODIEÆ, *Bartling.*—CHENOPO-DIACEÆ, *Lindl.*—SALSOLACEÆ, *Moquin-Tandon.*)

FLOWERS ☿ or *diclinous.* PERIANTH *herbaceous, regular,* 5-3-2-*phyllous, persistent.* STAMENS *sub-perigynous or hypogynous, equal and opposite to the sepals, or fewer.* OVARY 1-*celled,* 1-*ovuled;* OVULE *curved.* EMBRYO *annular or semi-annular, or spirally coiled.* ALBUMEN *usually farinaceous, sometimes* 0.

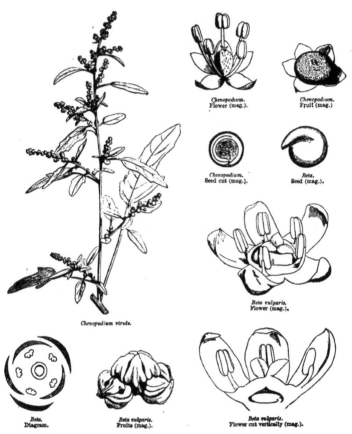

Chenopodium.
Flower (mag.).

Chenopodium.
Fruit (mag.).

Chenopodium.
Seed cut (mag.).

Beta.
Seed (mag.).

Beta vulgaris.
Flower (mag.).

Chenopodium viride.

Beta.
Diagram.

Beta vulgaris.
Fruits (mag.).

Beta vulgaris.
Flower cut vertically (mag.).

Herbaceous, suffrutescent, or rarely frutescent PLANTS, glabrous, pubescent or woody. STEMS cylindric or angular, erect or ascending, continuous and leafy, or articulate and often aphyllous. LEAVES alternate, rarely opposite, simple, sessile or petioled, usually flat, entire, toothed, sinuate or cut, sometimes fleshy, semi-cylindric or cylindric; stipules 0. FLOWERS ⚥, small, regular, often dimorphous (Atriplex); sometimes diclinous, or polygamous, sessile or pedicelled, solitary or variously agglomerated, axillary or terminal, bracteate and 2-bracteolate or naked. CALYX of 5, 3, 2 sepals (rarely 4), more or less coherent at their base, herbaceous, greenish, imbricate in æstivation, sometimes becoming fleshy and bacciform after flowering, or dry and capsular. COROLLA 0. STAMENS usually hypogynous, or at the bottom of the calyx, 5, or rarely fewer by arrest, all fertile and opposite to the sepals; *filaments* filiform or subulate, usually free, sometimes shortly united into a cup; *staminodes* (in some genera) very small, placed between the filaments (*Anabasis*, &c.); *anthers* 2-celled, ovoid or oblong, introrse, dehiscence longitudinal, connective sometimes vesicular above the cells (*Physogeton*). OVARY ovoid, or depressed-globose, usually free, very rarely adherent by the base, 1-celled; *stigmas* 2-4, distinct, or coherent at the base; *ovule* campylotropous, sessile at the bottom of the cell, or fixed laterally, or pendulous to a short funicle, micropyle facing the base, circumference or top of the ovary. FRUIT a utricle or caryopsis, or rarely a berry, included in the variously modified or unchanged calyx. SEED horizontal or vertical, erect or inverted, lenticular or reniform, integument double, of a crustaceous testa and membranous endopleura, or simple and membranous; *albumen* copious or scanty or 0 (*Anabasis*, *Salsola*, &c.), usually farinaceous, very rarely sub-fleshy. EMBRYO curved or annular, and surrounding the albumen, or coiled into a flat spiral, and dividing the albumen into 2 parts, or exalbuminous, and forming a conical spiral (*Salsola*); *radicle* facing the hilum; *cotyledons* plano-convex, slender.

[The following is Moquin-Tandon's classification in De Candolle's 'Prodromus':—

Sub-order I. CYCLOLOBEÆ.—Embryo annular or horse shoe-shaped, surrounding the albumen.

TRIBE I. CHENOPODEÆ. — Inflorescence normal. Flowers usually ⚥, homomorphous. Seed usually free, integument double.—Stem not jointed. Leaves membranous, more or less rhomboid. *Rhagodia*, *Beta, *Chenopodium, *Blitum, &c.

TRIBE II. SPINACIEÆ.—Inflorescence normal. Flowers ♂ ♀, dimorphous. Seed free or adnate to the pericarp.—Stem not jointed. Leaves membranous, more or less hastate. *Atriplex, Obione, *Spinacia, Eurotia, &c.

TRIBE III. CAMPHOROSMEÆ.—Inflorescence normal. Flowers ⚥ or ♂ ♀, homomorphous. Seed free, integument single.—Stem not jointed. Leaves fleshy or membranous, terete or linear. *Anisacantha*, *Sclerolæna*, *Camphorosma*, *Kochia, *Echinopsilum*, &c.

TRIBE IV. CORISPERMEÆ. — Inflorescence normal. Flowers ⚥, homomorphous. Seed usually adnate to the pericarp.—Stem not jointed. Leaves linear, coriaceous. *Agriophyllum*, *Corispermum*, &c.

TRIBE V. SALICORNIEÆ.—Inflorescence anomalous. Flower ⚥, homomorphous.—Stem jointed. Leaves succulent, scalelike or 0. *Salicornia*, *Halocnemum*, *Arthrocnemum*, &c.

Sub-order II. SPIROLOBEÆ.—Embryo spiral. Albumen scanty or 0.

TRIBE VI. SUÆDEÆ.—Flowers bracteolate. Seed with a double integument. Embryo in a flat spiral. *Schanginia*, *Suæda*, *Chenopodina*, *Schoberia*, &c.

TRIBE VII. SALSOLEÆ.—Flowers bracteate. Seed with a single integument. Embryo in a conical spiral.—Stem jointed or not. Leaves succulent. *Caroxylon*, *Salsola, *Halimocnemis*, *Halogeton*, *Noœa*, *Anabasis*, &c.—ED.]

Chenopodeæ approach *Baselleæ, Amarantaceæ, Phytolacceæ*, and *Tetragonieæ* in various characters, the most obvious of which is the curved embryo surrounding a more or less abundant farinaceous albumen. Most *Chenopodeæ* inhabit the shores of oceans and salt lakes, and deserts formerly covered by the sea; they are principally met with in the Mediteranean region and Asiatic Russia. Other species seem to prefer the neighbourhood of man, being abundant amongst ruins, along roads and in cultivated places where the soil is impregnated with azotized matters; presenting numerous varieties. They are generally rare in the tropics, where they are replaced by *Amarantaceæ*, and rarer still in the southern hemisphere; Australia, however, possesses several species remarkable for the singularity of their structure. Some follow the footsteps of man (*Chenopodium hybridum, leiospermum*, &c.).

Amongst *Chenopodeæ* some species contain mucilage, starch, or sugar, which renders them edible; others are medicinal; some furnish carbonate of soda by burning. Spinach (*Spinacia oleracea*), a potherb unknown to the ancients, was introduced into Spain by the Arabs, and thence spread over Europe; it is also cultivated in India. The leaves of the Orach (*Atriplex hortensis*), commonly called Bonne Dame, are edible and refreshing; the seeds possess emetic and purgative properties. *Chenopodium album, viride, ficifolium*, and *Blitum rubrum* and *Bonus-Henricus* are also used as Spinach. Beet, the origin of which is uncertain, has been cultivated for ages in kitchen gardens and fields. The leaves of the White Beet or Mangold Wurzel (*Beta Cycla*) are edible when young, and used in soup for their laxative qualities. Beet-root (*Beta Rapa*) has fleshy yellow red or white roots containing an abundance of crystallizable sugar, similar to that of sugar-cane, which is the object of a large European commerce. This plant is also cultivated as forage for cattle [and is also largely eaten].

The starchy seeds of *Chenopodium* can in times of scarcity be mixed with those of cereals. *C. Quinoa* is an annual, the seeds of which, made into broth, serve as food to the Peruvians. Other *Chenopodieæ* possess aromatic properties which act powerfully on the nervous system; *C. Botrys* is a herb growing in middle and southern Europe, the scent of which is pleasant; it is used as a bechic. *Ambrina ambrosioides*, Mexican Tea, is now cultivated in all gardens; its odour is sweet and penetrating, and an infusion of it is a useful stomachic. The seeds of *Chenopodium anthelminthicum*, a South American plant, are an excellent vermifuge. *Camphorosma monspeliaca* smells strongly of camphor; a tea is prepared from it which is praised as a diuretic, cephalic, and antispasmodic. *C. fœtidum* is an indigenous plant with branching and spreading stems, common in uncultivated places on calcareous soil; all parts of it exhale a smell of rotten fish, due to the carbonate of ammonia which it contains. It is used in injections and fomentations, as an antispasmodic, emmenagogue and antihysteric.

Salsola, Suæda, and *Salicornia* grow abundantly on seashores and in salt soil; they contain a large quantity of alkaline salts, which are obtained by incineration, and which are converted by means of charcoal and lime into carbonate of soda. The young shoots of *Salicornia*, gathered on the sea-shore, are eaten as Purslane by the inhabitants of the maritime provinces of Holland [and are pickled as Samphire in parts of England].

CLXXXII. *BASELLEÆ*, Ad. Brongniart.

(BASELLACEÆ, *Moquin-Tandon.*—ATRIPLICUM *genera, Jussieu.*)

Herbaceous rarely suffruticose glabrous PLANTS. STEMS often climbing or twining, sub-angular, sparingly leafy. LEAVES alternate, or very rarely opposite, petioled, simple, entire or sub-sinuate, fleshy, or rarely sub-coriaceous, exstipulate. FLOWERS ☿, regular, small, solitary, or in axillary spikes furnished with often winged or keeled bracts. CALYX often coloured, persistent, 5-fid, -partite or -phyllous, æstivation imbricate. COROLLA 0. STAMENS perigynous, opposite to the sepals; *filaments* subulate, dilated at the base; *anthers* 2-celled, introrse, dorsifixed, dehiscence longitudinal. OVARY free, 1-celled; *style* terminal, simple; *stigmas* 3, sub-divaricate, sometimes solitary, 3-lobed (*Tandonia*), rarely simple (*Ullucus*); *ovule*

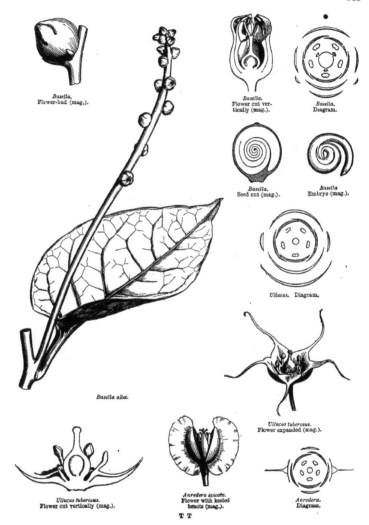

Basella,
Flower-bud (mag.).

Basella.
Flower cut ver-
tically (mag.).

Basella.
Diagram.

Basella.
Seed cut (mag.).

Basella
Embryo (mag.).

Ullucus. Diagram.

Basella alba.

Ullucus tuberosus.
Flower expanded (mag.).

Ullucus tuberosus.
Flower cut vertically (mag.).

Anredera spicata.
Flower with keeled
bracts (mag.).

Anredera,
Diagram.

T T

basal, solitary, curved, micropyle downwards. FRUIT indehiscent, enveloped in the
dry or succulent (*Basella*) calyx. SEED ovoid or sub-globose, *testa* membranous,
albumen farinaceous. EMBRYO either coiled into a flat spiral between two layers of
albumen, or annular or horseshoe-shaped and surrounding a copious albumen;
cotyledons plano-convex, sometimes sub-foliaceous; *radicle* descending.

PRINCIPAL GENERA.

[* *Embryo spiral, albumen excentric.*]
*Basella.   *Ullucus.

[** *Embryo annular, albumen central.*]
Tandonia.   Boussingaultia.   *Anredera.

Basellæ are intermediate between *Chenopodieæ* and *Amarantaceæ*; they somewhat resemble *Portulaceæ* and certain *Polygoneæ* in their climbing stem, but they are separated from all these families by their
habit, their stem twining to the right, their persistent and often winged bracts, sagittate anthers, filaments
dilated below, &c.

Basellæ are natives of America and tropical Asia. Some species of *Basella*, which are eaten in
China and India, have been introduced into Europe, and there cultivated as potherbs under the name of
Red Spinach (*B. rubra*), and White Spinach (*B. alba*); the berries of the first yield a fine red but fugitive
colour.

The starchy root of *Ullucus tuberosus*, which is used for food in Peru, has been introduced into France,
and recommended as a substitute for the Potato.

CLXXXIII. *PARONYCHIEÆ*, Saint-Hilaire.

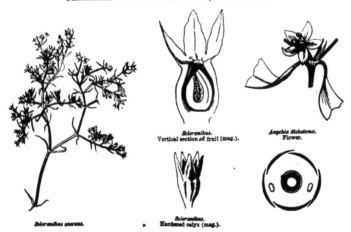

Scleranthus.
Vertical section of fruit (mag.).

Anychia dichotoma.
Flower.

Scleranthus annuus.

Scleranthus.
Hardened calyx (mag.).

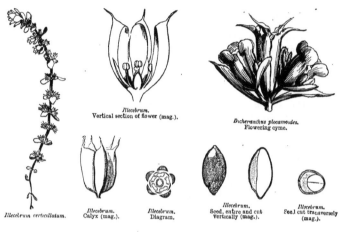

Illecebrum.
Vertical section of flower (mag.).

Dicheranthus plocamoides.
Flowering cyme.

Illecebrum verticillatum.

Illecebrum.
Calyx (mag.).

Illecebrum.
Diagram.

Illecebrum.
Seed, entire and cut vertically (mag.).

Illecebrum.
Seed cut transversely (mag.).

ALLIED TRIBE. *TELEPHIEÆ.*

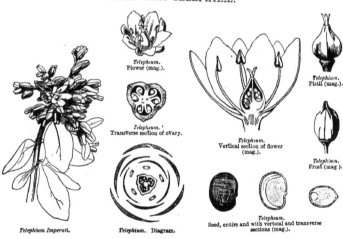

Telephium.
Flower (mag.).

Telephium.
Transverse section of ovary.

Telephium.
Vertical section of flower (mag.).

Telephium.
Pistil (mag.).

Telephium.
Fruit (mag.).

Telephium Imperati.

Telephium. Diagram.

Telephium.
Seed, entire and with vertical and transverse sections (mag.).

FLOWERS *small, æstivation imbricate.* CALYX *5-4-merous.* PETALS *squamiform* [*or* 0], *perigynous, alternate with the calyx-lobes, sometimes* 0. STAMENS *perigynous, equal or double the number of the sepals.* OVARY *free*, 1-*celled.* FRUIT *dry*, 1-*seeded.* ALBUMEN *farinaceous.* EMBRYO *curved or peripheric.*

Herbaceous or suffruticose PLANTS, very much branched, usually prostrate. LEAVES usually opposite, sometimes fascicled and in false whorls, sessile, small, entire; *stipules* scarious, rarely obsolete (*Scleranthus, Mniarum*). FLOWERS small, usually whitish or greenish, sessile or in axillary or terminal cymes often accompanied by arrested sometimes plumose branchlets; *bracts* stipular. CALYX of 5–4 more or less coherent sepals, imbricate in æstivation. PETALS small, squamiform, resembling staminodes, imbricate in æstivation, inserted on the calyx, alternate with its lobes, and usually of the same number, rarely 0 (*Scleranthus, Pteranthus,* &c.), or transformed into supernumerary stamens. STAMENS inserted on the calyx, opposite to its lobes, and equal or rarely fewer in number (*Pollichia, Mniarum*), or twice as many from the metamorphosis of the petals (*Scleranthus*); *filaments* distinct; *anthers* 2-celled, introrse. OVARY free, 1-celled; *style* usually 2-partite or -fid; *ovule* solitary, rarely 2 (*Pollichia*), semi-anatropous, basilar, erect, or pendulous from a basal funicle. FRUIT dry, small, usually membranous, 1-seeded. SEED with farinaceous albumen. EMBRYO cylindric, lateral, curved or annular; *cotyledons* small; *radicle* turned towards the hilum.

[The following grouping of *Paronychieæ* is that prepared for the 'Genera Plantarum':—

TRIBE I. PARONYCHIEÆ.—Flowers axillary or cymose, usually ☿, homomorphous. Embryo terete.—Leaves opposite, very rarely alternate; stipules scarious. *Paronychia, Anychia, Gymnocarpus, Herniaria, Illecebrum, Pollichia, Corrigiola,* &c.

TRIBE II. PTERANTHEÆ.—Flowers ternate on the top of a peduncle, lateral imperfect, surrounded by rigid simple or pinnatipartite bracts.—Leaves opposite; stipules scarious or 0. *Pteranthus, Cometes (Saltia), Dicheranthus.*

TRIBE III. SCLERANTHEÆ.—Flowers axillary, panicled, or 2–4 at the top of a peduncle. —Leaves opposite, connate at the base, exstipulate. *Scleranthus, Habrosia, Mniarum.*—ED.]

ALLIED TRIBE. *TELEPHIEÆ.*

Calyx, androecium, placentation, ovules and seeds as in *Paronychieæ.* Style 3-partite, or stigmas 3, recurved. Fruit a 1-seeded indehiscent utricle, enveloped in the calyx (*Corrigiola*), or a many-seeded capsule with 3 valves (*Telephium*). Leaves alternate, glaucous; stipules scarious. Flowers in racemes or terminal cymes.

PRINCIPAL GENERA.

Corrigiola.[1]   Telephium.[2]

Paronychieæ are closely allied to *Caryophylleæ* in æstivation, insertion, placentation, and albumen; they differ in habit, scale-like petals or 0, scarious stipules, &c. They are equally near *Amarantaceæ* in the 1-celled ovary, curved basilar ovule, and farinaceous albumen; but *Amarantaceæ* are more obviously apetalous, and are hypogynous, exstipulate, &c.

[1] *Corrigiola* should be retained in *Paronychieæ* proper, on account of its indehiscent fruit.—ED.

[2] *Telephium* is referable to *Molluginea* (pp. 261, 461), having obvious petals and dehiscent fruit.—ED.

Paronychieæ are also connected with *Portulaceæ* by their curved ovule and farinaceous albumen; but the latter are sufficiently distinguished by their fugacious petals, usually several-celled ovary with several-ovuled cells, habit, &c. They have some connection with *Baselleæ*, which differ especially in their twining stem, double calyx, dilated filaments, &c. Finally, they are linked with *Polygoneæ* by the 1-celled and 1-ovuled ovary and the nature of the seed, but, besides other differences, in *Polygoneæ* the ovule is orthotropous.

Paronychieæ are dispersed over the temperate regions of the northern hemisphere [often inhabiting deserts and very dry places]. Few are of use to man: the Rupture-wort (*Herniaria glabra*), an indigenous plant, living in sandy soil, was formerly esteemed as a diuretic and vulnerary; it is now fallen into disuse.

Scleranthus perennis, which grows in siliceous or granitic fields, is the food of the Polish Cochineal, which long supplied the place of the Mexican Cochineal as a red dye.

CLXXXIV. *CYNOCRAMBEÆ, Endlicher.*

An annual sub-succulent HERB. LEAVES petioled, the lower opposite, the upper alternate, entire, penni-tri-nerved; *stipules* cut, uniting the bases of the petioles. FLOWERS monœcious, sexes springing from different axils; ♂ *bracts* 2–3, sessile, ebracteate. PERIANTH of 2 antero-posterior leaflets, juxtaposed in æstivation, revolute after flowering. STAMENS 2–20, inserted at the base of the perianth leaflets; *filaments* capillary, free; *anthers* at first linear, then sagittate, versatile, 2-celled, dehiscence longitudinal. ♀ FLOWERS generally 3, rarely more (the intermediate one usually largest, the lateral sometimes imperfect), sessile, furnished with a posticous bract, 2

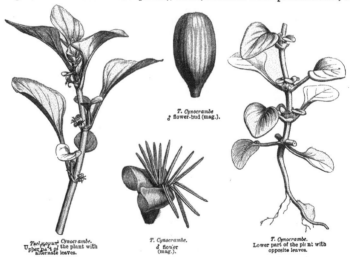

T. Cynocrambe
♂ flower-bud (mag.).

Theligonum Cynocrambe.
Upper part of the plant with alternate leaves.

T. Cynocrambe.
♂ flower
(mag.).

T. Cynocrambe.
Lower part of the plant with opposite leaves.

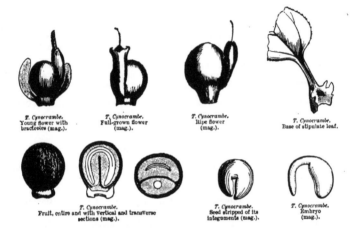

T. Cynocrambe.
Young flower with
bracteoles (mag.).

T. Cynocrambe.
Full-grown flower
(mag.).

T. Cynocrambe.
Ripe flower
(mag.).

T. Cynocrambe.
Base of stipulate leaf.

T. Cynocrambe.
Fruit, entire and with vertical and transverse
sections (mag.).

T. Cynocrambe.
Seed stripped of its
integuments (mag.).

T. Cynocrambe.
Embryo
(mag.).

anterior and 2 lateral bracteoles. PERIANTH excentric, tubular, sub-clavate, 3-lobulate, traversed by the style, becoming lateral by the enlargement of the ovary. OVARY 1-celled; *style* lateral; *stigma* clavate, undivided; *ovule* solitary, basilar, campylotropous. FRUIT a drupe. SEED horseshoe-shaped; *testa* membranous. EMBRYO hooked, in the axis of a sub-cartilaginous albumen; *cotyledons* linear, incumbent; *radicle* cylindric, inferior.

ONLY GENUS.
Thelygonum.

This genus approaches *Urticeæ* and *Cannabineæ* in diclinism, its single perianth, 1-celled 1-ovuled ovary, basilar ovule, chalaza corresponding to the hilum, fleshy albumen, and stipulate leaves; but in *Urticeæ* the ovary is free and the fruit an achene. It also recalls the apetalous *Cyclospermeæ*, and especially the *Tetragoniæ*, the ovary of which is inferior; but it differs in the number of stamens, linear anthers, basilar ovule, not farinaceous albumen, &c.

Thelygonum is a plant of the Mediterranean region; it is slightly acrid and purgative, and might be used as an inferior potherb.

CLXXXV. *MONIMIACEÆ.*

(URTICARUM *genera, Jussieu.*—MONIMIEÆ ET ATHEROSPERMEÆ, *Br.*—MONIMIACEÆ, *Endl., Tulasne.*)

FLOWERS *diclinous, apetalous.* SEPALS 4 *or more, imbricate.* STAMENS ∞ , *inserted on an open or urceolate receptacular cupule.* CARPELS ∞ , 1-*celled,* 1-*ovuled, sunk in the cupule, or seated on it;* OVULE *anatropous.* FRUIT *a drupe or nut.* SEED

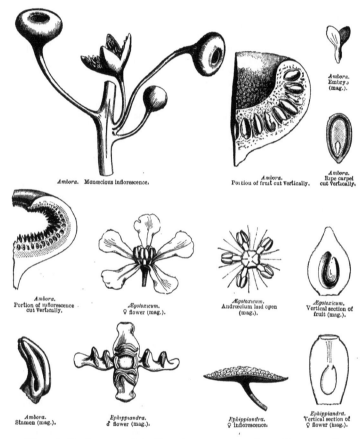

Ambora. Monœcious inflorescence.

Ambora. Portion of fruit cut vertically.

Ambora. Embryo (mag.).

Ambora. Ripe carpel cut vertically.

Ambora. Portion of inflorescence cut vertically.

Ægotoxicum. ♀ flower (mag.).

Ægotoxicum. Androecium laid open (mag.).

Ægotoxicum. Vertical section of fruit (mag.).

Ambora. Stamen (mag.).

Ephippiandra. ♂ flower (mag.).

Ephippiandra. ♀ inflorescence.

Ephippiandra. Vertical section of ♀ flower (mag.).

pendulous in the drupes, erect in the nuts; ALBUMEN copious. EMBRYO axile or basilar.—STEM woody. LEAVES opposite or whorled, exstipulate.

TREES or aromatic SHRUBS. LEAVES persistent, opposite or 3-4-nately whorled, very rarely alternate, entire or toothed, petioled, exstipulate, very often pellucid-punctate, glabrous, silky, cottony or scaly. FLOWERS apetalous, usually monœcious, very rarely ☿ (Hortonia) or polygamous (Doryphora, Atherosperma, &c.), solitary,

geminate, or in racemes cymes or panicles, bracteate; *bracteoles* caducous. RECEP-
TACULAR CUPULE discoid or urceolate, rarely capsular, usually accrescent. SEPALS
4, decussate, or 5–8–∞, many-seriate, imbricate in æstivation. STAMENS usually
indefinite, rarely 8 (*Ephippiandra*) or 5 (*Ægotoxicum*), lining the wall of the cupule
in the ♂ flowers, occupying the throat only in the ☿, free, all fertile or some reduced to staminodes; *filaments* linear and filiform, or dilated into a petaloid elongated membrane, or very short and nearly 0, naked or 2-appendiculate near the base
or towards the centre; *anthers* introrse or extrorse, usually adnate to and shorter
than the connective, cells opposite, dehiscing either by separate or confluent slits, or
transversely by 2 ascending valvules; *staminodes* situated within the fertile stamens,
and sometimes bearing half an anther. CARPELS numerous, 1-celled, free, sessile on
the surface of the receptacular cupule, rarely sunk in its thickened walls (*Ambora*);
ovule anatropous, sometimes pendulous, and then style terminal; sometimes erect, and
then style lateral or basilar. DRUPES or NUTS, superficial or sunk in the receptacular
cupule : *drupes* dry, or pulpy and odoriferous, with a pendulous or erect seed; *nuts*
with an erect seed, sometimes (*Doryphora*) terminated by a plumose style. SEED
free in the drupes, adnate to the pericarp in the nuts; *albumen* copious, fleshy,
sometimes oily. EMBRYO straight, axile in the pendulous seeds, basilar and minute
in the erect; *cotyledons* divaricate.

[The following disposition of the genera of *Monimiaceæ* is that of A. De Candolle
in the ' Prodromus ' :—

TRIBE I. SIPARUNEÆ.—Perianth fig-like, indehiscent. Ovule erect. *Siparuna, Palmeria.*

TRIBE II. TAMBOURISSEÆ.—Perianth fig-like, indehiscent. Ovule pendulous. *Tam-
bourissa.*

TRIBE III. MONIMIEÆ.—Perianth fig-like or sub-globose, at length dehiscent. Ovule
pendulous. *Monimia, Ephippiandra, Mathæa, Kibara,* &c.

TRIBE IV. HEDYCARTEÆ.—Perianth spreading. Fruit drupaceous. Ovule pendulous.
Hortonia, Hedycarya, Peumus (*Boldoa).*

TRIBE V. ATHEROSPERMEÆ.—Perianth spreading. Fruit an achene. Ovule erect.
Laurelia, Doryphora.

DOUBTFUL GENUS. *Ægotoxicon.*[1] (*Æxtoxicon.*)—ED.]

The affinities of *Monimiaceæ* have given rise to very different opinions. A. L. de Jussieu, who, when
his 'Genera Plantarum' was published, knew only *Ambora*, placed it near *Ficus*, in *Urticeæ*; when, later,
he knew *Monimia, Citrosma, Atherosperma,* &c., he united them with *Ambora* into the family of *Monimieæ*,
which he divided into two tribes, according to the drupaceous or nut-like character of the fruit, and which
he still placed near *Urticeæ.*

He had also noticed the affinity of *Monimieæ* with *Calycantheæ*, without however overlooking the
connection of the latter with *Rosaceæ* (see page 102). R. Brown separated *Monimieæ* from *Atherospermeæ*; he placed the former near *Urticeæ*, as Jussieu had done, and brought *Atherospermeæ* near
Laurineæ on account of their aromatic properties, and especially the structure of their anthers. Endlicher,
like Jussieu, placed *Atherospermeæ* and *Monimieæ* together, and like R. Brown he placed them before
Laurineæ. Tulasne, the author of a learned monograph of *Monimiaceæ*, thinks that they are nearest

---

[1] This remarkable genus, appended to *Monimiaceæ* by
some authors, to *Euphorbiaceæ* by others, and to *Ilicineæ*

by others, differs wholly from the first named in the geminate ovules, double perianth, and lepidote scales.—ED.

to *Rosaceæ*, through *Calycantheæ*, notwithstanding the alternate leaves, stipules, and absence of albumen. In most *Rosaceæ* the leaves are simple, as in *Monimiaceæ* ; their inflorescence, the numerical type of the floral envelopes, and perigynous stamens are the same. With regard to the pistil, the carpels of *Roseæ* and *Pomaceæ* are included in a receptacular cupule, like those of *Citrosma*, *Monimia*, and *Atherosperma* ; those of *Geum* and *Spiræa* are seated on a convex receptacle, as in *Mollinedia* and *Hedycarya*, and the same relation exists between *Sanguisorbeæ* and *Boldoa* with regard to the fewness and position of the carpels. Besides this, the dry or fleshy fruit and the usually solitary pendulous or erect ovule, strengthen this relationship, which, according to Tulasne, is much closer than that which exists between *Monimiaceæ* and *Laurineæ* and *Urticeæ*. He recognizes in *Laurineæ* no structural likeness except in their anthers. As to the relations with *Urticeæ* established by Jussieu principally on the genera *Ficus* and *Dorstenia*, he rejects it, on account of the great development of the stipules in the latter, the absence of aroma, the small number of stamens, the form of the perianth, the orthotropous or campylotropous ovule, and the constantly superior radicle. Endlicher was the first who sagaciously compared *Mollinedia* with the free-carpelled *Anonaceæ*.

Finally, Hooker fil. and Thomson place *Monimiaceæ* in the neighbourhood of *Myristiceæ* and of the second tribe of *Magnoliaceæ* (*Illicium*); this affinity, founded on the aromatic properties, pellucid-punctate leaves, diclinism, number of stamens, solitary anatropous ovule, albuminous seed, divaricate cotyledons, &c., appears to us the most natural.

Most *Monimiaceæ* live in south tropical, and several in south temperate latitudes. *Citrosma* and *Mollinedia* are purely tropical American. *Laurelia* inhabits Chili and New Zealand. *Boldoa*, a monotypic genus, belongs to Chili. *Hedycarya* and *Atherosperma* are dispersed over East Australia, Tasmania, and New Zealand. *Ambora* and *Monimia* grow in Madagascar, Comoro and the Mascarene Islands. *Kibara* is a native of tropical Asia. *Doryphora* is confined to East Australia.

Monimiaceæ possess a tonic and stimulating volatile oil in all their parts. The leaves of *Boldoa* are used to promote digestion, like tea and coffee ; its drupes are sugary, and its seeds contain a fixed oil [its bark is used for tanning]. The fruit of *Laurelia sempervirens* also is edible.

Atherosperma moschatum is a gigantic tree, much sought for ship-building ; a decoction of its highly aromatic bark, mixed with milk, is a substitute for tea. The drupes of *Ambora* yield a red juice analogous to arnotto, but they are only eaten by birds.

CLXXXVI. *MYRISTICEÆ.*

(MYRISTICEÆ, *Br.*—MYRISTICACEÆ, *Lindl.*)

TREES or SHRUBS, with a styptic juice reddening in contact with the air, bark of the branches very often reticulated, young shoots usually furfuraceous. LEAVES alternate, nearly distichous, shortly petioled, coriaceous, simple, entire, penninerved, folded lengthwise when young, pubescent or scaly, exstipulate. FLOWERS diœcious, usually axillary, in racemes, glomerules, heads or panicles, inconspicuous, white or yellow, often rusty-pubescent, glabrous within; *bract* usually solitary, concave. PERIANTH simple, coriaceous, tubular or urceolate or sub-campanulate, 3-2-4-fid, æstivation valvate.—FLOWERS ♂ : STAMENS 3-15, monadelphous; *filaments* united into a compact column, cylindric or turbinate or dilated into a sort of denticulate disk ; *anthers* extrorse, 2-celled, adnate to the column or to the teeth of the disk, rarely inserted on the teeth and free, cells parallel, opening longitudinally.—FLOWERS ♀ : CARPEL solitary (very rarely 2, of which one is minute and sterile), free, 1-celled ; *style* terminal, very short or 0 ; *stigma* undivided or sub-lobed ; *ovule* solitary, basal, erect, anatropous. CAPSULE fleshy, with 2 undivided or 2-fid valves. SEED erect,

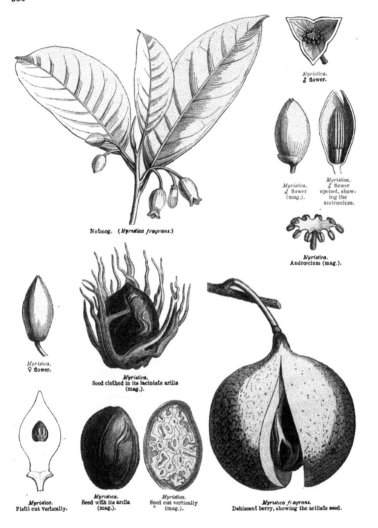

Myristica.
♂ flower.

Nutmeg. (*Myristica fragrans.*)

Myristica.
♂ flower
(mag.).

Myristica.
♂ flower
opened, show-
ing the
androecium.

Myristica.
Androecium (mag.).

Myristica.
♀ flower.

Myristica.
Seed clothed in its laciniate arilla
(mag.).

Myristica.
Pistil cut vertically.

Myristica.
Seed with its arilla
(mag.).

Myristica.
Seed cut vertically
(mag.).

Myristica fragrans.
Dehiscent berry, showing the arillate seed.

Myristica.
Ovary.

Myristica.
Ovule and arilla.

Myristica.
Ovule cut vertically.

Myristica.
Embryo.

M. fatua.
Transverse section of
andrœcium (mag.).

wholly or partially enveloped in a fleshy, laciniate, often aromatic aril, springing from the base of a short and thick funicle; *testa* hard; *albumen* scented, copious, sebaceous, deeply ruminate transversely from irregular processes of the inner membrane of the seed. EMBRYO minute, basilar, straight; *cotyledons* sub-foliaceous, divaricate, plane or rugose and folded; *radicle* short, cylindric, inferior.

GENUS.

Myristica.

Myristiceæ, long placed among *Monochlamydeæ* near *Laurineæ*, on account of their 3-partite inconspicuous diclinous and apetalous flowers, 1-celled 1-ovuled ovary, berried fruit, woody stem, coriaceous leaves, and aromatic properties, ought rather to be placed near *Anonaceæ* or *Monimieæ*, which resemble them in the 3-partite valvate perianth, extrorse anthers, solitary erect anatropous ovule, copious ruminate albumen, minute basilar embryo with [divaricate cotyledons and] inferior radicle, woody aromatic stem, and alternate nearly sub-distichous leaves folded lengthwise when young; but in *Anonaceæ* the flowers are usually hermaphrodite and petaled, the stamens are indefinite, multiseriate, and free, the carpels are more or less numerous, and the seed has no aril. *Myristiceæ* also approach *Magnoliaceæ* in the extrorse hypogynous anthers, the berried 2-partite fruit, copious albumen, basilar embryo, woody stem, and alternate leaves; they differ in the valvate æstivation of the calyx, the absence of corolla, the monadelphism, solitary carpel, solitary erect ovule, and the arillate not ruminate seed. We prefer to keep the name *aril* for the organ which envelops the Nutmeg, because, after an examination of two ovules, we have concluded that this organ springs rather from the base of the ovule than from the exostome, as is admitted by A. De Candolle and Planchon.

This little family is tropical; it inhabits the tropical eastern peninsula and islands of Asia, especially the Moluccas, but also America, Madagascar, and the Pacific Islands.

All parts of *Myristiceæ* are aromatic; their styptic juice contains a very tenacious acrid colouring matter, which reddens on contact with the air. The seed and its aril possess hydrocarbons, of which the aroma is more or less strong according to the species. The most remarkable is the Nutmeg-tree (*Myristica fragrans*), a fine tree of the Moluccas, especially cultivated in the Banda Islands, and introduced in 1770 into the Mauritius and Bourbon, whence it passed into America. Its seeds (Nutmeg), and its aril (Mace), are used as spices and as stimulating medicines. Nutmegs yield an essence by distillation, and a fixed solid oil by expression under heat, mixed with a volatile oil; this mixed oil is named Nutmeg-butter, on account of its consistence and yellow colour. The volatile oil, separated by distillation, is used in perfumery, but the aroma of the Nutmeg is narcotic, and it has been frequently stated that the exhalations from heaps of Nutmegs have proved injurious to persons sleeping near them. Nutmegs, if eaten in considerable quantities, are really poisonous; they excite thirst, cause oppression, dyspepsia, intoxication, delirium, and even fatal apoplexy.

Many other species of *Myristica* produce aromatic seeds, but less valuable than those of *M. fragrans*. Those of *M. spuria*, a native of the Philippines, are covered with a Mace which is at first yellow, and afterwards of a bright red; they lose their aroma at the end of a year. The red juice obtained by incision of the trunk is substituted in commerce for Dragon's-blood. The seed of *M. tomentosa* is considered an aphrodisiac by the inhabitants of Amboyna.

Among American species *M. bicuiba* and *officinalis* deserve to be mentioned; the aroma of their seed is faint, changed by some bitter principle; a fatty oil, and even a waxy substance, are obtained from them. The fruits of *M. Otoba*, a native of the mountains of Colombia, have a penetrating disagreeable odour; an antipsoric unguent is prepared from their whitish aril. *M. fatua* and *sebifera* inhabit the Antilles and Guiana; the aroma of their seed is very fugacious. The seed of the latter, treated with hot water, yields a grease of which candles are made; and the reddish and acrid juice which flows from an incision of the trunk is employed for ulcers and decayed teeth. The Mace of *M.* (*Pyrrosa*) *tingens*, a native of Amboyna, contains a red colouring matter, which the natives chew mixed with lime. [The pulpy exocarp of *M. fragrans* is preserved and eaten by the Dutch.]

CLXXXVII. *LAURINEÆ.*

(LAURI, *Jussieu.*—LAURINÆ, *Ventenat.*—LAURINEÆ, *D.C.*)

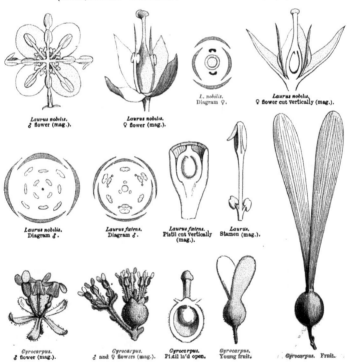

Laurus nobilis.
♂ flower (mag.).

Laurus nobilis.
♀ flower (mag.).

L. nobilis.
Diagram ♀.

Laurus nobilis,
♀ flower cut vertically (mag.).

Laurus nobilis.
Diagram ♂.

Laurus foetens.
Diagram ♂.

Laurus foetens.
Pistil cut vertically (mag.).

Laurus.
Stamen (mag.).

Gyrocarpus.
♂ flower (mag.).

Gyrocarpus.
♂ and ♀ flowers (mag.).

Gyrocarpus.
Pistil laid open.

Gyrocarpus.
Young fruit.

Gyrocarpus. Fruit.

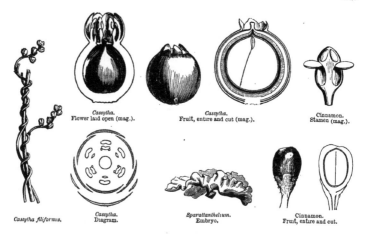

Cassytha.
Flower laid open (mag.).

Cassytha.
Fruit, entire and cut (mag.).

Cinnamon.
Stamen (mag.).

Cassytha filiformis.

Cassytha.
Diagram.

Sparattanthelium.
Embryo.

Cinnamon.
Fruit, entire and cut.

FLOWERS ☿ or *polygamous.* PERIANTH *simple, monosepalous,* 6–4–9-*fid.* STAMENS *perigynous, as many as the perianth-lobes, or a multiple, alternately fertile and sterile;* ANTHERS 2-*celled or* 4-*locellate, opening from below upwards by valvules.* OVARY 1-*celled;* OVULE *pendulous, anatropous.* FRUIT *a drupe or berry.* SEED *solitary, inverse, exalbuminous.* EMBRYO *straight.* RADICLE *superior.*

TREES or SHRUBS, rarely UNDERSHRUBS, aromatic, sometimes fœtid, very rarely aphyllous parasitic twining HERBS (*Cassytha*). LEAVES alternate, sometimes close together, sub-opposite or sub-whorled, simple, entire, penninerved, or palminerved, finely reticulate, coriaceous and persistent, or rarely soft and deciduous, punctate with glands filled with a volatile oil, and often pellucid; *stipules* 0. CYMES 3–∞ - (sometimes 1-) flowered, racemed or panicled or umbelled, capitate, bracteate at the base, rarely with a 4-6-phyllous or scaly-imbricate involucre. FLOWERS ☿ or diclinous, regular, small, white or yellow or green, usually scented. PERIANTH simple (CALYX), monosepalous, herbaceous or petaloid, rarely coriaceous, sometimes fleshy, or hardened when ripe, rotate, infundibuliform or urceolate, 6- (rarely 4- very rarely 9-)fid; *lobes* 2-seriate, imbricate, equal, or the outer smaller; *tube* usually persistent, often more or less accrescent, and changed into a cup surrounding the base of the fruit. STAMENS inserted at the base or throat of the calyx, definite, very rarely sub-indefinite (*Tetranthera*), 3-4-seriate, reduced in ♀ to glands, scales, or staminodes; ☿ and ♂ flowers with the outer stamens introrse, 4–6 or 9 usually fertile, eglandular at the base; the inner extrorse, 2-3 usually fertile and 2-glandular at the base; *filaments* free, or very rarely monadelphous, usually dilated at the top; *anthers* ovoid or oblong, sometimes acuminate from the connective being prolonged beyond the cells; cells 2, parallel, often transversely 2-locellate, and opening from the base upwards by a longitudinal persistent valvule. OVARY free (or very rarely adherent), 1-celled; *style* simple, rather stout, short; *stigma* obtuse, sub-capitate or discoid, obscurely 2–3-lobed; *ovule* solitary (very rarely 2), pendulous from the top of the cell, anatropous. FRUIT a berry, rarely a drupe, or dry, globose or ellipsoid, usually seated on the thickened pedicel, or girt at its base by or included in the calyx-tube. SEED inverted, exalbuminous; *testa* membranous. EMBRYO straight; *cotyledons* large, plano-convex (coiled in *Gyrocarpeæ*), peltate near the base, fleshy, oily; *radicle* very short, superior.

[The following sub-orders and tribes are those of Meissner in De Candolle's ‘Prodromus:’—

Sub-order I. LAURINEÆ VERÆ. Frutescent or arborescent leafy plants. Fruit superior, or very rarely inferior (*Agathophyllum*). Cotyledons plano-convex. Anthers 2-celled or 4-locellate.

TRIBE I. PERSEACEÆ.—Flowers usually ☿, 3-merous. Calyx 6-lobed; stamens 9, 3 inner extrorse, 2-glandular; staminodes 3, stipitate, rarely 0.—Leaves evergreen; buds usually naked. *Cinnamomum, Alseodaphne, Phœbe, Machilus, *Persea, Haasia, Beilschmiedia, Boldoa, Nesodaphne,* &c.

TRIBE II. CRYPTOCARYÆ.—Flowers usually 3- rarely 4-merous. Stamens 9, 6, 3, rarely 4; staminodes frequently obsolete. Fruit enclosed in the calyx-tube.—Leaves evergreen, buds not scaly. *Cryptocarya, Aiouea, Acrodiclidium, Aydendron, Mespilodaphne,* &c.

TRIBE III. OREODAPHNEÆ.—Flowers usually diœcious, 3-merous. Stamens 9, rarely 3, 3 innermost 2-glandular; staminodes 0 or obsolete. Berry naked or girt at the base by the calyx-tube. *Oreodaphne, Nectandra, *Sassafras, Gœppertia,* &c.

TRIBE IV. LITSÆACEÆ. — Flowers usually diœcious, 6- rarely 4-merous. Anthers all introrse; staminodes 0 or obsolete. *Tetranthera, Cylicodaphne, Actinodaphne, Litsæa, Daphnidium, *Laurus, Aperula, Lindera,* &c.

Sub-order II. CASSYTHEÆ. Parasitic herbs with filiform twining stem, adhering by suckers to living plants. Leaves replaced by scales. Ovary included in the calyx-tube. Anthers 2-celled. *Cassytha.*

Sub-order III. GYROCARPEÆ.[1]—Frutescent or arborescent erect or climbing leafy plants. Fruit inferior. Cotyledons spirally coiled around the gemmule. Anthers 2-celled. *Gyrocarpus, Sparattanthelium, Illigera.*—ED.]

Laurineæ form a very natural family, which, in its woody stem, its usually coriaceous evergreen and exstipulate leaves, its often unisexual flowers, simple perianth, perigynous stamens, 1-celled and 1-ovuled ovary, presents an affinity with *Atherospermeæ* and *Thymeleæ*; the latter also resemble it in the pendulous ovule, the exalbuminous seed, and the fleshy cotyledons, but differ in the dehiscence of the anthers. *Atherospermeæ* have, like *Laurineæ*, anthers opening by ascending valvules, but the pistil is polycarpellary, the ovule is erect, and the embryo minute at the base of a fleshy albumen.

Laurineæ grow especially in tropical regions, where they form forests on cool mountains [and in hot valleys, plains, &c.], but a few are found in North America [South Africa, Australia], the Canaries, and Mediterranean Europe; they are absent from North Asia [except Japan and China] and the tropical African continent. *Cassytha* is met with in South Africa [India], and all hot regions of the southern hemisphere.

Laurineæ secrete a pungent volatile oil in the bark and glands of the leaves and flowers. This oil

[1] See *Combretaceæ*, p. 420.—ED.

possesses, in different species, stimulating or sedative properties, represented in their maximum intensity, the one by Cinnamon, the other by Camphor, so that the specific virtues of the family may be considered as included in these two types.   We shall only here quote the most widely-spread species.

The Laurel (*Laurus nobilis*), a tree of South Europe, is cultivated in France, but does not attain a great size.   Its leaves have a pleasant scent and an acrid and aromatic taste; they are used as a flavouring.   Their chromule and their essence, soluble in fatty bodies, are ingredients in several ointments and other external medicaments, as are also the berries, which contain a fixed and a volatile oil.   Its leaves are invariably laid in layers over Smyrna figs as imported.

. *Sassafras officinalis* is a large forest tree, inhabiting the edges of streams from Canada to Florida; the aroma of its wood and of the bark of its root resembles that of Fennel and Camphor; both the wood and root-bark are employed as sudorifics.

Ocotea (*Puchury*) *major*, a Brazilian tree, yields the Pichurim bean, a seed containing a volatile acid, and a butyraceous oil whose taste and smell are between those of Nutmeg and Sassafras.   The Brazilians use it largely in cases of weakness of the bowels.

Persea gratissima, the Avocado or Alligator Pear, a large South American tree, has no aromatic principle, and is only useful for the thick and butyraceous flesh of its berry, which tastes like pistachio, and is eaten with spices and meat, and animals of all kinds feed upon it.   [It further yields an abundant oil for illuminating, soap-making, and its seeds a black dye used for marking linen.]

Cinnamomum officinale, which yields Cinnamon bark, is cultivated in Ceylon and tropical colonies; this bark is pale brown and sweet-scented, hot, aromatic and sugary; it is used as a condiment, tonic, and stimulant.   ·

The Chinese Cinnamon (*C. Cassia*) grows in Malabar, China, and the Moluccas; its bark [as exported] is thicker than that of the Ceylon Cinnamon, and not rolled, and its colour is darker; its taste is hot and pungent, and its scent recalls that of bugs; it is therefore less valued.   [This, *C. malabathricum*, and allied species, are used to adulterate true Cinnamon.]

Camphora officinarum grows wild in China and Japan, and is cultivated in tropical and sub-tropical colonies; its wood and leaves contain Camphor, a concrete volatile colourless oil, lighter than water, with a penetrating odour and an acrid but cooling taste; it is very soluble in fixed and volatile oils, in alcohol and ether, evaporating completely in the air, and is very inflammable.   This principle exists in several other *Laurineæ*, as also in several plants not connected with this family, and notably in *Labiatæ*.   Camphor is much used as a sedative, antispasmodic, and antiseptic; it is largely applied externally, dissolved in alcohol, oil, and vinegar; taken in overdoses it may produce complete insensibility, and even death.

[*Biri biri* or *bebeeru*, of Guiana, an alkaloid procured from *Nectandra Rodiæi* of Guiana, is a most powerful medicine, largely used as a febrifuge, and supposed to be the principal ingredient in Warburg's Drops, a medicine in use throughout British India.   Various species of Sassafras are used as stimulants, tonics, and vermifuges.]

The woods of most *Laurineæ* are of a fine and solid tissue, and are peculiarly useful to cabinet-makers and turners.   Those most used in France are the Anise or Sassafras of Orinoco (*Ocotea cymbarum*); the Bebeeru (*Nectandra Rodiæi*), a hard heavy greenish-yellow wood from Guiana and the famous Greenheart of Demerara; and the Licari (Clove Cassia of Brazil, or Rose of Cayenne, *Licaria guyanensis*[1]), which the French workmen call Pepper-wood, from the pungency of its dust.   [To these must be added the Vinatico or Madeira Mahogany, the produce of *Persea Indica*; the fœtid Til of the Canaries (*Oreodaphne fœtens*); the Sweetwood of Jamaica (*Oreodaphne exalbata*); and the Stinkwood of South Africa (*O. bullata*).]

---

[1] *Dicypellium caryophyllatum*, Nees, according to Steudel.—ED.

CLXXXVIII. *THYMELEÆ.*

(THYMELEÆ, *Adanson.*—DAPHNOIDEÆ, *Ventenat.*—DAPHNACEÆ, *Meyer.*—
AQUILARINEÆ, *Br.*—AQUILARIACEÆ, *Lindl.*)

FLOWERS ☿ or *polygamous.* PERIANTH *simple,* 4–5-*fid.* STAMENS ·*equal to the
perianth-lobes, fewer, or double the number, inserted on the tube or throat of the perianth.*

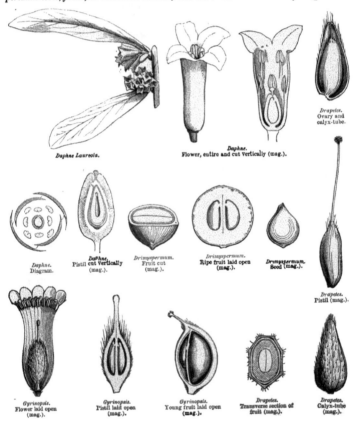

Daphne Laureola.

Daphne.
Flower, entire and cut vertically (mag.).

Drapetes.
Ovary and
calyx-tube.

Daphne.
Diagram.

Daphne.
Pistil cut vertically
(mag.).

Drimyspermum.
Fruit cut
(mag.).

Drimyspermum.
Ripe fruit laid open
(mag.).

Drimyspermum.
Seed (mag.).

Drapetes.
Pistil (mag.).

Gyrinopsis.
Flower laid open
(mag.).

Gyrinopsis.
Pistil laid open
(mag.).

Gyrinopsis.
Young fruit laid open
(mag.).

Drapetes.
Transverse section of
fruit (mag.).

Drapetes.
Calyx-tube
(mag.).

OVARY *free*, 1-2-*celled*; OVULE *pendulous, anatropous.* SEED *exalbuminous or nearly so.* EMBRYO *straight*; RADICLE *superior.*

SHRUBS or SMALL TREES, rarely annual HERBS, bark furnished with a fibrous and tenacious liber, juice acrid and caustic. LEAVES scattered or opposite, simple, entire, usually coriaceous and shining, jointed at the base, penninerved, exstipulate. FLOWERS ☿ or polygamous, axillary or terminal, solitary or in a spike, raceme, umbel, fascicle or head, sometimes involucrate, purple, white, yellow or greenish, very rarely blue, often handsome and scented, usually with silky pubescence. PERIANTH simple (CALYX), petaloid or rarely herbaceous, monosepalous, tubular, in-fundibuliform or urceolate, deciduous or marcescent; *tube* sometimes jointed above the ovary, or above its base, which persists when ripe; *limb* of 4–5 equal lobes, or the outer a little the largest, imbricate in æstivation; *throat* naked or furnished with scales, perigynous glands, or short filaments equal and alternate with the perianth-lobes, or multiple in number; *hypogynous scales* 4–8, small, membranous, fleshy or filiform, distinct or united into a cup or ring, sometimes 0. STAMENS equal and alternate with the perianth-lobes (*Drapetes, Struthiola*), or reduced to 2, or twice as many and 2-seriate, inserted on the tube or throat of the perianth, the upper series opposite to the perianth-lobes, the lower alternate; *filaments* filiform or flattened; *anthers* 2-celled, ovoid oblong or linear, basi- or dorsi-fixed, or adnate, dehiscence longitudinal, introrse, connective sometimes apiculate. OVARY free, 1 (rarely 2-) celled; *style* simple, sub-terminal or ventral, sometimes very short, or 0; *stigma* capitate or sub-clavate, papillose or hispidulous, rarely smooth; *ovules* anatropous, solitary, pendulous near the top of the cell; rarely 2–3, collateral or superimposed. FRUIT indehiscent, a nut drupe or berry, very rarely a capsule with 2 semi-placen-tiferous valves (*Aquilaria, Gyrinops*). SEEDS pendulous, *raphe* lateral, *testa* thin or crustaceous; *albumen* scanty and fleshy, or 0. EMBRYO straight; *cotyledons* plano-convex; *radicle* short, superior.

[In the original this order is simply divided into two tribes, *Daphneæ* and *Aquilarineæ*; the following is the more detailed arrangement of Meissner in De Candolle's 'Prodromus':—

Sub-order I.—THYMELEÆ. Ovary 1-celled, 1- (or very rarely 2-3-?) ovuled. Ovule appended near the top of the cell. Stem woody or herbaceous.

TRIBE I. DAPHNEÆ.—Scales or perigynous glands absent. *Pimelea, Drapetes, Canajera, Daphnopsis, *Lagetta, *Dais, *Daphne, *Wikstrœmia, Stellera, Thymelæa, *Passerina.*

TRIBE II. GNIDIEÆ. Scales or perigynous glands present in the tube or throat of the perianth. *Struthiola, Lachnœa, *Gnidia, *Lasiosiphon*, &c.

Sub-order II.—AQUILARIEÆ. Ovary 2-celled, cells 1-ovuled, or 1-celled with 2-3 1-ovuled parietal placentas.

TRIBE III. GYRINOPIDEÆ.—Perigynous scales various. Flowers 5-merous. *Aquilaria, Gyrinops*, &c.

TRIBE IV. DRIMYSPERMEÆ. — Perigynous scales absent. Flowers usually 4-merous. *Pseudais, Drimyspermum.*—ED.]

Thymeleæ present an affinity with *Santalaceæ, Elæagneæ,* and *Proteaceæ*. *Santalaceæ* are easily distinguished by the valvate perianth, the inferior ovary, the ovules reduced to a nucleus, pendulous from a free central placenta, and the abundant albumen; *Elæagneæ* by the erect basilar ovule, the often spinescent branches, and the peltate scales; *Proteaceæ* differ from *Thymeleæ* in the valvate æstivation and inferior radicle. *Thymeleæ* are scarcely distinguishable from *Rosaceæ*, except in habit, their often opposite and exstipulate leaves, the tenacious liber, and acrid vesicant juice. As in *Rosaceæ*, the flower is coloured, and if the petals are absent they are represented by scales accompanying the calyx, the stamens are perigynous, the ovary is free, the ovule pendulous and anatropous, the embryo straight, exalbuminous, &c. In a word, a flower of *Dais* or *Daphne* exactly resembles that of some *Amygdaleæ*.

Thymeleæ mostly inhabit the warm extra-tropical regions of the southern hemisphere, especially Africa and Australia; they are less abundant in the northern temperate hemisphere and between the tropics; they are rarer in America. *Daphne* belongs to the Old World, *Pimelea* is spread over the Australian continent [and New Zealand], *Gnidia* inhabits South Africa. *Lagetta* and many other genera are tropical, *Dirca* is North American, *Drapetes* temperate South American. Several genera, and especially the suborder *Aquilarieæ*, inhabit tropical Asia.

This family, a very natural one in its botanical characters, is also so in respect of the properties of its species. The bark and fruit of many contain, besides a bitter extractive, a peculiar green very acrid and active sebaceous matter. The roots of several furnish a yellow dye (*Passerina tinctoria*); others have tenacious cortical fibres, which are variously used in hot countries.

The extremely acrid seeds of the Spurge-flax (*Daphne Gnidium*), a native of South Europe, were formerly used as a purgative, but are dangerously powerful in their action; a decoction of the leaves was also used, the effect of which is less violent. The bark has a slightly nauseous smell and a corrosive taste; it acts as a caustic when applied to the skin, either entire, powdered, or as an ointment. The Mezereon (*D. Mezereum*), the bark of which is extensively used as a medicine in Germany, together with *D. alpina* and *Cneorum*, all indigenous in France, have the same virulent properties as the Spurge-flax. The leaves and bark of *D. Laureola*, a native of woods throughout France, are often used when fresh as an issue by the peasants. *Dirca palustris* in North America, *Lagetta lintenria* in South America, *Daphne cannabina* in India, are similarly used. The leaves of *Daphne Tarton-raira* in Sardinia, those of *Gnidia* at the Cape of Good Hope, and the berries of *Drimyspermum* in Java, are used in the popular pharmacy as purgatives and emetics.

In India a paper is made from the liber of *Daphne cannabina*, and cord is manufactured from that of *Lagetta funifera* and *lintenria* in South America [the latter of which produces the beautiful substance called Lace-bark].

CLXXXIX. *HERNANDIEÆ.*[1]

(HERNANDIEÆ, *Blume.*—HERNANDIACEÆ, *Dumortier.*)

[TREES. LEAVES scattered, coriaceous, petioled, ovate or peltate, quite entire, exstipulate. FLOWERS monœcious, in peduncled axillary and terminal cymes, ternate in a 4-leaved involucre; central flower ♀, 4-merous, sessile in an urceolate persistent involucel; lateral ♂, not involucellate, pedicelled, usually 3-merous. CALYX herbaceous; *tube* short, narrow, jointed in the ♀, upper part deciduous; *limb* of ♂ 6-, of ♀ 8–10-partite; lobes 2-seriate, valvate in æstivation, inner rather narrower. FLOWERS ♂ : STAMENS 3 (very rarely 4), inserted on the calyx-throat, opposite its outer lobes, connivent; *filaments* short, usually 1–2-glandular at the base; *anthers* didymous, large, erect, adnate to the large connective, 2-celled, dehiscing longitudinally by a deciduous valve. FLOWERS ♀ : STAMINODES glandular, opposite the outer calyx-lobes. OVARY

---

[1] This Order is omitted in the original.—ED.

included in the calyx-tube, free, sessile, 1-celled; *style* terminal, filiform, deciduous; *stigma* fleshy, dilated, 3–4-crenate or -lobed; *ovule* 1, pendulous from the top of the cell, anatropous. DRUPE large, spongy, ovoid, dry, 8-furrowed, or smooth, included in the membranous calyx-tube. SEED sub-globose, inverted; *testa* crustaceous, *raphe* annular, *albumen* 0. EMBRYO orthotropous; *cotyledons* large, fleshy, lobed, torulose; *radicle* short, superior.

ONLY GENUS.

Hernandia.

Hernandieæ are very nearly related to *Laurineæ* (of which they may be regarded as a tribe) and *Thymeleæ*, differing from both conspicuously in habit, from *Laurineæ* in the structure of the anther, from *Thymeleæ* in this, in wanting the tough bark, in the position of the style, and the 2-seriate perianth. The few species are natives of tropical coasts.

The wood of *Hernandia* is spongy, and used for floats and tinder; the bark, leaves, &c., are purgatives, and the juice is an active depilatory.—ED.]

CXC. ELÆAGNEÆ.

(ELÆAGNORUM pars, *Jussieu.*—ELÆAGNEÆ, *Br.*—ELÆAGNACEÆ, *Lindl.*)

FLOWERS ☿ or *diœcious or polygamous.* PERIANTH *simple, herbaceous, or coloured within.* STAMENS *perigynous, double the number of the perianth-lobes, or equal and alternate with them.* OVARY *free,* 1-*celled,* 1-*ovuled;* OVULE *ascending, anatropous.* FRUIT *included in the hardened perianth-tube.* SEED *erect;* ALBUMEN *scanty.* EMBRYO *straight, axile;* RADICLE *inferior.*

TREES or SHRUBS, covered with scarious discoid silvery or brown scales which are furnished at the margin with stellate hairs; branches sometimes spinescent, branchlets annual and deciduous, buds naked. LEAVES alternate (*Elæagnus, Hippophaë*) or opposite (*Shepherdia, Conuleum*), simple, penninerved, entire, shortly petioled; *stipules* 0. FLOWERS regular, ☿ or diœcious or polygamous, solitary in the axil of the leaves, or in spikes racemes or cymes, yellow or white, often scented; ♂ FL.: PERIANTH simple, herbaceous, composed of 2 antero-posterior sepals (*Hippophaë*), or of 4 sepals united at the base into a short tube (*Shepherdia*); ☿, ♀ FL.: PERIANTH tubular, scaly outside, often coloured within; *limb* 2-fid (*Hippophaë*) or 4-partite (*Shepherdia*), or 4–6-fid (*Elæagnus*), imbricate or valvate in æstivation, rarely conical and entire (*Conuleum*). DISK lining the perianth-tube, and forming a glandular ring at the throat (*Elæagnus, Shepherdia*), sometimes dilated into a fleshy cone traversed by the style and longer than the perianth (*Conuleum*), sometimes 0 (*Hippophaë*). STAMENS inserted on the torus, 4–8 (double the sepals) in the ♂ flowers, and opposite and alternate with them; equal and alternate with the perianth-lobes in the ☿ and polygamous flowers (*Elæagnus*); *anthers* erect, basi- or dorsi-fixed, cells sub-opposite, parallel, introrse, dehiscence longitudinal; *pollen* compressed, obscurely 3-gonous. OVARY sessile, free within the accrescent perianth-tube, 1-celled; *style* simple, erect, elongated, stigmatiferous on one side; *ovule* solitary, anatropous,

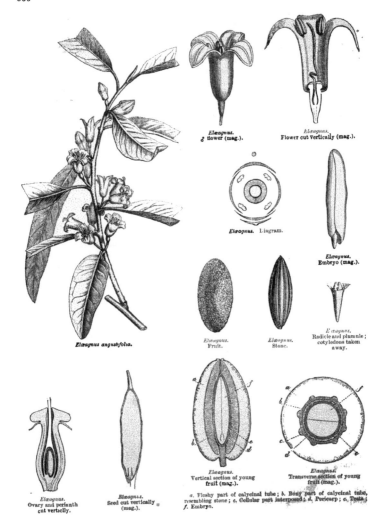

Elæagnus.
♂ flower (mag.).

Elæagnus.
Flower cut vertically (mag.).

Elæagnus. Diagram.

Elæagnus.
Embryo (mag.).

Elæagnus angustifolia.

Elæagnus.
Fruit.

Elæagnus.
Stone.

Elæagnus.
Radicle and plumule;
cotyledons taken
away.

Elæagnus.
Ovary and perianth
cut vertically.

Elæagnus.
Seed cut vertically
(mag.).

Elæagnus.
Vertical section of young
fruit (mag.).

Elæagnus.
Transverse section of young
fruit (mag.).

a. Fleshy part of calycinal tube; *b.* Bony part of calycinal tube,
resembling stone; *c.* Cellular part interposed; *d.* Pericarp; *e.* Testa;
f. Embryo.

nearly basal, ascending, sessile or shortly funicled. FRUIT indehiscent, enclosed in the drupe-like calyx-tube, which is fleshy outside and bony within. SEED ascending, *testa* membranous or cartilaginous, *hilum* basilar, *raphe* projecting, *chalaza* apical; *albumen* 0, or very thin. EMBRYO straight, axile; *cotyledons* thick; *radicle* superior.

PRINCIPAL GENERA.

* Hippophaè.   Shepherdia.   * Elæagnus.   Conuleum.

Elæagneæ are very near *Proteaceæ* (which see). They approach *Santalaceæ*, but these differ in their really adherent ovary and the ovules. We have indicated the affinity between *Elæagneæ* and *Thymeleæ* under that family.

Elæagneæ form a small family, chiefly natives of the mountains of tropical and sub-tropical Asia; a few species inhabit Europe, the Mediterranean region, and North America. They are very rare in tropical America, and entirely absent in south temperate latitudes.

The fleshy base of the perianth, enveloping the fruit of *Elæagnus*, contains free malic acid, which renders the fruit of some species edible, as the Zinzeyd (*E. hortensis* and *orientalis*) in Persia, and *E. arborea* and *conferta* in India. The fruit of *Hippophae rhamnoides*, an indigenous [French] shrub, is acid and resinous tasted; though described as very poisonous, the Finns are said to use it as a seasoning for fish. It is cultivated for its running roots and thorny close and interlaced branches, which form hedges and bind the sands. The balsamic flower of *Elæagnus angustifolia*, commonly called Bohemian Olive, is prescribed in many parts of South Europe for malignant fevers.

CXCI. *PROTEACEÆ.*

(PROTEÆ, *Jussieu.*—PROTEACEÆ, *Br.*)

Banksia quercifolia. Fructiferous cone.

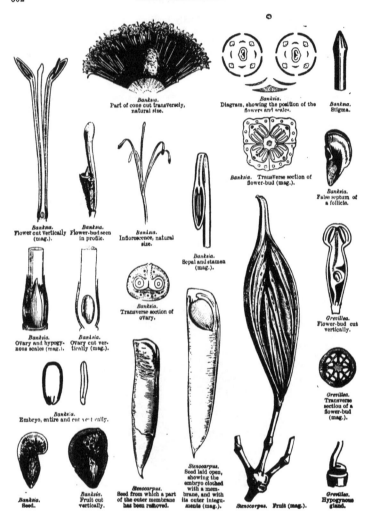

Banksia.
Part of cone cut transversely,
natural size.

Banksia.
Diagram, showing the position of the
flowers and scales.

Banksia.
Stigma.

Banksia. Transverse section of
flower-bud (mag.).

Banksia.
False septum of
a follicle.

Banksia.
Flower cut vertically
(mag.).

Banksia.
Flower-bud seen
in profile.

Banksia.
Inflorescence, natural
size.

Banksia.
Sepal and stamen
(mag.).

Grevillea.
Flower-bud cut
vertically.

Banksia.
Ovary and hypogy-
nous scales (mag.).

Banksia.
Ovary cut ver-
tically (mag.).

Banksia.
Transverse section of
ovary.

Grevillea.
Transverse
section of a
flower-bud
(mag.).

Banksia.
Embryo, entire and cut vertically.

Banksia.
Seed.

Banksia.
Fruit cut
vertically.

Stenocarpus.
Seed from which a part
of the outer membrane
has been removed.

Stenocarpus.
Seed laid open,
showing the
embryo clothed
with a mem-
brane, and with
its outer integu-
ments (mag.).

Stenocarpus. Fruit (mag.).

Grevillea.
Hypogynous
gland.

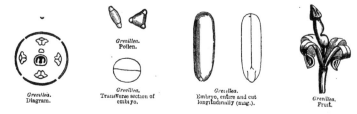

Grevillea.
Diagram.

Grevillea.
Pollen.

Grevillea.
Transverse section of
embryo.

Grevillea.
Embryo, entire and cut
longitudinally (mag.).

Grevillea.
Fruit.

Flowers *usually* ☿, 4-*merous.* Perianth *simple, æstivation valvate.* Stamens *perigynous, equal and opposite to the sepals.* Ovary *free,* 1-celled, 1-∞ -*ovuled* ; ovules *anatropous or orthotropous, micropyle always inferior.* Fruit *a nut or follicle,* 1-∞ -*seeded.* Seed *exalbuminous* ; radicle *inferior.*

Shrubs, trees, very rarely herbs. Leaves scattered, rarely opposite or whorled, usually coriaceous, persistent, simple, entire or often toothed or variously laciniate, or pinnatisect or pinnate, sometimes polymorphous on the same plant; *stipules* 0. Flowers ☿, rarely diclinous, terminal or axillary, in a head, spike, fascicle, umbel, raceme or panicle, rarely axillary and solitary, white yellow or red, very rarely blue or green, some abundantly nectariferous ; capitula and spikes furnished with imbricate bracts, sometimes with a general involucre ; *rachis* or *receptacle* usually thick, conical spheroidal or cylindric, alveolate, furfuraceous or velvety, rarely naked ; *pedicels* geminate or solitary in the axil of a bract. Perianth simple (calyx), coriaceous, coloured or herbaceous, regular or irregular, marcescent or deciduous, usually pubescent outside ; *sepals* 4, linear or spathulate, valvate in æstivation, or imbricate at the top, spreading or connivent, or united into a tube usually split on one side ; *limb* sometimes closed and retaining the stigma, sometimes 4-fid, regular or 1-2-labiate, lobes reflexed plane or concave. Stamens 4 (sometimes 1 arrested), opposite to the sepals and inserted on their limb or claw, very rarely hypogynous (*Bellendena*) ; *filaments* filiform, short, or completely adnate to the calyx ; *anthers* dorsi- or basi-fixed, linear, oblong, ovoid or cordate, 2-celled, introrse, distinct, or rarely syngenesious by the contiguous cells, the other sometimes imperfect ; *pollen* triangular, ellipsoid or lunate, rarely spherical ; *glands* or *scales* hypogynous (rarely obsolete or 0), sometimes 4, alternate with the sepals, either distinct or united into a cup or ring or adherent to the bottom of the calyx ; sometimes less than 4, or reduced to one anterior. Ovary free, sessile or stipitate, 1-celled (rarely pseudo-2-celled) ; *style* terminal, filiform, persistent or deciduous, either equalling the calyx and straight, or long, exserted, and curved ; *stigma* terminal or lateral, undivided, emarginate or 2-fid ; *ovules* solitary or geminate, or several 2-seriate ; *micropyle* always inferior, sometimes fixed to the base of the cell and anatropous, sometimes to the top of the cell and orthotropous. Fruit compressed, ventricose or gibbous, smooth or rugose or verrucose or bristling with points ; sometimes indehiscent, 1-celled, 1-2-seeded, a nut samara or drupe ; sometimes a capsule or follicle, 1-2-valved, 1-2-many-seeded, 1-celled, or 2-celled by a false septum formed by membranes detached from the testa of the contiguous seeds and separable into 2 plates. SEEDS usually ovoid or globose in the nuts, compressed and winged in the follicles, exalbuminous; *hilum* basilar or lateral. EMBRYO straight; *radicle* sometimes near the hilum, sometimes diametrically opposite to it, always inferior.

[Tribes of *Proteaceæ* according to Meissner in De Candolle's 'Prodromus' :—

Sub-order I. NUCUMENTACEÆ.—Fruit an indehiscent nut or drupe. Flowers usually solitary in the axil of each bract.

TRIBE I. PROTEÆ.—Anthers inserted at the base of the short spreading perianth-lobes, all perfect (upper rarely imperfect); cells 2, parallel, adnate. 1.—(Australia). *Petrophila, Isopogon, Adenanthos, Stirlingia.* 2.—(South Africa). *Aulax, *Leucodendron, *Protea, Leucospermum, Mimetes, Serruria, Nivenia, Sorocephalus, Spatalia.*

TRIBE II. CONOSPERMEÆ.—Anthers at first cohering by the adjacent cells, then free. Hypogynous scales 0. Ovary obconic, 1-celled, 1-ovuled. (Australia.) *Synaphea, Conospermum.*

TRIBE III. FRANKLANDIEÆ.—Anthers perfect, cells adnate to the perianth-tube. Ovule 1. Nut dry, with a pappus-like coma. (Australia.) *Franklandia.*

TRIBE IV. PERSOONIEÆ.—Anthers perfect, inserted on the perianth-segments. Ovules 2, rarely 1. Drupe usually fleshy. 1.—(Australia). *Bellendena, *Persoonia, &c.* 2.—(South Africa). *Fauria, Brabejum.* 3.—(America). *Andropetalum, *Guevina.*

Sub-order II. FOLLICULARES.—Fruit dehiscent, 1-2-valved, 1-∞-seeded.

TRIBE V. GREVILLEÆ.—Ovules 2-4, collateral. Seeds without an intervening membrane or substance. 1.—(Australia). *Helicia, Macadamia, Xylomelum, Orites, Lambertia, Grevillea, Hakea, Knightia,* &c. 2.—(Asia). *Helicia.* 3.—(America). *Rhopala, Adenostephanum.*

TRIBE VI. EMBOTHRIEÆ.—Ovules several, imbricate, in 2 rows. Seeds usually separated by an intervening substance. 1.—(Australia). *Telopea, Lomatia, Cardwellia, Stenocarpus.* 2.—(America). *Orocallis, Embothrium, Lomatia.*

TRIBE VII. BANKSIEÆ.—Ovules geminate. Seeds usually separated by a woody or membranous plate. Flowers in dense cones or heads. (Australia.) *Banksia, Dryandra.*—ED.]

Proteaceæ, placed by Endlicher in the same class with *Elæagneæ, Thymeleæ, Santalaceæ,* and *Laurineæ,* have been separated from them, with *Elæagneæ,* by Brongniart, and thus form a well-defined and much more natural group. These two families, in fact, are closely allied in habit, the simple perianth with valvate æstivation, perigynous stamens, free 1-celled ovary, ovule with inferior micropyle, exalbuminous seed, &c. *Elæagneæ* only differ in the always regular flower, the stamens alternate with the perianth-lobes in the isostemonous flowers, and the fruit included in the perianth-tube, which is fleshy outside and bony within.

Proteaceæ approach *Santalaceæ* in the valvate æstivation, isostemony, and stamens opposite to the perianth-lobes; but *Santalaceæ* differ in the inferior ovary, ovule, albuminous seed, and superior radicle. *Proteaceæ* have also some analogy with *Thymeleæ,* founded on the absence of petals, the 1-celled ovary, and the exalbuminous embryo; but *Thymeleæ* differ in the imbricate æstivation, the usually diplostemonous flower, the alternation of the stamens with the calyx-lobes in the isostemonous flowers, and the superior radicle. Beyond these, the most important character separating *Proteaceæ* from the families above-named resides in the micropyle invariably facing the base of the ovary, whatever may be the structure of the ovule and the situation of the hilum.

Proteaceæ almost exclusively belong to south temperate regions, being especially abundant at the Cape of Good Hope and in Australia. They are much rarer in New Zealand and in South America. A few are found in tropical Australia and equatorial Asia; some inhabit equatorial America; very few have been observed in equatorial Africa, and none have been found in the north temperate zone [except at the foot of the Himalayas and in Japan].

This family is more noticeable for the richness and elegance of its flowers than for its useful properties; it has therefore been long cultivated by gardeners. The bark of several species is astringent, the seeds of some are edible; and notably those of the Queensland Nut, *Macadamia ternifolia*. *Protea grandiflora*, from South Africa, is used by the natives for the cure of diarrhœa. The seeds of *Brabejum stellatum*, roasted like chestnuts, are edible, and its pericarp forms a substitute for coffee; those of *Guevina Avellana* are collected by the inhabitants of Chili, who like their mild and somewhat oily taste; its pericarp is there substituted for that of the pomegranate.

The nectaries of *Banksia* and *Protea* secrete an abundant nectar, eagerly sought by bees; that yielded by *Protea mellifera, lepidocarpos* and *speciosa*, is used, under the name of Protea Juice (*Boschjes stroop*), as a bechic at the Cape of Good Hope. The aborigines of Australia feed on the nectar of *Banksia*. [The wood of some Australian species is useful for cabinet work. *Protea grandiflora* is the Wagen-boom, whose wood is used for waggon wheels.]

CXCII. *URTICEÆ.*

(URTICARUM *genera, Jussieu.*—URTICEARUM *genera, D.C.*—URTICEÆ, *Br., Weddell.*—URTICACEÆ, *Endlicher.*)

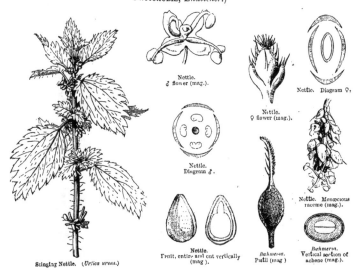

Stinging Nettle. (*Urtica urens.*)

Nettle.
♂ flower (mag.).

Nettle. Diagram ♀.

Nettle.
♀ flower (mag.).

Nettle.
Diagram ♂.

Nettle. Monœcious raceme (mag.).

Nettle.
Fruit, entire and cut vertically (mag.).

Bœhmeria.
Pistil (mag.)

Bœhmeria.
Vertical section of achene (mag.).

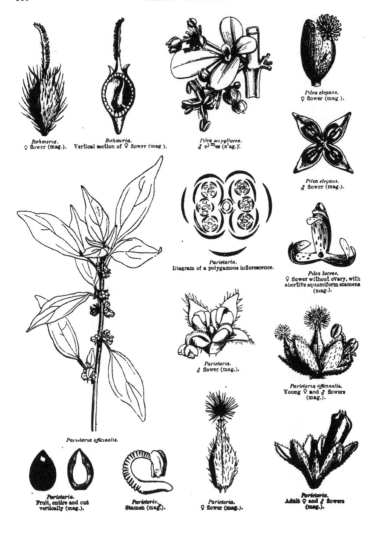

Bœhmeria.
♀ flower (mag.).

Bœhmeria.
Vertical section of ♀ flower (mag.).

Pilea serpyllacea.
♂ ⚥ flowers (n'ag.).

Pilea elegans.
♀ flower (mag.).

Pilea elegans.
♂ flower (mag.).

Parietaria.
Diagram of a polygamous inflorescence.

Pilea lucens.
♀ flower without ovary, with abortive squamiform stamens (mag.).

Parietaria.
♂ flower (mag.).

Parietaria officinalis.
Young ♀ and ♂ flowers (mag.).

Parietaria officinalis.

Parietaria.
Fruit, entire and cut vertically (mag.).

Parietaria.
Stamen (mag.).

Parietaria.
♀ flower (mag.).

Parietaria.
Adult ♀ and ♂ flowers (mag.).

Parietaria.
Pistil (mag.).

Helxine Soleirolii.
♂ flower, involucred.

Helxine.
♂ and ♀ flower (mag.).

Helxine.
♀ flower, involucred (mag.).

FLOWERS *diclinous or polygamous.*—♂ : PERIANTH *calyciform, isostemonous* ; FILAMENTS *uncoiling elastically.*—♀ : PERIANTH *calyciform, sometimes* 0. OVARY *1-celled* ; OVULE *solitary, erect, orthotropous.* FRUIT *dry or fleshy.* ALBUMEN *more or less copious.* EMBRYO *straight, axile, antitropous* ; RADICLE *superior.*—LEAVES *stipulate.*

HERBS, UNDERSHRUBS or SHRUBS, very rarely climbing (*Urera*), or TREES ; juice watery, very rarely milky (*Neraudia*) ; whole plant often armed with hairs, some simple, others furnished at the base with an outer layer of cellules containing an acrid and burning juice ; epidermal cells often containing cystoliths (see p. 121). STEM often angular ; *bark* thin, with very tenacious fibres. LEAVES alternate or opposite, petioled, penninerved, entire, toothed or serrate, rarely palmi-nerved and -lobed ; *stipules* adnate to the stem or petiole, lateral or axillary, free or joined to those of the opposite leaf [rarely 0]. FLOWERS monœcious, diœcious or polygamous, usually in cymes, very rarely solitary and axillary (*Helxine*), sometimes crowded upon a common receptacle, which is convex (*Pipturus, Procris,* &c.) or concave (*Elatostemma*) ; *cymes* loose or glomerate or capitate, solitary or geminate when the inflorescence is definite ; sometimes spiked racemed or panicled when it is mixed and indefinite ; *pedicels* frequently jointed ; *bracts* usually small and persistent, often forming an involucre, and distinct or coherent, sometimes 0. FLOWERS ♂ : PERIANTH single (CALYX), usually green, rarely coloured, gamosepalous, 4–5- (rarely 2-3-) partite, sometimes monosepalous (*Forsköhlea*) ; segments equal, concave, imbricate in æstivation (*Urtica,* &c.), or valvate (*Parietaria,* &c.). STAMENS equal and opposite to the segments, and inserted at their base ; *filaments* filiform, subulate or dilated, usually transversely wrinkled, inflexed in æstivation and uncoiling elastically in flower ; *anthers* 2-celled, introrse, dorsifixed, cells contiguous, usually oblong, globose or reniform, dehiscence longitudinal. OVARY rudimentary, sessile or stipitate, glabrous or hairy. FLOWERS ♀ : PERIANTH (CALYX) tubular or 3–5-partite or -lobed, often accrescent, rarely 0 (*Forsköhlea,* &c.). STAMENS rudimentary, scale-like, opposite to the sepals (*Pilea, Procris,* &c.), usually 0. OVARY free or adnate to the perianth (*Pipturus,* &c.), sessile or shortly stipitate, ovoid, 1-celled ; *style* termi-nal or sub-lateral, simple, or with a capitate or penicillate stigma, or stigmatiferous on one side, or very short, or obsolete ; *stigma* sessile, laciniate-multipartite ; *ovule* solitary, erect, sessile or funicled, orthotropous. FRUIT dry (*achene*) or fleshy (*drupe*), naked or enclosed in or adnate to the sometimes accrescent perianth; *achenes* compressed, ovoid or spherical, smooth, dotted or tubercled, sometimes · winged or cottony ; *drupes* ovoid, usually aggregated into heads like the fruit of the Mulberry ; *endocarp* crustaceous ; *sarcocarp* thin, sometimes adnate to the perianth. SEED erect, usually free within the endocarp ; *testa* thin ; *chalaza* broad, brownish ; *albumen* fleshy and oily, rarely abundant, very rarely 0 (*Elatostemma*). EMBRYO straight, axile, antitropous ; *cotyledons* fleshy, oval or sub-orbicular, plano-convex ; *radicle* cylindric or conical, superior.

[The following is Weddell's classification of *Urticeæ* in De Candolle's ' Prodromus ':—

TRIBE I. URERÆ.—Clothed with stinging hairs. Leaves decussately opposite, or spirally alternate. Perianth of ♀ 4-partite or -lobed, free, rarely 2-lobed or tubular. *Urtica, Obetia, Fleurya, Laportea, Urera, Girardinia*, &c.

TRIBE II. PROCRIDEÆ.—Unarmed. Leaves opposite, or alternate by arrest and often distichous. Perianth of ♀ free, 3- rarely 4-partite. Stigma penicillate. *Pilea, Pellionia, Elatostemma, Procris*, &c.

TRIBE III. BŒHMERIEÆ.—Unarmed. Leaves opposite or alternate. Perianth of ♀ free or adnate, usually tubular, rarely short or 0. *Bœhmeria, Pouzolzia, Memorialis, Pipturus, Villobrunea, Maoutia, Phenax*, &c.

TRIBE IV. PARIETARIEÆ.—Unarmed. Leaves alternate, quite entire. Flowers unisexual or polygamous ; ♀ inflorescence involucrate. Perianth tubular, free. *Parietaria, Hemistylis*, &c.

TRIBE V. FORSKOHLIEÆ.—Unarmed or prickly. Leaves alternate or opposite. Flowers unisexual, often involucrate. Perianth of ♀ tubular or 0. *Forskóhlea, Droguetia, Australina,* &c.—ED.]

Urticeæ are so closely allied to *Moreæ, Ulmaceæ, Celtideæ*, and *Cannabineæ* that all botanists place these families in one same class (URTICINEÆ, *Brongn.*), with the following common characters: apetalism, isostemony, stamens opposite the sepals, 1-celled and 1-ovuled ovary, usually orthotropous or campylotropous ovule, albumen fleshy or 0, and radicle superior.

We have indicated the analogies of *Urticeæ* with *Cynocrambeæ, Piperaceæ, Saururceæ, Ceratophylleæ* (see these families). Weddell, in his monograph of *Urticeæ*, has combined them with *Tiliaceæ*, chiefly on account of the tenacity of the liber-bundles of the bark, the stipules, definite inflorescence, valvate æstivation, 2-lobed anthers, smooth pollen, &c., whence he regards *Urticeæ* as a degradation from the type of *Malvaceæ*, through *Tiliaceæ*.

Urticeæ are mainly tropical ; Europe is of all parts of the world the poorest in species ; but, as Weddell observes, what it loses in variety and number of species is partly compensated for by the multitude of individuals, so that there is perhaps no exaggeration in saying that the five or six species of Nettles and Pellitories which swarm around our habitations cover nearly as much ground as the numerous species scattered through equatorial regions. *Urtica* is represented in many parts of the globe, but is confined to temperate and cold regions.

Except *Urtica* and *Parietaria*, all the genera (about thirty-six) are essentially tropical or sub-tropical, and it is very exceptionally that a few of their species have wandered into the extratropical regions of one or other hemisphere, as *Bœhmeria, Elatostemma, Pilea, Laportea*, &c. America possesses a third of the known species, Asia and Malacca another third, the Pacific Islands and Africa another, from which a dozen species inhabiting Europe must be deducted.

The medicinal properties of *Urticeæ* are confined to the Pellitories (*Parietaria*) and some species of Nettle. *Parietaria diffusa* and *erecta* contain Nitre, whence their use as a diuretic and in external appli-

. cations. Flogging with fresh Nettles (*Urtica urens* and *dioica*), was formerly frequently resorted to by doctors to produce a healthy counter-irritation of the skin. This practice, named *urtication*, is still suc‚ cessfully adopted, both in civilized countries and among savages, and especially by the Malays. The large Nettle (*U. dioica*) is good forage for milch cows; its leaves are eaten as spinach, and given as food to young turkeys.

From an industrial point of view *Urticeæ* deserve notice; their cortical fibres rival in tenacity those of Hemp. Such is the case with our *U. dioica*, with *U. cannabina* of North-east Asia and Persia, *Laportea canadensis* of North America, and especially the China Grass, *Bœhmeria nivea* (*Tchou-ma* of the Chinese, *Ramie* of the Isles of Sunda), of which there are perhaps two species; its fibres are as remarkable for whiteness and silkiness as for tenacity. [*B. Puya*, of the Himalayas, is similarly used.] Though but lately used extensively, the China Grass has been in use ever since the sixteenth century in the Netherlands. [Its culture is now extended into India and various British colonies. *Laportea oreₜulata*, of North India, at certain seasons emits when bruised so irritant an effluvia as to cause a copious flow of saliva and mucus from the nose and eyes for many hours, and its stinging hairs have produced violent fevers. A Timor species of *Urtica* is said to have caused death. The leaves of many *Urticeæ* are eaten cooked in the Himalayas, as are the tuberous roots of *Pouzolzia tuberosa* in India.]

CXCIII. *MOREÆ.*

(MOREÆ ET ARTOCARPEÆ, *Endliche*r.)

FLOWERS *diclinous.* PERIANTH *single, imbricate, sometimes* 0. OVARY 1-*celled*; STYLES 1–2; OVULE *solitary, basilar and orthotropous, or parietal and campylotropous, or anatropous.* ACHENE, DRUPE, *or* UTRICLE. ALBUMEN *fleshy, or* 0. EMBRYO *curved or straight, axile*; RADICLE *superior.—*LEAVES *alternate*; STIPULES *fugacious.* JUICE *milky.*

TREES or SHRUBS, sometimes climbing, with milky juice, rarely stemless HERBS (*Dorstenia*). LEAVES alternate, undivided or lobed, often polymorphous; *stipules* usually convolute and enveloping the terminal bud, persistent or deciduous, and usually leaving a semi-annular scar. FLOWERS diclinous; either diœcious, the ♂ in small spicate cymes, the ♀ capitate on a globose receptacle (*Broussonetia, Maclura*), or monœcious; sometimes in spikes, ♂ and ♀ distinct (Mulberry); sometimes lining the inner surface of a hollow pyriform fleshy receptacle which is furnished at the base with scaly bracteoles, and at the top with an orifice closed by scales, the ♂ above, ♀ below (Fig); sometimes covering a flat or cupped receptacle, the ♂ and ♀ intermixed and sunk in pits with laciniate edges (*Dorstenia*).— ♂ : PERIANTH single (CALYX), imbricate, 4-partite (*Morus, Maclura, Broussonetia,* &c.) or 3-partite (*Ficus, Artocarpus,* &c.) or 0 (*Dorstenia, Brosimum,* &c.), sometimes more or less tubular (*Pourouma, Cecropia*). STAMENS usually isostemonous, opposite to the calyx-lobes and inserted at their base, usually 4, rarely 3 (*Ficus*), sometimes 2 or more (*Dorstenia*); *filaments* filiform or subulate, smooth or transversely wrinkled, usually inflexed in æstivation, then spreading, a little longer than the sepals, usually free, rarely connate (*Pourouma*); *anthers* introrse or extrorse, usually 2-celled, ovoid or sub-globose, dorsifixed, erect or incumbent, dehiscence longitudinal; rarely peltate

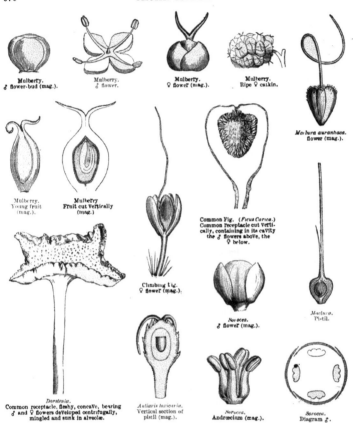

Mulberry.
♂ flower-bud (mag.).

Mulberry.
♂ flower.

Mulberry.
♀ flower (mag.).

Mulberry.
Ripe ♀ catkin.

Maclura auranthaca.
flower (mag.).

Mulberry.
Young fruit
(mag.).

Mulberry
Fruit cut vertically
(mag.)

Common Fig. (Ficus Carica.)
Common receptacle cut vertically, containing in its cavity
the ♂ flowers above, the
♀ below.

Climbing Fig.
♀ flower (mag.).

Sorocea.
♂ flower (mag.).

Maclura.
Pistil.

Dorstenia.
Common receptacle, fleshy, concave, bearing
♂ and ♀ flowers developed centrifugally,
mingled and sunk in alveolæ.

Antiaris toxicaria.
Vertical section of
pistil (mag.).

Sorocea.
Andrœcium (mag.).

Sorocea.
Diagram ♂.

and 1-celled (*Brosimum*). OVARY rudimentary or obsolete.— ♀ : PERIANTH imbricate, persistent, sometimes of 3–4 free sepals (*Morus, Maclura, Treculia*), sometimes
4–5-fid (*Conocephalus, Ficus*), or 4-toothed (*Olmedia*), or tubular or urceolate
(*Cecropia, Artocarpus,* &c.), or 0 (*Dorstenia*). OVARY sessile (*Morus, Maclura*) or
stipitate (*Ficus, Broussonetia, Dorstenia*), usually 1-celled, sometimes of 2 unequal
cells; *style* sometimes terminal, with 2 nearly free often unequal **branches, stigma-**

tiferous on their inner face or throughóut their length; sometimes lateral, filiform, more or less divided at the top (*Ficus, Dorstenia*), or undivided (*Broussonetia, Cecropia*, &c.) ; *ovule* solitary, either inserted in the middle of the wall, usually campylotropous or anatropous (*Artocarpus*), or basilar and orthotropous (*Cecropia*), micropyle superior. FRUIT: 1. An ACHENE (*Maclura*) or DRUPE (*Morus*), enveloped by the accrescent succulent calyx, sometimes on a fleshy gynophore (*Broussonetia*, &c.) ; 2. A UTRICLE sunk in a sub-succulent (*Dorstenia*) or succulent (*Ficus*) receptacle. SEED with a crustaceous fragile or finely membranous integument, *hilum* ventral. EMBRYO in the axis of a fleshy more or less copious albumen, sometimes 0 ; *cotyledons* oblong, plane, incumbent; r*adicle* superior.

TRIBE I. ARTOCARPEÆ.—Filaments usually inflexed in bud. Ovule ventral, basal or apical. *Brosimum, Antiaris, Olmedia, Cecropia, Trophis, Galactodendron, Bleekrodia, Streblus, *Artocarpus*, &c.

TRIBE II. MOREÆ.—Filaments usually straight in bud. Ovule ventral or apical. *Ficus, *Dorstenia, *Morus, *Maclura, Epicarpurus, *Broussonetia, Treculia, &c.—ED.]

Moreæ belong to the class *Urticineæ*. They are so closely allied to *Artocarpeæ* that Trécul, who has published a learned memoir on the latter, indicates as the only diagnostic character the inflexion of the filaments in the bud of *Moreæ*, while in all *Artocarpeæ*, except *Trophis*, the filaments are erect before as well as after flowering ; but this diagnosis fails in *Ficus*, so that if *Ficus* be retained in *Moreæ*, and *Trophis* in *Artocarpeæ*, there is no line of demarcation between the two families. *Moreæ* are also very near *Celtideæ* (which see). The true *Urticeæ*, with which they were formerly united, only differ in their watery juice and habit. The tenacity of the cortical fibres is common to both families.

Moreæ inhabit the tropical and sub-tropical regions of both hemispheres. A few occur in temperate North America. The common Fig, a native of Asia Minor, is now spread over the Mediterranean basin.

Moreæ possess a milky juice, scanty and nearly colourless in some, very copious in others, acrid and corrosive in most, containing various matters—such as mannite and succinic acid in Mulberries, a colouring principle in *Maclura*, an elastic resin (india-rubber) in many Figs. Astringent principles are combined with these in the bark, and mucilaginous and aromatic qualities in the herbaceous parts, which give to different species of this large family very various properties ; some are lenitive medicines, others stimulants, others poisonous. The inflorescence, at first charged with an acrid juice, acquires when ripe quite opposite qualities. Mucilage and sugar are thén developed in it in such proportions, that the fruit becomes a very nourish- ing food, or an acidulous cooling medicine which is successfully employed in a great many dis- orders.

The Black Mulberry (*Morus nigra*), a native of Persia, has been cultivated in Europe from the earliest times. Its drupaceous fruit (due to the flowers being densely spiked) owes its acidulous sugary taste in great part to the succulent calyces. It is nutritious and refreshing, and a slightly astringent syrup is prepared from it before it is quite ripe. The bark of its root is acrid, bitter, purgative and vermifuge ; and Dioscorides states that it destroys tænia. Its leaves may be employed to feed silkworms, but for this important purpose the White Mulberry (*M. alba*) is especially used. Both the White Mul- berry and silkworm are natives of China, where the former is cultivated for rearing this insect. Thence this culture passed into Persia, and on to Constantinople, in the reign of Justinian ; thence into Spain, and later, in the time of Roger, into Sicily and Calabria ; finally, at the conclusion of the French war for the conquest of the kingdom of Naples, both tree and insect were introduced into the south of France, where, thanks to Charles VIII., Henry IV., and especially to Colbert, *M. alba* is now almost naturalized. The Chinese attribute diuretic and anthelminthic qualities to the root of this species, and febrifugal to its leaves. The fruit and young leaves of *M. indica* are eaten in India, as are those of *M. rubra* in North America, and *M. celtidifolia* and *corylifolia* in South America. *M. pabularia* is cultivated for forage in

Tibet, Kashmir, &c. The Paper Mulberry (*Broussonetia papyrifera*), a diœcious tree, cultivated in China and Japan, has been introduced into European gardens. Its leaves are remarkable for their polymorphism; the fleshy gynophore of the fruit is insipid; the fibrous bark of its stem is used to make Chinese paper. From the bark of another species of *Broussonetia* the South Sea Islanders fabricate their clothing (Tappa).

The fruit of *Maclura tinctoria*, a native of the Antilles and Mexico, is used like that of our Mulberry. Its hard, compact wood takes a fine polish, and might be very useful in cabinet-making, but it is exclusively used for dyeing, as Fustic. *M. aurantiaca* is a small North American tree, which yields the flexible and very elastic Bow-wood. Its fruit, the Osage Orange, contains a yellow fœtid juice, with which the Indians paint their faces when they go to war.

Figs are trees or climbing shrubs; the principal species is *Ficus Carica*, cultivated throughout the Mediterranean region. Its fleshy receptacle, the Fig, whether fresh or dried, is an agreeable and very nutritious food, and an emollient medicine. By incision its bark yields an acrid and caustic milk, containing a considerable quantity of India-rubber. Certain tropical Figs contain much larger quantities of this milk; as the India-rubber Fig (*F. elastica*), the Banyan or Bo-tree (*F. indica*), and the Sacred Fig (*F. religiosa*) [all Indian species]. Of these the Banyan is a large evergreen tree, the adventitious roots of which descend to the earth, root there, and form arcades which extend on all sides, far from the trunk. A famous Banyan on the Nerbudda thus has a circumference of 2,000 feet, and 320 columns formed by adventitious roots. Some Chinese [and Indian] Figs are the food of the Lac, an hemipterous insect (*Coccus Lacca*). The females are crowded so densely on the young branches as to leave no spaces, and exude from their bodies a resinous substance (Lac) which is used in the manufacture of sealing-wax and certain varnishes. *F. Sycomorus* is a widely spread Egyptian tree. Its wood is very light, and supposed to be incorruptible, whence the ancient Egyptians used it for mummy-cases. *Dorstenia* differs from *Ficus* in the receptacle of the flowers being but slightly concave. The root of *D. braziliensis* has a faint and agreeable aroma; it is employed in Brazil against the bite of venomous serpents, whence its Spanish name *Contrayerva*.

Artocarpeæ, which we have annexed to *Moreæ*, are spread over all equatorial regions.

Most *Artocarpeæ* contain a milky juice, which exhibits the most opposite properties and even in the same genus; thus it is innocuous in *Antiaris innocua*, but acrid, caustic and poisonous in *A. toxicaria*, the viscid juice of which, the *Antiar* of the Javanese, is obtained by incisions in the bark, and hardens into a gum-rasin, with which the Malays prepare the *poison-upas*, to poison their arrows and cris. This *upas* differs from the equally deadly *Strychnos Ticuté* of the Javanese, and the *curare* of Guiana (also yielded by a Strychnos; see page 557), in acting on the organs of circulation, while the *Strychnos* acts on the nervous system. [Another species of *Antiaris* (*A. saccidora*, of Malabar), has so tough a bark that sacks (used for rice, &c.) are made by removing it from truncheons of the trunk or branches, merely leaving a thin section of the wood to serve as the bottom of the sack.]

The Breadfruit-tree (*Artocarpus incisa*) provides the South Sea Islanders with an abundant, wholesome and agreeable food. Its entire fruit, composed of carpels agglomerated on a fleshy receptacle, is as large as a man's head. It is gathered before it is perfectly ripe, when it consists of a white and farinaceous flesh (produced by the united calyces), and is sliced, slightly baked on cinders, and eaten like bread, of which it has somewhat the taste. The French and English have propagated this precious fruit over the torrid zone. The fruit of *A. integrifolia* is not less esteemed [it is a very inferior fruit, not eaten by Europeans]; its firm and sugary pulp is much sought by the Creoles, in spite of its disagreeable smell. Its seeds are eaten fried or boiled, as are also those of *Brosimum Alicastrum*, which in Jamaica takes the place of bread amongst the poor. Colombia possesses another tree of the same family, which is one of the most wonderful vegetable productions. This, the Cow-tree, or Palo de Leche (*Galactodendron utile*), the inhabitants of the Cordilleras of Venezuela regularly milk, procuring by incision an enormous quantity of a white and somewhat viscous liquid, presenting all the physical properties of the best milk, with a slight balsamic odour. It contains, besides sugar and albumine, a large proportion of a fatty waxy matter to which it appears to owe its principal properties; by evaporation in a water-bath an extract similar to frangipani is obtained.

The wood of *Artocarpus* is used in India and Cochin China for furniture and cabinet work. That of

Cudrania javanensis, a shrub from the Isles of Sunda, furnishes a dye. The leaves of the Breadfruit-tree, which are about three feet long and one and a half broad, are used as cloths and mats. The bark of the Guarumo (*Cecropia peltata*), a tall slender tropical American tree, contains india-rubber; it is given as an astringent in diarrhœa and blennorrhœa; its ashes are rich in alkaline salts. [The beautiful Snake-wood of Guiana is the produce of *Piratinera guianensis*.]

CXCIV. *CELTIDEÆ*, Endlicher.

FLOWERS *polygamous.* PERIANTH *single.* OVARY *free, 1-celled*; STYLES 2; OVULE *solitary, parietal or sub-apical, campylotropous.* FRUIT *a drupe.* ALBUMEN *scanty or* 0. EMBRYO *curved*; RADICLE *superior.*—STEM *woody.* LEAVES *alternate*; STIPULES *fugacious.*

SHRUBS or TREES with alternate branches, often armed with spinescent axillary

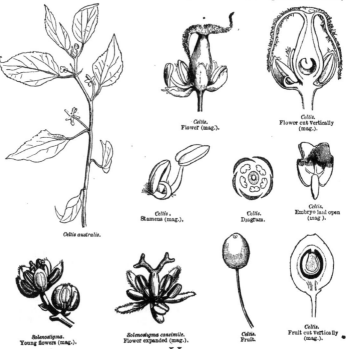

Celtis.
Flower (mag.).

Celtis.
Flower cut vertically
(mag.).

Celtis.
Stamens (mag.).

Celtis.
Diagram.

Celtis.
Embryo laid open
(mag.).

Celtis australis.

Solenostigma.
Young flowers (mag.).

Solenostigma consimile.
Flower expanded (mag.).

Celtis.
Fruit.

Celtis.
Fruit cut vertically
(mag.).

branchlets. LEAVES alternate, petioled, entire or serrate, usually 3-nerved, pubescent or scabrous, rarely glabrous; *stipules* 2, at the base of the petioles, caducous. FLOWERS ☿, or often ♂ by arrest of the ovary, solitary and peduncled, or in racemes or panicles. PERIANTH calycine, 5-phyllous or -partite, imbricate in æstivation, spreading after flowering, persistent. STAMENS 5, inserted at the bottom of the perianth, and opposite to the sepals; *filaments* cylindric, usually short, inflexed in bud, rising elastically as in Nettles (*Celtis tetrandra*); *anthers* dorsifixed, rarely reflexed in bud, and becoming introrse later; cells 2, opening by a longitudinal sometimes very short slit. OVARY free, ovoid, usually inequilateral, 1-celled; *stigmas* 2, terminal, usually elongate-subulate, undivided or 2-fid; *ovule* solitary, parietal, sub-apical, campylotropous or semi-anatropous; *micropyle* superior. DRUPE moderately fleshy. SEED pendulous, curved above; *testa* membranous; *albumen* fleshy, scanty, sometimes nearly 0. EMBRYO curved; *cotyledons* plane, or folded on each other, incumbent; *radicle* elongate, superior.

PRINCIPAL GENERA

* Celtis.	Sponia.	Solenostigma.	Homoioceltis.	Gironniera.
Mertensia.	Parasponia.	Trema.	Monisia.	

Celtideæ approach *Moreæ* in the structure of the ovary, ovule, seed, and embryo, but differ in their inflorescence and the nature of their fruit. They are closely allied to *Ulmaceæ* (see that family).

Celtideæ inhabit tropical and temperate Asia, America, and the Mediterranean region of Europe.

The bark and leaves have sometimes a sub-aromatic odour and a bitter acrid taste; the flesh of the fruit is astringent. Several species have oily seeds; some possess a compact very durable wood, some others a light almost spongy wood, or flexible and very tenacious. The shoots of *Celtis australis* are used to make forks or whip-handles; its slightly styptic fruit is edible; its seeds yield, by expression, an oil similar to oil of almonds. *C. occidentalis* is common in the hot parts of North America; its drupe is prescribed as an astringent. The root, bark, and leaves of *C. orientalis*, a West Asiatic species, are regarded as a specific against epilepsy.

CXCV. *CANNABINEÆ*, *Endlicher*.

FLOWERS *diclinous*. PERIANTH ♂ *calyciform*; ♀ *scalelike*. OVARY 1-*celled*, 1-*ovuled*; STYLES 2; OVULE *pendulous, campylotropous*. FRUIT *dry*. SEED *pendulous, exalbuminous*. EMBRYO *hooked or spirally coiled*. RADICLE *superior*.—STEM *herbaceous, with watery juice*. LEAVES *stipulate*.

Annual and erect HERBS, or perennial and twining, with watery juice. LEAVES opposite, or the upper alternate, petioled, cut, lobed, toothed, often glandular (Hemp), stipulate. FLOWERS diœcious.—♂ racemed or panicled. PERIANTH herbaceous, of 5 free sub-equal sepals, imbricate in æstivation. STAMENS 5, opposite to the sepals, and inserted on their base; *filaments* filiform, short; *anthers* terminal, 2-celled, oblong; cells opposite, marked with 4 furrows, muticous or apiculate by the connective which exceeds them, dehiscence longitudinal.—FLOWERS ♀ in a strobiloid spike (Hop) or in glomerulæ (Hemp), with 2-flowered bracts (Hop) or 1-flowered (Hemp), flowers bracteolate. PERIANTH urceolate, reduced to one bract-like mem-

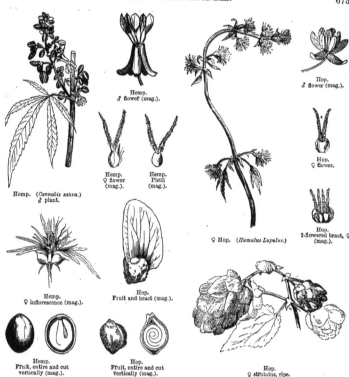

Hemp.
♂ flower (mag.).

Hop.
♂ flower (mag.).

Hemp.
♀ flower
(mag.).

Hemp.
Pistil
(mag.).

Hop.
♀ flower.

Hemp. (*Cannabis sativa.*)
♂ plant.

♀ Hop. (*Humulus Lupulus.*)

Hop.
2-flowered bract, ♀
(mag.).

Hemp.
♀ inflorescence (mag.).

Hop.
Fruit and bract (mag.).

Hemp.
Fruit, entire and cut
vertically (mag.).

Hop.
Fruit, entire and cut
vertically (mag.).

Hop.
♀ strobilus, ripe.

branous sepal, closely enveloping the ovary. OVARY free, ovoid or sub-globose, sub-compressed, 1-celled; *style* terminal, very short or 0; *stigmas* 2, elongate-filiform or subulate, pubescent; *ovule* solitary, pendulous from the top of the cell, campylotropous; *micropyle* superior. ACHENE glandular, embraced by the accrescent perianth (Hop), or a smooth CARYOPSIS, included in the perianth and 2-valved by pressure (Hemp). SEED pendulous; *testa* free or adnate to the pericarp; *endopleura* fleshy, simulating an albumen; *hilum* marked by a brown spot. EMBRYO exalbuminous, hooked, or spirally rolled; *cotyledons* incumbent; *radicle* superior.

GENERA.

Cannabis.   * Humulus.

x x 2

This little family was formerly placed in *Urticeæ*, with which it agrees in its diclinous greenish monochlamydeous flowers, imbricate æstivation, stamens opposite to the sepals, 1-celled and 1-ovuled ovary, dry fruit, superior radicle, and stipulate leaves ; but *Urticeæ* differ in the long coiled and elastic filaments, the globose anthers, erect orthotropous ovule, and the more or less abundant albumen. *Celtideæ* present the same affinities, besides which the seed is pendulous and the ovule campylotropous.

Cannabineæ have been cultivated from the remotest period, and are now spread over all northern temperate regions. Hemp is said to grow wild on the hills of Central and South Asia ; the Hop is a native of Europe, America, and West Asia.

The Hop and the Hemp, although very different in appearance, are similar not only in floral structure, but in the tenacity of their fibres and the bitter narcotic juice which they contain. The achene and bract of the Hop (*Humulus Lupulus*) are covered with yellow globose glands, containing a bitter aromatic principle (*lupuline*), which imparts to beer a slightly narcotic quality and an agreeable bitterness ; the young white underground shoots are edible. The seed of the Hemp (*Cannabis sativa*) contains a fixed oil, which is used for making soft soap and for lighting. The stem yields textile cortical fibres. The fibre of its bark, separated from the wood of the stem by prolonged maceration, is called Hemp. Hemp yields a gelatinous resin, contained in little glands on the surface of the stem and leaves, which possesses very intoxicating properties ; it is called *cherris* in Arabia, North Africa, and the East ; this principle, more narcotic and dangerous even than opium, is the basis of a thick green preparation, called *hachish* or *hasheesh* by the Arabs, and extensively smoked by them.

[In India the resin collected on the body or clothes by running amongst the Hemp plants in the fields is called *churras*, and taken in a solid state it yields the Tincture of *Cannabis* or Indian Hemp, of the Pharmacopœia. The bruised twigs are smoked as *gunja*, and the leaves, flowers, &c., yield the infusion called *bhang*. The *Cannabis* is grown at every cottage door in tropical Africa, and smoked under the name of *diamba*. Hemp-seed is imported in enormous quantities for feeding birds. Hemp appears to have been known to the Greeks and northern races from time immemorial, but not to the Hebrews and Egyptians.]

CXCVI. *ULMACEÆ, Mirbel.*

Elm.
Flower (mag.).

Elm.
Flower laid open
(mag.).

Elm. Fruit.

Elm.
Seed (mag.).

Elm.
Seed cut
vertically
(mag.).

Elm.
(*Ulmus campestris*.)
Flowering branch.

FLOWERS ☿ . PERIANTH *single, campanulate, persistent.* OVARY *free, 1-2-celled* ; STYLES 2 ; OVULES *solitary, pendulous, anatropous.* FRUIT *a samara or nut, 1-seeded.* SEED *inverted.* EMBRYO *exalbuminous, straight.* RADICLE *superior.*—STEM *woody.* LEAVES *alternate* ; STIPULES *fugacious.*

TREES or SHRUBS with alternate branches ·and watery juice. LEAVES alternate, distichous, simple, petioled, penninerved, usually inequilateral, scabrous ; *stipules* 2, caducous. FLOWERS lateral on the branchlets, fascicled, ☿ or 1-sexual. PERIANTH herbaceous or slightly coloured, sub·campanulate, *limb* 4-5-8-fid, imbricate in æstivation, erect after flowering, persistent. STAMENS inserted at the bottom of the perianth, equal and opposite to its lobes, rarely more numerous; *filaments* filiform, distinct; *anthers* 2-celled, dorsifixed; cells somewhat oblique, introrse, dehiscence longitudinal. OVARY free, of 2 connate carpels, 2-celled (*Ulmus*) or 1-celled (*Planera*); *styles* 2, divergent, stigmatiferous along their inner face; *ovules* solitary in each cell, pendulous near the top of the septum, or from the top of the single cell, anatropous. FRUIT a membranous samara (*Ulmus*), or a coriaceous 1-celled and 1-seeded indehiscent nut (*Planera*). SEED inverted, *testa* membranous, *raphe* longitudinal. EMBRYO exalbuminous, straight; *cotyledons* plane (*Ulmus*) or sub-sinuous (*Planera*), or conduplicate (*Holoptelea*); *radicle* short, superior.

GENERA.

* Ulmus.   * Planera.   Holoptelea.

By many botanists *Ulmaceæ* are united with *Celtideæ*, from which they only differ in their inflorescence, extrorse anthers and anatropous ovule; both are included in the class URTICINEÆ of Brongniart and adopted by Planchon; for in *Ulmaceæ*, though the ovule is anatropous, the micropyle faces the top of the ovary, as in the other families of *Urticineæ*.

Ulmaceæ are spread over the temperate northern regions [and sub-tropical India]. Their bark contains an astringent and tonic bitter mucilage and tannin. The inner bark of the common Elm (*Ulmus campestris*) has been considered a remedy for dropsy and eruptions. That of the American Elms (*U. fulva* and *americana*) is so rich in mucilage that poultices and a nutritious jelly are made of it. The Americans reduce it to a powder as fine as flour, and employ it largely in inflammatory diseases. *Planera Abelicea*, a native of Crete, produces an aromatic wood, formerly exported under the name of False Sandalwood. Elm wood is rather hard, reddish, and used especially for wheelwright's work, shafts, axle-trees, screws for presses, &c. Exostoses or wens are often developed on Elm trunks, which acquire great hardness, and are much sought by cabinet-makers from the different patterns the twisted arrangement of their wood-fibres exhibits.

CXCVII. *BETULACEÆ.*

(AMENTACEARUM genera, *Jussieu.*—BETULINÆ, *L.-C. Richard.*—BETULACEÆ, *Bartling.*)

FLOWERS *monœcious, in catkins, the* ♂ *with a calyciform scale-like perianth, the* ♀ *achlamydeous, with accrescent scales.* STAMENS *4 or 2.* OVARY *with two 2-ovuled cells;* OVULES *pendulous, anatropous.* NUTS *usually winged, 2-celled, 1-seeded.* EMBRYO *exalbuminous;* RADICLE *superior.* STEM *woody.* LEAVES *alternate, stipulate.*

TREES or SHRUBS, *branches* scattered, *buds* scaly. LEAVES often sprinkled with resinous glands, alternate, simple, toothed, the pinnate nerves terminating in the teeth; *stipules* free, caducous. FLOWERS monœcious, sessile, at the base of scaly bracts, in terminal or lateral catkins.—CATKINS ♂ : *Scales* bearing 2–3 flowers, each accompanied within by 2 or 4 squamules. PERIANTH calyciform, regular, 4-lobed (*Alnus*), or reduced to a scale (*Betula*). STAMENS 4, inserted at the base of the perianth-lobes, and opposite to them (*Alnus*); or 2, inserted at the base of the scale-like perianth; *filaments* bifurcate at the top (*Betula*); *anthers* basifixed, with cells juxtaposed (*Alnus*), or separate on the branches of the filament (*Betula*), dehiscence longitudinal.—CATKINS ♀ sometimes pendulous, solitary, with membranous or

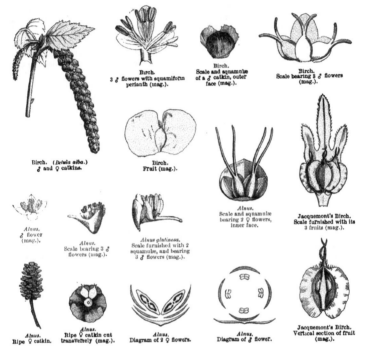

Birch.
3 ♂ flowers with squamiform perianth (mag.).

Birch.
Scale and squamule of a ♂ catkin, outer face (mag.).

Birch.
Scale bearing 3 ♂ flowers (mag.).

Birch. (*Betula alba.*)
♂ and ♀ catkins.

Birch.
Fruit (mag.).

Alnus.
Scale and squamule bearing 2 ♀ flowers, inner face.

Jacquemont's Birch.
Scale furnished with its 3 fruits (mag.).

Alnus.
♂ flower (mag.).

Alnus.
Scale bearing 3 ♂ flowers (mag.).

Alnus glutinosa.
Scale furnished with 2 squamule, and bearing 3 ♂ flowers (mag.).

Alnus.
Ripe ♀ catkin.

Alnus.
Ripe ♀ catkin cut transversely (mag.).

Alnus.
Diagram of 2 ♀ flowers.

Alnus.
Diagram of ♂ flower.

Jacquemont's Birch.
Vertical section of fruit (mag.).

coriaceous scales caducous when full grown (*Betula*); sometimes erect in a corymbose raceme, with the scales at first fleshy, then woody and persistent (*Alnus*); *scales* sessile, 3-lobed and flowered (*Betula*), or entire and each accompanied by 4 lateral squamules, and 2-flowered (*Alnus*). PERIANTH 0. OVARY sessile, with two 1-ovuled cells, of which one is often empty; *stigmas* 2, terminal, sessile, elongate, filiform; *ovules* pendulous from the septum a little below its top, anatropous. STROBILUS formed by the enlarged scales. NUTS winged or annular, 1-celled and 1-seeded. SEED inverted; *testa* membranous, thin. EMBRYO exalbuminous, straight; *cotyledons* plane, foliaceous in germination; *radicle* superior.

GENERA.

* Betula.   * Alnus.   Clethropsis.

We shall indicate the affinities of *Betulaceæ* with *Cupuliferæ* and *Myriceæ* under those families. They approach *Ulmaceæ* in the stamens being opposite to the calyx-lobes, in the 2-celled ovary and

1-ovuled cells, the pendulous and anatropous ovule, winged fruit, exalbuminous seed, and alternate leaves with caducous stipules; but their diclinism and amentaceous inflorescence decidedly separate them.

Betulaceæ are spread over the temperate and cold climates of the northern hemisphere. Birches form woods or vast forests in Northern Europe, Asia, and America. Some stunted shrubs inhabit the polar regions up to the limits of perpetual snow. Alders are dispersed over the temperate countries of the northern hemisphere, and over sub-tropical America.

The *Betulaceæ*, without rivalling *Cupuliferæ* in their use to mankind, furnish him with several useful species, at the head of which must be placed the White Birch (*Betula alba*), which grows nearly throughout Europe, Siberia, Asia Minor, and North America; it approaches nearer to the pole than any other European tree, and grows in alpine districts the least favourable to vegetation. Its wood is too spongy to be employed in buildings, but cartwrights, upholsterers, and turners value it for its tenacity; as firewood it is nearly as useful as beech, and its charcoal is in demand for forges. Its flexible branches are made into brooms. Its bark is impermeable to water, and hence used in the north for various utensils, shoes, cords, boxes, snuff-boxes, and for preserving roofs from moisture. The bark of a variety of the White Birch (the Canoe Birch or Paper Birch) is used in Canada for the construction of light and portable canoes; these are formed of sheets of bark bound together with the root-fibres of the White Fir, and smeared with the resin of the Balsam Fir. The bark of the White Birch contains, besides an astringent principle used in tanning leather, a resinous balsamic oil, which becomes empyreumatic by distillation, and is used in the preparation of Russia leather. Finally, the cellular part of this bark, which contains starch, is a valuable resource to the Samoiedes and the Kamtschatkans, who bruise it, and mix it with their food. The sap, which is sugary before the sprouting of the leaves, is considered an excellent antiscorbutic in North America, and both vinegar and beer are made from it. Other species of Birch (*B. lenta, nigra, lutea*, &c.) are similarly prized in North America for their wood, tannin and sugary sap. The common Alder (*Alnus glutinosa*) is a nearly aquatic tree; its wood is little used for buildings, because it does not withstand alternate dryness and damp; but if constantly submerged, it becomes as incorruptible as oak, whence its use for piles. Alder wood is used by upholsterers, cabinet-makers, turners, and sabot-makers, and is esteemed for firewood, because it burns with a bright and almost smokeless flame; its charcoal is usually employed in gunpowder making. The bark, leaves, and cones may be used for tanning and dyeing black.

CXCVIII. *PLATANEÆ.*

(PLATANEÆ, *Lestiboudois.*—PLATANACEÆ, *Lindl.*)

FLOWERS *diclinous, in 1-sexual capitula.* PERIANTH *replaced by clavate scales.* STAMENS *equal and alternate with the scales.* OVARIES *1-celled, 1-2-ovuled*; OVULE *pendulous, orthotropous.* FRUIT *a nut.* SEED *exalbuminous or not*; RADICLE *inferior.* —STEM *woody.* BUDS *long, concealed by the concave base of the petiole, and only appearing after the fall of the leaf.* LEAVES *stipulate, alternate, palminerved.*

Usually tall TREES, bark flaking. LEAVES alternate, petioled, palminerved, large, few-lobed, furnished with stellate hairs, fugacious; *stipules* leaf-opposed, tubular at the base, often crowned with a blade at the top, caducous. FLOWERS[1] monœcious, agglomerated in 1-sexual globose capitula, occupying different branches. —FLOWERS ♂ surrounded by scale-like bracts, minute, hairy at the top, and within by sepals (arrested stamens?) longer than the bracts, linear-clavate, furrowed, truncate at the top. STAMENS equal and alternate with the scales; *filaments* very short;

[1] [This description of the ♂ and ♀ flowers of *Plan-tanæ* does not accord with the usually received view, which is, that the stamens in the ♂ and the ovaries in the ♀ are mixed without definite order with scales which may be bracts, perianth-segments, or staminodes, or arrested ovaries.—ED]

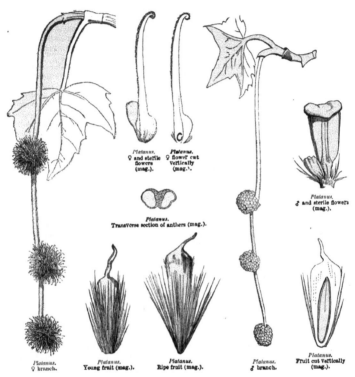

Platanus.
♀ and sterile
flowers
(mag.).

Platanus.
♀ flower cut
vertically
(mag.).

Platanus.
♂ and sterile flowers
(mag.).

Platanus.
Transverse section of anthers (mag.).

Platanus.
♀ branch.

Platanus.
Young fruit (mag.).

Platanus.
Ripe fruit (mag.).

Platanus.
♂ branch.

Platanus.
Fruit cut vertically
(mag.).

anthers elongate-clavate, cells lateral, *connective* longer than the cells, peltate, pubescent. —FLOWERS ♀ surrounded by 3–4–0 bracts; *sepals* 3–4, clavate, with a truncate top; *scales* (arrested stamens?) alternate with the sepals, much smaller, often 0, irregularly obovate, pointed. CARPELS 5–8–4–2, sub-whorled, opposite to the lobes, to which they adhere at the base, ovoid, 1-celled; *style* terminal, elongate-linear, tip recurved, stigmatiferous on the inner side; *ovule* solitary, pendulous from the top of the cell (rarely 2?), orthotropous. *NUTS* 1-seeded, coriaceous, terminated by the persistent style, surrounded at the base by stiff hairs. SEED pendulous; *testa* thin, membranous; *albumen* 0, or very scanty; *radicle* cylindric, elongated, inferior.

ONLY GENUS.
* Platanus.

Plataneæ are near *Balsamifluæ*, which are, however, distinguished by their balsamic juice, conical involucrate catkins, 2-styled ovary with 2-many-ovuled cells, sub-anatropous ovules and capsular fruit. *Plataneæ* also approach *Hamamelideæ* (see page 410), and have some affinity with *Garryaceæ* in inflores. cence, diclinism, 1 or 2 pendulous ovules, styles stigmatic throughout their length, and straight axile embryo; but *Garryaceæ* differ in the clearly tetrandrous flower with 4 sepals, the anatropous ovule, abundant albumen, berried fruit, opposite penninerved leaves, &c. [The affinities of *Plataneæ* are rather with the Amentaceous families, not with the families enumerated above.]

Plataneæ are natives of Mediterranean Asia and North America, but are cultivated as ornamental trees in temperate regions throughout the world. They possess a slightly astringent property [but are chiefly valuable for their timber and ornamental appearance. *P. orientalis* attains a gigantic size and immense age].

CXCIX. *MYRICEÆ.*

(AMENTACEARUM *pars, Jussieu.*—MYRICEÆ, *L.-C. Richard.*—MYRICACEÆ, *Lindl.*)

FLOWERS *diclinous, spicate, achlamydeous, sessile in the axil of a bract.* STAMENS 2–16. OVARY *with 2–4 scales at the base, 1-celled, 1-ovuled;* OVULE *sessile, basilar, orthotropous.* FRUIT *a drupe, succulent, waxy.* SEED *erect, exalbuminous.* EMBRYO *straight.*—STEM *woody.* LEAVES *alternate.*

UNDERSHRUBS, SHRUBS or TREES with scattered cylindrical unjointed branches. BUDS scaly, solitary. LEAVES alternate, simple, usually serrated, sometimes cut or pinnatifid, rarely entire, stiff or coriaceous when dry, usually sprinkled with globose or discoid ceraceous odoriferous particles, midrib giving off anastomosing veins; *stipules* 0 (except in *Comptonia*). FLOWERS monœcious or diœcious, in axillary simple or compound catkins, and forming close [open?] spikes; *catkins* of the monœcious species 2-sexual, and bearing the ♂ at the base, and ♀ at the top.— FLOWERS ♂ : PERIANTH 0. STAMENS 2–16, sessile in the axil of a bract, naked or furnished with 2 lateral bracteoles, often accompanied by several flowerless bracteoles; *filaments* free, connate at the base ; *anthers* extrorse, basifixed, sub-didymous, dehis- cence longitudinal.—FLOWERS ♀ in ovoid or cylindric catkins, ebracteate. OVARY sessile in the axil of the bract, furnished at the base with, and more or less adnate to

Myrica californica.

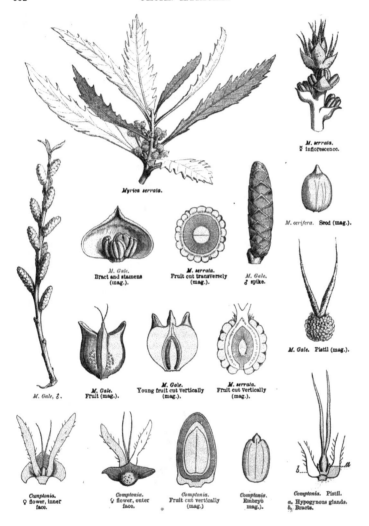

Myrica serrata.

M. serrata.
♀ Inflorescence.

M. cerifera. Seed (mag.).

M. Gale.
Bract and stamens
(mag.).

M. serrata.
Fruit cut transversely
(mag.).

M. Gale.
♂ spike.

M. Gale. Pistil (mag.).

M. Gale, ♂.

M. Gale.
Fruit (mag.).

M. Gale.
Young fruit cut vertically
(mag.).

M. serrata.
Fruit cut vertically
(mag.).

Camptonia.
♀ flower, inner
face.

Comptonia.
♀ flower, outer
face.

Comptonia.
Fruit cut vertically
(mag.)

Comptonia.
Embryo
mag.).

Comptonia. Pistil.
a, Hypogynous glands.
b, Bracts.

2–4 sterile scales, 1-celled; *stigmas* 2, lateral, filiform, sessile; *ovule* solitary, ortho-tropous, basal, sessile. DRUPE covered either with long fleshy papillæ, succulent, edible (*M. sapida*), or with superficial wax-secreting odorous glands; *endocarp* bony. SEED erect; *testa* thin, membranous. EMBRYO exalbuminous, antitropous; *cotyledons* fleshy, plano-convex; *radicle* cylindric, superior.

PRINCIPAL GENERA.

Myrica.   Comptonia.

Myriceæ approach *Juglandeæ* in diclinism, inflorescence, the 1-celled and 1-ovuled ovary, the erect orthotropous ovule, drupaceous fruit, exalbuminous seed, woody stem, and alternate aromatic leaves. They are distinguished by their achlamydeous flowers, free ovary, waxy fruit, habit, and simple coria-ceous 1-nerved leaves sprinkled with resinous glands. They have also some affinity with *Casuarineæ* (which see) and *Betulaceæ*, from which they are distinguished by their 1-celled ovary and erect ortho-tropous ovule, and by their properties. Mr. Chapman has joined to *Myriceæ* the genus *Leitneria*, which differs in its campylotropous ovule.

Myriceæ belong to both worlds, but are nowhere abundant [excepting that some species are emi-nently social]. They chiefly inhabit North America, South Africa, and the mountains of Asia and Java. *M. Faya* is found in the Azores and Canaries. *M. Gale* inhabits boggy places in the north-west of Europe [covering vast tracts].

The bark of several *Myriceæ* contains benzoic acid and tannin, united with a resinous substance, which gives it astringent tonic properties; such is *Comptonia aspleniifolia*, a decoction of which is used in North America for obstinate diarrhœa. From the fruits of *M. Gale* is obtained a volatile oil, of a faint scent, which is almost tasteless at first, but becomes acrid in the mouth. Those of *M. sapida*, of Nepal [and China], contain an acid sugary and very agreeable taste, due to clavate fleshy hairs which are loaded with a red juice. The wax of *M. cerifera* was long used extensively in America for lighting; its root is an emetic and violent purgative. The leaves of *M. Gale* were formerly praised as antipsoric.

CC. *CASUARINEÆ*, Mirbel.

FLOWERS *diclinous, achlamydeous, the ♂ with 2–4 bracteoles, monandrous; the ♀ with 2–4 scarious bracteoles.* OVARY *1-celled;* STYLES 2; OVULES 2, *semi-anatropous.* FRUIT *a samara.* SEED *exalbuminous.* RADICLE *superior.*—STEM *woody, branches jointed, leafless.*

SHRUBS or very much branched TREES, with the habit of *Ephedra* or *Equisetum*; *branches* whorled, jointed, knotty, striate or furrowed. LEAVES replaced by striate many-toothed sheaths surrounding the nodes of the branches. FLOWERS monœcious or diœcious, ♂ in spikes, ♀ in capitula:—FLOWERS ♂ seated within the last sheaths of the branches, each with 4 bracteoles, 2 lateral and 2 antero-posterior, all connate into a deciduous cap which is detached by the elongation of the stamen. STAMEN solitary, central; *filament* at first short, thick at the base, lengthening while flowering; *anther* incumbent; cells sub-opposite, separated at the top and bottom, dehiscence longitu-dinal. FLOWERS ♀ capitate at the tips of the branches, each sessile in the axil of a persistent bract, furnished with 2 boat-shaped bracteoles, at first open, then closing on the young fruit, and re-opening when ripe. PERIANTH 0. OVARY compressed-lenticular, 1-celled; *style* terminal, very short; *stigmas* 2, elongate-filiform; *ovules*

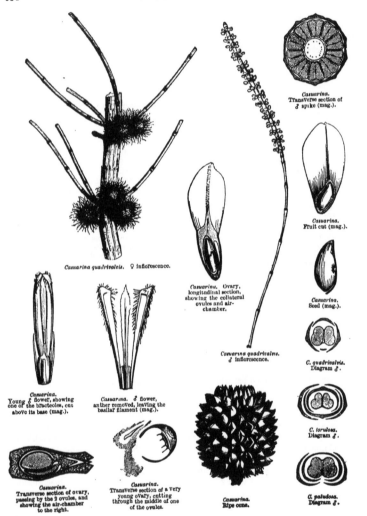

Casuarina quadrivalvis. ♀ inflorescence.

Casuarina. Ovary, longitudinal section, showing the collateral ovules and air-chamber.

Casuarina. Transverse section of ♂ spike (mag.).

Casuarina. Fruit cut (mag.).

Casuarina. Seed (mag.).

Casuarina quadrivalvis. ♂ inflorescence.

C. quadrivalvis. Diagram ♂.

Casuarina. Young ♂ flower, showing one of the bracteoles, cut above its base (mag.).

Casuarina. ♂ flower, anther removed, leaving the basilar filament (mag.).

Casuarina. Transverse section of ovary, passing by the 2 ovules, and showing the air-chamber to the right.

Casuarina. Transverse section of a very young ovary, cutting through the middle of one of the ovules.

Casuarina. Ripe cone.

C. torulosa. Diagram ♂.

C. paludosa. Diagram ♂.

2, collateral, fixed above the base of the cell, semi-anatropous. FRUIT a strobilus of woody bracts and bracteoles, each pair of bracts opening like a capsule into 2 spread_ ing valves, and containing a samaroid caryopsis membranous at the top, crustaceous at the base, and filled with spiral vessels. SEEDS solitary, erect, funicle attached to the middle of the testa and backed by the arrested ovule; *testa* membranous, nearly transparent. EMBRYO exalbuminous; *cotyledons* large, oblong, compressed; *radicle* minute, superior.

ONLY GENUS.

* Casuarina.

Casuarineæ are distinguishable at the first glance, having scarcely any affinity except with *Myriceæ*, which differ, besides their characteristic habit, in the nature of their drupaceous fruit.

Dr. Bornet, who has examined fresh flowers of *Casuarina*, cultivated by M. Thuret at Antibes, informs us that 'in *C. quadrivalvis* each stamen is surrounded by a 3-valved perianth (?); two valves are lateral, the third faces the axis; in very rare instances a fourth valve has been found pressed against the anterior face of the stamen. The posterior valve is linear and does not adhere to the lateral valves; the latter are much larger, plicate, enlarged at the top, recurved into a hood, and adhere by the interlacing of marginal hooked hairs. The fourth valve, when present, is narrow and linear. The ♂ flower of *C. torulosa* is the most complicated of all, and the only one which constantly presented a 4-nary perianth, composed of 2 lateral, a posterior and an anterior segment. This arrangement is easily seen in the young flowers, in which the filaments are still very short. As the filament lengthens, the valves are torn across a little above their bases, which persist as small brown truncate scales, surrounding the base of the filament, the upper part remaining pressed against the anther, capping its top until the anther is ready to open, when this latter becomes oval, greatly enlarged, and drives off, like a wedge, the lateral valves.'

As far as Bornet could judge of the ♀ flowers, the funicle in *C. quadrivalvis* is not normally free, and the placental vessels are curiously arranged; a section of the ovary shows that the cavity of the pericarp to be divided by a cellular mass of the placenta, one of the cavities containing the collateral ovules, the other being empty. This empty chamber is not accidental, for it is seen in very young ovaries, and is perhaps itself divided in two by a further prolongation of the placenta.

Casuarineæ are chiefly natives of Australia, where they have been found in a fossil state; they are also met with in India, the Indian Archipelago and Madagascar, where they are called *Filao*.

Casuarineæ are of little use to man; their hard and heavy wood may be used in ship-building, and is made into war-clubs by the Australians [and Pacific Islanders]. The bark of *C. equisetifolia* is astringent; it is used in powder in dressing wounds, and in a decoction to stop chronic and choleraic diarrhœa; it may also prove useful as a colouring matter, or mordant. That of *C. muricata* furnishes the Indians with a nerve-tonic medicine.

CCI. *SALICINEÆ, L.-C. Richard.*

FLOWERS *diœcious.* PERIANTH 0. STAMENS 2-∞. OVARY 1-*celled;* STYLES 2; PLACENTAS *parietal,* 2, ∞-*ovuled;* OVULES *anatropous.* CAPSULE *with 2 semi-placen- tiferous valves.* SEEDS *erect, bearded at the base, exalbuminous.* RADICLE *inferior.—* STEM *woody.* LEAVES *alternate, stipulate.*

TREES or SHRUBS, or dwarf creeping UNDERSHRUBS; *branches* cylindric, alternate. LEAVES alternate, simple, penninerved, entire or angular-toothed, petioled; *stipules* scaly and deciduous, or foliaceous and persistent. FLOWERS diœcious, in terminal sessile or pedicelled catkins, each furnished with a membranous entire or lobed

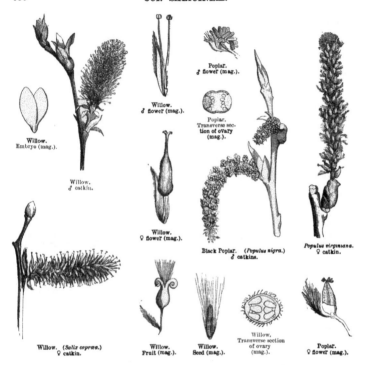

Willow.
Embryo (mag.).

Willow.
♂ flower (mag.).

Willow.
♂ catkin.

Poplar.
♂ flower (mag.).

Poplar.
Transverse section of ovary
(mag.).

Willow.
♀ flower (mag.).

Black Poplar.   (*Populus nigra.*)
♂ catkins.

Populus virginiana.
♀ catkin.

Willow.   (*Salix caprœa.*)
♀ catkin.

Willow.
Fruit (mag.).

Willow.
Seed (mag.).

Willow,
Transverse section
of ovary
(mag.).

Poplar.
♀ flower (mag.).

bract.—FLOWERS ♂ : PERIANTH replaced by a glandular annular or urceolate and obliquely truncate disk. STAMENS 2–∞, inserted on the centre of the torus ; *filaments* filiform, free, or more or less connate ; *anthers* basifixed ; cells parallel, contiguous, dehiscence longitudinal.—FLOWERS ♀ : PERIANTH 0, disk glandular or urceolate. OVARY sessile [or pedicelled], 1-celled ; *placentas* 2, parietal ; *styles* 2, very short, more or less connate, each terminated by a 2–3-lobed stigma ; *ovules* many, ascending, anatropous. CAPSULE 1-celled, valves 2, revolute when ripe, semi-seminiferous. SEEDS numerous, erect, minute ; *testa* membranous ; *hilum* basilar, truncate ; *funicle* short, thick, expanding into a woolly tuft which envelops the seed. EMBRYO exalbuminous, erect ; *cotyledons* plano-convex, elliptic ; *radicle* very short, inferior.

GENERA.

* Salix.   * Populus.

The affinities of *Salicineæ* are obscure; they approach *Balsamifluæ* (see that family), and resemble the Amentaceous families in their inflorescence, diclinism, the absence of perianth (at least in the ♀), the connate carpels, dry fruit, and straight exalbuminous embryo, but there the resemblance ceases. Some botanists have compared them with *Tamariscineæ* on account of the parietal placentation and bearded seeds, but this comparison appears unnatural.

Willows principally affect damp and swampy places throughout the northern hemisphere. [A few species inhabit temperate South America and South Africa, none Australia or the Pacific Islands.] Poplars grow in Central and South Europe, North Africa and North America. They have not been met with elsewhere.

Salicineæ possess astringent and bitter principles in their bark, which are used for the cure of inter-mittent fevers, especially in the form of the active principle *salicine*, which is obtained from *Salix alba, vitellina, amygdalina, viminalis, Helix, purpurea*, &c. Some Poplars (*Populus alba* and *tremula*) also con-tain salicine, but it is accompanied by another alkaloid named *populine*. Their bark also contains a yellow dye, which is often utilized. The resinous and balsamic buds of the Black Poplar (*P. nigra*) and Aspen (*P. tremula*) yield the base of the unguent *populeum*, a preparation used in cases of hæmorrhage; these buds are also taken internally in chronic affections of the lungs. *P. balsamifera*, of North America, yields the resin *tacamahaca*, a resolvent and vulnerary. The ♂ catkins of *Salix ægyptiaca* are very odori-ferous, and a medicinal water is prepared from them in the East which is regarded as a sudorific and cordial. Poplar wood, although soft, is valued for its lightness; that of Willows, and especially of *Salix capræa*, and of Osiers (*S. vitellina, viminalis, purpurea*), is universally used by basket-makers, coopers, and gardeners. Several species of Willows and Poplars contribute to the beauty of our gardens.

CCII. *EUPHORBIACEÆ.*

(TRICOCCÆ, *L.*—TITHYMALOIDEÆ, *Ventenat.*—EUPHORBIÆ, *Jussieu.*—EUPHORBIACEÆ, *R. Br.*—ANTIDESMEÆ, PUTRANJIVEÆ *adjunctæ.*—BUXINEÆ *exclusæ.*)

FLOWERS *diclinous, usually mono- rarely di-chlamydeous, or achlamydeous.* OVARY *free, 3-celled, rarely of 2 or several 1-2-ovuled cells*; OVULES *pendulous.* FRUIT *usually of 3 cocci, rarely a berry.* SEEDS *solitary or geminate, pendulous.* EMBRYO *straight, in the axis of an abundant fleshy albumen.*

Large or small TREES, or UNDERSHRUBS, or HERBS, of very various habit; juice milky, acrid, opaline or watery. LEAVES alternate, rarely opposite or whorled, sometimes much reduced, never wholly absent, petioled or sessile, usually 2-stipulate at the base, nearly always simple, rarely 3-foliolate [or pinnate]; *limb* entire, toothed or lobed, penni- or palmi-nerved, very various in size, shape, consistence and clothing. INFLORESCENCE axillary (rarely extra-axillary) or terminal, definite or indefinite, very polymorphous. FLOWERS monœcious or diœcious, ♂ with often a rudimentary pistil, ♀ central (i.e. terminal) when the inflorescence is androgynous and definite, peripheric (i.e. lowest on the axis) when it is indefinite. CALYX gamosepalous, usually 3–5-merous, æstivation various (very rarely 0); lobes equal or unequal, entire, toothed or cut. COROLLA polypetalous, or very rarely gamopetalous, hypogynous or perigynous, varying like the calyx in the number of parts, colour and æstivation; *petals* alternate with the calyx-lobes when they are isomerous, imbricate or twisted in æstivation. DISK varying in situation and form (or 0). STAMENS 1–∞, central in the ♂, or inserted at the bottom or on the base of the calyx; *filaments* free or mono-poly-adelphous, erect or incurved; *anthers* [usually absent] variously inserted, globose or didymous, free or coherent, opening longitudinally, obliquely or horizontally by 2 (or 1) extrorse or introrse slits or oblong pores, connective polymorphous. OVARY superior, 3- (rarely 1-many-) celled, whorled around a central column which is persistent after the fall of the fruit when it is a capsule; *style* various, usually short, with as many branches as there are cells, often again divided, more or less reflexed and papillose above; *ovules* solitary or geminate, pendulous from the inner angle of the cells, sessile, usually operculate by a cellular development from the placenta, usually anatropous or semi-anatropous; *raphe* ventral, rarely dorsal. FRUIT various, usually a capsule, opening in 3, or 2-∞ usually 2-valved cocci which separate from a persistent axis [or drupaceous]. SEEDS pendulous; *testa* crustaceous, very often arillate or carunculate; *albumen* fleshy, more or less copious. EMBRYO various; *cotyledons* large or small; *radicle* superior.

[In the classification of this family I have followed J. M. Müller's learned monograph in De Candolle's 'Prodromus.' Each of the below-mentioned tribes subdivides naturally into sub-tribes, the characters of which are given, with the most important genera in each.]

SERIES I. STENOLOBEÆ.

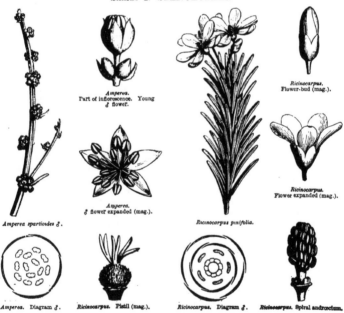

Amperea.
Part of inflorescence. Young ♂ flower.

Ricinocarpus.
Flower-bud (mag.).

Amperea.
♂ flower expanded (mag.).

Ricinocarpus.
Flower expanded (mag.).

Amperea spartioides ♂.

Ricinocarpus pinifolia.

Amperea. Diagram ♂.

Ricinocarpus. Pistil (mag.).

Ricinocarpus. Diagram ♂.

Ricinocarpus. Spiral androecium.

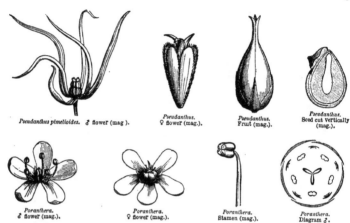

Pseudanthus pimelioides. ♂ flower (mag).

Pseudanthus. ♀ flower (mag.).

Pseudanthus. Fruit (mag.).

Pseudanthus. Seed cut vertically (mag.).

Poranthera. ♂ flower (mag.).

Poranthera. ♀ flower (mag.).

Poranthera. Stamen (mag.).

Poranthera. Diagram ♂.

Cotyledons semi-cylindric, not sensibly broader than the radicle, and much narrower than the albumen.—Stem sub-woody. Leaves narrow.

TRIBE I. *CALETIEÆ.*

Ovarian cells 2-ovuled. Calyx of ♂ quincuncial in æstivation. (Australia.)

Sub-tribe I. PORANTHEREÆ.—Flowers petalous. Stamens inserted around a rudimentary ovary; anthers 4-porose. *Poranthera.*

Sub-tribe II. EUCALETIEÆ.—Flowers apetalous. Stamens inserted around a rudimentary ovary; anthers with 2 slits. *Caletia, Micrantheum.*

Sub-tribe III. PSEUDANTHEÆ.—Stamens central. *Pseudanthus, Stachystemon.*

TRIBE II. *RICINOCARPEÆ.*

Ovarian cells 1-ovuled. Calyx of ♂ quincuncial in æstivation. (Australia.)

Sub-tribe I. EURICINOCARPEÆ.—Flowers petaloid, exinvolucrate. *Beyeria, Ricinocarpus,* &c.

Sub-tribe II. BERTYEÆ.—Flowers apetalous, involucrate. *Bertya.*

TRIBE III. *AMPEREÆ.*

Ovarian cells 1-ovuled. Calyx of ♂ valvate or sub-valvate in æstivation. (Australia.)

Sub-tribe I. MONOTAXIDEÆ.—Flowers petalous, exinvolucrate. *Monotaxis.*

Sub-tribe II. EUAMPEREÆ. Flowers apetalous, exinvolucrate. *Amperea.*

SERIES II. PLATYLOBEÆ.

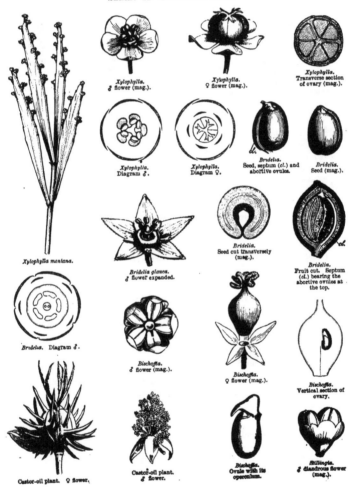

Xylophylla.
♂ flower (mag.).

Xylophylla.
♀ flower (mag.).

Xylophylla.
Transverse section
of ovary (mag.).

Xylophylla.
Diagram ♂.

Xylophylla.
Diagram ♀.

Bridelia.
Seed, septum (cl.) and
abortive ovules.

Bridelia.
Seed (mag.).

Xylophylla montana.

Bridelia glauca.
♂ flower expanded.

Bridelia.
Seed cut transversely
(mag.).

Bridelia.
Fruit cut. Septum
(cl.) bearing the
abortive ovules at
the top.

Bridelia. Diagram ♂.

Bischofia.
♂ flower (mag.).

Bischofia.
♀ flower (mag.).

Bischofia.
Vertical section of
ovary.

Castor-oil plant. ♀ flower.

Castor-oil plant.
♂ flower.

Bischofia.
Ovule with its
operculum.

Stillingia.
♂ diandrous flower
(mag.).

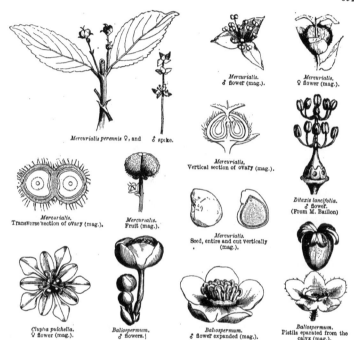

Mercurialis perennis ♀, and ♂ spike.

Mercurialis.
♂ flower (mag.).

Mercurialis.
♀ flower (mag.).

Mercurialis.
Vertical section of ovary (mag.).

Mercurialis.
Transverse section of ovary (mag.).

Mercurialis.
Fruit (mag.).

Mercurialis.
Seed, entire and cut vertically (mag.).

Ditaxis lancifolia.
♂ flower.
(From M. Baillon)

Clutia pulchella.
♀ flower (mag.).

Baliospermum.
♂ flowers.

Baliospermum.
♂ flower expanded (mag.).

Baliospermum.
Pistils separated from the calyx (mag.).

Cotyledons plane, much broader than the radicle, nearly as broad as the albumen.

TRIBE IV. *PHYLLANTHEÆ.*

Ovarian cells 2-ovuled. Calyx of ♂ quincuncial in æstivation.

* *Flower exinvolucrate, apetalous or petalous.*

Sub-tribe I. SAVIEÆ.—Stamens inserted around a rudimentary ovary. Glands of the extrastaminal disk alternating with the petals. *Actephila, Savia, Moacurra.*

Sub-tribe II. ANDRACHNEÆ.—Stamens inserted round a rudimentary ovary. Glands of the extrastaminal disk opposite to the petals. *Andrachne.*

Sub-tribe III. SAUROPIDEÆ.—Stamens central. Petals 0. Extrastaminal glands opposite the sepals. *Sauropus,* &c.

Sub-tribe IV. FREIREODENDREÆ.—Stamens excentric, inserted around a central disk; anthers always erect. *Freireodendron.*

YY2

Sub-tribe V. ANTIDESMEÆ.—Stamens inserted around the base of a rudimentary ovary. Anther-cells pendulous in bud. *Thecacoris, Antidesma.*

Sub-tribe VI. HIERONYMEÆ.—Stamens inserted around the base of a rudimentary ovary; anther-cells erect in bud, then versatile. *Hieronyma.*

Sub-tribe VII. EUPHYLLANTHEÆ. — Stamens central; anther-base always inferior. *Phyllanthus, Breynia, Putranjiva,* &c.

Sub-tribe VIII. LEPTONEMEÆ. — Stamens central; anther-base at length superior. *Leptonema.*

Sub-tribe IX. SECURINEGEÆ.—Stamens inserted around the base of a rudimentary ovary; anther-base always inferior. Calyx-lobes of ♂ with flat (not induplicate) margins. *Securinega, Baccaurea, Aporosa.*

Sub-tribe X. BISCHOFFIEÆ.—Stamens inserted around the base of a rudimentary ovary; anther-base always inferior. Calyx-lobes of ♂ induplicative. *Bischoffia.*

Sub-tribe XI. HYÆNANCHEÆ.—Stamens inserted around a central vacant space. Anther-base always inferior. *Hyœnanche.*

Sub-tribe XII. CYCLOSTEMONEÆ.—Stamens inserted around a central disk. Anther-base always inferior. *Cyclostemon, Hemicyclia,* &c.

* * *Male flowers involucrate.*

Sub-tribe XIII. UAPACEÆ.—Flowers apetalous. Stamens inserted around a rudimentary ovary. Anthers erect. *Uapaca.*

TRIBE V. *BRIDELIEÆ.*

Ovarian cells 2-ovuled. Calyx of ♂ valvate in æstivation.

GENERA.

Bridelia.   Cleistanthus, &c.

TRIBE VI. *CROTONEÆ.*

Ovarian cells 1-ovuled. Anthers inflexed in bud. Calyx of ♂ quincuncial in æstivation.

Sub-tribe I. EUCROTONEÆ.—Flowers ♂ with no rudimentary ovary. **Croton, Julocroton, Crotonopsis.*

Sub-tribe II. MICRANDREÆ.—Flowers ♂ with a rudimentary ovary. *Micrandra.*

TRIBE VII. *ACALYPHEÆ.*

Ovarian cells 1-ovuled. Anthers erect in bud. Flowers in the axils of the bracts, or involucrate; involucres 1-sexual. Calyx of ♂ valvate in æstivation.

* *Flowers not involucrate.*

Sub-tribe I. JOHANNESIEÆ.—Calyx of ♂ very short, induplicative. Stamens central. *Johannesia.*

Sub-tribe II. HEVEEÆ.—Calyx of ♂ induplicative. Stamens inserted around a central column. *Hevea.*

Sub-tribe III. GARCIEÆ.—Calyx of ♂ irregularly bursting. Petals (at least of the ♂) more numerous than the sepals. Stamens central. *Manniophyton, Garcia, Aleurites,* &c.

Sub-tribe IV. AGROSTOSTACHYDEÆ.—Calyx regularly valvate. Petals of ♂ double the sepals in number. Stamens central. *Agrostostachys.*

Sub-tribe V. CROZOPHOREÆ.—Calyx regularly valvate. Petals equalling the sepals in number, or 0. Stamens central. *Sarcoclinium, Ricinella, Argyrothamnium, Crozophora,* &c.

Sub-tribe VI. CAPERONIEÆ.—Calyx regularly valvate. Petals as many as the sepals, or obsolete. Stamens inserted around a sculptured rudimentary ovary. *Caperonia, Leucocroton.*

Sub-tribe VII. CŒLODISCEÆ.—Flowers apetalous. Stamens inserted around a broad central disk. *Cœlodiscus.*

Sub-tribe VIII. CEPHALOCROTONEÆ.—Flowers apetalous. Stamens inserted around the base of a rudimentary ovary. *Cephalocroton,* &c.

Sub-tribe IX. EUACALYPHEÆ.—Flower apetalous. Stamens central, not polyadelphous. *Phillenetia, Claoxylon, Mercurialis, Acalypha, Alchornea,* [*Cœlebogyne,*] *Bernardia, Tragia, Trewia, Mallotus, Cleidion, Macaranga,* &c.

Sub-tribe X. RICINEÆ.—Flowers apetalous. Stamens polyadelphous, with no rudiment of an ovary. **Ricinus, Homonoya.*

** *Flowers ♀ only involucrate. Involucre calyciform, 1-flowered.*

Sub-tribe XI. DIPLOCHLAMYDEÆ. — Stamens central, without a rudimentary ovary. *Diplochlamys.*

Sub-tribe XII. EPIPRINEÆ.—Stamens inserted around a rudimentary ovary. *Epiprinus.*

*** *Flowers ♂ and ♀ involucrate. Involucre bud-like, many-flowered.*

Sub-tribe XIII. PEREÆ.—Stamens central, without a rudiment of an ovary. *Pera.*

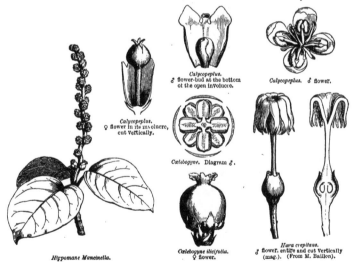

Calycopeplus.
♂ flower-bud at the bottom of the open involucre.

Calycopeplus. ♂ flower.

Calycopeplus.
♀ flower in its involucre, cut vertically.

Cœlebogyne. Diagram ♂.

Hippomane Mancinella.

Cœlebogyne ilicifolia.
♀ flower.

Hura crepitans.
♂ flower, entire and cut vertically (mag.). (From M. Baillon).

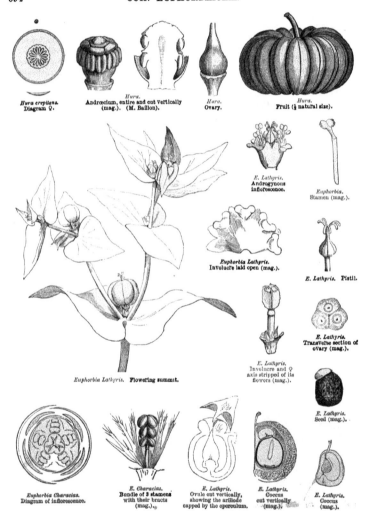

Hura crepitans.
Diagram ♀.

Hura.
Andrœcium, entire and cut vertically
(mag.). (M. Baillon).

Hura.
Ovary.

Hura.
Fruit (⅓ natural size).

E. Lathyris.
Androgynous
inflorescence.

Euphorbia.
Stamen (mag.).

Euphorbia Lathyris.
Involucre laid open (mag.).

E. Lathyris. Pistil.

E. Lathyris.
Transverse section of
ovary (mag.).

E. Lathyris.
Involucre and ♀
axis stripped of its
flowers (mag.).

E. Lathyris.
Seed (mag.).

Euphorbia Lathyris. Flowering summit.

Euphorbia Characias.
Diagram of inflorescence.

E. Characias.
Bundle of 3 stamens
with their bracts
(mag.).

E. Lathyris.
Ovule cut vertically,
showing the arillode
capped by the operculum.

E. Lathyris.
Coccus
cut vertically
(mag.).

E. Lathyris.
Coccus
(mag.).

Tribe VIII. *HIPPOMANEÆ.*

Ovarian cells 1-ovuled. Stamens sometimes inserted around a central disk. Anthers erect in bud. Flowers in the axils of the bracts, or involucrate. Involucres 1-sexual. Calyx of ♂ quincuncial in æstivation.

* *Flowers exinvolucrate, bracts never connate.*

† *Flower with petals, or with the glands of the extrastaminal disk opposite the sepals.*

Sub-tribe I. BENNETTIEÆ.—Petals induplicatively imbricate. Stamens inserted around a rudimentary ovary. *Bennettia.*

Sub-tribe II. POGONOPHOREÆ.—Stamens inserted around a rudimentary ovary. Stamens equalling the sepals, or if more the outer opposite the sepals. *Pogonophora, Microdesmis.*

Sub-tribe III. ACIDOCROTONEÆ.—Stamens central, rarely surrounding the rudiment of an ovary. Outer stamens alternating with the petals. *Acidocroton.*

Sub-tribe IV. CLUYTIEÆ.—Stamens inserted around a rudimentary ovary, alternate with the sepals. Petals perigynous. *Cluytia.*

Sub-tribe V. MANIHOTEÆ.—Stamens inserted around a central disk. **Manihot.*

Sub-tribe VI. JATROPHEÆ.—Stamens inserted around a central receptacle or a rudimentary ovary, equalling the sepals in number, or the outer alternating with them. **Jatropha, Trigonostemon, Ostodes, Codiæum,* &c.

Sub-tribe VII. CHÆTOCARPEÆ. Stamens inserted around a rudimentary ovary, outer alternating with the sepals. Petals suppressed. *Chætocarpus.*

† † *Flowers absolutely apetalous. Disk glands if extrastaminal and isomerous never opposite the sepals.*

Sub-tribe VIII. CHEILOSEÆ.—Stamens inserted around a rudimentary ovary, outer alternating with the sepals. *Cheilosa, Cunuria.*

Sub-tribe IX. MISCHODONTEÆ. Stamens inserted around a rudimentary ovary, opposite the sepals. Disk suppressed. *Mischodon.*

Sub-tribe X. GELONIEÆ.—Stamens central, or surrounding a rudimentary ovary, as many as the sepals, or the outer series opposite the sepals. *Baliospermum, Gelonium, Omphalia,* &c.

Sub-tribe XI. *Adenoclineæ.*—Flower not compressed. Stamens inserted around a conical glandular receptacle. *Adenocline,* &c.

Sub-tribe XII. STOMATOCALYCEÆ. —Flower compressed. Stamens inserted around a central area. *Stomatocalyx.*

Sub-tribe XIII. CARUMBIEÆ. —Flower compressed. Stamens central. *Wartmannia, Carumbium.*

Sub-tribe XIV. EUHIPPOMANEÆ.—Flower not compressed. Stamens central, equalling the sepals in number, or the outer alternating with the sepals. *Mabea, *Stillingia, Sebastiania, Hippomane, Excæcaria, Colliguaya,* &c.

* * *Flowers in the axils of bracts, not involucrate; bracts at first connate, then splitting and resembling an involucre.*

Sub-tribe XV. HUREÆ.—Flower not compressed. Stamens central. *Hura,* &c.

* * * *Flower enclosed in a calyciform involucre.*

Sub-tribe XVI. PHYLLOBOTRYEÆ.—Stamens central, disk 0. *Phyllobotryum.*

TRIBE IX. *DALECHAMPIEÆ.*

Ovarian cells 1-ovuled. Anthers erect in bud. Flowers involucrate. Involucres 2-sexual, i.e. bearing ♂ and ♀ flowers. Calyx of ♂ valvate in æstivation. Involucre compressed, 2-phyllous. Flowers ♂ polyandrous.

GENUS.

Dalechampia.

TRIBE X. *EUPHORBIEÆ.*

Ovarian cells 1-ovuled. Anthers erect in bud. Flowers involucrate; involucres 2-sexual. Calyx of ♂ (rarely developed) quincuncial in æstivation. Involucre calyciform, not compressed. Flowers ♂ monandrous.

PRINCIPAL GENERA.

Anthostema.   Synadenium.   * Euphorbia.   Calycopeplus.   Pedilanthus.

[The genus *Daphniphyllum,* separated from *Euphorbiaceæ* by J. Müller, and erected into a natural order, differs from the latter only in the small embryo, and the ovule with a ventral raphe. They are Indian shrubs of mountainous districts, extending from the Himalayas to Java, China, and Japan, and there is also a West Indian species.—ED.]

The great family of *Euphorbiaceæ* presents numerous affinities. Thus they approach *Malvaceæ* in habit, pubescence, stipulate leaves, often monadelphous stamens and loculicidal fruit; they differ in the nature of their secretions, in diclinism, the position of the ovules, and the structure of the seed. They are connected with *Urticeæ,* but the latter are separated by their 1-celled ovary, simple style, solitary orthotropous ovule, 1-celled indehiscent fruit, persistent calyx, &c. They approach *Rhamneæ* in habit (*Bridelia*), but these are easily distinguished by the hermaphrodite flowers, the position and direction of the ovules, and the situation of the stamens relatively to the petals. *Menispermeæ* also resemble in their flower some *Euphorbiaceæ,* but are distinguished by the arrangement of their free carpels.

About half the species inhabit equatorial America; they are much rarer in temperate America. In the Old World they are more abundant in the Mediterranean region and in temperate Asia than between the tropics. The vast genus *Euphorbia,* comprising more than 700 species, is dispersed over the whole world, except in alpine and cold localities.

The properties of *Euphorbiaceæ* are as distinctive as their botanical characters, and the ancients had so accurately recognized this, that all plants with a 3-coccous fruit were regarded by them as hurtful and suspicious. *Euphorbiaceæ* secrete a very acrid milky juice, varying in strength with the species, the climate, and the organ of the plant. In some this juice is one of the most deleterious of poisons, in others its acridity is so far neutralized by mucilage and resins as to reduce it to a simple purgative and diuretic. Some species are slightly narcotic-acrid, others aromatic. The albumen usually contains a fixed oil, without the acridity which is found in the embryo and integuments of the seed. It is to a liquid resin and a volatile principle that all the properties of *Euphorbiaceæ* are due; they are thus strongest in alcoholic tinctures, but are dissipated or weakened by the application of heat. The root of the Manihot offers a remarkable example: there is scarcely a more poisonous juice than this, yet the action of fire converts the plant into very wholesome food.

Euphorbia, the typical genus, is very various in habit: some species have a fleshy angular stem, furnished with spines, resembling that of a Cactus; these species (*E. antiquorum, canariensis, officinarum, abyssinica*) yield by incision a resinous juice which is strongly drastic, and used externally as a vesicant. [The true Euphorbium is the *E. resinifera* of Morocco.] The other Euphorbias, which have a normal stem and leaves, have a purgative milky juice; such are *E. Esula, Cyparissias, amygdaloides, Helioscopia, Peplus, palustris, Lathyris*, &c., all indigenous to France. Some are considered efficacious in syphilitic cachexia, as *E. parviflora* and *hirta*, of India, and *linearis*, of America. English doctors, before the introduction of mercury, prescribed *E. hiberna* in similar cases; and in Spain *E. canescens* is still administered with the same object. Others are emetic, and in North America *E. corollata* and *Ipecacuanha* are thus used. *E. thymifolia*, slightly astringent and aromatic, is given in India as a vermifuge to infants. *E. hypericifolia*, of tropical America, the juice of which is both astringent and slightly narcotic, furnishes a useful medicine in dysentery. The milky juice of *E. balsamifera*, which grows in the Canaries, is so little acrid, that the natives cook it, and convert it, they say, into a nourishing jelly; that of *E. cotinifolia*, on the contrary, is so poisonous, that the Caribbeans poison their arrows with it. Finally, a *Euphorbia* from the forests of Brazil (*E. phosphorea*) is described by Von Martius, from the stem of which distils a phosphorescent juice.

It is especially in the arborescent *Euphorbiaceæ* that the juice is abundant and caustic. The Blindingtree (*Excœcaria Agallocha*) of the Moluccas contains so acrid a juice that, if a drop falls into the eye, it is nearly blinded: this has happened to sailors sent on shore to cut wood. The smoke even of this wood, when burned, is dangerous. The Manchineel (*Hippomanes Mancinella*) is a fine tree of tropical America, which, according to some travellers, possesses such poisonous properties, that rain falling on the skin, after flowing over its leaves, produces a blister, and that to sleep beneath its shade is death. But J. Jacquin, who resided long in the Antilles, treats this tradition as a fable; he stood under a Manchineel tree naked for some hours, whilst the rain fell through it upon him, and yet he remained unhurt. It is, however, true that a drop of its milk will raise a blister full of serous matter on the skin. The fleshy fruit, of the shape, colour, and smell of an apple, is a very active poison, but so caustic that it would be impossible to eat it.

Another species of tropical America (*Hura crepitans*) contains an extremely deleterious principle, thus described by Boussingault in his 'Cours d'Agriculture':—'When M. Rivero and I analyzed some milk of the *Hura*, sent to us from Guaduas by Dr. Roulin, we were attacked with erysipelas. The courier who brought it was seriously injured, and the inhabitants of the houses where he had lodged on the road experienced the same effects.' This milk perfectly resembled that of the Cow-tree (see *Artocarpeæ*). The fruit of *Hura crepitans* [called Sandbox] is a woody capsule, composed of 12–18 cocci, which, in drying, break open suddenly down the back into two valves, at the same time separating elastically from the axis, with a noise like a pistol-shot. This capsule, boiled in oil to prevent dehiscence, and then emptied, is used as a sand-box in the colonies.

Siphonia elastica is a tree of Guiana and Brazil, 30 to 60 feet in height, whose milky juice is obtained by incisions in the trunk as a tenacious and very elastic mass, known as India-rubber; this is a hydrocarbon, softening in boiling water, insoluble in alcohol, but soluble in ether, sulphuret of carbon and volatile oils. Thanks to this solubility, India-rubber has become a great article of commerce, in the fabrication of elastic tissues, clothing, waterproof shoes, and various utensils. This milky juice is found in other families; several Figs of Asia and America, and especially *Ficus elastica*, of India, possess it; also *Castilloa elastica*, of Sumatra; *Vahea gummifera*, of Madagascar; and *Hancornia speciosa*, of Brazil; but none of these yield so copious a supply as the *Siphonia*.

Amongst the many *Euphorbiaceæ* prescribed for syphilis amongst foreign nations, we may mention *Stillingia sylvatica*, of Carolina and Florida; *Jatropha officinalis, Croton perdiceps* and *campestris*, of Brazil. *Tragia* and *Acalypha*, American and Asiatic plants, are praised as resolvents, diaphoretics, and diuretics. *Mercurialis*, of which two species (*M. annua* and *perennis*) are natives of France, is a moderate purge.

Omphalea triandra, a tree of Guiana, yields a juice, white at first, which blackens in the air, and is used as ink. Some Crotons of America and Africa yield, by incision, a balsamic scented resin. The bark [called Cascarilla bark] of *Croton Eleuteria*, a shrub of the Antilles, contains a volatile oil and a bitter resinous principle, to which it owes its stimulating tonic and slightly astringent qualities. Other congeners of tropical America (*C. nitens, micans, suberosus, Pseudo-china*) have similar properties, and the seeds of several of them contain a strong-smelling essential oil, but soluble in colonial perfumery; such is that of *C. gratissimum*, which recalls the scent of Mint. *C. Tiglium* is a small tree of the Moluccas, all parts of which are purgative. Its seeds contain a fixed oil [Croton-oil], combined with a resin and a peculiar acid, the action of which is so powerful that one or two drops, taken internally, purge violently, as does rubbing it on the stomach, and which further produces an eruption on the skin which may be useful to the patient. The seeds of *Jatropha Curcas* [Physic-nut], a shrub growing throughout the hot countries of America, yield a plentiful supply of an oil of which soap is made. But the most celebrated of the oleiferous *Euphorbiaceæ* is certainly the Palma Christi, or Castor-oil plant, *Ricinus communis*, the seeds of which yield by expression under cold a fixed oil, called Castor-oil, soluble in alcohol (which distinguishes it from all other oils), and much used as a purgative [and when fresh as a hair-oil in India. A virulent principle resides in the seed-coats and embryo, which are not crushed in the process of extracting the oil]. The seeds of the Spurge (*Euphorbia Lathyris*), an indigenous herb, rival those of the Castor-oil and *Croton Tiglium*. Its capsules, and those of its congeners, are used to stupefy fish. Those of *Phyllanthus*, a tropical plant, are similarly employed. The powdered seeds of *Hyænanche globosa*, a South African tree, are by the Cape colonists sprinkled over mutton to poison hyenas. The seeds of *Stillingia sebifera*, called Chinese Wax-tree, besides containing a fixed oil, are covered with a very white wax, used in China for making candles. The very poisonous seeds of the Japanese Oil-tree, (*Elæococca verrucosa*)· yield by expression an oil used for lighting.

The fruits or seeds of some *Euphorbiaceæ* may be eaten with impunity. The almond of *Aleurites triloba*, a small tree of the Moluccas, is very well tasted, and is considered an excitant. The seeds of *Conceveiba guianensis*, the juice of which is green, have a delicious taste. The kernel of the American Omphalens is edible when the embryo has been removed. The acidulous-sugary berries of *Cicca disticha* are eaten in India. *Emblica officinalis*, which also grows in tropical Asia, produces a fleshy fruit of a taste at first harsh, and afterwards sweet. This fruit, dried, was used as an astringent against dysentery, and as a tanning material by the natives. But of all edible *Euphorbiaceæ*, the most valuable (owing to the abundance of starch in their roots) are two kinds of Manioc (*Manihot*) cultivated throughout tropical Africa and America. That of the sweet Manioc (*Manihot Aipè*) is eaten cooked in ashes or in water, like the Potato, and animals eat it raw without injury. This is not the case with the Bitter Manioc (*M. utilissima*), the root of which contains a juice laden with a strongly poisonous principle, analogous to hydrocyanic acid ; but the volatility of this principle, and the facility with which it is destroyed by fermentation, explain the facility with which an abundant and wholesome food is obtained from the root. These roots are grated, pressed, dried, sifted, and then slightly baked on an iron plate ; thus prepared, it swells considerably in water or broth ; this food is called *couaque*. If, instead of drying the grated pulp, it is spread upon a hot iron plate, the starch and mucilage, by mixing together, consolidate the pulp, and form a biscuit called Cassava-bread. The Cipipa is the pure starch of the Manioc, which has been removed along with the expressed juice of the root, and which has been washed and dried in the air. This same starch, heated on iron plates, is partially cooked, and clusters into hard and irregular lumps, called Tapioca. Tapioca is partially soluble in cold water, and forms with boiling water a sort of transparent jelly frequently used in soup.

The Turnsole (*Crozophora tinctoria*), which grows in the Mediterranean region, possesses, like most *Euphorbiaceæ*, an acrid juice and purgative seeds, but its colouring principle is its most useful property. Woollen rags are dipped into the juice expressed from the tops of the plant, and are afterwards exposed to the ammoniacal vapour of urine ; the rags thus acquire a dark blue colour, and are called Flags of Girasol. This matter is used for the colouring of Dutch cheeses, which are dipped into water dyed blue by the Girasol, and immediately dried. The red tinge of the cheese-rind is probably due to the action of the lactic acid contained in the cheese. *Mercurialis* also contains a blue colouring principle, analogous to that of the Girasol. Certain Indian *Euphorbiaceæ*, like *Bischoffia*, are used for dyeing red. [African Teak (*Oldfieldia africana*), a little-known plant, has been referred to *Euphorbiaceæ*, but doubtfully.]

CCIII. *BUXINEÆ.*

(EUPHORBIACEARUM genera, *Jussieu.*—BUXINEÆ, *Franç. Plée.*—BUXACEÆ, *Baillon.*)

TREES or SHRUBS or perennial HERBS. LEAVES opposite or alternate, simple, entire or lobed, coriaceous, persistent, exstipulate. FLOWERS monœcious, axillary or

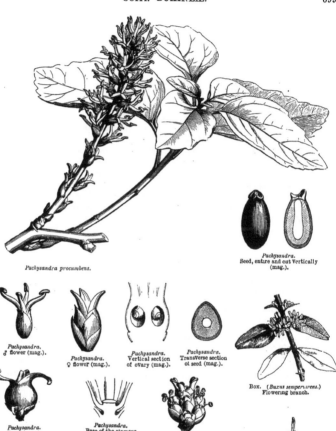

Pachysandra procumbens.

Pachysandra.
Seed, entire and cut vertically
(mag.).

Pachysandra.
♂ flower (mag.).

Pachysandra.
♀ flower (mag.).

Pachysandra.
Vertical section
of ovary (mag.).

Pachysandra.
Transverse section
of seed (mag.).

Box. (*Buxus sempervirens.*)
Flowering branch.

Pachysandra.
Fruit (mag.).

Pachysandra.
Base of the stamens
with abortive ovary
(mag.).

Box.
Inflorescence.

Pachysandra.
Transverse section of
ovary (mag.).

Pachysandra.
Embryo (mag.).

Box.
♂ flower (mag.).

Box.
♀ flower (mag.).

Box.
Dehiscent capsule.

Box.
Diagram ♂.

Box.
Diagram ♀.

Box.
Transverse section of
ovary (mag.).

Box.
Pistil (mag.).

Box.
Seed, entire and cut
vertically (mag.).

terminal, in a spike or raceme, one ♀ terminal (*Buxus*), or some ♀ lowest (*Sarcococca*, *Pachysandra*), the others ♂. FLOWERS ♂: CALYX deeply 4-partite, lobes decussate, 2 lateral outer, enveloping the 2 antero-posterior, imbricate in æstivation. STAMENS 4, opposite to the calyx-lobes; *filaments* hypogynous, erect, exserted when mature; *anthers* 2-celled, introrse, dehiscence longitudinal. OVARY rudimentary, central.— FLOWERS ♀: CALYX deeply 4–12-partite; lobes several-seriate, usually whorled in threes, imbricate in æstivation. OVARY superior, 2–3-celled; *styles* 2–3, excentric, divergent, stigmatic on their inner face, channelled; *ovules* geminate in the cells, suspended from the top of the inner angle, anatropous; *raphe* dorsal; *micropyle* superior and ventral. FRUIT 2–3-celled, or 1-celled by arrest, capsular or fleshy (*Sarcococca*), loculicidal, or indehiscent, crowned by the persistent styles; cells 1–2-seeded. SEEDS pendulous; *testa* crustaceous, black, brilliant, carunculate. EMBRYO curved, in a fleshy albumen; *radicle* superior.

GENERA.

* Buxus.   * Sarcococca.   * Pachysandra.   Styloceras.   Simmondeia.

Buxineæ have hitherto been placed amongst *Euphorbiaceæ* on account of their fruit with three cells or cocci opening elastically. M. Plée, in 1853, separated *Buxus* as the type of a small family which only differs from *Euphorbiaceæ* in the absence of milky juice, the peripheric styles leaving the top of the ovary naked, the placentas which are distinct in their upper portion instead of forming a central common axis, ovules constantly with exterior raphe and interior micropyle; but we have seen in *Celastrineæ*, &c., that the exterior raphe is a character of small value. *Buxineæ* also approach *Hamamelideæ* in their opposite or alternate leaves, inflorescence, dehiscent fruits, seed with exterior raphe and slightly curved embryo in the middle of a copious albumen.

The Box (*Buxus sempervirens*) inhabits the Mediterranean regions, whence it spreads to the north of Europe. Another species grows in the Balearic Islands; three or four others inhabit Asia. The Box with pedicelled ♂ flowers, forming the sub-genus *Tricera*, is American, as well as a species of *Pachysandra*. The other species of *Pachysandra* and *Sarcococca* are Asiatic.

The common Box is a shrub attaining 15 to 20 feet in height, of which a dwarf variety is cultivated for edgings to borders. The close and homogeneous tissue of its wood renders it valuable for wood-engraving, and it is on Box that the illustrations to this work are engraved. A decoction of the grated wood was formerly used as a sudorific and febrifuge; its leaves and seeds are purgative. It is often substituted for Hops to give bitterness to beer, but this adulteration is dangerous, as it induces intestinal inflammation.

CCIV. *PENÆACEÆ, Brown.*[1]

[Flowers ☿. Perianth *tubular, 4-lobed, valvate.* Stamens 4, *on the throat, alternate with the lobes, connective thick.* Ovary *free, 4-celled;* stigma *4-lobed, or* stigmas 4; ovules *2-4 in each cell.* Capsule *loculicidal.* Embryo *minute, in a fleshy albumen.*—Shrubs. Leaves *opposite.*

Evergreen shrubs or undershrubs, with the habit of *Ericeæ* or *Epacrideæ.* Leaves opposite, often imbricate, flat or ericoid, quite entire, penninerved; *stipules* minute, setiform or glandular. Flowers ☿, solitary in the axils of the upper leaves, or of coloured bracts, 2-4-bracteolate, sessile. Perianth coloured, accrescent; *tube* cylindric; *lobes* 4, valvate or reduplicate in æstivation. Stamens 4, inserted on the perianth-throat, alternate with its lobes; *filaments* usually very short, connective thickened; *anther-cells* adnate, introrse, dehiscence longitudinal; *pollen-grains* ovoid, 6-8-sulcate. Ovary free, 4-celled, cells opposite the perianth-lobes; *style* terete or with 4 angles that alternate with the cells; *stigmas* 4, or one 4-lobed; *ovules* 2 in each cell and erect, or 4, 2 erect and 2.pendulous from the inner angle, anatropous. Capsule 4-celled, loculicidally 4-valved; *valves* acuminate, breaking away from the style-base. Seeds oblong; *testa* shining; *hilum* excavated; *funicle* very short, white, thick; *albumen* 0. Embryo fleshy, ovoid; *cotyledons* 2, minute; *radicle* next the hilum, obtuse or concave.

Tribe I. Penæeæ.—Ovules 2 in each cell, erect. Bracteoles 2 or 0. *Penæa, Stylapterus, Brachysiphon, Sarcocolla.*

Tribe II. Endonemeæ.—Ovules 4 in each cell, 2 ascending, 2 pendulous. Bracteoles 2-4. *Glischrocolla, Endonema.*

A family of obscure affinity, most closely allied, according to Lindley, to *Rhamneæ,* but differing signally in the anthers, want of petals and disk, and in the stigma, which is indusiate, as in *Ericeæ.* It has been also compared with *Epacrideæ, Bruniaceæ,* and *Proteaceæ,* but with little reason.

Natives of South Africa, of no known use; some produce a nauseous viscid gum, called Sarcocoll, but which is not the Sarcocoll of the Greeks, whatever that was.—Ed.]

CCV. *GEISSOLOMEÆ.*[1]

(Geissolomeæ, *Endlicher.*— Geissolomaceæ, *Sonder.*)

[Flowers ☿, *subtended by imbricating bracts.* Perianth *4-partite, segments imbricate.* Stamens 8, *at the bottom of the perianth-tube;* anthers *versatile.* Ovary *free, 4-celled;* styles 4; stigmas *minute;* ovules 2, *pendulous in each cell.* Capsule *loculicidal.* Embryo *in the axis of a fleshy albumen;* cotyledons *linear, fleshy.*—Shrub. Leaves *opposite.*

A shrub. Leaves opposite, quite entire. Buds terminal, acute, silky, with

---

[1] This order is omitted in the original.—Ed.

the habit of *Buxus*, cordate, margins thickened, penninerved; *stipules* 0. FLOWERS
axillary, solitary, many-bracteate; *bracts* bifariously imbricate, in unequal pairs,
membranous, ovate, acute, penninerved. PERIANTH-segments 4, ovate, mucronate,
nerves parallel, æstivation imbricate. STAMENS 8, inserted at the bottom of the
perianth, the 4 opposite the lobes rather the longest; *filaments* elongate, included;
anthers erect, ovoid, versatile, cells mucronate, dehiscence longitudinal. OVARY free,
4-lobed, 4-celled, narrowed into the styles; *styles* 4, cohering into one conical
4-grooved style, but at length free; *stigmas* minute; *ovules* 2 in each cell, collateral,
pendulous from the apex, anatropous. CAPSULE 4-celled, loculicidally 4-valved at
the top, cells 1-seeded. SEEDS ovate, sub-compressed; *funicle* short; *testa* thin,
2-lobed between the hilum and micropyle; *albumen* fleshy. EMBRYO straight,
central, as long as the albumen; *cotyledons* linear, fleshy; *radicle* cylindric, obtuse,
superior.

ONLY GENUS.
Geissoloma.

A monotypic family, native of the mountains of South-west Africa, usually combined with *Penæaceæ*,
but differing wholly in the bracts, perianth, stamens, ovary, ovules, and seed. It has no known properties.—ED.]

CCVI. *LACISTEMACEÆ*.[1]

(LACISTEMACEÆ, *Martius*.—LACISTEMEÆ, *Meissner*.)

[FLOWERS ☿, *spiked, bracteate and 2-bracteolate.* SEPALS *0–6 or* 4, *petaloid,
imbricate.* DISK *large, fleshy.* STAMEN *1, 2-fid.* OVARY *free, 1-celled*; STIGMAS *3;*
OVULES *3–6, pendulous, apical.* DRUPE *with a 3-valved endocarp.* SEED *solitary,
pendulous, arillate.* EMBRYO *straight, in a fleshy albumen.*—LEAVES *alternate,
stipulate.*

SHRUBS OT TREES. LEAVES alternate, distichous, persistent, penninerved, often
pellucid-punctate; *stipules* lanceolate, caducous. INFLORESCENCE of axillary solitary
or fascicled spikes. FLOWERS ☿, minute, crowded; *bracts* erose, lower empty;
bracteoles 2, linear. PERIANTH of 2–6, usually 4 petaloid segments, persistent, the
anticous largest, posticous smallest, lateral exterior in æstivation. DISK fleshy,
shorter than the perianth, often unequal, margin inflexed and lobed. STAMEN 1,
anticous, inserted on the disk, persistent; *filaments* 2-fid, each arm 1-celled, cells
dehiscing transversely; *pollen* ovoid, smooth, 3-furrowed. OVARY free, 1-celled,
narrowed into the style; *stigmas* 3, minute; *ovules* 3–6, pendulous from near the top
of the cell, anatropous, funicle thickened. FRUIT a drupe, 1-celled, with a 3-valved
dry endocarp, each valve bearing a placenta on its axis, one only with a seed.
SEED pendulous; *aril* complete, fleshy; *testa* crustaceous; *albumen* copious, fleshy.
EMBRYO central, straight; *cotyledons* ovate, flat, appressed, nerved; *radicle* long,
cylindric.

ONLY GENUS.
Lacistema.

[1] This order is omitted in the original.—ED.

The affinities of *Lacistemeæ* are obscure; but most obvious with *Samydeæ* in the alternate often pellucid-dotted leaves, stipules, disk, 1-celled ovary with parietal placentas, style, stigmas, thin septiferous valves of the capsule, albuminous seed and embryo. I think that they can hardly be regarded otherwise than as reduced *Samydeæ*; and, in fact, Lindley observes that they may when not in flower be easily mistaken for Casearias. The species, of which there are about sixteen, are all tropical American, and have no known uses.—ED.]

CCVII. *NEPENTHEÆ.*

(NEPENTHINÆ, *Link.*—NEPENTHEÆ, *Blume.*—NEPENTHACEÆ, *Lindl.*)

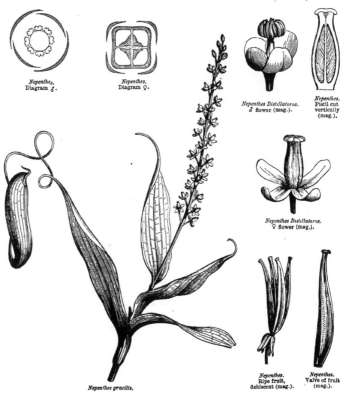

Nepenthes.
Diagram ♂.

Nepenthes.
Diagram ♀.

Nepenthes Distillatoria.
♂ flower (mag.).

Nepenthes.
Pistil cut vertically (mag.).

Nepenthes Distillatoria.
♀ flower (mag.).

Nepenthes gracilis.

Nepenthes.
Ripe fruit, dehiscent (mag.).

Nepenthes.
Valve of fruit (mag.).

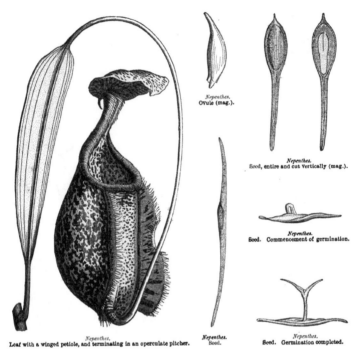

Nepenthes.
Ovule (mag.).

Nepenthes.
Seed, entire and cut vertically (mag.).

Nepenthes.
Seed. Commencement of germination.

Nepenthes.
Leaf with a winged petiole, and terminating in an operculate pitcher.

Nepenthes.
Seed.

Nepenthes.
Seed. Germination completed.

FLOWERS *diœcious*. PERIANTH *single*. STAMENS *united above into an antheri-ferous column*. OVARY *several-celled, many-ovuled*. CAPSULE *with loculicidal semi-septiferous valves*. SEEDS *scobiform, many-seriate on the 2 faces of the septa* ; ALBUMEN *fleshy*. EMBRYO *straight, axile.*—LEAVES *terminated by an operculate pitcher.*

Suffrutescent PLANTS with prostrate or sarmentose stem; *wood* without concentric zones, but with numerous bundles of tracheæ dispersed across the pith and liber, and surrounding the woody axis. LEAVES alternate; *petiole* winged at the base, the midrib prolonged at the top and curved or spirally twisted, and terminating in a second foliaceous expansion, which is hollowed like an urn (the pitcher), to the opening of which is fitted a sort of lid attached as by a hinge, and capable of being lowered or raised, so that the pitcher is sometimes closed, sometimes open; it is often found to contain a watery liquid before the raising of the lid. FLOWERS diœcious, numerous, in a raceme or sub-terminal panicle, which becomes lateral by the growth of the stem. FLOWERS ♂ : PERIANTH single (CALYX), 4-partite; lobes sub-oval, hairy on the outside, pitted within, imbricate in æstivation, the 2 outer somewhat the largest. STAMENS united into a central column; *anthers* about 16, extrorse, united into a spherical head, with 2 apposed and contiguous cells, debiscence longitudinal. FLOWERS ♀ : PERIANTH similar to that of ♂. OVARY free, 3–4-gonous, 3–4-celled, of 3–4 carpels opposite to the perianth-lobes; *stigma* sessile, discoid, obscurely 4-lobed, *lobes* answering to the septa; *ovules* numerous, inserted on the septa, many-seriate, ascending, anatropous. CAPSULE coriaceous, oblong, truncate, crowned by the stigma, 4-celled, with 4 semi-septiferous valves. SEEDS elongate, fusiform, imbricate; *testa* membranous, loose, produced far beyond the nucleus, tubular; *hilum* lateral, near the base; *raphe* filiform, free under the integument in the lower half, and united with it above the middle, and ending in a chalaza which supports a globular nucleus; *albumen* fleshy. EMBRYO straight, axile, subcylindric, or fusiform; *cotyledons* linear, plano-convex; *radicle* short, inferior.

ONLY GENUS.
Nepenthes.

Nepentheæ, which have some affinity with *Aristolochieæ* (p. 708), differ in diclinism, monadelphism, free. ovary, loculicidal capsule, and especially by the petioles dilated into a pitcher.[1] They also offer more than one analogy with *Droseraceæ* and *Parnassieæ*, while their leaves recall those of *Sarracenia* (p. 214).

Nepentheæ are natives of swamps in [the Malay Islands, Australia, and New Caledonia] tropical Asia, the Seychelle Islands, and Madagascar; their seeds, often held in a loose cellular integument, float at first on the surface of the water, which they imbibe by degrees, when they sink to the bottom to germinate there.

CCVIII. *ARISTOLOCHIEÆ.*

(ARISTOLOCHIÆ, *Adanson.*—ARISTOLOCHIEÆ, *Endlicher.*—ARISTOLOCHIACEÆ, *Lindl.*— ASARINEÆ, *Bartling.*)

PERIANTH *single, superior, regular or irregular, usually coloured.* STAMENS *epigynous and gynandrous, inserted at the base of the style.* OVARY *inferior, several-celled and ovuled;* OVULES *anatropous.* SEEDS *albuminous.* EMBRYO *minute, basilar, axile.*

Herbaceous PLANTS with creeping rhizomes, or tuberous, suffrutescent or frutescent. STEM often twining, simple or branched, often thickened at the nodes; *wood* scented, sometimes without concentric zones and liber fibres. LEAVES alternate, simple, all green, or some scale-like and the others green, various in form, usually cordate, penni- or pedati-nerved, veins reticulate; *petiole* very often dilated at the base and semi-amplexicaul, protecting the buds; *stipules* 0, but sometimes replaced by the axillary leaf (rarely 2) of an undeveloped branch. FLOWERS ☿, axillary or terminal, solitary, rarely in a spike or cymose raceme, sometimes fur-

---

[1] The pitcher is not the dilated petiole, but a special organ, represented by a gland at the top of the costa of the young leaf. See *Linn. Trans.*, vol. xxii., p. 415.—ED.

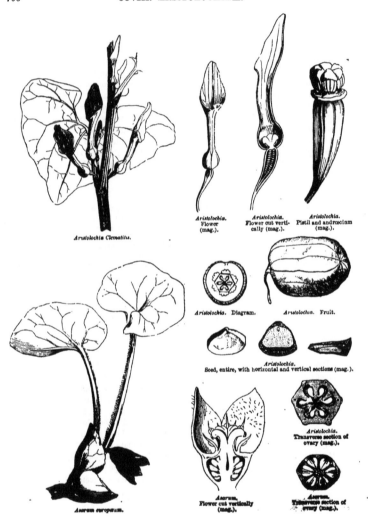

Aristolochia Clematitis.

Aristolochia.
Flower
(mag.).

Aristolochia.
Flower cut verti-
cally (mag.).

Aristolochia.
Pistil and androecium
(mag.).

Aristolochia. Diagram.

Aristolochia. Fruit.

Aristolochia.
Seed, entire, with horizontal and vertical sections (mag.).

Aristolochia.
Transverse section of
ovary (mag.).

Asarum europæum.

Asarum.
Flower cut vertically
(mag.).

Asarum.
Transverse section of
ovary (mag.).

Asarum.
Diagram.

Asarum.
Andrœcium and pistil (mag.).

Asarum.
Pistil (mag.).

Asarum.
Seed, entire and cut vertically (mag.).

nished with bracts, usually large, sometimes small, brown or blackish-purple, often fœtid. PERIANTH single (CALYX), sometimes regular, 3-lobed, campanulate or rotate, sometimes irregular, polymorphous; *tube* swollen above the summit of the ovary into a utricle enclosing the stamens, and spreading into a cup-shaped 1–2-labiate or peripheric limb, valvate or induplicate in æstivation, persistent or marcescent. STAMENS 6, rarely 5 or 12 (rarely 18–36), sometimes furnished with short filaments, free, or coherent below, inserted on an epigynous annular disk, or at the base of the stylary column, sometimes sessile and adnate to the stylary column by the whole inner surface of the anthers; *anthers* 2-celled, extrorse, or rarely sub-introrse, or some extrorse, and the others introrse in the same flower (*Heterotropa*), cells parallel, apposed, dehiscence longitudinal, connective sometimes prolonged into a point (*Asarum*). OVARY more or less completely inferior, slender (except *Asarum*), 6- or 4-celled; *styles* usually 6, rarely 3 or more, united at the base into a column (nearly always staminiferous), divided at the top into stigmatiferous lobes; *ovules* anatropous, *raphe* thick, 2-seriate at the inner angles of the 6-celled ovaries, and 1-seriate down the centre of the septum in the 4-celled ovaries (*Bragantia*). FRUIT sometimes crowned by the persistent calyx-limb or by its marcescent base, sometimes umbilicate by its scar, a capsule, rarely a berry, sometimes sub-globular or 4-gonous, usually 6-gonous, with 6 or 4 cells, sometimes irregularly dehiscent, usually septicidally 6–4 valved at the base, rarely at the top, rarely apical. SEEDS more or less numerous, horizontal, usually flattened, lower face convex, upper face concave and occupied by a prominent suberose fungoid raphe, which separates from the testa; *albumen* copious, fleshy or sub-horny. EMBRYO minute, basilar, axile; *cotyledons* very short, scarcely visible before germination; *radicle* near the hilum, centripetal or inferior.

TRIBE I. *ASAREÆ.*

Ovary 6-celled, more or less inferior, short and rather broad. Stamens 12, free; filaments distinct from the stylary column, 6 outer shortest, opposite to the styles; anthers introrse or extrorse. Calyx persistent; limb regular, 3-lobed. Capsule opening irregularly.—Herbs with perennial rhizomes, lower leaves scale-like, cauline normal, reniform. Flower terminal, solitary.

GENERA.

Asarum.　　　Heterotropa.

z z 2

Tribe II. *BRAGANTIEÆ.*

Ovary completely inferior, elongate, slender, stipitate, 4-gonous, 4-celled; ovules numerous, 1-seriate on the middle of the septa. Stamens 6–36, equal, furnished with filaments. Calyx deciduous, closely appressed to the top of the ovary, and 3-lobed. Capsule siliquose, 4-valved.—Shrubs or undershrubs. Leaves reniform, oval or oblong-lanceolate, reticulate. Flowers in spikes or racemes, small (*Bragantia*), or very large, campanulate (*Thottea*).

GENERA.

Bragantia.   Thottea.

Tribe III. *ARISTOLOCHIÆ.*

Ovary completely inferior, elongated, slender, stipitate, 6-gonous, 6- (rarely 5-) celled; ovules numerous, inserted at the central angle of the cells and 2-seriate. Stamens 6 (rarely 5); anthers sessile, extrorse, adnate by their whole dorsal surface to the stylary column. Calyx deciduous, constricted above the top of the ovary, irregular, tubular, limb various. Capsule oblong or globose, 6-angled, 6-valved, opening at the bottom or top of the fruit.

GENERA.

Holostylis.   * Aristolochia.

The affinities of *Aristolochieæ* are rather obscure; some botanists have placed them near *Cucurbitaceæ*, which they resemble in their twining stem, alternate leaves, inferior ovary and extrorse stamens; but *Cucurbitaceæ* differ in their diclinous double-perianthed flowers, imbricate æstivation, in the form and number of the stamens, the mode of placentation, the absence of albumen, &c. They might with better reason be placed near *Nepentheæ* and *Cytineæ*; they have, like *Cytineæ*, a monoperianthed isostemonous or diplostemonous flower, extrorse anthers, inferior often 1-celled ovary; but *Cytineæ* are parasitic, aphyllous and diclinous? *Nepentheæ* are allied to *Aristolochieæ*, and especially to the tribe *Bragantieæ*, by their single perianth, extrorse anthers, 4-gonous several-celled and many-ovuled ovary, and especially by the exceptional structure of the leaves (see p. 705).

Aristolochieæ mostly inhabit tropical America, they are rarer in northern temperate countries and in tropical Asia, and somewhat more frequent in the Mediterranean region. None have been met with in southern temperate latitudes [except *Aristolochia* in America].

Most *Aristolochieæ* contain in their root a volatile oil, a bitter resin, and an extractable acrid substance, which have been celebrated in all times and countries as stimulants of the glandular organs and the functions of the skin. Other species, in which the bitter resinous principle predominates, have been from the most ancient times administered as anti-hysterics, emmenagogues, &c., whence their name.

At the present time the species most in use are *Aristolochia serpentaria* and *A. officinalis*, designated in North America Virginian Snakeroot, Viperine, Colubrine, &c., and especially prescribed for the bite of the rattlesnake. It was not until the seventeenth century that European practitioners became aware of their properties, and employed them instead of the indigenous species. Their congeners of the Antilles, Peru, Brazil, and India, are equally praised as alexipharmics. A decoction of *A. fœtida* is used in Mexico for washing ulcers. The Aristolochias of Europe and the Mediterranean region, to which the exotic species are now preferred, are *A. rotunda, longa, pallida,* and *crenulata,* which grow in the south of Europe; *A. Mauritania,* of Syria, and *A. Clematitis,* dispersed throughout France. ᐧ *Asarum europæum (Asarabacca)* is an inconspicuous plant, growing in cool and shady parts of Europe; its roots are bitter, nauseous and strong-smelling; they were formerly used as emetics, but have fallen into disuse since the discovery of Ipecacuanha. The leaves reduced to powder furnish a good sternutatory. *A. asarifolium* is similarly used in America. *A. canadense* is also frequently used here as an emmenagogue, and from its ginger-like smell is used to flavour wines and food.

CCIX. *RAFFLESIAOEÆ*, Br.

(RAFFLESIÆ, HYDNOREÆ, CYTINEÆ, APODANTHEÆ.)

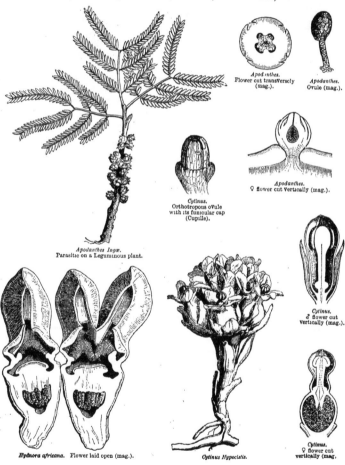

Apodanthes.
Flower cut transversely
(mag.).

Apodanthes.
Ovule (mag.).

Cytinus.
Orthotropous ovule
with its funicular cap
(Cupule).

Apodanthes.
♀ flower cut vertically (mag.).

Apodanthes Iagœ.
Parasitic on a Leguminous plant.

Cytinus.
♂ flower cut
vertically (mag.).

Hydnora africana. Flower laid open (mag.).

Cytinus Hypocistis.

Cytinus.
♀ flower cut
vertically (mag,

Cytinus.
Part of ovary cut trans-
versely (mag.).

Cytinus.
Diagram ♀.

Rafflesia Arnoldi.
Seed (mag.).

Rafflesia.
Partial section of the
seed, showing the
undivided embryo
(R. Brown) (mag.).

PARASITES *on the roots and sometimes on the branches of dicotyledonous plants.*
PERIANTH *monophyllous, regular.* COROLLA 0, *or rarely 4-petalous* (Apodantheæ).
ANTHERS ∞, 1- *(rarely 2–3-) seriate* (Apodantheæ). OVARY 1-*celled, with several many-
ovuled placentas*; OVULES *orthotropous, sometimes sub-anatropous.* FRUIT *indehiscent,
many-seeded.* EMBRYO *undivided, albuminous or not.*

TRIBE I. *RAFFLESIEÆ.*

Flowers ☿ or diœcious. Perianth 5–10-fid, æstivation imbricate (*Rafflesia,
Sapria*), or valvate (*Brugmansia*); anthers 1-seriate, adnate under the dilated top of
a staminal column (*synema*), and opening by a single or double pore. Ovary with
confluent or distinct many-ovuled placentas. Pericarp semi-adherent or free,
fleshy. Seeds recurved, funicle dilated at the top. Embryo albuminous, axile,
shorter than the albumen.—Parasites on the roots of Vines. Flower sub-sessile,
bracts imbricate.

GENERA.

Rafflesia.   Sapria.   Brugmansia.

TRIBE II. *HYDNOREÆ.*

Flowers   .   Perianth 3-fid, æstivation valvate. Stamens inserted on the
perianth-tube; anthers numerous, dehiscence longitudinal, united into a 3-lobed
ring, of which each lobe is opposite to the perianth-segments. Ovary inferior;
stigma sessile, depressed, of 3 lobes formed by apposed lamellæ, distinct to the cavity
of the ovary, where they become placentiferous; placentas pendulous from the top
of the ovary, sub-cylindric or branching, everywhere covered by very numerous
orthotropous ovules. Fruit fleshy. Embryo globose in the axis of a cartilaginous
albumen.—Parasitic on the rhizomes of *Euphorbia* in South Africa.

GENUS.
Hydnora.

TRIBE III. *CYTINEÆ.*

Flowers diclinous. Perianth 4-8-fid, æstivation imbricate.— ♂ : Stamens double
the number of the calyx-lobes, united into one bundle; anthers 1-seriate at the top
of the synema, with 2 apposed parallel? cells, dehiscence longitudinal. ♀ : Ovary
inferior, 8-16 celled above, but 1-celled below; placentas distinct, parietal, in

pairs, lobed; style solitary; stigma with radiating lobes. Fruit a berry, or sub-coriaceous, pulpy within. Embryo exalbuminous, undivided, homogeneous. Parasitic on *Cistus* in the Mediterranean region, and on the roots of other plants in America and South Africa.

GENUS.
Cytinus.

TRIBE IV. *APODANTHEÆ.*

Flowers diœcious. Calyx 4-fid or -partite, imbricate in æstivation and persistent. Corolla of 4 deciduous petals. ♂ : Anthers placed below the dilated top of the staminal column, 2–3-seriate, sessile, 1-celled. Ovary adherent, many-ovuled; ovules orthotropous, scattered through the cavity; stigma capitate. Fruit a berry, inferior. Embryo exalbuminous, undivided, homogeneous.—Parasitic on the stem and branches of dicotyledonous plants, never on the roots.

GENERA.
Apodanthes.   Pilostyles.

Rafflesia and *Brugmansia* belong to the Indian Archipelago. *Sapria* inhabits the shady forests of the extreme eastern Himalaya. *Hydnora* grows in Africa and South America. *Cytinus* principally inhabits South Africa and tropical North America; one species (*C. Hypocistis*) belongs to the Mediterranean region. *Apodanthes* and *Pilostyles* occur in America on the stems and branches of several *Leguminosæ* (*Adesmia, Bauhinia, Calliandra*). Some species are remarkable for their gigantic flowers; that of *Rafflesia Arnoldi* springs directly from the roots of *Cissus angustifolia*, expands on the surface of the earth, and attains nearly 3 feet in diameter. The perianth is 5-fid, spreading, and the throat bears an annular crown; its pink colour and scent of meat attract the flies, which thus become aids to its fertilization.

Cytinus contains, besides gallic acid and tannin, two colouring principles and a matter analogous to *ulmine*. From the herbage and fruit is obtained an extract called Hypocistis juice; it is blackish, acidulated, astringent and harsh in taste; it was known to the ancients, and is still used in the south of Europe for bloody flux and dysentery. The buds of *R. Patma* are used in Java for uterine hæmorrhage. *Brugmansia* also possesses powerful styptic properties. [The rhizome of *Hydnora* is used by the Hottentots for tanning their fishing nets, &c.]

CCX. *JUGLANDEÆ.*

(JUGLANDEÆ, *D.C.*—JUGLANDINEÆ, *Dumortier.*—JUGLANDACEÆ, *Lindl., Casim. D.C.*)

FLOWERS *diclinous, spiked.* ♂ : PERIANTH *single, 6–2–3-lobed, or* 0. STAMENS 3–∞, *inserted at the base of the perianth, or of the bract.* ♀ : PERIANTH *superior,* 4–2-*toothed.* OVARY *inferior,* 1-*celled;* OVULE *solitary, erect, orthotropous.* FRUIT *fleshy, dehiscent or not.* NUT *septate.* SEED *exalbuminous;* COTYLEDONS *fleshy, oily,* 2-*lobed.*—STEM *woody.* LEAVES *alternate, pinnate, exstipulate.*

TREES or SHRUBS with watery or resinous juice; buds 2–3, superimposed in the same axil, foliaceous or scaly, sessile, or stipitate before the leaves unfold. LEAVES alternate, exstipulate, impari- rarely pari-pinnate, glabrous pubescent tomentose or with scattered discoid hairs; *leaflets* membranous or coriaceous, not punctate.

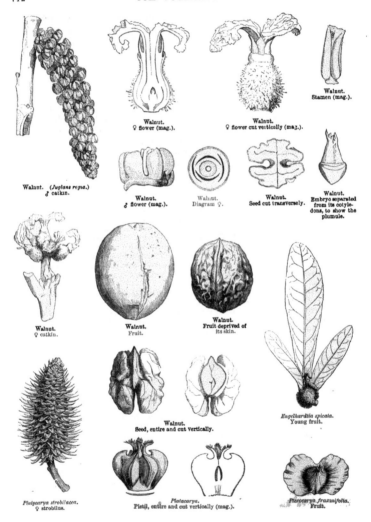

Walnut. (*Juglans regia*.)
♂ catkin.

Walnut.
♀ flower (mag.).

Walnut.
♀ flower cut vertically (mag.).

Walnut.
Stamen (mag.).

Walnut.
♂ flower (mag.).

Walnut.
Diagram ♀.

Walnut.
Seed cut transversely.

Walnut.
Embryo separated
from its cotyle-
dons, to show the
plumule.

Walnut.
♀ catkin.

Walnut.
Fruit.

Walnut.
Fruit deprived of
its skin.

Engelhardtia spicata.
Young fruit.

Walnut.
Seed, entire and cut vertically.

Platycarya strobilacea.
♀ strobilus.

Platycarya.
Pistil, entire and cut vertically (mag.).

Pterocarya fraxinifolia.
Fruit.

INFLORESCENCE indefinite, 1-sexual (when the ♂ are in axillary catkins, and the ♀ in terminal or axillary spikes), or 2-sexual (when they are in a catkin terminated by the ♂). FLOWERS monœcious, ♂ small: PERIANTH single, adnate to the inner face of a bract, which is 6-lobed, or 2–3-lobed, or obsolete. STAMENS ·3-36, inserted at the base of the perianth or bract, 2–several-seriate; *filaments* very short, free or coherent at the base; *anthers* 2-celled, glabrous or pubescent, dehiscence longitu.. dinal, connective usually prolonged beyond the cells. OVARY rudimentary or 0.— FLOWERS ♀: *Bract* more or less united to the flower, or free. RECEPTACULAR CUPULE (CALYX of authors) more or less adnate to the ovary, 3–∞ -toothed at the top, or forming a bracteal involucre. PERIANTH (COROLLA?) sometimes with 4 teeth, of which the 2 antero-posterior are exterior in æstivation, the anterior often largest and bracteiform; sometimes with 2 lateral teeth adnate to the ovary. OVARY inferior, 1-celled, at length imperfectly 2–4-celled at the base and top; *style* short; *stigmas* usually 2, rarely 4; *placenta* central, very short, bearing an orthotropous erect sessile ovule. FRUIT sometimes fleshy or membranous, indehiscent or bursting irregularly or 4-valved; *endocarp* free or united to the pericarp, indehiscent or 2–3-valved, with cartilaginous septa forming 2–4 imperfect cells at the base and top; endocarp and septa chambered. SEED exalbuminous; *testa* membranous; *endopleura* very thin. EMBRYO fleshy and oily, 2-lobed, cerebriform or cordate at the base; *radicle* very short, superior; *plumule* often 2-phyllous, usually showing the rudiments of small buds.

GENERA.

* Juglans.   * Carya.   * Pterocarya.   Engelhardtia.   * Platycarya.

Juglandeæ, which only comprise about thirty species, are very near *Myriceæ* (which see); they are also connected with *Terebinthaceæ*, through *Pistacia*, but *Terebinthaceæ* differ in their inflorescence, in having petals, a free ovary, and curved ovule. *Juglandeæ* approach *Cupuliferæ* and *Betulaceæ* in their amentaceous inflorescence, diclinism, apetalism, exalbuminous embryo, woody stem, and alternate leaves. They are separated by the structure of the fruit and ovule, the pinnate exstipulate leaves, and the aromatic principle.

Juglans and *Carya* are North American, but the most remarkable species (*J. regia*), as well as *Pterocarya*, inhabits the southern provinces of the Caucasus [and *Juglans* the Himalayas]. *Engelhardtia* is especially tropical Asiatic, *Platycarya* Chinese. The common Walnut (*J. regia*), a native of Persia [and the Himalayas], and introduced into Greece and Italy some centuries before our era, is now culti- vated throughout temperate Europe. Its wood is much sought by cabinet-makers and armourers for gun-stocks. Dyers also obtain a blackish-brown dye from it. All parts of the plant possess a peculiar scent, tolerably pleasant, but giving headache to those who remain long in its shade in hot weather. The pericarp contains a volatile oil (from which a tincture is prepared), associated with tannin and citric and malic acids, whence its use as an astringent, tonic, and stimulant. The leaves possess analogous qualities. The seed is edible; it contains an agreeable fixed oil, which quickly becomes rancid. The wood of *J. nigra* is more highly valued than that of *J. regia*, on account of its violet-black colour. The bark of *J. cinerea* is used as a purgative in America. The seeds of *Carya* are edible, except those of *C. amara*; but the latter (which is intensely bitter), mixed with oil of camomile, is considered efficacious in obstinate colics. *J. cinerea* yields the Butter-nut; *J. nigra*, the Black Walnut; *Carya oliveæformis*, the Pekan- nut; *C. alba* and *nigra*, the Hickory-nut; and *C. glabra*, the Pig-nut—all of North America. *Engelhardtia* contains an abundance of resinous juices. *E. spicata* attains a height of 160 to 280 feet, and its trunk becomes so large that three men with arms extended can scarcely touch around it. Its russet-coloured wood, hard and heavy, is used for cart-wheels in Java, and for vases the diameter of which is sometimes enormous.

CCXI. *CUPULIFERÆ.*

(CASTANEÆ, *Adanson.*—AMENTACEARUM *pars, Jussieu.*—CUPULIFERÆ, *Richard, Endlicher.*—QUERCINÆ, *Jussieu.*—QUERCINEÆ ET FAGINEÆ, *Dumortier.*)

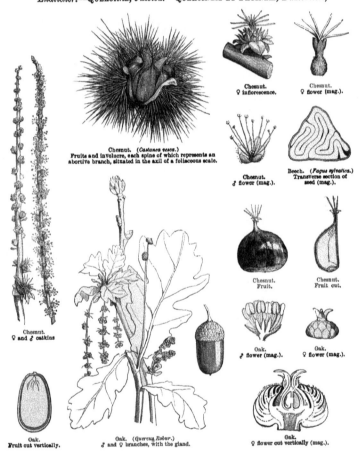

Chesnut.   (*Castanea vesca.*)
Fruits and involucre, each spine of which represents an
abortive branch, situated in the axil of a foliaceous scale.

Chesnut.
♀ inflorescence.

Chesnut.
♀ flower (mag.).

Chesnut.
♂ flower (mag.).

Beech.   (*Fagus sylvatica.*)
Transverse section of
seed (mag.).

Chesnut.
Fruit.

Chesnut.
Fruit cut.

Oak.
♂ flower (mag.).

Oak.
♀ flower (mag.).

Chesnut.
♀ and ♂ catkins

Oak.
Fruit cut vertically.

Oak.   (*Quercus Robur.*)
♂ and ♀ branches, with the gland.

Oak.
♀ flower cut vertically (mag.).

Q. Cerris.
♀ flower cut transversely
(mag.).

Q. costata.
Transverse section of
fruit.

Q. coccifera.
Ovule with outer flexuous
membrane and large
exostome (mag.).

Q. Ægilops. Abortive
ovules at the base of
the seed.

FLOWERS *diclinous, in spikes,* ♂ *with a single perianth.* STAMENS 5–∞. FLOWERS ♀ *sessile in a cupuliform involucre.* PERIANTH *single, calyciform.* OVARY *inferior, of 2–3–6 2-ovuled cells;* OVULES *anatropous, pendulous or erect.* NUTS *involucrate, usually 1-seeded.* SEED *exalbuminous.* EMBRYO *straight.*—STEM *woody.* LEAVES *alternate, stipulate.*

TREES, rarely SHRUBS. LEAVES alternate, simple, penninerved, caducous or persistent, stipulate. FLOWERS monœcious, usually in 1-sexual spikes, sometimes ♀ at the base, and ♂ at the apex; ♂ in cylindric or globose catkins, naked or bracteate. PERIANTH single, lobes often unequal. STAMENS 5–20, inserted at the base of the perianth, free; *filaments* weak; *anthers* 2-celled. OVARY rudimentary or 0.—FLOWERS ♀ 1–3–5, sessile in a common cupuliform involucre externally clothed with scaly or spinescent or accrescent scales. PERIANTH superior, regular, usually 6-lobed. OVARY inferior, 2–3–6-celled by centripetal septa which are early absorbed; *styles* as many as cells, undivided, stigmatiferous at the top; *ovule* geminate in each cell, erect, basal or pendulous from the apex, anatropous, furnished with a double integument. FRUIT composed of nuts contained in an involucre or a dehiscent cupule. SEED usually solitary in each nut, the arrested ovules persisting in their original position. EMBRYO straight, exalbuminous; *radicle* small, superior; *cotyledons* usually fleshy, folded or sinuous, and their inner faces pressed together.

GENERA.

* Quercus.   Lithocarpus.   * Castanea.   Castanopsis.   * Fagus.

Cupuliferæ approach *Betulineæ* in the alternate stipulate leaves, inflorescence, diclinism, several-celled ovary, anatropous ovule, dry 1-celled fruit, and exalbuminous seed; but they differ in their inferior ovary and cupulate fruit. They are also connected with *Juglandieæ*, for besides the affinities indicated (at p. 713), the acorn is sometimes divided into four by false septa, a character on which Lindley established his genus *Synædrys.* Finally, their affinity is obvious with *Pomaceæ* (which see), as indicated by Brongniart.

Cupuliferæ principally inhabit northern temperate regions. They especially abound in America; they are very rare in the north of Asia [except China and Japan], but form vast forests in South and Central Europe [the Himalayas and East Asiatic mountains]. Some attain the snow limit in the Antarctic regions [others inhabit the mountains of Australia, Tasmania, and New Zealand]. They become rare as they approach the equator, and only grow in elevated localities on the large islands of the Indian Archipelago. Chesnuts and Oaks are numerous on the high mountains of cisequatorial Asia, and species of the latter genus are not rare in tropical North America. In southern tropical countries they are almost wholly wanting. Africa produces none, except in the Mediterranean region, where a few Oaks are met with. The Beech is represented on the Andes of Chili by very tall trees (*Fagus procera*), which are only less lofty than the *Araucaria*; and on the mountains of the same country *F. Pumilio* marks the limit of arboreal vegetation. Other species of Beech have been observed in [Fuegia] Tasmania and New Zealand.

Cupuliferæ, besides the beauty of their habit and foliage, are amongst the most useful of plants. Not only do they furnish us with a valuable fuel, but their wood, being [often] close-grained, nearly imperishable, and easily worked, is used in the manufacture of agricultural and other implements, furniture, and utensils of all sorts, as well as in the construction of machines, buildings, ships, &c. These trees often attain a size which indicates a prodigious longevity. There are in Italy Chesnuts with the trunk of 40, 75, and even 160 feet in circumference, and which must certainly have been in existence many thousand years. [This is a disputed statement.] There were formerly in France many Oaks which were certainly cotemporaries of the Druids, i.e. the Oak of Autrage, in Alsace, whose trunk at the ground measured 45 feet in circumference when it was cut down in 1858 and sold by auction. That of Allouville, in Normandy, in which, 200 years ago, a chapel was hollowed out, is of about the same size. The largest Oak still existing in France is that of Montravail, near Saintes, 27 feet in diameter and upwards of 80 feet in circumference. It is very advisable that these venerable monuments of the Vegetable Kingdom should be placed under State protection, like the historical monuments erected by the hand of man.

Cupuliferæ possess, amongst other principles, tannin and gallic acid, which give them astringent properties, useful in medicine and manufactures. The bark of *Quercus tinctoria*, a large species, of the forests of Pennsylvania, is exported to Europe [as Quercitron] on account of the richness of its yellow colouring principle; it is also used in America for the tanning of leather. The bark of the European species (*Q. Robur, pedunculata, pubescens, Cerris*) is dried and pulverized as tan, and similarly used. *Q. coccifera*, a Mediterranean shrub, is the food of the Kermes, an hemipterous insect of the cochineal tribe, which is collected for dyeing silk and wool crimson. *Q. Suber* grows in the south of France and Spain; the outer spongy part of its bark is the elastic substance known as Cork. The acorns of most Oaks contain a large quantity of starch, a fixed oil, and a bitter astringent substance; baked and treated with boiling water they yield a highly tonic drink, which is successfully administered in the shape of coffee to children of a lymphatic temperament. The acorns of *Q. Ilex, Ballota, Æsculus*, and *Ægilops* have no bitter or harsh principle, and to this day are used as food by the inhabitants of some parts of the Mediterranean region, and especially of Algeria. The leaves of *Q. mannifera*, a Kurdistan species, secrete a sugary matter. The cups of *Q. Ægilops* are the object of a considerable commerce for dyeing black and for tanning leather.

Various species of Oak, and principally *Q. Ægilops*, yield Galls, formed by a hymenopterous insect which pierces the peticle to deposit its eggs; the vegetable juices are extravasated at the spot, and form an excrescence containing gallic acid and tannin. Our writing ink is made by an infusion of gall-nuts in a solution of a salt of iron (green copperas). The Beech (*Fagus sylvatica*) bears angular fruits, called *mast*, the seed of which is oily and of a pleasant taste; but if too largely eaten, they cause headache and vertigo. The Chesnut (*Castanea vesca*) produces farinaceous seeds which, eaten raw, are astringent, but furnish, if cooked or baked, an agreeable and wholesome food. The so-called Lyons Chesnut is only an improved variety of the common Chesnut.

CCXII. *CORYLACEÆ.*

(CASTANEARUM pars, *Adanson.* —AMENTACEARUM pars, *Jussieu.*—CUPULIFERARUM pars, *Richard.*—CORYLACEÆ, *Hartig, Alph. D.C.*)

FLOWERS *diclinous, in spikes, the ♂ achlamydeous, furnished with a staminiferous bract.* FLOWERS ♀ *geminate on a bract, furnished with very accrescent bracteoles.* PERIANTH *single, irregularly lobed.* OVARY *inferior, partially 2-celled, 2-ovuled;* OVULES *pendulous, anatropous.* Nut *involucrate by foliaceous bracteoles.* Seed *solitary, exalbuminous.* Embryo *straight.--*Stem *woody.* Leaves *alternate, stipulate.*

Shrubs or small trees. Leaves alternate, penninerved, doubly-toothed,

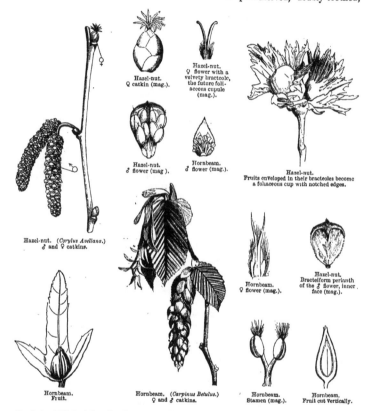

Hazel-nut.
♀ catkin (mag.).

Hazel-nut.
♀ flower with a
velvety bracteole,
the future foli-
aceous cupule
(mag.).

Hazel-nut.
♂ flower (mag.).

Hornbeam.
♂ flower (mag.).

Hazel-nut.
Fruits enveloped in their bracteoles become
a foliaceous cup with notched edges.

Hazel-nut. (*Corylus Avellana.*)
♂ and ♀ catkins.

Hornbeam.
♀ flower (mag.).

Hazel-nut.
Bracteiform perianth
of the ♂ flower, inner
face (mag.).

Hornbeam.
Fruit.

Hornbeam. (*Carpinus Betulus.*)
♀ and ♂ catkins.

Hornbeam.
Stamen (mag.).

Hornbeam.
Fruit cut vertically.

stipulate, folded obliquely along their lateral nerves, and facing the axis either by their inner spreading face (*Ostrya, Carpinus*), or by one of their sides pressed against the other (*Corylus*). Flowers monœcious, in 1-sexual spikes; ♂ in cylindric catkins, accompanied by a naked bract, or folded within 2 juxtaposed bracteoles. Perianth 0. Stamens several, inserted at the base or middle of the bract, and included; *filaments* often divided or bifid; *anthers* with separated cells, usually hairy at the top. Rudimentary OVARY 0.—FLOWERS ♀ in a short spike, geminate in each bract, and each furnished with very accrescent bracteoles. PERIANTH superior, irregularly lobed at the top. OVARY inferior, imperfectly-2-celled by 2 prominent placentas, of which one only bears at the top 2 pendulous anatropous ovules, each with a single integument; *style* very short, divided into 2 elongated linear stigmas. NUT largely umbilicate at the base, enclosed in a foliaceous lobed or laciniate involucre. SEED solitary. EMBRYO straight, exalbuminous; *cotyledons* fleshy, their inner faces appressed, longer than the small superior radicle.

GENERA.

<table>
<tr><td>* Ostrya.</td><td>* Carpinus.</td><td>Distegocarpus.</td><td>* Corylus.</td></tr>
</table>

Corylaceæ can only be distinguished from *Cupuliferæ* by their achlamydeous male flowers consisting of a staminiferous bract, and by the foliaceous or tubular laciniate involucre of their fruit, which is acid in taste. They inhabit cold or temperate regions of the northern hemisphere.

The Filbert or Hazel (*Corylus Avellana*) is a shrub spread over Europe and northern Asia. Its seed is of an agreeable taste, and yields by expression a bland undrying oil; its bark is astringent and considered a febrifuge. *C. Colurna* and *tubulosa*, which grow in southern Europe, *C. rostrata* and *americana*, of North America, bear edible fruits like the Filbert. The Hornbeam (*Carpinus Betulus*) is an indigenous tree with elegant foliage, cultivated for hedges. Its wood is white, very fine and close, and becomes exceedingly hard when dry. It is used for wheelwright's work, screws of presses, and handles of tools, and is besides a very good fuel. The scales of the fruit of *Ostrya* are covered with prurient hairs.

CCXIII. *LORANTHACEÆ.*

(LORANTHEÆ, *Jussieu.*—LORANTHACEÆ, *Lindl.*—VISCOIDEÆ, *Richard.*)

FLOWERS *diclinous or* ☿. PERIANTH *simple, often girt by a calycule.* SEPALS *4-6-8, rarely 3, inserted round an epigynous disk, distinct or coherent, æstivation valvate.* STAMENS *as many as the sepals, inserted on them and opposite.* OVARY *inferior, 1-celled, 1-ovuled;* OVULE *sessile, erect, orthotropous;* STYLE *simple.* FRUIT *a berry.* ALBUMEN *fleshy.* EMBRYO *straight;* RADICLE *superior.—Parasitic* SHRUBS. LEAVES *entire.*

Evergreen SHRUBS, parasitic on the wood of other Dicotyledons, sometimes appearing epiphytal, and emitting roots which creep over the branches of the infested tree; *branches* knotted, very often jointed, cylindric, 4-gonous or compressed. LEAVES opposite, rarely alternate or whorled, thick, coriaceous, entire, penni- or palmi-nerved, nerves inconspicuous, sometimes reduced to stipuliform scales, or 0; *stipules* 0. FLOWERS sometimes imperfect, small, inconspicuous, whitish or greenish-yellow, sometimes perfect, brightly coloured, variously arranged, usually furnished with 1 or several bracts, and with a calycule simulating an outer perianth, which is sometimes obsolete. PERIANTH single (CALYX), superior in the ☿, inserted in the ♀ around a disk crowning the top of the ovary; *sepals* 4-6-8, rarely 3, distinct, or connate into a tube often split on one side, æstivation valvate. STAMENS equal and

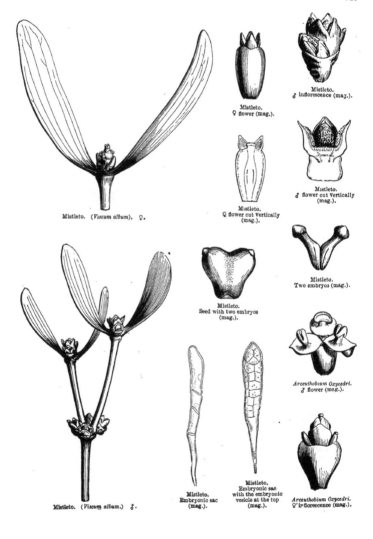

Mistleto. (*Viscum album*). ♀.

Mistleto.
♀ flower (mag.).

Mistleto.
♀ flower cut Vertically
(mag.).

Mistleto.
♂ inflorescence (mag.).

Mistleto.
♂ flower cut Vertically
(mag.).

Mistleto.
Two embryos (mag.).

Mistleto.
Seed with two embryos
(mag.).

Arceuthobium Oxycedri.
♂ flower (mag.).

Mistleto. (*Viscum album.*) ♂.

Mistleto.
Embryonic sac
(mag.).

Mistleto.
Embryonic sac
with the embryonic
vesicle at the top
(mag.).

Arceuthobium Oxycedri.
♀ inflorescence (mag.).

Loranthus.
Young embryo with
its suspensor (mag.).

Loranthus albus.
♂ flower (mag.).

Loranthus.
Embryo (mag.).

Mistleto.
Fruit cut vertically.

Mistleto. Fruits.

Loranthus albus.
Ovary cut vertically (mag.).

Loranthus albus.
♀ flower (mag.).

Loranthus albus.
Diagram ♀. showing
the bracts the caly-
code, and calyx.

Loranthus albus.
Fruit cut vertically (mag.).

opposite to the perianth-lobes, and inserted on them; *filaments* adnate at the base, free above, or very rarely coherent, variable in length, sometimes 0; *anthers* introrse, 2-celled, erect and adnate, or incumbent and versatile, dehiscence longitudinal, sometimes many-celled and opening by numerous pores (*Viscum*), rarely 1-celled dehiscing transversely (*Arceuthobium*). OVARY inferior, usually crowned by an annular disk, 1-celled; *style* terminal, simple, sometimes 0; *stigma* terminal, more or less thick (sometimes obsolete), undivided or emarginate; *ovule* sessile, orthotropous, often reduced to the nucleus or the embryonic sac, erect, solitary, or accompanied by 2 rudimentary ovules. BERRY 1-seeded. SEED erect; *albumen* fleshy, copious. EMBRYO (often several) axile, or inserted in an excentric cavity of the albumen, peripheric or lateral, clavate, straight or arched; *cotyledons* somewhat fleshy, obtuse, sometimes coherent; *radicle* thick, superior.

PRINCIPAL GENERA.

Arceuthobium.	Loranthus.	.Tupeia.	Viscum.
Lepidoceras.	Nuytsia.	Loranthus.	Erythranthus.

Loranthaceæ are closely allied to *Santalaceæ* : in both, besides the important analogy founded on parasitism, the æstivation is valvate, the flower is isostemonous, the stamens are epigynous and opposite to the sepals, the ovary is inferior and 1-celled, the ovule is reduced to the embryonic sac, the albumen is fleshy, and the leaves are entire, coriaceous and exstipulate. *Loranthaceæ* also approach *Proteaceæ* in the valvate æstivation, isostemony, 1-celled ovary, &c.

Loranthaceæ are mostly tropical, but some inhabit temperate and cool regions of the northern, and still more of southern temperate latitudes.' Three genera are represented in Europe : the Mistleto (*Viscum album*) lives principally on Apple-trees and Poplars, although it scarcely objects to. any tree or shrub, and even attaches itself to *Loranthus europæus*, which is itself parasitic on Oaks and Chesnuts. *Arceuthobium* grows on the Juniper. The dissemination of *Loranthaceæ* is mostly effected by birds, which feed on their berries, and drop on the trees the undigested seeds. In *Arceuthobium* the seed is projected from the fruit by a peculiar contractile force.

The fruit of *Loranthaceæ* contains Birdlime, a peculiar viscous, tenacious and elastic substance, between resin and india-rubber ; it exists in other plants (Holly), but that of *Loranthaceæ*, and especially of *Viscum album* and *Loranthus albus*, is the best. Its abundance depends on the stock on which the Mistleto grows. The most suitable are the Maple and Elm, next the Birch and Service, and then the Apple and Pear, &c.

Many species of *Loranthus* are used medicinally by the Brazilians, who prepare with the young shoots and leaves of *L. citrocolus* an ointment much praised as a cure of œdematous tumours ; *L. globosus*, *elasticus*, and *longiflorus* are similarly used in India. *L. bicolor* is a reputed antisyphilitic. The leaves of *L. rotundifolius*, cooked in milk, are recommended in Brazil for diseases of the chest. The leaves of the Mistleto, formerly used as antispasmodics and antiepileptics, have long fallen into disuse. This plant was worshipped by the ancient Gauls, who saw a mysterious emblem in a shrub vegetating and propagating its kind without touching the earth. When gathered from the Oak it was held sacred by the Druids.

INTERMEDIATE GENUS.

MYSODENDRON, Banks.

Mysodendron punctulatum. ♂ plant.

Mysodendron.
Flower, entire and cut vertically (mag.).

Mysodendron. ♀ plant.

3 A

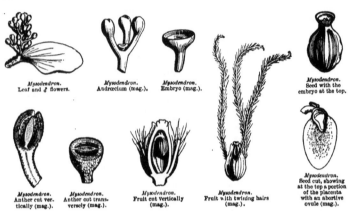

Mysodendron.   Mysodendron.   Mysodendron.   Mysodendron.
Leaf and ♂ flowers.   Andrœcium (mag.),   Embryo (mag.).   Seed with the
                                                                                embryo at the top.

Mysodendron.   Mysodendron.   Mysodendron.   Mysodendron.   Mysodendron.
Anther cut ver-   Anther cut trans-   Fruit cut vertically Fruit with twining hairs   Seed cut, showing
tically (mag.).   versely (mag.).   (mag.).   (mag.).   at the top a portion
                                                                                of the placenta
                                                                                with an abortive
                                                                                ovule (mag.).

Mysodendron forms a group intermediate between *Loranthaceæ* and *Santalaceæ*. They are small diœcious parasitic shrubs, from antarctic America, living especially upon Beech. Characters: FLOWERS ♂ : PERIANTH 0. STAMENS in the axil of a bract; *anthers* 1-celled, opening at the top by a transverse slit.—FLOWERS ♀ : PERIANTH superior, entire, and contracted. OVARY inferior [trigonous], with feathery setæ rising from the base [from slits at the 3 angles], 1-celled ; *placenta* cylindric, surmounted by 3 naked pendulous ovules. [FRUIT ovoid, 3-gonous, 1-celled, 1-seeded. SEED pendulous from the top of the placental column, which bears the 2 arrested ovules at its summit, obovoid ; *testa* membranous ; *albumen* fleshy. EMBRYO apical ; *cotyledons* obscure] ; *radicle* superior, [apex discoid, exserted]. The plumose bristles which escape from the three longitudinal slits at the angles of the ovary of *Mysodendron*, by twining round the branches of trees to which the fruit is wafted, fulfil the functions of the viscous substance enclosed in the fruits of *Loranthus* and *Viscum*. *M. punctulatum* is so abundant on the Beeches of Tierra del Fuego that it may be recognized at a distance by its yellow colour, which contrasts with the dark green of the trees on which it parasitically lives.

CCXIV. *SANTALACEÆ, Br.*

FLOWERS *usually* ☿. PERIANTH *single*. STAMENS *equal and opposite to the perianth-lobes*. OVARY *inferior, or adherent by the base only, 1-celled ;* OVULES *2-3-5, reduced to the nucleus, pendulous from the top of a central free placenta.* FRUIT *dry or fleshy, 1-seeded.* ALBUMEN *fleshy.* EMBRYO *straight, axile ;* RADICLE *superior.*

HERBS, SHRUBS or TREES, often (always?) parasitic on the roots or branches (*Henslovia*) of other plants ; *branches* frequently angular. LEAVES usually alternate, sometimes opposite, entire, penninerved, or with 3-5 lateral oblique nerves, usually

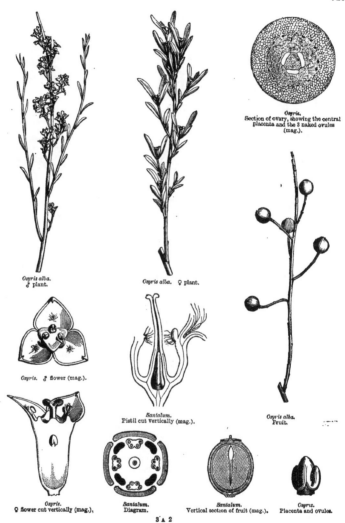

Osyris.
Section of ovary, showing the central placenta and the 3 naked ovules (mag.).

Osyris alba.
♂ plant.

Osyris alba. ♀ plant.

Osyris. ♂ flower (mag.).

Santalum.
Pistil cut vertically (mag.).

Osyris alba.
Fruit.

Osyris.
♀ flower cut vertically (mag.).

Santalum.
Diagram.

Santalum.
Vertical section of fruit (mag.).

Osyris.
Placenta and ovules.

3 A 2

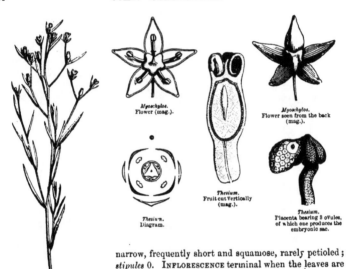

Thesium intermedium.

Myoschylos.
Flower (mag.).

Myoschylos.
Flower seen from the back
(mag.).

Thesium.
Fruit cut Vertically
(mag.).

Thesium.
Diagram.

Thesium.
Placenta bearing 3 ovules,
of which one produces the
embryonic sac.

narrow, frequently short and squamose, rarely petioled; *stipules* 0. INFLORESCENCE terminal when the leaves are opposite, usually indefinite when the leaves are alternate, and then flowers in a spike or head, or in small extra-axillary cymes with the peduncle united with the floral leaf, or sometimes solitary. FLOWERS ☿ or polygamous or diclinous, white green dirty yellow or red, often minute; *bracteoles* usually 2, lateral, situated within the bract or the floral leaf accompanying the solitary flower, or the lateral flowers of the cyme; *pedicels* 0, or short, and continuous with the perianth. PERIANTH single, tube often merging into the receptacular cup (*calycode*), which is often prolonged beyond the ovary; *limb* 5–4–3-lobed, valvate in æstivation, lobes caducous or persistent, often bearded in the centre of the inner surface. STAMENS opposite to the perianth-lobes, and inserted at their base or in the middle; *filaments* short; *anthers* basi- or dorsi-fixed, introrse, 2-celled, dehiscence longitudinal, sometimes 4-locellate and opening above by a large orifice (*Choretrum*). DISK epigynous, often apparent, sometimes dilating into a lobed plate, the lobes alternate with the stamens. OVARY inferior, 1-celled, or free when young, and finally united with the receptacular cup or calycode (*Santalum*), sometimes adherent by the base only to the receptacular cup (*Anthobolus*); *style* included, entire, 2–3–4–5-lobed, lobes alternate or opposite to the stamens; *placenta* basal, central, erect, cylindric; *ovules* 2–3–5, pendulous from the top of the placenta, naked, i.e. without coats; embryonic sac protruding from the nucleus, recurved, developing the embryo and albumen outside the nucleus. FRUIT a nut, rarely a berry, indehiscent, epicarp thin, mesocarp usually hardened; endocarp pulpy when young, then drying, separating from the mesocarp and enveloping the seed. SEED solitary by arrest, inverted, covered with the remains of the endocarp and placenta, and accompanied by the arrested ovules; *albumen* fleshy. EMBRYO straight, axile; *cotyledons* linear or oblong, convex on their dorsal face, and shorter than the radicle, which is superior.

TRIBE I. *SANTALEÆ.*

Flowers ☿, or rarely diœcious. Ovary inferior. Stamens inserted on the middle of the perianth-lobes.

PRINCIPAL GENERA.

Quinchamalium.	Osyris.	Choretrum.	Arjoona.
Pyrularia.	Comandra.	Leptomeria.	Nanodea.
Henslovia.	Thesium.	Santalum.	Myoschylos.

TRIBE II. *ANTHOBOLEÆ.*

Flowers ☿, polygamous or diœcious. Ovary adherent at the base only. Stamens inserted at the base of the perianth-lobes. [All Australian].

PRINCIPAL GENERA.

Anthobolus.   Exocarpus.

[To these tribes should be added that of *Buckleyeæ*, Alph. D.C., which is founded on the United States genus *Buckleya*, a plant with a double perianth to the ♀ flowers, and a single one to the ♂.]

We have indicated the affinity of *Santalaceæ* with *Loranthaceæ*, *Proteaceæ*, *Elæagneæ*, and *Thymeleæ* (see these families). They also approach the apetalous *Combretaceæ* in the simple perianth with valvate æstivation, the stamens opposite the perianth-lobes, the inferior ovary 1-celled and crowned by a disk, &c. [They approach still nearer to *Olacineæ* and *Corneæ*, differing from the former in the superior ovary, and from the latter in being monochlamydeous and 3-ovulate.]

Santalaceæ are dispersed over the temperate and tropical regions of the whole world, especially in Asia, Europe, South Africa, and Australia; they seem to be absent from tropical America and tropical Africa. They are herbaceous in Europe, Central Asia, and South America (where *Nanodea* attains the height of only 6 or 9 feet), suffruticose in the Mediterranean region, generally arborescent in Asia and Australia as well as in the northern temperate regions of the New World, and sometimes parasitic in Asia (*Henslovia*).

Very little is known about the properties of *Santalaceæ*. The most remarkable are the species of *Santalum*, especially *album*, a large tree of South Asia, the aromatic and scented wood of which was formerly celebrated medicinally, and is still used [as a medicine in India and] in perfumery and cabinet making [as are various Pacific Island species of this genus]. The roots and fruits of *Osyris* and *Thesium* are astringent. The leaves of *O. nepalensis* are used as tea. An infusion of the leaves of *Myoschilos oblongus*, the Senna of the Chilians, is purgative. The Peruvians eat the seeds of *Cervantesia tomentosa* like filberts; those of *Pyrularia pubera*, which grows on the mountains of Carolina and Virginia, yield a fixed and edible oil.

CCXV. *GRUBBIACEÆ.*[1]

(OPHIRACEÆ, *Arnott.*—GRUBBIACEÆ, *Alph. D.C.*)

[FLOWERS ☿, *in a cone, involucrate.* PERIANTH-SEGMENTS 4, *superior, valvate.* STAMENS 8, *inserted at the bases of the segments;* ANTHERS *opening by valves.* OVARY

---
[1] This order is omitted in the original.—ED.

inferior, at first 2, then 1-*celled*; STYLE *short*; STIGMA *simple or* 2-*lobed*; OVULES *solitary in each cell, pendulous*; NUCULES 4, *laterally connate,* 1-*seeded.* EMBRYO *straight, in a fleshy albumen*; RADICLE *superior.*—SHRUBS. LEAVES *opposite, exstipulate.*

SHRUBS with the habit of *Phylica* and *Bruniaceæ.* LEAVES opposite, linear-lanceolate, quite entire, margin revolute; *stipules* 0. INFLORESCENCE in cones in the axils of the upper leaves, subtended by 2 lateral bracts. FLOWERS ☿. PERIANTH superior, of 4 caducous ovate-acute segments, hairy externally, valvate in æstivation. STAMENS 8, slightly adherent to the bases of the segments, 4 alternate rather longer than the others; *filaments* ligulate; *anthers* adnate, erect, 2-celled, cells dehiscing longitudinally by valves. OVARY inferior, capped by an annular disk, 2-celled when young, 1-celled afterwards by the rupture of the septum, which remains on a central placenta; *style* short; *stigma* truncate or 2-lobed; *ovules* 2, ovoid, compressed, one in each cell or on each side of the central free septum, pendulous from its summit, anatropous. NUCULE crowned by the style and alveolar disk, 1-seeded. SEED inverse, sub-spherical, bearing the remains of the septum and undeveloped ovule on one side; *albumen* fleshy. EMBRYO straight, cylindric, almost as long as the seed; *cotyledons* short, appressed; *radicle* long, cylindric, superior.

ONLY GENUS.

Grubbia.

A family of one genus, whose affinity is rather obscure. It has been appended to *Bruniaceæ* by Decaisne, Lindley, and Arnott, and to *Hamamelideæ* by Gardner; but is probably nearer *Santalaceæ*, as Endlicher has pointed out, a point that cannot be settled till the structure of the ovule is known. Brongniart and A. de Candolle regard it as intermediate between *Bruniaceæ* and *Santalaceæ*. *Grubbia* is a native of South Africa, where three species have been discovered; they are of no known use. —ED.]

CCXVI. *BALANOPHOREÆ, L.-C. Richard.*

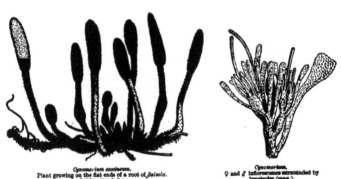

Cynomorium coccineum.
Plant growing on the flat ends of a root of *Salsola.*
1-5th natural size.

Cynomorium.
♀ and ♂ inflorescence surrounded by
bracteoles (mag.).

Cynomorium.
♂ flower with 5-phyllous perianth, laid open, showing an abortive style (mag.).

Cynomorium.
♀ flower, nearly ripe, with 4-phyllous perianth (mag.).

Cynomorium.
♀ flower, with 3-phyllous perianth (mag.).

Cynomorium.
Fruit cut longitudinally, showing the embryo at the base of the albumen, the inferior micropyle, and the chalaza at the opposite end.

Parasitic HERBS, *fleshy, aphyllous, monœcious or diœcious.* SCAPES *naked or scaly.* FLOWERS *in a capitulum.* PERIANTH *usually 3-lobed.* STAMENS 3 (*rarely more or less*), *inserted on the perianth.* OVARY *inferior,* 1-*celled;* OVULE *pendulous, orthotropous.* ALBUMEN *fleshy.* EMBRYO *undivided.*

Fleshy HERBS; *rhizome* hypogeous, sub-globose, or branching and creeping, parasites on the roots of other plants. SCAPES simple or branched, bearing flowers throughout, or sterile at the base, naked, or furnished with scales replacing the leaves, and often bearing at the base bracts, a ring, or a volva or cup, which at an early stage encloses the inflorescence. FLOWERS monœcious or diœcious, rarely polygamous (*Cynomorium*), sessile or sub-sessile, in a globose oblong or cylindric [simple or branched] capitulum; the ♂ and ♀ sometimes in different inflorescences, sometimes in the same, often mingled with peltate scales and rudiments of arrested flowers. PERIANTH single, 3-6-phyllous, or 3-lobed, or tubular, campanulate, or 2-labiate, æstivation valvate or induplicate [sometimes 0].—FLOWERS ♂: STAMENS usually 3, rarely more, sometimes 1 (*Cynomorium*), opposite to the sepals, inserted at their base when they are free, and then distinct, or on their tube when monadelphous; *anthers* 1-2-∞-celled, introrse or extrorse, dehiscence longitudinal or apical.—FLOWERS ♀: OVARY inferior, 1- (rarely 2-) celled (*Helosis*); *style* filiform; *stigma* terminal, sometimes sessile; *ovules* solitary, erect or pendulous from the top of the cell, and orthotropous. FRUIT dry, coriaceous. SEED inverted; *testa* crustaceous; *albumen* fleshy. EMBRYO minute, globose, undivided, near to or distant from the hilum.

[The following is a slight modification of Eichler's arrangement of the genera in Martius' ' Flora of Brazil.' Tribe *Cynomorieæ*, which I have added, he considers as not belonging to *Balanophoreæ*, but to be near *Gunneraceæ*; whilst the rest of the order he refers to the neighbourhood of *Santalaceæ.*

Tribe I. Cynomorieæ.—Flowers ☿ or unisexual by arrest, with a distinct perianth, superior in the ♀, sometimes 0 in the ♂. Stamens free. Anthers 2-celled, dehiscence longitudinal. Ovary 1-celled; style single; ovule solitary, pendulous. *Cynomorium, Mystropetalum, Dactylanthus.*

Tribe II. Eubalanophoreæ.—Perianth of the ♂ 3–6-lobed, of the ♀ 0. Stamens monadelphous; anthers extrorse. Ovary 1-celled; style 1; ovule pendulous, anatropous. *Balanophora.*

Tribe III. Langsdorffieæ.—Perianth of the ♂ 3-lobed or of a few scales, of the ♀ tubular. Stamens monadelphous; anthers extrorse. Ovary 1-celled; style 1; ovule 1, erect. *Langsdorffia, Thonningia.*

Tribe IV. Helosideæ.—Perianth of the ♂ 3-lobed or tubular or campanulate, of the ♀ 0. Stamens monadelphous; anthers connate, bursting at the top. Ovary 1-celled; styles 2; ovule 1, erect. *Helosis, Phyllocoryne, Sphærorhizon, Corynæa, Rhopalocnemis.*

Tribe V. Scybalieæ.—Perianth of the ♂ 3-lobed, of the ♀ 0. Stamens monadelphous; anthers extrorse. Ovary at first 1-celled; styles 2; ovules 2, pendulous from an apical placenta which descends and divides the cell into two. *Scybalium.*

Tribe VI. Lophophyteæ.—Perianth of ♂ and ♀ 0. Stamens 2, free. Ovary and ovules as in *Scybalieæ. Lophophytum, Ombrophytum, Lathrophytum.*

Tribe VII. Sarcophyteæ.—Perianth of ♂ 3-lobed, of ♀ 0. Stamens 3, free; anthers many-celled. Ovary at first 1-celled; stigma sessile; ovules 3, pendulous from an apical placenta that descends and divides the cell into three. *Sarcophyte.*—Ed.]

The parasitism of *Balanophoreæ*, the anatomical structure of their tissue, composed of cells crossed by rayed scalariform vessels, and the nature of their seeds, connect them with *Cytineæ* and *Rafflesiaceæ* ; [1] but they differ in habit, inflorescence, and the composition of their ovary. They have also some analogy with *Gunneraceæ* in diclinism, apetalism, inflorescence, oligandry, the inferior 1-celled and 1-ovuled ovary, the pendulous ovule, albuminous seed, undivided embryo, and astringent property. They differ in parasitism, the absence of leaves, &c.

Balanophoreæ principally inhabit the intertropical region of both worlds, but are not abundant anywhere; one species alone (*Cynomorium coccineum*), the analysis of which we have taken from the learned memoir of Weddell, grows on various plants of the Mediterranean shores. Another (*Dactylanthus*) inhabits New Zealand. *Mystropetalum* and *Sarcophyte* inhabit South Africa. The third, fourth, fifth, and sixth tribes are all American, except *Rhopalocnemis*, which is tropical Asiatic, and *Thonningia*, which is tropical African.

The properties of *Balanophoreæ* are more or less astringent. *Cynomorium coccineum*, the *Fungus melitensis* [of the Crusaders], has an astringent and slightly acid taste. Its reddish juice was formerly prescribed as an infallible styptic for hæmorrhage and diarrhœa. In Jamaica *Helosis* has a similar reputation. *Sarcophyte*, a Cape species, exhales a fœtid odour, as do several others. *Ombrophytum*, which grows in Peru with marvellous rapidity after rain, is named by the inhabitants Mountain Maize; they cook the scape, and eat it like mushrooms. [Candles are made from a peculiar hydrocarbon contained in a Javanese *Balanophora.*]

CCXVII. PIPERACEÆ, L.-C. Richard.

Flowers ☿ or diœcious, achlamydeous. Stamens 2-3-6-∞. Ovary 1-*celled*, 1-*ovuled*; ovule *sessile, basilar, erect, orthotropous.* Berry *nearly dry.* Albumen *fleshy, dense.* Embryo *antitropous, apical, included in the embryonic sac;* radicle *superior.*

Annual or perennial herbs, usually succulent, or shrubs. Stems simple or

[1] I regard them as having no affinity whatever with *Rafflesiaceæ.*—Ed.

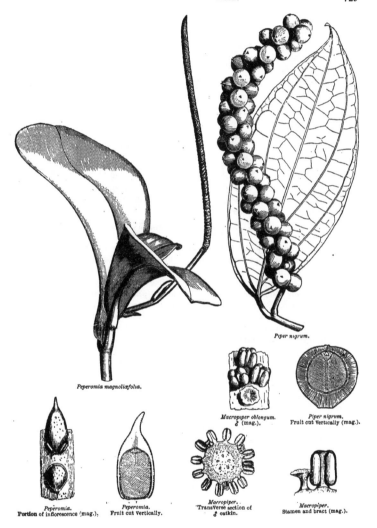

Piper nigrum.

Peperomia magnoliæfolia.

Macropiper oblongum.
♂ (mag.).

Piper nigrum.
Fruit cut vertically (mag.).

Peperomia.
Portion of inflorescence (mag.).

Peperomia.
Fruit cut vertically.

Macropiper.
Transverse section of
♂ catkin.

Macropiper.
Stamen and bract (mag.).

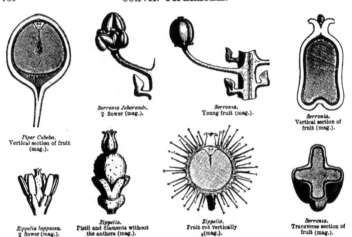

Piper Cubeba.
Vertical section of fruit
(mag.).

Serronia Jaborandi.
♀ flower (mag.).

Serronia.
Young fruit (mag.).

Serronia.
Vertical section of
fruit (mag.).

Zippelia lappacea.
♀ flower (mag.).

Zippelia.
Pistil and filaments without
the anthers (mag.).

Zippelia.
Fruit cut vertically
(mag.).

Serronia.
Transverse section of
fruit (mag.).

branched, cylindric, jointed at the nodes, with axillary branches solitary or leaf-opposed, never whorled, furnished with fibro-vascular bundles scattered through their pith, which gives them the appearance of Monocotyledonous plants. LEAVES often succulent, opposite or whorled, sometimes alternate by arrest, simple, entire, nerves scarcely visible, or prominent, reticulate; *petiole* very short, sheathing at the base; *stipules* 0. SPADICES solitary or fascicled, terminal or leaf-opposed, naked or with a short foliaceous spathe. FLOWERS ☿, or diœcious by arrest of the stamens, achlamydeous, furnished with a peltate or decurrent bract sessile on an often fleshy and sub-cylindrical spadix, or half-buried in its pits, rarely pedicelled. STAMENS sometimes 2, occupying the right and left sides of the ovary, sometimes 3, of which one is posterior; often more numerous (6–∞), of which several are arrested; *fila-ments* very short, linear, adnate at the base to the ovary; *anthers* extrorse, ovoid, adnate, 2- rarely 1-celled (by confluence of the cells), and reniform, dehiscence longi-tudinal; *pollen-grains* smooth, sub-globose, pellucid. OVARY sessile, sub-globose, 1-celled, 1-ovuled; *stigma* usually sessile, terminal or sub-oblique, short or elongate-subulate or orbicular, undivided or 3–4-lobed, glabrous or hispid; *ovule* basilar, sessile, orthotropous. BERRY dry or fleshy. SEED erect, basilar, sub-globose; *testa* cartilaginous, thin; *albumen* fleshy-farinaceous, or sub-cartilaginous, usually hollow in the middle. EMBRYO antitropous, at the top of the seed, occupying a superficial cavity in the albumen, and included in the persistent embryonic sac, small, turbinate or lenticular; *cotyledons* very short, thick; *radicle* superior.

PRINCIPAL GENERA.

Piper.	Chavica.	Peperomia.	Verhuelia.	Macropiper.
Cubeba.	Artanthe.	Serronia.	Zippelia.	

Piperaceæ are closely allied to *Saurureæ*, from which they are distinguished by their 1-celled ovary and 1-seeded fruit (see this family). They are equally near *Chloranthaceæ* in the achlamydeous flowers, the filaments joined to the ovary, the 1-celled 1-ovuled ovary, orthotropous ovule, sessile stigma, and albuminous seed; but in *Chloranthaceæ* the seed is pendulous, the embryonic sac is not persistent, the leaves are always opposite, and their petioles are united into an amplexicaul 2-stipulate sheath. The orthotropy and basilar position of the ovule also bring *Piperaceæ* near *Urticeæ*, *Loranthaceæ*, and even *Polygoneæ*.

Piperaceæ, contained between 35° north latitude and 42° south latitude, abound especially in tropical America. They are less numerous in the Indian Archipelago and the Isles of Sunda; whence they spread north and south over the Asiatic continent. Africa is poor in species; but some grow in South Africa; and few more are met with in the African islands. The woody species are especially Asiatic, the herbaceous American.

Piperaceæ possess an acrid resin, an aromatic volatile oil, and a crystallizable proximate principle (*piperine*), which is sometimes present in all parts of the plant, sometimes principally in the root or fruit. The Pepper (*Piper nigrum*) grows wild and is largely cultivated in Java, Sumatra and Malabar. Its fruit, gathered before it is ripe, dried and pulverized, is the Black Pepper, a spice known in Europe since the conquests of Alexander; the ripe fruit, macerated in water, then dried and cleared from its pericarp by friction, is White Pepper. The latter is preferred for use at table, but the other is the better stimulant. *P. trioicum*, which grows in Asia, is as much esteemed as a condiment as *P. nigrum*. The fruit of the American species, *P. citrifolium, crocatum, Amalago*, is similarly used. The Long Pepper is the entire spike, gathered before it is ripe, of *P. longum*, a shrub of the mountains of India. Its young fruits have a still more burning taste than those of Black Pepper. Cubebs (*P. Cubeba*) grows wild in Java; its properties are as powerful as those of Black Pepper. Its berries are administered in affections of the urethra. Betel (*P. Betel*) has bitter aromatic leaves, which the inhabitants of equatorial Asia mix with Areca-nut and lime to form a masticatory, which they use constantly, and which stimulates the digestive organs in hot and damp climates; but the abuse of it makes the teeth as black as ebony, and the gums vascular.

The Ava, or Kava (*P. methysticum*), is cultivated in the tropical islands of the Pacific; its root, when bruised, chewed, impregnated with saliva, and mixed with Coco juice, is used to prepare a very intoxicating and narcotic liquor, the frequent use of which is not less pernicious than that of Betel. This root is used as a sudorific by English doctors. Several American species are renowned for their diaphoretic and antispasmodic qualities, as *P. crystallinum, rotundifolium, heterophyllum, churumaya*, &c., the leaves of which are used in an infusion. A decoction of *P. elongatum* is administered as an anti-syphilitic in Peru, where also the ripe spikes of *P. crocatum* yield a saffron-yellow dye.

CCXVIII. *SAURUREÆ, L.-C. Richard.*

FLOWERS ☿. PERIANTH 0. OVARY 1–*several-celled*; OVULES *ascending, orthotropous.* SEEDS *with farinaceous or horny albumen.* EMBRYO *antitropous, included on the top of the albumen in the embryonic sac*; RADICLE *superior.*

Aquatic or land HERBS, perennial, with a creeping scaly or tuberous rhizome. STEMS either simple or slightly branching, jointed-knotty, cylindric and leafy, sometimes short or obsolete and scapigerous. LEAVES radical or alternate, petioled, entire, reticulate; *petiole* sheathing by its dilated base, or adnate to an intrapetiolar sheath split along one side. FLOWERS ☿, leaf-opposed, arranged on a spadix in racemes or dense spikes, terminal, solitary or sometimes geminate, naked or furnished at the base with several coloured spathes, each flower furnished with one or 2 collateral bracts. PERIANTH 0. STAMENS 3–6, or more, whorled around the ovary, free or united below to its base, or inserted on its top; *filaments* subulate or clavate, longer than the bracts; *anthers* introrse, 2-celled, opposite, contiguous, or disjoined by the connective, dehiscence longitudinal. OVARY superior or inferior, composed of 3–5 more or less coherent carpels, and 3–5-celled with central placentation, or 1-celled with parietal placentation; *stigmas* terminating the free and narrowed tips of the carpels, papillose on their inner face; *ovules* inserted at the central angle of the cells, 2–4–8, 2-seriate, ascending or horizontal, orthotropous. FRUIT of follicles or a lobed berry. SEEDS few or solitary in each cell, sub-basilar, ovoid, sub-globose or cylindric; *testa* coriaceous, thick; *albumen* farinaceous or horny.

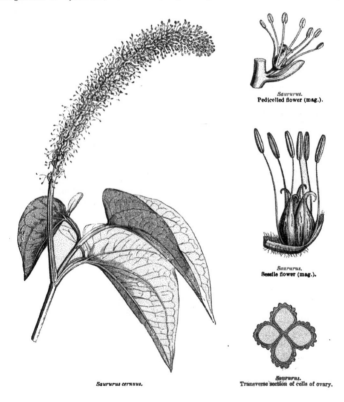

Saururus.
Pedicelled flower (mag.).

Saururus.
Sessile flower (mag.).

Saururus cernuus.

Saururus.
Transverse section of cells of ovary.

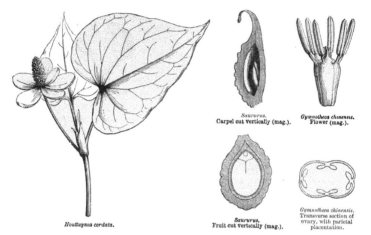

Saururus.
Carpel cut vertically (mag.).

Gymnotheca chinensis.
Flower (mag.).

Houttuynia cordata.

Saururus.
Fruit cut vertically (mag.).

Gymnotheca chinensis.
Transverse section of ovary, with parietal placentation.

EMBRYO apical, situated in a superficial cavity of the albumen, and enclosed in the persistent embryonic sac; *cotyledons* very short; *radicle* superior.

GENERA.

Saururus.   Houttuynia.   Anemiopsis.   Gymnotheca.   Saururopsis.

Saurureæ are near *Piperaceæ* in the intrapetiolar sheath, achlamydeous flowers, and antitropous embryo enclosed, at the top of the albumen, within the persistent embryonic sac; but *Piperaceæ* are distinguished from them by their 1-carpelled 1-celled and 1-ovuled ovary.

This family has been found in tropical Asia, equatorial and South Africa, Japan, and extra-tropical North America. *Saurureæ* possess a somewhat acrid aroma, which confirms their affinity with *Piperaceæ*. The root of *Saururus cernuus*, which grows in America, is boiled and applied externally in pleurodynia. The herbage of *Houttuynia* is used as an emmenagogue in Cochin China.

CCXIX. *CHLORANTHACEÆ.*

(CHLORANTHEÆ, *Br.*—CHLORANTHACEÆ, *Lindl.*)

FLOWERS ☿ or *diclinous, achlamydeous.* OVARY *1-celled, 1-ovuled;* OVULE *pendulous, orthotropous.* DRUPE *fleshy.* ALBUMEN *fleshy, copious.* EMBRYO *small, antitropous, apical;* RADICLE *inferior.*

Small TREES or UNDERSHRUBS, rarely annual HERBS, aromatic, with opposite jointed-knotty branches. LEAVES opposite, petioled, simple, penninerved, dentate or rarely entire; *petioles* united into a short amplexicaul sheath, 2-stipulate on each

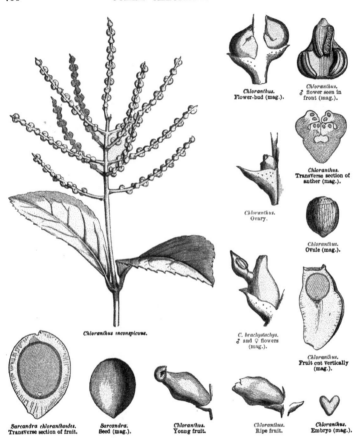

Chloranthus.
Flower-bud (mag.).

Chloranthus.
♂ flower seen in front (mag.).

Chloranthus.
Transverse section of anther (mag.).

Chloranthus.
Ovary.

Chloranthus.
Ovule (mag.).

Chloranthus inconspicuus.

C. brachystachys.
♂ and ♀ flowers (mag.).

Chloranthus.
Fruit cut vertically (mag.).

Sarcandra chloranthoïdes.
Transverse section of fruit.

Sarcandra.
Seed (mag.).

Chloranthus.
Young fruit.

Chloranthus.
Ripe fruit.

Chloranthus.
Embryo (mag.).

side. FLOWERS small, terminal, or rarely axillary, each sunk in an avicular bract, or rarely naked. PERIANTH 0. STAMENS in the ♂ inserted on a common axis and forming a spike, sometimes scattered and bracteate, sometimes close, imbricate and ebracteate; *filaments* short; *anther-cells* marginal on the connective, dehiscence longitudinal. STAMENS in the ♀ 1–3, united with the dorsal face of the ovary, the 2 lateral with 1-celled anthers, the intermediate one with a 2-celled anther; cells introrse, opposite, dehiscence longitudinal; *filaments* keeled, united by their base. OVARY sessile, 3-gonous or sub-globose, 1-celled; *stigma* terminal, sessile, obtuse or depressed, furrowed or sub-lobed, deciduous; *ovule* solitary, suspended near the top of the cavity, orthotropous. DRUPE fleshy, sub-globose or 3-gonous; *epicarp* thick; *endocarp* thin, fragile. SEED pendulous, integument finely membranous. EMBRYO antitropous, minute, included in the top of a fleshy copious albumen; *cotyledons* very short, divaricate; *radicle* inferior.

PRINCIPAL GENERA.

| Hedyosmum. | * Chloranthus. | Sarcandra. | Ascarina. |

Chloranthaceæ are very near *Piperaceæ* (p. 731). They also approach *Saurureæ* in the achlamydeous flower, the filaments adnate to the ovary, orthotropous ovule, and copious albumen. They are separated by the fewer stamens, 1-celled ovary, solitary pendulous ovule, woody stem, and opposite leaves. Payer has erroneously described and figured as anatropous the ovule of *Chloranthus* (Organ., p. 423, tab. 90).

This little family is [chiefly] tropical. *Chloranthus* and *Sarcandra* grow in hot climates and especially in the Asiatic islands. *Ascarina* inhabits the Society Islands [and New Zealand], *Hedyosmum* America.

Chloranthaceæ are aromatic, and classed among excitants. The root of *C. officinalis*, which grows in the damp forests of Java, has a very penetrating odour of camphor, and a sub-bitter aromatic taste; it is renowned in Java as an antispasmodic and febrifuge. The celebrated botanist Blume found it efficacious in the treatment of the fatal intermittent fever which, in 1824, killed 30,000 natives, and of the typhus fever which, in 1825, devastated several provinces of that island. The young shoots and the leaves of *Hedyosmum nutans* and *arborescens* are used in Jamaica in popular medicine as antispasmodics and digestives.

CCXX. *CERATOPHYLLEÆ,* S. Fr. Cray.

FLOWERS *monœcious, sessile in the axils of the leaves, involucrate, achlamydeous.* ANTHERS ∞. OVARY *1-celled, 1-ovuled*; OVULE *pendulous, orthotropous.* ALBUMEN 0. EMBRYO *antitropous, thick*; RADICLE *inferior*; PLUMULE *polyphyllous.*

Aquatic submerged HERBS, much branched; *stem* and *branches* cylindric, jointed-knotty. LEAVES whorled, sessile, exstipulate, dissected; *segments* 2–3-chotomous, filiform, acute, finely toothed. FLOWERS achlamydeous, monœcious, sessile in the axils of the leaves, those of each sex enclosed separately in a 10–12-partite involucre with linear equal entire or cut segments.—FLOWERS ♂: ANTHERS ∞, crowded in the centre of the involucre, sessile, ovoid-oblong, 2–3-cuspidate at the top; cells 2, collateral, buried in a cellular mass, rupturing irregularly.—FLOWERS ♀: OVARY solitary, sessile, sides cuspidate a little above the base, 1-celled; *style* terminal, continuous with the ovary, attenuated, tip with unilateral stigmatic papillæ; *ovule* solitary, pendulous from the top of the cell, orthotropous. NUT coriaceous, enclosed in the persistent involucre, armed with 2 basilar accrescent points, and terminated by the persistent style. SEED pendulous; *testa* finely membranous, thickened around the hilum, exalbuminous. EMBRYO antitropous; *cotyledons* oval, thick; *radicle* very short, inferior; *plumule*-herbaceous, sub-stipitate, polyphyllous, equalling the cotyledons.

ONLY GENUS.
Ceratophyllum.

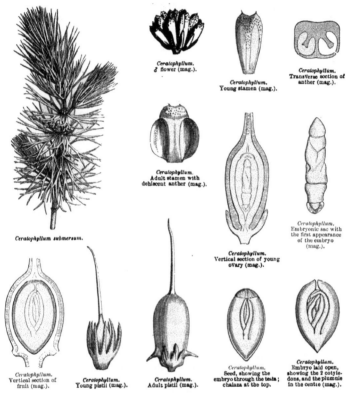

Ceratophyllum.
♂ flower (mag.).

Ceratophyllum.
Young stamen (mag.).

Ceratophyllum.
Transverse section of
anther (mag.).

Ceratophyllum.
Adult stamen with
dehiscent anther (mag.).

Ceratophyllum.
Embryonic sac with
the first appearance
of the embryo
(mag.).

Ceratophyllum submersum.

Ceratophyllum.
Vertical section of young
ovary (mag.).

Ceratophyllum.
Vertical section of
fruit (mag.).

Ceratophyllum.
Young pistil (mag.).

Ceratophyllum.
Adult pistil (mag.).

Ceratophyllum.
Seed, showing the
embryo through the testa;
chalaza at the top.

Ceratophyllum.
Embryo laid open,
showing the 2 cotyle-
dons, and the plumule
in the centre (mag.).

The affinities of *Ceratophylleæ* are not obvious; they have, like *Chloranthaceæ*, opposite leaves, diclinous achlamydeous flowers, a 1-celled 1-ovuled ovary, and a pendulous orthotropous ovule; but they are aquatic, their leaves are capillary, and their seed is exalbuminous. They have the same analogies with *Urticeæ*, but in this family the flower has a perianth, and the ovule is basilar. They resemble *Callitrichineæ* only in being aquatic and achlamydeous, and having involucrate flowers and a 1-celled and 1-ovuled ovary. Brongniart was the first to point out the similarity of structure between the seeds of *Ceratophyllum* and of *Nelumbium* (page 210).

This family is composed of a small number of species inhabiting stagnant water in Europe, Asia, and North America. They possess no known useful quality.

CCXXI. *PODOSTEMACEÆ.*[1]

(PODOSTEMEÆ, *Richard.*—PODOSTEMACEÆ, *Lindl.*)

[FLOWERS ☿ or *diclinous.* PERIANTH 0 *or simple.* ANDRŒCIUM *hypogynous,* 1–3-*seriate.* STAMENS *definite or indefinite.* OVARY 1–3-*celled;* STIGMAS 1–3, *sessile or on a style;* OVULES *numerous in the cells.* CAPSULE *septicidal.* SEEDS *minute, albuminous.—Aquatic* HERBS, *sometimes frondose, often resembling* Algæ *or* Hepaticeæ.

WATER-PLANTS, with a distinct simple or branched stem and leaves, or with these confluent into broad or narrow Alga-like fronds. INFLORESCENCE various, often scapose; *scapes* 1- or many-flowered, arising from a tubular sheath or involucre, or naked. FLOWERS unisexual or ☿, naked or monochlamydeous, usually enclosed in a spathaceous marcescent involucre, which is at first closed, then bursts; mouth 2- or more- lobed. PERIANTH 0, or calycine or petaloid, 3-lobed or -partite, membranous, marcescent. STAMENS definite or indefinite, free or monadelphous, erect; *filaments* linear flat or membranous, marcescent; *anthers* ovate-oblong or linear, 2-lobed, dorsifixed, dehiscence longitudinal; *pollen* powdery, globose didymous or 3-gonous. STAMINODES filiform or subulate, as many as and alternate with the stamens, or inserted with them or outside them, or more than them, or 0. OVARY free, sessile or stipitate, central or excentric, smooth or costate, 1–3-celled; *style* cylindric or 0; *stigmas* 2–3, simple or dilated or laciniate or toothed, rarely 1 capitate; *ovules* numerous in each cell, attached to thick axile or slender parietal placentas, anatropous. CAPSULE 1–3-celled, septicidally and septifragally 2–3-valved, many-seeded; valves equal and persistent, or unequal, the smaller deciduous. SEEDS microscopic, compressed; *testa* mucilaginous; *tegmen* membranous; *albumen* 0. EMBRYO straight, oily; *cotyledons* erect; *radicle* very short, obtuse.

TRIBE I. HYDROSTACHYEÆ.—Flowers naked, diœcious. Ovary 1-celled; carpels 2, alternating with bracts; placentas linear, parietal. *Hydrostachys.*

TRIBE II. EUPODOSTEMACEÆ.—Flowers ☿ without perianth and enclosed in an involucre. Ovary 2–3-celled, with axile placentas, or 1-celled with a central placenta. *Mourera, Apinagia, Dicræa, Podostemon, Hydrobryum, Castelnavia, Tristicha.*

The affinities of *Podostemaceæ* are most obscure. I have suggested that they are reduced forms of *Lentibularineæ* or *Scrophularineæ.* The decidedly dicotyledonous embryo removes them from all the Monocotyledons with which they have been compared. Lindley has suggested *Piperaceæ* and *Callitriche,* Meissner *Ceratophylleæ*; more lately Lindley allies them undoubtedly to *Elatineæ.*

Podostemaceæ are natives of rocky river-beds in the tropics; *Hydrostachys* is Madagascarian; *Dicræa* and *Tristicha* are Asiatic, Madagascarian, and American. One *Podostemon* inhabits temperate North America; the rest of the family are chiefly American, except *Hydrobryum,* which is Indian.

The only known use of the family is that the South American Indians make some use of their saline ashes.—ED.]

[1] This order is omitted in the original. Its characters are taken from Tulasne's monograph.—ED.

CCXXII. *BATIDEÆ. Lindl.*

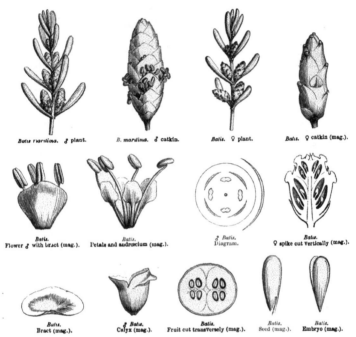

Batis maritima. ♂ plant.　　*B. maritima.* ♂ catkin.　　*Batis.* ♀ plant.　　*Batis.* ♀ catkin (mag.).

Batis.
Flower ♂ with bract (mag.).　　*Batis.* Petals and androecium (mag.).　　*Batis.* Diagram.　　*Batis.* ♀ spike cut vertically (mag.).

Batis. Bract (mag.).　　♂ *Batis.* Calyx (mag.).　　*Batis.* Fruit cut transversely (mag.).　　*Batis.* Seed (mag.).　　*Batis.* Embryo (mag.).

Littoral saline greyish-coloured PLANTS. STEMS branched, straggling, weak. LEAVES opposite, oblong-linear or obovate-oblong, sessile or sub-sessile, flat above, convex below, fleshy, exstipulate. FLOWERS diœcious, arranged in 4 rows, in conical oblong spikes, opposite, sessile, green.— ♂ : FLOWERS distinct; *bracts* cochleariform, obtuse or very shortly acuminate, concave, entire, persistent, close together. CALYX membranous, campanulate, or in a compressed cup, truncate, bilabiate. PETALS 4, claws united; *limb* rhomboidal. STAMENS 4, alternate with the petals, and projecting; *filaments* subulate, glabrous; *anthers* 2-celled, introrse, oblong, didymous, incumbent, versatile, dehiscence longitudinal. OVARY rudimentary or 0.— ♀ : FLOWERS united into a fleshy spike; *bracts* as in the ♂, deciduous, the 2 lower united. CALYX and COROLLA 0. OVARIES 8–12, coherent, and adherent to the base of the bracts, 4-celled; *style* 0; *stigma* capitate, sub-bilobed; *ovules* solitary, erect, anatropous; *pericarps* 4-celled, united, and forming a fleshy ovoid conical fruit; *endocarp* coriaceous. SEEDS oblong, erect, straight, exalbuminous; *testa* membranous. EMBRYO conformable to the seed; *cotyledons* fleshy, oblong, compressed; *radicle* short, near the hilum.

ONLY GENUS.

Batis.

The affinities of *Batis* are very obscure; in habit it resembles *Chenopodiaceæ*, but the structure of its flowers would appear to indicate a closer analogy with *Reaumuriaceæ, Tamariscineæ*, and some *Zygophylleæ* through *Tribulus.*

Batis inhabits the seashores of tropical America.

CCXXIII. *CONIFERÆ, Jussieu.*

(ABIETINÆ, CUPRESSINÆ, TAXINÆ, *L.-C. Richard,* ET GNETEÆ, *Blume.—* GYMNOSPERMARUM *pars, Brongniart.*)

FLOWERS *diclinous, amentaceous, achlamydeous.* OVARY, STYLE *and* STIGMA 0; OVULES *naked;* MICROPYLE *gaping, receiving directly the pollen-grains.* SEED *albuminous.* EMBRYO *with 2 or more cotyledons.*—STEM *woody.* LEAVES *scattered, opposite, whorled or fascicled.*

TREES, or UNDERSHRUBS or SHRUBS, usually resinous, wood without proper vessels, and composed of fibres with one or more series of concave disks; *buds* naked or protected by scales. LEAVES exstipulate, usually persistent, scattered, distichous, opposite, ternate, imbricate, or fascicled on shortened branches, simple, entire, or rarely denticulate, or very rarely lobed, usually linear or acicular, often boat-shaped or scale-like, rarely elliptic or flabellate, always with simple nerves, very rarely reduced to teeth in the axils of which spring branches dilated into phyllodes [*Phyllocladus*], sometimes dimorphous (some acicular, the others scale-like) on the same individual (Juniper). FLOWERS in catkins, monœcious or diœcious, achlamydeous; *catkins* ♂ composed of antheriferous scales; *anthers* of 1–20 parallel or radiating contiguous or distant cells, adnate to the scale (dilated or peltate), which serves as connective and filament, sometimes pendulous (*Araucaria*), dehiscence longitudinal, rarely transverse; *pollen* sometimes of 2 rather turgid vesicles, united by an intermediate membrane (*Pinus, Abies, Dacrydium, Podocarpus,* &c.); sometimes of very small smooth globose grains (*Araucaria, Sequoia, Cunninghamia, Cupressineæ, Taxineæ*). FLOWERS ♀ reduced to naked ovules,[1] usually orthotropous, springing from spreading scales, or a cupuliform disk; *micropyle* gaping, endostome often prolonged into a styloid tube. FRUIT sometimes forming a dry strobilus, or fleshy by the union of the thickened and often hardened scales, sometimes drupe-like with a fleshy coriaceous or crustaceous testa; sometimes surrounded by a fleshy cupule. SEED naked, often winged; *albumen* horny or fleshy-farinaceous, or oily, rarely ruminate (*Torreya*), originally containing several rudimentary embryos, of which one only is developed. EMBRYO axile, usually antitropous, often furnished with a very

---

[1] [The outer coat of the so-called naked ovule of *Coniferæ* and *Cycadeæ* is by various botanists considered to be a true but incomplete ovary; a disputed point which has given rise to much discussion and volumes of literature. · Under a third theory the outer coat is regarded as of the nature of a disk.—ED.]

long suspensor formed of a skein of filaments; *cotyledons* 2 or several-whorled (or, according to Duchartre, 2, each many-partite and opposite), epigeous or hypogeous in germination; *radicle* superior or inferior.

Coniferæ, so remarkable for the exceptional structure of their woody fibres, for their seed containing originally several embryos (all abortive but one, which has often several cotyledons), and for the long interval between the fertilization of the ovules and the ripening of the seeds, are distinctly characterized by the extreme simplicity of their reproductive organs, and form with *Cycadeæ* (the flowers of which are equally reduced to naked ovules) an isolated group in the Vegetable Kingdom [known as *Gymnosperms*]. They might be considered as intermediate between *Phænogams* and *Cryptogams*, if a few external resemblances only were noted, such as those which exist between *Ephedra* and *Equisetum*, *Cycadeæ* and *Filices*, &c.

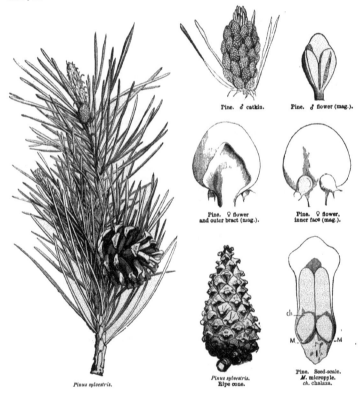

Pine. ♂ catkin.

Pine. ♂ flower (mag.).

Pine. ♀ flower
and outer bract (mag.).

Pine. ♀ flower,
inner face (mag.).

Pinus sylvestris.

Pinus sylvestris.
Ripe cone.

Pine. Seed-scale.
M. micropyle.
ch. chalaza.

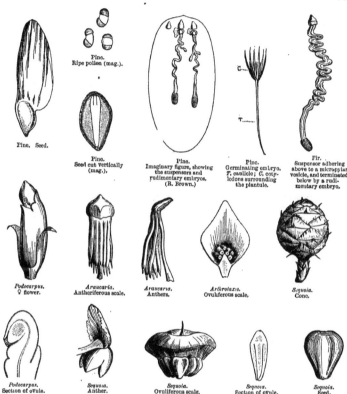

Pine. Seed.

Pine.
Ripe pollen (mag.).

Pine.
Seed cut vertically
(mag.),

Pine.
Imaginary figure, showing
the suspensors and
rudimentary embryos.
(R. Brown.)

Pine.
Germinating embryo.
T. caulicle; C. coty-
ledons surrounding
the plantule.

Fir.
Suspensor adhering
above to a micropylar
vesicle, and terminated
below by a rudi-
mentary embryo.

Podocarpus.
♀ flower.

Araucaria.
Antheriferous scale.

Araucaria.
Anthers.

Arthrotaxis.
Ovuliferous scale,

Sequoia.
Cone.

Podocarpus.
Section of ovule.

Sequoia.
Anther.

Sequoia.
Ovuliferous scale.

Sequoia.
Section of ovule.

Sequoia.
Seed.

Coniferæ, which have played so considerable a part in all the geological epochs of our planet, are to this day one of the most numerous and widely-spread families in the world. They form a class rather than a family, and their tribes may be considered as so many distinct orders, themselves capable of sub-division. These tribes or sub-orders are the following:—

Tribe I. *ABIETINÆ, L.-C. Rich.*

TREES, usually tall, often gigantic, resinous, trunk conical, branches numerous, most frequently whorled; or SHRUBS with divaricate branches; BUDS naked or scaly. LEAVES usually persistent, stiff, narrow-linear, subulate, lanceolate or elliptic, scattered, or united in 1–7-foliate fascicles, girt at the base with a scarious sheath (*Pinus*). FLOWERS monœcious, or rarely diœcious, stamens and ovuliferous scales spirally arranged around a common axis, and forming terminal or lateral catkins. CATKINS ♂: STAMENS numerous, naked, more or less compact; *filaments* very short, thick, usually prolonged at the top into a scale-like straight or inflexed connective; *anthers* sometimes 2-celled, with ovoid-oblong apposed cells, separated by a more or less developed connective which extends beyond them; sometimes 3–many-celled; cells cylindric, 1–2-seriate below the connective, dehiscence longitudinal or transverse; *pollen* composed of 2 vesicles united by an intermediate membrane.—CATKINS ♀: OVULIFEROUS SCALES usually numerous, sessile on their axis, or shortly unguiculate, never peltate, imbricate, accrescent, naked, or inserted in the axil of a bract which is early arrested or accrescent, exceeding the scale; *ovules* with inferior micropyle, 2 collateral (*Pinus, Abies, Picea, Larix, Cedrus*) or 3–5 (*Cunninghamia, Arthrotaxis*), or 5–9 (*Sequoia, Sciadopitys*), or solitary (*Araucaria, Eutassa, Dammara, Dacrydium, Podocarpus*), inserted by their base towards the middle of the scale, or sometimes adnate to it throughout their length, near its top (*Araucaria, Podocarpus*), orthotropous, or very rarely anatropous (*Podocarpus, Dacrydium*). CONE usually composed of persistent or deciduous seminiferous scales, woody or coriaceous, thickened, or thin at the top. SEEDS as many as the ovules, inverted, adhering to the scale, or caducous; *testa* coriaceous or bony, rarely soft (*Podocarpus*), often terminating in a membranous wing above (*Pinus, Abies*, &c.), or unilateral (*Araucaria*); *albumen* fleshy, oily. EMBRYO with oblong-linear cotyledons; *radicle* cylindric, inferior.

[Under the tribes of this important family I have thought it right to introduce the sub-tribes and all the genera, as given in the latest work on the subject, by Parlatore, in De Candolle's 'Prodromus':—

Sub-tribe I. ARAUCARIEÆ.—Anthers 6 or more, 2-seriate, linear. Scales of cone spirally arranged, deciduous. Seeds solitary, pendulous, wingless, or unequally winged; cotyledons 2, entire or 2-partite.—Leaves flat, often broad, or 4-gonous, sub-opposite or spirally arranged. *Araucaria*, (*Eutassa*), *Dammara*.

Sub-tribe II. PINEÆ.—Anthers 2, sub-globose or oblong. Scales of cone spirally arranged, usually persistent. Seeds 2, collateral (rarely 1 arrested), pendulous, usually winged; cotyledons 2, 2–∞-partite.—Leaves various. *Pinus*, (*Larix*, *Cedrus*, *Picea*, *Abies*, *Tsuga*, *Pseudolarix*).

Sub-tribe III. TAXODIEÆ.—Anthers 2–5, rarely 9, 1-seriate. Scales of cone spirally arranged (whorled in *Widdringtonia*), usually persistent. Seeds 3–9 (very rarely fewer), winged or not; cotyledons 2, usually entire.—Leaves usually linear, rarely scale-like. *Cunninghamia*, *Arthrotaxis*, *Sciadopitys*, *Sequoia*, *Cryptomeria*, *Glyptostrobus*, *Taxodium*, *Widdringtonia*.

Sub-tribe IV. CUPRESSINEÆ. (See Tribe II. *Cupressineæ*, p. 744.) Anthers 3–5, rarely 2. Scales of cone 4 or more, decussately opposite, or 3–4 in a whorl, persistent, free or connate (*Juniperus*). Seeds erect, 2–3-winged, rarely wingless; cotyledons 2, usually entire.—Leaves opposite or whorled. *Actinostrobus, Frenela, Callitris, Libocedrus, Thuja, Thuyopsis, Biota, Diselma, Fitzroya, Chamæcyparis, Cupressus, Juniperus.* •

Pines, Firs, Larches, and Cedars, which composed the genus *Pinus* of Linnæus, cover a vast area of the northern hemisphere; they are gregarious on the mountains of temperate regions, and descend towards the plains as they approach the pole; in the Alps they mark the limit of arborescent vegetation (*P. Pumilio, Larix*). Many species of *Pinus*, L., occur in North America, from the lofty mountains of Mexico to the Frozen Ocean; there are fewer in Europe. Asia possesses the Cedar of Lebanon [and of the Taurus], and that of the Himalayas (*Deodara*) [and North Africa the Cedar of the Atlas]. In China and in Japan are found the singular *Sciadopitys* and *Cunninghamia*, besides all our European genera. *Sequoia semper-virens* and *gigantea* are Californian and Mexican trees which attain a height of 300 feet, and the trunk a circumference of 80. As Dr. Hooker has remarked, it is the southern hemisphere which possesses the largest number of genera which are not found elsewhere, or but rarely. Such are, amongst *Abietineæ*, the genera *Dammara, Eutassa, Araucaria, Arthrotaxis, Dacrydium, Podocarpus*, &c., and of the other tribes the genera *Callitris, Actinostrobus, Pachylepis, Thuja, Phyllocladus*, &c.

Araucaria forms vast forests on the mountains of Brazil and Chili; *Dammara* grows in the Moluccas [Pacific Islands] and New Zealand; *Eutassa* is Australian, Norfolk Island and New Caledonian; *Arthro-taxis*, which is remarkable for having the habit of *Lycopodium*, is confined to Tasmania. *Dacrydium* belongs principally to New Zealand [but extends to the Malayan Peninsula, Tasmania, and New Cale-donia]. *Podocarpus* is cosmopolitan, inhabiting Australia, New Zealand, tropical Asia, Africa, Chili, Japan and the Antilles.

Abietineæ, besides their elegant habit, gigantic stature, their persistent [except *Larix, Pseudolarix*, and *Glyptostrobus*] and singular leaves and fruits, which give so marked a character to the landscape, take the first rank amongst plants useful to man; this depends principally on the nature of their wood, which is flexible, light, and so saturated with a resin as to resist humidity and avert decay. This durability of *Abietineæ* renders them useful for building purposes, both on land and water. The resins which they contain are very important both in the arts and medicinally. The trunks of Pines, Firs, and Larches especially either exude or yield by incision Turpentine, a semi-liquid substance, acrid and of a penetrating odour, essentially composed of a fixed resin dissolved in a volatile oil, in combination with a certain quantity of succinic acid. Several natural and artificial products are obtained from turpentine, which, according to its consistency and the season of the year, is designated as 'en pâte,' 'de barras,' 'de galipot.' Exposed to the sun it is termed Yellow or White or False Venetian Turpentine. Oil of turpentine is that refined by filtration. Distilled over a gentle fire, it yields the spirit of turpentine so extensively used in the arts, and which, when mixed with alcohol, forms the liquid hydrogen, used for lighting. The residuum of the distillation is rosin (*arcanson* or *colophane* or *colophony*). The galipot, triturated in water, is the yellow resin of commerce. Pitch is prepared by burning in a copper the resinous refuse. Tar (*poix noire* or *brais gras*) is half-liquid pitch, obtained by burning the refuse of the preceding products in a covered vessel. Lampblack is the product (soot) of all the above-named materials, after being burnt in a furnace leading to a chamber in which the smoke is deposited as an impalpable powder. The turpentine of the Larch (*Larix europæa*) is most esteemed of any, and is known as Venetian Turpentine.

The Balm of Gilead Fir (*Abies balsamea*), a North American tree, yields Canada Balsam, a sweet-scented turpentine, administered in affections of the urethra; in North America antisyphilitic qualities are attributed to a decoction of its root. The Canadians use the cones of *Pinus Banksiana* as a sudorific. The Silver Fir (*Abies pectinata*) is one of the most useful species for ship-building, timber-work, planks and furniture; its buds, which are resinous in smell and taste, are used medicinally. The Larch yields, besides turpentine, a white substance, sugary and laxative, named Manna of Briançon, and which is analogous to gum arabic. From the trunk of *Pinus Sabiniana*, of North America, exudes under heat a substance (*pinite*) analogous to the preceding. The Cedar of Lebanon (*Cedrus Libani*) is one of the most majestic species of the family; the Jews looked upon its wood as incorruptible. We read in the Bible that the Temple of Solomon was built of Cedar cut on Mount Lebanon, but it is probable that Larch[1] or Cypress wood was used, these being much more durable and compact, and less likely to split. The Deodar (*C. Deodara*), a species as beautiful as that of Lebanon, inhabits the Eastern Himalayas and Affghanistan: it is considered sacred by the Hindoos, and yields an oil efficacious in certain cutaneous disorders. The

[1] Larch is found nowhere in Syria or Asia Minor. The wood of the Lebanon Cedar is very hard and closely-grained in its native forests, though soft and loose in cultivation.—ED.

Stone-Pine (*Pinus Pinea*) is a very picturesque tree of the Mediterranean region, of which the seeds, called 'pignons doux,' are of an oily, mild and agreeable taste, as is also the case with the seeds of *P. Cembra*, an alpine species, which are used as food by the Siberians.

Linnæus informs us that the Lapps and Eskimos, for want of cereals, find material for bread in the inner layer of the bark of *Pinus sylvestris* and *Abies alba*, which they bake slightly and reduce to flour ; with this flour they make thin cakes, which they keep a long time, and consider excellent. From the young shoots of several kinds of Fir may be prepared an antiscorbutic beer [as Spruce beer from the tops of *Abies excelsa*] ; those of *Dacrydium cupressinum*, a fine New Zealand tree, contain a slightly bitter resinous matter, of which Captain Cook availed himself to prepare a drink with which he cured his sailors of scurvy. The bark of *Pinus Pinea, Cembra, maritima*, &c., was formerly valued for its astringency, and is used in tanning leather. On the old roots of *P. Massoniana* a peculiar brown scabrid fungus grows, which is waxy and whitish within, and of which a decoction is used by the Chinese and Japanese in diseases of the lungs and bladder. *Dammara orientalis* is a large-leaved tree of the Malayan Archipelago, and yields Dammar, a white and hard resin, analogous to copal [used throughout India to render fabrics waterproof]. The Kauri (*D. australis*) is one of the tallest trees in New Zealand ; a resin exudes from its trunk which is also found in semi-fossilized blocks, which bear the name of Kapia. It resembles elemi ; the natives pound and burn it to obtain a soot, with which they tattoo their faces [and it is largely imported into England for the manufacture of varnish]. The seeds of *Araucaria brasiliensis* and *imbricata* are edible, like our Chesnuts [those of *A. Bidwillii*, in Australia, form the chief food of whole tribes of natives during certain seasons], as are those of some species of *Podocarpus*, and especially of *P. neriifolia*. The hard light and durable wood of *P. Totara* is much sought by the New Zealanders for the construction of canoes [it is the best wood in those islands]. [The wood of *Dacrydium Franklinii*, a magnificent tree, confined to a very limited area on the west coast of Tasmania, is a beautiful furniture wood of a golden colour.] Amber is a fossil resin, procured from the lignites of the Baltic shores; the most transparent pieces are worked into ornaments; varnish and medicine are also prepared from it. Petroleum, a liquid bitumen, of which abundant springs are found in some countries, has the same origin as amber.

TRIBE II. *CUPRESSINEÆ, L.-C. Rich.*

TREES or SHRUBS, resinous, branched ; *branches* mostly scattered, cylindric or sometimes angular; *buds* naked, or rarely scaly. LEAVES persistent, opposite or whorled in threes, or scattered, very often adnate and decurrent, narrowly linear or scale-like, usually small, stiff, imbricate, rarely caducous (*Taxodium*). FLOWERS monœcious or diœcious, stamens and ovuliferous scales inserted on a common axis, usually ebracteate, imbricate, and forming terminal or lateral catkins. CATKINS ♂ : STAMENS numerous, naked, nearly horizontal ; *filaments* short, thick, prolonged into a scale-like connective, and peltate excentrically ; *anthers* 2-3-celled or more, separate, adnate, ovoid or oblong, dehiscence longitudinal; *pollen* globose.—CATKINS ♀ : OVULIFEROUS SCALES few, peltate, very often mucronate at the back, below the tip, whorled in one or several series around a more or less shortened axis; *ovules* solitary geminate or numerous, sessile, inserted at the base or towards the middle of the scale, orthotropous; *micropyle* superior. FRUIT a cone with woody or fleshy scales, closely connivent, or sometimes bony within (*Juniperus drupacea*). SEEDS solitary or geminate, rarely numerous ; *testa* thin, woody or bony, angular or with a membranous margin. EMBRYO antitropous, in the axis of a fleshy scanty albumen ; *cotyledons* 2, rarely 3-9, oblong, obtuse ; *radicle* cylindric, superior.

[See Sub-tribe IV. *Cupressineæ*, p. 742].

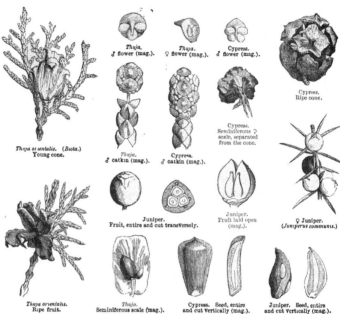

Thuja.
♂ flower (mag.).

Thuja.
♀ flower (mag.).

Cypress.
♂ flower (mag.).

Cypress.
Ripe cone.

Cypress.
Seminiferous ♀
scale, separated
from the cone.

Thuja orientalis. (Biota.)
Young cone.

Thuja.
♂ catkin (mag.).

Cypress.
♂ catkin (mag.).

Juniper.
Fruit, entire and cut transversely.

Juniper.
Fruit laid open
(mag.).

♀ Juniper.
(Juniperus communis.)

Thuja orientalis.
Ripe fruit.

Thuja.
Seminiferous scale (mag.).

Cypress. Seed, entire
and cut vertically (mag.).

Juniper. Seed, entire
and cut vertically (mag.).

Cupressineæ prefer a temperate climate; they extend from Central Europe to the eastern extremity of Asia; they are spread over North America, South Africa, and Australia. Junipers and Cypresses inhabit all the northern temperate zone; the common Juniper (*Juniperus communis*) ascends to the snow line. *Cryptomeria* and *Biota* belong to China and Japan, as do *Thuyopsis* and *Chamæcyparis*, several species of which are also American. *Taxodium* and *Thuja* belong to North America [and Japan], *Libocedrus* to Chili and New Zealand, *Widdringtonia* to South Africa and Madagascar, *Frenela* [and *Actinostrobus*] to Australia, *Callitris* to North Africa, [*Diselma* to Tasmania; *Fitzroya* to Chili].

Cupressineæ possess volatile resinous matters with properties analogous to those of *Abietineæ*. The essence contained in their herbaceous parts and fruit is of the same nature as the essence of turpentine, but the resin which exudes from their trunk contains only traces of volatile oil, and no succinic acid. This resin, combined with an astringent principle, is in some species stimulating and tonic. Other *Cupressineæ* are useful in manufactures, from the hardness and aroma of their wood.

The common Juniper (*Juniperus communis*), a diœcious tree, like all its congeners, is a native of Europe and Siberia; it produces fleshy fruits, improperly called berries, which contain sugar, and are thus fermentable; they also yield a medicinal extract or *rob*, which is sugary, resinous and very stomachic, and enters into the composition of gin. The aromatic wood of the Juniper is used for fumigating.

The fruits of *J. Oxycedrus*, a shrub of the Mediterranean region, may replace those of the preceding species; its wood, when burnt in a closed vessel, gives off an oily liquid, with a very strong empyreumatic odour, used in veterinary medicine. The leaves of *J. Sabina*, a European shrub, contain a fœtid volatile oil, which is anthelminthic and an emmenagogue. *J. Virginiana*, commonly called Red Cedar, has leaves with a resinous scent, but not fœtid, which may be substituted for those of *J. Sabina*. Its reddish sweet-scented light wood is easily worked, and used in making lead pencils [as is its variety *J. Bermudiana*, of which the celebrated Bermuda sailing vessels are also made].

Taxodium distichum, the Deciduous Cypress, is a tree of the Louisianian swamps, now naturalized in several parts of Europe. Its cones are used as diuretics in Anglo-American medicine, and its resin is praised as efficacious in arthritic pains. The roots produce conical hollow excrescences [several feet high], of which the Americans make beehives. The Cypress (*Cupressus sempervirens*) grows wild in the Levant; its sombre hue has caused it to be consecrated to the dead, but the variety *pyramidale* is more particularly planted in cemeteries; its wood is hard, reddish, aromatic and nearly indestructible. The Thuya (*Biota orientalis*), commonly called Tree of Life, a native of China, was introduced into France in the reign of Francis the First. *T. occidentalis*, an American species whose branches exhale a strong smell of treacle, was formerly recommended for its diuretic qualities. *Callitris quadrivalvis*, a Mauritanian tree, secretes a resin known as Sandarac. At the base of its trunk enormous knots are produced, called by the ancient Romans Citron-wood, of which they made tables, which were bought for their weight in gold; Pliny mentions one purchased by Cicero for 8,750*l.*, and another which was sold by auction for 12,250*l.*

Tribe III. *TAXINEÆ, L.-C. Rich.*

TREES or SHRUBS, not resinous; *branches* scattered, rarely whorled; *buds* scaly. LEAVES persistent or annual (*Salisburya*), scattered or distichous, rarely fascicled (*Salisburya*), simple, entire, rigid, linear, sometimes flabelliform, lobed, or reduced to a scale which bears in its axil a branch dilated into a phyllode (*Phyllocladus*). FLOWERS diœcious, the ♂ in sub-globose or elongated catkins, the ♀ solitary, or united into a short spike, often surrounded by imbricate bracts at their base.— CATKINS ♂ naked or with scales at the base. STAMENS numerous, naked, arranged along the axis of the catkin; *filaments* very short, prolonged into a laciniate connective (*Salisburya, Phyllocladus*), or peltate (*Taxus*); *anthers* 2–3–8-celled, dehiscence longitudinal; *pollen* globose.—FLOWERS ♀ naked or bracteate, each inserted on a cupuliform disk, at first short, then accrescent; *ovule* solitary, sessile in the centre of the disk, erect, orthotropous; *micropyle* superior. DRUPE composed of the thickened and fleshy disk, surrounding an erect seed, with bony testa, or sometimes fleshy (*Salisburya*). EMBRYO antitropous, in the axis of a fleshy dense albumen, sometimes farinaceous (*Taxus*), or ruminate (*Torreya*); *cotyledons* 2, semi-cylindric; *radicle* cylindric, superior.

PRINCIPAL GENERA.

* Taxus.	* Phyllocladus.	* Salisburya.	* Cephalotaxus.
Torreya.	Pherosphæra.	Lepidothamnus.	Saxe-Gothea.[1]

Taxineæ are separated from the preceding tribe, both by the succulent cup which surrounds their seeds, and by their fleshy testa. They are met with in all temperate regions, as well as on tropical mountains in Asia and America. Central and Mediterranean Europe possess the common Yew (*Taxus baccata*), which is also spread over North Asia [and America]. Of *Torreya* one species is Japanese, the other Floridan; *Phyllocladus* inhabits Tasmania, New Zealand [New Caledonia and Borneo]; *Cephalotaxus* and *Salisburya* are natives of Japan and China [*Pherosphæra* of Tasmania, and *Lepidothamnus* and *Saxe-Gothea* of Chili].

Taxineæ secrete, like the preceding Conifers, but much less abundantly, resinous juices combined with a volatile oil, and astringent bitter, sometimes narcotic-acrid principles. The common Yew formerly formed forests in some parts of Europe; its longevity is greater than that of any other tree; its red wood,

---

[1] To these Parlatore adds *Dacrydium* and *Podocarpus*, both included under *Abietineæ* by others.—ED.

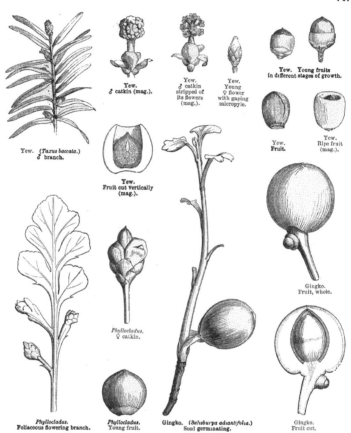

Yew. (*Taxus baccata.*)
♂ branch.

Yew.
♂ catkin (mag.).

Yew.
♂ catkin
stripped of
its flowers
(mag.).

Yew.
Young
♀ flower
with gaping
micropyle.

Yew. Young fruits
in different stages of growth.

Yew.
Fruit cut vertically
(mag.).

Yew.
Fruit.

Yew.
Ripe fruit
(mag.).

Gingko.
Fruit, whole.

Phyllocladus.
♀ catkin.

Phyllocladus.
Foliaceous flowering branch.

Phyllocladus.
Young fruit.

Gingko. (*Salisburya adiantifolia.*)
Seed germinating.

Gingko.
Fruit cut.

hard and capable of receiving a high polish, was much valued in cabinet-work [and for making bows]; the fleshy cup of its fruit contains a mild mucilaginous juice, and may be eaten harmlessly; but its seeds, and especially its leaves, are said to be very poisonous [its wood-fibres are remarkable in presenting a spiral line]. The Gingko (*Salisburya adiantifolia*) is considered sacred in China and Japan, and is planted around temples. Its seed, with a fleshy and oily testa, exhales when ripe a strong smell of rancid butter; its kernel tastes like a filbert, though slightly harsh; in Japan it is considered a digestive, and always served at banquets.

Tribe IV. *GNETACEÆ, Lindl.*

Large or small TREES, or UNDERSHRUBS, not resinous, often sarmentose; *branches* jointed-knotty, opposite or fascicled, sometimes aphyllous, sheathed at the joints, or with very small setaceous leaves (*Ephedra*), or with oval penninerved entire leaves (*Gnetum*), or with only 2 large permanent radical leaves (*Welwitschia*). FLOWERS monœcious or diœcious, with sheaths or laciniate scales, the ♀ with a membranous tubular bifid calyciform sheath. STAMEN solitary (*Gnetum*), or 6–∞ united into a

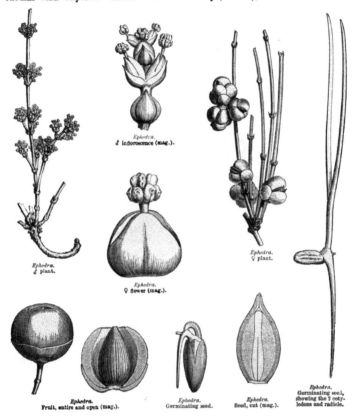

Ephedra.
♂ inflorescence (mag.).

Ephedra.
♀ plant.

Ephedra.
♂ plant.

Ephedra.
♀ flower (mag.).

Ephedra.
Fruit, entire and open (mag.).

Ephedra.
Germinating seed.

Ephedra.
Seed, cut (mag.).

Ephedra.
Germinating seed,
showing the 2 coty-
ledons and radicle.

Welwitschia.
Entire plant, 1·18th natural size; furnished with its cotyledons, which take the place of leaves.

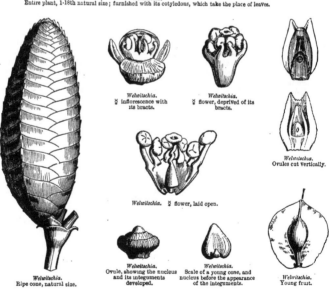

Welwitschia.
Ripe cone, natural size.

Welwitschia.
☿ inflorescence with its bracts.

Welwitschia.
☿ flower, deprived of its bracts.

Welwitschia.
Ovules cut vertically.

Welwitschia. ☿ flower, laid open.

Welwitschia.
Ovule, showing the nucleus and its integuments developed.

Welwitschia.
Scale of a young cone, and nucleus before the appearance of the integuments.

Welwitschia.
Young fruit.

column; *anthers* with 2-4 cells, opening at the top by as many pores or valvules; *ovule* solitary, sometimes in the centre of the stamens (*Welwitschia*) and then abortive, sessile, erect, orthotropous; *primine* membranous, open or notched at the top (*exostome*); *secundine* exserted as a styliform tube which expands into a perforated stigma-like more or less persistent disk (*endostome*). SEED with coriaceous or fleshy testa. EMBRYO antitropous, in the apex of a fleshy albumen; *radicle* superior.

GENERA.

Gnetum.   Ephedra.   Welwitschia.

Some *Gnetaceæ*, by the pinnate and anastomosing nerves of their leaves (*Gnetum*), and by the structure of their wood-fibres, which, like those of *Coniferæ*, are marked with a series of disks, but are associated with large punctate vessels, establish the passage from *Coniferæ* to the other Dicotyledons, and in particular to *Casuarineæ* through *Ephedra*, and to *Chlorantheæ* through *Gnetum*.

Gnetum Gnemon, *funiculare*, *edule*, and *Brunonianum* are all natives of tropical Asia; *G. urens* and *nodiflorum*, &c., of Guiana. *Ephedra* inhabits temperate regions of both hemispheres, affecting salt deserts and the edges of torrents; some species ascend into the alpine region, both in Europe (*E. helvetica*) and America (*E. andina*) [and *E. vulgaris* in the Himalayas]. [*E. distachya*, *alata*, *altissima*, *fragilis*, *campylopoda*, and *vulgaris* inhabit the Mediterranean region. *E. andina*, *americana*, *triandra*, &c., are American.]

The most curious of the *Gnetaceæ*, and perhaps of all Dicotyledons, is that which was discovered a few years ago on the west coast of Africa, near Cape Negro, by Dr. Welwitsch. It never exceeds a foot in height, but the stem is often more than four feet in diameter. It bears no appendages but its two cotyledons, which last throughout its life, i.e. more than a century, and in time grow to an extraordinary size, attaining six feet in length and two to three in width; they are green, very coriaceous, and torn by the wind into numerous segments which spread out upon the earth. Along the margin of its enormous platform-like stem, marked with concentric circles, rise short dichotomous floral peduncles, the branchlets of which bear terminal catkins or young cones, with brilliant scarlet bracts, imbricate in four rows, each containing a flower. After flowering the cones enlarge, and attain about two inches in length and one in diameter. This bizarre plant is named *toumbo* by the natives. We have taken all the dissections from the valuable work of Dr. Hooker.

The *Ephedra* are of little use to man; the flowering branches of the Mediterranean species were formerly used as styptics. *Gnetum* yield textile fibres more tenacious than those of hemp. The leaves and fruit of *G. Gnemon*, which is cultivated in Amboyna and Java, are eaten as a vegetable. The branches of *G. urens* contain a limpid somewhat mucilaginous potable juice; its seeds, when cooked and baked, are edible.

CCXXIV. *CYCADEÆ.*

(CYCADEÆ, *Persoon*, *Br.*, *L.-C. Richard.*—CYCADACEÆ, *Lindl.*—CYCADEACEÆ, *Endlicher.*
—TYMPANOCHETÆ, *Martius.*)

FLOWERS *diœcious*, *achlamydeous*. FLOWERS ☿, *in terminal cones*, *and formed of scales bearing on their dorsal face numerous 1-celled anthers.* FLOWERS ♀ *reduced to naked orthotropous ovules*, *sometimes solitary*, *erect*, *inserted in the crenatures of velvety foliiform appendages*, *sometimes geminate*, *inverted*, *on the inner face of peltate scales.* SEED *albuminous.*—STEM *woody.* LEAVES *pinnate*, *crowning the stem.*

Large or small TREES, elegant, very long-lived. STEM usually simple, straight,

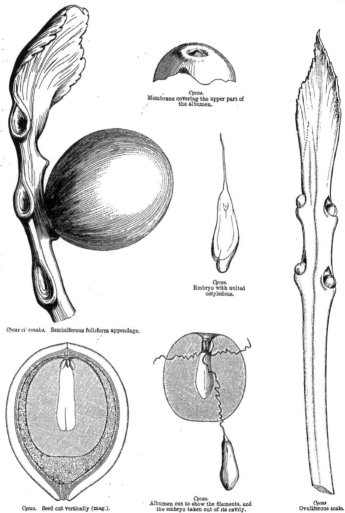

Cycas.
Membrane covering the upper part of
the albumen.

Cycas.
Embryo with united
cotyledons.

Cycas ci cinalis. Seminiferous foliiform appendage.

Cycas. Seed cut vertically (mag.).

Cycas.
Albumen cut to show the filaments, and
the embryo taken out of its cavity.

Cycas
Ovuliferous scale.

Cycas circinalis.

globose ovoid or cylindric, and sometimes 10 feet in circumference (thicker in the ♀), covered with the persistent bases of the petioles, or marked with circular scars; *pith* voluminous, surrounded with one or more zones of wood, each the result of several years' growth, and composed of woody fibres and punctate rayed or reticulate vessels arranged in radiating lines separated by medullary rays, and enveloped in a thick layer of cortical parenchyma. LEAVES of 2 forms: the one short, hard, scaly (*perules*), enclosing the terminal bud; the others normal (as in Palms), and pinnate [or 2–3-pinnate, *Bowenia*]; *leaflets* entire or denticulate, coriaceous, plane or waved; nerves slender, parallel and equal (*Zamia*), or reduced to a prominent

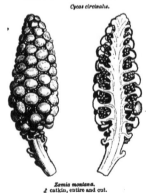

Zamia montana.
♂ catkin, entire and cut.

Zamia.
Antheriferous scale,
anterior face.

Zamia.
Antheriferous scale,
posterior face.

Zamia.
1-celled dehiscent anther
(mag.).

Zamia.
Pollen-grains.

ridge (*Cycas*), or pinnate from a midrib, and then simple or bifurcate (*Stangeria*). VERNATION various : (1) petiole and leaflets rolled into a crozier, as in Ferns ; (2) petiole alone involute, and leaflets imbricate ; (3) petiole straight, and leaflets folded along their midrib and juxtaposed. FLOWERS achlamydeous, diœcious, collected into strobili or terminal cones ; ♂ forming large ovoid or oblong cones ; *scales* thickly coriaceous, oblong, or dilated at the apex, which is plane (*Zamia*), or cuspidate (*Cycas*), or bidentate (*Ceratozamia*), bearing on their dorsal face numerous 1-celled coriaceous anthers, which cover the face of the scale (*Cycas*), or form two groups along the midrib (*Zamia*), dehiscence longitudinal; *pollen* hyaline, globose or ellipsoid.— FLOWERS ♀ : OVARY, *style*, and *stigma* 0 ; *scales* foliiform, imbricate, forming a sort of cone at the top of the stem, crenellated and bearing in each crenature an erect ovule (*Cycas*) ; or forming a true peduncled cone (*Zamia*), composed of stipitate peltate scales, under which are placed 2 ovules ; *ovules* naked, sessile, orthotropous. SEED drupe-like, presenting several openings corresponding to the embryonic vesicles, from which descend folded cords, terminated by embryos of which one only is developed ; *testa* fleshy without, crustaceous within; *albumen* fleshy, thick, in the centre of which is the cavity containing the perfect embryo. EMBRYO appearing undivided, owing to the cohesion of the cotyledons ; *radicle* superior (*Cycas*), or inferior, or obliquely directed towards the rachis (*Zamia*, &c.); *cotyledons* unequal, often hypogeous in germination.

PRINCIPAL GENERA.

* Cycas.	* Macrozamia.	* Dioon.	Bowenia.	* Zamia.
* Ceratozamia.	* Stangeria.	Encephalartos.	Microcycas.	

Cycadeæ, which the earlier botanists, relying on habit and vernation, placed near either the Palms or Tree-ferns, and other Cryptogamic families, are evidently Dicotyledons, and closely allied to *Coniferæ*; the anatomy of the stem, inflorescence, structure of the stamens, ovules, seeds and embryo, are almost identical in the two families; besides which, in *Cycadeæ* the ovules are sometimes geminate, with an inferior micropyle as in *Abietineæ*; or solitary with a superior micropyle, as in *Taxineæ* (*Salisburya*). The only important difference is in the habit and foliation of *Cycadeæ*. *Cycas* more frequently inhabits tropical Asia and its large islands, but also Madagascar and equatorial Australia. *Macrozamia* [and *Bowenia*] is peculiar to Australia. *Encephalartos* and *Stangeria* are South African. *Zamia, Microcycas, Ceratozamia,* and *Dioon* are tropical and sub-tropical American.

The central and cortical pith of *Cycadeæ* abounds in nutritious starch. The *Cycas* of the Moluccas and of Japan yields a sort of sago, with which the natives make bread. The Hottentots feed on tho [pith of the] *Encephalartos*, called by the Dutch colonists Broodboom (Bread-tree). The seeds of *Cycas* and *Zamia* are edible, containing starch combined with a gummy matter, but they are astringent in the raw state; those of an Australian species are reputed to be violently emetic.

CLASS II.—**MONOCOTYLEDONS.**

I. *HYDROCHARIDEÆ.*

(HYDROCHARIDÉES, *Jussieu, L.-C. Richard.*—HYDROCHARIDEÆ, *D.C.*)

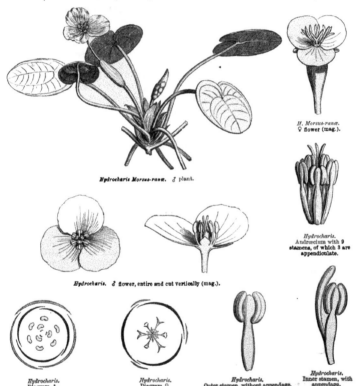

H. Morsus-ranæ.
♀ flower (mag.).

Hydrocharis Morsus-ranæ. ♂ plant.

Hydrocharis.
Andrœcium with 9
stamens, of which 3 are
appendiculate.

Hydrocharis. ♂ flower, entire and cut vertically (mag.).

Hydrocharis.
Diagram ♂.

Hydrocharis.
Diagram ♀.

Hydrocharis.
Outer stamen, without appendage.

Hydrocharis.
Inner stamen, with
appendage.

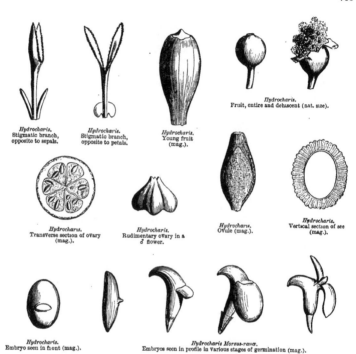

Hydrocharis.
Stigmatic branch,
opposite to sepals.

Hydrocharis.
Stigmatic branch,
opposite to petals.

Hydrocharis.
Young fruit
(mag.).

Hydrocharis.
Fruit, entire and dehiscent (nat. size).

Hydrocharis.
Transverse section of ovary
(mag.).

Hydrocharis.
Rudimentary ovary in a
♂ flower.

Hydrocharis.
Ovule (mag.).

Hydrocharis.
Vertical section of see
(mag.).

Hydrocharis.
Embryo seen in front (mag.).

Hydrocharis Morsus-ranæ.
Embryos seen in profile in various stages of germination (mag.).

FLOWERS *usually diclinous, enclosed in a membranous spathe.* PERIANTH 6-*merous,* 2-*seriate* (*calyx and corolla*). STAMENS 3-6-9-12, *inserted at the base of the perianth, several often sterile.* OVARY *inferior,* 1- *or more-celled;* OVULES *several in each cell, ascending or orthotropous, placentation parietal.* FRUIT *a utricle or berry.* SEEDS *exalbuminous.—Aquatic* PLANTS. LEAVES *usually radical.*

Aquatic HERBS, usually perennial, submerged or floating, stoloniferous, sometimes gemmiparous at the axils (*Hydrocharis*). ROOTSTOCK short and creeping, or elongated, jointed and knotted, cylindric. LEAVES usually all radical, rarely cauline, opposite or whorled (*Udora, Anacharis,* &c.), floating or submerged, sometimes emerged, petiolate; *blade* entire, vernation convolute; *petiole* sometimes sheathing at the base, often reduced to a phyllode by the arrest of the blade; nerves longitudinal, margins denticulate (*Vallisneria, Blyxa*). FLOWERS diœcious, rarely ☿ (*Udora, Ottelia*); *buds* enclosed in a membranous or herbaceous spathe, which is sessile or

petioled, sometimes 1-2-phyllous, smooth, or fringed on the dorsal nerve (*Enhalus*) ;
sometimes tubular, or longitudinally split on one side.—FLOWERS ♂ usually numerous
in a 1-2- phyllous spathe, rarely solitary (*Hydrilla*, &c.), usually pedicelled, sometimes
furnished with a spathella or true bract. PERIANTH 6-phyllous, 2-seriate ; outer
leaflets calycine, tubular, or sub-coherent at the base, imbricate or valvate in æstivation ; inner leaflets petaloid, larger, contorted and folded in æstivation, very rarely
0 (*Vallisneria*). STAMENS inserted at the bottom of the perianth, 3 opposite to the
sepals (*Hydrilla, Vallisneria*), or 6-9-12, several-seriate, some often imperfect ; *filaments* free, or sub-monadelphous at the base, short, cylindric, compressed or clavate,
sometimes appendaged (*Hydrocharis*); *anthers* introrse, rarely extrorse (*Hydrocharis*),
2-celled, ovoid-globose or linear, adnate to the connective, dehiscence longitudinal ;
pollen smooth or.papillose. OVARY rudimentary, occupying the centre of the flower.
—FLOWERS ♀ and ☿ usually solitary ; *spathe* tubular, or split longitudinally, very
often sessile. PERIANTH superior ; limb 6-partite, 2-seriate ; outer segments calycine, inner petaloid. STAMENS inserted at the bottom of the perianth. OVARY
inferior, 1-celled, placentation parietal (*Udora, Anacharis, Hydrilla, Vallisneria,
Blyxa*), or 6-8-9-celled (*Stratiotes, Enhalus, Ottelia, Boottia, Limnobium, Hydrocharis*);
style very short, or long and adnate to the perianth-tube ; *stigmas* 3 in the 1-celled
ovary, 6 in the several-celled ovary, more or less deeply 2-fid, glandular-papillose on
the ventral side ; *ovules* numerous, ascending, orthotropous or anatropous. FRUIT
submerged, various in form, usually longitudinally ridged, naked at the top, or
crowned by the persistent perianth, coriaceous, sub-fleshy, rupturing by decay in
the water, 1-celled, or more or less completely several-celled ; *septa* membranous,
opposite to the stigmas, and projected from the periphery towards the axis. SEEDS
numerous, on pulpy parietal placentas which spread partially over the septa ; *testa*
membranous, tough, usually clothed with cylindric cells of a very elegant and often
spiral structure. EMBRYO exalbuminous, straight; *radicle* reaching to the hilum;
plumule usually very conspicuous, more or less lateral.

[The following tribes are from Endlicher's 'Genera' :—

TRIBE I. ANACHARIDEÆ.—Caulescent. Leaves opposite or whorled. *Udora, Anacharis,
Hydrilla, Apalanthe.*

TRIBE II. VALLISNERIEÆ.—Stemless, scapigerous. Leaves all radical, linear. Ovary 1-
celled ; stigmas 3. *Vallisneria, Blyxa.*

TRIBE II. STRATIOTIDEÆ.—Stemless, scapigerous. Ovary several-celled ; stigmas 6.
Stratiotes, Enhalus, Ottelia, Boottia, Limnobium, Hydrocharis.—ED.]

Hydrocharideæ, placed by Brongniart in the class of *Fluriales*, with *Butomeæ, Alismaceæ, Juncagineæ,
Naiadeæ*, &c., are separated from these families mainly by their inferior ovary. Though recognizing the
trivial value of this character, we have used it to distinguish *Hydrocharideæ*, to avoid the alternative of
uniting in one family all the exalbuminous Monocotyledons, which, as sagaciously remarked by A. de
Jussieu, present a continuous series from the most simple type of flower, reduced to a stamen and a carpel
(*Najas*), to the most complex, represented by *Hydrocharis.*

The few known species of *Hydrocharis* mostly inhabit fresh and still waters in temperate regions of
both worlds, or are maritime (*Enhalus*). *Ottelia* inhabits the Nile and rivers of tropical Asia and Australia. *Vallisneria* is European, Asiatic, African, North American, and Australian ; (for its mode of ferti-

lization see p. 156). *Anacharis*, a native of America, is now abundant in North Europe, where it has increased so rapidly as to impede navigation in many parts of England. *Hydrocharis* and *Stratiotes* also abound in North Europe, where they are utilized as manures. *Blyxa* is a native of India and Madagascar. The herbage of *Hydrocharideæ* is mucilaginous and moderately astringent; *Hydrocharis Morsusranæ* was formerly employed with *Nymphæa*. *Ottelia* and *Boottia* are eaten by the Indians, but form a poor vegetable [the starchy rootstock is the part eaten]. The tubers and fruits of *Enhalus*, n Indian and Celebes plant, are edible; the fibre of its leaves is textile.

II. *CANNACEÆ.*

(CANNEÆ, *Br.*—CANNACEÆ, *Agardh.*—MARANTACEÆ, *Lindl.*)

FLOWERS ☿. PERIANTH *superior, double*; *outer herbaceous, 3-phyllous*; *inner petaloid, irregular, composed of petals and staminodes.* STAMEN *solitary, lateral*; ANTHER *1-celled.* OVARY *inferior, 3-1-celled*; OVULES *campylotropous or anatropous.*

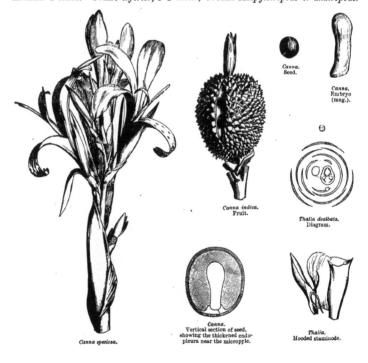

Canna.
Seed.

Canna.
Embryo
(mag.).

Canna indica.
Fruit.

Thalia dealbata.
Diagram.

Canna.
Vertical section of seed,
showing the thickened endopleura near the micropyle.

Thalia.
Hooded staminode.

Canna speciosa.

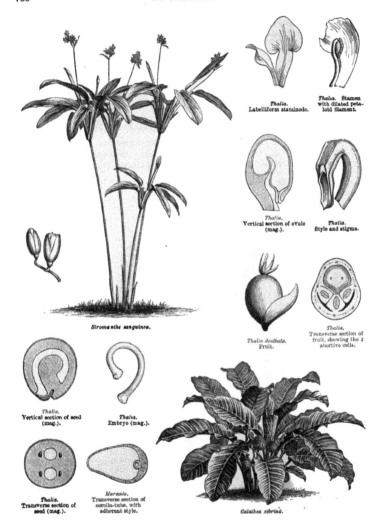

Stromanthe sanguinea.

Thalia.
Labelliform staminode.

Thalia. Stamen
with dilated peta-
loid filament.

Thalia.
Vertical section of ovule
(mag.).

Thalia.
Style and stigma.

Thalia dealbata.
Fruit.

Thalia.
Transverse section of
fruit, showing the 2
abortive cells.

Thalia.
Vertical section of seed
(mag.).

Thalia.
Embryo (mag.).

Thalia.
Transverse section of
seed (mag.).

Maranta.
Transverse section of
corolla-tube, with
adherent style.

Calathea zebrina.

Calathea zebrina. Flower. *Calathea.* Young anther (mag.). *Calathea.* Hooded staminode.

Calathea.
Transverse section of fruit.
a. Septal glands; *b.* Septa
c. Testa; *d.* Raphe;
e. Albumen; *f.* Embryo.

CAPSULE *3-valved.* SEED *albuminous.* EMBRYO *straight or hooked.*—HERBS. LEAVES *alternate, sheathing, midrib giving off laterally simple and parallel secondary nerves.*

Perennial HERBS with fibrous roots or fleshy creeping rhizomes. STEM simple, or branched above, enveloped by the sheathing petioles. LEAVES alternate, simple; *petiole* sheathing at the base, often thickened at the top; *limb* plane, large, entire; *midrib* thick, giving off laterally simple parallel oblique or horizontal secondary nerves which are incurved at the tips. FLOWERS ☿, irregular, in a raceme or terminal or lateral panicle, bracteate. PERIANTH superior, apparently formed of 3–4 irregular petaloid whorls: (1) (CALYX) herbaceous or scarious, 3-phyllous, imbricate; (2) (COROLLA) of 3 imbricate divisions, alternating with the calyx, coloured, tubular at the base, equal or sub-equal; (3) OUTER STAMINODES petaloid, imbricate, inserted on the corolla and alternate with it, the interior one bilobed or ringent; (4) INNER STAMINODES petaloid, alternate with the outer, one labelliform, the other antheriferous; *anther* 1-celled, introrse, dehiscence longitudinal; *pollen* globose, smooth, or tubercled (*Canna*). OVARY inferior, 3-celled, or 1-celled by arrest; *style* sometimes dilated, petaloid, straight or curved; sometimes slender, free, or adnate below to the corolla-tube and margin of the antheriferous segment; *stigma* terminal or sub-lateral, entire, sub-labiate or concave; *ovules* sometimes solitary, basilar, campylotropous or semi-anatropous, sometimes numerous, 2-seriate at the inner angle of the cells, horizontal and anatropous. CAPSULE 1-celled, sometimes fleshy, or 3-celled and loculicidally 3-valved. SEEDS globose or angular, when immature sometimes furnished with a filamentous aril (*Canna*); *testa* coriaceous; *endopleura* thickened around the micropyle; *albumen* horny. EMBRYO of the anatropous seeds straight, or slightly curved at the top; *radicle* turned towards the hilum; embryo of the campylotropous seeds hooked, or bent like a crozier, accompanied by two chalazal canals which cross the albumen.

PRINCIPAL GENERA.

* Thalia.	Calathea.	Ichnosiphon.	Phrynium.
* Canna.	Maranta.	Stromanthe.	

A. L. de Jussieu united *Cannaceæ* and *Zingiberaceæ* in one family, under the name of *Balisiers.* They differ only in the stamen of *Zingiberaceæ* belonging to the first whorl of the staminodes, and having a 2-celled anther, and in the double albumen.

Cannaceæ, properly so called, are chiefly natives of tropical and sub-tropical America, where they replace *Zingiberaceæ*, and whence they have spread throughout the hot parts of the Old World.

Cannaceæ have in contrast with *Zingiberaceæ* no aromatic principles, but their rhizome abounds with a nutritious starch. That of *Maranta arundinacea* (Arrowroot), cultivated in the West Indies, is recommended for its digestibility. Its uncooked rhizome is acrid, rubefacient, salivatory, and is considered an antidote to the poison from the juice of the Manchineel, when applied to the wounded surface. The leaves of *M. lutea* are covered on their under face with a resinous excretion, supposed to be efficacious in dysuria. The tubers of *M. Allouya*, cooked and seasoned with pepper, are eaten in the Antilles. The root of *Canna* is reputed to be diuretic and diaphoretic. The seeds of several species are considered substitutes for coffee, and yield a purple dye.

III. *ZINGIBERACEÆ.*

(ZINGIBERACEÆ, *L.-C. Richard.*—SCITAMINEÆ, *Br.*—AMOMEÆ, *Jussieu.*—
ALPINIACEÆ, *Link.*)

FLOWERS ☿. PERIANTH *and* STAMINODES *as in* Cannaceæ. STAMEN *solitary, anterior;* ANTHER *2-celled.* OVARY *inferior, usually 3-celled;* OVULES *anatropous.* FRUIT *usually a capsule.* SEEDS *with 2 albumens, a farinaceous and a horny (vitellus).* EMBRYO *with the cotyledonary end sheathed by the vitellus, the radicular free and touching the hilum.*—HERBS, *with creeping or tuberous rhizome.* LEAVES *as in* Cannaceæ.

Perennial HERBS with a creeping or tuberous rhizome, rarely with fibrous roots, stemless, or stem simple, enveloped by the leaf-sheaths. LEAVES all radical, or alternate, simple; *petiole* forming a split (very rarely closed) sheath, sometimes ligulate; *limb* flat, entire; *midrib* thick, giving off laterally numerous secondary simple parallel oblique or transverse nerves. FLOWERS ☿, irregular, naked or bracteolate, spiked, racemed or panicled, radical or terminal, often accompanied by spathaceous bracts. PERIANTH double, superior; outer (*calyx*) coloured or herbaceous, tubular, entire or split like a spathe, 3-toothed or -fid; inner (*corolla*) with a long or short tube, 3-partite, divisions more or less unequal, the upper usually largest, cucullate; *staminodes* petaloid, dissimilar, forming a 2-lipped tube adnate to the corolla-tube, lower lip the largest. STAMEN solitary, inserted at the base of the corolla-tube; *filament* free, usually dilated and petaloid, often prolonged beyond the anther; *anther* erect or incumbent, introrse; cells distant, marginal. OVARY inferior, 3- (rarely 1-2-) celled, often surmounted by 1 or more staminodes; *ovules* 1 or more in each cell, 2–several-seriate, inserted at the central angle of the cells, horizontal, anatropous. FRUIT crowned by the remains of the perianth, usually a loculicidally 3-valved capsule, rarely irregularly ruptured or dehiscing by longitudinal slits. SEEDS usually numerous, sub-spherical or angular, arillate or not, *testa* cartilaginous; *albumen* farinaceous, absent near the hilum, interposed between the seed-coats and a second horny albumen (*vitellus*), which is closed at the top opposite to the hilum, and perforated at the base to allow of the passage of the radicle. EMBRYO straight, sub-cylindric, axile, capped at the cotyledonary end by the vitellus; *radicle*

Alpinia.
Flower.

Alpinia nutans. Entire plant and flower, reduced.

*Renealmia
sanguinolenta.*
Dehiscent fruit.

Alpinia.
Flower cut
vertically.

Alpinia.
Transverse section of
fruit.

*Amomum Granum-
paradisi.* Vertical
section of seed.

Amomum.
Embryo
(mag.).

Renealmia.
Embryo
(mag.).

Renealmia.
Arillate seed.

Renealmia.
Vertical section of seed.

Renealmia.
Transverse section of
seed, on a level
with the embryo.

Hedychium Gardnerianum.
Diagram.

A. Axis; *B.* Outer bract; *C.* Inner bract;
D. Shoot; *e.* Sepals; *f.* Petals; *g, g', g',*
Outer staminodes; *h, h,* Inner staminodes;
i. Fertile stamen.

protruded through the vitellus, prolonged beyond the albumen, and reaching to the hilum.

PRINCIPAL GENERA.

* Globba.	Trilophus.	* Alpinia.	* Roscoëa.	Diracodes.
Colebrookia.	Achasma.	* Hedychium.	Ceranthera.	Stenochasma.
* Renealmia.	Zingiber.	* Amomum.	Piperidium.	Curcuma.
* Elettaria.	* Gastrochilus.	Kæmpferia.	Donacodes.	* Costus.
		Hitchenia.		

We have indicated the close affinity between *Zingiberaceæ, Cannaceæ,* and *Musaceæ* under the latter families. *Zingiberaceæ* are for the most part tropical, and especially Asiatic ; they are rare in the sub-tropical regions of Japan, as well as in equinoctial Africa and America [but very abundant in tropical Africa].

The root of *Zingiberaceæ* contains various volatile oils, an aromatic resin, a bitter principle, a more or less abundance of starch, and sometimes a yellow colouring matter (*curcumine*). The odoriferous principles which abound in the roots are also found in the fruits, but are [usually] scarcely perceptible in the herbage. The root of Ginger (*Zingiber officinale*), introduced from India into the Antilles by the Spaniards, has an acrid pungent taste and a strongly aromatic scent ; it is considered in India to be anti-scorbutic and aphrodisiac. Many European doctors recommend it as a powerful stimulant, for which reason it enters into the composition of an English beer (Ginger beer) much used in North Europe. It is equally valued as a condiment, preserved in sugar. The roots of Galanga, the origin of which is some-what obscure, and the use of which is nearly given up, are furnished in India by different species of *Alpinia.* Zodoary is the produce of *Curcuma Zedoaria* and *Zerumbet.* From the root of *C. leucorhiza* and *angustifolia* an arrowroot is obtained, but it is charged with a yellow matter, and much inferior to that of *Maranta.* The roots of *Costus,* formerly renowned in Europe, have now fallen into disuse, although their great bitterness causes them still to be used in India as a tonic. It is the same with the roots of several *Curcumæ* and *Kæmpferiæ,* commonly called Terra-merita [Turmeric], Indian Saffron, &c., which contain a very abundant yellow colouring matter, more useful in dyeing than in medicine [and which form a principal ingredient in Curries]. The fruits of *Amomum,* called Cardamoms, are employed as a condiment, and esteemed for their stomachic properties. Maniguette [Meliguetta], or Grains of Paradise (*A. Granum-paradisi*), a Guinea species, is used, with several of its congeners (*A. citriodorum,* &c.), to add strength to vinegar, and to adulterate pepper [and spirits]. The Peruvians, according to Pöppig, apply the odoriferous leaves of *Renealmia* as a topic for rheumatic pains.

IV. *MUSACEÆ.*

(Musæ, *Jussieu.*—Musaceæ, *Agardh.*)

FLOWERS ☿. PERIANTH *superior, petaloid, irregular, 6-merous, 2-seriate.* STAMENS 6, *of which 1 or more are usually imperfect.* OVARY *inferior, 3-celled, 1–many-ovuled ;* OVULES *anatropous.* FRUIT *fleshy, indehiscent, or sub-drupaceous with loculicidal or septicidal dehiscence.* SEED *albuminous ;* RADICLE *inferior or centripetal.* —HERBS. LEAVES *alternate, sheathing, midrib giving off laterally simple and parallel secondary nerves.*

HERBS, often gigantic. STEM or SCAPE enveloped by the thick and persistent sheathing bases of the petioles, simple, sometimes appearing as an arborescent trunk, sometimes very short or 0. LEAVES alternate, petioled, simple, entire, con-volute in vernation ; *limb* usually elongated, sometimes arrested ; *midrib* thick, giving off laterally transverse or oblique secondary nerves, parallel, very close, a

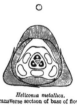

Heliconia metallica.
Transverse section of base of flower.

Heliconia Bihai. Entire plant.

Heliconia formosa.
Lower portion of the
flower cut vertically.

Heliconia metallica.
Abortive stamen.

Heliconia formosa.
Ovule (mag.).

Urania guianensis.
Seed, with hairy aril (mag.).

Heliconia metallica. Flowers.

Urania guianensis.
Seed, with hairy aril, cut
vertically (mag.).

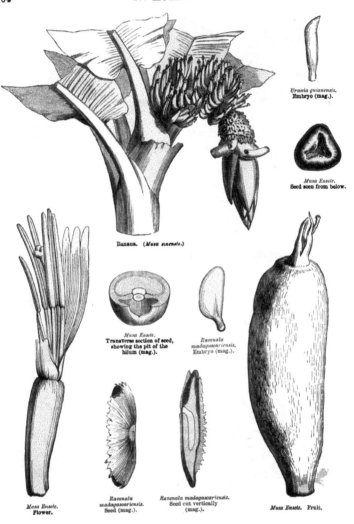

Urania guianensis.
Embryo (mag.).

Musa Ensete.
Seed seen from below.

Banana. (*Musa sinensis.*)

Musa Ensete.
**Transverse section of seed,
showing the pit of the
hilum (mag.).**

*Ravenala
madagascariensis.*
Embryo (mag.).

Musa Ensete.
Flower.

*Ravenala
madagascariensis.*
Seed (mag.).

Ravenala madagascariensis.
Seed cut vertically
(mag.).

Musa Ensete. Fruit.

little incurved at the tips. FLOWERS ☿, irregular, sessile, or pedicelled, in the axil of a spathe; *peduncles* radical. or axillary, furnished with inflated spathelike coloured distichous or alternate bracts. PERIANTH epigynous, petaloid, of six 2-seriate often dissimilar segments, one of the outer anterior usually very large, often keeled, 2 of the inner lateral often smaller, the third posterior, always very minute, labelliform; *segments* all distinct (*Ravenala, Heliconia*), or variously coherent; some-times the 2 inner lateral united into a tube split behind, and enclosing the stamens (*Strelitzia*); sometimes the 3 outer and the 2 lateral inner form a tube split behind, and 5-lobed at the top (*Musa*). STAMENS 6, inserted at the base of the perianth-segments, the posterior opposite to the labelliform segment usually imperfect, as are sometimes several of the others; *filaments* plane, free; *anthers* introrse; cells 2, sub-opposed, elongated and adnate to a connective which is prolonged into a point or membranous appendage, dehiscence longitudinal; *pollen* globose. OVARY inferior, cells 3, opposite the inner perianth-segments; *style* simple, cylindric; *stigma* with 3 linear lobes, papillose on their inner face, or concave, obscurely 6-lobed; *ovules* solitary and basilar in each cell, or numerous and 2-several-seriate at the central angle of the cell, anatropous. FRUIT umbilicate, with 3 1-many-seeded cells; sometimes fleshy, indehiscent, with numerous seeds buried in the pulp (*Musa*), some-times sub-drupaceous, with a fleshy coriaceous epicarp and bony endocarp; dehis-cence either by loculicidal many-seeded valves (*Strelitzia, Ravenala*), or into 3 septicidal 1-seeded cocci (*Heliconia*). SEEDS ovoid [or cubical or angled], fixed by one end or by the centre; *funicle* obsolete, or dilated into a fleshy membranous laciniate or hairy aril; *testa* coriaceous, hard, smooth or rugose; *albumen* fleshy, farinaceous. EMBRYO straight, oblong-linear or fungiform; radicular end perfora-ting the albumen, reaching the hilum, inferior or centripetal.

PRINCIPAL GENERA.

| Heliconia. | Musa. | Strelitzia. | Ravenala. |

Musaceæ approach *Cannaceæ* and *Zingiberaceæ* in the structure of their stem, the nervation of their leaves, the 3-celled inferior ovary, and the albuminous seeds; they are distinguished by their 2-seriate perianth without staminodes, the number of their normal stamens, and the absence of aromatic principles. They are especially distinguished from other epigynous Monocotyledons by their habit, irregular flower, and the nature of their sometimes spathaceous bracts.

Heliconia inhabits tropical America, *Urania* the tropics of the Old World, *Strelitzia* South Africa, *Ravenala* Madagascar; *Musa*, natives of the Old World, were transported to America before its discovery by Europeans, and are now dispersed throughout the tropical and sub-tropical zone.

Musaceæ, which, by their elegant habit and the beauty of their flowers and foliage, are amongst the greatest ornaments of the tropical Flora, are further eminently useful to the inhabitants of those climates. The Banana and Plantain fruits (*Musa paradisiaca* and *sapientum*) afford an agreeable sweet farinaceous food and a refreshing drink. The pith of the stem, the top of the floral spike, and even the shoots of several species are eaten as vegetables. The culture of these valuable plants is not less important in the tropics than that of cereals and farinaceous tubers in temperate regions. The variety of food furnished by Bananas in different stages of ripeness is the admiration of travellers. Cultivation has produced nume-rous varieties of form, colour, and taste. Humboldt and Boussingault have estimated that, under good cultivation, a Banana plant will produce on an average in one year three bunches of fruit, each weighing 44 lb.; these would yield for every 3 acres, in hot climates, 404,800 lb. of Bananas; and in countries on the limits of its culture 140,800 lb., an amount far exceeding the maximum yield of our tuberous plants, which are, besides, much less nutritious than an equal weight of Bananas.

The petioles of Bananas, and especially those of the Abaca (*Musa textilis*), are formed of very tenacious fibres, of which the natives make thread and textile fabrics; they also use the blade of the leaves to cover their huts. *Ravenala madagascariensis* is the finest species of the family; its popular name of Traveller's-tree is due to the reservoir formed by the leaf-sheaths, in which a limpid and fresh water collects, which may be obtained by piercing the base of the petiole [probably the rain-water which falls on the blade and is conducted by the grooved midrib to the sheathing petiole]. The inhabitants of Madagascar cook its bruised seeds with milk, and prepare a broth from them; the pulpy aril of the seed, remarkable for its magnificent blue colour, yields an abundant volatile oil. The juice of *Musa Ensete* is considered in Abyssinia a strong diaphoretic [the succulent interior of the stem is eaten; not the fruit, which is small, dry, and full of very large seeds].

V. *BROMELIACEÆ.*

(Bromelieæ, *A.-L. de Jussieu.*—Bromeliaceæ, *Lindl.*—Bromellæ et Tillandsiæ, *Adr. Jussieu.*)

Flowers ☿. Perianth 6-*merous, 2-seriate, the exterior calycoid, the inner* petaloid. Stamens 6, *epigynous, perigynous or hypogynous.* Ovary *inferior or semi-inferior or superior,* 3-*celled.* Berry *indehiscent, or* Capsule 3-*valved.* Seeds *albu-*

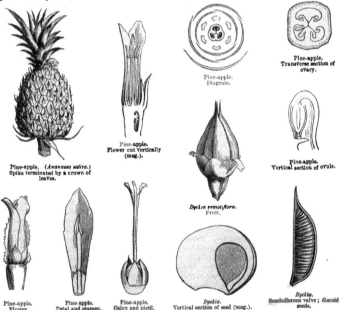

Pine-apple. (*Ananassa sativa*.)
Spike terminated by a crown of
leaves.

Pine-apple.
Flower cut Vertically
(mag.).

Pine-apple.
Diagram.

Pine-apple.
Transverse section of
ovary.

Pine-apple.
Vertical section of ovule.

Dyckia remotiflora.
Fruit.

Pine-apple.
Flower.

Pine-apple.
Petal and stamen.

Pine-apple.
Calyx and pistil.

Dyckia.
Vertical section of seed (mag.).

Dyckia.
Seminiferous valve; discoid
seeds.

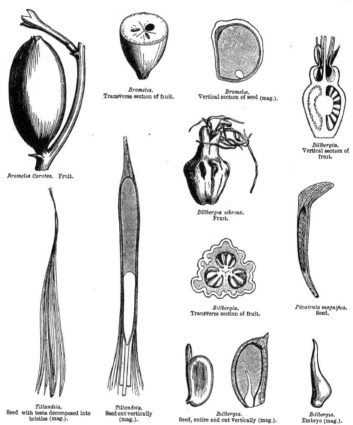

Bromelia Carotas. Fruit.

Bromelia.
Transverse section of fruit.

Bromelia.
Vertical section of seed (mag.).

Billbergia.
Vertical section of fruit.

Billbergia zebrina.
Fruit.

Billbergia.
Transverse section of fruit.

Pitcairnia magnifica.
Seed.

Tillandsia.
Seed with testa decomposed into bristles (mag.).

Tillandsia.
Seed cut vertically (mag.).

Billbergia.
Seed, entire and cut vertically (mag.).

Billbergia.
Embryo (mag.).

minous, often bearded. EMBRYO small, axile, outside the albumen.—Usually herbaceous PLANTS. LEAVES sheathing.

Herbaceous, sometimes woody PLANTS, generally stemless, with perennial stock and fibrous roots, mostly epiphytes. LEAVES usually all at the base of the stem or scape, sheathing, stiff, channelled, margin often toothed or spinous, epidermis clothed with scaly hairs. FLOWERS ☿, regular or sub-irregular, spiked, racemed or panicled, and each with a scarious or coloured bract. PERIANTH quite inferior, semi-superior or superior, 6-partite, 2-seriate; outer segments (*calyx*) herbaceous, the 2 posterior usually coherent, the third anterior, sometimes shorter, imbricate or rarely valvate in æstivation; inner segments petaloid (*corolla*), more or less coherent, usually furnished within at the base with a scale or nectariferous crest, spirally twisted in æstivation, or rarely valvate, marcescent, and again twisted in age. STAMENS 6, epigynous, perigynous, or hypogynous; *filaments* subulate, usually dilated at the base, free or connate, and more or less adnate to the inner perianth-segments; *anthers* introrse, 2-celled, basi- or dorsi-fixed, erect or incumbent, dehiscence longitudinal. OVARY either completely superior (*Dyckia*), or semi-inferior (*Pitcairnia*), or inferior (*Ananassa, Billbergia*, &c.), 3-celled; *style* simple, 3-gonous, sometimes 3-partite; *stigmas* 3, simple, or rarely 2-fid, sometimes fleshy or petaloid, straight, or spirally twisted; *ovules* anatropous, numerous, 2-seriate at the inner angle of the cells, horizontal or ascending, rarely definite and pendulous from the top of the central angle (*Ananassa*). FRUIT a 3-celled berry or a septicidally 3-valved capsule, rarely loculicidal, endocarp usually separable. SEEDS usually numerous, oblong (*Guzmannia, Brocchinia*), or linear (*Pitcairnia, Tillandsia*), or ovoid (*Bromelia, Billbergia*), or discoid (*Dyckia*); *testa* cellular (*Pitcairnia*), or fleshy (*Ananassa, Billbergia*), or suberose (*Dyckia*), or silky (*Tillandsia*), often obtusely acuminate at both ends; *albumen* farinaceous. EMBRYO placed outside the albumen, straight or hooked, radicular end near the hilum.

PRINCIPAL GENERA.

* Ananassa.	* Billbergia.	* Tillandsia.	* Guzmannia.
* Bromelia.	* Acanthostachys.	* Quesnelia.	* Dyckia.
* Æchmea.	* Pitcairnia.	* Caraguata.	Pourretia.

Bromeliaceæ, in being epigynous, perigynous or hypogynous, are intermediate between Monocotyledons with a free, and those with an adherent ovary. Brongniart has placed them in the class of *Pontederiaceæ*, which they approach in the herbaceous stem, sheathing radical leaves, flowers in a spike or raceme, bracteate 2-seriate perianth, superior or semi-adherent 3-celled ovary, loculicidally 3-valved capsule, and farinaceous albumen; but *Pontederiaceæ* are separated by their completely petaloid perianth, the ovary-cells either unequal or reduced to one, the solitary ovule pendulous from the top of the fertile cell, and the axile and included embryo. On the other hand, *Bromeliaceæ* are near *Hæmodoraceæ*, which differ in their equitant distichous leaves, their wholly petaloid perianth, their stamens, of which three only are fertile, their undivided stigma, their not farinaceous albumen, &c.

Bromeliaceæ are all American, where most are epiphytes in tropical forests; they are much rarer in hot extra-tropical regions.

The fruit of the baccate *Bromeliaceæ* contains citric and malic acids, to which it owes astringent medicinal properties. The ripe berries of some abound in sugar, which gives them an exquisite flavour. The Pine-apple (*Ananassa*), the most important species, has been introduced into Asia and Africa. The fruit consists of a dense spike of connate fleshy berries and bracts, forming together an ovoid or sub-globose syncarpous compound fruit, which is seedless through cultivation, and crowned by a tuft of leaves. When ripe it is full of an acidulous perfumed sugary juice, and is considered one of the most delicate of fruits; but when unripe the juice is acid and acrid, and much esteemed in the Antilles as a vermifuge and diuretic. *Bromelia Pinguin* and several other species possess the same properties. *Tillandsia usneoides* [a very slender filamentous, much-branched species that hangs in hair-like masses from trees in the North-west Indies and southern United States] is used in America in the preparation of an ointment used in cases of hæmorrhage; its fragile and very long stems, deprived of their outer parenchyma, are employed for stuffing mattresses, under the name of 'vegetable hair;' they are also remarkable for having no spiral vessels. *Billbergia tinctoria* yields a yellow colouring matter, and the Pine-apple leaf an extremely beautiful silky fibre.

VI. *ORCHIDEÆ.*

(ORCHIDES, *Jussieu.*—ORCHIDEÆ, *Br.*—ORCHIDACEÆ, *Lindl.*)

PERIANTH *superior, irregular, 2-seriate.* STAMENS 1-2, *gynandrous;* POLLEN-GRAINS *variously agglomerated.* OVARY *inferior, 1-celled, with 3 parietal placentas;* OVULES *numerous, anatropous.* SEEDS *numerous, scobiform, exalbuminous.* EMBRYO *minute.*—STEM *herbaceous.* ROOTS *fibrous, often tubercled.* LEAVES *radical or alternate, sheathing, sometimes scale-like.* FLOWERS *usually in a spike or raceme.*

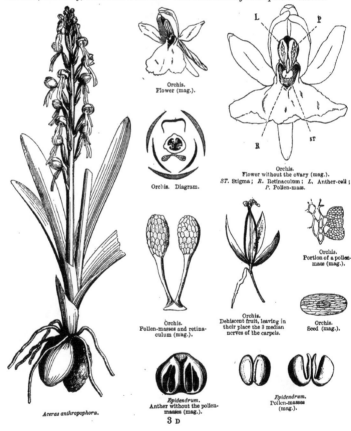

Orchis.
Flower (mag.).

Orchis. Diagram.

Orchis.
Flower without the ovary (mag.).
ST. Stigma; *R.* Retinaculum; *L.* Anther-cell;
P. Pollen-mass.

Orchis.
Portion of a pollen-mass (mag.).

Orchis.
Pollen-masses and retina-culum (mag.).

Orchis.
Dehiscent fruit, leaving in their place the 3 median nerves of the carpels.

Orchis.
Seed (mag.).

Aceras anthropophora.

Epidendrum.
Anther without the pollen-masses (mag.).

Epidendrum.
Pollen-masses (mag.).

3 D

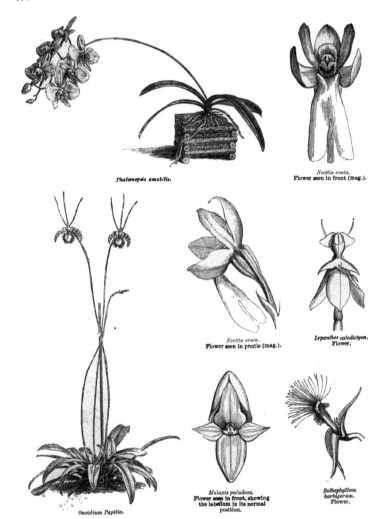

Phalænopsis amabilis.

Neottia ovata.
Flower seen in front (mag.).

Neottia ovata.
Flower seen in profile (mag.).

Lepanthes calodictyon.
Flower.

Malaxis paludosa.
Flower seen in front, showing
the labellum in its normal
position.

*Bolbophyllum
barbigerum.*
Flower.

Oncidium Papilio.

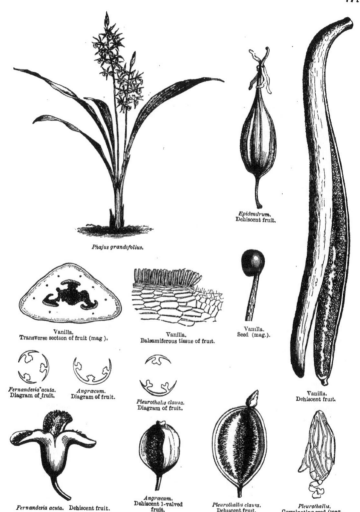

Phajus grandifolius.

Epidendrum.
Dehiscent fruit.

Vanilla.
Transverse section of fruit (mag).

Vanilla.
Balsamiferous tissue of fruit.

Vanilla.
Seed (mag.).

Fernanderia acuta.
Diagram of fruit.

Angræcum.
Diagram of fruit.

Pleurothalis clausa.
Diagram of fruit.

Vanilla.
Dehiscent fruit.

Fernanderia acuta. Dehiscent fruit.

Angræcum.
Dehiscent 1-valved
fruit.

Pleurothallis clausa.
Dehiscent fruit.

Pleurothallis.
Germinating seed (mag.

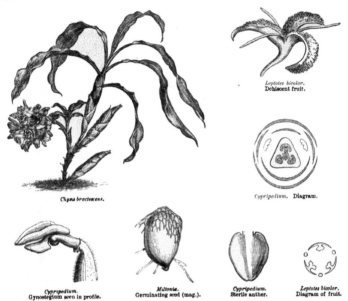

Leptotes bicolor.
Dehiscent fruit.

Chysis bractescens.

Cypripedium. Diagram.

Cypripedium.
Gynostegium seen in profile.

Miltonia.
Germinating seed (mag.).

Cypripedium.
Sterile anther.

Leptotes bicolor.
Diagram of fruit.

Perennial herbaceous terrestrial PLANTS, or epiphytes, or parasites? (*Epipo-gium, Corallorhiza, Neottia Nidus-avis*), sometimes sarmentose and furnished with adventitious roots (*Vanilla*), sometimes marsh plants (*Liparis, Malaxis*); *rhizome* creeping, or with fascicled fibrous roots, often accompanied with ovoid or palmate tubercles, caulescent or stemless; leaves often connate at the base, and forming with the thickened stem an oblong swollen or flattened organ (*pseudo-bulb*). STEM or SCAPE usually simple, cylindric or angular, often aphyllous, or furnished with scales. LEAVES: the radical and lower cauline close together, the upper equitant, alternate or opposite, sheathing, glabrous, rarely velvety (*Eria*), fleshy or membranous, cylindric, linear or linear-lanceolate, usually entire or emarginate (*Vanda*), or flabelliform (*Pogonia*), or cordate (*Neottia*); nerves parallel, rarely reticulate (*Anæctochilus*), sometimes gemmiparous (*Malaxis paludosa, Spiranthes gemmipara*). FLOWERS ☿, or imperfect by arrest, terminal, solitary or in a spike raceme or panicle, and bracteate, sometimes springing from the middle of a leaf (*Pleurothallis*). PERIANTH superior, usually petaloid, irregular, formed of 6 2-seriate free or coherent persistent or caducous leaflets; the outer (*sepals*) 3, of which 2 are lateral and 1 inferior, usually becoming superior by the torsion of the pedicel or ovary; the inner (*petals*) 3, alternating with the sepals, the 2 lateral similar, and the third (*labellum*) originally superior, then becoming inferior, usually dissimilar, larger, very various in shape and colouring, frequently hollowed at the base into a sac or spur; *limb* of the labellum usually 3-lobed, sometimes entire; *disk* naked, callous, glandular or laminate. ANDRŒCIUM and STYLE adnate into a column (*gynostegium*), of which the anterior face, opposite to the labellum, and terminated by the stigma, belongs to the substance of the style, while the dorsal face, terminated by the anthers, belongs to the androecium. STAMENS usually normally one, opposite the upper sepal, and accompanied by two rudimentary stamens, reduced to inconspicuous or obsolete prominences, very rarely there are 2 normal stamens opposite to the 2 lateral petals (*Cypripedium*); *anther* 2-celled (or 1-celled by absorption of the septum) or 4-celled by more or less perfect secondary septa, sometimes divided or multilocellate by transverse septa, erect, or inclined and covered by the concave top of the gynostegium (*clinandrium*); *pollen* agglomerated into 2–4–8 masses (*pollinia*), lodged in the membranous pockets of the anther, and composed of grains usually collected by fours into numerous groups (*masses*), cohering by means of elastic filaments, or gathered round a cellular axis; *grains* sub-pulverulent, easily separable or agglutinated into a solid compact waxy tissue; *pollen-masses* sometimes free, usually fixed, either directly or by means of a cellular pedicel (*caudicle*), to a viscous gland (*retinaculum*) below the anther, naked, or enclosed in a membranous fold of the stylary surface (*bursicula*). OVARY inferior, 1-celled (more rarely 3-celled), of 3 connate carpels with parietal placentation, sometimes surmounted by a cupule (analogous to that of *Santalaceæ*); *style* confluent with the stamens, occupying the face opposite to the labellum, produced at the top into a prominence or fleshy beak (*rostellum*); *stigmatic surface* (*gynixus*) oblique, concave, viscous, composed, according to R. Brown, of 3 stigmas, usually confluent, but sometimes distinct and opposite to the sepals; *ovules* numerous, shortly funicled, anatropous. CAPSULE membranous or coriaceous, cylindric, ovoid or winged, 1-celled, dehiscence very various, mostly by 3 semi-placentiferous valves, which separate from the 3 persistent midribs of the carpels, the latter remaining united at the base and apex. SEEDS very numerous, very minute; *testa* loose, reticulate, sometimes crustaceous and black (*Vanilla, Cyrtosia*). EMBRYO exalbuminous, fleshy.

TRIBE I. *MALAXIDEÆ.*

Pollen coherent in waxy masses, applied directly to the stigma, without accessory cellular tissue. Anther terminal or opercular.—Epiphytes, or rarely terrestrial plants; pseudo-bulbs formed by the connate leaf-bases and thickened stem.

PRINCIPAL CULTIVATED GENERA.

Pleurothallis.	Masdevallia.	Dendrochylum.	Pedilonum.	Stelis.
Octomeria.	Malaxis.	Aporum.	Lepanthes.	Liparis.
Dendrobium.	Eria.	Physosiphon.	Bolbophyllum.	Cirrhopetalum.
Polystachia.	Oberonia.	Calypso.		

TRIBE II. *EPIDENDREÆ.*

Pollen cohering in defined waxy masses. Cellular membrane prolonged into elastic caudicles, often folded, without true glands. Anther terminal, opercular.—

Usually epiphytes, caulescent or pseudo-bulbous, rarely furnished with fleshy roots.

PRINCIPAL CULTIVATED GENERA.

Cœlogyne.	Ponera.	Barkeria.	Cattleya.	Evelina.
Pholidota.	Hexadesmia.	Broughtonia.	Schomburgkia.	Isochilus.
Dinema.	Chysis.	Leptotes.	Bletia.	Diothonea.
Sophronitis.	Lælia.	Spathoglottis.	Epidendrum.	Phajus.

TRIBE III. *VANDEÆ.*

Pollen cohering in defined waxy masses, fixed after flowering to a caudicle and retinaculum. Anther terminal, rarely dorsal, opercular.—Epiphytal, rarely terrestrial, caulescent (especially the American species), or pseudo-bulbous (especially the Asiatic). Leaves often emarginate at the tip.

PRINCIPAL CULTIVATED GENERA.

Eulophia.	Pilumna.	Grobya.	Mormodes.	Galeandra.
Dichæa.	Huntleya.	Cychnoches.	Cyrtopera.	Fernandezia.
Zygopetalum.	Cyrtopodium.	Lissochilus.	Oncidium.	Warrea.
Notylia.	Vanda.	Ornithidium.	Cirrhea.	Renanthera.
Odontoglossum.	Maxillaria.	Camarotis.	Brassia.	Dicrypta.
Ornithocephalus.	Saccolabium.	Miltonia.	Lycaste.	Rodriguezia.
Sarcanthus.	Stanhopea.	Camaridium.	Burlingtonia.	Œceoclades.
Houlletia.	Scaphiglottis.	Jonopsis.	Angræcum.	Peristeria.
Colax.	Calanthe.	Sarcadenia.	Govenia.	Galeottia.
Phalænopsis.	Acanthophipplum.	Gongora.	Catesetum.	Ansellia.
Acropera.	Aeriopsis.	Cœlia.	Trichopilia.	Trigonidium.

TRIBE IV. *OPHRYDEÆ.*

Pollen composed of indefinite masses, united into 2 pollinia by an elastic cobweb-like axis, agglutinated to a retinaculum. Anther terminal, erect or reclinate, persistent, with perfect cells.—Terrestrial plants with tuberous roots.

PRINCIPAL GENERA.

Orchis.	Aceras.	Ophrys.	Gymnadenia.	Habenaria.	Anacamptis.
Serapias.	Satyrium.	Platanthera.	Bonatea.	Nigritella.	Peristylus.
Holothrix.	Bartholina.	Disa.	Disperis.	Pterygodium.	Schizodium.
Penthea.	Herschelia.	Corycium.			

TRIBE V. *ARETHUSEÆ.*

Pollinia sub-pulverulent or formed of angular lobules, fixed by their base, or by a point below the top. Anther terminal, opercular.—Usually terrestrial stemless plants, or caulescent and sarmentose, some aphyllous and parasitic.

PRINCIPAL GENERA.

Limodorum.	Cephalanthera.	Cyathoglottis.	Guebina.	Sobralia.	Vanilla.[1]
Caladenia.	Corysanthes.	Glossodia.	Pterostylis.	Microtis.	Caleya.
Asianthus.	Pogonia.	Chiloglottis.	Chloræa.		

[1] *Vanilla*, with probably *Gastrodia, Epipogium, Erythorchus, Cyrtosia,* and some others, forms a very distinct tribe (*Vanilleæ*) of *Orchideæ.*—ED.

Tribe VI. *NEOTTIEÆ.*

Pollinia sub-pulverulent; granules loosely coherent, fixed to a retinaculum. Anther parallel to the stigma, persistent, cells close together.—Terrestrial plants, with fascicled fibrous or tuberous roots, sometimes epiphytes, sometimes aphyllous, or parasites? resembling *Orobancheæ.*

PRINCIPAL GENERA.

Ponthiæva.	Prescottia.	Listera.	Neottia.	Epipactis.	Spiranthes.
Stenorhynchus.	Pelexia.	Goodyera.	Anæctochilus.	Physurus.	Zeuxine.
Diuris.	Orthoceras.	Prasophyllum.	Thelymitra.		

Tribe VII. *CYPRIPEDIEÆ.*

Anthers 2, lateral, both fertile, the intermediate one petaloid. Pollen granular, softening during fertilization. Stigma divided into 3 areolæ opposite to the stamens.

CULTIVATED GENERA.

Cypripedium.   Uropedium.
Selenipedium.

Orchideæ, which form one of the most natural families of the Vegetable Kingdom, have taxed the sagacity of our most eminent botanists—Dupetit-Thouars, R. Brown, L. C. Richard, Blume, Lindley, &c. They are especially remarkable for the curiously varied shapes and colours of their perianth, which resembles most dissimilar objects—as a helmet, slipper, fly, bee, beetle, a little monkey, &c.—and the relative sizes of which are sometimes extraordinarily different (*Uropedium*). The andrœcium, which is gynandrous, like that of *Aristolochieæ* (page 705), the pollen agglomerated into masses, as in *Asclepiadeæ* (page 553), and the undivided embryo, are all exceptional characters, which might render their position in the system doubtful, were it not that the structure of their stem, the nervation of their leaves, and the arrangement of their hexaphyllous and 2-seriate perianth, evidently place them among Mono-

Cypripedium spectabile.

cotyledons. Incomplete as their andrœcium appears, the ternary type of most Monocotyledonous families may yet be traced in it. According to the sagacious observations of R. Brown, it is composed, sometimes of an outer whorl of three stamens represented by one normal anther and two rudimentary ones opposite to the sepals; sometimes of an inner whorl, equally triandrous, of which one stamen is arrested, and two are normal, opposite to each lateral petal, and alternating with the carpels. This ternary arrangement is confirmed by *Apostasiæ*, which are closely allied to *Orchideæ* in their petaloid hexaphyllous 2-seriate irregular perianth, gynandrous andrœcium composed of three stamens one of which is often arrested, and which are only distinguished by their granular pollen and their 3-celled ovary. *Orchideæ* also approach *Burmanniaceæ* in epigyny, 1-celled ovary, 3-valved capsule, and scobiform seeds; and *Canneæ* in the inferior ovary, the perianth, and the andrœcium reduced to a single stamen.

The floral structure of *Orchideæ* sometimes presents a singularity equally remarkable and rare in the Vegetable Kingdom : we find on the same inflorescence dimorphous flowers (*Cychnoches ventricosum, Vanda Lowii, Spiculæa, Drakæa,* &c.); or even three different forms (*Catasetum, Myanthus, Cychnoches*).

In *Orchideæ*, owing to the consistence of their pollen, extraneous agency is required to ensure fertilization, which, as in *Asclepiadeæ*, is effected by insects; and in our hothouses, where these auxiliaries are wanting, fertilization must be artificially secured. In some species the lip is irritable; it oscillates opposite the column (*Megaclinium*), or turns round it (*Caleana*); on an insect settling on the surface of the lip, the latter quickly approaches the column, and presses the insect against it, which in its efforts to disengage itself breaks up and crushes the pollen-masses, and spreads them over the stigma.[1]

Orchideæ mostly inhabit tropical forests; they abound especially in the New World, where their numerous species generally grow on the trunks of trees, to which they attach themselves by their long adventitious roots; but they are terrestrial in the temperate regions of the northern hemisphere; they become rare near the pole, and *Calypso borealis* is the only one which reaches 68° north latitude. *Malaxideæ* inhabit the Indian continent and islands, and principally the Malay Archipelago; they are less numerous in tropical America, and the islands of South Africa; they are rather frequent in Australia and the Pacific Islands, but rare in the northern hemisphere, and entirely absent from the Mediterranean region, temperate America, and the Cape of Good Hope. [*Liparis, Malaxis,* and *Calypso* are all European; *Liparis, Microstylis,* and *Calypso* are temperate American; *Liparis* and *Polystachya* are both South African.] *Epidendreæ* nearly all belong to the tropical regions of the New World; some, however, inhabit the same zone in Asia; a very few [many species of *Cœlogyne*] are found in North India and near China; one species only advances as far as South Carolina. *Vandeæ* are found in equal numbers in tropical Asia and America; they are common in Madagascar, rare in Africa, and very rare beyond the tropics. *Ophrydeæ* inhabit all temperate and subtropical regions, especially Central and Mediterranean Europe, and South Africa; they are rarer in the tropics. *Neottieæ* principally grow in temperate Asia and Australia; they are much less numerous within the tropic of Cancer, and very rare in Africa. *Arethuseæ* abound in south temperate regions, and especially Australia [and South Africa]; they become rarer in the tropics and north temperate zone. *Chloreæ* extend as far south as the Straits of Magellan. *Cypripedieæ* inhabit the temperate and cool regions of the northern hemisphere; they are somewhat frequent in America; [and are spread over tropical Asia and its islands].

Orchideæ are much admired for the singularity, beauty and scent of their flowers. Their cultivation, which usually requires a hothouse and extreme care, has during the last forty years become an absolute passion in Europe. Linnæus, in the middle of the last century, knew but a dozen exotic Orchids, whereas at the present day about 2,500 are known to English horticulturists.

Of the few Orchids which are of use to man, the Vanillas (*Vanilla claviculata, planifolia,* &c.) hold the first rank. They are sarmentose plants, natives of the hot and damp regions of Mexico, Colombia, and Guiana [and tropical Africa]. Their fruit is a fleshy long capsule, and the black globose seeds are enveloped in a special tissue which secretes a balsamic oil; if kept in a dry place the capsule becomes covered with pointed and brilliant crystals of benzoic acid, and imparts its delicious perfume to various delicate dishes, chocolates, liqueurs, &c. The Faham (*Angræcum fragrans*) is a native of Bourbon; its leaves, known as Bourbon Tea, taste of bitter almonds and smell like Tonquin beans; they are used to stimulate digestion, and in pulmonary consumption. Salep, which is imported from Asia Minor and Persia, is produced by the tubers of several species of *Orchis*, which are equally natives of Europe

---

[1] For an account of the phenomena of Orchid fertilization, see Darwin 'On the Fertilization of Orchids.'—ED.

(*O. mascula, Morio, militaris, maculata*, &c.). Salep contains in a small volume an abundance of nutritive starch, associated with a peculiar gum, analogous to Bassorine; it was formerly considered a powerful analeptic; it is now used as a sweet, scented jelly, or mixed with chocolate. The root of Helleborine (*Epipactis latifolia*) is employed for arthritic pains; those of *Himantoglossum hircinum, Spiranthes autumnalis*, and *Platanthera bifolia*, are reputed to be aphrodisiac. The flowers of *Gymnadenia conopsea* are administered for dysentery; and in North America the tubers of *Arethusa bulbosa* are used to stimulate indolent tumours and in toothache. The root of *Spiranthes diuretica* is renowned in Chili. The rhizome of *Cypripedium pubescens* replaces the Valerian as an antispasmodic in the estimation of the Anglo-Americans.

VII. *APOSTASIACEÆ*.[1]

[Perennial HERBS, ROOTS fibrous. STEM rigid, simple or slightly branched, slender. LEAVES cauline, sheathing, rather rigid, alternate, lanceolate, strongly nerved, nerves parallel. FLOWERS ☿, in simple or compound terminal nodding racemes, fragrant; *pedicels* bracteate at the base, sometimes bracteolate. PERIANTH superior, of 6 sub-equal segments in two series, rather oblique, deciduous; *segments* linear-oblong; of the 3 outer 1 is anticous and 2 lateral; inner narrow, posticous, and more or less labelliform. STAMENS 3, 2 fertile opposite

Apostasia. Diagram.   *Neuwiedia.* Diagram.

the lateral inner segments of the perianth; third, if present, rarely fertile, opposite the anticous segment; *filaments* short, adnate to the base of the style; *anthers* basifixed, erect, 2-celled, introrse; *pollen* of free grains. OVARY inferior, elongate, 3-celled; *style* slender, terete; *stigma* obscurely 3-lobed; *ovules* numerous, attached to the inner angles of the cells. CAPSULE membranous, 3-celled, loculicidally 3-valved; *valves* cohering at the base and top. SEEDS numerous, very minute, ovoid or scobiform; *testa* membranous, lax at both ends.

GENERA.

Apostasia.   Neuwiedia.

A very small order, closely allied to *Orchideæ*, distinguished by the style and the 3-celled ovary, which, however, occur in the *Selenipedium* section of *Cypripedieæ*. The species, which are very few, are natives of Trans-Gangetic India and the Malay peninsula and islands. They have no known properties.—ED.]

VIII. *BURMANNIACEÆ*.

(BURMANNIÆ, *Sprengel*.—BURMANNIACEÆ, *Bl*.—TRIPTERELLEÆ, *Nuttall*.—THISMIEÆ ET TRIURIDEÆ, *Miers*.)

FLOWERS ☿. PERIANTH *superior, 6-partite, 2-seriate*. STAMENS 3–6. OVARY *inferior, 1-3-celled*; STIGMAS 3. SEEDS *with cellular testa, exalbuminous*. EMBRYO *undivided.—Weak* HERBS *with linear leaves, or aphyllous*.

[1] This Order is omitted in the original.

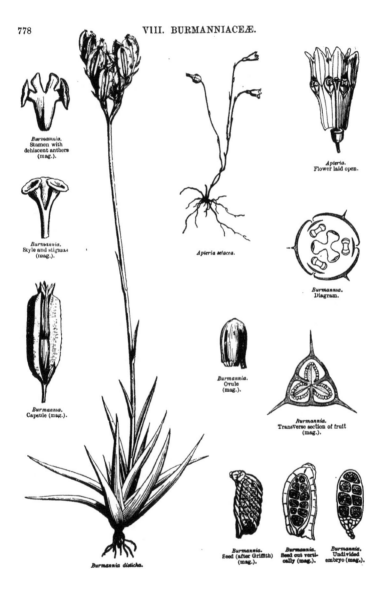

Burmannia.
Stamen with
dehiscent anthers
(mag.).

Burmannia.
Style and stigmas
(mag.).

Burmannia.
Capsule (mag.).

Burmannia disticha.

Apteria setacea.

Burmannia.
Ovule
(mag.).

Apteria.
Flower laid open.

Burmannia.
Diagram.

Burmannia.
Transverse section of fruit
(mag.).

Burmannia.
Seed (after Griffith)
(mag.).

Burmannia.
Seed cut verti-
cally (mag.).

Burmannia.
Undivided
embryo (mag.).

Annual or perennial HERBS, terrestrial or parasitic (?), very weak, green whitish or pink, aphyllous, rarely sarmentose and bearing leaves. FLOWERS ☿, in a 2-fid many-flowered cyme, or 1–2-flowered, very various in colour, and bracteate. PERIANTH superior, petaloid, tubular, tube regular or gibbous; *limb* of 6 2-seriate segments, the inner smallest, the outer sometimes very long (*Ophiomeris*). STAMENS inserted at the top of the tube or on the throat of the perianth; *filaments* distinct or monadelphous (*Thismia*); *anthers* with 2 disconnected cells, introrse; *connective* dilated, various in form. OVARY inferior, 1-celled with 3 parietal placentas, or 3-celled with 3 axile placentas; *style* simple, springing from the thickened top of the ovary; *stigmas* 3, 2–3-fid; *ovules* numerous. CAPSULE crowned by the marcescent perianth, terete or 3-angled, 3-winged, membranous; sometimes 1-celled, opening at the top or on one side into three semi-placentiferous valves; sometimes 3-celled, opening laterally between its angles by transverse slits, or a pyxidium. SEEDS numerous, small, oblong, some truncate and umbilicate at the end, the others pointed; *testa* lax. EMBRYO minute, undivided, cellular.

Burmanniaceæ form a small well-marked group, from their parasitism and floral structure, and are connected on the one hand with *Taccaceæ*, and on the other with *Aristolochieæ*, as well as with the Rhizanthous plants allied to the latter. Around *Burmannieæ* proper, various other genera may be grouped as sections of the Order as follows:—

BURMANNIEÆ proper.—Terrestrial plants, green and leafy, or discoloured and aphyllous. Perianth of 6 segments, the 3 outer winged. Stamens 3, opposite to the inner segments. Ovary 3-celled; stigmas 3. *Burmannia, Gonyanthes, Nephrocodum.*

APTERANTHEÆ.—Discoloured aphyllous plants. Perianth persistent or caducous, wingless. Stamens 3. Ovary 1-celled. *Apteria, Dictyostegia, Gymnosiphon, Benitzia, Cymbocarpus.*

THISMIEÆ.—Discoloured aphyllous plants. Perianth regular or gibbous, wingless. Stamens 6, monadelphous or free. Ovary 1-celled. Capsule opening transversely. *Thismia, Ophiomeris.*

STENOMERIDEÆ.—Green leafy sarmentose plants. Leaves cordate, resembling those of *Dioscoreæ* or *Smilax*. Perianth with 6 divisions. Stamens 6. Ovary 3-celled. Capsule linear, elongated, membranous, triquetrous. *Stenomeris.*

ALLIED TRIBE. TRIURIDEÆ.[1]—Pale monœcious plants. Perianth 6-merous. Stamens 6. Ovaries numerous, free on a rounded receptacle, 1-ovuled; style lateral and basilar, resembling the carpels and gynophore of a Strawberry. *Sciaphila, Hexuris, Triuris.*

Burmanniaceæ have been placed near *Orchideæ* by several botanists on account of their undivided embryo, reduced to a little cellular mass, which appears to be entirely formed of the tigellus. They are allied to *Irideæ* by their 6-merous 2-seriate perianth, their triandrous andrœcium, anthers with longi-

---

[1] *Triurideæ* have little affinity with *Burmanniaceæ*, and belong to the Apocarpous series of Monocotyledon* (which see).—ED.

tudinal dehiscence, inferior 3-celled ovary, and 3 dilated stigmas. They have also some connection with *Hæmodoraceæ* (which see).

Burmanniæ grow in damp and grassy soils, but most of the other genera live in the shade of large forests, on vegetable detritus, and are perhaps parasitical. They inhabit the tropical regions of Asia and America, extending in the New World to 37° north latitude. They are also met with in Madagascar.

The herbs of this family are slightly bitter-astringent; they are not known to possess any useful property.

IX. *TACCACEÆ.*

(TACCEÆ, *Presl.*— TACCACEÆ, *Lindl.*)

FLOWERS ☿. PERIANTH *superior, petaloid, 6-merous, 2-seriate.* STAMENS 6; FILAMENTS *concave*; ANTHERS *adnate to their concave face.* OVARY *inferior, 1-celled, with 3 parietal placentas*; OVULES *numerous, anatropous or semi-anatropous.* BERRY. SEEDS *numerous, albuminous.*—HERBS *with radical leaves; veins reticulate.*

Perennial stemless HERBS with a tuberous rhizome. LEAVES all radical; *petiole* semi-sheathing; *limb* sometimes entire, sometimes palmiseet or bipinnatifid, with prominent nerves. FLOWERS ☿, regular, in an involucrate umbel on the top of a simple cylindric or angular scape; *involucre* foliaceous, 4-phyllous; *pedicels* 1-flowered, long, naked, several flowerless, elongated-filiform, or mixed with the fertile ones. PERIANTH superior, petaloid; *segments* 6, 2-seriate, equal or the inner a little

Tacca pinnatifida. Fructiferous branch.

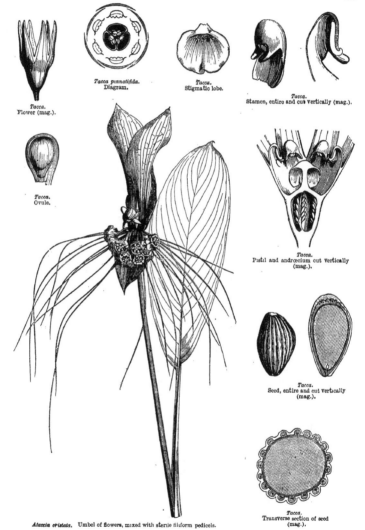

Tacca.
Flower (mag.).

Tacca pinnatifida.
Diagram.

Tacca.
Stigmatic lobe.

Tacca.
Stamen, entire and cut vertically (mag.).

Tacca.
Ovule.

Tacca.
Pistil and androecium cut vertically
(mag.).

Tacca.
Seed, entire and cut vertically
(mag.).

Ataxia cristata. Umbel of flowers, mixed with sterile filiform pedicels.

Tacca.
Transverse section of seed
(mag.).

the largest, persistent. STAMENS 6, inserted at the base of the segments; *filaments* dilated, vaulted or cucullate at the top; *anthers* introrse; cells 2, distant, parallel, adnate to the cavity of the filaments, free at the top, straight or incurved, dehiscence longitudinal. OVARY inferior, 1-celled or incompletely 3-celled; *placentas* 3, parietal, nerviform or 2-lobed; *style* short, thick; *stigma* orbicular or depressed, with 3 radiating emarginate or bifid lobes; *ovules* numerous, several-seriate, subascending and anatropous, or horizontal and semi-anatropous. BERRY umbilicate by the persistent limb of the perianth, 1-celled, or incompletely 3-celled. SEEDS ovoid, angular or lunate; *testa* coriaceous, striate, easily separable from the membranous endopleura; *albumen* fleshy. EMBRYO excessively small, ovoid, included in the albumen, near the basilar hilum, or distant from the ventral hilum.

GENERA.

* Tacca.　　　* Ataccia.

Lindley has placed *Taccaceæ*, *Dioscoreæ*, and *Smilaceæ* in his class *Dictyogens*, so called on account of the reticulated nerves of the leaves, which recall the nervation of Dicotyledons, which they besides resemble in the structure of the stem, which presents fibro-vascular bundles arranged with tolerable regularity around a central pith. *Taccaceæ* differ from *Dioscoreæ* in habit, conformation of the stamens, and 1-celled ovary. They have an affinity with *Aroideæ* (*Dieffenbachia*, *Dracunculus*, *Amorphophallus*), principally founded on the nature of the leaves, and R. Brown considers them intermediate between that family and *Aristolochieæ*.[1]

Taccaceæ inhabit mountain forests of Asia, Africa, Oceania (and Guiana? according to Planchon). *Tacca pinnatifida* is principally found at the mouth of damp and shady valleys of the Oceanic Islands; it is cultivated for its starchy tubers, which furnish the islanders with a sort of Arrowroot. The Tahitians prepare from the floral scapes of the *Tacca* a very white and shining straw, with which they form hats and coronets with much skill and taste.

X. *IRIDEÆ.*

(ENSATÆ, *L. Ker.*—IRIDES, *Jussieu.*—IRIDEÆ, *Br.*—IRIDACEÆ, *Lindl.*)

FLOWERS ☿. PERIANTH *superior, petaloid, 6-merous, 2-seriate.* STAMENS 3, *opposite to the outer perianth-segments;* ANTHERS *extrorse.* OVARY *inferior, of 3 many-ovuled cells;* OVULES *anatropous.* CAPSULE *loculicidally 3-valved.* SEEDS *albuminous.* —STEM *herbaceous.* LEAVES *equitant or sheathing, ensiform or linear.*

Perennial HERBS with a tuberous or bulbous rhizome, rarely with fibrous roots, very rarely suffrutescent (*Witsenia*), glabrous, sometimes pubescent or velvety. SCAPE central, jointed or not, simple or branched, sometimes nearly 0. LEAVES usually all radical, equitant, distichous, ensiform or linear, angular, entire, flat, or folded longitudinally, the cauline alternate, sheathing. FLOWERS ☿, regular or irregular, terminal, in a spike corymb or loose panicle, rarely solitary, each furnished with 2 (rarely more) spathaceous bracts, usually scarious; *inflorescence* with a double sub-foliaceous bract. PERIANTH superior, petaloid, tubular, 6-fid or -partite, regular or sub-2-labiate; *segments* 2-seriate, equal, or the inner smallest,

---

[1] They are hardly separable as an order from *Burmanniaceæ.*—ED.

Iris.
Rhizome and equitant leaves.

Iris. Seed cut vertically
(mag.).

Iris.
Flower deprived of the perianth.
limb.

Pardanthus.
Diagram, showing the
stigmas alternate with the
stamens.

Iris germanica. Flower.

Iris.
Diagram, showing the
stigmas opposite to the
stamens.

Hydrokæna Meleagris.
Andrœcium and stigmas
seen in profile.

Iris. Flower cut vertically.

Iris.
Capsule with loculicidal
valves.

Hydrokæna.
Andrœcium and stigmas
seen in front.

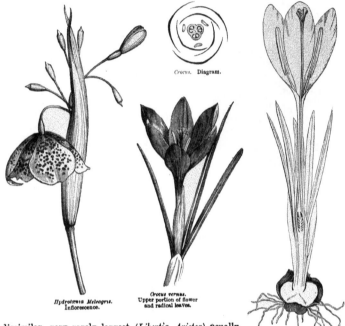

Crocus. Diagram.

Hydrokœnia Meleagris.
Inflorescence.

Crocus vernus.
Upper portion of flower
and radical leaves.

Crocus.
Entire plant cut vertically.

dissimilar, very rarely largest (*Libertia, Aristea*), usually fugacious, sometimes twisted spirally after flowering, and persistent (*Moræa, Pardanthus, Aristea, Galaxia,* &c.), æstivation twisted. STAMENS 3, epigynous, or inserted either on the tube or at the base of the outer perianth-segments ; *filaments* distinct, or more or less monadelphous (*Tigridia, Ferraria, Vieusseuxia,* &c.); *anthers* extrorse, 2-celled, basi- or dorsi-fixed and versatile, oblong or ovoid, or sagittate, dehiscence longitudinal. OVARY inferior, or rarely semi-inferior (*Witsenia*) ; cells 3, many- (rarely few-) ovuled (*Aristea*); *style* simple; *stigmas* 3, either opposite to the stamens (*Iris, Moræa, Vieusseuxia*), or alternate (*Pardanthus, Sisyrinchium, Libertia,* &c.), very often dilated, petaloid, or lamellate, gyrose, contorted (*Patersonia, Galaxia, Libertia,* &c.), entire, or 2-3-fid, or 2-labiate (*Diplarrhena, Iris*); *ovules* few or many, in 1–2-several series at the inner angle of each cell, usually horizontal, sometimes ascending (*Patersonia, Galaxia, Crocus,* &c.) ; or pendulous (*Gladiolus, Watsonia,* &c.), anatropous. CAPSULE 3-gonous, or lobed, or gibbous, 3-celled, loculicidally 3-valved ; *valves* membranous, coriaceous or cartilaginous, semi-septiferous ; *placentas* nerviform, adnate to the edge of the septum, sometimes coherent in a persistent central free column (*Pardanthus*). SEEDS usually numerous, sub-globose, or [usually horizontally] compressed, sometimes margined or winged; *testa* membranous, loose or papery, sometimes coriaceous or fleshy; *raphe* usually free, or easily separable; *albumen* fleshy or cartilaginous, sometimes sub-horny. EMBRYO axile or concentric, usually half the length of the albumen; *radicle* reaching to the hilum, situation variable.

[*Irideæ* have been divided into the following sub-orders by Klatt :—

SUB-ORDER I. GLADIOLEÆ.—Spathe 2-valved. Filaments free, unequal; stigmas filiform. *Gladiolus, *Watsonia, *Sparaxis, Antholyza, Diasia, Tritonia, Babiana, *Galaxia, *Witsenia, *Anomatheca, *Aristea, &c.

SUB-ORDER II. IRIDEÆ PROPER.—Spathe many-valved, outer herbaceous, inner membranous. Filaments equal, free or connate at the base. Stigmas petaloid. *Moræa, *Cipura, Xiphium, Dietes, Diplarrhena, *Iris, &c.

SUB-ORDER III. CYPELLEÆ.—Spathe 2- or many-valved. Perianth-segments unequal. Filaments connate at the base. Stigmas dilated. *Pardanthus, Cypella, Libertia,* &c.

SUB-ORDER IV. SISYRINCHIEÆ.—Spathe 2-valved. Filaments connate throughout their length. Stigmas simple, or involute-filiform. *Herbertia, *Vieusseuxia, *Tigridia, *Ferraria, Patersonia, *Sisyrinchium, *Gelasine, *Hydrotænia,* &c.

SUB-ORDER V. IXIEÆ.—Spathe 2-valved. Filaments equal, free. Stigmas linear (except *Crocus*). *Ixia, Hesperantha, Geissorhiza, Trichonema, *Crocus,* &c.—ED.]

Irideæ are distinguished from other Monocotyledons with inferior ovaries by their trinary andrœcium, extrorse anthers, and the petaloid stigmas of most of the genera. They have some affinity with *Burmanninceæ* and *Hæmodoraceæ* (which see).

Irideæ are much more extra-tropical in both hemispheres than tropical; South Africa contains a great number and variety of species, as does Mexico, but they are rare in Asia. Many genera are exclusively African (*Sparaxis, Vieusseuxia,* &c.), or American (*Sisyrinchium, Hydrotænia*), or Australian (*Patersonia*), whilst several others are dispersed over Australasia and the American continent. *Iris* inhabits the north temperate regions. *Gladiolus* and *Trichonema,* which abound in South Africa, advance as far as the Mediterranean region and Central Europe. *Crocus* inhabits sub-alpine regions and the plains of Europe and temperate Asia.

The tuberous or bulbous rhizomes of *Irideæ* contain a small proportion of a fatty and acrid matter, and a large quantity of starch, combined with a peculiar volatile oil, which gives them stimulating properties. Some species lose their acridity by drying or boiling, and their tubers may be used as emollients, or even as food; such are several South African species, which are eaten by the Hottentots. The rhizome of *Iris florentina* is medicinally the most important of the family; when fresh it is a strong purgative; dried, it stimulates moderately the pulmonary and gastro-intestinal mucous membranes; it enters into several pharmaceutical preparations, and its violet scent is a well-known perfume [Orris-root]; little balls of it, called Iris peas, are used to maintain suppuration after cautery. *I. germanica* and *pallida* were formerly used as diuretics and purgatives. The tubers of the Flag or Bog Iris (*I. Pseud-acorus*), the taste of which is acrid and astringent, are still administered by some country doctors in dropsy and chronic diarrhœa. Those of *I. virginica* and *versicolor* are similarly prescribed in North America. *I. sibirica* is considered an antisyphilitic in North Asia. The rhizome of *I. fœtidissima* was renowned among the ancients for the cure of hysteria and scrofula. The bulbs of *Sisyrinchium galaxioides, Ferraria purgans* and *cathartica,* and *Libertia ixioides,* are used in South America as purgatives and diuretics. *Pardanthus chinensis* has a high repute in India as an aperient. The root of *Gladiolus communis* is made into an amulet by the superstitious peasants of Germany; that of *G. segetum* was anciently considered an emmenagogue and aphrodisiac. The bulbs of *Moræa collina,* of the Cape, are very poisonous, and have the same effects as Fungi.

3 E

The stigmas of the Saffron (*Crocus sativus*), the origin of which is unknown, and the cultivation of which dates from very ancient times, contain a strong-scented volatile oil and a rich yellow dye; they are greatly esteemed as an emmenagogue, and as an excitant of the gastric and cerebral functions; it is still cultivated in France and Spain, and largely used by dyers and liqueur manufacturers, and also as a condiment in some countries. The stigmas of the other species of Crocus, although containing a colouring matter, are useless. The blue perianth of *Iris germanica*, crushed and mixed with lime, yields the Iris green of painters. Finally, the seeds of *I. Pseud-acorus* are a well-known substitute for coffee.

XI. *AMARYLLIDEÆ.*

(NARCISSORUM *sectio, Jussieu.*—NARCISSEÆ, *Agardh.*—AMARYLLIDEÆ, *Br.*—
AMARYLLIDACEÆ, *Lindl.*)

FLOWERS ☿ . PERIANTH *superior, petaloid, 6-fid or -partite, 2-seriate, sometimes with a crown simulating a supplementary perianth,* STAMENS 6, *very rarely 12–18, inserted on the perianth.* OVARY *inferior,* 3·1-*celled;* STYLE *simple;* OVULES *anatropous.* FRUIT *a loculicidally 3-valved capsule, or fleshy and indehiscent.* SEEDS *albuminous;* TESTA *membranous or thick;* RAPHE *lateral, immersed.* EMBRYO *short,*

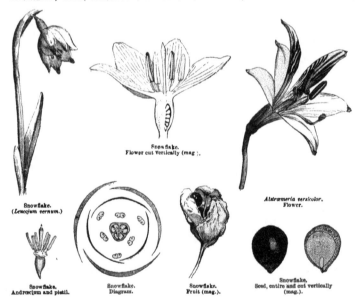

Snowflake.
Flower cut vertically (mag.).

Snowflake.
(*Leucojum vernum.*)

Alstræmeria versicolor.
Flower.

Snowflake.
Androecium and pistil.

Snowflake.
Diagram.

Snowflake.
Fruit (mag.).

Snowflake,
Seed, entire and cut vertically
(mag.).

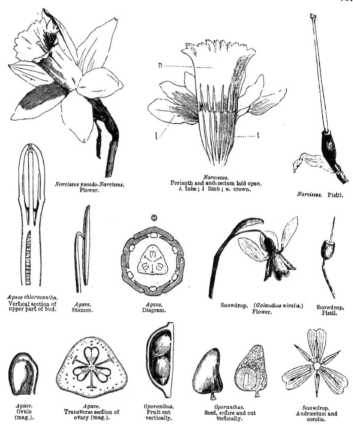

Narcissus pseudo-Narcissus.
Flower.

Narcissus.
Perianth and andrœcium laid open.
t. tube; *l* limb; *n.* crown.

Narcissus. Pistil.

Agave chloracantha.
Vertical section of
upper part of bud.

Agave.
Stamen.

Agave.
Diagram.

Snowdrop. (*Galanthus nivalis.*)
Flower.

Snowdrop.
Pistil.

Agave.
Ovule
(mag.).

Agave.
Transverse section of
ovary (mag.).

Oporanthus.
Fruit cut
vertically.

Oporanthus.
Seed, entire and cut
vertically.

Snowdrop.
Andrœcium and
corolla.

axile.—Perennial HERBS, *usually bulbous, stemless.* LEAVES *radical, elongated, entire.*
SCAPE *terminated by one or several flowers furnished with spathaceous bracts.*

Perennial HERBS, usually stemless, bulbous and with fibrous roots; rarely
caulescent with fascicled roots and alternate cauline leaves (*Alstrœmeria, Doryanthes*).
LEAVES radical, in 2 or several rows, sometimes 2 spreading (*Hæmanthus*), entire,
sheathing at the base, nerves parallel. SCAPE cylindric or angular, solid or fistular,
sometimes very short, or nearly 0; rarely stem erect or twining (*Bomarea*). FLOWERS

☿, elegant, regular or irregular, solitary or umbellate, or rarely in aggregated spikes (*Doryanthes*), enclosed in spathaceous bracts. PERIANTH superior, petaloid, 6-phyllous, or tubular-infundibuliform ; *limb* 6-partite, 2-seriate, regular or ringent, imbricate, deciduous or marcescent, often bearing at the throat a petaloid crown simulating an accessory corolla (*Narcissus, Pancratium,* &c.). STAMENS inserted either on an epigynous disk or on the tube or throat of the perianth, 6, opposite to the divisions of the perianth, or sometimes 12–18 (*Gethyllis*), then very rarely all fertile ; *filaments* cohering by their dilated bases, equal and erect, or unequal and inclined ; *anthers* introrse, 2-celled, basi- or dorsi-fixed, erect or incumbent, very rarely adnate within to a thick connective (*Chlidanthus*), opening by 2 longitudinal slits, or at their tips. OVARY inferior, 3-celled, rarely sub-1-celled (*Calostemma*); *style* simple, erect, or inclined with the stamens; *stigma* undivided or 3-lobed ; *ovules* numerous, rarely definite (*Griffinia, Hæmanthus, Calostemma,* &c.), 2-seriate at the central angle of the cells, parietal in the 1-celled ovary, usually horizontal or pendulous, rarely ascending (*Griffinia, Hæmanthus, Gethyllis,* &c.), always anatropous. FRUIT a loculicidally 3-valved capsule, or rupturing irregularly, rarely 1–2-celled by arrest, sometimes an indehiscent berry (*Gethyllis, Hæmanthus, Sternbergia, Clivia,* &c.). SEEDS shortly funicled, rarely solitary, sub-globose, angular or flat ; *testa* sometimes membranous or papery, often margined or winged ; sometimes thick and fleshy, or even enormously hypertrophied (*Pancratium, Calostemma,* &c.); *raphe* longitudinal, deep seated, sometimes fleshy ; *chalaza* apical; *albumen* fleshy. EMBRYO straight, axile, shorter than the albumen ; *radicle* reaching to the hilum, centripetal or superior, rarely inferior.

PRINCIPAL GENERA.

* Galanthus.	* Sternbergia.	* Crinum.	* Calostemma.	* Leucojum.
* Oporanthus.	* Hæmanthus.	* Pancratium.	* Amaryllis.	* Griffinia.
Eustephia.	* Narcissus.	Gethyllis.	* Alstrœmeria.	* Doryanthes.
* Clivia.	* Bomarea.			

AGAVEÆ. (CLOSELY ALLIED GENERA.)

    * Agave.   Fourcroya.

Amaryllideæ only differ from *Liliaceæ* (which see) in their inferior ovary. They approach *Irideæ*, *Hypoxideæ*, and *Hæmodoraceæ* : *Irideæ* are separated by triandry and extrorse anthers ; *Hypoxideæ* by habit, the texture of their flower, and their black crustaceous testa ; *Hæmodoraceæ* by their stamens, which are often reduced to three, their not bulbous roots, &c. *Agaveæ* are true *Amaryllideæ*, without bulbs, with valvate perianth-segments and a fistular style perforated at the top, and they are further remarkable for their spiny fleshy leaves and their often gigantic scape, which flowers but once, and terminates in a large panicle.

Amaryllideæ mostly grow in temperate or tropical regions; the remarkable fact in their geographical distribution is that the genera without a corona to the perianth are very rare in Europe and North America, but abound in South Africa and trans-equatorial America. Several genera are confined to Europe, South Africa, America, and Australia respectively. The Snowdrop (*Galanthus nivalis*) alone reaches high latitudes. *Crinum* and *Pancratium* prefer seashores in temperate and hot regions. *Agave americana* is now spread throughout the tropics, and even into Mediterranean Europe and Africa, where it is used for fences.

Amaryllideæ are much sought as ornamental plants, and rival *Liliaceæ* in the magnificence of their flowers and the sweet smell of several species, which are therefore used in perfumery: Their properties are also analogous to those of *Liliaceæ*; the mucilage of their bulbs is more abundant and less acrid, but it is combined with a bitter gum-resin, which is a violent emetic. This property induced the ancients to class among medicinal plants *Narcissus pseudo-Narcissus* and the Snowflake (*Leucojum vernum*), which both flower in the spring. The bulb of *Sternbergia lutea*, which grows in the East, was formerly employed to hasten the ripening of indolent tumours; those of *Amaryllis, Crinum,* and *Pancratium* are still thus used in Asia and America. *Pancratium maritimum* possesses properties similar to those of *Scilla*, and is sometimes substituted for it. *Amaryllis Belladonna*, of the Antilles, and *Hæmanthus toxicaria*, of South Africa, are eminently poisonous; the Kaffirs make use of the latter to poison their arrows. *Crinum zeylanicum* is also considered in the Moluccas a violent poison. Finally, the flowers of *Narcissus pseudo-Narcissus* are narcotic in small doses, and dangerous in larger ones. *Alstræmeriæ*, from South America, which are noticeable for their habit and the beauty of their flowers, bear farinaceous tubers, which may serve as food. *A. Salsilla* is used in Chili as a substitute for Sarsaparilla.

The *Agave americana*, cultivated in our gardens under the incorrect name of Aloe, is greatly esteemed in Mexico, on account of the various uses which can be made of it. When its central bud is removed previous to the lengthening of the scape, it yields an abundance of sugary liquid, which, when fermented, becomes a spirituous drink, called *pulque*, greatly esteemed by the Mexicans, and which by distillation yields an alcohol analogous to rum, named *mescal*. The expressed juice of the leaves is prescribed by American doctors as a resolvent and alterative, very efficacious in syphilis, scrofula, and even cancers. The woody fibres which form the framework of the leaves afford a very tenacious thread, the Vegetable Silk of commerce, from which the ancient Mexicans made paper. The scape, dried and cut in pieces of a varying thickness, is used for razor-strops, and as a substitute for cork.

XII. *HÆMODORACEÆ, Br.*

FLOWERS ☿. PERIANTH *petaloid, 6-merous, 2-seriate, regular or sub-irregular, usually superior.* STAMENS 6, *of which 3 are often sterile, or 0, inserted on the perianth-segments.* OVARY *inferior, or rarely superior, 3-celled, or sub-1-celled;* OVULES *usually semi-anatropous.* FRUIT *usually a loculicidally 3-valved capsule.* SEEDS *albuminous.* EMBRYO *with radicle near or far from the hilum.*—*Perennial* HERBS. LEAVES *ensiform, equitant.* FLOWERS *in a panicle or corymb.*

Perennial HERBS; *roots* fibrous, fascicled. STEM simple or nearly so, sometimes shortened, or a rhizome. LEAVES alternate, usually distichous, ensiform, sheathing at the base, equitant. FLOWERS ☿, regular or sub-irregular, in racemes or a corymb, bracteolate. PERIANTH petaloid, tubular or sub-campanulate, usually hairy or woolly outside, glabrous within, usually superior, 6-partite; divisions 2-seriate, either free to the base, or joined below into a tube, sometimes sub-irregular, and unilateral above (*Anigosanthus*). STAMENS 6, inserted at the base of the perianth-segments, of which 3 are opposite to the outer segments, often imperfect or 0, the 3 others fertile, 1 sometimes deformed; *filaments* filiform or subulate, rarely dilated and petaloid, free, or partially adnate to the perianth-segments; *anthers* introrse, 2-celled, basi- or dorsi-fixed, dehiscence longitudinal. OVARY inferior or rarely superior (*Xiphidium, Wachendorfia,* &c.), with 3 cells opposite to the inner segments of the perianth, rarely sub-1-celled by failure of the septa (*Phlebocarya*); *style* terminal, simple, base sometimes dilated and hollow; *stigma* undivided; *ovules* inserted at the inner angle of the cells, solitary or geminate, or indefinite, peltate,

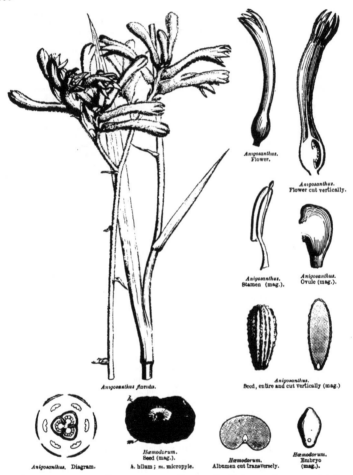

Anigosanthus.
Flower.

Anigosanthus.
Flower cut vertically.

Anigosanthus.
Stamen (mag.).

Anigosanthus.
Ovule (mag.).

Anigosanthus flavidus.

Anigosanthus.
Seed, entire and cut vertically (mag.)

Anigosanthus. Diagram.

Hæmodorum.
Seed (mag.).
h. hilum; m. micropyle.

Hæmodorum.
Albumen cut transversely.

Hæmodorum.
Embryo
(mag.).

semi-anatropous, or rarely anatropous. FRUIT a 3-celled capsule, accompanied or crowned by the marcescent perianth, loculicidally 3-valved; valves **septiferous**, or septa remaining attached to an axile column, rarely **a 1-seeded nut** (*Phlebocarya*).

SEEDS solitary or geminate or numerous, oblong, peltate or basifixed; *testa* coriaceous, glabrous or hairy; *albumen* cartilaginous, hard. EMBRYO straight, short;
radicle usually distant from the hilum, and placed almost outside the albumen.

PRINCIPAL GENERA.

| Lachnanthes. | Xiphidium. | * Hæmodorum. | * Anigosanthus. |
| Blancoa. | * Wachendorfia. | Conostylis. | |

Hæmodoraceæ are near *Amaryllideæ* and *Irideæ* in the 6-merous 2-seriate petaloid perianth, the 6
or 3 stamens, the usually inferior and 3-celled ovary, albuminous seeds, &c.; they differ from *Amaryllideæ*
in their usually hairy or woolly perianth, equitant leaves, stamens often reduced to 3, ovary sometimes
superior, and root never bulbous; they are separated from *Irideæ* by their introrse anthers. *Anigosanthus*
approaches *Bromeliaceæ* in the perigynous androecium.

Hæmodoraceæ are principally North American, South African, and South-west Australian. *Xiphidium* and *Hugenbachia* are tropical American.

The roots and seeds of several species contain a red colouring principle; such is *Lachnanthes
tinctoria*, of North America; but this principle, analogous to Madder, is much less solid, and is little
used.

XIII. *HYPOXIDEÆ*, Br.

FLOWERS ☿, *regular*. PERIANTH *superior, petaloid*, 6-*merous*, 2-*seriate*. STAMENS
6; ANTHERS *introrse*. OVARY *inferior, with 3 many-ovuled cells*; OVULES *semi-anatropous*. FRUIT *a capsule or berry*. SEEDS *strophiolate, albuminous*. EMBRYO *axile*;
RADICLE *distant from the hilum, superior.—Stemless* HERBS. LEAVES linear.

Herbaceous stemless perennials; *root* tuberous or fibrous. LEAVES all radical,
linear, entire, folded, nerves parallel. SCAPES simple, or branched at the top,
cylindric, sometimes very short, or 0 (*Curculigo*). FLOWERS ☿, yellow, rarely
diclinous by arrest, regular, either sessile and radical, or terminating the scape,
solitary whorled or panicled, 1-2-bracteolate. PERIANTH petaloid, superior, 6-par-
tite, persistent or deciduous; *segments* 2-seriate, the outermost velvety. STAMENS
6, inserted at the base of the perianth-segments; *filaments* free; *anthers* introrse,
2-celled, basifixed, erect, sagittate, dehiscence longitudinal, sometimes cohering into
a tube. OVARY inferior, 3-celled, or 1-celled with 3 parietal placentas (*Curculigo*);
style terminal, simple; *stigmas* 3, free or connate; *ovules* numerous, 2-several-
seriate at the inner angle of the cells (*Hypoxis*), anatropous. FRUIT a capsule dehiscing
longitudinally, or a berry, 3-celled, or 1-2-celled by abortion. SEEDS numerous, sub-
globose; *testa* black, crustaceous, wrinkled; *funicle* sometimes persistent; *albumen*
fleshy. EMBRYO straight, axile, nearly as long as the albumen; *radicle* distant from
the hilum, superior.

PRINCIPAL GENERA.

* Hypoxis.　　　　* Curculigo.

Hypoxideæ are near *Amaryllideæ* in their perianth, inferior ovary, &c.; they are separated by their
habit, their seeds with black and crustaceous testa, &c.; they especially approach *Asteliæ* in habit,
hairiness, the number of stamens and stigmas, the 1-3-celled ovary, &c. Their linear-folded leaves, with
parallel nerves, recall those of *Gagea*, a Liliaceous genus.

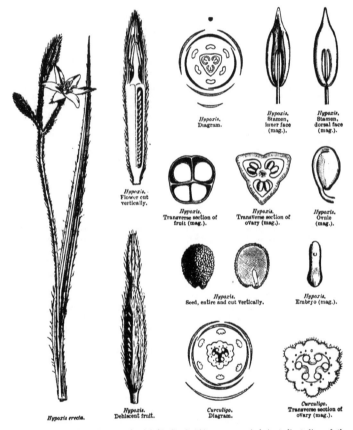

Hypoxis erecta.

Hypoxis.
Flower cut
vertically.

Hypoxis.
Diagram.

Hypoxis,
Stamen,
inner face
(mag.).

Hypoxis,
Stamen,
dorsal face
(mag.).

Hypoxis.
Transverse section of
fruit (mag.).

Hypoxis.
Transverse section of
ovary (mag.).

Hypoxis.
Ovule
(mag.).

Hypoxis.
Seed, entire and cut vertically.

Hypoxis.
Embryo (mag.).

Hypoxis.
Dehiscent fruit.

Curculigo.
Diagram.

Curculigo.
Transverse section of
ovary (mag.).

Hypoxideæ abound nowhere; a few inhabit South Africa, extra-tropical Australia, India, and the tropical and hot extra-tropical regions of America.

Little is known of their properties. The tubers of *Curculigo orchioides,* which resemble those of *Orchideæ,* become when dry transparent like amber; their sub-aromatic bitterness leads to their employment in affections of the urethra. The roots of *C. stans,* which grows in the Marianne Islands, are edible. The tubers of *Hypoxis erecta* are prescribed by the natives of North America for the cure of ulcers, and they are used internally for intermittent fevers.

XIV. *VELLOSIEÆ, Don.*

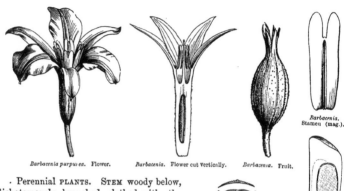

Barbacenia purpurea. Flower.   Barbacenia. Flower cut vertically.   Barbacenia. Fruit.

Barbacenia.
Stamen (mag.).

. Perennial PLANTS. STEM woody below, dichotomously branched, clothed with the leaf-bases, which are agglutinated by a resinous viscous juice. LEAVES collected at the top of the stem and branches, grass-like, spinescent or very stiff. PERIANTH superior, petaloid, 6-partite, 2-seriate, regular. STAMENS inserted at the base of the perianth, either 6 free, or indefinite and united into several

Barbacenia. Diagram.

Barbacenia.
Vertical section
of seed
(mag.).

bundles, naked or with a scale at their base (*Vellosia*); *filaments* filiform, or plane and 2-fid at the top; *anthers* linear, dorsi- or basi-fixed, 2-celled, introrse. OVARY inferior, 3-celled; *ovules* numerous, horizontal, anatropous or semi-anatropous. CAPSULE opening at the top into 3 incomplete loculicidal semi-placentiferous valves. SEEDS numerous, cuneate or angular; *testa* coriaceous or suberose; *albumen* fleshy. EMBRYO minute, placed laterally and outside the albumen.

GENERA.

Vellosia.   Barbacenia.

Vellosieæ are closely allied to *Bromeliaceæ* in the perianth, style, ovary, fruit, and embryo, and in the leaves crowning the top of the stem; *Bromeliaceæ* are separated by their exterior calycoid perianth and farinaceous albumen. *Vellosieæ* also approach *Hæmodoraceæ*, from which they only differ in the number of stamens, sometimes indefinite, the 3-gonous and 3-partite style, the usually arborescent stem, clothed with the persistent leaf-bases, and leafy at the top [and the situation of the embryo]. They abound in Brazil, and are exceptionally met with in Madagascar, Arabia, and Abyssinia.

XV. *DIOSCOREÆ.*

(DIOSCOREÆ, *Br.*—DIOSCOREACEÆ, *Lindl.*)

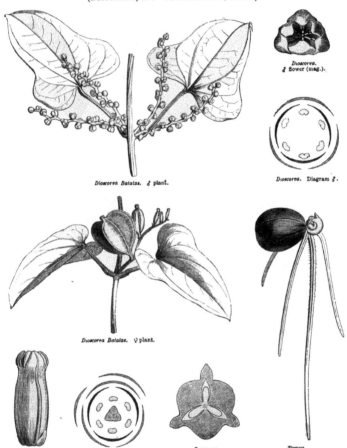

Dioscorea Batatas. ♂ plant.

Dioscorea.
♂ flower (mag.).

Dioscorea. Diagram ♂.

Dioscorea Batatas. ♀ plant.

Dioscorea.
♀ flower (mag.).

Dioscorea.
Diagram ♀.

Dioscorea.
Transverse section of ovary
(mag.).

Tamus.
Germinating seed, the plumule
raising the operculum.

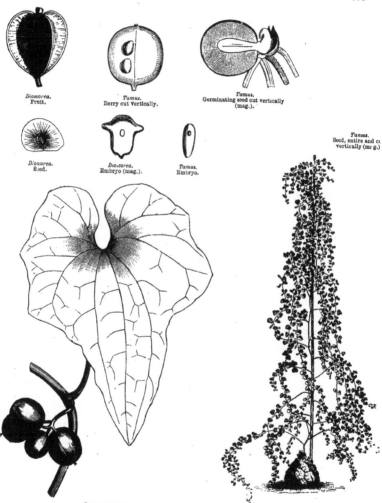

Dioscorea.
Fruit.

Tamus.
Berry cut vertically.

Tamus.
Germinating seed cut vertically
(mag.).

Dioscorea.
Seed.

Dioscorea.
Embryo (mag.).

Tamus.
Embryo.

Tamus.
Seed, entire and cut
vertically (mag.).

Tamus communis.
Fructiferous branch.

Testudinaria elephantipes (reduced in size).

FLOWERS *diœcious*. PERIANTH *superior, 6-merous, 2-seriate*. STAMENS 6. OVARY *inferior, with three 2-1-ovuled cells*; OVULES *pendulous, superimposed, anatropous.* CAPSULE *or* BERRY. SEEDS *compressed, winged, or globose, albuminous.—Twining or sarmentose* HERBS, *with tuberous rhizome.* LEAVES *reticulate-veined.*

Perennial HERBS, or UNDERSHRUBS, twining from right to left; *rhizome* subterranean, tuberous, fleshy, or epigeal, and covered with a thick and regularly cleft suberose bark (*Testudinaria*), giving off annual branches at the top. LEAVES alternate or sub-opposite, petioled, simple, palminerved, nerves reticulate, entire or palmisect; *petioles* often biglandular at the base, and often producing bulbils or large tubers at their axils. FLOWERS diœcious by arrest, small, inconspicuous, regular, in axillary racemes or spikes. PERIANTH herbaceous or sub-petaloid, superior in the ♀ flowers; *limb* with 6 segments, 2-seriate, equal, persistent. STAMENS 6, inserted at the base of the perianth-segments; in the ♀ 0, or rudimentary; *filaments* short, free; *anthers* introrse, 2-celled, shortly ovoid or globose, dorsifixed, dehiscence longitudinal. OVARY inferior, 3-celled; *styles* 3, short, often coherent at the base; *stigmas* obtuse, or rarely emarginate-bilobed; *ovules* solitary or geminate, pendulous, superimposed at the central angle, anatropous. FRUIT sometimes membranous, capsular, 3-gonous, 3-celled, opening at the projecting angles loculicidally (*Dioscorea*); sometimes 1-celled by the arrest of 2 cells, the third fertile, winged (*Rajania*); sometimes an indehiscent berry, 3-celled, or 1-celled by obliteration of the septa (*Tamus*). SEEDS compressed, and often winged in the capsular fruits, globose in the berried; *albumen* fleshy and dense, or cartilaginous. EMBRYO small, included, near the hilum, thinner and auricled at the upper end (*Dioscorea, Rajania*), or oblong-cylindric (*Tamus*); *radicle* near the hilum.

PRINCIPAL GENERA.

| * Dioscorea. | Rajania. | Tamus. | * Testudinaria. |

Dioscoreæ are very near *Smilax* in the nervation of the leaves, the perianth, androecium, fleshy fruit, &c., but are distinguished by the inferior ovary. They differ from *Taccaceæ* (see p. 782) in habit, 3-celled 1-2-ovuled ovary, and the internal structure of the seed; like *Taccaceæ*, they have some points of resemblance with *Aristolochiæ*.

Dioscoreæ inhabit especially southern tropical and extra-tropical regions; they are much rarer in northern temperate latitudes. *Rajania* is peculiar to tropical America. *Tamus* inhabits woods in temperate Europe and Asia. *Dioscorea* is met with in the tropics, and in temperate Australia; one small species has recently been discovered in the Pyrenees (*D. pyrenaica*). *Testudinaria* is peculiar to South Africa.

The root-tuber of *Dioscoreæ*, often called Ubi, Ufi, or Papa (names given by the Americans to the Potato), is filled with an abundant starch, mixed with an acrid and bitter principle. *D. sativa, alata, penta-phylla, bulbifera, Batatas*, &c. are cultivated throughout the tropics, and contribute largely to the sustenance of the Malays and Chinese, and the natives of Oceania and West Africa. The leaves of some species are used in intermittent fevers. The tuber of *Tamus communis* was formerly used as a purgative and diuretic; resolvent qualities were also attributed to it, and it was rasped, and applied as a plaster on arthritic strumous tumours, and on bruises—whence its name of Beaten Woman's Herb. The shoots, deprived of their acridity by boiling, are eaten like Asparagus.

XVI. *TRIURIDEÆ.*[1]

(Triuriaceæ, *Miers.*—Triuraceæ, *Gardn.*—Triuridaceæ, *Lindl.*—Triurideæ, *Dcne. and Le Maout.*)

[Very slender white or discoloured rarely green HERBS ; *roots* fibrous. STEM simple, rarely divided, filiform, straight or flexuous, erect. LEAVES 0, or bract-like, alternate, nerveless. FLOWERS minute, racemose or spiked, monœcious or diœcious, rarely unisexual; *pedicels* bracteate. PERIANTH 3-4-6-8-partite, hyaline ; *segments* connate at the base, valvate, tips often caudate. STAMENS few, various in number, sessile in the base of the perianth, usually seated on an androphore; *anthers* 4-celled, 2-valved, lobes rarely separated. CARPELS many, on a central receptacle, 1-celled; *style* excentric, lateral or basal, smooth or feathery; *stigma* obsolete or truncate or clavate; *ovule* 1, basal, erect. Ripe CARPELS obovoid, coriaceous and indehiscent, or 2-valved, 1-seeded. SEED ovoid; *testa* reticulate; *nucleus* cellular.

SECTION I. TRIURIEÆ.—Perianth-lobes with twisted tails that are inflexed in bud. Anther-cells separate, each 2-locellate. Ovary gibbous ; style ventral. *Triuris, Hexuris.*

SECTION II. SCIAPHILEÆ.—Perianth-lobes without tails. Anther-cells confluent. Style almost basilar. *Soridium, Sciaphila, Hyalisma.*

A very singular little order, well defined and illustrated by Miers in the Linnæan Transactions, from which work the above descriptions are taken. According to him they are allied to *Alismaceæ*; in the neighbourhood of which the late R. Brown also informed me they must in his opinion be placed.

Triurideæ are natives of tropical forests in America and Asia, growing on mossy banks and dead leaves, with hardly any attachment to the ground. *Hyalisma* is a native of Ceylon (it is referred to *Sciaphila* by Thwaites) ; *Sciaphila* of both Asia and America ; all the other genera are American. —ED.]

XVII. *BUTOMEÆ.*

(Butomeæ, *L.-C. Richard.*—Butomaceæ, *Endl., Lindl*)

FLOWERS ☿. PERIANTH 6-*merous, 2-seriate (calyx and corolla)*. STAMENS *hypogynous, 9-∞*. OVARIES *6-∞, whorled, more or less distinct, 1-celled, many-ovuled*; OVULES *erect, anatropous or campylotropous, placentation parietal*. FRUIT *follicular*. SEEDS *numerous*. EMBRYO *straight or hooked, exalbuminous*; RADICLE *inferior.*— *Marsh* HERBS, *perennial, stemless*. FLOWERS *solitary or umbelled.*

Perennial marsh or aquatic HERBS, stemless, glâbrous, sometimes milky. LEAVES all radical ; *petiole* semi-sheathing at the base; *blade* linear or oval, large, nerved, sometimes arrested. SCAPES simple, 1-many-flowered. FLOWERS ☿, regular, solitary (*Hydrocleis*), or umbelled (*Butomus, Limnocharis*) ; *pedicels* with membranous

---

[1] See tribe *Triurideæ* of *Burmanniaceæ*, p. 779.—ED.

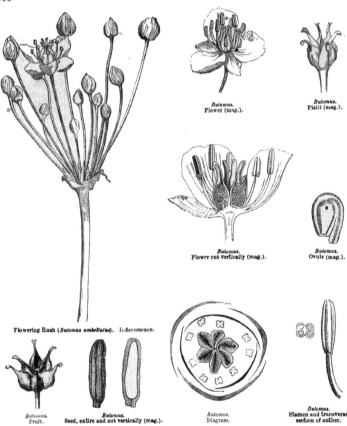

Butomus.
Flower (mag.).

Butomus.
Pistil (mag.).

Butomus.
Flower cut vertically (mag.).

Butomus.
Ovule (mag.).

Flowering Rush (*Butomus umbellatus*). Inflorescence.

Butomus.
Fruit.

Butomus.
Seed, entire and cut vertically (mag.).

Butomus.
Diagram.

Butomus.
Stamen and transverse
section of anther.

bracts. PERIANTH 6-phyllous, leaflets 2-seriate; outer herbaceous or sub-coloured; inner petaloid, imbricate, usually deciduous. STAMENS hypogynous, sometimes 9, of which 6 are in pairs opposite to the sepals, and 3 opposite to the petals (*Butomus*), sometimes indefinite, the outer often imperfect (*Limnocharis, Hydrocleis*); *filaments* filiform, free; *anthers* introrse, 2-celled, linear, dehiscence longitudinal. OVARIES 6 or more, whorled, free, or slightly coherent by their ventral suture, 1-celled, many-ovuled; *styles* continuous with the ovaries, stigmatiferous on their ventral face;

ovules many, covering the surface of the cell, or attached to a reticulate parietal placenta, erect, anatropous (*Butomus*) or campylotropous (*Limnocharis*). CARPELS distinct, coriaceous, usually beaked by the persistent styles, dehiscing ventrally (*Butomus*) or dorsally (*Limnocharis*), many-seeded. SEEDS erect, sometimes shortly funicled, straight, with membranous testa (*Butomus*); sometimes sessile, hooked, with a crustaceous transversely wrinkled testa (*Limnocharis*). EMBRYO exalbuminous, straight or hooked; *radicle* inferior.

PRINCIPAL GENERA.

     * Butomus.   Butomopsis.   * Limnocharis.   * Hydrocleis.

Butomeæ are closely allied to *Alismaceæ*, through *Limnocharis*, only differing in their singular placentation and the number of their ovules.

This family is not numerous; *Butomus* inhabits the north temperate zone, *Limnocharis* and *Hydrocleis* tropical America, *Butomopsis* Africa. The roots and seeds of *Butomus umbellatus* (Flowering Rush) were formerly recommended as emollients and refrigerants. The baked root is still eaten in North Asia. *Hydrocleis* is remarkable for its milky juice, and *Limnocharis* for the structure of its leaves, which have a large terminal pore, by which the plant appears to relieve its tissues when gorged with liquid. This phenomenon is identical with that described by Schmidt, Duchartre, and C. Musset, as occurring in several *Aroideæ* (*Colocasia*), and which consists in an intermittent more or less abundant emission of pure water, to the extent of more than half an ounce in a hot summer's night, a phenomenon which has been observed in the leaves of *Gramineæ*, and several other Monocotyledons.

XVIII. *ALISMACEÆ*, Br.

FLOWERS ☿, or *monœcious*. PERIANTH 6-*merous*, 2-*seriate* (*calyx and corolla*). STAMENS *hypogynous or perigynous, equal or multiple in number with the perianth leaflets*. OVARIES *more or less numerous, whorled or capitate, distinct,* 1-*celled,* 1-2-*ovuled*; OVULES *campylotropous.* FRUIT *a follicle.* SEEDS *recurved, exalbuminous.* EMBRYO *hooked.*—STEM *herbaceous.* LEAVES *radical, strongly nerved.*

Aquatic or marsh HERBS, perennial, sometimes producing subterranean tuber-like buds (*Sagittaria*). LEAVES usually radical, rosulate or fascicled; *petiole* with a dilated sheathing base; *blade* entire, nerves prominent, converging towards the top and united by secondary transverse nerves, cordate or sagittate or oval-oblong, arrested when the leaf is submerged, and then replaced by the petiole changed into a linear or spathulate phyllode. FLOWERS regular, ☿, or rarely monœcious (*Sagittaria*), in a raceme or panicle with whorled pedicels. PERIANTH 6-phyllous, leaflets 2-seriate, the 3 outer calycinal, the 3 inner petaloid, æstivation imbricate or convolute, caducous. STAMENS inserted on the receptacle, or at the base and on the sides of the inner perianth leaflets, equalling them or double or multiple in number; *filaments* filiform; *anthers* 2-celled, introrse, dorsifixed in the ☿ flowers, extrorse and basifixed in the ♂ (*Sagittaria*), dehiscence longitudinal. OVARIES 6-8-∞, whorled or capitate, quite distinct (*Alisma, Sagittaria*), or coherent by their ventral suture (*Damasonium*); *style* ventral, very short; *stigma* simple; *ovules* campylotropous, solitary, basilar, erect (*Alisma, Sagittaria*), or 2-3 superimposed, the one

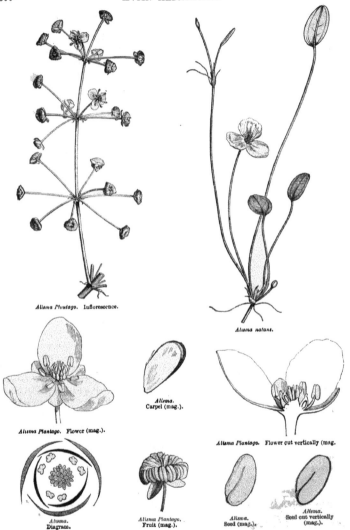

Alisma Plantago. Inflorescence.

Alisma natans.

Alisma Plantago. Flower (mag.).

Alisma. Carpel (mag.).

Alisma Plantago. Flower cut vertically (mag.

Alisma. Diagram.

Alisma Plantago. Fruit (mag.).

Alisma. Seed (mag.).

Alisma. Seed cut vertically (mag.).

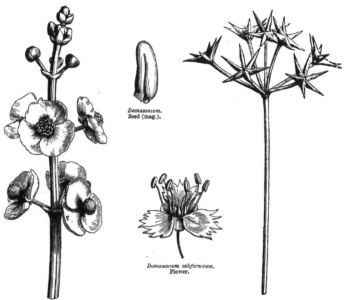

Damasonium.
Seed (mag.).

Damasonium californicum.
Flower.

Sagittaria sagittæfolia. Inflorescence.

Damasonium stellatum.

basilar and erect, the others horizontal. Ripe CARPELS indehiscent, or dehiscent by their ventral suture. SEEDS recurved, exalbuminous; *testa* membranous. EMBRYO hooked, sub-cylindric; *radicle* inferior or centripetal.

PRINCIPAL GENERA.

*Alisma.   * Sagittaria.   Damasonium.

Alismaceæ have by a great many botanists been united with *Juncagineæ,* which only differ in their always extrorse anthers, anatropous ovules, and straight embryo; *Alismaceæ* are also connected on the other hand with *Butomeæ,* which are separated by their placentation and the number of their ovules.

They are found, though not abundantly, in the temperate and tropical regions of both worlds. *Alisma* grows in the temperate zone of the northern hemisphere and the tropics of the New World. *Sagittaria* inhabits the same countries, but is rarer in the tropics. *Damasonium* inhabits certain parts of Europe, North Africa, North-west United States, and East Australia.

Most *Alismaceæ* possess an acrid juice, which led formerly to their use in medicine. The Water Plantain (*Alisma Plantago*) and *Sagittaria sagittæfolia* have been prescribed, but without good reason, for hydrophobia; the feculent rhizomes of the latter lose their acridity by desiccation, and serve as food to the Tartar Kalmucks; the same is the case with *S. sinensis,* cultivated in China, and *S. obtusifolia,* of North America.

XIX. *JUNCAGINEÆ, L.-C. Richard.*

FLOWERS ⚥, or *diclinous.* PERIANTH 6-*merous,* 2-*seriate, calycinal, sometimes* 0. STAMENS 6, *perigynous or hypogynous, sometimes* 1 *only;* ANTHERS *extrorse.* OVARIES 3 *or more, distinct, or more or less coherent,* 1-2-*ovuled;* OVULES *basilar, anatropous.*

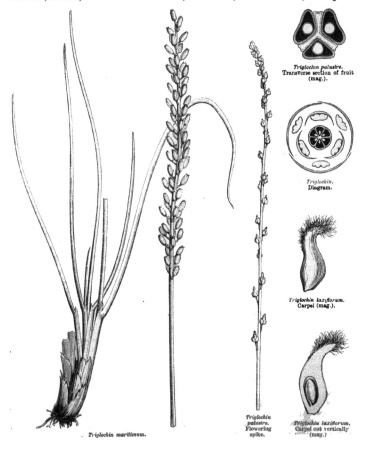

Triglochin palustre.
Transverse section of fruit
(mag.).

Triglochin.
Diagram.

Triglochin laxiflorum.
Carpel (mag.).

*Triglochin
palustre.*
Flowering
spike.

Triglochin laxiflorum.
Carpel cut vertically
(mag.)

Triglochin maritimum.

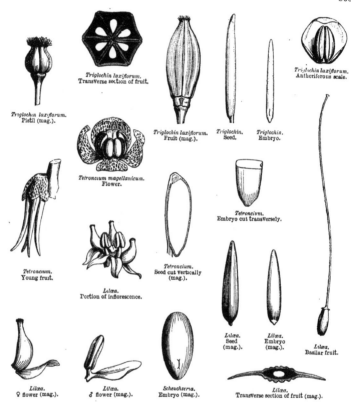

Triglochin laxiflorum.
Transverse section of fruit.

Triglochin laxiflorum.
Pistil (mag.).

Triglochin laxiflorum.
Antheriferous scale.

Tetroncium magellanicum.
Flower.

Triglochin laxiflorum.
Fruit (mag.).

Triglochin.
Seed.

Triglochin.
Embryo.

Tetroncium.
Embryo cut transversely.

Tetroncium.
Young fruit.

Tetroncium.
Seed cut vertically
(mag.).

Lilæa.
Portion of inflorescence.

Lilæa.
Seed
(mag.).

Lilæa.
Embryo
(mag.).

Lilæa.
Basilar fruit.

Lilæa.
♀ flower (mag.).

Lilæa.
♂ flower (mag.).

Scheuchzeria.
Embryo (mag.).

Lilæa.
Transverse section of fruit (mag.).

FRUIT *a capsule or follicular, or indehiscent.* SEEDS *erect, exalbuminous*; EMBRYO *straight.*—STEM *or* SCAPE *herbaceous.* LEAVES *all radical, or cauline alternate.*

Marsh HERBS. LEAVES sheathing at the base, semi-cylindric or linear-ensiform, sometimes scented. FLOWERS ☿ (*Triglochin, Scheuchzeria*), or diœcious (*Tetroncium*), sometimes 0 (*Lilæa*). STAMENS 5, inserted at the base of the perianth-leaflets (*Tetroncium, Triglochin*), or hypogynous (*Scheuchzeria*), rarely 1 only (*Lilæa*); *filaments* very short; *anthers* 2-celled, extrorse, dehiscence longitudinal. CARPELS 3, distinct, 1-celled (*Scheuchzeria*), or 6 united into a 6-celled ovary, of which 3 cells are often imperfect (*Triglochin*), rarely solitary or spiked (*Lilæa*); *styles* as many as

3 F 2

carpels, elongated or very short; *stigmas* simple (*Tetroncium*), or capitate (*Lilæa*), or plumose (*Triglochin*), or papillose (*Scheuchzeria*); *ovules* 2, collateral, erect, or solitary and basilar, anatropous. FRUIT of distinct spreading follicles, opening by their ventral suture (*Scheuchzeria*), or a 4–6–3-celled capsule, opening by the ventral sutures of the carpels (*Triglochin, Tetroncium*), or indehiscent (*Lilæa*). SEEDS erect, exalbuminous. EMBRYO straight; *radicle* inferior.

PRINCIPAL GENERA.

Scheuchzeria.   Triglochin.   Tetroncium.   Lilæa.

Juncagineæ, which are closely allied to *Alismaceæ,* also approach *Naiadeæ* (see these families).

Triglochin is a widely-dispersed genus, growing in marshes or brackish grounds of all temperate regions. *Scheuchzeria* grows in the turfy swamps of Europe and North America. *Tetroncium* belongs to the Magellanic lands, and *Lilæa* to New Grenada and Chili.

XX. *POTAMEÆ, Jussieu.*

Annual or perennial PLANTS; *rhizome* sometimes with swollen joints, growing in fresh, brackish or salt water. STEMS knotty-jointed, usually branched, radicant.

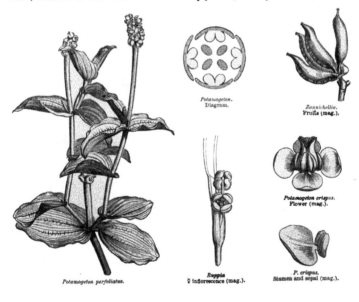

Potamogeton.
Diagram.

Zannichellia.
Fruits (mag.).

Potamogeton crispus.
Flower (mag.).

Potamogeton perfoliatus.

Ruppia
☿ inflorescence (mag.).

P. crispus.
Stamen and sepal (mag.).

Potamogeton crispus.
Young fruit (mag.).

P. crispus. Fruit cut
vertically and transversely (mag.).

P. crispus.
Seed (mag.).

P. crispus.
Embryo.

LEAVES all submerged, or the upper floating, alternate or distichous or close together, or rarely opposite, sessile or petioled, entire, filiform, linear, oval or oblong-lanceolate, all similar, or the submerged narrower and deprived of a stomatiferous epidermis; *stipules* free, or joined to the base of the petiole, membranous, intrafoliar, entire or emarginate. FLOWERS ☿ or monœcious or polygamous, in a spike or glomerate or solitary. PERIANTH of 4 herbaceous valvate sepals (*Potamogeton*); or forming a membranous 3-toothed cup in the ♂ flowers, 0 in the ♀ (*Zannichellia*); or totally absent in the ☿ flowers (*Ruppia*). STAMENS 4, sub-sessile, inserted on the claw of the sepals (*Potamogeton*); or 2, sessile, hypogynous (*Ruppia*); or 1, stipitate (*Zannichellia, Althenia*); *anthers* rounded, obtuse or apiculate and 3-celled, or oblong and 1-celled (*Althenia*); *pollen* globose or oblong or arched, and very finely granular. OVARIES 1–4–6, 1-celled, 1-ovuled; *style* elongate; *stigma* peltate or unilateral, sessile; *ovule* pendent, orthotropous or campylotropous. FRUIT sessile or stipitate, indehiscent, coriaceous; *epicarp* membranous; *endocarp* hard, or opening into 2 valves at germination. SEED oblong; *testa* membranous; *albumen* 0. EMBRYO macropodous, antitropous or amphitropous, cotyledonary end arched or spirally coiled.

GENERA.

Potamogeton.	Grœnlandia.	Zannichellia.	Spirillus.
Althenia.	Ruppia.	.	

Potameæ inhabit either stagnant water, sluggish streams, or brackish estuaries, or the shallow seas of the cold and temperate regions of the globe. They are rare in the tropics. *Zannichellia* lives in ditches in Europe and North America; *Althenia* in the lagoons of the South of France and of Algeria; *Ruppia* in the muddy salt marshes in both continents. They are of no use to man.

XXI. *APONOGETEÆ, Planchon*

Aquatic stemless HERBS with tuberous starchy feculent rhizomes. LEAVES submerged or floating, petioled; *petiole* enlarged and membranous at the base; *blade* linear, oval or oblong; *nerves* 1–5, parallel, often united by transverse venules, between which the cellular tissue is sometimes wanting, when the leaf is elegantly latticed (*Ouvirandra fenestralis*). FLOWERS white or pink, in a unilateral simple or

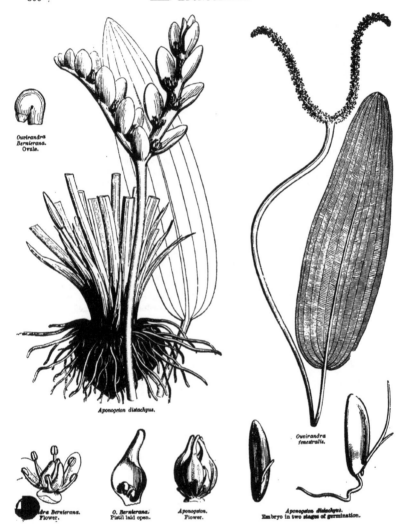

Ouvirandra
Bernierana.
Ovule.

Aponogeton distachyus.

Ouvirandra
fenestralis.

dra Bernierana.
Flower.

O. Bernierana.
Pistil laid open.

Aponogeton.
Flower.

Aponogeton distachyus.
Embryo in two stages of germination.

2-3-fid spike, at the top of an axillary scape, enclosed at first in a membranous or coloured conical spathe. PERIANTH 0, or 2-3-phyllous, caducous or persistent, sometimes accompanied by 10-15 distichous white thick and accrescent bracts (*Aponogeton*). STAMENS 6-18-20, hypogynous, sub-equal; *filaments* subulate, persistent; *anthers* ovoid, basifixed, 2-celled; *pollen* globose, or acutely elliptic. OVARIES 3-5, flagon-shaped, sessile, 1-celled; *style* continuous with the ovary, oblique, and stigmatiferous on the inner face; *ovules* 2-4-6, inserted a little to one side at the bottom of the cell, ascending, anatropous. FRUIT a 1-2-seeded follicle. SEEDS with membranous or sub-spongy testa; *albumen* 0. EMBRYO oval or elliptic, thick, compressed, radicular end inferior.

GENERA.

* Aponogeton.   * Ouvirandra.

[*Aponogeton* is very closely allied to *Potameæ*, and is distinguished chiefly by the bracts, hypogynous stamens, and sub-basal ovules.]

These genera belong to tropical Africa, India, and Madagascar. The name *Ouvirandra* is taken from the Madagascar language, in which the word *ouvirandrou* signifies edible root, and is applied equally to several species of *Dioscorea*; the root *oui* is found bearing the same meaning throughout the islands of the Southern Ocean. We have noticed elsewhere that the genus *Spathium*, of Loureiro, is a synonym of *Saururus*.

XXII. *NAIADEÆ.*

Marine or fresh-water annual or perennial HERBS. STEMS creeping or rooting, branched. LEAVES alternate, distichous or opposite, often close together at the top of the internodes, linear, 1-3-nerved, entire, or tip denticulate, base sheathing, persistent, or jointed and caducous; *sheath* furnished with intravaginal free or connate membranous stipules, sometimes accompanied by scales (*Phucagrostis*); *spadices* usually solitary, sometimes several united in a common spathe, each furnished with a 2-valved spathella, and bearing 3-6 ☿ flowers (*Posidonia*). FLOWERS ☿ (*Posidonia*), or monœcious (*Zostera*), or diœcious (*Najas, Phucagrostis*); solitary (*Phucagrostis*), or sub-solitary (*Najas*), or in axillary fascicles (*Caulinia*), or united, 2 or more, on a spadix contained in a foliaceous spathe. PERIANTH 0 (*Zostera, Phyllospadix, Phucagrostis, Posidonia*, &c.); or tubular, membranous, 4-lobed (*Caulinia, Najas*), or tubular, denticulate (*Halophila*). STAMENS 1 (*Najas, Caulinia, Zostera*, &c.), or 2 (*Phucagrostis*), or 4-3 (*Posidonia*); *filaments* 0 (*Najas*), or very short, scale-like (*Zostera*), or dilated and aristate (*Posidonia*), or geminate and coherent (*Phucagrostis*); *anthers* 1-celled (*Zostera*), or 2-celled (*Posidonia, Phucagrostis*, &c.), or 4-celled (*Caulinia, Najas*), dehiscence longitudinal; *pollen* usually confervoid (*Zostera, Phucagrostis*, &c.), or globose (*Najas, Caulinia*). OVARIES 1-2-4, distinct, 1-celled, usually 1-ovuled, sometimes with 3 several-ovuled parietal placentas (*Halophila, Lemnopsis*); *stigmas* 2 apiculate, or 3 filiform, sometimes jointed at the top of the style (*Halophila*, &c.), rarely discoid and sharply branching (*Posidonia*); *ovule* pendulous and orthotropous (*Zostera*), or campylotropous (*Phucagrostis*, &c.), or ascending and anatropous (*Caulinia, Halophila*, &c.). FRUIT usually a nut or utricle, sometimes a berry (*Posidonia*), indehiscent, or opening more or less irregularly at germination. SEED

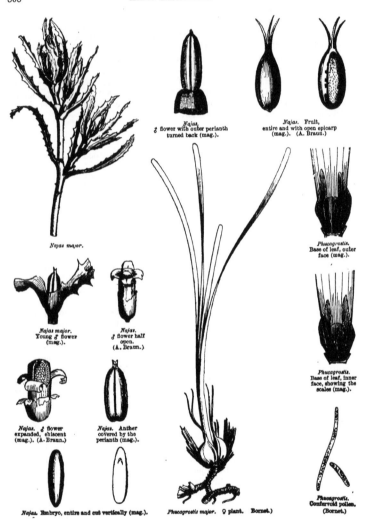

Najas,
♂ flower with outer perianth
turned back (mag.).

Najas. Fruit,
entire and with open epicarp
(mag.). (A. Braun.)

Najas major.

Phucagrostis.
Base of leaf, outer
face (mag.).

Najas major.
Young ♂ flower
(mag.).

Najas.
♂ flower half
open.
(A. Braun.)

Najas.
♂ flower
expanded, ehiscent
(mag.). (A. Braun.)

Najas. Anther
covered by the
perianth (mag.).

Phucagrostis.
Base of leaf, inner
face, showing the
scales (mag.).

Najas. Embryo, entire and cut vertically (mag.).

Phucagrostis major. ♀ plant. Bornet.)

Phucagrostis.
Confervoid pollen.
(Bornet.)

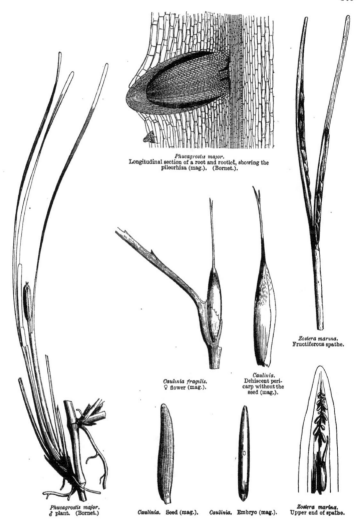

Phucagrostis major.
Longitudinal section of a root and rootlet, showing the pileorhiza (mag.). (Bornet.).

Zostera marina.
Fructiferous spathe.

Caulinia fragilis.
♀ flower (mag.).

Caulinia.
Dehiscent peri-
carp without the
seed (mag.).

Phucagrostis major.
♂ plant. (Bornet.)

Caulinia. Seed (mag.).   *Caulinia.* Embryo (mag.).

Zostera marina.
Upper end of spathe.

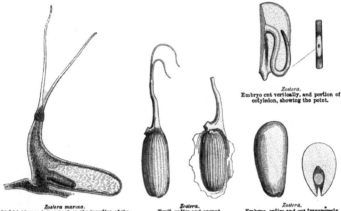

Zostera.
Embryo cut vertically, and portion of cotyledon, showing the point.

Zostera marina.
Pistil laid open below to show the insertion of the ovule (mag.).

Zostera.
Fruit, entire and opened (mag.).

Zostera.
Embryo, entire and cut transversely (mag.).

sub-globular or ovoid; *testa* thin or membranous, smooth or reticulate. EMBRYO macropodous.

GENERA.

Zostera.   Phucagrostis.   Lemnopsis.   Phyllospadix.   Caulinia.
Halophila.   Posidonia.   Najas.

Zostera inhabits the estuaries of the North Sea and Atlantic and Indian Oceans, *Posidonia* and *Phuca-grostis* the Mediterranean, and *Phyllospadix* the western shores of North America. *Caulinia* and *Najas* inhabit fresh still waters in Europe and America. In Holland the leaves of *Zostera* are used in the con-struction of dykes. For some years they have been used in France for stuffing mattresses and for packing.

Adrien de Jussieu, who studied the classification of Monocotyledons, divided them into albuminous and exalbuminous, and the latter again into terrestrial (*Orchideæ*) and aquatic. The exalbuminous aquatics have been divided into two sections, according to the presence or absence of a true perianth. The chlamydeous section includes *Alismaceæ, Butomeæ,* and *Hydrocharideæ,* which have six perianth divisions, the three inner petaloid; the other section comprises *Juncagineæ, Naiadeæ, Potameæ,* and *Zosteraceæ,* which have a scaly, membranous, or herbaceous perianth, or are achlamydeous. The three families of the first section are distinguished by free or coherent ovaries, and by the placentation; those of the second section by the embryo, which is brachypodous and homotropous in *Juncagineæ,* macropodous and antitropous in *Zosteraceæ,* macropodous and amphitropous in *Potameæ,* macropodous and homotropous in *Naiadeæ.*

With modifications we have adopted this classification,[1] and after many endeavours we have suc-ceeded in uniting in what appear to us homogeneous groups, the exalbuminous aquatic Monocotyledonous genera, placed in one family by most botanists. Without overlooking the close affinity between *Junca-gineæ, Potameæ, Naiadeæ,* &c., we think that the form of the stigmas—entire and peltate, or divided and pointed—may serve to group very naturally the different genera of *Naiadeæ* and *Potameæ,* the latter being

[1] After many attempts by many botanists, it is pretty clear that any linear arrangement of the Mono-cotyledonous families is quite impossible, and that there is little choice between several of them; under which cir-stances the simplest and most practical is that adopted in this work. See Synopsis of Orders at the end.—ED.

connected with *Juncagineæ*. It is thus that we have united to *Potameæ*, *Ruppia*, which has hitherto been placed near *Posidonia* and *Zostera*. On the other hand, it is probable that when the fruits and seeds of *Halophila*, *Lemnopsis*, &c. are known, a family will be made of these genera, which, by its many-ovuled ovaries with parietal placentation, will stand in the same relation to *Naiadeæ* that *Butomeæ* do to *Alismaceæ*, which *Aponogeton* and *Ouvirandra* approach.

It appears superfluous to discuss the modern view, based on that of Adanson, who considers 'as very rational' the affinity between *Alismaceæ* and *Ranunculaceæ*, and we shall retain our opinion until we find, on examining their seeds, with or without a microscope, an albumen and a dicotyledonous embryo in *Sagittaria*, which Adanson believed that he had seen, just as he fancied he saw two cotyledons in the seed of Reeds.

If, in spite of the conscientious work and the sagacious observations which during the last hundred and fifty years have so greatly advanced Botany, it is allowable to revive paradoxes that have been absolutely condemned by science ; if mere superficial resemblance is sufficient to establish natural affinity, we do not see why we should hesitate to follow Adanson in uniting, as he has done, *Cycadeæ* with Palms, *Aristolochieæ* with *Vallisneria*, *Polygala* with *Tithymaleæ*, and so forth.

XXIII. *PALMÆ, L., Juss., Martius, Blume,* &c.

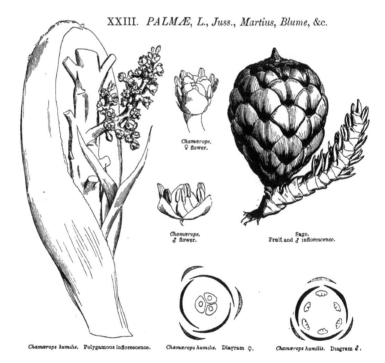

Chamærops,
♀ flower.

Chamærops,
♂ flower.

Sago.
Fruit and ♂ inflorescence.

Chamærops humilis. Polygamous inflorescence.   *Chamærops humilis.* Diagram ♀.   *Chamærops humilis.* Diagram ♂.

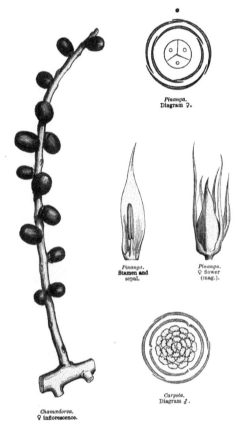

Pinanga.
Diagram ♀.

Pinanga.
Stamen and
sepal.

Pinanga.
♀ flower
(mag.).

Caryota.
Diagram ♂.

Chamædorea.
♀ inflorescence.

Rhapis.
Diagram ♂.

Pinanga.
Vertical section of ovary
(mag.).

Rhapis.
Half androecium and
sepal.

Monœcious inflorescence, ♀ flowers

FLOWERS *usually diclinous, sessile, or pedicelled on a simple or branched spadix.* CALYX *and* COROLLA *3-merous.* STAMENS *usually 6, hypogynous or perigynous.* OVARY *free, with 1–3 coherent or free carpels;* OVULES *solitary in each cell, rarely geminate.* FRUIT *a berry or drupe.* SEED *albuminous.* EMBRYO *peripheric.*—STEM *woody.* LEAVES *alternate, petiole sheathing, blade usually laciniate.*

Perennial woody PLANTS, elegant or majestic in habit. PRIMARY ROOT decaying early, and replaced by numerous adventitious roots, which are developed at the base of the trunk, and form a compact conical mass, often very voluminous, and rising more or less above the soil, and in certain cases raising the trunk, and supporting it like the shrouds of a ship. TRUNK (*stipe*) usually tall and slender, sometimes short and tumid (*Geonoma* and *Phœnix acaulis, Astrocaryum acaule,* &c.), or a short and creeping inclined stock, or forming underground a branched rhizome, the top of which, crowned by leaves, is on the surface of the soil (*Sabal, Rhapis*); simple, or very rarely dichotomous (*Hyphœne,* &c.), sub-cylindric, or rarely swollen towards the middle (*Iriartea, Acrocomia, Jubœa*), with or without nodes, smooth or armed with hairs, which are thickened and elongated into spines (*Martinezia, Bactris*), usually rough and annulated by the persistent bases of the leaves, sometimes marked with spiral scars (*Corypha elata*). LEAVES springing from the terminal bud, alternate, base sheathing the stem; *sheath* sometimes with a ligulate prolongation at the upper part (*Sabal, Copernicia,* &c.), and usually decomposing into a fibrous network after the decomposition of the leaf; *petiole* convex below; *blade* pinnatisect or flabellate, or peltate (*Licuala peltata*), or simply split; *segments* or *pinnules* callous at the base, quite distinct, or coherent below, folded longitudinally in vernation, with margins recurved [or erect or depressed], often split along the secondary nerves; *nerves* sometimes persistent and resembling filaments, sometimes much prolonged in tendril-like appendages (*Calamus*). INFLORESCENCE axillary. SPADIX (*régime*) furnished with an herbaceous or almost woody spathe, monophyllous, or composed of several distichous bracts, wholly or partially enveloping the inflorescence, or considerably shorter than it. FLOWERS small, usually diœcious or monœcious, rarely ☿ (*Corypha, Sabal,* &c.), shortly pedicelled or sessile, often sunk in the pits of the spadix, furnished with a bract and 2 opposite bracteoles, free or coherent, sometimes reduced to a callosity, or 0. PERIANTH double, persistent, coriaceous, formed of a calyx and a calycoid corolla. SEPALS 3, distinct or more or less coherent, often keeled. PETALS 3, more or less distinct, æstivation valvate in the ♂ flowers, imbricate-convolute in the ♀. STAMENS hypogynous on a sub-fleshy disk, or perigynous at the base of the perianth, usually 6, 2-seriate, opposite to the sepals and petals, rarely 3 (sp. *Areca,* sp. *Phœnix*), or multiples of 3 (15–30 in *Borassus,* 24–36 in *Lodoicea*), sometimes rudimentary in the ♀ flowers; *filaments* distinct, or united at the base into a tube or cup; *anthers* introrse, or sometimes extrorse, 2-celled, linear, dorsifixed, dehiscence longitudinal. CARPELS 3 (rarely 2–1), distinct, or coherent into a sub-globose or 3-lobed ovary, with 1–3 cells, of which 2 are very often arrested, usually rudimentary in the ♂ flowers; *styles* continuous with the back of the carpels, coherent, or rarely sub-distinct; *stigmas* simple; *ovules* rarely geminate and collateral in each cell, usually solitary, fixed to the central angle a little above the base, sometimes orthotropous with the micropyle superior, sometimes more or less anatropous with the micropyle inferior, or facing the wall of the ovary. FRUIT sometimes 3-2-1-celled, 3-1-seeded, sometimes 3-lobed, sometimes composed of 3 distinct carpels, accompanied at the base with the persistent and usually hardened perianth. BERRY or DRUPE with smooth or scaly epicarp; *sarcocarp* fleshy, and sometimes oily or fibrous; *endocarp* membranous, fibrous, woody, or bony. SEED oblong, ovoid or spherical, erect or pendulous laterally; *testa* often adhering to the endocarp; *albumen* copious, cartilaginous, horny or sub-woody, dry or oily, homogeneous or ruminate. EMBRYO pressed against the periphery of the seed, and covered with a thin layer of albumen, turbinate or conical or cylindric.

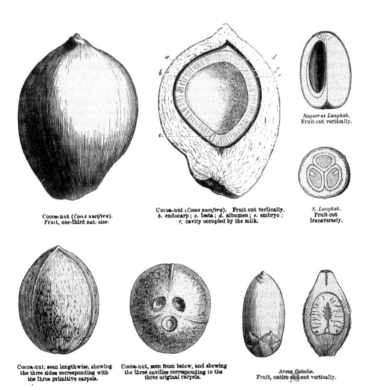

Cocoa-nut (*Cocos nucifera*).
Fruit, one-third nat. size.

Cocoa-nut (*Cocos nucifera*). Fruit cut vertically.
b. endocarp; *c.* testa; *d.* albumen; *e.* embryo;
f. cavity occupied by the milk.

Saguerus Langkab.
Fruit cut vertically.

S. Langkab.
Fruit cut
transversely.

Cocoa-nut, seen lengthwise, showing
the three sides corresponding with
the three primitive carpels.

Cocoa-nut, seen from below, and showing
the three cavities corresponding to the
three original carpels.

Areca Catechu.
Fruit, entire and cut vertically.

Cucifera thebaica.
Fruit cut vertically.

Chamærops.
Fruit.

Chamærops.
Seed, entire and cut vertically.

Chamærops humilis.
Cluster of fruits, natural size.

Seaforthia elegans.

Chamædorea latifolia.

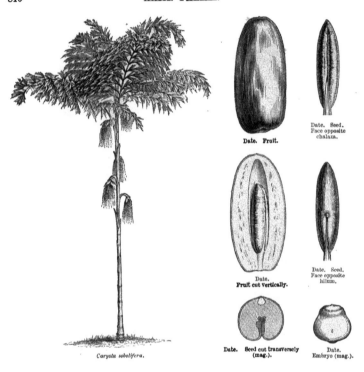

Caryota sobolifera.

Date. Fruit.

Date. Seed.
Face opposite
chalaza.

Date.
Fruit cut vertically.

Date. Seed.
Face opposite
hilum.

Date. Seed cut transversely
(mag.).

Date.
Embryo (mag.).

TRIBE I. *ARECINEÆ.*

Trees or shrubs with pinnate, pinnatifid or bipinnate leaves; pinnules with
curved margins. Spathe polyphyllous, rarely monophyllous, very rarely 0. Flowers
monœcious or diœcious, sessile on a smooth foveolate or bracteolate rachis.
Stamens hypogynous. Fruit deeply 3-lobed, a berry or drupe. Albumen homogeneous or ruminate. Embryo usually basilar.

PRINCIPAL GENERA.

* Chamædorea.	* Areca.	Harina.	Hyospathe.	Pinanga.
Iriartea.	Hyophorbe.	Kentia.	Ceroxylon.	Œnocarpus.
Seaforthia.	* Arenga.	Oreodoxa.	Oranin.	* Caryota.

Tribe II. *CALAMEÆ.*

Sarmentose or arborescent plants. Leaves pinnate or palmate-flabellate, often terminating in a long appendage armed with hooks; pinnules with decurved margins. Spathe usually polyphyllous, rarely monophyllous. Spadices branched. Flowers usually diclinous, sessile; bracts and bracteoles enveloping the flowers, and simulating an amentaceous inflorescence. Stamens hypogynous or perigynous. Fruit a berry covered with imbricate quadrate scales, which are at first erect, then recurved; albumen homogeneous or ruminate. Embryo lateral or sub-basilar.

PRINCIPAL GENERA.

   * Calamus.   Plectocomia.   Zalacca.   Dæmonorops.   * Sagus.   * Mauritia.

Tribe III. *BORASSINEÆ.*

Trees with palmate-flabellate or pinnate leaves; pinnules of the flabellate with erect margins. Spathes woody or fibrous, reticulate (*Manicaria*), imperfect and sheathing the base of the spadices, or perfect, and completely enveloping them. Flowers usually diœcious, the ♂ nearly glumaceous in texture, sunk in the pits formed on the spadix by the union of the bracts, and presenting an amentaceous appearance. Stamens hypogynous. Fruit a drupe, rarely a berry; albumen homogeneous. Embryo usually apical.

PRINCIPAL GENERA.

Borassus.   Hyphæne.   Manicaria.
* Latania.   Geonoma.   Bentinckia.

Tribe IV. *CORYPHINEÆ.*

Trees or stemless plants. Leaves usually palmate-flabellate, very rarely pinnate (*Phœnix*); pinnules with erect

Rhapis flabelliformis.

3 G

Livistonia australis.   *Phœnix dactylifera.*

margins. Spathes imperfect, or rarely perfect. Flowers sessile, usually ⚥, or polygamo-diœcious. Stamens hypogynous or perigynous. Fruit a berry; albumen homogeneous or ruminate. Embryo dorsal.

PRINCIPAL GENERA.

Corypha.	Brahea.	* Trachycarpus.	* Livistona.
Copernicia.	* Rhapis.	* Licuala.	Sabal.
Thrinax.	Saribus.	* Chamærops.	* Phœnix.

Tribe V. *COCOINEÆ.*

Large or small trees. Leaves pinnate; pinnules with decurved margins. Trunk thorny or not. Flowers at first enclosed in the spathe, diclinous, bracteolate, sessile, or sunk in pits formed by the union of the bracts. Stamens hypogynous, filaments confluent at the base. Fruit a drupe, sarcocarp fibrous or oily (*Elaeis*), endocarp thick, woody, marked with 3 scars, 1 of which corresponds to the embryo. Seed oily; albumen homogeneous. Embryo basilar.

PRINCIPAL GENERA.

Desmoncus.	Acrocomia.	* Cocos.	Bactris.
Astrocaryum.	Diplothemium.	Gulielma.	Attalea.
Maximiliana.	Martinezia.	* Elaeis.	* Jubæa.

Palms are not closely allied to any of the families in the division to which they belong; although *Cyclantheæ, Nipaceæ* and *Phytelephasieæ* approach them in their sheathing flabellate or pinnate leaves and inflorescence. Endlicher notices some resemblance in habit between the Palms and the arundinaceous Grasses, but R. Brown does not admit this analogy, and rather considers Palms as near *Junceæ* through *Xerotes* and *Flagellaria.*

Palms belong [almost] exclusively to the torrid zone and to the hottest regions of the temperate zone. The species which extend furthest from the equator do not advance higher than 44° north latitude, and 39° south latitude (*Areca sapida,* New Zealand), and these are but few in number. Even in the tropics they are very unequally distributed, and are most abundant where great heat is accompanied by great humidity. They are numerous in India and its Archipelago: they abound in Central America; but they are comparatively rare in Africa on account of the long droughts. One alone, the Dwarf Palm (*Chamærops humilis*), is a native of South Europe; it is more abundant on the neighbouring shores of Africa, forming a link between the Mediterranean and the sub-tropical regions of Africa. The Date (*Phœnix dactylifera*) is a native of Arabia and Africa. Certain Palms are social, and occupy immense tracts; some grow in inundated savannahs (*Iriartea*), others in arid plains.

The trunk of Palms is very various in size; that of *Oreodoxa frigida* is scarcely equal to a small reed, while that of *Jubæa* measures more than three feet in diameter. Certain species are stemless; others rise to 250 feet in height. The number of flowers borne by a Palm is sometimes prodigious; 12,000 male flowers have been counted in a single Date spathe; 207,000 in a spathe, and 600,000 in a single individual of *Alfonsia amygdalina.*

The family of Palms, of which nearly 1,000 species are now known, ranks next to *Gramineæ* in utility. Perhaps there is no species for which a use is not found in domestic or industrial economy. All yield textile fibres, useful especially in the manufacture of paper; their large leaves serve to thatch houses, and cut in strips are made into cords, mats, baskets, hats and various utensils. The wood of many is used in house building. Some contain an edible starch in their trunks, others have a sugary fermentable sap; certain species are valuable for their fruit, others for the oil in their seed or pericarp. In many Palms the central bud is, when young, a highly esteemed vegetable. Finally, Palms yield several interesting medicinal products. *Sagus Rumphii, lævis* and *genuina,* which grow in the Moluccas, contain a very nourishing farinaceous pith, known as Sago; and several *Mauritiæ,* from tropical America, rival the Sago Palm in this respect. *Arenga saccharifera, Corypha umbraculifera, Borassus flabelliformis, Cocos nucifera, Sagus Rumphii, Raphia, Mauritia vinifera* [and some species of *Caryota* and *Phœnix*], possess an abundant sap, from which sugar is extracted, or which is converted by fermentation into an alcoholic drink known as Toddy, Palm Wine, Laymi and Arrack.

The Date Palm (*Phœnix dactylifera*) is a diœcious tree. Each female flower produces three berries, of which two are usually arrested; their solid flesh, of a vinous, sugary taste, somewhat viscous, serves as food to the Negros and Arabs who live in the Belad-el-Djerid, or Date country, situated to the south of the Atlas, and extending from Morocco to Tunis.

The Cocoa-nut (*Cocos nucifera*), which affects the coasts throughout the tropics, is called by some travellers the King of Plants, a name expressive of its great value. Its stem and leaves, and the fibres which accompany them, and its fruit, suffice for all the wants of the inhabitants of the torrid zone; it yields sugar, milk, solid cream, wine, vinegar, oil, cordage, cloth, cups, wood for building, thatch, &c.

The Cabbage Palm is an *Areca,* the 'cabbage' being its central bud; but many other Palms yield a cabbage much larger and more sapid; these are *Cocos nucifera, Arenga saccharifera, Maximiliana regia,* and all the species of *Attalea*. A sort of cabbage is also obtained from *Chamærops humilis*. *Elaeïs guineensis*, a large monœcious Palm of West Africa, which is cultivated in America, bears a drupe, the sarcocarp of which contains a yellow scented oil, called Palm Oil, used in Africa and Guiana as olive oil; the kernel also yields a white solid oil, used as butter: this latter, much less abundant than the other, is not imported into Europe; but the first, which always remains liquid in the tropics, is imported into France and England, where it arrives congealed, and is used in making soap [and candles]. *Ceroxylon andicola*, a magnificent species growing in Peru, and *Corypha cerifera*, named in Brazil *Carnaüba*, produce a true wax, which exudes from the leaves, and especially from the trunk, at the rings. The Double Cocoa-nut (*Lodoicea sechellarum*) is a very tall tree [confined to the Seychelles Islands], the enormous 2-lobed fruit of which was formerly in great repute as a universal antidote; it is now only an object of curiosity.

Areca Catechu, a large Palm of India, Ceylon and the Moluccas, produces the Areca-nut, from the seed of which is prepared a much esteemed astringent juice, which is chewed, mixed with quicklime and the leaves of Betel Pepper, by the inhabitants of tropical Asia (see p. 731).

From the leaves of all Palms are made more or less coarse hats, for which the young leaves are used, being carefully cut before they unfold, and while still whitish and supple; the leaves of *Corypha* are preferred for this purpose. The fibrous husk of the Cocoa-nut is used for making cords; and the other parts of several Palms also yield fibres with which cordage is manufactured. The Piaçaba[1] is the most important for ships' cables, as it does not decay in water; mattresses, brushes and brooms are also made of it. The species which produce the Piaçaba are *Leopoldinia Piaçaba* and *Attalea funifera*. In Brazil they obtain from the leaves of several species of *Bactris*, especially *B. setosa*, a textile matter named *tecum*, finer and more tenacious than hemp, of which fine hammocks and fishing-nets are made. M. Marius Porte, in a notice of the uses of some Palms, tells us that this thread is not used for garments, on account of a sort of rasping property, which causes it to cut like a file or sandpaper, excoriating the skin, and if worn with other clothes quickly rubbing them to pieces. With a thread of tecum and patience, says M. Porte, a bar of iron may be cut.

The Rattans (*Calamus*), or Cane Palms, have a very slender stem, scarcely as thick as the thumb; this stem, in some species, climbs up trees, sometimes attaining a length of 12,000 to 18,000 feet[2] (*Rumphia*, vol. ii. p. 158). The flexible stems are sent to Europe, where they are used for various light and solid articles, trellised furniture, switches, canes (known as Dutch canes), &c. The fruit of *Calamus Draco* is impregnated with a red astringent resin named Dragon's Blood, much more used by druggists than the Dragon's Blood of the *Pterocarpus* or the *Dracœna*. The roots of *Sabal Palmetto* are very rich in tannin.

The sap of *Corypha umbraculifera* and *sylvestris*, Asiatic species, is an emetic, and considered an alexipharmic. *Hyphæne cucifera*, an Egyptian Palm, remarkable for the dichotomy of its stem, yields a gum-resin (Egyptian bdellium), formerly classed among diuretics, and the sarcocarp of its fruit tastes like gingerbread.

This elegant family forms the principal ornament of our southern gardens, and with care some species may even be grown in the climate of Paris. The Dwarf or Fan Palm (*Chamærops humilis*), mentioned above, is a small polygamous tree, stemless or caulescent, abundantly spread over Sicily, Italy, Spain, and Algeria, and which lives in the open air in south-eastern France. The Chusan Palm (*Trachycarpus* or *Chamærops excelsa*), a diœcious tree a foot to 18 inches high, is less picturesque; but hardier than the preceding; its trunk is furnished with a sort of tow or hair, resulting from the decomposed bases of the petioles; this tow, used by the Chinese in the manufacture of cordage and of coarse stuffs, forms a natural clothing to this Palm-tree, and protects it so much from the cold as to enable it to stand the winter in the gardens of Provence and Languedoc, as well as on the coast from Bordeaux to Cherbourg, and even in the Isle of Wight.

Some North American dwarf Palms are also cultivated; the best known is the *Sabal Adansonii*, a stemless species, hardy in the south of France. Another, and a greatly preferable one, and equally hardy, is the *Chamærops Hystrix*, a caulescent species, of which the stem, bristling with sharp points, rarely attains the height of three feet. The species first introduced into Europe, probably by the Arabs, is the Date, *the* tree of the African oases, without which the Sahara would be uninhabitable. It was

---

[1] Coci, or Cocoa-nut fibre, is probably here referred to.—ED.

[2] This has not been verified. Three hundred feet is not an uncommon length in Ceylon and the Malay islands.—ED.

cultivated in biblical times, and its origin is unknown, although it may be reasonably supposed to be indigenous to Arabia. But even in ancient times it was cultivated in South Persia, Egypt, and North Africa, whence it was much later introduced into the south of Europe. Its fruits only acquire all their qualities under the torrid and dry sky of desert regions. The best come to us from the oases of the Central Sahara; those of second quality from the northernmost oases of Algeria and Tunis. Dates little inferior to these last are still gathered in the environs of the town of Elche, in Spain, between 38° and 39° north latitude; but this is the extreme north limit of Date culture, considered as a fruit tree. Above this point the pulp remains more or less acid, and the Date is only an ornamental tree. On the coast of Liguria it is cultivated for the sake of the leaves, which are used by Roman Catholics in the ceremonies of Palm Sunday, as well as at the Jewish Passover. It is common on the shores of Provence, between Toulon and Nice; it appears to be about as hardy as the Orange, for it dies wherever the latter is killed by the cold.

XXIV. PHYTELEPHASIEÆ, Brongniart.

Palm-like stemless or caulescent PLANTS. LEAVES very long, pinnate, crowded at the top of the [very short inclined] stem; *leaflets* decurved at the base. FLOWERS monœcious or polygamo-diœcious; *spathe* monophyllous (*Phytelephas*) or diphyllous (*Wettinia*); *spadices* simple, clavate or cylindric, densely covered with flowers. PERIANTH-LEAFLETS 2-seriate, unequal, æstivation imbricate or valvate. STAMENS ∞, inserted at the base of the perianth; *anthers* linear or oblong, apiculate by the connective, 2-celled, opening longitudinally. OVARY of 4 1-ovuled cells (*Phytelephas*), or 1-celled and 1-ovuled (*Wettinia*); *style* terminal, divided into 5–6 stigmatic branches (*Phytelephas*), or basilar, lateral and 3-fid (*Wettinia*); *ovule* basilar, ascending, anatropous. DRUPES aggregated, angular, muricate, 4-celled; *endocarp* crustaceous, simulating a rounded cone (*Phytelephas*), or berry, coriaceous, 1-celled, 1-seeded (*Wettinia*). SEEDS with coriaceous or membranous testa; *albumen* copious, ivory-like; *radicle* near the hilum.

GENERA.

Phytelephas.   Wettinia.

Phytelephasieæ are near *Pandaneæ* and *Cyclantheæ*. *Wettinia*, by its one-celled ovary and anatropous ovule, forms the passage to Palms.

This little group belongs to Peru [Ecuador and New Grenada]. The ivory-like albumen of *Phytelephas* is edible when young; it hardens so much when ripe that it is used as a substitute for elephants' tusks, whence its common name of Vegetable Ivory.

XXV. NIPACEÆ, Brongniart.

A Palm-like PLANT. TRUNK unarmed, thick, short, spongy within. LEAVES terminal, large, pinnatisect; *pinnules* narrow, erect, stiff, plaited. SPADIX monœcious, terminal, sheathed in a polyphyllous spathe, persistent, at first erect, afterwards drooping. FLOWERS ♂ minute, yellowish, each with a bract, and united into lateral cylindric catkins; the ♀ agglomerated into a terminal capitulum.—FLOWERS ♂: SEPALS 3. PETALS 3, valvate in æstivation. STAMENS 3, with coherent filaments;

anthers adnate, extrorse, sub-didymous.—FLOWERS ♀: PERIANTH 0. PISTIL composed of 3 distinct carpels, obliquely truncate, angular, with scales at the base; *stigmas* 3, sessile, excentric, marked with a lateral slit; [*ovule* ascending, solitary, anatropous]. DRUPES of a chesnut brown, forming by their aggregation a voluminous capitulum, turbinate, angular, 1-seeded; *sarcocarp* thick, dry, fibrous; *endocarp* fibrous-woody, perforated at the base. SEED furrowed longitudinally by a projection of the kernel; *albumen* homogeneous, cartilaginous, hollow in the centre. EMBRYO basilar.

ONLY GENUS.

Nipa.

Nipa, like *Phytelephasieæ*, is near *Pandaneæ*, *Cyclantheæ* and Palms. The only species nhabits the swampy estuaries of the large rivers of India and the Moluccas.

The seed germinates within the fruit, which falls into the sea, by which it is carried away; but it only detaches from the spadix after several years, when the germination of the seed is so far advanced that the salt water cannot hurt the embryo. The seeds are edible before they are quite ripe, their insipidity being corrected by sugar. The inhabitants of the Philippines and of Cochin China obtain from the spadix a juice which yields by fermentation a spirituous liquor and acetic acid. The fronds are used for covering huts in the Indian Archipelago; hats and cigar-cases are also made of them. [The fruits of *Nipa* abound in the Eocene formations at the mouth of the Thames.]

XXVI. *PANDANEÆ, Br.*

Frutescent or arborescent PLANTS. STEM simple or branched, ringed, supported on strong adventitious roots; *branches* leafy at the extremity. LEAVES imbricate in 3 rows, in very close spirals, linear-lanceolate, amplexicaul, edges often spinous. FLOWERS diœcious, achlamydeous, completely covering simple or branched spadices, accompanied by herbaceous or coloured caducous spathes.—FLOWERS ♂: SPADIX branched, thyrsoid, in tufts or large catkins. STAMENS numerous, very dense; *filaments* filiform, isolated or in bundles; *anthers* 2-celled, dehiscence longitudinal.— FLOWERS ♀: SPADIX simple. OVARIES numerous, 1-celled, coherent in bundles, or rarely isolated; *stigmas* sessile, distinct; *ovule* solitary, anatropous, adnate to a parietal placenta (*Pandanus*); or ovules 3, orthotropous (?), inserted at the base of the cell (*Souleyetia*). FRUIT formed of fibrous drupes united in closely cohering groups; *endocarp* bony. SEED ovoid; *testa* membranous; *raphe* filiform, scarcely projecting, *chalaza* very conspicuous, reaching the top of the cell; *albumen* fleshy, dense. EMBRYO straight, basilar; *radicle* directed towards the bottom of the cell.

PRINCIPAL GENERA.

Pandanus.   Souleyetia.   Heterostigma.

Pandaneæ, properly so called, which are joined by several botanists to *Freycinetieæ*, *Cyclantheæ*, *Phytelephasieæ* and *Nipaceæ*, form with these small families and with *Typhaceæ* a group approaching Palms and Aroids. *Nipaceæ*, *Phytelephasieæ*, *Cyclantheæ* and *Freycinetieæ* resemble Palms in habit. *Pandaneæ* inhabit the shores [and interior] of Asia, Arabia Felix, the large islands of the Pacific Ocean, Madagascar, West Africa, North Australia, &c. [The number of species in Mauritius, where they are

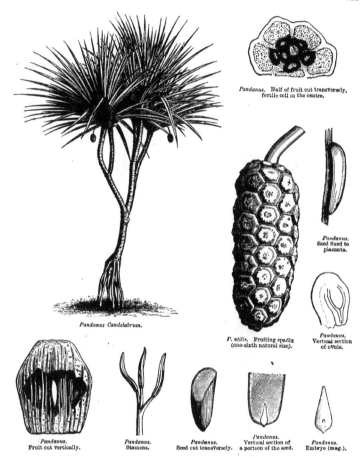

Pandanus. Half of fruit cut transversely, fertile cell in the centre.

Pandanus. Seed fixed to placenta.

P. utilis. Fruiting spadix (one-sixth natural size).

Pandanus. Vertical section of ovule.

Pandanus. Fruit cut vertically.

Pandanus. Stamens.

Pandanus. Seed cut transversely.

Pandanus. Vertical section of a portion of the seed.

Pandanus. Embryo (mag.).

Pandanus Candelabrum.

called Vacoas, is a remarkable botanical feature of that island.] Their habit is similar to that of gigantic *Bromeliaceæ*; one of the most remarkable is *Pandanus Candelabrum*, which owes its specific name to the elegance of its dichotomous ramification. Mungo Park relates a singular phenomenon presented by another species named *Fang-Jani* (*Heterostigma Heudelotianum*), which grows at Goree, on the Gambia, and which, according to this celebrated traveller, burns spontaneously when ripe; it is now known that this apparent combustion is due to a disease arising from the growth of a parasitic Fungus (*Fumago*), which covers the leaves with a black powder analogous to charcoal.

Gaudichaud has divided *Pandanus* into several genera, the characters of which appear to depend on the form of the stigmas ; these genera not being described, but only figured, we shall enumerate here only *Souleyetia* and *Heterostigma*, which are more distinctly characterized.

The male flowers of *Pandanus* have a sweet but very penetrating odour. The leaves are used for making sacks, in which Bourbon coffee is sent to Europe. The juice of certain *Pandani* is recommended as an astringent in dysentery ; the young fruit is considered an emmenagogue.

XXVII. *TYPHACEÆ*.

(TYPHÆ, *Jussieu.*—TYPHINÆ, *Agardh.*—TYPHACEÆ, *D.C.*)

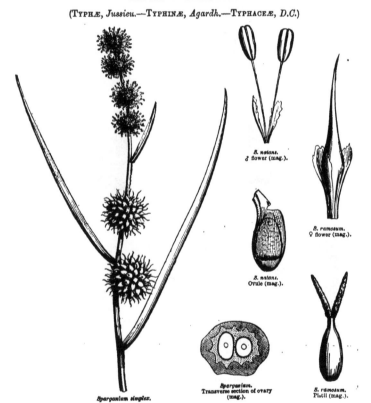

S. natans.
♂ flower (mag.).

S. ramosum.
♀ flower (mag.).

S. natans.
Ovule (mag.).

Sparganium.
Transverse section of ovary
(mag.).

S. ramosum.
Pistil (mag.).

Sparganium simplex.

Sparganium.
Ovary cut verti-
cally (mag.).

Sparganium. Ovarian cell
opened, showing the seed with
its outer raphe (mag.).

S. natans.
Fruit cut, with the
four scales (mag.).

S. natans.
Seed
(mag.).

S. natans.
Embryo
(mag.).

FLOWERS *monœcious, in a spike or heads, seated on a spadix.* PERIANTH 0.
STAMENS *accompanied by bristles or scales.* OVARIES *1-2-celled, 1-ovuled;* OVULE
pendulous, anatropous. FRUIT *dry or fleshy.* SEED *albuminous.* EMBRYO *axile; or*
RADICLE *superior.*—STEM *herbaceous.* LEAVES *alternate.*

Aquatic or marsh HERBS, perennial, with creeping rhizome. STEMS cylindric,
not knotty, solid, simple or branched. LEAVES alternate, linear, striate, entire,
gathered at the base of the stem, sheathing; the cauline subtending the spadices,
or forming an involucre before flowering. FLOWERS imperfect, seated on a monœcious spadix, in heads (*Sparganium*), or dense spikes, sometimes continuous and
furnished at intervals with foliaceous very caducous spathes, sometimes interrupted
(*Typha*); the upper flowers staminiferous, the lower pistilliferous.—FLOWERS ♂:
PERIANTH 0. STAMENS numerous, accompanied by bristles or membranous scales;
anthers basifixed, oblong; cells 2, often shorter than the connective (*Typha*), opening
longitudinally.—FLOWERS ♀: PERIANTH 0. OVARIES accompanied by bristles or
scales, sessile (*Sparganium*), or on long stalks when ripe (*Typha*), with 1-2 1-ovuled
cells; *style* continuous with the ovary, simple; *stigma* unilateral, papillose, lingui-
form, elongated; *ovule* pendulous from the top of the cell, anatropous. FRUITS sub-
drupaceous or dry, angular, tipped by the style; *epicarp* membranous or sub-spongy;
endocarp sub-woody, indehiscent (*Sparganium*), or split on one side when ripe, and
endocarp coriaceous (*Typha*). SEED inverted; *albumen* farinaceous or fleshy, copious.
EMBRYO straight, axile; *radicle superior.*

TRIBE I. *SPARGANIEÆ.*

Flowers agglomerated on a hemispheric receptacle in capitula with foliaceous
bracts. Spadix branched or simple. Stamens accompanied by membranous scales;
ovaries 1-2-celled, each with 3-4 imbricate and persistent scales. Fruit an
indehiscent drupe, 1-2-celled, epicarp spongy. Seed ovoid, testa smooth, raphe
external.

ONLY GENUS.

Sparganium.

Tribe II. *TYPHEÆ.*

Flowers in a compact cylindric spike. Spadix simple. Stamens springing from the spadix, accompanied by numerous bristles. Ovaries 1-celled, inserted on small

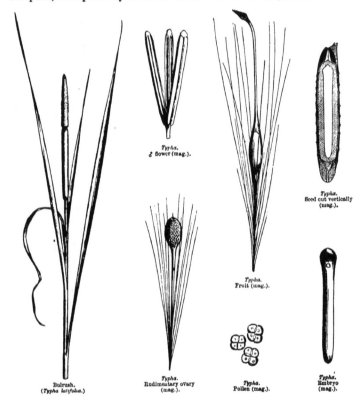

Typha.
♂ flower (mag.).

Typha.
Seed cut vertically
(mag.).

Typha.
Fruit (mag.).

Bulrush.
(*Typha latifolia.*)

Typha.
Rudimentary ovary
(mag.).

Typha.
Pollen (mag.).

Typha.
Embryo
(mag.).

protuberances of the rachis, on long stalks when ripe, accompanied by numerous bristles, and by clavate rudimentary ovaries. Fruit dry, epicarp split on one side. Seed linear, testa striate.

ONLY GENUS.
Typha.

We have thought it best to divide *Typhaceæ* into two tribes, distinguished by habit, inflorescence, and the structure of the male and female flowers, and of the fruit and seed. The ovule of *Typha* has been figured as orthotropous, but it is evidently anatropous; the micropyle faces the top of the cell, and the chalaza is turned towards the bottom.

Typhaceæ are allied on the one hand to *Aroideæ* and *Cyperaceæ*, and on the other to *Pandaneæ*, from which *Sparganium* only differs in its small stature, more simple fruit, and pendulous seed; this resemblance is so striking that one is tempted to regard *Pandaneæ* as gigantic Sparganiums. This family contains few species: *Typha* is dispersed over the tropical and temperate regions of the whole world, and principally of the northern hemisphere, where its species inhabit stagnant water and the sides of streams. *Sparganium* prefers cold or temperate regions.

The starchy rhizome of *Typha* possesses slightly astringent and diuretic properties, which lead to its use in East Asia for the cure of dysentery, urethritis and aphthæ. The stems and leaves are used for thatching cottages. It has been vainly tried to utilize the bristles of the spike in the manufacture of a sort of velvet. The seeds of *Sparganium* are eaten by water-birds. [The pollen of *Typha* is made into bread by the natives of Scind and New Zealand.]

XXVIII. *CYCLANTHEÆ*, Poiteau.

Stemless or caulescent PLANTS. STEM sub-woody, often climbing by means of adventitious epiphytic roots. LEAVES cauline or radical, petioled, alternate or

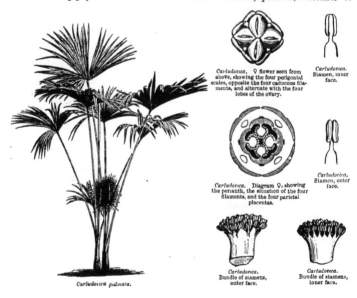

Carludovica. ♀ flower seen from above, showing the four perigonial scales, opposite the four caducous filaments, and alternate with the four lobes of the ovary.

Carludovica. Stamen, inner face.

Carludovica. Diagram ♀, showing the perianth, the situation of the four filaments, and the four parietal placentas.

Carludovica. Stamen, outer face.

Carludovica. Bundle of stamens, outer face.

Carludovica. Bundle of stamens, inner face.

Carludovica palmata.

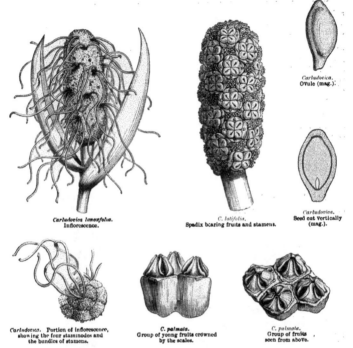

Carludovica lanexfolia.
Inflorescence.

C. latifolia.
Spadix bearing fruits and stamens.

Carludovica.
Ovule (mag.).

Carludovica.
Seed out vertically
(mag.).

Carludovica. Portion of inflorescence,
showing the four staminodes and
the bundles of stamens.

C. palmata.
Group of young fruits crowned
by the scales.

C. palmata.
Group of fruits
seen from above.

alternate-distichous, coriaceous, entire or 2–3–5-partite; *nerves* parallel or oblique, flabellate. SPATHES 4- to 3-phyllous, imbricate. SPADIX monœcious, cylindric. FLOWERS densely covering the spadix; the ♂ grouped in 4 bundles accompanying the ♀ (*Carludovica*), or ♀ and ♂ in alternating spirals (*Cyclanthus*). FLOWERS ♂: PERIANTH multifid; *lobes* very short, irregularly 2-seriate, imbricate in æstivation (*Carludovica*), or 0 (*Cyclanthus*). STAMENS in 4 bundles opposite to the lobes of the perianth; *filaments* short, slightly dilated (*Carludovica*), or fragile (*Cyclanthus*); *anthers* oblong or linear, 2-celled, dehiscence longitudinal.—FLOWERS ♀: PERIANTH 0 (*Cyclanthus*), or of 4 fleshy valvate herbaceous or coloured scales, each with a long caducous filament (*staminode*) at the base, the remains of which are inconspicuous (*Carludovica*). OVARY 2-4-lobed at the top, 1-celled, ∞-ovuled, with 4 parietal placentas; *stigmas* small, sessile, with 2 antero-posterior lobes (*Cyclanthus*), or linear and answering to the placentas (*Carludovica*); *ovules* anatropous, sessile (*Carludovica*), or on long funicles (*Cyclanthus*). FRUIT a syncarpous berry, of fleshy ♀ flowers; bark of the spadix fructiferous, bursting at the base into 3–4 irregular fleshy strips, which roll up by degrees towards the top of the spadix, and retain the berries fixed in their pulp; these soon deliquesce and leave the seeds (*Carludovica palmata*). SEEDS numerous; *testa* soft or coriaceous, filled with raphides; *raphe* often thickened; *albumen* horny. EMBRYO small, straight, cylindric, basilar; *radicle* near the hilum.

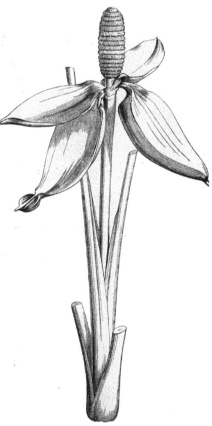

Cyclanthus bipartitus. Inflorescence.

GENERA.

Cyclanthus.　　*Carludovica.

Cyclantheæ, which are closely allied to *Pandaneæ* and *Freycinetieæ*, are equally near *Aroideæ* and Palms. They are exclusively tropical American.

The flowering spadices of several *Cyclanthi*, and especially of *C. bipartitus*, cultivated by the natives of the province of Maynas, in Brazil, have a sweet scent, between that of vanilla and cinnamon. The Indians cook them with meat as an aphrodisiac. Pœppig has observed that these spadices are never attacked by fructivorous animals, not even by the numerous species of ants, usually so fond of succulent fruits.

Carludovica palmata, which grows in the damp forests of Ecuador, Peru, and New Grenada, yields a much-valued straw, from which are manufactured Guayaquil or Panama hats. Weddell remarks that the young leaves are gathered in bud, while still scarcely tinged with green; the fan-shaped blade is so cut into strips as only to preserve the middle part of the divisions of the blade, which remains attached to the petiole, and the size of which varies according to the fineness of the work for which it is intended; the leaf thus prepared is steeped successively in boiling water, in water acidulated with lemon-juice, and in very cold water, and allowed to dry; the bleaching is then perfect. In drying, the edges of each strip become revolute, giving it a cylindrical shape, which greatly increases its solidity. The price of these hats varies singularly; the commonest are sold for 1s. 6d., those of medium quality are worth 5s.; a fine one will fetch 3l. 2s. 6d. to 5l. 4s. 2d.; but some are made of so fine a tissue that they are sold at the enormous price of 20l. 16s. 8d. The straw of *Carludovica* is also used in the manufacture of cigar-cases.

XXIX. *FREYCINETIEÆ*, Brongniart.

Freycinetia Banksii.
Vertical section of a group of stamens enclosing the abortive pistil.

Freycinetia imbricata.
Transverse section of fruit (mag.).

F. Banksii.
Ovary accompanied by sterile stamens.

Freycinetia.
Seed, entire and cut vertically (mag.).

STEM often rooting or sarmentose, rarely arborescent, with the habit of *Pandanus*. LEAVES narrow, sheathing, amplexicaul below, with parallel nerves, denticulate or sub-spinous on their edge and dorsal face, equitant in æstivation. INFLORESCENCE terminal, rarely lateral; *spathes* usually yellow or red. SPADIX polygamo-diœcious, simple, entirely covered with naked flowers.—FLOWERS ♂ in a tuft, often grouped around an abortive ovary. STAMENS ∞; *filaments* filiform, isolated or in groups of 2–3; *anthers* 2-celled.—FLOWERS ♀: OVARIES numerous, 1-celled, many-ovuled, accompanied at the base with sterile stamens, which are isolated or agglomerated in bundles of 3–4–∞; *stigmas* sessile, distinct; *ovules* anatropous, ascending, 2-seriate on 3 parietal placentas, linear, alternating with the stigmas. BERRIES aggregated, 1-celled, several-seeded. SEEDS minute, sunk in a colourless pulp, erect on short funicles; *testa* membranous, smooth or striate; *raphe* lateral, more or less developed, and strophiolate; *albumen* fleshy, dense. EMBRYO minute, straight; *radicle* near the hilum, and inferior.

ONLY GENUS.
Freycinetia.

Freycinetieæ are distinguished from *Pandaneæ* by their ovaries with three many-ovuled placentas, and the baccate lower part of the fruit. Like *Pandaneæ* they inhabit the large islands of the [Malayan Archipelago,] Pacific Ocean, Norfolk Island, New Zealand and North Australia.

XXX. *AROIDEÆ.*

(AROIDEÆ, *Jussieu.*—ARACEÆ, *Schott.*—CALLACEÆ, *Bartling.*)

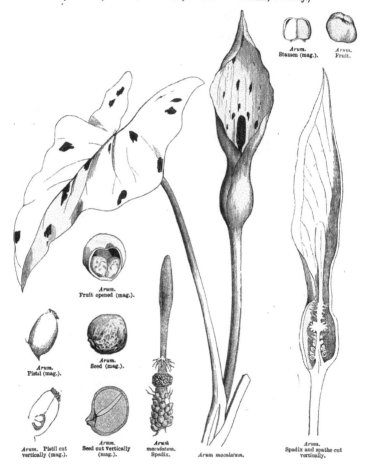

Arum.
Stamen (mag.).

Arum.
Fruit.

Arum.
Fruit opened (mag.).

Arum.
Seed (mag.).

Arum.
Pistil (mag.).

Arum. Pistil cut
vertically (mag.).

Arum.
Seed cut vertically
(mag.).

*Arum.
maculatum.*
Spadix.

Arum maculatum.

Arum.
Spadix and spathe cut
vertically.

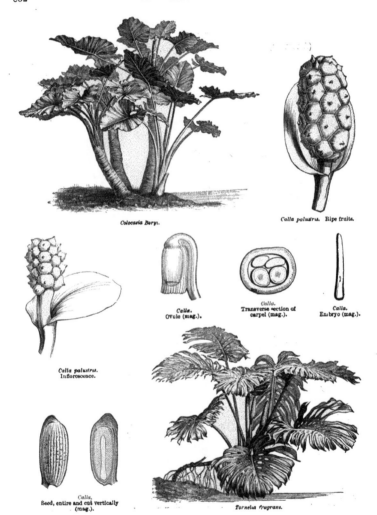

Colocasia Boryi.

Calla palustris. Ripe fruits.

Calla.
Ovule (mag.).

Calla.
Transverse section of carpel (mag.).

Calla.
Embryo (mag.).

Calla palustris.
Inflorescence.

Calla.
Seed, entire and cut vertically (mag.).

Tornelia fragrans.

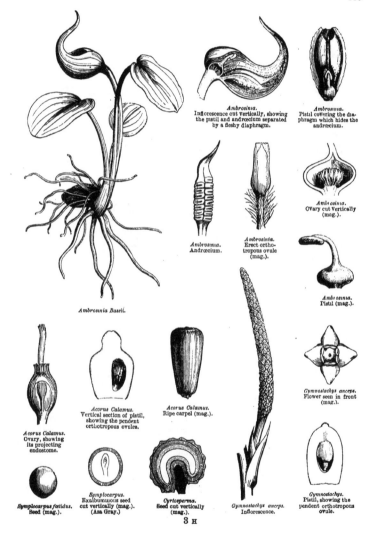

Ambrosinia.
Inflorescence cut vertically, showing the pistil and androecium separated by a fleshy diaphragm.

Ambrosinia.
Pistil covering the diaphragm which hides the androecium.

Ambrosinia.
Androecium.

Ambrosinia.
Erect orthotropous ovule (mag.).

Ambrosinia.
Ovary cut vertically (mag.).

Ambrosinia.
Pistil (mag.).

Ambrosinia Bassii.

Gymnostachys anceps.
Flower seen in front (mag.).

Acorus Calamus.
Ovary, showing its projecting endostome.

Acorus Calamus.
Vertical section of pistil, showing the pendent orthotropous ovules.

Acorus Calamus.
Ripe carpel (mag.).

Symplocarpus fœtidus.
Seed (mag.).

Symplocarpus.
Exalbuminous seed cut vertically (mag.). (Asa Gray.)

Cyrtosperma.
Seed cut vertically (mag.).

Gymnostachys anceps.
Inflorescence.

Gymnostachys.
Pistil, showing the pendent orthotropous ovule.

3 H

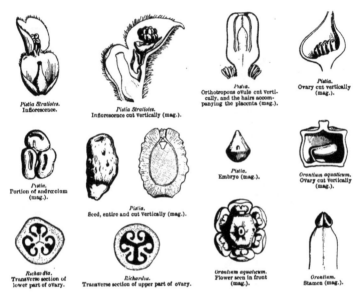

Pistia Stratiotes.
Inflorescence.

Pistia Stratiotes.
Inflorescence cut vertically (mag.).

Pistia.
Orthotropous ovule cut vertically, and the hairs accompanying the placenta (mag.).

Pistia.
Ovary cut vertically
(mag.).

Pistia.
Portion of andrœcium
(mag.).

Pistia.
Seed, entire and cut vertically (mag.).

Pistia.
Embryo (mag.).

Orontium aquaticum.
Ovary cut vertically
(mag.).

Richardia.
Transverse section of
lower part of ovary.

Richardia.
Transverse section of upper part of ovary.

Orontium aquaticum.
Flower seen in front
(mag.).

Orontium.
Stamen (mag.).

FLOWERS *monœcious, or more rarely diœcious, or* ☿ *, inserted on a simple spadix, furnished with a spathe, with or without a perianth.* OVARY *1–several-celled;* OVULES *basilar or parietal, erect, ascending or pendulous, orthotropous, campylotropous or anatropous.* FRUIT *a berry.* SEED *albuminous, or very rarely exalbuminous.* EMBRYO *axile.—Stemless or caulescent* PLANTS. LEAVES *radical or alternate, blade dilated or linear, nerves prominent, reticulate.*

Usually herbaceous PLANTS, with colourless or milky sap, perennial, sometimes with a rhizome or tubers, and then stemless; sometimes caulescent, with straight, branched and arborescent stems marked with petiolar scars, sometimes sarmentose, or climbing by means of adventitious roots; sometimes viviparous (*Remusatia vivipara*), very rarely floating (*Pistia*). LEAVES sometimes solitary, usually terminating an epigeal rhizome or stem, alternate, glabrous; *petioles* sheathing at the base; *blade* usually dilated, strongly palmi-pedati-pelti-nerved, cordate or hastate, entire or variously cut, sometimes perforated or bulbilliferous, vernation convolute. SCAPE or STEM terminated by a spadix. SPATHE monophyllous, variously involute, persistent or deciduous. SPADIX simple, springing from the axil of the spathe, free, or adhering to its midrib, sessile or stipitate, entirely covered with flowers, or terminated by a sterile appendage, very various in form. FLOWERS usually imperfect, rarely ☿, sessile on the spadix, the ♀ usually below, the ♂ above, contiguous, or separated by a naked space, or by rudimentary ovaries, or by staminodes which sometimes succeed the ♂ flowers, and clothe the top of the spadix. PERIANTH 0, or (in the ☿ flowers) 4–5–6–8-phyllous, or 5–8-fid. STAMENS numerous, free or variously coherent; filaments short or obsolete; anthers extrorse, 2-celled, opening either by a longitudinal or transverse slit, or by an apical or sub-apical pore; pollen-grains sometimes agglutinated. OVARIES generally aggregated, distinct or coherent, 1-celled, or 2–3–4–many-celled by prolongation of the parietal placentas, furnished internally with hairs that secrete an abundant mucilage; styles 0 or simple; stigma capitate or discoid, undivided or sometimes lobed (Asterostigma); ovules solitary or more or less numerous, basilar parietal or apical, erect, ascending, sub-horizontal or pendulous, orthotropous or campylotropous, rarely anatropous and with external raphe (Amorphophallus variabilis). FRUIT an indehiscent berry, 1–several-celled, 1–many-seeded. SEEDS sub-globose, oblong or angular; testa coriaceous, thick; albumen farinaceous or fleshy, copious, disappearing at germination, or rarely 0 (Symplocarpus). EMBRYO white or green, axile, turbinate or cylindric, or sometimes slightly angular (Acorus), or very rarely curved (Cyrtosperma).

TRIBE I. CALLACEÆ.

Flowers ☿, or ♂ and ♀ on the same spadix, achlamydeous or not.

SECTION I. ACOROIDEÆ.—Spathe leaf-like, adnate to the peduncle. Flowers ☿, covering the spadix. Perianth 4–6-phyllous. Stamens as many as, and opposite to the sepals. Ovaries 1–3-celled; ovules pendulous, orthotropous. Seeds albuminous. —Herbs, usually containing an aromatic oil (Acorus), rhizome jointed. Leaves ensiform, equitant, sheathing in vernation.

PRINCIPAL GENERA.

* Acorus. Gymnostachys.

SECTION II. ORONTIACEÆ. — Spathe persistent, herbaceous, or sometimes coloured (Dracontium, Symplocarpus), rarely 0 (Orontium). Spadix covered with ☿ flowers. Perianth 4–5–6–8-phyllous, or sometimes 5–8-fid (Dracontium). Stamens as many as, and opposite to the perianth-lobes. Ovaries 1–few-celled; style usually 0, rarely subulate-elongate (Dracontium), or tetragonal-pyramidal (Symplocarpus); ovules sometimes basilar and campylotropous (Pothos), or horizontal and semi-anatropous (Orontium), sometimes pendent and anatropous (Anthurium), or campylotropous (Cyrtosperma, Lasia). Seed usually exalbuminous.—Herbs, rarely aquatic (Orontium), stemless or caulescent, and often sarmentose or climbing (Pothos, Anthurium). Leaves alternate, sometimes jointed (Pothos, &c.), simple, entire (Orontium, Symplocarpus, Pothos), or pinnatisect (Lasia), or palmisect (Anthurium), or pedate (Dracontium). Stipulary sheaths adnate to the petiole, or opposite to the leaves, sometimes alternating with the petiole (Anthurium). Spathe longer or shorter than the spadix.

PRINCIPAL GENERA.

Orontium.	Symplocarpus.	Cyrtosperma.	Arctiodracon.	* Spathiphyllum.
Lasia.	Dracontium.	* Anthurium.	Pothos.	

Section III. Calleæ.—Spathe coloured, persistent (*Calla*), or deciduous (*Monstera, Scindapsus*). Flowers achlamydeous. Spadix sometimes ☿ below, sometimes ♀ below and ☿ above (*Monstera, Scindapsus*); filaments flat; anthers 2-celled, opening longitudinally. Ovaries 1-few-celled; stigma sessile or sub-sessile; ovules erect, anatropous (*Calla, Monstera*), or campylotropous (*Scindapsus*). Seed sometimes exalbuminous (*Scindapsus*).—Terrestrial herbs, sometimes aquatic (*Calla*). Stem elongate, often stoloniform, branched or climbing, and more or less furnished with adventitious roots. Leaves sub-distichous, blade entire or perforated (*Monstera, &c.*). Stipulary sheaths adnate to the petiole, or free and opposite to the leaves.

PRINCIPAL GENERA.

* Calla.   * Monstera.   * Scindapsus.   * Tornelia.

Tribe II. *ARACEÆ.*

Flowers diclinous, achlamydeous, the ♀ on the lower, the ♂ on the upper part of the spadix.

Section IV. Anaporeæ.—Spadix free (*Aglaonema, Richardia*), or adnate to the spathe (*Spathicarpa, Dieffenbachia, &c.*), rarely terminated by a sterile appendage (*Pinellia*). Flowers ♀ and ♂ contiguous, the ♀ usually mingled with staminodes. Anthers free or coherent, sunk in a thick connective, opening by pores. Ovaries 1-3-celled; style short or 0; ovules solitary or numerous, erect or ascending, rarely pendulous (*Richardia*), orthotropous, or rarely anatropous (*Richardia, Aglaonema*).—Herbs with knotted rhizome, stemless or caulescent. Petiolar sheaths elongated, the stipulary 0.

PRINCIPAL GENERA.

* Richardia.   * Pinellia.   Spathicarpa.   * Aglaonema.   * Dieffenbachia.

Section V. Colocasieæ.—Spadix free, terminated by a naked and sterile appendage (*Colocasia, Peltandra, &c.*), or with no appendage (*Caladium, Xanthosoma, Acontias, Syngonium, Philodendron, &c.*). Flowers ♀ and ♂, numerous, usually separated by rudimentary organs. Anthers free or coherent, cells sunk in a thick and peltate connective. Ovaries numerous, free, 1-many-celled, several-ovuled; style short or 0; ovules orthotropous or semi-anatropous. Seeds albuminous.—Herbs with tuberous rhizome, stemless or caulescent, sometimes climbing. Leaf-blade pelti-palmi-nerved. Petiolar sheaths long or short, the stipular 0, or elongated and opposite to the leaves. Spathes usually sweet-scented.

PRINCIPAL GENERA.

* Colocasia.   Peltandra.   Syngonium.   * Alocasia.
* Xanthosoma.   * Philodendron.   * Caladium.   * Acontias.

Section VI. Dracunculineæ.—Spadix free, or rarely adnate to the base of the spathe, monœcious, or very rarely diœcious (*Arisæma*), terminated by a naked clavate appendage (*Arum*), or flagellate (*Arisæma*), or globose and irregular (*Amor-*

phophallus). Flowers ♀ and ♂ numerous, sometimes separated by rudimentary organs. Anthers free or rarely coherent, cells larger than the connective. Ovaries free, 1-celled, 1-several-ovuled; stigma sessile or sub-sessile; ovules orthotropous or very rarely anatropous (*Amorphophallus*), pendulous and erect in the same cell. Seeds albuminous, or very rarely exalbuminous (*Amorphophallus*).—Herbs with a usually tuberous or thick rhizome. Leaves strongly palmi-pelti-nerved, entire, cordate, hastate, sagittate, or palmi-pedati-partite. Spathe coloured, usually violet, glabrous or hairy within and fœtid.

PRINCIPAL GENERA.

| Arisarum. | * Arum. | * Dracunculus. | * Arisæma. | Typhonium. |
| Pythonium. | Biarum. | Sauromatum. | * Amorphophallus. | |

SECTION VII. CRYPTOCORYNEÆ.—Spadix included and jointed to the spathe by its top (*Cryptocoryne*), or projecting and free (*Stylochæton*). Flowers ♀ numerous, the lower separated from the ♂. Anthers sessile or sub-sessile at the top of the spadix. Carpels numerous, whorled around the base of the spadix, and united into a several-celled ovary; styles as many as carpels; ovules ascending, orthotropous. Rudimentary organs 0 or indistinct. Seeds albuminous.—Marsh herbs (*Cryptocoryne*), or growing in sand (*Stylochæton*), rhizome stoloniferous. Leaves sub-1-nerved, or palmi-nerved, entire, lanceolate, emarginate at the base or sagittate.

PRINCIPAL GENERA.

| Cryptocoryne. | Stylochæton. | Lagenandra. |

SECTION VIII. PISTIACEÆ.—Spadix adnate to the spathe. Flower ♀ solitary, separate from the ♂ flowers. Anthers sessile at the top or side of the spadix. Ovary 1-celled, many-ovuled; styles distinct; ovules basilar or sub-lateral, erect, orthotropous. Rudimentary organs 0. Seeds albuminous.—Aquatic floating herbs, stoloniferous, or terrestrial with tuberous rhizome. Leaves entire, several-nerved.

PRINCIPAL GENERA.

| * Pistia. | Ambrosinia. |

Aroideæ, in spite of their polymorphism, form a very homogeneous group; they have an obvious affinity with *Typhaceæ* and *Pandaneæ*; the former differing in the structure of the anthers, the latter in the conformation of the fruit. *Pistia* approaches *Lemnaceæ*, the anthers and seeds of which present so close an analogy with *Aroideæ* that certain authors have placed them in the same family. Lindley, in fact, joins *Lemnaceæ* to *Pistiaceæ*, which he separates from *Aroideæ*; but he equally separates *Orontiaceæ*, which he classes (we think wrongly) between *Liliaceæ* and *Junceæ*; *Orontiaceæ* are inseparable from true *Aroideæ*, and this affinity is confirmed by an observation of Gasparrini, who has seen monstrous flowers of *Arum italicum* with a perianth analogous to that of *Acorus* and *Orontium*.

The leaves of *Aroideæ*, which are very variable in shape, texture and nervation, recall sometimes *Sparganieæ* (*Acorus*), sometimes *Marantaceæ* (*Aglaonema marantæfolium*), sometimes *Smilaceæ* (*Goniurus*), sometimes *Taccaceæ* (*Dracunculus, Amorphophallus*), and sometimes even some Dicotyledons, as *Aquilarineæ* (*Heteropsis salicifolia*), or *Cycadeæ* (*Zamioculcas*). They are sometimes jointed like those of Oranges (*Pothos*), or stipulate like those of *Piperaceæ*; but, with the exception of *Anthurium violaceum*, which bears some peltate epidermal scales, all known *Aroideæ* have glabrous leaves. The fruits of some Anthuriums detach from the spadix by a peculiar mechanism, and remain suspended by elongated fibrous cords, similar to those which retain the seeds of *Magnolia* at the moment when the fruit bursts.

Most *Aroideæ* are tropical, and especially abound in the large American forests and in the temperate regions of the Andes. Tropical Asia possesses fewer, but this is there compensated by the elegance and variety of the species. Few are met with north of the tropic of Cancer. *Orontiaceæ* and *Callaceæ* are the most arctic; one (*Calla palustris*) extends in North Europe to 64°. The true Arums are principally met with in the east of the Mediterranean region. The number of *Aroideæ* hitherto observed in Africa is not considerable; the United States of America possess at most six species. *Symplocarpus* grows in North Asia and America, as well as *Arctiodracon*, which extends beyond 49°. *Richardia* belongs to South Africa; *Cryptocoryneæ* are met with in the swamps of Asia and on the sandy hills of tropical Africa; *Arisæma* inhabits the mountains of sub-equatorial Asia and North America. *Acorus* is a native of North Asia, and has been introduced into Europe. *Gymnostachys* grows in east and extra-tropical Australia. *Pistia* is common in the waters of the whole tropical zone.

The spadix of several Arums gives off when in flower more or less heat. This phenomenon, observed by Lamarck, Bory, Hubert, Brongniart, Van Beck, &c., is especially remarkable in tropical species; the maximum of heat developed by our *Arum maculatum* is 7° to 0° above that of the surrounding air; but *Colocasia cordifolia* and *odora* emit a heat more than 10°, 12°, and even 22° above that of the atmosphere, according to Van Beck and Bergsma.

Some *Aroideæ* exhale during flowering a repulsive odour; as, amongst others, *Dracunculus crinitus*, the cadaverous exhalations from which attract flies, which descend to the bottom of the spathe, where they are entangled in the long hairs; but if some species are fœtid, others on the contrary are sweet-scented, as *Richardia æthiopica*, of South Africa, cultivated in Europe for the beauty and perfume of its white spathe, which encloses a golden spadix. Many Colocasias and Caladiums are now cultivated in our hothouses and public gardens on account of the elegance and size of their leaves.

Aroideæ are remarkable for the abundance of crystals throughout their tissue. Delile has discovered them even in the anthers, where they are mixed with the pollen-grains.

The rhizome and leaves of *Aroideæ* contain a very acrid juice, which may occasion serious accidents; *Layemandra toxicaria*, quoted by Lindley, is considered a most violent poison; the stem and leaves of *Dieffenbachia Seguina* produce, when chewed, a violent inflammation of the mucous membrane, and a swelling of the tongue which renders it impossible to speak; the leaves of *Colocasia* and of *Arum* are also extremely irritating, but this acridity is removed by desiccation or cooking, and almost entirely disappears at the flowering season. This acrid principle accompanies, in the rhizome, an abundance of starch, useful as food.

Arum maculatum, a European plant, was prescribed by the ancients as an excitant in affections of the mucous membrane; it was also applied externally as a rubefacient and epispastic; but its qualities vary so much with the age of the plant that it has fallen into disuse. *Calla palustris* was formerly classed among alexipharmics, on account of its violent diaphoretic properties. Many other *Aroideæ* are used as medicinal plants, such as, among the genera of the Mediterranean region, *Arum, Arisarum, Dracunculus*, and *Biarum*. The principal species renowned in Indian medicine are: *Amorphophallus campanulatus, Typhonium trilobatum, Arisæma triphyllum, pentaphyllum, Dracontium, Scindapsus officinalis*, &c. The root of *Symplocarpus fœtidus*, remarkable for its fœtid odour, resembling that of the polecat, is much employed by the Americans against asthma and chronic coughs. The dried root of *Orontium aquaticum* is considered edible in the United States. The leaves of *Monstera pertusa*, full of raphides, are slightly caustic, and are used, bruised, as a topic, for anasarca in tropical America.

Those *Aroideæ* which have starchy and edible rhizomes principally belong to the section *Colocasieæ*. The most celebrated is *Colocasia antiquorum*, a native of India, cultivated in Egypt from time immemorial, and spread all over the tropics. *C. himalaiensis*[1] supports, with *Arisæma utile*, the inhabitants of the Indian mountains. Other congeneric species are cultivated in Bengal. The Taro (*Alocasia macrorrhiza*) abounds in the Pacific Islands. The rhizome and fructiferous spadix of *Peltandra virginica*, of North America, are equally edible. The fleshy spadices, bearing perfumed and well-tasted fruits, of *Tornelia fragrans* (*Monstera deliciosa*) are habitually sold in Mexican markets, where they rival the Pine-apple in estimation. The shoots of *Xanthosoma sagittæfolium*, known under the name of Caraïbe Cabbage, are

---

[1] Probably *C. antiquorum*, or *Alocasia indica*, is here referred to. The Himalayan Arisæmas are only resorted to in times of scarcity.—ED.

used as vegetables in the Antilles. The rhizomes of our *Arum maculatum* and *Calla palustris*, bruised, washed, and mixed with the farina of cereals, serve, according to Pallas, as food for the poor populations of Lapland and Finland. *Arum* starch is sold in London under the name of Portland Arrowroot.

Acorus Calamus, now naturalized in various parts of Europe, yields an aromatic acrid and bitter rhizome, used as a tonic and excitant, and entering into the composition of some compound medicines; the same is the case with *A. gramineus*, a native of China. The herbage of *Pistia*, brought each year from Central Africa to Egypt, was formerly prescribed as an emollient and refrigerant.

The long adventitious roots of several *Aroideæ*, and particularly those of *Phyllodendron*, designated in Central America under the names of *Imbe, Oumbe*, &c., are used as cords to tie up the bundles of Sarsaparilla which are sent to Europe.

XXXI. *LEMNACEÆ*.

(LEMNACEÆ, *D.C. et Duby, Link, Schleiden.*—PISTIACEÆ, *Lindl. in part.*)

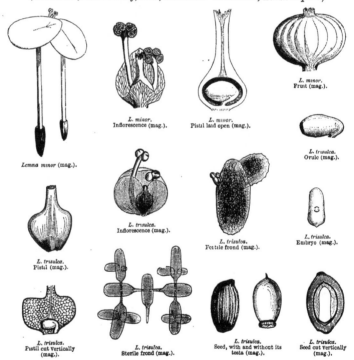

L. minor.
Fruit (mag.).

L. minor.
Inflorescence (mag.).

L. minor.
Pistil laid open (mag.).

L. trisulca.
Ovule (mag.).

Lemna minor (mag.).

L. trisulca.
Inflorescence (mag.).

L. trisulca.
Fertile frond (mag.).

L. trisulca.
Embryo (mag.).

L. trisulca.
Pistil (mag.).

L. trisulca.
Pistil cut vertically
(mag.).

L. trisulca.
Sterile frond (mag.).

L. trisulca.
Seed, with and without its
testa (mag.).

L. trisulca.
Seed cut vertically
(mag.).

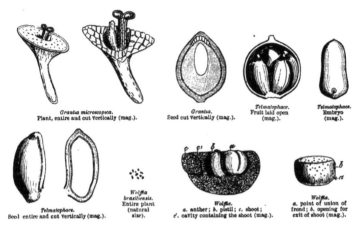

Grantia microscopica.
Plant, entire and cut vertically (mag.).

Grantia.
Seed cut vertically (mag.).

Telmatophace.
Fruit laid open
(mag.).

Telmatophace.
Embryo
(mag.).

Telmatophace.
Seed entire and cut vertically (mag.).

*Wolffia
brasiliensis.*
Entire plant
(natural
size).

Wolffia.
a. anther; *b.* pistil; *c.* shoot;
c'. cavity containing the shoot (mag.).

Wolffia.
a. point of union of
frond; *b.* opening for
exit of shoot (mag.).

Herbaceous PLANTS, very small, floating freely on the surface of stagnant water, stemless and reduced either to lenticular or obovate disks (*fronds*), flattened above, sometimes more or less convex below; or to membranous productions springing at right angles to each other; *fronds* emitting by 2 lateral slits, or sometimes by 1 basilar slit (*Wolffia*) young fronds similar to the first, with no vessels, or with rudimentary tracheæ, transient in the pistil, sometimes apparent throughout the tissue of the plant (*Spirodela*); lower surface of the fronds usually bearing at the middle rootlets terminated by a membranous cap (*pileorrhiza*), fascicled (*Spirodela*), or reduced to one (*Lemna, Grantia, Telmatophace*); sometimes with no rootlets (*Wolffia*). INFLORESCENCE imbedded in the frond. FLOWERS achlamydeous, naked, or enclosed in a spathe, reduced to 1–2 stamens, accompanied by a sessile pistil; *spathe* urceolate, membranous, bursting irregularly by the evolution of the stamens; *filaments* filiform; *anthers* with 2 sub-globose cells, opening by a transverse slit; *pollen* muricate, opening by a single slit; *style* continuous with the top of the ovary; *stigma* infundibuliform. OVARY 1-celled, 1-many-ovuled; *ovule* anatropous (*Spirodela, Telmatophace*), or semi-anatropous (*Lemna*), or orthotropous (*Wolffia*). FRUIT 1-many-seeded, indehiscent (*Lemna, Wolffia*), or dehiscing transversely (*Telmatophace*). SEED with coriaceous corky fleshy testa, *endopleura* membranous, forming an embryotegium at the micropylar point; *albumen* fleshy or scanty. EMBRYO axile, straight; *radicle* superior, inferior or vague.

PRINCIPAL GENERA.

Lemna.   Telmatophace.   Spirodela.   Grantia.   Wolffia.

At the beginning of winter Lemnas sink into the mud, where they perhaps all undergo transformations analogous to those exhibited by *L. trisulca*, the sterile membranous portions of which give rise to an orbicular fertile plant, which is similar in all points to *L. minor*.

Lemnaceæ, which are the smallest known Phanerogams, are intermediate between *Naiadeæ* and *Aroideæ*; they are closely allied to the latter through the genus *Pistia*, which approaches them in inflorescence and the structure of its seeds. They are found in stagnant water in all climates, but especially in temperate regions; they are rarer in the tropics on account of the heat which dries up the swamps, and the torrents of rain which violently agitate the water. Their part in nature seems to be the protection from the solar light of the inferior organisms of the Animal Kingdom which live in swamps, and at the same time to serve them for food.

XXXII. *ASPIDISTREÆ.*

Glabrous stemless HERBS; *rhizome* usually creeping. LEAVES radical, sheathing, oblong-lanceolate, coriaceous, nerves prominent. FLOWERS ☿, solitary, spiked, rising out of the ground (*Aspidistra*), or scape terminated by a dense spike (*Rhodea*, &c.). PERIANTH petaloid, valvate in æstivation, sub-globose (*Rhodea*), or campanulate (*Tupistra, Aspidistra*), 6–8-fid. STAMENS 6–8, inserted on the perianth. OVARY free, with 3 2-ovuled cells; *stigma* sessile, rayed, 3-fid (*Rhodea*), or stipitate, 3–6-lobed (*Tupistra, Aspidistra*); *ovules* semi-anatropous. BERRY (*Rhodea*) 1-celled, 1-seeded [or several-celled and -seeded in *Aspidistra* and *Tupistra*. SEED large, sub-globose; *testa* very thin; *albumen* copious, dense. EMBRYO cylindric]. Japanese and Asiatic plants.

GENUS.

| Aspidistra. | Tupistra. | Rhodea. |

XXXIII. *OPHIOPOGONEÆ.*

Stemless [tufted] HERBS. LEAVES sheathing, linear-ensiform, or oblong-lanceolate. SCAPE simple, terminated by a raceme of ☿ flowers. PERIANTH petaloid, rotate, 6-fid or -partite; *throat* naked, or furnished with an annular crown (*Peliosanthes*). STAMENS 6; *filaments* dilated, almost 0; *anthers* basifixed, sagittate, mucronate (*Ophiopogon*), or adnate to the annular crown (*Peliosanthes*). OVARY adherent to the base of the perianth, of 3 2-ovuled cells; *style* 3-gonous, thick; *stigma* shortly 3-fid (*Ophiopogon*), or 3-fid and radiating (*Peliosanthes*); *ovules* anatropous.

Rhodea japonica. Flowering spike.

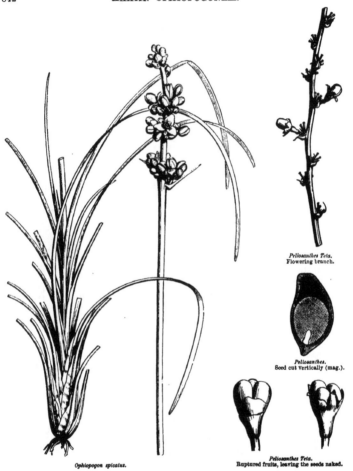

Peliosanthes Teta.
Flowering branch.

Peliosanthes.
Seed cut vertically (mag.).

Ophiopogon spicatus.

Peliosanthes Teta.
Ruptured fruits, leaving the seeds naked.

FRUIT breaking up and leaving the seeds exposed while still unripe. SEEDS with fleshy herbaceous testa.—Indian and Japanese plants.

GENERA.

Ophiopogon.   Peliosanthes.

Ophiopogon japonicus, called by the natives Serpent's Beard, produces mucilaginous sugary tubers, frequently used in China and Japan for diseases of the abdomen.

XXXIV. *LILIACEÆ.*

(LILIACEÆ ET NARCISSORUM *pars, Jussieu.*—HEMEROCALLIDEÆ ET ASPHODELEÆ, *Br.*— LILIACEÆ, TULIPACEÆ ET ASPHODELEÆ, *D.C.*).

FLOWERS ☿. PERIANTH *inferior, petaloid, 6-merous, 2-seriate.* STAMENS 6, *hypogynous or perigynous.* OVARY *superior, with 3 several–many-ovuled-cells;* STYLE *simple.* FRUIT *capsular.* SEEDS *with membranous or crustaceous testa;* ALBUMEN *fleshy.*—STEM *or* SCAPE *with a bulbous, tuberous or fibrous-fascicled root.*

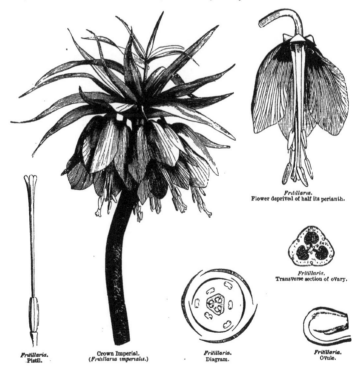

Fritillaria.
Flower deprived of half its perianth.

Fritillaria.
Transverse section of ovary.

Fritillaria.
Pistil.

Crown Imperial.
(*Fritillaria imperialis.*)

Fritillaria.
Diagram.

Fritillaria.
Ovule.

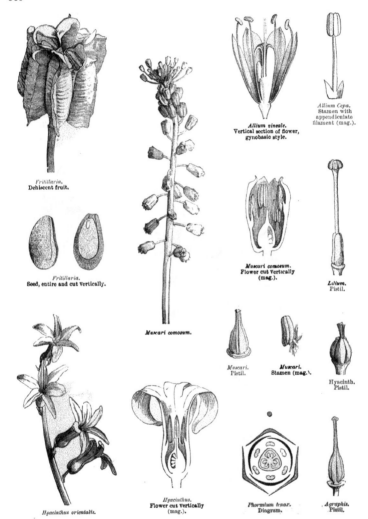

Fritillaria.
Dehiscent fruit.

Fritillaria.
Seed, entire and cut vertically.

Muscari comosum.

Hyacinthus orientalis.

Allium vineale.
Vertical section of flower,
gynobasic style.

Muscari comosum.
Flower cut vertically
(mag.).

Muscari.
Pistil.

Muscari.
Stamen (mag.).

Hyacinthus.
Flower cut vertically
(mag.).

Phormium tenax.
Diagram.

Allium Cepa.
Stamen with
appendiculate
filament (mag.).

Lilium.
Pistil.

Hyacinth.
Pistil.

Agraphis.
Pistil.

Herbaceous perennials, very rarely annuals, sometimes frutescent or arborescent; *root* bulbous, tuberous, fibrous-fascicled, or with a creeping rhizome. STEM simple, or branching above, straight or flexuous; or SCAPE aphyllous, erect, very rarely twining (*Streptolirion*) [*Bowiesa*]. LEAVES simple, entire, sheathing or amplexicaul, the radical fascicled, the cauline usually sessile, generally linear, flat or channelled, sometimes cylindric, rarely terminated by a tendril (*Methonica*). FLOWERS ⚥, mostly terminal, solitary, racemose, spiked, umbellate or capitate, rarely panicled (*Yucca*), furnished with scarious or spathaceous bracts, regular or very rarely 2-labiate (*Daubenya*). PERIANTH inferior, petaloid, of 6 2-seriate divisions, distinct, or forming a tube 6-fid at the top, sometimes nectariferous at the base (*Fritillaria*); æstivation imbricate. STAMENS 6, inserted on the receptacle or at the base of the perianth; *filaments* distinct, filiform or flat, sometimes appen-

Aloe vulgaris.

diculate or 3-toothed, the intermediate tooth antheriferous; *anthers* introrse, 2-celled, basi- or dorsi-fixed, or versatile, dehiscence longitudinal. OVARY free, with 3 several-many-ovuled cells; *style* simple, terminal, or rarely gynobasic (*Allium vineale*, &c.); *stigmas* 3, more or less distinct; *ovules* inserted at the central angle of the cells, anatropous or semi-anatropous, FRUIT dry, very rarely a berry (*Sanseviera, Lomatophyllum*), or sub-berried (*Yucca*, sp. *Asphodelus*). CAPSULE 3-celled, loculicidally 3-valved, rarely septicidal (*Calochortus, Agapanthus, Kniphofia,* &c.). SEEDS usually numerous; *testa* various, sometimes membranous or suberose, pale, sometimes margined; sometimes crustaceous, fragile, black; *albumen* fleshy. EMBRYO axile or excentric, variable in length, straight, or variously bent; *radicle* near the hilum.

TRIBE I. *TULIPACEÆ.*

Perianth-leaflets distinct and coherent at the base, sometimes nectariferous. Stamens hypogynous or obscurely perigynous. Ovules anatropous. Fruit a capsule, rarely a berry. Seeds usually compressed; testa pale brown, spongy or hard. Embryo small, straight or sub-arched, basilar.—Herbs with a usually bulbous root, sometimes tuberous (*Methonica*), or frutescent and annulated (*Yucca*).

PRINCIPAL GENERA.

* Tulipa.	* Methonica.[1]	* Erythronium.	* Lilium.	Gagea.
* Yucca.	* Fritillaria.	Lloydia.	* Calochortus.	

TRIBE II. *HEMEROCALLIDEÆ.*

Perianth tubular, limb 6-fid. Stamens inserted on the perianth. Fruit capsular. Ovules anatropous. Seeds more or less compressed; testa membranous, usually pale. Embryo axile, straight.—Perennial herbs, with tuberous or fibrous root.

PRINCIPAL GENERA.

* Funkia.	* Polianthes.	Brodiæa.	* Phormium.	* Blandfordia.
* Triteleia.	* Agapanthus.	Leucocoryne.	* Hemerocallis.	

TRIBE III. *ALOINEÆ.*

Perianth tubular, 6-fid, -toothed, or -partite. Stamens inserted on the receptacle or perianth-tube. Ovules anatropous. Fruit capsular, rarely a berry. Seeds compressed or angular, or winged; testa membranous and pale, or crustaceous and black. Embryo axile, straight.—Perennial herbs, sometimes frutescent or arborescent, and with fleshy leaves (*Aloe*); roots fibrous-fascicled, often swollen.

PRINCIPAL GENERA.

* Sanseviera.	* Tritoma.	Lomatophyllum.	Kniphofia.
* Aloe.	* Asphodelus.	* Eremurus.	

TRIBE IV. *HYACINTHINEÆ.*

Perianth tubular or 6-partite. Stamens inserted on the receptacle or perianth-tube. Ovules anatropous or semi-anatropous. Fruit capsular. Seeds globose or angular; testa crustaceous, black. Embryo straight or bent; radicle facing the hilum.—Herbs with bulbous or fibrous-fascicled root.

PRINCIPAL GENERA.

* Muscari.	* Lachenalia.	Myogalum.	Bellevalia.
* Eucomis.	* Allium.	* Phalangium.	* Hyacinthus.
* Scilla.	Thysanotus.	* Cyanella.	* Weltheimia.
* Urginea.	* Arthropodium.	Uropetalum.	* Ornithogalum.
* Bulbine.	* Agraphis.	* Albuca.	* Anthericum.

We have indicated the extremely close affinities between *Liliaceæ*, *Asparageæ* and *Smilaceæ*; families which together form a group to which most other Monocotyledonous families may be linked, directly or by intermediates. Thus *Junceæ*, which are near certain *Melanthaceæ* and *Liliaceæ*, connect these with other families with a free ovary; and on the other hand, those *Amaryllideæ* and *Dioscoreæ* which

---

[1] *Methonica* belongs to a very different tribe. See Tribe III., *Methoniceæ*, of *Melanthaceæ*, p. 853.—ED.

belong, the one to the *Liliaceæ*, the other to the *Smilaceæ* with an inferior ovary, connect them with the epigynous Monocotyledons.

Liliaceæ are spread over all the world, except the Arctic zone; they principally inhabit the tempe‑ rate and sub-tropical regions of the Old World. *Tulipaceæ*, with the exception of *Methonica* [referred to *Melanthaceæ*], belong to the northern hemisphere. *Hemerocallideæ* are most frequently met with in south temperate latitudes, rarely in North America and Japan. *Aloineæ* principally inhabit South Africa; *Asphodeleæ*, Europe. *Hyacinthineæ*, which is the most numerous tribe, inhabit the temperate regions of both hemispheres; they are especially abundant in the Mediterranean region, and are met with in Australia. Most of the genera are confined to certain countries: thus *Drimia*, *Eucomis*, *Lachenalia*, inhabit South Africa; *Arthropodium* is Australian; while *Scilla*, *Urginea*, &c., are dispersed over Europe, Africa and Japan. *Allium* is spread over all East Europe and North and Mediterranean [and Western] Asia.

Liliaceæ are remarkable for the beauty of their flowers; at their head must be placed the Tulip (*Tulipa Gesneriana*), the varieties of which are eagerly sought by amateurs, especially in Belgium and Holland. Among the principal genera cultivated for ornament, after Tulips, must be mentioned the Hyacinth, Lily, Tuberose, *Yucca*, *Agapanthus*, *Tritoma*, *Hemerocallis*, *Funkia*, *Fritillaria*, *Methonica*, &c.

Liliaceæ contain an abundant mucilage often rich in sugar and starch, a resinous bitter substance, a volatile acrid oil, and an extractable principle, combined in very different proportions, and thus exhibiting a variety of properties; some being condiments and alimentary, others yielding more or less powerful medicines, and some even being poisonous.

The bulbs of several of the tribe of *Tulipaceæ* were formerly employed in pharmacy on account of their bitter acrid principle, analogous to that of *Scilla*. The root of *Erythronium Dens-canis* was once prescribed in Europe as an anthelminthic and aphrodisiac; the North American species are used there to induce vomiting. The roots of *Methonica* are considered very poisonous; the bulbs of *Gagea*, more mucilaginous and less acrid, are also used as an emetic. Those of Lilies (*Lilium candidum*, &c.), which are very rich in starch, are employed as an emollient plaister. The fruits of *Yucca* are purgative; its root is used as soap.

In the tribe of *Hemerocallideæ* we can only cite as useful species the Tuberose (*Polianthes tuberosa*), of which the flowers are used in perfumery, and the New Zealand Flax (*Phormium tenax*), cultivated in some parts of western France for the sake of the fibres of its leaves, from which cordage is made. The flower of *Hemerocallis* was formerly used as a cordial.

The tribe of *Aloineæ* consists almost exclusively of Aloes, plants with thick fleshy brittle leaves, which possess under their epidermis peculiar vessels filled with an extremely bitter resinous juice, much used in medicine as a tonic, purgative, drastic and emmenagogue; the principal species which yield it are *Aloe ferox*, *spicata*, *plicatilis*, *arborescens*, from the Cape of Good Hope; *A. socotrina*, which grows at Socotra and in North-east Africa; *A. vulgaris*, a native of the Cape, and now naturalized in India and America, and cultivated in some parts of the Mediterranean region. *Sanseviera* grows in India; the root of some of its species is administered for affections of the urethra and lungs, and in rheumatism. The leaves of others (*S. cylindrica*, &c.) abound in extremely fine textile fibres.

The most important species of the tribe of *Hyacinthineæ* is the Squill (*Scilla maritima* or *Urginea Scilla*), which grows principally in Algeria; its very large bulb is composed of numerous coats filled with a viscous juice, very bitter and acrid, and even corrosive, which contains a peculiar principle (*scillitine*), to which some of the qualities of the Squill are due. It is a powerful diuretic and an efficacious expec‑ torant and emetic. It is also used in tanning leather. The bulbs of *Camassia esculenta* are edible, and sought by some Indian tribes of North America.

The numerous species of *Allium*, which mostly contain nutritious matters, joined to a sulphurous volatile oil, and have an acrid taste and a pungent odour, owe to these principles alimentary and medicinal properties. The bulb of the cultivated Garlic (*Allium sativum*), used in cookery as a seasoning, is also used in medicine, externally as a rubefacient, and internally as a vermifuge; and it enters, with camphor, into the composition of a celebrated prophylactic, known as Marseilles or Thieves' Vinegar. The other species of *Allium* cultivated as condiments are the Common Onion (*A. Cepa*), the Winter Onion (*A. fistulosum*), the Eschalot (*A. ascalonicum*), the Leek (*A. Porrum*), the False Leek (*A. Ampeloprasum*), the Rocambole (*A. Scorodoprasum*), and the Chive (*A. Schœnoprasum*). Several other species (*A. Moly*, *nigrum*, *Dioscoridis*, *Victorialis*, *ursinum*, &c.), formerly classed among the officinal species, are now fallen into disuse. It is the same with the genera *Hyacinthus, Muscari, Ornithogalum,* the bulbs of which were formerly used as purgatives and diuretics; those of *Ornithogalum altissimum* are still in use at the Cape for asthma and pulmonary catarrhs [those of *O. pyrenaicum* are sold and eaten at Bath as French asparagus]. The tuberous roots of *Anthericum* and *Asphodelus* lose their acridity by desiccation or boiling; they were formerly considered diuretics and emmenagogues; the *Asphodel* (*A. ramosus*) was used as a substitute for the Squill. Some endeavours have recently been made to extract an alcoholic spirit from its tuberous roots.

Tulbaghia alliacea, cepacea, &c., with an alliaceous odour, have thick fibrous roots, which are cooked in milk and administered at the Cape for phthisis and worms.

XXXV. *ERIOSPERMEÆ.*

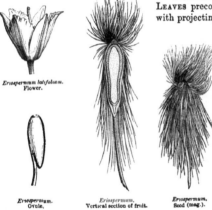

Eriospermum latifolium.
Flower.

Eriospermum.
Ovule.

Eriospermum.
Vertical section of fruit.

Eriospermum.
Seed (mag.).

Perennial HERBS with tuberous roots. LEAVES precocious, petioled, rounded, thick, with projecting reticulate nerves; *blade* bulbilliferous or gemmiparous below. SCAPE growing after the decay of the leaves, simple, cylindric. FLOWERS ☿, racemed or panicled; *pedicels* with a scarious bract at the base. PERIANTH petaloid, 6-partite, campanulate, persistent. STAMENS 6, inserted at the bottom of the perianth; *filaments* flat, dilated at the base; *anthers* sagittate-didymous, incumbent. OVARY free, with 3 few-ovuled cells; *style* filiform, trigonous; *stigma* sub-capitate, obscurely 3-fid; *ovules* inserted at the bottom of the cell, ascending, anatropous. CAPSULE membranous, ovoid, trilobed-trigonous, with 3 loculicidal semi-septiferous valves. SEEDS few or solitary, erect, lanceolate; *testa* thin, covered with long silky hairs bent back to the chalaza, and longer than the nucleus.

GENUS.

Eriospermum.

Stemless plants of South Africa. The villosity of the seeds is the only character which distinguishes *Eriospermeæ* from *Liliaceæ,* a difference which, moreover, is found between the genera of *Malvaceæ, Ternstrœmiaceæ* and *Convolvulaceæ,* but which has not therefore dissociated them. The tubers of one species, which are scarlet, are used as a topic for the cure of ulcers.

XXXVI. *CONANTHEREÆ, Don.*

Stemless HERBS, natives of Peru and Chili. ROOT tubero-bulbous, with fibrous coats; *scape* branched, bracteate. FLOWERS ☿, blue, panicled. PERIANTH petaloid, adhering to the ovary by its base; *limb* 6-partite, spirally twisted after flowering, and then detaching transversely above its base. STAMENS 6, inserted on the perianth; *filaments* compressed, short, glabrous; *anthers* basifixed, connivent in a cone, opening by a pore at the top. OVARY sub-adherent, of 3 many-ovuled cells; *style* filiform ; *stigma* simple. CAPSULE loculicidally 3-valved. SEEDS globose.

GENERA.

| Conanthera. | Cummingia. | Zephyra, &c. |

XXXVII. *GILLIESIEÆ, Lindley.*

PLANTS of Chili. Bulbous glabrous HERBS. LEAVES radical, linear. FLOWERS ☿, inconspicuous, arranged in an umbel with a double coloured involucre. PERIANTH greenish, fleshy, sometimes with 3 2-labiate leaflets (*Gilliesia*); sometimes regular, urceolate, 6-toothed (*Miersia*). STAMENS 6, sometimes adnate to the base of the perianth, and united into a cup, the 3 posterior sterile (*Gilliesia*); sometimes minute, inserted at the throat of the perianth (*Miersia*). OVARY of 3 many-ovuled cells; *style* filiform; *stigma* capitate. CAPSULE with 3 semi-septiferous valves. SEEDS with a crustaceous black testa.

GENERA.

| Gilliesia. | Miersia. |

XXXVIII. *MELANTHACEÆ.*

(MELANTHEÆ, *Batsch.*—MELANTHACEÆ, *Br.*—COLCHICACEÆ, *D.C.*—
VERATREÆ, *Salisbury.*)

FLOWERS ☿. PERIANTH *sub-herbaceous or petaloid, 6-merous, 2-seriate, æstivation imbricate or valvate.* STAMENS 6, *inserted at the base or throat of the perianth ;* ANTHERS *dorsifixed, usually extrorse.* OVARY *superior, or rarely semi-inferior, of 3 many-ovuled cells;* STYLES 3. FRUIT *of three follicles, rarely a capsule, very rarely a berry.* SEEDS *albuminous.* EMBRYO *small, included.*—STEM or SCAPE *herbaceous.* LEAVES *radical or alternate.*

Perennial HERBS with bulbous or tuberous roots, rarely of fascicled fibres, some-times with a horizontal rhizome. STEM or SCAPE annual, simple, or very rarely branched, often shortened, or subterranean. LEAVES all radical, or the cauline alternate, sometimes Grass-like or setaceous; sometimes large, nerved, entire, sessile,

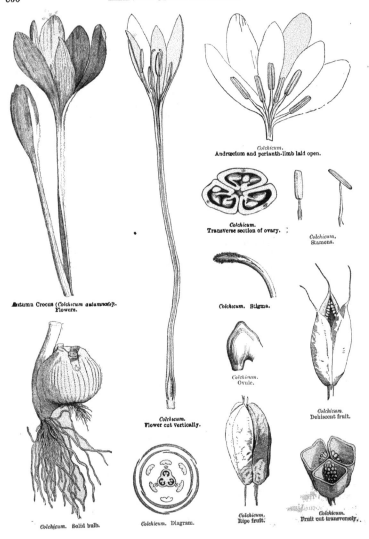

Colchicum.
Andrœcium and perianth-limb laid open.

Colchicum.
Transverse section of ovary.

Colchicum.
Stamens.

Autumn Crocus (*Colchicum autumnale*)-
Flowers.

Colchicum. Stigma.

Colchicum.
Ovule.

Colchicum.
Deliscent fruit.

Colchicum.
Flower cut vertically.

Colchicum. Solid bulb.

Colchicum. Diagram.

Colchicum.
Ripe fruit.

Colchicum.
Fruit cut transversely.

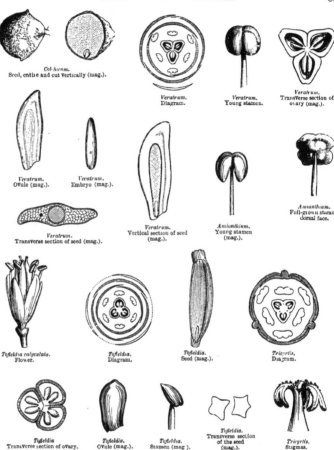

Colchicum.
Seed, entire and cut vertically (mag.).

Veratrum.
Diagram.

Veratrum.
Young stamen.

Veratrum.
Transverse section of ovary (mag.).

Veratrum.
Ovule (mag.).

Veratrum.
Embryo (mag.).

Veratrum.
Transverse section of seed (mag.).

Veratrum.
Vertical section of seed (mag.).

Amianthium.
Young stamen (mag.).

Amianthium.
Full-grown stamen, dorsal face.

Tofieldia calyculata.
Flower.

Tofieldia.
Diagram.

Tofieldia.
Seed (mag.).

Tricyrtis.
Diagram.

Tofieldia
Transverse section of ovary.

Tofieldia.
Ovule (mag.).

Tofieldia.
Stamen (mag).

Tofieldia.
Transverse section of the seed (mag.).

Tricyrtis.
Stigmas.

and more or less sheathing. FLOWERS ☿, or very rarely imperfect by arrest (*Veratrum*, &c.), regular, radical, or axillary or terminal on the scape or stem, in a racemed spike or panicle, naked or bracteate, exceptionally calyculate (*Tofieldia*). PERIANTH inferior, very rarely semi-superior, petaloid, tubular, with 6 often persistent segments, or with 6 2-seriate leaflets, all distinct, sessile or unguiculate, deciduous, the inner conformed to the outer, rarely hollowed into a nectariferous sac (*Tricyrtis, Melan-thium, Androcymbium*); the claw often furnished with pores or glandular pits (*Uvularia, Burchardia, Ornithoglossum*); æstivation imbricate (*Colchicum, Tofieldia, Veratrum*), valvate (*Tricyrtis*), or sub-valvate (*Tofieldia nepalensis*). STAMENS 6, or very rarely 3 (*Scoliopus*), inserted at the base, or above the claw of the perianth-leaflets, very rarely hypogynous (*Uvularia*), sometimes mingled with staminodes (*Kreysigia*); *filaments* filiform, free, often persistent; *anthers* 2-celled, or falsely 1-celled (*Veratrum*), usually dorsifixed, very rarely basifixed (*Helonias*), extrorse in bud, and versatile after flowering, or introrse (*Tofieldia, Bulbocodium, Colchicum, Amianthium*); cells linear (*Colchicum, Uvularia, Tricyrtis, Ledebouria, &c.*), or reni-form, or didymous (*Amianthium, Veratrum, Melanthium, Helonias, &c.*); dehiscence longitudinal, rarely transverse (*Amianthium*). OVARY superior, or very rarely semi-inferior (*Zygadenus, ? Tofieldia, ? Veratrum*), 3-celled and formed of 3 carpels, rarely simple and 1-celled (*Monocaryum*); *styles* 3, opposite to the ovary-cells, distinct and papillose-stigmatiferous along their inner edge, or more or less con-nate below, and terminated by three stigmas; *ovules* inserted at the inner angle of the cells, 1-2-many-seriate, usually numerous, anatropous or semi-anatropous. FRUIT usually capsular, membranous or coriaceous, 3-celled, sometimes of 3 follicles, sometimes opening in three semi-placentiferous loculicidal valves (*Ornithoglossum, Anguillaria, Tricyrtis, &c.*), rarely sub-baccate, and opening at the top (*Uvularia*), very rarely an indehiscent berry (*Drapiezia*). SEEDS usually numerous, globose, angular, oblong or compressed; *testa* thin, membranous, soft and suberose, or black and brilliant (*Disporum*); *albumen* copious, coriaceous or cartilaginous. EMBRYO included, small; *radicle* near or distant from the hilum.

TRIBE I. *VERATREÆ.*

Stem or scape leafy. Flowers axillary, solitary, or racemed or spiked. Styles short, usually distinct. Perianth leaves free, sessile, or very shortly clawed, some-times united at the base. Ovary free or semi-inferior.

PRINCIPAL GENERA.

Tofieldia.	Amianthium.	Schelhammera.	* Veratrum.	Zygadenus.
* Uvularia.	Asagræa.	Burchardia.	* Tricyrtis.	* Xerophyllum.
Ornithoglossum.	* Disporum.	* Helonias.	Anguillaria.	Scoliopus.
Schænocaulon.	Melanthium.			

TRIBE II. *COLCHICEÆ.*

Stemless. Flowers springing from a subterranean bulb. Perianth tubular, or with long clawed segments, naked or crested. Ovaries free; styles fragile, distinct, or more or less coherent.

PRINCIPAL GENERA.

* Colchicum.   * Merendera.   *** Bulbocodium.**

[Tribe III. *METHONICEÆ*.

Perianth tubular or 6-partite. Stamens inserted on the base of the perianth-segments or on its tube. Ovules anatropous; stigmas 3. Fruit capsular. Seeds turgid; testa thick or fleshy, white or red. Embryo straight.—Herbs, erect or with climbing elongate stems. Root a fleshy lobed perennial tuber. Leaves alternate, or in whorls of 3 or more, sessile, tips sometimes curled. Peduncles 1-flowered, axillary or leaf-opposed or extra-axillary; flowers yellow or red.

GENERA.

Methonica (Gloriosa). Sandersonia. Littonia.

A very distinct group, of doubtful affinity, upon the whole, I think, nearer *Uvularia, Tricyrtis*, and other *Melanthaceæ* than *Liliaceæ*. The root is very peculiar, and unlike that of *Liliaceæ* proper. *Methoniceæ* are natives of eastern South Africa, and one genus (*Methonica*) has species in tropical Africa and India: the root of the latter is reputed to be a virulent poison.—Ed.]

Melanthaceæ are in some respects near *Junceæ* and especially *Liliaceæ*, but are separated from the first by their petaloid perianth, and from both by the direction of the anthers; they are distinguished from *Liliaceæ*, tribe *Asphodeleæ*, by the nature of their testa. The tribe *Colchiceæ* is a very natural one; *Veratreæ* contain several less closely allied genera. Some, in fact, are berried, and have an analogy with *Smilaceæ*, in spite of the difference arising from the direction of the anthers; on the other hand, the genera with loculicidal capsules approach the tribe *Tulipaceæ* of *Liliaceæ*. The tribe of *Veratreæ*, characterized by its perianth with free very often clawed and nectariferous leaflets, is the nucleus of the family. One-third of the species inhabit North America; a second third belong to South Africa; the others are dispersed over the north shores of Africa, in India, Australia, Central Asia, north alpine and sub-alpine Europe; very few are met with in South America. *Tofieldia* inhabits Europe, North America [the Himalayas] and New Granada. *Pleea* is met with in one of the Antilles. Some species of *Uvularia* grow in the New World, others in India, China and Japan. *Tricyrtis* is Asiatic. *Veratrum* belongs to both worlds. *Colchicum* inhabits Central Europe and the Mediterranean and Caucasian regions. *Burchardia, Anguillaria, Schelhammera, Kreysigia* are Australian.

Melanthaceæ occupy an important place among medicinal plants; they are acrid, drastic, emetic, and their employment demands the greatest caution; they are principally recommended in gouty and rheumatic affections; their properties depend on various alkaloids (*veratrine, colchicine, sabadilline*), which modern chemists have separated. The officinal species are:—the White Hellebore (*Veratrum album*), a native of the alpine and sub-alpine districts of Europe; its root, which is a violent drastic and emetic, and dangerous even to powder, is now scarcely used, except externally in cutaneous diseases, and to kill vermin. *V. nigrum*, which is collected in forest clearances and mountain pastures in Central and South Europe, possesses similar but less powerful properties. *V. viride* is similarly used by the Anglo-Americans. The Cevadille, or Sabadilla, inhabits Mexico; its capsules and seeds alone are brought to Europe; they were for a long time attributed to *V. Sabadilla*, a Chinese species, and later to several species of *Schœnocaulon* (*S. officinale, caricifolium*, &c.); they are now known to be produced by *Asagraya officinalis* [a plant also found in Caraccas]. They are used externally in a powder or liniment to remove vermin; inwardly as pills, injections, or infusions, as a vermifuge. *Colchicum autumnale* is common in meadows throughout Europe; its flowers, which spring from an underground bulb, like those of the Crocus, appear in autumn; the leaves are developed in the following spring, and the fruit along with them. The bulb, flowers and seed are all used medicinally.

. . The tuber of *Hermodacte*, prescribed from the sixth century by Greek doctors, and later by the Arabians, in the treatment of arthritic affections, has now nearly fallen into disuse; its origin was long

obscure, as much from the spurious drugs which usurped its name, as from the error of botanists. Linnæus attributed it to the *Iris tuberosa*, but the researches of modern pharmacists, and especially Guibourt
and Planchon, have shown that *Hermodacte* is the produce of *Colchicum variegatum*, a native of the
Mediterranean region.

In North America the root of *Helonias dioica* is used as a vermifuge; steeped in wine it is a bitter
tonic: a decoction of the root of *H. bullata* is administered there for obstructions of the abdominal
viscera. The seeds of *Amianthium muscætoxicum* are narcotic, and used to destroy flies. The bulb of
Ledebouria hyacinthioides replaces in India that of *Scilla*. *Uvularia* is distinguished from the other
Melanthaceæ by its medicinal properties as well as by its botanical characters, which place it near
Streptopus, belonging to *Asparageæ*. The roots of *U. latifolia* and *flava* are, in fact, mucilaginous and
slightly astringent; the Anglo-American doctors prescribe it in an infusion for gargling. A decoction of
the leaves and root of *U. grandiflora* is considered by the natives of North America a cure for the bite
of the rattle-snake.

XXXIX. *SMILACEÆ.*

(ASPARAGORUM *genera, Jussieu.*—SMILACEÆ, *Br.*—LILIACEARUM *pars,* SMILACEÆ ET
PHILESIACEÆ, *Lindl.*)

FLOWERS *usually* ☿. PERIANTH *inferior, petaloid, mostly 6-merous, 2-seriate,
isostemonous.* STAMENS *hypogynous or perigynous.* OVARY *superior, 3-celled, rarely
1-2-4-celled;* OVULES *more or less numerous, anatropous or semi-anatropous or orthotropous.* BERRY *few-seeded.* SEEDS *globose;* TESTA *membranous;* ALBUMEN *very
dense.* EMBRYO *small, included.*—HERBS *or sarmentose* UNDERSHRUBS, *sometimes
with tendrils or thorns.* LEAVES *all radical, or cauline alternate or whorled.*

Perennials, with a usually creeping rhizome, or herbaceous or sarmentose
UNDERSHRUBS; *branches* unarmed or thorny. LEAVES all radical, or the cauline
alternate or whorled, sessile, sheathing or petioled, sometimes with stipular tendrils
(*Smilax*); *nerves* parallel or 3-5-7-palmate and anastomosing (*Smilax*); sometimes
scale-like, and then accompanied by branches dilated into a phyllode (or cladode)
(*Ruscus*). FLOWERS regular, ☿, or diclinous by arrest, terminal or axillary, solitary
(*Paris, Trillium*), or sub-solitary (*Streptopus*); sometimes racemose (*Convallaria,
Polygonatum, Smilacina,* &c.), or umbellate (*Smilax, Medeola*); *pedicels* often jointed
and bracteolate. PERIANTH petaloid, with 6 leaflets, rarely 4 (*Majanthemum*), or
5-∞ (*Paris*), 2-seriate, all alike, or the inner narrowest or largest (*Paris, Trillium*);
distinct, or forming a tubular or campanulate perianth (*Polygonatum, Convallaria*);
æstivation imbricate. STAMENS equal in number with the perianth leaflets [or ∞ in
Paris], inserted on them or on the receptacle; *filaments* free, rarely more or less
monadelphous (*Ruscus, Paris,* &c.); *anthers* introrse, 2-celled, connective apiculate
(*Trillium, Paris*). OVARY free, sessile, usually 3-celled, sometimes 1-celled with 3-5
parietal placentas (*Lapageria, Paris,* &c.), or 2-celled (*Majanthemum*); cells many-
ovuled (*Paris, Trillium, Medeola, Drymophila, Streptopus*); or 1-2-ovuled (*Convallaria, Polygonatum, Smilacina, Smilax, Ruscus,* &c.); *styles* as many as the carpels,
distinct or coherent; *stigmas* distinct, simple; *ovules* inserted at the inner angle of
the cells, anatropous or semi-anatropous or orthotropous. BERRY 1-2-3-4-celled,
1-few-seeded. SEEDS sub-globose; *testa* membranous, thin; *albumen* cartilaginous
or sub-horny. EMBRYO small, included, often remote from the hilum.

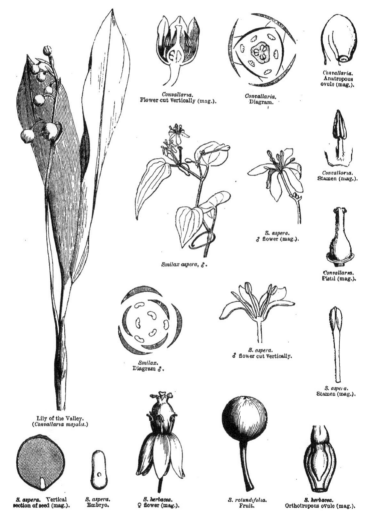

Convallaria.
Flower cut vertically (mag.).

Convallaria.
Diagram.

Convallaria.
Anatropous
ovule (mag.).

Smilax aspera, ♂.

S. aspera.
♂ flower (mag.).

Convallaria.
Stamen (mag.).

Convallaria.
Pistil (mag.).

Smilax.
Diagram ♂.

S. aspera.
♂ flower cut vertically.

S. aspera.
Stamen (mag.).

Lily of the Valley.
(*Convallaria majalis.*)

S. aspera. Vertical
section of seed (mag.).

S. aspera.
Embryo.

S. herbacea,
♀ flower (mag.).

S. rotundifolia.
Fruit.

S. herbacea.
Orthotropous ovule (mag.).

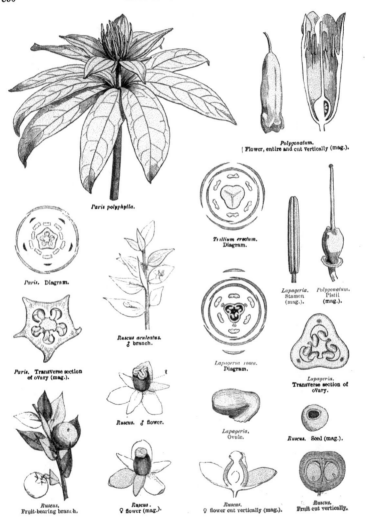

Paris polyphylla.

Polygonatum.
Flower, entire and cut vertically (mag.).

Paris. Diagram.

Trillium erectum.
Diagram.

Ruscus aculeatus.
♂ branch.

Lapageria.
Stamen
(mag.).

Polygonatum.
Pistil
(mag.).

Paris. Transverse section
of ovary (mag.).

Lapageria rosea.
Diagram.

Lapageria.
Transverse section of
ovary.

Ruscus. ♂ flower.

Lapageria.
Ovule.

Ruscus. Seed (mag.).

Ruscus.
Fruit-bearing branch.

Ruscus.
♀ flower (mag.).

Ruscus.
♀ flower cut vertically (mag.).

Ruscus.
Fruit cut vertically.

TRIBE I. *CONVALLARIEÆ.*

Flowers axillary ☿ or polygamous. Styles united; ovules semi-anatropous.—
Leaves radical, or cauline and alternate.

PRINCIPAL GENERA.

Drymophila.	Clintonia.	Ruscus.	* Streptopus.
Majanthemum.	Callixene.	* Polygonatum.	Smilax.
Philesia.	* Convallaria.	Luzuriaga.	Lapageria.
* Smilacina.	Ripogonum.		

TRIBE II. *PARIDEÆ.*

Flowers terminal ☿. Styles distinct; ovules anatropous.—Leaves whorled.

PRINCIPAL GENERA.

Paris.	* Trillium.	Medeola.

Smilaceæ approach *Melanthaceæ* in habit, if we compare *Streptopus* and *Uvularia*, &c., the fruit of which is fleshy when young, as is that of *Asphodeleæ*. They are closely allied to the berried *Asparageæ*, and are only distinguished by their membranous testa, cartilaginous albumen, and the habit of some genera, which recalls that of Dicotyledons. These differences, although slight, nevertheless led R. Brown to restrict *Smilaceæ* by separating them from the true *Asparageæ*, which he joined to *Liliaceæ.*

Lapageria and its two allied genera (*Callixene* and *Philesia*) approach *Smilax* in their vegetation, organs, fruits, and depurative properties (whence their roots are employed as a substitute for Sarsaparilla), and bear flowers which but for size and colour resemble in plan of structure those of *Asparagus.* According to Planchon, they form an intermediate link connecting the genera on the confines of the group of *Smilaceæ* with *Liliaceæ* properly so called.

Smilaceæ mostly grow in tropical and extra-tropical regions of the New World, from Canada to the Straits of Magellan. Half of the species occur north of the tropic of Cancer, a quarter inhabit the same latitudes of Europe and Asia, the other quarter are dispersed over tropical Asia and Australasia; South Africa appears to contain none. The genera *Polygonatum, Convallaria, Smilacina, Majanthemum, Streptopus*, belong to the temperate and cold regions of the northern hemisphere; *Ruscus* to the south of Europe and the Canaries; *Medeola* to North America; *Trillium* inhabits cool and shady places in America and North Asia. *Paris* grows in Central Europe and Asia; *Smilax* is spread over the temperate and tropical regions of both hemispheres.

Smilaceæ, divided into two tribes by their botanical characters, are so also by their properties. *Paris quadrifolia* and its congeners are regarded as poisonous, narcotic and acrid; its leaves, root and berries were formerly employed medicinally. The root of *Medeola virginica* (Indian Cucumber-root), is used by Anglo-American doctors as a diuretic and emetic.

The root of our *Polygonatum*, called Solomon's Seal, from the circular marks left on the rhizome by the flowering stems, is inodorous, sugary, mucilaginous, astringent, and was formerly considered a vulnerary. From its abundant starch, it has been mixed in bread in some parts of North Europe; its shoots are edible, like Asparagus, but its berries are nauseous, emetic and purgative. The berries of *Smilacina racemosa* are considered as a tonic for the nerves, as are those of the Lily of the Valley (*Convallaria majalis*), well known for its odorous flowers; its root is a sternutatory, and yields a drastic extract. The leaves of *Streptopus amplexifolius* are used in popular medicine for astringent gargles; the young root is eaten as a salad. The mucilaginous, slightly acrid and bitter roots of *Ruscus* share the qualities of *Asparagus*, and were formerly valued as an aperient, diuretic and emmenagogue. Their seeds, when roasted, have an agreeable aroma, whence, like those of *Asparagus*, they have been used as a substitute for coffee. The roots of Sarsaparilla, so valuable in the treatment of syphilis, belong to different species of *Smilax* (*S. Sarsaparilla, officinalis, papyracea, syphilitica*), natives of tropical America. The South European

S. aspera, nigra, mauritanica and *alpina*, yield Italian Sarsaparilla, the properties of which are analogous, but much inferior.   The China-root, procured from Asiatic species (*S. China, zeylanica, perfoliata*), possesses the same qualities as the American Sarsaparilla.   The bulky roots of some species of the same genus, and of *Ripogonum* of Asia and Australia, are full of starch, and hence edible.   That of *Luzuriaga radicans* is used in Peru and Chili as a substitute for Sarsaparilla.

XL. *ASTELIEÆ, Brongniart.*

Astelia hemichrysa. Fruit.	*A. hemichrysa.* Seed, entire and cut vertically.	*A. hemichrysa.* Albumen cut vertically (mag.).	*A. Solandri.* Seed cut vertically.	*A. Solandri.* Embryo (mag.).

Perennial tufted HERBS, often epiphytal on old trees.   ROOTS fibrous.   LEAVES radical, imbricate, linear-lanceolate or ensiform, keeled, covered below, or on both sides, with long silky or silvery hairs.   FLOWERS polygamo-diœcious, racemed or panicled, or rarely sub-solitary; *pedicels* not jointed, 1-bracteolate at the base. PERIANTH sub-glumaceous [rather membranous or sub-coriaceous], silky outside, 6-partite, imbricate, persistent.   STAMENS 6, inserted at the base of the perianth; *anthers* introrse.   OVARY 3-celled (*Astelia Solandri, nervosa*, &c.), or 1-celled by absorption of the septa, and with 3 parietal placentas (*A. linearis, Cunninghamii*, &c.); *style* 1 or 0; *stigmas* 3; *ovules* numerous, anatropous.   FRUIT a berry, or a loculicidally 3-valved capsule.   SEEDS more or less numerous, appendiculate at the top or at the two ends; *testa* black, crustaceous; *endopleura* membranous; *albumen* thick.   EMBRYO straight, cylindric, axile.

PRINCIPAL GENERA.

Astelia.   Milligania.

Asteliæ are not closely allied to any family, but they most nearly approach *Hypoxideæ* in their radical Grass-like and velvety leaves, their perianth, andrœcium, ovary, &c.   The habit of most of the species recalls that of *Tillandsia* amongst *Bromeliaceæ*; like these they are often epiphytes, and live on large trees, where they resemble birds' nests; others inhabit swamps.   They are met with in New Zealand, Bourbon, Tasmania, the Sandwich Islands and South America.

Blume has described a Javanese diœcious undershrub, with fibrous root, lanceolate leaves tomentose beneath, and panicled flowers with a six-partite persistent perianth, three-celled ovary and one-seeded berry, of which he made his genus *Hanguana*, which he considers near *Asteliæ*.

XLI. *ASPARAGEÆ.*

(ASPARAGI, *partim, Jussieu.*—ASPHODELEARUM *genera, Br.*—ASPARAGINEARUM *genera,*
A. Richard.—LILIACEARUM *genera, Endlicher, Lindl., Brongn.*)

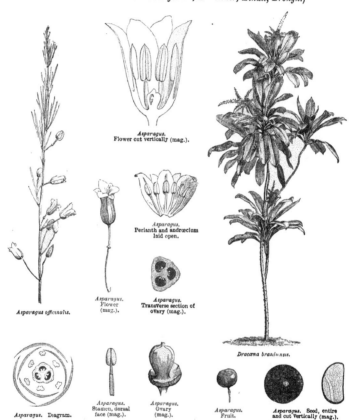

Asparagus.
Flower cut vertically (mag.).

Asparagus.
Perianth and androecium
laid open.

Asparagus.
Flower
(mag.).

Asparagus.
Transverse section of
ovary (mag.).

Asparagus officinalis.

Asparagus. Diagram.

Asparagus.
Stamen, dorsal
face (mag.).

Asparagus.
Ovary
(mag.).

Dracæna brasiliensis.

Asparagus.
Fruit.

Asparagus. Seed, entire
and cut vertically (mag.).

FLOWERS *usually* ☿, *regular, pedicels jointed.* PERIANTH *inferior, petaloid, 6-fid*
or -partite, 2-seriate. STAMENS 6, *perigynous or hypogynous.* OVARY *superior,*

3-celled; STYLE *simple*. FRUIT *a berry*. SEEDS *with black crustaceous testa* ; ALBUMEN *fleshy* ; RADICLE *variable in direction*.

HERBS, SHRUBS or TREES, and then marked with annular scars. ROOT tuberous or fibrous. STEM sometimes sarmentose (*Eustrephus, Myrsiphyllum*, &c.). LEAVES distichous or alternate, distant, or fascicled at the top of the branches, often sheathing, sessile or petioled, linear, ensiform, ovate-lanceolate, or elliptic, nerves parallel (*Dracæna*), or divergent (*Cordyline*), sometimes reduced to membranous scales, and bearing in their axils fascicled filiform simple green branches, which replace leaves (*Asparagus*). FLOWERS ☿, or rarely diclinous (*Asparagus*), solitary, or variously disposed ; *pedicels* jointed. PERIANTH inferior, petaloid, 6-partite or -fid ; *segments* campanulate, connivent or spreading. STAMENS 6, inserted on the receptacle, or base of the perianth, rarely on its throat (*Cordyline*); *filaments* filiform, sometimes swollen at the top (*Dianella*); *anthers* 2-celled, basifixed, linear or sagittate, or rarely dorsifixed (*Asparagus*), and versatile (*Cordyline*). OVARY 3-celled ; *style* simple ; *stigma* 3-lobed ; *ovules* 1-2-several, on the inner angle of the cells, semi-anatropous or anatropous. BERRY globose or 3-lobed, 3-celled, often 1-seeded by arrest. SEEDS sub-globose, ovoid or angular ; *testa* black, brilliant, crustaceous or coriaceous; *hilum* usually ventral, sometimes strophiolate (*Cordyline*); *albumen* fleshy. EMBRYO axile or excentric, straight or curved ; *radicle* centripetal, inferior or vague.

PRINCIPAL GENERA.

* Dianella.	Myrsiphyllum.	Eustrephus.
* Cordyline.	* Asparagus.	* Dracæna.

Asparageæ are intermediate between *Liliaceæ* and *Smilaceæ*, from which they are with difficulty distinguished. They differ from *Liliaceæ* only in their berried fruit ; and their black and crustaceous testa brings them especially near the tribes *Hyacinthineæ* and *Asphodeleæ*. They are still nearer *Smilaceæ* in their fleshy fruit and habit, the testa being their only differential mark ; this difference was disregarded by Jussieu and A. Richard, and most botanists now unite *Asparageæ* and *Smilaceæ* with *Liliaceæ* proper ; but by this combination the characters of *Liliaceæ* become so indefinite that various genera which do not belong to it must be technically included, and the limitation of one of the most natural groups of the Vegetable Kingdom be rendered very uncertain. It is with the view of clearly defining these families, and thus simplifying their study, that we have isolated *Liliaceæ, Asparageæ* and *Smilaceæ*, and placed the nearly-related *Aphyllanthes, Xerotes, Abama, Ophiopogon, Aspidistra*, &c., in close proximity to them. The species of *Asparagus* are dispersed over the temperate and hot regions of the Old World, but are absent from the New. *Cordyline* inhabits the tropics [and temperate regions] of the southern hemisphere. *Dianella* is scattered over Madagascar, tropical India, Malacca, Polynesia, Australia and New Zealand ; *Eustrephus* is Australian ; *Myrsiphyllum* is confined to South Africa. The true *Dracænas* are met with in Brazil [?], India, Africa, and their neighbouring islands.

Asparageæ are not less distinguished from the true *Liliaceæ* by their botanical characters than by their properties, especially those residing in their organs of vegetation. The roots of the Asparagus (*A. officinalis*) were formerly reckoned among the five principal aperient roots ; its berries and seeds were also prescribed as diuretics and aphrodisiacs ; it is cultivated throughout Europe for its shoots, from which a sedative syrup is prepared, recommended in affections of the heart. Its properties are due to a peculiar principle (*asparagine*), which abounds most in a Mediterranean species (*A. acutifolius*). The roots of *Cordyline* are used in tropical Asia for dysentery ; the flowers of *C. reflexa* are considered emmenagogues. *C. australis*, from the South Sea Islands, produces a fleshy root called *ti* by the New Zealanders, who use it as food, and also prepare from it a spirit sought by European sailors for its antiscorbutic qualities.[1] The root of *Dianella odorata* is used in Java, according to Blume, with other aromatics, in the composition of pastilles for burning. The Dragon Tree (*Dracæna Draco*), remarkable for its monstrous dimensions and prodigious longevity, exudes from its trunk a red resinous juice (one of the many kinds of Dragon's Blood), used medicinally as an astringent. This resin has its analogue amongst *Xanthorrhœas*, which we have placed near *Xerotideœ*. The Dragon Tree of Orotava is visited by all travellers in the Canary Islands. Its trunk below the lowest branches is 80 feet in height, and ten men holding hands can scarcely encircle it. When Teneriffe was discovered in 1402, tradition affirms that it was already as large as it is now; a tradition confirmed by the slow growth of the young Dragon Trees in the Canaries, of which the age is exactly known; whence it has been calculated that the Dragon Tree of Orotava is the oldest plant now existing on the globe.

XLII. *ROXBURGHIACEÆ, Wallich.*[2]

[Twining half-shrubby PLANTS. LEAVES alternate, petioled, ovate, coriaceous; *nerves* parallel, with cross venules. FLOWERS large, solitary, axillary, fœtid. PERIANTH of 4 lanceolate petaloid segments. STAMENS 4, hypogynous; *filaments* very short; *anthers* large, basifixed, 2-celled, dehiscence longitudinal, produced into a very long appendage. OVARY 1-celled; *style* 0; *stigma* pointed; *ovules* many, on 2 basal placentas, anatropous. CAPSULE 1-celled, 2-valved, many-seeded. SEEDS narrow, obovoid, appendiculate, on very long funicles, which form two fascicles, and have a whorl of vesicles round the apex below the seed; *testa* thick, furrowed; *albumen* fleshy. EMBRYO slender, axile.

ONLY GENUS.

Roxburghia.

A genus of very few species, natives of Trans-Gangetic India and the Malayan Islands. The tuberous root is candied in India.—ED.]

XLIII. *XEROTIDEÆ, Endlicher.*

Perennial HERBS. STEM 0 or very short. LEAVES Grass-like or filiform, dilated at the base, or reduced to radical sheaths (*Aphyllanthes*). FLOWERS ☿, or diœcious (*Xerotes, Dasylirion*), in a raceme, spike, head or umbel. PERIANTH petaloid or sub-coloured (*Xerotes, Abama*), with 6 equal leaflets or segments. STAMENS 6, hypogynous or perigynous; *anthers* 2-celled, oblong or ovoid, sometimes peltate (*Xerotes*). OVARY free, with 3 1–2- rarely several-ovuled cells (*Abama, Xanthorrhœa*), sometimes 1-celled (*Calectasia*); *styles* 3, usually united; *ovules* attached near the base of the cells, erect. CAPSULE with 3 loculicidal semi-septiferous valves, sometimes 1-seeded and indehiscent (*Kingia, Calectasia*); *albumen* fleshy or cartilaginous. EMBRYO straight, basilar or axile.

PRINCIPAL GENERA.

| Abama. | Dasylirion. | Sowerbæa. | Aphyllanthes. |
| Xerotes. | Xanthorrhœa. | Kingia. | Calectasia. |

[1] I am unaware of the authority for this statement.—ED.   [2] This Order is omitted in the original.—ED.

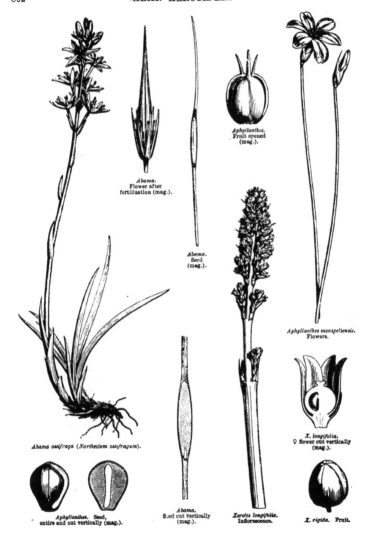

Abama.
Flower after
fertilization (mag.).

Aphyllanthes.
Fruit opened
(mag.).

Abama.
Seed
(mag.).

Aphyllanthes monspeliensis.
Flowers.

Abama ossifraga (Narthecium ossifragum).

X. longifolia.
♀ flower cut vertically
(mag.).

Aphyllanthes. Seed,
entire and cut vertically (mag.).

Abama.
Seed cut vertically
(mag.).

Xerotes longifolia.
Inflorescence.

X. rigida. Fruit.

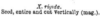

Xerotes longifolia.
Diagram ♀.

X. longifolia.
Diagram ♂.

X. rigida.
Seed, entire and cut vertically (mag.).

Dasylirion.
Young fruit (mag.).

These genera were formerly placed among *Junceæ*, but have now been withdrawn and grouped by themselves. *Abama* (*Narthecium*) is European and North American; *Aphyllanthes* South European; *Xerotes*, *Xanthorrhœa* and *Sowerbœa* Australian; *Dasylirion* Mexican.

The stem of *Xanthorrhœa arborea* exudes a resinous yellow juice, commonly termed Botany Bay Gum, of an acrid taste, and having when burnt the smell of benzoin; it is used by the Australian doctors for lienteria (a form of diarrhœa), and the different diseases of the thoracic cavity. *Abama ossifraga* was formerly classed among vulneraries; it inhabits the swamps of North and Central Europe, and was supposed to soften the bones of the sheep and oxen which fed on it; to which prejudice it owes its specific name.

XLIV. *JUNCEÆ.*

(JUNCI, *Jussieu.*—JUNCEÆ, *D.C.*—JUNCACEÆ, *Agardh.*)

FLOWERS *usually* ☿. PERIANTH *inferior, 6-phyllous, glumaceous, 2-seriate.* STAMENS 6, *or rarely 3, inserted at the base of the perianth-segments.* OVARY *superior, 3- or 1-celled, 1-many-ovuled;* OVULES *erect, anatropous.* CAPSULE *1-3-celled, loculicidal or septifragal.* SEEDS *albuminous.* EMBRYO *basilar;* RADICLE *inferior.*— STEM *herbaceous.* LEAVES *alternate, sheathing.*

Annual or perennial HERBS, cæspitose, or with a creeping rhizome. STEMS cylindric, spongy, or sometimes chambered by medullary septa, simple or rarely branched, leafy, or sometimes shortened and giving off flowering scapes. LEAVES alternate, base sheathing, blade linear, pointed, nerves striate, entire or denticulate, glabrous or hairy, flat channelled or cylindric, sometimes compressed and chambered like the stems, finally sometimes obsolete. FLOWERS ☿, or diclinous by arrest, regular, in a cyme spike or head, rarely solitary, each furnished with a bracteole. PERIANTH of 6 2-seriate leaflets, equal, glumaceous-scarious, persistent, æstivation imbricate. STAMENS usually 6, opposite the perianth-segments, and inserted at their base, or hypogynous, rarely 3, opposite to the outer perianth-segments; *filaments* filiform, free or united by their base; *anthers* introrse, 2-celled, dehiscence longitudinal. OVARY free, 3-celled, or 1-celled (*Luzula, Rostkovia*); *style* terminal, simple; *stigmas* 3, filiform, papillose; *ovules* 3, basilar, or numerous and with central or parietal placentation, erect or ascending, anatropous. CAPSULE 1-3-celled, with 3 semi-septiferous valves sometimes coherent at their base, rarely septifragal. SEEDS

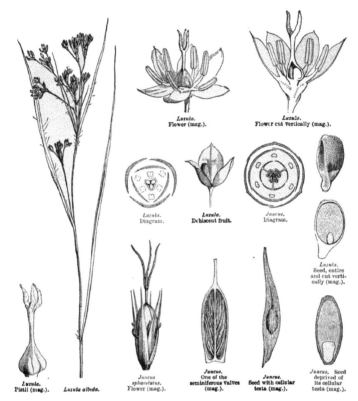

Luzula.
Flower (mag.).

Luzula.
Flower cut vertically (mag.).

Luzula.
Diagram.

Luzula.
Dehiscent fruit.

Juncus.
Diagram.

Luzula.
Seed, entire
and cut verti-
cally (mag.).

Luzula.
Pistil (mag.).

Luzula albida.

*Juncus
sphacelatus.*
Flower (mag.).

Juncus.
One of the
seminiferous valves
(mag.).

Juncus.
Seed with cellular
testa (mag.).

Juncus. Seed
deprived of
its cellular
testa (mag.).

3–∞, small, rounded or angular, with a membranous testa, or fusiform with a cellular loose testa; *albumen* fleshy, dense, rarely farinaceous (*Prionium*). EMBRYO included, basilar; *radicle* thick, near the hilum, and inferior.

PRINCIPAL GENERA.

Luzula.   Prionium.   *Juncus.

Junceæ, which approach *Liliaceæ* in several points of structure, are especially distinguished by their glumaceous perianth and habit, which recalls that of *Cyperaceæ* and *Gramineæ*, while the triandrous species approach *Restiaceæ*; but in the latter the three stamens are opposite to the inner sepals, the ovules are orthotropous and pendulous, &c.

Junceæ inhabit damp meadows and swamps, or grassy and woody mountain regions; they are scarce in dry ground. They mostly grow in north temperate latitudes, and some species advance to both polar regions; they become rarer as we approach the equator. *Juncus* and *Luzula* are cosmopolitan, and are met with on the highest mountains of both worlds. *Prionium* is South African; *Rostkovia* inhabits the Maganellic lands [and New Zealand].

The properties of *Junceæ* are uninteresting. The fruit of *Juncus acutus*, baked and steeped in wine, is said to be a diuretic, and to stop menorrhagia, but it gives headaches. The rhizomes of *J. conglomeratus*, *effusus, glaucus*, and especially of *Luzula vernalis*, are popular diuretics in Central Europe. *J. glaucus* is cultivated by gardeners to make bands of. The Chinese use the pith of certain species for candle-wicks. [The farthing rushlight of England had a wick of rush pith.]

XLV. *RAPATEÆ*,[1] *Endlicher.*

[Perennial usually tall marsh HERBS. ROOTSTOCK short. LEAVES equitant, ensiform, rigid, base sheathing. SCAPES erect. SPATHE 1–2-leaved. FLOWERS ☿, in a dense terminal spathaceous capitulum. Yellow or rose-coloured, sessile or pedicelled, with many imbricate bracts. PERIANTH 6-partite; 3 outer leaflets navicular, rigid, imbricate; 3 inner petaloid, fugacious, often cohering in a tube, contorted in æstivation. STAMENS 6, inserted in pairs opposite the inner perianth-segments; *anthers* basifixed, 2-celled; apex tubular, opening by a terminal single or double pore, or terminated by a polliniferous or sub-glandular appendage. OVARY 3-celled; *style* simple, filiform; *stigma* capitate, papillose; *ovules* 1–2 in each cell, erect, anatropous. CAPSULE membranous or coriaceous, 1–3-celled, loculicidally 2–3-valved. SEEDS oblong or sub-globose; *hilum* basilar; *albumen* fleshy. EMBRYO immersed in the albumen, minute; *radicle* next the hilum.

GENERA.

Rapatea.	Spathanthus.	Schœnocephalum.

A small order allied to *Junceæ*, with the habit of *Cyperaceæ*, very peculiar anthers, and the perianth of *Xyrideæ*. Natives of Brazil; of no known use.—ED.]

XLVI. *PONTEDERIACEÆ.*

(PONTEDERIACEÆ, *A. Richard.*—PONTEDEREÆ, *Kunth.*)

FLOWERS ☿. PERIANTH *inferior, petaloid, 6-partite, irregular, persistent.* STAMENS *inserted on the perianth,* 6, *or* 3 *opposite to the inner segments.* OVARY *superior, of* 3 *many-ovuled cells, or* 2 *sterile, and* 1 *fertile and* 1-*ovuled.* FRUIT *a capsule, enveloped by the fleshy perianth.* ALBUMEN *farinaceous.* EMBRYO *straight, axile.*—*Marsh* PLANTS. STEM *herbaceous.* LEAVES *alternate, with sheathing petiole.*

Perennial aquatic or marsh HERBS, with a short rhizome or rooting stem. LEAVES alternate; *petiole* cylindric or vesicular, largely sheathing at the base; *blade*

---

[1] This Order is omitted in the original.—ED.

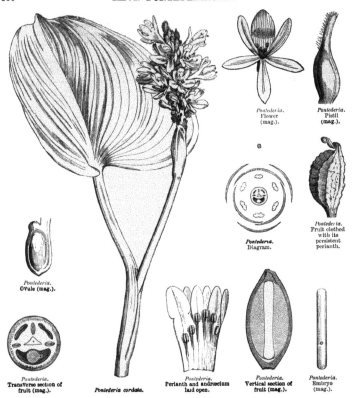

Pontederia.
Flower
(mag.).

Pontederia.
Pistil
(mag.).

Pontederia.
Diagram.

Pontederia.
Fruit clothed
with its
persistent
perianth.

Pontederia.
Ovule (mag.).

Pontederia.
Transverse section of
fruit (mag.).

Pontederia cordata.

Pontederia.
Perianth and andrœcium
laid open.

Pontederia.
Vertical section of
fruit (mag.).

Pontederia.
Embryo
(mag.).

oval, orbicular, or sub-cordate, with arched nerves; sometimes replaced by a petiolar phyllode representing a leaf with parallel nerves (*Heteranthera*). INFLORESCENCE axillary or sub-terminal. FLOWERS in a spike or panicle, ebracteate, usually sessile. PERIANTH inferior, petaloid, glandular, or hairy outside, usually white, blue or violet, infundibular or hypocrateriform; *tube* more or less elongated; *limb* 6-partite; *segments* unequal, obscurely 2-seriate, the inner median superior and larger; 2-labiate-ringent, spirally rolled in æstivation, marcescent or succulent. STAMENS inserted on the tube or throat of the perianth, sometimes 6, opposite to the 6 perianth-segments; sometimes 3, opposite to the inner segments, the anterior usually dissimilar; *filaments* filiform, arched, sometimes thickened in the middle; *anthers* introrse, basi-dorsi-fixed; cells 2, parellel, opening longitudinally. Ovary free, of 3 several-ovuled cells, sometimes sub-1-celled with 3 incomplete cells, of which 2 are smallest and empty, and one fertile 1-ovuled; *style* terminal, simple; *stigma* thick, unilateral, or obscurely 3-lobed; *ovules* anatropous, sometimes several horizontal or erect, sometimes one pendulous. Capsule enveloped by the persistent base of the perianth, with 3 many-seeded cells, and 3 loculicidal semi-septiferous valves, or indehiscent, 1-celled, 1-seeded. Seeds numerous, inserted at the inner angle of the cells, or solitary, pendulous from the top of the fertile cell, oblong-cylindric; *testa* parchment-like, striate or tessellated; *raphe* filiform, inconspicuous; *chalaza* apical, thickened; *albumen* farinaceous. Embryo straight, axile, clavate or cylindric; *radicle* near the hilum.

PRINCIPAL GENERA.

| Heteranthera. | * Pontederia. | Reussia. |

Pontederiaceæ have been placed by Brongniart in the same class with *Bromeliaceæ*, *Vellosieæ* and *Hæmodoraceæ*; they differ from the two latter in their free ovary and farinaceous albumen, and from *Bromeliaceæ* in their completely petaloid perianth, often triandrous andrœcium, ovary with unequal cells, and axile embryo as long as the albumen. They approach *Asphodeleæ* in their perianth and andrœcium; but are separated by habit, æstivation and the nature of the albumen. *Pontederiaceæ* especially inhabit America between 40° north latitude, and 30° south latitude; they are rare in tropical Africa and Asia.

Pontederia vaginalis is esteemed as a medicinal plant in Japan, Java, and on the Coromandel coast, where a decoction of its root is used in diseases of the liver and stomach; pulverized and mixed with sugar, it is administered for asthma; it is chewed for toothache; its leaves, bruised and mixed with milk, are administered in cholera; and its young shoots are edible.

XLVII. *PHILYDREÆ.*[1]

(Philydreæ, *Brown.*—Philydraceæ, *Lindl.*)

[Herbs with fibrous roots. Stems erect, simple, often woolly. Leaves ensiform; bases sheathing, equitant, narrow. Flowers spiked or racemose, enclosed in spathaceous persistent bracts, inodorous. Perianth of 2 (antero-posterior) membranous marcescent segments, yellow. Stamen opposite the posticous perianth-segment; *filament* dilated, with broad lateral wings and petaloid staminodes; *anther* 2-celled. Ovary superior, 3-celled; *style* excentric; *stigma* cupulate; *ovules* numerous, on the inner angles of the cells. Capsule 3-celled, loculicidally 3-valved, many-seeded. Seeds minute, horizontal; *testa* thick; *albumen* fleshy. Embryo axile, large.

GENERA.

| Philydrum. | Heteria. |

A small order of two genera, each monotypic, usually associated with *Xyrideæ*, but from which they differ in their three-celled ovary, and embryo included in the albumen. To me they appear more closely allied to *Commelyneæ*, from which they differ chiefly in the reduced perianth, stamens and large embryo. They are natives of eastern tropical Asia and Australia.—Ed.]

---

[1] This Order is omitted in the original.—Ed.

XLVIII. *COMMELYNEÆ.*

(JUNCORUM *genera, Jussieu.*—EPHEMERÆ, *Batsch.*—COMMELINEÆ, *Br.*—
COMMELYNACEÆ, *Lindl.*)

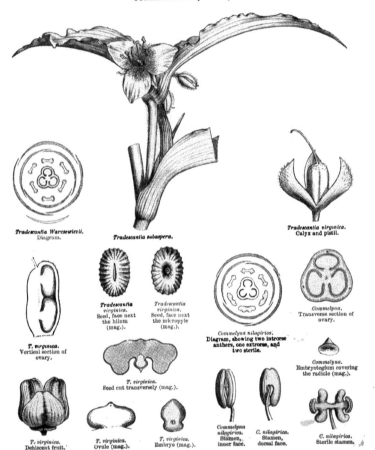

Tradescantia Warscewiczii.
Diagram.

Tradescantia subaspera.

Tradescantia virginica.
Calyx and pistil.

T. virginica.
Vertical section of ovary.

Tradescantia virginica.
Seed, face next the hilum (mag.).

Tradescantia virginica.
Seed, face next the micropyle (mag.).

Commelyna nilagirica.
Diagram, showing two introrse anthers, one extrorse, and two sterile.

Commelyna.
Transverse section of ovary.

T. virginica.
Seed cut transversely (mag.).

Commelyna.
Embryotegium covering the radicle (mag.).

T. virginica.
Dehiscent fruit.

T. virginica.
Ovule (mag.).

T. virginica.
Embryo (mag.).

Commelyna nilagirica.
Stamen, inner face.

C. nilagirica.
Stamen, dorsal face.

C. nilagirica.
Sterile stamen.

FLOWERS ☿. PERIANTH *inferior, double.* SEPALS 3. PETALS 3. STAMENS 6, *hypogynous, all fertile, or some sterile.* OVARY *superior, with 3 few-ovuled cells;* OVULES *orthotropous.* CAPSULE *with* 3–2 *cells, loculicidal.* SEEDS *albuminous.* EMBRYO *antitropous.*—STEM *herbaceous.* LEAVES *alternate.*

Succulent HERBS, annual with fibrous root, or perennial with a tuberous rhizome. STEMS cylindric, knotty. LEAVES alternate, simple, entire, sheathing at the base, flat or channelled, soft, nerved, sheath entire. FLOWERS ☿, or incomplete by arrest of the ovary, regular or sub-irregular, usually blue, solitary, fascicled, umbellate or racemed, furnished either with bracts or with spathaceous or cucullate 1–2-phyllous involucres. PERIANTH double; the outer calycine of 3 persistent sepals, the interior corolline of 3 distinct sessile or clawed petals, very rarely united at the base into a short tube (*Cyanotis*); caducous or marcescent, sometimes becoming fleshy after flowering (*Campelia*), one often dissimilar or obsolete; æstivation imbricate. STAMENS 6, hypogynous, opposite to the sepals and petals, sometimes in two groups (*Dichorisandra*), rarely 3–5 by arrest, some often antherless; *filaments* filiform, usually furnished with jointed hairs; *connective* dilated; *anthers* introrse, or rarely one extrorse and 2 introrse in the same flower, 2-celled, cells divergent, marginal on the connective, very rarely adnate to its anterior face, and contiguous parallel (*Dichorisandra*), dehiscence longitudinal, all fertile, or some sterile and deformed. OVARY free, 3-celled; *style* simple; *stigma* undivided, or obscurely 3-lobed, or sometimes concave (*Cyanotis*); *ovules* inserted at the inner angle of the cells, either more than two, peltate, 2-seriate on nerviform placentas; or 2 basifixed and collateral; or superimposed, 1 pendulous, the other erect. CAPSULE usually accompanied by the persistent perianth, 3-celled, or 2-celled by arrest, of 3–2 loculicidal semi-septiferous valves. SEEDS few or solitary, ovoid, angular, peltate, or nearly square; *testa* membranous or rigid, rugose or foveolate, closely adherent to the fleshy dense albumen; *hilum* ventral and depressed, or at one end of the seed, and large. EMBRYO pulley-shaped, sunk in a pit of the albumen, diametrically opposite to the hilum; *radicle* covered by a hood (*embryotegium*).

PRINCIPAL GENERA.

* Commelyna.	* Aneilema.	* Cyanotis.	* Dichorisandra.
* Tradescantia.	Spironema.	Campelia.	Cochliostemma.

Commelyneæ are distinguished from other Monocotyledonous families by their double perianth, clearly separable, like that of *Alismaceæ*, into calyx and corolla. Their habit and structure separate them from *Junceæ*, to which they were formerly united; they approach *Restiaceæ*, and especially *Xyrideæ*, in their antitropous embryo and sheathing leaves; but these differ in several other characters, and notably in the form and situation of their embryo, which is lenticular and pressed against the outside of the albumen. *Commelyneæ* grow in the tropics of both worlds and especially the New; a few are Australian, where they extend to 35° south latitude. Some extend to 40° north latitude.

Little is known of the properties of *Commelyneæ*. Many species possess an abundant mucilage, which is alimentary after being cooked. The tuberous rhizomes of some contain, besides mucilage, starch, which adds to their nutritive qualities; such are *Commelyna tuberosa, cœlestis, angustifolia, stricta,* &c.; other Mexican species are administered in diseases of the liver. The rhizome of *C. Rumphii* is praised as an emmenagogue. The tubers of *C. medica* are in use in China for coughs, asthma, pleurisy, and strangury. The herbage of *Tradescantia malabarica*, cooked in oil, is employed in the treatment of leprosy and ringworm. The Indians drink a decoction of *Cyanotis axillaris* for dropsy. *Tradescantia diuretica* is prescribed in Brazil.

XLIX. *MAYACEÆ*,[1] *Kunth.*

[Moss-like slender creeping pellucid PLANTS, growing in damp places. STEMS branched. LEAVES alternate, crowded, linear, emarginate, flaccid. FLOWERS ☿, axillary, solitary, peduncled, white pink or violet. PERIANTH of 6 pieces in two series, marcescent; outer of 3 herbaceous sepals; inner of 3 membranous petals. STAMENS 3, inserted on the base of the sepals; *anthers* basifixed, spathulate, imperfectly 2-celled, introrse. OVARY 1-celled; *carpels* opposite the petals; *style* filiform, persistent; *stigma* simple, minutely 3-lobed; *ovules* sessile, horizontal, on 3 parietal placentas, orthotropous. CAPSULE membranous, 1-celled, loculicidally 3-valved, many-seeded. SEEDS sub-globose, terminated by a conical tubercle; *testa* ribbed; *albumen* dense, of radiating cells. EMBRYO minute, half immersed in the albumen at the end opposite the hilum.

GENUS.

Mayaca.

A small order allied either to *Commelyneæ* or to *Xyrideæ* and *Restiaceæ*, according to the value given to the position of the embryo with regard to the albumen.

The species are all American, extending from Virginia to Brazil.—ED.]

L. *XYRIDEÆ*,[2] *Kunth.*

[Herbaceous Rush-like or Sedge-like tufted erect usually rigid PLANTS, often growing in watery places. ROOTS fibrous. STEMS simple. LEAVES radical, ensiform or filiform, base sheathing, often equitant. FLOWERS ☿, in terminal solitary heads of densely imbricating 1-flowered rigid scarious bracts. PERIANTH of 6 segments in two series; 3 outer segments calycine, lateral, navicular, persistent, rigid; anticous more membranous, larger, sheathing the lateral, caducous; 3 inner segments petaloid, clawed, claws more or less connate. STAMENS 3–6, inserted on the inner perianth-segments, 3 opposite fertile, the others sterile penicillate or obsolete; *filaments* filiform; *anthers* extrorse, 2-celled, dehiscence longitudinal. OVARY 1- or incompletely 3-celled; *style* terminal, 3-fid; *stigmas* 3, 3-lobed; *ovules* numerous, basal, orthotropous, funicle elongate. CAPSULE 1-celled, loculicidally 3-valved; or 3-celled and fenestrate at the base, operculate above. SEEDS numerous, angled or globose; *testa* coriaceous, striate or costate; *albumen* fleshy. EMBRYO minute, lenticular, placed outside the albumen at the end opposite the hilum.

PRINCIPAL GENERA.

Xyris. Abolboda.

Xyrideæ are closely allied to *Commelyneæ* and *Eriocauloneæ*, as also to *Rapateaceæ*, from which they differ in the embryo not being immersed in the albumen. They are natives chiefly of swamps in tropical and warm countries The leaves and root are used in the cure of itch and leprosy in India and South America.—ED.]

[1] This Order is omitted in the original.—ED.

[2] This Order is only alluded to in the original at the end of *Eriocauloneæ* (p. 873).—ED.

LI. *FLAGELLARIEÆ*, *Endlicher*.

Brongniart places near *Restiaceæ Flagellarieæ*, composed of the genera *Flagellaria* and *Joinvillea*, of which the following are the characters :—FLOWERS ☿. SEPALS 3, distinct. PETALS 3, distinct, scarious, like the sepals. STAMENS 6, hypogynous, free; *anthers* introrse. OVARY of 3 1-ovuled cells; *stigmas* 3, apical, divergent, filiform, papillose within from bottom to top; *ovules* pendulous by a short funicle from the top of each cell, orthotropous, micropyle inferior. FRUIT a 1-2-seeded berry. SEED with crustaceous testa; *albumen* farinaceous. EMBRYO minute, antitropous, lenticular, outside the albumen, covered with an embryotegium.—Reedy or sarmentose HERBS. LEAVES with long sheaths and parallel nerves.

Herbs of tropical Asia, Australia and New Caledonia.

LII. *ERIOCAULONEÆ*.

(JUNCORUM *genera*, *Jussieu*.—ERIOCAULONEÆ, *L.-C. Richard*.)

FLOWERS *monœcious or diœcious*. PERIANTH *inferior, double, the outer 2-3-phyllous, the inner sub-tubular, 3-2-fid*. STAMENS *double the number of the perigonial leaflets, inserted on the inner, the alternate often sterile*. OVARY *superior, of 2-3 1-ovuled cells*; OVULES *pendulous, orthotropous*. CAPSULE *2-3-celled, loculicidal*. SEEDS *albuminous*. EMBRYO *globose or sub-lenticular, antitropous, outside the albumen.* —STEM *or* SCAPE. LEAVES *cauline or radical, semi-sheathing*. FLOWERS *in a capitulum*.

HERBS inhabiting swamps or flooded grounds, perennial, stemless, rarely caulescent, very rarely suffrutescent. LEAVES linear, sub-fleshy, entire, sometimes fistular, nerves striate, semi-sheathing at the base, the radical crowded, the cauline alternate. FLOWERS minute, united into an involucrate capitulum, on a usually fleshy receptacle, incomplete, monœcious in the same capitulum, or rarely diœcious, each furnished with a bract, and accompanied by hairs or scales. PERIANTH double, the inner usually discoloured. — FLOWERS ☿: Outer PERIANTH of 2 lateral sepals, or 3, of which 1 is posterior ; inner tubular, sub-campanulate ; *limb* 2-fid or 3-toothed or 3-fid; *segments* imbricate, equal, or the anterior largest, æstivation imbricate. STAMENS inserted on the tube of the inner perianth, equal and opposite to its divisions, or twice as many ; some larger, opposite to these divisions ; the others smaller, alternate, often antherless or rudimentary ; *filaments* subulate, inflexed in æstivation ; *anthers* 2-celled, very rarely 1-celled, dorsifixed, dehiscence longitudinal. Rudimentary OVARIES 3-2, glanduliform or tuberculiform.—FLOWERS ♀: Inner and outer PERIANTHS 3- rarely 2-phyllous, the inner leaflets most delicate, sometimes replaced by 3 bundles of hairs (*Lachnocaulon, Tonina*), sometimes the claws distinct and blades coherent (*Philodice*); *staminodes* 0. OVARY free, of 3-2 carpels, sometimes with a second series of superimposed and sterile carpels, simulating inner stigmas (*Pæpalanthus*); *style* terminal, very short; *stigmas* as many as carpels, simple or 2-fid,

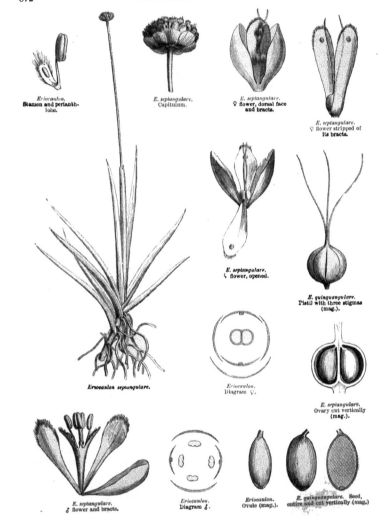

Eriocaulon.
**Stamen and perianth-
lobe.**

E. septangulare.
Capitulum.

E. septangulare.
♀ flower, dorsal face
and bracts.

E. septangulare.
♀ flower stripped of
its bracts.

E. septangulare.
♂ flower, opened.

E. quinquangulare.
Pistil with three stigmas
(mag.).

Eriocaulon septangulare.

Eriocaulon.
Diagram ♀.

E. septangulare.
Ovary cut vertically
(mag.).

E. septangulare.
♂ flower and bracts.

Eriocaulon.
Diagram ♂.

Eriocaulon.
Ovule (mag.).

E. quinquangulare. Seed,
entire and cut vertically (mag.)

surrounding the sterile carpels; *ovules* solitary in each carpel, pendulous near the top of the inner angle, orthotropous. CAPSULE crowned by the style, surrounded by the perianth, 2-3-celled, loculicidal. SEEDS pendulous, ovoid or sub-cylindric, longitudinally ribbed ; ribs membranous, hyaline, breaking up into fine hairs; *testa* coriaceous, shining ; *endopleura* 0 ; *albumen* farinaceous. EMBRYO diametrically opposite to the hilum, pressed to the outside of the albumen, antitropous, sub-globose or lenticular.

PRINCIPAL GENERA.

| Eriocaulon. | Tonina. | Philodice. | Pæpalanthus. | Lachnocaulon. |

Eriocauloneæ, with *Commelyneæ, Xyrideæ, Restiaceæ*, and *Centrolepideæ*, form Von Martius' class of *Enantioblasteæ*, so named on account of the invariable position of the embryo at the opposite end of the seed to the hilum. *Eriocauloneæ* approach *Restiaceæ* in the ovary with 2-3 1-ovuled cells, the pendulous and orthotropous ovule, the structure of the seed and the direction of the embryo ; but *Restiaceæ* are separated by their inflorescence, the completely glumaceous perianth, the 1-seriate stamens, 1-celled anthers, and the smooth testa with naked or strophiolate hilum, &c.

Eriocauloneæ are tolerably rich in species ; two-thirds of the family are tropical American, and half of the remainder are North Australian. A few species are found in tropical Asia, Madagascar, and the islands of South Africa. They are less rare in North America, where they extend to 44° N. latitude ; one alone (*Eriocaulon septangulare*) inhabits North America and Scotland, where it has been met with in Skye [and in the west of Ireland].

Respecting the properties of this family, nothing is known except as regards *Eriocaulon setaceum*, of which the herbage, cooked in oil, is used as a popular antipsoric in India.

LIII. *RESTIACEÆ.*

(JUNCORUM *genera, Jussieu.*—RESTIACEÆ, *Br.* et CENTROLEPIDEÆ, *Desvaux.*)

FLOWERS *diclinous.* PERIANTH *inferior, calyciform, of 2-6 2-seriate glumes, or imperfect.* STAMENS 3-2. OVARY *3-2-1-celled* ; OVULES *solitary, orthotropous, pendulous.* FRUIT *a 3-celled capsule or a nut.* SEED *albuminous.* EMBRYO *antitropous, outside the albumen.*—STEM *or* SCAPE. LEAVES *all radical or cauline, sheathing.* FLOWERS *spiked or racemed.*

HERBS or UNDERSHRUBS with a creeping rhizome. STEMS branched-knotty or simple. LEAVES either all radical, crowded, or cauline alternate, sheathing at the base, sheath split, blade entire, narrow-linear or arrested. FLOWERS regular, spiked racemed or panicled, mixed with scarious bracts, usually diclinous, rarely ☿ (*Lepyrodia*). PERIANTH glumaceous, of 4-6 2-seriate glumes, 2 of the outer lateral, and 1 posterior—the inner larger or smaller, persistent in the ♀ flowers ; sometimes of a single scale. STAMENS 2-3, opposite to the inner glumes, and inserted at their base, sterile or absent in the ♀ flowers ; *filaments* filiform, usually free ; *anthers* 1-celled, dorsifixed, peltate, rarely 2-celled (*Lyginia, Lepidanthus, Anarthria*, &c.), dehiscence longitudinal, introrse. OVARY free, 3-2-celled, rarely 1-celled (*Chætanthus, Leptocarpus, Loxocarya*) ; *styles* 1-3, continuous with the back of the carpels, distinct, or jointed at the base ; *stigmas* 1-3, plumose, usually (?) introrse ; *ovules* solitary in each cell, pendulous, orthotropous. FRUIT a loculicidal capsule, or a follicle or nut.

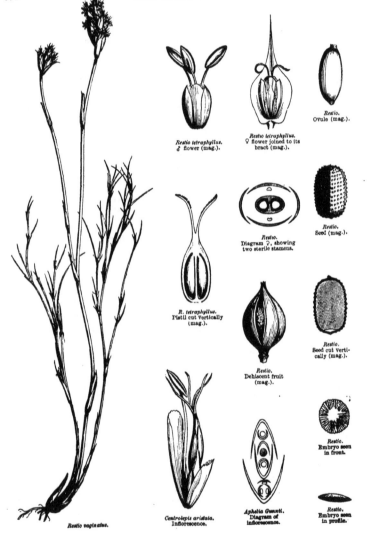

Restio tetraphyllus.
♂ flower (mag.).

Restio tetraphyllus.
♀ flower joined to its
bract (mag.).

Restio.
Ovule (mag.).

Restio.
Diagram ♀, showing
two sterile stamens.

Restio.
Seed (mag.).

R. tetraphyllus.
Pistil cut vertically
(mag.).

Restio.
Seed cut verti-
cally (mag.).

Restio.
Dehiscent fruit
(mag.).

Restio.
Embryo seen
in front.

Restio vaginatus.

Centrolepis aristata.
Inflorescence.

Aphelia Gunnii.
Diagram of
inflorescence.

Restio.
Embryo seen
in profile.

SEEDS pendulous from the top of the cell; *testa* coriaceous, hard, crustaceous, smooth or tubercled, rarely membranous; *hilum* naked or strophiolate; *albumen* fleshy, copious. EMBRYO at the end of the seed opposite to the hilum, and pressed against the outside of the albumen, lenticular, antitropous.

PRINCIPAL GENERA.

Restio.	Thamnochortus.	Lyginia.	Anthochortus.	Calopsis.
Elegia.	Lepyrodia.	Leucoplocus.	Willdenovia.	

Restiaceæ are closely allied to *Eriocauloneæ* (which see). They approach *Junceæ* in their rhizome, knotty stem, alternate sheathing leaves, glumaceous 3-seriate perianth, 3-1-celled ovary, capsular fruit and fleshy albumen; and are distinguished by their split sheath, 3-2-stamens, orthotropous ovule and lenticular embryo outside the albumen. They have also some affinity with *Cyperaceæ* in habit, diclinism, number of stamens, &c.; but *Cyperaceæ* differ in the leaf-sheaths not being split, the perianth being replaced by bristles or scales, the 2-celled and basifixed anthers, the erect anatropous ovule, farinaceous albumen, more or less included embryo, &c.

Restiaceæ all live south of the equator; the greater number are South African [and Australian]; some inhabit Madagascar and Australia. None have yet been observed in the New World.

Restiaceæ possess no known property, or other use than that made by savages of their stems to thatch their huts.

CENTROLEPIDEÆ, Desvaux [*Desvauxiaceæ*, Lindl.], originally annexed to *Restiaceæ*, then separated, and again united to this family, differ only in the perianth being reduced to one or two sub-opposite glumes; their andrœcium is monandrous; the ovary consists of one or more irregularly connate 1-celled carpels, each with a filiform style; and the fruit is a membranous utricle, opening laterally by a longitudinal slit.

GENERA.

Centrolepis.	Aphelia.	Alepyrum.	Gaimardia.

[*Centrolepideæ* inhabit sandy places and swamps in Australia, and the mountains of New Zealand and Tierra del Fuego.]

LIV. *CYPERAOEÆ.*

(CYPEROIDEÆ, *Jussieu.*—CYPERACEÆ, *Br., D.C.*)

FLOWERS *glumaceous*, ☿ *or diclinous.* PERIANTH 0, *or replaced by bristles.* STAMENS *hypogynous, usually 3–2;* ANTHERS *basifixed.* OVARY 1-*celled,* 1-*ovuled;* STYLES 3–2; OVULE *basilar, anatropous.* ACHENE. SEED *albuminous.* EMBRYO *minute, included or exserted.*—STEM *usually angular.* LEAVES *Grass-like, sheath very rarely split.* FLOWERS *in spikes.*

Usually Grass-like HERBS, with rhizome shortened or creeping, stoloniferous, sheathed by foliar scales, sometimes tuberous at its extremities. STEMS angular or cylindric, without nodes, or septate within (*Eleocharis geniculata,· articulata,* &c.), often hypogeal, last internode elongated and epigeal, simple, or very rarely branched, solid when young, fistular when full grown. LEAVES alternate, springing from the nodes, equitant in 2–3 rows; *petiole* in a closed or very rarely split sheath, sometimes with no blade, but elongated and mucronate; *blade* linear or ribbon-like, or channelled, with parallel nerves and transverse venules, margin entire, often scabrid;

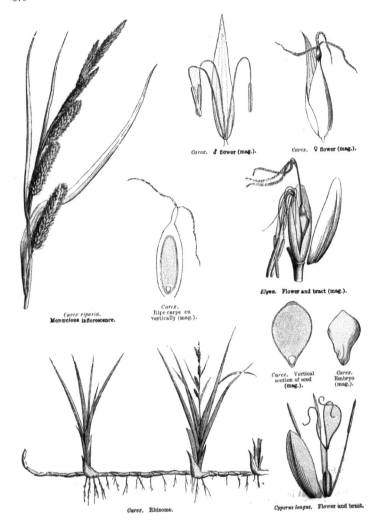

Carex. ♂ flower (mag.).

Carex. ♀ flower (mag.).

Elyna. Flower and bract (mag.).

Carex.
Ripe carpe cu
vertically (mag.).

Carex riporia.
Monœcious inflorescence.

Carex. Vertical
section of seed
(mag.).

Carex.
Embryo
(mag.).

Carex. Rhizome.

Cyperus longus. Flower and bract.

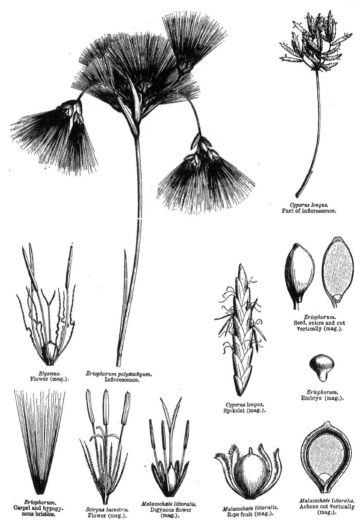

Cyperus longus.
Part of inflorescence.

Blysmus.
Flower (mag.).

Eriophorum polystachyum.
Inflorescence.

Cyperus longus.
Spikelet (mag.).

Eriophorum.
Seed, entire and cut
vertically (mag.).

Eriophorum.
Embryo (mag.).

Eriophorum.
Carpel and hypogynous bristles.

Scirpus lacustris.
Flower (mag.).

Malacochæte littoralis.
Digynous flower
(mag.).

Malacochæte littoralis.
Ripe fruit (mag.).

Malacochæte littoralis.
Achene cut vertically
(mag.).

stipule axillary, membranous, joined throughout its length by its dorsal face to the leaf-sheath, not longer than the sheath, or produced into a cushion or membrane (*ligule*), free only at the top. FLOWERS ☿ or monœcious or diœcious, in spikelets which are rarely solitary, usually spiked panicled or glomerate, and furnished with bracts or polymorphous involucres. FLOWERS each with 1-2 scarious bracts (*glumes*), solitary, or inserted in 2 or several rows on a common pedicel, and forming a 1-few-many-flowered spikelet; *bracts* below the spikelet often flowerless, sometimes heteromorphous, and serving as a common spathe to the spikelet. PERIANTH 0 or formed of 6, sometimes 3–∞ bristles, sometimes united into a ring at their base; [of 6 2-seriate regular coriaceous segments in *Oreobolus*]. STAMENS hypogynous, usually 3, of which 1 is anterior and 2 posterior, sometimes 2–1, rarely 4 (*Hypolytrum*) or 5 (*Scleria, Caustis*), or 6–8 (*Lepironia, Diplasia*), or 12 (*Evandra, Chrysithrix*); *filaments* filiform or flattened, free; *anthers* basifixed, 2-celled, linear, introrse, dehiscence longitudinal. OVARY free, sessile or stipitate, often surrounded at its base by a cupuliform disk (*Scleria, Ficinia, Melancranis*, &c.), or by a membranous adherent ring (*Fimbristylis*), or by 3 spathulate staminodes (*Fuirena*); compressed or plano-convex, or more usually with 3 angles answering to the 3 stamens, 1-celled, 1-ovuled; *styles* 3, rarely 2, stigmatiferous on their inner edge, more or less coherent below, base continuous with or thickened and jointed to the top of the ovary; *ovule* inserted at the inner base of the ovary, erect, anatropous. FRUIT 1-seeded, indehiscent, lenticular, plano-convex, 3-gonous, cylindric or globose, smooth, spotted, scabrid, bristly, tubercular, rugose or striate, usually terminated by the base of the style, which sometimes completely caps it (*Cladium*, &c.); *pericarp* membranous, crustaceous or bony, very rarely a drupe (*Diplasia*). SEED erect; *testa* thin; *albumen* farinaceous or sometimes fleshy. EMBRYO near the hilum, minute; extruded, or rarely surrounded by a thin layer of albumen, and included (*Carex*, &c.); *cotyledon* lenticular, fleshy, undivided; *plumule* inconspicuous; *radicle* inferior, obtuse.

TRIBE I. *CYPEREÆ.*

Spikelets usually many-flowered. Glumes distichous, imbricate, some of the lower often empty. Flowers ☿. Perianth 0, or represented by hispid bristles. Style very rarely swollen at the base, deciduous.

PRINCIPAL GENERA.

| * Papyrus. | * Cyperus. | Kyllingia. | Mariscus. |

TRIBE II. *SCIRPEÆ.*

Spikelets usually many-flowered. Glumes imbricate in several rows, very rarely distichous (*Androtrichum, Abilgaardia*), some of the lower often empty. Flowers ☿. Perianth 0, or represented by scaly or hairy bristles. Achene usually pointed or beaked by the persistent base of the style.

PRINCIPAL GENERA.

| Eleocharis. | Fuirena. | Androtrichum. | Scirpus. | Isolepis. |
| Ficinia. | Eriophorum. | Fimbristylis. | Melancranis. | |

Tribe III. *HYPOLYTREÆ.*

Spikelets 1-flowered, agglomerated in capitate heads or cymose panicles. Flowers ☿, each with 2–4–6 closely imbricate glumes. Perianth 0. Stamens 2–3, or 6–8. Style 2–3-fid, deciduous, or the base persistent.

PRINCIPAL GENERA.

| Hypolytrum. | Lipocarpha. | Hemicarpha. | Diplasia. |

Tribe IV. *RHYNCHOSPOREÆ.*

Spikelets usually few-flowered [flower solitary and axillary in *Oreobolus*]. Glumes imbricate in 2 or several rows, the lower empty. Flowers usually polygamous. Perianth 0, or composed of 6 bristles, rarely less, very rarely more (8–10) [or of 6 2-seriate pieces in *Oreobolus*]. Stamens 3, sometimes 6. Achene often beaked by the persistent base of the style.

PRINCIPAL GENERA.

Arthrostylis.	Caustis.	Blysmus.	Pleurostachys.
Lepidosperma.	Dulichium.	Rhyncospora.	Carpha.
Schœnus.	Cladium.	Chætospora.	Oreobolus.

Tribe V. *SCLERIEÆ.*

Spikelets diclinous, the ♂ many-flowered, with glumes imbricate in 2 or several rows, the lower ones sometimes empty. Perianth 0. Stamens 1–3, very rarely 5. ♀ spikelets 1-flowered, with glumes imbricate in several rows. Perianth 0. Style 3-fid, equal at the base. Achene bony or crustaceous, usually seated on a 3-lobed disk.

PRINCIPAL GENERA.

| Scleria. | Diplacrum. |

Tribe VI. *CARICINEÆ.*

Flowers monœcious or diœcious, in spikes, with glumes imbricate in several rows. Perianth 0. ♂ spikes simple. Stamens 3–2. ♀ spikes simple or compound. Pistil embraced by an inner scale with its back to the axis, 2-keeled (analogous to the upper glumelle of *Gramineæ*), with edges usually joined and thus forming an envelope or utricle (*urceole, perianth, perigynium*), persistent and accrescent, and enclosing either the ovary only, or the ovary accompanied by a sterile setiform pedicel.

PRINCIPAL GENERA.

| Carex. | Elyna. |

Cyperaceæ are closely allied to *Gramineæ*; the latter being distinguished by their split leaf-sheath, rounded culm with projecting nodes, dorsifixed anthers, usually plumose stigmas, fruit with pericarp adhering to the testa, and always extruded embryo. The habit of *Cyperaceæ* is that of *Restiaceæ*, but the latter essentially differ in their 3-celled fruit, and their orthotropous ovule suspended from the top of the cell.

Cyperaceæ are spread over the world, and especially in the cold regions of the northern hemisphere; they are gregarious in marshy plains, damp meadows, and the dry slopes of high mountains. They are less frequent in maritime estuaries. *Carex* and *Scirpus* diminish in number as we approach the equator; the contrary is the case with *Cyperus*, which abounds on the shores of large tropical rivers, and in the clearings of virgin forests. *Cyperaceæ* are less abundant in the southern hemisphere, where they are replaced by *Restiaceæ*.

Cyperaceæ differ from *Gramineæ* in possessing but little sugar and starch; their leaves and stem contain no juice, and form but poor pasturage. The rhizomes of some species contain a bitter principle and a volatile oil, together with a little starch, whence they rank as resolvents and diuretics. The bitter and slightly camphorate rhizomes of our *Carices* were formerly used (especially those of *C. arenaria*, which sometimes attain a very great length) in herpetic and syphilitic cachexy, as a substitute for Sarsaparilla. The stock of *Scirpus lacustris* is astringent and diuretic. *Remirea maritima*, common in tropical America, possesses the same qualities in a high degree. The herbage of *Eriophorum* was formerly administered for dysentery, and the spongy pith of its stem is considered by the German peasants efficacious against tænia. The tubers of *Cyperus longus, rotundus*, and some of their congeners, growing in South Europe and the hot parts of Asia, are aromatic, bitter, tonic and stimulating. The root of *Kyllingia triceps* is prescribed in India for diabetes. The tubers of *Cyperus esculentus*, a native of Africa, cultivated by the ancient Egyptians, and mentioned by Theophrastes, contain, besides starch, a notable quantity of sweet oil, which seldom occurs in subterranean organs; these tubers yield a nourishing food, and are reputed aphrodisiac. The tubers are moniliform in *C. articulatus*, a native of the tropics of both worlds; hence its common name of Paternoster. Our maritime *Scirpi* (as *S. tuberosus*) bear starchy and edible tubers at the tips of their rootlets. In Egypt the stems of *Cyperus dives* and *alopecuroides* are used in the fabrication of very fine mats, preferable to those of straw on account of their freshness. In France the best mattresses are made with the stems of *Scirpus lacustris*; this plant undergoes a singular modification in running water, being changed into a floating ribbon-like phyllode. In the south the long stems of *Carex nervosa* are used in chair-making. Finally, this family includes the Papyrus (*Papyrus antiquorum*), which grows in the swamps of Upper Egypt [and other parts of tropical Africa], and from which the ancients made paper by slicing the culm horizontally, pressing and hammering, and thus flattening the slices, so as to form a sheet, which was then smoothed with an ivory instrument.

LV. *GRAMINEÆ*, *Jussieu*.

[GRAMINA, *Juss.*—GRAMINEÆ, *R. Br.*—GRAMINACEÆ, *Lindl.*]

FLOWERS *glumaceous, in spikelets, usually* ☿. PERIANTH *imperfect or* 0. STAMENS *hypogynous, usually* 3, *rarely less or more;* ANTHERS *dorsifixed*. OVARY *free*, 1-*celled*, 1-*ovuled;* OVULE *parietal, semi-anatropous*. CARYOPSIS. ALBUMEN *farinaceous*. EMBRYO *basilar, outside the albumen.*—STEMS *usually knotted*. LEAVES *with a split sheath, usually ligulate*.

Annuals or perennials, usually herbaceous, cæspitose, rarely suffrutescent, frutescent or arborescent; with fibrous roots or a creeping rhizome, often stoloniferous at the radical nodes. STEM (*culm*) cylindric, rarely compressed, fistular, or sometimes solid, usually jointed at the insertion of the leaves; nodes annular, solid, swollen, rarely contracted (*Molinia*), simple, or branched from the evolution of an axillary bud with its primary leaf next the stem. LEAVES alternate, distichous, springing from the nodes; *petiole* dilated, convolute, sheathing the stem;

margins free, or very rarely more or less united ; *blade* entire, usually narrow, linear, sometimes oblong or oval, margins very often scabrid, nerves parallel [united by transverse venules] ; *stipule* axillary, adnate by its dorsal face to the sheath, and produced as a membranous tongue (*ligule*). INFLORESCENCE of spikelets arranged along an axis (*rachis*), sometimes sessile on the rachis (*spiked*), sometimes borne on branched peduncles and diffuse (*panicled*) or shortly branched (a *spicate panicle*), rarely fascicled and enclosed in a common spathe ; *spikelets* 1-several-flowered, often containing sterile flowers, each with an involucre of two scaly opposite bracts (*glumes*),[1] nearly on a level, one embracing the other, sometimes absent. FLOWERS ☿, rarely diclinous monœcious or diœcious, sometimes polygamous, each with 2 sub-opposite bracts (*paleæ* or *glumelles*), of which the lower and outer is largest, unequally nerved or keeled, furnished with a terminal or dorsal or basilar awn, or muticous ; upper and inner glumelle sheathed by the lower, emarginate or bifid, rarely obsolete or arrested, usually with no midrib, and having 2 lateral nerves. PERIANTH imperfect, very rarely 0, composed of whorled hypogynous membranous fleshy irregular scales (*squamules*), which are free or connate, normally 3, the 2 outer alternate with the paleæ, the inner opposite to the upper palea, often heteromorphous and narrower, usually obsolete. STAMENS hypogynous, definite, usually 3, sometimes 6 (*Oryza, Potamophila, Hydrochloa, Zizania, Pharus, Nastus, Bambusa,* &c.), rarely 4 (*Microlœna, Anomochloa, Tetrarrhena*), or 2 (*Anthoxanthum,* &c.), or 1 (*Uniola,* &c.), very rarely indefinite, when the ovary is arrested (*Luziola, Pariana*) ; in the hexandrous flowers whorled around the ovary ; in the triandrous, 2 opposite to the lateral nerves of the upper palea, and 1 to the lower glumelle ; in the diandrous the outer is wanting ; in the monandrous the outer only is present ; *filaments* capillary, free, or sometimes cohering at the base ; *anthers* dorsifixed, 2-celled, linear, usually 2-fid at the two ends, dehiscence lateral, longitudinal, or very rarely apical. OVARY free, 1-celled, 1-ovuled ; *styles* 2, very rarely 3, free or connate at the base, sometimes united into an undivided style ; *stigmas* with simple or branched hairs ; *ovule* adnate to the posterior part of the ovary throughout its length, or by its base, very rarely suspended below the top. FRUIT free or adnate to one or both of the glumelles, dry, indehiscent ; *pericarp* usually thin, membranous or coriaceous, and adhering to the seed (*caryopsis*), rarely membranous and dehiscent (*Sporobolus*), usually presenting a dark mark at the level of the hilum, where the testa is attached to the pericarp ; *albumen* farinaceous, or between farinaceous and horny, very thick. EMBRYO outside of the albumen, in a pit at the base of its anterior face ; *cotyledon* scutellate, often split along its outer face, and showing the radicle and plumule ; *plumule* terminal, conical, composed of 1–4 primary convolute leaves ; *radicle* basilar, thick, obtuse, often with several tubercles which are perforated at germination by radical fibres, each springing from one of these tubercles, and surrounded at their base by a small sheath (*coleorrhiza*), the remains of the perforated portion of the embryo.

---

[1] By recent authors the term *glume* is, as here, confined to these two bracts ; by other authors these are called *empty glumes* ; and the lower or outer glumelle is called *flowering glume.*—ED.

TRIBE I. *ANDROPOGONEÆ.*

Spikelets usually geminate or in threes, polygamous, the middle fertile, the lateral ♂ or neuter, very rarely all fertile, in a spicate or branched, or sometimes digitate panicle, more rarely in a spicate raceme. Fertile spikelets composed of a ☿ flower accompanied by a lower ♂ or neuter one. Glumes sub-equal, often longer than the ☿ flower, or rarely unequal, the lowest largest. Glumelles membranous, rarely cartilaginous; lower glumelle of the ☿ flower facing the upper glume. Stamens 3. Stigmas usually long, protruding under or at the top of the flower. Caryopsis with a punctiform hilary spot.

PRINCIPAL GENERA.

* Andropogon.	* Erianthus.	* Saccharum.	Ischœmum.
Imperata.	Tripsacum.	* Sorghum.	

TRIBE II. *PANICEÆ.*

Spikelets all fertile, in a spicate or branched, sometimes digitate panicle, composed of one ☿ flower accompanied by one lower ♂ or neuter flower. Lower glume smaller than the upper, often minute or arrested. Glumelles usually cartilaginous, shining; lower glumelle of the ☿ flower facing the upper glume. Stamens 3. Stigmas usually long, protruding at or under the top of the flower. Caryopsis with a punctiform hilary spot.

PRINCIPAL GENERA.

Reimaria.	Oplismenus.	* Pennisetum.	Paspalum.	* Setaria.
* Penicillaria.	* Panicum.	Digitaria.	Tragus.	

TRIBE III. *ORYZEÆ.*

Spikelets all fertile, in a raceme or panicle, 1-flowered, often with arrested glumes, or 2–3-flowered, the lower flowers with 1 glumelle, neuter, the terminal only fertile. Glumelles parchment-like, stiff. Stamens usually 6, often 3 (*Hygroryza, Ehrharta, Leersia*), or 4 (*Microlæna, Tetrarrhena*), rarely 1 (*Leersia*). Stigmas divergent, protruding at the sides of the flower. Caryopsis with a linear hilary spot.

PRINCIPAL GENERA.

Pharus.	Zizania.	Leersia.
Ehrharta.	* Oryza.	Anomochloa.

TRIBE IV. *PHALARIDEÆ.*

Spikelets ☿, monœcious or polygamous, in a spicate panicle or in **spikes**, sometimes with 2 flowers, ☿ ♀ or ♂; sometimes with 2–3 flowers, **the upper only** fertile. Glumes usually equal, longer than or as long as the flower. Glumelles more or less hardened after flowering; lower glumelle of the fertile flower facing the lower glume. Stamens 3–2. Stigmas usually elongated or filiform, protruding at the top or sides of the flower. Caryopsis with a linear or punctiform spot.

PRINCIPAL GENERA.

* Anthoxanthum.	* Phalaris.	* Zea.
Hierochloa.	Lygeum.	* Coix.

Tribe V. *PHLEINEÆ.*

Spikelets all fertile, laterally compressed, in a spicate panicle or spike, with one ☿ flower, with or without the pedicellate rudiment of a second flower. Glumes sub-equal or unequal, as long as or longer than the flower. Glumelles membranous; lower glumelle facing the lower glume. Stamens 3–2. Stigmas elongated, protruding at the top of the flower or spikelet. Caryopsis with a punctiform hilary spot.

PRINCIPAL GENERA.

* Phleum.	Alopecurus.	Crypsis.
Beckmannia.	Mibora.	Cornucopia.

Tribe VI. *AGROSTIDEÆ.*

Spikelets all fertile, more or less laterally compressed, in a branched or spiked panicle, with a single ☿ flower, rarely accompanied by the pedicellate rudiment of a second upper flower. Glumes sub-equal or unequal, usually longer than the flower. Glumelles between membranous and herbaceous, as the glumes, the lower muticous or aristate; awn usually dorsal, and facing the lower glume. Stamens 3, rarely 1–2. Stigmas usually sessile, protruding laterally at the base of the spikelet.

PRINCIPAL GENERA.

Chæturus.	Gastridium.	Cinna.	Polypogon.
Sporobolus.	Muhlenbergia.	* Agrostis.	

Tribe VII. *STIPEÆ.*

Spikelets all fertile, sub-cylindric or compressed, in panicles containing one ☿ flower. Glumes sub-equal or unequal, equalling or longer than the flower. Glumelles becoming coriaceous when ripe, the lower answering to the lower glume, often convolute, awned at the tip; awn simple or 3-fid, very rarely muticous. Stamens 3. Stigmas protruding laterally towards the base of the spikelet. Caryopsis with a linear hilary spot towards its middle or near its top.

PRINCIPAL GENERA.

* Milium.	Lasiagrostis.	* Stipa.
Piptatherum.	Machrochloa.	Aristida.

Tribe VIII. *ARUNDINEÆ.*

Spikelets all fertile, in a branched or spicate panicle, sometimes with 1 ☿ flower, with or without the pedicellate rudiment of an upper flower; sometimes many-flowered. Glumes equalling or longer than the flowers. Glumelles usually surrounded at the base with long hairs, membranous-herbaceous, as are the glumes; the lower awned or muticous, and facing the lower glume. Stamens 3, or rarely 2. Stigmas usually sessile or sub-sessile, protruding from the sides or towards the base of the spikelet. Caryopsis with a punctiform or linear hilary spot.

PRINCIPAL GENERA.

* Calamagrostis.	* Arundo.	Phragmites.	Deyeuxia.
Ampelodesmos.	* Gynerium.	* Ammophila.	

Tribe IX. *CHLORIDEÆ.*

Spikelets all fertile, in unilateral digitate or panicled spikes, sessile on the inner face of a continuous rachis, laterally compressed; sometimes with several flowers, the 1-3 lowest ☿, the upper rudimentary; sometimes with 1 ☿ flower, with or without the rudiment of a second flower. Glumes more or less unequal, usually shorter than the flowers. Glumelles membranous, the lower answering to the lower glume. Stamens 3. Stigmas usually elongated, erect, protruding towards the top or above the middle of the flower. Caryopsis with a punctiform hilary spot.

PRINCIPAL GENERA.

Cynodon.	Chloris.	Leptochloa.
Dactylotenium.	* Eleusine.	Spartina.

Tribe X. *PAPPOPHOREÆ.*

Spikelets all fertile, in cylindric-globose spikes, or in a panicle; more or less laterally compressed, with 2 or several flowers, the lower 1-5 ☿, the upper usually imperfect. Glumes more or less unequal. Glumelles membranous or sub-coriaceous, the lower with 9-13 nerves, often prolonged into bristles or teeth; lower glumelle of the base of the spikelet answering to the lower glume. Stamens 3, rarely 2. Stigmas erect, protruding at the top of the flower. Caryopsis with a punctiform or oblong hilary spot.

PRINCIPAL GENERA.

Echinaria.	Sesleria.

Tribe XI. *AVENEÆ.*

Spikelets all fertile, pedicelled or sub-sessile, in a branched **spreading or spicate** panicle, more rarely in a raceme or spike, 2-many-flowered, **the upper or lower** flower often ♂ or rudimentary. Glumes large, **sub-equal or unequal, usually almost** completely embracing the flowers. Glumelles membranous or somewhat coriaceous, the lower usually awned; awn usually dorsal, geniculate and bent below; lower glumelle of the flower at the base of the spikelet answering to the lower glume. Stamens 3, rarely 2. Stigmas sessile or sub-sessile, divergent, protruding from the sides of the flower. Caryopsis with a linear or punctiform hilary spot.

PRINCIPAL GENERA.

* Aira.	* Lagurus.	Gaudinia.	- Corynephorus.
Trisetum.	Arrhenatherum.	Deschampsia.	* Holcus.
Danthonia.	Airopsis.	* Avena.	Uralepis.
Monandraira.			

Tribe XII. *FESTUCEÆ.*

Spikelets all fertile, pedicelled, or more rarely sub-sessile, in a branched spreading or spicate panicle, more rarely in a raceme or spike, 2–many-flowered, the upper or lower flower often rudimentary or ♂. Glumes 2, often shorter than the contiguous flower. Glumelles 2, membranous or somewhat coriaceous, the lower awned at or below the top, awn not twisted or muticous; lower glumelle of the flower at the base of the spikelet answering to the lower glume. Stamens 3, rarely 2–1. Stigmas usually sessile or sub-sessile, divergent, protruding at the sides, and usually towards the base of the flower. Caryopsis with a linear or punctiform hilary spot.

PRINCIPAL GENERA.

* Poa.	Catabrosa.	Kœleria.	Lamarckia.
Eragrostis.	* Briza.	Schismus.	* Festuca.
Glyceria.	Melica.	* Dactylis.	* Bromus.
Oreochloa.	Molinia.	* Cynosurus.	Uniola.
Diarrhena.	Nastus.	* Arundinaria.	* Bambusa.

Tribe XIII. *TRITICEÆ.*

Spikelets all fertile, or rarely polygamous, spicate, sessile or sub-sessile on the notches of the usually waved rachis; 1–2–many-flowered, the upper flower usually arrested. Glumes 2, rarely 1, variable in length. Glumelles herbaceous or sub-coriaceous, rarely membranous, the lower awned at or below the top, or muticous; lower glumelle of the base of the spikelet answering to the lower glume. Stamens 3, rarely 1. Stigmas sessile or sub-sessile, divergent, protruding from the sides and often towards the base of the flower. Caryopsis with a linear hilary spot.

PRINCIPAL GENERA.

* Lolium.	* Triticum.	Psilurus.	* Hordeum.	Ægilops.
Lepturus.	* Elymus.	Nardus.	Rottbœllia.	* Secale.

Gramineæ form one of the most natural groups of plants; it is principally to their numerous species that the name of Grass is given; but in tropical Asia we find *Gramineæ* of great height and even forming true trees. Like all clearly defined families, they have few affinities with other orders, and they are really only related to *Cyperaceæ*, called by the ancients ' spurious Grasses ' (*Gramineæ spuriæ*), from which they differ by their parietal ovule, their seed with abundant farinaceous albumen, their leaves with a split

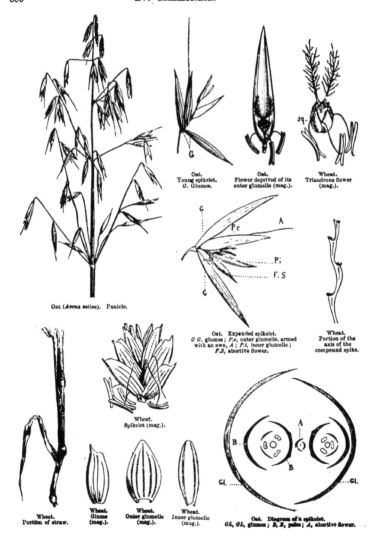

Oat. Young spikelet. *G.* Glumes.

Oat. Flower deprived of its outer glumelle (mag.).

Wheat. Triandrous flower (mag.).

Oat (*Avena sativa*). Panicle.

Oat. Expanded spikelet. *G G*, glumes; *P.e*, outer glumelle, armed with an awn, *A* ; *P.i*, inner glumelle ; *F.S*, abortive flower.

Wheat, Portion of the axis of the compound spike.

Wheat. Spikelet (mag.).

Wheat. Portion of straw.

Wheat. Glume (mag.).

Wheat. Outer glumelle (mag.).

Wheat. Inner glumelle (mag.).

Oat. Diagram of a spikelet. *GL, GL*, glumes ; *B, B*, pales ; *A*, abortive flower.

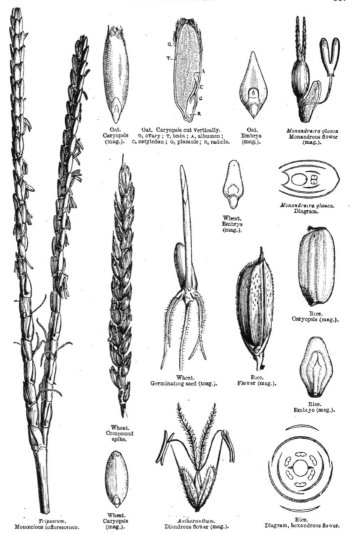

Oat.
Caryopsis
(mag.).

Oat. Caryopsis cut vertically.
o, ovary; T, testa; A, albumen;
C, cotyledon; G, plumule; R, radicle.

Oat.
Embryo
(mag.).

Monandraira glauca.
Monandrous flower
(mag.).

Wheat.
Embryo
(mag.).

Monandraira glauca.
Diagram.

Rice.
Caryopsis (mag.).

Wheat.
Germinating seed (mag.).

Rice.
Flower (mag.).

Rice.
Embryo (mag.).

Wheat.
Compound
spike.

Tripsacum.
Monœcious inflorescence.

Wheat.
Caryopsis
(mag.).

Anthoxanthum.
Diandrous flower (mag.).

Rice.
Diagram, hexandrous flower.

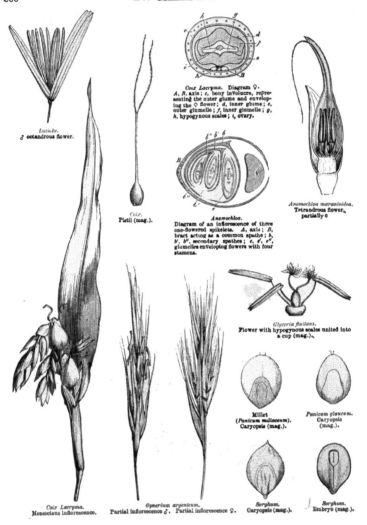

Luziola.
♂ octandrous flower.

Coix Lacryma. Diagram ♀.
A, B, axis ; *c,* bony involucre, representing the outer glume and enveloping the ♀ flower ; *d,* inner glume ; *e,* outer glumelle ; *f,* inner glumelle ; *g,* *h,* hypogynous scales ; *i,* ovary.

Coix.
Pistil (mag.).

Anomochloa.
Diagram of an inflorescence of three one-flowered spikelets. *A,* axis ; *B,* bract acting as a common spathe ; *b,* *b', b'',* secondary spathes ; *e, e', e'',* glumelles enveloping flowers with four stamens.

Anomochloa marantoidea.
Tetrandrous flower,
partially ☿

Glyceria fluitans.
Flower with hypogynous scales united into a cup (mag.).

Coix Lacryma.
Monœcious inflorescence.

Gynerium argenteum.
Partial inflorescence ♂.   Partial inflorescence ♀.

Millet
(*Panicum miliaceum*).
Caryopsis (mag.).

Panicum glaucum.
Caryopsis
(mag.).

Sorghum.
Caryopsis (mag.).

Sorghum.
Embryo (mag.).

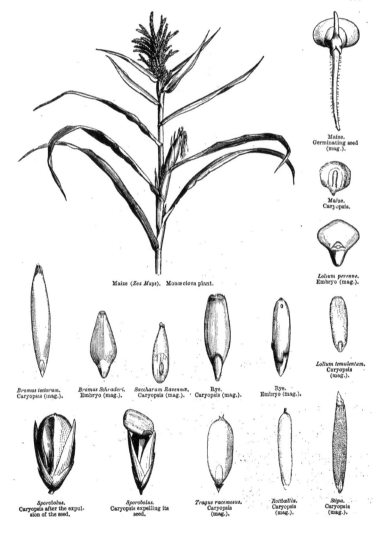

Maize (*Zea Mays*). Monœcious plant.

Maize.
Germinating seed
(mag.).

Maize.
Caryopsis.

Lolium perenne.
Embryo (mag.).

Lolium temulentum.
Caryopsis
(mag.).

Bromus tectorum.
Caryopsis (mag.).

Bromus Schraderi.
Embryo (mag.).

Saccharum Ravennæ.
Caryopsis (mag.).

Rye.
Caryopsis (mag.).

Rye.
Embryo (mag.).

Sporobolus.
Caryopsis after the expulsion of the seed.

Sporobolus.
Caryopsis expelling its seed.

Tragus racemosus.
Caryopsis
(mag.).

Rottbœllia.
Caryopsis
(mag.).

Stipa.
Caryopsis
(mag.).

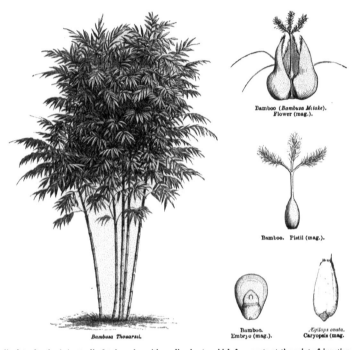

Bamboo (*Bambusa Mitake*).
Flower (mag.).

Bamboo. Pistil (mag.).

Bambusa Thouarsii.

Bamboo.
Embryo (mag.).

Ægilops ovata.
Caryopsis (mag.

ligulate sheath, their usually fistular culm, with swollen knots, which form septa at the point of junction
with the leaves. The arborescent *Gramineæ*, and notably the Bamboos, of which the flower is hexandrous
and furnished with a whorl of three glumelles, offer some points of resemblance with Palms.

Brongniart has recently observed the phenomena of sleep in *Strephium guyanense*, belonging to the
tribe *Paniceæ*; its leaves, which spread during the day, are erect and close over each other at night, like
those of the Sensitive-plant and some other *Mimoseæ*.

This immense family is distributed over all the globe, from the tropics to the frozen regions; the
majority inhabit the temperate zone, but *Paniceæ, Chlorideæ, Oryzeæ, Andropogoneæ* and Bamboos are
especially tropical. The native country of the cultivated cereals is still unknown.

Gramineæ contain in their herbage, and especially in their seeds, nutritious principles, which
entitle them to the first rank amongst plants useful to man, and which are of the greatest importance in
an economic and political point of view. Besides starch, sugar and mucilage, they yield sulpho-azotized
matters (*fibrine, casein, albumine*), elements essential to the formation of flesh in animals, and especially
phosphate of lime, which is the basis of their bony framework. The Cerealia, whose seeds abound in
starch, azotized matter and phosphates, are:—Wheat (*Triticum sativum*), Rye (*Secale cereale*), Barley
(*Hordeum vulgare, distichum*, &c.), Oats (*Avena sativa*), all cultivated by the Caucasian race in the northern
and temperate regions. Rice (*Oryza sativa*) and Millet (*Panicum miliaceum*) originated amongst the Asiatic

races, as well as *Eleusine coracana*, which is a great resource in India when the Rice crop fails. Maize (*Zea Mays*), which formerly served as food to the American races, is now spread over the whole world. *Bromus Mango*, a species near our *B. secalinus*, was cultivated in South Chili before the arrival of Europeans; the Araucanians have now abandoned it for the cereals of the Old World. *Sorghum vulgare* and *Penicillaria spicata* are the support of the negro race. The inhabitants of East Africa cultivate *Poa abyssinica*, *Eleusine*, and our European cereals, which are often infested by various cryptogamic parasites (Rust, Smut, Ergot), of which we shall treat under *Fungi*.

The Sugar-cane (*Saccharum officinarum*) is, in all probability, a native of tropical Asia; it has been cultivated from very ancient times in the East Indies. After the conquests of Alexander it became known in Europe; and towards the end of the thirteenth century it was introduced from India into Arabia and the Mediterranean region. At the beginning of the fifteenth century the Portuguese planted it in Madeira, where it prospered, and whence it passed to the Canaries and to St. Thomas. In 1506 the Spaniards introduced it into St. Domingo; it rapidly increased there, and soon spread over all tropical America, where it has produced numerous varieties. It is especially the lower part of its culm which yields the sap from which is extracted the readily crystallizable principle so universally used as a food, condiment and medicine. Cane-sugar is fermentable, like that of many other vegetables; and it is from the non-crystallizable syrup (treacle), which remains after the crystallization of the sugar, and which is submitted to spirituous fermentation, that rum is obtained by distillation. *Sorghum saccharatum*, the stem of which is very rich in sugar, is cultivated in China, Africa, &c.

A considerable number of *Gramineæ* are medicinal. The rhizome of the Dog's-tooth Grass (*Triticum repens*), which infests cultivated ground in Europe, is used as an emollient and aperient tisane; other European species (*T. glaucum* and *junceum*, and *Cynodon Dactylon*) possess similar properties; as also does the *Cynodon lineare*, of India, and *Andropogon bicornis*, of tropical America. *Arundo Donax* is a large reed, of which the root is diuretic and sudorific. Formerly that of *Phragmites communis* was prescribed as a depurative and anti-syphilitic. *Calamagrostis* is considered diuretic by the French peasants. *Perotis latifolia* has the same reputation in India.

The mucilaginous seeds of Barley are still used, as in the time of Hippocrates, in the preparation of a diluent and cooling drink; they are also used in the making of beer. Under the influence of moist heat they are allowed to germinate, a process which converts the starch into sugar; this sugary matter is dried and pulverized, and its decoction, flavoured with hops, is submitted to spirituous fermentation. The seeds of Rice are emollient, like those of Barley, and slightly astringent. They are equally fermentable, and yield by distillation an alcohol called *arrack*.

Coix Lachryma (Job's Tears), a native of tropical Asia, and cultivated in China [and India], is a monœcious Grass, remarkable for its ♀ spikelets enveloped in an involucre which becomes stony when ripe; its seeds are considered in China to be tonic and diuretic, and are administered in a tisane in phthisis and dropsy. The root of *Manisuris granularis* is prescribed in India for obstructions of the bowels. A decoction of the seeds of *Dactyloctenium ægyptiacum* is renowned in Africa as an alleviator of nephritic pains, and its herbaceous parts are applied externally for the cure of ulcers.

Andropogons have aromatic roots, whence some species are used in India as stimulants. Such are *A. Nardus*, or False Spikenard; *A. Iwarunkusa*, *Purancura*, and *citratus* (Lemon Grass). The leaves of the Sweet Rush (*A. laniger* and *Schœnanthus*), of Africa and Arabia, are prescribed in the East for their stimulating, antispasmodic, diaphoretic qualities. The Vetiver or Viti-Vayr [or Kus-Kus], is the very sweet-scented fibrous root of *Andropogon muricatus*, first imported into Europe about fifty years ago, which is used in India to perfume rooms and to preserve stuffs and clothes from insects; according to Vauquelin, it contains an aromatic principle, analogous to Myrrh, and possesses the stimulating properties of its congeners. [The roots are made into fans, and worked into slips of bamboo to form the screens used to mitigate the heat in India.] The Bamboo (*Bambusa arundinacea* and *verticillata* [and many other species]) is used in building in China and Japan. The young shoots of these two trees contain a sugary pith, which the Indians seek eagerly; when they have acquired more solidity a liquid flows spontaneously from their nodes, and is converted by the action of the sun into drops of true sugar. The internodes of the stems often contain siliceous concretions, of an opaline nature, named *tabasheer* [a substance presenting remarkable optical properties]. Several American Bamboos contain a very fresh drinkable water, sought by Indians and travellers.

Most of our native *Gramineæ* form a pasturage for flocks, and, when dried, become hay, which has an agreeable scent, especially when mixed with *Anthoxanthum odoratum*, the roots of which contain benzoic acid. Some species are too siliceous, or are armed with awns which may penetrate the skin, or irritate the intestines of the animals which have eaten them (*Calamagrostis, Stipa,* &c.); others are purgative (*Bromus catharticus,* &c.); others poisonous, as the Darnel (*Lolium temulentum*), the seeds of which, if mixed with those of cereals, cause vomiting, giddiness and intoxication. The *Molinia cærulea,* a native of damp meadows, becomes dangerous to horses towards the flowering season. *Festuca quadri-dentata,* common in Peru, is eminently poisonous and mortal to cattle.

The straw of our cereals, besides its use in agriculture, is also used in the manufacture of hats and bonnets, and especially of Leghorn hats, which rival those of Panama in fineness and high price. *Lygeum Spartium* [Alpha Grass] and *Macrochloa tenacissima* [Esparto] are used in the manufacture of paper, baskets, &c.

Various *Gramineæ* are ornamental garden plants. The Provence Cane (*Arundo Donax*), of South Europe, is cultivated as an economic and medicinal plant; we have indicated the properties of its root; its long solid and light stems are of various uses; they are made into fishing-rods, trellis-work, &c. It flowers but seldom even in its native climate, and never in the north of France, where it does not flourish; this peculiarity, which is shared by its congener, the common Reed, is probably owing to its rapid propagation by rhizomes, which renders its reproduction by seeds superfluous. The Mauritanian Reed (*Arundo mauritanica*) differs from the preceding in its lower growth (6 to 10 feet), and especially in flowering abundantly, even in the climate of Paris. The Pampas Grass (*Gynerium argenteum*), a native of the temperate regions of South America, and introduced a few years ago into European gardens, is universally admired; it is diœcious, and the female plants are distinguished by the larger size and greater spread of the panicles. Only two or three species of Bamboos are hardy in our northern climates; these are all natives of Central China or of the Himalaya. They are the Black Bamboo (*B. Metake,* and *B. glaucescens*). The Large Bamboo (*B. arundinacea*), a tree of South China and India, is one of the most ornamental species, but it only succeeds in the hottest parts of Mediterranean Europe. The Arundinarias of the Himalaya are true Bamboos in habit, foliage, and the woody consistency of the stem. One species only, *A. falcata,* has been introduced into European gardens; it stands the winters as far north as 43°, and is even cultivated with some success in the west of France. Here and there in the gardens of the Mediterranean region the Ravenna Cane (*Saccharum Ravennæ*) is met with, attaining nearly the stature of the Sugar-cane; and the *Panicum plicatum,* a plant suitable for the ornamentation of lawns, also flourishes in the same locality. The Canary Grass, or Gardener's Garters (*Phalaris arundinacea*) [which produces Canary-seed], presents a variety with white striped leaves, of some value as an ornamental plant, [as does the common *Dactylis glomerata*]. *Briza, Agrostis, Festuca, Lolium,* and *Aira* are principally used to form lawns and edgings.

CLASS III.—ACOTYLEDONS.

I. *FILICES, L.*

Acotyledonous PLANTS, *very generally perennial and terrestrial, stemless, caulescent, or arborescent.* FRONDS *springing from the upper surface of the creeping rhizomes, or forming regular crowns which terminate erect stems; blade leafy, crozier-shaped in vernation, stomatiferous, simple, pinnatifid or pinnatisect.*

REPRODUCTIVE ORGANS *composed of capsules* (sporangia) *collected into groups* (sori), *situated on the nerves, at the back or margin of the frond.* SORI *usually covered with a pellicle* (indusium *or* involucre). SPORANGIA *opening lengthwise, or girt by an elastic ring which unrolling tears them irregularly.* SPORES *numerous, at first collected in fours in the cells* (mother-cells), *filling the sporangium, then freed by the decay of these cells, and developing on the damp soil a cellular expansion* (prothallus), *on the lower surface of which are developed:—*1. *Cellular bodies* (antheridia) *containing flattened threads, coiled in a helix, furnished with cilia, and moving actively* (antherozoids) ;— 2. *Cellular sacs, open at one end* (archegonia), *into which the antherozoids enter to fertilize a contained vesicle which is destined to reproduce the plant.*

Perennial, very rarely annual (*Gymnogramme leptophylla*), terrestrial, or very rarely aquatic (*Ceratopteris*). STEM sometimes forming a rhizome, which is tuberous and fleshy (*Angiopteris*), or creeping on soil or rocks or trees, sometimes vertical and arborescent, or rarely twining (*Lygodium*), or sub-sarmentose and dichotomous (*Gleichenia*). It is composed of fibro-vascular bundles, disposed in a more or less regular circle around a copious cellular tissue ; each bundle presents at its circumference a black zone formed of woody fibres (*prosenchyma*), and a white centre formed of annular and rayed vessels. The central cellular tissue of the stem communicates, through spaces between the vascular bundles, with an outer zone of similar tissue. The whole is surrounded by a bark formed of the persistent bases of the branches.

Foliiferous BRANCHES (*fronds*), springing sometimes from the upper surface of the rhizome, at greater or less distances, and becoming partially or wholly disjointed as the stem lengthens, and new fronds are developed, sometimes crowded and covering the rhizome [then called *caudex*], the tip of which turns up and emits a crown of fresh fronds (*Struthiopteris germanica*, &c.). This arrangement forms the passage to Tree-ferns, in which the stem (*caudex*) rises vertically, in some species attaining a height of 50 to 65 feet. This stem [often called trunk] grows not only in diameter, but in length, as is shown by the scars of the fronds being at first close together, and afterwards sundered, the spaces between them increasing. Ferns, whether creeping or erect, give off numerous roots, which in the arborescent species extend

I. FILICES.

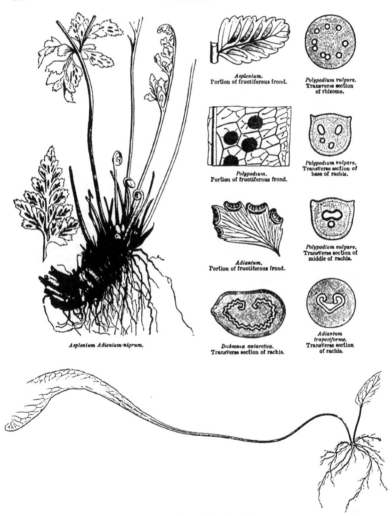

Asplenium.
Portion of fructiferous frond.

Polypodium vulgare.
Transverse section
of rhizome.

Polypodium.
Portion of fructiferous frond.

Polypodium vulgare.
Transverse section of
base of rachis.

Adiantum.
Portion of fructiferous frond.

Polypodium vulgare.
Transverse section of
middle of rachis.

Asplenium Adiantum-nigrum.

Dicksonia antarctica.
Transverse section of rachis.

*Adiantum
trapeziforme.*
Transverse section
of rachis.

Asplenium rhizophyllum. Frond rooting at its extremity.

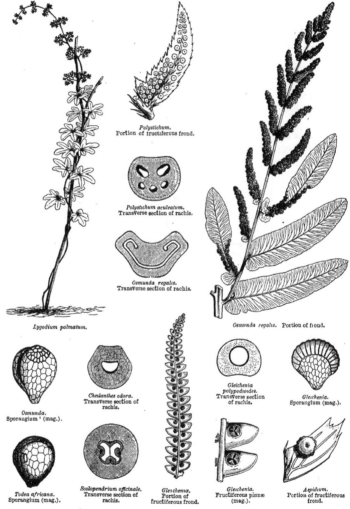

Polystichum.
Portion of fructiferous frond.

Polystichum aculeatum.
Transverse section of rachis.

Osmunda regalis.
Transverse section of rachis.

Lygodium palmatum.

Osmunda regalis. Portion of frond.

Osmunda.
Sporangium [1] (mag.).

Cheilanthes odora.
Transverse section of rachis.

Gleichenia polypodioides.
Transverse section of rachis.

Gleichenia.
Sporangium (mag.).

Todea africana.
Sporangium (mag.).

Scolopendrium officinale.
Transverse section of rachis.

Gleichenia.
Portion of fructiferous frond.

Gleichenia.
Fructiferous pinnæ (mag.).

Aspidium.
Portion of fructiferous frond.

[1] The annulus should have been represented as more oblique.—ED.

I. FILICES.

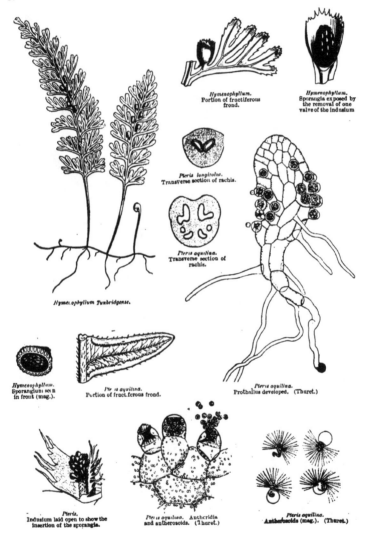

Hymenophyllum.
Portion of fructiferous
frond.

Hymenophyllum.
Sporangia exposed by
the removal of one
valve of the indusium

Pteris longifolia.
Transverse section of rachis.

Pteris aquilina.
Transverse section of
rachis.

Hymenophyllum Tunbridgense.

Hymenophyllum.
Sporangium seen
in front (mag.).

Pteris aquilina.
Portion of fructiferous frond.

Pteris aquilina.
Prothallus developed. (Thuret.)

Pteris.
Indusium laid open to show the
insertion of the sporangia.

Pteris aquilina. Antheridia
and antherozoids. (Thuret.)

Pteris aquilina.
Antherozoids (mag.). (Thuret.)

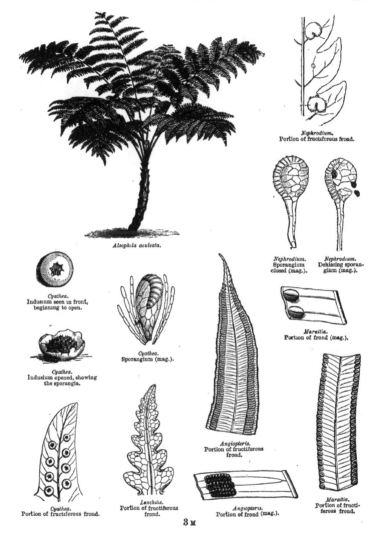

Alsophila aculeata.

Nephrodium.
Portion of fructiferous frond.

Nephrodium.
Sporangium
closed (mag.).

Nephrodium.
Dehiscing sporangium (mag.).

Cyathea.
Indusium seen in front,
beginning to open.

Cyathea.
Sporangium (mag.).

Marattia.
Portion of frond (mag.).

Cyathea.
Indusium opened, showing
the sporangia.

Cyathea.
Portion of fructiferous frond.

Lonchitis.
Portion of fructiferous
frond.

Angiopteris.
Portion of fructiferous
frond.

Angiopteris.
Portion of frond (mag.).

Marattia.
Portion of fructiferous frond.

3 M

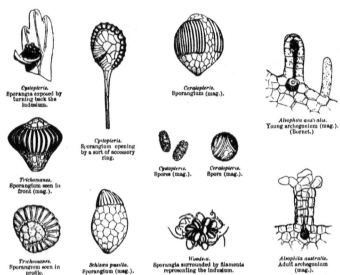

Cystopteris.
Sporangia exposed by
turning back the
indusium.

Cystopteris.
Sporangium opening
by a sort of accessory
ring.

Trichomanes.
Sporangium seen in
front (mag.).

Ceratopteris.
Sporangium (mag.).

Cystopteris.
Spores (mag.).

Ceratopteris.
Spore (mag.).

Alsophila australis.
Young archegonium (mag.).
(Bornet.)

Trichomanes.
Sporangium seen in
profile.

Schizæa pusilla.
Sporangium (mag.).

Woodsia.
Sporangia surrounded by filaments
representing the indusium.

Alsophila australis.
Adult archegonium
(mag.).

over the whole circumference; these root-fibres are blackish glabrous or velvety, fragile, cylindric, usually mingled with russet scaly hairs, which are often also found on the aërial stem, on the rachis, and even on the principal nerves of the fronds. The fronds are rolled into a crozier before they expand, in such a way that their tip forms the centre of the crozier, and that their lower surface is outside. The petiole (*rachis*) is cylindric, compressed, or hexagonal at the base; the blade, usually covered with a stomatiferous epidermis, is sometimes simple and entire, more frequently pinnatisect, or 2–3-pinnatisect; pinnules sometimes extremely fine (*Trichomanes Pluma*), nearly always continuous with the midrib of the secondary pinnæ, sometimes caducous (*Drynaria*), rarely membranous, pellucid and deprived of stomata (*Hymenophyllum*). The nerves of the fronds are slender and well-defined, sometimes simple and springing laterally from the median nerve, sometimes forked and dichotomous, and often, in consequence of this dichotomy, forming a network with more or less regular and hexagonal meshes. In some genera they form transverse and regular arches, or large irregular meshes, whence spring short nerves which terminate in the middle of these cellular spaces. Further, the nerves often anastomose in arches near the median nerve, and give off on the outer side simple or forked or anastomosing and reticulate venules.

The fronds are sometimes furnished with bulbils (*Hemionitis*, &c.), or are radicant (*Asplenium rhizophyllum, Woodwardia*, &c.); they are often very dissimilar in the same species (*Platycerium*), some being sterile, others fructiferous.

REPRODUCTIVE ORGANS composed of capsules (*sporangia*). SPORANGIA springing from the nerves, on the lower surface or near the margin of the fronds, and collected in groups (*sori*). SORI naked, or covered either with a fold of the margin of the frond, or with a prolongation of the epidermis (*indusium*). Sometimes their abundance induces the atrophy and more or less complete disappearance of the leaf-like blade of the frond, and they then form panicles or spikes isolated at the end of the general frond (*Osmunda, Aneimia, Lygodium,* &c.). Each sporangium is pedicelled or sessile, variously dehiscent, and usually furnished with a variously shaped elastic ring ; the sporangium contains numerous free reproductive spherical or angular corpuscules (*spores*), with a smooth warted or reticulate surface. These spores were originally enclosed by fours in cells, which afterwards decay. Under the influence of humidity [rather of drought] the sporangium opens or bursts, and the spores are elastically expelled.

The spores when placed on damp earth at once begin to germinate by emitting a filament which develops into a small foliaceous cellular expansion, emarginate at the tip (*proembryo, prothallus*). On the lower surface of the prothallus are soon developed small cellular protuberances, usually resulting from the superposition of three cells, of which the lower acts as a support, and the upper as a cover to the middle one ; this latter contains a mucilaginous tissue, the cells of which contain flattened threads, coiled in a helix, furnished with a series of numerous short cilia, accompanied by a small vesicle ; these moveable bodies have been termed *antherozoids*, and the organ which contains them *antheridia*.

In the vicinity of the antheridia appear, a little later, somewhat larger cellular ovoid or rounded organs, terminated by a sort of style, which is open at the period of fertilization. These cellular sacs, analogous to the ovules of Phanerogams, are named *archegonia* ; at the bottom of their cavity is seen a globose utricle which has been compared to the embryonic sac. In this utricle a vesicle soon appears from which the new plant will be developed.

All the conditions for fertilization being thus arranged, the antherozoids break the wall of the antheridia, drawing after them the mucilaginous vesicles, and escape, executing, by means of the vibratory hairs at one of their extremities, rapid movements, which are assisted by the rain or dew which moistens the mucilage expelled with them from the mother-cell. They thus reach the canal of the archegonium, and fertilization is secured ; a small cellular mass is then developed in the fertilized archegonium, which lengthens into an erect axis, on the top of which fronds will be developed, and from the base lateral roots. The prothallus soon disappears.

Some species, notably those which grow on rocks or on walls exposed to the heat of the sun, and the fronds of which are fragile, have the power of reviving after being almost entirely dried up.

TRIBE I. *POLYPODIACEÆ.*

Elastic ring generally narrow, prolonged from one side of the rather long pedicel, interrupted at the top or the opposite side near the pedicel.

Acrostichum.	Olfersia.	Platycerium.	Hemionitis.	Niphobolus.
Doodia.	Gymnogramme.	Cheilanthes.	Scolopendrium.	Ceterach.
Lonchitis.	Nephrodium.	Grammitis.	Davallia.	Aspidium.
Notochlæna.	Blechnum.	Cystopteris.	Polypodium.	Struthiopteris.
Woodsia.	Dictyopteris.	Asplenium.	Cibotium.	Phymatodes.
Adiantum.	Dicksonia.	Drynaria.	Woodwardia.	

Tribe II. *CYATHEACEÆ.*

Elastic ring usually oblique, and completely surrounding the sporangium ; sporangium often compressed, sessile, or shortly pedicelled, not continuous with the ring.

PRINCIPAL GENERA.

Hemitelia.	Matonia.	Alsophila.	Cyathea.

Tribe III. *HYMENOPHYLLEÆ.*

Elastic ring nearly as in *Cyatheaceæ*, but the sporangia are nearly globose, and the elastic ring is situated on a plane nearly perpendicular to the point of attachment of the sporangium.

PRINCIPAL GENERA.

Hymenophyllum.	Trichomanes.	Loxsoma.

Tribe IV. *CERATOPTERIDEÆ.*

Elastic ring large, formed of vertical cells, not completely surrounding the sporangium, which is sessile.

GENERA.

Ceratopteris.	Parkeria.

Tribe V. *GLEICHENIEÆ.*

Sporangia solitary or grouped in definite numbers (2–3), sessile, globose ; elastic ring perfect, but not corresponding to the point of attachment of the sporangium.

PRINCIPAL GENERA.

Gleichenia.	Mertensia.	Platysoma.

Tribe VI. *LYGODIEÆ.*

Sporangia sessile, ovoid or turbinate ; elastic ring replaced by a sort of cap with radiating striæ, occupying the end of the sporangium, opposite to the point of attachment.

PRINCIPAL GENERA.

Aneimia.	Schizæa.	Lygodium.	Mohria.

I. FILICES.

Tribe VII. *OSMUNDEÆ.*

Elastic ring embracing a part of the circumference of the sporangium, or reduced to a small disk of cells with thick walls.

GENERA.

Osmunda. Todea.

Tribe VIII. *MARATTIEÆ.*

Sporangia free, appressed, in 2 rows, or in a circle, or confluent, and together resembling a several-celled capsule, deprived of rings, each opening by a slit or pore.

GENERA.

Kaulfussia. Angiopteris. Marattia. Danæa.

Ferns present such marked characters that in all classifications they form a distinct group. Brongniart, from whom we have taken most of the details relating to this important family, places near them *Marsileaceæ, Lycopodiaceæ, Equisetaceæ* and *Characeæ* to form his class of *Filicineæ.* He has divided Ferns into several very natural tribes, founded on the structure of the sporangia and their mode of insertion.

The numerous genera of this vast family have been classed according to the arrangement of the sori and indusia; but it must be observed that, in certain cases, the sori of the same species appear with and without indusium; thus *Polypodium rugulosum* and *Hypolepis tenella* have been separated, though they are, in fact, the same species; the same may be said of *Polystichum venustum* and *Polypodium sylvaticum*, &c.

'Ferns inhabit the most different climates, from the polar regions (*Woodsia hyperborea, Pteris argentea*, &c.), where, however, few species are found, to the tropics, where they are abundant and varied. Many genera are indeed limited to equatorial regions, or extend but little beyond, and then especially in the southern hemisphere. Few genera, on the contrary, are confined to a single continent, and those which are have generally but few species. Most genera have a very wide range, a fact not only true of large genera as limited by Swartz and Willdenow, but generally of those into which they have been subdivided. Some tribes are entirely or almost entirely confined to hot regions; as *Cyatheaceæ, Ceratopterideæ* and *Hymenophylleæ*, of which three species only (*Trichomanes radicans* and *Hymenophyllum tunbridgense* and *Wilsoni*) grow in Europe. All the Tree-ferns, and particularly those of the tribe of *Cyatheaceæ*, are tropical, or extend but a little way into some islands situated far from the equator (*Alsophila Colensoi, Cyathea Smithii*, of New Zealand). The arborescent *Dicksoniæ* (*D. antarctica, lanata*, &c.), extend farther south in New Zealand; and the Lomarias with an erect but short stem are found in the Magellanic lands.

'The family of Ferns comprises at least 3,000[1] described species (the proportion to Phanerogams being as one to thirty), of which about 150 to 200 belong to each of the temperate zones, and 2,600 to the tropical regions of both continents, and to the islands included in this zone. In each of these zones the number of Ferns varies much, according to locality. A peculiar combination of climatic conditions is almost always essential to their existence, dry regions producing very few species; damp, cool and shady places suit them better, and the number of species is so much the greater the more these conditions are fulfilled; insular climates are therefore very favourable to them, and the predominance of Ferns in such has long been noticed. We know, in fact, that the smaller and more distant from continents islands are, the more maritime is their climate, owing to the habitual humidity of the air and its uniform temperature, and the larger is the proportion which Ferns bear to Phænogamic plants.

'The family of Ferns, together with *Coniferæ*, has more fossil representatives than any other

[1] These numerical estimates need revision.—ED.

throughout the series of geological formations, and Ferns are without doubt one of the most interesting of orders if looked at from this point of view. In fact, this family, so numerous and so widely spread over the surface of the globe at the present time, presents, in the most ancient of those strata which contain vegetable remains, species which appear almost identical, in many cases, with those now living. It predominates even in those ancient strata composing the coal formation; upwards of two hundred species being now known to be scattered for the most part through the coal measures of Europe and of some parts of North America.' [1]

The numerous species of the tribe of *Polypodiaceæ* possess similar principles; their frond is mucilaginous, slightly astringent, and sometimes sub-aromatic;' the rhizome is usually bitter, astringent, and somewhat acrid. Some species contain an adipose waxy matter, fixed and volatile oils; others yield by analysis a principle analogous to manna. The rhizome and stem of a large number abound in starch.

The rhizome of the Male Fern (*Nephrodium Filix-mas*), which grows in the forests of Central and South Europe, is much used as an anthelminthic. The various European species of *Aspidium* have the same property but in a less degree. Several American and tropical Asiatic species of *Asplenium, Polypodium, Diplazium*, &c., are used in the same way as our Male Fern. The rhizome of *Polypodium Calaguala* is much valued in Peru as an astringent and diaphoretic. From the Maidenhair (*Adiantum Capillus-Veneris*), which grows in the south of Europe, a bechic syrup is prepared; *Asplenium Trichomanes, Adiantum-nigrum, Ruta-muraria*, from North Europe, possess analogous properties, as do the Canadian Maidenhair (*Adiantum pedatum*) and other exotic congeners. *Scolopendrium officinale, Hemionitis* and *Ceterach officinarum*, all European plants, are employed as astringents and mucilages.

The herbage of *Aspidium fragrans*, which has the scent of raspberries, is much esteemed in the North of Asia as an antiscorbutic, and the Mongols use it as tea. The rhizome of *Aspidium Baromez*, commonly named Scythian Lamb, is clothed with golden yellow hairs, whence perhaps was obtained the famous *byssus* of the ancients, with which they manufactured stuffs which fetched an extraordinary price; its red and viscous juice is an esteemed astringent in China. The hairs which clothe the stem of some Polypodiums are renowned in the Antilles as a styptic, and English doctors use them as hæmostatics, as well as the hairs of several Cyatheas from the Moluccas, which have been of late years prescribed under the name of *Penjuar Yambi*. [Those of several Polynesian species of Tree-fern (*Cibotium*) are extensively imported into Australia, and used to stuff pillows, &c.]

The young mucilaginous shoots of several Ferns are eaten as a salad, especially in North Europe. *Ceratopteris thalictroides* is in tropical Asia considered a pot-herb. The rhizome of *Pteris esculenta*, a New Zealand species very near our Common Fern, serves as food to the natives, as do the tubers of *Nephrodium esculentum* in Nepal. *Cyathea medullaris*, of New Zealand, contains in the lower part of its stem a reddish glutinous pith, which when baked acquires the taste of the radish, and is much liked by the natives. The stem of *Gleichenia Hermanni* is starchy, somewhat bitter, sub-aromatic and edible.

Aneimia tomentosa, with the odour of myrrh, *Mohria thurifraga*, of the Cape, which smells of benzoin, as well as *Lygodium microphyllum* and *circinatum*, have incisive and bechic properties.

Marattieæ inhabit America, Asia and Polynesia, but are not numerous; they are very rare in south temperate latitudes. Some species are arborescent. The bruised fronds of *Angiopteris evecta*, a species spread over the Pacific Islands, communicate an agreeable scent to cocoa-nut oil; its young shoots are edible.

II. *OPHIOGLOSSEÆ, Br.*

Ophioglosseæ are separated from true Ferns by the nature of their rhizome, and the development and texture of their fronds, which are not rolled into a crozier in vernation, and by their sporangia being arranged in a longitudinal series on a sort

[1] Ad. Brongniart, *Dictionnaire universel d'Histoire naturelle.*

of scape, at the end of which they form a simple spike (*Ophioglossum*), or a raceme (*Botrychium*). These sporangia have no ring, and contain smooth triangular spores, which ally *Ophioglosseæ* to *Lycopodiaceæ*, through the genus *Phylloglossum*.

Ophioglossum.
Portion of fructiferous spike (mag.).

Ophioglossum.
Spores (mag.).

GENERA.

Botrychium.   Ophioglossum.
Helminthostachys.

Most *Ophioglosseæ* are exotic, nearly all are terrestrial, except *O. pendulum*, which lives on trees like some Lycopods; some inhabit the Mascarene Islands and extra-tropical Australia; they are rarer in the West Indies and in America between the tropic of Cancer and the equator; the number of European species is even fewer; one of these also inhabits North America, another the north of Asia. One or two species are more frequent at the Cape. None have been observed in North Africa.

All are mucilaginous, and a decoction of them is alimentary. *O. vulgatum* was formerly esteemed as a vulnerary; it is still used as an astringent in angina. *Helminthostachys dulcis*, from the Moluccas, is succulent and laxative; its young shoots are edible. The herbage of *Botrychium cicutarium* is a reputed alexipharmic in St. Domingo.

Ophioglossum vulgatum.

III. *EQUISETACEÆ, D.C.*

(FILICUM *genus, L.*—GONOPTERIDES, *Willdenow.*—PELTATA, *Hoffmann.*)

Perennial Acotyledons, inhabiting moist places; rhizome subterranean. STEMS *straight, cylindric, channelled, stiff, simple or branched, jointed, fistular; joints accompanied by sheaths denticulate at the top, from the base of which spring branches, which are whorled, and resemble the stems.* REPRODUCTIVE ORGANS *in catkins formed of numerous whorled peltate polygonal scales, arranged perpendicularly to the axis, bearing on their lower face 5-6-9 sacs (sporangia), which open longitudinally to emit free globose smooth spores, accompanied by 2 spathulate filaments* (elaters); *spores in germination forming a prothallus, like that of Ferns, on which are developed antheridia containing antherozoids or archegonia, into which the antherozoids enter for the purpose of fertilization.*

Equisetum. Ripe spore and elaters.

E. fluviatile.
Rhizome and tubers.

Equisetum.
Spore surrounded by its elaters.

Equisetum. Clypeole
furnished with
sporangia.

Equisetum.
Dehiscing
sporangium.

Equisetum sylvaticum.

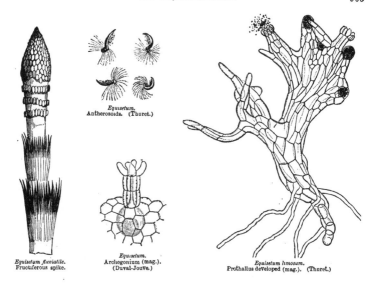

Equisetum.
Antherozoids. (Thuret.)

Equisetum fluviatile.
Fructiferous spike.

Equisetum.
Archegonium (mag.).
(Duval-Jouve.)

Equisetum limosum.
Prothallus developed (mag.). (Thuret.)

Perennial terrestrial or aquatic PLANTS; rhizome subterranean, running, often branched, covered with brown hairs, joints sometimes bulbous. STEMS straight in evolution, jointed, formed of cylindric regularly furrowed internodes, and terminated above in a ring, of which the free edge is elongated as a toothed sheath. A similar sheath terminates also the joints of the rhizome. Internodes of the stem fistular, closed above by a cellular diaphragm opposite the base of the sheath. Walls of the internodes composed of 2 concentric cylinders; outer or cortical cylinder entirely fibro-cellular, and usually presenting large longitudinal cavities opposite the furrows of the internode; inner cylinder formed of bundles of annular or spiral vessels, and presenting small longitudinal cavities opposite the ridges of the internode, and consequently alternate with the cavities in the cortical cylinder. The number and arrangement of these cavities (best seen in a transverse section of the stem) are a certain guide to the determination of the species, which are divided into *vernal* and *æstival*, according to the period at which fructification takes place. The stems are simple, or furnished with regularly whorled branches, invariably placed below the nodes and the commencement of the sheaths; these branches, and the whorled branchlets which they sometimes bear, resemble the stem in structure, but in some species they want the central cavity and the cortical spaces; the cavities and fibro-vascular bundles of the inner cylinder, are however, always present; the same is the case with the internodes of the rhizome. The epidermis of the stem, branches, and sheaths is furnished with stomata, of which the position is always limited to those parts which cover a parenchyma filled with chlorophyll. The siliceous layer which covers the epidermis is looked upon by M. Duval-Jouve as a secretion from that part of the cells which is in contact with the air, and not as entering into the composition of their membranes, as several authors maintain.

FRUCTIFICATION in spikes or conical catkins, either æstival and terminal, or vernal and springing directly from the rhizome. SPIKE formed of several whorls of horizontal pedicels, vertically dilated at their free end into a peltate body (*clypeola*), bearing on its inner face 6-9 sporangia, whorled around the pedicel, and parallel to it. SPORANGIA, at the period of the emission of the spores, dehiscing longitudinally on the side facing the pedicel.

SPORES very numerous, free, uniform, spherical, and bearing two filiform appendages (*elaters*) dilated at each end into a flattened very hygroscopic and spirally coiled spathula, which uncoils under moisture. Before expansion the elaters form, according to M. Duval-Jouve, a hollow sphere around the spore, their common point of attachment being on the equator, and their spathulate ends on the poles of the spores. The spores germinate as in Ferns, developing an irregularly lobulate diœcious or monœcious prothallus, which bears antheridia at the end of its lobes, and archegonia on the upper surface of the fleshy tissue of its base.

The ANTHERIDIA appear as ovoid swellings, composed of large cells enveloping a central group of prismatic cells, which soon multiply into a number of cellules, each containing a flattened ellipsoid globule. Soon after, the walls of these cellules disappear, and the globules (the future antherozoids) become isolated. After some days there appears within the globules a colourless imperfect ring, with unequally swollen ends, fixed against the circumference of the disk, of which it occupies three-fourths, the remaining space being occupied by a mucilaginous mass. Soon the terminal cells of the antheridia separate at the top, spread out so as to form a crown, and leave a passage for the globules; these are no sooner set at liberty than they are seen to quiver, and oscillate like a pendulum. These oscillations continue for a few seconds only, after which the antherozoids are found to have replaced the globules, not the smallest trace of the latter remaining. M. Duval-Jouve, who has published a very beautiful and complete monograph of this family, thinks that the globule is absorbed simultaneously with the formation of the antherozoid. The antherozoids of *Equisetaceæ* are formed like those of Ferns, and have the same faculty of translation.

The ARCHEGONIA are found near the lobed branches of the prothallus; these branches are nearly always deprived of antheridia at their extremity; their lower part is thicker, and composed of cells smaller than those of the antheridiferous branches. This fleshy region bears several small bright red cellular cushion-like organs, with a globose base (*ventricle*), and a long neck terminated by a 4-lobed bell-mouth; the ventricle is immersed in the small-celled tissue, and is nearly filled by a more or less spherical cell, from which, after fertilization, the new plant will start. It is into the cavity of these little cushions that the antherozoids are said to penetrate to effect fertilization.

Numerous observations have shown that, in most cases, Equisetums are diœcious; the prothallus furnished with numerous and well-developed archegonia rarely bears antheridia; and if a few archegonia occur at the base of the antheridiferous pro. thallus, they are nearly always sterile. This tendency to become diœcious is not an obstacle to reproduction; for the proximity or overlapping of the prothalli of different sexes, resulting from the entanglement of the spores by their elaters, favours fertilization. Owing to this proximity, a drop of rain or dew enables the antherozoids by a swimming movement to reach the orifice of the archegonium which they are to fertilize.

ONLY GENUS.

Equisetum.

Equisetaceæ approach Ferns in the structure of their antheridia and archegonia, as well as in their mode of germination; but as regards habit, they have no affinity with any family, unless it may be with *Casuarineæ*, with which we have compared them, or *Calamites*, which resemble gigantic Horsetails.

The *Equisetaceæ* now living are in general of small stature and not numerous; the largest have been met with near Caraccas by M. Ernst; these attain about 30 feet in height.

Horsetails principally inhabit temperate regions; they decrease towards the pole, and are rare in the tropics, and are absent throughout nearly the whole of the southern hemisphere. A remarkable species (*E. ramosissimum*) is found in Algeria, of which the branches spread for considerable distances over hedges or bushes.

The stems of several species, being encrusted with silica, are used in the arts to polish wood and even metals [and called Dutch reeds]. The rhizomes of some are starchy. *E. arvense, fluviatile, limosum,* and *hyemale* were formerly used in medicine as astringents and diuretics.

IV. *MARSILEACEÆ*, Br., Brongn.

Acotyledonous marsh HERBS, *with a creeping rhizome.* FRONDS *rolled into a crozier in vernation.* FRUCTIFICATION *rhizocarpian, enclosed in 2–several-celled 2–4-valved sporocarps.* ANTHERIDIA *and* SPORANGIA *enclosed in the same sporocarp, but occupying separate cells, sessile or sub-sessile on a sub-gelatinous placenta. Sporangia emitting a single spore, from which springs a prothallus bearing a single archegonium.*

Perennial PLANTS, inhabiting swamps or inundated places. RHIZOME filiform, creeping and rooting; axis composed of rayed and annular vessels, and elongated cells. FRONDS radical, epidermis furnished with stomata, the extremity rolled into a crozier before expansion; rachis naked (*Pillularia*), or terminated by 2 pairs of leaflets placed crosswise. LEAFLETS cuneate, entire or lobed, traversed by fan-shaped dichotomous nerves, and presenting, as in *Oxalis*, the phenomenon of sleep (*Marsilea*).

FRUCTIFICATION of two sorts, enclosed in capsular receptacles (*sporocarps*), solitary, axillary, situated near the rhizome, or geminate, or inserted near the base or along the fronds, globose or reniform, glabrous or velvety. SPOROCARPS 2–4-valved, emitting either a mucilaginous mass which contains micro- and macro-spores mixed (*Pillularia*), or a mucous cylinder bearing at intervals oblong sporangia which each

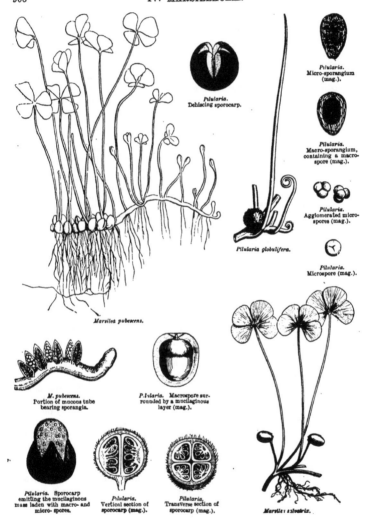

Pilularia.
Dehiscing sporocarp.

Pilularia.
Micro-sporangium
(mag.).

Pilularia.
Macro-sporangium,
containing a macro-
spore (mag.).

Pilularia.
Agglomerated micro-
spores (mag.).

Pilularia.
Microspore (mag.).

Pilularia globulifera.

Marsilea pubescens.

M. pubescens.
Portion of mucous tube
bearing sporangia.

Pilularia. Macrospore surrounded by a mucilaginous
layer (mag.).

Pilularia. Sporocarp
emitting the mucilaginous
mass laden with macro- and
micro- spores.

Pilularia.
Vertical section of
sporocarp (mag.).

Pilularia.
Transverse section of
sporocarp (mag.).

Marsilea salvatrix.

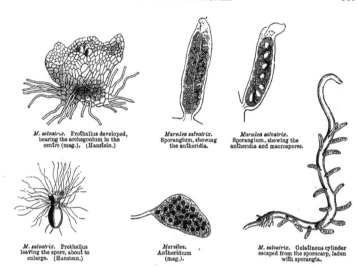

M. salvatrix. Prothallus developed, bearing the archegonium in the centre (mag.). (Hanstein.)

Marsilea salvatrix. Sporangium, showing the antheridia.

Marsilea salvatrix. Sporangium, showing the antheridia and macrospores.

M. salvatrix. Prothallus leaving the spore, about to enlarge. (Hanstein.)

Marsilea. Antheridium (mag.).

M. salvatrix. Gelatinous cylinder escaped from the sporocarp, laden with sporangia.

contain, but on opposite sides, micro- and macro-spores (*Marsilea*). MACRO-SPORANGIA (*oophoridia*) and MICRO-SPORANGIA (*antheridia*) formed originally of cells, containing a cellular mass, which afterwards divides into granules agglomerated in fours. The antheridia give birth to fragile vermiform many-haired antherozoids, similar to those of Ferns. In the macro-sporangia all the cells are absorbed with their 4 contained granules, except one, in which one of the 4 granules is further developed into a spore; this spore, enveloped in a mucous layer, rapidly emits a prothallus, which bears at its centre a large cell surmounted by a hollow papilla, formed of 4 rows of superimposed cells; this is the archegonium, on which the antherozoids act to produce fertilization.

GENERA.

Marsilea.   Pilularia.

Marsileaceæ resemble Ferns in their coiled fronds, but the structure of their spores or reproductive organs brings them nearer to *Lycopodiaceæ*.

This family inhabits the temperate and hot regions of both worlds and of New Holland. It contains but few species, of which however one (*M. salvatrix*) has of late acquired celebrity from having been the means of saving the lives of a party of intrepid exploring naturalists, who, lost in the immense deserts of Australia, had no other food but its sporangia. [These after the drying up of the pools in which the *Marsilea* grows, are found in vast abundance on the surface of the soil, and contain a certain proportion of starch and probably other nutritious matters.]

V. *SALVINIEÆ*, *Bartling*.

Acotyledonous aquatic PLANTS. FRONDS *with edges reflexed in vernation.* FRUC-
TIFICATION *rhizocarpian, enclosed in distinct 1-celled sporocarps.* ANTHERIDIA *and*
SPORANGIA *springing from a branched placenta, situated at the bottom of the sporocarp.*
SPORE *solitary.* PROTHALLUS *with several archegonia.*

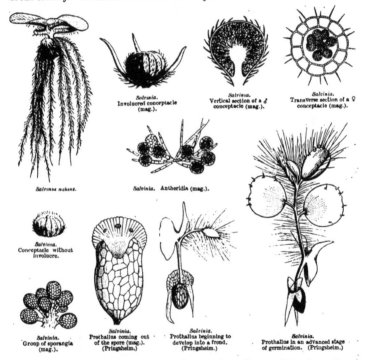

Salvinia.
Involucred conceptacle
(mag.).

Salvinia.
Vertical section of a ♂
conceptacle (mag.).

Salvinia.
Transverse section of a ♀
conceptacle (mag.).

Salvinia natans.

Salvinia. Antheridia (mag.).

Salvinia.
Conceptacle without
involucre.

Salvinia.
Group of sporangia
(mag.).

Salvinia.
Prothallus coming out
of the spore (mag.).
(Pringsheim.)

Salvinia.
Prothallus beginning to
develop into a frond.
(Pringsheim.)

Salvinia.
Prothallus in an advanced stage
of germination. (Pringsheim.)

Annual floating HERBS, not attached to the soil, resembling large *Lemnæ*
(*Salvinia*), or a *Jungermannia* (*Azolla*), with no true stem. FRONDS with margins
reflexed before expansion, usually claret-coloured on the under surface, sometimes
composed of cellular tissue, without nerves and stomata (*Salvinia*); sometimes with
a stomatiferous epidermis (*Azolla*), rounded or lobed, sessile or sub-sessile, alternate

or distichous, imbricate. ·REPRODUCTIVE ORGANS of two inds, similar to those óf *Marsileaceæ*, inserted at the base of the fronds. Antheridia and sporangia enclosed in distinct sporocarps (*conceptacles*), globose, 1-celled, composed of 2 laminæ with intervening empty laminæ. ANTHERIDIA spherical, springing from a much-branched basilar column. SPORANGIA ovoid, pedicelled, at the top of a central clavate column. The only spore developed in the sporangium emits a prothallus, on which several archegonia are usually developed.

GENERA.

Azolla.   Salvinia.

Salviniæ, which formerly composed, with *Marsileaceæ*, the family of *Rhizocarpeæ*, differ from *Marsileaceæ* and *Pilulariæ* in habit, and in reproductive organs, which occupy distinct one-celled sporocarps with central placentation; also in the sporocarp not producing mucous bodies, and in its decaying when ripe.

These plants float on still water like *Lemna*.

Salvinia is met with throughout the northern hemisphere, as well as in tropical and South America. *Azolla* inhabits Asia, Africa, Australia, and America from Canada to the Straits of Magellan.

VI. *LYCOPODIACEÆ.*

(LYCOPODINEÆ, *Swartz, Br.*—LYCOPODIACEÆ, *L.-C. Richard, Brongn., Spring.*)

Acotyledonous terrestrial moss-like PLANTS. STEM *herbaceous, rooting or creeping, simple or branched.* LEAVES *opposite or whorled, persistent, small, 1-nerved.* FRUCTIFICATION *epiphyllous at the base of the leaves, and dispersed over the entire stem, or arranged in catkins at the ends of the branches.* SPORANGIA *sessile, or shortly stipitate, reniform or cordate, coriaceous, 2–3-valved, containing either powdery quaternary granules, smooth or papillose, or larger granules, marked on one side with 3 prominent lines.*

Terrestrial perennials, or very rarely annuals, resembling Mosses or Jungermannias. ROOTS filiform, at first simple, then dichotomous, exceptionally fusiform (*Phylloglossum*). STEM herbaceous, leafy, rooting, creeping or erect or sometimes sarmentose (*Selaginella scandens, Lycopodium volubile*), simple or dichotomously branched. Axis formed of scalariform vessels, arranged in bundles, the number of which is a multiple of 2 (4, 8, 16, 32, 64), and the division of which produces a dichotomous ramification of the stem, as in Ferns.

Lycopodiaceæ are divided by their habit into two natural groups: the one has branches developed in all directions, or at least in undetermined directions (true *Lycopodia, Psilotum, Tmesipteris*); in the other the branches spread symmetrically, forming a sort of frond analogous to that of some Ferns (*Selaginella*). Some species have a compressed stem (*Lycopodium complanatum*), or square (*L. tetragonum*); several resemble large Mosses (*L. fontinaloides*), or long cords (*L. funiforme*). *L. clavatum* has often creeping stems sometimes more than 13 feet long, and

Pœppig has observed stems of *Selaginella exaltata* which measured more than 60 feet.

LEAVES simple, sessile, more or less decurrent, never jointed, regular or falcate, with a single stomatiferous nerve, sometimes whorled around the axis, and all, at a given height, of the same shape and size (*Lycopodium*, &c.); sometimes in 4 regular series, arranged on the plan of the branches, and presenting two different forms; the one largest, occupying the side of the axis, and called *lateral*; the others smaller, called *intermediate* or *stipuliform* (*Selaginella*); sometimes auricled; glabrous or very rarely sub-pubescent, green or dark or light red (*Lycopodium rubrum, rubescens,* &c.), sometimes shot (*Selaginella cæsia, atro-viridis*),

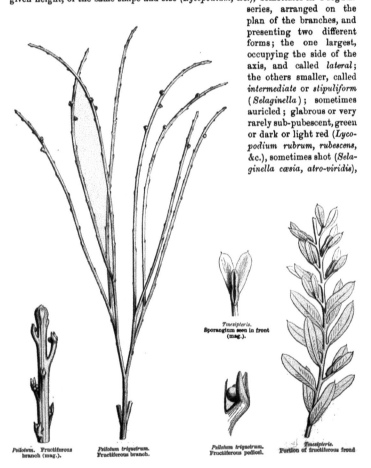

Psilotum. Fructiferous branch (mag.).

Psilotum triquetrum. Fructiferous branch.

Tmesipteris. Sporangium seen in front (mag.).

Psilotum triquetrum. Fructiferous pedicel.

Tmesipteris. Portion of fructiferous frond

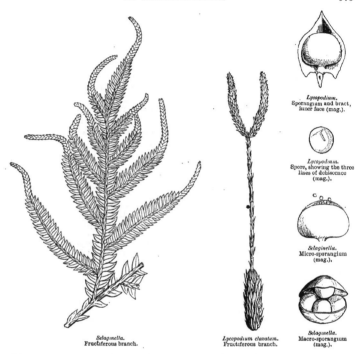

Lycopodium.
Sporangium and bract,
inner face (mag.).

Lycopodium.
Spore, showing the three
lines of dehiscence
(mag.).

Selaginella.
Micro-sporangium
(mag.).

Selaginella.
Macro-sporangium
(mag.).

Selaginella.
Fructiferous branch.

Lycopodium clavatum.
Fructiferous branch.

either acute or terminating in a scarious white tongue (*L. vestitum*); axillary buds never present.

REPRODUCTIVE ORGANS springing either near the base of the leaves, over all the branches, or only at their upper part, sometimes towards the base of bracteal leaves, and forming spikes, catkins, or terminal cones; rarely borne at the end of a sort of naked scape, which rises from the middle of a rosette of small subulate leaves (*Phylloglossum*). SPORANGIA sometimes all alike, and then 1-celled (*Lycopodium, Phylloglossum*), or 2-celled (*Tmesipteris*), or 3-celled (*Psilotum*), containing only very small homogeneous granules, sometimes dimorphous (*Selaginella*). The *dimorphous sporangia* consist of :—1st, MICRO-SPORANGIA (*goniotheca*), which are 2–3-valved, and contain numerous smooth or papillose granules (*microspores, antheridia*), which are developed in groups of 4 in cells that are soon re-absorbed; these antheridia, when placed in suitable conditions of moisture, break, and project beyond the cells, whence antherozoids emerge, similar to those of Ferns and Equisetums:—2nd, the MACRO-

3 N

SPORANGIA (*oophoridia, spherotheca*), less numerous, solitary and very large, or 4–5–6 at the base of the spike, or indeterminate in number, and mixed with the microsporangia, sometimes placed on distinct spikes, and as it were monœcious, 2–3-lobed and valved, and containing 3–4–8 sub-globose bodies (*macrospores*). The MACRO-SPORES are the true spores, which are much larger than the microspores, and alone have the power of germinating; they are rounded on the outer side, and present where they are in contact within the sporangium 3–4 triangular faces, somewhat flattened, and separated by more or less projecting lines; the prothallus springs from the point where these lines meet. PROTHALLUS orbicular, composed of 3–4 layers of cells; the archegonia are developed on its surface, and are somewhat similar to those of Ferns. No one has yet seen the germination of the microspores, which are only sexual organs of all Lycopods but *Selaginella*.

GENERA.

Psilotum.　　Tmesipteris.　　Lycopodium.　　Selaginella.　　Phylloglossum.

The fertilization of Lycopods is still very obscure. In true Lycopods, in which the sporocarps are all alike (*Lycopodium, Psilotum,* &c.), we only find microspores, apparently homogeneous, but perhaps of two different natures, mixed in the same sporangium. It is different in *Selaginella,* where we can trace the development of the embryonic vesicle and of the embryo, precisely as in Ferns. One archegonium alone is fertile, and it is not rare to find at the base of the young plant the remains of the prothallus laden with sterile archegonia. One peculiarity of the formation of the embryo in *Selaginella* is, that at the moment when fertilization takes place, the cavity of the spore partially fills with a very delicate cellular tissue. The embryonic vesicle grows, becomes divided, and lengthens into a filament which buries itself in the middle of this fresh tissue. Then the lower cells of the filament multiply and form a small cellular mass, from which spring the first bud and the radicle. Thus formed, the embryo quickly rends the prothallus, and gives birth to a young plant.

Lycopodiaceæ, as at present defined, are closely allied to *Isoeteæ* only; they differ in the regular and valvate dehiscence of the sporangia and the number of the macrospores, which are four and rarely eight, while they are numerous in *Isoeteæ.*

The creeping species grow at one end, while the other end decays, so that they travel, like the rhizomes of many Phænogams (*Carex, Iris, Polygonatum,* &c.) recall by their dichotomous and leafy stems certain fossils of the coal formation (*Calamites*); their cone-shaped fructification resembles the structure of another fossil, the *Triplosporites,* described by R. Brown, and placed by him between *Lycopodiaceæ* and *Ophioglosseæ.* There is also a resemblance between *Lycopodiaceæ* and Gymnosperms.

About 350 *Lycopodiaceæ* are now known, of which 100 are Lycopods and 200 Selaginellas. This group is represented up to the polar regions by *Lycopodium Selago, alpinum, complanatum, magellanicum,* as well as by *Selaginella spinulosa, helvetica* and *denticulata,* which advance to the snow limit, where *S. sanguinolenta* is met with on the Altai and towards the mouth of the Amoor. *Phylloglossum (P. Sanguisorba,* Spg.), the smallest known species, resembling in habit a very small *Plantago pusilla* (an inch or two in height), is found on the west coast of Australia, and in New Zealand. *Tmesipteris* belongs to Australia, *Psilotum* to Madagascar, the Mascarene, the Moluccas and the Sandwich Islands.

We have no very precise knowledge of the properties of *Lycopodiaceæ.* The Club Moss (*L. clavatum*), which grows in the mountainous woods of Europe, is an insipid herb, still administered in Russia for hydrophobia. The granules which fill the sporangia are extremely inflammable, and form 'vegetable sulphur,' which is valuable for theatrical purposes [to produce lightning]. This dust is also used for rolling pills in, and medicinally as a desiccator, to cure excoriations in the skin of young infants. A decoction of *L. Selago* is emetic, drastic, vermifuge and emmenagogue; *L. myrsinitis* and *catharticum* are also considered purgative. The root of *L. Phlegmaria* is slightly salt; the Indians attribute to it marvellous virtues in checking vomiting, curing pulmonary affections, dropsy, &c.; they also use it in the composition of philters.

Some species of *Lycopodiaceæ* are cultivated; the true Lycopods are very difficult to rear, but this is not the case with *Selaginellæ*, several of which play a considerable part in the ornamentation of our hot-houses, as a covering for damp walls or borders: such are *S. apoda, denticulata, Hugelii, cæsia*, and *cuspidata* as a sarmentose plant. Two or three species possess the property of drying up and reviving when moistened (*S. convoluta, involvens*, &c.); thus recalling the Rose of Jericho (see p. 232).

VII. *ISOETEÆ, Bartling.*

Aquatic and submerged Acotyledons, or terrestrial. RHIZOME *very short, furrowed, emitting dichotomous roots and subulate cæspitose fronds, erect in evolution, enlarged and membranous at the base.* SPORANGIA *situated at the lower part of the fronds, those containing macrospores attached to the outer fronds; those containing microspores attached to the central fronds of the rosette; the macrospores are marked on one hemisphere with a tricrural line, the microspores are marked with a furrow.*

Perennial grass-like PLANTS, aquatic or amphibious or terrestrial, stemless. RHIZOME sub-globose or depressed, formed of a fleshy utricular tissue, often oily, bearing below 2–3–4 furrows or fissures, along which it divides, by a sort of budding process, into distinct individuals (*Isoetes setacea*, &c.). ROOTS often springing in longitudinal series in each of the furrows of the stock, dichotomous, brown, almost glabrous in the lacustrine species, velvety in the terrestrial. FRONDS fascicled, straight in vernation, their base more or less amplexicaul, and pressed together like a bulb, terminated by a foliar linear-filiform or subulate blade, resembling the leaves of some Phænogams (*Littorella, Lobelia Dortmanna*, &c.), along with which *Isoetes* is often found.

Following J. Gay, A. Braun distinguishes in the frond the phyllopode, the pouch (*le voile*), the border (*l'aréa*), the sporangium, the ligule, and the blade. The phyllopode is the dilated or sheathing base of the frond, an organ analogous to the petiole of a leaf. In the terrestrial species the phyllopodes persist for several years, on the outside of the rhizome, as brown, hard, 2–3-toothed scales. The phyllopode is hollowed into a pouch (*le voile*), which occupies almost its whole surface; this pouch either opens towards the axis, or is perforated at the base. A narrow border of a peculiar tissue (*l'aréa*) circumscribes it. The pouch encloses a membranous sac (*sporangium*), closed on all sides, which is divided transversely into several compartments by membranous septa. Above the sporangium is a small smooth scale, formed of a delicate tissue (*ligule*). The rest of the frond, of a more or less deep green, forms the blade; it is usually subulate, flat on the inner surface, convex on the outer; it is traversed longitudinally by transversely septate tubes, which surround a central bundle of annular and spiral vessels. The epidermis of the terrestrial species bears stomata, which are absent in the lacustrine.

The SPORANGIA, although alike in form, structure, and insertion, contain 2

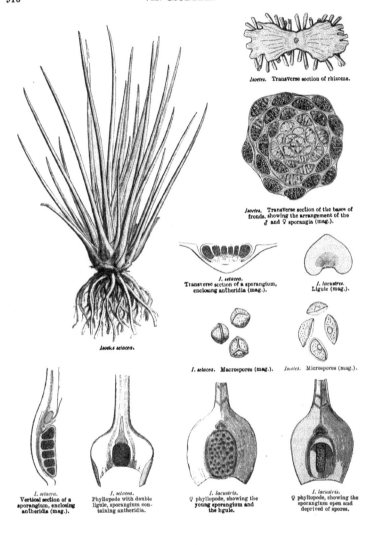

Isoetes. Transverse section of rhizome.

Isoetes. Transverse section of the bases of fronds, showing the arrangement of the ♂ and ♀ sporangia (mag.).

Isoetes setacea.

I. setacea.
Transverse section of a sporangium, enclosing antheridia (mag.).

I. lacustris.
Ligule (mag.).

I. setacea. Macrospores (mag.).   *Isoetes.* Microspores (mag.).

I. setacea.
Vertical section of a sporangium, enclosing antheridia (mag.).

I. setacea.
Phyllopode with double ligule, sporangium containing antheridia.

I. lacustris.
♀ phyllopode, showing the young sporangium and the ligule.

I. lacustris.
♀ phyllopode, showing the sporangium open and deprived of spores.

kinds of reproductive organs, namely, macrospores in the outer fronds, and microspores in the central fronds.

The MACROSPORES, to the number of 40 to 200 in each sporangium, are at first
united in fours, and later separate; they are divided by a circular ridge into 2 hemispheres, one of which is smooth, and the other, which is a little produced, is marked by
3 ridges, dividing its surface into 3 triangular spaces. The membrane of the macrospores is double; the inner very thin and very smooth; the outer thicker, crustaceous,
granulated, foveolate, or muricate, white or opaline. The macrospores open at
germination into 3 valves along the connivent ridges.

The MICROSPORES, which resemble fine flour, are whitish when fresh, brownish
later; there are above a million in each sporangium. They are at first united in
fours, then free, oblong, convex on the back, furrowed, furnished with a double membrane like the macrospores, often granulated or papillose on the surface, and contain
a minute drop of oil.

 · ˙ The development of the prothallus in *Isoetes* is absolutely the same as in
Selaginella, only the macrospores are developed in greater numbers in the sporangia,
and the archegonia are less abundant on the prothallus. The antherozoids are
somewhat similar to those of Ferns.

ONLY GENUS.

Ísoeteæ.

Isoeteæ, regarded as a section of *Lycopodiaceæ* by most botanists, and having decided affinities with
them, seem to us nevertheless to form either a small family or a distinct tribe, whether we consider the
nature of their vegetative organs, or the structure of their sporangia and macrospores.

The species of *Isoetes* are spread over the whole world. Some form a sort of close lawn, scarcely
more than an inch in height; while others, which live in deep water, often attain two feet in length
(*I. Malinverniana*).

VIII. *CHARACEÆ*, L.-C. Richard.

(CHAREÆ, *Kütz.*)

Acotyledonous PLANTS, *cellular, aquatic.* STEMS *tubular, jointed, naked or surrounded by several parallel elongated cells.* BRANCHES *whorled on a level with the
joints.* REPRODUCTIVE ORGANS *formed of antheridia and sporangia, borne on the
branches, and often accompanied by branchlets or bracteoles.* ANTHERIDIA *composed of
a spherical sac, containing oblong vesicles, whence spring numerous chambered tubes,
containing antherozoids.* SPORANGIA *oblong or ovoid, formed of spiral tubes, and
crowned by five protuberances* (corona), *and containing a single starchy uni-embryonic
spore.*

Characeæ are aquatic submerged vegetables, often exhaling a fœtid alliaceous
odour, with transparent rhizome, composed of joints always formed of a single tube,
and fixed in the mud of stagnant and running water by tubular filiform very fine
colourless rootlets. The plant is sometimes reproduced by the lower nodes of the

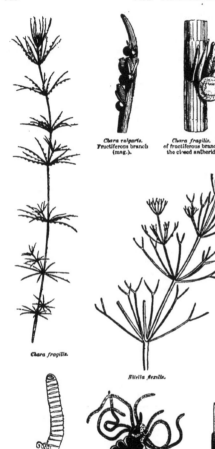

Chara fragilis.

Chara vulgaris.
Fructiferous branch
(mag.).

Chara fragilis. Portion
of fructiferous branch, showing
the closed antheridia (mag.).

Chara fragilis. Portion of fructiferous branch, showing the dehiscent
antheridium (mag.).

C. fragilis. Top of sporangium, to
show the corona formed of four non-
chambered protuberances (mag.).

Nitella flexilis.

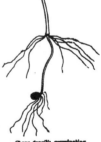

Chara.
End of a chambered filament,
of which each cell contains
an antherozoid (mag.).

Chara fragilis.
Valve of antheridium, showing the
central antheridiferous cell (mag.).

Chara.
Portion of joint,
showing the cylin-
dric inner cell,
and its sheath
formed of tubes.

Chara fragilis, germinating.

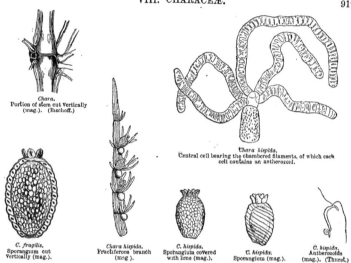

Chara.
Portion of stem cut vertically
(mag.). (Bischoff.)

Chara hispida.
Central cell bearing the chambered filaments, of which each
cell contains an antherozoid.

C. fragilis.
Sporangium cut
vertically (mag.).

Chara hispida.
Fructiferous branch
(mag.).

C. hispida.
Sporangium covered
with lime (mag.).

C. hispida.
Sporangium (mag.).

C. hispida.
Antherozoïds
(mag.). (Thuret.)

stem being converted into starchy tubers of various shapes, or by whitish crustaceous bulbils, developed at the joints of the stems.

STEMS tubular, cylindric, leafless, jointed, sometimes transparent and flexible, even after being dried (*Nitella*); sometimes opaque and fragile, even before being dried (*Chara*); often covered with calcareous salts, usually branched ; *joints* composed of a cylindric tubular cell, naked (*Nitella*), or clothed with a sort of sheath formed of smaller tubes united together, and giving rise on the outer surface to longitudinal and oblique striæ (*Chara*); the joints or tubes are filled with a colourless liquid, in which pale green granules float ; their inner wall is covered with green uniform granules, arranged in a chaplet or longitudinal parallel very regular series, and more or less pressed together ; these series are oblique with relation to the axis of the tube, an obliquity which is the result of the greater or less twisting of the tube. The series of green granules covers the whole inner surface of the tube, with the exception of 2 parallel opposite bands, which are quite free from them. This arrangement of the granules obtains both in the simple tube of *Nitella*, and in the central and peripheric tubes of *Chara*; but the intracellular circulation which is described at p. 147, and which has exercised the sagacity of many physiologists (Corti, Slack, Sachs, Schumacher, &c.), is best observed in the central tube of *Chara* when deprived of its envelope of cortical tubes. This circulation does not extend along the 2 bands without granules, which proves, as supposed by Amici, and confirmed by Dutrochet, that the intracellular currents take place under the influence of these series of globules fixed to the walls of the tube, and are determined by an action of these

granules on this fluid. Amici attributed them to electric action. Becquerel and
Dutrochet opposed this view. Donné, having observed that the green granules, if
detached from the tube which they line, and removed to one of the bands where the
current is not felt, execute a rapid rotatory motion, attributes the movement im-
parted to the liquid to the rotation of the granules; and as no trace of vibrating
hairs has been discovered in these granules, Brongniart was led to suppose that the
ambient fluid is set in motion by successive contractions of the various parts of each
granule, that is to say, by a change of form analogous to a sort of peristaltic motion.

The BRANCHES are whorled on a level with the joints; they may be simple and bear
along their inner surface the reproductive organs, furnished with an involucre of
branchlets or bracts; or may be more or less branched, often dichotomously, and
bear the reproductive organs at the top or on a level with the forks of the branches.

The REPRODUCTIVE ORGANS (*antheridia* and *sporangia*) are borne on the same
individual, and then usually close together (*monœcious*); or on different individuals
(*diœcious*).

The ANTHERIDIA appear before the sporangia, and are situated immediately
below them (*Chara*), or above them (*Nitella*); their wall is composed of 8 flattened
triangular crenulate convex valves, the crenatures of which so interlock, that they
together form a sphere (*globule*). Each valve consists of 12–20 cells radiating from
a common centre, and each crenulation answers to an imperfect septum radiating
from the centre of the valve. The inner surface of the valve is lined with a layer of
red granules; the cavity of the globule is occupied by a colourless liquid, the thickness of which gives the appearance of a whitish ring surrounding the antheridium.
At the centre of [the inner surface of] each valve is fixed perpendicularly an oblong
vesicle, which contains orange-coloured granules placed in a line, and presenting a
circulation observed by Thuret, and analogous to what occurs in the stems. The 8
vesicles emanating from the 8 valves converge towards the centre of the antheridium,
where their ends unite in a small cellular mass. A larger and flask-shaped ninth
vesicle attaches the antheridium to the plant, its enlarged base rising from a branch
of the *Chara*, while its opposite end penetrates the globule through a space between
the 4 lower valves, and reaches the central cellular mass. From this point emanate
a large number of flexuous hyaline chambered tubes, each joint of which contains
a filiform spirally-coiled antherozoid, furnished with 2 very long and fine setæ,
which finally escapes from the joint in which it was imprisoned.

The SPORANGIA are terminated by a *corona*, composed of 5 simple cells, whorled
and more or less persistent (*Chara*), or of 5 chambered cells simulating a corona, of 2
whorls, and very caducous (*Nitella*). The corona surrounds an orifice, which may
be regarded as a sort of micropyle. The walls of the sporangium are composed of 2
coats; the outer membranous, colourless, transparent; the inner formed of compressed
thick spiral tubes, containing colouring matters. The solitary spore is amylaceous,
and fills the sporangium; it is clothed with a membranous fawn-coloured coat,
marked, as well as the outer membrane, with spiral striæ due to the impression of
the tubes of the sporangium; it germinates without developing a prothallus.

GENERA.

Chara.   Nitella.

The affinities of *Characeæ* are very obscure. A. L. de Jussieu placed them among Monocotyledonous Phænogams, near *Naiadeæ*; R. Brown near *Hydrocharideæ*; some botanists have placed them near *Myriophyllum* and *Ceratophyllum*, from having the same habitat, and resembling them in habit. Several modern authors—Wallroth, Von Martius, Agardh, Endlicher—have classed them after *Algæ*, where Adr. de Jussieu retained them, whilst observing that, in spite of their purely cellular structure, they recall axiferous plants. Brongniart having regard to the fact that the stems and leaves of submerged Phænogams often are simple in structure as compared with terrestrial plants of the same families, thinks that the structure of the stem of *Characeæ* ought not to determine their affinity, and that it is rather the nature of their reproductive organs which should be considered; whence he places these plants amongst the highest Cryptogams, near Ferns and *Marsileaceæ*, or between these and Mosses and *Hepaticæ*. *Characeæ* all grow in fresh water, or sometimes in the brackish waters of sea-coasts, and appear to extend nearly over the globe.

This family is nearly useless to man; some species of *Chara* are covered with calcareous salts, and are used for polishing plate, whence their common name of *Herbe à écurer* and *Lustre d'eau*.

IX. MUSCI, *Dillenius, Hedwig*.

Cellular terrestrial or aquatic annual or perennial Acotyledons. STEMS *erect or prostrate, leafy.* LEAVES *alternate, or rarely distichous, sessile, simple.* REPRODUCTIVE ORGANS *formed of antheridia and archegonia, in the axils of bracteoles, or from the centre of a common involucre* (perigonium, perigynium). INFLORESCENCE *polygamous, monœcious, or diœcious.* ANTHERIDIA *formed of shortly pedicelled cellular sacs, opening at the top, and emitting a semi-fluid parenchyma, composed of cells each containing an antherozoid.* ARCHEGONIA *flask-shaped, containing an embryonic nucleus suspended in mucus.* NUCLEUS *formed of a cellular sac, which is transversely ruptured; the upper part is then raised by the rapid elongation of the embryo, whilst the lower part forms at the base of the embryo a membranous sheath* (vaginula), *and surmounts the fruit with a sort of cap* (calyptra). FRUIT *capsular* (urn or theca) *borne on a pedicel* (seta); *seta sheathed at the base by the vaginula, and sometimes terminated* [*below the urn*] *by a swelling* (apophysis). URN *spherical, ovoid, cylindric or prismatic, sometimes indehiscent, mostly furnished with a cover* (operculum), *which detaches circularly when ripe, and leaves an orifice* (peristome); *axis of the urn occupied by a cellular column* (columella) *rising into the operculum.* PERISTOME *naked, or furnished with a separable border* (annulus), *and usually crowned with 4, or a multiple of 4 teeth or hairs.* SPORANGIUM *enclosed in the urn, and formed of a double sac lining the walls of the urn and columella.* SPORES *in fours in the mother-cells, contained between the 2 sacs of the sporangium.*

STEM elongating in one direction, i.e. from below upwards, sometimes scarcely perceptible (*Buxbaumia, Phascum, Discelium,* &c.), sometimes several feet long (*Neckera,* &c.), simple or branched, of equal diameter throughout, cylindric or trigonous, variable in consistency and colour; sometimes soft, watery and nearly transparent (*Schistostega,* &c.), or green or reddish (*Bryum, Splachnum, Mnium,* &c.); sometimes nearly woody and black (*Polytrichum, Hypnum, Neckera,* &c.). Axis of the stem solely composed of more or less elongated and narrow fibrous cells; the outer superficial cells, which take the place of the epidermis, are thicker and of a more or less intense red; they are often prolonged into aërial rootlets or appendages

IX. MUSCI.

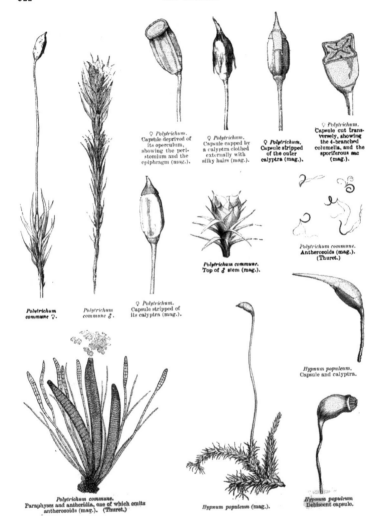

♀ *Polytrichum.*
Capsule deprived of its operculum, showing the peristomium and the epiphragm (mag.).

♀ *Polytrichum.*
Capsule capped by a calyptra clothed externally with silky hairs (mag.).

♀ *Polytrichum.*
Capsule stripped of the outer calyptra (mag.).

♀ *Polytrichum.*
Capsule cut transversely, showing the 4-branched columella, and the sporiferous sac (mag.).

Polytrichum commune.
Antherozoids (mag.). (Thuret.)

Polytrichum commune.
Top of ♂ stem (mag.).

Polytrichum commune ♀.

Polytrichum commune ♂.

♀ *Polytrichum.*
Capsule stripped of its calyptra (mag.).

Hypnum populeum.
Capsule and calyptra.

Polytrichum commune.
Paraphyses and antheridia, one of which emits antherozoids (mag.). (Thuret.)

Hypnum populeum (mag.).

Hypnum populeum.
Dehiscent capsule.

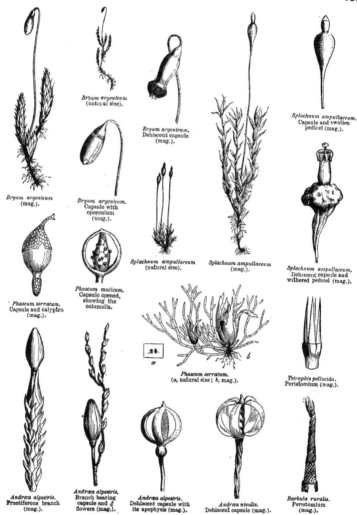

Bryum argenteum
(natural size).

Bryum argenteum.
Dehiscent capsule
(mag.).

Bryum argenteum.
(mag.).

Bryum argenteum.
Capsule with
operculum
(mag.).

Splachnum ampullaceum.
Capsule and swollen
pedicel (mag.).

Splachnum ampullaceum
(natural size).

Splachnum ampullaceum.
(mag.).

Splachnum ampullaceum.
Dehiscent capsule and
withered pedicel (mag.).

Phascum serratum.
Capsule and calyptra
(mag.).

Phascum muticum.
Capsule opened,
showing the
columella.

Phascum serratum.
(a, natural size; b, mag.).

Tetraphis pellucida.
Peristomium (mag.).

Andræa alpestris.
Fructiferous branch
(mag.).

Andræa alpestris.
Branch bearing
capsule and ♂
flowers (mag.).

Andræa alpestris.
Dehiscent capsule with
its apophysis (mag.).

Andræa nivalis.
Dehiscent capsule (mag.).

Barbula ruralis.
Peristomium
(mag.).

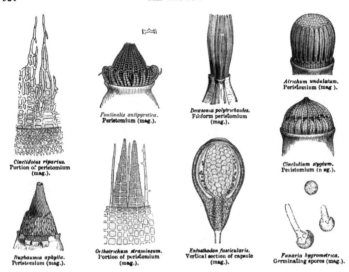

Cinclidotus riparius.
Portion of peristomium
(mag.).

Fontinalis antipyretica.
Peristomium (mag.).

Dawsonia polytrichoides.
Filiform peristomium
(mag.).

Atrichum undulatum.
Peristomium (mag.).

Buxbaumia aphylla.
Peristomium (mag.).

Orthotrichum stramineum.
Portion of peristomium
(mag.).

Entosthodon fascicularis.
Vertical section of capsule
(mag.).

Cinclidium stygium.
Peristomium (n ag.).

Funaria hygrometrica.
Germinating spores (mag.).

of very various form ; the fibrous cells are larger, very thin, hyaline, filled with a watery liquid and amylaceous granules ; the medullary cells are narrower and thicker, but soft and brownish ; they contain no granules, but colour by iodine.

Roots.—The roots spring either from basilar cells or from peripheric cells of the stem ; they are always composed of a single series of cells united by oblique walls ; they are more or less branched, the principal branches of a reddish brown, the branchlets white or hyaline.   In many species a resinous exudation forms a granular deposit on their surface, which is of great importance in attaching certain species to hard bodies ; in other cases contributing to the agglutination of sands and the binding of the dunes on sea-coasts by tufted species (*Polytrichum piliferum, nanum, Barbula ruralis, Rhacomitrium canescens*, &c.).   Besides their subterranean roots, most Mosses possess aërial or adventitious roots, which are developed over the whole surface of the stem, but more especially in the axils of the leaves of the branches.

Leaves.—Blade usually simple, nerveless, or traversed longitudinally by one, rarely by two, cellular bundles (*Hylocomium*), commonly termed nerves, which may be shorter than the blade, or may reach its apex, or extend beyond it as an awn, bristle or hair-point.   In some species the nerves present on the upper or lower surface more or less regular excrescences (*Fissidens, Pottia*), or plates with thickened margins (*Barbula, Polytrichum*).   The leaves are always sessile, and inserted more or less horizontally [or obliquely or longitudinally], often decurrent, with usually symmetrical wings ; and are distichous or alternate, and arranged in regular spirals.

The cells of the tissue of the leaves form regular dodecahedrons, or a dense parenchyma (*Mnium, Orthotrichum,* &c.), or elongated dodecahedrons, or rhomboids (*Bryum*), or approach vermicular fibrous cells (*Hypnum,* &c.); in *Dichelyma,* and in the midrib and marginal edge of many other species, the cells lengthen so much as to resemble vessels. The cells often vary in form and size in the same leaf, being usually larger and more elongated towards the base of the blade than towards the top, and formed of a thinner membrane, and deprived of chlorophyll ; the marginal series is always composed of narrow cells which often project under the form of tubercles or teeth (*Polytrichum,* &c.), or branched hairs (*Buxbaumia,* &c.). The cell-membrane itself is smooth or covered with papillæ, whence the name of papillose cells (*Barbula*). In certain Mosses the foliar parenchyma is composed of several similar or dissimilar cellular layers (*Leucobryum*); in the latter, each layer is homogeneous; in some the cells are small, elongated, nearly cylindric, full of chlorophyll; in others they are large, nearly octahedrons and tubular, with hyaline walls pierced with holes, and deprived of green granules. These layers are alternate, and the tabular cells always cover both surfaces of the leaves.

REPRODUCTIVE ORGANS.—Mosses are 2-sexual; both sexes may occur in the same involucre, or in separate ones, and the plant is accordingly monœcious or diœcious. In the 2-sexual Mosses the two reproductive organs are intermixed in the centre of the involucre, or arranged in two groups, or contained in separate special involucral leaves. The inflorescence differs in external appearance in the two sexes; the involucre of the male flowers is termed a *perigonium,* that of the female a *perigynium,* and that of the 2-sexual *perigamium.* The two latter form an elongated, nearly closed bud, composed of leaves resembling the cauline leaves, of which, in fact, they are a modification ; the perigonium is stouter, and its leaves are larger and more concave. At the time of the development of the fruit, a new cycle of leaves, which are in a rudimentary state before fertilization, is simultaneously evolved ; it bears the name of *perichœtium,* and varies in form in different species.

The *paraphyses* are jointed filaments found in the inflorescence of most Mosses: in the ♀ they are simple filiform bodies, composed of a single series of cells ; in the ♂ filiform, clavate or spathulate, and terminated by several series of cells.

The *perigonium* is often traversed by the stem, which is sometimes continuous (*Polytrichum*) through several superimposed series of perigonia, a character peculiar to Polytrichum, as is also the presence of an *epiphragm,* closing the urn.

The ANTHERIDIA, or male organs, are little elongated cylindric or sub-spherical (*Buxbaumia*) usually very shortly pedicelled sacs formed of tabular cellular tissue filled with green granules, enveloped in an extra-cellular thick and hyaline substance, and filled with a granular mucilage, destined to be expelled by jets through an apical opening in the antheridium when the latter is mature. The mucilaginous mass filling the antheridium is composed of spherical hyaline cells ·00026 to ·00039 in. in diameter, each containing a filiform antherozoid, and furnished anteriorly with two extremely slender vibrating hairs, which equal it in length. The antherozoid describes a spiral of two turns within the mother-cell, and presents, either in the middle, or at the posterior part, a heap of 12–20 amylaceous granules; water dissolves the mother-cell, and frees the antherozoid, which by means of its vibrating hairs reaches the gaping mouth of the archegonium.[1]

The ARCHEGONIUM or female organ originates, like the antheridium, as a cell, which by subdivision develops a cylindrical cellular body with a rounded top, traversed by a canal. As its walls elongate, the lower part of the archegonium dilates, till, when mature, it becomes flagon-shaped, and at the same time its tubular apex has expanded into a funnel-shaped terminal orifice. The cavity of the archegonium contains a nucleus (a free cell) immersed in mucilage. The nucleus, after fertilization of the archegonium by an antherozoid, elongates and becomes cylindric, the archegonium itself at the same time increasing both in length and breadth; but the nucleus being developed far more rapidly than the archegonium, finally ruptures this transversely, carrying up its upper part as a conical cap (*calyptra*), and leaving its lower part as a sheath (*vaginule*) surrounding the base of the nucleus, which now forms a slender bristle. This bristle goes on elongating, still capped by the calyptra, till it has attained its full height, when the upper part within the calyptra swells and forms the capsule (*theca* or *urn*), commonly called the fruit of the Moss. The rest of the bristle is the *pedicel* or *seta* of the capsule.

The CAPSULE is usually ovoid or cylindric, sometimes spherical (*Phascum*), or angular (*Polytrichum*), rarely compressed on one side and unequal. It may remain entire (*Phascum*), or split into four segments united at the top (*Andræa*); but in most other Mosses it dehisces transversely above the middle. This detached portion is named the *operculum* of the capsule. The base of the capsule is usually narrowed equally into the seta; but sometimes there is a symmetrical oblique or gibbous swelling at the junction, called *apophysis*. In *Andræa* and *Archidium* the apex of the branch at the base of the seta is swollen and fleshy, forming a pseudopodium, as in *Sphagneæ*. Between the margins of the operculum and of the capsule there often exists an intermediate organ, called the *annulus*, composed of one or several rows of very hygrometrical cells, the rapid growth of which facilitates the detachment of the operculum. The mouth of the capsule is naked, or furnished with a peristome consisting of 1-2 rows of lanceolate or filiform (*Dawsonia*) appendages or teeth, which are definite in number (4, 8, 16, 32, 64). When the peristome is simple, it usually originates in the loose tissue which lines the inner surface of the urn; when it is double, the inner is a prolongation of the sac (*sporangium*) containing the spores. Sometimes (*Polytrichum*) the inner peristomium extends horizontally from the circumference towards the centre, to form a membrane called the *epiphragm*. The walls of the capsule are composed of an epidermal layer formed of tabular small and thick cells, and of 2-3 layers of large thin and hyaline parenchymatous cells. The epidermal layer is often pierced with stomata, especially at its lower part, as well as at the neck and apophysis.

The SPORANGIUM is enclosed in the capsule; it is composed of a membranous sac which lines the cavity of the capsule, and whose base is drawn up like the finger of a glove over a central axis (*columella*). A loose tissue, or of jointed and sometimes

[1] I am unable to make the structure of the organs of   from which I have widely departed in the description Mosses clear by a literal translation of the original, of the development of the archegonium.—ED.

anastomosing filaments, unites the outer wall of the sac to the inner wall of the capsule; the columella is fitted with lax cellular tissue, which rises into the oper. culum, and is continuous below with the tissue of the seta. The columella is wanting in some Phascums.

The SPORES are developed in fours in *mother-cells*, which constitute a very soft tissue between the columella and inner walls of the sporangium, and which tissue is rapidly absorbed. *Archidium* alone presents an exception to this rule, each mother-cell containing only a single spore. The *prothallus* resulting from the germination of the spores is a cellular confervoid filament, which branches dichotomously or in tufts, on several points of which buds appear; these buds become leafy stems, of which some bear archegonia or antheridia, or both. The prothallus is persistent in some minute Mosses with very slender stems (*Schistostega, Ephemerum,* &c.).

Tribe I. *BRYACEÆ*.

Mosses properly so called, *stegocarpous* or *cleistocarpous* [capsule dehiscing or not]. Capsule sessile or pedicelled, indehiscent or with a separable operculum; mouth with or without an annulus, naked, or with a simple or double peristome.

PRINCIPAL GENERA.

Ephemerum.	Dicranodontium.	Calymperes.	Discelium.	Leptostomum.
Physcomitrella.	Campylopus.			Fontinalis.
	Angstrœmia.	Zygodon.	Funaria.	Dichelyma.
Phascum.	Trematodon.	Amphidium.	Entosthodon.	Climacium.
Acaulon.			Physcomitrium.	
Bruchia.	Leucobryum.	Orthotrichum.	Pyramidula.	Cryphæa.
	Octoblepharum.	Ulota.		Leucodon.
Voitia.		Ptychomitrium.	Meesia.	Leptodon.
	Fissidens.	Glyphomitrium.	Amblyodon.	Cladomnium.
Archidium.	Conomitrium.	Macromitium.		Neckera.
		Coscinodon.	Bartramia.	Trachyloma.
Pleuridium.	Seligeria.	Sclotheimia.	Oreades.	Homalia.
Sporledera.	Campylostelium.		Glyphocarpus.	
	Blindia.	Grimmia.	Cryptopodium.	Hookeria.
Schistostega.	Brachyodus.	Scouleria.	Conostomum.	Cyathophorum.
		Schistidium.		Pterygophyllum.
Astomum.	Pottia.	Dryptodon.	Mielichoferia.	Daltonia.
Rhabdoweisia.	Anacalypta.	Rhacomitrium.		Mniadelphus.
Hymenostomum.	Trichostomum.		Bryum.	
Weisia.	Desmatodon.	Hedwigia.	Orthodontium.	Fabronia.
Gymnostomum.	Barbula.	Braunia.	Mnium.	Aulacopilum.
	Leptotrichum.		Aulacomnium.	Anisodon.
Anœctangium.	Didymodon.	Cinclidotus.	Timmia.	Anacamptodon.
	Ceratodon.		Paludella.	Thedenia.
Dicranum.	Distichium.	Splachnum.	Webera.	
Dicranella.		Tayloria.	Hymenodon.	Hypnum.
Cynodontium.	Tetraphis.	Dissodon.	Cinclidium.	Hypopterygium.
Stylostegium.	Tetrodontium.	Œdipodium.	Buxbaumia.	Rhizogonium.
Holomitrium.	Encalypta.	Tetraplodon.	Diphyscium.	Hymenodon.

Plagiothecium.	Lyellia.	Pterogonium.	Anomodon.
Rhynchostegium.		Lescurea.	Heterocladium.
Thamnium.	Orthothecium.	Pterigynandrum.	Peeudoleskea.
Eurbynchium.	Pylaisæa.		Thuidium.
	Homalothecium.	Antitrichia.	Hylocomium.
Polytrichum.	Platygyrium.		
Dawsonia.	Cylindrothecium.	Leskea.	

Tribe II. *ANDRÆACEÆ.*

Schistocarpous Mosses. Capsule borne on a pseudopodium, not operculate, opening by 4 longitudinal fissures, and forming 4 valves cohering by their tips (*Andrœa*), or free (*Acroschisma*).

This little tribe, which was formerly placed near *Jungermanniece* on account of its habit and the valvate capsule, is separated by its columella, the absence of elaters, and the coherence of the valves at the top, or towards the middle of the urn; but it approaches both these and *Sphagnece* in the development of the fruit.

GENERA.

Andrœa.   Acroschisma.

Mosses, like Phænogams, possess other modes of reproduction than that by spores. In nearly all tubercles are formed on the subterranean and even aërial roots, which when exposed to the air germinate like the buds formed on the prothallus, some of the peripheric cells elongating to form the roots, others furnishing new cells for the stem and leaves.

In some Mosses reproductive tubercles are also developed in the axils of the leaves (*Phascum nitidum*, *Bryum erythrocarpum*). Sometimes they are developed into buds before detachment, and root in the soil as soon as they reach it (*Bryum annotinum*). In still other cases these buds root before separating from the mother-plant (*Conomitrium Julianum, Cinclidotus aquaticus*). In some cases even a detached leaf can, according to Schimper, produce a prothallus by cell-multiplication (*Funaria hygrometrica*). Finally, excrescences or *propagula* form at the extremity of the stem and branches of some Mosses (*Aulacomnium*, *Tetraphis*, &c.), or even on the leaves (*Orthotrichum*), which when detached form new individuals. Mosses, so distinct from *Hepaticæ* in their organs of vegetation and fructification, evidently approach them in their sexual organs. *Sphagnee* connect the two families, resembling Mosses in habit, foliage and formation of fruit, and *Hepaticæ* in the evolution of the prothallus, and in the structure of the antheridia and antherozoids, &c.

Mosses inhabit all climates and most opposite localities, from the equator to the poles; they abound in temperate regions, ascending the highest mountains and descending into the deepest valleys. They clothe with perpetual verdure the trunks of trees, rocks, and often old walls and roofs. Wherever there is moisture they are found; some are submerged in running water (*Fontinalis*), others in stagnant water (*Hypnum*). A large number, after being totally dried up during the summer, recover their verdure in the cool and rain of autumn. Their part in the economy of nature as contributing to the formation of soil, is not less important than that of Lichens, whose work they carry on, adding their own detritus to that of these latter, and forming on sandy lands, by their decay and reproduction, a layer of soil suitable for the agriculturist.

Certain species were formerly used in medicine as astringents and diuretics. *Leskea sericea* is still applied externally in some countries for its hæmostatic properties. Several Mosses are used in the arts and domestic economy; in Sweden and Norway the peasants utilize *Hypnum parietinum* and *Fontinalis* to fill the crevices in the walls of their huts. The common *Polytrichum* is made into brushes much used to give a dressing to stuffs. Finally, *Hypnum triquetrum* is used, on account of its great elasticity, in packing Phænogamous plants.

X. *SPHAGNA, Schimper.*

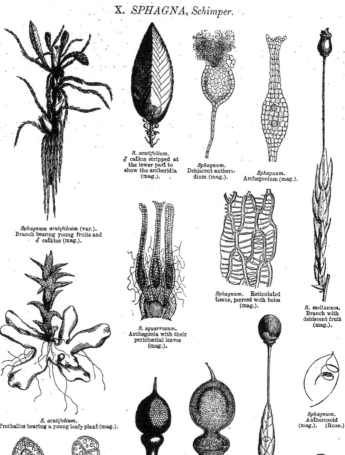

S. acutifolium.
♂ catkin stripped at
the lower part to
show the antheridia
(mag.).

Sphagnum.
Dehiscent antheri-
dium (mag.).

Sphagnum.
Archegonium (mag.).

Sphagnum acutifolium (var.).
Branch bearing young fruits and
♂ catkins (mag.).

Sphagnum. Reticulated
tissue, pierced with holes
(mag.).

S. squarrosum.
Archegonia with their
perichætial leaves
(mag.).

S. molluscum.
Branch with
dehiscent fruit
(mag.).

S. acutifolium.
Prothallus bearing a young leafy plant (mag.).

Sphagnum.
Antherozoid
(mag.).   (Roze.)

Sphagnum.
Spore (mag.).
(Schimper.)

Sphagnum.
Spores aggregated
(mag.).

S. squarrosum.
Young fruit enveloped
in its cap
(mag.).

S. squarrosum.
Young fruit
cut vertically
(mag.).

*Sphagnum
squarrosum.*
Fruit (mag.).

Sphagnum.
Microspore.

3 o

Moss-like soft flaccid spong Acotyledons. STEMS *erect in bogs, floating in swamps;* BRANCHES *regularly ʾfascicled, attached to the stem laterally to the insertion of the leaves.* LEAVES *imbricate, concave, nerveless, uncoloured and nearly transparent.* REPRODUCTIVE ORGANS *composed of antheridia and archegonia.* ANTHERIDIA *situated on long-pedicelled amentiform branchlets, opening elastically at the top, and emitting antherozoids.* ARCHEGONIA *terminal, in bud-like involucres.* FRUIT *solitary, capsular, globose or ovoid, sessile on a hemispherical or sub-discoid vaginula.* Calyptra ʾforming *a very imperfect cap, the largest part remaining attached to the vaginula.* CAPSULE *operculate; peristome naked, without annulus.* SPORANGIUM *hemispheric, seated on a very short thick columella, which disappears when ripe.* SPORES *dimorphous, some larger, pyramidal, the others polyhedral.* PROTHALLUS *filamentous, knotty or lobed, analogous to that of* Hepaticæ.

STEM.—Composed of a woody cortical cylinder of fibrous cells, surrounding a medullary or axile bundle. The cortical cylinder formed of 1–4 layers of sometimes fibrous cells, nearly always pierced with annular holes, which by facilitating capillary action in the stem and branches, enable the plant to grow without roots, and to rise to a considerable height, often many feet above the soil, without loss of vegetative power; this result is principally due to the reflexed branches, which always remain sterile, and which, acting like adventitious roots, contribute, with the spongy tissue of the bark of the stem, to raise the water from the base of the plant to its top.

ROOTS or rootlets only exist in young plants, completely disappearing afterwards; they are composed, as in true Mosses, of a single series of cylindric hyaline cells.

LEAVES.—The leaves are not developed from the cortical layer of the stem, but, like the branches, from the outer cellular layer of the woody cylinder. Cellular tissue uniform, but subsequently dimorphic, owing to the unequal segmentation of the cells, of which some become elongated, and contain chlorophyll; others tabular, diaphanous and porous. This structure is analogous to that of true Mosses, but in *Sphagna* the two kinds of cells are distributed in the same layer, and alternate with each other, so that the cylindrical green cells form the threads of the foliar network, and the large porous cells form the interstices. The cells pierced with holes are evidently intended, like those surrounding the stem, to render the plant more hygrometric. With the exception of *Leucophanes*, no Moss possesses a greater power than *Sphagna* do, of conveying water to considerable heights; they are true siphons, contributing to the draining of marshes, converting swamps into bogs, and bogs into dry land.

REPRODUCTIVE ORGANS.—*Antheridia* and *archegonia* (like true Mosses and *Hepaticæ*), usually on different individuals, never in the same involucre, as is so often the case with Mosses. *Antheridia* collected in inflorescences (the only analogue to which is in the leafy *Hepaticæ*), forming catkins or small cones, situated on secondary axes. Each inflorescence is composed of an involucral leaf, and an antheridium inserted laterally to this leaf. The antheridia greatly resemble in form those of *Hepaticæ*; they are globose or ovoid sacs, formed of one layer of hyaline chlorophyll cells, with a pedicel formed of 4 series of cylindric cells. The antherozoids escape by a rupture of the top of the antheridium (as in *Hepaticæ* and Mosses), which then quickly decays; and this distinguishes it from the antheridia of Mosses, which sometimes persist for years. Thuret has observed that the cells in the antheridia are lenticular, flatter on one side than on the other, as in *Hepaticæ*, whilst in Mosses they are perfectly symmetrical; each contains an antherozoid, which is set free by the rapid dissolution of the cell-wall under water. The antherozoid is a 2-haired filament, spirally coiled twice, adhering to a vesicle, which, according to Roze, contains an amylaceous granule. Schimper says that the antherozoid, as long as it is enclosed in its cell, revolves rather quickly upon its axis; but, as the spiral tends to become excentric, it soon leaves its prison, and once free it adds a movement of progression to that of rotation; the latter appears to be produced by the rapid oscillation of the 2 filaments, which thus perform the function of vibrating hairs. The spiral itself has no motion, either of contraction or dilatation, and the antherozoid appears to be propelled upon the principle of the Archimedean screw, its oscillating filaments causing it to revolve on its axis, and its helicoid form determining its advance in the liquid; the more rapid the rotations are, the quicker is its progression. The antheridia are accompanied by numerous *paraphyses*, which are branched very fine succulent filaments, evidently adapted to retain moisture around the antheridium, and to facilitate its dehiscence.

The FEMALE INFLORESCENCE is an elongated bud, composed of variously formed leaves that form an involucre to the archegonia; in the centre is a whorl of very small leaves, constituting the *perichætium* of the fruit. The archegonia, numbering 1–3 (rarely 4), occupy the rounded top of the fertile branchlet; one alone ripens. Their structure is nearly identical with that of Mosses and *Hepaticæ*. Like the antheridia, they are accompanied by numerous filaments (*paraphyses*), much branched and very flaccid, interlaced so as to resemble a loose arachnoid tissue. As soon as a nucleus in the cavity of the archegonium is fertilized, as in Mosses, its cells rapidly multiply, especially at the lower part; soon it opens a passage across the foot of the archegonium, perforates the top of the stem, and advances, continually multiplying, to the place where the involucral leaves begin. This cellular tissue, once established within the receptacle, that is to say, on that portion of the stem which is to form the *vaginula*, develops so fast, that it becomes in a few days a hemisphere, or almost a bulb, bearing the abortive archegonia on its sides. The dilated top of the stem becomes the vaginula. A cushion is then developed in this protuberance, sometimes at the base even of the archegonium itself, which is the rudiment of the future capsule. This enlarges without the smallest change occurring in the body of the archegonium, which, in Mosses, at first enlarges simultaneously with the evolution of its contents. As the young fruit rises above the vaginula, its cellular layers become modified; the outer, which is to form the capsular walls, is still simple; the following layer, destined to form the sporangium, is equally simple; the central cells, which represent the commencement of the columella, are loose and transparent; soon the cells of the capsule and of the sporangium become doubled and quadrupled by vertical division. The walls of the outer layer thicken and darken, and the sporangial layer, which rests on the central or columellar tissue, divides into three concentric parts in the middle, on one of which the spores are developed.

The organ which forms the *cap* or *calyptra* in true Mosses, and which is merely the top of the archegonium, does not take a definite shape in *Sphagna*. The archegonial walls persist until the fruit is nearly ripe, when the primordial envelope of the latter ruptures irregularly, a portion remains attached to the base, and the rest, which slightly adheres to the top of the capsule, usually falls away with the operculum.

The capsule then rests immediately on the vaginula, which is usually discoid; it is separated from the perichætium by a prolongation of the branch, so that the capsule rises above the involucre; this false pedicel, which is not (as in Mosses) an integral part of the capsule, is named *pseudopodium*.

The capsule is spherical or ovoid, often becoming oblong by the desiccation of its outer tissue. Dehiscence takes place by an operculum, which consists of a horizontal section of the upper part of the capsule. When the capsule, in drying, elongates and contracts, the operculum is detached with a slight crackling, and the spores are expelled by the compressed air which had penetrated through the stomata into the capsule. The mouth of the capsule is smooth, with no trace of a *ring* or *peristomium*.

When the fruit is perfectly ripe, the columella contracts to the bottom of the capsule, and leaves the sporangium attached to the inner walls of the latter. The sporiferous layer of the sporangium produces four successive generations [of mother-cells?] before the spores appear; these then belong to the fifth generation. They are of two shapes and dimensions: some are tetrahedral, with a convex base, and reproduce the plant; the others, infinitely smaller, form regular polyhedrons $\frac{1}{3750}$ in. in diameter; these are always sterile.

The *prothallus*, developed in water or on damp earth, is at first filamentous and branched; then there appear, at the extremities of one or more of its ramifications, swellings, consisting of vesicular cells containing a mucilage, in which green granules float; these are rudimentary plants. Under other circumstances these tubercles become transformed into thalliform lobed expansions, in the sinus of which young plants are also developed (*Sphagnum acutifolium*); in both cases recalling the mode of germination of *Hepaticæ*, properly so-called, and of *Jungermannieæ*.

ONLY GENUS.

Sphagnum.

Sphagna, which formerly constituted a tribe of Mosses, and of which the celebrated muscologist, Schimper, now makes a separate family, are, in fact, intermediate between *Hepaticæ* and Mosses. They approach the former in their mode of germination, and in the early stages of their growth in the form of the male inflorescence, the structure of the antheridia and the antherozoids, and in the absence of a normal calyptra; while they are connected with *Bryaceæ* or true Mosses by the structure of the stem and leaves, the mode of ramification, the discoid vaginula, the imperfect calyptra, and the dimorphous spores.

Sphagna prefer temperate and cold countries; they occupy an immense extent of marsh land in the northern hemisphere, where their remains, accumulated for centuries, form turf, a valuable fuel in countries where wood is wanting. It is in the marshy tracts inhabited by *Sphagnum* that the elements of a higher vegetation are gradually established; *Cyperaceæ* and *Gramineæ* at first, then *Vaccinium*, *Myrica*, dwarf Willows, and at last Pines, Birches, &c.

Some species of *Sphagnum* which abound in the Polar swamps serve as food to the reindeer. Mingled with reindeer's hair they are used for stuffing mattresses. *Sphagnum* is frequently used in our hot-houses in the cultivation of certain epiphytal *Orchideæ*, to which it serves as it were for soil by keeping their roots always moist.

XI. *HEPATICÆ, Jussieu, Bischoff, Nees v. Esenb.*

Cellular Acotyledons, either stemless with creeping often dichotomous fronds, furnished or not with nerves; or with leafy stems. REPRODUCTIVE ORGANS *antheridia and archegonia.* ENVELOPE *of the fruit (calyptra) rupturing at the top, and persistent at the base of the pedicel.* SPORANGIA *capsular, usually 4-valved, nearly always accompanied by elaters.* SPORES *very numerous.*

Small annual or perennial, usually very delicate PLANTS, composed of a cellular tissue, generally prostrate, rooting, inhabiting shady and damp places. FRONDS green, violet or brown, sometimes expanded as membranous plates, foliaceous, lobed, stomatiferous, nerveless, or traversed by a nerve formed of elongated cells; sometimes furnished with a simple or branched leafy axis. LEAVES membranous, usually distichous, often entire, lobed or dentate, sometimes deeply divided, frequently accompanied by accessory stipulary leaves named stipules or amphigastria. ROOTS of simple fibres, tubular, white brown or purplish, transparent, scattered or in small tufts, and springing from the lower surface of the fronds, sometimes from the base of the amphigastria.

The REPRODUCTIVE ORGANS are monœcious or diœcious, sometimes sunk in the thickness of the frond (*Riccia*, &c.); sometimes projecting above the frond, and often pedicelled (*Marchantia*). ANTHERIDIA oblong or spherical, formed of a single layer of transparent cells; they contain a mucilage which becomes organized into discoid very small and delicate cells; when ripe, the antheridia rupture, or dehisce transversely at the top, and expel a large number of these discoid cells, whence come out, after a few seconds, filiform antherozoids, coiled in a spiral, and bearing at their extremity two excessively slender threads.

The ARCHEGONIA precede the appearance of the *perianth*, and are usually found collected at the top of the principal axis, or of the lateral branches, or at the axils of the amphigastria (*Calypogeia, Mastigobryum*); they are usually enclosed in a peculiar organ resembling a wide-mouthed cup; they are cellular sacs, usually narrowed into a styloid process, channelled throughout their length, which, at a certain period, ruptures at the top to admit the action of the antherozoids. The broadest part of the sac contains one cell larger than the rest, the *nucleus*; soon the nucleus is divided in two by a horizontal septum; the lower half forms the pedicel (*seta*), which, when the capsule is ripe, is usually white and transparent, often very short, and which by pushing the sporangium before it, ruptures irregularly the

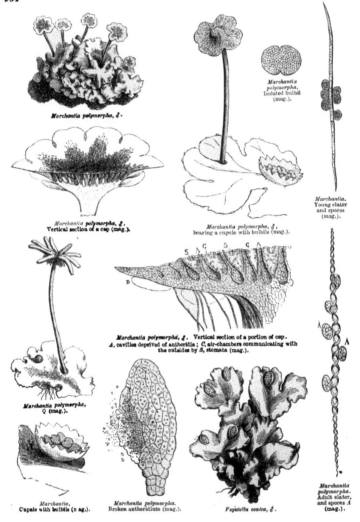

Marchantia polymorpha, ♂ .

Marchantia polymorpha, ♀ .
Vertical section of a cap (mag.).

Marchantia polymorpha.
Isolated bulbil
(mag.).

Marchantia polymorpha, ♂ ,
bearing a cupule with bulbils (mag.).

Marchantia.
Young elater
and spores
(mag.).

Marchantia polymorpha, ♂ . Vertical section of a portion of cap.
A, cavities deprived of antheridia ; *C*, air-chambers communicating with
the outsides by *S*, stomata (mag.).

Marchantia polymorpha,
♀ (mag.).

Marchantia.
Cupule with bulbils (mag.).

Marchantia polymorpha.
Broken antheridium (mag.).

Fegatella conica, ♂ .

*Marchantia
polymorpha.*
Adult elater,
and spores *A*
(mag.).

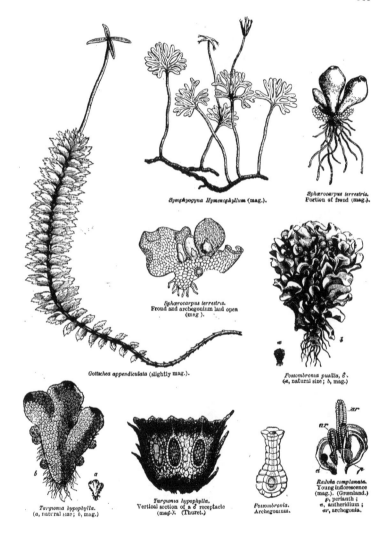

Symphyogyna Hymenophyllum (mag.).

Sphærocarpus terrestris.
Portion of frond (mag.).

Sphærocarpus terrestris.
Frond and archegonium laid open
(mag.).

Gottschea appendiculata (slightly mag.).

Fossombronia pusilla, ♂.
(*a*, natural size; *b*, mag.)

Targionia hypophylla.
(*a*, natural size; *b*, mag.)

Targionia hypophylla.
Vertical section of a ♂ receptacle
(mag.). (Thuret.)

Fossombronia.
Archegonium.

Radula complanata.
Young inflorescence
(mag.). (Grœnland.)
p, perianth ;
a, antheridium ;
ar, archegonia.

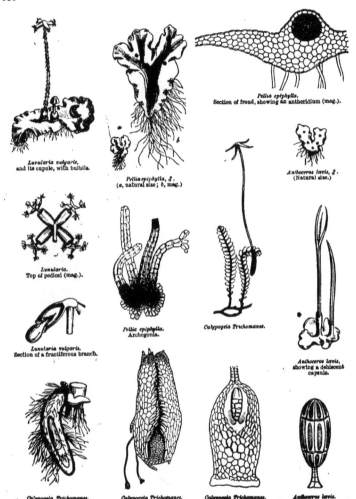

Lunularia vulgaris,
and its cupule, with bulbils.

Pellia epiphylla, ♂.
(*a,* natural size; *b,* mag.)

Pellia epiphylla.
Section of frond, showing an antheridium (mag.).

Anthoceros lævis, ♂.
(Natural size.)

Lunularia.
Top of pedicel (mag.).

Pellia epiphylla.
Archegonia.

Calypogeia Trichomanes.

Lunularia vulgaris.
Section of a fructiferous branch.

Anthoceros lævis,
showing a dehiscent
capsule.

Calypogeia Trichomanes.
Sac containing an archegonium
(mag.).

Calypogeia Trichomanes.
Vertical section of sac, showing
the archegonium.

Calypogeia Trichomanes.
Archegonium (mag.).
(Gottsche.)

Anthoceros lævis.
Young antheridium (mag.).
(Rose.)

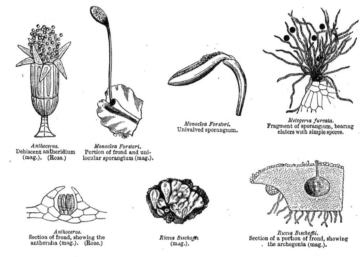

Anthoceros.
Dehiscent antheridium
(mag.). (Roze.)

Monoclea Forsteri.
Portion of frond and uni-
locular sporangium (mag.).

Monoclea Forsteri.
Univalved sporangium.

Metzgeria furcata.
Fragment of sporangium, bearing
elaters with simple spores.

Anthoceros.
Section of frond, showing the
antheridia (mag.). (Roze.)

Riccia Bischoffii
(mag.).

Riccia Bischoffii.
Section of a portion of frond, showing
the archegonia (mag.).

upper part of the membranous sac or *calyptra*; the base of the sac persists, and forms a sheath (*vaginula*) around the base of the pedicel.

The SPORANGIUM usually opens in four valves, very rarely more, or breaks irregularly (*Fossombronia*). In *Anthoceros* it is traversed by a central *columella*. It contains :—1st, very elongated cells, usually fusiform, sometimes truncate (*Frullania, Lejeunia*), which are usually free at the dehiscence of the sporangium, but may be attached to the tips of the valves (*Frullania, Lejeunia, Aneura, Metzgeria*), or partially adhere to the bottom of the sporangium (*Pellia*); these cells, called elaters, contain, pressed against their wall, a small dark-coloured ribbon in a single (*Aneura*) or double spiral; the elaters, by their twisting movement, serve to disseminate the spores:—2nd, spheroid mother-cells, within each of which are developed four spores, which are set free by the absorption of the mother-cells. In germination the spore emits a small cellular prothallus, which reproduces the plant.

Hepaticeæ, besides sexual reproduction, can be propagated, like Mosses, by buds, or *propagula*, which appear on the surface or margins of the frond. They are cellular, rounded, polymorphous, sometimes tolerably large, and analogous to bulbils. In some cases (*Blasia*), these propagula occupy ovoid pouches, hollowed in the nerve and top of the frond. One rather curious example of this mode of propagation is met with in the tribe of *Marchantieæ*. On the surface of the fronds of some species of *Marchantia* are seen foliaceous membranous cups, with entire or elegantly fringed margins; at the bottom of these cups are ovoid or lenticular bodies, composed of granular cells; these little bodies are at first fixed to the bottom of the cup by a small cell which serves as a pedicel. In *Lunularia*, which covers the damp earth of our gardens, this cup resembles an arched vase, or a semicircular basket, whence its name of *lunule*.

Hepaticæ are very naturally grouped, according to their vegetative organs. Some present flat dichotomous membranous fronds appressed to the soil by means of rootlets, and usually nerveless, but furnished with stomata; more rarely the membranous frond twists spirally around a sort of axis formed by the nerve (*Riellia*). Other *Hepaticæ* have a simple or branched prostrate or erect stem, and are furnished with true leaves. The stem sometimes creeps over the soil, emitting fronds at different distances, and then simulates a rhizome or runner (*flagella*). The ramification is usually dichotomous, as in Lycopods and Ferns; but one of the branches is sometimes arrested, or changed into a secondary flagelliform branchlet.

The leaves of the caulescent *Hepaticæ* are extremely varied in form, but are nevertheless so characteristic in each genus as to enable botanists to describe the species without seeing the reproductive organs. These little leaves, without nerves or stomata, are opposite or alternate, and their insertion resembles that of *Selaginella*. Usually they are nearer the upper than the lower surface of the stem, and in the intervals are seen, turned towards the soil, a third series of smaller leaves, obliquely inserted, named *amphigastria* (or stipules). When the leaves are alternate the spiral arrangement is from right to left (*Frullania*), or from left to right (*Lophocolea*). The leaves and amphigastria are always sessile; but the leaves are not always flat and spreading; they have often on their upper surface a crest or appendage which forms a sort of wing (*Gottschea*), or they have a small basal raised fold (*Lejeunia*), or a sort of pouch open at both ends (*Frullania*). These modifications, though apparently very slight, affect all the species of the genus, and the form of these pouches is constant in the same species.

The leaves of *Jungermanniæ* are so arranged that the lines of insertion of two consecutive leaves converge, and represent an erect or reversed V. In *Anthoceros* the vegetative system consists of a simple membranous expansion spread upon the earth.

The form and position of the perianth have supplied a methodical division of the caulescent *Hepaticæ* of the group of *Jungermanniæ*. This perianth, which is identical in structure with the leaves, forms a cup or small urn, sometimes contracted at the top, which in this case is ruptured to allow the passage of the sporangium. But this dehiscence presents many modifications: it results in four nearly equal segments in *Marchantia*; in numerous strips in *Fimbriaria*. The perianth remains nearly entire and campanulate in *Lejeunia*; it is cylindric in many true *Jungermanniæ*; it is absent in some genera, and replaced by leaves in *Gymnomitrium*. With the exception of *Marchantia*, *Fimbriaria* and *Preissia*, it is absent in the tribe of *Marchantieæ*. The pedicel which bears the sporangium is always cylindric, cellular, very delicate, usually transparent and colourless; its growth is often very rapid.

In most *Hepaticæ* the perigonium [inflorescence] contains but one archegonium (*Marchantia*, &c.). In *Jungermanniæ* it always contains several, all but one of which are abortive; in *Sarcoscyphus* two or three are developed. The name *perichætium* or *perichætial leaves* is given to the outer envelope or involucre of the perianth. The perichætium originally envelops the archegonium with its styliform appendage; the perigonium is formed afterwards. In *Calypogeia*, *Harpanthus*, &c., there is only a perigonium and no perichætium; while in *Gymnomitrium*, *Schisma*, &c., there is a perichætium and no perigonium.

The antheridia of *Hepaticæ* were first observed by Schmidel on *Jungermannia pusilla*, L. (*Fossombronia*). In this little plant the antheridia are free, shortly pedicelled, and planted on the central nerve of the frond. The cells of its walls contain bright yellow granules, which cause the antheridium to resemble a grain of pollen. When the antheridia are fully developed, the top cells become markedly turgid; this indicates the moment of dehiscence. Thuret, from whom we have taken most of these details, has observed the following phenomena: the cells which form the upper half of the antheridium suddenly bend in the opposite direction to that previously occupied; a complete discoloration is the result, and the contents of the antheridium are set free, when a membrane or cuticle covering or connecting the cells becomes visible. The antherozoids are flexible, furnished with two hairs, and the archegonia are in juxtaposition with the antheridia. In the tribe of *Marchantieæ* the antheridia occupy peculiar receptacles of very varied form; sometimes they are little pedicelled disks with waved edges (*Marchantia polymorpha*), or are sessile on the margin of the fronds (*Feyatella conica*), or they form small appendages along both margins of the frond (*Targionia*). But whatever their form, the receptacles all agree in presenting a tissue with a mamillated surface, the mamillæ corresponding to an ovoid antheridium immersed in their parenchyma, and communicating with the outside by a small canal which abuts against the top of the mamilla. These antheridia contain minute cells which again contain antherozoids; these differ from those of *Fossombronia* in their minute size, the shortness of their body, and the extreme tenuity of their vibrating hairs.

The sporangia, like the antheridia, are immersed in a common receptacle (*Marchantia*); sometimes this receptacle is conical (*Fegatella*) or hemispherical (*Reboulia*). In *Lunularia*, the vegetation of which' is identical with that of the preceding genera, the sporangia are tubular, four in number, arranged in a cross at the top of the pedicel. In *Jungermannia* the sporangia, at first ovoid or spherical, open when ripe into four valves, bearing the elaters in the middle. The elaters are wanting in the tribe of *Ricciea*, and are very rudimentary in *Anthocerea*.

The spores vary in size and appearance; they are ovoid or spheroidal, smooth or granular, or echinulate.

Tribe I. *JUNGERMANNIEÆ.*

Leafless, or furnished with stem and leaves. Archegonia and antheridia developed at the extremity of the stem. Sporangium furnished with elaters, deprived of columella.

PRINCIPAL GENERA.

§ 1.—*Plants with Leaves.*

Jungermannia.	Frullania.	Geocalyx.	Schisma.
Gottschea.	Sarcoscyphus.	Tricholea.	Sendtnera.
Lophocolea.	Calypogeia.	Gymnomitrium.	Pleuranthe.
Lejeunia.	Saccogyna.		

§ II.—*Plants without Leaves.*

Pellia.	Fossombronia.	Aneura.	Blasia.
Blytia.	Metzgeria.	Symphyogyna.	

Tribe II. *MONOCLEEÆ.*

Frond an irregular thallus or leafy stem. Sporangium solitary, opening lengthwise, deprived of columella, elaters carried away with the spores.

GENERA.

Monoclea.	Calobryum.

Tribe III. *RICCIEÆ.*

Frond or thallus dichotomous, traversed by a median nerve. Archegonia and antheridia immersed, or superficial and sessile. Sporangium without columella or elaters.

PRINCIPAL GENERA.

Riccia.	Ricciella.	Riellia.	Oxymitra.	Corsinia.	Sphærocarpus.

Tribe IV. *MARCHANTIEÆ.*

Frond or thallus irregular, with emarginate divisions, and no median nerve. Archegonia and antheridia borne on the lower surface of a peltate pedicelled disk, arising in a notch of the thallus. Sporangium furnished with elaters, but with no columella.

PRINCIPAL GENERA.

Marchantia.	Athalamia.	Reboulia.	Duvalia.	Monosolenium.
Dumortiera.	Plagiochasma.	Preissia.	Cymatodium.	Askepos.
Lunularia.	Grimaldi.	Fegatella.	Targionia.	Antrocephalus.
Sauteria.	Fimbriaria.			

Tribe V. *ANTHOCEREÆ.*

Frond or thallus irregular, without median nerves. Archegonia and antheridia dispersed over the thallus. Sporangium siliculose, 2-valved, furnished with a central columella covered with elaters.

PRINCIPAL GENERA.

Anthoceros.   Nothotylus.

Hepaticæ differ from true Mosses in habit, in the calyptra breaking at the top and sheathing the base of the pedicels, and in the structure of their sporangium without operculum, and usually accompanied by elaters. They are connected with *Sphagna* by their mode of germination, the form of the antheridia, and the absence of a normal cap or calyptra; but *Sphagna* are distinguished by their habit, the form of their sporangium, which is always furnished with a columella that disappears at maturity, their dimorphous spores, &c.

Most of the genera are cosmopolitan; some prefer temperate or cold regions; but *Gottschea, Thysananthus, Polyotus,* &c., are principally tropical; *Targionia* is exclusively temperate; *Cymatodium* grows in the Antilles. Two-thirds of *Ricciea* have been observed in Europe. Most *Anthocereæ,* as well as the rest of the family, are dispersed over the world. Some *Hepaticæ* have a very strong peculiar odour, confined to certain species, of which they are characteristic, and a somewhat acrid taste. The ancients looked upon *Marchantia polymorpha* as a resolvent medicine, and employed it in diseases of the liver. *M. chenopoda* has a similar reputation in tropical America.

XII. *LICHENES, Jussieu.*

Cellular perennial Acotyledons, growing on the ground, stones, bark, the leaves of other plants, and even on other Lichens. ORGANS OF VEGETATION (thallus) *polymorphous, irregular, spreading or erect, varied in substance and colour.* REPRODUCTIVE ORGANS *of two sorts:* (1) (apothecia), *situated on the surface or margin, or imbedded in the thallus, varied in form and colour, composed of sporangia* (thecæ) *containing* 2-∞ *spores;* (2) *spermogonia, formed of spherical conceptacles imbedded in the thallus near the apothecia, lined with filaments* (sterigmata), *which bear corpuscules of an extreme tenuity, transparent, polymorphous* (spermatia), *considered as analogues of the antherozoids, but not endowed with motion.*

A perfect *Lichen* usually consists of: 1. A THALLUS, or vegetative apparatus, 2. APOTHECIA, or organs of fructification; 3. SPERMOGONIA, or organs of fertilization. The thallus varies much in form, texture, and colour; it never has stomata; its texture is usually dry and coriaceous, sometimes gelatinous. It is *foliaceous* when it presents lobed laciniate peltate expansions, &c.; *fruticulose*, when it assumes an erect cylindric form, and branches (*Roccella, Cladonia,* &c.); *filamentous,* when its ramifications are soft and prostrate (*Ephebe, Evernia, Cornicularia,* &c.); *crustaceous,* when it forms, either on the surface of the soil, or on that of organic or inorganic bodies which support it, a more or less friable crust (*Opegrapha, Endocarpon*). It is *hypophleous* when concealed under the epidermis or between the fibres of trees (*Verrucaria, Xylographa,* &c.). It is grey, white, yellow, red or black, and usually becomes greenish when moistened.

Anatomically the substance of the *thallus* consists of 3–4 layers of different elements :—1, *cortical*; 2, *gonidial*; 3, *medullary*; 4, sometimes a lower layer whence spring the root filaments, and called *hypothallus*.

1. The *cortical* layer is usually formed of colourless cells with thicker or thinner walls; its surface also presents a sort of amorphous crust, variously coloured, and called *epithallus.*

2. The *gonidial* layer is placed immediately below the cortical layer; the elements composing it are continuous or disconnected, and appear under the form of bright or olive-green granules (*gonidia*). The presence of these gonidia distinguishes the tissue of Lichens from that of Fungi, in which they do not exist.

3. The *medullary* layer presents three principal modifications : it is *felted,* i.e. composed of closely interlaced filaments; *crustaceous,* when the filaments, fewer in number, are accompanied by white molecules mixed with numerous crystals of oxalate of lime; it is *cellular* when it is composed of rounded or angular utricles, associated with the filaments.

4. The lower layer, or *hypothallus,* is usually of a darker colour than the upper it is also covered with rootlike hairs, which have been called *rhizines.*

The thallus of some species (*Collema,* &c.), presents a more simple structure, and appears to be reduced to two membranes separated by a mucilaginous mass, in the middle of which filaments are suspended.

The APOTHECIA are sometimes superficial and sessile or stipitate, and at others buried in the tissue of the thallus. In the first case, they are discoid, scutellate or patelliform (*Usnea,* &c.), or linear-elongated (*Opegrapha*), or globose (*Roccella*); in the second, they form a sort of pouch or conceptacle, occupying the thickness of the thallus. The apothecia are rarely of the same colour as the thallus; they are usually black or brown, or present all shades between bright yellow and red; no blue ones have been seen. Their size is very variable; the smallest measure at most $\frac{1}{50}$ in., the largest sometimes attain more than $\frac{3}{25}$ in. (*Nephroma*). The apothecia are composed of sporangia (*thecæ*), which contain the spores. These thecæ, more or less pressed together, are usually accompanied by filaments thickened at the top (*paraphyses*). The sporangia or thecæ are large oblong cylindric or ovoid vesicles, with an attenuated base, and fixed to a layer of a special tissue denser than that of

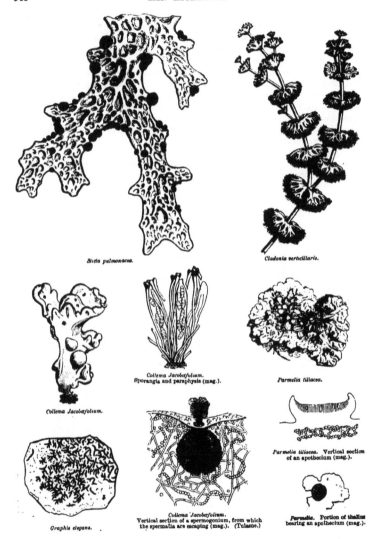

Sticta pulmonacea.

Cladonia verticillaris.

Collema Jacobæfolium.
Sporangia and paraphysis (mag.).

Parmelia tiliacea.

Collema Jacobæfolium.

Parmelia tiliacea. Vertical section
of an apothecium (mag.).

Graphis elegans.

Collema Jacobæfolium.
Vertical section of a spermogonium, from which
the spermatia are escaping (mag.). (Tulasne.)

Parmelia. Portion of thallus
bearing an apothecium (mag.).

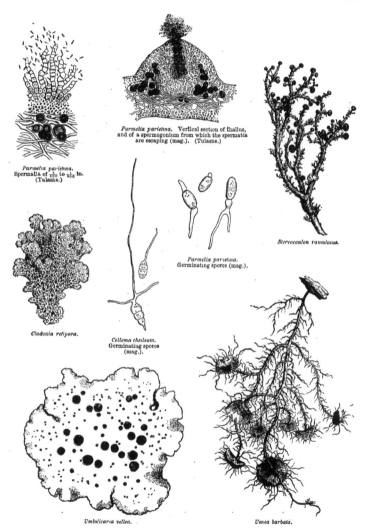

Parmelia parietina. Vertical section of thallus, and of a spermogonium from which the spermatia are escaping (mag.). (Tulasne.)

Parmelia parietina. Spermatia of 1/100 to 1/166 in. (Tulasne.)

Stereocaulon ramulosus.

Parmelia parietina. Germinating spores (mag.).

Cladonia retipora.

Collema cheileum. Germinating spores (mag.).

Umbilicaria vellea.

Usnea barbata.

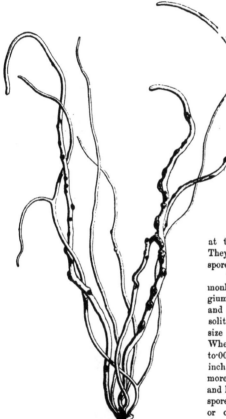

Roccella tinctoria.

the middle layer of the thallus (*hypothecium*). The dimensions of the sporangia vary much according to age, to genera and species, and to the number of spores which they contain. Their wall is formed of a rather thick membrane, especially when young, of an extreme tenuity in certain genera; in others it attains ·00039 in., and their greatest thickness is usually at the top of the sporangium. They persist after emitting the spores.

There are usually 8, less commonly 6–4–2 spores in each sporangium, but sometimes 20–100 spores, and even more, occur; as does a solitary spore. The spores vary in size according to their number. When 8, they measure from ·00027 to ·00157, by ·00007 to ·00006 of an inch. The smallest known are not more than ·00003 in. in length, and half as much in breadth. The spores are ellipsoid, ovoid, fusiform or oblong-cylindrical; they are simple, or chambered and 2–3–4–∞-celled. Two layers may be distinguished in their walls—an outer or *epispore*, and an inner or *endospore*. The epispore is usually extremely thin, and scarcely perceptible. The colour of the spores is always determined by that of the epispore; white is most usual. Iodine often colours them blue; but in all cases the epispore alone becomes coloured. The spores are expelled by a peculiar contraction of the sporangium, in the same way as in *Helvelleæ*, *Pezizæ* and most Fungi of the group of *Ascophoreæ*. At germination the spores of Lichens produce, like Fungi, a net-work of filaments.

The *spermogonia* are small conceptacles immersed in the superficial layers of the thallus, rarely placed in distinct tubercles, communicating with the outside by a small orifice, and containing simple or jointed filaments (*sterigmata*), which produce small arched oblong linear or acicular corpuscules (*spermatia*), the supposed fertilizing agents. The spermatia are of extreme tenuity; the largest are ·00043 to ·00067 in., the smallest ·00003 in. in length, with half that breadth; they do not appear to have any power of motion, and possess no locomotive organ. Although lichenists are agreed in considering the spermatia as the analogues of the antherozoids, it is difficult to describe their action on the spores. It has been observed in certain genera that some individuals of a species have spermogonia, others not; whence it would appear that such species are diœcious, and not monœcious, like most Lichens. The genus *Ephebe* is an example, the species of which may be considered as monœcious or diœcious, according as the spermogonia and apothecia are found on the same or on different individuals.

Pycnidia are small conceptacular protuberances, resembling spermogonia, but differing in their less abundant, more bulky contents, and in their germinating power. Their origin and functions are still quite obscure. Tulasne considers them to be a supplementary means of propagation.

Nylander, from whom we have adopted most of the preceding views, has divided Lichens into several families, which, in our opinion, represent tribes, and form the three following very unequal groups as regards the number of genera which they contain: 1. *Collemaceæ*, with fifteen genera; 2. *Myriangiaceæ*, with one, *Myriangium*; 3. *Lichinaceæ* proper, which contains a hundred.

Tribe I. *COLLEMACEÆ*.

Gonionema.	Synalyssis.	Leptogium.	Spilonema.
Paulia.	Obryzum.	Ephebe.	Collema, &c.

Tribe II. *MYRIANGIACEÆ*.

Myriangium.

Tribe III. *LICHINACEÆ* proper.

Section I. Epiconiodeæ.

Calycium.	Coniocybe.	Sphærophoron.	Acroscyphus.

Section II. Cladoniodeæ.

Bæomyces.	Cladonia.	Stereocaulon.

Section III. Ramalodeæ.

Roccella.	Siphula.	Usnea.	Alectoria.
Evernia.	Ramalina.	Cetraria.	

Section IV. Phyllodeæ.

Nephroma.	Peltigera.	Solorina.	Sticta.
Parmelia.	Physcia.	Umbilicaria.	

Section V. Placodeæ.

Squamaria.	Placodium.	Lecanora.	Urceolaria.	Pertusaria.
Thelotrema.	Leciden.	Opegrapha.	Chiodecton.	

Section VI. Pyrenodeæ.

Endocarpon.	Verrucaria.	Endococcus.	Trypethelium, &c.

Several botanists are, however, now inclined to unite with Lichens the entire group of *Thecasporeæ*, Fungi which differ from them in no important character but the absence of a gonidial layer. The absence of oxalate of lime in the tissue of Fungi, on which the separation of the two groups was founded, cannot be relied on, since long ago Dawson Turner, Tripier and Steinheil proved chemically the presence of this salt in certain *Boleti* (*Boletus sulphureus*, &c.). The action of iodine, which tinges with blue the sporangia of most Lichens, does not appear to us sufficient to separate the Lichens and thecasporal Fungi. Our opinion is corroborated by that of a learned botanist, Dr. Léveillé, an authority on all questions relative to this branch of cryptogamy, who has sent us a letter from which we extract the following passages :—

'You enquire, my dear Decaisne, what is the difference between the Lichens and the thecasporal Fungi: the question is definite, and cannot be evaded. I answer that I have frequently examined into it, and I find the distinctions to be so trifling that I have always regretted that these vegetables should not be placed under one head. The paraphyses, the thecæ and the spores are identical. The *hypothallus* of Lichens corresponds to the *mycelium* of Fungi; like it it spreads over the surface of bodies, develops under the epidermis of plants, in the thickness of tissues, and in the earth. The receptacle (*apothecium*) of Lichens varies as much in form as that of Fungi; it is sometimes superficial, sometimes sunk in the tissues; but in Lichens it is always, like the thallus, coriaceous. In Fungi it is often similar, and in *Peziza* it is fleshy, watery, or friable like wax. The surface of the receptacle (*epithecium*) is naked in Fungi; the extremity of the paraphyses, which often projects and colours the disk, is fugitive, and disappears in the Fungus. In Lichens, on the contrary, the *epithecium* is a normal state; it is formed, not only by the swollen projecting extremity of the paraphyses, but by a granular and persistent matter. Again, the receptacle of Fungi is fugacious; but in *Sphæriæ*, which continue for a long time, without being perennial, the conceptacles only last a year at the most. *Sphæriæ*, once developed and fructified, have completed their existence; they do not vegetate afresh. With Lichens it is different; their receptacle is perennial; it may remain for several years, and always be in a perfect state of fructification. This perennial condition of the receptacle has been noticed by Meyen, and I have verified it in *Parmelia parietina*, and in *P. Lagascæ* in Corsica. Also I have seen at Montmorency *Lecanora sulphurea*, of which the disk (*thecium*) had been destroyed by snails, reproduce a new thecium. This phenomenon of the reproductive organs, added to the character of the thallus of Lichens, and the (at most) annual existence of the most compound *Sphæriæ*, I consider as one of the most curious and important of biological characters. Unfortunately it is I who have brought it forward and established it: it does not pass current on the Exchange of Science. How can we distinguish a Lichen from an ascosporous Fungus, since the reproductive organs are similar? The only difference we can detect is in the thallus; but if it is crustaceous, scaly, filamentose, &c., in Lichens and *Sphæriæ*, let us not forget that the thallus is the characteristic mark of Lichens, the more so as it is confined to them alone. The principal point then is to establish the difference between the thallus of Lichens and the *stroma* of Fungi. That the one resembles branches, the other a simple cushion, is a matter of indifference. The thallus of Lichens always presents three layers: the cortical, the gonidial, and the medullary. The *Sphæriæ*, on the contrary, have but two; the gonidial layer is always wanting: this is an anatomical truth which must perforce be admitted. These gonidia,[1] by their presence, remove every difficulty when in doubt as to the nature of a Fungus or

---

[1] Sachs has adopted in his 'Lehrbuch' the views entertained by Schwendener and Reess, that Lichens are not autonomous organisms, but colonies of Algæ held as it were in captivity by the meshes of an ascomycetous Fungus. This is an extension, in fact, to all Lichens of a suggestion of De Bary's that *Collema*, *Ephebe*, &c., are

Lichen. If it had occurred to me to ascertain whether it was present, I should not have described *Ascro-scyphus* as a Fungus; I should have recognized it as a Lichen. You will be forced to regard the gonidia as peculiar to Lichens; they play too important a part to be overlooked; their nearly constant presence, their form and green colour certainly demand that they should not be passed over in silence. What is their part in the economy of Lichens? Are they of no use in the respiration of these vegetables, in the formation of the colouring principles which they yield, and in the power of reproducing Lichens? Their bursting through the epidermis in the form of soridia, the change of colour which they sometimes undergo in contact with the air, are so many points which demand attention; we have nothing similar in the thecasporal Fungi. The thallus of Lichens is very hygrometric, as long as it is alive; it dilates or contracts according to the degree of moisture, and becomes green again; but if it be moistened after it is dead, it changes colour on almost every occasion. It is probably owing to this hygrometric action, which is particularly exercised at night, that Lichens remain alive in countries where the drought is uninterrupted, or where they have taken root on hard stones impermeable to water, or on iron, or glass, as you saw with me in the crypt of the church at Jouarre. The thallus of Lichens is never viscous, which is very commonly the case in true Fungi; but I ought to say that I have seen very few viscous *Ascophoreæ* (*Geoglossum viscosum, glutinosum*). Lichens, especially those with a crustaceous thallus, or which grow upon stones, have a strong tendency to become coloured if the stones contain oxide of iron or manganese; they then acquire a ferruginous tint, which has led to many bad species having been made. The thecasporal Fungi develop nearly everywhere, in damp or dark places. Lichens like plenty of light, and appear indifferent as to what they grow on: the thecasporal Fungi are more particular; they especially like wood, they live parasitically on insects, on Fungi, on ergoted Rye: hence habitat, without being a character, is of some importance. Lichens draw their nourishment from the air, from the dew, as the *Lecanora* of Pallas proves indubitably.

'Nylander indicates as the characters of Lichens the colouring of blue or red by iodine; the presence of oxalate of lime, of discoid or lenticular starch-grains, and of gonidia, which are often absent. The colouring blue or red is to me a phenomenon more curious than characteristic. As the thallus of Lichens contains starch, and Fungi never, this blueness readily distinguishes the thallus of a Lichen; but with regard to the parts of fructification, we see in Lichens, as in Fungi, that the influence of the iodine is shown, sometimes on the inner substance of the conceptacle, sometimes at the extremity of the paraphyses, sometimes on all the thecæ, or only on their extremity. The spores also sometimes exhibit it. The presence of oxalate of lime in Lichens may be of importance; but as you tell me it exists also in Fungi, the latter must have been very superficially examined for it not to have been met with. *Clavariæ* especially possess a large quantity.

' One word on the Lichens and Fungi which live parasitically on the thallus of Lichens. These productions have neither mycelium nor thallus. Tulasne considers them as Lichens, probably because they become blue with iodine; others think them Fungi, because they have no thallus. How shall we reconcile these views? It is true we find gonidia, but these belong to the thallus of the Lichen on which they live.

' Now we must open another chapter, that of *spermogonia*, or rather of *spermatogonia*. I do not require you to adopt this name, because a conceptacle, whether masculine or feminine, is always a conceptacle. In the same way as we say *flores masculi, flores fœminei*, we may say *conceptaculum masculinum*, and *fœmineum*. I think, however, that you may adopt the name of *spermatia* (although they are not found on all Lichens, and although botanists persist in considering them as parasitic Fungi), because they appear to play a very real part in the economy of Lichens, and because in a complete description they must be noticed; but I should wish a mark of interrogation (?) to accompany *conceptaculum masculinum*.

' I would also advise you to use the word *thecæ* preferably to *asci* (*ascospores*). Fries, the *summus magister*, has said somewhere that he would never use the word *thecæ*, but always *asci*. See how easy it is to express oneself when the following adjectives are added to these words:—*asci, ascelli, ascidii*

either conditions of Nostochineous or Chroococcous Algæ modified by the parasitism of ascomycetous Fungi, or that their immature states have hitherto been confounded with Algal forms belonging to these groups. Thwaites indeed, many years ago, pointed out their parallelism. Archer considers that the hypothesis of Schwendener is confuted by the fact that the assumed parasitic Fungus does not destroy, or live upon, its assumed Algal host.—ED.

*primarii, constitutivi, reproductorii, liberi, fixi, diffluentes, persistentes, emersi, semi-emersi, immersi, inclusivi,
suffultorii, sporophori,* &c. I spare you the rest: all this implies that it is but a cell infinitely varied. . . .'

Lichens are found in all climates, but their numbers increase as we recede from the equator;
they usually grow, as we have said, on earth, stones, leaves, bark, and other Lichens; also on Mosses,
dead wood, bones, leather, old iron, the old windows of country churches, which they decompose in
process of time, under the influence of damp, by extracting a little potash ; such is especially the case
with *Parmelia parietina,* which grows on almost anything.   Some prefer calcareous rocks, others
granitic; some inhabit rocks moistened by the sea (*Lichina, Roccella*). The edible Lichen (*Lecanora
esculenta*) is quoted in M. Léveillé's letter in support of the opinion that Lichens derive their nourishment
from the atmosphere. The thallus of this species occurs in little rounded masses the size of a filbert: the
interior is white and crustaceous ; the surface is grey, uneven, wrinkled, with warts enlarged into lobes ;
these lobes overlap irregularly, but have evidently been developed centrifugally, and in consequence of
the early interlacing of their ramifications, or rather of their destruction, have formed a body solid
within, and imperfectly foliaceous without. This Lichen, which has been found in Algeria, is frequently
met with in the most arid mountains of the deserts of Tartary, where the soil is chalk and gypsum, and it
grows on the soil amongst the flints, from which it is only to be distinguished by practised eyes. Large
quantities are found in the Kirghis deserts, to the south of the river Jaïk, at the foot of gypsum hills
which surround the salt lakes. The traveller Parrot brought home specimens of this Lichen, which, at
the beginning of 1828, had fallen like rain in several parts of Persia. He was assured that the ground
was covered with it to a height of eight inches, that cattle eagerly ate it, and that the natives
gathered it as a manna fallen from heaven, and made bread of it. The naturalists Pallas and
Professor Eversmann, who observed it growing, never found a single specimen attached to any object;
they gathered some of the size of a pin's head; all were absolutely free of support. Eversmann con-
jectured that this Lichen had originally germinated around a grain of sand, which it had then entirely
surrounded ; but observation not having confirmed this hypothesis, he has been led to admit that the
germ of this Lichen develops in all directions, and derives its nourishment from the surrounding air.

Lichens, which mostly contain starch, may, like *Lecanora esculenta,* contribute more or less to
the food of men or animals; such is the Reindeer Lichen (*Cenomyce rangiferina*), which serves as
pasturage, in the northern regions, to the herds of reindeer and some other herbivorous mammalia.
The Iceland Moss (*Cetraria islandica*) and the 'Pulmonaire de Chêne' (*Sticta pulmonacea*) contain
a bitter and mucilaginous principle, which cause them to be used medicinally in diseases of the lungs.
Variolaria amara and several *Parmeliæ* are used in certain countries as febrifuges and anthelminthics.
Peltigera canina formerly entered into the composition of a remedy for hydrophobia. Some Lichens also
yield useful dyes; as *Roccella, Lecanora tartarea* and *Purella,* and *Parmelia saxatilis,* which yield the
Orchil and Cudbear of commerce.

Lichens also play an important part in the economy of nature. It may be said that they, with
Mosses, have been the first cultivators of the soil; or rather, that is they who have created the soil that
covers the great mineral masses of the globe. It is of their detritus that are formed even now on the
most arid rocks the first layers of *humus,* or earth, in which plants of a higher order speedily take root,
and their débris, accumulating during centuries, finally form a soil capable of sustaining and affording
nourishment to the largest vegetables. It is generally thought that Lichens, as well as Mosses, are
injurious to the trees on which they grow : this opinion does not rest on any solid foundation.[1]

We find in a memoir recently published by Dr. Lortet some very interesting details regarding the
action of electricity on the spermatia of Lichens and Fungi. Neither static nor voltaic electricity exercises
any influence on these organs, but electricity developed by the induction coil presents a most curious
phenomenon. The spermatia, according to M. Lortet, who adopts in this the opinion of M. Itzigsohn,
are endowed with very active movements; contrary to the ideas of most botanists, who look upon these
movements as a *Brownian oscillation,* M. Lortet perceives no difference between these movements and
those of antherozoids, although the strongest magnifying powers have not enabled him to detect any
vibrating hairs on the spermatia. However this may be, these bodies, placed in water, execute two
extremely quick movements—the one of oscillation, which consists of a tremulous motion of the organ

---

[1] Lichens and Mosses are not parasitic; but when
they clothe trees they impede the circulation of air, and
hasten decay. They further intercept light when enveloping young shoots, and interfere with the development of cambium and the evolution of the foliage.—
ED.

itself; the other of translation, which enables it in a short time to traverse a somewhat considerable space. To observe these motions the glass plate for the object should be traversed by two grooves crossing at right angles; in each groove a metallic thread should be firmly cemented, and these threads leave in the middle of the glass a free space where the corpuscules swim; the induction apparatus is a reel, the generator being a simple element of bichromate of potash. All things being thus arranged, it is easy to pass the inductive currents through the prepared liquid between the supporting plate and the film of glass which covers it. The antherozoids of *Hepaticæ* and Mosses are not influenced by the induced currents; their movements are not modified, and their relative positions remain the same, although they may be in the path of a strong current. But it is different with the spermatia of Lichens and Fungi: the moment that the small imbedded threads on the object-glass are brought into contact with the points of the induction coil, the thousands of spermatia visible in the field of the microscope place themselves parallel to the current, i.e. with their longest diameter in a straight line between the points; their movements of translation are then entirely arrested; their trepidation continues, but feebly. If, by means of the two other threads cemented to the object-glass, the electricity is passed in a perpendicular direction to the first, the spermatia immediately move, and place themselves in this direction. Instead of touching end to end, under the influence of the current, as if they attracted each other, they arrange themselves parallel to each other and to the current. If the current be weakened by degrees, its influence is no longer felt in the centre of the arrangement; there the spermatia resume their movements and irregular positions; whilst towards the two points the action of the current continues to manifest itself, and the arrangement in line continues. If the current be entirely arrested, the corpuscules disperse in every direction; as soon as it recommences, they again fall into line, and may remain for hours without change. While the induction current is passing, there can be no movement of the liquid, since, there being no progression of the spermatia at each change of direction of the current, they remain motionless in the water which contains them, and only revolve upon themselves. This singular arrangement in line can only arise from a polarization similar to that which is produced by induction in several metallic conductors placed close together.

XIII. *FUNGI, Jussieu, Persoon,* &c.

(HYMENOMYCETES, *Fries.*—DISCOMYCETES, *Fries.*—GASTEROMYCETES, *Fries.* — PYRENOMYCETES, *Fries.*—HYPHOMYCETES, *Link.*—GYMNOMYCETES, *Link.*)

Cellular Acotyledons, very various in duration and texture, epigeal or hypogeal, generally parasitic on decaying vegetable or animal substances, on the bark of trees, on the surface or interior of leaves, and even on other Fungi, very rarely living on stones or in water, lovers of shade, always deprived of fronds, stomata and green parts. ORGAN OF VEGETATION (mycelium) *mostly subterranean, composed of elongated cells, which are isolated, or form a web or membranes.* ORGANS OF FRUCTIFICATION *borne on the mycelium, sessile or pedicelled, naked, or contained in a peculiar envelope, very various in form, and bearing the spores on the interior or exterior.* SPORES, *sometimes borne upon basidia, sometimes contained in sporangia* (thecæ), *mostly formed of two membranes, always* (?) *motionless.*

Fungi, together with Lichens, form a group of Cryptogams possessing no archegonia; they are polymorphous, [often] ephemeral, annual, or perennial, never green; composed either of filaments, or of a loose or close tissue, pulpy or fleshy, rarely woody; sometimes furnished with peculiar vessels containing a white, yellow or orange milky juice.

They grow above or under ground, on decomposing vegetable or animal matter, or are parasites on vast numbers of Phænogamous plants, and even on other Fungi; some are developed in living animal tissues, and are the reputed causes of various diseases. They are very rarely found on stones, or in water. In no particular can they be compared with Phænogams, having no organs comparable with leaves and flowers. Among Acotyledons, they approach *Algæ* in their vegetation, and Lichens in their fructification, but they have no fronds; and no known Fungus resembles a unicellular Alga.

In a Mushroom or Toadstool the following organs are distinguished:—the *mycelium*, the *volva*, the *stipe* (or *pedicel*), the *receptacle* or *pileus*, the *conceptacle*, the *basidia*, the *thecæ*, and the *spores*.

The *mycelium* fulfils at once the functions of root and stem; this, of which the Mushroom spawn is an instance, is the result of the vegetation of the spores. It is composed of cells, originally free, variable in colour, more or less elongated, and is sometimes so scanty as to escape observation. It appears, when completely developed, under four different forms : 1. The *filamentous* or *nematoid* is composed of elongated branched cells, isolated, or collected in threads. 2. The *membranous* or *hymenoid* is composed of cells united into a membrane of diverse texture. 3. The *pulpy* or *malacoid* presents a soft and pulpy, branched or grumous mass; in this case the granules which compose it present a remarkable analogy with the protozoid animalcules called *Amœbæ*, and with animal *sarcode*; this form of mycelium, placed in water, vegetates, but does not fructify. 4. The *tubercular* or *scleroid* consists of globose or flattened regular or irregular tubercles, of a firm texture, homogeneous structure, and composed of extremely minute cells. This scleroid form, which plays a very important part in the vegetation of Fungi, is only a transitory one; it always proceeds from the filamentous state, and may be compared with the tubers of the Potato, and not with a true subterranean stem; its life is truly latent, and only preserved by the hygrometric nature of its tissue. In a favourable season the scleroid mycelium becomes saturated with moisture, and either develops a perfect Fungus itself, or a nematoid mycelium, which again develops perfect individuals. Its function is analogous to the albumen of an amylaceous seed, or of a tuber, the substance of which is exhausted as the plant it nourishes grows, and leaves only a cortical membrane. It is thus that the *Agaricus tuberosus* is developed from *Sclerotium cornutum, A. racemosus* from *S. lacunosum, Clavaria phacorrhiza* from some other *Sclerotium*; it is the same with certain *Pezizæ* (*P. tuberosa, Candolleana,* &c.), and with *Botrytis cinerea*, which grows indifferently on various forms of *Sclerotia* (*S. durum, compactum, medullosum,* &c.). The mycelium is remarkable for its power of retaining vitality long after it has been collected, starting into life as soon as placed under favourable conditions, reproducing its filaments, and extending indefinitely and consuming the organic substances it meets with, until exposed to the influence of light, which enables it to produce its organs of fructification.

The *volva* is the more or less firm and membranous envelope or pouch, which contains the young Fungus, and which the latter bursts through as it develops (*Amanita*, &c.). The *pedicel* or *stipes* is the stem-like portion which supports the *receptacle*; it is often surrounded by a *ring* (*annulus*), or a *cortina*, membranous or filamentous veils, which extend from the stipes to the margin of the pileus, thus protecting when young the organs of fructification (*Agaricus, Amanita*, &c.).

The term *pileus* is usually given to the dilated portion of a Toadstool, above the pedicel, which always bears on its lower surface the organs of fructification and their appendages, which consist of *gills* or *tubes* or *processes*. The *gills* (*lamellæ*) are radiating or flabellate appendicular and membranous plates (*Agaricus*). The *tubes* form small cylindric or angular pipes (*Boletus, Polyporus*, &c.). The *processes* are teeth or points (*Hydnum*). These appendages are clothed with a special fructiferous layer, the *hymenium*. The name of *receptacle* is indifferently applied to the entire [reproductive system of the] Fungus (*Agaricus*, &c.), or to the part which bears the organs of fructification (*basidia, thecæ, spores*, &c.); which part may be filamentous (Moulds), or membranous (*Thelephora*), or alveolate (*Morchella*). The *clinode* is an organ analogous to the hymenium, springing from the inner wall of the conceptacle [a closed sporiferous cavity], or from the surface of the receptacle, and which terminates in simple or branched filaments bearing an isolated spore at their extremities. It might in strictness be called a hymenium, for it fulfils the functions of that organ.

The *spores* (seeds or reproductive bodies) are free, or borne on the extremity of a filament, or inserted on special organs called *basidia*, surmounted by 2–4 points or *sterigmata*, or enclosed in cells (*sporangia, thecæ*, &c.). They are formed of two membranes; the outer (*epispore*) is smooth, areolate, or verrucose, &c.; the inner (*endospore*) is thin, colourless, apparently structureless, and contains granules, and sometimes particles of oil. The spores germinate by emitting 1 or 2 filaments, the first rudiments of the mycelium. The fertilization of Fungi was, till lately, involved in complete obscurity; but the researches of MM. de Bary and Woronin encourage the hope that it will shortly be as well known as that of other Cryptogams. It is supposed that Fungi possess, besides the spores, other sporomorphic reproductive organs, which are :—

FEMALE ORGANS.—*Oogonia*, globose bodies, at first filled with a granular mass which divides into several reproductive globules, named *oospores*. *Gonospheria* only differ from *oogonia* in the condensation of the protoplasm at the centre of the cell, consequently leaving an empty space between the cell and the protoplasm; this entails a slight modification of structure, but little differing from that of the antheridia. The *scolecite* is a vermiform body, composed of cells somewhat resembling *oogonia*, arranged either in little groups or in linear series.

MALE ORGANS.—The *antheridia* are composed of simple cells, springing from the mycelium under or around the female organs. They are at first filiform, then they swell at the top, separate from the mycelium by a septum, become filled with protoplasm, but never with antherozoids, and rest on the female organs to effect fertilization. The *spermatia* are simple cells, ovoid, straight or bent, never globose, and are enclosed in a conceptacle (*spermogonium*), whence they issue, mixed with mucilage, as threads and globules, which harden in the air, but which in water are reduced to cells, without leaving any trace of the parts with which they were connected. These, which have been regarded as male sexual organs, are intended, according to some modern botanists, for fertilization. They do not emit germ filaments, as they ought were they to be considered as *stylospores* or reproductive organs. *Zygospores* are round or oval cells, terminating a filamentous receptacle, or developed on the sides of two branchlets from one branch, which approach and unite so as to form a single body (*zygosporangium*) containing a single spore (*zygospore*). This mode of fertilization has hitherto only been observed on *Syzygites megalocarpus*, *Ascophora rhizopus*, and *Mucor fusipes*.

SECONDARY REPRODUCTIVE ORGANS.—The *conidia* are simple cells, globose or ovoid, naked, pulverulent, isolated, or agglomerated in a compact mass. In the first case they are joined end to end, or arranged in racemes, or situated at the extremity of simple or branched filaments; in the second case, they resemble variously coloured pulpy or fleshy tubercles, which soften, and are almost entirely dissolved in water. The *stylospores* are ovoid, spheroidal or elliptic cells, straight or curved, simple or chambered, variously coloured, always pedicelled and included in a conceptacle (*pycnide*). The *zoospores* are absolutely identical with those of some *Algæ* (see p. 976); they are furnished with two hairs, by the help of which they move readily; if placed on a slightly moistened leaf they germinate by emitting filaments which penetrate the stomata, or pierce the epidermis, and ramify in the parenchyma; they have been observed in the joints of *Cystopus* and in *Peronospora*. Whatever view may be taken of the nature of *conidia, stylospores, spermatia,* &c., we may arrange the immense class of Fungi into the following six distinct groups, by the spores properly so called.

TRIBE I. *BASIDIOSPOREÆ.*

Spores simple, borne on rounded semi-elliptic or conical cells, named *basidia*, which terminate in 2–4 points (*sterigmata*), each bearing a spore; the basidia are

Amanita rubescens. v, Volva; *s,* stipe; *a,* ring.   *Agaricus campestris,* in different stages of growth.

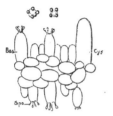

Agaricus campestris.
Portion of gill cut transversely,
and showing the two lateral faces.
Bas, basidia ; *Spo,* spores ;
Cys, cystide (mag.).

Agaricus campestris.
Portion of gill cut
transversely, and showing
one face (mag.).
H, cells ; *S.H,* hymenium ;
B, basidia ; *Sp,* spores.

Geastrum tenuipes.
Adult plant : volva divided into a star ;
globular conceptacle.

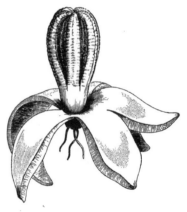

Clavaria Ligula.

Lysurus pentactinus.
Adult plant : volva divided into a star ; receptacle with
five conniving branches.

Gymnosporangium aurantiacum.
On a branch of Juniper.

Gymnosporangium aurantiacum.
Portion of young tissue.

Gymnosporangium aurantiacum.
Basidia, sterigmata,
and spores.

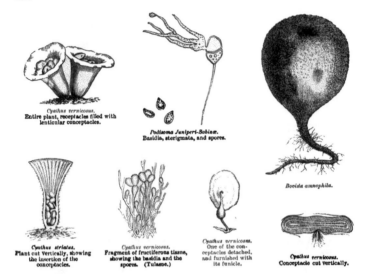

Cyathus vernicosus.
Entire plant, receptacles filled with lenticular conceptacles.

Podisoma Juniperi-Sabinæ.
Basidia, sterigmata, and spores.

Bovista amnophila.

Cyathus striatus.
Plant cut vertically, showing the insertion of the conceptacles.

Cyathus vernicosus.
Fragment of fructiferous tissue, showing the basidia and the spores. (Tulasne.)

Cyathus vernicosus.
One of the conceptacles detached, and furnished with its funicle.

Cyathus vernicosus.
Conceptacle cut vertically.

often accompanied by other large projecting cells, transparent, acute or obtuse, always deprived of sterigmata, to which have been given the name of *cystides*. The basidia are borne on the gills, folds, veins, and processes of the receptacle, or in the tubes (in which case these are exterior); sometimes in conceptacles, the cavities of which they line.

SECTION I.—Basidia external, placed on the surface of a smooth veined lamellose porous, &c., receptacle (*ectobasidia*).

PRINCIPAL GENERA.

Amanita.	Boletus.		Clavaria.	
Agaricus.	Favolus.	Phlebophora.	Lachnocladium.	Phallus.
Lentinus.	Hexagona.	Phlebia.	Merisma.	Dictyophora.
Cantharellus.		Fistulina.		Sophronia.
Lenzites.	Craterellus.		Tremella.	Clathrus.
Cyclomyces.	Thelephora.	Hydnum.	Dacrymyces.	Ileodictyon.
	Leptochæte.	Hericium.	Exidia.	Laternea.
Dædalea.	Stereum.		Gymnosporangium.	
Polyporus.	Schizophyllum.	Gomphus.	Podisoma.	Lysurus.
			Aservë.	Calathiscus.

SECTION II.—Basidia internal, enclosed in a dehiscent or indehiscent conceptacle, usually presenting cavities lined by basidia (*endobasidia*).

PRINCIPAL GENERA.

Podaxon.	Myriostoma.	Scleroderma.		Crucibulum.
Cauloglossum.	Plecostoma.	Sclerangium.	Trichia.	
Secotium.		Trichoderma.	Arcyria.	Carpobolus.
Cycloderma.	Broomeia.	Pilacre.		Atractobolus.
Polyplocium.			Licea.	Thelebolus.
Gyrophragmium.	Lycoperdon.	Reticularia.	Perichæna.	
Hyperhiza.	Mycenastrum.	Lignidium .	Tipularia.	Gautiera.
Stemonitis.	Phellorina.	Diphtherium.		Hymenangium.
		Spumaria.	Polygaster.	Octaviana.
Battarrea.	Bovista.	Physarum.	Endogone.	Melanogaster.
	Lycogala.	Didymium.	Myriococcum.	
Tylostoma.		Diderma.	Polyangium.	Hymenogaster.
Schizostoma.	Polysaccum.	Cenangium.		Hysterangium.
Geaster.		Triptotrichia.	Cyathus.	Hydnangium.

The *Basidiosporeæ*, especially those of the first section, comprise the Fungi commonly so called. The principal genus, *Agaricus*, contains a large number of species, very difficult to distinguish, notwithstanding the divisions and subdivisions established by mycologists. Most of them are inodorous and insipid; but others are odoriferous, and others acrid and even burning. One entire group of Agarics contains a peculiar milky juice, white, yellow, or reddish, insipid or caustic. The Mushroom, a variety of *A. campestris*, abounds in meadows, especially where horses feed; it is the only species which is cultivated, and which is a rather important object of commerce. It may be obtained at all seasons by cultivating it in beds in cellars or underground caves. *A. neapolitanus* was formerly cultivated by the nuns of a convent near Naples, who raised it on a bed of coffee-grounds. The Poplar Agaric (*A. Ægerita*), and that of the Hazel-nut (*A. Avellanus*), may also be obtained from slabs of poplar or hazel covered with cinders, and this again with a light layer of earth which is occasionally watered. The genus *Amanita*, separated from *Agaricus*, yields the delicious Orange Agaric (*A. Cæsarea*), so much sought by gastronomes, together with some very poisonous species, such as *A. bulbosa, phalloides, muscaria*, &c., which contain a narcotic acrid principle that acts like Indian Hemp or hashish. Hitherto no antidote to this poison has been found. Salt, in which Fungi are preserved in Russia, is far from being such : witness the death of the wife of the Czar Alexis I. from eating Mushrooms preserved in salt during Lent. *Cantharellus cibarius*, which is distinguished from all other Mushrooms by its form and colour, is found abundantly from June till October in oak and chesnut woods, and yields an excellent food to the country people. *Boletus edulis* is eaten fresh or dried; cut in slices or dried in the sun or by a stove, it is stored and sold, as might be *B. castaneus* were it sufficiently plentiful. Amadou, or German tinder, is prepared from *Polyporus igniarius* and *fomentarius*; it was used in the last century as a hæmostatic. Its ashes, as well as those of some other *Polypori*, are used by the Ostiaks and Kamtschatkans as snuff, probably to keep up an irritation of the mucous membrane, and thus to preserve the nose from being frost-bitten. *P. officinalis*, improperly termed White Agaric, is a violent purgative, now fallen into disuse. The *Thelephoræ* are all useless; they are membranous coriaceous Fungi, marked on their upper surface with various, sometimes bright-coloured zones; the most remarkable species is the *T. princeps* of Java, which is the largest known Fungus, attaining often one foot to one and a half. The genera *Hydnum* (*repandum*, &c.) and *Clavaria* furnish edible species; the latter form small bushy tufts, white, yellow, orange, rose, or blue; but those of the latter colour are suspicious, and *C. amethystea* is possibly dangerous, since it brings on violent colic.

Tremella violacea, which gives a bluish colour, as well as Jew's Ears (*Exidia Auricula-Judæ*), formerly used for dropsy, are quite given up. The latter is eaten by the inhabitants of the Ukraine.

The group of Endobasidial Fungi comprises both small and almost gigantic Fungi (*Lycoperdon giganteum*). To this group belong the Puff-balls, which alter remarkably in appearance : the young or adult are white and firm; then they turn brown, and the interior softens till it appears decayed, after which it dries up and is converted into dust, which escapes by an apical opening, leaving behind filaments, and at the base some spongy cells. This spongy substance, saturated with a solution of nitre, formerly served as amadou and was used as a hæmostatic; it was also burnt in hives to stifle bees. The *Scleroderma* outwardly resemble Lycoperdons, and their interior recalls the colour and texture of Truffles; but their sulphurous alliaceous smell prevents their being used as food, pigs even rejecting them.

Geastrum hygrometricum, a hypogeal globose plant, presents a curious phenomenon: when mature, and still underground, if the season be dry, the outer envelope, which is hard, tough, and hygrometric, divides into strips from the crown to the base; these strips spread horizontally, raising the plant above its former position in the ground; on rain or damp weather supervening, the strips return to their former position; on the return of the drought, this process is repeated, until the Fungus reaches the surface, becomes epigeal, and spreads out there; then the membrane of the conceptacle opens to emit the spores in the form of dust.

Podisoma Juniperi-Sabinæ also belongs to the group of *Basidiosporeæ*. This plant is confounded with *Gymnosporangium aurantiacum*, and to it is attributed the production of *Rœstelia cancellata*, a disease which first appears in the form of orange patches sprinkled with little black spots, on the leaves of the Pear-tree. The experiments which we have carried on to verify this transformation of genus and species not having confirmed it, we wait for fresh proofs before admitting a theory of metamorphoses and trans-formations which tend to upset all the ideas we have acquired in mycology, by transferring to the Vegetable Kingdom the series of phenomena which zoologists call 'alternation of generations,' or 'digenesis.' It is now admitted that the loose comparison drawn by the anatomists of the seventeenth century between the animal egg and its appendages, and the vegetable egg, long retarded our knowledge of the fertilization of Phænogamic plants: let us then beware of bequeathing a similar stumbling-block to our successors by introducing into science theories regarding the specific identity of productions so dissimilar in appearance, and of which their author even confesses that we can scarcely hope to obtain a direct proof.

TRIBE II. *THECASPOREÆ.*

Spores usually contained by eights in cells (*thecæ, sporangia*), covering wholly or partially the surface of a receptacle, or the interior of a conceptacle. Thecæ accompanied or not by paraphyses, and opening at the top by an inconspicuous operculum, for the emission of simple or chambered spores.

SECTION I.—Thecæ elongated, covering the surface of a receptacle (*Ectothecæ*).

PRINCIPAL GENERA.

			Stictis.	Glonium.
Geoglossum.	Helvella.		Cryptodiscus.	Schizothecium.
Spathularia.		Agyrium.	Godronia.	Rhytisma.
Mitrula.	Peziza.	Pyronema.	Mellitiosporium.	Excipula.
	Ascobolus.	Cryptomyces.		
Morchella.	Bulgaria.			
Eromitra.	Cyttaria.	Cenangium.	Hysterium.	Cliostomum.
Verpa.	Helotium.	Tympanis.	Stegilla.	Actidium.
Gyrocephalus.	Rhizina.		Lophium.	Phacidium.

SECTION II.—Thecæ rounded, ovoid, clavate or cylindric, enclosed in a concep-tacle (*Endothecæ*).

PRINCIPAL GENERA.

			Tuber.	
Sphæria.	Asterina.	Hydnobolites.	Picoa.	Onygena.
Hypoxylon.	Elaphomyces.	Hydnotria.	Chæromyces.	
Cordyceps.	Hydnocystis.	Genabea.	Terfesia.j	Erysiphe.
Thamnomyces.	Genea.	Stephensia.	Delastria.	
Dothidea.	Balsamia.	Pachyphlœus.		

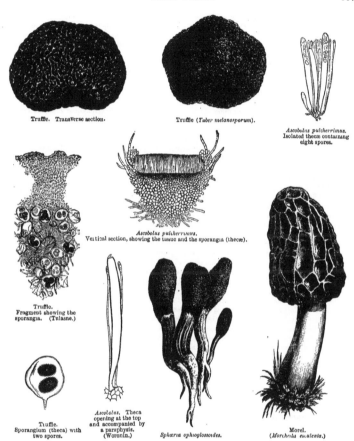

Truffle. Transverse section.

Truffle (*Tuber melanosporum*).

Ascobolus pulcherrimus.
Isolated theca containing
eight spores.

Ascobolus pulcherrimus.
Vertical section, showing the tissue and the sporangia (thecæ).

Truffle.
Fragment showing the
sporangia. (Tulasne.)

Truffle.
Sporangium (theca) with
two spores.

Ascobolus. Theca
opening at the top
and accompanied by
a paraphysis.
(Woronin.)

Sphæria ophioglossoides.

Morel.
(*Morchella esculenta.*)

The Ectothecal *Thecasporeæ* approach Lichens in the organization of their operculate reproductive organs (*thecæ*), associated with paraphyses. *Mitrula paludosa* is a small orange Fungus, of which the habitat is curious and exceptional; it grows at the bottom of watery swamps, fixed to leaves and small twigs. The common Morel (*Morchella esculenta*) may be looked on as the harbinger of spring, appearing [in France] with tolerable regularity in April, if this month is rainy. It is eaten fresh or dried, as are all its congeners. It is the same with *Helvellæ*, though one species of this genus has been noted by Krombhotz as suspicious; it is *H. suspecta*, which he refers to a plant discovered at Fontainebleau by Paulet, and described under the name of *M. Pleopos.* *Cyttaria Gunnii*, *Berteroi* and *Hookeri* grow on the small branches of several species of evergreen Beeches (*Fagus Cunninghamii*, &c.) in the southern hemisphere, where they are found in immense quantities in the form of little fleshy cartilaginous masses pierced with holes; they serve as food to the natives during part of the year. The *Pezizæ* are sometimes most brilliantly coloured; they form a large genus, the species of which are difficult to define: some are edible, as *P. cochleata*, &c. It was the beauty of the scarlet and orange *Pezizæ* which attracted two illustrious mycologists, Persoon and Battarra, to the study of Fungi; of whom Persoon was the first to methodize the order. Amongst the most curious genera of ectothecal *Thecasporeæ* are *Phacidium*, *Hysterium* and *Stegilla*, the receptacles of which open either transversely or by a longitudinal slit, or by strips resembling those of a *Geastrum.*

The Endothecal *Thecasporeæ* differ much in appearance: from the epiphyllous *Erysiphe*, formed of cobwebby white filaments mixed with small blackish globose conceptacles, surrounded by extremely elegant organs, to the Truffles properly so called, every transition and every degree of complication of structure is found. The genus *Sphæria*, in spite of its dismemberment, is the most numerous in species and the most singular of the group, being represented throughout the world and on all plants. The spores of several species germinate within the body of certain caterpillars, whence they emerge, still growing. We know the history of *S. militaris*, which has been cited as an example of the transformation of an animal into a vegetable; *S. Robertsii*, of New Zealand, and *S. sinensis* have a similar origin; the latter species is in much repute in China, where it is sold in small bundles as a marvellous medicine. Under the name of Truffle (*Tuber cibarium*) three species are confounded; they are black and rugged externally, and composed of a mass of tissue, the interior of which is black, and traversed by white veins. The thecæ, which contain 4-8 spores, give its black colour to this Fungus. Young Truffles are white, because they are composed of a homogeneous tissue; they become black with age, owing to the presence of the reproductive bodies, at which period they have acquired their full taste and smell. The great profits which would result from the cultivation of the edible Truffles have often stimulated efforts to grow them; but every such attempt has failed. If under some circumstances these valuable Fungi have appeared in consequence of sowing acorns, it was soon observed that their appearance was very ephemeral, and that the culture yielded irregular profits. The Black Truffle is not the only edible species of the genus. *T. magnatum*, *griseum*, *album*, &c., are much sought in Hungary, Italy and Algeria, where *T. album* is known under the name of *Terfez*. The *Onygenæ*, which partake of the characters of *Tuber* and *Sphæria*, grow on all epidermal animal substances, such as the hoofs of horses, the horns of oxen, feathers, hairs, and even old cloth rags. The species of *Erysiphe* are most curious in point of organization; they are called Mildews [*pollard*, Fr.], in allusion to the leaves on which they are found looking as if powdered with flour. Erysiphes are in general innocuous; but when they completely overrun certain plants, they arrest the vegetation or the flowering, as may be seen in Hop plants, which they damage considerably. The leaves of our large *Cucurbitaceæ* are sometimes whitened by an *Erysiphe*, which, however, does not seem to hurt the plant much.

Tribe III. *CLINOSPOREÆ.*

Spores springing from a clinodium covering wholly or partially the surface of the receptacle, or enclosed in a conceptacle.

Section I.—Receptacle fleshy, sessile or pedicelled, convex or concave, covered by the clinodium (*Ectoclinal Clinosporeæ*).

PRINCIPAL GENERA.

Tubercularia.	Stilbum.	Melanconium.	Uredo.	Puccinia.
Ægerita.	Graphium.	Stilbospora.	Uromyces.	Phragmidium.
Fusarium.		Dictyosporium.	Polycystis.	Triphragmium.
Selenosporium.	Dinemasporium.		Ustilago.	Coryneum.
Sphacelia.	Asterosporium.	Myrothecium.	Thecaphora.	

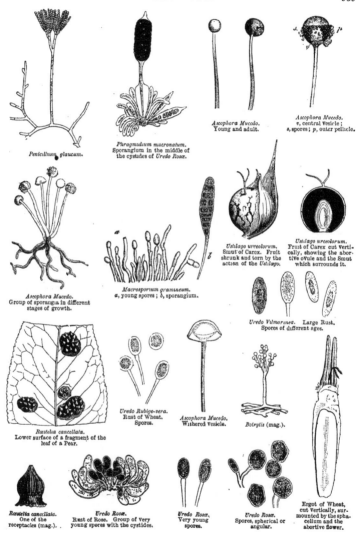

Penicillium glaucum.

Phragmidium mucronatum.
Sporangium in the middle of
the cystides of *Uredo Rosæ.*

Ascophora Mucedo.
Young and adult.

Ascophora Mucedo.
v, central vesicle;
s, spores; p, outer pellicle.

Ascophora Mucedo.
Group of sporangia in different
stages of growth.

Macrosporium gramineum.
a, young spores; b, sporangium.

Ustilago urceolorum.
Smut of Carex. Fruit
shrunk and torn by the
action of the Ustilago.

Ustilago urceolorum.
Fruit of Carex cut verti-
cally, showing the abor-
tive ovule and the Smut
which surrounds it.

Uredo Vilmorinea. Large Rust.
Spores of different ages.

Ræstelia cancellata.
Lower surface of a fragment of the
leaf of a Pear.

Uredo Rubigo-vera.
Rust of Wheat.
Spores.

Ascophora Mucedo.
Withered vesicle.

Botrytis (mag.).

Ræstelia cancellata.
One of the
receptacles (mag.).

Uredo Rosæ.
Rust of Rose. Group of very
young spores with the cystides.

Uredo Rosæ.
Very young
spores.

Uredo Rosæ.
Spores, spherical or
angular.

Ergot of Wheat,
cut vertically, sur-
mounted by the spha-
celium and the
abortive flower.

Section II.—Conceptacle membranous, more or less thick, fleshy, coriaceous or horny, sessile or pedicelled, opening variously and enclosing the clinodium (*Endoclinal Clinosporeæ*).

PRINCIPAL GENERA.

Œcidium.	Parmu.aria.	Prosthemium.	Hendersonia.	Microthecium.
Rœstelia.	Asteroma.	Sphæronema.		Angiospoma.
Peridermium.	Pestalozzia.	Hercospora.	Polychæton.	Ravenalia.
Endophyllum.	Discosia.	Septaria.	Phylacia.	
Actinothyrium.	Dilophospora.	Phoma.	Piptostomum.	
Leptothyrium.	Neottiospora.	Melasmia.	Scopinula.	

The *Tubercularia* are extremely common Fungi; they affect the bark of branches, and are remarkable for their intense red colour. Several species of *Sphæria* live parasitically upon them, and this parasitism is so frequent, that it has led some botanists to look upon them as a peculiar (*conidial*) condition of the *Sphæria* on which they grow; that is to say, a supplementary mode by which these *Sphæria* are reproduced; but when we find the same *Tubercularia* giving birth simultaneously to two perfectly distinct *Sphæriæ* (*S. parasitans* and *S. cinnabarina*), and the common *Tubercularia* of the Gooseberry sometimes supporting *S. cinnabarina* and *appendiculata* (which is certainly not a variety),—*Fusarium aurantiacum* bearing at the same time *S. pulicaris* and *olerum*, and *F. tremellosum* producing at the same time *Peziza Tulasnorum* and *Sphæria coccinea*—it may be doubted whether the theory of this conidial state rests on a firm base. To support it, it has been found necessary to unite different species of *Fusarium* and *Selenosporium* in one, as has been done with *Tubercularia*. *Sphacelia vegetum* vegetates between the pericarp and the ovule of *Graminea* and *Cyperaceæ*; during development it rends the pericarp and suppresses it, and affects the growth of the ovule, and is then termed *ergot*; this ergot somewhat resembles the seed in shape, but has no envelopes; it has a fœtid smell, and a very deep violet or black fissured surface; if sown it does not germinate, but if one of its ends be placed in the earth, and it be covered by a bell-glass, it produces two elegant *Sphæriæ* (*S. purpurea* and *mici ucephala*). The same ergot has been seen to bear both these at the same time. The *Stillæ* have the same structure as *Tubercularia*, only their pedicel is longer; they are also regarded as the conidial state of some *Sphæriæ*. The spores of *Asterosporium*, placed between two plates of glass, resemble a star, and still more what was formerly called a Crow-foot. The spores of *Dictyosporium* are oval, compressed, and latticed like the leaves of *Ouvirandra*.

Most of the diseases of our cereals must be attributed to the group of *Clinosporeæ*, whose spores pierce the tissues of their leaves and roots. The Rust of Wheat, which comprises two badly-described species (*Uredo linearis* and *U. Rubigo-vera*)—appears as a yellow or orange dust on the leaves and straw of *Gramineæ*; it is composed of spherical or slightly ovoid globules. The Greater Rust (*U. Vilmorinea*) is easily distinguished by its large elliptic spores, covered with very small, usually shortly pedicelled spiculæ, and by its dark orange colour; it appears principally on the haulm; when very prevalent, the farmers say that the Wheat is turning red. It has been believed, on their authority, that this Rust is the first state of Mildew (*Puccinia Graminis*); but to prove the contrary, it is only necessary to place a mark on the stalks under observation. The Rust of the Glumes (*U. glumarum*) is developed on the floral envelopes, and often on the seed itself. It does not really exist, for we find on the same glumes and bolls the three species in question, and with them the *Puccinia Graminis*. Bunt (*Ustilago Caries*) is very common, and attacks Wheat, occupying the interior of the pericarp, and leaving no trace of the ovule; the diseased seed nearly retains its shape, but when pressed emits an unctuous soft pulp or black dust, which smells like decaying fish. Smut (*Ustilago segetum*) is a Fungus which usurps the place of the ovule of cereals, or renders them abortive, attacking the pericarp, floral envelopes, and even the spikelets, and reducing them to a black powder which is wafted to a distance by the wind. It is observed in Wheat, Barley, Oats, Millet, and Sorghum, but very rarely in Rye. The Maize Smut (*U. Maydis*) is remarkable from its attacking all parts of the plant above ground, forming larger or smaller irregular tubercles, which finally break and emit a black sanious matter that stains all parts of the plant. If the organs of fructification are attacked, no fruit is to be expected. *Pucciniæ* are brown or blackish parasitic Fungi, the sporangia of which present two superimposed cells; they are developed on an infinity of Phænogams, principally on the lower surface of their leaves. *Phragmidia* much resemble *Puccinia*; but their sporangia are many-celled; they are parasites on the *Uredines* which inhabit the leaves of several *Rosaceæ*. Here again the frequent occurrence of this parasitism has led to the belief that *Uredo* is only a form of *Phragmidium*.

Actinothyrium, Asteroma, Pestalozzia, and the neighbouring genera have led to various statements of the same nature; some botanists have described them as *Sphæriæ*, and have left to their successors the task of characterizing and classing them methodically, which is extremely difficult. MM. Tulasne have singularly simplified the question by considering them as conidia or stylospores of various species. If this theory be admitted, it will in future be necessary to ascertain for every species of *Sphæria* whether it is hermaphrodite, monœcious or diœcious, and to describe its spermatia, conidia, stylospores, &c. To add to the difficulty, great enough of itself, considering that these plants are microscopic, we must add others, consequent on the various forms being only met with either at great distances apart or mingled with other organs of the same nature which belong to distinct species. How, it may be asked, are these organs or forms to be referred to the species they really belong to? and are we not in constant fear of referring them incorrectly? And it is all the easier to fall into error, when it is remembered that a spermatium, a conidium, or a stylospore has no character indicating that it belongs to one species rather than to another. The new theory may be attractive, but when different isolated forms are met with, they ought to be described, and their descriptions placed among those of autonomous species with a certainty which is based on incontestable characters, and not, as has been done, on the authority of others. Lastly, we must admit that the mode of fertilization described by MM. de Bary and Woronin as occurring in *Peziza confluens, Melaloma*, and *Ascobolus furfuraceus*, does not modify our ideas with regard to conidia and stylospores; it still leaves much to be desired with regard to spermatia. For ourselves, all the ectoclinal or endoclinal Fungi, although very simple in their composition, will always be regarded as complete Fungi, and as worthy of attention as those whose organization is more complicated.

The *Æcidia*, which grow on the lower surface of the leaves of some *Euphorbiæ*, alter the appearance of the latter so much, that individuals of *E. Cyparissias* attacked by these Fungi have been described by old botanists as a different species. The *Æcidium* of the *Berberis vulgaris* is accused by farmers of producing Rust and even *Puccinia* on *Gramineæ*, which nevertheless does not prevent their frequently using this shrub to enclose their fields.[1] *Ræstelia cancellata* grows on the Pear and allied genera, always avoiding the Apple; it is first perceived towards June, in the shape of orange-red patches sprinkled in the centre with black spots, on the upper surface of the leaves; on the corresponding part of the lower surface the *Ræstelia* appears as a small cone which opens laterally by several longitudinal slits. This Fungus possesses spores and spermatia, and is therefore capable of reproduction; nevertheless it has been regarded as a form of *Gymnosporangium*, a Tremella-like plant which grows upon Junipers (*Juniperus Sabina, Oxycedrus*, &c.). It has been further stated that if a spray of Savine laden with *Gymnosporangium* be placed near a Pear-tree, *Ræstelia* will appear upon the leaves of the latter; but as the experiment does not always succeed, one of the promulgators of this opinion was told that 'a lucky hand was necessary.'

TRIBE IV. *CYSTOSPOREÆ*.

Receptacles flocculent, continuous or chambered, simple or branched, terminated by a vesicular sporangium containing the spores.

PRINCIPAL GENERA.

Didymocrater.	Mucor.	Pilobolus.	Diamphora.	Hydrophora.
Syzygites.	Ascophora.	Melidium.	Rhizopus.	Azygites.

[1] The history of this matter is curious and instructive. The so-called Rusts of the Barberry and Wheat were popularly supposed to be identical, and to be communicated from the Barberry to the Wheat. The microscope, however, showed them to be essentially distinct, and that one was an *Æcidium* (*Æ. berberidis*), the other a *Puccinia* (*P. graminis*). Recent observations, however, on the development of *Puccinia* and *Æcidium* have been regarded as proving them to be states of one dimorphic genus, and as thus establishing the popular view as the true one.—ED.

Ascophora Mucedo develops principally upon neglected vegetable substances, bread, sweetmeats, &c. ; its spores germinate in ten to twelve hours. *Mucor caninus* forms tufts on dog's excrement ; its sporangium tears irregularly. *Mucor nitens*, better known under the name of *Phycomyces*, vegetates on fat, and bodies steeped in oil, as linen, wood, and earth. It is the giant of the *Mucedineæ* ; its filaments attain four inches in height ; they are as lustrous as silk, preserve well, and do not stick to paper. Agardh first described it under the name of *Ulva nitens*. *Syzygites megalocarpus*, which grows only on Fungi, is remarkable for its mode of fertilization, which is analogous to that which takes place in the conjugate *Algæ*. The same phenomenon is observable in *Ascophora rhizopus*. *Philobus* is a small Fungus which grows, especially in autumn, on the excrement of nearly all animals ; its life is very brief ; it grows during the night, and disappears in the middle of the day. It resembles a small pedicelled urn covered with an operculum ; the sporangium is sunk in the cavity of the receptacle, whence it emerges by throwing off the operculum to a distance, and as it usually ruptures, its vestiges only are found, which has led to the belief that the operculum was the sporangium itself, and that it contained the spores. These spores move and swim, so to speak, in the sporangium before their emission, a phenomenon which is perhaps unique, and which ought to be followed throughout all its stages.

TRIBE V. *TRICHOSPOREÆ.*

Receptacles filamentous, simple or branched, fistular, continuous or chambered. Spores very various in form, simple or compound, clustered at the extremity of the branches, or around the receptacle.

PRINCIPAL GENERA.

Ceratium.	Menispora.	Polyactis.	Periconia.	Arthrinium.
Botrytis.	Sporocybe.	Gonatotrichum.	Peronospora.	Pachnocybe.
Psilonia.	Verticillium.	Asterophora.	Haplaria.	Helicotrichum.
Mycogone.	Desmotrichum.	Helicoma.	Sepedonium.	Gonytrichum.
Helminthosporium.	Acrothamium.	Rhopalomyces.		

This tribe is certainly one of the most curious to study ; the numerous genera which compose it grow on decomposing vegetable matter, and even in the tissue of living leaves ; hitherto they have not been found to have any useful quality, and unfortunately two species are notorious for the mischief they have caused to agriculture and industry. *Ceratium hydnoides*, of which the structure is extremely delicate, in damp weather sometimes covers old rotten trunks with its white tufts.

Sepedonium mycophilum is remarkable for converting the whole substance of *Boleti* into a brilliant golden yellow dust. *Botrytis Bassiana* is the cause of the *Muscardine*, a disease which for the last twenty years has devastated the silkworm nurseries ; it is developed on several caterpillars, but especially on the silkworm. Its mycelium attacks the interior of the living caterpillar, which it finally kills ; twenty-four hours after its death the Fungus appears like a little forest on the surface of the worm, which looks as though covered with flour or plaster. All means devised against the ravages of the Muscardine have proved fruitless ; and just as it was disappearing, it was replaced by the *spot* or *pébrine*, a still more fatal disease, which some naturalists consider to be of vegetable origin. *Peronospora infestans* has since 1845 shown itself in all countries where Potatos are cultivated, and produces the rot. The disease first shows itself on the leaves, which curl, turn black, and dry up. The spores of the Fungus which occupies the lower surface of the leaves, become detached, are washed into the earth by rain, and reach the tubers, on which they begin to grow, forming superficial patches which each day increase in diameter and depth, till the tuber finally decays.[1] As in the case of the Muscardine, the remedies prescribed for this disease have proved absolutely useless.

---

[1] This is not always the case, as the disease frequently commences within the tuber ; the spores which fall on the stem also germinate there, and quickly destroy the haulm. The disease is propagated still more rapidly by zoospores, which are occasionally produced within the spores, as first observed by De Bary.—ED.

TRIBE VI. *ARTHROSPOREÆ.*

Receptacle filamentous, fistular, simple or branched, or almost obsolete, continuous or chambered. Spores naked, terminal, jointed end to end, continuous or chambered, separating more or less easily.

PRINCIPAL GENERA.

Antennaria.	Aspergillus.	Torula.	Fumago.	Penicillium.
Tetracotium.	Phragmotrichum.	Coremium.	Hormiscium.	Dematium.
Isaria.	Monilia.	Oidium.		

Of the genus *Antennaria*, which we place at the head of the *Arthrosporeæ*, the fructification is scarcely known. It is parasitic on [Ferns and] woody vegetables; the branches and branchlets which compose it are very numerous and formed of joints united end to end, very unequal and cohering strongly. All the species are black, and spread over leaves, branches, and even trunks. They absolutely suffocate the plants which they attack, by obstructing respiration, as is seen in Pines, Cisti, arborescent Heaths, Olives, &c. *Fumago* appears on herbaceous and woody plants, and on the inorganic objects which they overshadow. It is this which, mixed with dust, begrimes the statues in public walks with a sort of coat which resembles a layer of soot. Its filaments are very fine and branched, formed of unequal joints, without fructification, and can be carried off by the wind or by rubbing. In this state it constitutes the *Fumago vagans*, which comprises several species of Fungi, as *Cladosporium Fumago*, and various species of *Polychæton* and *Triposporium*. *Polychæton* belongs to the tribe of the Endothecal *Thecasporeæ*: that of the Lemon-tree (*P. Citri*) commits great ravages on the *Aurantiaceæ* cultivated in Italy, Spain, Portugal, the Azores, &c. *Fumago* is not parasitic like *Antennaria*; it vegetates on the honey-like excretions of the *Aphis*, the excreta of Cochineal and other insects. It is, therefore, upon these that we must make war to preserve our trees from the black which soils their leaves. *Penicillium glaucum* is the most common Mould; it is found on all animal or vegetable substances when they begin to decay, only needing moisture to induce its development. *Coremium glaucum* appears principally upon fruit which is spoiling, and on that which has been preserved. It differs from *Penicillium* in its yellow pedicel, formed by the union of several. *Isaria* grows on decomposing wood, and principally on dead insects, entomological collections not being always free from it. MM. Tulasne consider it as the conidial state of several *Sphæriæ*. *Oidium Tuckeri* (the Vine disease) appeared in 1847 in English vineries, but it is doubtless the same as that recorded in the sixteenth century by M. Mizauld, and it has long been observed in America in its perfect state of an *Erysiphe*. The mischief which it has caused in Europe is incalculable. Amongst the various means which have been suggested for destroying it, we need only mention sulphur. Hot and damp seasons are particularly favourable to it. The Yeast of beer (*Torula cerevisiæ*) is a production of which the nature is still far from being determined. It is certain that it is the cause and effect of fermentation, and that the globules composing it are jointed end to end, and are mostly free or separate, and that their multiplication is due to cell-division. Mr. Berkeley thinks that it is a modification of *Penicillium glaucum*, due to the medium in which it is developed, since this Fungus is found to spring from these globules when they are exposed to the air. According to M. E. Hallier, the Yeast of beer would be the conidial state of *Leptothrix*, which arrives at the perfect state in the beer after fermentation. All this requires fresh researches.

Fungi have nearly the same geographical distribution as Lichens; they are met with in the tropics, and in the coldest regions of both hemispheres at the top of the highest mountains beyond Phæno-gamic vegetation. M. Martins gathered on the top of the Faulhorn, at the height of 9,000 feet, two Lycoperdons, a *Peziza*, and several Agarics. In Java they occupy a zone comprised between 5,000 and 8,500 feet, an elevation about equal, as Junghuhn has remarked, to that we have indicated on the Alps.[1] Their northern limit appears to be Melville Island, 74° 47' north latitude. *Lanosa nivalis* spreads its mucous filaments over the surface of the snow, and several of our European species have been found in the southern hemisphere, beyond 50° south latitude, in Auckland and Campbell Islands. Our Common Agaric, the Amadou Boletus, and *Ascophora Mucedo*, exist in all countries of the world. Certain genera, poor in species, have widely dispersed representatives: thus, *Montagnites Pallasii*, from

---

[1] Several species occur up to 18,000 feet in the Himalaya.—ED.

the shores of the Irtisch, in 61° north latitude, scarcely differs from *M. Candollei*, which inhabits the dunes near Montpellier; *Mitremyces lutescens*, of Carolina, is represented in Tasmania by *M. fuscus*; *Cyclomyces fuscus*, from the Isle of France, has its analogue in the United States in *C. Greenii*; finally, the genus *Secotium*, which had previously only been found at the Cape and in New Zealand, has been observed in Algeria, the Ukraine, and even in France. These examples might be indefinitely multiplied.

Fungi are sometimes met with under very singular conditions: thus, the common *Schizophyllum* has been found growing on the remains of a whale's jaw abandoned on the sea-shore; Réaumur saw in Poitou five or six specimens of *Clathrus cancellatus*, which had grown between the stones of a wall; Tode found *Pyrenium metallorum* in the barrel of a pistol; *Polyporus terrestris*, *Agaricus epigæus*, and *Thamnomyces Chamissonii* have been gathered upon rocks; and we have ourselves found at Montmorency *Lycogala parietinum* on a large millstone which had been taken out of the earth not more than a week.

Certain Fungi, like Mosses of the genus *Splachnum*, have for their constant habitat the excrement of herbivorous animals; most of the *Ascoboli* vegetate upon cow-dung; *Mucor murinus*, *caninus*, &c., on the dung of rats, dogs, &c.; the names of *Hormospora stercoris* and *Sordaria coprophila* indicate the places usually affected by *Pilobolus*.

A very few Fungi live in water: we have mentioned *Mitrula paludosa*. In other groups, as fresh-water species may be quoted *Peziza rivularis* and *Clavus,' Helotium Sphagnorum*, &c. *Sphæria Posidoniæ* and *Corallinarum* live parasitically at the bottom of the sea on the leaves of *Zosteraceæ* and on the calcareous fronds of *Corallina officinalis*.

The growth of certain Fungi is proverbially rapid. This may be readily conceived when it is considered that they are developed from an invisible underground mycelium, and that they only await favourable conditions for appearing and expanding. The *Mucedineæ*, or Moulds, appear in a few hours, and disappear as rapidly. When the organs of fructification are enclosed in a volva, they seem rather to dilate than to grow by the production of fresh tissue; the pedicel lengthens and swells, just as certain bodies increase in volume by imbibing water. The growth of some suberose *Boleti* is very slow, and some-times occupies several years. Certain Fungi are only known to us by their sclerotioid mycelium, which attains a considerable size, and serves as food to the inhabitants of New Zealand and China. These white masses, covered with a brown or blackish bark, which are often as large as the head, have been described under the names of *Mylitta*, *Pachyma*, &c. One of them, *P. pinetorum*, is found in China on Pine-roots. *Bromicolla aleutica*, which so much resembles *Sclerotium muscorum*, is used as food by the inhabitants of the Aleutian Isles.

The smell of Fungi is not generally strong, and might be termed *fungous* when it is mild and pleasant, like that of the Mousseron (*Agaricus albellus*), &c. The smell of the Truffle is somewhat peculiar, being observed elsewhere only in a genus of Madrepores (*Astroita*), which is hence called Truffle-stone. Others have the smell of a goat, of old cheese, &c. In *Phallus* and *Clathrus* the cadaverous odour is so strong that these Fungi are smelt at a great distance, and they are eaten by insects as if they were really dead bodies; but here the odour is localized, and confined to decomposed fructiferous parts. Certain species have a sweet smell when fresh; as *Polyporus suaveolens*, sought by the Lapps; *Uredo suaveolens*, &c. Others, on the contrary, like *Agaricus camphoratus*, *Hydnum graveolens*, &c., are only scented when dry. The Moulds have a peculiar and very characteristic smell.

The taste of Fungi is usually mild, and not very pleasant. Some are so extremely acrid that it would be dangerous to retain much of them in the mouth; however, this acridity disappears when they are properly cooked. Many species, as Truffles, Morels, and certain Agarics, are edible and much sought after. Many others, which strongly resemble the preceding, and which nearly all belong to the genera *Agaricus* and *Amanita*, are poisonous. How can the edible be distinguished from the poisonous? The answer is very difficult, especially if we compare the contradictory statements which declare the same species to be innocuous and hurtful. The usual advice is to reject those Fungi of which the smell and taste are disagreeable, and the flesh soft and watery; those which grow in shady and damp places, which quickly spoil, rapidly change colour when their tissues are torn; those which tinge silver brown, or which blacken onions. But these various indications are not certain: the surest way is to analyze the botanical character of the Fungus, or to be guided by the popular traditions of the country. In all cases, suspicious Fungi should be boiled after being cut up, and the water in which they have been cooked should be thrown away. It is still better to slice and macerate them in vinegar and water, which should be thrown away, for it is now averred that vinegar neutralizes the poisonous principle of Fungi; but after this preparation they are never pleasant tasted.

Their colour is most variable, and often depends on the age and degree of moisture of the species. In general it is not brilliant: if we find yellow, red, blue, violet, white, and black Fungi, the tints are usually subdued, and the green one (*Peziza æruginosa*) cannot be compared as to tint with the green of chlorophyll. The peculiar tissues of Fungi present the same variety of shades; although white is the chief colour of Agarics, and russet that of *Boleti*, yet some change quickly to indigo the moment they are broken and their tissue exposed to the air. Certain Fungi are phosphorescent: several exotic Agarics, and those of the Olive (*A. olearius*), of the south of Europe, present this singular phenomenon.

If plunged in water, and exposed to the action of light, Fungi give off hydrogen, azote, and carbonic acid; nevertheless, according to some experimentalists, *Tremellæ* behave differently, and give off oxygen when they are placed in the same conditions as Phænogams provided with green matter.

A Fungus, if broken or cut in one of its parts, continues to live; but the wound does not heal, and the exposed surfaces wither or dry.

Properly speaking, only one species of Fungus is cultivated; but *A. Ægerita* is sometimes produced by burying round slabs of Poplar, and *A. albellus* by moving earth from one place to another. The young Truffle-oaks do not give rise to Truffles; but it often happens that the presence of young plants in suitable soil tends to the production of these valuable tubers.

The chemical composition of Fungi is somewhat complicated: they contain, in addition to water, which sometimes forms nine-tenths of their substance, *cellulose*, associated with other peculiar elements, which together constitute what chemists have termed *fungine*; general principles, as *agaricine, viscosine, mycetide*, a certain quantity of *albumine*, as well as an azotized fatty matter; some colouring, resinous or hydrogenized principles; fixed or volatile oils; an amyloid substance, turning blue with iodine; *fermentable sugar*; *mannite*; *glucose*, more abundant in old Fungi than in young ones (probably the result of the transformation of other hydro-carbons); *propylamine* is also extracted from them, which gives the smell of rotten fish to Bunt and the Ergot of Rye; finally, *amanitine*, a principle of which the poisonous action is only too certain, but the atomic composition of which is not yet known. Besides these substances, Fungi contain phosphates, malates, oxalates, and citrates of lime, magnesia, and alumina, and even free oxalic acid and chloride of potassium. The spores of Fungi are so minute, says Fries, that a single *Reticularia* possesses millions; so subtle as to be individually imperceptible, so light that they are carried by the atmospheric vapours, and so numerous that it is difficult to imagine a space too small to contain them, as is shown by the learned and ingenious experiments of M. Pasteur, and the difficulty of guarding against the introduction of *Penicillium glaucum* in a multitude of experiments upon the spontaneous generation of Fungi. Fermentation and the mysterious part which it plays in the decomposition of organic matter is closely connected with the existence of the plants belonging to the genera *Cryptococcus, Hormiscium*, &c., which consist of free oblong or ovoid microscopic cells.

Fungi are of scarcely any use in the arts, with the exception of *amadou* and some colouring principles, among which we shall only quote that of *Dothidea tinctoria*, which lives on the leaves of a *Baccharis* of New Granada, and yields a very solid green dye. In medicine none are now used but the Ergot of Rye or Wheat, to arrest certain hæmorrhages, and to assist in childbirth. The Ergot of *Paspalum ciliare*, which has no resemblance in form or colour to that of Rye, is similarly used in North America. Dr. Roulin saw, in New Granada, mules, deer and parrots seriously injured, and even die after eating ergoted Maize, the taste of which was disguised by that of the *Sphacelia*, which is rather sweet; this Ergot is called in the country *peladero*, on account of its causing to animals the loss of hair, nails, claws and beak. It is known that flour which contains a certain quantity of Ergot becomes extremely poisonous, and causes very serious diseases, described by doctors as *ergotism* and *dry caries* of the joints; diseases which in certain rainy years have raged in an endemic form, like cholera, in various places.   J. H. Léveillé.

XIV. *ALGÆ, Jussieu, Agardh, Lamouroux, Kützing*, &c.

Cellular Acotyledons, aquatic, or growing on damp ground, always exposed to the light; very various in form, texture and colour; free, or fixed by roots or fulcra; sometimes microscopic, unicellular; sometimes furnished with a simple or branched stem, terminated by fronds, always deprived of stomata. REPRODUCTIVE ORGANS *sometimes of one kind only, resulting from the concentration of the green matter, and becoming spores furnished with vibrating hairs and gifted with motion (zoospores); sometimes of antheridia and sporangia, monœcious or diœcious, and mostly producing motionless spores, solitary or quaternary in the same sporangium (perispore).*

TRIBE I. *FLORIDEÆ, Lamouroux, Thuret.—RHODOSPERMEÆ, Harv., CHORISTOSPOREÆ, Dcne.*

Marine, or very rarely fresh-water *Algæ*, rose, violet, purple, red-brown, or rarely greenish, often mucilaginous, formed of simple or branched filaments (*Dasya*), or of

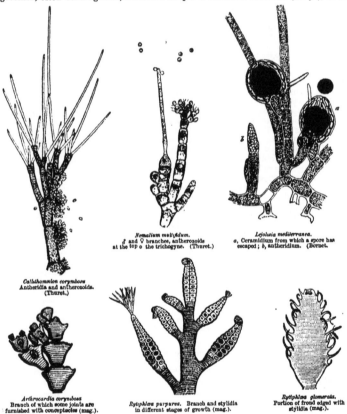

Nemalium multifidum.
♂ and ♀ branches, antherozoids
at the top o the trichogyne. (Thuret.)

Lejolisia mediterranea.
a, Ceramidium from which a spore has
escaped; b, antheridium. (Bornet.

Callithamnion corymbosu
Antheridia and antherozoids.
(Thuret.)

Arthrocardia corymbosa
Branch of which some joints are
furnished with conceptacles (mag.).

Rytiphlœa purpurea. Branch and stylidia
in different stages of growth (mag.).

Rytiphlœa glomerata.
Portion of frond edged with
stylidia (mag.).

Delesseria ruscifolia. Branch bearing a coccidium.

Plocamium vulgare.
Portion of sterile branch.

Delesseria ruscifolia.
Coccidium cut lengthwise (mag.).

Delesseria ruscifolia.
Portion of frond bearing
tetraspores along the
median nerve (mag.).

Delesseria ruscifolia.
Tetraspore (mag.).

Plocamium.
Portion of branch, bearing a
favella.

Plocamium.
Stichidium contain-
ing quaternary
spores (mag.).

Plocamium.
Favella cut,
containing globular
spores (mag.).

Corallina officinalis.
Fructiferous branch
(mag.).

Corallina cupressina.
Section of a branch
terminated by a
ceramidium (mag.).

several tubes composing a simple filamentous stem (*Polysiphonia*, &c.), or of membranous irregular *fronds* (*Porphyra*), or apparently foliaceous (*Delesseria*) or cartilaginous (*Iridea*), with or without nerves, entire, or latticed (*Hemitrema*, *Thuretia*), or umbellate (*Constantinia*), or lomentaceous (*Catenella*), or Jungermannoid (*Leveillea*, *Polyzonia*), sometimes encrusted with lime and fragile (*Corallina*, &c.). REPRODUCTIVE ORGANS of two kinds, monœcious or diœcious. *Sporangia* superficial, or sunk in the frond and contained in conceptacles of various forms. *Spores* rounded or oblong, solitary or quaternary. *Antheridia* of various form, or making a part of the tissue of the frond, composed of colourless cells, each containing an antherozoid with no vibrating hairs or power of motion. The antherozoids fertilize the sporangium by means of a special tube (*trichogyne*).

PRINCIPAL GENERA.

Porphyra.	Cruoria.	Thamnidium.	Dumontia.	Chondrus.
Bangia.		Callithamnion.	Catenella.	Gigartina.
	Wrangelia.	Griffithsia.		Callymenia.
Nemalium.	Bornetia.	Crouania.	Grateloupia.	Gymnogongrus.
Batrachospermum.	Naccaria.	Dudresnaya.	Fastigiaria.	Phyllophora.
Liagora.	Spermothamnion.	Ptilothamnion.	Schizymenia.	
Helminthora.	Ceramium.	Ptilota.	Halymenia.	Peyssonnelia.
	Microcladia.			Rhododermis.

Petrocelis.	Gracilaria.	Polyides.	Chondria.	
Hildenbrandtia.	Nitophyllum.		Rytiphlæa.	Amansia.
Champia.	Delesseria.	Chylocladia.		Leveillea.
Rhodymenia.	Thuretia.		Dasya.	Polyzonia.
Lomentaria.	Hemitrema.	Rhodomela.	Melobesia.	
Plocamium.		Polysiphonia.	Jania.	Polyphacum.
	Constantinia.	Bonnemaisonia.	Corallina.	Osmundaria
Sphærococcus.	Gelidium.	Laurencia.	Lithotamnium.	

The reproductive organs of *Floridæ* have received various names. *Ceramidia* are oval conceptacles, presenting an aperture (*ostiole*) at the top, and containing quaternary spores (*spherospores*). *Kalidia*, *capsules*, and *cystocarps* are bodies of the same form as the preceding, but containing undivided spores. These organs and the following terminate, when young and unfertilized, in a sort of long hair on which the antherozoids rest, and which MM. Bornet and Thuret have termed *trichogyne*; this hair is intended to transmit the fertilizing fluid. *Favellæ* are spherical conceptacles, axillary or terminal, with a smooth wall, sometimes surrounded by a sort of involucre, formed of small branchlets (*Ceramium*, &c.). *Coccidia* are coriaceous conceptacles, usually open at the top, and containing reproductive corpuscules (*Delesseria*). *Stychidia* are little spikes containing quaternary spores regularly arranged. The fronds of some *Floridæ* sometimes bear on their surface isolated cells, forming a sort of sporangium, containing four spores (*Delesseria*); sometimes each of the cells of the membranous frond gives birth to spores (*Porphyra*).

The antheridia of *Floridæ* form a thin layer, transparent at the surface of the frond, or little axillary tufts either naked or involucred (*Griffithsia setacea*), or appear as a small distorted palette (*Chondria*), or as a small conical mass (*Lejolisia*). In all cases the antheridia, whether united in little racemes, or in an assemblage of small utricles, contain an oblong or globose antherozoid, with neither hairs nor motion. In *Porphyra*, the antheridia originate in the transformation of the marginal cells of the frond, which divide into a large number of small colourless cells, becoming so many antherozoids.

TRIBE II. *PHÆOSPOREÆ* et *FUCACEÆ*, *Thuret.—APLOSPOREÆ*, *Dcne.—*
MELANOSPOREÆ, *Harv.*

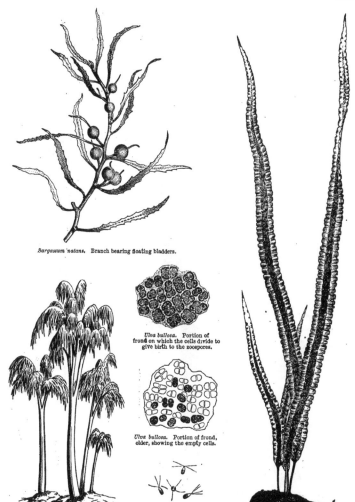

Sargassum 'natans. Branch bearing floating bladders.

Lessonia fuscescens (1-30th natural size).

Ulva bullosa. Portion of frond on which the cells divide to give birth to the zoospores.

Ulva bullosa. Portion of frond, older, showing the empty cells.

Ulva bullosa. Zoospores, bearing four hairs.

Laminaria saccharina (1-30th natural size).

Chætophora elegans.
Zoospore with four hairs.

Œdogonium vesicatum.
Zoospores furnished with
a crown of hairs.   (Thuret.)

Chætophora elegans,
Branches producing a zoospor
n each joint

Œdogonium ciliatum.
Two cells or sporangia, in
each of which is formed a
zoospore (mag. 200).
(Pringsheim).

ss, two sporangia, on which
are fixed the antheridia, *ad* ;
a, antheridium, of which the
lid is removed ; *st*, terminal
bristle.

Œdogonium vesicatum.
Filament of which each
joint presents annular
striæ near the septa
before the exit of the
spore.   (Thuret.)

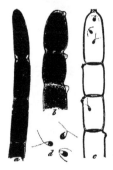

Œdogonium.
Joint opening to give
passage to a zoospore.

Cladophora glomerata.
a, end of young filament ; *b*, older, before
the exit of the zoospores ; *c*, the same
empty ; *d*, zoospores.   (Thuret.)

Fucus vesiculosus.
fr, frond ; *F*, fructiferous tubercle ;
V, aërin vesicle.

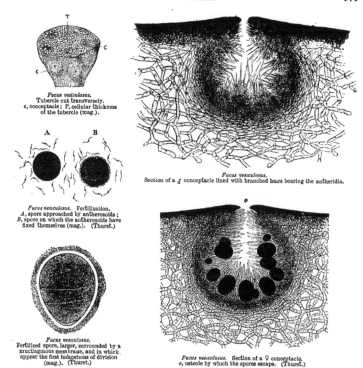

Fucus vesiculosus.
Tubercle cut transversely.
c, conceptacle; *T*, cellular thickness
of the tubercle (mag.).

Fucus vesiculosus. Fertilization.
A, spore approached by antherozoids;
B, spore on which the antherozoids have
fixed themselves (mag.). (Thuret.)

Fucus vesiculosus.
Fertilized spore, larger, surrounded by a
mucilaginous membrane, and in which
appear the first indications of division
(mag.). (Thuret.)

Fucus vesiculosus.
Section of a ♂ conceptacle lined with branched hairs bearing the antheridia.

Fucus vesiculosus. Section of a ♀ conceptacle.
o, osteole by which the spores escape. (Thuret.)

Marine *Algæ*, brown or olive, mucilaginous, very various in form: stemless or caulescent, rounded or elongated, formed into a cup (*Himanthalia*), or cord (*Chorda*), or plate (*Laminaria*), or fan (*Padina*); frond with or without nerves, entire or variously cut, sometimes pierced with holes (*Agarum*), or twisted into a spiral (*Thalassiophyllum*), or furnished with floating bladders (*Sargassum, Macrocystis, &c.*), or with a fistular stem (*Ecklonia buccinalis, Nereocystis, &c.*). REPRODUCTIVE ORGANS, superficial sporangia under the form of sori; *spores*, sometimes proceeding from the condensation of the colouring matter of the fronds (*trichosporangia*), very small, furnished with vibrating hairs and endowed with the power of motion, sometimes resulting from a true bisexual fertilization, and then monœcious or diœcious; the ♀ (*oosporangia*) either external or conceptacular, containing 1, 2, 4, 8 large ovoid or globose motionless spores; *antheridia* sessile or in branches, oblong or linear;

antherozoids containing a red globule having the power of motion, and furnished with 2 vibrating hairs, the one anterior, the other posterior.

SECTION I. LAMINARIEÆ.—Reproductive organs superficial (*sori*); spores usually mobile, germinating without apparent previous fertilization.

PRINCIPAL GENERA.

Scytosiphon.	Ectocarpus.	Petrospongium.		Alaria.
Phyllitis.	Streblonema.	Leathesia.	Sporochnus.	Costaria.
	Myriotrichia.		Stilophora.	Agarum.
Punctaria.	Giraudia.	Castagnea.	Carpomitra.	Thalassiophyllum.
Litosiphon.		Liebmannia.		
	Sphacelaria.	Mesogloia.	Laminaria.	Dictyota.
Desmarestia.	Cladostephus.	Chordaria.	Haligenia.	Taonia. ;
	Arthrocladia.	Chorda.	Lessonia.	Padina.
Dictyosiphon.			Nereocystis.	Dictyopteris.
Aglaozonia.	Myrionema.	Asperococcus.	Macrocystis.	
	Elachista.	Ralfsia.	Ecklonia.	

SECTION II. FUCACEÆ.—Reproductive organs ♂ and ♀ contained in conceptacles; antherozoids furnished with hairs; spores motionless.

PRINCIPAL GENERA.

Fucus.	Cystophora.	Halidrys.	Himanthalia.
Pelvetia.	Landsburgia.	Sargassum.	Splachnidium.
Urvillea.	Cytoseira.	Myriodesma.	Blosvillea.

In the first section of *Phæosporeæ* the sporangia are superficially and irregularly distributed over the surface of the frond, like sori; from these sporangia escape ovoid *zoospores*, terminated by hairs and moving actively, which germinate immediately they become fixed. In the second section of *Phæosporeæ* (*Fucaceæ*, properly called, or Wrack) the fructification usually corresponds to tubercles, dispersed over the frond, or united on special organs, terminal, or arranged in axillary racemes. Each tubercle answers to a fructiferous cavity or *conceptacle* in the thickness of the frond; this cavity contains a gelatinous matter, and bears on its inner wall when young a sort of hairs or transparent filiform cells. At the period of reproduction, such of these hairs (*sporangia, perispores*) as are to fructify, swell, and become filled with brownish matter. This becomes organized into reproductive bodies, which escape by a small orifice in the centre of the tubercle, and soon divide into 2–4–8 spores, which quickly germinate. There remains nothing in the cavity but the torn mother-cells (*perispores* or *sporangia*), and the other sterile cellular hairs, which lengthen into little tufts, and also issue in succession through the same orifice. Sometimes the *antheridia* exist in the same conceptacle with the *sporangia*, and the plant is monœcious; sometimes the antheridia and sporangia are borne upon distinct individuals, when the plant is diœcious. The conceptacles containing the antheridia are generally recognizable by their orange colour. The antheridia consist of ovoid vesicles, containing a whitish mass, sprinkled with red granules: they are borne upon branched chambered hairs. When the plant is exposed to the air, the antheridia are expelled in a mass through the orifice of the conceptacle, and at their extremities issue numerous transparent lageniform antherozoids, moving rapidly; each usually contains a red granule, forming a dorsal protuberance; the locomotive organs consist of two unequal very mobile hairs, the shortest in front, the other dragging behind the body of the antherozoid. In the conceptacles where the sporangia and antheridia are united, the latter line the upper half, near the opening, and the sporangia occupy the bottom of the conceptacle.

TRIBE III. *CHLOROSPOREÆ*, *Thuret.*—*CONFERVÆ*, *Agardh.*

Marine or fresh-water *Algæ*, green, sometimes reduced to a single microscopic cell (*Hydrocytium*), or composed of simple or branched capillary filaments (*Conferva*, &c.), or arranged in a network with horizontal meshes (*Hydrodictyon*, *Trypethallus*), or interlaced and forming spongy balls (*Codium*), sometimes dilated into cellular foliaceous plates (*Ulva*, *Udotea*, *Anadyomene*), sometimes tubular (*Enteromorpha*), sometimes umbrella-shaped (*Acetabularia*), or resembling a Moss, a Lycopod, or the branch of a Conifer (*Caulerpa*). REPRODUCTIVE ORGANS formed by the concentration of the green matter into motile *spores*, furnished with vibrating hairs, or the result of fertilization by antherozoids.

SECTION I. CONFERVÆ.—Tubes or cells containing ovoid spores furnished with 2–4 vibrating hairs.

PRINCIPAL GENERA.

Conferva.	Ulva.	Anadyomene.	Caulerpa.	Penicillus.
	Ulothrix.		Dasycladus.	Cymopolia.
Hydrodyction.	Coleochæte.	Microdyction.		
	Chætophora.		Acetabularia.	Halymeda.
Microspora.	Draparnaldia.	Bryopsis.	Neomeris.	Udotea.
	Cladophora.	Codium.	Bellotia.	

SECTION II. UNICELLULARES.—Each cell producing several spores furnished with vibrating hairs.

PRINCIPAL GENERA.

Codiolum.	Characium.	Hydrocytium,	Sciadium.

SUB-TRIBE. *ŒDOGONIEÆ.*

Green *Algæ*, very simple in structure, composed of a series of simple or branched cells; spores formed by a concentration of the green matter, and escaping by a peculiar division of the mother-cell, ovoid, mobile, furnished with a crown of vibrating hairs, or resulting from impregnation; *antheridia* formed of filaments composed of a series of small cells containing 1–2 *antherozoids*, which issue by an operculum to fertilize the spores contained in the sporangial cell.

GENERA.

Œdogonium.	Bolbochæte.	? Derbesia.

TRIBE IV. *VAUCHERIEÆ.*

Green fragile *Algæ*, formed of simple not septate filaments, presenting two sorts of reproductive organs; the one resulting from the concentration of the green matter at the extremity of the filaments, which they burst, and whence they issue in the form of a large oval mobile *spore*, entirely covered by a ciliate epithelium; the others the result of fertilization.

The *antheridium* first appears as a small horn (*cornicula*), filled with mucilage, and placed by the side of a small rounded organ which acts as a *sporangium*; the antheridium contains in its apex *antherozoids* of $\frac{1}{4000}$ in., furnished with two

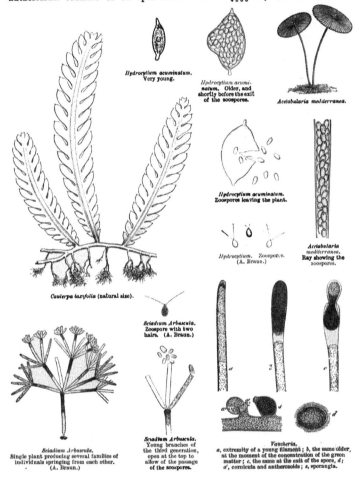

Hydrocytium acuminatum. Very young.

Hydrocytium acuminatum. Older, and shortly before the exit of the zoospores.

Acetabularia mediterranea.

Hydrocytium acuminatum. Zoospores leaving the plant.

Hydrocytium. Zoospores. (A. Braun.)

Acetabularia mediterranea. Ray showing the zoospores.

Caulerpa taxifolia (natural size).

Sciadium Arbuscula. Zoospore with two hairs. (A. Braun.)

Sciadium Arbuscula. Single plant producing several families of individuals springing from each other. (A. Braun.)

Sciadium Arbuscula. Young branches of the third generation, open at the top to allow of the passage of the zoospores.

Vaucheria. *a*, extremity of a young filament; *b*, the same older, at the moment of the concentration of the green matter; *c*, the same at the exit of the spore, *d*; *a'*, cornicula and antherozoids; *e*, sporangia.

vibrating hairs, and resembling the antherozoids of *Fucaceæ*; the *sporangium* contains green granules; a septum forms in each organ, below the antherozoids in the antheridium, and below the granular mass in the sporangium. When fertilization is about to take place, these two organs bend towards each other, the antheridium bursts open at the tip, the antherozoids escape, seek the extremity of the sporangium, the top of which has also opened, enter it, and thus effect fertilization.[1] The membrane of the sporangium thereupon thickens, and encloses a mass of green granules; finally, this sporangium becomes detached from the mother-plant, and sinks into the mud, to give birth sooner or later to a fresh individual.

ONLY GENUS.

Vaucheria.

Some *Algæ* of the group of *Chlorosporeæ* have, as we have just seen, two sorts of spores; the one germinating immediately, the others sinking into the mud, where they remain buried during a longer or shorter time before germinating; these spores are clothed in a tolerably thick membrane, and resemble the Infusorial animals termed *Enkystæ*. The strangest phenomenon is that of *Hydrodictyon*, where the green matter is concentrated within a joint, and becomes converted into motile corpuscules, which arrange themselves within the tube, so as to form a complete network, at the same time that other corpuscules escape, and form large germs with a much slower development, named *chroniospores*, usually $\frac{1}{500}$ to $\frac{1}{500}$ of an inch in diameter; these increase for some time in size, then give birth to other corpuscules furnished with vibrating hairs, which develop into a new individual in the form of a net (*Hydrodictyon, Colæochæte*, &c.). During their torpid state these chroniospores resemble a *Protococcus* of $\frac{1}{2000}$ in. in diameter, and have often been described as very different species from their parents. The knowledge of these curious phenomena is due to MM. Pringsheim and Hofmeister.

SAPROLEGNIEÆ.—MYCOPHYCEÆ, Kützing.

Colourless aquatic filamentous *Algæ* (?), resembling *Vaucheria* in structure, growing on decomposing organic matter, presenting rounded motile zoospores, furnished with hairs, resembling the spores of *Confervæ* or *Vaucheria*; and also sporangia containing spherical oogonia.

GENERA.

Saprolegnia.   Achlya.   Pythium.

These singular vegetables are considered to be Fungi by some botanists; they live, in fact, on organic matters in a state of decomposition in water, where they act upon oxide of iron by decomposing the carbonic acid, absorbing the oxygen, and thus setting free the sulphuretted hydrogen, which destroys the vegetables or animals near it. Notwithstanding the significance of these biological phenomena, several physiologists who have carefully studied *Saprolegnieæ* do not hesitate to class them amongst *Algæ*. '*Saprolegnia ferax*,' says Thuret, 'is usually found on the bodies of drowned animals, which it covers with a whitish down; it even attacks live fish. Nothing is easier than to procure this singular *Alga*. Let a vase be filled with water from a garden tub, and some flies be thrown into it, and it will usually be developed in a few days. The body of the fly becomes covered with hyaline filaments, which radiate around it, enveloping it with a whitish zone. Under a microscope, these filaments are seen to be continuous, simple or scarcely branched, and to contain minute granules which show a motion resembling that which is seen in the hairs of Phænogams. These granules are very numerous, especially towards the upper extremity of the tube, to which they give a grey, somewhat russet tint. This portion soon becomes isolated from the rest

---

[1] I have not adhered literally to the description of this operation in the original, which would not be clear in English.—ED.

of the filament by the formation of a diaphragm. Then the contained matter coagulates in small masses, which become more and more sharply defined, and end by forming so many zoospores. These phenomena succeed each other very rapidly; often in less than an hour the granular matter becomes condensed at the top of a filament, the septum forms, and the zoospores appear. Finally the tube, which has a small

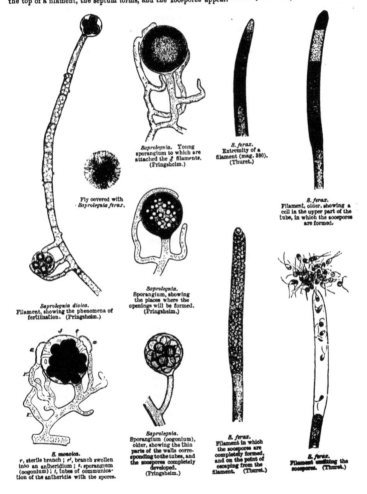

Saprolegnia. Young sporangium to which are attached the ♂ filaments. (Pringsheim.)

S. ferax. Extremity of a filament (mag. 380). (Thuret.)

Fly covered with Saprolegnia ferax.

S. ferax. Filament, older, showing a cell in the upper part of the tube, in which the zoospores are formed.

Saprolegnia. Sporangium, showing the places where the openings will be formed. (Pringsheim.)

Saprolegnia dioica. Filament, showing the phenomena of fertilization. (Pringsheim.)

Saprolegnia. Sporangium (oogonium), older, showing the thin parts of the walls corresponding to the tubes, and the zoospores completely developed. (Pringsheim.)

S. ferax. Filament in which the zoospores are completely formed, and on the point of escaping from the filament. (Thuret.)

S. monoica. r, sterile branch; r′, branch swollen into an antheridium; s, sporangium (oogonium); t, tubes of communication of the antheridia with the spores.

S. ferax. Filament emitting the zoospores. (Thuret.)

protuberance at its extremity, bursts there, and the zoospores escape, the first with impetuosity, the others more slowly; they are turbinate in shape, and furnished with two hairs. This is not the only mode of reproduction possessed by *Saprolegnia*; another phenomenon succeeds. The filaments emit small lateral branches, the extremities of which swell into sacs of a blackish hue, due to the condensation of their granular contents. Soon a septum forms, isolating the sacs from the little tubes which serve as pedicels to them. After some time the granular matter divides into several masses, which at first adhere to the walls of the sac, but which later become free and spherical. Sometimes there is only one of these masses; sometimes the same sac contains fifteen or twenty. I have fancied that I could recognize on their periphery little *mamillæ* resembling regularly arranged opercula.' The sacs have been termed by Pringsheim *oosporangia*. The oosporangia require fertilization to enable them to produce fertile spores. It is obvious, therefore, that *Saprolegnieæ* have a double mode of reproduction, similar to that of *Vaucheria*: the one asexual by means of zoospores; the other sexual, producing *ovyonia* arising from the fertilization of a sporangium (*oosporangium*).

In retaining *Saprolegnieæ* amongst *Algæ*, we are necessarily led to connect with them some plants which appear to approach them in their mode of reproduction by zoospores, but which some distinguished botanists also place amongst Fungi; these are *Cystopus*, *Peronospora*, &c.; as well as the small family of *Chytridiaceæ*, comprising *Rhizidium*, *Chytridium*, *Synchytrium*, &c. The great differences of opinion as to the nature of these plants which exist, show how obscure is the whole history of those Cryptogams that possess no archegonia.

TRIBE V. *SYNSPOREÆ*, Decaisne.—*CONJUGATÆ*, Link.

Fresh-water *Algæ*, consisting of cells of very various form, or of chambered tubes containing green matter, either granular or disposed in spiral plates. At the period of reproduction, the cells forming each tube swell or mamillate laterally, and the mamillæ of two contiguous tubes then unite, and their walls disappear at the surface of contact. Communication being thus established between the two cells, the green matter of the one passes into the cavity of the other, and mixes with its contents, and from this fusion simple or compound spores result, which produce one or several new plants.

PRINCIPAL GENERA.

Zygnema.	Mougeotia.	Craterospermum.	Spirogyra.
Zygonium.	Staurospermum.	Sirogonium.	Mesocarpus.

SUB-TRIBE. *DESMIDIEÆ*.

Microscopic green *Algæ*, presenting the appearance of corpuscules composed of two opposite hemispheres, joined base to base, free, isolated, or associated in flat or spiral bands, enveloped in mucilage, very varied and elegant in form, always symmetrical, entire or lobed, surface smooth or sculptured, reproduced either by conjugation (as in *Synsporeæ*) and the passage of the endochrome from one half to the other, resulting in a reproductive spore, or by the division of an individual, or by means of sporangia of very various forms. The green matter of *Desmidieæ* possesses, according to M. de Brébisson, a circulation analogous to that of *Chara*.

PRINCIPAL GENERA.

Micrasterias.	Penium.	Mesotænium.	Desmidium.	Binatella.
Euastrum.	Closterium.	Pleurotænium.	Spondylorium.	Trochiscia.
Staurastrum.	Tetmemorus.	Spirotænia.	Helierella.	Scenedesmus.
Xanthidium.	Cylindrocystis.	Sphærosoma.	Heterocarpella.	Pediastrum.
Cosmarium.				

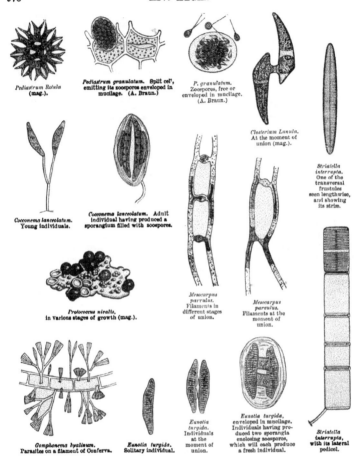

Pediastrum Rotula
(mag.).

Pediastrum granulatum. Split cell,
emitting its zoospores enveloped in
mucilage. (A. Braun.)

P. granulatum.
Zoospores, free or
enveloped in mucilage.
(A. Braun.)

Closterium Lunula.
At the moment of
union (mag.).

*Striatella
interrupta.*
One of the
transversal
frustules
seen lengthwise,
and showing
its striæ.

Cocconema lanceolatum.
Young individuals.

Cocconema lanceolatum. Adult
individual having produced a
sporangium filled with zoospores.

Protococcus nivalis,
in various stages of growth (mag.).

*Mesocarpus
parvulus.*
Filaments in
different stages
of union.

*Mesocarpus
parvulus.*
Filaments at the
moment of
union.

Gomphonema hyalinum.
Parasites on a filament of Conferva.

Eunotia turgida.
Solitary individual.

*Eunotia
turgida.*
Individuals
at the
moment of
union.

Eunotia turgida,
enveloped in mucilage.
Individuals having pro-
duced two sporangia
enclosing zoospores,
which will each produce
a fresh individual.

*Striatella
interrupta,*
with its lateral
pedicel.

TRIBE VI. *DIATOMEÆ* et *BACILLARIEÆ, Ehrenberg, Brébisson,* &c.

Microscopic *Algæ* (?) living in fresh, brackish, or salt water, generally prismatic
and rectangular, free sessile or pedicelled, naked or immersed in mucilage, and

dividing into polymorphic fragments [*frustules*]. *Diatomeæ* have a rigid envelope, marked with extremely fine striæ, fragile, siliceous, 2-valved, remaining unchanged when dry, containing a brown or yellowish nitrogenized matter, sometimes with a rather rapid creeping movement.—Some authors refer *Diatomeæ* to the Animal Kingdom. Certain species are parasitic; others form flakes, or gelatinous masses upon rocks; others live in fresh and pure spring-water; others again cover the soil with a more or less thick brown sticky layer, often occupying a considerable space. They abound in public fountains, the walls of which they stain brown.

PRINCIPAL GENERA.

Micromega.	Encyonema.	Diatoma.	Exilaria.	Cyclotella.
Schizonema.	Gaillonella.	Achnanthes.	Stigmatella.	Asteromphalus.
Homœocladia.	Fragillaria.	Cymbophora.	Surirella.	Spatangidium.
Berkeleya.	Meridion.	Gomphonema.	Frustulia.	

Besides their multiplication by spores, resembling that of *Desmidieæ*, *Diatomeæ* are multiplied by deduplication. On the centre of each frustule in the solitary species, and of each segment or joint in the species of which the frustules are aggregated, there is frequently visible in the young Diatom a line or stria dividing it into two or several frustules, which become distinct individuals, similar to the parents. In the Diatoms with united frustules the effect of this deduplication is to multiply the number of its segments.

Many *Diatomeæ* have been found in a fossil state. Ehrenberg discovered that certain 'rotten-stones,' employed in the arts, were entirely composed of their siliceous shells. M. de Brébisson has shown that calcining living species neither destroys nor distorts their envelope, and that by this process an artificial rotten-stone can be made. The so-called 'mineral flour,' to which geologists have given the name of *Diatomepelite*, has the same origin. *Diatomeæ* abound in guano.

ALGÆ SPURIÆ.

We have classed as Doubtful *Algæ* (*Algæ Spuriæ*) certain imperfectly known genera, and which are probably only degraded forms of the preceding families: such are the families *Rivularieæ, Oscillarieæ, Nostochineæ, Palmelleæ,* and *Volvocineæ.*

They are gelatinous *Algæ* of a whitish green, brackish, red or brown colour, living on earth or damp stones, or in fresh either cold or thermal water, rarely in the sea; composed either of globules, or of simple or branched filaments, continuous or chambered, nearly always enveloped in mucilage.

[Many of their genera are now so familiar to us through the researches of microscopists, that a brief notice of them seems to be required, under the tribes to which they have been referred.

I. OSCILLATORIEÆ, including RIVULARIEÆ.—Plants growing in fresh and brackish pools, hot springs, rivers, vegetable infusions, &c., formed of transversely striated or moniliform filaments, sometimes spirally curled or sheathed in mucus, exhibiting a serpentine motion. Reproduction by transverse division.

PRINCIPAL GENERA.

Oscillatoria.	Spirulina [? Spirularia].	Microcoleus.	Lyngbya.	Scytonema.
Ulothrix.	Calothrix.	Sclerothrix.	Rivularia.	Glæotrichia.
Euactis.	Leptothrix.	Bacterium.	Vibrio.	

3 R 2

Very obscure plants, abounding in damp places, and in various situations; some form woolly patches, or gelatinous strata of a green, olive, brown or æruginous colour; some emit a strong odour; some are almost invariably found on dripping rocks. *Vibrios* are minute colourless active jointed bodies, that abound in decomposing infusions, and, like the still simpler *Bacteriæ*, which are mere inflexible rods of excessive minuteness, are probably rudimentary states of other *Algæ*.

II. NOSTOCHINEÆ.—Plants growing on damp moss or earth, and on stones in fresh water, consisting of slender moniliform tranquil or oscillating filaments, composed of cells placed end to end, immersed in a dense gelatinous matter, formed by the fusion of the gelatinous sheaths of the filaments. Cells of the filaments of three kinds:—1. Ordinary cells; 2. Vesicular cells, large, without granular matter, but often with erect hairs; 3. Globose, elliptic or cylindric sporangia.—Reproduction by cell-division.

PRINCIPAL GENERA.

Nostoc (Hormosiphon).	Sphærozyga.	Spermoseira.	Monormia.
Aphanizomenon.	Anabaina.	Trichodesmium.	

A group of obscure plants, resembling *Collema* amongst Lichens externally, found all over the globe, even on ice and snow; often occurring in detached masses. *Nostoc edule* is sold in China, dried, and forms an ingredient in soups. Several species inhabit salt lakes in Tibet and elsewhere. *Monormia* forms floating jelly-like masses on brackish water, sometimes of great extent. *Trichodesmium*, which floats on the surface of the great oceans and Red Sea, and resembles chopped straw, has been referred both to this group and to *Oscillatorieæ*.

III. PALMELLACEÆ.—Gelatinous or powdery crusts found on damp surfaces, and in fresh or salt water, composed of globular and elliptic cells, aggregated in a gelatinous matrix, increasing by cell-division and by ciliated zoospores.

PRINCIPAL GENERA.

Chlorococcus.	Protococcus.	Glœocapsa.	Hormospora.
Palmella.	Trypothallus.		

Palmella cruenta forms the well-known rose-coloured gelatinous patches on damp walls, which flake off when dry. *P. prodigiosa*, which spreads over meat and boiled vegetables, under certain atmospheric and other conditions, with alarming rapidity, is probably not an *Alga*. *Protococcus* includes various uni-cellular *Palmellaceæ*, which increase by division into two or four parts, which separate, but are connected by a semi-gelatinous layer. Sometimes its cells give rise to four ciliated zoospores, of two sizes, the larger of which settle down and develop a cellulose coat; whilst of the further development of the smaller nothing is known. The famous Red Snow (*P. nivalis*) of the arctic regions and Alps, which is also found on stones in fresh-water streams, belongs to this genus, and is probably a rudimentary *Palmella*.

IV. VOLVOCINEÆ.—Minute fresh-water *Algæ*, consisting of a number of permanently active zoospore-like bodies, associated in various forms, and surrounded by a gelatinous coat, with or without an enveloping membrane.

PRINCIPAL GENERA.

Volvox.	Stephanosphæra.	Gonium.

Volvox is a pale green globule, one-fiftieth of an inch in diameter, common in ponds, in constant rolling motion. It consists of a membranous sac, studded with green points, and clothed with innumer-able cilia. The green points consist of layers of zoospore-like bodies coating the inside of the sac, with two cilia, which project through holes in the sac, and are further provided with delicate filaments, that extend from their sides and meet similar filaments from the adjoining bodies. The zoospores are pyriform, have a reddish eye-spot and transparent contractile vacuole. Their mode of reproduction is not satisfactorily ascertained, but young Volvoxes occupy the centre of the sphere. *Stephanosphæra* has eight biciliated green cells, placed at equal distances along the equator of a spherical cell. *Gonium* presents a flat frond of about sixteen cells. These organisms present two forms of cells, an active and a motionless. The active cells have each a pair of vibratile cilia projecting through their hyaline envelope.

V. CRYPTOCOCCEÆ (including CRYPTOCOCCUS, ULVINA, and SPHÆROTILUS) consist of minute colourless globules, floating in vinegar, aromatic waters, &c., and are probably mycelia of Mildew Fungi.—ED.]

The roots of *Algæ* are sometimes absent, sometimes reduced to fulcra or claspers fixed to solid bodies, as stones, shells, wood, &c. ; sometimes filamentous and descending into the sand (*Caulerpa, Penicillus,* &c.); sometimes discoid and attached to other *Algæ* like suckers (*Lejolisia, Leveillea,* &c.).

The stems of *Phæosporeæ* are usually cylindric, formed of elongated cells, with thick walls, united by means of a firm intercellular substance ; the stems are simple or branched, of various lengths, sometimes erect, and simulating a small tree of 10 to 13 feet in height (*Lessodia*), sometimes swollen and bulbous at the base (*Haligenia*), always deprived of stomata. The stems (stipes) of several *Algæ* periodically lose their fronds, which are annually renewed (*Laminaria Cloustoni*). They sometimes present distinct zones on a transverse section, due to the relative size of the cells composing them : the axis is formed of large cells, and called pith (*stratum medullare*) ; to this succeeds a median zone (*stratum intermedium*), covered by an outer (*stratum corticale*) ; this latter is formed of rather dense cells, filled with a colouring matter, which determines the colour of the species.

The dimensions of *Algæ* are extremely various. *Protococcus* measures scarcely the $\frac{1}{700}$ of an inch, while *Macrocystis* extends to nearly 1,700 feet. The frond of *Haligenia bulbosa* measures sometimes 13 feet in width. The microscopic plant described by Montagne in 1844 as *Trichodesmium Ehrenbergii* covered the Red Sea farther than the eye could reach, over a surface of more than 200 miles, with its little filaments of a brick-red colour; hence the name [1] given to the sea which this *Alga* sometimes colours, but which several navigators have observed at different parts of the Atlantic.

Algæ are usually surrounded by a thicker or thinner layer of mucilage, the origin of which is very obscure, and this whether they belong to the groups of microscopic unicellular *Algæ* or to *Confervæ, Fucaceæ, Florideæ,* &c. There are some, however, which are covered with a calcareous coat, like certain *Charæ*; as *Acetabularia, Neomeris, Corallina,* and several other genera of the group *Chlorosporeæ*. Several *Phæosporeæ* (*Laminaria,* &c.), after having been washed in fresh water and dried, become covered with a white efflorescence of a sweet taste (*mannite*). Others again appear to secrete a liquid, acid enough to decompose the hardest limestone; as the singular *Euactis calcivora*, observed by A. Braun on the pebbles of the Lake of Neufchatel, which this little microscopic *Alga* ($\frac{1}{1818}$ of an inch) destroys, furrowing them with worm-like lines, often of great depth.

Several species exhale a peculiar smell, especially some time after being taken out of the water, or when moistened after having been dried; as *Polysiphonia,* which smells of chlorine, *Desmarestia* and *Stilophora,* which exhale a sharp unwholesome smell. Certain *Florideæ,* when moistened, smell something like violets. Finally, some *Oscillariæ* have a slight musky scent. The majority of *Algæ* respire in the same way as Phænogams; they decompose carbonic acid in light, and disengage oxygen even when deprived of green matter, as M. Rosanoff has stated. *Florideæ* owe this remarkable property to their red pigment, as De Candolle noticed when submitting *Porphyra vulgaris* to the action of light. The *Saprolegnieæ,* however, behave differently, as we have seen ; they absorb the oxygen of the water in which they live, and respire like animals.

Many *Algæ* grow exclusively on others; as *Lithosiphon, Streblonema, Myriotrichia, Myrionema, Elachista, Erythrotrichia, Polysiphonia,* &c. ; but these plants are not really parasites, they do not appear to draw their sustenance from the *Algæ* on which they grow; the colour which certain botanists have

[1] This etymology is not admissible.—ED.

thought they could perceive in those *Floridæ* which are parasitic on *Chlorosporeæ* is foreign to these latter, as may easily be ascertained by transverse slices, which show two very distinct rings, the central one green, the outer red.   It is, however, certain that some *Algæ* can only grow on the tissues of other species, and some even only on a particular species; as *Elachista scutulata* on *Himanthalia*, *Ectocarpus brachiatus* on *Rhodymenia palmata*, *Rhizophyllis dentata* on *Peyssonnelia Squamaria*, *Polysiphonia fastigiata* on *Fucus nodosus*, &c., &c.   Some *Confervæ* and *Saprolegnieæ* develop on Fishes, which they finally kill.   The shell of some marine Tortoises of Japan are sometimes quite covered with *Confervæ*, giving them a most singular aspect.

The marine *Algæ* inhabit the most varied localities; some affect the shore, others the open sea. Everyone has heard of the immense bank of Seaweed (*Sargassum natans*) floating in the middle of the Atlantic Ocean, which Christopher Columbus crossed in 1492, in 38° 50′, and again in 1493, in 37°, between 40° and 43° W. long.   Now, as this enormous bank of *Sargassum* still exists, we may conclude that it has not changed its position during 350 years.   The Polar seas support the largest *Algæ* of the groups of *Phæosporeæ*, *Laminariæ*, or *Fucaceæ*.   *Fucus vesiculosus* grows on anything.   The nature of the rock on which they grow does not appear to be of importance to *Algæ*; the quality of the water, more or less muddy, more or less tranquil, the exposure to the shock of the waves, the level of the beach, &c., appear much more necessary conditions for their development.   On bottoms of pure mud scarcely anything is found but *Vaucheria*, *Diatomeæ*, and some *Oscillarieæ*; but on muddy rocks, and shores with *Zostera*, the algological flora is often very rich, and contains a large number of peculiar species.

The *Algæ* of the open sea often live at very considerable depths: *Udotea vitifolia* has been brought up from a depth of 250 feet, near the Canaries; Péron and Maugé have dredged up brilliant living *Algæ* from 500 feet; and Bory declares that he fished up *Sargassum turbinatum* from a depth of nearly 600 feet off the Isle of France.[1]   *Anadyomene stellata* has been gathered at twenty fathoms in the Gulf of Mexico ; *Constantinia*, at fifty fathoms in the Polar seas.   Immense masses of *Diatomeæ* are met with in the deepest abysses of ocean.   Although, according to divers, light does not appear to penetrate below fifteen fathoms, nevertheless the above-mentioned *Udotea vitifolia* was of a rich green colour, comparable, according to Humboldt, with that of the leaves of the Vine and of Grass.   The temperature suitable to the growth of *Algæ* also presents remarkable differences.   While *Protococcus nivalis* lives on snow, and the largest *Laminariæ* inhabit the polar regions, we meet with *Algæ* also in thermal springs: such as the *Oscillarieæ*, which in Tuscany inhabit boracic pools ; and *Anabaina thermalis*, which grows in mineral waters of a temperature above 104° Fahr.

As to the geographical distribution of *Algæ*, *Phæosporeæ* or *Fucaceæ* prefer the cold regions of both hemispheres; nevertheless *Sargassum* abounds in tropical and subtropical regions.   *Rhodosporeæ* or *Florideæ* principally inhabit temperate zones; *Chlorosporeæ* or *Confervæ* abound everywhere.

The genera and species of *Algæ* are infinitely less numerous than those of other Cryptogamic families, but the number of individuals is much more considerable.   The sounding lead brings them up in masses from all depths, and *Diatomeæ* are at the present time aiding in the formation of siliceous deposits similar to the fossil deposits which yield rotten-stone.   Fresh-water *Algæ* are much more abundant than marine ; *Confervæ* sometimes multiply enormously in winter in submerged meadows, and form a sort of felt, which, on the retreat of the waters, remains dry, whitens, and is known by the name of *natural paper* or *water-flannel*.

The marine *Algæ* which are medicinal agents owe their property as vermifuges to iodine, a powerful medicine, and to a very sweet volatile oil ; as the Corsican Moss (*Gigartina Helminthochorton*) and the common Coralline (*Corallina officinalis*).   From the ashes of Wrack, iodine and soda are obtained.   The ashes of many species were employed as antiscrofulous and antiscorbutic before Courtois and Gay-Lussac discovered and separated the simple substance iodine, now so much used in medicine.   *Ulva Lactuca* was formerly considered a resolvent and vulnerary ; *Conferva rivularis*, when distended with water, is used as a topic for burns.   Several species contain a mucilage, which, when not altered by iodine, renders them edible ; as in the Carrageen Moss (*Chondrus polymorphus*), which supports the poor natives of the sea-shores of northern seas ; and as in *Alaria esculenta*, *Rhodymenia palmata* [Dulse], *Ulva*

---

[1] This I may venture to assure the reader is an error ; *Sargassa* are eminently littoral *Algæ*, with the exception of the *Sargassum natans*, which floats, in an abnormal condition, in the Great Ocean.—ED.

Lactuca [Green Laver], *Porphyra purpurea, Halymenia edulis,* &c. The inhabitants of South Chili use as food the large mucilaginous fronds of *Urvillea utilis. Porphyra vulgaris,* cooked with lemon-juice, forms a sort of condiment [called Laver, much used by persons of a scrofulous habit]; the Chinese make cakes of *Porphyra,* which they dry, and then soak, to obtain a nutritious jelly. They also prepare from *Gracilaria lichenoides,* or Ceylon Moss, an alimentary substance analogous to isinglass.

The nests of the Salangane Swallow, a species belonging to the Sonda Islands, the Moluccas, &c., and of which the Chinese make great use [as bird's-nest soup], were long supposed to be made of certain marine *Aiyæ* of the group of *Florideæ* (*Gelidium, Gracilaria, Sphærococcus,* &c.); but M. Trécul has clearly shown that these nests are of animal origin, and that the Salangane constructs them by means of a mucus which flows abundantly from its beak at the pairing season, and which it arranges in thin concentric layers.

Besides the iodine and soda yielded by marine *Algæ,* the farmers on the coast sometimes use them as cattle-food, besides which they are a valuable manure, enriching the soil with an abundance of organic matter. They are collected at the end of winter, when their fertilization is over, and the annual shoots have ceased growing by the sprouting of the cellular tissue of the stipe or frond. The *Phæosporeæ* are particularly used for improving the soil, with *Zostera,* which abounds in estuaries. The peasants of Brittany carry thousands of cartloads of *Fucus* and *Laminaria* to places situated from twelve to fifteen miles from the shore. The stipes of several *Laminarieæ* become horny when dry, and the peasants of the North of Scotland make knife handles of them, as the inhabitants of the Magellanic lands [and sealers and whalers] do with the stipes of *Lessonia.* [After undergoing a process, these stipes are manufactured in England into whip and knife handles, and especially into bougies, which swell when moistened, and thus distend openings.—ED.]

The structure, and especially, the mode of reproduction of Cryptogams without archegonia, are as yet so imperfectly known, that it is often doubtful in what class to place certain of them, and to decide whether they belong to *Algæ, Fungi,* or *Lichens.* It is this imperfect state of Cryptogamic science to which A. L. de Jussieu alluded in his 'Genera,' when, speaking of *Fungi,* he says: 'Incerta adhucdum Fungorum generatio, licet ab auctoribus descripta.' Amongst the causes which have contributed to obscure this branch of Botany, we must especially allude to the prevalence of the habit indulged in by many authors of creating a special nomenclature without regard to that of their predecessors; the slightest structural modification is no sooner recognized than a new term is invented for it, so that the same organ has received several names; and, to add to the complication, the same name has been on several occasions applied to different organs. This redundant glossology, which even Linnæus termed a calamity ('Verbositas præsente seculo calamitas scientiæ') has always proved an obstacle to the progress of science. Even amongst Phænogams, it is enough to quote the classification of fruits proposed at the commencement of this century by eminent naturalists, and which is now forgotten. Botany has long needed a reform and a simplification of Cryptogamic glossology, and all sensible people recognize its necessity. Léveillé has already taken the work in hand in his excellent article, entitled, 'Considérations Mycologiques, &c.'; let us hope that some botanical philosopher will relieve Cryptogamy of a nomenclature which encumbers it, and which renders it so difficult to beginners.

The most remarkable example of this abuse of a term (by applying it to different organs) is shown in the word *spore,* used in Cryptogams as an equivalent for the *seed* of Phænogams. In *Ferns, Equisetaceæ, Marsiliaceæ,* &c., this pretended seed can only be compared with a bulbil, or rather with a flower-bud, which contains the germs of the reproductive organs, but which will only develop and *flower* after being separated from the mother-plant. This separation of the reproductive organs, by detachment from the mother-plant before fertilization, offers some analogy with the extra-uterine fecundation of Fishes and Batrachians.

Charæ, in their mode of fecundation and the very complicated structure of their antheridia, approach Phænogams; the antherozoid penetrates into the sporangium by the orifice of a coroniform organ that fulfils the part of a stigma, and it fecundates a simple amylaceous spore, which really germinates without forming a prothallus, as do the seeds of many Monocotyledons; but this analogy is not carried out in the vegetative organs, the extreme simplicity of which resembles that of *Confervæ.* These latter are reproduced, without having been fecundated, merely by the concentration of the green matter into spores, which, in spite of their powers of motion, may also be regarded as bulbils.

In those *Fucaceæ* which, like *Chara*, are monœcious or diœcious, the antherozoids act directly on a *naked* spore, after it leaves the sporangium, and which is thus analogous to the germinal vesicle of the superior animals. In *Florideæ*, which are monœcious or diœcious, the antherozoids, which have no power of motion, empty their contents on a tubular organ (*trichogyne*), and this matter assists in the formation of several spores in a sort of capsule (*cystocarp*), which may be compared with an archegonium. Finally, in certain *Fungi* the spores are fecundated by the action of a peculiar fluid, proceeding from simple cells, which rest on the female organ, but which have no antherozoids. This imperfect sexuality, joined to the simplicity of their structure, the absence of green matter, and the phenomena of their respiration, would authorize us in considering them as terminating the vegetable series, if the presence of vessels containing peculiar juices, and their very varied proximate principles, did not indicate an organization higher than that of *Algæ*, with which, nevertheless, they are connected through *Saprolegniea*; these being true *Vaucheriæ* with no green matter, which have been considered by eminent botanists now as *Fungi*, now as *Algæ*.

This diversity in the structure and functions of the reproductive organs of Cryptogams, which we have explained as accurately as we could, clearly separates them from Phænogams, in which the action of the pollen and the development of the embryo are, in spite of the polymorphism of the flower, so strikingly uniform.

When, in studying those Acotyledonous families of apparently a higher order, which the brilliant discoveries of some modern observers tend from time to time to separate from Cryptogams, we bear in mind the physiological rôle of the antherozoids and zoospores, which seem borrowed from animals, we cannot but recognize here the mysterious link between the two Kingdoms, which are drawn still closer together by the intimate connection that exists between the Spurious *Algæ* (*Diatomeæ, Volvocineæ, Palmelleæ*) and animals of the simplest organization. Hence the ingenious comparison of the Animal and Vegetable Kingdoms to two trees of which the tops are far apart, while their roots interlace; or to two cones, the tops of which are occupied by the most perfect beings, while the juxtaposed bases are represented by a commingling of inferior organisms. Linnæus entertained the same view, and thus gave expression to it in his ' Philosophia Botanica':—Nature connects animals and plants by their most imperfect species; ' Natura sociat plantas et animalia; hoc faciendo non connectit perfectissimas plantas cum animalibus maxime imperfectis, sed imperfecta animalia et imperfectas plantas consociat. Naturæ regna conjunguntur in minimis.'

APPENDIX.

ON THE CLASSIFICATION OF PLANTS BY THE NATURAL
METHOD; WITH AN ANALYSIS OF THEIR CLASSES,
COHORTS, AND ORDERS, AS ARRANGED IN THIS
WORK.

BY THE EDITOR.

CLASSIFICATION OF PLANTS.

The originator of the Natural System of Plants, as at present understood, was the English naturalist, John Ray, who, in his 'Methodus Plantarum emendata et aucta,'[1] published in 1703, established the great primary divisions of Flowering and Flowerless Plants, and divided the former into Monocotyledons and Dicotyledons, and this on a scientific basis. In the same work Ray further classified all plants known to him, whether from specimens or books (about 18,000), under genera, which, though imperfectly defined and limited, are for the most part natural groups, and answer in many cases (as *Cruciferæ, Umbelliferæ, Coniferæ*, &c.) to the Natural Orders of Jussieu's method. Ray thus showed that he had a very clear and correct appreciation of the subordination of all other characters to those of the seed, in respect of the primary groups of the Vegetable Kingdom; but he very imperfectly appreciated the subordination of the characters afforded by the floral whorls, in the classification of his genera, and hence failed to complete even a rudimentary Natural System.

Antoine de Jussieu, when alluding to the crude efforts of Tournefort's predecessors (in 1719), thus records his opinion of Ray's philosophical views of classification:— 'Johannes Raius hos inter celeberrimos, ac optime de re Herbaria in Anglia meritus, hujus incommodi cavendi causa, non a floribus tantum fructibusve, sed etiam a foliorum, caulium, radicumque partium organicarum figura earumque colore, odore, sapore, et totius plantæ facie exteriori sumenda esse genuinæ methodi principia affirmabat.'—*Judicium de Tournefortii Methodo*, p. xiv.; *Institutiones*, ed. iii. (posthuma), v. 1.

In 1693 J. P. Tournefort, Professor of Botany at the Jardin des Plantes, Paris, the friendly rival and correspondent of Ray, first defined genera as now accepted. He did not, however, recognize the great divisions of Ray, and his twenty-four classes are, like so many of his genera, purely artificial combinations (see p. 162).

Profiting by the labours of Ray and Tournefort, Linnæus, in 1735, established genera on a scientific basis, with as clear a conception of the subordination of characters for this purpose as Ray had for the establishment of higher combinations. He, however, like Tournefort, failed to apply his principles to the formation of groups of a higher value than modern genera, and to appreciate the importance of Ray's primary divisions of the Vegetable Kingdom. As is well known, Linnæus never regarded his sexual system as other than artificial. His attempt to form a natural one resulted in his indicating, without

---

[1] In the first edition of the 'Methodus' (1682) Ray enunciated his views in the following terms: 'Ex hac seminum divisione sumi potest generalis plantarum distinctio, eaque meo jure omnium prima et longe optima.' p. 9.

characters, 65 families of very unequal value. At the commencement of his 'Fragmenta Methodi Naturalis' (1738) he says: 'Diu et ego circa methodum naturalem inveniendam laboravi, bene multa quæ adderem obtinui, perficere non potui, continuaturus dum vixero; interea quæ novi proponam.'

It was left to the sagacity of Bernard de Jussieu, when arranging the plants in the gardens of the Trianon (in 1758), to seize upon the characters whereby the genera of Linnæus should be methodized naturally under the primary divisions of Ray, and to the genius of his nephew Antoine-Laurent de Jussieu so to characterize these subdivisions as that, with but slight modifications, his 'Genera Plantarum secundum Ordines Naturales disposita,' has, ever since its publication in 1774, retained its position as the basis of a complete scientific classification of plants, and secured for its author the well-earned reputation of the expositor of the Natural System. A.-L. de Jussieu's system comprised 15 classes, 100 Natural Orders, and about 1,740 genera. His classes are to some extent artificial, and being founded upon modifications of the floral whorls alone, have been undervalued by many botanists; these modifications, though inconstant in each class, are excellent guides to affinity, and (with the solitary exception to be noticed immediately) no better means than these afford of subdividing either Monocotyledons or Dicotyledons into primary groups, has hitherto been recognized.

In 1763 Michel Adanson, who had studied in his youth under Bernard de Jussieu, published his 'Familles des Plantes,' a work of great originality and research, and which is not only remarkable on this account, but as being the first complete system of Natural Orders that ever appeared in print. These, however, though founded on natural characters, being essentially artificial in construction, have had little influence in developing the Natural System.

In 1827 Robert Brown published his discovery of the direct action of the pollen-tube on the nucleus of the ovule in two Dicotyledonous orders, *Coniferæ* and *Cycadeæ*, thus affording a sure basis for a subdivision of the class Dicotyledons into two primary groups, the Gymnospermous and Angiospermous. Brown was also the first to propose the combining of Jussieu's Orders into groups of a higher value, but subordinate to his classes; and in his various works he so improved both Jussieu's Orders and his Method, that he ranks second only to that author and to Ray as the expositor of the Natural System of Plants.

In 1813 Augustin Pyrame de Candolle, then Professor of Botany at Montpellier, published the first edition of his 'Théorie Elémentaire,' in which he proposed a modification of Jussieu's classes, and of his sequence of the Orders, which is in some respects an improvement, and in a few the reverse; as where he breaks up the Acotyledons, and places the vascular Orders of these amongst Phænogams, calling them Cryptogamous Phænogams. A great improvement was his combining the Diclinous and Apetalous Classes of Jussieu. The first edition of the 'Théorie' contains 145 Orders; the second, which appeared in 1819, contains 161; and the last, edited by his son, A. de Candolle, in 1844, and which represents his latest views, contains 213 Orders: in it the vascular Cryptogams are removed from the Phænogams, but the Gymnospermous division of Dicotyledons is not recognized.

The sequence of Orders given in this work is for the most part that proposed by De Candolle, and is that followed in the universities and schools of England and its dependencies, in America, and throughout a great part of the continent of Europe. The arrangement of De Candolle, given at p. 165 of this work, is that of the first edition of the 'Théorie Elémentaire.' The following is his matured plan, as given by his son:—

DIVISION I. VASCULAR (or PHÆNOGAMIC) PLANTS.
    CLASS I. EXOGENS (or DICOTYLEDONS).
        Sub-class 1. Thalamifloral.
           „   2. Calycifloral.
           „   3. Corollifloral.
           „   4. Monochlamydal.
    CLASS II. ENDOGENS (or MONOCOTYLEDONS).
DIVISION II. CELLULAR PLANTS (or CRYPTOGAMS).
        Sub-class 1. Æthiogams (or Vascular Cryptogams).
           „   2. Amphigams (or Cellular Cryptogams).

In 1830, Dr. Lindley, Professor of Botany at University College, London, published his 'Introduction to the Natural Orders of Plants,' a very original and valuable work, and the first that discussed fully the characters, affinities, geographical distribution, medical and economic properties of the Natural Orders. It follows what the author designates as a slight modification of De Candolle's plan, but which is in truth a great improvement of it. The Orders are there arranged as follows :—

    CLASS I. VASCULAR or FLOWERING PLANTS.
        SUB-CLASS I. EXOGENS or DICOTYLEDONS.
           Tribe I. Angiospermous.
               § 1. Polypetalous (inclusive of Incompletæ).
               § 2. Monopetalous.
           Tribe II. Gymnospermous.
        SUB-CLASS II. ENDOGENS or MONOCOTYLEDONS.
           Tribe I. Petaloideæ.
             „ II. Glumaceæ.
    CLASS II. CELLULAR or FLOWERLESS PLANTS.
           Tribe I. Filicoideæ.
             „ II. Muscoideæ.
             „ III. Aphylleæ.

In 1833 Lindley published his 'Nixus Plantarum,' which is remarkable as being an attempt to carry out Brown's suggestion of throwing the Natural Orders into groups, which the author designated 'Nixus.' The result was an arrangement of the Orders into Cohorts and minor groups, so badly constructed and artificially limited, that it is not needful to reproduce them here, even in outline. It was subsequently somewhat modified in the second edition of the author's 'Natural System.'

Nor is it necessary to dwell upon Lindley's further efforts to group systematically, and place naturally, the Orders of Plants, which he attempted to do in the article on Exogens in the 'Penny Cyclopædia' (1838), and again in 1839 in the 'Botanical Register,' and lastly, in 1845, in his 'Vegetable Kingdom.' This latter work is a copiously illustrated edition of the original 'Introduction to the Natural System,' enlarged and improved in every particular, except that of the classification adopted, which, as it has not been followed, need not here be discussed. It is sufficient to state that undue weight is given in it to the absence or presence of albumen, and to uni- or bi-sexuality, and too little to the floral envelopes, fruit, &c.

In 1835 Elias Fries, Professor of Botany at Upsala, in his ' Corpus Florarum Provincialium Suæcica,' proposed a subdivision of De Candolle's four divisions (Thalamifloral, &c.), upon what approaches to a ternary plan ; each subdivision being characterized by floral characters, and being again divided into others characterized by the ovary being syncarpous with axile placentation, apocarpous, or syncarpous with central placentation. Monocotyledons and Cryptogams are each subdivided into four principal branches.　This system has been adopted to some extent in Scandinavia.

Professor Stephan Endlicher, of Vienna, commenced in 1836 his ' Genera Plantarum secundum Ordines Naturales disposita,' a truly great work, brought to a conclusion (exclusive of three supplements) in 1850.　His arrangement differs from that of both Jussieu and De Candolle, is less simple, and is burdened with a cumbrous nomenclature.　The work is in universal use, and its arrangement much followed in Germany and elsewhere, both because of its intrinsic merits, and of its comprehensiveness, it being the only good and complete account of the genera of plants published since that of A. L. de Jussieu (in 1774).　Endlicher divides plants primarily into two Regions, viz. *Thallophyta* (*Algæ*, *Fungi*, &c.), having no definite ax's f growth, or opposition of root and stem ; and *Cormophyta*, which have such an axis.　The following are the outlines of the system :—

REGION I.　THALLOPHYTA.

 SECTION I.　PROTOPHYTA (2 classes, *Algæ* and *Lichenes*).
  „　II.　HYSTEROPHYTA (class *Fungi*).

REGION II.　CORMOPHYTA.

 SECTION III.　ACROBRYA.

  Cohort 1.　Anophyta (2 classes, *Hepaticæ* and *Musci*).
   „　2.　Protophyta (5 classes, *Filices, Equisetaceæ, Cycadem*, &c.).
   „　3.　Hysterophyta (3 classes, *Balanophoreæ, Cytineæ*, &c.).

 SECTION IV.　AMPHIBRYA (all Monocotyledons, 11 classes).

  „　V.　ACRAMPHIBRYA.

  Cohort 1.　Gymnospermeæ (1 class, *Coniferæ*).
   „　2.　Apetala (6 classes).
   „　3.　Gamopetala (10 classes = *Monopetalæ* of Jussieu).
   „　4.　Dialypetala (22 classes = *Polypetalæ* of Jussieu).

Endlicher conceived that a series is thus obtained, ascending from the simplest to the most specialized forms, regarding the *Leguminosæ* as the latter.　The chief errors in it are the placing *Cycadeæ* amongst Acotyledons, far apart from the other Gymnospermous Orders ; and *Balanophoreæ* and *Cytineæ* together, and both amongst the Cryptogams.

In 1843 Adolphe Brongniart, Professor of Botany at the Jardin des Plantes, Paris, published his ' Enumération des Genres des Plantes cultivées au Muséum d'Histoire Naturelle de Paris, suivant l'ordre établi dans l'Ecole de Botanique.'　The principal feature of this system, which is given in detail at p. 165, which in other respects closely follows Endlicher's, is the intercalation of the Incomplete and Diclinous Orders of Jussieu with the *Polypetalæ* (as proposed by Lindley), whereby the series of the latter is much deranged.

C. F. Meisner (now written Meissner), Professor of Botany at Basle, published in 1843 an excellent synoptical disposition of the Orders in his ' Plantarum Vascularum

Genera secundum Ordines Naturales digesta, eorumque differentiæ et affinitates tabulis diagnosticis expositæ,' of which the following is a sketch :—

 A. Vasculares.

  I. Dicotyledons.

   † Diplochlamydal.

    * Dialypetalous *or* Polypetalous.

     1. Thalamifloral (13 classes and 75 orders).

     2. Calycifloral (7 classes and 36 orders).

    ** Monopetalous.

      α. Fruit inferior (3 classes and 13 orders).

      β. Fruit superior (8 classes and 49 orders).

   †† Monochlamydal (7 classes and 64 orders).

  II. Monocotyledons (9 classes and 38 orders).

 B. Cellulares.

  III. Acotyledons (not described).

   This is much the same as the Candollean arrangement; the principal defects are : the not recognizing the Gymnospermous division except as a division of Monochlamyds ; and the retention of Lindley's faulty class of Rhizanths, which he further places in Monocotyledons.

   Adrien de Jussieu, Professor of Plant Culture at the Jardin des Plantes, son of Antoine-Laurent, published in 1844 a modification of his father's arrangement, which is much used in France, and is further modified at p. 167 of this work. It regards the Gymnosperms as a primary division of Dicotyledons, retains Rhizanths, classing them under Angiospermous Dicotyledons, and brings the Diclines into juxtaposition with the Apetalæ.

   Various other methods and systems have been proposed, most of which have never been adopted by any one, and but few even by their proposers. None are of sufficient importance to merit discussion here, though many good suggestions may be obtained from their study. A list of them will be found in Lindley's ' Vegetable Kingdom,' 2nd edit., published in 1847.

   A review of the foregoing schemes establishes the following propositions :—The primary division of Vegetables is into Phænogams and Cryptogams ;—that of Phænogams is into Dicotyledons and Monocotyledons ;—that of Dicotyledons is into Angiosperms and Gymnosperms ;—that of Cryptogams is into Acrogens (for the most part vascular) and Thallogens (or purely cellular) ; also that the perianth must be resorted to for the purpose of further grouping the Angiospermous Dicotyledons and the Monocotyledons. Beyond this, except in the case of Cryptogams, it is difficult to establish any subdivisions higher than that of Orders; and of the Phænogamous Orders themselves, it is astonishing how few are absolutely limited. Of the 278 described in this work, excluding those containing only one or two genera, a large proportion either are connected with one or more others by a series of intermediate genera, or contain genera which present so many of the characters of other Orders, that it is altogether uncertain in which of them they should be placed : such do not oblige us to unite the Orders between which they oscillate, but they render vague the definition of that in which they are placed. With Cryptogams the case is very different ; the principal Orders and higher groups of the Vascular division are comparatively well-limited, and it is only the Cellular whose ordinal limits are undefined. And not only are the Orders of Cryptogams far better limited, but many of them are far more comprehensive, as *Fungi* and *Algæ*, each of which is divisible into tribes more distinct on the whole (whether strictly limited or not) than are most allied Orders of Phænogams from one another. In other words, the Orders of Cryptogams and Phænogams have very different absolute and relative values.

MONOCOTYLEDONS.—In the classification of these, the three following structural characters have been regarded as of primary importance. 1. The relations of the floral envelopes to one another and to the ovary, by Jussieu (followed by Fries), who divided them into epigynous, hypogynous and perigynous, in conformity with his division of Dicotyledons (see p. 164). 2. The presence or absence of floral envelopes, and their nature; whence originated the Petaloid and Glumaceous divisions of Lindley (subsequently modified in his 'Nixus'). 3. The presence or absence and nature of the albumen, as proposed by Brongniart. An important subsidiary character, derived from the position of the embryo, whether enclosed in or external to the albumen, is also of great importance.

In the classification I have proposed, the greatest importance is given to Jussieu's character of the adhesion or not of the perianth to the ovary; this appearing to afford the means of subdividing the Monocotyledons into two series better than does any other single character, or combination of characters. The exceptions are few, and the typical Orders with inferior perianth are, on the whole, more closely allied to one another than to any in the series with an inferior perianth; but the character of perigynism and hypogynism does not answer for primary divisions.

This disposition of the Monocotyledons appears to follow a simple series, descending from the Orders most complex in structure (*Orchideæ*, &c.) to those most simple (*Gramineæ*); the principal exceptions are the Incomplete-flowered Orders, *Aponogetæ* and *Naiadeæ*, which are obviously reduced *Alismaceæ*, and *Lemnaceæ* and *Typhaceæ*, which are obviously closely allied to *Aroideæ*.

ANGIOSPERMOUS DICOTYLEDONS.—A practicable linear classification of the Orders of this division, which shall also be natural, has not hitherto been proposed, nor is it likely that it is attainable. In the first place, it is not so obvious in what a high type of structure consists; and in the next place, so many of the Orders contain both high and low types, and in such various proportions, that it is impossible to establish a table of precedence for them. De Candolle suggested that plants whose flowers had the greatest number of distinct and separate organs should head the series, namely, those with hermaphrodite polypetalous flowers, having apocarpous ovaries; and that unisexual plants with incomplete perianths or none should terminate it.

Succeeding authorities have considered departure from the primitive condition of the floral whorls (regarded as foliar organs) as evidence of high type, and consequently that plants with consolidated floral organs should head the series; placing Monopetalæ before Polypetalæ, and Monopetalæ with the calyx-tube and ovary consolidated (epigynous Mono-petalæ) before those in which the ovary is free. It has further been held that a flower in which many such consolidated organs are subsidiary to the perfecting of one seed is of a higher type than one in which as many organs are subsidiary to the perfecting of many seeds, and that *Compositæ*, by satisfying the first of these conditions more nearly than any other Order, should hence head the series.

Whatever the difference of opinion as to the value of cohesion of the petals, that is, of monopetalism in contradistinction to polypetalism, as evidence of high or low type of structure, there is none as to its value as an index of affinity; the Monopetalous Orders being indisputably more nearly related on the whole to one another than they are to the Polypetalous. It is very different with the Orders of Jussieu's and De Candolle's third great group of Angiospermous Dicotyledons, the Incomplete-flowered or Diclinous. These consist in great part of Orders which are manifestly very near allies of Polypetalous Orders, and in part of Orders or groups of Orders that have no recognized close affinity with any in either Monopetalæ or Polypetalæ; and the fact of the presence or absence of floral envelopes being no positive proof of affinity, has led to the abandonment of the Incomplete-flowered division by Brongniart and other systematists, and the dispersion of its members amongst the Polypetalæ and Monopetalæ, guided by affinity, when that is known, and by analogy in other cases. To me it appears that, under a classificatory point of view, the adoption of an Incomplete division cannot be avoided: firstly, because it contains so many natural and well-established cohorts that have no recognized affinity with any amongst either Polypetalæ or Monopetalæ; and secondly, because it contains so many which, though closely allied to others amongst Polypetalæ, could not be intercalated amongst these without disturbing their sequence, thus sundering Orders that should stand in contiguity.

The subdivision of the Polypetalous Dicotyledons by Jussieu and De Candolle into Hypogynous, Perigynous, and Epigynous, has been by many authors objected to as artificial; but I think it cannot be doubted that the characters indicated by these terms are guides to affinity, and that, exceptions notwithstanding, there is a recognizable, though often distorted, sequence of Orders from hypogynous *Ranunculaceæ* to perigynous *Leguminosæ*, and from these again to epigynous *Araliaceæ*, which indisputably establish a direct passage to the epigynous Monopetalæ. The great obstacle to the recognition of the Thalamifloral and Calycifloral series, lies in the fact that (putting aside the many cases of hypogynous Orders containing perigynous genera) there are many Orders of which it is difficult to say to which they belong. Thus Brongniart regards as hypogynous *Anacardieæ*, *Connaraceæ*, *Burseraceæ*, and *Celastrineæ*, all of which are regarded as perigynous by De Candolle; and as perigynous *Caryophylleæ*, *Elatineæ*, and *Olacineæ*, which De Candolle and Lindley regard as hypogynous. To reduce this difficulty, Mr. Bentham and I, observing that a highly developed staminiferous disk prevailed in the Orders that intervened between the manifestly perigynous and hypogynous Orders, collected them into a division of Polypetalæ, called Series Discifloræ. In doing this, we did not look on the disk as a proof of affinity, but as a guide to that amount of affinity which certainly exists between the Orders included under that Series.

It remains to say a few words on the sequence of Dicotyledonous Orders here adopted. That a linear, and at the same time natural sequence is unattainable, is conceded by all philosophical systematists. It is further, I think, established that the Monopetalæ exhibit the greatest departure from the imaginary primitive type of the Dicotyledonous flower, and that they should, in preference to Polypetalæ, head a system founded on this consideration.[1]

But if Monopetalæ be chosen to head the Dicotyledons, *Compositæ* should lead off, followed by *Valerianeæ*, *Rubiaceæ*, and *Caprifoliaceæ*, Orders that establish the passage to epigynous Polypetalæ: under which arrangement, *Caprifoliaceæ*, &c., instead of leading to *Corneæ* and the other Polypetalæ, would be followed by the Campanal Cohort, a group more closely allied to *Compositæ* than are *Rubiaceæ*, &c. This suggests the expediency of following De Candolle in placing the epigynous Monopetalæ in the middle of the Dicotyledons, beginning with *Valerianeæ*, *Rubiaceæ*, and *Caprifoliaceæ*, which lead to the

[1] Separation of the sexes, which in the Animal Kingdom is universal amongst the higher groups, is in the Vegetable Kingdom no indication of a high or low development; and is a fallacious guide to affinity. This is, no doubt, connected with the fact, that in plants the transmission of the sperm-cell or antherozoid to the pistil or germ-cell is directly brought about by external agencies, and not by the spontaneous action of the individual of either sex.

epigynous Polypetalæ, and ending with *Goodenieæ*, *Campanulaceæ*, and *Vaccinieæ*, which lead to the hypogynous Monopetalæ.

Continuing the Monopetalous Series, its successive Cohorts seem to follow a fairly natural sequence, terminating abruptly with *Labiatæ*, and followed by the Incompletæ, with *Nyctagineæ*, whose perianth is petaloid and gamophyllous, at their head. The Incomplete Series again exhibits a succession of less and less specialized floral organs, finally approaching in simplicity to those of Gymnosperms, which undoubtedly close the Dicotyledonous Class.

Commencing again with *Compositæ*, and proceeding in the other direction, through the epigynous Polypetalæ, to the perigynous and hypogynous Orders, the succession of Cohorts terminating with the Ranal is not unnatural.

On these grounds I am disposed to approve of the sequence adopted by De Candolle, which places Monopetalæ in the centre of the series, flanked on either hand by Polypetalæ and Incompletæ, which two latter, as remarked above, have many cross affinities, but have few affinities of consequence with Monopetalæ. The Cohorts may thus be fancifully likened to the parti-coloured beads of a necklace, joined by a clasp, the beads touching at similarly coloured points of their surfaces. The position of each bead in the necklace is determined by the predominance of colours common to itself and those nearest to it; whilst the number and proportion of the other colours which each bead presents, indicates its claims to be placed elsewhere in the necklace; in other words, such colours represent the cross affinities which the Cohorts display with others remote from the position they occupy.

In conclusion, the student must remember that the above sketches of systems and their foundations are of a very superficial description, and that rapidly accumulating discoveries in the development of organs and of species, in dichogamy, in Fossil Botany, and in the distribution of Plants in time and area, will no doubt one day throw a new light on the whole subject, and teach us why it is that such and such modifications of form and of structure are more or less faithful guides to affinity. It is one thing to perceive affinities, the power to do which is intuitive and possessed in very different degrees by different persons, the child often detecting a consanguinity where the sage fails to see it when pointed out to him; it is another thing to seize the clues to such affinities, and, like Jussieu, to generalize them, and develop them into principles of classification; and it is a third and widely different thing to prove that such affinities are genetic and real, not chance-produced or imitative:—this is one of the deepest problems of nature, the solution of which is to be arrived at through the patient labours of the anatomist and experimenter, which alone can reveal the philosophy of classification.

·SYNOPSIS

CLASSES, SUB-CLASSES, COHORTS AND ORDERS.

(BY THE EDITOR.)

———••◦◦◦◦•◦——

THE following synopsis of the Orders is intended to facilitate reference to the detailed descriptions contained in the body of this work, and to afford the student the means of studying the subordination of characters for systematic purposes. The principal exceptions are given to the diagnosis of the higher groups ; but the beginner must not expect to be able to refer a Plant to its Order with much facility or any certainty, till he has laboriously acquired a general insight into the principles and practice of classification.

SUB-KINGDOM I..

PHÆNOGAMOUS, COTYLEDONOUS, OR FLOWERING PLANTS.

Plants furnished with flowers, normally composed of whorls of foliar organs, enclosing stamens or ovules or both. Fertilization effected by the action of pollen-cells on the nucleus of an ovule. Propagated by seeds.

CLASS I. DICOTYLEDONS.

Stem, when perennial, furnished with a pith, surrounded by concentric layers of wood, and that by a separable bark. Leaves with usually netted venation. Floral whorls usually in fours or fives, or multiples of those numbers. Embryo with two (sometimes connate) cotyledons. In germination the radicle lengthens, and forks or branches.

Sub-class I. Angiospermous.

Ovules produced in a closed ovary, fertilized by the pollen-tube traversing a stigmatic tissue to reach the cavity of the ovary, and hence the embyro-sac of the ovule.

DIVISION I. POLYPETALOUS.

Flowers with both a calyx and a corolla, the latter of separate petals. (See also the exceptions to Monopetalous, p. 1005, and Apetalous, p. 1012, Divisions.)

Exceptions.—In the following Orders both apetalous flowers and flowers with connate petals occur :—
Menispermeæ, Caryophylleæ, Malvaceæ, Sterculiaceæ, Tiliaceæ, Rutaceæ, Simarubeæ, Burseraceæ, Olacineæ, Celastrineæ, Saxifrageæ, Crassulaceæ, Myrtaceæ, Passifloreæ.

Apetalous flowers also occur in *Ranunculaceæ, Magnoliaceæ, Berberideæ, Sarraceniaceæ, Papaveraceæ, Cruciferæ, Canellaceæ, Bixineæ, Violarieæ, Zygophylleæ, Geraniaceæ, Rhamneæ, Sapindaceæ, Terebinthaceæ,*

Rosaceæ, Hamamelideæ, Balsamifluæ, Halorageæ, Gunneraceæ, Callitrichineæ, Rhizophoreæ, Combretaceæ, Lythrarieæ, Onagrarieæ, Samydaceæ, Loaseæ, Datisceæ, Ficoideæ, Tetragonieæ, Corneæ, Garryaceæ.

Plants with connate petals also occur in *Anonaceæ, Pittosporeæ, Polygaleæ, Portulaceæ, Tamariscineæ, Camelliaceæ, Dipterocarpeæ, Humiriaceæ, Diosmeæ, Balsamineæ, Meliaceæ, Ilicineæ, Stackhousieæ, Droseraceæ, Bruniaceæ, Napoleoneæ, Melastomaceæ, Turneraceæ, Cucurbitaceæ, Cacteæ.*

Series I. *THALAMIFLORAL.*

Sepals usually distinct and separate, free from the ovary.   Petals 1-2-∞ -seriate, hypogynous.   Stamens hypogynous, rarely inserted on a short or long torus, or on a disk. Ovary superior.

Exceptions.—Connate sepals occur in a few orders.   The calyx is adnate to the ovary, or to a fleshy torus embracing the ovary, in *Pæonia* of *Ranunculaceæ, Calycanthaceæ,* and in a few *Anonaceæ, Nymphæaceæ, Portulaceæ, Capparideæ, Bixineæ, Polygaleæ, Camelliaceæ, Vochysiaceæ, Tiliaceæ,* and *Dipterocarpeæ;* petaloid sepals occur in apetalous *Ranunculaceæ,* in *Berberideæ, Tiliaceæ,* and others; the stamens are manifestly perigynous in a few *Dilleniaceæ, Papaveraceæ, Capparideæ, Moringeæ, Resedaceæ, Violarieæ, Caryophylleæ, Portulaceæ, Malvaceæ,* and *Sterculiaceæ.*

Cohort I. Ranales.—Stamens very rarely definite.   Carpels free or immersed in the torus, very rarely connate.   Micropyle usually inferior.   Embryo minute in a fleshy albumen.

Cohort II. Parietales.—Stamens ∞ or definite.   Carpels connate into a 1-celled ovary with parietal placentas, rarely spuriously 2- or more- celled by the prolongation of the placentas.   (Carpels free in a few *Papaveraceæ* and *Resedaceæ;* ovary regularly 3- or more- celled in some *Sarraceniaceæ, Papaveraceæ, Capparideæ* and *Bixineæ.*)

Stamens ∞, free. Ovary 1-celled; placentas parietal.—Herbs, rarely shrubs, with alternate leaves and milky juice 214

13. FUMARIACEÆ. Flowers irregular. Stamens 6, in two bundles, rarely 4, free. Ovary 1-celled; placentas parietal.—Herbs, erect or climbing; leaves alternate, usually multifid 218

** *Embryo large, curved (except in Moringeæ); albumen* 0.

14. CRUCIFERÆ. Sepals and petals 4 each. Stamens 6, 4 longer. Capsule usually. spuriously 2-celled, or 2-multi-locellate.—Herbs, very rarely shrubs; leaves alternate . . 221

15. CAPPARIDEÆ. Sepals 4-8. Petals 4-8 or 0. Stamens 4 or ∞, often perigynous. Ovary 1-celled, usually stalked. Fruit a capsule or berry.—Shrubs, trees or herbs; leaves alternate . . . . '. . . . 232

? 16. MORINGEÆ. Sepals and petals 5. Stamens 8-10, perigynous. Capsule elongate, 3-valved. — Trees; leaves alternate, compound 235

17. RESEDACEÆ. Calyx 4-8-partite. Petals 4, 8, 2, 0, often cut. Stamens 3-40. Fruit various.—Herbs or shrubs; leaves alternate 236

*** *Embryo large, in fleshy albumen.*

18. CISTINEÆ. Flowers regular. Sepals many. Ovules orthotropous. Embryo curved.—Leaves opposite or alternate, usually stipulate 238

19. VIOLARIEÆ. Flowers usually irregular. Sepals 5. Stamens 5; anthers usually curved; connective usually dilated above the cells. Fruit usually a capsule. Embryo straight.—Leaves opposite or alternate, stipulate 240

20. CANELLACEÆ. Flowers regular. Sepals 4-5. Petals 4-5-0. Stamens 20-30, on the outside of a tube formed by the filaments. Berry 2-∞ -seeded.—Aromatic trees; leaves exstipulate 243

21. BIXINEÆ. Sepals 2-6. Petals various or 0. Stamens ∞, free, inserted on a disciform torus.—Shrubs or trees; leaves stipulate or not. (See also 102, *Samydaceæ*) . 243

Cohort III. Polygalales.—Sepals and petals 5 each,. rarely 4 or 3. Stamens as many or twice as many as the petals. Ovary 2- (rarely 1- or more-) celled. Albumen fleshy, rarely absent.—Herbs or shrubs; leaves exstipulate. (Petals and stamens 1-3 or more in *Vochysiaceæ*:)

22. PITTOSPOREÆ. Flowers regular. Sepals and petals 5 each. Stamens 5, free.— Trees or erect or twining shrubs; leaves alternate, simple, exstipulate . . 247

23. POLYGALEÆ. Flowers irregular. Sepals 5. Petals 3-5. Stamens usually 8, monadelphous. — Herbs, shrubs or trees; leaves usually alternate, simple, exstipulate . 249

23b. TREMANDREÆ. Flowers regular. Sepals and petals 3, 4, 5 each. Stamens twice

as many, free.—Small shrubs, often Heath-like; leaves alternate, opposite or whorled, exstipulate 249

24. VOCHYSIACEÆ. Sepals 5, sometimes adnate to the ovary, unequal, posterior spurred or gibbous. Petals 1, 3, or 5, hypogynous or perigynous. Stamens several, usually one fertile. — Shrubs or trees; leaves usually opposite or whorled, simple; stipules 0 or small 251

Cohort IV. Caryophyllales.—Flowers regular. Sepals 2-5, rarely 6. Petals usually as many. Stamens as many or twice as many, rarely more or fewer. Ovary 1-celled, or imperfectly 2-5-celled; placenta central, free, rarely parietal. Embryo usually curved, and in a floury albumen. (Petals connate in some *Portulaceæ* and *Tamariscineæ*.) This Cohort is closely allied to XIV., *Ficoidales*, and XXVII., *Chenopodiales*.

25. FRANKENIACEÆ. Sepals connate. Petals 4-5, long-clawed. Stamens as many. Placentas 2-4, parietal. Albumen fleshy; embryo straight. — Herbs or undershrubs; leaves small, opposite, simple, exstipulate . 252

26. CARYOPHYLLEÆ. Sepals 3-5, free or connate. Petals 3-5, clawed or not. Stamens 3-5 or 6-10. Placenta usually central and free.—Herbs, rarely shrubby; leaves opposite, simple, stipulate or not . . 254

Cohort V. Guttiferales.—Flowers regular. Sepals and petals each usually 4-5, imbricate in bud. Stamens usually ∞. Ovary 3-∞-celled, rarely 2-celled, or of one carpel; placentas on the inner angles of the cells.

Cohort VI. Malvales.—Flowers rarely irregular. Sepals 5, rarely 2-4, free or connate, valvate or imbricate. Petals as many or 0. Stamens usually ∞ and monadelphous. Ovary 3-∞-celled (rarely of 1 carpel); ovules in the inner angles of the cells.—Shrubs, rarely trees; leaves alternate, usually stipulate, simple or compound.

SERIES II. *DISCIFLORAL.*

Sepals distinct or connate, free or adnate to the ovary. Disk usually conspicuous, as a ring or cushion, or spread over the base of the calyx-tube, or confluent with the base of the ovary, or broken up into glands. Stamens usually definite, inserted upon or at the outer or inner base of the disk. Ovary superior.

Exceptions.—Disk inconspicuous or flat in *Malpighiaceæ*; often reduced to minute glands or absent in *Lineæ, Oxalideæ, Geraniaceæ, Simarubeæ, Celastrineæ, Coriarieæ, Olacineæ, Ilicineæ*, and a few other Orders. Stamens sometimes perigynous in *Chailletiaceæ, Rhamneæ, Anacardiaceæ, Celastrineæ, Stackhousieæ, Burseraceæ*; indefinite in a few genera of various Orders. Ovary more or less inferior in some *Olacineæ, Rhamneæ*, and *Anacardiaceæ*.

Cohort VII. Geraniales.—Flowers often irregular. Disk usually annular, adnate to the stamens, or reduced to glands, rarely 0. Ovary of several carpels, syncarpous or subapocarpous ; ovules 1–2, rarely ∞, ascending or pendulous ; raphe usually ventral.

Cohort VIII. Olacales.—Flowers regular, ☿ or unisexual. Calyx small. Disk free, cupular or annular, rarely glandular or 0. Ovary entire, 1-∞ -celled; ovules 1-3 in each cell, pendulous; raphe dorsal, integuments confluent with the nucleus. Albumen usually copious, fleshy; embryo small.—Shrubs or trees; leaves alternate, simple, exstipulate. (This Cohort is closely allied to XXXVII., *Santalales*.)

Cohort IX. Celastrales.—Flowers ☿, regular; corolla hypogynous or perigynous. Disk tumid, adnate to the base of the calyx-tube or lining it. Stamens as many as the petals or fewer, rarely twice as many, perigynous or inserted outside the disk or on its edge. Ovary usually entire; ovules 1-2 in each cell, erect; raphe ventral.—Leaves undivided, except in *Ampelideæ* and *Staphyleaceæ*.

PAGE

Ovary lobed.—Herbs; leaves alternate, simple, narrow; stipules minute or 0 . . 346

66. RHAMNEÆ. Calyx-lobes valvate. Petals imbricate, small, concave, or 0. Stamens usually perigynous, inserted on the edge of the disk, opposite the petals. Ovary free or adnate to the disk or inferior.—Trees or shrubs; leaves alternate, stipulate . . 346

PAGE

67. AMPELIDEÆ. Calyx small, lobes imbricate. Petals valvate, caducous. Stamens inserted outside the disk and opposite to the petals. Ovary free.—Shrubs, usually climbers; leaves simple or compound, stipulate or not 349

Cohort X. Sapindales.—Flowers often irregular and unisexual. Disk tumid, adnate to the base of the calyx or lining its tube. Stamens perigynous or inserted upon the disk, or between it and the ovary, usually definite. Ovary entire lobed or apocarpous; ovules 1-2 in each cell, usually ascending with a ventral raphe, or reversed, or pendulous from a basal funicle, rarely ∞ horizontal. Seed usually exalbuminous. Embryo often curved or crumpled.—Shrubs or trees; leaves usually compound.

68. SAPINDACEÆ. Flowers often irregular. Petals 4–5 or 0. Stamens fewer or more than the petals, rarely as many. Ovary often excentric. Embryo exalbuminous, usually curved or spiral.—Leaves usually alternate, simple or compound, stipulate or not . . 351

69. ACERINEÆ. Flowers regular. Petals 4–5 or 0. Stamens as many as the petals or more. Ovary central. Embryo albuminous; cotyledons folded or convolute.—Leaves opposite, usually simple, exstipulate . 353, 354

70. HIPPOCASTANEÆ. Flowers irregular. Petals 4–5. Stamens 5–8, declinate. Embryo large, exalbuminous; cotyledons fleshy, often connate.—Trees; leaves opposite, compound, exstipulate . . . 353, 356

71. MELIANTHEÆ. Flowers irregular. Petals 4 or 5, the fifth minute. Stamens 4, inclined. Embryo small, green, in copious albumen.—Herbs or shrubs; leaves simple or compound, usually stipulate . 353, 358

72. SABIACEÆ. Flowers regular, usually ☿. Bracts, sepals, petals and stamens usually so whorled as to be more or less regularly opposite to one another. Stamens free or adnate to the petals, some often sterile. Embryo with thick or rugose or membranous and contorted cotyledons.—Trees or shrubs; leaves alternate, often simple, exstipulate . 359

73. TEREBINTHACEÆ. Flowers regular or irregular, ☿ or unisexual. Petals as many as the sepals or twice as many, or 0. Ovary 1-celled, with a 2–3-fid style, or 2–5-celled; cells 1-ovuled; ovules pendulous from a basal funicle, or attached to the inner angle of the cell. Fruit usually a drupe. Seeds exalbuminous. — Trees or shrubs; leaves various, exstipulate . . . 360

SERIES III. *CALYCIFLORAL.*

Sepals connate (rarely free), often adnate to the ovary. Petals 1-seriate, perigynous or epigynous. Disk adnate to the base of the calyx, rarely tumid or raised into a torus or gynophore. Stamens perigynous, usually inserted on or beneath the outer margin of the disk. Ovary frequently inferior.

Cohort XI. Rosales.—Flowers usually ☿, regular or irregular. Carpels 1 or more, usually quite free in bud, sometimes variously united afterwards with the calyx-tube, or enclosed in the swollen top of the peduncle; styles usually distinct.

* *Ovules inserted on the inner angle of the carpel, rarely basal or parietal.*

74. CONNARACEÆ. Flowers regular. Stamens definite. Carpels 1-5, free; ovules 2 in each carpel, basal, orthotropous. Fruit a follicle. Seeds often albuminous.—Trees or shrubs; leaves alternate, compound, exstipulate 364

[1] This division should perhaps form a separate Cohort, but the affinity between *Saxifrageæ* and *Hamamelideæ* is very close indeed.

88. BRUNIACEÆ. Flowers regular, ☿, in spikes or heads. Petals 4–5, perigynous. Stamens 4–5, free or adnate to the petals. Ovary free or inferior or semi-inferior, 1–3-celled; styles more or less connate; ovules pendulous, in the 1-celled ovaries 1 or 10, in the 2–3-celled 1 or 2 collateral from the inner angle. Fruit capsular or indehiscent. Seeds albuminous. — Shrubs; leaves alternate, Heath-like, exstipulate 413

89. HALORAGEÆ. Flowers regular, ☿ or unisexual. Petals 2–4, epigynous, or 0. Stamens definite. Ovary 1-4-celled; styles 1–4; ovules solitary, pendulous in each cell. Fruit small, indehiscent. Seed albuminous.—Herbs or small shrubs; leaves alternate, rarely opposite or whorled, simple or much divided; flowers usually small, axillary . . 414

90. GUNNERACEÆ. Flowers ☿ or unisexual, in dense or lax spikes, ebracteate. Calyx-teeth 2–3 or 0. Petals 2, epigynous, concave or 0. Stamens 2, opposite the petals when very large; anthers basifixed. Ovary 1-celled; styles 2; ovule 1, pendulous. Fruit indehiscent. Seed albuminous; embryo minute.—Large or small herbs; leaves alternate, simple, stipulate; flowers minute.

416 (and *Gunnera*) 414

? 91. CALLITRICHINEÆ. Flowers ☿ or unisexual. Petals 0 or 2 scales. Stamens 1 or 2 at the base of the ovary; anthers basifixed. Ovary 4-celled, 4-lobed; styles 2, filiform, distinct; ovules solitary, pendulous in each cell. Fruit indehiscent, 4-lobed. Seed albuminous.—Tender water-herbs; leaves opposite, simple, exstipulate 417

Cohort XII. Myrtales.—Flowers regular or sub-regular, usually ☿. Ovary syncarpous, usually inferior; style undivided (very rarely styles free); placentas axile or apical, rarely basal.—Leaves simple, usually quite entire (rarely 3-foliolate in *Combretaceæ*).

* *Ovules pendulous from the top of the ovarian cell.*

92. RHIZOPHOREÆ. Calyx-lobes valvate. Petals perigynous or epigynous, usually toothed or cut. Stamens 2–4 times as many as the petals. Ovary 2-6-celled, superior or inferior. Albumen 0 or fleshy.—Trees or shrubs; leaves simple, opposite, stipulate, rarely alternate and exstipulate . . . 418

93. COMBRETACEÆ. Calyx-lobes valvate. Petals epigynous. Stamens usually definite. Ovary 1-celled; ovules 2–5, funicles long. Drupe 1-seeded. Embryo convolute or folded, exalbuminous.—Shrubs or trees; leaves opposite or alternate, simple, very rarely 3-foliolate, exstipulate 420

* * *Ovules usually numerous, and axile (solitary or few, and apical or basal, in some Myrtaceæ, Melastomaceæ, Onagrarieæ, and Trapeæ).*

94. MYRTACEÆ. Calyx-lobes valvate or imbricate. Petals epigynous. Stamens ∞, rarely definite. Ovary inferior, or semi-inferior, rarely free, 1-∞-celled. Fruit a capsule or berry. Seeds exalbuminous.—Trees or shrubs; leaves opposite or alternate, simple, often 3-plinerved, usually exstipulate and gland-dotted 422

95. NAPOLEONEÆ. Calyx-lobes valvate. Corolla epigynous, simple and multifid, or double, the outer plaited, the inner multifid. Stamens many; filaments united into a petaloid cup, or free. Ovary 5-6-celled; stigma discoid; ovules several in each cell.—Shrubs; leaves alternate, simple, exstipulate . . 426

96. MELASTOMACEÆ. Calyx-lobes usually imbricate. Petals epigynous or perigynous, contorted in bud. Stamens 3–12, usually declinate and of 2 forms; anthers usually opening by pores, inflexed in bud. Ovary free or adnate to the calyx-tube, 1-2-∞-celled. Fruit capsular or berried. Seeds numerous, minute, rarely few large, exalbuminous. — Shrubs or trees, rarely herbs; leaves opposite or whorled, simple, exstipulate, usually 3-5-plinerved 428

97. LYTHRARIEÆ. Calyx-lobes valvate. Petals perigynous, rarely epigynous, usually crumpled in bud. Stamens usually definite. Ovary free, rarely adnate to the calyx-tube, 2-∞-celled; cells ∞-ovuled. Seeds small, exalbuminous. — Herbs trees or shrubs; leaves usually opposite, simple, exstipulate . 432

98. OLINIEÆ. Calyx 4-5-toothed. Petals epigynous, with sometimes interposed scales. Stamens 4-5-∞; filaments flexuous in bud. Ovary 2-4-5-celled; cells 2-3-∞-ovuled.

Berry or drupe 3–4-celled. Embryo curved or spiral, exalbuminous. — Shrubs; leaves opposite, simple, exstipulate, not gland-dotted 434

99. GRANATEÆ. Calyx-lobes valvate. Petals 5–7, epigynous, imbricate in bud. Stamens ∞, many-seriate. Ovary with 2 superimposed tiers of cells, upper tier with parietal, lower with central placentation. Berry traversed by membranous septa. Seeds with fleshy testa, exalbuminous; cotyledons convolute. — A shrub; leaves sub-opposite, simple, exstipulate, not gland-dotted . . 435

100. ONAGRARIEÆ. Calyx-lobes valvate.

Petals 2–4, epigynous, contorted. Stamens 2, 4, or 8. Ovary inferior, 2-4- rarely 1-celled; cells 1-∞-ovuled. Embryo exalbuminous, straight. — Herbs shrubs or trees; leaves opposite or alternate, simple, exstipulate 436

101. TRAPEÆ. Calyx-lobes valvate. Petals 4, epigynous. Stamens 4. Ovary semi-inferior, 2-celled; cells 1-ovuled; stigma flattened. Fruit indehiscent, 2–4-spined or horned, 1-seeded. Embryo exalbuminous; cotyledons very unequal, one very large, the other minute. — Floating herbs; leaves rosulate, simple, exstipulate 439

Cohort XIII. Passiflorales.—Flowers usually regular, ☿ or unisexual. Ovary usually inferior, syncarpous, 1-celled; placentas parietal; sometimes 3- or more- celled by the produced placentas; styles free or connate.

* *Flowers ☿, except some Samydaceæ and Passifloreæ.*

102. SAMYDACEÆ. Petals like the sepals or 0. Stamens definite, alternating with scales, or ∞. Ovary inferior or superior, and inserted by a broad base; style rarely 3–5-fid. Embryo straight, albuminous.—Trees or shrubs; leaves usually alternate, simple, stipulate or 0. (See also 21 *Bixineæ*) . . 441

103. LOASEÆ. Petals epigynous. Stamens ∞, rarely definite, many usually imperfect. Ovary inferior; ovules many. Capsule many-seeded. Embryo albuminous.—Herbs, often clothed with hooked or stinging hairs; leaves opposite or alternate, simple lobed or pinnatifid 442

104. TURNERACEÆ. Calyx-lobes imbricate. Petals 5, perigynous, contorted in bud. Stamens 5, sub-hypogynous. Ovary free; styles 3. Capsule 3-valved. Embryo albuminous. — Herbs or shrubs; leaves alternate, simple or pinnatifid; stipules minute or 0 445

105. PASSIFLOREÆ. Petals perigynous, often like the petals, or 0. Corona at the base of the petals single, treble, or absent, or reduced to scales. Ovary free; style simple, or styles 3–5. Capsule or berry usually stipitate. Embryo exalbuminous.—Shrubs,

usually scandent; leaves alternate, simple or compound, stipulate or not, with or without lateral tendrils 446

* * *Flowers unisexual.*

106. CUCURBITACEÆ. Petals epigynous, usually confluent with the calyx. Stamens 3, rarely 5; anthers extrorse. Corona 0. Ovary inferior; placentas produced to the axis and revolute. Seeds exalbuminous.— Herbs or undershrubs, climbing or trailing; leaves alternate, often lobed, rarely compound, exstipulate; tendrils 0 or lateral . 440

107. BEGONIACEÆ. Perianth-segments 2-∞, often 4, epigynous, outer sepaloid or all petaloid, rarely perianth tubular. Stamens ∞, rarely definite; anthers adnate. Ovary usually 3-angled, 3-celled; styles often 2-fid; placentas axile, many-ovuled. Capsule or berry many-seeded. Seeds minute, albumen scanty or 0. — Herbs, succulent; leaves alternate, often unequal-sided, entire toothed lobed or digitate, stipulate 453

108. DATISCEÆ. Perianth-segments small, epigynous. Stamens 4-∞; anthers dorsifixed. Ovary usually gaping at the top. Capsule membranous, many-seeded. Albumen scanty. —Herbs or trees; leaves alternate, simple or pinnate, exstipulate 453

Cohort XIV. Ficoidales.—Flowers regular or sub-regular. Ovary syncarpous, inferior semi-inferior or superior, 1-celled with parietal placentas, or 2-∞-celled with basilar or axile placentas. Embryo albuminous and curved, or cyclical, or exalbuminous and oblique. (This Cohort is allied to IV. *Caryophyllineæ*, and XXVII., *Chenopodiales*.)

100. Cacteæ. Flowers ☿. Sepals petals and stamens usually ∞. Ovary inferior, 1-celled, placentas parietal; stigmas usually radiating.—Spinous and leafless (rarely leafy) succulent plants, stem often fleshy, angled or ribbed 457

110. Ficoideæ. Flowers ☿ or unisexual. Calyx-lobes 4-5. Petals ∞, or small or 0. Stamens few or ∞. Ovary 2-∞-celled; styles free or connate. — Herbs or undershrubs; leaves quite entire 461

111. Mesembryantemeæ. Flowers ☿. Calyx superior, 2-8-parted. Petals and stamens ∞, ∞-seriate, epigynous. Ovary in-ferior, several-celled, many-seeded; embryo curved, in floury albumen.—Succulent herbs or undershrubs; leaves opposite or alternate, simple, often 3-gonous or cylindric, exstipulate 461, 462

112. Tetragonieæ. Flowers ☿ or unisexual. Calyx-lobes induplicate-valvate. Stamens 1-5-0, epigynous. Ovary inferior, 1-9-celled; styles 1-9; ovules 1, pendulous in each cell. Drupe with the accrescent calyx. Embryo annular, in floury albumen.—Herbs or shrubs; leaves sub-opposite or alternate, simple, succulent, exstipulate . . 461, 464

Cohort XV. Umbellales.—Flowers regular, usually ☿. Stamens usually definite. Ovary inferior, 1-2-∞-celled; ovules solitary, pendulous in each cell from its top; styles free or connate at the base; ovules with the coats confluent with the nucleus. Seeds albuminous; embryo usually minute.

113. Umbelliferæ. Flowers ☿. Petals imbricate, rarely valvate. Ovary 2-celled; styles 2. Fruit of dry indehiscent mericarps.—Herbs; stems often fistular; leaves alternate, usually much dissected, exstipulate; flowers umbelled, rarely capitate . 465

114. Araliaceæ. Flowers ☿. Petals valvate, rarely imbricate. Ovary 1-∞-celled; styles 2-∞, or stigmas sessile. Fruit usually a berry or drupe. Seed with a ventral raphe.—Trees or shrubs, rarely herbs; leaves alternate, rarely opposite, usually compound, often stipulate; flowers usually umbelled 471

115. Corneæ. Flowers ☿ or unisexual. Petals valvate or imbricate. Ovary 1-∞-celled; style 1. Fruit a drupe. Seed with a dorsal raphe. — Shrubs or trees; leaves usually opposite, simple, exstipulate; flowers capitate or corymbose 475

116. Garryaceæ. Flowers unisexual. Sepals of ♂ 4, valvate in bud; of ♀ obsolete or 2-lobed. Stamens 4; anthers basifixed. Ovary 1-celled; styles 2; ovules 2, collateral, pendulous. Berry 1-2-seeded. — Shrubs; leaves opposite, simple, exstipulate; flowers in drooping catkins 477

DIVISION II. MONOPETALOUS.

Flowers furnished with both sepals and petals, the latter connate.

Exceptions.—Apetalous and polypetalous species occur in the orders *Primulaceæ, Oleineæ,* and *Plantagineæ*; and polypetalous ones also occur in *Lobeliaceæ, Ericaceæ, Monotropeæ, Pyrolaceæ, Plumbagineæ, Myrsineæ, Sapoteæ, Cyrilleæ, Styraceæ, Ebenaceæ,* and *Jasmineæ.* (See also the exceptions to Polypetalæ, p. 995.)

Series I. *EPIGYNOUS.*

Ovary inferior (superior in some *Goodeniaceæ* and in *Brunoniaceæ*). (See also 131, *Vaccinieæ,* and 164, *Gesneraceæ.*)

Cohort XVI. Caprifoliales.—Flowers regular or irregular. Stamens as many as the corolla-lobes, inserted on the corolla. Ovary inferior, 2-∞-celled; cells 1-∞-ovuled. Seeds albuminous or very rarely exalbuminous.—Shrubs or trees, rarely herbs; leaves opposite or whorled, often stipulate; calyx never pappose.

Cohort XVII. Asterales.—Flowers regular or irregular; if unisexual, usually collected in involucrate heads. Stamens as many as the corolla-lobes, rarely fewer, inserted on the corolla. Ovary inferior, 1-celled, 1-ovuled (or if 2-3-celled, with one cell only ovuliferous). —Herbs or shrubs, rarely trees; leaves exstipulate; limb of the calyx usually reduced to a pappus or 0.

Cohort XVIII. Campanales.—Flowers most often irregular, rarely unisexual or collected into involucrate heads. Stamens as many as the corolla-lobes or fewer. Ovary 2-6-celled, rarely 1-celled; style simple, stigma often indusiate; ovules numerous in the cells, rarely solitary.

lobes valvate. Stamens epigynous or epipetalous, cohering in a tube. Ovary 1-3-
celled; stigma not indusiate; cells many-
ovuled. Fruit a capsule or berry. Seeds

albuminous.—Herbs, rarely shrubby, usually
milky; leaves alternate or radical, exstipulate 512

SERIES II. *HYPOGYNOUS* or *PERIGYNOUS.*

Ovary superior, inferior in *Vacciniæ* and some *Primulaceæ, Myrsineæ, Styraceæ,* and *Gesneraceæ.* (See also exceptions to Series I.)

* Flowers usually regular.

Cohort XIX. Ericales.—Corolla hypogynous. Stamens as many or twice as many as the corolla-lobes, epipetalous or hypogynous. Ovary 1-∞ -celled; cells 1-∞ -ovuled; stigma simple, entire or lobed. Seeds minute.

128. ERICINEÆ. Corolla hypogynous,
mono- rarely poly-petalous; lobes 4-5, contorted or imbricate. Stamens 8-10, rarely
4-5; anthers 2-celled, opening by terminal
pores, rarely by slits. Disk glandular. Fruit
a capsule berry or drupe. Seeds albuminous.
—Shrubs or trees; leaves opposite or alternate, usually evergreen, exstipulate . . 514

129. MONOTROPEÆ. Corolla hypogynous,
mono- or poly-petalous; lobes 4-5, imbricate.
Stamens 8-10, hypogynous; anthers 1-2-
celled, opening variously. Disk glandular.
Capsule loculicidally 4-5-valved. Seed with
a loose testa and undivided minute exalbuminous embryo.—Parasitic fleshy brown leafless herbs 517

130. PYROLACEÆ. Corolla hypogynous,
polypetalous or nearly so; petals 5, imbricate.
Stamens 10, hypogynous; anthers 2-celled,
opening by pores or slits. Disk glandular.
Capsule loculicidally 3-5-valved. Seeds with
a loose testa and minute undivided exalbuminous embryo. — Perennial scapigerous
herbs; leaves alternate, evergreen, exstipulate 519

131. VACCINIEÆ. Corolla epigynous, monopetalous; lobes 4-5-6, imbricate. Stamens 8-10-12, epigynous; anthers 2-celled,
opening by pores. Berry 4-10-celled, many-
seeded. Seeds albuminous.—Shrubs or trees;
leaves alternate, exstipulate 520

132. EPACRIDEÆ. Corolla hypogynous,
monopetalous; lobes 4-5, valvate or imbricate. Stamens hypogynous or epipetalous,
4-5, rarely 8-10; anthers 1-celled, opening
longitudinally. Disk cup-shaped or glandular. Ovary 2-∞ -celled; cells 1-∞ -ovuled.
Fruit a capsule or drupe. Seeds albuminous.
— Shrubs or small trees, leaves alternate,
rarely opposite, often Heath-like, sometimes
sheathing, exstipulate 522

133. DIAPENSIACEÆ. Corolla monopetalous, hypogynous; lobes 5, imbricate. Stamens 5, epipetalous, often alternating with
staminodes; anther-cells 2-valved. Disk 0.
Capsule 3-4-celled, loculicidally 3-4-valved.
Seeds albuminous, testa firm or lax.—Small
herbs; leaves alternate, crowded or scattered,
evergreen, exstipulate 524

Cohort XX. Primulales.—Corolla regular, hypogynous, rarely .epigynous, mono- rarely poly-petalous. Stamens equalling the corolla-lobes and opposite to them, or if more, one series always opposite them, hypogynous or epipetalous. Ovary 1-celled, with free basal placentation.—Herbs or shrubs, rarely trees; leaves rarely opposite, exstipulate.

134. PLUMBAGINEÆ. Corolla mono–polypetalous, hypogynous; lobes 5, contorted
or imbricate. Stamens 5, hypogynous or
epipetalous. Ovary 1-celled; styles 3-5;
ovule 1, pendulous from a basal funicle.
Fruit membranous, included in the calyx.

Embryo straight, in floury albumen.—Herbs,
sometimes shrubby below; leaves alternate,
exstipulate, base often amplexicaul . . 525

135. PRIMULACEÆ. Corolla monopetalous,
hypogynous, rarely epigynous, rarely 0;
lobes 4-5, contorted or imbricate. Stamens

4-5, opposite the corolla-lobes. Ovary rarely inferior, 1-celled ; placenta central, globose, many-ovuled. Capsule 3-5-valved. Seeds albuminous.—Herbs ; leaves radical alternate opposite or whorled   528

136. MYRSINEÆ. Corolla mono- rarely poly-petalous, hypogynous or epigynous ; lobes 4-5, usually contorted. Stamens 4-5, epipetalous, opposite the corolla-lobes or petals, sometimes alternating with staminodes. Ovary 1-celled ; placenta central, globose. Fruit a 1- or few-seeded drupe or berry, rarely a follicle. Seeds albuminous.— Trees or shrubs ; leaves usually alternate, gland-dotted, exstipulate   .   .   . 531

Cohort XXI. Ebenales. Corolla mono-poly-petalous, hypogynous or epigynous, rarely perigynous. Stamens usually many more than the corolla-lobes, if equalling them alternate with them (except *Sapoteæ*). Ovary 2-∞-celled ; cells usually few-ovuled. Fruit rarely capsular.—Shrubs or trees ; leaves alternate, exstipulate.

137. SAPOTEÆ. Corolla monopetalous, hypogynous ; lobes 4-8, imbricate in 1 or 2 series. Stamens equalling the corolla-lobes and opposite them, with alternating staminodes, or many and 2-seriate ; anthers usually extrorse. Ovary many-celled ; cells 1-ovuled. Fruit a 1-∞-seeded berry. Seeds with a thick often hard shining testa, long broad hilum, scanty fleshy albumen, and large embryo.—Trees or shrubs, with milky juice ; leaves alternate, quite entire, exstipulate   . 535

138. EBENACEÆ. Corolla monopetalous, hypogynous ; lobes 3-7, contorted-imbricate. Stamens hypogynous or epipetalous,2-4 times as many as the corolla-lobes, or if equal alternating with them, or ∞ ; anthers introrse. Ovary 3-∞-celled ; ovules 1-2 in each cell, pendulous. Berry globose or ovoid, few-seeded. Embryo short, in copious cartilaginous albumen.—Trees or shrubs ; leaves alternate, coriaceous, exstipulate, quite entire 537

139. CYRILLEÆ. Corolla sub-polypetalous, hypogynous ; petals 5, contorted or convolute. Stamens 5 or 10, hypogynous ; anthers introrse. Disk 0. Ovary 2-4-celled ; ovules 1-∞ in each cell, pendulous. Fruit a capsule or drupe. Embryo straight, in fleshy albumen.—Shrubs ; leaves alternate, membranous, quite entire, exstipulate   . 539

140. STYRACEÆ. Corolla mono-poly-petalous, perigynous or epigynous ; lobes or petals 5-7, imbricate valvate or contorted. Stamens inserted at the base of the corolla, 5-∞ , 1-∞ -seriate, free, scattered or fascicled or 1-∞ -adelphous ; anthers introrse. Ovary 2-5-celled ; cells 1-∞ -ovuled ; ovules various in insertion and direction, upper or all pendulous. Fruit fleshy or hard, 1-celled, few-seeded. Embryo straight, in fleshy albumen.—Shrubs or trees ; leaves alternate, exstipulate   541

Cohort XXII. Gentianales.—Corolla mono- rarely sub-poly-petalous, hypogynous Stamens equalling the corolla-lobes or fewer, always inserted on the corolla, and usually included in its tube. Ovary usually syncarpous and 2-celled.—Herbs shrubs or trees ; leaves very rarely alternate, or stipulate.

141. JASMINEÆ. Corolla-lobes 5-8, imbricate. Stamens 2. Ovary 2-celled ; cells 1-2-ovuled. Fruit a berry or capsule. Seeds exalbuminous.—Shrubs, often climbing, or trees ; leaves opposite or alternate, simple or 1-7-foliolate, exstipulate   .   .   . 543

142. OLEINEÆ. Corolla-lobes 4 (or corolla 0), valvate. Stamens 2, inserted on the corolla or hypogynous. Ovary 2-celled ; cells 2-3-ovuled. Fruit various. Seed albuminous.—Shrubs or trees ; leaves opposite, simple or compound, exstipulate   .   . 545

143. SALVADORACEÆ. Corolla-lobes 4, imbricate. Stamens 4, alternate with the corolla-lobes. Disk 4-lobed. Ovary 2-celled ; ovules geminate in each cell. Fruit indehiscent.—Shrubs or trees ; leaves opposite ; stipules minute   .   .   . 547

144. APOCYNEÆ. Corolla-lobes contorted or valvate. Stamens as many as the corolla-lobes ; anthers often sagittate ; pollen granular. Carpels 2, distinct or cohering ; style

1, often dilated below the usually 2-fid stigma; ovules many. Fruit and seeds various.—Trees or shrubs, rarely herbs, often climbing; juice milky; leaves opposite or whorled (rarely alternate); stipules 0 or of cilia 548

145. ASCLEPIADEÆ. Corolla-lobes 5, usually contorted. Stamens usually cohering in a crown, and clasping the top of the short style; anther-cells 2–4; pollen agglutinated in waxy masses, those of the adjacent anthers often united in pairs. Ovary of 2 free carpels; stigmas 2, connate; ovules many. Fruit of 1 or 2 follicles. Seeds often comose. —Trees or shrubs, often climbing; leaves opposite, rarely whorled alternate or obsolete, exstipulate 551

146. LOGANIACEÆ. Corolla-lobes 5–10, valvate imbricate or contorted. Stamens as many as the corolla-lobes, and opposite them when 10, alternate when 5. Ovary 2–4-celled; cells 1–∞-ovuled. Fruit various. Seed albuminous.—Trees or shrubs; leaves opposite, stipulate or with dilated petioles . 555

147. GENTIANEÆ. Corolla - lobes contorted valvate or induplicate. Stamens as many as the corolla-lobes, alternate with them. Ovary 1- rarely sub-2-celled; placentas parietal; ovules many, horizontal. Capsule septicidal. Seeds small, albuminous. —Glabrous herbs, rarely shrubby below; leaves opposite or whorled, quite entire, exstipulate (alternate and 3-foliolate in *Menyantheæ*) 558

Cohort XXIII. Pelemoniales.—Corolla hypogynous, monopetalous, regular. Stamens as many as the corolla-lobes, and inserted on the tube; filaments usually exserted. Ovary 1–5-celled, syncarpous (except *Dichondreæ* and *Nolaneæ*); cells 1–2- very rarely ∞ -ovuled. Embryo albuminous.—Herbs, rarely shrubby below; leaves alternate or 0, exstipulate.

- 148. HYDROPHYLLEÆ. Corolla on a hypogynous disk; lobes 5, imbricate. Stamens 5, inserted at the base of the corolla-tube; filaments slender, exserted, often appendaged at the base. Ovary 1-2-celled; style slender, 2-fid; ovules 2 or more on each of 2 placentas. Capsule 2-valved; embryo straight, in cartilaginous albumen.— Herbs; juice watery 560

149. HYDROLEACEÆ. Corolla hypogynous; lobes 5, imbricate. Stamens inserted on the corolla-tube; filaments slender, exserted. Disk 0. Ovary 1-2-celled; styles 2, distinct; ovules many. Capsule 2-valved. Seeds minute; embryo straight in scanty albumen. —Herbs 561

150. POLEMONIACEÆ. Corolla hypogynous; lobes 5, contorted. Stamens inserted on the middle or top of the corolla-tube; filaments exserted, sometimes unequal. Disk glandular. Ovary 3-5-celled; style 3-5-fid; ovules solitary and erect, or many 2-seriate and ascending. Capsule 3-5-valved. Embryo straight, in fleshy albumen.—Herbs, rarely undershrubs; leaves alternate, or the lower opposite 563

151. CONVOLVULACEÆ. Corolla-lobes 5, contorted. Stamens inserted at the base of

the corolla-tube; filaments included or exserted, equal or unequal. Disk usually annular. Ovary 1–4-celled, syncarpous; style usually slender; ovules 1-2 in each cell, basal, erect. Capsule (rarely a berry) 1-4-celled. Embryo curved, with crumpled cotyledons and scanty mucilaginous albumen. —Herbs, usually twining, rarely shrubs; juice often milky 565

152. DICHONDREÆ. Corolla-lobes 5, valvate. Stamens inserted on the corolla. Ovary apocarpous; carpels 2-4, 2-ovuled; styles basal; ovules erect. Utricles 2, 1-seeded. Embryo curved; cotyledons crumpled, in scanty albumen.—Small herbs, erect or prostrate 567

153. CUSCUTEÆ. Corolla-lobes 4-5. Stamens inserted on the corolla-tube, with often as many fimbriate scales below their insertion, included. Disk 0. Ovary 2-celled; styles 2, free or connate; ovules 2, erect in each cell. Capsule 2-celled, circumscissa at the base. Embryo spiral, in copious fleshy albumen.—Leafless parasitic filiform twining herbs 508

154. BORRAGINEÆ. Corolla-lobes 4–5, imbricate. Stamens often conniving into a tube around the style, sometimes cohering

at the tips. Disk annular or cupular. Ovary of 2 bipartite 2-celled carpels; style gyno-basic; stigma simple; ovules pendulous. Fruit of 4 nucules or a 2-4 pyrened drupe. Embryo usually exalbuminous; cotyledons flat.—Herbs, rarely shrubs, usually hispid or prickly; leaves quite entire; inflorescence scorpioid 509

155. CORDIACEÆ. Corolla-lobes 4-5, con-torted. Stamens 5; filaments usually long, exserted; anthers distant. Disk annular or capsular. Ovary 4-8-celled; style forked or twice forked at the top; ovules pendulous or erect. Drupe with a 1- or 4-8-celled stone. Embryo exalbuminous; cotyledons longitu-dinally folded.—Trees or shrubs; leaves scabrous 573

156. NOLANEÆ. Corolla-lobes 5-10, folded. Stamens 5; filaments slender, ex-serted. Disk fleshy. Carpels numerous, distinct, crowded on the receptacle, 1-celled, 1-ovuled. Style simple; stigma capitate; ovule erect. Drupes distinct; embryo curved in fleshy albumen.—Herbs; leaves geminate, quite entire; peduncles solitary, axillary 574, 580

Cohort XXIV. Solanales.—Corolla monopetalous, hypogynous, regular or oblique. Stamens as many as the corolla-lobes, epipetalous, equal or unequal. Ovary 2-celled, syn-carpous; cells very numerous, very many-ovuled.—Herbs, rarely shrubs or trees; leaves alternate or geminate, rarely opposite, exstipulate.

157. SOLANEÆ. Corolla-lobes 4 or 5, in-duplicate or contorted. Stamens 4-5, in-cluded or exserted; filaments short or long. Disk annular, cupular or 0. Ovary 2-celled, placentas on the septum; style simple; ovules very numerous. Fruit a 2-celled capsule or berry. Embryo curved or annular in fleshy albumen. — Herbs, rarely shrubs; upper leaves often geminate 575

158. CESTRINEÆ. Corolla-lobes 4-5, in-duplicate. Stamens 4-5, included. Disk annular or cupular. Ovary 2-celled; pla-centas on the septum; style simple; stigma capitate; ovules few or many. Fruit a capsule or berry. Embryo straight, albu-minous.—Shrubs or trees 582

** Flowers very irregular, rarely regular.

Cohort XXV. Personales. — Corolla monopetalous, hypogynous, often 2-labiate. Stamens fewer than the corolla-lobes, rarely as many, unequal, most' often 4 didynamous, or 2. Ovary 1-2- very rarely 4-celled; style simple; stigmas 1-2; ovules usually very numerous. Fruit usually capsular.—Herbs, rarely shrubs or trees; leaves exstipulate.

159. SCROPHULARINEÆ. Corolla often 2-lipped; lobes imbricate or folded. Stamens 4 didynamous, or 2. Ovary 2-celled; pla-centas on the septum; style simple; stigmas 1-2; ovules definite or ∞. Fruit a capsule, rarely a berry. Embryo straight or curved, albuminous.—Herbs, very rarely shrubs or trees; leaves opposite, alternate or whorled, simple 583

160. VERBASCEÆ. Corolla sub-regular; lobes 5, imbricate. Stamens 5, unequal, exserted; anthers sub-1-celled. Ovary 2-celled; placentas on the septum, many-ovuled. Capsule 2-valved. Embryo straight, albuminous.—Herbs; leaves alternate . . 588

161. UTRICULARIEÆ. Corolla 2-lipped; lobes imbricate. Stamens 2, included; anthers 1-celled. Ovary 1-celled; placenta globose, basal, many-ovuled; stigma 2-lipped. Fruit capsular. Embryo straight, undivided, exalbuminous. — Scapigerous herbs, often floating; leaves radical, entire or capillary and multifid 589

162. OROBANCHEÆ. Corolla 2-lipped; lobes imbricate. Stamens 4 didynamous. Disk fleshy. Ovary 1- rarely 2-celled; placentas parietal, many-ovuled. Capsule 2-valved. Embryo minute, in fleshy albumen. —Parasitic leafless herbs 591

163. COLUMELLIACEÆ. See 123 a . 593

164. GESNERACEÆ. Corolla more or less 2-lipped, hypogynous perigynous or epigy-nous; lobes 5, imbricate. Stamens usually 4 didynamous; anthers often cohering.

Disk annular or unilateral. Ovary 1-celled, superior; placentas 2, parietal, many-ovuled. Fruit a berry or capsule. Seeds minute. Embryo straight, albuminous or not.—Herbs, rarely shrubs; leaves usually opposite or whorled, exstipulate 595

165. RAMONDIEÆ. Corolla nearly regular, rotate; lobes 5, imbricate. Stamens 5, sub-equal; anthers with short terminal slits. Ovary 1-celled; placentas parietal, many-ovuled. Capsule septicidally 2-valved. Seeds minute, hispidulous. Embryo straight, albuminous—A scapigerous herb . . . 599

166. BIGNONIACEÆ. Corolla usually 2-lipped; lobes 5, imbricate. Stamens usually 4 didynamous, or 2. Disk glandular. Ovary 1-celled with parietal placentas, or 2-celled with placentas along the edges of the septum; ovules ∞. Capsule 1-2-valved. Seeds large, imbricate, usually broadly winged. Embryo straight, exalbuminous; cotyledons broad, flat.—Trees or shrubs, rarely herbs, erect or climbing; leaves usually opposite, often compound and cirrhose, exstipulate . 600

167. CRESCENTIEÆ. Corolla sub-2-lipped; lobes imbricate. Stamens 4 didynamous.

Disk annular. Ovary 1- or spuriously 2-4-celled; placentas 2-4, parietal, many-ovuled. Fruit woody or fleshy, indehiscent. Seeds large, buried in pulp; embryo exalbuminous; cotyledons fleshy.—Trees or shrubs; leaves alternate opposite or whorled, exstipulate . 603

168. ACANTHACEÆ. Corolla usually 2-lipped; lobes imbricate or twisted. Stamens 4 didynamous, or 2. Disk cupular or annular. Ovary 2-celled; cells 2-∞-ovuled; placentas usually on the septum; ovules often inserted on processes of the placenta. Capsule 2-valved. Embryo exalbuminous; cotyledons large, sometimes crumpled.—Herbs, rarely shrubs; nodes tumid; leaves opposite or whorled, exstipulate 604

169. SESAMEÆ. Corolla 2-lipped; lobes imbricate. Stamens 4 didynamous, or 2; anther-cells shorter than the connective; tip glandular. Disk annular or cupular. Ovary 1-4-celled; stigma 2-lamellate; placentas axile or parietal; ovules few or many. Fruit a capsule or drupe, often of remarkable form. Embryo straight, exalbuminous or nearly so.—Herbs with vesicular glands; leaves opposite or alternate, exstipulate . 607

Cohort XXVI. Lamiales.—Corolla usually 2-lipped, rarely sub-regular or quite regular, hypogynous. Stamens fewer than the corolla-lobes, rarely as many, unequal, most often 4 didynamous, or 2. Ovary 2-4-celled; style simple; ovules solitary in the cells (rarely 2 or more in some *Myoporineæ* and *Verbenaceæ*). Fruit an indehiscent 2-4-celled drupe or of 2-4 nucules.—Leaves exstipulate.

170. MYOPORINEÆ. Corolla sub-regular or 2-lipped; lobes 5, imbricate. Stamens 4, sub-equal or didynamous; anther-cells confluent. Ovary of 2 carpels, usually 4-celled, with 2 ovules in each cell, separated by a septum. Fruit a 2-4-celled drupe. Embryo cylindric in scanty albumen.—Shrubs or undershrubs, leaves alternate, often gland-dotted 610

171. SELAGINEÆ. Corolla sub-regular or 2-labiate; lobes 4-5, imbricate. Stamens 4 sub-equal, or 2; anthers 1-celled. Ovary 2-celled; stigma simple; ovules 1, pendulous in each cell. Fruit of 1 or 2 unequal achenes or utricles. Embryo straight in fleshy albumen.—Herbs or undershrubs; leaves alternate fascicled or sub-opposite, narrow 611

172. STILBINEÆ. Corolla sub-regular;

lobes 5, imbricate. Stamens 4, equal; filaments slender, exserted; anthers 2-celled. Ovary dicarpellary, 1-2-celled; cells 1-ovuled or one cell empty; stigma simple; ovule 1, erect. Fruit a 1-seeded utricle, or 2-celled 4-valved loculicidal capsule. Embryo sub-cylindric in fleshy albumen.—Heath-like shrubs; leaves whorled . . . 613

173. VERBENACEÆ. Corolla regular or 1-2-lipped; lobes 4-5, imbricate. Stamens 4 didynamous, or 2, very rarely 5; anthers 2-celled. Ovary 2-4-8-celled; stigma simple or 2-fid; ovules 1 or geminate in each cell. Fruit a drupe, with 2-4 1-2-celled pyrenes; or a 1-4-celled berry. Embryo straight in scanty albumen; cotyledons foliaceous.—Herbs, shrubs, or trees; leaves opposite or whorled, rarely alternate, simple or compound 615

174. GLOBULARIEÆ. Corolla 1-2-lipped;

lobes imbricate. Stamens 4, didynamous; anther-cells confluent. Ovary 1-celled; style slender, stigma simple or 2-lobed; ovule solitary, pendulous. Caryopsis invested by the calyx, mucronate. Embryo straight, in fleshy albumen.—Herbs or shrubs; leaves alternate, quite entire; flowers in an involucrate head 618

175. LABIATÆ. Corolla 2-lipped, rarely sub-regular; lobes 2–5, imbricate. Stamens 4, didynamous, rarely sub-equal, or 2; anther-cells often confluent at the tip, sometimes separated. Disk thick. Ovary of 2 2-partite carpels; lobes 1-ovuled; style slender, simple, unequally 2-lobed, gynobasic; ovule solitary, erect in each lobe. Fruit of 4 nucules. Embryo straight, rarely curved, exalbuminous or with thin albumen.—Herbs, rarely shrubs, often strong-scented; leaves

opposite or whorled; flowers usually in false whorls 620

ANOMALOUS ORDER.

176. PLANTAGINEÆ. Flowers ☿ or unisexual. Corolla scarious, monopetalous, hypogynous, 3–4-lobed, lobes imbricate, or 0. Stamens 4, inserted on the corolla, or hypogynous; filaments very slender, exserted, persistent; anthers large, versatile, deciduous. Ovary free, with 2–4 1-8-ovuled cells, or 1-celled and 1-ovuled; style filiform with 2 lines of stigmatic papillæ. Fruit a circumsciss capsule or bony nucule. Seeds peltate; embryo straight or curved; albumen fleshy.—Scapigerous herbs, rarely shrubby below; leaves alternate or radical, rarely opposite; scapes axillary from the lower leaves; flowers inconspicuous . . 623

DIVISION III.   APETALOUS or INCOMPLETE-FLOWERED.

Flowers with a single floral envelope (the calyx), or 0.

Exceptions.—A double floral envelope occurs in some *Paronychieæ, Euphorbiaceæ, Rafflesiaceæ, Loranthaceæ, Santalaceæ,* and *Podostemaceæ.* See various incomplete genera under the exceptions to Divisions I. and II.

SUB-DIVISION I.   Ovary superior (inferior in *Cynocrambeæ* and *Gyrocarpeæ*). Perianth usually distinct.

Cohort XXVII.   Chenopodiales.—Flowers usually ☿. Perianth green or coloured, usually regular; tube short or 0; segments imbricate in bud. Ovary superior (except *Cynocrambeæ*), of 1 (rarely several) carpels. Ovules solitary (2 or more in some *Amaranthaceæ* and *Paronychieæ*), basal. Embryo usually coiled or curved.—Herbs or shrubs. (This Cohort is closely allied to Nos. IV. and XIV.)

177. NYCTAGINEÆ. Perianth petaloid, gamophyllous; tube slender; base persistent, enclosing the 1-seeded achene or utricle. Stamens hypogynous. Embryo large, convolute or conduplicate; albumen scanty.—Herbs and shrubs; nodes tumid; leaves opposite . 625

178. PHYTOLACCEÆ. Perianth green or petaloid; tube short or 0. Stamens hypogynous or nearly so. Ovary of several free or connate 1-ovuled carpels. Embryo usually curved or coiled; albumen floury or 0.—Herbs, shrubs, or trees; leaves usually alternate, stipulate or not 628

179. POLYGONEÆ. Calyx green or coloured. Stamens 1 or 5–9, perigynous; anthers 2-celled. Ovary 1-celled; styles 2–4; ovule basal. Achene usually enclosed in the often

accrescent perianth. Embryo straight or curved, sometimes foliaceous; albumen floury. —Herbs or shrubs, rarely trees; leaves alternate; stipules 0, or membranous and sheathing 631

180. AMARANTHACEÆ. Sepals 3–5, 1-seriate, more or less scarious. Stamens hypogynous, opposite the sepals; anthers often 1-celled. Ovary 1-celled; stigmas 2–3; ovules 1 or more, basal. Embryo annular or curved; albumen floury.—Herbs, rarely shrubs; leaves opposite or alternate, exstipulate 635

181. CHENOPODIEÆ. Sepals herbaceous, 3–5, 1-seriate, or perianth utricular or 0. Stamens 1–5, opposite the sepals, hypogynous or perigynous; anthers 2-celled. Ovary 1-

celled; stigmas 2–4; ovule 1, basal. Embryo curved or spiral; albumen fleshy or floury or 0.—Herbs, shrubs, or rarely trees; leaves various, exstipulate 637

182. BASELLEÆ. Sepals 5, 1–2-seriate, green or coloured. Stamens 5, perigynous, opposite the sepals; anthers 2-celled. Ovary 1-celled; stigmas usually 3 or 3-lobed; ovule 1, basal. Embryo spiral; albumen floury. — Herbs, fleshy, often trailing; leaves scattered, rarely opposite, fleshy, exstipulate . . 640

183. PARONYCHIEÆ. Sepals 4–5, green or white. Stamens 1–5, opposite the sepals, or

0, perigynous or hypogynous, sometimes alternating with petals or staminodes. Ovary 1-celled; styles 2–3; ovule 1, rarely 2, basal. Embryo curved or annular; albumen floury. —Herbs, rarely shrubby; leaves opposite or fascicled; stipules scarious or 0 . . . 642

184. CYNOCRAMBEÆ. Flowers monœcious. Perianth ♂ of 2 juxtaposed leaflets; of ♀ tubular, excentric, superior, 3 - lobulate. Stamens ∞; anthers linear, versatile. Ovary inferior; style simple; ovule 1, basal. Embryo hooked; albumen subcartilaginous. — An annual herb; leaves opposite below, stipulate 645

Cohort XXVIII. Laurales.—Flowers usually unisexual. Perianth green or coloured, usually regular. Ovary superior (inferior in *Gyrocarpeæ*), 1-celled; stigma 1; ovule solitary. Embryo straight, albuminous or not.

185. MONIMIACEÆ. Flowers usually monœcious, in a receptacular cup. Sepals 4–∞, many-seriate, imbricate. Stamens usually many, free, inserted on the cup; anthers opening by slits or valves, alternating with staminodes. Ovule pendulous, lateral or basilar. Fruit a nut or drupe. Embryo albuminous, cotyledons divaricate.—Aromatic trees or shrubs; leaves exstipulate . . 646

186. MYRISTICEÆ. Flowers diœcious. Perianth 2–4-fid, valvate. Stamens monadelphous; anthers extrorse. Ovule basal, erect. Seed with ruminate albumen; cotyledons divaricate; radicle inferior.—Aromatic

trees; leaves alternate, exstipulate. (Should be placed next to *Anonaceæ*) . . . 649

187. LAURINEÆ. Flowers ☿ or polygamous. Perianth 4–9-fid, imbricate. Stamens free; anthers extrorse or introrse, opening by valves. Ovule pendulous. Seed exalbuminous; radicle superior.—Aromatic trees; rarely scentless twining leafless herbs; leaves alternate sub-opposite or whorled, exstipulate 652

187a. GYROCARPEÆ. Flowers of *Laurineæ*, but ovary inferior and cotyledons spirally coiled round the plumule 654

Cohort XXIX. Daphnales.—Flowers usually hermaphrodite. Perianth green or coloured, regular or irregular, often tubular. Ovary 1- rarely 2-celled, superior; stigma 1; ovule usually solitary, pendulous or sub-erect. Albumen 0, rarely scanty; embryo straight.

188. THYMELEÆ. Flowers ☿ or polygamous. Perianth 4–5-fid, imbricate. Stamens inserted on the tube. Ovary 1- rarely 2-celled; style ventral or sub-terminal; ovule pendulous. Seed exalbuminous or nearly so; radicle superior.—Shrubs or trees, rarely herbs; bark tough; leaves opposite or scattered, exstipulate 656

189. HERNANDIEÆ. Flowers monœcious. Perianth 6–10-partite; segments 2-seriate, valvate. Stamens 3–4; anthers opening by a deciduous valve. Ovary 1-celled; style terminal; ovule pendulous. Seed exalbuminous; radicle superior.—Trees; leaves scattered, exstipulate 658

190. ELÆAGNEÆ. Flowers ☿ or diclinous. Perianth tubular or 4–6-fid or -partite, imbricate or valvate. Stamens perigynous or inserted on the perianth-tube, free. Ovary 1-celled; ovule sub-basal. Seed with scanty albumen; radicle inferior.—Shrubs or trees with silvery scales; leaves alternate or opposite, exstipulate 659

191. PROTEACEÆ. Flowers usually ☿. Perianth of 4 usually spathulate, free or partially free, valvate or sub-valvate segments. Anthers inserted on the perianth-segments. Ovary 1-celled; ovules 1–∞. Seed exalbuminous; radicle inferior.—Shrubs or trees; leaves rarely opposite, exstipulate . . 661

Cohort XXX. Urticales. — Flowers diclinous, ⚥ in *Ulmaceæ*. Perianth green, usua'ly regular, rarely 0. Stamens opposite the perianth-lobes or sepals. Ovary superior, 1-celled (2-celled in *Ulmaceæ*) ; stigmas 1–2 ; ovule solitary, micropyle always superior. Fruit usually an achene or samara. Embryo straight, albuminous or not.

102. URTICEÆ. Flowers diclinous or polygamous. Perianth various, imbricate or valvate, rarely 0. Stamens usually equal to the perianth-lobes; filaments uncoiling elastically. Style simple or multifid. Ovule erect, orthotropous. Embryo straight, albuminous. —Herbs, rarely trees; juice limpid; leaves stipulate 665

103. MOREÆ. Flowers diclinous, minute, often on an open or closed receptacle. Perianth tubular or 3-4-partite or 0. Stamens as in *Urticeæ*, but filaments sometimes straight. Styles 1 or 2; ovule various. Embryo straight or curved, albuminous or not. —Trees or shrubs; juice milky; leaves alternate; stipules large, fugacious . . . 669

104. CELTIDEÆ. Flowers polygamous. Perianth 5-partite, persistent, imbricate. Stamens as in *Urticeæ*. Styles 2; ovule basilar, campylotropous. Embryo curved; albumen scanty or 0. —Trees; juice watery; leaves alternate; stipules fugacious . . . 673

195. CANNABINEÆ. Flowers diclinous. Perianth ♂, sepals 5, free, imbricate; ♀, various. Stamens 4, opposite the sepals; filaments short. Styles 2; ovule pendulous, campylotropous. Embryo hooked or coiled, exalbuminous.—Herbs; juice watery; leaves stipulate 674

196. ULMACEÆ. Flowers ⚥. Perianth campanulate, persistent, 4-8-fid. Stamens opposite the perianth-lobes; anthers extrorse. Ovary 1-2-celled; styles 2; ovules 1 pendulous in each cell. Embryo straight, exalbuminous.—Trees; juice watery; leaves alternate; stipules fugacious 676

Cohort XXXI. Amentales.—Flowers diclinous, in catkins, cones or heads. Perianth 0, or calyciform, or of one or more bristles, bracts, bracteoles or scales. Ovary superior, 1-2-celled. Seeds exalbuminous.—Trees or shrubs; leaves alternate, simple.

107. BETULACEÆ. Flowers monœcious, in slender catkins. Perianth ♂ a 4-lobed cup or scale; ♀ of accrescent scales. Ovary with 2 2-ovuled cells; styles 2; ovules pendulous, anatropous. Radicle superior. —Leaves stipulate 677

198. PLATANEÆ. Flowers monœcious, in globose catkins. Perianth ♂ and ♀ of slender scales or bristles. Ovary 1-celled; style 1; ovule 1, pendulous, orthotropous. Radicle inferior. —Leaves alternate; stipules leaf-opposed, sheathing 679

199. MYRICEÆ. Flowers mon- or diœcious, in long or short catkins, sessile in the axils of scales. Perianth ♀ 0, or of scales or bracts. Ovary 1-celled; styles 2; ovule 1, erect, orthotropous. Radicle superior.— Leaves alternate, stipulate or not . . . 681

200. CASUARINEÆ. Flowers mon- or diœcious; ♂ in catkins, monandrous, 4-bracteate; ♀ in heads, bracteate and 2-bracteolate. Ovary 1-celled; ovules geminate, semi-anatropous. Radicle superior.—Leafless trees, with slender jointed branches; joints ending in toothed sheaths 683

201. SALICINEÆ. Flowers diœcious, in the scales of catkins; ♂ of a 1-celled ovary, with several basal anatropous ovules; stigma 2-3-lobed. Seeds comose. Radicle inferior. —Leaves stipulate 685

Cohort XXXII. Euphorbiales.—Flowers ⚥ or diclinous. Perianth various or 0. Ovary superior, 2-∞-celled; ovules 1-∞ in each cell, pendulous, anatropous. Fruit usually capsular, 1-∞-celled; cells 1-∞-seeded.

202. EUPHORBIACEÆ. Flowers diclinous. Perianth various or 0, sometimes dichlamydeous. Anthers usually globose or didymous. Ovary 3- (rarely 1-2- or ∞-) celled; ovules solitary or geminate in each cell, usually operculate. Fruit usually 3-coccous, rarely a berry. Embryo straight in copious albumen.—Herbs, shrubs or trees; leaves usually alternate and 2-stipulate 687

203. BUXINEÆ. Flowers monœcious. Se-

pals or bracts 4-12, imbricate, rarely 0. Stamens 4, opposite the sepals. Ovary 2-3-celled; ovules solitary or geminate, pendulous, not operculate. Fruit capsular or fleshy. Seed albuminous.—Trees or shrubs; leaves opposite or alternate, exstipulate . . 608

204. PENÆACEÆ. Flowers ☿. Perianth tubular, 4-lobed, valvate. Stamens 4, on the perianth. Ovary 4-celled; cells 2-4-ovuled. Capsule loculicidal. Embryo minute, albuminous.—Shrubs; leaves opposite 701

205. GEISSOLOMEÆ. Flowers ☿. Sepals 4, imbricate. Stamens 8. Ovary 4-celled. Capsule loculicidal; cells 2-ovuled. Embryo albuminous.—Shrubs; leaves opposite . . 701'

206. LACISTEMACEÆ. Flowers ☿, spiked. Sepals 0 or 4-6, petaloid, imbricate. Stamen 1, 2-fid. Ovary 1-celled; stigmas 3; ovules 3-6. Drupe with a 3-valved endocarp. Seed arillate, albuminous.—Shrubs or trees; leaves alternate, stipulate 702

Cohort XXXIII. Piperales.—Flowers ☿ or diclinous, usually in spikes or catkins. Perianth rudimentary or 0. Ovary superior, of one 1-celled 1-ovuled carpel, or of several free 2-∞ -celled.

206a. (217.) PIPERACEÆ. Flowers crowded on a spadix, very minute, ☿ or diœcious. Ovary 1-celled; ovule 1, basilar, orthotropous. Embryo albuminous, included in the amniotic sac.—Herbs or shrubs; leaves opposite alternate or whorled, exstipulate . 728

206b. (218.) SAURUREÆ. Flowers crowded on a spadix, ☿. Ovary superior or inferior, 1-celled with parietal placentas, or 3-5-celled with axile placentas; ovules ascending, 2-∞ in each cell. Embryo albuminous, imbedded in the amniotic sac.—Herbs; leaves alternate, exstipulate 731

206c. (219.) CHLORANTHACEÆ. Flowers ☿, or diclinous; ♂ spicate, ♀ cymose or panicled. Ovary 1-celled; ovule 1, pendulous, orthotropous. Embryo small, albuminous.—Shrubs; leaves opposite, stipulate . 733

206d. (220.) CERATOPHYLLEÆ. Flowers monœcious, axillary, sessile. Ovary 1-celled; ovule 1, pendulous, orthotropous. Embryo exalbuminous; plumule polyphyllous.—Submerged herbs, with whorled dissected exstipulate leaves 735

Cohort XXXIV. Nepenthales.—Flowers diœcious. Perianth 4-partite, imbricate. Stamens monadelphous. Ovary superior, 3-4 celled; ovules very many, on the septa. Capsule loculicidal. Seed scobiform, albuminous.—Scandent shrubs; leaves alternate, terminated by pitchers.

207. NEPENTHEÆ 703

SUB-DIVISION II. Ovary inferior. Perianth more or less distinct in the ♂ or ♀ or both (very obscure in some *Balanophoreæ*).

Cohort XXXV. Asarales.—Flowers ☿ or diclinous. Perianth usually coloured. Stamens epigynous in the ☿ flowers. Ovary inferior, 1-many-celled. Fruit a capsule or berry.

208. ARISTOLOCHIEÆ. Flowers ☿. Perianth regular or irregular, valvate.—Herbs or undershrubs; leaves alternate, exstipulate . 705

209. RAFFLESIACEÆ. Flowers diœcious, rarely ☿. Perianth regular, valvate or imbricate.—Leafless root-parasites . . . 700

Cohort XXXVI. Quernales.—Flowers diclinous, ♂ in catkins, ♀ solitary or in spikes. Perianth green, of ♂ lobed or reduced to a scale; of ♀ minute, 2-6-lobed or -toothed. Ovary inferior, 1-6-celled; ovule 1 basal, or 1 or more pendulous. Fruit 1-seeded. Albumen 0.—Trees; leaves simple or compound.

210. JUGLANDEÆ. Perianth of ♂ a scale or bract; of ♀ 2-4-toothed. Ovary 1-celled; ovule 1, erect, orthotropous. Fruit a drupe with a 2-valved endocarp.—Leaves pinnate, stipulate 711

211. CUPULIFERÆ. Perianth of ♂ unequally lobed; of ♀ 6-lobed. Ovary 2-6-celled. Fruit of 1-3 1-seeded nuts in an involucre.—Leaves simple, alternate, stipulate 714

212. CORYLACEÆ. Perianth of ♂ a staminiferous scale; ♀ irregularly lobed. Ovary 1-celled, with 2 ovules, pendulous from prominent placentas. Fruit a 1-seeded nut, enclosed in the accrescent foliaceous tubular bracteoles.—Leaves simple, stipulate . . 716

Cohort XXXVII. Santalales.

—Flowers ☿ or diclinous. Perianth usually conspicuous, coloured, polymorphous and valvate. Ovary 1-∞-celled; cells 1-∞-ovuled; ovules usually reduced to a naked nucleus. Fruit a 1-seeded berry or drupe.—Parasitic herbs or shrubs. This Cohort is closely allied to VIII., *Olacales*.

213. LORANTHACEÆ. Flowers ☿, rarely diclinous. Perianth often coloured, gamo- or poly-sepalous, valvate. Stamens inserted on the perianth-lobes or segments. Ovary 1-celled; ovule solitary, erect, adnate to the cell-wall. Embryo in fleshy albumen.—Shrubs, parasitic on branches; leafless, or leaves thickly coriaceous, exstipulate . . 718

214. SANTALACEÆ. Perianth and stamens of *Loranthaceæ*. Ovary 1-celled; ovules 2-∞, pendulous from a free central column. Embryo in fleshy or dry albumen.—Shrubs or herbs, parasitic on roots or branches; leaves opposite or alternate, exstipulate . . 722

215. GRUBBIACEÆ. Flowers ☿, in small axillary cones. Perianth of 4 valvate segments. Stamens 8. Ovary 2-celled at first, with one pendulous ovule in each cell. Embryo in fleshy albumen.—Shrubs; leaves opposite, ericoid, exstipulate . . . 725

216. BALANOPHOREÆ. Flowers 1-sexual, sometimes ☿ (*Cynomorium*). Perianth ♂ usually 3-lobed, valvate; ♀ various or 0. Stamens usually 3, monadelphous (1 in *Cynomorium*). Ovary 1-2-celled; ovules 1 in each cell, pendulous, often adnate to the cell-wall. Embryo undivided, in fleshy or granular albumen.—Fleshy scapigerous leafless root-parasites 726

Incomplete-flowered Orders of unknown or very dubious Affinity.

221. PODOSTEMACEÆ. Flowers ☿ or diclinous. Perianth 0 or simple. Stamens definite or indefinite, hypogynous. Ovary superior, 1-3-celled; cells many-ovuled; placentas axile or parietal; ovules anatropous. Fruit capsular. Embryo exalbuminous; radicle superior.—Aquatic herbs often resembling *Hepaticæ*, or *Musci*, or *Algæ* . . 737

Sub-class II. Gymnospermous.

Ovules produced superficially on a scale (bract or open ovary); fertilized by the direc application of the pollen to the apex of the nucleus, which the pollen-tube penetrates. —Flowers unisexual (except in *Welwitschia*).

223. CONIFERÆ. Stem branched, exogenous, with slender pith and no proper vessels (vascular tissue proper). Leaves opposite whorled or fascicled, simple; nerves simple. Perianth 0 739

223a. GNETACEÆ. Stem branched, exogenous, or tissues confused, with slender pith and proper vessels. Leaves 0 or opposite; nerves branched (except *Welwitschia*). Perianth of bracts or scales 748

224. CYCADEÆ. Stem simple, rarely branched or 0; pith large and proper vessels abundant. Leaves compound; nerves straight. Perianth 0 750

CLASS II. MONOCOTYLEDONS.

Stem without distinct layers of wood surrounding a column of pith, or a separable bark. Leaves with usually parallel venation. Floral whorls when present usually in threes or multiples of three. Embryo with one cotyledon. In germination adventitious roots usually at once proceed from the radicular end of the embryo. (Obscure rings of wood and a bark are distinguishable in some arborescent *Asparagi.* Net-veined leaves occur in *Dioscoreæ, Smilaceæ,* and a few other Orders.)

DIVISION I.

Ovary inferior (superior in some *Bromeliaceæ* and *Hæmodoraceæ*). Perianth usually distinct, 2-seriate and coloured.

* Albumen 0 ; embryo distinct.

Cohort I.   Hydrales.—Flowers usually diclinous, regular. Perianth 6-partite, 3 outer segments herbaceous, 3 inner petaloid or 0. Stamens 3 or more, epigynous or inserted on the base of the perianth-segments. Ovary 1, 3, or 6-celled. Fruit a berry. Embryo distinct, exalbuminous. Aquatic herbs.

* * Albumen floury, embryo distinct.

Cohort II.   Amomales.—Flowers usually ☿ and very irregular (regular in *Bromeliaceæ*). Perianth of 5-6 segments. Stamens 6, 1 or 5 antheriferous, the rest petaloid, all antheriferous in *Bromeliaceæ.* Ovary usually 3-celled. Fruit a berry or capsule.

* * * Albumen 0 or cellular ; embryo very obscure. Seeds very minute (except Taccaceæ*).*

Cohort III.   Orchidales. — Flowers ☿ and very irregular. Perianth of 6 rarely 3 segments. Stamens 1, 2, or 3, confluent with the style. Fruit capsular. Embryo very minute.

Cohort IV.   Taccades.—Flowers ☿, regular. Perianth 6-lobed. Stamens 3 or 6, inserted on the perianth-tube; anthers peculiar. Ovary 1- or 3-celled. Fruit capsular or berried. Seeds minute, exalbuminous, or larger and albuminous.

* * * * *Albumen fleshy or horny ; embryo distinct.* (*See also* Taccaceæ).

Cohort V. Narcissales.—Flowers ☿, regular or irregular. Perianth usually petaloid. Stamens 3 or 6, inserted on the perianth-tube. Ovary 3-celled. Seeds with copious fleshy or horny albumen and a distinct embryo.—Leaves parallel-veined.

<table>
<tr><td>

10. IRIDEÆ. Perianth regular or irregular. Stamens 3 ; anthers extrorse . . 782

11. AMARYLLIDEÆ. Perianth glabrous, regular or irregular. Stamens 6, inserted on the perianth-tube. Ovary 1- or 3-celled . 786

12. HÆMODORACEÆ. Perianth hairy without, regular or nearly so. Stamens 3 or 6, inserted on the base of the perianth-segments; anthers introrse. Ovary 1- or 3-celled . 780

</td><td>

13. HYPOXIDEÆ. Perianth hairy without, . regular. Stamens 6, inserted on the perianth-segments; anthers introrse, sagittate. Seeds strophiolate 791

14. VELLOSIEÆ. Perianth regular. Stamens 6 or more, inserted on the perianth-tube; anthers linear. Embryo outside the albumen 793

</td></tr>
</table>

Cohort VI. Dioscorales.—Flowers diœcious, regular. Perianth herbaceous. Stamens 6, inserted at the base of the perianth-segments. Ovary 3-celled. Fruit a berry or capsule. Seeds with copious fleshy dense albumen, and a distinct included embryo.—Climbing herbs or undershrubs. Leaves net-veined.

15. DIOSCOREÆ 704

DIVISION II.

Ovary superior. (See also *Bromeliaceæ* and *Hæmodoraceæ*).

SUB-DIVISION I. Ovary apocarpous (reduced to 1 carpel in some *Naiadeæ*. See also some *Palmæ*).

Cohort VII. Triurales.—Flowers unisexual. Perianth 6-lobed or -partite. Stamens 6. Carpels 1-ovuled ; style basal or lateral. Seed minute, with very dense albumen and obscure embryo.—Minute leafless slender herbs.

16. TRIURIDEÆ 707

Cohort VIII. Potamales.—Flowers ☿ or unisexual. Perianth of 3, 4, or 6 segments, or 0. Stamens 1-6. Albumen 0; embryo conspicuous.—Usually water-plants.

<table>
<tr><td>

17. BUTOMEÆ. Flowers ☿. Perianth 6-leaved, 3 outer herbaceous. Stamens 9-∞, hypogynous. Carpels many-ovuled, follicular when ripe 707

18. ALISMACEÆ. Flowers ☿ or monœcious. Perianth 6-leaved, 3 outer herbaceous. Stamens 6 or 12, hypogynous or perigynous. Carpels 1-2-ovuled, follicular when ripe . 790

19. JUNCAGINEÆ. Flowers ☿ or diclinous. Perianth 6-leaved, all herbaceous. Stamens 6, rarely 1 ; anthers extrorse. Carpels 3, 1-2-ovuled, follicular or capsular when ripe 802

20. POTAMEÆ. Flowers ☿ or unisexual. Perianth of 4 herbaceous valvate leaflets or

</td><td>

0, on a 3-toothed cup. Stamens 1, 2 or 4. Carpels 1-6, 1-ovuled, indehiscent or 2-valved when ripe ; ovules pendulous . 804

21. APONOGETEÆ. Flowers ☿. Perianth 0, or of 2-3 caducous leaflets, inserted on a spadix with often white bracts. Stamens 6-18-20, hypogynous. Carpels 3-5, 2-4-6-ovuled, follicular when ripe; ovules ascending 805

22. NAIADEÆ. Flowers ☿ or unisexual, obscure. Perianth 0 or tubular or membranous. **Stamens 1-2-4 ; anthers 1-2-4-celled. Carpels 1-4, 1-∞-ovuled ; when ripe an indehiscent berry or utricle, or dehiscent; stigmas usually 2-3, filiform** . . . 807

</td></tr>
</table>

SUB-DIVISION II. Ovary syncarpous (rarely apocarpous in some *Palmæ*).

Cohort IX. Palmales.—Flowers unisexual, in a simple or branched spadix, enclosed or not in a spathe. Perianth of distinct 2-seriate coriaceous segments, green, rarely coloured, or 0. Fruit a 1- rarely 2-seeded drupe or berry. Seed albuminous.—Shrubs or trees, with flabellate or pinnately divided, rarely simple leaves.

23. PALMÆ. Flowers usually diclinous on a branched spadix. Perianth of 6 herbaceous 2-seriate segments. Stamens 6, rarely more or fewer, hypogynous or perigynous. Ovary 3- rarely 1-celled, or of 3 separate carpels; styles short, free or connate; cells 1- rarely 2-ovuled. Fruit various, 1–3-celled. Seeds large; embryo minute, sunk in a pit of the fleshy or horny albumen. —Trees or shrubs 811

24. PHYTELEPHASIEÆ. Flowers monœcious or polygamo-diœcious, on a simple club-shaped spadix. Perianth-segments 2-seriate, unequal. Stamens ∞, inserted at the base of the perianth. Ovary 1-celled and 1-ovuled, or 4-celled and 4-ovuled; style 3-6-fid; ovule basal, anatropous. Fruit a muricated syncarpous or coriaceous berry. Seed large. Albumen like ivory.—Dwarf Palm-like plants; leaves pinnate   .   . 821

25. NIPACEÆ. Flowers monœcious, ♂ in cylindric catkins, ♀ capitate. Perianth ♂ of 6 2-seriate segments, ♀ 0. Stamens 3, filaments connate; anthers extrorse. Carpels 3, obliquely truncate; stigmas sessile, excentric; ovule 1, ascending. Fruit an aggregation of dry fibrous coriaceous drupes. Seed large, furrowed; albumen cartilaginous   . 821

Cohort X. Arales.—Flowers ☿ or unisexual, arranged in a spadix or spike, with or without a spathe, or sunk in pits of a minute scale-like frond. Perianth of distinct pieces, white or green, or of minute scales, or 0. Fruit a drupe or berry with one, few, or many small or minute albuminous seeds.—Herbs, often gigantic, rarely trees; leaves simple or pinnatifid, very rarely pinnately divided.

26. PANDANEÆ. Flowers diœcious, in oblong or globose spadices. Perianth 0. Stamens numerous, crowded. Ovaries numerous, 1-celled; stigmas sessile; ovules 1 or 3. Fruit a head of fibrous 1-seeded drupes. Seed small; albumen fleshy. — Trees or shrubs; leaves simple, narrow, margins and keel spinous 822

27. TYPHACEÆ. Flowers monœcious, in simple or branched heads or catkins. Perianth 0 or of slender scales or bristles. Stamens crowded; anthers basifixed. Ovaries crowded, sessile or stalked, 1-celled; style slender; stigma unilateral; ovule 1, pendulous, anatropous. Fruit dry or drupaceous. Albumen floury or fleshy. — Herbs; leaves narrow, linear or ligulate, quite entire   . 824

28. CYCLANTHEÆ. Flowers diclinous, in oblong or cylindric spadices. Perianth of ♂ multifid, of ♀ 0 or of 4 long-tailed segments. Stamens in 4 bundles, opposite the perianth-lobes. Ovary 1-celled; stigmas 4; ovules many, anatropous, on 4 parietal placentas. Fruit syncarpous, berried. Seeds small. Albumen horny.—Large stemless or climbing herbs; leaves large, entire or 2-5-partite . 827

29. FREYCINETIEÆ. Flowers polygamo-diœcious, in globose or oblong or cylindric spadices. Perianth 0. Stamens ∞; filaments isolated or grouped. Ovaries many, 1-celled; stigmas sessile; ovules many, anatropous, on parietal placentas. Fruit syncarpous, berried. Seed minute; albumen fleshy.—Large herbs or trees; stem erect or climbing; leaves narrow; margins and keel spinous   .   . 830

30. AROIDEÆ. Flowers ☿ or diclinous, in cylindric or oblong spadices. Perianth 0, or of 4-8 scales. Stamens few or many, filaments short or 0; anthers extrorse or opening by pores. Ovaries aggregated, 1-4-celled; style distinct or 0; ovules 1 or more, basal or parietal. Berry 1- or more- seeded. Seeds minute; albumen fleshy or floury, or 0.—Herbs, stemless or with erect or scandent stems; leaves usually large, simple or pinnatifidly divided, usually net-veined . 831

31. LEMNACEÆ. Flowers ☿. Perianth 0. Stamens 1-2; anther-cells sub-globose. Ovary 1-celled; style terminal; stigma funnel-shaped; ovules 1 or more. Fruit 1-many-seeded. Seed most minute. Albumen fleshy. —Minute floating green scales .   .   . 839

Cohort XI. Liliales.—Flowers ⚥, very rarely unisexual, spiked, racemed, panicled or solitary, rarely capitate. Perianth of 6 (very rarely 4) sub-similar pieces, or monopetalous and 6-lobed, regular (except *Gilliesieæ*), usually all coloured and petaloid (coriaceous or subglumaceous in *Junceæ*). Embryo immersed in copious albumen (not external to, nor in a lateral cavity of the albumen).

Cohort XII. Pontederales.—Flowers ☿, spiked, panicled or capitate. Perianth of 2 segments or of 6 2-seriate segments, all petaloid, or 3 outer herbaceous or coriaceous, and 3 inner petaloid. Style single; stigma sub-entire. Embryo immersed in copious albumen (not external to or in a lateral cavity of the albumen).—Marsh or water herbs.

Cohort XIII. Commelynales.—Flowers ☿, spiked, panicled, solitary or capitate. Perianth regular or irregular, of 6 segments in 2 series, 3 outer herbaceous, 3 inner very different, petaloid, coloured. Style usually 3-fid. Embryo outside the albumen, or in a distinct cavity in its side.

Cohort XIV. Restiales.—Flowers ☿ or unisexual, regular or irregular. Perianth of 4-6 glumaceous or scarious or membranous segments in 1-2 series, or reduced to scales, or 0. Stamens 1-3, free or united in a cup. Ovary usually 3-celled; ovules solitary, pendulous in each cell, orthotropous. Fruit capsular, rigid or membranous. Embryo outside the base of the albumen.

Cohort XV. Glumales.—Flowers in the axils of scales, which are arranged in spikelets. Perianth 0, or of minute scales or hairs or bristles. Stamens 1–3, rarely more. Ovary 1-celled, 1-ovuled. Fruit a caryopsis. Albumen fleshy or floury; embryo immersed or not. —Grasses or grass-like herbs.

SUB-KINGDOM II.

CRYPTOGAMOUS, ACOTYLEDONOUS, OR FLOWERLESS PLANTS.

Plants destitute of stamens and ovaries. Fertilization effected by the action of antherozoids on the nucleus of an archegonium, or by the union of the contents of two special cells. Propagation by means of homogeneous spores, consisting of a single cell.

CLASS III. ACROGENS.

Axis of growth distinct, growing from the apex, with usually no provision for subsequent increase in diameter, and with frequently distinct foliage. Reproduction by the action of antherozoids on archegonia.

Cohort I. Filicales.—Plants with both cellular and vascular tissue. Antheridia, or archegonia, or both, formed on a prothallus that is developed from the spore on its germination.

* *Spores of one kind: antheridia and archegonia both produced upon a prothallus.*

** *Spores of two kinds, one containing antherozoids, the other developing a prothallus with archegonia.*

Sporangia solitary, placed at the base of the leaves, or in the scales of terminal cones, 2-3-valved, containing either minute quaternary microspores full of antherozoids, or large sub-globose macrospores with a tricrural mark on one hemisphere. Macrospores developing a prothallus in germination, on the surface of which archegonia are produced . . . 911

7. ISOÊTEÆ. Submerged or terrestrial plants, with a tumid caudex clothed with the sheathing bases of elongate fronds. Sporangia enclosed in the bases of the fronds; those of the outer frond bearing macrospores, of the inner microspores. Germination, &c., as in *Lycopodiaceæ* 915

Cohort II. Muscales.—Plants composed of cellular tissue only. Archegonia or antheridia, or both, formed on the stem or branches of a new plant that is developed from the spore on its germination.

8. CHARACEÆ. Aquatic branched plants, with whorled branches, consisting of a series of long superimposed fascicles (internodes) of inarticulate tubes. Antheridia consisting of spherical vesicles containing articulate tubes, each joint (cell) of which contains an antherozoid. Archegonia consisting of a single spore, covered with spirally arranged tubes, and fertilized in situ 917

9. MUSCI. Stems leafy; leaves alternate or distichous. Antheridia consisting of delicate open sacs full of cells containing an antherozoid. Archegonia consisting of a flask-shaped body enclosing a vesicle which developes, after fertilization, a stalked urn-shaped sporangium, full of spores . . 921

10. SPHAGNA. Moss-like plants, differing from Mosses in their regular fascicled branches arising from the stem by the sides of the leaves, by some peculiarities in the structure of the stem and leaves, and stalk of the sporangium, and by having dimorphic spores 929

11. HEPATICÆ. Stems leafy, with alternate or distichous leaves, or frondose. Antheridia and archegonia as in *Musci*, but these are in some tribes buried in the substance of the frond, in others borne on the under surface of a stalked disk, in others axillary or terminal. Archegonia after fertilization giving rise to a sporangium that usually bursts into 4 spreading horizontal valves, but is sometimes 1-2-valved, or consists of a simple sac sunk in the frond. Spores usually mixed with spiral filaments (elaters) 933

CLASS IV. THALLOGENS.

Axis of growth indeterminate, growth taking place chiefly peripherically and horizontally. Plants wholly composed of cellular tissue. Reproductive organs various. Spores not developing a prothallus in germination.

12. LICHENES. Terrestrial plants. Thallus coriaceous and irregularly lobed, or erect, or a mere crust, various in colour and consistence. Fructification of two sorts: 1. *Apothecia*, which are superficial, marginal, or sunk in the frond, and contain, or consist of vertical densely packed tubes or sacs (sporangia) containing 2-8 spores; 2. *Spermogonia*, which are spherical bodies sunk in the substance of the frond, whose inner surface is lined with filaments (sterigmata) which support slender transparent corpuscles, called spermatia (supposed to represent antherozoids) . . . 940

13 FUNGI. Usually terrestrial polymorphous plants, sometimes subterranean, often parasitic, destitute of chlorophyll or starch, of most varied form, colour, and consistence, sometimes reduced to a few filaments or cells. Vegetative organs consisting of a mycelium (or tissue of slender simple threads). Spores most minute, sometimes superficial, at others borne upon projections called basidia, at others enclosed in cells or sacs 949

14. ALGÆ. Usually highly coloured plants, aquatic or natives of damp rocks, walls, &c., sometimes frondose, at others reduced to a few cells or a single cell. Fructification monœcious or diœcious, sometimes of special cells of two sexes, sometimes of simple mobile spores, sometimes of antheridia and sporangia, which are free or enclosed in capsules. 965

INDEX OF TECHNICAL TERMS.

—•◦✳◦•—

INDEX OF NAMES.

———o✦o✦o——

3 x

3 x 2

LONDON   PRINTED BY

SPOTTISWOODE AND CO., NEW-STREET SQUARE

AND PARLIAMENT STREET

Errata.

Page 24, line 1, for *apillary* read *capillary*.

" 152, " 4 and 16, for *Cœlebogyne* read *Cælebogyne*.

" 188, " 52, for *Tecta* read *Teeta*.

" 243, Order XX. to be enclosed in [ ].

" 251, Order XXIV. to be enclosed in [ ].

" 349, line 9, for *elatus* read *elata*.

" 433, " 39, for *Lafœnsia* read *Lafoensia*.

, 448, dele note.

" 536, line 3, *for* nearly *read* very.

" 547, " 40, *after* anatropous *insert* [BERRY 1–2-celled.]

" 571, " 3, for *appendiculate* read *pendulous*.

" 574, " 2 and 12, *for* appendiculate *read*, pendulous.

" 597, " 16, *under* Tribe II. BESLERIEÆ *dele* Mitraria.

" 604, " 33, *after* æstivation imbricate *insert* [or twisted].

" 630, " 11, *under* Tribe II. PHYTOLACCEÆ *dele* Rivina.

" 721, " 8, *dele* Loranthus.

" 926, " 13, for *vaginule* read *vaginula*.